The following sections feature extra practice problems posted on the Internet. They can be downloaded to your disk or hard drive from the *Elementary Algebra: Concepts and Applications*, Fifth Edition, home page, which is located at **http://hepg.awl.com/be/elem_5**. For your convenience, these exercises are also printed in the *Instructor's Resource Guide*. Please contact your instructor for assistance.

Section 1.5	Addition of Real Numbers
Section 1.6	Subtraction of Real Numbers
Section 1.7	Multiplication and Division of Real Numbers
Section 1.8	Exponential Notation and Order of Operations
Section 2.5	Problem Solving
Section 2.6	Solving Inequalities
Section 3.2	Graphing Linear Equations
Section 3.3	Graphing and Intercepts
Section 4.7	Division of Polynomials
Section 4.8	Negative Exponents and Scientific Notation
Section 5.1	Introduction to Factoring
Section 5.2	Factoring Trinomials of the Type $x^2 + bx + c$
Section 5.3	Factoring Trinomials of the Type $ax^2 + bx + c$
Section 5.4	Factoring Perfect-Square Trinomials and Differences of Squares
Section 5.5	Factoring: A General Strategy
Section 5.7	Solving Applications
Section 6.3	Addition, Subtraction, and Least Common Denominators
Section 6.4	Addition and Subtraction with Unlike Denominators
Section 6.5	Complex Rational Expressions
Section 6.6	Solving Rational Equations
Section 6.7	Applications Using Rational Equations and Proportions
Section 6.8	Formulas
Section 7.1	Slope–Intercept Form
Section 7.2	Point–Slope Form
Section 7.3	Linear Inequalities in Two Variables
Section 8.2	Systems of Equations and Substitution
Section 8.3	Systems of Equations and Elimination
Section 8.4	More Applications Using Systems
Section 8.5	Systems of Linear Inequalities
Section 9.2	Multiplying and Simplifying Radical Expressions
Section 9.3	Quotients Involving Square Roots
Section 9.4	More Operations with Radicals
Section 9.5	Radical Equations
Section 9.6	Applications Using Right Triangles
Section 9.7	Higher Roots and Rational Exponents
Section 10.3	The Quadratic Formula and Applications
Section 10.5	Graphs of Quadratic Equations
Section 10.6	Functions

ELEMENTARY
Algebra
CONCEPTS AND APPLICATIONS

ELEMENTARY
Algebra

CONCEPTS AND APPLICATIONS

Marvin L. Bittinger
Indiana University–Purdue University at Indianapolis

David J. Ellenbogen
Community College of Vermont

FIFTH EDITION

ADDISON-WESLEY

An imprint of Addison Wesley Longman, Inc.

Reading, Massachusetts • Menlo Park, California • New York • Harlow, England
Don Mills, Ontario • Sydney • Mexico City • Madrid • Amsterdam

Publisher	Jason A. Jordan
Associate Editor	Christine Poolos
Production Supervisors	Kathleen Manley and Ron Hampton
Text Design	Geri Davis/The Davis Group, Inc.
Editorial and Production Services	Martha Morong/Quadrata, Inc.
Art Editor	Janet Theurer
Marketing Managers	Ben Rivera and Liz O'Neil
Illustrators	Scientific Illustrators and Jim Bryant
Compositor	The Beacon Group
Cover Designer	Barbara Atkinson
Cover Photograph	Eric Neurath/Stock, Boston/PNI
Manufacturing Supervisor	Ralph Mattivello

PHOTO CREDITS

1, Douglas Peebles **63,** Nancy Sheehan/PhotoEdit **119,** Pamela Brackett/VisionQuest **158,** AP/Wide World Photos **167,** Mary Clay/Tom Stack & Associates **175,** Joseph McBride/Tony Stone Images **243,** Alison Langley/Stock Newport **297,** Denver Bryan/Comstock **339,** Todd Phillips/Third Coast Stock Source, Inc. **361,** Mike King/Tony Stone Images **387,** Steve Krongard/The Image Bank **425,** Erick Meola/The Image Bank **452,** Comstock **471,** Comstock **489,** The Toronto Star/B. Spremo **502,** NASA

Library of Congress Cataloging-in-Publication Data
Bittinger, Marvin L.
 Elementary algebra: concepts and applications/Marvin L. Bittinger,
 David Ellenbogen.—5th ed.
  p. cm.
  Includes index.
  ISBN 0-201-84749-3—ISBN 0-201-41731-6 (teacher's edition)
  1. Algebra. I. Ellenbogen, David. II. Title.
QA152.2.B5797 1997
512.9—dc21   97-8476
                                                                            CIP

Copyright © 1998 Addison Wesley Longman, Inc.
All rights reserved. No part of this publication may be reproduced, stored in a retrieval system, or transmitted, in any form or by any means, electronic, mechanical, photocopying, recording, or otherwise, without the prior written permission of the publisher. Printed in the United States of America.
2 3 4 5 6 7 8 9 10—RNT—009998

To Peggy

Contents

1 Introduction to Algebraic Expressions 1

- **1.1** Introduction to Algebra 2
- **1.2** The Commutative, Associative, and Distributive Laws 9
- **1.3** Fractional Notation 16
- **1.4** Positive and Negative Real Numbers 24
- **1.5** Addition of Real Numbers 31
- **1.6** Subtraction of Real Numbers 37
- **1.7** Multiplication and Division of Real Numbers 44
- **1.8** Exponential Notation and Order of Operations 50

  SUMMARY AND REVIEW 58
  TEST 61

2 Equations, Inequalities, and Problem Solving 63

- **2.1** Solving Equations 64
- **2.2** Using the Principles Together 71
- **2.3** Formulas 78
- **2.4** Applications with Percent 84
- **2.5** Problem Solving 90
- **2.6** Solving Inequalities 100
- **2.7** Solving Applications with Inequalities 108

  SUMMARY AND REVIEW 115
  TEST 117

3 Introduction to Graphing 119

- **3.1** Interpreting Graphs and Ordered Pairs 120
- **3.2** Graphing Linear Equations 129

- **3.3** Graphing and Intercepts 138
- **3.4** More Applications Using Graphs 145
- **3.5** Slope 157

SUMMARY AND REVIEW 168
TEST 171

CUMULATIVE REVIEW: CHAPTERS 1–3 173

4 Polynomials 175

- **4.1** Exponents and Their Properties 176
- **4.2** Polynomials 184
- **4.3** Addition and Subtraction of Polynomials 192
- **4.4** Multiplication of Polynomials 200
- **4.5** Special Products 207
- **4.6** Polynomials in Several Variables 215
- **4.7** Division of Polynomials 223
- **4.8** Negative Exponents and Scientific Notation 228

SUMMARY AND REVIEW 238
TEST 241

5 Polynomials and Factoring 243

- **5.1** Introduction to Factoring 244
- **5.2** Factoring Trinomials of the Type $x^2 + bx + c$ 250
- **5.3** Factoring Trinomials of the Type $ax^2 + bx + c$ 257
- **5.4** Factoring Perfect-Square Trinomials and Differences of Squares 265
- **5.5** Factoring: A General Strategy 273
- **5.6** Solving Quadratic Equations by Factoring 277
- **5.7** Solving Applications 284

SUMMARY AND REVIEW 294
TEST 296

6 Rational Expressions and Equations — 297

- **6.1** Rational Expressions   298
- **6.2** Multiplication and Division   305
- **6.3** Addition, Subtraction, and Least Common Denominators   310
- **6.4** Addition and Subtraction with Unlike Denominators   319
- **6.5** Complex Rational Expressions   327
- **6.6** Solving Rational Equations   332
- **6.7** Applications Using Rational Equations and Proportions   337
- **6.8** Formulas   350

SUMMARY AND REVIEW   355
TEST   358

CUMULATIVE REVIEW: CHAPTERS 1–6   359

7 More with Graphing — 361

- **7.1** Slope–Intercept Form   362
- **7.2** Point–Slope Form   370
- **7.3** Linear Inequalities in Two Variables   373
- **7.4** Direct and Inverse Variation   378

SUMMARY AND REVIEW   384
TEST   386

8 Systems of Equations and Problem Solving — 387

- **8.1** Systems of Equations and Graphing   388
- **8.2** Systems of Equations and Substitution   395
- **8.3** Systems of Equations and Elimination   401
- **8.4** More Applications Using Systems   410
- **8.5** Systems of Linear Inequalities   419

SUMMARY AND REVIEW   421
TEST   423

9 Radical Expressions and Equations — 425

- **9.1** Introduction to Square Roots and Radical Expressions 426
- **9.2** Multiplying and Simplifying Radical Expressions 433
- **9.3** Quotients Involving Square Roots 438
- **9.4** More Operations with Radicals 443
- **9.5** Radical Equations 447
- **9.6** Applications Using Right Triangles 454
- **9.7** Higher Roots and Rational Exponents 461

SUMMARY AND REVIEW 466
TEST 469

10 Quadratic Equations — 471

- **10.1** Solving Quadratic Equations: The Principle of Square Roots 472
- **10.2** Solving Quadratic Equations: Completing the Square 476
- **10.3** The Quadratic Formula and Applications 482
- **10.4** Complex Numbers as Solutions of Quadratic Equations 491
- **10.5** Graphs of Quadratic Equations 494
- **10.6** Functions 502

SUMMARY AND REVIEW 511
TEST 513

CUMULATIVE REVIEW: CHAPTERS 1–10 515

Appendixes
- A Sets 519
- B Factoring Sums or Differences of Cubes 523

Tables
1. Fractional and Decimal Equivalents 529
2. Squares and Square Roots 530
3. Geometric Formulas 531

Answers A-1

Index I-1

Preface

*A*ppropriate for a one-term course in elementary algebra, this text is intended for those students who have a firm background in arithmetic. It is the first of three texts in an algebra series that also includes *Intermediate Algebra*: *Concepts and Applications*, Fifth Edition, by Bittinger/Ellenbogen and *Elementary and Intermediate Algebra*: *Concepts and Applications*, *A Combined Approach*, Second Edition, by Bittinger/Ellenbogen/Johnson. *Elementary Algebra*: *Concepts and Applications*, Fifth Edition, is a significant revision of the Fourth Edition with respect to design, contents, pedagogy, and an expanded supplements package. This series is designed to prepare students for any mathematics course at the college algebra level.

Approach

Our approach is designed to help today's students both learn and retain mathematical concepts. The goal of this revision is to address three major challenges for teachers of developmental mathematics courses that we have seen emerging during the 1990s. One challenge is to prepare students of developmental mathematics for the transition from "skills-oriented" elementary and intermediate algebra courses to more "concept-oriented" college-level mathematics courses. A second challenge is to teach these same students critical thinking skills: to reason mathematically, to communicate mathematically, and to solve mathematical problems. The third challenge is to reduce the amount of content overlap between elementary algebra and intermediate algebra texts.

Following are some aspects of the approach that we have used in this revision to help meet the challenges we all face teaching developmental mathematics.

Problem Solving

One distinguishing feature of our approach is our treatment of and emphasis on problem solving. We use problem solving and applications to motivate the material wherever possible, and we include real-life applications and problem-solving techniques throughout the text. Problem solving not only encourages students to think about how mathematics can be used, it helps to prepare them for more advanced material in later courses.

- In Chapter 2, we introduce the five-step process for solving problems: (1) Familiarize, (2) Translate, (3) Carry out, (4) Check, and (5) State the answer. These steps are then used consistently throughout the text whenever we encounter a problem-solving situation. Repeated use of this problem-solving strategy gives students a sense that they have a starting point for any type of problem they encounter, and frees them to focus on the mathematics necessary to successfully translate the problem situation. In this edition (see pages 92, 341, and 410), estimation and carefully-checked guesses are frequently used to help with the Familiarize and Check steps.

Applications

Interesting applications of mathematics help motivate both students and instructors. Solving applied problems gives students the opportunity to see their conceptual understanding put to use in a real way. In the Fifth Edition of *Elementary Algebra: Concepts and Applications*, not only have we increased the number of applications by more than 74 percent, but we have also included a wide variety of real-data applications throughout the text. Over 80 percent of our applications are new. Art has also been integrated into the applications and exercises to aid the student in visualizing the mathematics. (See pages 121, 291, 318, and 437.)

Content Notes (see also Content Changes on p. xvi)

- Chapter 1 includes a brief review of arithmetic topics and then moves quickly into elementary algebra topics. This allows instructors sufficient time to cover the topics necessary to prepare students for intermediate algebra.
- Chapter 3 contains an intuitive introduction to graphing, a topic that is integrated throughout the text. This helps students to visualize the mathematics of many concepts while allowing them to develop facility with graphing throughout the course.

Pedagogy

Skill Maintenance Exercises. Retention of skills is critical to the future success of our students. To this end, nearly every exercise set includes carefully chosen exercises that review skills and concepts from preceding chapters of the text, often in preparation for the next section. In this edition, we have increased the number of skill maintenance exercises by about 50%. (See pages 127, 199, 255, and 377.)

Cumulative Review. After Chapters 3, 6, and 10, we have also included a Cumulative Review, which reviews skills and concepts from all preceding chapters of the text. (See pages 173, 359, and 517.)

Synthesis Exercises. Each exercise set ends with a group of synthesis exercises. These problems can offer opportunities for students to synthesize skills and concepts from earlier sections with the present material, or they can provide students with deeper insights into the current topic. Synthesis exercises are generally more challenging than those in the main body of the exercise set. We have increased the number of synthesis exercises by 10% for the Fifth Edition. (See pages 50, 113–114, 214, and 292–293.)

Writing Exercises. Nearly every set of synthesis exercises begins with four writing exercises—almost twice the number found in the Fourth Edition. All writing exercises are marked with a maze icon (◆). These exercises are usually not as difficult as other synthesis exercises, but require written answers that aid in student comprehension, critical thinking, and conceptualization. Because some instructors may collect answers to writing exercises, and because more than one answer may be correct, answers to writing exercises are not listed at the back of the text. (See pages 89, 182, 199, and 249.)

What's New in the Fifth Edition?

We have rewritten many key topics in response to user and reviewer feedback and have made significant improvements in design, art, and pedagogy. Detailed information about the content changes is available in the form of a Conversion Guide. Please ask your local Addison Wesley Longman sales consultant for more information. Following is a list of the major changes in this edition.

New Design

- The new design is more open and readable. We have increased the type size in many features to emphasize key topics. Pedagogical use of color has been increased to make it easier to see how examples unfold.
- The entire art program is new for this edition. We have ensured the accuracy of the graphical art through the use of computer-generated graphs. Color in the graphical art is used pedagogically and precisely to help the student visualize the mathematics. (See pages 32, 131, 134, 163, and 316.)

Technology Connections

- The Technology Connections, which appear throughout the text, integrate technology, increase the understanding of concepts through visualization, encourage exploration, and motivate discovery learning. Due to user demand, we have more than doubled the number of these strictly optional features. Optional Technology Connection exercises appear in many exercise sets and are marked with a grapher icon (▱). (See pages 148, 188, and 282.)

Collaborative Corners

- In today's professional world, teamwork is essential. We have included optional Collaborative Corner features throughout the text that require students to work in groups to solve problems. There is an average of three Collaborative Corner activities per chapter, each one appearing after the appropriate section's exercise set. (See pages 128, 319, 394, and 418.) Additional Collaborative Corner activities and suggestions for directing collaborative learning appear in the *Printed Test Bank/Instructor's Resource Guide*.

World Wide Web Integration

- The World Wide Web is a powerful resource that reaches more and more people every day. In an effort to get students more involved in using this resource, we have enhanced every chapter opener to include a World Wide Web address **http://hepg.awl.com/be/elem_5**. Students can go to this page on the World Wide Web to further explore the subject matter of the chapter opening application. Selected exercise sets also have additional practice-problem worksheets that can be downloaded over the Internet. These exercise sets are listed in the front of the text.

Content Changes

A variety of content changes have been made throughout the text. Some of the more significant changes are listed below.

- In Chapter 1, Section 1.2 is now condensed for better flow. The coverage of rational numbers in Section 1.4 has been improved. The examples in Section 1.8 are improved to prepare students for factoring in Chapter 5.
- In Chapter 2, we have increased our use of real data in examples and exercises to convey the universality of problem solving. Section 2.5 spends more time on estimating in the "Familiarize" step of the five-step problem-solving process.
- In Chapter 3, we have merged subsections in Section 3.2 for better flow, and we illustrate in greater depth the use of graphs to represent solutions. Section 3.4 introduces the student to the tremendously important concept of rate. In Section 3.5, the study of rate is linked to the concept of slope.
- In Chapter 4, Section 4.2 has been reorganized for better flow and graphs are used to estimate the value of a polynomial for selected values. New geometric applications in Sections 4.3 and 4.6 help students to better visualize the study of polynomials.
- In Chapter 5, factoring is introduced more intuitively in Section 5.1, and the concept of checking by evaluating is utilized.
- In Chapter 6, Section 6.3 streamlines material that was formerly spread out in Sections 6.3, 6.4, and 6.5. Denominators that are opposites are now in Section 6.4 so they can be regarded as special types of "unlike" denominators. More emphasis is placed on listing restrictions at the start of the problem in Section 6.6, and we have included more exercises with geometry in Section 6.7.

- Chapter 7 now includes a new section of applications in the exercise sets to help the student review slope, a topic introduced in Chapter 3. We also feature more graphs to help the student visualize direct and inverse variation in Section 7.4.
- In Chapter 10, we have included new synthesis exercises involving graphs in Section 10.1. We have also reorganized examples and subsections in Section 10.5 to better prepare the student for future coursework.

Supplements for the Instructor

Instructor's Edition

The *Instructor's Edition* is a specially bound version of the student text with worked-out solutions to the even-numbered exercises and answers to the odd-numbered exercises at the back of the text.

Instructor's Solutions Manual

The *Instructor's Solutions Manual* contains worked-out solutions to all exercises in the exercise sets.

Printed Test Bank/Instructor's Resource Guide
by Donna DeSpain

This supplement contains the following:

- Extra practice problems with more sections covered than in the Fourth Edition
- Black-line masters of grids and number lines for transparency masters or test preparation
- A videotape index and section cross references to the tutorial software packages available with this text
- Additional collaborative learning activities and suggestions
- A syllabus conversion guide from the Fourth Edition to the Fifth Edition

The test bank portion contains the following:

- Six alternative free-response test forms for each chapter
- Two multiple-choice versions of each chapter test
- Eight final examinations: three with questions organized by chapter, three with questions organized by type, and two with multiple-choice questions

All test forms have been completely rewritten.

TestGen—EQ

TestGen—EQ is a computerized test generator that allows instructors to select test questions manually or randomly from selected topics. The test questions are algorithm-driven so that regenerated number values maintain problem types and provide a large number of test items in both multiple-choice and open-ended formats for one or more test forms. Test items can be viewed on screen, and the built-in question editor lets instructors modify existing questions or add new questions that include pictures, graphs, math symbols, and variable text and numbers. Instructors can also customize both the look and content of test banks and tests. Test questions are easily transferred from the test bank to a test and can be sorted, searched, and displayed in various ways. Available in both Windows and Macintosh versions, TestGen—EQ is free to qualifying adopters.

QuizMaster—EQ

QuizMaster—EQ enables instructors to create and save tests and quizzes using TestGen—EQ so students can take them on a computer network. Instructors can set preferences for how and when tests are administered. QuizMaster—EQ automatically grades the exams and allows the instructor to view or print a variety of reports for individual students, classes, or courses. This software is available for both Windows and Macintosh and is fully networkable. QuizMaster—EQ is free to qualifying adopters.

InterAct Math Plus

This software, available for both Windows and Macintosh, combines course management and on-line testing with the features of InterAct Math Tutorial Software (see *InterAct Math Tutorial* under Supplements for the Student) to create an invaluable teaching resource. Contact your local Addison Wesley Longman sales consultant for a demonstration.

Supplements for the Student

Student's Solutions Manual by Judith A. Penna

This manual contains completely worked-out solutions with step-by-step annotations for all the odd-numbered exercises in the exercise sets. Solution processes match those illustrated in the text. The manual also contains answers for all the even-numbered exercises in the text.

Videotapes

Developed especially for the Bittinger/Ellenbogen texts, these videotapes feature an engaging team of lecturers presenting material from every section of the text in an interactive format. The lecturers' presentations support an approach that emphasizes visualization and problem solving. The videotapes are free to qualifying adopters.

Video Manual by Janina Udrys

Designed to be used with the Bittinger/Ellenbogen videotapes in distance-learning situations, this manual includes many examples and art pieces from the video series. The manual also includes additional problems specifically developed for distant learners.

InterAct Math Tutorial Software

InterAct Math Tutorial Software includes exercises that are linked one-to-one with the odd-numbered exercises in the textbook. Every exercise is accompanied by an example and an interactive guided solution designed to involve students in the solution process and to help them identify precisely where they are having trouble. In addition, the software recognizes common student errors and provides students with appropriately customized feedback. Available for both Windows or Macintosh and fully networkable, Interact Math Tutorial Software is free to qualifying adopters.

World Wide Web Supplement http://hepg.awl.com/be/elem_5

This on-line supplement contains links to web sites related to the chapter openers in the text. It also contains extra practice-problem worksheets for selected text sections. Students can download these worksheets and use them for practice, as needed. Text sections that correlate with on-line worksheets are listed in the front of the text.

Acknowledgments

No book can be produced without a team of professionals who take pride in their work and are willing to put in long hours. Barbara Johnson, in particular, deserves special thanks for her work as development editor. Barbara's tireless attention to detail and her many fine suggestions have contributed immeasurably to the quality of this text. Laurie A. Hurley and Irene Doo also deserve deeply felt thank you's for their careful accuracy checks and well-thought-out suggestions. Judy Penna's work in preparing the *Student's Solution Manual*, the *Instructor's Solution Manual*, and the indexes, amounts to an inspection of the text that goes beyond the call of duty, and for which we are extremely grateful. We are also indebted to Chris Burditt for his many fine ideas that appear in our Collaborative Corners.

Jason Jordan, the sponsoring editor, displayed both wisdom and restraint in providing direction when it was needed and allowing us to work independently as much as possible. Martha Morong, of Quadrata, Inc., provided editorial and production services that set the standard for the industry. Janet Theurer, of Theurer Briggs Design, performed outstanding work as art editor. George and Brian Morris of Scientific Illustrators generated the graphs, charts, and many of the illustrations. Their work is always precise and attractive. The many hand-drawn illustrations appear thanks to Jim Bryant, an artist with true mathematical sensibilities. Finally, a special thank you to Christine Poolos for coordinating reviews, tracking down information, and managing so many of the day-to-day details—always with a pleasant demeanor.

In addition, we thank the students at Community College of Vermont and the following professors for their thoughtful reviews and insightful comments.

Prerevision Diary Reviewers (Fourth Edition)

Mazie Akana, *University of Hawaii—Leeward Community College*
Irene Doo, *Austin Community College*
Andrea Seavitt, *Henry Ford Community College*
Kathy Struve, *Columbus State Community College*

Manuscript Reviewers

Julie Bonds, *Sonoma State University*
Beverly Broomell, *Suffolk Community College*
Michael Butler, *College of the Redwoods*
Richard Butterworth, *Massasoit Community College*
Virginia Crisonino, *Union City College*
Florence Chambers, *Mohawk Valley Community College*
Irene Doo, *Austin Community College*
Elizabeth Farber, *Bucks County Community College*
Jim Fryxell, *College of Lake County*
Donna Goldstein, *Mercer County Community College*
Lonnie Hass, *North Dakota State University*
Dale Hughes, *Johnson County Community College*
Paulette Kirkpatrick, *Wharton County Community College*
Allan Robert Marshall, *Cuesta College*
Carol Metz, *Westchester Community College*
Eveline Robbins, *Gainesville College*
Julia Schroeder, *John A. Logan College*
Diane Tesar, *South Suburban College*
Jane Theiling, *Dyersburg State Community College*

Finally, a special thank you to all those who so generously agreed to discuss their professional use of mathematics in our chapter openers. These dedicated people, none of whom we knew prior to writing this text, all share a desire to make math more meaningful to students. We cannot imagine a finer set of role models.

M.L.B.
D.J.E.

Feature Walkthrough

The following six pages show you how to use *Elementary Algebra* to maximize understanding while making studying easier.

- **Chapter Openers:** Each chapter opens with a list of the sections covered and a real-life application that helps show how mathematics is needed when solving real problems. Real data are often used in these applications, as well as in many other exercises and examples, to increase student interest. For those students with access to the Internet, we have equipped every chapter opener with a World Wide Web address, **http://hepg.awl.com/be/elem_5**. At this site, students can further explore the subject matter of the chapter opening application.

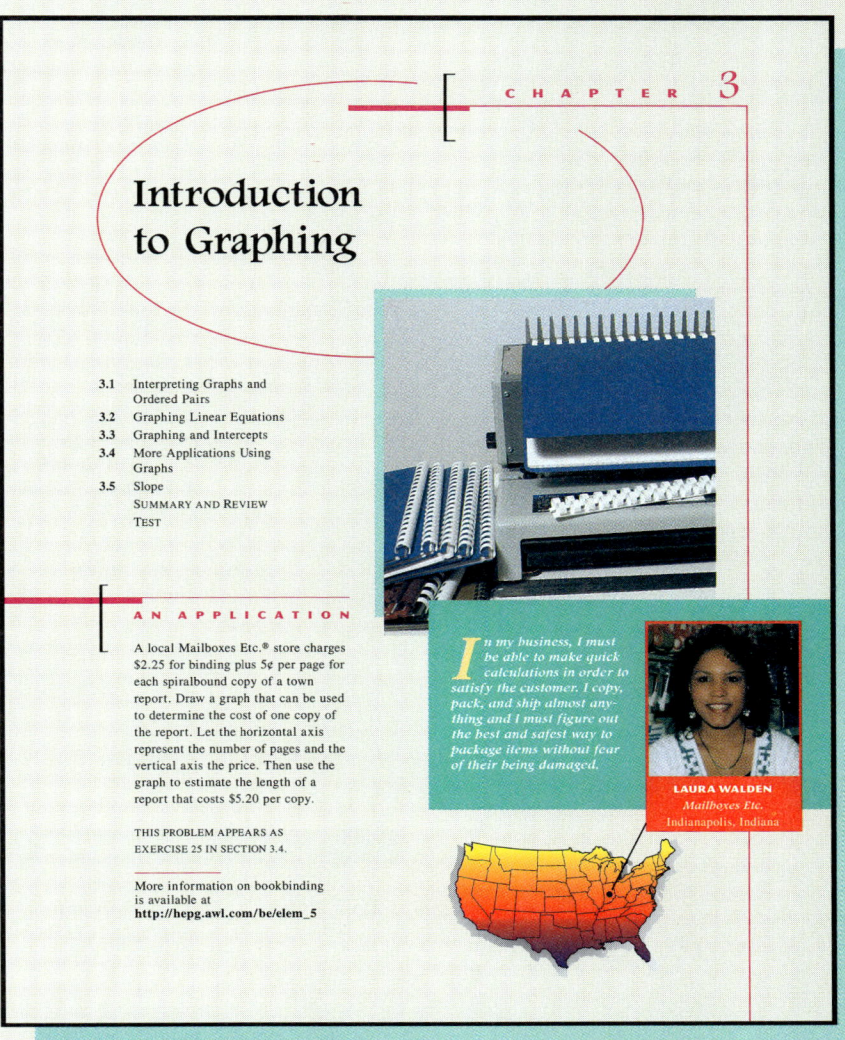

- **Problem Solving:** A distinguishing feature of the authors' approach is their **Five-Step Process for Solving Problems:** (1) Familiarize, (2) Translate, (3) Carry Out, (4) Check, and (5) State the Answer. These steps are used throughout the text whenever a problem-solving situation occurs. This problem-solving process provides a structure for students to use with any type of problem they encounter and allows them to focus on all aspects of problem solving.

5.7 SOLVING APPLICATIONS **285**

EXAMPLE 1 **Page Numbers.** The product of the page numbers on two consecutive pages of a book is 156. Find the page numbers.

SOLUTION

1. **Familiarize.** Recall that consecutive page numbers are one apart, like 49 and 50. Let x = the first page number; then $x + 1$ = the next page number.

2. **Translate.** We reword the problem before translating:

   Rewording: The first page number times the next page number is 156.

   Translating: $x \cdot (x + 1) = 156$

3. **Carry out.** We solve the equation as follows:

   $x(x + 1) = 156$
   $x^2 + x = 156$ Multiplying
   $x^2 + x - 156 = 0$ Subtracting 156 to get 0 on one side
   $(x - 12)(x + 13) = 0$ Factoring
   $x - 12 = 0$ *or* $x + 13 = 0$ Using the principle of zero products
   $x = 12$ *or* $x = -13$.

4. **Check.** The solutions of the equation are 12 and -13. Since page numbers cannot be negative, -13 is not a solution. On the other hand, if x is 12, then $x + 1$ is 13 and $12 \cdot 13 = 156$. Thus, 12 checks.

5. **State.** The page numbers are 12 and 13.

EXAMPLE 2 **Dimensions of a Garden.** A vacant square-shaped lot is being turned into a community garden. Because a path 2 m wide is needed at one end, only 48 m² of the lot will be garden space. Find the dimensions of the lot.

SOLUTION

1. **Familiarize.** We first make a drawing. Recall that the area of any rectangle, including a square, is length · width. We let x = the length (or width) of the square lot. Note that the path is a rectangle 2 m wide and x m long.

2. **Translate.** It helps to reword this problem before translating.

- **New Design:** The new design is more open and readable. Throughout the text, generous amounts of color-coded technical and situational art are used to enhance student understanding of examples and exercises. This feature will also aid in the visualization of mathematical concepts.

3.5 SLOPE 163

steepness. For example, numbers like 2%, 3%, and 6% are often used to represent the **grade** of a road, a measure of a road's steepness. For example, a 3% grade means that for every horizontal distance of 100 ft, the road rises or drops 3 ft. The concept of grade also occurs in skiing or snowboarding, where a 4% grade is considered very tame, but a 40% grade is considered steep.

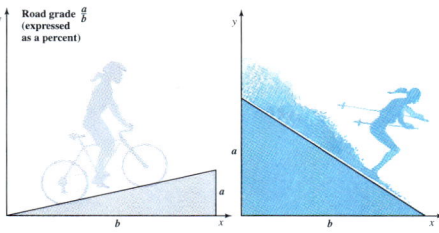

Road grade $\frac{a}{b}$ (expressed as a percent)

EXAMPLE 5 *Skiing.* Among the steepest skiable terrain in North America, the Headwall on Mount Washington, in New Hampshire, drops 720 ft over a horizontal distance of 900 ft. Find the grade of the Headwall.

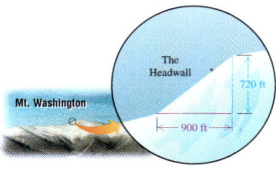

SOLUTION The grade of the Headwall is its slope, expressed as a percent:

$$= \frac{720}{900}$$
$$= \frac{8}{10}$$
$$= 80\%.$$

Carpenters use slope when designing stairs, ramps, or roof pitches. An application occurs in the engineering of a dam—the force or strength of depends on how much the river drops over a specified distance. Ratios ge also appear in areas of research ranging from the rate at which a cooling (in astronomy) to the growth rate of an animal (in zoology).

---

134 CHAPTER 3 • INTRODUCTION TO GRAPHING

Since $(0, b)$ is the point at which the graph crosses the y-axis, it is called the **y-intercept.** Sometimes, for convenience, we simply refer to b as the y-intercept. In Sections 3.3 and 7.1, we will make extensive use of y-intercepts.

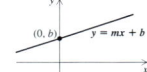

Linear equations appear in many real-life situations.

EXAMPLE 7 *Value of an Office Machine.* The value of Dupligraphix's color copier is given by $v = -\frac{1}{2}t + 3$. Here v represents the value, in thousands of dollars, t years from the date of purchase. Graph the equation and then use the graph to estimate the value of the copier $2\frac{1}{2}$ yr after the date of purchase.

SOLUTION We graph $v = -\frac{1}{2}t + 3$ by selecting values for t and then calculating the associated value v. Since time cannot be negative in this case, we select nonnegative values for t.

If $t = 0$, then $v = -\frac{1}{2} \cdot 0 + 3 = 3$.
If $t = 4$, then $v = -\frac{1}{2} \cdot 4 + 3 = 1$.
If $t = 8$, then $v = -\frac{1}{2} \cdot 8 + 3 = -1$.

t	v
0	3
4	1
8	-1

We label the axes and plot the points (see the figure on the left below). Since they line up, our calculations are probably correct. However, including any points below the horizontal axis seems unrealistic since (at least for tax purposes) the value of the copier cannot be negative. Thus, when drawing the graph, we end the solid line at the horizontal axis.

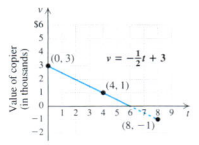 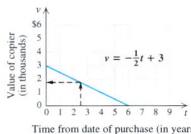

To estimate the value of the copier after $2\frac{1}{2}$ yr, we need to determine what second coordinate is paired with $2\frac{1}{2}$. To do this, we locate the point on the line that is above $2\frac{1}{2}$ and then find the value on the vertical axis that corresponds to that point (see the figure on the right above). It appears that after $2\frac{1}{2}$ yr, the copier is worth about $1800.

CAUTION! When the coordinates of a point are read from a graph, as in Example 7, values should not be considered exact.

- **Applications:** In this edition, the authors have increased the number of applications and have also included real-data applications to reinforce the relevance of mathematics to students. Over 80% of the applications are new and are taken from disciplines as varied as the natural and social sciences, business and economics, and health care. Art has also been integrated to aid the student in visualizing the many applications that appear in the examples and exercises.

COLLABORATIVE CORNER

Focus: Least common multiples and proportions
Time: 20–30 minutes
Group size: 2

Travel between different countries usually necessitates an exchange of currencies. Recently, one Canadian dollar was worth 72 cents in U.S. funds. Use this exchange rate for the activity that follows.

ACTIVITY

1. Within each group of two students, one student should play the role of a U.S. citizen planning a visit to Canada. The other student should play the role of a Canadian citizen planning a visit to the United States.

   a) Determine how much Canadian money the U.S. citizen would receive in exchange for $72 of U.S. funds. Remember: The U.S. dollar is worth more ($1.00 vs. 72¢) than the Canadian dollar. Then determine how much U.S. money the Canadian would receive in exchange for $72 of Canadian funds. (*Hint*: Use a proportion.)

   b) Determine how much U.S. money the Canadian citizen would receive in exchange for $100. Then determine how much the U.S. citizen would receive in exchange for $100 of U.S. funds.

2. The answers to parts (a) and (b) above should indicate that change (coins) is needed to exchange $72 of Canadian money for U.S. funds, or to exchange $100 of U.S. funds for Canadian money.

   a) What is the smallest amount of Canadian dollars that can be exchanged for U.S. dollars without requiring coins?

   b) What is the smallest amount of U.S. dollars that can be exchanged for Canadian dollars without requiring coins? (*Hint*: See part a.)

   c) Use your results from parts (a) and (b) of this question to find two other amounts of U.S. currency and two other amounts of Canadian currency that can be exchanged without requiring coins. Answers may vary.

   d) Use your results from parts (a) and (b) of this question to find the smallest number x for which neither conversion—from x Canadian dollars to U.S. funds or from x U.S. dollars to Canadian funds—will require coins.

3. At one time in 1997, one New Zealand dollar was worth about 76 cents in U.S. funds. Find the smallest number x for which neither conversion—from x New Zealand dollars to U.S. funds or from x U.S. dollars to New Zealand funds—will require coins. (*Hint*: See part d above.)

6.4 Addition and Subtraction with Unlike Denominators

Adding and Subtracting with LCDs • When Factors Are Opposites

Adding and Subtracting with LCDs

We now know how to rewrite two rational expressions in an equivalent form that uses the LCD. Once rational expressions share a common denominator,

- **Collaborative Corner Features:** The authors now include optional Collaborative Corner features throughout the text. These require students to work as a group to solve problems or to perform specially designed activities. There is an average of three Collaborative Corners per chapter, each one appearing after the appropriate exercise set.

---

point (4, 3) appears to line up with the intercepts, so we draw the graph.

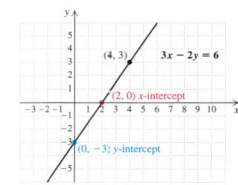

TECHNOLOGY CONNECTION 3.3

Most graphers require that an equation have y alone on one side before the equation can be entered. Once the equation has been entered, depending on the dimensions of the window, we may not be able to see both intercepts. For example, if $y = -0.8x + 17$ is graphed in the standard $[-10, 10, -10, 10]$ window, neither intercept is visible.

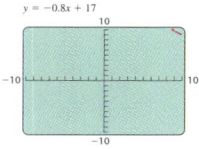

To better view the intercepts, we can try different window dimensions or we can zoom out. The ZOOM fea-

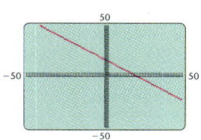

ture allows us to reduce or magnify a graph or a portion of a graph. Before zooming, the ZOOM *factors* must be set in the memory of the ZOOM key. If we zoom out with factors set at 5, both intercepts are visible but the axes are heavily drawn, as shown in the preceding figure.

This suggests that the *scales* of the axes should be changed. To do this, we use the WINDOW menu and set Xscl to 5 and Yscl to 5. The resulting graph has tick marks 5 units apart and clearly shows both intercepts.

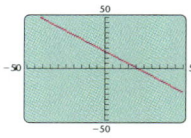

Graph each equation so that both intercepts can be easily viewed. Zoom or adjust the window settings so that tick marks can be clearly seen on both axes.

1. $y = -0.72x - 15$
2. $y = 2.13x = 27$
3. $5x + 6y = 84$
4. $2x - 7y = 150$

- **Technology Connection Features:** Due to user demand, the authors have significantly increased the number of optional Technology Connection features in this edition. Optional Technology Connection exercises also appear in many exercise sets and are marked with a grapher icon ( ).

228   CHAPTER 4   •   POLYNOMIALS

32. $(t^3 - t^2 + t - 1) \div (t - 1)$
33. $(6x^4 - 3x^2 + x - 4) \div (2x^2 + 1)$
34. $(4x^4 - 4x^2 - x - 3) \div (2x^2 - 3)$

SKILL MAINTENANCE

35. The perimeter of a rectangle is 640 ft. The length is 15 ft greater than the width. Find the area of the rectangle.
36. Solve: $3(2x - 1) = 7x - 5$.
37. Solve: $2x > 12 + 7x$.
38. Plot the points $(4, -1), (0, 5), (-2, 3)$, and $(-3, 0)$.
39. In which quadrant are both coordinates negative?
40. Graph: $3x - 2y = 12$.

SYNTHESIS

41. ◆ Explain how the equation
$$(2x + 3)(3x - 1) = 6x^2 + 7x - 3$$
can be used to devise two division problems.

42. ◆ On an assignment, a student *incorrectly* writes
$$\frac{12x^3 - 6x}{3x} = 4x^2 - 6x.$$
What mistake do you think the student is making and how might you convince the student that a mistake has been made?

43. ◆ Can the quotient of two binomials be a trinomial? Why or why not?
44. ◆ How is the distributive law used when dividing a polynomial by a polynomial?

Divide.

45. $(45x^{8k} + 30x^{6k} - 60x^{4k}) \div (3x^{2k})$
46. $(10a^{9k} - 32a^{6k} + 28a^{3k}) \div (2a^{3k})$
47. $(y^4 + a^2) \div (y + a)$
48. $(5a^3 + 8a^2 - 23a - 1) \div (5a^2 - 7a - 2)$
49. $(15y^3 - 30y + 7 - 19y^2) \div (3y^2 - 2 - 5y)$
50. Divide the sum of $4x^5 - 14x^3 - x^2 + 3$ and $2x^5 + 3x^4 + x^3 - 3x^2 + 5x$ by $3x^3 - 2x - 1$.
51. Divide $5x^7 - 3x^4 + 2x^2 - 10x + 2$ by the sum of $(x - 3)^2$ and $5x - 8$.
52. Divide $6a^{3h} + 13a^{2b} - 4a^h - 15$ by $2a^h + 3$.

If the remainder is 0 when one polynomial is divided by another, the divisor is a factor of the dividend. Find the value(s) of c for which $x - 1$ is a factor of each polynomial.

53. $x^2 - 4x + c$
54. $2x^2 - 3cx - 8$
55. $c^2x^2 + 2cx + 1$

4.8   Negative Exponents and Scientific Notation

Negative Integers as Exponents • S
Multiplying and Dividing Using Scie
Problem Solving with Scientific Nota

We now attach a meaning to negativ
positive and negative exponents, we
known as *scientific notation*.

Negative Integers as Expo

If we define negative exponents carefu
exponents will hold for all integer ex

5.1   INTRODUCTION TO FACTORING   249

EXERCISE SET 5.1

Find three factorizations for each monomial.

1. $10x^3$
2. $6x^3$
3. $-6x^5$
4. $-15x^6$
5. $26x^5$
6. $25x^4$

Factor. Remember to use the largest common factor and to check by multiplying.

7. $x^2 - 6x$
8. $x^2 + 8x$
9. $8x^2 + 4x$
10. $5x^2 - 10x$
11. $x^3 + 6x^2$
12. $4x^4 + x^2$
13. $8x^4 - 24x^2$
14. $5x^5 + 10x^3$
15. $2x^2 + 2x - 8$
16. $6x^2 + 3x - 15$
17. $5x^6 - 10x^3 + 8x^2$
18. $10x^5 + 6x^4 - 3x^3$
19. $2x^8 + 4x^6 - 8x^4 + 10x^2$
20. $5x^4 - 15x^3 - 25x - 10$
21. $x^5y^5 + x^4y^3 + x^3y^3 - x^2y^2$
22. $x^9y^6 - x^7y^5 + x^4y^4 + x^3y^3$
23. $\frac{3}{5}x^6 + \frac{4}{5}x^5 + \frac{1}{5}x^4 + \frac{1}{5}x^3$
24. $\frac{5}{7}x^7 + \frac{3}{7}x^5 - \frac{6}{7}x^3 - \frac{1}{7}x$

Factor.

25. $y(y + 3) + 7(y + 3)$
26. $b(b - 5) + 3(b - 5)$
27. $x^2(x + 3) - 7(x + 3)$
28. $3z^2(2z + 9) + (2z + 9)$
29. $y^2(y + 8) + (y + 8)$
30. $x^2(x - 7) - 3(x - 7)$

Factor by grouping, if possible, and check.

31. $x^3 + 3x^2 + 4x + 12$
32. $6z^3 + 3z^2 + 2z + 1$
33. $2x^3 + 6x^2 + 3x + 9$
34. $3x^3 + 2x^2 + 3x + 2$
35. $9x^3 - 12x^2 + 3x - 4$
36. $10x^3 - 25x^2 + 4x - 10$
37. $5x^3 - 2x^2 + 5x - 2$
38. $18x^3 - 21x^2 + 30x - 35$
39. $x^3 + 8x^2 - 3x - 24$
40. $2x^3 + 12x^2 - 5x - 30$
41. $w^3 - 7w^2 + 4w - 28$
42. $y^3 + 8y^2 - 2y - 16$
43. $x^3 - x^2 - 2x + 5$
44. $p^3 + p^2 - 3p + 10$

45. $2x^3 - 8x^2 - 9x + 36$
46. $20g^3 - 4g^2 - 25g + 5$

SKILL MAINTENANCE

47. Graph: $y = x - 6$.
48. Solve: $4x - 8x + 16 \geq 6(x - 2)$.
49. Subtract: $-13 - (-25)$.
50. Solve $A = \dfrac{p + q}{2}$ for p.

Multiply.

51. $(y + 5)(y + 7)$
52. $(y + 7)^2$
53. $(y + 7)(y - 7)$
54. $(y - 7)^2$

SYNTHESIS

55. ◆ Write a two-sentence paragraph in which the word "factor" is used at least once as a noun and once as a verb.
56. ◆ Josh says that for Exercises 1–46 there is no need to print answers in the back of the text. Is he correct in saying this? Why or why not?
57. ◆ Explain how someone could construct a polynomial with four terms that can be factored by grouping.
58. ◆ In answering a factoring problem, Taylor says the largest common factor is $-5x^2$ and Natasha says the largest common factor is $5x^2$. Can they both be correct? Why or why not?

Factor, if possible.

59. $4x^5 + 6x^3 + 6x^2 + 9$
60. $x^6 + x^4 + x^2 + 1$
61. $x^{12} + x^7 + x^5 + 1$
62. $x^3 + x^2 - 2x + 2$
63. $5x^5 - 5x^4 + x^3 - x^2 + 3x - 3$
64. $ax^2 + 2ax + 3a + x^2 + 2x + 3$

65. ◆ Explain how to construct a polynomial of degree 9 for which $5x^3y^2$ is the largest common factor.
66. ◆ Marlene recognizes that evaluating provides only a partial check of her factoring. Because of this, she often performs a second check with a different replacement value. Is this a good idea? Why or why not?

238 CHAPTER 4 • POLYNOMIALS

SUMMARY AND REVIEW 4

KEY TERMS

Polynomial, p. 184
Term, p. 184
Monomial, p. 184
Binomial, p. 185
Trinomial, p. 185
Degree of a term, pp. 185, 216
Coefficient, p. 185
Leading term, p. 186
Leading coefficient, p. 186
Degree of a polynomial, p. 186
Descending order, p. 187
Opposite of a polynomial, p. 193
FOIL, p. 208
Difference of squares, p. 209
Polynomial in several variables, p. 215
Like terms, p. 217
Scientific notation, p. 232

IMPORTANT PROPERTIES AND FORMULAS

Definitions and Properties of Exponents

Assuming that no denominator is 0 and that 0^0 is not considered, for any integers m and n,

1 as an exponent:	$a^1 = a$
0 as an exponent:	$a^0 = 1$
Negative exponents:	$a^{-n} = \dfrac{1}{a^n}$
The Product Rule:	$a^m \cdot a^n = a^{m+n}$
The Quotient Rule:	$\dfrac{a^m}{a^n} = a^{m-n}$
The Power Rule:	$(a^m)^n = a^{mn}$
Raising a product to a power:	$(ab)^n = a^n b^n$
Raising a quotient to a power:	$\left(\dfrac{a}{b}\right)^n = \dfrac{a^n}{b^n}$

Special Products of Polynomials

$(A + B)(A - B) = A^2 - B^2$
$(A + B)(A + B) = A^2 + 2AB + B^2$
$(A - B)(A - B) = A^2 - 2AB + B^2$

Scientific notation: $N \times 10^n$, where $1 \leq N < 10$ and n is an i[nteger]

- **Summary and Review:** Each chapter ends with a Summary and Review that lists key terms and important properties and formulas. The review exercises (skill maintenance, synthesis, and writing exercises) and a Chapter Test can be used to confirm and solidify student understanding of the key concepts from the chapter.

CUMULATIVE REVIEW: CHAPTERS 1-6   359

CUMULATIVE REVIEW 1-6

1. Use the commutative law of addition to write an expression equivalent to $a + 2b$.
2. Write a true sentence using either < or >:
   -3.1 ☐ -3.15.
3. Evaluate $(y - 1)^2$ for $y = -6$.
4. Simplify: $-4[2(x - 3) - 1]$.

Simplify.

5. $-\dfrac{1}{2} + \dfrac{3}{8} + (-6) + \dfrac{3}{4}$
6. $-\dfrac{72}{108} \div \left(-\dfrac{2}{3}\right)$
7. $-6.262 \div 1.01$
8. $4 \div (-2) \cdot 2 + 3 \cdot 4$

Solve.

9. $3(x - 2) = 24$
10. $49 = x^2$
11. $-4x = -18$
12. $5x + 7 = -3x - 9$
13. $4(y - 5) = -2(y + 2)$
14. $x^2 + 11x + 10 = 0$
15. $\dfrac{1}{3}x - \dfrac{2}{9} = \dfrac{2}{3} + \dfrac{4}{9}x$
16. $\dfrac{4}{x} + x = 5$
17. $3 - y \geq 2y + 5$
18. $\dfrac{2}{x-3} = \dfrac{5}{3x+1}$
19. $2x^2 + 7x = 4$
20. $4(x + 7) < 5(x - 3)$
21. $\dfrac{x^2}{x+2} = \dfrac{4}{x+2}$
22. $(2x + 7)(x - 5) = 0$
23. $\dfrac{2}{x^2 - 9} + \dfrac{5}{x - 3} = \dfrac{3}{x + 3}$

Solve each formula.

24. $A = \dfrac{4b}{t}$, for t
25. $\dfrac{1}{t} = \dfrac{1}{m} - \dfrac{1}{n}$, for n
26. $r = \dfrac{a-b}{c}$, for c

Combine like terms.

27. $x + 2y - 2z + \dfrac{1}{2}x - z$
28. $2x^3 - 7 + \dfrac{3}{7}x^2 - 6x^3 - \dfrac{4}{7}x^2 + 5$

Graph.

29. $y = 1 - \dfrac{1}{2}x$
30. $x = -3$
31. $x - 6y = 6$

32. Find the slope of the line containing the points (1, 5) and (2, 3).

Simplify.

33. $\dfrac{x^{-5}}{x^{-3}}$
34. $y^2 \cdot y^{-10}$
35. $-(2a^2b^7)^2$
36. Subtract:
   $(-8y^2 - y + 2) - (y^3 - 6y^2 + y - 5)$.

Multiply.

37. $4(3x + 4y + z)$
38. $(2x^2 - 1)(x^3 + x - 3)$
39. $(6x - 5y)^2$
40. $(x + 3)(2x - 7)$
41. $(2x^3 + 1)(2x^3 - 1)$

Factor.

42. $6x - 2x^2 - 24x^4$
43. $16x^2 - 81$
44. $x^2 - 10x + 24$
45. $8x^2 + 10x + 3$
46. $6x^2 - 28x + 16$
47. $2x^2 - 18$
48. $16x^2 + 40x + 25$
49. $3x^2 + 10x - 8$
50. $x^4 + 2x^3 - 3x - 6$

Simplify.

51. $\dfrac{y^2 - 36}{2y + 8} \cdot \dfrac{y + 4}{y + 6}$
52. $\dfrac{x^2 - 1}{x^2 - x - 2} \div \dfrac{x - 1}{x - 2}$
53. $\dfrac{5ab}{a^2 - b^2} + \dfrac{a + b}{a - b}$
54. $\dfrac{x+2}{4-x} - \dfrac{x+3}{x-4}$
55. $\dfrac{1 + \dfrac{2}{x}}{1 - \dfrac{4}{x^2}}$
56. $\dfrac{\dfrac{1}{t} + 2t}{t - \dfrac{2}{t^2}}$

Divide.

57. $\dfrac{15x^4 - 12x^3 + 6x^2 + 2x + 18}{3x^2}$
58. $(15x^4 - 12x^3 + 6x^2 + 2x + 18) \div (x + 3)$

Solve.

59. Linnae has $36 budgeted for stationery. Engraved stationery costs $20 for the first 25 sheets and $0.08 for each additional sheet. How many sheets of stationery can Linnae order and still stay within her budget?
60. The price of a box of cereal increased 15% to $4.14. What was the price of the cereal before the increase?

- **Cumulative Review:** A Cumulative Review appears after Chapters 3 and 6 and at the end of the text. These Cumulative Reviews require students to use skills and concepts from all preceding chapters of the text. Using the keyed answers provided in the back of the book, students can identify any areas that need special attention.

ELEMENTARY
Algebra
CONCEPTS AND APPLICATIONS

CHAPTER 1

Introduction to Algebraic Expressions

1.1 Introduction to Algebra
1.2 The Commutative, Associative, and Distributive Laws
1.3 Fractional Notation
1.4 Positive and Negative Real Numbers
1.5 Addition of Real Numbers
1.6 Subtraction of Real Numbers
1.7 Multiplication and Division of Real Numbers
1.8 Exponential Notation and Order of Operations
SUMMARY AND REVIEW
TEST

AN APPLICATION

The tallest mountain in the world, as measured from base to peak, is Mauna Kea in Hawaii. From a base 19,684 ft below sea level, it rises 33,480 ft. What is the elevation of its peak?

THIS PROBLEM APPEARS AS EXERCISE 57 IN SECTION 1.5.

More information on Hawaiian volcanoes is available at
http://hepg.awl.com/be/elem_5

*M*ath is the backbone of the surveying field. From algebra to trigonometry to geometry, it is impossible to do surveying without using math.

CLYDE MATSUNAGA
Land Surveyor
Hilo, Hawaii

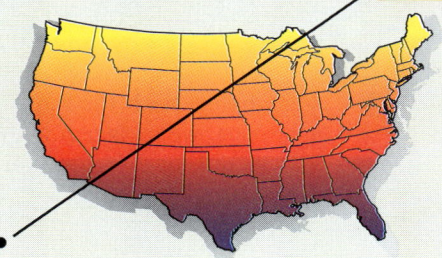

CHAPTER 1 • **INTRODUCTION TO ALGEBRAIC EXPRESSIONS**

Problem solving is the focus of this text. Chapter 1 presents some preliminaries that are needed for the problem-solving approach that is developed in Chapter 2 and used throughout the rest of the book. More specifically, Chapter 1 reviews some arithmetic, discusses real numbers and their properties, and examines how real numbers are added, subtracted, multiplied, divided, and raised to powers.

1.1 Introduction to Algebra

Algebraic Expressions • Translating to Algebraic Expressions • Translating to Equations

This section introduces some basic concepts of algebra. Since equation solving is central to the study of algebra, we concentrate on the expressions that appear in equations and some important words for translating English to mathematics.

Algebraic Expressions

One major difference between arithmetic and algebra is the use of *variables* in algebra. When a letter can represent a variety of different numbers, that letter is called a **variable**. For example, if n represents the number of students enrolled in a college's 8 A.M. section of Elementary Algebra, the number n will vary, depending on how many students drop or add the class. Thus the number n is a variable. If each student pays $600 for the course, the college collects a total of $600 \cdot n$ dollars from the students. Since the cost to each student is consistently $600, the number 600 is called a **constant**.

Cost to the Student (in dollars)	Number of Students Registered	Total Collected (in dollars)
600	n	$600 \cdot n$

The expression $600 \cdot n$ is a **variable expression** because its value varies with the choice of n. In this case, the total amount collected, $600 \cdot n$, will change with the number of students enrolled. In the chart on the following page, we replace n with a variety of values and compute the total amount collected. In doing so, we are **evaluating the expression** $600 \cdot n$.

Cost to the Student (in dollars), 600	Number of Students Enrolled, n	Total Collected (in dollars), $600 \cdot n$
600	20	12,000
600	25	15,000
600	30	18,000

Variable expressions are examples of *algebraic expressions*. An **algebraic expression** consists of variables and/or numerals, often with operation signs and grouping symbols. Examples are

$$a + 1900, \quad 5 \cdot x, \quad 12t - r, \quad 18 \div y, \quad \frac{9}{7}, \quad \text{and} \quad 3a(b + c).$$

Recall that a fraction bar is a division symbol: $\frac{9}{7}$ means $9 \div 7$. Similarly, multiplication can be written in several ways. For example, "5 times x" can be written as $5 \cdot x$, $5 \times x$, $5(x)$, or simply $5x$.

To evaluate an algebraic expression, we **substitute** a number for each variable in the expression.

EXAMPLE 1   Evaluate each expression for the given values.

a) $x + y$ for $x = 37$ and $y = 28$
b) $5ab$ for $a = 2$ and $b = 3$

SOLUTION

a) We substitute 37 for x and 28 for y and carry out the addition:

$$x + y = 37 + 28 = 65.$$

The number 65 is called the **value** of the expression.

b) We substitute 2 for a and 3 for b and multiply:

$$5ab = 5 \cdot 2 \cdot 3 = 10 \cdot 3 = 30.$$

EXAMPLE 2   The area A of a rectangle of length l and width w is given by the formula $A = lw$. Find the area when l is 24.5 in. and w is 10 in.

SOLUTION   We evaluate, using 24.5 in. for l and 10 in. for w and carry out the multiplication:

$$A = lw = (24.5 \text{ in.})(10 \text{ in.})$$
$$= (24.5)(10)(\text{in.})(\text{in.})$$
$$= 245 \text{ in}^2, \text{ or } 245 \text{ square inches.}$$

Note that $(\text{in.})(\text{in.}) = \text{in}^2$. Exponents are discussed in detail in Section 1.8.

EXAMPLE 3   The area of a triangle with a base of length b and a height of length h is given by the formula $A = \frac{1}{2}bh$. Find the area when b is 8 m and h is 6.4 m.

SOLUTION   We substitute 8 m for b and 6.4 m for h and then multiply:

$$A = \tfrac{1}{2}bh = \tfrac{1}{2}(8 \text{ m})(6.4 \text{ m})$$
$$= \tfrac{1}{2}(8)(6.4)(\text{m})(\text{m})$$
$$= 4(6.4) \text{ m}^2$$
$$= 25.6 \text{ m}^2, \text{ or } 25.6 \text{ square meters.}$$

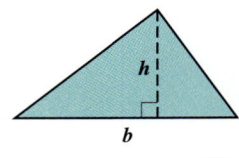

Translating to Algebraic Expressions

Before attempting to translate problems to equations, we need to be able to translate certain phrases to algebraic expressions.

KEY WORDS

Addition (+)	Subtraction (−)	Multiplication (·)	Division (÷)
add	subtract	multiply	divide
sum	difference	product	quotient
plus	minus	times	divided by
more than	less than	twice	ratio of
increased by	decreased by	of	per

EXAMPLE 4   Translate each phrase to an algebraic expression.

a) Seven less than some number
b) Eighteen more than a number
c) A number divided by five

SOLUTION   To help think through a translation, we sometimes begin with a specific number in place of a variable.

a) If the number were 19, then seven less than 19 would mean $19 - 7$. If the number were 50, the translation would be $50 - 7$. If we use x to represent "some number," the translation of "Seven less than some number" is

$x - 7.$

b) If we knew the number to be 10, the translation would be $10 + 18$, or $18 + 10$. We let t represent "a number," so the translation of "Eighteen more than a number" is

$t + 18,$ or $18 + t.$

c) We let m represent "a number." If the number were 9, the translation would be $9 \div 5$, or $\tfrac{9}{5}$. Thus our translation of "A number divided by five" is

$m \div 5,$ or $\dfrac{m}{5}.$

> **CAUTION!** Because the order in which we subtract and divide affects the answer, answering $7 - x$ or $5 \div m$ in Examples 4(a) and 4(c) is incorrect.

EXAMPLE 5   Translate each of the following.

a) Some number, increased by five
b) Half of a number
c) Five more than twice some number
d) Six less than the product of two numbers
e) Seventy-six percent of some number

SOLUTION

Phrase	*Algebraic Expression*
a) Some number, increased by five	$n + 5$, or $5 + n$
b) Half of a number	$\frac{1}{2}t$, or $\frac{t}{2}$
c) Five more than twice some number	$2x + 5$
d) Six less than the product of two numbers	$mn - 6$
e) Seventy-six percent of some number	76% of z, or $0.76z$

Translating to Equations

The symbol = ("equals") indicates that the expressions on either side of the equals sign represent the same number. An **equation** is a number sentence with the verb =. Equations may be true, false, or neither true nor false.

EXAMPLE 6   Determine whether each equation is true, false, or neither.

a) $3 + 2 = 5$ b) $7 - 2 = 4$ c) $x + 6 = 13$

SOLUTION

a) $3 + 2 = 5$ The equation is *true*.
b) $7 - 2 = 4$ The equation is *false*.
c) $x + 6 = 13$ The equation is *neither* true nor false, because we do not know what number x represents.

SOLUTION

> A replacement or substitution that makes an equation true is called a *solution*. Some equations have more than one solution, and some have no solution. When all solutions have been found, we have *solved* the equation.

To determine whether a number is a solution of an equation, we evaluate the expressions on each side of the equation. If the values are the same, then the number is a solution.

EXAMPLE 7 Determine whether 7 is a solution of $x + 6 = 13$.

SOLUTION

$x + 6 = 13$ 　　Writing the equation
$7 + 6 \;?\; 13$ 　　Substituting 7 for x
$13 \;|\; 13$ 　　$13 = 13$ is TRUE.

Since the left-hand and the right-hand sides are the same, we have a solution. No other number makes the equation true, so the only solution is the number 7.

Although we do not study solving equations until Chapter 2, we can translate certain problem situations to equations now.

EXAMPLE 8 Translate the following problem to an equation.

What number plus 478 is 1019?

SOLUTION We let y represent the unknown number. The translation then comes almost directly from the English sentence.

What number plus 478 is 1019?

$y \quad + \quad 478 = 1019$

Note that "is" translates to "=" and "plus" translates to "+."

Sometimes it helps to reword a problem before translating.

EXAMPLE 9 Translate the following problem to an equation.

The elevation of Denver, 5280 ft, is 88 times the elevation of Houston. What is the elevation of Houston?

SOLUTION We let h represent the elevation of Houston. The rewording and translation follow:

Rewording: 　88 times the elevation of Houston is 5280.

Translating: 　88 　·　 　　　h　　　 　= 5280

EXERCISE SET 1.1

Evaluate.
1. $5a$, for $a = 9$
2. $9x$, for $x = 7$
3. $t + 6$, for $t = 2$
4. $13 - r$, for $r = 9$
5. $\dfrac{x + y}{2}$, for $x = 2$ and $y = 14$
6. $\dfrac{p + q}{5}$, for $p = 15$ and $q = 20$
7. $\dfrac{m - n}{7}$, for $m = 20$ and $n = 6$
8. $\dfrac{x - y}{6}$, for $x = 23$ and $y = 5$
9. $\dfrac{a}{b}$, for $a = 45$ and $b = 9$
10. $\dfrac{m}{n}$, for $m = 54$ and $n = 9$
11. $\dfrac{9m}{q}$, for $m = 6$ and $q = 18$
12. $\dfrac{5z}{y}$, for $z = 9$ and $y = 15$

Substitute to find the value of each expression.

13. *Distance driven.* Traveling r miles per hour for t hours, a driver will travel rt miles. How far will Yvonne travel in 3 hr at a speed of 55 mph?

14. *Orbit time.* A communications satellite orbiting 300 mi above the earth travels about 27,000 mi in one orbit. The time, in hours, for an orbit is
$$\dfrac{27{,}000}{v},$$
where v is the velocity, in miles per hour. How long will an orbit take at a velocity of 1125 mph?

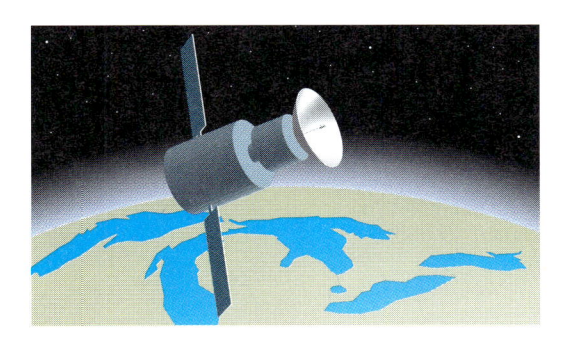

15. *Area of a parallelogram.* The area of a parallelogram with base b and height h is bh. Find the area of the parallelogram when the height is 15.4 cm (centimeters) and the base is 6.5 cm.

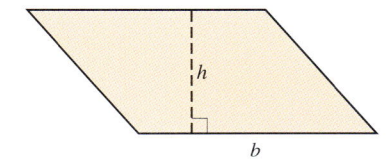

16. *Olympic softball.* A softball player's batting average is h/a, where h is the number of hits and a is the number of "at bats." In the 1996 Summer Olympics, Dot Richardson had 9 hits in 33 at bats. What was her batting average?

17. *Zoology.* A great white shark has triangular teeth. Each tooth measures about 5 cm across the base and has a height of 6 cm. Find the surface area of the front side of one such tooth. (See Example 3.)

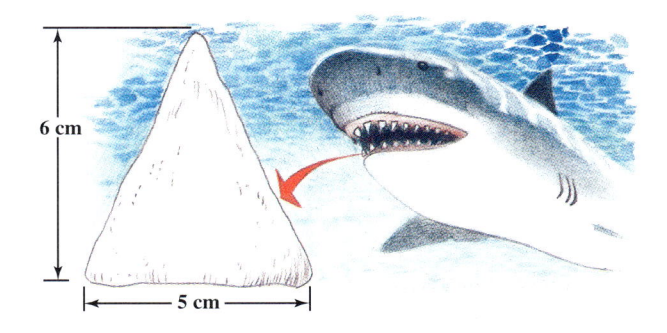

18. *Work time.* Enrico takes five times as long to do a job as Rosa does. Suppose t represents the time it takes Rosa to do the job. Then $5t$ represents the time it takes Enrico. How long does it take Enrico if Rosa takes 30 sec? 90 sec? 2 min?

Translate to an algebraic expression.

19. Seven times x
20. The product of 4 and a
21. 6 more than b
22. 8 more than t
23. 9 less than c
24. 4 less than d
25. 6 increased by q
26. 11 increased by z
27. 9 more than a number
28. c more than d
29. x less than y
30. 2 less than a number
31. x divided by w
32. The quotient of two numbers
33. m subtracted from n
34. p subtracted from q
35. The sum of two numbers
36. The sum of d and f
37. The product of 9 and twice m
38. t minus twice r
39. One quarter of some number
40. One third of the product of two numbers
41. 64% of some number
42. 38% of a number
43. Lita had $50 before paying x dollars for a pizza. How much remains?
44. Dino drove his pickup truck at 65 mph for t hours. How far did he go?

Determine whether the given number is a solution of the given equation.

45. 15; $x + 17 = 32$
46. 75; $y + 28 = 93$
47. 93; $a - 28 = 75$
48. 12; $8t = 96$
49. 63; $\dfrac{t}{7} = 9$
50. 52; $\dfrac{x}{8} = 6$
51. 3; $\dfrac{108}{x} = 36$
52. 7; $\dfrac{94}{y} = 12$

Translate each problem to an equation. Do not solve.

53. What number added to 73 is 201?
54. Seven times what number is 2303?
55. When 42 is multiplied by a number, the result is 2352. Find the number.
56. When 345 is added to a number, the result is 987. Find the number.
57. *Games.* A game board has 64 squares. If you win 35 squares and your opponent wins the rest, how many does your opponent get?
58. *Hours worked.* A carpenter charges $25 an hour. How many hours did she work in order to make $53,400?
59. *Travel to work.* In the Northeast, the average commute to work is 24.5 min. The average commuting time in the West is 1.8 min less. How long is the average commute in the West?
60. *Fuel consumption.* Between 1975 and 1995, the amount of fuel used by trucks in the United States doubled. If 31.4 billion gallons were used in 1975, how much was used in 1995?

SYNTHESIS

To the student and the instructor: Synthesis exercises *are designed to challenge students to extend the concepts or skills studied in each section. Some synthesis exercises will require the assimilation of skills and concepts from several sections.*

Writing exercises, denoted by ◆, *should be answered using one or more complete English sentences. In most sections, four writing exercises appear as the first synthesis exercises. These exercises are not as challenging as those exercises appearing later in the exercise set and can be assigned to students who might not otherwise attempt synthesis exercises. Because answers to many writing exercises will vary, solutions are not listed in the answer section.*

61. ◆ If the length of a rectangle is doubled, does its area double? Why or why not?
62. ◆ How does a variable expression differ from a variable, and how does it differ from an equation?
63. ◆ Write a problem, in words, that translates to the equation $35x = 840$.
64. ◆ Write a problem, in words, that translates to the equation $1989 + t = 2003$.

Translate to an algebraic expression.

65. 5 less than three times a number
66. One third of one half of the product of two numbers
67. The perimeter of a rectangle with length l and width w (perimeter means distance around)

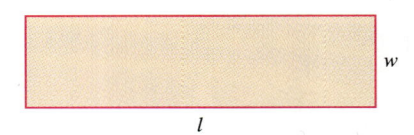

68. The perimeter of a square with side s (perimeter means distance around)

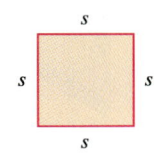

69. Ray's age in 7 yr if he is 2 yr older than Monique and Monique is a years old

70. Signs of Distinction charges $90 per square foot for handpainted signs. The town of Belmar commissioned a triangular sign with a base of 3 ft and a height of 2.5 ft. How much will the sign cost?

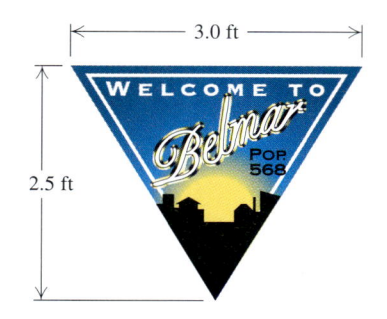

71. Find the area that is shaded.

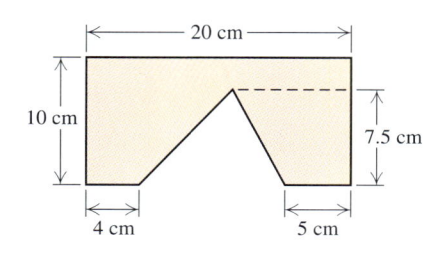

72. Evaluate $\dfrac{x+y}{4}$ when $y = 8$ and x is twice y.

73. Evaluate $\dfrac{x-y}{7}$ when $y = 35$ and x is twice y.

74. Evaluate $\dfrac{y-x}{3}$ when $x = 9$ and y is three times x.

75. Evaluate $\dfrac{y+x}{2} + \dfrac{3 \cdot y}{x}$ for $x = 2$ and $y = 4$.

Answer each question with an algebraic expression.

76. If $w + 3$ is a whole number, what is the next whole number after it?

77. If $d + 2$ is an odd number, what is the preceding odd number?

78. If $d + 2$ is an even number, what is the preceding even number?

1.2 The Commutative, Associative, and Distributive Laws

Equivalent Expressions • The Commutative Laws • The Associative Laws • The Distributive Law • The Distributive Law and Factoring

In order to solve equations, it is important to be able to manipulate algebraic expressions. The commutative, associative, and distributive laws discussed in this section enable us to write *equivalent expressions* that can streamline our work.

Equivalent Expressions

The expressions $4 + 4 + 4$, $3 \cdot 4$, and $4 \cdot 3$ all represent the same number, 12. Expressions that represent the same number are said to be **equivalent**.

The equivalent expressions $t + 18$ and $18 + t$ were used in Section 1.1 when we translated "eighteen more than a number." To check that these expressions are equivalent, we make some choices for t:

When $t = 3$, $t + 18 = 3 + 18 = 21$ and $18 + t = 18 + 3 = 21$.
When $t = 40$, $t + 18 = 40 + 18 = 58$ and $18 + t = 18 + 40 = 58$.

The Commutative Laws

We have seen that changing the order in addition or multiplication does not change the result. Equations like $3 + 18 = 18 + 3$ and $5 \cdot 4 = 4 \cdot 5$ illustrate this idea and show that addition and multiplication are **commutative**.

THE COMMUTATIVE LAWS

For Addition. For any numbers a and b,

$a + b = b + a$.

(We can change the order when adding without affecting the answer.)

For Multiplication. For any numbers a and b,

$ab = ba$.

(We can change the order when multiplying without affecting the answer.)

EXAMPLE 1   Use the commutative laws to write an expression equivalent to each of the following: **(a)** $y + 5$; **(b)** $9x$; **(c)** $7 + ab$.

SOLUTION

a) An expression equivalent to $y + 5$ is $5 + y$ by the commutative law of addition.
b) An expression equivalent to $9x$ is $x9$ by the commutative law of multiplication.
c) An expression equivalent to $7 + ab$ is $ab + 7$ by the commutative law of *addition*.

A second expression equivalent to $7 + ab$ is $7 + ba$ by the commutative law of *multiplication*.

Also equivalent to $7 + ab$ is $ba + 7$ by both commutative laws.

The Associative Laws

Parentheses are used to indicate groupings. We normally simplify within the parentheses first. For example,

$3 + (8 + 4) = 3 + 12 = 15$

1.2 THE COMMUTATIVE, ASSOCIATIVE, AND DISTRIBUTIVE LAWS

and

$$(3 + 8) + 4 = 11 + 4 = 15.$$

Similarly,

$$4(2 \cdot 3) = 4(6) = 24$$

and

$$(4 \cdot 2)3 = (8)3 = 24.$$

Note that, so long as only addition or only multiplication appears in an expression, changing the grouping does not change the result. Equations such as $3 + (8 + 4) = (3 + 8) + 4$ and $4(2 \cdot 3) = (4 \cdot 2)3$ illustrate that addition and multiplication are **associative**.

THE ASSOCIATIVE LAWS

For Addition. For any numbers a, b, and c,

$$a + (b + c) = (a + b) + c.$$

(Numbers can be grouped in any manner for addition.)

For Multiplication. For any numbers a, b, and c,

$$a \cdot (b \cdot c) = (a \cdot b) \cdot c.$$

(Numbers can be grouped in any manner for multiplication.)

EXAMPLE 2 Use an associative law to write an expression equivalent to each of the following: **(a)** $y + (z + 3)$; **(b)** $(8x)y$.

SOLUTION

a) An expression equivalent to $y + (z + 3)$ is $(y + z) + 3$ by the associative law of addition.
b) An expression equivalent to $(8x)y$ is $8(xy)$ by the associative law of multiplication.

When only additions or only multiplications are involved, parentheses can be placed any way we please. For that reason, we sometimes omit them altogether. Thus,

$$x + (y + 7) = x + y + 7, \quad \text{and} \quad l(wh) = lwh.$$

A sum such as $(5 + 1) + (3 + 5) + 9$ can be simplified by pairing numbers that add to 10. The associative and commutative laws allow us to do this:

$$(5 + 1) + (3 + 5) + 9 = 5 + 5 + 9 + 1 + 3$$
$$= 10 + 10 + 3 = 23.$$

EXAMPLE 3 Use the commutative and/or associative laws of addition to write at least two expressions equivalent to $(x + 5) + y$.

SOLUTION

a) $(x + 5) + y = x + (5 + y)$ Using the associative law; $x + (5 + y)$ is one equivalent expression.
$= x + (y + 5)$ Using the commutative law

b) $(x + 5) + y = y + (x + 5)$ Using the commutative law; $y + (x + 5)$ is one equivalent expression.
$= y + (5 + x)$ Using the commutative law again

EXAMPLE 4 Use the commutative and/or associative laws of multiplication to rewrite $2(x3)$ as $6x$. Show and give reasons for each step.

SOLUTION

$2(x3) = 2(3x)$ Using the commutative law
$= (2 \cdot 3)x$ Using the associative law
$= 6x$ Simplifying

The Distributive Law

The *distributive law* is probably the single most important law for manipulating algebraic expressions. Unlike the commutative and associative laws, the distributive law uses multiplication together with addition.

You have already used the distributive law although you may not have realized it at the time. To illustrate, try to multiply $3 \cdot 21$ mentally. Many people find the product, 63, by thinking of 21 as $20 + 1$ and then multiplying 20 by 3 and 1 by 3. The sum of the two products, $60 + 3$, is 63. Note that if the 3 does not multiply both 20 and 1, the result will not be correct.

EXAMPLE 5 Compute in two ways: $4(3 + 2)$.

SOLUTION

a) As in the multiplication of $3(20 + 1)$ above, to compute $4(3 + 2)$, we can multiply both 3 and 2 by 4 and add the results:

$4(3 + 2) = 4 \cdot 3 + 4 \cdot 2$ Multiplying both 3 and 2 by 4
$= 12 + 8 = 20.$ Adding

b) By first adding inside the parentheses, we get the same result in a different way:

$4(3 + 2) = 4(5)$ Adding; $3 + 2 = 5$
$= 20.$ Multiplying

THE DISTRIBUTIVE LAW

For any numbers a, b, and c,

$a(b + c) = ab + ac.$

(The product of a number and a sum can be written as the sum of two products.)

1.2 THE COMMUTATIVE, ASSOCIATIVE, AND DISTRIBUTIVE LAWS

EXAMPLE 6 Multiply: $3(x + 2)$.

SOLUTION Since $x + 2$ cannot be simplified unless a value for x is given, we use the distributive law:

$$3(x + 2) = 3x + 3 \cdot 2 \quad \text{Using the distributive law}$$
$$= 3x + 6.$$

In the expression $x + 2$, the parts separated by the plus sign are called **terms**.* The distributive law can also be used when more than two terms are inside the parentheses.

EXAMPLE 7 Multiply: $6(s + 2 + 5w)$.

SOLUTION

$$6(s + 2 + 5w) = 6s + 6 \cdot 2 + 6 \cdot 5w \quad \text{Using the distributive law}$$
$$= 6s + 12 + (6 \cdot 5)w \quad \text{Using the associative law for multiplication}$$
$$= 6s + 12 + 30w$$

Because of the commutative law of multiplication, the distributive law can be used on the "right": $(b + c)a = ba + ca$.

EXAMPLE 8 Multiply: $(c + 4)5$.

SOLUTION

$$(c + 4)5 = c \cdot 5 + 4 \cdot 5 \quad \text{Using the distributive law on the right}$$
$$= 5c + 20$$

> **CAUTION!** To use the distributive law for removing parentheses, multiply *each* term inside the parentheses by the multiplier outside:
> $$a(b + c) \neq ab + c.$$

The Distributive Law and Factoring

If we reverse the statement of the distributive law, we have the basis of a process called **factoring**: $ab + ac = a(b + c)$. To **factor** an expression means to write an equivalent expression that is a product. The parts of the product are called **factors**. Note that "factor" can be used as either a verb or a noun.

EXAMPLE 9 Use the distributive law to factor each of the following.

a) $3x + 3y$ **b)** $7x + 21y + 7$

---

*Terms are discussed in greater detail in Sections 1.5, 1.6, and 1.8.

14 CHAPTER 1 • INTRODUCTION TO ALGEBRAIC EXPRESSIONS

SOLUTION

a) By the distributive law,

$$3x + 3y = 3(x + y).$$ The *common factor* is 3.

b) $7x + 21y + 7 = 7 \cdot x + 7 \cdot 3y + 7 \cdot 1$ The common factor is 7.
$$= 7(x + 3y + 1)$$ Using the distributive law

Be sure not to omit the 1 or the common factor, 7.

To check our factoring, we multiply to see if the original expression is obtained. For example, to check the **factorization** in Example 9(b), note that

$$7(x + 3y + 1) = 7x + 7 \cdot 3y + 7 \cdot 1$$
$$= 7x + 21y + 7.$$

EXERCISE SET 1.2

Use the commutative law of addition to write an equivalent expression.

1. $3 + x$
2. $a + 2$
3. $a + bc$
4. $x + 3y$
5. $9x + 3y$
6. $3a + 7b$
7. $2(a + 3)$
8. $9(x + 5)$

Use the commutative law of multiplication to write an equivalent expression.

9. $8 \cdot a$
10. xy
11. st
12. $7b$
13. $5 + ab$
14. $x + 3y$
15. $2(a + 3)$
16. $9(x + 5)$

Use the associative law of addition to write an equivalent expression.

17. $(a + 3) + b$
18. $(5 + m) + r$
19. $r + (t + 9)$
20. $x + (2 + y)$
21. $(ab + c) + d$
22. $(m + np) + r$

Use the associative law of multiplication to write an equivalent expression.

23. $(3x)y$
24. $(9a)b$
25. $4(xy)$
26. $9(rp)$
27. $3[2(a + b)]$
28. $5[x(2 + y)]$

Use the commutative and/or associative laws to write two equivalent expressions.

29. $r + (t + 6)$
30. $5 + (v + w)$
31. $(5a)b$
32. $x(3y)$

Use the commutative and/or associative laws to rewrite each of the following. Label each step with a reason, as in Example 4.

33. $(7 + x) + 2$ as $x + 9$
34. $(2a)4$ as $8a$
35. $(m3)7$ as $21m$
36. $4 + (9 + x)$ as $x + 13$

Multiply.

37. $3(a + 4)$
38. $4(x + 3)$
39. $6(1 + x)$
40. $6(v + 4)$
41. $3(x + 1)$
42. $9(x + 3)$
43. $4(1 + y)$
44. $7(s + 5)$
45. $9(2x + 6)$
46. $9(6m + 7)$
47. $5(r + 2 + 3t)$
48. $4(5x + 8 + 3p)$
49. $(a + b)2$
50. $(x + 2)7$
51. $(x + y + 2)5$
52. $(2 + a + b)6$

Use the distributive law to factor each of the following. Check by multiplying.

53. $7x + 7z$
54. $5y + 5z$
55. $5 + 5y$
56. $13 + 13x$
57. $18x + 3y$
58. $5x + 20y$
59. $5x + 10 + 15y$
60. $3 + 27b + 6c$
61. $12x + 9$
62. $6x + 6$

63. $9x + 3y$
64. $15x + 5y$
65. $2a + 16b + 64$
66. $5 + 20x + 35y$
67. $11x + 44y + 121$
68. $7 + 14b + 56w$

SKILL MAINTENANCE

To the student and the instructor: Skill maintenance exercises *review skills studied in earlier sections. Often these exercises serve as preparation for a following section.*

Translate to an algebraic expression.

69. 9 less than t
70. Half of m

SYNTHESIS

71. ◈ Are subtraction and division associative? Why or why not?

72. ◈ Are subtraction and division commutative? Why or why not?

73. ◈ Are terms and factors the same thing? Why or why not?

74. ◈ When Kara multiplied $7(1 + a + 2c)$, she got $7 + 7a + 14c$. Kara claimed she never used the commutative or associative laws to get this result. Is this possible? Why or why not?

Tell whether the expressions in each pairing are equivalent. Then explain why or why not.

75. $8 + 4(a + b)$ and $4(2 + a + b)$
76. $7 \div 3m$ and $m3 \div 7$
77. $(rt + st)5$ and $5t(r + s)$
78. $yax + ax$ and $xa(1 + y)$
79. $30y + x15$ and $5[2(x + 3y)]$
80. $[c(2 + 3b)]5$ and $10c + 15bc$

81. ◈ Evaluate the expressions $3(2 + x)$ and $6 + x$ for $x = 0$. Do your results indicate that $3(2 + x)$ and $6 + x$ are equivalent? Why or why not?

82. ◈ Factor $15x + 40$. Then evaluate both $15x + 40$ and the factorization for $x = 4$. Do your results *guarantee* that the factorization is correct? Why or why not? (*Hint:* See Exercise 81.)

COLLABORATIVE CORNER

Focus: Application of commutative and associative laws

Time: 15–20 minutes

Group size: 2–3

Legend has it that while still in grade school, the mathematician Carl Friedrich Gauss (1777–1855) was able to add the numbers from 1 to 100 mentally. Gauss did not add them sequentially, but rather paired 1 with 99, 2 with 98, and so on.

ACTIVITY

1. Use a method similar to Gauss's to simplify the following:

   $1 + 2 + 3 + 4 + 5 + 6 + 7 + 8 + 9 + 10.$

   At least one group member should add from left to right as a check.

2. How were the associative and commutative laws applied in part (1) above?

3. Use Gauss's method to find the sum of the first 100 counting numbers:

   $1 + 2 + 3 + \cdots + 48 + 49 + 50 + 51 + 52 + \cdots + 97 + 98 + 99 + 100.$

   How many pairs of numbers adding to 100 are there? Is any number left over? What is the total sum?

4. Find the sum of the first 200 counting numbers. What will the pairs of numbers add to? How many pairs will there be?

5. Compare your group's answers with those of another group.

1.3 Fractional Notation

Factors and Prime Factorizations • Fractional Notation • Multiplication and Simplification • Canceling • Addition, Subtraction, and Division

This section covers multiplication, addition, subtraction, and division with fractional notation. Although much of this may be review, note that fractional expressions that contain variables are also included.

Factors and Prime Factorizations

In order to be able to study addition and subtraction using fractional notation, we first review how *natural numbers* are factored. **Natural numbers** can be thought of as the counting numbers:

1, 2, 3, 4, 5, ...*

(The dots indicate that the established pattern continues without ending.) To factor a number, we simply express it as a product of two or more numbers.

EXAMPLE 1 Write several factorizations of 12. Then list all factors of 12.

SOLUTION   The number 12 can be factored in several ways:

1 · 12,   2 · 6,   3 · 4,   2 · 2 · 3.

The factors of 12 are 1, 2, 3, 4, 6, and 12.

Some numbers have only two factors, the number itself and 1. Such numbers are called **prime**.

PRIME NUMBER

A *prime number* is a natural number that has exactly two different factors: the number itself and 1.

EXAMPLE 2 Which of these numbers are prime?   7, 4, 1

SOLUTION

7 is prime. It has exactly two different factors, 7 and 1.

4 is not prime. It has three different factors, 1, 2, and 4.

1 is not prime. It does not have two *different* factors.

---

*A similar collection of numbers, the **whole numbers,** includes 0: 0, 1, 2, 3, ...

If a natural number, other than 1, is not prime, we call it **composite**. Every composite number can be factored into a product of prime numbers. Such a factorization is called the **prime factorization** of that composite number.

EXAMPLE 3 Find the prime factorization of 36.

SOLUTION We first factor 36 in any way that we can. One way is like this:

$$36 = 4 \cdot 9.$$

The factors 4 and 9 are not prime, so we factor them:

$$36 = 4 \cdot 9 = 2 \cdot 2 \cdot 3 \cdot 3. \quad \text{2 and 3 are both prime.}$$

The prime factorization of 36 is $2 \cdot 2 \cdot 3 \cdot 3$.

Fractional Notation

An example of **fractional notation** for a number is

$$\frac{2}{3} \begin{matrix} \longleftarrow \text{Numerator} \\ \longleftarrow \text{Denominator} \end{matrix}$$

The top number is called the **numerator**, and the bottom number is called the **denominator**. When the numerator and the denominator are the same nonzero number, we have fractional notation for the number 1.

FRACTIONAL NOTATION FOR 1

For any number a, except 0,

$$\frac{a}{a} = 1.$$

(Any nonzero number divided by itself is 1.)

Multiplication and Simplification

Recall from arithmetic that fractions are multiplied as follows.

MULTIPLICATION OF FRACTIONS

For any two fractions a/b and c/d,

$$\frac{a}{b} \cdot \frac{c}{d} = \frac{ac}{bd}.$$

(The numerator of the product is the product of the two numerators. The denominator of the product is the product of the two denominators.)

CHAPTER 1 • INTRODUCTION TO ALGEBRAIC EXPRESSIONS

EXAMPLE 4 Multiply: (a) $\frac{2}{3} \cdot \frac{7}{5}$; (b) $\frac{4}{x} \cdot \frac{8}{y}$.

SOLUTION We multiply numerators as well as denominators.

a) $\frac{2}{3} \cdot \frac{7}{5} = \frac{2 \cdot 7}{3 \cdot 5} = \frac{14}{15}$

b) $\frac{4}{x} \cdot \frac{8}{y} = \frac{4 \cdot 8}{x \cdot y} = \frac{32}{xy}$

When one of the fractions being multiplied is 1, multiplying yields an equivalent expression because of the *identity property of* 1.

THE IDENTITY PROPERTY OF 1

For any number a,

$$a \cdot 1 = a.$$

(Multiplying a number by 1 gives that same number.)

EXAMPLE 5 Multiply: $\frac{4}{5} \cdot \frac{6}{6}$.

SOLUTION We have

$$\frac{4}{5} \cdot \frac{6}{6} = \frac{4 \cdot 6}{5 \cdot 6} = \frac{24}{30}.$$

Since $\frac{6}{6} = 1$, the expression $\frac{4}{5} \cdot \frac{6}{6}$ is equivalent to $\frac{4}{5} \cdot 1$, or simply $\frac{4}{5}$. Thus, $\frac{24}{30}$ is equivalent to $\frac{4}{5}$.

The steps of Example 5 are reversed by "removing a factor equal to 1" —in this case, $\frac{6}{6}$. By removing a factor that equals 1, we can *simplify* an expression like $\frac{24}{30}$ to an equivalent expression like $\frac{4}{5}$.

To simplify, we factor the numerator and the denominator, looking for the largest factor common to both. This is sometimes made easier by writing the prime factorizations. After identifying common factors, we can express the fraction as a product of two fractions, one of which is in the form a/a.

EXAMPLE 6 Simplify: (a) $\frac{15}{40}$; (b) $\frac{36}{24}$.

SOLUTION

a) Note that 5 is a factor of both 15 and 40:

$\frac{15}{40} = \frac{3 \cdot 5}{8 \cdot 5}$ Factoring the numerator and the denominator, using the common factor, 5

$= \frac{3}{8} \cdot \frac{5}{5}$ Rewriting as a product of two fractions; $\frac{5}{5} = 1$

$= \frac{3}{8} \cdot 1 = \frac{3}{8}.$ Using the identity property of 1 (removing a factor equal to 1)

b) $\dfrac{36}{24} = \dfrac{2 \cdot 2 \cdot 3 \cdot 3}{2 \cdot 2 \cdot 2 \cdot 3}$ Writing the prime factorizations and identifying common factors

$= \dfrac{3}{2} \cdot \dfrac{2 \cdot 2 \cdot 3}{2 \cdot 2 \cdot 3}$ Rewriting as a product of two fractions; $\dfrac{2 \cdot 2 \cdot 3}{2 \cdot 2 \cdot 3} = 1$

$= \dfrac{3}{2} \cdot 1 = \dfrac{3}{2}$ Using the identity property of 1

It is always wise to check your result to see if any common factors of the numerator and the denominator remain. (This will never happen if prime factorizations are used correctly.) If common factors remain, repeat the process and simplify your result.

Canceling

Canceling is a shortcut that you may have used for removing a factor equal to 1 when working with fractional notation. With *great* concern, we mention it as a possible way to speed up your work. You should use canceling only when removing common factors in numerators and denominators. Canceling should *not* be used in sums or differences. Our concern is that "canceling" be done with care and understanding. Example 6(b) might have been done faster as follows:

$$\dfrac{36}{24} = \dfrac{\cancel{2} \cdot \cancel{2} \cdot 3 \cdot 3}{\cancel{2} \cdot \cancel{2} \cdot 2 \cdot 3} = \dfrac{3}{2}, \quad \text{or} \quad \dfrac{36}{24} = \dfrac{3 \cdot \cancel{12}}{2 \cdot \cancel{12}} = \dfrac{3}{2}, \quad \text{or} \quad \dfrac{\overset{3}{\cancel{\underset{18}{36}}}}{\underset{2}{\cancel{\underset{12}{24}}}} = \dfrac{3}{2}$$

CAUTION! Unfortunately, canceling is often performed incorrectly:

$$\dfrac{2 + \cancel{3}}{\cancel{2}} = 3, \quad \dfrac{4 - \cancel{1}}{4 - \cancel{2}} = \dfrac{1}{2}, \quad \dfrac{1\cancel{5}}{\cancel{5}4} = \dfrac{1}{4}.$$

  Wrong!   Wrong!   Wrong!

$$\dfrac{2 + 3}{2} = \dfrac{5}{2}, \quad \dfrac{4 - 1}{4 - 2} = \dfrac{3}{2}, \quad \dfrac{15}{54} = \dfrac{5 \cdot 3}{18 \cdot 3} = \dfrac{5}{18}.$$

In each of these situations, the expressions canceled out were *not* factors. Factors are parts of products. For example, in 2 · 3, the numbers 2 and 3 are factors, but in 2 + 3, 2 and 3 are *not* factors. **If you can't factor, you can't cancel! If in doubt, don't cancel!**

The identity property of 1 allows us to use 1 as a factor in the numerator and the denominator whenever it is convenient.

EXAMPLE 7 Simplify: $\dfrac{9}{72}$.

SOLUTION

$$\frac{9}{72} = \frac{1 \cdot 9}{8 \cdot 9}$$ Factoring and using the identity property of 1 to write 9 as $1 \cdot 9$

$$= \frac{1 \cdot 9}{8 \cdot 9} = \frac{1}{8}$$ Simplifying by removing a factor equal to 1: $\frac{9}{9} = 1$

Addition, Subtraction, and Division

When denominators are the same, fractions are added or subtracted by adding or subtracting numerators and keeping the same denominator.

ADDITION AND SUBTRACTION OF FRACTIONS

For any two fractions a/d and b/d,

$$\frac{a}{d} + \frac{b}{d} = \frac{a+b}{d} \quad \text{and} \quad \frac{a}{d} - \frac{b}{d} = \frac{a-b}{d}.$$

EXAMPLE 8   Add and simplify: $\frac{4}{8} + \frac{5}{8}$.

SOLUTION   The common denominator is 8. We add the numerators and keep the common denominator:

$$\frac{4}{8} + \frac{5}{8} = \frac{4+5}{8} = \frac{9}{8}.$$

In arithmetic, we often write $1\frac{1}{8}$ rather than the "improper" fraction $\frac{9}{8}$. In algebra, the notation $\frac{9}{8}$ is more useful and is quite "proper" for our purposes.

When denominators are different, we use the identity property of 1 and multiply to find a common denominator.

EXAMPLE 9   Add or subtract as indicated: **(a)** $\frac{7}{8} + \frac{5}{12}$; **(b)** $\frac{9}{8} - \frac{4}{5}$.

SOLUTION

a) The number 24 is divisible by both 8 and 12. We multiply both $\frac{7}{8}$ and $\frac{5}{12}$ by suitable forms of 1 to obtain two fractions with denominators of 24:

$$\frac{7}{8} + \frac{5}{12} = \frac{7}{8} \cdot \frac{3}{3} + \frac{5}{12} \cdot \frac{2}{2}$$ Multiplying by 1. Since $3 \cdot 8 = 24$, we multiply $\frac{7}{8}$ by $\frac{3}{3}$. Since $2 \cdot 12 = 24$, we multiply $\frac{5}{12}$ by $\frac{2}{2}$.

$$= \frac{21}{24} + \frac{10}{24}$$ Performing the multiplications

$$= \frac{31}{24}.$$ Adding fractions

b) $\dfrac{9}{8} - \dfrac{4}{5} = \dfrac{9}{8} \cdot \dfrac{5}{5} - \dfrac{4}{5} \cdot \dfrac{8}{8}$ Using 40 as a common denominator

$\phantom{b) \dfrac{9}{8} - \dfrac{4}{5}} = \dfrac{45}{40} - \dfrac{32}{40} = \dfrac{13}{40}$ Subtracting fractions

Two numbers whose product is 1 are called **reciprocals**, or **multiplicative inverses,** of each other. All numbers, except zero, have reciprocals. For example,

the reciprocal of $\tfrac{2}{3}$ is $\tfrac{3}{2}$ because $\tfrac{2}{3} \cdot \tfrac{3}{2} = \tfrac{6}{6} = 1$;

the reciprocal of 9 is $\tfrac{1}{9}$ because $9 \cdot \tfrac{1}{9} = \tfrac{9}{9} = 1$; and

the reciprocal of $\tfrac{1}{4}$ is 4 because $\tfrac{1}{4} \cdot 4 = 1$.

Reciprocals are used to rewrite division as multiplication.

DIVISION OF FRACTIONS

To divide two fractions, multiply by the reciprocal of the divisor:

$$\dfrac{a}{b} \div \dfrac{c}{d} = \dfrac{a}{b} \cdot \dfrac{d}{c}.$$

EXAMPLE 10   Divide: $\dfrac{1}{2} \div \dfrac{3}{5}$.

SOLUTION

$\dfrac{1}{2} \div \dfrac{3}{5} = \dfrac{1}{2} \cdot \dfrac{5}{3}$ $\dfrac{5}{3}$ is the reciprocal of $\dfrac{3}{5}$

$\phantom{\dfrac{1}{2} \div \dfrac{3}{5}} = \dfrac{5}{6}$

After adding, subtracting, multiplying, or dividing, we may still need to simplify the answer.

EXAMPLE 11   Perform the indicated operation and simplify.

a) $\dfrac{7}{10} - \dfrac{1}{5}$ b) $\dfrac{5}{6} \cdot \dfrac{9}{25}$ c) $\dfrac{2}{3} \div 4$

SOLUTION

a) $\dfrac{7}{10} - \dfrac{1}{5} = \dfrac{7}{10} - \dfrac{1}{5} \cdot \dfrac{2}{2}$ Using 10 as the common denominator

$\phantom{\dfrac{7}{10} - \dfrac{1}{5}} = \dfrac{7}{10} - \dfrac{2}{10}$

$\phantom{\dfrac{7}{10} - \dfrac{1}{5}} = \dfrac{5}{10} = \dfrac{1 \cdot 5}{2 \cdot 5} = \dfrac{1}{2}$ Removing a factor equal to 1: $\dfrac{5}{5} = 1$

b) $\dfrac{5}{6} \cdot \dfrac{9}{25} = \dfrac{5 \cdot 9}{6 \cdot 25}$ Multiplying numerators and denominators

$= \dfrac{5 \cdot 3 \cdot 3}{2 \cdot 3 \cdot 5 \cdot 5}$ Factoring

$= \dfrac{3 \cdot 3 \cdot \cancel{5}}{2 \cdot 5 \cdot \cancel{3} \cdot \cancel{5}}$ Removing a factor equal to 1: $\dfrac{3 \cdot 5}{3 \cdot 5} = 1$

$= \dfrac{3}{10}$ Simplifying

c) $\dfrac{2}{3} \div 4 = \dfrac{2}{3} \cdot \dfrac{1}{4}$ Multiplying by the reciprocal of the divisor

$= \dfrac{\cancel{2} \cdot 1}{3 \cdot \cancel{2} \cdot 2}$ Factoring and removing a factor equal to 1: $\dfrac{2}{2} = 1$

$= \dfrac{1}{6}$

EXERCISE SET 1.3

Write at least two factorizations of each number. Then list all of the factors of the number.

1. 30
2. 70
3. 42
4. 60

Find the prime factorization of each number. If the number is prime, state this.

5. 22
6. 15
7. 30
8. 55
9. 50
10. 20
11. 27
12. 98
13. 18
14. 24
15. 40
16. 56
17. 43
18. 120
19. 210
20. 79
21. 115
22. 143

Simplify.

23. $\dfrac{10}{14}$
24. $\dfrac{16}{56}$
25. $\dfrac{49}{14}$
26. $\dfrac{72}{27}$
27. $\dfrac{6}{48}$
28. $\dfrac{12}{70}$
29. $\dfrac{56}{7}$
30. $\dfrac{132}{11}$
31. $\dfrac{19}{76}$
32. $\dfrac{17}{51}$
33. $\dfrac{100}{20}$
34. $\dfrac{150}{25}$
35. $\dfrac{75}{80}$
36. $\dfrac{42}{50}$
37. $\dfrac{120}{82}$
38. $\dfrac{75}{45}$
39. $\dfrac{210}{98}$
40. $\dfrac{140}{350}$

Perform the indicated operation and, if possible, simplify.

41. $\dfrac{3}{5} \cdot \dfrac{1}{2}$
42. $\dfrac{11}{10} \cdot \dfrac{8}{5}$
43. $\dfrac{17}{2} \cdot \dfrac{3}{4}$
44. $\dfrac{11}{12} \cdot \dfrac{12}{11}$
45. $\dfrac{1}{8} + \dfrac{3}{8}$
46. $\dfrac{1}{2} + \dfrac{1}{8}$
47. $\dfrac{4}{9} + \dfrac{13}{18}$
48. $\dfrac{4}{5} + \dfrac{8}{15}$
49. $\dfrac{3}{a} \cdot \dfrac{b}{7}$
50. $\dfrac{x}{5} \cdot \dfrac{y}{z}$
51. $\dfrac{4}{a} + \dfrac{3}{a}$
52. $\dfrac{7}{a} - \dfrac{5}{a}$
53. $\dfrac{3}{10} + \dfrac{8}{15}$
54. $\dfrac{9}{8} + \dfrac{7}{12}$
55. $\dfrac{9}{7} - \dfrac{2}{7}$
56. $\dfrac{12}{5} - \dfrac{2}{5}$
57. $\dfrac{13}{18} - \dfrac{4}{9}$
58. $\dfrac{13}{15} - \dfrac{8}{45}$
59. $\dfrac{5}{6} - \dfrac{5}{18}$
60. $\dfrac{15}{16} - \dfrac{2}{3}$
61. $\dfrac{7}{6} \div \dfrac{3}{5}$
62. $\dfrac{7}{5} \div \dfrac{3}{4}$
63. $\dfrac{8}{9} \div \dfrac{4}{15}$
64. $\dfrac{3}{4} \div 3$
65. $6 \div \dfrac{3}{7}$
66. $\dfrac{1}{10} \div \dfrac{1}{5}$
67. $\dfrac{13}{12} \div \dfrac{39}{5}$

68. $\dfrac{17}{6} \div \dfrac{3}{8}$ **69.** $24 \div \dfrac{2}{5}$ **70.** $78 \div \dfrac{1}{6}$

71. $\dfrac{3x}{4} \div 6$ **72.** $\dfrac{5}{6} \div 15a$ **73.** $\dfrac{5}{3} \div \dfrac{a}{b}$

74. $\dfrac{x}{7} \div \dfrac{4}{y}$ **75.** $\dfrac{x}{6} - \dfrac{1}{3}$ **76.** $\dfrac{9}{10} + \dfrac{x}{2}$

SKILL MAINTENANCE

Use a commutative law to write an equivalent expression. There can be more than one correct answer.

77. $5(x + 3)$ **78.** $7 + (a + b)$

SYNTHESIS

79. ◈ Under what circumstances would the sum of two fractions be easier to compute than the product of the same two fractions?

80. ◈ Under what circumstances would the product of two fractions be easier to compute than the sum of the same two fractions?

81. ◈ Use the word *factor* in two sentences—once as a noun and once as a verb.

82. ◈ Is multiplication of fractions commutative? Why or why not?

Simplify.

83. $\dfrac{256}{192}$ **84.** $\dfrac{pqrs}{qrst}$

85. $\dfrac{9 \cdot 4 \cdot 16}{8 \cdot 15 \cdot 12}$ **86.** $\dfrac{8 \cdot 9xy}{2xy \cdot 36}$

87. $\dfrac{15 \cdot 4xy \cdot 9}{6 \cdot 25x \cdot 15y}$ **88.** $\dfrac{10x \cdot 12 \cdot 25y}{2 \cdot 30x \cdot 20y}$

89. *Packaging.* Tritan Candies uses two sizes of boxes, 6 in. and 8 in. long. These are packed end to end in bigger cartons to be shipped. What is the shortest-length carton that will accommodate boxes of either size without any room left over? (Each carton must contain boxes of only one size; no mixing is allowed.)

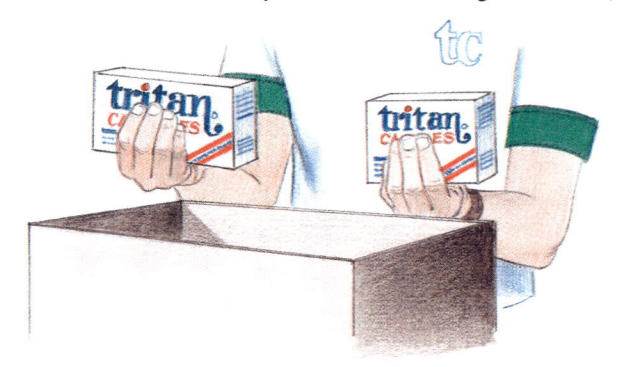

90. In the following tables, the top number should be factored in such a way that the sum of the factors is the bottom number. For example, in the first column, 56 has been factored as $7 \cdot 8$, since $7 + 8 = 15$, the bottom number. Find the missing numbers in each table.

Product	56	63	36	72	140	96
Factor	7					
Factor	8					
Sum	15	16	20	38	24	20

Product		168	110			
Factor				9	24	3
Factor	8	8		10	18	
Sum	14		21			24

Find the area of each figure.

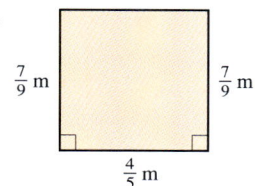

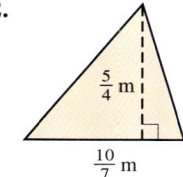

91. square with sides $\dfrac{7}{9}$ m and $\dfrac{4}{5}$ m

92. triangle with height $\dfrac{5}{4}$ m and base $\dfrac{10}{7}$ m

93. Find the perimeter of a square with sides of length $\dfrac{5}{9}$ m.

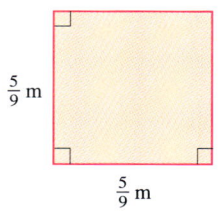

94. Find the perimeter of the rectangle in Exercise 91.

95. ◈ Make use of the properties and laws discussed in Sections 1.2 and 1.3 to explain why $x + y$ is equivalent to $(2y + 2x)/2$.

1.4 Positive and Negative Real Numbers

The Integers • The Rational Numbers • Real Numbers and Order • Absolute Value

A **set** is a collection of objects. The set containing the numbers 1, 3, and 7 is generally written {1, 3, 7}. In this section, we examine some important sets of numbers. More on sets can be found in Appendix A.

The Integers

Two sets of numbers were mentioned in Section 1.3. We represent these sets using dots on a number line.

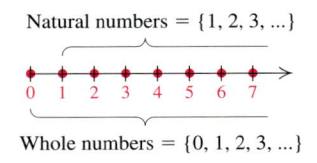

Natural numbers = {1, 2, 3, ...}

Whole numbers = {0, 1, 2, 3, ...}

To create a new set, the *integers,* we include all whole numbers, along with their *opposites*. To find the opposite of a number, we locate the number that is the same distance from 0 but on the other side of the number line. For example,

the opposite of 1 is negative 1, written -1;

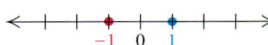

and

the opposite of 3 is negative 3, written -3.

The **integers** consist of all whole numbers and their opposites.

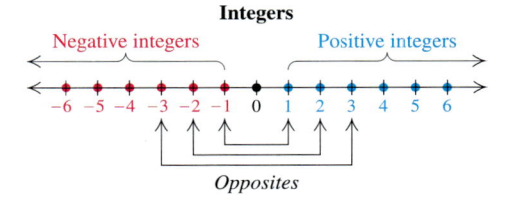

Opposites

Opposites are discussed in more detail in Section 1.6. Note that, except for 0, opposites occur in pairs. Thus, 5 is the opposite of -5, just as -5 is the opposite of 5. Note that 0 acts as its own opposite.

SET OF INTEGERS

The set of integers = {..., $-4, -3, -2, -1, 0, 1, 2, 3, 4, ...$}.

Integers are associated with many real-world problems and situations.

EXAMPLE 1 State which integer(s) corresponds to each situation.

a) Noreen lost 21 points in one round of a card game.
b) Death Valley is 280 ft below sea level.
c) Jaco's Bistro made $329 on Sunday, but lost $53 on Monday.

SOLUTION

a) Losing 21 points in a card game gives you -21 points.

b) The integer -280 corresponds to the situation (see the figure below). The elevation is -280 ft.

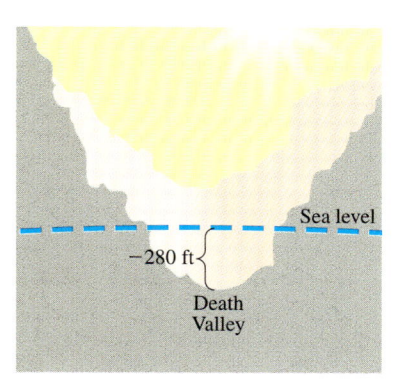

c) The integer 329 corresponds to the profit on Sunday and -53 corresponds to the loss on Monday.

The Rational Numbers

Although numbers like $\frac{4}{7}$ are built out of integers, these numbers are not themselves integers. Another set, the **rational numbers,** contains fractions and decimals, in addition to the integers. Some examples of rational numbers are

$$\frac{4}{7}, \quad -\frac{4}{7}, \quad 5, \quad -3, \quad 0, \quad \frac{-35}{8}, \quad 2.4, \quad -0.31.$$

The number $-\frac{4}{7}$ (read "negative four-sevenths") can also be written $\frac{-4}{7}$ or $\frac{4}{-7}$. In fact, every one of the numbers above can be written as an integer over another integer. For example, 5 can be written as $\frac{5}{1}$ and 2.4 can be written as $\frac{24}{10}$. In this manner, any *ratio*nal number can be expressed as the *ratio* of two integers. Because we cannot even begin to list all rational numbers, this idea of ratio is used to describe the set as follows.

26   CHAPTER 1   •   INTRODUCTION TO ALGEBRAIC EXPRESSIONS

SET OF RATIONAL NUMBERS

The set of rational numbers $= \left\{ \dfrac{a}{b} \,\middle|\, a \text{ and } b \text{ are integers and } b \neq 0 \right\}.$

This is read "the set of all numbers $\dfrac{a}{b}$, where a and b are integers and $b \neq 0$."

To *graph* a number is to mark its location on a number line.

EXAMPLE 2 Graph each of the following rational numbers.

a) $\dfrac{5}{2}$ b) -3.2 c) $\dfrac{11}{8}$

SOLUTION

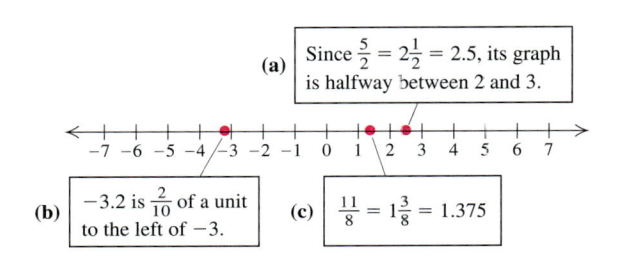

(a) Since $\dfrac{5}{2} = 2\dfrac{1}{2} = 2.5$, its graph is halfway between 2 and 3.

(b) -3.2 is $\dfrac{2}{10}$ of a unit to the left of -3.

(c) $\dfrac{11}{8} = 1\dfrac{3}{8} = 1.375$

Every rational number can be written as a fraction or a decimal.

EXAMPLE 3 Convert to decimal notation: $-\dfrac{5}{8}$.

SOLUTION We first find decimal notation for $\dfrac{5}{8}$. Since $\dfrac{5}{8}$ means $5 \div 8$, we divide.

$$\begin{array}{r} 0.625 \\ 8\overline{)5.000} \\ \underline{4\,8} \\ 20 \\ \underline{1\,6} \\ 40 \\ \underline{40} \\ 0 \end{array}$$

Thus, $\dfrac{5}{8} = 0.625$, so $-\dfrac{5}{8} = -0.625$.

Because the division in Example 3 ends with the remainder 0, we consider -0.625 a **terminating decimal.** When a remainder reappears, we have a **repeating decimal.**

1.4 POSITIVE AND NEGATIVE REAL NUMBERS

EXAMPLE 4 Convert to decimal notation: $\frac{7}{11}$.

SOLUTION We divide:

$$
\begin{array}{r}
0.6\,3\,6\,3\ldots \\
11\,\overline{\smash{)}7.0\,0\,0\,0} \\
\underline{6\,6} \\
4\,0 \\
\underline{3\,3} \\
7\,0 \\
\underline{6\,6} \\
4\,0
\end{array}
$$

4 reappears as a remainder.

We abbreviate repeating decimals by writing a bar over the repeating part, in this case, $0.\overline{63}$. Thus, $\frac{7}{11} = 0.\overline{63}$.

Although we do not prove it here, every rational number can be expressed as either a terminating or repeating decimal, and every terminating or repeating decimal can be expressed as a ratio of two integers.

Real Numbers and Order

Some numbers, when written in decimal form, neither terminate nor repeat. Such numbers are called **irrational numbers.**

What kinds of numbers are irrational numbers? One example is π (the Greek letter *pi*, read "pie"), which is used to find the area and the circumference of a circle: $A = \pi r^2$ and $C = 2\pi r$.

Another irrational number, $\sqrt{2}$ (read "the square root of 2"), is the length of the diagonal of a square with sides of length 1. It is also the number that, when multiplied by itself, gives 2. No rational number can be multiplied by itself to get 2, although some approximations come close:

1.4 is an *approximation* of $\sqrt{2}$ because $(1.4)^2 = 1.96$;

1.41 is a better approximation because $(1.41)^2 = 1.9881$;

1.4142 is an even better approximation because $(1.4142)^2 = 1.99996164$.

To approximate $\sqrt{2}$ on some calculators, simply press $\boxed{2}$ and then $\boxed{\sqrt{}}$. With other calculators, press $\boxed{\sqrt{}}$, $\boxed{2}$, and $\boxed{\text{ENTER}}$, or consult a manual.

EXAMPLE 5 Graph the real number $\sqrt{3}$ on a number line.

SOLUTION We use a calculator and approximate: $\sqrt{3} \approx 1.732$ ("$\approx$" means "approximately equals"). Then we locate this number on a number line.

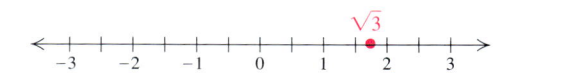

The rational numbers and the irrational numbers together correspond to all the points on a number line and make up what is called the **real-number system.**

28 CHAPTER 1 • INTRODUCTION TO ALGEBRAIC EXPRESSIONS

SET OF REAL NUMBERS

The set of real numbers = The set of all numbers corresponding to points on the number line.

The following figure shows the relationships among various kinds of numbers.

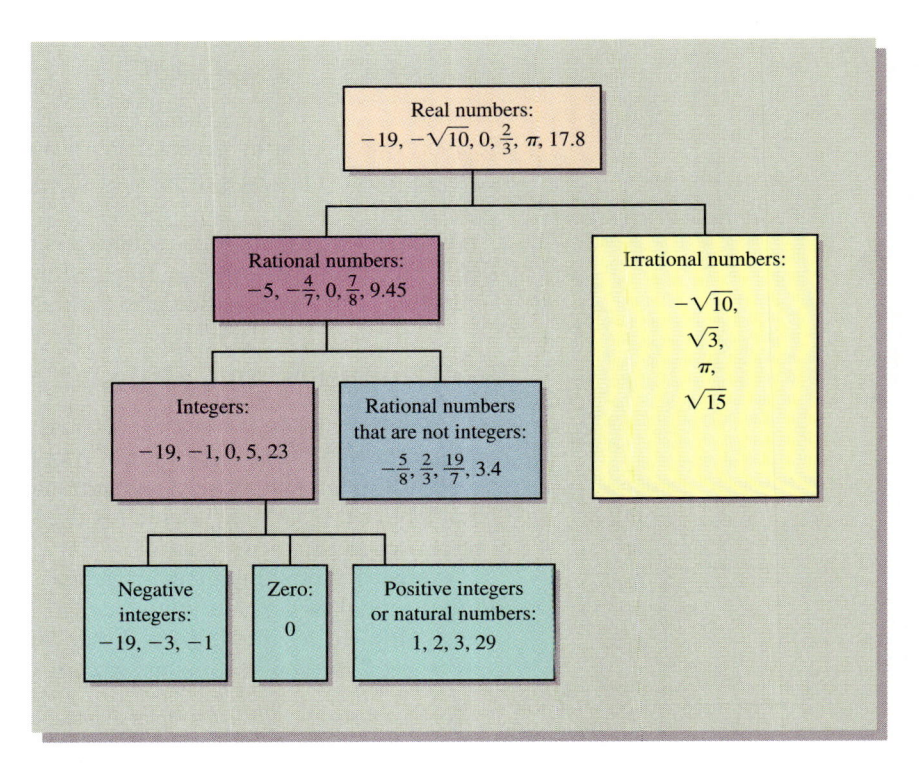

```
Real numbers:
−19, −√10, 0, 2/3, π, 17.8
├── Rational numbers:
│   −5, −4/7, 0, 7/8, 9.45
│   ├── Integers:
│   │   −19, −1, 0, 5, 23
│   │   ├── Negative integers: −19, −3, −1
│   │   ├── Zero: 0
│   │   └── Positive integers or natural numbers: 1, 2, 3, 29
│   └── Rational numbers that are not integers:
│       −5/8, 2/3, 19/7, 3.4
└── Irrational numbers:
    −√10, √3, π, √15
```

Real numbers are named in order on the number line, with larger numbers further to the right. For any two numbers, the one to the left is less than the one to the right. We use the symbol < to mean "**is less than.**" The sentence −8 < 6 means "−8 is less than 6." The symbol > means "**is greater than.**" The sentence −3 > −7 means "−3 is greater than −7."

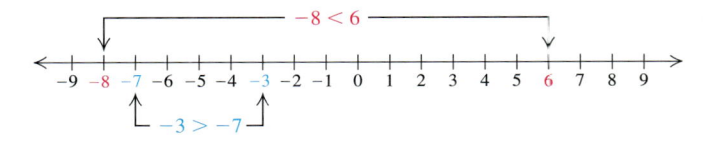

$$-8 < 6$$
$$\leftarrow | -9 | -8 | -7 | -6 | -5 | -4 | -3 | -2 | -1 | 0 | 1 | 2 | 3 | 4 | 5 | 6 | 7 | 8 | 9 | \rightarrow$$
$$-3 > -7$$

EXAMPLE 6 Use either < or > for ▨ to write a true sentence.
a) 2 ▨ 9
b) −3.45 ▨ 1.32
c) 6 ▨ −12
d) −18 ▨ −5
e) $\frac{7}{11}$ ▨ $\frac{5}{8}$

SOLUTION

a) Since 2 is to the left of 9, we know that 2 is less than 9, so $2 < 9$.
b) Since -3.45 is to the left of 1.32, we have $-3.45 < 1.32$.
c) Since 6 is to the right of -12, we have $6 > -12$.
d) Since -18 is to the left of -5, we have $-18 < -5$.
e) We convert to decimal notation: $\frac{7}{11} = 0.\overline{63}$ and $\frac{5}{8} = 0.625$. Thus, $\frac{7}{11} > \frac{5}{8}$.

We also could have used a common denominator: $\frac{7}{11} = \frac{56}{88} > \frac{55}{88} = \frac{5}{8}$.

Sentences like "$a < -5$" and "$-3 > -8$" are called **inequalities**. It is useful to remember that every inequality can be written two ways. For instance,

$-3 > -8$ has the same meaning as $-8 < -3$.

It may be helpful to think of an inequality sign as an "arrow" with the smaller side pointing to the smaller number.

Note that $a > 0$ means that a represents a positive real number and $a < 0$ means that a represents a negative real number.

Statements like $a \leq b$ and $b \geq a$ are also inequalities. We read $a \leq b$ as "*a* **is less than or equal to** *b*." We read $a \geq b$ as "*a* **is greater than or equal to** *b*."

EXAMPLE 7 Write true or false for each inequality.

a) $-3 \leq 5$ b) $-3 \leq -3$ c) $-5 \geq 4$

SOLUTION

a) $-3 \leq 5$ is *true* because $-3 < 5$ is true.
b) $-3 \leq -3$ is *true* because $-3 = -3$ is true.
c) $-5 \geq 4$ is *false* since neither $-5 > 4$ nor $-5 = 4$ is true.

Absolute Value

There is a convenient terminology and notation for the distance a number is from 0 on a number line. It is called the **absolute value** of the number.

ABSOLUTE VALUE

We write $|a|$, read "the absolute value of a," to represent the number of units that a is from zero.

EXAMPLE 8 Find each absolute value: (a) $|-3|$; (b) $|7.2|$; (c) $|0|$.

SOLUTION

a) $|-3| = 3$ since -3 is 3 units from 0.
b) $|7.2| = 7.2$ since 7.2 is 7.2 units from 0.
c) $|0| = 0$ since 0 is 0 units from itself.

Distance is never negative, so numbers that are opposites have the same absolute value. If a number is nonnegative, its absolute value is the number itself. If a number is negative, its absolute value is its opposite.

EXERCISE SET 1.4

Tell which real numbers correspond to each situation.

1. Aimee's golf score was 3 under par, while Jackson's score was 7 over par.
2. The temperature on Wednesday was 18° above zero. On Thursday, it was 2° below zero.
3. The largest daily drop in the Dow Jones Average was 508 points; the largest increase was 186.84. (*Source: The Guinness Book of Records,* 1996)
4. A printer earned $1200 one week and lost $560 the next.
5. The Dead Sea is 1286 feet below sea level, whereas Mt. Everest is 29,029 feet above sea level.
6. In bowling, the Jets are 34 pins behind the Strikers after one game. Describe the situation in two ways.
7. Janice deposited $750 in a savings account. Two weeks later, she withdrew $125.
8. During the late 1980s, the world birthrate, per thousand, was 27. The death rate, per thousand, was 9.7. (*Source: The Guinness Book of Records,* 1996)
9. During a video game, Cindy intercepted a missile worth 20 points, lost a starship worth 150 points, and captured a base worth 300 points.
10. Ignition occurs 10 seconds before liftoff. A spent fuel tank is detached 235 seconds after liftoff.

Graph each rational number on a number line.

11. $\frac{10}{3}$ 12. $-\frac{17}{5}$ 13. -4.3
14. 3.87 15. -2 16. 5

Find decimal notation.

17. $\frac{7}{8}$ 18. $-\frac{1}{8}$ 19. $-\frac{3}{4}$
20. $\frac{5}{6}$ 21. $\frac{7}{6}$ 22. $\frac{5}{12}$
23. $\frac{2}{3}$ 24. $\frac{1}{4}$ 25. $-\frac{1}{2}$
26. $-\frac{3}{8}$ 27. $\frac{1}{10}$ 28. $-\frac{7}{20}$

Write a true sentence using either < or >.

29. 5 ☐ 0   30. 9 ☐ 0
31. -9 ☐ 5   32. 8 ☐ -8
33. -6 ☐ 6   34. 0 ☐ -7
35. -8 ☐ -5 36. -4 ☐ -3
37. -5 ☐ -11 38. -3 ☐ -4
39. -12.5 ☐ -9.4 40. -10.3 ☐ -14.5
41. $\frac{5}{12}$ ☐ $\frac{11}{25}$ 42. $-\frac{14}{17}$ ☐ $-\frac{27}{35}$

Write an inequality with the same meaning as each of the following.

43. $-2 > a$ 44. $a > 9$
45. $-10 \leq y$ 46. $12 \geq t$

Classify each inequality as true or false.

47. $-3 \geq -11$ 48. $5 \leq -5$
49. $0 \geq 8$ 50. $-5 \leq 7$
51. $-8 \leq -8$ 52. $8 \geq 8$

Find each absolute value.

53. $|-4|$ 54. $|-9|$ 55. $|17|$
56. $|3.1|$ 57. $|5.6|$ 58. $\left|-\frac{2}{5}\right|$
59. $|329|$ 60. $|-456|$ 61. $\left|-\frac{9}{7}\right|$
62. $|8.02|$ 63. $|0|$ 64. $|-1.07|$

65. $|x|$, for $x = -8$ **66.** $|a|$, for $a = -5$

For Exercises 67–72, consider the following list of numbers:

$$-23, -4.7, 0, \tfrac{5}{9}, \pi, \sqrt{17}, 8.31, 62.$$

67. List all rational numbers.
68. List all natural numbers.
69. List all integers.
70. List all irrational numbers.
71. List all real numbers.
72. List all nonnegative integers.

SKILL MAINTENANCE

73. Multiply and simplify: $\tfrac{21}{5} \cdot \tfrac{1}{7}$.
74. Evaluate $3xy$ for $x = 2$ and $y = 7$.
75. Use a commutative law to write an expression equivalent to $ab + 5$.
76. Factor: $3x + 9 + 12y$.

SYNTHESIS

77. ◆ How many rational numbers are between 0 and 1? Why?
78. ◆ Is every nonnegative integer a whole number? Why or why not?
79. ◆ Why is it impossible for the absolute value of a number to be negative?
80. ◆ Classes at Stockard Community College must have at least 10 students in order to run. Campus policy says no more than 35 students can enroll in one class. What is the range of class sizes at the college? Why?

List in order from least to greatest.

81. $13, -12, 5, -17$
82. $-23, 4, 0, -17$
83. $\tfrac{4}{5}, \tfrac{4}{3}, \tfrac{4}{8}, \tfrac{4}{6}, \tfrac{4}{9}, \tfrac{4}{2}, -\tfrac{4}{3}$
84. $-\tfrac{2}{3}, \tfrac{1}{2}, -\tfrac{3}{4}, -\tfrac{5}{6}, \tfrac{3}{8}, \tfrac{1}{6}$

Write a true sentence using either $<$, $>$, or $=$.

85. $|-5|$ ▪ $|-2|$ **86.** $|4|$ ▪ $|-7|$
87. $|-8|$ ▪ $|8|$ **88.** $|23|$ ▪ $|-23|$
89. $|-3|$ ▪ $|5|$ **90.** $|-19|$ ▪ $|-27|$

Solve. Consider only integer replacements.

91. $|x| = 7$ **92.** $-3 < |x| < 3$
93. $2 < |x| < 5$

Given that $0.3\overline{3} = \tfrac{1}{3}$ and $0.6\overline{6} = \tfrac{2}{3}$, express each of the following as a ratio of two integers.

94. $0.1\overline{1}$ **95.** $0.9\overline{9}$ **96.** $5.5\overline{5}$

To the student and instructor: *The calculator icon, ▦, is used to indicate those exercises designed to be solved with a calculator.*

97. ◆ ▦ When Helga's calculator gives a decimal value for $\sqrt{2}$ and that value is promptly squared, the result is 2. Yet when that same decimal approximation is entered by hand and then squared, the result is not exactly 2. Why do you suppose this is?

1.5 Addition of Real Numbers

Adding with a Number Line • Adding without a Number Line • Problem Solving • Combining Like Terms

We now consider addition of real numbers. To gain understanding, we will use a number line first. After observing the principles involved, we will develop rules that allow us to work more quickly.

Adding with a Number Line

To perform the addition $a + b$ on a number line, we start at a and then move according to b.

a) If b is positive, we move to the right (the positive direction).
b) If b is negative, we move to the left (the negative direction).
c) If b is 0, we stay at a.

EXAMPLE 1   Add: $-4 + 9$.

SOLUTION   To add on a number line, we locate the first number, -4, and then move 9 units to the right. Note that it requires 4 units to reach 0. The difference between 9 and 4 is where we finish.

$$-4 + 9 = 5$$

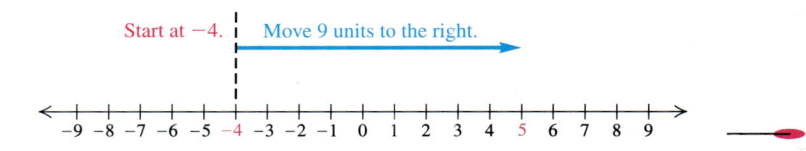

EXAMPLE 2   Add: $3 + (-5)$.

SOLUTION   We locate the first number, 3, and then move 5 units to the left. Note that it requires 3 units to reach 0. The difference between 5 and 3 is 2, so we finish 2 units to the left of 0.

$$3 + (-5) = -2$$

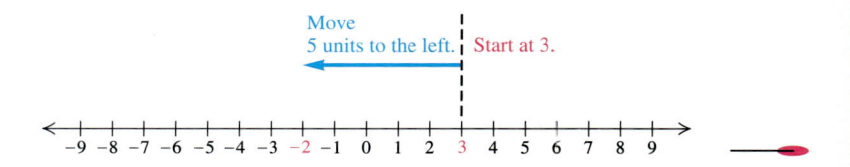

EXAMPLE 3   Add: $-4 + (-3)$.

SOLUTION   After locating -4, we move 3 units to the left. We finish a total of 7 units to the left of 0.

$$-4 + (-3) = -7$$

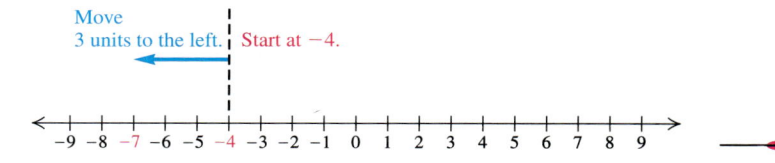

1.5 ADDITION OF REAL NUMBERS

EXAMPLE 4 Add: $-5.2 + 0$.

SOLUTION We locate -5.2 and move 0 units. Thus we finish where we started.

$$-5.2 + 0 = -5.2$$

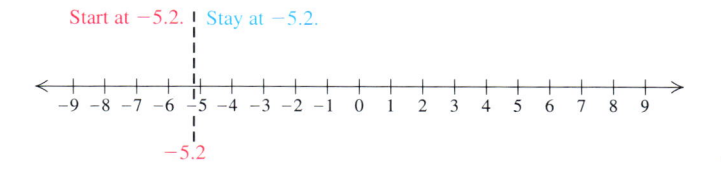

Start at -5.2. Stay at -5.2.

From Examples 1–4, we develop the following rules.

RULES FOR ADDITION OF REAL NUMBERS

1. *Positive numbers*: Add as usual. The answer is positive.
2. *Negative numbers*: Add absolute values and make the answer negative (see Example 3).
3. *A positive number and a negative number*: Subtract the smaller absolute value from the greater absolute value. Then:
   a) If the positive number has the greater absolute value, the answer is positive (see Example 1).
   b) If the negative number has the greater absolute value, the answer is negative (see Example 2).
   c) If the numbers have the same absolute value, the answer is 0.
4. *One number is zero*: The sum is the other number (see Example 4).

Rule 4 is known as the **identity property of 0**. It says that for any real number a, $a + 0 = a$.

Adding without a Number Line

The rules listed above can be used without drawing a number line.

EXAMPLE 5 Add without using a number line.

a) $-12 + (-7)$ b) $-1.4 + 8.5$
c) $-36 + 21$ d) $1.5 + (-1.5)$
e) $-\frac{7}{8} + 0$ f) $\frac{2}{3} + \left(-\frac{5}{8}\right)$

SOLUTION

a) $-12 + (-7) = -19$ Two negatives. *Think*: Add the absolute values, 12 and 7, to get 19. Make the answer *negative*, -19.

b) $-1.4 + 8.5 = 7.1$ A negative and a positive. *Think*: The difference of absolute values is $8.5 - 1.4$, or 7.1. The positive number has the larger absolute value, so the answer is *positive*, 7.1.

c) $-36 + 21 = -15$ A negative and a positive. *Think*: The difference of absolute values is $36 - 21$, or 15. The negative number has the larger absolute value, so the answer is *negative*, -15.

d) $1.5 + (-1.5) = 0$ A negative and a positive. *Think*: Since the numbers are opposites, they have the same absolute value and the answer is 0.

e) $-\dfrac{7}{8} + 0 = -\dfrac{7}{8}$ One number is zero. The sum is the other number, $-\dfrac{7}{8}$.

f) $\dfrac{2}{3} + \left(-\dfrac{5}{8}\right) = \dfrac{16}{24} + \left(-\dfrac{15}{24}\right) = \dfrac{1}{24}$ This is similar to part (b) above.

If we are adding several numbers, some positive and some negative, the commutative and associative laws allow us to add all the positives, then add all the negatives, and then add the results. Of course, we can also add from left to right, if we prefer.

EXAMPLE 6   Add: $15 + (-2) + 7 + 14 + (-5) + (-12)$.

SOLUTION

$15 + (-2) + 7 + 14 + (-5) + (-12)$

$= 15 + 7 + 14 + (-2) + (-5) + (-12)$ Using the commutative law of addition

$= (15 + 7 + 14) + [(-2) + (-5) + (-12)]$ Using the associative law of addition

$= 36 + (-19)$ Adding the positives; adding the negatives

$= 17$ Adding a positive and a negative

Problem Solving

Addition of real numbers occurs in many real-world applications.

EXAMPLE 7   *Lake Level.* In the course of one four-month period, the water level of Lake Champlain went down 2 ft, up 1 ft, down 5 ft, and up 3 ft. How much had the lake level changed at the end of the four months?

SOLUTION The problem translates to a sum:

Rewording: The 1st change plus the 2nd change plus the 3rd change plus the 4th change is the total change.

Translating: $-2 + 1 + (-5) + 3 =$ Total change.

Adding from left to right, we have

$$-2 + 1 + (-5) + 3 = -1 + (-5) + 3$$
$$= -6 + 3$$
$$= -3.$$

The lake level has dropped 3 ft at the end of the four-month period.

Combining Like Terms

When two terms have variable factors that are exactly the same, like $5ab$ and $7ab$, the terms are called **like**, or **similar, terms.*** The distributive law enables us to **combine**, or **collect, like terms.** The above rules for addition will again apply.

EXAMPLE 8 Combine like terms.

a) $-7x + 9x$ b) $2a + (-3b) + (-5a) + 9b$
c) $7 + y + (-3.5y) + 2$

SOLUTION

a) $-7x + 9x = (-7 + 9)x$ Using the distributive law
$ = 2x$ Adding -7 and 9

b) $2a + (-3b) + (-5a) + 9b = 2a + (-5a) + (-3b) + 9b$
                                            Using the commutative law of addition
$ = (2 + (-5))a + (-3 + 9)b$
                                            Using the distributive law
$ = -3a + 6b$ Adding

c) $7 + y + (-3.5y) + 2 = y + (-3.5y) + 7 + 2$ Using the commutative law of addition
$ = (1 + (-3.5))y + 7 + 2$ Using the distributive law
$ = -2.5y + 9$ Adding

With practice we can leave out some steps, combining like terms mentally. Numbers like 7 and 2 in the expression $7 + y + (-3.5y) + 2$ are constants and are also considered to be like terms.

---

*Like terms are discussed in greater detail in Section 1.8.

EXERCISE SET 1.5

Add using a number line.

1. $-7 + 3$
2. $2 + (-5)$
3. $-5 + 9$
4. $8 + (-3)$
5. $-8 + 8$
6. $6 + (-6)$
7. $-3 + (-5)$
8. $-4 + (-6)$

Add. Do not use a number line except as a check.

9. $-15 + 0$
10. $-6 + 0$
11. $0 + (-8)$
12. $0 + (-2)$
13. $-15 + 15$
14. $17 + (-17)$
15. $-24 + (-17)$
16. $-17 + (-25)$
17. $11 + (-11)$
18. $-18 + 18$
19. $-7 + 8$
20. $8 + (-5)$
21. $10 + (-12)$
22. $-4 + (-5)$
23. $-3 + 14$
24. $13 + (-6)$
25. $-14 + (-19)$
26. $11 + (-9)$
27. $19 + (-19)$
28. $-20 + (-6)$
29. $23 + (-5)$
30. $-15 + (-7)$
31. $-23 + (-9)$
32. $40 + (-8)$
33. $40 + (-40)$
34. $-25 + 25$
35. $85 + (-65)$
36. $63 + (-18)$
37. $-3.6 + 1.9$
38. $-6.5 + 4.7$
39. $-5.4 + (-3.7)$
40. $-3.8 + (-9.4)$
41. $-\frac{4}{3} + \frac{2}{3}$
42. $-\frac{3}{5} + \frac{2}{5}$
43. $-\frac{4}{9} + \left(-\frac{6}{9}\right)$
44. $-\frac{3}{7} + \left(-\frac{5}{7}\right)$
45. $-\frac{5}{6} + \frac{2}{3}$
46. $-\frac{5}{8} + \frac{1}{4}$
47. $-\frac{5}{8} + \left(-\frac{1}{3}\right)$
48. $-\frac{3}{7} + \left(-\frac{2}{5}\right)$
49. $35 + (-14) + (-19) + (-5)$
50. $28 + (-44) + 17 + 31 + (-94)$
51. $-44 + \left(-\frac{3}{8}\right) + 95 + \left(-\frac{5}{8}\right)$
52. $24 + 3.1 + (-44) + (-8.2) + 63$

Solve.

53. *Telephone bills.* Maya's telephone bill for July was $82. She sent a check for $50 and then made $37 worth of calls in August. How much did she owe at the end of August?

54. *Yardage gained.* In a college football game, the quarterback attempted passes with the following results.

First try	13-yd gain
Second try	incomplete
Third try	12-yd loss
Fourth try	21-yd gain
Fifth try	14-yd loss

Find the total gain (or loss).

55. *Profits and losses.* The following table shows the profits and losses of Fax City over a 5-yr period. Find the profit or loss after this period of time.

Year	Profit or loss
1992	−$28,375
1993	+$37,425
1994	+$95,485
1995	−$19,365
1996	+$98,245

56. *Barometric pressure.* The barometric pressure at Omaha dropped 6 millibars (mb); then it rose 3 mb. After that, it dropped 14 mb and then rose 4 mb. What was the total change in pressure?

57. *Peak elevation.* The tallest mountain in the world, as measured from base to peak, is Mauna Kea in Hawaii. From a base 19,684 ft below sea level, it rises 33,480 ft (*Source: The Guinness Book of Records,* 1996). What is the elevation of its peak?

58. *Credit card bills.* Kyle's credit card bill is $470. She sends a check to the credit card company for $45, charges another $160 in merchandise, and then pays off another $500 of her bill. How much does either Kyle owe the company or the company owe Kyle?

59. *Stock growth.* Monday the value of a share of IBM stock dropped $\$\frac{1}{4}$. Tuesday it rose in value $\$\frac{5}{8}$ and Wednesday it lost $\$\frac{3}{8}$. How much did the stock's value rise or fall at the end of the three-day period?

60. *Account balance.* Tony has $460 in a checking account. He writes a check for $530, makes a deposit of $75, and then writes a check for $90. What is the balance in the account?

Combine like terms.

61. $4a + 9a$
62. $3x + 8x$
63. $-3x + 12x$
64. $2m + (-7m)$
65. $4x + 7x$
66. $5a + 9a$

67. $7m + (-9m)$
68. $-4x + 9x$
69. $-5a + (-2a)$
70. $10n + (-17n)$
71. $-3 + 8x + 4 + (-10x)$
72. $8a + 5 + (-a) + (-3)$

Find the perimeter of each figure.

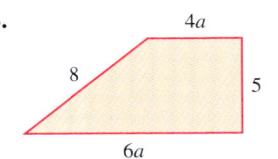

73. (figure with sides 8, 7x, 5x, 9)
74. (figure with sides 4a, 8, 5, 6a)
75. (figure with sides 9, 4n, 6n, 8n, 7)
76. (figure with sides 2, 6, 7n, 5n, 7n, 3)

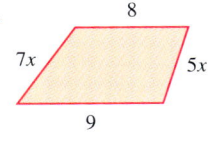

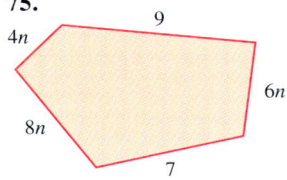

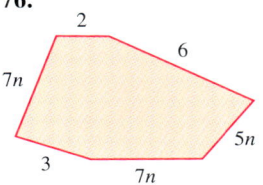

SKILL MAINTENANCE

77. Multiply: $7(3z + y + 2)$.
78. Divide and simplify: $\frac{7}{2} \div \frac{3}{8}$.

SYNTHESIS

79. ◈ Without performing the actual addition, explain why the sum of all integers from -50 to 50 is 0.

80. ◈ Write a problem for a classmate to solve. Devise the problem so that it translates to a sum of negative and positive integers.

81. ◈ Under what circumstances will the sum of one positive number and several negative numbers be positive?

82. ◈ Describe in your own words why the sum of two negative numbers is always negative.

83. *Stock prices.* The value of EKB stock rose $\$2\frac{3}{8}$ and then dropped $\$3\frac{1}{4}$ before finishing at $\$64\frac{3}{8}$. What was the stock's original value?

84. *Sports card values.* The value of a sports card dropped $12 and then rose $17.50 before settling at $61. What was the original value of the card?

Find the missing term.

85. $4x + \underline{\quad} + (-9x) + (-2y) = -5x - 7y$
86. $-3a + 9b + \underline{\quad} + 5a = 2a - 6b$
87. $3m + 2n + \underline{\quad} + (-2m) = 2n + (-6m)$
88. $\underline{\quad} + 9x + (-4y) + x = 10x - 7y$

89. *Geometry.* The perimeter of a rectangle is $7x + 10$. If the length of the rectangle is 5, express the width in terms of x.

90. *Golfing.* After five rounds of golf, a golf pro was 3 under par twice, 2 over par once, 2 under par once, and 1 over par once. On average, how far above or below par was the golfer?

1.6 Subtraction of Real Numbers

Opposites and Additive Inverses • Subtraction • Problem Solving

In arithmetic, when a number b is subtracted from another number a, the difference, $a - b$, is the number that when added to b gives a. For example, $45 - 17 = 28$ because $28 + 17 = 45$. In this section, this approach is used to develop an efficient way of finding the value of $a - b$ for any real numbers a and b. Before doing so, however, we must develop some terminology.

Opposites and Additive Inverses

Numbers such as 6 and -6 are opposites, or *additive inverses,* of each other. Whenever opposites are added, the result is 0; and whenever two numbers add to 0, those numbers are opposites.

EXAMPLE 1 Find the opposite of each number: **(a)** 34; **(b)** -8.3; **(c)** 0.

SOLUTION

a) The opposite of 34 is -34: $34 + (-34) = 0$.
b) The opposite of -8.3 is 8.3: $-8.3 + 8.3 = 0$.
c) The opposite of 0 is 0: $0 + 0 = 0$.

To name the opposite, we use the symbol $-$, as follows.

OPPOSITE

The *opposite,* or *additive inverse,* of a number a is written $-a$ (read "the opposite of a" or "the additive inverse of a").

Note that if we take a number, say 8, and find its opposite, -8, and then find the opposite of the result, we will have the original number, 8, again. Thus, for any number a,

$$-(-a) = a.$$

EXAMPLE 2 Find $-x$ and $-(-x)$ when $x = 16$.

SOLUTION

If $x = 16$, then $-x = -16$.   The opposite of 16 is -16.
If $x = 16$, then $-(-x) = -(-16) = 16$.   The opposite of the opposite of 16 is 16.

EXAMPLE 3 Find $-x$ and $-(-x)$ when $x = -3$.

SOLUTION

If $x = -3$, then $-x = -(-3) = 3$.
If $x = -3$, then $-(-x) = -(-(-3)) = -(\ 3\ ) = -3$.

Note in Example 3 that an extra set of parentheses is used to show that we are substituting the negative number -3 for x. The notation $-\ -x$ is not used.

A symbol such as -8 is usually read "negative 8." It could be read "the additive inverse of 8," because the additive inverse of 8 is negative 8. It could also be read "the opposite of 8," because the opposite of 8 is -8. Thus a symbol like -8 can be read in more than one way. A symbol like $-x$, which has a variable, should be read "the opposite of x" or "the additive inverse of x" and *not* "negative x," since to do so suggests that $-x$ represents a negative number. As we saw in Example 3, $-x$ can represent a positive number. This notation can be used to restate a result from Section 1.5 as *the law of opposites*:

THE LAW OF OPPOSITES

For any two numbers a and $-a$,

$a + (-a) = 0$.

(When opposites are added, their sum is 0.)

1.6 SUBTRACTION OF REAL NUMBERS

A negative number is said to have a "negative *sign*." A positive number is said to have a "positive *sign*." When we replace a number by its opposite, or additive inverse, we can say that we have "changed or reversed its sign."

EXAMPLE 4 Change the sign (find the opposite) of each number: **(a)** -3; **(b)** -10; **(c)** 14.

SOLUTION

a) When we change the sign of -3, we obtain 3.
b) When we change the sign of -10, we obtain 10.
c) When we change the sign of 14, we obtain -14.

Subtraction

Opposites are helpful when subtraction involves negative numbers. To see why, look for a pattern in the following:

Subtracting		*Adding an Opposite*
$5 - 8 = -3$	since $-3 + 8 = 5$	$5 + (-8) = -3$
$-6 - 4 = -10$	since $-10 + 4 = -6$	$-6 + (-4) = -10$
$-7 - (-10) = 3$	since $3 + (-10) = -7$	$-7 + 10 = 3$
$-7 - (-2) = -5$	since $-5 + (-2) = -7$	$-7 + 2 = -5$

The matching results suggest that we can subtract by adding the opposite of the number being subtracted. This can always be done and often provides the easiest way to subtract real numbers.

SUBTRACTION OF REAL NUMBERS

For any real numbers a and b,
$$a - b = a + (-b).$$
(To subtract, add the opposite, or additive inverse, of the number being subtracted.)

EXAMPLE 5 Subtract each of the following and then check by addition.

a) $2 - 6$ **b)** $4 - (-9)$ **c)** $-4.2 - (-3.6)$

SOLUTION

a) $2 - 6 = 2 + (-6) = -4$ The opposite of 6 is -6. We change the subtraction to addition and add the opposite. *Check*: $-4 + 6 = 2$.

b) $4 - (-9) = 4 + 9 = 13$ The opposite of -9 is 9. We change the subtraction to addition and add the opposite. *Check*: $13 + (-9) = 4$.

c) $-4.2 - (-3.6) = -4.2 + 3.6$
$ = -0.6$ Adding the opposite. *Check*: $-0.6 + (-3.6) = -4.2$.

EXAMPLE 6   Subtract $-\frac{3}{5}$ from $-\frac{2}{5}$.

SOLUTION   A common denominator already exists, so we subtract:

$$-\frac{2}{5} - \left(-\frac{3}{5}\right) = -\frac{2}{5} + \frac{3}{5} \quad \text{Adding the opposite}$$

$$= \frac{-2+3}{5} = \frac{1}{5}.$$

Check: $\frac{1}{5} + \left(-\frac{3}{5}\right) = \frac{1+(-3)}{5} = \frac{-2}{5}.$

The symbol "−" is read differently depending on where it appears. For instance, $-5 - (-x)$ is read "negative five minus the opposite of x."

EXAMPLE 7   Read each of the following and then subtract.

a) $3 - 5$ b) $-4.6 - (-9.8)$ c) $-\frac{3}{4} - \frac{7}{5}$

SOLUTION

a) $3 - 5$;   Read "three minus five"
$3 - 5 = 3 + (-5) = -2$ Adding the opposite

b) $-4.6 - (-9.8)$;   Read "negative four point six minus negative nine point eight"
$-4.6 - (-9.8) = -4.6 + 9.8 = 5.2$ Adding the opposite

c) $-\frac{3}{4} - \frac{7}{5}$;   Read "negative three-fourths minus seven-fifths"
$-\frac{3}{4} - \frac{7}{5} = -\frac{15}{20} + \left(-\frac{28}{20}\right) = -\frac{43}{20}$ Finding a common denominator and adding the opposite

When several additions and subtractions occur together, we can convert them all to additions.

EXAMPLE 8   Simplify: $8 - (-4) - 2 - (-5) + 3$.

SOLUTION

$8 - (-4) - 2 - (-5) + 3 = 8 + 4 + (-2) + 5 + 3$ To subtract, we add the opposite.

$= 18$

Recall that the terms of an algebraic expression are separated by plus signs. For instance, the terms of the expression $5x - 7y - 9$ are $5x$, $-7y$, and -9, since $5x - 7y - 9 = 5x + (-7y) + (-9)$.

EXAMPLE 9   Identify the terms of the expression $4 - 2ab + 7a - 9$.

SOLUTION   We have

$4 - 2ab + 7a - 9 = 4 + (-2ab) + 7a + (-9)$,   Rewriting as addition

so the terms are 4, $-2ab$, $7a$, and -9.

1.6 SUBTRACTION OF REAL NUMBERS

EXAMPLE 10

Combine like terms.

a) $1 + 3x - 7x$
b) $-5a - 7b - 4a + 10b$
c) $9 - 3m - 14 + 7m$

SOLUTION

a) $1 + 3x - 7x = 1 + 3x + (-7x)$ Adding the opposite
$= 1 + (3 + (-7))x$ Using the distributive law.
$= 1 + (-4x)$ Try to do this mentally.
$= 1 - 4x$ Rewriting as subtraction to be more concise

b) $-5a - 7b - 4a + 10b = -5a + (-7b) + (-4a) + 10b$ Rewriting as addition
$= -5a + (-4a) + (-7b) + 10b$ Using the commutative law of addition
$= -9a + 3b$ Combining like terms mentally

c) $9 - 3m - 14 + 7m = 9 + (-3m) + (-14) + 7m$ Rewriting as addition
$= 9 + (-14) + (-3m) + 7m$ Using a commutative law
$= -5 + 4m$

Problem Solving

Subtraction is used to solve problems involving differences.

EXAMPLE 11

Changes in Elevation. The lowest point in Asia is the Dead Sea, which is 400 m below sea level. The lowest point in the United States is Death Valley, which is 86 m below sea level. What is the difference in elevation between the Dead Sea and Death Valley?

SOLUTION It is helpful to draw a picture of the situation.

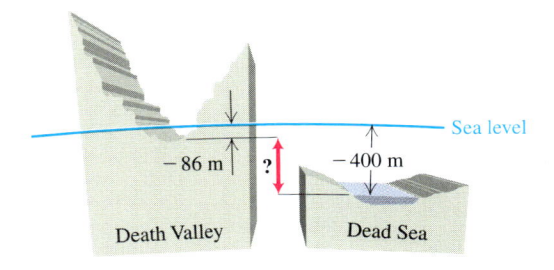

To find the difference in elevation, we always subtract the lower elevation—in this case, -400 m—from the higher elevation—here, -86 m:

$$-86 - (-400) = -86 + 400 = 314.$$

Death Valley is 314 m higher than the Dead Sea.

EXERCISE SET 1.6

Find the opposite, or additive inverse.

1. 39
2. −17
3. −9
4. $\frac{7}{2}$
5. −3.14
6. 48.2

Find −x when x is each of the following.

7. 23
8. −26
9. $-\frac{14}{3}$
10. $\frac{1}{328}$
11. 0.101
12. 0

Find −(−x) when x is each of the following.

13. 72
14. 29
15. $-\frac{2}{5}$
16. −9.1

Change the sign. (Find the opposite.)

17. −1
18. −7
19. 7
20. 10

Subtract.

21. 4 − 13
22. 4 − 9
23. 0 − 7
24. 0 − 10
25. −7 − (−9)
26. −9 − (−3)
27. −10 − (−10)
28. −8 − (−8)
29. 12 − 16
30. 14 − 19
31. 20 − 27
32. 30 − 4
33. −8 − (−3)
34. −7 − (−9)
35. −40 − (−40)
36. −9 − (−9)
37. 7 − 7
38. 5 − 5
39. 6 − (−6)
40. 4 − (−4)
41. 8 − (−3)
42. −7 − 4
43. −6 − 8
44. 6 − (−10)
45. −4 − (−9)
46. −14 − 2
47. −6 − (−5)
48. −4 − (−3)
49. 3 − (−12)
50. 5 − (−6)
51. 0 − 5
52. 0 − 6
53. −5 − (−2)
54. −3 − (−1)
55. −7 − 14
56. −9 − 16
57. 0 − (−5)
58. 0 − (−1)
59. −8 − 0
60. −9 − 0
61. 3 − (−7)
62. 12 − (−5)
63. 2 − 25
64. 18 − 63
65. −42 − 26
66. −18 − 63
67. −71 − 2
68. −49 − 3
69. 24 − (−92)
70. 48 − (−73)
71. −50 − (−50)
72. −70 − (−70)
73. $\frac{3}{8} - \frac{5}{8}$
74. $\frac{3}{9} - \frac{9}{9}$
75. $\frac{4}{5} - \frac{2}{3}$
76. $\frac{5}{8} - \frac{3}{4}$
77. $-\frac{3}{4} - \frac{2}{3}$
78. $-\frac{5}{8} - \frac{3}{4}$
79. −2.8 − 0
80. 6.04 − 1.1
81. 0.09 − 1
82. 0.089 − 1
83. $\frac{1}{6} - \frac{2}{3}$
84. $-\frac{3}{8} - \left(-\frac{1}{2}\right)$
85. $-\frac{4}{7} - \left(-\frac{10}{7}\right)$
86. $\frac{12}{5} - \frac{12}{5}$

Translate each phrase to mathematical language and simplify. See Example 11.

87. The difference between 3.8 and −5.2
88. The difference between −2.1 and −5.9
89. The difference between 114 and −79
90. The difference between 23 and −17
91. Subtract 37 from −21.
92. Subtract 19 from −7.
93. Subtract −25 from 9.
94. Subtract −31 from −5.

Write words for each of the following and then perform the subtraction.

95. −1.8 − 2.7
96. −2.7 − 5.9
97. −250 − (−425)
98. −350 − (−1000)

Simplify.

99. 25 − (−12) − 7 − (−2) + 9
100. 22 − (−18) + 7 + (−42) − 27
101. −31 + (−28) − (−14) − 17
102. −43 − (−19) − (−21) + 25
103. −34 − 28 + (−33) − 44
104. 39 + (−88) − 29 − (−83)
105. −93 − (−84) − 41 − (−56)
106. 84 + (−99) + 44 − (−18) − 43

Identify the terms in each expression.

107. −7x − 4y
108. 7a − 9b

109. $-5 + 3m - 6mn$ **110.** $-9 - 4t + 10rt$

Combine like terms.

111. $4x - 7x$ **112.** $3a - 14a$
113. $7a - 12a + 4$ **114.** $-9x - 13x + 7$
115. $-8n - 9 + n$ **116.** $-7 + 9n - 8$
117. $3x + 5 - 9x$ **118.** $2 + 3a - 7$
119. $2 - 6t - 9 - 2t$ **120.** $-5 + 3b - 7 - 5b$
121. $7 + (-3x) - 9x + 1$
122. $14 - (-5x) + 2x - (-32)$
123. $13x - (-2x) + 45 - (-21)$
124. $8x - (-2x) - 14 - (-5x) + 53$

Solve.

125. *Record temperature drop.* The greatest recorded temperature change in one day occurred in Browning, Montana, when the temperature fell from 44°F to −56°F (*Source: The Guinness Book of Records,* 1996). How much did the temperature drop?

126. *Loan repayment.* Gisela owed Ramon $290. Ramon decides to "forgive" $125 of the debt. How much does Gisela owe?

127. *Elevation extremes.* The elevation of Mount Whitney, the highest peak in California, is 14,776 ft more than the elevation of Death Valley, California (*Source:* 1996 *Information Please Almanac*). If Death Valley is 282 ft below sea level, find the elevation of Mount Whitney.

128. *Elevation extremes.* The lowest elevation in Asia, the Dead Sea, is 1312 ft below sea level. The highest elevation in Asia, Mount Everest, is 29,028 ft (*Source: The World Almanac and Book of Facts* 1996). Find the difference in elevation.

129. *Underwater elevation.* The deepest point in the Pacific Ocean is the Marianas Trench, with a depth of 10,415 m. The deepest point in the Atlantic Ocean is the Puerto Rico Trench, with a depth of 8648 m. What is the difference in elevation of the two trenches?

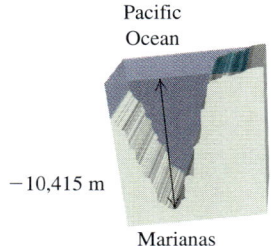

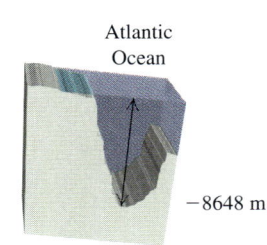

130. *Changes in elevation.* The lowest point in Africa is Lake Assal, which is 156 m below sea level. The lowest point in South America is the Valdes Peninsula, which is 40 m below sea level. How much lower is Lake Assal than the Valdes Peninsula?

SKILL MAINTENANCE

131. Find the area of a rectangle when the length is 36 ft and the width is 12 ft.

132. Find the prime factorization of 864.

SYNTHESIS

133. ◆ Why might it be advantageous to rewrite a long series of additions and subtractions as all additions?

134. ◆ Explain why $-a + b$ is the opposite of $a + (-b)$.

135. ◆ A student claims to be able to add real numbers but unable to subtract them. What advice would you offer this student?

136. ◆ Under what circumstances will the difference of two negative numbers be negative?

Tell whether each statement is true or false for all real numbers m and n. Use various replacements for m and n to support your answer.

137. If $m > n$, then $m + n > 0$.
138. If $m > n$, then $m - n > 0$.
139. If m and n are opposites, then $m - n = 0$.
140. If $m = -n$, then $m + n = 0$.

141. ◆ A gambler loses a wager and then loses "double or nothing" (meaning the gambler owes twice as much) twice more. After the three losses, the gambler's assets are −$20. Explain how much the gambler originally bet and how the $20 debt occurred.

142. ◆ If n is positive and m is negative, what is the sign of $n + (-m)$? Why?

1.7 Multiplication and Division of Real Numbers

Multiplication • Division

We now develop rules for multiplication and division of real numbers. Because multiplication and division are closely related, the rules are quite similar.

Multiplication

We already know how to multiply two nonnegative numbers. To see how to multiply a positive number and a negative number, consider the following pattern in which multiplication is regarded as repeated addition:

This number → $4(-5) = (-5) + (-5) + (-5) + (-5) = -20$ ← This number
decreases $\quad\ 3(-5) = \qquad\quad (-5) + (-5) + (-5) = -15 \quad$ increases
by 1 each $\quad\ 2(-5) = \qquad\qquad\qquad\ \ (-5) + (-5) = -10 \quad$ by 5 each
time. $\qquad\ 1(-5) = \qquad\qquad\qquad\qquad\quad\ \ (-5) = -5 \qquad$ time.
$\qquad\qquad\ 0(-5) = \qquad\qquad\qquad\qquad\qquad\qquad 0 = 0$

This pattern illustrates that the product of a negative number and a positive number is negative.

THE PRODUCT OF A NEGATIVE NUMBER AND A POSITIVE NUMBER

To multiply a positive number and a negative number, multiply their absolute values. The answer is negative.

EXAMPLE 1   Multiply: **(a)** $8(-5)$; **(b)** $-\frac{1}{3} \cdot \frac{5}{7}$.

SOLUTION

a) $8(-5) = -40$ *Think*: $8 \cdot 5 = 40$; make the answer negative.
b) $-\frac{1}{3} \cdot \frac{5}{7} = -\frac{5}{21}$ *Think*: $\frac{1}{3} \cdot \frac{5}{7} = \frac{5}{21}$; make the answer negative.

The pattern developed above includes not just products of positive and negative numbers, but a product involving zero as well.

THE MULTIPLICATIVE PROPERTY OF ZERO

For any real number a,
$$0 \cdot a = a \cdot 0 = 0.$$
(The product of 0 and any real number is 0.)

EXAMPLE 2 Multiply: $173(-452)0$.

SOLUTION We have

$173(-452)0 = 173[(-452)0]$ Using the associative law of multiplication
$ = 173[0]$ Using the multiplicative property of zero
$ = 0.$ Using the multiplicative property of zero again

Note that whenever 0 appears as a factor, the product will be 0.

We can extend the above pattern still further to examine the product of two negative numbers.

This number decreases by 1 each time. →
$2(-5) = (-5) + (-5) = -10$ ← This number increases by 5 each time.
$1(-5) = (-5) = -5$
$0(-5) = 0 = 0$
$-1(-5) = -(-5) = 5$
$-2(-5) = -(-5) - (-5) = 10$

According to the pattern, the product of two negative numbers is positive.

THE PRODUCT OF TWO NEGATIVE NUMBERS

To multiply two negative numbers, multiply their absolute values. The answer is positive.

EXAMPLE 3 Multiply: **(a)** $(-5)(-7)$; **(b)** $(-1.2)(-3)$.

SOLUTION

a) The absolute value of -5 is 5 and the absolute value of -7 is 7. Thus,

$(-5)(-7) = 5 \cdot 7$ Multiplying absolute values
$ = 35.$

b) $(-1.2)(-3) = (1.2)(3)$ Multiplying absolute values
$ = 3.6$ Try to go directly to this step.

When three or more numbers are multiplied, we can order and group the numbers as we please, because of the commutative and associative laws.

EXAMPLE 4 Multiply: **(a)** $-3(-2)(-5)$; **(b)** $-4(-6)(-1)(-2)$.

SOLUTION

a) $-3(-2)(-5) = 6(-5)$ Multiplying the first two numbers. The product of two negatives is positive.
$ = -30$ The product of a positive and a negative is negative.

b) $-4(-6)(-1)(-2) = 24 \cdot 2$ Multiplying the first two numbers and the last two numbers
$ = 48$

We can see the following pattern in the results of Example 4.

The product of an even number of negative numbers is positive.
The product of an odd number of negative numbers is negative.

Division

Recall that $a \div b$, or $\frac{a}{b}$, is the number, if one exists, that when multiplied by b gives a. For example, to show that $10 \div 2$ is 5, we need only note that $5 \cdot 2 = 10$. Thus division can always be checked with multiplication.

EXAMPLE 5 Divide, if possible, and check your answer.

a) $14 \div (-7)$ **b)** $\dfrac{-32}{-4}$ **c)** $\dfrac{-10}{7}$ **d)** $\dfrac{-17}{0}$

SOLUTION

a) $14 \div (-7) = -2$ We look for a number that when multiplied by -7 gives 14. That number is -2. *Check*: $(-2)(-7) = 14$.

b) $\dfrac{-32}{-4} = 8$ We look for a number that when multiplied by -4 gives -32. That number is 8. *Check*: $8(-4) = -32$.

c) $\dfrac{-10}{7} = -\dfrac{10}{7}$ We look for a number that when multiplied by 7 gives -10. That number is $-\frac{10}{7}$. *Check*: $-\frac{10}{7} \cdot 7 = -10$.

d) $\dfrac{-17}{0}$ is **undefined**.   We look for a number that when multiplied by 0 gives -17. There is no such number because the product of 0 and *any* number is 0, not -17.

The sign rules for division are the same as those for multiplication: The quotient of a positive number and a negative number is negative; the quotient of two negative numbers is positive.

RULES FOR MULTIPLICATION AND DIVISION

To multiply or divide two real numbers:

1. Using the absolute values, multiply or divide, as indicated.
2. If the signs are the same, the answer is positive.
3. If the signs are different, the answer is negative.

Had Example 5(a) been written as $-14 \div 7$ or $-\frac{14}{7}$, rather than $14 \div (-7)$, the result would still have been -2. Thus from Examples 5(a) and 5(b), we have the following:

$$\frac{-a}{b} = \frac{a}{-b} = -\frac{a}{b}$$

and

$$\frac{-a}{-b} = \frac{a}{b}.$$

1.7 MULTIPLICATION AND DIVISION OF REAL NUMBERS

EXAMPLE 6 Rewrite each of the following in two equivalent forms: (a) $\frac{5}{-2}$; (b) $-\frac{3}{10}$.

SOLUTION We use one of the properties just listed.

a) $\frac{5}{-2} = \frac{-5}{2}$ and $\frac{5}{-2} = -\frac{5}{2}$

b) $-\frac{3}{10} = \frac{-3}{10}$ and $-\frac{3}{10} = \frac{3}{-10}$

Since $\frac{-a}{b} = \frac{a}{-b} = -\frac{a}{b}$

When a fraction has a negative sign, it may help to rewrite (or simply visualize) the fraction in an equivalent form.

EXAMPLE 7 Perform the indicated operation: (a) $\left(-\frac{4}{5}\right)\left(\frac{-7}{3}\right)$; (b) $-\frac{2}{7} + \frac{9}{-7}$.

SOLUTION

a) $\left(-\frac{4}{5}\right)\left(\frac{-7}{3}\right) = \left(-\frac{4}{5}\right)\left(-\frac{7}{3}\right)$ Rewriting $\frac{-7}{3}$ as $-\frac{7}{3}$

$= \frac{28}{15}$ Try to go directly to this step.

b) Given a choice, we generally choose a positive denominator:

$-\frac{2}{7} + \frac{9}{-7} = \frac{-2}{7} + \frac{-9}{7}$ Rewriting both fractions with a common denominator of 7

$= \frac{-11}{7}$, or $-\frac{11}{7}$.

EXAMPLE 8 Find the reciprocal: (a) -27; (b) $\frac{-3}{4}$; (c) $-\frac{1}{5}$.

SOLUTION

a) The reciprocal of -27 is $\frac{1}{-27}$. More often, this number is written as $-\frac{1}{27}$.
b) The reciprocal of $\frac{-3}{4}$ is $\frac{4}{-3}$, or, equivalently, $-\frac{4}{3}$.
c) The reciprocal of $-\frac{1}{5}$ is -5.

Keep in mind that the opposite, or additive inverse, of a number is what we add to the number to get 0, whereas a reciprocal is what we multiply the number by to get 1. Compare the following.

Number	Opposite (Change the sign.)	Reciprocal (Invert but do not change the sign.)
$-\frac{3}{8}$	$\frac{3}{8}$	$-\frac{8}{3}$
19	-19	$\frac{1}{19}$
0	0	Undefined

$\left(-\frac{3}{8}\right)\left(-\frac{8}{3}\right) = 1$

$-\frac{3}{8} + \frac{3}{8} = 0$

When dividing with fractional notation, it is usually easier to multiply by a reciprocal. With decimal notation, it is usually easier to carry out division.

EXAMPLE 9   Divide: (a) $-\frac{2}{3} \div \left(-\frac{5}{4}\right)$; (b) $-\frac{3}{4} \div \frac{3}{10}$; (c) $27.9 \div (-3)$.

SOLUTION

a) $-\dfrac{2}{3} \div \left(-\dfrac{5}{4}\right) = -\dfrac{2}{3} \cdot \left(-\dfrac{4}{5}\right) = \dfrac{8}{15}$ *Multiplying by the reciprocal*

> Be careful not to change the sign when taking a reciprocal!

b) $-\dfrac{3}{4} \div \dfrac{3}{10} = -\dfrac{3}{4} \cdot \left(\dfrac{10}{3}\right) = -\dfrac{30}{12} = -\dfrac{5}{2} \cdot \dfrac{6}{6} = -\dfrac{5}{2}$ *Removing a factor equal to 1: $\frac{6}{6} = 1$*

c) $27.9 \div (-3) = \dfrac{27.9}{-3} = -9.3$ *Do the division $3\overline{)27.9}$ = 9.3. The answer is negative.*

In Example 5(d), we explained why we cannot divide -17 by 0. This also explains why *no* nonzero number b can be divided by 0: Consider $b \div 0$. Is there a number that when multiplied by 0 gives b? No, because the product of 0 and any number is 0, not b. We say that $b \div 0$ is **undefined** for $b \neq 0$.

On the other hand, if we divide 0 by 0, we look for a number r such that $0 \div 0 = r$ and $r \cdot 0 = 0$. But, $r \cdot 0 = 0$ for *any* number r. Thus it appears that $0 \div 0$ could be any number we choose. Getting any answer we want when we divide 0 by 0 would lead to contradictions. Thus we say that $0 \div 0$ is **indeterminate**.

Finally, note that $0 \div 7 = 0$ since $0 \cdot 7 = 0$. This can be written $0/7 = 0$.

EXAMPLE 10   Divide, if possible: (a) $\frac{0}{-2}$; (b) $\frac{5}{0}$; (c) $\frac{0}{0}$.

SOLUTION

a) $\dfrac{0}{-2} = 0$ *Check: $0(-2) = 0$.*

b) $\dfrac{5}{0}$ is undefined.

c) $\dfrac{0}{0}$ is indeterminate.

DIVISION INVOLVING ZERO

For any nonzero real number a,

$$\dfrac{0}{a} = 0 \quad \text{and} \quad \dfrac{a}{0} \text{ is undefined.}$$

The expression $\frac{0}{0}$ is indeterminate.

EXERCISE SET 1.7

Multiply.

1. $-9 \cdot 3$
2. $-3 \cdot 7$
3. $-8 \cdot 7$
4. $-9 \cdot 2$
5. $8 \cdot (-3)$
6. $9 \cdot (-5)$
7. $-9 \cdot 8$
8. $-10 \cdot 3$
9. $-8 \cdot (-2)$
10. $-2 \cdot (-5)$
11. $-5 \cdot (-9)$
12. $-9 \cdot (-2)$
13. $15 \cdot (-8)$
14. $-12 \cdot (-10)$
15. $-14 \cdot 17$
16. $-13 \cdot (-15)$
17. $-25 \cdot (-48)$
18. $39 \cdot (-43)$
19. $-3.5 \cdot (-28)$
20. $97 \cdot (-2.1)$
21. $6 \cdot (-13)$
22. $7 \cdot (-9)$
23. $-7 \cdot (-3.1)$
24. $-4 \cdot (-3.2)$
25. $\frac{2}{3} \cdot \left(-\frac{3}{5}\right)$
26. $\frac{5}{7} \cdot \left(-\frac{2}{3}\right)$
27. $-\frac{3}{8} \cdot \left(-\frac{2}{9}\right)$
28. $-\frac{5}{8} \cdot \left(-\frac{2}{5}\right)$
29. -6.3×2.7
30. -4.1×9.5
31. $-\frac{5}{9} \cdot \frac{3}{4}$
32. $-\frac{8}{3} \cdot \frac{9}{4}$
33. $3 \cdot (-7) \cdot (-2) \cdot 6$
34. $9 \cdot (-2) \cdot (-6) \cdot 7$
35. $-2 \cdot (-5) \cdot (-9)$
36. $-4 \cdot (-3) \cdot (-5)$
37. $-\frac{1}{3} \cdot \frac{1}{4} \cdot \left(-\frac{3}{7}\right)$
38. $-\frac{1}{2} \cdot \frac{3}{5} \cdot \left(-\frac{2}{7}\right)$
39. $-2 \cdot (-5) \cdot (-3) \cdot (-5)$
40. $-3 \cdot (-5) \cdot (-2) \cdot (-1)$
41. $(-14) \cdot (-27) \cdot 0$
42. $7 \cdot (-6) \cdot 5 \cdot (-4) \cdot 3 \cdot (-2) \cdot 1 \cdot 0$
43. $(-8)(-9)(-10)$
44. $(-7)(-8)(-9)(-10)$
45. $(-6)(-7)(-8)(-9)(-10)$
46. $(-5)(-6)(-7)(-8)(-9)(-10)$

Divide, if possible, and check. If a quotient is undefined or indeterminate, state this.

47. $24 \div (-4)$
48. $\frac{28}{-7}$
49. $\frac{36}{-9}$
50. $26 \div (-13)$
51. $\frac{-16}{8}$
52. $-32 \div (-4)$
53. $\frac{-48}{-12}$
54. $-63 \div (-9)$
55. $\frac{-72}{9}$
56. $\frac{-50}{25}$
57. $-100 \div (-50)$
58. $\frac{-200}{8}$
59. $-108 \div 9$
60. $\frac{-64}{-7}$
61. $\frac{400}{-50}$
62. $-300 \div (-13)$
63. $\frac{28}{0}$
64. $\frac{0}{-5}$
65. $\frac{88}{-9}$
66. $\frac{0}{0}$
67. $\frac{0}{-9}$
68. $\frac{-35}{0}$
69. $0 \div 0$
70. $0 \div (-47)$

Write each number in two equivalent forms, as in Example 6.

71. $\frac{-8}{3}$
72. $\frac{-12}{7}$
73. $\frac{29}{-35}$
74. $\frac{9}{-14}$
75. $-\frac{7}{3}$
76. $-\frac{4}{15}$
77. $\frac{-x}{2}$
78. $\frac{9}{-a}$

Find the reciprocal of each number.

79. $\frac{4}{-5}$
80. $\frac{2}{-9}$
81. $-\frac{47}{13}$
82. $-\frac{31}{12}$
83. -10
84. 13
85. 4.3
86. -8.5
87. $\frac{-9}{4}$
88. $\frac{-6}{11}$
89. -1
90. $\frac{1}{1/2}$

Perform the indicated operation and simplify, if possible. If a quotient is undefined or indeterminate, state this.

91. $\left(\frac{-7}{4}\right)\left(-\frac{3}{5}\right)$
92. $\left(-\frac{5}{6}\right)\left(\frac{-1}{3}\right)$
93. $\left(\frac{-6}{5}\right)\left(\frac{2}{-11}\right)$
94. $\left(\frac{7}{-2}\right)\left(\frac{-5}{6}\right)$
95. $\frac{3}{-8} + \frac{-5}{8}$
96. $\frac{-4}{5} + \frac{7}{-5}$
97. $\left(\frac{-9}{5}\right)\left(-\frac{10}{7}\right)$
98. $\left(-\frac{2}{7}\right)\left(\frac{5}{-8}\right)$

50 CHAPTER 1 • INTRODUCTION TO ALGEBRAIC EXPRESSIONS

99. $\left(-\frac{3}{11}\right) + \frac{5}{-11}$
100. $\frac{-9}{7} + \left(-\frac{4}{7}\right)$
101. $\frac{7}{8} \div \left(-\frac{1}{2}\right)$
102. $\frac{3}{4} \div \left(-\frac{2}{3}\right)$
103. $\frac{9}{5} \cdot \frac{-20}{3}$
104. $\frac{-5}{12} \cdot \frac{7}{15}$
105. $\left(-\frac{18}{7}\right) + \left(-\frac{3}{7}\right)$
106. $\left(-\frac{12}{5}\right) + \left(-\frac{3}{5}\right)$
107. $-\frac{5}{9} \div \left(-\frac{5}{6}\right)$
108. $-\frac{5}{4} \div \left(-\frac{3}{4}\right)$
109. $-44.1 \div (-6.3)$
110. $-6.6 \div 3.3$
111. $\frac{-5}{9} - \frac{2}{9}$
112. $\frac{-3}{7} - \frac{2}{7}$
113. $\frac{-3}{10} + \frac{2}{-5}$
114. $\frac{-5}{9} + \frac{2}{-3}$
115. $\frac{7}{10} \div \left(\frac{-3}{5}\right)$
116. $\left(\frac{-3}{5}\right) \div \frac{6}{15}$
117. $\frac{5}{7} - \frac{1}{-7}$
118. $\frac{4}{9} - \frac{1}{-9}$
119. $\frac{-4}{15} + \frac{2}{-3}$
120. $\frac{3}{-10} + \frac{-1}{5}$

SKILL MAINTENANCE

121. Simplify: $\frac{264}{468}$.

122. Combine like terms: $x + 12y + 11x - 14y - 9$.

SYNTHESIS

123. ◆ Multiplication can be regarded as repeated addition. Using this idea and a number line, explain why $3 \cdot (-5) = -15$.

124. ◆ Most calculators have a key, often appearing as $\boxed{1/x}$, for finding reciprocals. To use this key, enter a number and then press $\boxed{1/x}$ to find its reciprocal. What should happen if you enter a number and then press the reciprocal key twice? Why?

125. ◆ What advice would you offer a student who claims to be able to multiply, but not divide, any two real numbers?

126. ◆ If two numbers are opposites of each other, are their reciprocals opposites of each other? Why or why not?

127. Show that the reciprocal of a sum is *not* the sum of the two reciprocals.

128. Which real numbers are their own reciprocals?

Tell whether the expression represents a positive number or a negative number when m and n are negative.

129. $\frac{m}{-n}$
130. $\frac{-n}{-m}$
131. $-m \cdot \left(\frac{-n}{m}\right)$
132. $-\left(\frac{n}{-m}\right)$
133. $(m + n) \cdot \frac{m}{n}$
134. $(-n - m)\frac{n}{m}$

135. What must be true of m and n if $-mn$ is to be (a) positive? (b) zero? (c) negative?

136. The following is a proof that a positive number times a negative number is negative. Provide a reason for each step. Assume that $a > 0$ and $b > 0$.
$$a(-b) + ab = a[-b + b]$$
$$= a(0)$$
$$= 0$$
Therefore, $a(-b) = -ab$.

137. ◆ Is it true that for any numbers a and b, if a is larger than b, then the reciprocal of a is smaller than the reciprocal of b? Why or why not?

1.8 Exponential Notation and Order of Operations

Exponential Notation • Order of Operations • Simplifying and the Distributive Law • The Opposite of a Sum

Algebraic expressions often contain *exponential notation*. In this section, we learn how to use exponential notation as well as rules for the *order of operations*, in performing certain algebraic manipulations.

Exponential Notation

A product like $3 \cdot 3 \cdot 3 \cdot 3$, in which the factors are the same, is called a **power**. Powers occur often enough that a simpler notation called **exponential notation** is used. For

$$\underbrace{3 \cdot 3 \cdot 3 \cdot 3}_{4 \text{ factors}}, \quad \text{we write} \quad 3^4.$$

This is read "three to the fourth power," or, simply, "three to the fourth." The number 4 is called an **exponent** and the number 3 a **base**.

Expressions like s^2 and s^3 are usually read "s squared" and "s cubed," respectively. This comes from the fact that a square of side s has an area A given by $A = s^2$ and a cube of side s has a volume V given by $V = s^3$.

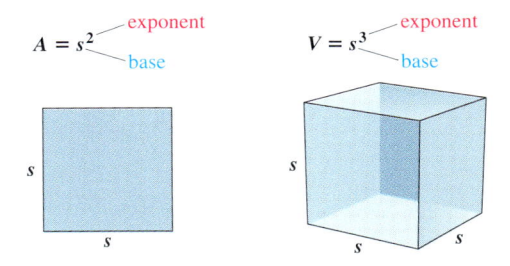

$A = s^2$ — exponent, base

$V = s^3$ — exponent, base

EXAMPLE 1 Write exponential notation for $10 \cdot 10 \cdot 10 \cdot 10 \cdot 10$.

SOLUTION

Exponential notation is 10^5.   5 is the exponent.
10 is the base.

EXAMPLE 2 Evaluate: **(a)** 5^2; **(b)** $(-5)^3$; **(c)** $(2n)^3$.

SOLUTION

a) $5^2 = 5 \cdot 5 = 25$ The second power indicates two factors of 5.

b) $(-5)^3 = (-5)(-5)(-5)$
$= 25(-5)$ Using the associative law of multiplication
$= -125$

c) $(2n)^3 = (2n)(2n)(2n)$
$= 2 \cdot 2 \cdot 2 \cdot n \cdot n \cdot n$ Using the associative and commutative laws of multiplication
$= 8n^3$

To determine what the exponent 1 will mean, look for a pattern in the following:

$$7 \cdot 7 \cdot 7 \cdot 7 = 7^4$$
$$7 \cdot 7 \cdot 7 = 7^3$$
$$7 \cdot 7 = 7^2$$
$$7 = 7^?$$

We divide by 7 each time.

The exponents decrease by 1 each time. To continue the pattern, we say that $7 = 7^1$.

EXPONENTIAL NOTATION

For any natural number n,
$$b^n \text{ means } \underbrace{b \cdot b \cdot b \cdot b \cdots b}_{n \text{ factors}}.$$

Order of Operations

How should $4 + 2 \times 5$ be computed? If we multiply 2 by 5 and add 4, the result is 14. If we add 2 and 4 and then multiply by 5, the result is 30. Since these results differ, the order in which we perform operations is clearly important. Sometimes grouping symbols such as parentheses ( ), brackets [ ], braces { }, absolute-value symbols | |, or fraction bars tell us what to do first. For example,

$$(4 + 2) \times 5 = 6 \times 5 = 30$$

and

$$4 + (2 \times 5) = 4 + 10 = 14.$$

Besides grouping symbols, there are other rules for determining the order in which operations should be performed.

RULES FOR ORDER OF OPERATIONS

1. Perform all calculations within grouping symbols.
2. Simplify all exponential expressions.
3. Perform all multiplication and division, in order, from left to right.
4. Perform all addition and subtraction, in order, from left to right.

Thus the correct way to compute $4 + 2 \times 5$ is to first multiply 2 by 5 and then add 4. The result is 14.

EXAMPLE 3 Simplify: $15 - 2 \times 5 + 3$.

SOLUTION When no groupings or exponents appear, we always multiply or divide before adding or subtracting:

$$\begin{aligned} 15 - 2 \times 5 + 3 &= 15 - 10 + 3 \quad &\text{Multiplying} \\ &= 5 + 3 \\ &= 8. \end{aligned} \quad \text{Subtracting and adding from left to right}$$

Always calculate within parentheses first. When there are exponents and no parentheses, simplify powers before multiplying or dividing.

1.8 EXPONENTIAL NOTATION AND ORDER OF OPERATIONS 53

EXAMPLE 4 Simplify: **(a)** $(3 \times 4)^2$; **(b)** 3×4^2.

SOLUTION

a) $(3 \times 4)^2 = (12)^2$ Working within parentheses first
$= 144$

b) $3 \times 4^2 = 3 \times 16$ Simplifying the power
$= 48$ Multiplying

Note that $(3 \times 4)^2 \neq 3 \times 4^2$.

CAUTION! Example 4 illustrates that, in general, $(ab)^2 \neq ab^2$.

EXAMPLE 5 Evaluate for $x = 5$: **(a)** $(-x)^2$; **(b)** $-x^2$.

SOLUTION

a) $(-x)^2 = (-5)^2 = (-5)(-5) = 25$ We square the opposite of 5.

b) $-x^2 = -(5)^2 = -25$ We square 5 and then find the opposite.

CAUTION! Example 5 illustrates that, in general, $(-x)^2 \neq -x^2$.

EXAMPLE 6 Evaluate $-15 \div 3(6 - a)^3$ for $a = 4$.

SOLUTION

$-15 \div 3(6 - a)^3 = -15 \div 3(6 - 4)^3$ Substituting 4 for a
$= -15 \div 3(2)^3$ Working within parentheses first
$= -15 \div 3 \cdot 8$ Simplifying the exponential expression
$= -5 \cdot 8$
$= -40$ Dividing and multiplying from left to right

When combinations of grouping symbols are used, the rules still apply. We begin with the innermost grouping symbols and work to the outside.

EXAMPLE 7 Simplify: $8 \div 4 + 3[9 + 2(3 - 5)^3]$.

SOLUTION

$8 \div 4 + 3[9 + 2(3 - 5)^3] = 8 \div 4 + 3[9 + 2(-2)^3]$ Doing the calculations in the innermost parentheses first

$= 8 \div 4 + 3[9 + 2(-8)]$ $(-2)^3 = (-2)(-2)(-2) = -8$

$= 8 \div 4 + 3[9 + (-16)]$

$= 8 \div 4 + 3[-7]$ Completing the calculations within the brackets

$= 2 + (-21)$ Multiplying and dividing from left to right

$= -19$

54 CHAPTER 1 • INTRODUCTION TO ALGEBRAIC EXPRESSIONS

EXAMPLE 8 Calculate: $\dfrac{12(9 - 7) + 4 \cdot 5}{3^4 + 2^3}$.

SOLUTION An equivalent expression with brackets is

$$[12(9 - 7) + 4 \cdot 5] \div [3^4 + 2^3].$$

In effect, we need to simplify the numerator, simplify the denominator, and then divide the results:

$$\frac{12(9 - 7) + 4 \cdot 5}{3^4 + 2^3} = \frac{12(2) + 4 \cdot 5}{81 + 8}$$
$$= \frac{24 + 20}{89} = \frac{44}{89}.$$

Simplifying and the Distributive Law

Sometimes we cannot simplify within parentheses. When a sum or difference is within the parentheses, we can use the distributive law to help simplify.

EXAMPLE 9 Simplify: $5x - 9 + 2(4x + 5)$.

SOLUTION

$5x - 9 + 2(4x + 5) = 5x - 9 + 8x + 10$ Using the distributive law
$ = 13x + 1$ Combining like terms

Now that exponents have been introduced, we can make our definition of *like* or *similar terms* more precise. **Like,** or **similar, terms** are either constant terms or terms containing the same variable(s) raised to the same power(s). Thus, 5 and -7, $19xy$ and $2yx$, and $4a^3b$ and a^3b are all pairs of like terms.

EXAMPLE 10 Simplify: $7x^2 + 3(x^2 + 2x) - 5x$.

SOLUTION

$7x^2 + 3(x^2 + 2x) - 5x = 7x^2 + 3x^2 + 6x - 5x$ Using the distributive law
$ = 10x^2 + x$ Combining like terms

The Opposite of a Sum

When a number is multiplied by -1, the result is the opposite of that number. For example, $-1(7) = -7$ and $-1(-5) = 5$.

THE PROPERTY OF -1

For any real number a,
$$-1 \cdot a = -a.$$
(Negative one times a is the opposite of a.)

When grouping symbols are preceded by a "−" symbol, we can multiply the grouping by −1 and use the distributive law. In this manner, we can find the *opposite*, or *additive inverse, of a sum.*

EXAMPLE 11 Write an expression equivalent to $-(3x + 2y + 4)$ without using parentheses.

SOLUTION

$$-(3x + 2y + 4) = -1(3x + 2y + 4) \qquad \text{Using the property of } -1$$
$$= -1(3x) + (-1)(2y) + (-1)4 \qquad \text{Using the distributive law}$$
$$= -3x - 2y - 4 \qquad \text{Using the property of } -1$$

Example 11 illustrates an important property of real numbers.

THE OPPOSITE OF A SUM

For any real numbers a and b,
$$-(a + b) = -a + (-b).$$
(The opposite of a sum is the sum of the opposites.)

To remove parentheses from an expression like $-(x - 7y + 5)$, we can first rewrite the subtraction as addition:

$$-(x - 7y + 5) = -(x + (-7y) + 5) \qquad \text{Rewriting as addition}$$
$$= -x + 7y - 5. \qquad \text{Taking the opposite of a sum}$$

This procedure is normally streamlined to one step in which we find the opposite by "removing parentheses and changing the sign of every term":

$$-(x - 7y + 5) = -x + 7y - 5.$$

EXAMPLE 12 Simplify: $3x - (4x + 2)$.

SOLUTION

$$3x - (4x + 2) = 3x + [-(4x + 2)]$$
$$= 3x + [-4x - 2]$$
$$= 3x + (-4x) + (-2) \qquad \text{Adding the opposite of } 4x + 2$$
$$= 3x - 4x - 2 \qquad \text{Try to go directly to this step.}$$
$$= -x - 2 \qquad \text{Combining like terms}$$

In practice, the first three steps of Example 12 are generally skipped.

EXAMPLE 13 Simplify: $5t^2 - 2t - (4t^2 - 9t)$.

SOLUTION

$$5t^2 - 2t - (4t^2 - 9t) = 5t^2 - 2t - 4t^2 + 9t \qquad \text{Removing parentheses and changing the sign of each term inside}$$
$$= t^2 + 7t \qquad \text{Combining like terms}$$

Expressions such as $7 - 3(x + 2)$ can be simplified as follows:

$$7 - 3(x + 2) = 7 + [-3(x + 2)] \quad \text{Adding the opposite of } 3(x + 2)$$
$$= 7 + [-3x - 6] \quad \text{Multiplying } x + 2 \text{ by } -3$$
$$= 7 - 3x - 6 \quad \text{Try to go directly to this step.}$$
$$= 1 - 3x. \quad \text{Combining like terms}$$

EXAMPLE 14 Simplify: **(a)** $3n - 2(4n - 5)$; **(b)** $7x^3 + 2 - [5(x^3 - 1) + 8]$.

SOLUTION

a) $3n - 2(4n - 5) = 3n - 8n + 10$ ⟵ Multiplying each term in the parentheses by -2
$\qquad\qquad\qquad\quad = -5n + 10$ ⟵ Combining like terms

b) $7x^3 + 2 - [5(x^3 - 1) + 8] = 7x^3 + 2 - [5x^3 - 5 + 8]$ ⟵ Removing parentheses
$\qquad\qquad\qquad\qquad\qquad = 7x^3 + 2 - [5x^3 + 3]$
$\qquad\qquad\qquad\qquad\qquad = 7x^3 + 2 - 5x^3 - 3$ ⟵ Removing brackets
$\qquad\qquad\qquad\qquad\qquad = 2x^3 - 1$ ⟵ Combining like terms

EXERCISE SET 1.8

Write exponential notation.

1. $17 \times 17 \times 17$
2. $5 \times 5 \times 5 \times 5$
3. $x \cdot x \cdot x \cdot x \cdot x \cdot x \cdot x$
4. $y \cdot y \cdot y \cdot y \cdot y \cdot y$
5. $6y \cdot 6y \cdot 6y \cdot 6y$
6. $5m \cdot 5m \cdot 5m \cdot 5m \cdot 5m$

Simplify.

7. 3^4
8. 5^3
9. $(-3)^2$
10. $(-7)^2$
11. $(-1)^5$
12. $(-1)^8$
13. 4^3
14. 9^1
15. $(-5)^4$
16. 5^4
17. 7^1
18. $(-1)^7$
19. $(2x)^4$
20. $(3x)^2$
21. $(-7x)^3$
22. $(-5x)^4$
23. $5 + 3 \times 7$
24. $9 - 4 \times 2$
25. $8 \times 7 + 6 \times 5$
26. $10 \times 5 + 1 \times 1$
27. $19 - 5 \times 3 + 3$
28. $14 - 2 \times 6 + 7$
29. $9 \div 3 + 16 \div 8$
30. $32 - 8 \div 4 - 2$
31. $7 + 10 - 10 \div 2$
32. $(2 - 5)^2$
33. $2 \cdot 5^3$
34. $3 \cdot 2^3$
35. $8 - 2 \cdot 3 - 9$
36. $8 - (2 \cdot 3 - 9)$
37. $(8 - 2 \cdot 3) - 9$
38. $(8 - 2)(3 - 9)$
39. $(-24) \div (-3) \cdot \left(-\frac{1}{2}\right)$
40. $32 \div (-2) \cdot (-2)$
41. $13(-10) + 45$
42. $7 \cdot 8 - 9 \cdot 6$
43. $2^4 + 2^3 - 10$
44. $40 - 3^2 - 2^3$
45. $5^3 + 26 \cdot 71 - (16 + 25 \cdot 3)$
46. $4^3 + 10 \cdot 20 + 8^2 - 23$
47. $[2 \cdot (5 - 3)]^2$
48. $5^3 - 7^2$
49. $\dfrac{7 + 2}{5^2 - 4^2}$
50. $\dfrac{5^2 - 3^2}{2 \cdot 6 - 4}$
51. $8(-7) + |6(-5)|$
52. $|10(-5)| + 1(-1)$
53. $19 - 5(-3) + 3$
54. $14 - 2(-6) + 7$
55. $9 \div (-3) \cdot 16 \div 8$
56. $-32 - 8 \div 4 \cdot (-2)$
57. $20 + 4^3 \div (-8) \cdot 2$
58. $4(-10)^3 - 5000$
59. $3|7 - (9 - 14)|$
60. $5|8 - 3 \cdot 7|$

Evaluate.

61. $9 - 5x$, for $x = 3$
62. $7 + x^3$, for $x = -2$
63. $24 \div t^3$, for $t = -2$
64. $20 \div a \cdot 4$, for $a = 5$
65. $45 \div 3a$, for $a = 3$
66. $50 \div 2t$, for $t = 5$
67. $5x \div 15x^2$, for $x = 3$
68. $6a \div 12a^3$, for $a = 2$
69. $-20 \div t^2 - 3(t - 1)$, for $t = -4$
70. $-30 \div t(t + 4)^2$, for $t = -6$
71. $-x^2 - 5x$, for $x = -3$
72. $(-x)^2 - 5x$, for $x = -3$
73. $\dfrac{3a - 4a^2}{a^2 - 20}$, for $a = 5$
74. $\dfrac{a^3 - 4a}{a(a - 3)}$, for $a = -2$

Rename each expression without using parentheses.

75. $-(9x + 1)$
76. $-(3x + 5)$
77. $-(7 - 2x)$
78. $-(6x - 7)$
79. $-(4a - 3b + 7c)$
80. $-(5x - 2y - 3z)$
81. $-(3x^2 + 5x - 1)$
82. $-(8x^3 - 6x + 5)$

Remove parentheses and simplify.

83. $5x - (2x + 7)$
84. $7y - (2y + 9)$
85. $2a - (5a - 9)$
86. $11n - (3n - 7)$
87. $2x + 7x - (4x + 6)$
88. $3a + 2a - (4a + 7)$
89. $9t - 5r - 2(3r + 6t)$
90. $4m - 9n - 3(2m - n)$
91. $15x - y - 5(3x - 2y + 5z)$
92. $4a - b - 4(5a - 7b + 8c)$
93. $3x^2 + 7 - (2x^2 + 5)$
94. $7x^4 + 9x - (5x^4 + 3x)$
95. $5t^3 + t - 3(t + 2t^3)$
96. $8n^2 + n - 2(n + 3n^2)$
97. $12a^2 - 3ab + 5b^2 - 5(-5a^2 + 4ab - 6b^2)$
98. $-8a^2 + 5ab - 12b^2 - 6(2a^2 - 4ab - 10b^2)$
99. $-7t^3 - t^2 - 3(5t^3 - 3t)$
100. $9t^4 + 7t - 5(9t^3 - 2t)$

101. $7(x + 2) - 5(3x - 4)$
102. $5(x - 3) - 4(2x - 1)$
103. $6(a^3 + a) - 12 - 2(a^3 + 2a - 1)$
104. $2(t^2 - 2t) - 3(t^2 - 4t) + t^2$
105. $6(3x - 7) - [4(2x - 5) + 2]$
106. $4(7x - 5) - [3(1 - 8x) + 6]$

SKILL MAINTENANCE

Translate to an algebraic expression.

107. Nine more than twice a number
108. Half of the sum of two numbers

SYNTHESIS

109. ◆ Is $\dfrac{16}{2x}$ equal to 16/2x? Why or why not?
110. ◆ Write the sentence $(-x)^2 \neq -x^2$ in words. Explain why $(-x)^2$ and $-x^2$ are not equivalent.
111. ◆ Write the sentence $-|x| \neq -x$ in words. Explain why $-|x|$ and $-x$ are not equivalent.
112. ◆ Some students use the mnemonic device PEMDAS to help remember the rules for the order of operations. Explain how this can be done.

Simplify.

113. $5t - \{7t - [4r - 3(t - 7)] + 6r\} - 4r$
114. $z - \{2z - [3z - (4z - 5z) - 6z] - 7z\} - 8z$
115. $\{x - [f - (f - x)] + [x - f]\} - 3x$
116. ◆ Is it true that for any real numbers a and b,
$$-(ab) = (-a)b = a(-b)?$$
Why or why not?
117. ◆ Is it true that for any real numbers a and b,
$$ab = (-a)(-b)?$$
Why or why not?

If $n > 0$, $m > 0$, and $n \neq m$, determine whether each of the following is true.

118. $-n + m = -(n + m)$
119. $-n + m = m - n$
120. $m - n = -(n - m)$
121. $n^2 - mn = -(m - n)n$
122. $n(-n - m) = -n^2 + nm$
123. $-m(n - m) = -(mn + m^2)$
124. $-m(-n + m) = m(n - m)$
125. $-n(-n - m) = n(n + m)$

COLLABORATIVE C•O•R•N•E•R

Focus: Order of operations

Time: 10–15 minutes

Group size: 2

One exercise often used to help students learn the order of operations is to insert symbols within a display of numbers in order to obtain a predetermined result. For example, the display

1 2 3 4 5

can be used to obtain the result 21 as follows:

$(1 + 2) \div 3 + 4 \cdot 5.$

ACTIVITY

Each group member should prepare an exercise similar to the example shown at left. To do so, first select five numbers for display. Then, privately, select groupings and/or operations and calculate the result. Give your partner that result, along with the five numbers selected for display, and challenge him or her to insert symbols that will make the display equal the number given.

SUMMARY AND REVIEW 1

KEY TERMS

Variable, p. 2
Constant, p. 2
Variable expression, p. 2
Evaluate an expression, p. 2
Algebraic expression, p. 3
Substitute, p. 3
Value of an expression, p. 3
Equation, p. 5
Solution, p. 5
Equivalent expressions, p. 9
Term, p. 13
Factors, p. 13
Factorization, p. 14
Natural number, p. 16
Whole number, p. 16

Prime number, p. 16
Composite number, p. 17
Prime factorization, p. 17
Fractional notation, p. 17
Numerator, p. 17
Denominator, p. 17
Simplifying, p. 18
Reciprocal, p. 21
Multiplicative inverse, p. 21
Set, p. 24
Integer, p. 24
Rational number, p. 25
Terminating decimal, p. 26
Repeating decimal, p. 26
Irrational number, p. 27

Real-number system, p. 27
Less than, p. 28
Greater than, p. 28
Inequality, p. 29
Absolute value, p. 29
Combine like terms, p. 35
Opposite, p. 37
Additive inverse, p. 37
Undefined, p. 48
Indeterminate, p. 48
Power, p. 51
Exponential notation, p. 51
Exponent, p. 51
Base, p. 51
Like terms, p. 54

IMPORTANT PROPERTIES AND FORMULAS

Area of a rectangle: $A = lw$
Area of a triangle: $A = \frac{1}{2}bh$
Area of a parallelogram: $A = bh$
Commutative laws: $a + b = b + a, \quad ab = ba$
Associative laws: $a + (b + c) = (a + b) + c, \quad a(bc) = (ab)c$
Distributive law: $a(b + c) = ab + ac$
Identity property of 1: $1 \cdot a = a \cdot 1 = a$
Identity property of 0: $a + 0 = 0 + a = a$
Law of opposites: $a + (-a) = 0$
Multiplicative property of 0: $0 \cdot a = a \cdot 0 = 0$
Property of -1: $-1 \cdot a = -a$
Opposite of a sum: $-(a + b) = -a + (-b)$
Division involving 0: For $a \neq 0$, $\frac{0}{a} = 0$ and $\frac{a}{0}$ is undefined. $\frac{0}{0}$ is indeterminate.

$$\frac{-a}{b} = \frac{a}{-b} = -\frac{a}{b}, \quad \frac{-a}{-b} = \frac{a}{b}$$

Rules for Order of Operations

1. Perform all calculations within grouping symbols.
2. Simplify all exponential expressions.
3. Perform all multiplication and division, in order, from left to right.
4. Perform all addition and subtraction, in order, from left to right.

REVIEW EXERCISES

Evaluate.

1. $3a$, for $a = 5$
2. $\frac{x}{y}$, for $x = 12$ and $y = 2$
3. $\frac{2p}{q}$, for $p = 20$ and $q = 8$
4. $\frac{x - y}{3}$, for $x = 17$ and $y = 5$
5. $10 - y^2$, for $y = 5$
6. $-10 + a^2 \div (b + 1)$, for $a = 5$ and $b = 4$

Translate to an algebraic expression.

7. 8 less than z
8. The product of x and z
9. One more than the product of two numbers
10. Determine whether 35 is a solution of $x/5 = 8$.
11. Translate to an equation. Do not solve.

   A total of 197 countries participated in the 1996 Summer Olympics. This is 37 more than the number that participated in 1988. How many countries participated in the 1988 Olympics?

12. Use the commutative law of multiplication to write an expression equivalent to $2x + y$.
13. Use the associative law of addition to write an expression equivalent to $(2x + y) + z$.
14. Use the commutative and associative laws to write three expressions equivalent to $4(xy)$.

Multiply.
15. $6(3x + 5y)$
16. $8(5x + 3y + 2)$

Factor.
17. $21x + 15y$
18. $35x + 14 + 7y$
19. Find the prime factorization of 52.

Simplify.
20. $\dfrac{20}{48}$
21. $\dfrac{18}{8}$

Perform the indicated operation and, if possible, simplify.
22. $\dfrac{4}{9} + \dfrac{5}{12}$
23. $\dfrac{9}{16} \div 3$
24. $\dfrac{2}{3} - \dfrac{1}{15}$
25. $\dfrac{9}{10} \cdot \dfrac{16}{5}$

26. Tell which integers correspond to this situation: Renir has a debt of $45 and Raoul has $72 in his savings account.
27. Graph on a number line: $\dfrac{-1}{3}$.
28. Write an inequality with the same meaning as $-3 < x$.
29. Classify as true or false: $8 \geq 8$.
30. Classify as true or false: $0 \leq -1$.
31. Find decimal notation: $-\dfrac{7}{8}$.
32. Find the absolute value: $|-1|$.
33. Find $-(-x)$ when x is -5.

Simplify.
34. $4 + (-7)$
35. $-\dfrac{2}{3} + \dfrac{1}{12}$
36. $6 + (-9) + (-8) + 7$
37. $-3.8 + 5.1 + (-12) + (-4.3) + 10$
38. $-3 - (-7)$
39. $-\dfrac{9}{10} - \dfrac{1}{2}$
40. $-3.8 - 4.1$
41. $-9 \cdot (-6)$
42. $-2.7(3.4)$
43. $\dfrac{2}{3} \cdot \left(-\dfrac{3}{7}\right)$
44. $3 \cdot (-7) \cdot (-2) \cdot (-5)$
45. $35 \div (-5)$
46. $-5.1 \div 1.7$
47. $-\dfrac{3}{5} \div \left(-\dfrac{4}{5}\right)$
48. $|-3 \cdot 4 - 12 \cdot 2| - 8(-7)$
49. $|-12(-3) - 2^3 - (-9)(-10)|$
50. $120 - 6^2 \div 4 \cdot 8$
51. $(120 - 6^2) \div 4 \cdot 8$
52. $(120 - 6^2) \div (4 \cdot 8)$
53. $\dfrac{4(18 - 8) + 7 \cdot 9}{9^2 - 8^2}$

Combine like terms.
54. $11a + 2b + (-4a) + (-5b)$
55. $7x - 3y - 9x + 8y$

56. Find the opposite of -7.
57. Find the reciprocal of -7.
58. Write exponential notation for $2x \cdot 2x \cdot 2x \cdot 2x$.
59. Simplify: $(-3y)^3$.

Remove parentheses and simplify.
60. $2a - (5a - 9)$
61. $3(b + 7) - 5b$
62. $3[11x - 3(4x - 1)]$
63. $2[6(y - 4) + 7]$
64. $[8(x + 4) - 10] - [3(x - 2) + 4]$
65. $5\{[6(x - 1) + 7] - [3(3x - 4) + 8]\}$

SYNTHESIS

66. ◆ Explain the difference between a constant and a variable.
67. ◆ Explain the difference between a term and a factor.
68. ◆ Describe at least three ways in which the distributive law was used in this chapter.
69. ◆ Devise a rule for determining the sign of a negative quantity raised to a power.
70. Evaluate $a^{50} - 20a^{25}b^4 + 100b^8$ for $a = 1$ and $b = 2$.
71. If $0.090909\ldots = \dfrac{1}{11}$ and $0.181818\ldots = \dfrac{2}{11}$, what rational number is named by each of the following?
    a) $0.272727\ldots$
    b) $0.909090\ldots$

Simplify.
72. $-\left|\dfrac{7}{8} - \left(-\dfrac{1}{2}\right) - \dfrac{3}{4}\right|$
73. $(|2.7 - 3| + 3^2 - |-3|) \div (-3)$

CHAPTER TEST 1

1. Evaluate $\dfrac{3x}{y}$ for $x = 10$ and $y = 5$.
2. Write an algebraic expression: Nine less than some number.
3. Find the area of a triangle when the height h is 30 ft and the base b is 16 ft.
4. Use the commutative law of addition to write an expression equivalent to $3p + q$.
5. Use the associative law of multiplication to write an expression equivalent to $x \cdot (4 \cdot y)$.
6. Determine whether 3 is a solution of $96 - a = 93$.
7. Translate to an equation. Do not solve.

   On a hot summer day, a power company met a demand of 2518 megawatts. This is only 282 megawatts less than its maximum production capability. What is the maximum capability of production?

Multiply.

8. $3(6 - x)$
9. $-5(y - 1)$

Factor.

10. $11 - 44x$
11. $7x + 21 + 14y$
12. Find the prime factorization of 300.
13. Simplify: $\dfrac{10}{35}$.

Write a true sentence using either $<$ or $>$.

14. $-4 \;\square\; 0$
15. $-3 \;\square\; -8$

Find the absolute value.

16. $\left|\dfrac{9}{4}\right|$
17. $|-2.7|$
18. Find the opposite of $\dfrac{2}{3}$.
19. Find the reciprocal of $-\dfrac{4}{7}$.
20. Find $-x$ when x is -8.
21. Write an inequality with the same meaning as $x \leq -2$.

Compute and simplify.

22. $3.1 - (-4.7)$
23. $-8 + 4 + (-7) + 3$
24. $-\dfrac{1}{5} + \dfrac{3}{8}$
25. $2 - (-8)$
26. $3.2 - 5.7$
27. $\dfrac{1}{8} - \left(-\dfrac{3}{4}\right)$
28. $4 \cdot (-12)$
29. $-\dfrac{1}{2} \cdot \left(-\dfrac{3}{8}\right)$
30. $-45 \div 5$
31. $-\dfrac{3}{5} \div \left(-\dfrac{4}{5}\right)$
32. $4.864 \div (-0.5)$
33. $-2(16) - |2(-8) - 5^3|$
34. $6 + 7 - 4 - (-3)$
35. $256 \div (-16) \div 4$
36. $2^3 - 10[4 - (-2 + 18)3]$
37. Combine like terms: $18y + 30a - 9a + 4y$.
38. Simplify: $(-2x)^4$.

Remove parentheses and simplify.

39. $5x - (3x - 7)$
40. $4(2a - 3b) + a - 7$
41. $4\{3[5(y - 3) + 9] + 2(y + 8)\}$

SYNTHESIS

42. Evaluate $\dfrac{5y - x}{4}$ when $x = 20$ and y is 4 less than x.

Simplify.

43. $\dfrac{13{,}800}{42{,}000}$
44. $|-27 - 3(4)| - |-36| + |-12|$
45. $a - \{3a - [4a - (2a - 4a)]\}$

CHAPTER 2

Equations, Inequalities, and Problem Solving

2.1 Solving Equations
2.2 Using the Principles Together
2.3 Formulas
2.4 Applications with Percent
2.5 Problem Solving
2.6 Solving Inequalities
2.7 Solving Applications with Inequalities
SUMMARY AND REVIEW
TEST

AN APPLICATION

Chassman and Bem Booksellers offers a preferred-customer card for $25. The card entitles a customer to a 10% discount on all purchases for a period of one year. Under what circumstances would an individual save money by purchasing a card?

THIS PROBLEM APPEARS AS EXERCISE 65 IN SECTION 2.7.

More information on books is available at
http://hepg.awl.com/be/elem_5

*W*e use math every day to track the number of customers in the store and the number of books we've sold, as well as to determine whether we've made a profit.

ALICE OSBORNE
Bookstore Owner
Seattle, Washington

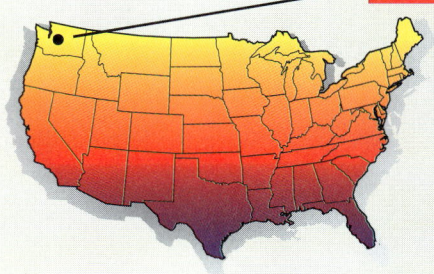

2.1 Solving Equations

Equations and Solutions • The Addition Principle • The Multiplication Principle

Solving equations is essential for problem solving in algebra. In this section, we study two of the most important principles used to solve equations.

Equations and Solutions

We have already seen that an equation is a number sentence stating that the expressions on either side of the equals sign represent the same number. Some equations, like $3 + 2 = 5$ or $2x + 6 = 2(x + 3)$, are *always* true and some, like $3 + 2 = 6$ or $x + 2 = x + 3$, are *never* true. In this text, we will concentrate on equations like $x + 6 = 13$ or $7x = 141$ that are true or false, depending on the replacement value.

SOLUTION OF AN EQUATION

> Any replacement for the variable that makes an equation true is called a *solution* of the equation. To *solve* an equation means to find all of its solutions.

To determine whether a number is a solution, we substitute that number for the variable throughout the equation. If the values on both sides of the equals sign are the same, then the number is a solution.

EXAMPLE 1   Determine whether 7 is a solution of $x + 6 = 13$.

SOLUTION   We have

$$\begin{array}{rl} x + 6 = 13 & \text{Writing the equation} \\ \overline{7 + 6 \;?\; 13} & \text{Substituting 7 for } x \\ 13 \;|\; 13 \;\; \text{TRUE} & \end{array}$$

Since the left-hand and the right-hand sides are the same, 7 is a solution.

EXAMPLE 2   Determine whether 19 is a solution of $7x = 141$.

SOLUTION   We have

$$7x = 141 \quad \text{Writing the equation}$$
$$7(19) \stackrel{?}{=} 141 \quad \text{Substituting 19 for } x$$
$$133 \mid 141 \quad \text{FALSE}$$

Since the left-hand and the right-hand sides differ, 19 is not a solution.

The Addition Principle

Consider the equation

$x = 7$.

We can easily see that the solution of this equation is 7. Replacing x with 7, we get

$7 = 7$, which is true.

Now consider the equation

$x + 6 = 13$.

In Example 1, we found that the solution of $x + 6 = 13$ is also 7. Although the solution of $x = 7$ may seem more obvious, the equations $x + 6 = 13$ and $x = 7$ are **equivalent**.

EQUIVALENT EQUATIONS

Equations with the same solutions are called *equivalent equations*.

There are principles that enable us to begin with one equation and end up with an equivalent equation, like $x = 7$, in which the solution is obvious.

One of the principles for solving equations concerns adding. An equation $a = b$ says that a and b stand for the same number. Suppose this is true, and some number c is added to a. We get the same result if we add c to b, because a and b are the same number.

THE ADDITION PRINCIPLE

For any real numbers a, b, and c,

$a = b$ is equivalent to $a + c = b + c$.

To visualize the addition principle, consider a balance similar to one a jeweler might use. (See the figure on the following page.) When the two sides of the balance hold quantities of equal weight, the balance is level. If weight is added or removed, equally, on both sides, the balance will remain level.

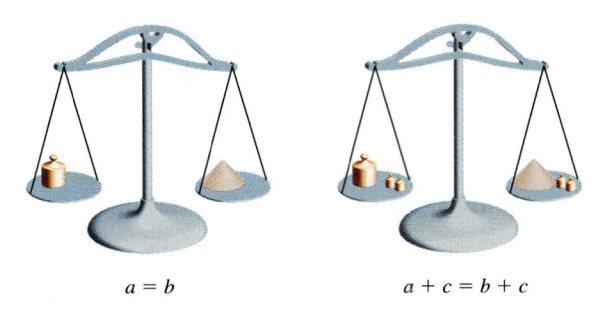

66 CHAPTER 2 • EQUATIONS, INEQUALITIES, AND PROBLEM SOLVING

$a = b$ $a + c = b + c$

When using the addition principle, we often say that we "add the same number on both sides of an equation." We can also "subtract the same number on both sides," since subtraction can be regarded as the addition of an opposite.

EXAMPLE 3

Solve: $x + 5 = -7$.

SOLUTION   We have

$x + 5 = -7$
$x + 5 - 5 = -7 - 5$ Using the addition principle: adding -5 on both sides or subtracting 5 on both sides
$x + 0 = -12$ Simplifying; $x + 5 - 5 = x + 5 + (-5) = x + 0$
$x = -12$.   Identity property of 0

It is obvious that the solution of $x = -12$ is the number -12. To check the answer in the original equation, we substitute.

Check:
$$\begin{array}{c|c} x + 5 = -7 \\ \hline -12 + 5 \; ? \; -7 \\ -7 \; | \; -7 \end{array} \text{ TRUE}$$

The solution of the original equation is -12.

In Example 3, to get x alone, we added the *opposite*, or *additive inverse*, of 5. We then simplified the left-hand side to x plus the *additive identity*, 0, or simply x. These steps effectively replaced the 5 on the left with a 0. When solving $x + a = b$ for x, we simply add $-a$ (or subtract a) on both sides.

The addition principle can also be used to solve subtraction problems.

EXAMPLE 4

Solve: $-6.5 = y - 8.4$.

SOLUTION   We have

$-6.5 = y - 8.4$ This can be regarded as $-6.5 = y + (-8.4)$.
$-6.5 + 8.4 = y - 8.4 + 8.4$ Using the addition principle: adding 8.4 on both sides eliminates -8.4 on the right
$1.9 = y$. $y - 8.4 + 8.4 = y + (-8.4) + 8.4$
$= y + 0 = y$

Check:
$$-6.5 = y - 8.4$$
$$-6.5 \; ? \; 1.9 - 8.4$$
$$-6.5 \;|\; -6.5 \quad \text{TRUE}$$

The solution is 1.9.

Note that the equations $a = b$ and $b = a$ have the same meaning. Thus, $-6.5 = y - 8.4$ could have been rewritten as $y - 8.4 = -6.5$.

EXAMPLE 5 Solve: $-\frac{2}{3} + x = \frac{5}{2}$.

SOLUTION We have

$$-\frac{2}{3} + x = \frac{5}{2}$$
$$-\frac{2}{3} + x + \frac{2}{3} = \frac{5}{2} + \frac{2}{3} \qquad \text{Adding } \tfrac{2}{3} \text{ on both sides}$$
$$x = \frac{5}{2} + \frac{2}{3}$$
$$= \frac{5}{2} \cdot \frac{3}{3} + \frac{2}{3} \cdot \frac{2}{2} \qquad \text{Multiplying by 1 to obtain a common denominator}$$
$$= \frac{15}{6} + \frac{4}{6}$$
$$= \frac{19}{6}.$$

The check is left to the student. The solution is $\frac{19}{6}$.

The Multiplication Principle

An equation like $\frac{5}{4}x = \frac{3}{8}$ says that $\frac{5}{4}x$ and $\frac{3}{8}$ represent the same number. Because of this, if $\frac{5}{4}x$ and $\frac{3}{8}$ are both multiplied by some number c, the products $c \cdot \frac{5}{4}x$ and $c \cdot \frac{3}{8}$ must also be equal.

THE MULTIPLICATION PRINCIPLE

For any real numbers a, b, and c, with $c \neq 0$,
$$a = b \quad \text{is equivalent to} \quad a \cdot c = b \cdot c.$$

EXAMPLE 6 Solve: $\frac{5}{4}x = \frac{3}{8}$.

SOLUTION We have

$$\frac{5}{4}x = \frac{3}{8} \qquad \text{Note that the reciprocal of } \tfrac{5}{4} \text{ is } \tfrac{4}{5} \text{ and that } \tfrac{4}{5} \cdot \tfrac{5}{4} = 1.$$
$$\frac{4}{5} \cdot \frac{5}{4}x = \frac{4}{5} \cdot \frac{3}{8} \qquad \text{Using the multiplication principle: multiplying on both sides by } \tfrac{4}{5} \text{ "eliminates" } \tfrac{5}{4} \text{ on the left}$$
$$1 \cdot x = \frac{3}{10} \qquad \text{Simplifying}$$
$$x = \frac{3}{10}. \qquad \text{Using the identity property of 1: } 1 \cdot x = x$$

Check:
$$\frac{\frac{5}{4}x = \frac{3}{8}}{\frac{5}{4}\left(\frac{3}{10}\right) \;?\; \frac{3}{8}}$$
$$\frac{3}{8} \;\bigg|\; \frac{3}{8} \quad \text{TRUE}$$

The solution is $\frac{3}{10}$.

In Example 6, to get x alone, we multiplied by the *reciprocal*, or *multiplicative inverse* of $\frac{5}{4}$. We then simplified the left-hand side to x times the *multiplicative identity*, 1, or simply x. These steps effectively replaced the $\frac{5}{4}$ on the left with 1.

Because division is the same as multiplying by a reciprocal, the multiplication principle also tells us that we can "divide on both sides by the same nonzero number." That is,

$$\text{if } a = b, \text{ then } \frac{1}{c} \cdot a = \frac{1}{c} \cdot b \text{ and } \frac{a}{c} = \frac{b}{c} \quad \text{(provided } c \neq 0\text{).}$$

In an expression like $3x$, the multiplier 3 is called the **coefficient**. When the coefficient of the variable is not written in fractional notation, it is usually more convenient to divide on both sides. When the coefficient *is* in fractional notation, it is usually easier to multiply by the reciprocal.

EXAMPLE 7 Solve: **(a)** $-4x = 92$; **(b)** $12.6 = 3x$; **(c)** $-x = 9$; **(d)** $\frac{2y}{9} = 10$.

SOLUTION

a) $-4x = 92$

$\dfrac{-4x}{-4} = \dfrac{92}{-4}$ Using the multiplication principle. Dividing by -4 on both sides is the same as multiplying by $-\frac{1}{4}$.

$1 \cdot x = -23$ Simplifying

$x = -23$ Using the identity property of 1

Check:
$$\frac{-4x = 92}{-4(-23) \;?\; 92}$$
$$92 \;\bigg|\; 92 \quad \text{TRUE}$$

The solution is -23.

b) $12.6 = 3x$

$\dfrac{12.6}{3} = \dfrac{3x}{3}$ Dividing by 3 on both sides or multiplying by $\frac{1}{3}$

$4.2 = 1x$

$4.2 = x$ Simplifying

Check:
$$\frac{12.6 = 3x}{12.6 \;?\; 3(4.2)}$$
$$12.6 \;\bigg|\; 12.6 \quad \text{TRUE}$$

The solution is 4.2.

c) To solve an equation like $-x = 9$, remember that when an expression is multiplied or divided by -1, its sign is changed. Here we multiply on both sides by -1 to change the sign of $-x$:

$$-x = 9$$
$$(-1)(-x) = (-1) \cdot 9 \quad \text{Multiplying by } -1 \text{ on both sides}$$
$$x = -9. \quad \text{Note that } (-1)(-x) \text{ is the same as } (-1)(-1)x.$$

Check:
$$\begin{array}{c|c} -x = 9 \\ \hline -(-9) \;?\; 9 \\ 9 \;|\; 9 \quad \text{TRUE} \end{array}$$

The solution is -9.

d) To solve an equation like $2y/9 = 10$, we rewrite the left-hand side as $\frac{2}{9} \cdot y$ and then use the multiplication principle:

$$\frac{2y}{9} = 10$$

$$\frac{2}{9} \cdot y = 10 \quad \text{Rewriting } \frac{2y}{9} \text{ as } \frac{2}{9} \cdot y$$

$$\frac{9}{2} \cdot \frac{2}{9} \cdot y = \frac{9}{2} \cdot 10 \quad \text{Multiplying by } \frac{9}{2} \text{ on both sides}$$

$$\left. \begin{array}{c} 1y = \dfrac{90}{2} \\ y = 45. \end{array} \right\} \quad \text{Simplifying}$$

Check:
$$\begin{array}{c|c} \dfrac{2y}{9} = 10 \\ \hline \dfrac{2 \cdot 45}{9} \;?\; 10 \\ \dfrac{90}{9} \\ 10 \;|\; 10 \quad \text{TRUE} \end{array}$$

The solution is 45.

EXERCISE SET 2.1

Solve using the addition principle. Don't forget to check!

1. $x + 7 = 20$
2. $x + 5 = 8$
3. $x + 15 = -5$
4. $y + 9 = 43$
5. $x + 6 = -8$
6. $t + 9 = -12$
7. $-5 = x + 8$
8. $-6 = y + 25$
9. $x - 9 = 6$
10. $x - 8 = 5$
11. $x - 7 = -21$
12. $x - 3 = -14$
13. $9 + t = 3$
14. $3 + t = 21$

15. $13 = -7 + y$
16. $15 = -9 + z$
17. $-3 + t = -9$
18. $-6 + y = -21$
19. $r + \frac{1}{3} = \frac{8}{3}$
20. $t + \frac{3}{8} = \frac{5}{8}$
21. $x + \frac{3}{5} = -\frac{7}{10}$
22. $x + \frac{2}{3} = -\frac{5}{6}$
23. $x - \frac{5}{6} = \frac{7}{8}$
24. $y - \frac{3}{4} = \frac{5}{6}$
25. $-\frac{1}{5} + z = -\frac{1}{4}$
26. $-\frac{1}{8} + y = -\frac{3}{4}$
27. $m + 3.9 = 5.4$
28. $y + 4.6 = 9.3$
29. $-9.7 = -4.7 + y$
30. $-7.8 = 2.8 + x$

Solve using the multiplication principle. Don't forget to check!

31. $5x = 80$
32. $3x = 39$
33. $5x = 45$
34. $6x = 72$
35. $84 = 7x$
36. $56 = 8x$
37. $-x = 23$
38. $100 = -x$
39. $-x = -8$
40. $-68 = -r$
41. $7x = -49$
42. $9x = -36$
43. $-12x = 72$
44. $-15x = 105$
45. $-3.4t = -20.4$
46. $-1.3a = -10.4$
47. $\frac{a}{4} = 12$
48. $\frac{y}{-8} = 11$
49. $\frac{3}{4}x = 27$
50. $\frac{4}{5}x = 16$
51. $\frac{-t}{3} = 7$
52. $\frac{-x}{6} = 9$
53. $\frac{2}{9} = -\frac{t}{4}$
54. $\frac{1}{9} = -\frac{z}{7}$
55. $-\frac{3}{5}r = -\frac{9}{10}$
56. $-\frac{2}{5}y = -\frac{4}{15}$
57. $\frac{-3r}{2} = -\frac{27}{4}$
58. $\frac{5x}{7} = -\frac{10}{14}$

Solve.

59. $2.8 + t = -3.1$
60. $\frac{3}{4}x = 18$
61. $-8.2x = 20.5$
62. $t - 7.4 = -12.9$
63. $17 = y + 29$
64. $96 = -\frac{3}{4}t$
65. $a - \frac{1}{6} = -\frac{2}{3}$
66. $-\frac{x}{7} = \frac{2}{9}$
67. $-24 = \frac{8x}{5}$
68. $\frac{1}{5} + y = -\frac{3}{10}$
69. $-16 = -\frac{2}{3}x$
70. $\frac{19}{23} = -x$

SKILL MAINTENANCE

Combine like terms.

71. $3x + 4x$
72. $6x + 5 - 7x$

Remove parentheses and simplify.

73. $3x - (4 + 2x)$
74. $2 - 5(x + 5)$

SYNTHESIS

75. ◆ What is the difference between equivalent expressions and equivalent equations?
76. ◆ Explain what is meant by the phrase "both sides of an equation."
77. ◆ Explain why it is not necessary to state a subtraction principle: For any real numbers a, b, and c, $a = b$ is equivalent to $a - c = b - c$.
78. ◆ When solving an equation, how do you determine what number to add, subtract, multiply, or divide by on both sides of that equation?

Solve, if possible. If no solution exists, state this. The icon ▓ indicates an exercise designed to give practice using a calculator.

79. ▓ $-356.788 = -699.034 + t$
80. ▓ $-0.2344m = 2028.732$
81. $5 + x = 5 + x$
82. $0 \cdot x = 7$
83. $4|x| = 48$
84. $2|x| = -12$
85. $0 \cdot x = 0$
86. $x + x = x$
87. $x + 4 = 5 + x$
88. $|3x| = 6$

Solve for x.

89. $mx = 9.4m$
90. $x - 4 + a = a$
91. $\frac{7cx}{2a} = \frac{21}{a} \cdot c$
92. $5c + cx = 7c$
93. $5a = ax - 3a$
94. $|x| + 6 = 19$
95. If $x - 4720 = 1634$, find $x + 4720$.
96. ▓ Lydia makes a calculation and gets an answer of 22.5. On the last step, she multiplies by 0.3 when she should have divided by 0.3. What should the correct answer be?
97. ◆ Are the equations $x = 5$ and $x^2 = 25$ equivalent? Why or why not?

2.2 Using the Principles Together

Applying Both Principles • Combining Like Terms • Clearing Fractions and Decimals • Equations Containing Parentheses

The equations in Section 2.1 required use of *either* the addition principle *or* the multiplication principle. Now we examine equations that require *both* principles.

Applying Both Principles

Consider the equation $3x + 5 = 17$. To solve such an equation, we first isolate the variable term, $3x$, using the addition principle. We then use the multiplication principle to get the variable by itself.

EXAMPLE 1 Solve: $3x + 5 = 17$.

SOLUTION We have

$$3x + 5 = 17$$
$$3x + 5 - 5 = 17 - 5 \quad \text{Using the addition principle: subtracting 5 on both sides (adding } -5)$$

First isolate the x-term.

$$3x = 12 \quad \text{Simplifying}$$
$$\frac{3x}{3} = \frac{12}{3} \quad \text{Using the multiplication principle: dividing by 3 on both sides } \left(\text{multiplying by } \tfrac{1}{3}\right)$$

Then isolate x.

$$x = 4. \quad \text{Simplifying}$$

Check:
$$\begin{array}{c|c} 3x + 5 = 17 \\ \hline 3 \cdot 4 + 5 \;?\; 17 \\ 12 + 5 \\ 17 \;|\; 17 \;\; \text{TRUE} \end{array}$$

We use the rules for order of operations: Find the product, $3 \cdot 4$, and then add.

The solution is 4.

EXAMPLE 2 Solve: $-5x - 6 = 16$.

SOLUTION We have

$$-5x - 6 = 16$$
$$-5x - 6 + 6 = 16 + 6 \quad \text{Adding 6 on both sides}$$
$$-5x = 22$$
$$\frac{-5x}{-5} = \frac{22}{-5} \quad \text{Dividing by } -5 \text{ on both sides}$$
$$x = -\frac{22}{5}, \text{ or } -4\frac{2}{5}. \quad \text{Simplifying}$$

Check:
$$\begin{array}{c|c} -5x - 6 = 16 \\ \hline -5\left(-\frac{22}{5}\right) - 6 \;?\; 16 \\ 22 - 6 \\ 16 \;\Big|\; 16 \quad \text{TRUE} \end{array}$$

The solution is $-\frac{22}{5}$.

EXAMPLE 3 Solve: $45 - t = 13$.

SOLUTION We have

$$45 - t = 13$$
$$\left.\begin{array}{r} 45 - t - 45 = 13 - 45 \\ -t = -32 \end{array}\right\}$$

Subtracting 45 on both sides:
$45 - t - 45 = 45 + (-t) - 45$
$ = -t + 45 - 45 = -t$

$$(-1)(-t) = (-1)(-32)$$

Multiplying by -1 on both sides (Dividing on both sides by -1 would also change the sign on both sides.)

$$t = 32.$$

We leave the check to the student. The solution is 32.

As our equation-solving skills improve, we will shorten some of our writing. Thus, instead of indicating the number being added, subtracted, multiplied, or divided on both sides, we simply write it on the opposite side, as in the following example.

EXAMPLE 4 Solve: $16.3 - 7.2y = -8.18$.

SOLUTION We have

$$16.3 - 7.2y = -8.18$$
$$-7.2y = -8.18 - 16.3$$

Subtracting 16.3 on both sides. We write the subtraction of 16.3 on the right side and remove 16.3 from the left side.

$$-7.2y = -24.48$$
$$y = \frac{-24.48}{-7.2}$$

Dividing by -7.2 on both sides. We write the division by -7.2 on the right side and remove the -7.2 from the left side.

$$y = 3.4.$$

Check:
$$\begin{array}{c|c} 16.3 - 7.2y = -8.18 \\ \hline 16.3 - 7.2(3.4) \;?\; -8.18 \\ 16.3 - 24.48 \\ -8.18 \;\Big|\; -8.18 \quad \text{TRUE} \end{array}$$

The solution is 3.4.

Combining Like Terms

If like terms appear in an equation, we combine them and then solve. Should like terms appear on both sides of an equation, we can use the addition principle to rewrite all like terms on one side.

2.2 USING THE PRINCIPLES TOGETHER 73

EXAMPLE 5 Solve.

a) $3x + 4x = -14$ **b)** $2x - 4 = -3x + 1$
c) $6x + 5 - 7x = 10 - 4x + 7$

SOLUTION

a) $3x + 4x = -14$

$\qquad 7x = -14$ Combining like terms

$\qquad x = \dfrac{-14}{7}$ Dividing by 7 on both sides

$\qquad x = -2$

The check is left to the student. The solution is -2.

b)

> Isolate variable terms on one side and constant terms on the other side.

$\qquad 2x - 4 = -3x + 1$

$\qquad 2x = -3x + 1 + 4$ Adding 4 on both sides

$\qquad 2x = -3x + 5$ Simplifying

$\qquad 2x + 3x = -3x + 3x + 5$ Adding $3x$ on both sides

$\qquad 5x = 5$ Combining like terms and simplifying

$\qquad x = \dfrac{5}{5}$ Dividing by 5 on both sides

$\qquad x = 1$ Simplifying

Check:
$$\begin{array}{c|c} 2x - 4 = -3x + 1 \\ \hline 2 \cdot 1 - 4 \;?\; -3 \cdot 1 + 1 \\ 2 - 4 \;\big|\; -3 + 1 \\ -2 \;\big|\; -2 \end{array}$$ TRUE

The solution is 1.

c) $\quad 6x + 5 - 7x = 10 - 4x + 7$

$\qquad -x + 5 = 17 - 4x$ Combining like terms

$\qquad -x + 5 + 4x = 17 - 4x + 4x$ Adding $4x$ on both sides

$\qquad 3x + 5 = 17$ Simplifying; this is identical to Example 1.

$\qquad 3x = 12$ Subtracting 5 on both sides and simplifying

$\qquad x = 4$ Dividing by 3 on both sides and simplifying

The student can confirm that 4 checks and is the solution.

Clearing Fractions and Decimals

Equations are usually easier to solve when they do not contain fractions or decimals. For example, if we multiply both sides of $\frac{1}{2}x + 5 = \frac{3}{4}$ by 4 and both sides of $2.3x + 7 = 5.4$ by 10, we have

$$4\left(\tfrac{1}{2}x + 5\right) = 4 \cdot \tfrac{3}{4} \quad \text{and} \quad 10(2.3x + 7) = 10 \cdot 5.4,$$

or

$$2x + 20 = 3 \quad \text{and} \quad 23x + 70 = 54.$$

The first equation has been "cleared of fractions" and the second equation has been "cleared of decimals." Each resulting equation is equivalent to its respective original equation.

The easiest way to clear an equation of fractions is to multiply *every term on both sides* of the equation by the smallest, or *least*, common denominator.

EXAMPLE 6

Solve: $\frac{2}{3}x - \frac{1}{6} = 2x$.

SOLUTION The number 6 is the least common denominator, so we multiply by 6 on both sides.

$$6\left(\frac{2}{3}x - \frac{1}{6}\right) = 6 \cdot 2x \quad \text{Multiplying by 6 on both sides}$$

$$6 \cdot \frac{2}{3}x - 6 \cdot \frac{1}{6} = 6 \cdot 2x$$

CAUTION! Be sure the distributive law is used to multiply *all* the terms by 6.

$$4x - 1 = 12x \quad \text{Simplifying; note that the fractions are cleared.}$$
$$-1 = 8x \quad \text{Subtracting } 4x \text{ on both sides}$$
$$-\frac{1}{8} = x \quad \text{Dividing by 8}$$

The number $-\frac{1}{8}$ checks and is the solution.

To clear an equation of decimals, we count the greatest number of decimal places in any one number. If the greatest number of decimal places is 1, we multiply on both sides by 10; if it is 2, we multiply by 100; and so on.

EXAMPLE 7

Solve: $16.3 - 7.2y = -8.18$.

SOLUTION The greatest number of decimal places in any one number is *two*. Multiplying by 100 will clear all decimals.

$$100(16.3 - 7.2y) = 100(-8.18) \quad \text{Multiplying by 100 on both sides}$$
$$100(16.3) - 100(7.2y) = 100(-8.18) \quad \text{Using the distributive law}$$
$$1630 - 720y = -818 \quad \text{Simplifying}$$
$$-720y = -818 - 1630 \quad \text{Subtracting 1630 on both sides}$$
$$-720y = -2448 \quad \text{Combining like terms}$$
$$y = \frac{-2448}{-720} \quad \text{Dividing by } -720 \text{ on both sides}$$
$$y = 3.4$$

In Example 4, the same solution was found without clearing decimals. Finding the same answer two ways is a good check. The solution is 3.4.

Equations Containing Parentheses

To solve equations that contain parentheses, we can use the distributive law to first remove the parentheses. Then we proceed as before.

2.2 USING THE PRINCIPLES TOGETHER **75**

EXAMPLE 8   Solve: **(a)** $2 - 5(x + 5) = 3(x - 2) - 1$; **(b)** $\frac{2}{5}(3x + 1) = \frac{7}{10}$.

SOLUTION

a)
$$2 - 5(x + 5) = 3(x - 2) - 1$$
$$2 - 5x - 25 = 3x - 6 - 1 \quad \text{Using the distributive law}$$
$$-5x - 23 = 3x - 7 \quad \text{Simplifying}$$
$$-5x - 23 + 7 = 3x$$
$$-23 + 7 = 3x + 5x \quad \left.\begin{array}{l}\text{Adding 7 and } 5x \text{ on both sides to}\\ \text{isolate the } x\text{-terms on the right side}\end{array}\right.$$
$$-16 = 8x \quad \text{Simplifying}$$
$$-2 = x \quad \text{Dividing by 8}$$

Check:
$$\begin{array}{c|c}
\multicolumn{2}{c}{2 - 5(x + 5) = 3(x - 2) - 1}\\ \hline
2 - 5(-2 + 5) \;?\; 3(-2 - 2) - 1\\
2 - 5(3) & 3(-4) - 1\\
2 - 15 & -12 - 1\\
-13 & -13 \quad \text{TRUE}
\end{array}$$

The solution is -2.

b)
$$\frac{2}{5}(3x + 1) = \frac{7}{10}$$
$$10 \cdot \frac{2}{5}(3x + 1) = 10 \cdot \frac{7}{10} \quad \text{Multiplying by the least common denominator, 10, on both sides}$$
$$4(3x + 1) = 7 \quad \text{Simplifying}$$
$$12x + 4 = 7 \quad \text{Using the distributive law}$$
$$12x = 3 \quad \text{Subtracting 4 on both sides}$$
$$x = \frac{3}{12}, \text{ or } \frac{1}{4} \quad \text{Dividing by 12 on both sides}$$

The student can confirm that $\frac{1}{4}$ checks and is the solution.

Here is a procedure for solving the types of equations discussed in this section.

AN EQUATION-SOLVING PROCEDURE

1. If necessary, use the distributive law to remove parentheses. Then combine like terms on each side.
2. Multiply on both sides to clear fractions or decimals. (This is optional, but it can ease computations.)
3. Get all terms with variables on one side and all constant terms on the other side, using the addition principle.
4. Combine like terms again, if necessary.
5. Multiply or divide to solve for the variable, using the multiplication principle.
6. Check all possible solutions in the original equation.

EXERCISE SET 2.2

Solve and check.

1. $4x + 5 = 41$
2. $3x + 6 = 30$
3. $8x + 4 = 68$
4. $6z + 3 = 57$
5. $5x - 8 = 27$
6. $6x - 3 = 15$
7. $3x - 9 = 33$
8. $5x - 7 = 48$
9. $7x + 2 = -54$
10. $4x + 3 = -21$
11. $-39 = 1 + 8x$
12. $-91 = 9t + 8$
13. $9 - 4x = 37$
14. $12 - 4x = 108$
15. $-7x - 24 = -129$
16. $-6z - 18 = -132$
17. $36 = 5x + 7x$
18. $4x + 5x = 45$
19. $27 - 6x = 99$
20. $32 - 7x = 11$
21. $4x + 3x = 42$
22. $6x + 19x = 100$
23. $-2a + 5a = 24$
24. $-4y - 8y = 48$
25. $-7y - 8y = -15$
26. $-10y - 3y = -39$
27. $10.2y - 7.3y = -58$
28. $3.4t - 1.2t = -44$
29. $x + \frac{1}{3}x = 8$
30. $x + \frac{1}{4}x = 10$
31. $8y - 35 = 3y$
32. $4x - 6 = 6x$
33. $6x - 5 = 7 + 2x$
34. $5y - 2 = 28 - y$
35. $6x + 3 = 2x + 11$
36. $5y + 3 = 2y + 15$
37. $5 - 2x = 3x - 7x + 25$
38. $10 - 3x = 2x - 8x + 40$
39. $4 + 3x - 6 = 3x + 2 - x$
40. $5 + 4x - 7 = 4x - 2 - x$
41. $4y - 4 + y + 24 = 6y + 20 - 4y$
42. $5y - 7 + y = 7y + 21 - 5y$

Solve and check. Clear fractions or decimals first.

43. $\frac{5}{4}x + \frac{1}{4}x = 2x + \frac{1}{2} + \frac{3}{4}x$
44. $\frac{7}{8}x - \frac{1}{4} + \frac{3}{4}x = \frac{1}{16} + x$
45. $\frac{2}{3} + \frac{1}{4}t = 6$
46. $-\frac{3}{2} + x = -\frac{5}{6} - \frac{4}{3}$
47. $\frac{2}{3} + 3y = 5y - \frac{2}{15}$
48. $\frac{1}{2} + 4m = 3m - \frac{5}{2}$
49. $\frac{1}{3}x + \frac{2}{5} = \frac{4}{15} + \frac{3}{5}x - \frac{2}{3}$
50. $1 - \frac{2}{3}y = \frac{9}{5} - \frac{1}{5}y + \frac{3}{5}$
51. $2.1x + 45.2 = 3.2 - 8.4x$
52. $0.96y - 0.79 = 0.21y + 0.46$
53. $1.03 - 0.6x = 0.71 - 0.2x$
54. $1.7t + 8 - 1.62t = 0.4t - 0.32 + 8$
55. $\frac{2}{5}x - \frac{3}{2}x = \frac{3}{4}x + 2$
56. $\frac{5}{16}y + \frac{3}{8}y = 2 + \frac{1}{4}y$

Solve and check.

57. $7(2a - 1) = 21$
58. $5(2t - 3) = 30$
59. $40 = 5(3x + 2)$
60. $9 = 3(5x - 2)$
61. $2(3 + 4m) - 9 = 45$
62. $3(5 + 3m) - 8 = 88$
63. $5r - (2r + 8) = 16$
64. $6b - (3b + 8) = 16$
65. $10 - 3(2x - 1) = 1$
66. $5(d + 4) = 7(d - 2)$
67. $3(t - 2) = 9(t + 2)$
68. $8(2t + 1) = 4(7t + 7)$
69. $7(5x - 2) = 6(6x - 1)$
70. $5(t + 3) + 9 = 3(t - 2) + 6$
71. $19 - (2x + 3) = 2(x + 3) + x$
72. $13 - (2c + 2) = 2(c + 2) + 3c$
73. $\frac{1}{3}(6x + 24) - 20 = -\frac{1}{4}(12x - 72)$
74. $\frac{1}{4}(8y + 4) - 17 = -\frac{1}{2}(4y - 8)$
75. $\frac{4}{5}(3x + 4) = 10$
76. $\frac{2}{3}(2x - 1) = 20$
77. $\frac{3}{2}(2x + 5) + \frac{1}{4} = -\frac{7}{2}$
78. $\frac{5}{6}\left(\frac{3}{4}x - 2\right) + \frac{1}{3} = -\frac{2}{3}$
79. $\frac{3}{4}\left(3x - \frac{1}{2}\right) - \frac{2}{3} = \frac{1}{3}$
80. $\frac{2}{3}\left(\frac{7}{8} - 4x\right) - \frac{5}{8} = \frac{3}{8}$
81. $0.7(3x + 6) = 1.1 - (x + 2)$
82. $0.9(2x + 8) = 20 - (x + 5)$
83. $a + (a - 3) = (a + 2) - (a + 1)$
84. $0.8 - 4(b - 1) = 0.2 + 3(4 - b)$

SKILL MAINTENANCE

85. Divide: $-22.1 \div 3.4$.
86. Factor: $7x - 21 - 14y$.
87. Use $<$ or $>$ for ▨ to write a true sentence:
    -15 ▨ -13.
88. Find $-(-x)$ when $x = -14$.

SYNTHESIS

89. ◆ Explain how the commutative and associative laws are used in Example 3.
90. ◆ What procedure would you follow for solving an equation like $0.23x + \frac{17}{3} = -0.8 + \frac{3}{4}x$? Could your procedure be streamlined? If so, how?
91. ◆ Dave is determined to solve the equation $3x + 4 = -11$ by first using the multiplication principle to "eliminate" the 3. How should Dave proceed and why?
92. ◆ Consider any equation of the form $ax + b = c$. Describe a method that can be used to solve for x.

Solve.

93. ▦ $8.43x - 2.5(3.2 - 0.7x) = -3.455x + 9.04$
94. ▦ $0.008 + 9.62x - 42.8 = 0.944x + 0.0083 - x$
95. $-2[3(x - 2) + 4] = 4(1 - x) + 8$
96. $0 = y - (-14) - (-3y)$
97. $3(x + 4) = 3(4 + x)$
98. $475(54x + 7856) + 9762 = 402(83x + 975)$
99. $2x(x + 5) - 3(x^2 + 2x - 1) = 9 - 5x - x^2$
100. $x(x - 4) = 3x(x + 1) - 2(x^2 + x - 5)$
101. $9 - 3x = 2(5 - 2x) - (1 - 5x)$
102. $-2y + 5y = 6y$
103. $\dfrac{x}{14} - \dfrac{5x + 2}{49} = \dfrac{3x - 4}{7}$
104. $\dfrac{5x + 3}{4} + \dfrac{25}{12} = \dfrac{5 + 2x}{3}$

COLLABORATIVE CORNER

Focus: Solving linear equations

Time: 10–20 minutes

Group size: 3

There is usually more than one correct sequence of steps for solving an equation. Thus it is important to be able to follow someone else's steps even if his or her approach is not what you might have selected.

ACTIVITY

1. Each group member should select a different one of the following equations and perform the first step of the solution.

    $4 - 3[2 + 5(x - 3)] = 7x + 6(2 - x)$,
    $4x - 7[2 + 3(x - 5) + x] = 4 - 9[2 - 3(x + 7)]$,
    $5 - 7[x - 2(x - 6) - 8] = 3x + 4(2x - 7) + 9$

2. Group members should then rotate the equations through the group so that the second and third steps of each solution are performed by the other two group members. Continue passing the problems around until all equations have been solved.

3. If an error has been made, the person who spots the error should not repair it. Instead, he or she should determine who performed the step in which the error appears and then consult with the third group member to confirm that, indeed, a mistake has been made. If a mistake really does exist, the person who made the error should then make the repair and the equation-solving process should resume.

4. Each group should form a consensus on what the three solutions are and then submit their answers to the entire class.

2.3 Formulas

Evaluating Formulas • Solving for a Letter

Many applications of mathematics involve relationships among two or more quantities. An equation that represents such a relationship will use two or more letters and is known as a **formula**. Although most of the letters in this book represent variables, some—like c in $E = mc^2$ or π in $C = \pi d$—represent constants.

Evaluating Formulas

EXAMPLE 1

Distance from a Storm. The formula $M = \frac{1}{5}n$ can be used to determine how far you are from a thunderstorm. Here n is the number of seconds that it takes the sound of thunder to reach you once lightning appears and M is the distance, in miles, that you are from the lightning. If it takes 10 sec for the sound of thunder to reach you after you have seen lightning, how far away is the storm?

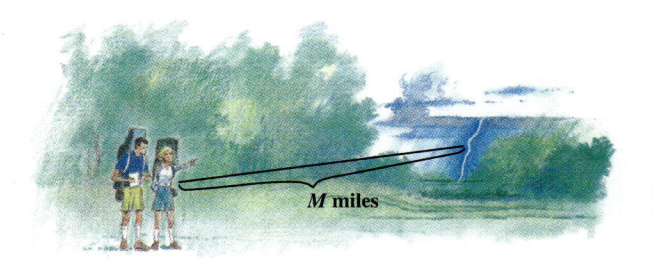

M miles

SOLUTION We substitute 10 for n and calculate M:
$$M = \tfrac{1}{5}n = \tfrac{1}{5}(10) = 2.$$
The storm is 2 mi away.

Solving for a Letter

Suppose that we are told how far away a storm is and we want to predict how long it will take the sound of thunder to reach us. We could substitute the distance, say 2, for M, and then solve for n:

$2 = \tfrac{1}{5}n$ Replacing M with 2

$10 = n.$ Multiplying by 5 on both sides

When this must be done for a variety of distances, it is easier to first solve for n and *then* substitute values for M.

2.3 FORMULAS

EXAMPLE 2   Solve for n: $M = \frac{1}{5}n$.

SOLUTION   We have

$M = \frac{1}{5}n$ We want this letter alone.

$5 \cdot M = 5 \cdot \frac{1}{5}n$ Multiplying by 5 on both sides

$5M = n$.

The equation $5M = n$ gives a quick, easy way to find the number of seconds it takes thunder to reach us when a storm is M miles away.

To see how the addition and multiplication principles apply to formulas, compare the following. In (A), we solve as usual; in (B), we do not simplify; and in (C), we cannot simplify since a, b, and c are unknown.

A. $5x + 2 = 12$ **B.** $5x + 2 = 12$ **C.** $ax + b = c$

$5x = 12 - 2$ $5x = 12 - 2$ $ax = c - b$

$5x = 10$ $x = \frac{12 - 2}{5}$ $x = \frac{c - b}{a}$

$x = \frac{10}{5} = 2$

EXAMPLE 3   *Circumference of a Circle.*   The formula $C = 2\pi r$ gives the *circumference* C of a circle with radius r. Solve for r.

SOLUTION   The **circumference** is the distance around a circle.

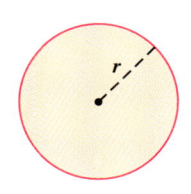

Given a radius r, we can use this equation to find a circle's circumference C.

$C = 2\pi r$ We want this letter alone.

$\dfrac{C}{2\pi} = \dfrac{2\pi r}{2\pi}$ Dividing by 2π on both sides

Given a circle's circumference C, we can use this equation to find the radius r.

$\dfrac{C}{2\pi} = r$

EXAMPLE 4   *Nutrition.*   The number of calories K needed each day by a moderately active woman who weighs w pounds, is h inches tall, and is a years old, can be estimated using the formula

$$K = 917 + 6(w + h - a).*$$

Solve for a.

---

*Based on information from M. Parker (ed.), *She Does Math!* (Washington DC: Mathematical Association of America, 1995), p. 96.

80 CHAPTER 2 • EQUATIONS, INEQUALITIES, AND PROBLEM SOLVING

SOLUTION We have

$$K = 917 + 6(w + h - a)$$ ←— We want the letter a alone.

$$K - 917 = 6(w + h - a)$$ Subtracting 917 on both sides

$$\frac{K - 917}{6} = w + h - a$$ Dividing by 6 on both sides

$$\frac{K - 917}{6} - w - h = -a$$ Subtracting $w + h$ (adding $-w - h$) on both sides

$$-\frac{K - 917}{6} + w + h = a.$$ Multiplying by -1 on both sides

We can also write this as

$$a = w + h - \frac{K - 917}{6}.$$

This formula can be used to estimate a woman's age, knowing her weight, height, and caloric needs.

The above steps are just like those used in Section 2.2 to solve equations.

TO SOLVE A FORMULA FOR A GIVEN LETTER:

1. Identify the letter being solved for and multiply on both sides to clear fractions or decimals, if necessary.
2. Get all terms with the letter to be solved for on one side of the equation and all other terms on the other side.
3. Combine like terms, if necessary. This may require factoring.
4. Multiply or divide to solve for the letter in question.

EXAMPLE 5 Solve for x: $y = ax + bx - 4$.

SOLUTION We solve as follows:

$$y = ax + bx - 4$$ We want this letter alone.

$$y + 4 = ax + bx$$ Adding 4 on both sides

$$y + 4 = (a + b)x$$ Combining like terms by factoring out x

$$\frac{y + 4}{a + b} = x.$$ Multiplying by $\frac{1}{a + b}$ on both sides

We can also write this as

$$x = \frac{y + 4}{a + b}.$$

CAUTION! Had we performed the following steps in Example 5, we would *not* have solved for *x*:

$$y = ax + bx - 4$$
$$y - ax + 4 = bx \quad \text{Subtracting } ax \text{ and adding 4 on both sides}$$

Two occurrences of *x*

$$\frac{y - ax + 4}{b} = x. \quad \text{Dividing by } b \text{ on both sides}$$

The mathematics of each step is correct, but since *x* occurs on both sides of the formula, *we have not solved the formula for x*. Remember that the letter being solved for should be alone on one side of the equation, with *no* occurrence of that letter on the other side!

EXERCISE SET 2.3

1. *Furnace output.* The formula
   $$b = 30a$$
   is used in New England to estimate the minimum furnace output *b*, in Btu's, for a modern house with *a* square feet of flooring. Determine the minimum furnace output for a 1900-ft^2 modern house.

2. *Furnace output.* The formula
   $$b = 50a$$
   is used in New England to estimate the minimum furnace output *b*, in Btu's, for an old, poorly insulated house with *a* square feet of flooring. Determine the minimum furnace output for a 2500-ft^2 older house.

3. *Surface area of a cube.* The surface area *A* of a cube with side *s* is given by
   $$A = 6s^2.$$
   Find the surface area of a cube with sides of 3 in.

4. *College size.* At many colleges, the number of "full-time-equivalent" students *f* is given by
   $$f = \frac{n}{15},$$
   where *n* is the total number of credits for which students enroll in a given semester. Determine the number of full-time-equivalent students on a campus in which students registered for a total of 21,345 credits.

5. *Electrical power.* The power rating *P*, in watts, of an electrical appliance is determined by
   $$P = I \cdot V,$$
   where *I* is the current, in amperes, and *V* is measured in volts. If a kitchen requires 30 amps of current and the voltage in the house is 115 volts, what is the wattage of the kitchen?

6. *Wavelength of a musical note.* The wavelength *w*, in meters per cycle, of a musical note is given by
   $$w = \frac{s}{f},$$
   where *s* is the speed of the sound, in meters per second, and *f* is the frequency, in cycles per second. The speed of sound in air is 344 m/sec. What is the wavelength of a note whose frequency in air is 24 cycles per second?

7. *Absorption of ibuprofen.* When 400 mg of the painkiller ibuprofen is swallowed, the number of milligrams *n* in the bloodstream *t* hours later (for $0 \le t \le 6$) is estimated by
   $$n = 0.5t^4 + 3.45t^3 - 96.65t^2 + 347.7t.$$
   How many milligrams of ibuprofen remain in the blood 1 hr after 400 mg has been swallowed?

8. *Size of a league schedule.* When all *n* teams in a league play every other team twice, a total of

N games are played, where
$$N = n^2 - n.$$
If a soccer league has 7 teams and every team plays every other team twice, how many games will be played?

Solve each formula for the indicated letter.

9. $A = bh$, for b
   (Area of parallelogram with base b and height h)

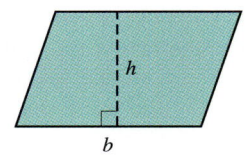

10. $A = bh$, for h
11. $d = rt$, for r
    (A distance formula, where d is distance, r is speed, and t is time)
12. $d = rt$, for t
13. $I = Prt$, for P
    (Simple-interest formula, where I is interest, P is principal, r is interest rate, and t is time)
14. $I = Prt$, for t
15. $H = 65 - m$, for m
    (To determine the number of heating degree days H for a day with $m°$ Fahrenheit as the average temperature)
16. $d = h - 64$, for h
    (To determine how many inches d above average an h-inch-tall woman is)
17. $P = 2l + 2w$, for l
    (Perimeter of a rectangle of length l and width w)

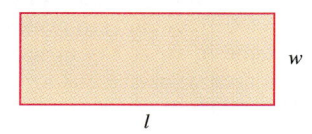

18. $P = 2l + 2w$, for w
19. $A = \pi r^2$, for π
    (Area of a circle with radius r) 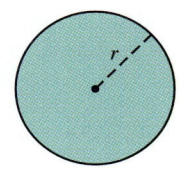
20. $A = \pi r^2$, for r^2
21. $A = \frac{1}{2}bh$, for h
    (Area of a triangle with base b and height h) 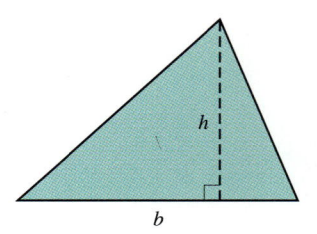
22. $A = \frac{1}{2}bh$, for b
23. $E = mc^2$, for m
    (A relativity formula)
24. $E = mc^2$, for c^2
25. $Q = \frac{c + d}{2}$, for d
26. $Q = \frac{p - q}{2}$, for p
27. $A = \frac{a + b + c}{3}$, for b
28. $A = \frac{a + b + c}{3}$, for c
29. $M = \frac{A}{s}$, for A
    (To compute the Mach number M for speed A and speed of sound s)
30. $P = \frac{ab}{c}$, for b
31. $Ax + By = C$, for y
32. $Ax + By = C$, for x
33. *Area of a sector.* The area of a sector of a circle is given by
    $$A = \frac{\pi r^2 S}{360},$$
    where r is the radius and S is the angle measure, in degrees, of the sector. Solve for S.

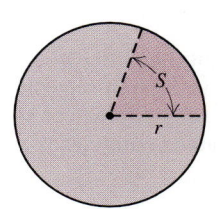

34. *Compounding interest.* The formula
$$A = P + Prt$$
is used to find the amount A in an account when simple interest is added to an investment of P dollars (see Exercise 13). Solve for P.

35. *Area of a trapezoid.* The formula
$$A = \tfrac{1}{2}ah + \tfrac{1}{2}bh$$
can be used to find the area A of a trapezoid with bases a and b and height h. Solve for h.

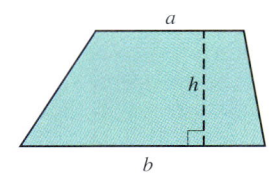

36. *Chess rating.* The formula
$$R = r + \frac{400(W - L)}{N}$$
is used to establish a chess player's rating R after that player has played N games, won W of them, and lost L of them. Here r is the average rating of the opponents (*Source*: The U.S. Chess Federation, 1996). Solve for L.

SKILL MAINTENANCE

Multiply.

37. $7(-3)2$ **38.** $-\tfrac{2}{3} \cdot \tfrac{9}{10}$

Simplify.

39. $10 \div (-2) \cdot 5 - 4$ **40.** $3|7 - (2 - 5)|$

SYNTHESIS

41. ◆ The equations
$$P = 2l + 2w \quad \text{and} \quad w = \frac{P}{2} - l$$
are equivalent formulas involving the perimeter P, length l, and width w of a rectangle. Devise a problem for which the second of the two formulas would be more useful.

42. ◆ Devise an application in which it would be useful to solve the equation $d = rt$ for r (see Exercise 11).

43. ◆ Amy has a formula that allows her to convert Celsius temperatures to Fahrenheit temperatures but not Fahrenheit to Celsius. What advice can you offer?

44. ◆ Is it correct to state that when the length and the width of a rectangle are doubled, the area is also doubled? Why or why not?

45. Revise the formula in Example 4 so that a woman's weight in kilograms (2.2046 lb = 1 kg) and her height in centimeters (0.3937 in. = 1 cm) are used.

46. The number of calories K needed each day by a moderately active man who weighs w kilograms, is h centimeters tall, and is a years old, can be determined by
$$K = 19.18w + 7h - 9.52a + 92.4.*$$
If Janos is moderately active, weighs 82 kg, is 185 cm tall, and needs to consume 2627 calories a day, how old is he?

47. *Dosage size.* Clark's rule for determining the size of a particular child's medicine dosage c is
$$c = \frac{w}{a} \cdot d,$$
where w is the child's weight, in pounds, a is the average adult weight, in pounds, and d is the usual adult dosage.[†] Solve for a.

48. ▣ *Weight of a fish.* An ancient fisherman's formula for estimating the weight of a fish is
$$w = \frac{lg^2}{800},$$
where w is the weight, in pounds, l is the length, in inches, and g is the girth (distance around the midsection), in inches. Estimate the girth of a 700-lb yellow tuna that is 8 ft long.

49. *Altitude and temperature.* Air temperature drops about 1° Celsius (C) for each 100-m rise above ground level, up to 12 km.[‡] If the ground level temperature is t°C, find a formula for the temperature T at an elevation of h meters.

Solve each formula for the given letter.

50. $\dfrac{y}{z} \div \dfrac{z}{t} = 1$, for y

51. $ac = bc + d$, for c

52. $qt = r(s + t)$, for t

53. $3a = c - a(b + d)$, for a

54. ◆ How might the answer to Exercise 34 be useful?

---

[*]Based on information from M. Parker (ed.), *She Does Math!* (Washington DC: Mathematical Association of America, 1995).

[†]Olsen, June Looby, Leon J. Ablon, and Anthony Patrick Giangrasso, *Medical Dosage Calculations,* 6th ed. (Redwood City CA: Addison-Wesley, 1995), p. A-31.

[‡]*A Sourcebook of Applications of School Mathematics,* National Council of Teachers of Mathematics, 1980, p. 93.

COLLABORATIVE CORNER

Focus: Formulas

Time: 15–25 minutes

Group size: 3–4

Materials: Calculators may be helpful.

The formula

$$N = d + 2m + \left[\!\left[\frac{3(m+1)}{5}\right]\!\right] + y + \left[\!\left[\frac{y}{4}\right]\!\right] - \left[\!\left[\frac{y}{100}\right]\!\right] + \left[\!\left[\frac{y}{400}\right]\!\right] + 2$$

can be used to determine the day of the week for any date that uses our current calendar.* To do so, d is replaced with the day of the month of the given date; m is replaced with the number of the month in the year, with January and February regarded as the 13th and 14th months of the previous year (thus 2/8/1958 becomes 14/8/1957) and March through December numbered as usual; and y is replaced with the year. The symbols $[\![\ ]\!]$ indicate use of only the integer portion of the enclosed quotient (round down). Once N has been determined, it should be divided by 7—the remainder is the desired day of the week, with Sunday as day one and Saturday as day seven, or zero.

ACTIVITY

1. Each group member should determine the day of the week on which he or she was born.
2. Ask a volunteer (possibly your instructor) to give the values of N, d, and m for his or her birthday. Then determine the age of the volunteer. (*Hint*: All group members can help where trial and error is involved.)

A Sourcebook of Applications of School Mathematics, National Council of Teachers of Mathematics, 1980.

2.4 Applications with Percent

Converting Between Percent Notation and Decimal Notation • Solving Percent Problems

Formulas like those in Section 2.3 appear in a wide variety of applications. In this section, we examine applications involving percent.

Suppose that Village Stationers installs a new cash register and the sales clerks inadvertently print out "totals" on each receipt without separating each transaction into "merchandise" and the five-percent "sales tax." For tax purposes, the shop needs a formula for separating each total into the amount spent on merchandise and the amount spent on tax. Before developing such a formula, we review the basics of percent problems.

Converting Between Percent Notation and Decimal Notation

The average family spends 26% of its income for food. What does this mean? It means that out of every $100 earned, $26 is spent for food. Thus 26% is a ratio of 26 to 100.

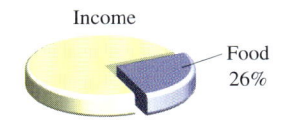

Income
Food 26%

The percent symbol % means "per hundred." We can regard the percent symbol as part of a name for a number. For example,

26% is defined to mean 26×0.01, or $26 \times \dfrac{1}{100}$, or $\dfrac{26}{100}$.

In general,

PERCENT NOTATION

$n\%$ means $n \times 0.01$, or $n \times \dfrac{1}{100}$, or $\dfrac{n}{100}$.

EXAMPLE 1 Convert to decimal notation: (a) 78%; (b) 1.3%.

SOLUTION

a) $78\% = 78 \times 0.01$ Replacing % with ×0.01
   $= 0.78$

b) $1.3\% = 1.3 \times 0.01$ Replacing % with ×0.01
   $= 0.013$

To convert from percent notation to decimal notation, move the decimal point two places to the left and drop the percent symbol.

EXAMPLE 2 Convert 43.67% to decimal notation.

SOLUTION

43.67%   0.43.67   43.67% = 0.4367

Move the decimal point two places to the left.

The procedure used in Example 2 can be reversed. Consider 0.38:

$0.38 = \dfrac{38}{100}$ Converting to fractional notation

$= 38\%.$ $\dfrac{n}{100}$ means $n\%$.

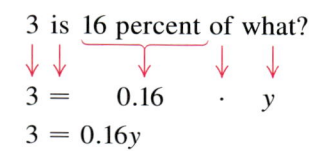

To convert from decimal notation to percent notation, move the decimal point two places to the right and write a percent symbol.

EXAMPLE 3 Convert to percent notation: (a) 1.27; (b) $\frac{1}{4}$; (c) 0.3.

SOLUTION

a) We first move the decimal point two places to the right: 1.27.

and then write a % symbol: 127%

b) Note that $\frac{1}{4} = 0.25$. We move the decimal point two places to the right: 0.25.

and then write a % symbol: 25%

c) We first move the decimal point two places to the right (recall that 0.3 = 0.30): 0.30.

and then write a % symbol: 30%

Solving Percent Problems

To solve problems involving percents, we translate to mathematical language and then solve an equation.

EXAMPLE 4 What is 11% of 49?

SOLUTION

Translate: What is 11% of 49?

$a = 0.11 \cdot 49$ 11% = 0.11; "of" means ·

$a = 5.39$

> A way of checking answers is by estimating as follows:
> $11\% \times 49 \approx 10\% \times 50$
> $= 0.10 \times 50 = 5.$
> Since 5 is close to 5.39, our answer is reasonable.

Thus, 5.39 is 11% of 49. The answer is 5.39.

EXAMPLE 5 3 is 16 percent of what?

SOLUTION

Translate: 3 is 16 percent of what?

$3 = 0.16 \cdot y$

$3 = 0.16y$

or

$$\frac{3}{0.16} = y \quad \text{Dividing on both sides by 0.16}$$

$$18.75 = y$$

Thus, 3 is 16 percent of 18.75. The answer is 18.75.

EXAMPLE 6 What percent of $50 is $16?

SOLUTION

Translate: What percent of $50 is $16?

$$n \cdot 50 = 16$$

$$n \cdot 50 = 16$$

$$n = \frac{16}{50} \quad \text{Dividing on both sides by 50}$$

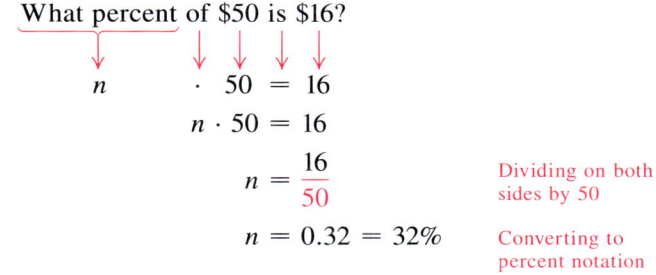

$$n = 0.32 = 32\% \quad \text{Converting to percent notation}$$

Thus, 32% of $50 is $16. The answer is 32%.

Examples 4–6 represent the three basic types of percent problems.

EXAMPLE 7 *Retail Sales.* A receipt from Village Stationers indicates the total amount paid (including tax), but not the price of the merchandise. If the sales tax is 5%, find the following:

a) The cost of the merchandise when the total is $31.50
b) A formula for the cost of the merchandise c when the total is T dollars

SOLUTION

a) When tax is added to the cost of an item, the customer actually pays more than 100% of the item's price. When sales tax is 5%, the total paid is 105% of the price of the merchandise. Thus if $c =$ the cost of the merchandise, we have

$31.50 is 105% of c

$$31.50 = 1.05 \cdot c$$

$$\frac{31.50}{1.05} = c \quad \text{Dividing by 1.05 on both sides}$$

$$30 = c. \quad \text{Simplifying}$$

The merchandise cost $30 before tax.

b) When the total is T dollars, we modify the approach used in part (a):

$$T = 1.05c$$

$$\frac{T}{1.05} = c. \quad \text{Dividing by 1.05 on both sides}$$

EXERCISE SET 2.4

Find decimal notation.
1. 32%
2. 49%
3. 7%
4. 91.3%
5. 24.1%
6. 2%
7. 0.46%
8. 4.8%

Find percent notation.
9. 4.54
10. 1
11. 0.998
12. 0.73
13. 2
14. 0.0057
15. 1.34
16. 9.2
17. 0.0068
18. 0.675
19. $\frac{3}{8}$
20. $\frac{3}{4}$
21. $\frac{7}{25}$
22. $\frac{4}{5}$
23. $\frac{2}{3}$
24. $\frac{5}{6}$

Solve.
25. What percent of 75 is 39?
26. What percent of 136 is 17?
27. What percent of 125 is 30?
28. What percent of 300 is 57?
29. 18 is 30% of what number?
30. 20.4 is 24% of what number?
31. 0.3 is 12% of what number?
32. 7 is 175% of what number?
33. What number is 65% of 420?
34. What number is 1% of a million?
35. What percent of 60 is 75?
36. What percent of 40 is 90?
37. What is 2% of 40?
38. What is 40% of 2?
39. 2 is what percent of 40?
40. 40 is 2% of what number?
41. *Infant health.* In a study of 300 pregnant women with "good-to-excellent" diets, 95% had babies in good or excellent health. How many women in this group had babies in good or excellent health?
42. *Infant health.* In a study of 200 pregnant women with "poor" diets, 8% had babies in good or excellent health. How many women in this group had babies in good or excellent health?
43. *Votes for president.* In 1996, Bill Clinton received 45.3 million votes. This accounted for 49% of all votes cast. How many people voted in the 1996 presidential election?
44. *Lotteries and education.* From 1980–1994, $47.1 billion of state lottery money was used for education. This accounted for 53% of all lottery proceeds for the period (*Source: Statistical Abstract of the United States*, 1995). What was the total amount of lottery proceeds from 1980–1994?
45. *Health care costs.* In 1985, the average U.S. citizen paid $1108 for health care. By 1993, that figure had increased by $668 (*Source: Statistical Abstract of the United States*, 1995). Determine the percentage by which the average citizen's health-care bill grew.
46. *FBI recruiting.* The FBI annually receives 16,000 applications for agents. It accepts 600 of these applicants. What percent does it accept?
47. *Junk mail.* The U.S. Postal Service reports that we open and read 78% of the junk mail that we receive. A business sends out 9500 advertising brochures. How many of them can the business expect to be opened and read?
48. *Left-handed bowlers.* It has been determined by sociologists that 17% of the population is left-handed. Each week 160 bowlers enter a tournament conducted by the Professional Bowlers Association. How many would you expect to be left-handed? Round to the nearest one.
49. *Kissing and colds.* In a medical study, it was determined that if 800 people kiss someone else who has a cold, only 56 will actually catch the cold. What percent is this?
50. On a test of 88 items, a student got 76 correct. What percent were correct?
51. A baseball player had 13 hits in 25 times at bat. What percent were hits?

52. A bill at Officeland totaled $37.80. How much did the merchandise cost if the sales tax is 5%?

53. Doreen's checkbook shows that she wrote a check for $987 for building materials. What was the price of the materials if the sales tax is 5%?

54. *Tipping.* Julian left a $4 tip for a meal that cost $25. What percentage of the cost of the meal was the tip?

55. *Deducting sales tax.* A tax-exempt charity received a bill of $145.90 for a sump pump. The bill incorrectly included sales tax of 5%. How much does the charity owe?

56. *Deducting sales tax.* A tax-exempt school group received a bill of $157.41 for educational software. The bill incorrectly included sales tax of 6%. How much should the school group pay?

57. *Cost of self-employment.* Because of additional taxes and fewer benefits, it has been estimated that a self-employed person must earn 20% more than a non–self-employed person performing the same task(s). If Trey earns $12 an hour working for The Typed Page, how much would he need to earn on his own for a comparable income?

58. Refer to Exercise 57. Clara earns $15 an hour working for Round Edge stairbuilders. How much would Clara need to earn on her own for a comparable income?

59. *Calorie content.* Pepperidge Farm Light Style 7 Grain Bread® has 140 calories in a 3-slice serving. This is 15% less than the number of calories in a serving of regular bread. How many calories are in a serving of regular bread?

60. *Fat content.* Peek Freans Shortbread Reduced Fat Cookies® contain 35 calories of fat in each serving. This is 40% less than the fat content in the leading imported shortbread cookie. How many calories of fat are in a serving of the leading shortbread cookie?

SKILL MAINTENANCE

61. Convert to decimal notation: $\frac{17}{25}$.

62. Add: $-23 + (-67)$.

63. Subtract: $-45.8 - (-32.6)$.

64. Remove parentheses and simplify:
$$4a - 8b - 5(5a - 4b).$$

SYNTHESIS

65. ◆ Would it be better to receive a 5% raise and then an 8% raise or the other way around? Why?

66. ◆ Erin is returning a tent that she bought during a 25%-off storewide sale that has ended. She is offered store credit for 125% of what she paid (not to be used on sale items). Is this fair to Erin? Why or why not?

67. ◆ Amber is in the 28% tax bracket. Would she be wiser to invest money in a CD (certificate of deposit) paying 5% (taxable) interest or in a (tax-free) municipal bond fund paying 4% interest? Why?

68. ◆ Does the following advertisement provide a convincing argument that summertime is when most burglaries occur? Why or why not?

69. If a is 130% of b, what percent of a is b?

70. The community of Bardville has 1332 left-handed females. If 48% of the community is female and 15% of the community is left-handed, how many people are in the community?

71. Claude pays 26% of his pretax earnings in taxes. What percentage of his *post*-tax earnings do his taxes comprise?

72. Rollie's Music charges $11.99 for a compact disc. Sound Warp charges $13.99 but you have a coupon for $2 off. In both cases, a 7% sales tax is charged on the *regular* price. How much does the disc cost at each store?

73. The new price of a car is 25% higher than the old price of $20,800. The old price is what percent lower than the new price?

COLLABORATIVE CORNER

Focus: Applications and models using percent
Time: 10–20 minutes
Group size: 3
Materials: Calculators are optional.

Often a store will reduce the price of an item by a fixed percentage and then return to the original price once the sale is over. Suppose a department store reduces all sporting goods 20%, all clothing 25%, and all electronics 15%.

ACTIVITY

1. Each group member should select one of the following three items: a $50 basketball, an $80 jacket, or a $240 portable sound system.
2. Each group member should apply the appropriate discount and determine the sale price of his or her item. Then he or she should find a multiplier that can be used to convert the discounted price back to the original price. Does this multiplier depend on the price of the item? By what percentage is the discounted price increased when returning to the original price?
3. Working together as a group, compare the results of part (2) for all three items. Then develop a formula that can be used to find a multiplier that will restore an $r\%$ discounted price to its original price. Check that your formula will duplicate the results of part (2).

2.5 Problem Solving

Five Steps for Problem Solving • Applying the Five Steps

One of the most important uses of algebra is as a tool for problem solving. In this section, we develop a problem-solving approach that will be used throughout the remainder of the text.

Five Steps for Problem Solving

In Section 2.4, we solved a problem in which Village Stationers needed a formula. To solve the problem, we *familiarized* ourselves with percent notation so that we could then *translate* a percent problem into an equation. We then *solved* the equation, *checked* the solution, and *stated* the answer at the end of the section.

FIVE STEPS FOR PROBLEM SOLVING IN ALGEBRA

1. *Familiarize* yourself with the problem situation.
2. *Translate* to mathematical language. (This often means writing an equation.)
3. *Carry out* some mathematical manipulation. (This often means *solving* an equation.)
4. *Check* your possible answer in the original problem.
5. *State* the answer clearly.

Of the five steps, the most important is probably the first one: becoming familiar with the problem situation. Here are some hints for familiarization.

TO BECOME FAMILIAR WITH A PROBLEM:

1. Read the problem carefully.
2. Reread the problem, perhaps aloud. Try to visualize the problem.
3. List the information given and the questions to be answered. Choose a variable (or variables) to represent the unknown and specify what the variable represents. For example, let L = length in centimeters, d = distance in miles, and so on.
4. Gather further information. Look up a formula in this book or in a reference book. Talk to a reference librarian or an expert in the field.
5. Make a table that uses all of the information you have available. Look for patterns that may help in the translation.
6. Make a drawing and label it with known and unknown information, using specific units if given.
7. Guess the answer and check the guess. Observe the manner in which the guess is checked.

Applying the Five Steps

EXAMPLE 1

Hiking. The Appalachian Trail stretches for 2100 mi, from Springer Mountain, Georgia, to Mount Katahdin, Maine. A hiker on Big Walker Mountain, in Virginia, is three times as far from the northern end as from the southern end. How far is the hiker from each end of the trail?

SOLUTION

1. **Familiarize.** It may be helpful to draw a picture. If we let

    d = the distance, in miles, to the southern end,

   then

    $3d$ = the distance, in miles, to the northern end.

(We could also let $x = $ the distance to the northern end and $\frac{1}{3}x = $ the distance to the southern end.)

To gain more familiarity, let's make a guess. Suppose that $d = 500$ mi. Then $3d = 1500$ mi and $500 + 1500 = 2000$. Since $2000 \neq 2100$, our guess was incorrect. Still, we have gained familiarity—we see that d must be more than 500 mi.

2. **Translate.** From the sketch, we see that the lengths of the two parts of the trail must add up to 2100 mi. This leads to our translation.

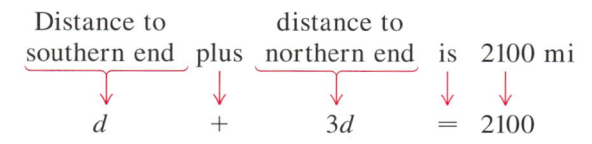

$$\underbrace{\text{Distance to southern end}}_{d} \underbrace{\text{plus}}_{+} \underbrace{\text{distance to northern end}}_{3d} \underbrace{\text{is}}_{=} \underbrace{2100 \text{ mi}}_{2100}$$

3. **Carry out.** We solve the equation:

$$d + 3d = 2100$$
$$4d = 2100 \quad \text{Combining like terms}$$
$$d = 525. \quad \text{Dividing by 4 on both sides}$$

4. **Check.** As predicted in the *Familiarize* step, d is more than 500 mi. If $d = 525$ mi, then $3d = 1575$ mi. Since $525 + 1575 = 2100$, we have a check.

5. **State.** It is 525 mi from Big Walker Mountain to Springer Mountain and 1575 mi from Big Walker Mountain to Mount Katahdin.

EXAMPLE 2

Page Numbers. The sum of two consecutive page numbers is 305. Find the page numbers.

SOLUTION

1. **Familiarize.** If the meaning of the word consecutive is unclear, we should consult a dictionary or someone who might know. Consecutive numbers are integers that are one unit apart. Thus, 18 and 19 are consecutive numbers, as are -24 and -23. If we let $x = $ the first page number, then $x + 1 = $ the next page number.

To become more familiar with the problem, we can make a table. How do we get the entries in the table? First, we just guess a value for x. Then we find $x + 1$. Finally, we add the two numbers to find their sum.

x	$x + 1$	Sum of x and $x + 1$
40	41	81	← This guess was much too small.
140	141	281	← This guess was a bit too small.

Our second guess leads us to suspect that the value of x is not much greater than 140. The problem could actually be solved with further guessing, but we need to practice using algebra.

2. **Translate.** We reword the problem and translate as follows.

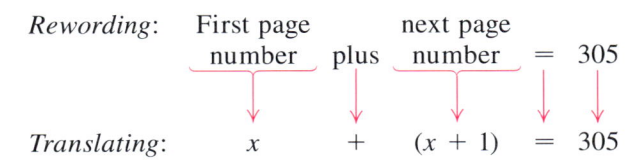

Rewording: First page number plus next page number = 305

Translating: x + $(x + 1)$ = 305

3. **Carry out.** We solve the equation:

$x + (x + 1) = 305$
$2x + 1 = 305$ Using an associative law and combining like terms
$2x = 304$ Subtracting 1
$x = 152.$ Dividing by 2

If x is 152, then $x + 1$ is 153.

4. **Check.** Our possible answers are 152 and 153. These are consecutive integers and their sum is 305, so the answers check in the original problem.

5. **State.** The page numbers are 152 and 153.

EXAMPLE 3

Truck Rentals. Truck-Rite Rentals rents trucks at a daily rate of $49.95 plus 39¢ per mile. Concert Productions has budgeted $100 for renting a truck to haul equipment to an upcoming concert. How many miles can a rental truck be driven on a $100 budget?

SOLUTION

1. **Familiarize.** Suppose that Concert Productions drives 75 mi. Then the cost is

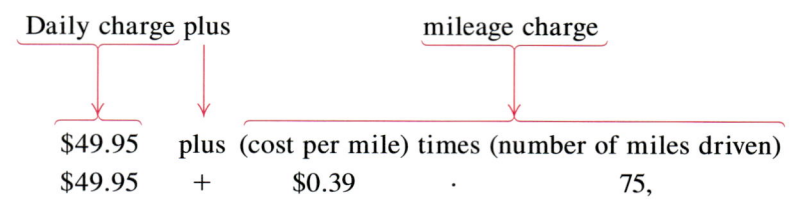

   Daily charge plus mileage charge

   $49.95 plus (cost per mile) times (number of miles driven)
   $49.95 + $0.39 · 75,

   which is $49.95 + $29.25, or $79.20. This familiarizes us with the way in which a calculation is made. Note that more than 75 mi can be driven without exceeding the budget.

   We let $m =$ the number of miles that can be driven for $100.

2. **Translate.** We reword the problem and translate as follows.

   Daily rate plus cost per mile times number of miles driven is total cost

   $49.95 + $0.39 · m = $100

3. **Carry out.** We solve the equation:

   $49.95 + 0.39m = 100$

   $0.39m = 50.05$ Subtracting 49.95 on both sides. This leaves $50.05 for mileage.

   $m = \dfrac{50.05}{0.39}$ Dividing by 0.39 on both sides

   $m \approx 128.3.$ Rounding to the nearest tenth

4. **Check.** To check, we multiply 128.3 by $0.39, getting $50.037. Then we add $50.037 to $49.95 and get $99.987, which is just about the $100 allotted. Note that $128.3 > 75$, as predicted in the *Familiarize* step.

5. **State.** The truck can be driven about 128.3 mi on the $100 budget.

EXAMPLE 4

Gardening. A rectangular community garden is to be enclosed with 92 m of fencing. In order to allow for compost storage, the garden must be 4 m longer than it is wide. Determine the dimensions of the garden.

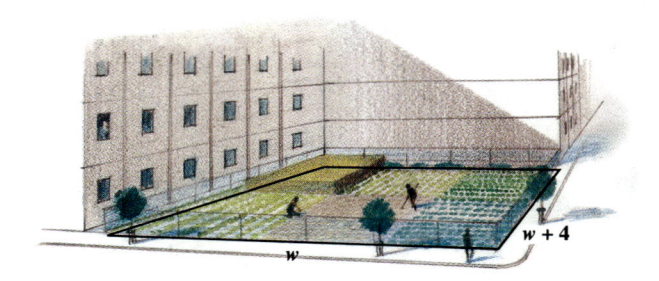

SOLUTION

1. **Familiarize.** Suppose the garden were 30 m wide. The length would then be 30 + 4 m, or 34 m, and the perimeter would be 2 · 30 + 2 · 34 m, or 128 m. Our guess was too big, but checking it has familiarized us with the problem.

   We let w = the width of the garden. Since the garden is "4 m longer than it is wide," we have $w + 4$ = the length. Recall that the perimeter P of a rectangle is the distance around it and is given by the formula $2l + 2w = P$, where l = the length and w = the width.

2. **Translate.** To translate, we substitute $w + 4$ for l and 92 for P:

   $$2l + 2w = P$$
   $$2(w + 4) + 2w = 92.$$

3. **Carry out.** We solve the equation:

   $$2(w + 4) + 2w = 92$$
   $$2w + 8 + 2w = 92 \quad \text{Using the distributive law}$$
   $$4w + 8 = 92 \quad \text{Combining like terms}$$
   $$4w = 84$$
   $$w = 21.$$

   The dimensions appear to be $w = 21$ m and l, or $w + 4$, = 25 m.

4. **Check.** If the width is 21 m and the length 25 m, then the garden is 4 m longer than it is wide. The perimeter is 2(25 m) + 2(21 m), or 92 m, and since 92 m of fencing is available, we have a check.

5. **State.** The garden should be 21 m wide and 25 m long.

EXAMPLE 5

Selling a Home. The Rosettis are planning to sell their home. If they want to clear $117,500 after having paid a 6% commission to a realtor, for how much must they sell the house?

SOLUTION

1. **Familiarize.** Suppose that the Rosettis sold the house for $120,000. A 6% commission can be determined by finding 6% of $120,000:

   6% of $120,000 = 0.06($120,000) = $7200.

Subtracting this commission from $120,000 would leave the Rosettis with

$120,000 − $7200 = $112,800.

This shows that in order for the Rosettis to clear $117,500, the house must be sold for more than $120,000. To determine exactly what the sale price must be, we could check more guesses. Instead, we let $x =$ the selling price. Because the commission is 6%, the realtor receives $0.06x$.

2. **Translate.** We reword the problem and translate as follows.

   Rewording: Selling price less commission is amount cleared.

   Translating: $\quad x \quad - \quad 0.06x \quad = \quad 117{,}500$

3. **Carry out.** We solve the equation:

   $x - 0.06x = 117{,}500$
   $1x - 0.06x = 117{,}500$
   $0.94x = 117{,}500$ Combining like terms. Had we noted that after the commission has been paid, 94% remains, we could have begun with this equation.

   $x = \dfrac{117{,}500}{0.94}$ Dividing by 0.94 on both sides

   $x = 125{,}000.$

4. **Check.** To check, we first find 6% of $125,000:

   6% of $125,000 = 0.06($125,000) = $7500.   This is the commission.

   Next, we subtract the commission to find the remaining amount:

   $125,000 − $7500 = $117,500.

   Since, after the commission, the Rosettis are left with $117,500, our answer checks. Note that the $125,000 sale price is greater than $120,000, as predicted in the *Familiarize* step.

5. **State.** The house must sell for $125,000.

EXAMPLE 6

Angles in a Triangle. The second angle of a triangle is 20° greater than the first. The third angle is twice as large as the first. How large are the angles?

SOLUTION

1. **Familiarize.** We draw a picture. In this case, the measure of the first angle $= x$, the measure of the second angle $= x + 20$, and the measure of the third angle $= 2x$.

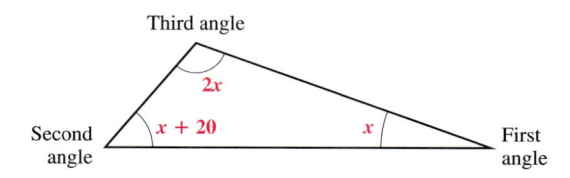

2. **Translate.** To translate, we need to recall a geometric fact (you might, as part of step 1, look it up in a geometry book): The measures of the angles of any triangle add up to 180°.

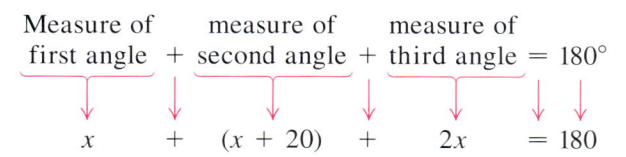

$$\underbrace{\text{Measure of first angle}}_{x} + \underbrace{\text{measure of second angle}}_{(x+20)} + \underbrace{\text{measure of third angle}}_{2x} = 180°$$

3. **Carry out.** We solve:

$$x + (x + 20) + 2x = 180$$
$$4x + 20 = 180$$
$$4x = 160$$
$$x = 40.$$

The measures for the angles appear to be:

First angle: $x = 40°$,

Second angle: $x + 20 = 40 + 20 = 60°$,

Third angle: $2x = 2(40) = 80°$.

4. **Check.** Consider 40°, 60°, and 80°. The second angle is 20° greater than the first, the third is twice the first, and the sum is 180°. These numbers check.

5. **State.** The measures of the angles are 40°, 60°, and 80°.

We close this section with some tips to aid you in problem solving.

PROBLEM-SOLVING TIPS

1. The more problems you solve, the more your skills will improve.
2. Look for patterns when solving problems. Each time you study an example in a text, you may observe a pattern for problems that you will encounter later in the exercise sets or in other practical situations.
3. When translating in mathematics, consider the dimensions of the variables and constants in the equation. The variables that represent length should all be in the same unit, those that represent money should all be in dollars or all in cents, and so on.
4. Make sure that units appear in the answer whenever appropriate.

EXERCISE SET 2.5

Solve. Even though you might find the answer quickly in some other way, practice using the five-step problem-solving process.

1. Three less than twice a number is 19. What is the number?
2. Two fewer than 10 times a number is 78. What is the number?
3. Five times the sum of 3 and some number is 70. What is the number?
4. Twice the sum of 4 and some number is 34. What is the number?
5. *Price of a CD player.* Doug paid $72 for a portable CD player during a 20%-off sale. What was the regular price?
6. *Price of a textbook.* Evelyn paid $50.40, including 5% tax, for a textbook. How much did the book itself cost?
7. The sum of three consecutive integers is 48. Find the numbers.
8. The sum of two consecutive odd numbers is 40. Find the numbers. (*Hint*: Odd numbers, like even numbers, are separated by 2 units.)
9. The sum of two consecutive even numbers is 50. Find the numbers. (*Hint*: See Exercise 8.)
10. The sum of two consecutive even integers is 106. What are the integers?
11. The sum of two consecutive odd integers is 128. What are the integers?
12. *Running.* Jenna is twice as far from the finish line as she is from the start of a 10-km race. How far has she run?
13. *Sled-dog racing.* The Iditarod sled-dog race extends for 1049 mi from Anchorage to Nome. If a musher is twice as far from Anchorage as from Nome, how many miles has the musher traveled?
14. A 480-m wire is cut into three pieces. The second piece is 3 times as long as the first. The third is 4 times as long as the second. How long is each piece?
15. *Home remodeling.* In 1995, Americans spent $35 billion to remodel bathrooms and kitchens. Twice as much was spent on kitchens as on bathrooms (*Source: Indianapolis Star*, 7/26/96). How much was spent on each?
16. *Angles of a triangle.* The second angle of a triangle is 3 times as large as the first. The third angle is 30° more than the first. Find the measure of each angle.
17. *Angles of a triangle.* The second angle of a triangle is 4 times as large as the first. The third angle is 45° less than the sum of the other two angles. Find the measure of each angle.
18. *Angles of a triangle.* The second angle of a triangle is 3 times as large as the first. The third angle is 10° more than the sum of the other two angles. Find the measure of the third angle.
19. *Angles of a triangle.* The second angle of a triangle is 4 times as large as the first. The third angle is 5° more than the sum of the other two angles. Find the measure of the second angle.
20. *Audio sales.* In the last quarter of the year, the Sound Meister tripled the total sales of the previous three quarters. Sales for the year totaled $480,000. How much was sold in the last quarter?
21. *Page numbers.* The sum of the page numbers on the facing pages of a book is 285. What are the page numbers?
22. *Page numbers.* The sum of the page numbers on the facing pages of a book is 281. What are the page numbers?
23. *Perimeter of a triangle.* The perimeter of a triangle is 195 mm. If the lengths of the sides are consecutive odd integers, find the length of each side.
24. *Perimeter of a triangle.* The perimeter of a triangle is 396 mm. If the lengths of the sides are consecutive even integers, find the length of each side.
25. *Hancock Building dimensions.* The ground floor of

the John Hancock Building in Chicago is a rectangle whose length is 100 ft more than the width. The perimeter is 860 ft. Find the width and the length of the rectangle. Find the area of the rectangle.

26. *Hancock Building dimensions.* The top of the John Hancock Building is a rectangle whose length is 60 ft more than the width. The perimeter is 520 ft. Find the width and the length of the rectangle. Find the area of the rectangle.

27. *Dimensions of a state.* The perimeter of the state of Wyoming is 1280 mi. The width is 90 mi less than the length. Find the width and the length.

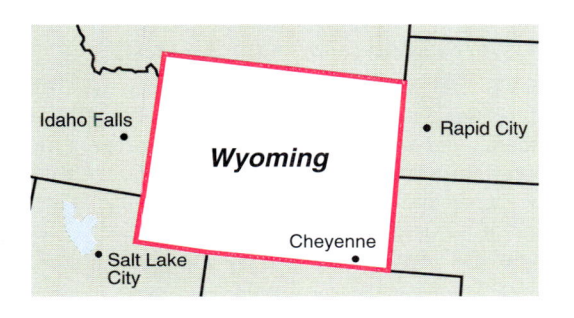

28. *Typing paper.* The perimeter of standard-size typewriter paper is 99 cm. The width is 6.3 cm less than the length. Find the length and the width.

29. *Savings interest.* Sharon invested money in a savings account at a rate of 6% simple interest. After 1 yr, she has $6996 in the account. How much did Sharon originally invest?

30. *Loan interest.* Alvin borrowed money from a cousin at a rate of 10% simple interest. After 1 yr, $7194 paid off the loan. How much did Alvin borrow?

31. *Car rentals.* Badger Rent-A-Car rents a compact car at a daily rate of $34.95 plus 10¢ per mile. A businessperson is allotted $80 for car rental. How many miles can she travel on the $80 budget?

32. *Car rentals.* Badger rents midsized cars at a rate of $43.95 plus 10¢ per mile. A tourist has a car-rental budget of $90. How many miles can he travel on the $90?

33. *Complementary angles.* The sum of the measures of two *complementary* angles is 90°. If one angle measures 15° more than twice the measure of its complement, find the measure of each angle.

34. *Supplementary angles.* The sum of the measures of two *supplementary* angles is 180°. If one angle measures 45° less than twice the measure of its supplement, find the measure of each angle.

35. *Race time.* The equation $R = -0.028t + 20.8$ can be used to predict the world record in the 200-m dash, where R represents the record in seconds and t represents the number of years since 1920. In what year will the record be 18.0 sec?

36. *Cricket chirps and temperature.* The equation $T = \frac{1}{4}N + 40$ can be used to determine the temperature T, in degrees Fahrenheit, given the number of times N a cricket chirps per minute. Determine the number of chirps per minute for a temperature of 80°F.

SKILL MAINTENANCE

Factor.

37. $5a + 10b - 45$
38. $3x - 12y + 60$

Simplify.

39. $7x - 3(8 - 2x) + 12$
40. $2x + 9 - 3(4x - 7)$

SYNTHESIS

41. ◆ While solving Exercise 14, Erin writes "$x =$ the first piece," Bob writes "$x =$ the length of the first piece," and Doris writes "$x =$ the length of the first piece, in meters." Is one of these approaches preferable to the others? Why or why not?

42. ◆ A student claims to be able to solve most of the problems of this section by guessing. Is there anything wrong with this approach? Why or why not?

43. ◆ Write a problem for a classmate to solve. Devise it so that the problem can be translated to the equation $x + (x + 2) + (x + 4) = 375$.

44. ◆ Write a problem for a classmate to solve. Devise it so that the solution is "Audrey can drive the rental truck for 50 mi without exceeding her budget."

45. *Basketball scoring.* Had Stacey's 3-pointer gone in, her team, the Blazers, would not have lost by a score of 79–77 and she would have scored half of the Blazers' points. How much did Stacey score in the loss?

46. *Car rentals.* If the daily rental for a car is $18.90 plus a certain price per mile and if Alanis must drive 190 mi on a $55 budget, what is the highest price per mile that Alanis can afford?

47. *Truck rentals.* Ed needs a truck for three days. He can rent either from Reston for $49 a day plus 20¢ per mile, or from Long Haul for $80 a day with unlimited mileage. What guidelines can you offer Ed to help him choose the less expensive rental?

48. *Test scores.* Pam scored 78 on a test that had 4 fill-ins worth 7 points each and 24 multiple-choice questions worth 3 points each. She had one fill-in wrong. How many multiple-choice questions did Pam get right?

49. *Gettysburg Address.* Abraham Lincoln's 1863 Gettysburg Address refers to the year 1776 as "four *score* and seven years ago." Write an equation and find what a score is.

50. One number is 25% of another. The larger number is 12 more than the smaller. What are the numbers?

51. *Angles in a quadrilateral.* The measures of the angles in a quadrilateral are consecutive odd numbers. Find the measure of each angle.

52. *Angles in a pentagon.* The measures of the angles in a pentagon are consecutive even numbers. Find the measure of each angle.

53. Apples are collected in a basket for six people. One third, one fourth, one eighth, and one fifth of the apples are given to four people, respectively. The fifth person gets ten apples, and one apple remains for the sixth person. Find the original number of apples in the basket.

54. *Winning percentage.* In a basketball league, the Falcons won 15 of their first 20 games. In order to win 60% of the total number of games, how many more games will they have to play, assuming they win only half the remaining games?

55. *Budgeting.* Luke spends 12% of his weekly salary on dining out. Of the money spent dining out, 55%, or $39.60, is spent on fast food. What is Luke's weekly salary?

56. *Perimeter of a rectangle.* The width of a rectangle is three fourths of the length. The perimeter of the rectangle becomes 50 cm when the length and the width are each increased by 2 cm. Find the length and the width.

57. *Area of a triangle.* The area of a triangle is 2.9047 in^2. Find the height of the triangle if the base is 8 in.

58. *Test scores.* Ella has an average score of 82 on three tests. Her average score on the first two tests is 85. What was the score on the third test?

59. *Music-club purchases.* During a recent sale, BMG Music Service® charged $8.49 for the first CD ordered and $3.99 for all others. For shipping and handling, BMG charged $2.47 for the first CD, $2.28 for the second CD, and $1.99 for all others. The total cost of a shipment (excluding tax) was $65.07. How many CD's were in the shipment?

60. A school purchases a piano and must choose between paying $2000 at the time of purchase or $2150 at the end of one year. Which option should the school select and why?

61. Sarah claims the following problem has no solution: "The sum of the page numbers on facing pages is 191. Find the page numbers." Is Sarah correct? Why or why not?

2.6 Solving Inequalities

Solutions of Inequalities • Graphs of Inequalities • Solving Inequalities Using the Addition Principle • Solving Inequalities Using the Multiplication Principle • Using the Principles Together

Many real-world situations translate to *inequalities*. For example, a student might be interested in what test scores will ensure *at least* a 90 average; an elevator might be designed to hold *at most* 2000 pounds; a tax credit might be allowable for families with incomes of *less than* $25,000; and so on. Before solving applications of this type, we must adapt our equation-solving principles to the solving of inequalities.

Solutions of Inequalities

In Section 1.4, we learned that an inequality is a number sentence containing $>$ (is greater than), $<$ (is less than), $\geq$ (is greater than or equal to), or $\leq$ (is less than or equal to). Inequalities like

$$-7 > x, \quad t < 5, \quad 5x - 2 \geq 9, \quad \text{and} \quad -3y + 8 \leq -7$$

are true for some replacements of the variable and false for others.

EXAMPLE 1 Determine whether the given number is a solution of $x < 2$: **(a)** -3; **(b)** 2.

SOLUTION

a) Since $-3 < 2$ is true, -3 is a solution.
b) Since $2 < 2$ is false, 2 is not a solution.

EXAMPLE 2 Determine whether the given number is a solution of $y \geq 6$: **(a)** 6; **(b)** -4.

SOLUTION

a) Since $6 \geq 6$ is true, 6 is a solution.
b) Since $-4 \geq 6$ is false, -4 is not a solution.

Graphs of Inequalities

Because the solutions of inequalities like $x < 2$ are too numerous to list, it is helpful to make a drawing that represents all the solutions. The **graph** of an inequality is such a drawing. Graphs of inequalities in one variable can be drawn on a number line by shading all points that are solutions. Open dots are used to indicate endpoints that are *not* solutions and closed dots indicate endpoints that *are* solutions.

EXAMPLE 3 Graph each inequality: **(a)** $x < 2$; **(b)** $y \geq -3$; **(c)** $-2 < x \leq 3$.

SOLUTION

a) The solutions of $x < 2$ are those numbers less than 2. They are shown on the graph by shading all points to the left of 2. The open dot at 2 indicates that 2 is *not* part of the graph, but numbers like 1.2 and 1.99 are.

b) The solutions of $y \geq -3$ are shown on the number line by shading the point for -3 and all points to the right of -3. The closed dot at -3 indicates that -3 *is* part of the graph.

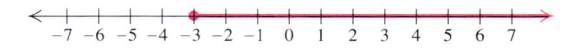

c) The inequality $-2 < x \leq 3$ is read "-2 is less than x and x is less than or equal to 3," or "x is greater than -2 and less than or equal to 3." To be a solution of $-2 < x \le 3$, a number must be a solution of both $-2 < x$ and $x \le 3$. The number 1 is a solution, as are -0.5, 1.9, and 3. An open dot indicates that -2 is not a solution, whereas a closed dot indicates that 3 is a solution. The other solutions are shaded.

Solving Inequalities Using the Addition Principle

Consider a balance similar to one that appears in Section 2.1. When one side of the balance weighs more than the other, the balance will tip in that direction. If equal amounts of weight are then added or subtracted on both sides of the balance, the balance will remain tipped in the same direction.

The balance illustrates the idea that when a number, such as 2, is added (or subtracted) on both sides of a true inequality, such as $3 < 7$, we get another true inequality:

$3 + 2 < 7 + 2$, or $5 < 9$.

Similarly, if we add -4 on both sides of $x + 4 < 10$, we get an *equivalent* inequality:

$x + 4 + (-4) < 10 + (-4)$, or $x < 6$.

We say that $x + 4 < 10$ and $x < 6$ are **equivalent**, which means that both inequalities have the same solution set.

THE ADDITION PRINCIPLE FOR INEQUALITIES	For any real numbers a, b, and c: $a < b$ is equivalent to $a + c < b + c$; $a \le b$ is equivalent to $a + c \le b + c$; $a > b$ is equivalent to $a + c > b + c$; $a \ge b$ is equivalent to $a + c \ge b + c$.

As with equation solving, when solving inequalities, our goal is to isolate the variable on one side.

2.6 SOLVING INEQUALITIES

EXAMPLE 4 Solve $x + 2 > 8$ and then graph the solution.

SOLUTION We use the addition principle, subtracting 2 on both sides:

$x + 2 - 2 > 8 - 2$ Subtracting 2 or adding -2 on both sides
$x > 6$.

From the inequality $x > 6$, we can determine the solutions easily. Any number greater than 6 makes $x > 6$ true and is a solution of that inequality as well as the inequality $x + 2 > 8$. The graph is as follows:

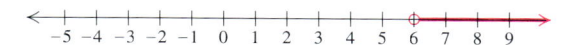

$$\leftarrow\!\!+\!\!\!+\!\!\!+\!\!\!+\!\!\!+\!\!\!+\!\!\!+\!\!\!+\!\!\!+\!\!\!+\!\!\!+\!\!\!\circ\!\!\!+\!\!\!+\!\!\!+\!\!\!\rightarrow$$
$$-5\ -4\ -3\ -2\ -1\ \ 0\ \ 1\ \ 2\ \ 3\ \ 4\ \ 5\ \ 6\ \ 7\ \ 8\ \ 9$$

Because most inequalities have an infinite number of solutions, we cannot possibly check them all. A partial check can be made using one of the possible solutions. For this example, we can substitute any number greater than 6, say 6.1, into the original inequality:

$$\begin{array}{r|l} x + 2 > 8 \\ \hline 6.1 + 2 \ ? \ 8 \\ 8.1 \ | \ 8 \ \ \text{TRUE} \end{array}$$

Since $8.1 > 8$ is true, 6.1 is a solution. Any number greater than 6 is a solution.

Although the inequality $x > 6$ is easy to solve (we merely replace x with numbers greater than 6), it is important to note that $x > 6$ is an *inequality*, not a *solution*. In fact, the solutions of $x > 6$ are numbers. To describe the set of all solutions, we will use **set-builder notation** to write the *solution set* of Example 4 as

$\{x|\ x > 6\}$.

This notation is read

"The set of all x such that x is greater than 6."

Thus a number is in $\{x|\ x > 6\}$ if that number is greater than 6. From now on, solutions of inequalities will be written using set-builder notation.

EXAMPLE 5 Solve $3x - 1 \leq 2x - 5$ and then graph the solution.

SOLUTION We have

$3x - 1 \leq 2x - 5$
$3x - 1 + 1 \leq 2x - 5 + 1$ Adding 1 on both sides
$3x \leq 2x - 4$ Simplifying
$3x - 2x \leq 2x - 4 - 2x$ Subtracting $2x$ on both sides
$x \leq -4$.   Simplifying

The graph is as follows:

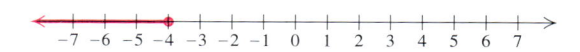

$$\leftarrow\!\!+\!\!\!+\!\!\!+\!\!\!\bullet\!\!\!+\!\!\!+\!\!\!+\!\!\!+\!\!\!+\!\!\!+\!\!\!+\!\!\!+\!\!\!+\!\!\!+\!\!\!+\!\!\!\rightarrow$$
$$-7\ -6\ -5\ -4\ -3\ -2\ -1\ \ 0\ \ 1\ \ 2\ \ 3\ \ 4\ \ 5\ \ 6\ \ 7$$

Any number less than or equal to -4 is a solution so the solution set is $\{x \mid x \leq -4\}$.

Solving Inequalities Using the Multiplication Principle

There is a multiplication principle for inequalities similar to that for equations, but it must be modified when multiplying on both sides by a negative number. Consider the inequality

$$3 < 7.$$

If we multiply on both sides by a *positive* number, say 2, we get another true inequality:

$$3 \cdot 2 < 7 \cdot 2, \quad \text{or} \quad 6 < 14. \qquad \text{TRUE}$$

If we multiply on both sides by a *negative* number, say -2, we get a *false* inequality:

$$3 \cdot (-2) < 7 \cdot (-2), \quad \text{or} \quad -6 < -14. \qquad \text{FALSE}$$

The fact that $6 < 14$ is true, but $-6 < -14$ is false, stems from the fact that the negative numbers, in a sense, mirror the positive numbers. Whereas 14 is to the *right* of 6, the number -14 is to the *left* of -6. Thus if we reverse the inequality symbol in $-6 < -14$, we get a true inequality:

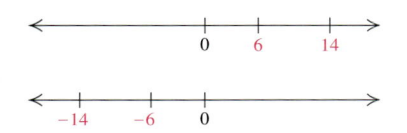

$$-6 > -14. \qquad \text{TRUE}$$

THE MULTIPLICATION PRINCIPLE FOR INEQUALITIES

For any real numbers a and b, and for any *positive* number c:

$a < b$ is equivalent to $ac < bc$, and
$a > b$ is equivalent to $ac > bc$.

For any real numbers a and b, and for any *negative* number c:

$a < b$ is equivalent to $ac > bc$, and
$a > b$ is equivalent to $ac < bc$.

Similar statements hold for $\leq$ and $\geq$.

EXAMPLE 6 Solve and graph each inequality: (a) $\frac{1}{4}x < 7$; (b) $-2y < 18$.

SOLUTION

a) $\quad \frac{1}{4}x < 7$

$\quad 4 \cdot \frac{1}{4}x < 4 \cdot 7 \qquad$ Multiplying by 4, the reciprocal of $\frac{1}{4}$, on both sides

$\qquad\qquad\qquad\qquad$ The symbol stays the same.

$\quad x < 28 \qquad$ Simplifying

The solution set is $\{x \mid x < 28\}$. The graph is as follows:

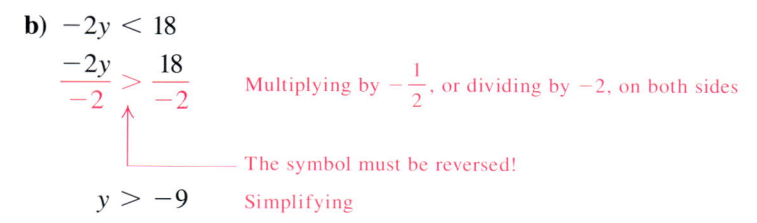

b) $-2y < 18$

$$\dfrac{-2y}{-2} > \dfrac{18}{-2} \qquad \text{Multiplying by } -\dfrac{1}{2}, \text{ or dividing by } -2, \text{ on both sides}$$

The symbol must be reversed!

$$y > -9 \qquad \text{Simplifying}$$

As a partial check, we substitute a number greater than -9, say -8, into the original inequality:

$$\dfrac{-2y < 18}{-2(-8) \ ? \ 18}$$
$$16 \ | \ 18 \quad \text{TRUE}$$

The solution set is $\{y \mid y > -9\}$. The graph is as follows:

Using the Principles Together

We use the addition and multiplication principles together to solve inequalities much as we did when solving equations.

EXAMPLE 7   Solve: **(a)** $6 - 5y > 7$; **(b)** $2x - 9 \leq 7x + 1$.

SOLUTION

a)
$$6 - 5y > 7$$
$$-6 + 6 - 5y > -6 + 7 \qquad \text{Adding } -6 \text{ on both sides}$$
$$-5y > 1 \qquad \text{Simplifying}$$
$$-\tfrac{1}{5} \cdot (-5y) < -\tfrac{1}{5} \cdot 1 \qquad \text{Multiplying by } -\tfrac{1}{5}, \text{ or dividing by } -5, \text{ on both sides}$$

The symbol must be reversed.

$$y < -\tfrac{1}{5} \qquad \text{Simplifying}$$

As a check, we substitute a number smaller than $-\tfrac{1}{5}$, say -1, into the original inequality:

$$\dfrac{6 - 5y > 7}{6 - 5(-1) \ ? \ 7}$$
$$6 - (-5) \ |$$
$$11 \ | \ 7 \quad \text{TRUE} \qquad 11 > 7 \text{ is a true statement.}$$

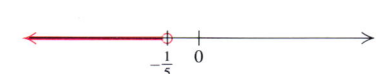

The solution set is $\left\{y \mid y < -\tfrac{1}{5}\right\}$.

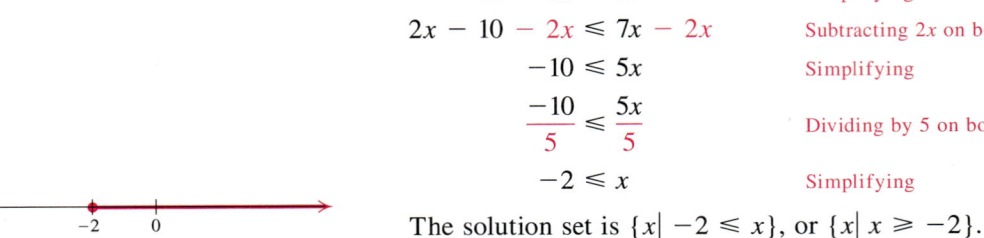

b)
$$2x - 9 \leq 7x + 1$$
$$2x - 9 - 1 \leq 7x + 1 - 1 \quad \text{Subtracting 1 on both sides}$$
$$2x - 10 \leq 7x \quad \text{Simplifying}$$
$$2x - 10 - 2x \leq 7x - 2x \quad \text{Subtracting } 2x \text{ on both sides}$$
$$-10 \leq 5x \quad \text{Simplifying}$$
$$\frac{-10}{5} \leq \frac{5x}{5} \quad \text{Dividing by 5 on both sides}$$
$$-2 \leq x \quad \text{Simplifying}$$

The solution set is $\{x \mid -2 \leq x\}$, or $\{x \mid x \geq -2\}$.

All of the equation-solving techniques used in Sections 2.1 and 2.2 can be used with inequalities provided we remember to reverse the inequality symbol when multiplying or dividing on both sides by a negative number.

EXAMPLE 8   Solve: **(a)** $16.3 - 7.2p \leq -8.18$; **(b)** $3(x - 2) - 1 < 2 - 5(x + 6)$.

SOLUTION

a) The greatest number of decimal places in any one number is *two*. Multiplying by 100 will clear decimals. Then we proceed as before.

$$16.3 - 7.2p \leq -8.18$$
$$100(16.3 - 7.2p) \leq 100(-8.18) \quad \text{Multiplying by 100 on both sides}$$
$$100(16.3) - 100(7.2p) \leq 100(-8.18) \quad \text{Using the distributive law}$$
$$1630 - 720p \leq -818 \quad \text{Simplifying}$$
$$-720p \leq -818 - 1630 \quad \text{Subtracting 1630 on both sides}$$
$$-720p \leq -2448 \quad \text{Simplifying}$$
$$p \geq \frac{-2448}{-720} \quad \text{Multiplying by } -\frac{1}{720}$$

The symbol must be reversed.

$$p \geq 3.4$$

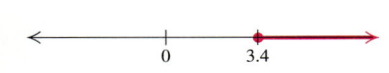

The solution set is $\{p \mid p \geq 3.4\}$.

b) $3(x - 2) - 1 < 2 - 5(x + 6)$
$$3x - 6 - 1 < 2 - 5x - 30 \quad \text{Using the distributive law to remove parentheses}$$
$$3x - 7 < -5x - 28 \quad \text{Simplifying}$$
$$3x + 5x < -28 + 7 \quad \text{Adding } 5x \text{ and also 7, to get all } x\text{-terms on one side and all other terms on the other side}$$
$$8x < -21 \quad \text{Simplifying}$$
$$x < -\frac{21}{8} \quad \text{Multiplying by } \frac{1}{8} \text{ on both sides}$$

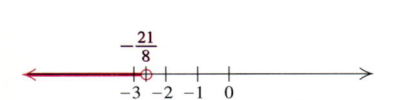

The solution set is $\{x \mid x < -\frac{21}{8}\}$.

EXERCISE SET 2.6

Determine whether each number is a solution of the given inequality.

1. $x > -3$
   a) 4
   b) 0
   c) -4.1
   d) -3.9
   e) 5.6

2. $y < 5$
   a) 0
   b) 5
   c) 4.99
   d) -13
   e) $7\frac{1}{4}$

3. $x \geq 6$
   a) -6
   b) 0
   c) 6
   d) 6.01
   e) $-3\frac{1}{2}$

4. $x \leq 10$
   a) 4
   b) -10
   c) 0
   d) 10.2
   e) -4.7

Graph on a number line.

5. $x \geq 6$
6. $y < 0$
7. $t < -3$
8. $y > 5$
9. $m > -4$
10. $p \leq 3$
11. $-3 < x \leq 5$
12. $-5 \leq x < 2$
13. $0 < x < 3$
14. $-5 \leq x \leq 0$

Describe each graph using set-builder notation.

15.
16.
17.
18.
19.
20.
21.
22.

Solve using the addition principle. Graph and write set-builder notation for the answers.

23. $y + 1 > 7$
24. $y + 6 > 9$
25. $x + 8 \leq -10$
26. $x + 9 \leq -12$
27. $x - 4 < 6$
28. $x - 3 < 14$
29. $x - 6 \geq 2$
30. $x - 9 \geq 4$
31. $y - 7 > -12$
32. $y - 10 > -16$
33. $2x + 4 \leq x + 9$
34. $2x + 4 \leq x + 1$

Solve using the addition principle. Write the answers in set-builder notation.

35. $4x - 6 \geq 3x - 1$
36. $3x - 9 \geq 2x + 11$
37. $y + \frac{1}{3} \leq \frac{5}{6}$
38. $x + \frac{1}{4} \leq \frac{1}{2}$
39. $x - \frac{1}{8} > \frac{1}{2}$
40. $y - \frac{1}{3} > \frac{1}{4}$
41. $-9x + 17 > 17 - 8x$
42. $-7x + 13 > 13 - 6x$

Solve using the multiplication principle. Graph and write set-builder notation for the answers.

43. $5x < 35$
44. $8x \geq 32$
45. $9y \leq 81$
46. $10x > 240$
47. $-7x < 13$
48. $8y < 17$
49. $12x > -36$
50. $-16x < -64$

Solve using the multiplication principle. Write the answers in set-builder notation.

51. $7y \geq -2$
52. $5x > -3$
53. $-5x < -17$
54. $-3y \leq -14$
55. $-4y \leq \frac{1}{3}$
56. $-2x \geq \frac{1}{5}$
57. $-\frac{8}{5} > -2x$
58. $-\frac{5}{8} < -10y$

Solve using the addition and multiplication principles.

59. $5 + 3x < 32$
60. $5 + 4y < 37$
61. $6 + 5y \geq 26$
62. $7 + 8x \geq 71$
63. $3x - 5 \leq 13$
64. $5y - 9 \leq 21$
65. $13x - 7 < -46$
66. $8y - 4 < -52$
67. $16 < 4 - 3y$
68. $22 < 6 - 8x$
69. $39 > 3 - 9x$
70. $40 > 5 - 7y$
71. $3 - 6y \geq 23$
72. $8 - 2y > 14$
73. $-3 < 8x + 7 - 7x$
74. $-5 < 9x + 8 - 8x$
75. $6 - 4y > 4 - 3y$
76. $7 - 8y > 5 - 7y$
77. $5 - 9y \leq 2 - 8y$
78. $6 - 13y \leq 4 - 12y$
79. $33 - 12x < 4x + 97$
80. $27 - 11x > 14x - 18$

108 CHAPTER 2 • EQUATIONS, INEQUALITIES, AND PROBLEM SOLVING

81. $2.1x + 43.2 > 1.2 - 8.4x$
82. $0.96y - 0.79 \leq 0.21y + 0.46$
83. $0.7n - 15 + n \geq 2n - 8 - 0.4n$
84. $1.7t + 8 - 1.62t < 0.4t - 0.32 + 8$
85. $\frac{x}{3} - 4 \leq 1$
86. $\frac{2}{3} - \frac{x}{5} < \frac{4}{15}$
87. $\frac{y}{5} + 2 \leq \frac{3}{5}$
88. $\frac{3x}{5} \geq -15$
89. $3(2y - 3) < 27$
90. $4(2y - 3) > 28$
91. $3(t - 2) \geq 9(t + 2)$
92. $8(2t + 1) > 4(7t + 7)$
93. $3(r - 6) + 2 < 4(r + 2) - 21$
94. $5(t + 3) + 9 > 3(t - 2) + 6$
95. $\frac{2}{3}(2x - 1) \geq 10$
96. $\frac{4}{5}(3x + 4) \leq 20$
97. $\frac{3}{4}\left(3x - \frac{1}{2}\right) - \frac{2}{3} < \frac{1}{3}$
98. $\frac{2}{3}\left(\frac{7}{8} - 4x\right) - \frac{5}{8} < \frac{3}{8}$

SKILL MAINTENANCE

Simplify.

99. $5 - 3^2 + (8 - 2)^2 \cdot 4$
100. $10 \div 2 \cdot 5 - 3^2 + (-5)^2$
101. $5(2x - 4) - 3(4x + 1)$
102. $9(3 + 5x) - 4(7 + 2x)$

SYNTHESIS

103. ◆ Are the inequalities $3x - 4 < 10 - 4x$ and $2(x - 5) > 3(2x - 6)$ equivalent? Why or why not?
104. ◆ Are all solutions of $x > -5$ solutions of $-x < 5$? Why or why not?
105. ◆ Explain in your own words why it is necessary to reverse the inequality symbol when multiplying both sides of an inequality by a negative number.
106. ◆ Explain how it is possible for the graph of an inequality to consist of just one number. (*Hint*: See Example 3c.)

Solve.

107. $6[4 - 2(6 + 3t)] > 5[3(7 - t) - 4(8 + 2t)] - 20$
108. $27 - 4[2(4x - 3) + 7] \geq 2[4 - 2(3 - x)] - 3$

Solve for x.

109. $-(x + 5) \geq 4a - 5$
110. $\frac{1}{2}(2x + 2b) > \frac{1}{3}(21 + 3b)$
111. $y < ax + b$ (Assume $a > 0$.)
112. $y < ax + b$ (Assume $a < 0$.)
113. Determine whether each number is a solution of the inequality $|x| < 3$.
    a) 3.2 b) -2 c) -3
    d) -2.9 e) 3 f) 1.7
114. Graph the solutions of $|x| < 3$ on a number line.

2.7 Solving Applications with Inequalities

Translating to Inequalities • Solving Problems

The five steps for problem solving can be used for problems involving inequalities.

Translating to Inequalities

Before solving problems that involve inequalities, we list some important phrases to look for. Sample translations are listed as well.

Important Words	Sample Sentence	Translation
is at least	Bill is at least 21 years old.	$b \geq 21$
is at most	At most 5 students dropped the course.	$n \leq 5$
cannot exceed	To qualify, earnings cannot exceed $12,000.	$r \leq 12{,}000$
must exceed	The speed must exceed 15 mph.	$s > 15$
is less than	Tucker's weight is less than 50 lb.	$w < 50$
is more than	Boston is more than 200 miles away.	$d > 200$
is between	The film was between 90 and 100 minutes long.	$90 < t < 100$

Solving Problems

EXAMPLE 1

Dietary Restrictions. The team trainer has urged Gerald to eat, on average, no more than 30 grams (g) of fat each day. During the course of a four-day weekend, Gerald estimates that he consumed 28, 37, and 46 g of fat on the first three days. Determine (in terms of an inequality) how many grams of fat Gerald can consume on the last day of the weekend if he is to satisfy the trainer's recommendation.

SOLUTION

1. **Familiarize.** Let us suppose that Gerald consumed 25 g of fat on the fourth day. His daily average for the weekend would then be

$$\frac{28 + 37 + 46 + 25}{4} = 34.$$

This shows that if Gerald is to average fewer than 30 g of fat per day for the weekend, he cannot consume 25 g of fat on the fourth day. Rather than guess again, let's translate to an inequality in which $x =$ the number of grams of fat that Gerald consumes on the fourth day.

2. **Translate.** We reword the problem and translate as follows:

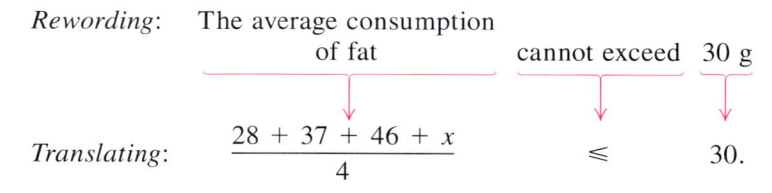

Rewording: The average consumption of fat   cannot exceed   30 g

Translating: $\dfrac{28 + 37 + 46 + x}{4} \leq 30.$

3. **Carry out.** Because of the fraction, this is a case in which the multiplication principle is best used first:

$$\frac{28 + 37 + 46 + x}{4} \leq 30$$

$$4\left(\frac{28 + 37 + 46 + x}{4}\right) \leq 4 \cdot 30 \qquad \text{Multiplying by 4 to clear the fraction}$$

$$28 + 37 + 46 + x \leq 120$$

$$111 + x \leq 120 \qquad \text{Simplifying}$$

$$x \leq 9. \qquad \text{Subtracting 111}$$

4. **Check.** As a partial check, we show that Gerald can consume 9 g of fat on the fourth day and not exceed a 30-g average for the four days:

$$\frac{28 + 37 + 46 + 9}{4} = \frac{120}{4} = 30.$$

5. **State.** In order for Gerald to average no more than 30 g of fat per day for the four days, he should consume no more than 9 g of fat on the fourth day.

EXAMPLE 2

Banquet Costs. The women's volleyball team can spend at most $450 for its awards banquet at a local restaurant. If the restaurant charges a $40 setup fee plus $16 per person, at most how many can attend?

SOLUTION

1. **Familiarize.** Let us suppose that 20 people were to attend the dinner. The cost would then be $40 + $16 · 20, or $360. This shows that more than 20 people could attend without exceeding $450. We could make another guess, or subtract $360 from $450 and "use up" the difference. Instead, we translate to an inequality and solve. We let $n =$ the number of people in attendance.

2. **Translate.** The cost of the banquet will be $40 for the setup fee plus $16 times the number of people attending. We can reword as follows:

Rewording: The setup fee plus the cost of the dinners cannot exceed $450

Translating: 40 + 16 · n ≤ 450.

3. **Carry out.** We solve for n:

$$40 + 16n \leq 450$$

$$16n \leq 410 \qquad \text{Subtracting 40 on both sides}$$

$$n \leq \frac{410}{16} \qquad \text{Dividing by 16 on both sides}$$

$$n \leq 25.625. \qquad \text{Simplifying}$$

4. **Check.** Although the solution set of the inequality is all numbers less than or equal to 25.625, since *n* represents the number of people in attendance, we round *down* to 25. If 25 people attend, the cost will be $40 + $16 · 25, or $440, and if 26 attend, the cost will exceed $450.

5. **State.** At most 25 people can attend the banquet.

An important point should be made regarding Example 2: Solutions of equations or inequalities do not always solve the problem from which the equation or inequality arises. In some cases, answers must be nonnegative, and in other cases, answers must be integers. Be sure to always check that the original problem has been solved.

EXERCISE SET 2.7

Translate to an inequality.

1. A number is less than 9.
2. A number is greater than or equal to 5.
3. The bag weighs at least 2 lb.
4. Between 75 and 100 people attended the concert.
5. The average speed was between 90 and 110 mph.
6. At least 400,000 people attended the Million Man March.
7. At most 1,200,000 people attended the Million Man March.
8. The amount of acid is not to exceed 40 liters (L).
9. The cost of gasoline is no less than $1.20 per gallon.
10. The temperature is at most −2°.

Use an inequality and the five-step process to solve each problem.

11. *Grade average.* Nadia is taking a literature course in which four tests are given. To get a B, a student must average at least 80 on the four tests. Nadia scored 82, 76, and 78 on the first three tests. What scores on the last test will earn her at least a B?

12. *Quiz average.* Rod's quiz grades are 73, 75, 89, and 91. What scores on a fifth quiz will make his average quiz grade at least 85?

13. *Van rentals.* Atlas rents a cargo van at a daily rate of $44.95 plus $0.39 per mile. A business has budgeted $250 for a one-day van rental. What mileages will allow the business to stay within budget? Round to the nearest tenth of a mile.

14. *Truck rentals.* Ridem rents trucks at a daily rate of $42.95 plus $0.46 per mile. The Letsons want a one-day truck rental, but must stay within a budget of $200. What mileages will allow them to stay within budget? Round to the nearest tenth of a mile.

15. *Phone costs.* Simon claims that it costs him at least $3.00 every time he calls a customer from a pay phone. If an average call costs 75¢ plus 45¢ for each minute, how long do his calls typically last?

16. *Parking costs.* Laura is certain that every time she parks in the municipal garage it costs her at least $2.20. If the garage charges 45¢ plus 25¢ for each half hour, for how long is Laura's car generally parked?

17. *Fruit servings.* Following the guidelines of the Food and Drug Administration, Dale tries to eat at least 5 servings of fruit each day. For the first six days of one week, she had 4, 6, 7, 4, 6, and 4 servings. How many servings of fruit should Dale eat on Saturday, in order to average at least 5 servings per day for the week?

18. *College course load.* To remain on financial aid, Millie needs to complete an average of at least 7 credits per quarter each year. In the first three quarters of 1997, Millie completed 5, 7, and 8 credits. How many credits of course work must Millie complete in the fourth quarter if she is to remain on financial aid?

19. *Well drilling.* All Seasons Well Drilling offers two plans. Under the "pay-as-you-go" plan, they charge $500 plus $8 a foot for a well of any depth. Under

their "guaranteed-water" plan, they charge a flat fee of $4000 for a well that is guaranteed to provide adequate water for a household. For what depths would it save a customer money to use the pay-as-you-go plan?

20. *Cost of road service.* Rick's Automotive charges $50 plus $15 for each (15-min) unit of time when making a road call. Twin City Repair charges $70 plus $10 for each unit of time. Under what circumstances would it be more economical for a motorist to call Rick's?

21. *Area of a rectangle.* The width of a rectangle is fixed at 4 cm. For what lengths will the area be less than 86 cm²?

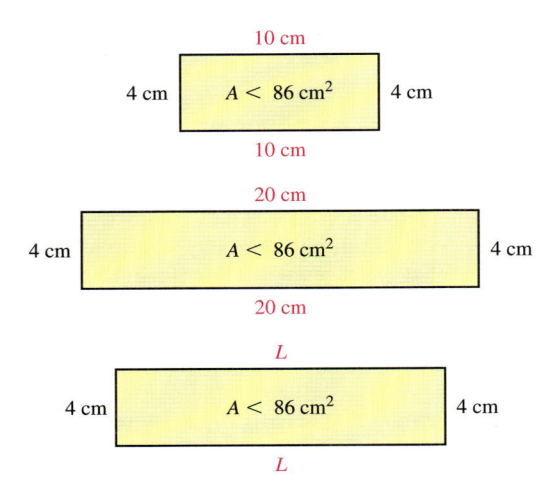

22. *Area of a rectangle.* The width of a rectangle is fixed at 16 yd. For what lengths will the area be at least 264 yd²?

23. *Insurance-covered repairs.* Most insurance companies will replace a vehicle if an estimated repair exceeds 80% of the "blue-book" value of the vehicle. Michelle's insurance company paid $8500 for repairs to her Subaru after an accident. What can be concluded about the blue-book value of the car?

24. *Insurance-covered repairs.* Following an accident, Jeff's Ford pickup was replaced by his insurance company because the damage was so extensive. Before the damage, the blue-book value of the truck was $21,000. How much would it have cost to repair the truck? (See Exercise 23.)

25. *Track records.* The formula $R = -0.075t + 3.85$ can be used to predict the world record in the 1500-m run t years after 1930. For what years will the world record be less than 3.5 min?

26. *Track records.* The formula $R = -0.028t + 20.8$ can be used to predict the world record in the 200-m dash t years after 1920. For what years will the world record be less than 19.0 sec?

27. *Weight gain.* A 9-lb puppy is gaining weight at a rate of $\frac{3}{4}$ lb per week. When will the puppy's weight exceed $22\frac{1}{2}$ lb?

28. *Pond depth.* On July 1, Garrett's Pond was 25 ft deep. Since that date, the water level has dropped $\frac{2}{3}$ ft per week. For what dates will the water level not exceed 21 ft?

29. *Body temperature.* A person is considered to be feverish when his or her temperature is higher than 98.6°F. The formula $F = \frac{9}{5}C + 32$ can be used to convert Celsius temperatures C to Fahrenheit temperatures F. For which Celsius temperatures is a person considered feverish?

30. *Melting butter.* Butter stays solid at Fahrenheit temperatures below 88°. Use the formula in Exercise 29 to determine those Celsius temperatures for which butter stays solid.

31. *Electrician visits.* Dot's Electric made 17 customer calls last week and 22 calls this week. How many calls must be made next week in order to maintain an average of at least 20 for the three-week period?

32. Find all numbers such that 3 times the number minus 10 times the number is greater than or equal to 8 times the number.

33. *Perimeter of a rectangle.* The length of a rectangle is fixed at 26 cm. What widths will make the perimeter greater than 80 cm?

34. *Perimeter of a rectangle.* The width of a rectangle is fixed at 8 ft. What lengths will make the perimeter at least 200 ft? at most 200 ft?

35. *Perimeter of a triangle.* One side of a triangle is 2 cm shorter than the base. The other side is 3 cm longer than the base. What lengths of the base will allow the perimeter to be greater than 19 cm?

36. *Perimeter of a pool.* The perimeter of a rectangular swimming pool is not to exceed 70 ft. The length is to be twice the width. What widths will meet these conditions?

37. *Volunteer work.* George and Joan do volunteer work at a hospital. Joan worked 3 more hr than George, and together they worked more than 27 hr. What possible numbers of hours did each work?

38. *Cost of clothes.* Angelo is shopping for a new pair of jeans and two sweaters of the same kind. He is determined to spend no more than $120.00 for the clothes. He buys jeans for $21.95. What is the most that Angelo can spend for each sweater?

39. *Net weight.* A medium-sized box of dog biscuits weighs 1 lb more than the small size. The large size weighs 2 lb more than the small size. The total weight of the three boxes is at most 30 lb. What are the possible weights of the small box?

40. *Area of a rectangle.* The width of a rectangle is 32 km. For what lengths is the area at least 2048 km^2?

41. *Fat content in foods.* Reduced Fat Hydrox® cookies contain 4 g of fat per serving. In order for a food to be labeled "reduced fat," it must have at least 25% less fat than the regular item. What can you conclude about how many grams of fat are in a serving of the regular Hydrox cookies?

42. *Fat content in foods.* Reduced Fat Skippy® peanut butter contains 12 g of fat per serving (see Exercise 41). What can you conclude about how many grams of fat are in regular Skippy peanut butter?

43. *Area of a triangle.* The base of a triangle is 16 cm. What lengths of the height will guarantee that the area of the triangle is at least 72 cm^2?

44. *Area of a triangle.* The height of a triangle is 20 cm. For what lengths of the base is the area at most 40 cm^2?

45. *Price of a movie ticket.* The average price of a movie ticket can be estimated by the equation $P = 0.1522Y - 298.592$, where Y is the year and P the average price, in dollars. The price is lower than what might be expected due to senior-citizen discounts, children's prices, and special volume discounts. For what years will the average price of a movie ticket be at least $6? (Include the year in which the $6 ticket first occurs.)

46. *Toll charges.* The equation $y = 0.027x + 0.19$ can be used to determine the approximate cost y, in dollars, of driving x miles on the Indiana toll road. For what mileages x will the cost be at most $6?

SKILL MAINTENANCE

Simplify.

47. $-3 + 2(-5)^2(-3) - 7$

48. $7 - a^2 - 9 + 5a^2$

49. $9x - 5 + 4x^2 - 2 - 13x$

50. $3x + 2[4 - 5(2x - 1)]$

51. $5ab + 9b - 8ab - 12a$

52. $3a^2b - 4ab + 2ab^2 - 7a^2b$

SYNTHESIS

53. ◆ Write a problem for a classmate to solve. Devise the problem so the answer is "The Rothmans can drive for 90 mi without exceeding their truck rental budget."

54. ◆ The symbols ≮, ≯, and ≰, and ≱ have not been discussed. Do you feel that this was an oversight? Why or why not?

55. ◆ Suppose that $t =$ Todd's age and $f =$ Frances's age. Write a sentence that would translate to the inequality
$$t \geq f + 10.$$

56. ◆ In Example 2, the number 25.625 was rounded down rather than up. Why?

57. The sum of two consecutive odd integers is less than 100. What is the largest possible pair of such integers?

58. The area of a square can be no more than 64 cm^2. What lengths of a side will allow this?

59. *Ski wax.* Green ski wax works best at Fahrenheit temperatures between 5° and 15°. Determine those Celsius temperatures for which green ski wax works best. (See Exercise 29.)

60. *Parking fees.* Mack's Parking Garage charges $4.00 for the first hour and $2.50 for each additional hour. For how long has a car been parked when the charge exceeds $16.50?

61. *Earnings.* Mariah can choose to be paid in one of the following two ways.

Plan A: A salary of $600 per month, plus a commission of 4% of gross sales

Plan B: A salary of $800 per month, plus a commission of 6% of gross sales over $10,000

For what gross sales is plan A better than plan B, assuming that gross sales are always more than $10,000?

62. *Parking fees.* When asked how much the parking charge is for a certain car (see Exercise 60), Mack replies "between 14 and 24 dollars." For how long has the car been parked?

63. *Nutritional standards.* In order for a food to be labeled "lowfat," it must have fewer than 3 g of fat per serving. Reduced fat Tortilla Pops® contain 60% less fat than regular nacho cheese tortilla chips, but still cannot be labeled lowfat. What can you conclude about the fat content of a serving of nacho cheese tortilla chips?

64. ◆ After 9 quizzes, Blythe's average is 84. Is it possible for Blythe to improve her average by two points with the next quiz? Why or why not?

65. ◆ Chassman and Bem Booksellers offers a preferred-customer card for $25. The card entitles a customer to a 10% discount on all purchases for a period of one year. Under what circumstances would an individual save money by purchasing a card?

COLLABORATIVE CORNER

Focus: Problem solving and inequalities

Time: 15–25 minutes

Group size: 4–5

Materials: Calculators are optional.

Residents in parts of Indiana can choose either MCI or Ameritech as their long-distance telephone company for long-distance calls within their area code. Each company's (1997) rates are outlined below.

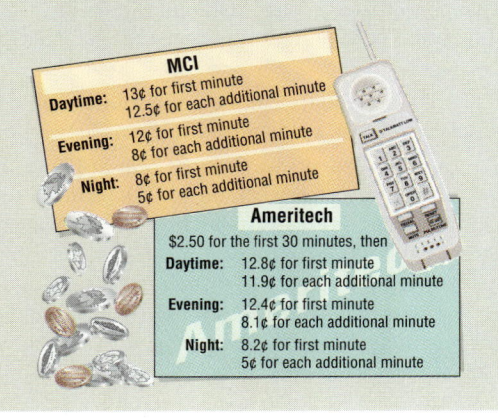

MCI
Daytime: 13¢ for first minute
12.5¢ for each additional minute
Evening: 12¢ for first minute
8¢ for each additional minute
Night: 8¢ for first minute
5¢ for each additional minute

Ameritech
$2.50 for the first 30 minutes, then
Daytime: 12.8¢ for first minute
11.9¢ for each additional minute
Evening: 12.4¢ for first minute
8.1¢ for each additional minute
Night: 8.2¢ for first minute
5¢ for each additional minute

ACTIVITY

1. Each group member should estimate how many long-distance calls within their area code he or she makes during each time category: daytime, evening, and night. Estimate the lengths of the calls and then estimate what the bill would be under each calling plan.

2. One pair of group members should then describe and list as many situations as possible for which MCI would be less expensive. The other group members should describe and list those situations for which Ameritech would be less expensive.

3. A "master" list should then be compiled, with all groups reporting their results from part (2). What general guidelines should the class give to someone who is about to choose between MCI and Ameritech?

SUMMARY AND REVIEW 2

KEY TERMS

Equivalent equations, p. 65
Coefficient, p. 68
Clearing fractions and decimals, p. 73
Formula, p. 78
Circumference, p. 79

Consecutive integers, p. 92
Graph of an inequality, p. 101
Equivalent inequalities, p. 102
Set-builder notation, p. 103
Solution set, p. 103

IMPORTANT PROPERTIES AND FORMULAS

Solving Equations

Addition principle:
For any real numbers a, b, and c,

$a = b$ is equivalent to $a + c = b + c$.

Multiplication principle:
For any real numbers a, b, and c, with $c \neq 0$,

$a = b$ is equivalent to $a \cdot c = b \cdot c$.

Solving Inequalities

Addition principle:
For any real numbers a, b, and c,

$a < b$ is equivalent to $a + c < b + c$;
$a > b$ is equivalent to $a + c > b + c$.

Multiplication principle:
For any real numbers a and b and any *positive* number c,

$a < b$ is equivalent to $ac < bc$;
$a > b$ is equivalent to $ac > bc$.

For any real numbers a and b and any *negative* number c,

$a < b$ is equivalent to $ac > bc$;
$a > b$ is equivalent to $ac < bc$.

Similar statements hold for $\leq$ and $\geq$.

An Equation-Solving Procedure

1. If necessary, use the distributive law to remove parentheses. Then combine like terms on each side.
2. Multiply on both sides to clear fractions or decimals. (This is optional, but it can ease computations.)
3. Get all terms with variables on one side and all constant terms on the other side, using the addition principle.
4. Combine like terms again, if necessary.
5. Multiply or divide to solve for the variable, using the multiplication principle.
6. Check all possible solutions in the original equation.

(continued)

To Solve a Formula for a Given Letter

1. Identify the letter being solved for and multiply on both sides to clear fractions or decimals, if necessary.
2. Get all terms with the letter to be solved for on one side of the equation and all other terms on the other side.
3. Combine like terms, if necessary. This may require factoring.
4. Multiply or divide to solve for the letter in question.

Percent Notation

$n\%$ means $n \times 0.01$, or $n \times \dfrac{1}{100}$, or $\dfrac{n}{100}$.

Five Steps for Problem Solving in Algebra

1. *Familiarize* yourself with the problem situation.
2. *Translate* to mathematical language. (This often means writing an equation.)
3. *Carry out* some mathematical manipulation. (This often means *solving* an equation.)
4. *Check* your possible answer in the original problem.
5. *State* the answer clearly.

REVIEW EXERCISES

Solve.

1. $x + 5 = -17$
2. $-8x = -56$
3. $-\dfrac{x}{4} = 48$
4. $n - 7 = -6$
5. $15x = -35$
6. $x - 0.1 = 1.01$
7. $-\dfrac{2}{3} + x = -\dfrac{1}{6}$
8. $\dfrac{4}{5}y = -\dfrac{3}{16}$
9. $5z + 3 = 41$
10. $5 - x = 13$
11. $5t + 9 = 3t - 1$
12. $7x - 6 = 25x$
13. $\dfrac{1}{4}x - \dfrac{5}{8} = \dfrac{3}{8}$
14. $14y = 23y - 17 - 10$
15. $0.22y - 0.6 = 0.12y + 3 - 0.8y$
16. $\dfrac{1}{4}x - \dfrac{1}{8}x = 3 - \dfrac{1}{16}x$
17. $4(x + 3) = 36$
18. $3(5x - 7) = -66$
19. $8(x - 2) = 5(x + 4)$
20. $-5x + 3(x + 8) = 16$

Solve each formula for the given letter.

21. $C = \pi d$, for d
22. $V = \dfrac{1}{3}Bh$, for B
23. $A = \dfrac{a + b}{2}$, for a

24. Find decimal notation: 0.1%.
25. Find percent notation: $\dfrac{11}{25}$.
26. What percent of 60 is 12?
27. 198 is 55% of what number?

Determine whether the given number is a solution of the inequality $x \leq 4$.

28. -3
29. 7
30. 4

Graph on a number line.

31. $4x - 6 < x + 3$
32. $-2 < x \leq 5$
33. $y > 0$

Solve. Write the answers in set-builder notation.

34. $y + \dfrac{2}{3} \geq \dfrac{1}{6}$
35. $9x \geq 63$
36. $2 + 6y > 14$
37. $7 - 3y \geq 27 + 2y$
38. $3x + 5 < 2x - 6$
39. $-4y < 28$
40. $3 - 4x < 27$
41. $4 - 8x < 13 + 3x$
42. $-3y \geq -21$
43. $-4x \leq \dfrac{1}{3}$

Solve.

44. An ink jet printer sold for $289 in July. This was $28 less than the cost in February. Find the cost in February.

45. A can of powdered infant formula makes 120 oz of formula. How many 6-oz bottles of formula will the can make?

46. An 8-m board is cut into two pieces. One piece is 2 m longer than the other. How long are the pieces?

47. About 30% of Hallmark's 19,600 full-time employees work in the company's headquarters in Kansas City, Missouri. How many people work in the headquarters?

48. The sum of two consecutive odd integers is 116. Find the integers.

49. The perimeter of a rectangle is 56 cm. The width is 6 cm less than the length. Find the width and the length.

50. After a 25% reduction, a picnic table is on sale for $120. What was the regular price?

51. Children from the ages of four through twelve in the United States had a combined income of $20.3 billion in 1996. This was a 16% increase from 1995. What was their combined income in 1995?

52. The measure of the second angle of a triangle is 50° more than that of the first. The measure of the third angle is 10° less than twice the first. Find the measures of the angles.

53. Jason has budgeted an average of $45 a month for entertainment. For the first five months of the year, he has spent $48, $39, $60, $35, and $53. How much can Jason spend in the sixth month without exceeding his average budget?

54. The length of a rectangle is 43 cm. For what widths is the perimeter greater than 120 cm?

SKILL MAINTENANCE

55. Evaluate $\dfrac{x - y}{5}$ for $x = 27$ and $y = 2$.

56. Multiply: $4(3t + 2 + s)$.

57. Divide: $12.42 \div (-5.4)$.

58. Remove parentheses and simplify:
$5x - 8(6x - y)$.

SYNTHESIS

59. ◆ What is the difference between using the multiplication principle for solving equations and using it for solving inequalities?

60. ◆ Explain how checking the solutions of an equation differs from checking the solutions of an inequality.

61. The combined length of the Nile and Amazon Rivers is 13,108 km. If the Amazon were 234 km longer, it would be as long as the Nile. Find the length of each river.

62. Consumer experts advise us never to pay the sticker price for a car. A rule of thumb is to pay the sticker price minus 20% of the sticker price, plus $200. A car is purchased for $11,520 using the rule. What was the sticker price?

Solve.

63. $2|n| + 4 = 50$ **64.** $|3n| = 60$

65. $y = 2a - ab + 3$, for a

CHAPTER TEST 2

Solve.

1. $x + 7 = 15$
2. $t - 9 = 17$
3. $3x = -18$
4. $-\frac{4}{7}x = -28$
5. $3t + 7 = 2t - 5$
6. $\frac{1}{2}x - \frac{3}{5} = \frac{2}{5}$
7. $8 - y = 16$
8. $-\frac{2}{5} + x = -\frac{3}{4}$
9. $3(x + 2) = 27$
10. $-3x + 6(x + 4) = 9$
11. $\frac{5}{6}(3x + 1) = 20$

Solve. Write the answers in set-builder notation.

12. $x + 6 \leq 2$
13. $14x + 9 > 13x - 4$
14. $\frac{1}{3}x < \frac{7}{8}$
15. $-2y \geq 26$
16. $4y \leq -32$
17. $-5x \geq \frac{1}{4}$
18. $4 - 6x > 40$
19. $5 - 9x \geq 19 + 5x$

Solve each formula for the given letter.

20. $A = 2\pi rh$, for r
21. $w = \dfrac{P - 2l}{2}$, for l

22. Find decimal notation: 500%.
23. Find percent notation: 0.054.
24. What number is 42% of 50?
25. What percent of 75 is 33?

Graph on a number line.

26. $y < 9$
27. $-2 \leq x \leq 2$

Solve.

28. The perimeter of a rectangle is 36 cm. The length is 4 cm greater than the width. Find the width and the length.
29. Kari is taking a 120-mi bicycle trip through Vermont. She has three times as many miles to go as she has already ridden. How many miles has she biked so far?
30. The perimeter of a triangle is 249 mm. If the sides are consecutive odd integers, find the length of each side.
31. Higher grain prices caused the value of Indiana cropland to rise 14% from 1995 to 1996 (*Source*: *Indianapolis Star*, 9/18/96). An acre of farmland was sold for $3078 in 1996. How much would it have been worth in 1995?
32. Find all numbers for which 6 times the number is greater than the number plus 30.
33. The width of a rectangle is 96 yd. Find all possible lengths so that the perimeter of the rectangle will be at least 540 yd.

SKILL MAINTENANCE

34. Translate to an algebraic expression:
    10 less than *x*.
35. Factor: $3a + 24b + 12$.
36. Multiply: $-\frac{3}{8} \cdot \left(-\frac{4}{5}\right)$.
37. Simplify: $8 + (-10) \div 5 \cdot 2 + 1$.

SYNTHESIS

38. Solve $c = \dfrac{1}{a - d}$ for *d*.
39. Solve: $3|w| - 8 = 37$.
40. A movie theater had a certain number of tickets to give away. Five people got the tickets. The first got one third of the tickets, the second got one fourth of the tickets, and the third got one fifth of the tickets. The fourth person got eight tickets, and there were five tickets left for the fifth person. Find the total number of tickets given away.

CHAPTER 3

Introduction to Graphing

3.1 Interpreting Graphs and Ordered Pairs
3.2 Graphing Linear Equations
3.3 Graphing and Intercepts
3.4 More Applications Using Graphs
3.5 Slope
SUMMARY AND REVIEW
TEST

AN APPLICATION

A local Mailboxes Etc.® store charges $2.25 for binding plus 5¢ per page for each spiralbound copy of a town report. Draw a graph that can be used to determine the cost of one copy of the report. Let the horizontal axis represent the number of pages and the vertical axis the price. Then use the graph to estimate the length of a report that costs $5.20 per copy.

THIS PROBLEM APPEARS AS EXERCISE 25 IN SECTION 3.4.

More information on bookbinding is available at
http://hepg.awl.com/be/elem_5

In my business, I must be able to make quick calculations in order to satisfy the customer. I copy, pack, and ship almost anything and I must figure out the best and safest way to package items without fear of their being damaged.

LAURA WALDEN
Mailboxes Etc.
Indianapolis, Indiana

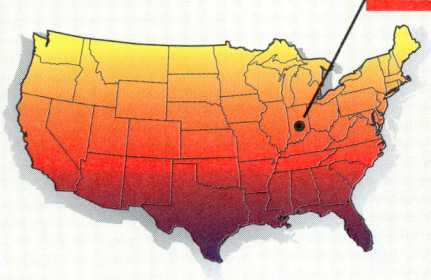

We now begin our study of graphing. First we will examine graphs as they commonly appear in newspapers or magazines and develop some terminology. Following that, we will study the graphs of certain equations. Finally, we will use graphs as a problem-solving tool for certain applications.

3.1 Interpreting Graphs and Ordered Pairs

Problem Solving with Graphs • Points and Ordered Pairs

Today's print and electronic media make almost constant use of graphs. This is due in part to the ease with which some graphs are prepared by computer, and in part to the large quantity of information that a graph can display. We first consider problem solving with bar graphs, line graphs, and circle graphs. Then we examine graphs that use a coordinate system.

Problem Solving with Graphs

A *bar graph* is a convenient way of showing comparisons. In every bar graph, certain categories, such as body weight in Example 1, are paired with certain numbers.

EXAMPLE 1

Driving under the Influence. Although some states use a cutoff rate of 0.08%, in *all* states, a blood-alcohol level of 0.10% or higher indicates that an individual has consumed too much alcohol to drive. The following bar graph shows the number of drinks that a person of a certain weight would consume to achieve a blood-alcohol level of 0.10% (*Source:* Adapted from Neighbor-

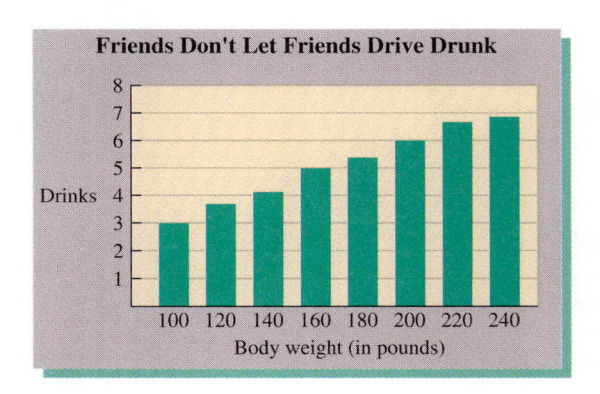

hood Digest, Vol. 7, No. 12). Note that a 12-oz beer, a 5-oz glass of wine, or a cocktail containing $1\frac{1}{2}$ oz of distilled liquor all count as one drink.

a) Approximately how many drinks would a 200-lb person have consumed if he or she had a blood-alcohol level of 0.10%?

b) What can be concluded about the weight of someone who can consume 4 drinks without reaching a blood-alcohol level of 0.10%?

SOLUTION

a) We go to the top of the bar that is above the body weight 200 lb. Then we move horizontally from the top of the bar to the vertical scale listing numbers of drinks. It appears that approximately 6 drinks will give a 200-lb person a blood-alcohol level of 0.10%.

b) By moving up the vertical scale to the number 4, and then moving horizontally, we see that the first bar to reach a height of 4 corresponds to a weight of 140 lb. Thus an individual should weigh over 140 lb if he or she wishes to consume 4 drinks without exceeding a blood-alcohol level of 0.10%.

Circle graphs, or *pie charts,* are often used to show what percent of the whole each particular item in a group represents.

EXAMPLE 2   *Use of Tax Dollars.*   The following pie chart shows how federal income tax dollars are spent.

Where Your Tax Dollars Are Spent

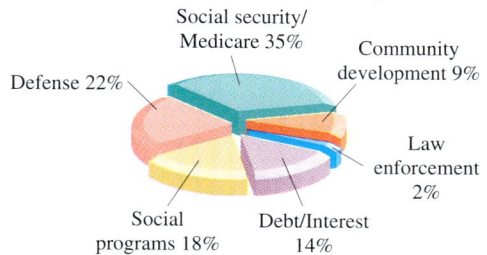

- Social security/Medicare 35%
- Community development 9%
- Defense 22%
- Law enforcement 2%
- Social programs 18%
- Debt/Interest 14%

As a freelance graphic artist, Jennifer has a taxable income* of $23,500, which puts her income in the 15% tax bracket. How much of Jennifer's income goes toward defense?

SOLUTION

1. Familiarize. If we did not already know, we would need to find out that "15% tax bracket" means that 15% of all taxable income is paid as federal taxes. Thus the total amount paid in federal taxes by Jennifer is

0.15 × $23,500,   or $3525.

The chart shows that 22% of all tax dollars are spent on defense. We let $y =$ the amount spent on defense.

---
*Taxable income is the income that remains after allowable deductions have been made.

2. **Translate.** We reword and translate the problem as follows:

   Rewording: What is 22% of $3525?

   Translating: $y = 22\% \cdot 3525$

3. **Carry out.** We solve the equation:

   $y = 0.22 \cdot \$3525 = \$775.50.$

4. **Check.** The check is left to the student.

5. **State.** $775.50 of Jennifer's income will be spent on defense.

Line graphs are often used to illustrate change over time. Certain points are first drawn to represent given information. When segments are drawn to connect the points, a line graph is formed.

Sometimes it is impractical to begin the listing of horizontal or vertical values with zero. When this occurs, as in Example 3, the symbol ⑤ is used to indicate a break in the listing of values.

EXAMPLE 3

Exercise and Pulse Rate. The following line graph shows the relationship between a person's resting pulse rate and months of regular exercise.*

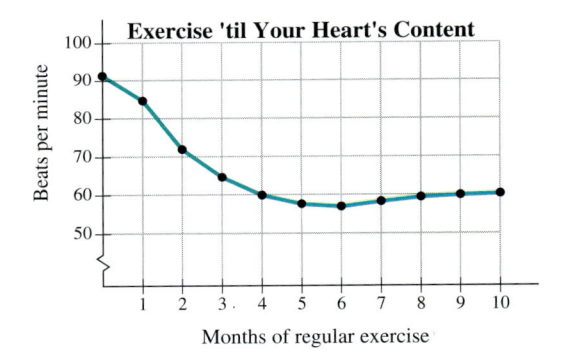

Exercise 'til Your Heart's Content

[Graph: Beats per minute vs Months of regular exercise]

a) How many months of regular exercise are required to lower the pulse rate as much as possible?

b) How many months of regular exercise are needed to achieve a pulse rate of 65 beats per minute?

SOLUTION

a) The lowest point on the graph occurs above the number 6. Thus, after 6 months of regular exercise, the pulse rate is lowered as much as possible.

b) We locate 65 on the vertical scale and then move right until the line is reached. At that point, we move down to the horizontal scale and read the information we are seeking.

---

*Data from *Body Clock* by Dr. Martin Hughes (New York: Facts on File, Inc.), p. 60.

3.1 INTERPRETING GRAPHS AND ORDERED PAIRS   123

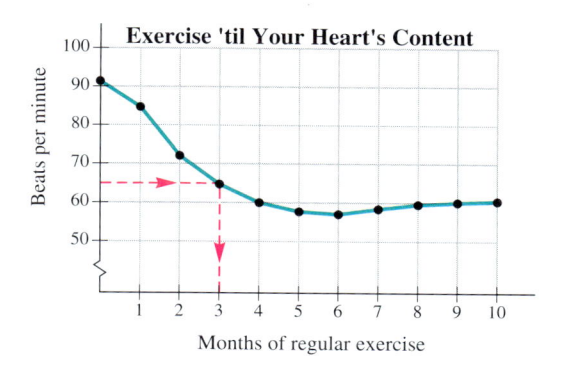

The pulse rate is 65 beats per minute after 3 months of regular exercise.

Points and Ordered Pairs

The line graph in Example 3 contains a collection of points. Each point pairs up a number of months of exercise with a pulse rate. To create such a graph, we **graph,** or **plot,** pairs of numbers on a plane. This is done using two perpendicular number lines called **axes** (singular, **axis**). The axes cross at a point called the **origin.** Arrows on the axes show the positive directions.

Consider the pair (3, 4). The numbers in such a pair are called **coordinates.** In (3, 4), the **first coordinate** is 3 and the **second coordinate** is 4. To plot (3, 4), we start at the origin and move horizontally to the 3. Then we move up vertically 4 units and make a "dot." Note that (3, 4) is located above 3 on the first axis and to the right of 4 on the second axis.

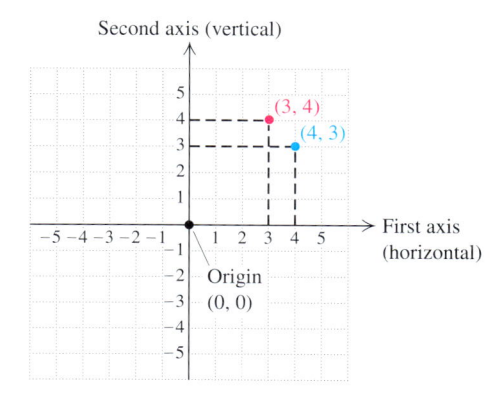

The point (4, 3) is also plotted in the figure above. Note that (3, 4) and (4, 3) are different points. They are called **ordered pairs** because the order in which the numbers appear is important.

EXAMPLE 4   Plot the point (−3, 4).

SOLUTION   The first number, −3, is negative. Starting at the origin, we move 3 units in the negative horizontal direction (3 units to the left). The second number, 4, is positive, so we move 4 units in the positive vertical direction (up). The point (−3, 4) is above −3 on the first axis and to the left of 4 on the second axis.

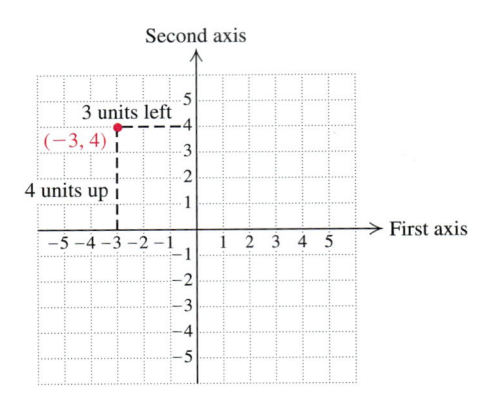

To find the coordinates of a point, we see how far to the right or left of the vertical axis the point is and how far above or below the horizontal axis it is.

EXAMPLE 5 Find the coordinates of points A, B, C, D, E, F, and G.

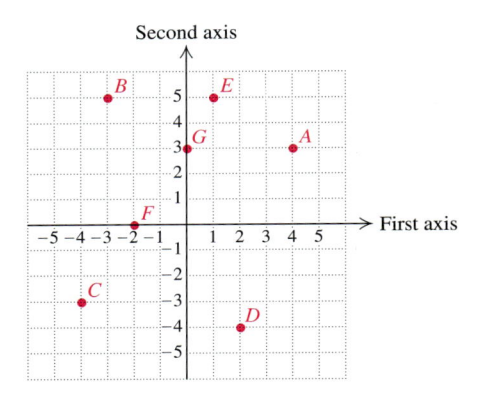

SOLUTION Point A is 4 units to the right (horizontal direction) and 3 units up (vertical direction). Its coordinates are (4, 3). The coordinates of the other points are as follows:

B: (−3, 5);   C: (−4, −3);   D: (2, −4);
E: (1, 5);   F: (−2, 0);   G: (0, 3).

The horizontal and vertical axes divide the plane into four regions, or **quadrants,** as indicated by Roman numerals in the figure at the top of the following page. In region I (the *first quadrant*), both coordinates of any point are positive. In region II (the *second quadrant*), the first coordinate is negative and the second is positive. In region III (the *third quadrant*), both coordinates are negative. In region IV (the *fourth quadrant*), the first coordinate is positive and the second is negative.

Note that the point (−4, 5) is in the second quadrant and the point (5, −5) is in the fourth quadrant. The points (3, 0) and (0, 1) are on the axes and are not considered to be in any quadrant.

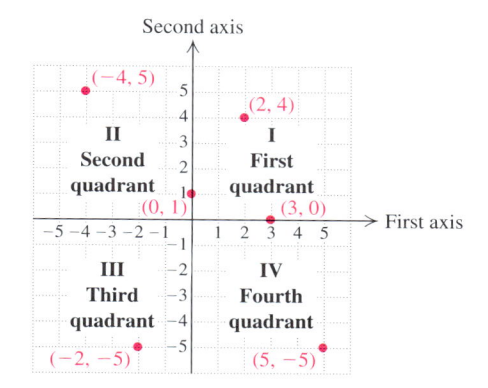

EXERCISE SET
3.1

Blood alcohol level. Use the bar graph in Example 1 to answer Exercises 1–4.

1. Approximately how many drinks would it take for a 160-lb person to reach a blood-alcohol level of 0.10%?
2. Approximately how many drinks would it take for a 100-lb person to reach a blood-alcohol level of 0.10%?
3. What can you conclude about the weight of someone who has consumed $3\frac{1}{2}$ drinks without reaching a blood-alcohol level of 0.10%?
4. What can you conclude about the weight of someone who has consumed 5 drinks without reaching a blood-alcohol level of 0.10%?

Use of tax dollars. Use the pie chart in Example 2 to answer Exercises 5–8.

5. Leila pays 18% of her taxable income of $31,200 in taxes. How much of her earnings will be spent on social programs?
6. Lionel pays 16% of his taxable income of $26,000 in taxes. How much of his earnings will be spent on community development?
7. The Caseys pay 23% of their taxable income of $101,500 in taxes. How much of their earnings will be spent on social security/medicare?
8. The Lunts pay 24% of their taxable income of $116,000 in taxes. How much of their earnings will be spent on law enforcement?

Sorting solid waste. Use the following pie chart to answer Exercises 9–12.

9. In 1993, Americans generated 206.9 million tons of waste. How much of the waste was plastic?
10. In 1996, the average American generated 4.5 lb of waste per day. How much of that was paper and cardboard?
11. Americans are recycling about 23% of all glass that is in the waste stream. How much glass does the average American recycle each day? (See Exercise 10.)
12. Americans are recycling about 5% of all plastic waste. How much plastic does the average family of four recycle each day? (See Exercise 10.)

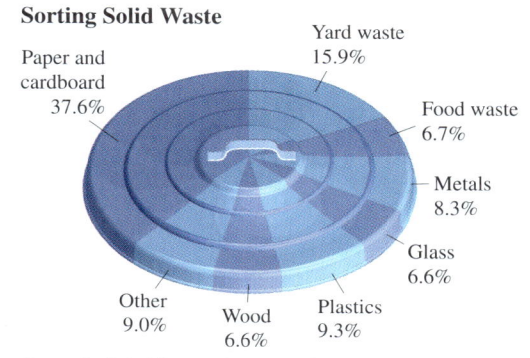

Sorting Solid Waste

Paper and cardboard 37.6%
Yard waste 15.9%
Food waste 6.7%
Metals 8.3%
Glass 6.6%
Plastics 9.3%
Wood 6.6%
Other 9.0%

Source: Statistical Abstract of the United States, 1995

Spending on health, education, and defense. *Use the following line graphs to answer Exercises 13–20. The graphs show the percentages of the Gross National Product (GNP) spent on health, education, and defense over a period of 45 years.*

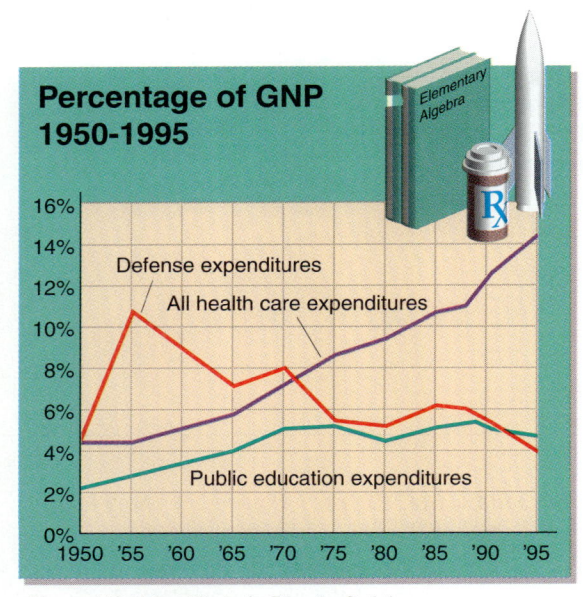

Percentage of GNP 1950–1995

Sources: U.S. National Center for Education Statistics, Health Care Financing Administration, U.S. Office of Management and Budget

13. Approximately what percent of the GNP was spent on public education in 1965?
14. Approximately what percent of the GNP was spent on defense in 1970?
15. Approximate the year in which health care expenditures were 10% of the GNP.
16. In what year did health care expenditures first exceed 12% of the GNP?
17. In 1980, the GNP was $2732 billion. Approximately how much was spent on public education?
18. In 1995, the GNP was $6977 billion. Approximately how much was spent on health care?
19. In 1970, about $50.8 billion was spent on education. What was the GNP?
20. In 1990, about $668 billion was spent on health care. What was the GNP?

Plot each group of points.

21. $(1, 2), (-2, 3), (4, -1), (-5, -3), (4, 0), (0, -2)$
22. $(-2, -4), (4, -3), (5, 4), (-1, 0), (-4, 4), (0, 5)$
23. $(4, 4), (-2, 4), (5, -3), (-5, -5), (0, 4), (0, -4),$ $(3, 0), (-4, 0)$
24. $(2, 5), (-1, 3), (3, -2), (-2, -4), (0, 4), (0, -5),$ $(5, 0), (-5, 0)$

In Exercises 25–28, find the coordinates of points A, B, C, D, and E.

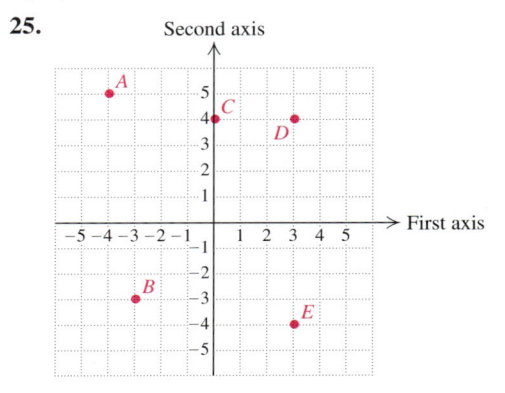

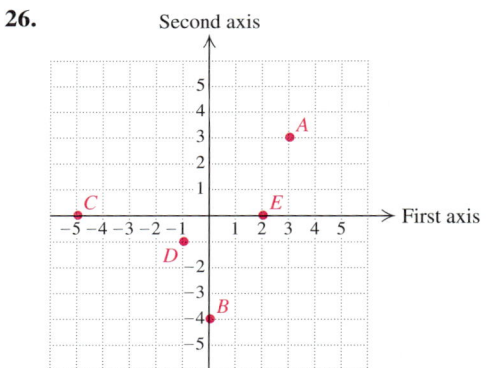

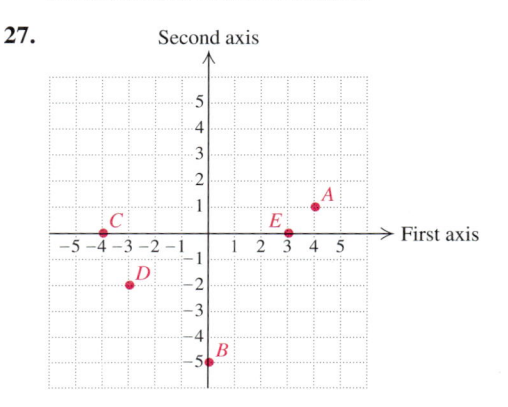

25.

26.

27.

---

*"Health vs. Education vs. Defense," Copyright 1992 by Consumers Union of U.S., Inc., Yonkers, NY 10703-1057. Adapted with permission from CONSUMER REPORTS, July 1992. Although this material originally appeared in CONSUMER REPORTS, the selective adaptation and resulting conclusions presented are those of the author(s) and are not sanctioned or endorsed in any way by Consumers Union, the publisher of CONSUMER REPORTS.

28.

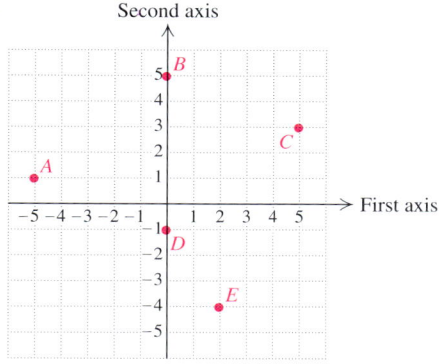

In which quadrant is each point located?

29. (−5, 3) **30.** (−12, 1)
31. (100, −1) **32.** (35.6, −2.5)
33. (−6, −29) **34.** (−3.6, −105.9)
35. (3.8, 9.2) **36.** (1895, 1492)

37. In quadrant III, first coordinates are always _____ and second coordinates are always _____.

38. In quadrant II, _____ coordinates are always positive and _____ coordinates are always negative.

In Exercises 39–42, display the given information as a line graph using the horizontal axis to represent years.

39. *Changing populations.* The percentage of female first-year college students has grown from 45% in 1970 to 51% in 1980 and 54% in 1990. (*Source: Statistical Abstract of the United States,* 1995, and *Information Please Almanac,* 1997)

40. *Changing attitudes.* The percentage of first-year college students who believe that "the activities of married women are best confined to the home and family" has declined from 48% in 1970 to 27% in 1980 and 25% in 1990. (*Source: Statistical Abstract of the United States,* 1995, and *Information Please Almanac,* 1997)

41. *Meat consumption.* Yearly consumption of red meat in the United States has declined from 125.8 lb per person in 1975 to 124.9 lb in 1985 and 114.8 lb in 1994. (*Source: Statistical Abstract of the United States,* 1995, and *Information Please Almanac,* 1997)

42. *Cheese consumption.* Yearly cheese consumption in the United States has risen from 14.3 lb per person in 1975 to 22.5 lb in 1985 and 26.8 lb in 1994. (*Source: Statistical Abstract of the United States,* 1995, and *Information Please Almanac,* 1997)

SKILL MAINTENANCE

Perform the indicated operation and simplify.

43. $\frac{2}{9} + \frac{2}{3}$ **44.** $\frac{5}{8} \cdot \frac{2}{15}$

45. $\frac{3}{5} \cdot \frac{10}{9}$ **46.** $\frac{2}{3} + \frac{1}{5}$

47. $\frac{3}{7} - \frac{4}{5}$ **48.** $-\frac{2}{3} \div 5$

SYNTHESIS

49. ◆ Would a circle graph be the best way to display the number of sport utility vehicles sold in each of the last five years? Why or why not?

50. ◆ The graph accompanying Example 3 flattens out. Why do you think this occurs?

51. ◆ What advantage(s) does the use of a line graph have over that of a bar graph?

52. ◆ Describe what the result would be if the first and second coordinates of every point in the following graph of an arrow were interchanged.

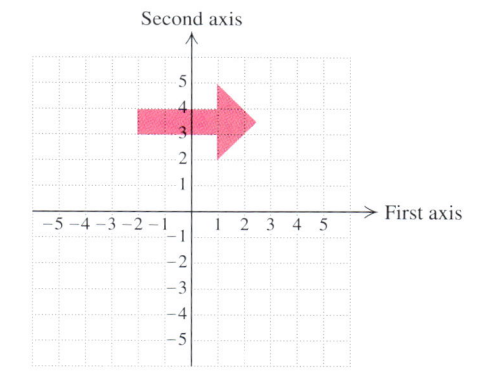

In Exercises 53–56, tell in which quadrant(s) the given point could be located.

53. The first coordinate is negative.

54. The second coordinate is positive.

55. The first and second coordinates are opposites.

56. The first coordinate is the reciprocal of the second coordinate.

57. The points (−1, 1), (4, 1), and (4, −5) are three vertices of a rectangle. Find the coordinates of the fourth vertex.

58. The pairs (−2, −3), (−1, 2), and (4, −3) can serve as three (of four) vertices for three different parallelograms. Find the fourth vertex of each parallelogram.

59. Graph eight points such that the sum of the coordinates in each pair is 7.

60. Graph eight points such that the first coordinate minus the second coordinate is 1.

61. Find the perimeter of a rectangle if three of its vertices are $(5, -2)$, $(-3, -2)$, and $(-3, 3)$.

62. Find the area of a triangle whose vertices have coordinates $(0, 9)$, $(0, -4)$, and $(5, -4)$.

Coordinates on the globe. Coordinates can also be used to describe the location on a sphere: 0° latitude is the equator and 0° longitude is a line from the North Pole to the South Pole through France and Algeria. In the figure shown here, hurricane Clara is at a point about 260 mi northwest of Bermuda near latitude 36.0° North, longitude 69.0° West.

63. Approximate the latitude and the longitude of Bermuda.

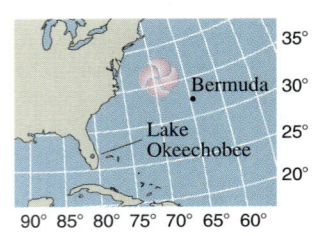

64. Approximate the latitude and the longitude of Lake Okeechobee.

65. ◆ In the *Star Trek* science-fiction series, a three-dimensional coordinate system is used to locate objects in space. If the center of a planet is used as the origin, how many "quadrants" will exist? Why? If possible, sketch a three-dimensional coordinate system and label each "quadrant."

COLLABORATIVE C•O•R•N•E•R

Focus: Graphing points

Time: 15–25 minutes

Group size: 3–5

Materials: Graph paper

In the game Battleship®, each player places miniature ships on a grid that only that player can see. An opponent guesses at coordinates that might "hit" one of the "hidden" ships. The following activity is similar to this game.

ACTIVITY

1. Using only integers from −10 to 10 (inclusive), one group member should secretly record the coordinates of a point on a slip of paper. (This point is the hidden "battleship.")

2. The other group members are then permitted to ask up to 10 "yes/no" questions in an effort to determine the coordinates of the hidden battleship. Be sure to phrase each question mathematically (for example, "Is the *x*-coordinate even?").

3. After answering each question, the group member who selected the point should use a graph to shade those points no longer under consideration. The other group members should confirm that the shading is consistent with each answer given.

4. If the hidden battleship has not been determined after 10 questions have been answered, the secret coordinates should be revealed to all group members.

5. Repeat parts (1)–(4) until each group member has had the opportunity to select the hidden point and answer questions.

3.2 Graphing Linear Equations

Solutions of Equations • Graphing Linear Equations

We have seen how bar, line, and circle graphs can represent information. Now we begin to learn how graphs can be used to represent solutions of equations.

Solutions of Equations

When an equation contains two variables, solutions must be ordered pairs in which each number in the pair replaces a letter in the equation. Unless stated otherwise, the first number in each pair replaces the variable that occurs first alphabetically.

EXAMPLE 1 Determine whether each of the following pairs is a solution of $4q - 3p = 22$:
(a) (2, 7); (b) (1, 6).

SOLUTION

a) We substitute 2 for p and 7 for q (alphabetical order of variables):

$$\begin{array}{c|c} 4q - 3p = 22 \\ \hline 4 \cdot 7 - 3 \cdot 2 \;?\; 22 \\ 28 - 6 \\ 22 \;\bigg|\; 22 \quad \text{TRUE} \end{array}$$

Since $22 = 22$ is *true*, the pair (2, 7) *is* a solution.

b) In this case, we replace p with 1 and q with 6:

$$\begin{array}{c|c} 4q - 3p = 22 \\ \hline 4 \cdot 6 - 3 \cdot 1 \;?\; 22 \\ 24 - 3 \\ 21 \;\bigg|\; 22 \quad \text{FALSE} \end{array}$$

Since $21 = 22$ is *false*, the pair (1, 6) is *not* a solution.

EXAMPLE 2 Show that the pairs (3, 7), (0, 1), and (−3, −5) are solutions of $y = 2x + 1$. Then graph the three points to determine another pair that is a solution.

SOLUTION To show that a pair is a solution, we substitute, replacing x with the first coordinate and y with the second coordinate of each pair:

$$\begin{array}{c|c} y = 2x + 1 \\ \hline 7 \;?\; 2 \cdot 3 + 1 \\ \;\bigg|\; 6 + 1 \\ 7 \;\bigg|\; 7 \quad \text{TRUE} \end{array} \qquad \begin{array}{c|c} y = 2x + 1 \\ \hline 1 \;?\; 2 \cdot 0 + 1 \\ \;\bigg|\; 0 + 1 \\ 1 \;\bigg|\; 1 \quad \text{TRUE} \end{array} \qquad \begin{array}{c|c} y = 2x + 1 \\ \hline -5 \;?\; 2(-3) + 1 \\ \;\bigg|\; -6 + 1 \\ -5 \;\bigg|\; -5 \quad \text{TRUE} \end{array}$$

In each of the three cases, the substitution results in a true equation. Thus the pairs (3, 7), (0, 1), and (−3, −5) are all solutions. We graph them below. Note that the three points appear to "line up." Will other points that line up with these points also represent solutions of $y = 2x + 1$? To find out, we use a ruler and lightly sketch a line passing through (−3, −5), (0, 1), and (3, 7).

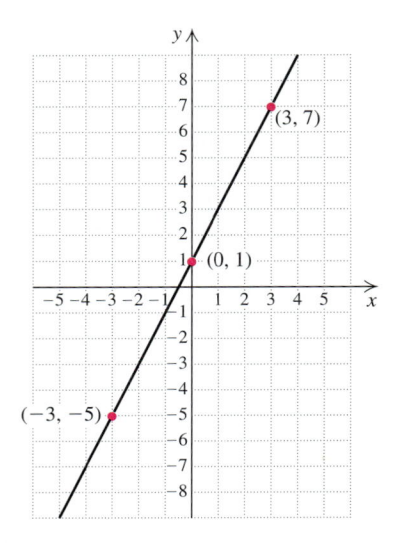

The line appears to pass through (2, 5). Let's check if this pair is a solution of $y = 2x + 1$:

$$\begin{array}{c|c} y = 2x + 1 \\ \hline 5 \;?\; 2 \cdot 2 + 1 \\ \;|\; 4 + 1 \\ 5 \;|\; 5 \qquad \text{TRUE} \end{array}$$

We see that (2, 5) *is* a solution. You should perform a similar check for at least one other point that appears to be on the line.

Example 2 leads us to suspect that *any* point on the line that passes through (3, 7), (0, 1), and (−3, −5) represents a solution of $y = 2x + 1$. In fact, every solution of $y = 2x + 1$ is represented by a point on this line and every point on this line represents a solution. The line is said to be the **graph** of the equation.

Graphing Linear Equations

Equations like $y = 2x + 1$ or $4q - 3p = 22$ are said to be **linear** because the graph of the solutions of each equation is a line. In general, any equation that can be written in the form $y = mx + b$ or $Ax + By = C$ (where m, b, A, B, and C are constants and A and B are not both 0) is linear.

To *graph* an equation is to make a drawing that represents its solutions. Linear equations can be graphed as follows.

TO GRAPH A LINEAR EQUATION:

1. Select a value for one coordinate and calculate the corresponding value of the other coordinate. Form an ordered pair. This pair is one solution of the equation.
2. Repeat step (1) to find at least one other ordered pair.
3. Plot the ordered pairs and draw a straight line passing through the points. The line represents all solutions of the equation.

EXAMPLE 3 Graph: $y = -3x + 1$.

SOLUTION We select a value for x, compute y, and form an ordered pair. Then we repeat the process for other choices of x.

If $x = 2$, then $y = -3 \cdot 2 + 1 = -5$, and $(2, -5)$ is a solution.
If $x = 0$, then $y = -3 \cdot 0 + 1 = 1$, and $(0, 1)$ is a solution.
If $x = -1$, then $y = -3(-1) + 1 = 4$, and $(-1, 4)$ is a solution.

Results are often listed in a table, as shown below. The points corresponding to each pair are then plotted.

x	y $y = -3x + 1$	(x, y)
2	-5	$(2, -5)$
0	1	$(0, 1)$
-1	4	$(-1, 4)$

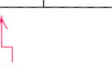

(1) Choose x.
(2) Compute y.
(3) Form the pair (x, y).
(4) Plot the points.

Note that all three points line up. If they didn't, we would know that we had made a mistake, because the equation is linear. When only two points are plotted, an error is more difficult to detect.

We now use a ruler or other straightedge to draw a line. Every point on the line represents a solution of $y = -3x + 1$.

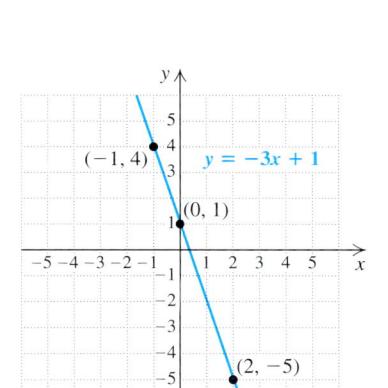

132 CHAPTER 3 • INTRODUCTION TO GRAPHING

Calculating ordered pairs is generally easiest when y is isolated on one side of the equation. To graph an equation in which y is not isolated, we can use the addition and multiplication principles (much as in Section 2.3) to solve for y.

EXAMPLE 4 Graph: $3y = 2x$.

SOLUTION To isolate y, we divide by 3, or multiply by $\frac{1}{3}$, on both sides:

$3y = 2x$

$\frac{1}{3} \cdot 3y = \frac{1}{3} \cdot 2x$ Using the multiplication principle to multiply by $\frac{1}{3}$ on both sides

$1y = \frac{2}{3} \cdot x$

$y = \frac{2}{3}x.$ Simplifying

Because all of the equations above are equivalent, we can use $y = \frac{2}{3}x$ to draw the graph of $3y = 2x$.

To graph $y = \frac{2}{3}x$, we select x-values that are multiples of 3. This allows us to avoid fractions when the corresponding y-values are computed.

If $x = 3$,   then $y = \frac{2}{3} \cdot 3 = 2.$
If $x = -3$, then $y = \frac{2}{3}(-3) = -2.$ Note that when multiples of 3 are substituted for x, the y-coordinates are not fractions.
If $x = 6$,   then $y = \frac{2}{3} \cdot 6 = 4.$

The following table lists these solutions. Next, we plot the points and see that they form a line. Finally, we draw and label the line.

x	$3y = 2x,$ or $y = \frac{2}{3}x$	(x, y)
3	2	(3, 2)
−3	−2	(−3, −2)
6	4	(6, 4)

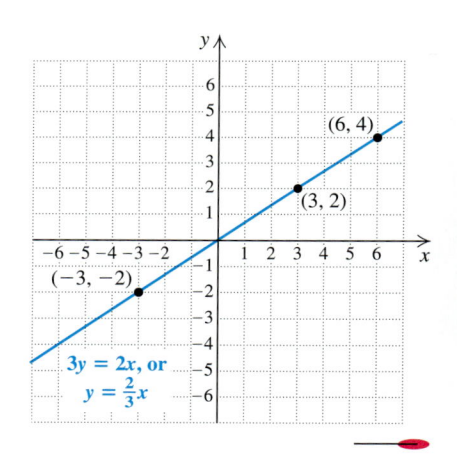

EXAMPLE 5 Graph: $x + 3y = -6$.

SOLUTION We first solve for y:

$x + 3y = -6$

$3y = -x - 6$ Adding $-x$ on both sides

$y = \frac{1}{3}(-x - 6)$ Multiplying by $\frac{1}{3}$ on both sides

$y = -\frac{1}{3}x - 2.$ Using the distributive law

Thus, $x + 3y = -6$ is equivalent to $y = -\frac{1}{3}x - 2$. We choose values for x and calculate the corresponding y-values.

If $x = 0$, then $y = -\frac{1}{3} \cdot 0 - 2 = 0 - 2 = -2$.
If $x = 6$, then $y = -\frac{1}{3} \cdot 6 - 2 = -2 - 2 = -4$.
If $x = -6$, then $y = -\frac{1}{3}(-6) - 2 = 2 - 2 = 0$.

The following table lists these solutions. Next, we plot the points and check that they line up. Finally, we draw and label the line.

x	$x + 3y = -6,$ or $y = -\frac{1}{3}x - 2$	(x, y)
0	-2	$(0, -2)$
6	-4	$(6, -4)$
-6	0	$(-6, 0)$

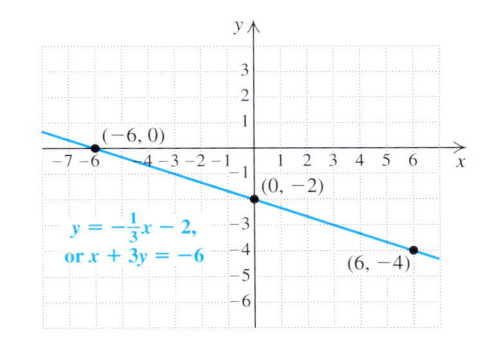

EXAMPLE 6   Graph: $y = 2x - 3$.

SOLUTION   We select some x-values and compute y-values.

If $x = 4$, then $y = 2 \cdot 4 - 3 = 5$, and $(4, 5)$ is a solution.
If $x = 1$, then $y = 2 \cdot 1 - 3 = -1$, and $(1, -1)$ is a solution.
If $x = 0$, then $y = 2 \cdot 0 - 3 = -3$, and $(0, -3)$ is a solution.

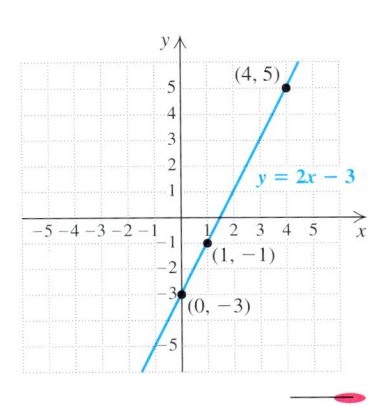

x	$y = 2x - 3$	(x, y)
4	5	$(4, 5)$
1	-1	$(1, -1)$
0	-3	$(0, -3)$

In Example 3, we saw that $(0, 1)$ is a solution of $y = -3x + 1$. Similarly, in Example 6, we found that $(0, -3)$ is a solution of $y = 2x - 3$. In general, for $y = mx + b$, if $x = 0$, then $y = m \cdot 0 + b = b$. Thus the graph of any equation of the form $y = mx + b$ contains the point $(0, b)$.

134 CHAPTER 3 • INTRODUCTION TO GRAPHING

Since $(0, b)$ is the point at which the graph crosses the y-axis, it is called the **y-intercept.** Sometimes, for convenience, we simply refer to b as the y-intercept. In Sections 3.3 and 7.1, we will make extensive use of y-intercepts.

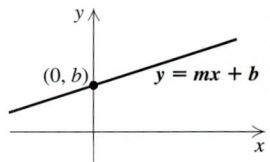

Linear equations appear in many real-life situations.

EXAMPLE 7   *Value of an Office Machine.* The value of Dupligraphix's color copier is given by $v = -\frac{1}{2}t + 3$. Here v represents the value, in thousands of dollars, t years from the date of purchase. Graph the equation and then use the graph to estimate the value of the copier $2\frac{1}{2}$ yr after the date of purchase.

SOLUTION   We graph $v = -\frac{1}{2}t + 3$ by selecting values for t and then calculating the associated value v. Since time cannot be negative in this case, we select nonnegative values for t.

If $t = 0$, then $v = -\frac{1}{2} \cdot 0 + 3 = 3$.
If $t = 4$, then $v = -\frac{1}{2} \cdot 4 + 3 = 1$.
If $t = 8$, then $v = -\frac{1}{2} \cdot 8 + 3 = -1$.

t	v
0	3
4	1
8	-1

We label the axes and plot the points (see the figure on the left below). Since they line up, our calculations are probably correct. However, including any points below the horizontal axis seems unrealistic since (at least for tax purposes) the value of the copier cannot be negative. Thus, when drawing the graph, we end the solid line at the horizontal axis.

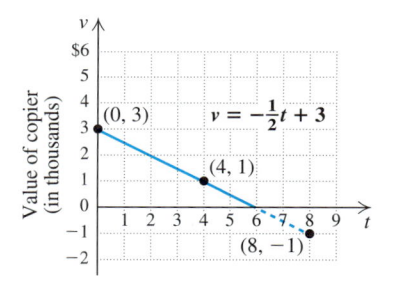

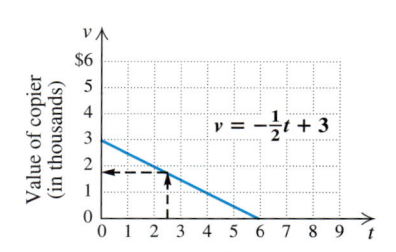

To estimate the value of the copier after $2\frac{1}{2}$ yr, we need to determine what second coordinate is paired with $2\frac{1}{2}$. To do this, we locate the point on the line that is above $2\frac{1}{2}$ and then find the value on the vertical axis that corresponds to that point (see the figure on the right above). It appears that after $2\frac{1}{2}$ yr, the copier is worth about $1800.

CAUTION!   When the coordinates of a point are read from a graph, as in Example 7, values should not be considered exact.

TECHNOLOGY CONNECTION 3.2

Beginning in this chapter, we will include activities that utilize graphing calculators or computer graphing software. Such calculators and software will be referred to simply as *graphers*. Most activities will use only basic features common to most graphers. All will be presented in a generic form—check with a user's manual or an instructor for more exact procedures.

All graphers have a *window*, the rectangular portion of the screen in which a graph appears. A RANGE feature is often used to determine what parts of the x- and y-axes will be shown. The minimum and maximum values for both x and y can be adjusted using Xmin, Xmax, Ymin, and Ymax. Sometimes we will describe a window using numbers of the form [L, R, B, T] to represent the Left and Right endpoints of the x-axis and the Bottom and Top endpoints of the y-axis.

Let's graph the equation $y = -\frac{4}{5}x + \frac{13}{5}$. Selecting the standard window, [-10, 10, -10, 10], results in the graph shown below.

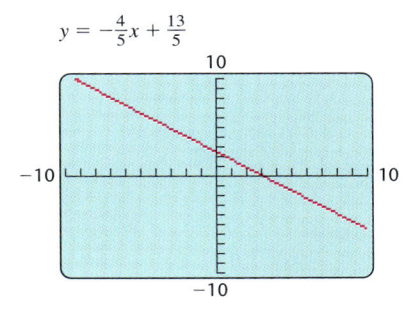

Use a grapher to graph each equation. Select the window [-10, 10, -10, 10] for each graph.

1. $y = -5x + 6.5$
2. $y = 3x - 4.5$
3. $y = \frac{4}{7}x - \frac{22}{7}$
4. $y = -\frac{11}{5}x - 4$
5. $y = 0.5x^2$
6. $y = 8 - x^2$

Many equations in two variables have graphs that are not straight lines. Three such graphs are shown below. As before, each graph represents the solutions of the given equation. Graphing calculators and computers are often helpful when drawing these *nonlinear* graphs. Nonlinear graphs are studied in Chapter 10 and in more advanced courses.

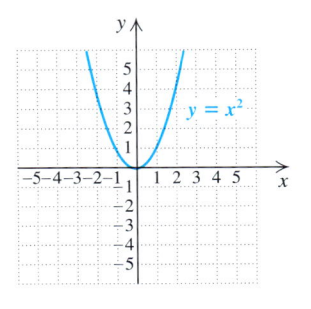

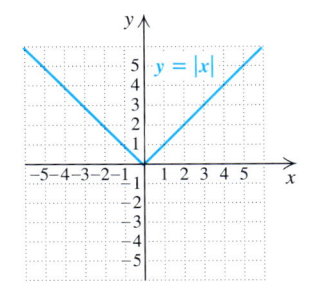

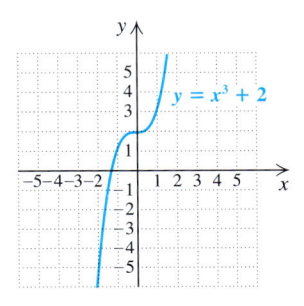

EXERCISE SET 3.2

Determine whether each equation has the given ordered pair as a solution.

1. $y = 4x + 3$; (2, 9)
2. $y = 2x + 5$; (1, 7)
3. $2x + 3y = 12$; (4, 2)
4. $5x - 3y = 15$; (0, 5)
5. $3a - 4b = 13$; (3, -1)
6. $2p - 3q = -13$; (-5, 1)

In Exercises 7–14, an equation and two ordered pairs are given. Show that each pair is a solution of the equation. Then graph the two pairs to determine another solution. Answers may vary.

7. $y = x - 5$; $(7, 2)$, $(1, -4)$
8. $y = x + 3$; $(-1, 2)$, $(4, 7)$
9. $y = \frac{1}{2}x + 3$; $(4, 5)$, $(-2, 2)$
10. $y = \frac{1}{2}x - 1$; $(6, 2)$, $(0, -1)$
11. $3x + y = 7$; $(2, 1)$, $(4, -5)$
12. $x + 2y = 5$; $(-1, 3)$, $(7, -1)$
13. $4x - 2y = 10$; $(0, -5)$, $(4, 3)$
14. $6x - 3y = 3$; $(1, 1)$, $(-1, -3)$

Graph each equation.

15. $y = x + 1$
16. $y = x - 1$
17. $y = x$
18. $y = -x$
19. $y = \frac{1}{2}x$
20. $y = \frac{1}{3}x$
21. $y = x - 3$
22. $y = x + 3$
23. $y = 3x - 2$
24. $y = 2x + 2$
25. $y = \frac{1}{2}x + 1$
26. $y = \frac{1}{3}x - 4$
27. $x + y = -5$
28. $x + y = 4$
29. $y = \frac{5}{3}x - 2$
30. $y = \frac{5}{2}x + 3$
31. $x + 2y = 8$
32. $x + 2y = -6$
33. $y = \frac{3}{2}x + 1$
34. $y = -\frac{2}{3}x + 4$
35. $8x - 4y = 12$
36. $6x - 3y = 9$
37. $8y + 2x = -4$
38. $6y + 2x = 8$

Solve by graphing. Label all axes, and show where each solution is located on the graph.

39. *Value of computer software.* The value v of a shopkeeper's inventory software program, in hundreds of dollars, is given by $v = -\frac{3}{4}t + 6$, where t is the number of years since the shopkeeper first bought the program. Graph the equation and use the graph to estimate what the program is worth 4 yr after it was first purchased.

40. *Vinyl phonograph singles.* The number of vinyl phonograph singles s, in millions, manufactured each year can be estimated by $s = -\frac{7}{2}t + 22$, where t is the number of years since 1991 (based on information in *Statistical Abstract of the United States,* 1995). Graph the equation and use the graph to estimate the number of singles that were produced in 1997.

41. *Increasing life expectancy.* A smoker is 15 times more likely to die from lung cancer than a nonsmoker. An exsmoker who stopped smoking t years ago is w times more likely to die from lung cancer than a nonsmoker, where

$$t + w = 15.*$$

Graph the equation and use the graph to estimate how much more likely it is for Sandy to die from lung cancer than Polly, if Polly never smoked and Sandy quit $2\frac{1}{2}$ years ago.

42. *Price of printing.* The price p, in cents, of a photocopied and bound lab manual is given by $p = \frac{7}{2}n + 20$, where n is the number of pages in the manual. Graph the equation and use the graph to estimate the cost of a 25-page manual. (*Hint*: Count by 5's on both axes.)

43. *Cost of college.* The cost T, in hundreds of dollars, of tuition and fees at many community colleges can be approximated by $T = \frac{6}{5}c + 1$, where c is the number of credits for which a student registers (based on information provided by the Community College of Vermont). Graph the equation and use the graph to estimate the cost of tuition and fees when a student registers for 4 three-credit courses.

---

*Source: Data from *Body Clock* by Dr. Martin Hughes, p. 60. New York: Facts on File, Inc.

44. *Cost of college.* The cost T, in thousands of dollars, of a year at a private four-year college (all expenses) can be approximated by $T = \frac{4}{5}d + 17$, where d is the number of years since 1992 (based on information in *Statistical Abstract of the United States,* 1995). Graph the equation and use the graph to estimate the cost of a year at a private four-year college in 2002.

45. *Tea consumption.* The number of gallons of tea n consumed each year by the average U.S. consumer can be approximated by $n = \frac{1}{10}d + 7$, where d is the number of years since 1991 (based on information in *Statistical Abstract of the United States,* 1995). Graph the equation and use the graph to estimate what the average tea consumption was in 1997.

46. *Record temperature drop.* On January 22, 1943, the temperature T, in degrees Fahrenheit, in Spearfish, South Dakota, could be approximated by $T = -2m + 54$, where m is the number of minutes since 9:00 A.M. that morning (*Source*: 1997 *Information Please Almanac*). Graph the equation and use the graph to estimate the temperature at 9:15 A.M.

SKILL MAINTENANCE

Solve and check.

47. $3x - 7 = -34$
48. $4(3 - 2x) = 7 - 3(5x - 1)$
49. $2(x - 9) + 4 = 2 - 3x$

Solve.

50. $pq + p = w$, for p
51. $Ax + By = C$, for y
52. $A = \dfrac{T + Q}{2}$, for Q

SYNTHESIS

53. ◆ The equations $3x + 4y = 8$ and $y = -\frac{3}{4}x + 2$ are equivalent. Which equation would be easier to graph and why?

54. ◆ Is it possible for the graph of a linear equation to have *no* y-intercept? Why or why not?

55. ◆ Suppose that a linear equation is graphed by plotting three points and that the three points line up with each other. Does this guarantee that the equation is being correctly graphed? Why or why not?

56. ◆ Explain how the graph in Example 7 can be used to determine when the value of the color copier has dropped to $1500.

57. Using only whole numbers for x and y, find as many solutions of $x + y = 7$ as possible. Then graph the equation.

58. Using only whole numbers for x and y, find as many solutions of $x + y = 9$ as possible. Then graph the equation.

59. ◆ Examine the graphs from Exercises 57 and 58. Then describe what the graph of $x + y = C$ will be for any constant C.

In Exercises 60–63, try to find an equation for the graph shown.

60.

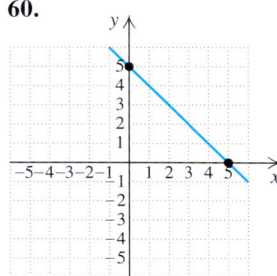

61.

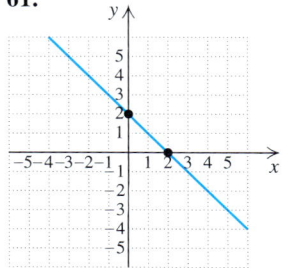

62.

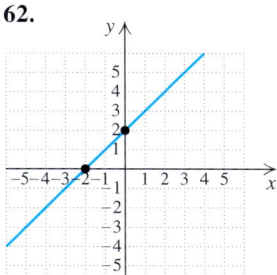

63.

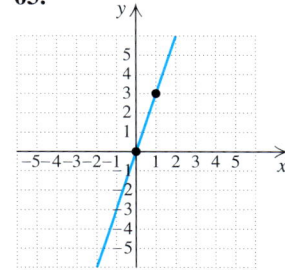

64. Translate to an equation:

d dimes and n nickels total $1.75.

Then graph the equation and use the graph to determine three different combinations of dimes and nickels that total $1.75 (see also Exercise 77).

65. Translate to an equation:

d $25 dinners and l $5 lunches total $225.

Then graph the equation and use the graph to determine three different combinations of lunches and dinners that total $225 (see also Exercise 77).

Using the x-values $-3, -2, -1, 0, 1, 2,$ and 3, graph each equation.

66. $y = 6 - x^2$
67. $y = \frac{1}{2}x^3$
68. $y = |x| + 2$
69. $y = |x| - 3$
70. $y = 3 - |x|$

138 CHAPTER 3 • INTRODUCTION TO GRAPHING

For Exercises 71–76, use a grapher to graph the equation. Use a $[-10, 10, -10, 10]$ *window.*

71. $y = -2.8x + 3.5$
72. $y = 4.5x + 2.1$
73. $y = 2.8x - 3.5$
74. $y = -4.5x - 2.1$
75. $y = x^2 + 4x + 1$
76. $y = -x^2 + 4x - 7$

77. ◆ Study the graph of Exercise 64 or 65. Does *every* point on the graph represent a solution of the associated problem? Why or why not?

COLLABORATIVE CORNER

Focus: Graphing and problem solving
Time: 25 minutes
Group size: 3
Materials: Each group will need a rubber ball, graph paper, and a tape measure.

ACTIVITY

Does a rubber ball always rebound a fixed percentage of the height from which it is dropped? To answer this, perform the following activity (read all steps before beginning).

1. One group member should hold a rubber ball at some height above the floor. A second group member should measure this height. The ball should then be dropped and caught at the peak of its bounce. The second group member should measure this rebound height and the third group member should record the two measurements.
2. Repeat part (1) two more times from the same height. Then find the average of the three rebound heights and use that number to form the ordered pair (original height, rebound height).
3. Repeat parts (1) and (2) four more times at different heights to find five ordered pairs. Graph the five pairs on graph paper and draw a straight line (starting at (0, 0)) that comes as close as possible to all six points.
4. Use the graph in part (3) to predict the rebound height for a ball dropped from a height that is 2 ft more than the greatest height used in parts (1)–(3). Then perform the drop and check your prediction. Repeat this experiment at another "new" height.
5. Does it appear that a rubber ball will always rebound a fixed percentage of the height from which it is dropped? Why or why not?

3.3 Graphing and Intercepts

Using Intercepts to Graph • Graphing Horizontal or Vertical Lines

Unless a line is horizontal or vertical, it will cross both axes. Locating the points at which the axes are crossed gives us another way of graphing linear equations.

Using Intercepts to Graph

Recall from Section 3.2 that the point at which a graph crosses the *y*-axis is called the *y*-intercept. To find the *y*-intercept, we replace *x* with 0 and then solve for *y*.

EXAMPLE 1   Find the *y*-intercept of the graph of $4x + 3y = 12$.

SOLUTION   To find the *y*-intercept, we let $x = 0$ and solve for *y*:

$4 \cdot 0 + 3y = 12$ Replacing *x* with 0
$3y = 12$
$y = 4.$

Thus, (0, 4) is the *y*-intercept.

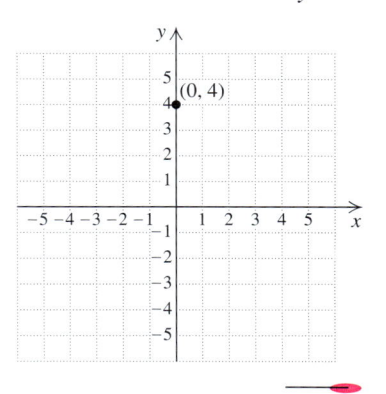

The **x-intercept** of a line is the point (if one exists) where the line crosses the *x*-axis. Since any point on the *x*-axis has a *y*-coordinate of 0, to find an *x*-intercept we replace *y* with 0 and solve for *x*.

EXAMPLE 2   Find the *x*-intercept of the graph of $4x + 3y = 12$.

SOLUTION   To find the *x*-intercept, we let $y = 0$ and solve for *x*:

$4x + 3 \cdot 0 = 12$ Replacing *y* with 0
$4x = 12$
$x = 3.$

Thus, (3, 0) is the *x*-intercept.

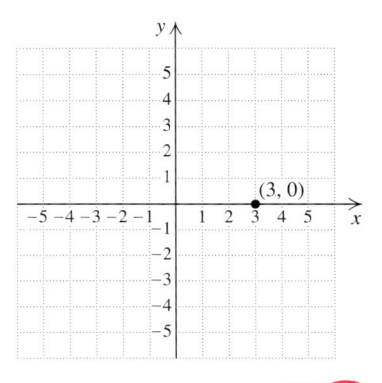

Although some graphs (see Example 4 in Section 3.2) have both intercepts at (0, 0), in most cases the two intercepts are two different points. Using those points, along with a third point as a check, we can quickly graph most linear equations.

140 CHAPTER 3 • INTRODUCTION TO GRAPHING

EXAMPLE 3

Graph $4x + 3y = 12$ using intercepts.

SOLUTION Examples 1 and 2 show that the y-intercept is $(0, 4)$ and the x-intercept is $(3, 0)$. Before drawing a line, we plot a third point as a check. We substitute any convenient value for x and solve for y.

If we let $x = 1$, then

$$4 \cdot 1 + 3y = 12 \quad \text{Substituting 1 for } x$$
$$4 + 3y = 12$$
$$3y = 12 - 4$$
$$3y = 8$$
$$y = \tfrac{8}{3}, \text{ or } 2\tfrac{2}{3}. \quad \text{Solving for } y$$

The point $\left(1, 2\tfrac{2}{3}\right)$ appears to line up with the intercepts, so our work is probably correct. To finish, we draw and label the line.

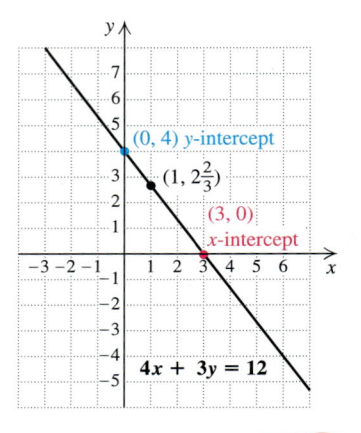

Note that the equation $4x + 3y = 12$ simplified to $3y = 12$ when we solved for the y-intercept. Thus, to find the y-intercept, we can momentarily ignore the x-term and solve the remaining equation.

In a similar manner, $4x + 3y = 12$ simplified to $4x = 12$ when we solved for the x-intercept. Thus, to find the x-intercept, we can momentarily ignore the y-term and then solve this remaining equation.

EXAMPLE 4

Graph $3x - 2y = 6$ using intercepts.

SOLUTION To find the y-intercept, we let $x = 0$. This amounts to temporarily ignoring the x-term and then solving:

$$-2y = 6 \quad \text{For } x = 0, \text{ we have } 3 \cdot 0 - 2y, \text{ or simply } -2y.$$
$$y = -3.$$

The y-intercept is $(0, -3)$.

To find the x-intercept, we let $y = 0$. This amounts to temporarily disregarding the y-term and then solving:

$$3x = 6 \quad \text{For } y = 0, \text{ we have } 3x - 2 \cdot 0, \text{ or simply } 3x.$$
$$x = 2.$$

The x-intercept is $(2, 0)$.

To find a third point, we replace x with 4 and solve for y:

$$3 \cdot 4 - 2y = 6 \quad \text{Numbers other than 4 can be used for } x.$$
$$12 - 2y = 6$$
$$-2y = -6$$
$$y = 3. \quad \text{This means that } (4, 3) \text{ is on the graph.}$$

The point (4, 3) appears to line up with the intercepts, so we draw the graph.

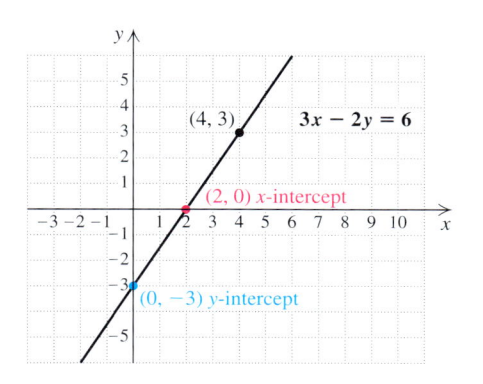

TECHNOLOGY CONNECTION 3.3

Most graphers require that an equation have y alone on one side before the equation can be entered. Once the equation has been entered, depending on the dimensions of the window, we may not be able to see both intercepts. For example, if $y = -0.8x + 17$ is graphed in the standard $[-10, 10, -10, 10]$ window, neither intercept is visible.

$y = -0.8x + 17$

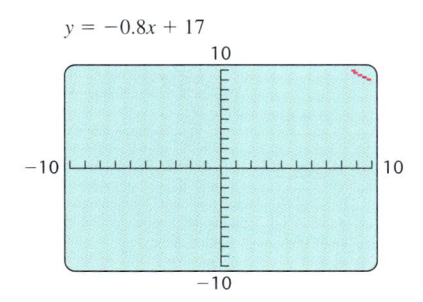

To better view the intercepts, we can try different window dimensions or we can zoom out. The ZOOM feature allows us to reduce or magnify a graph or a portion of a graph. Before zooming, the ZOOM *factors* must be set in the memory of the ZOOM key. If we zoom out with factors set at 5, both intercepts are visible but the axes are heavily drawn, as shown in the preceding figure.

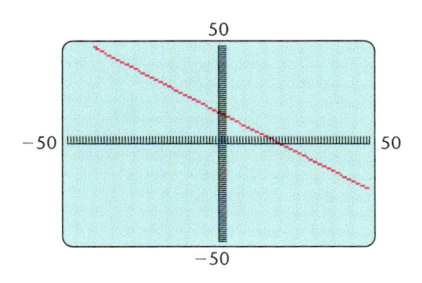

This suggests that the *scales* of the axes should be changed. To do this, we use the WINDOW menu and set Xscl to 5 and Yscl to 5. The resulting graph has tick marks 5 units apart and clearly shows both intercepts.

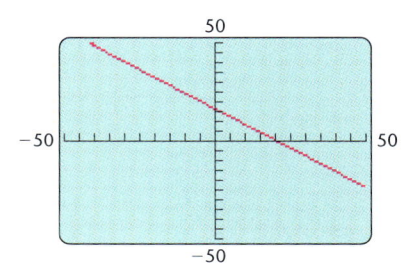

Graph each equation so that both intercepts can be easily viewed. Zoom or adjust the window settings so that tick marks can be clearly seen on both axes.

1. $y = -0.72x - 15$
2. $y - 2.13x = 27$
3. $5x + 6y = 84$
4. $2x - 7y = 150$

Graphing Horizontal or Vertical Lines

The equations graphed in Examples 3 and 4 are both in the form $Ax + By = C$. We have already stated that any equation in the form $Ax + By = C$ is linear, provided A and B are not both zero. What if A or B (but not both) is zero? We will find that when A is zero, the graph is a horizontal line, and when B is zero, the graph is a vertical line.

EXAMPLE 5

Graph: $y = 3$.

SOLUTION We can regard the equation $y = 3$ as $0 \cdot x + y = 3$. No matter what number we choose for x, we find that y must be 3 if the equation is to be solved. Consider the following table.

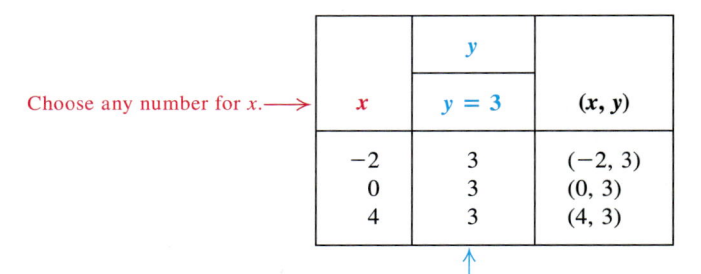

Choose any number for x. →

x	$y = 3$	(x, y)
-2	3	$(-2, 3)$
0	3	$(0, 3)$
4	3	$(4, 3)$

↑ y must be 3.

All pairs will have 3 as the y-coordinate.

When we plot the ordered pairs $(-2, 3)$, $(0, 3)$, and $(4, 3)$ and connect the points, we obtain a horizontal line. Any ordered pair $(x, 3)$ is a solution, so the line is parallel to the x-axis with y-intercept $(0, 3)$.

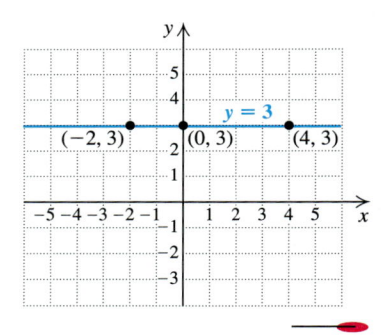

EXAMPLE 6

Graph: $x = -4$.

SOLUTION We can regard the equation $x = -4$ as $x + 0 \cdot y = -4$. We make up a table with all -4's in the x-column.

x must be -4. →

$x = -4$	y	(x, y)
-4	-5	$(-4, -5)$
-4	1	$(-4, 1)$
-4	3	$(-4, 3)$

↑ Choose any number for y.

All pairs will have -4 as the x-coordinate.

When we plot the ordered pairs (−4, −5), (−4, 1), and (−4, 3) and connect them, we obtain a vertical line. Any ordered pair (−4, y) is a solution. The line is parallel to the y-axis with x-intercept (−4, 0).

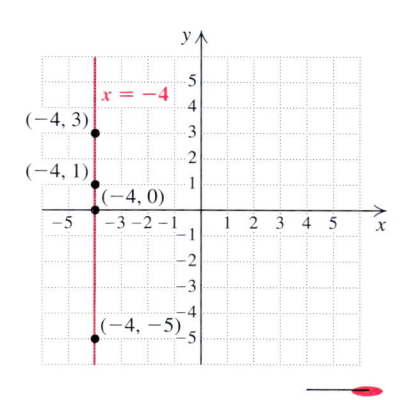

LINEAR EQUATIONS IN ONE VARIABLE

The graph of $y = b$ is a horizontal line, with y-intercept $(0, b)$.

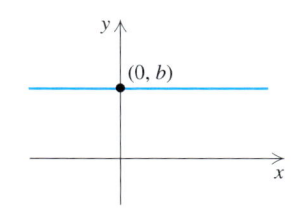

The graph of $x = a$ is a vertical line, with x-intercept $(a, 0)$.

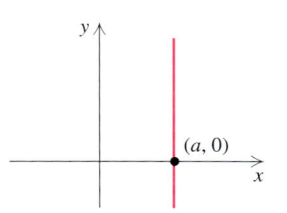

EXAMPLE 7 Write an equation for each graph.

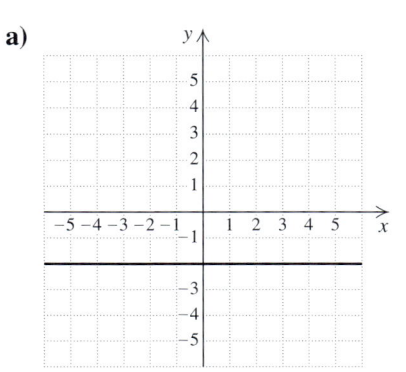

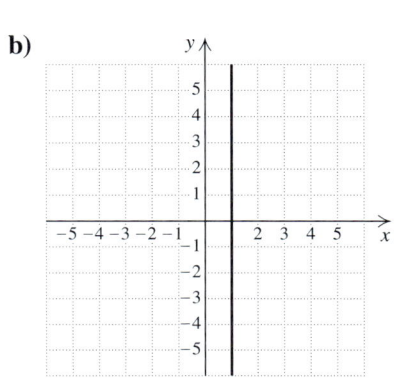

a)

b)

SOLUTION

a) Note that every point on the horizontal line passing through (0, −2) has −2 as the y-coordinate. Thus the equation of the line is $y = -2$.

b) Note that every point on the vertical line passing through (1, 0) has 1 as the x-coordinate. Thus the equation of the line is $x = 1$.

EXERCISE SET 3.3

For Exercises 1–4, find **(a)** *the coordinates of the y-intercept and* **(b)** *the coordinates of the x-intercept.*

1.

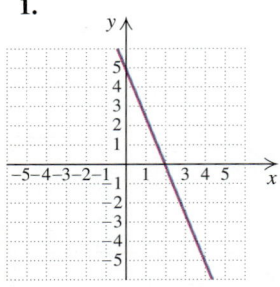

2.

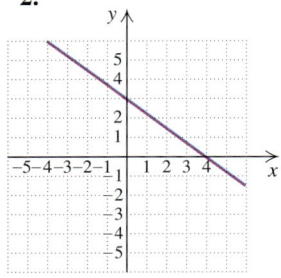

3.

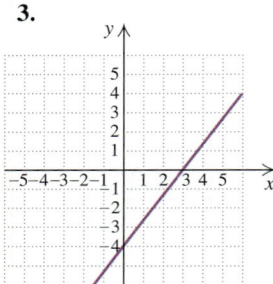

4.

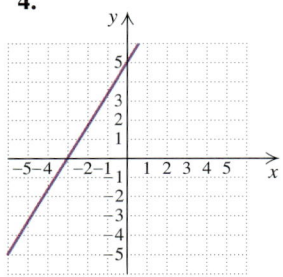

For Exercises 5–12, find **(a)** *the coordinates of the y-intercept and* **(b)** *the coordinates of the x-intercept. Do not graph.*

5. $3x + 5y = 15$
6. $5x + 2y = 20$
7. $7x - 2y = 28$
8. $3x - 4y = 24$
9. $-4x + 3y = 10$
10. $-2x + 3y = 7$
11. $6x - 3 = 9y$
12. $4y - 2 = 6x$

Find the intercepts. Then graph.

13. $2x + y = 6$
14. $3x + 2y = 12$
15. $6x + 9y = 18$
16. $x + 3y = 6$
17. $-x + 3y = 9$
18. $-x + 2y = 4$
19. $2x - y = 8$
20. $3x + y = 9$
21. $6 - 3y = 9x$
22. $2y - 2 = 6x$
23. $5x - 10 = 5y$
24. $3x - 9 = 3y$
25. $2x - 5y = 10$
26. $2x - 3y = 6$
27. $2x + 6y = 12$
28. $4x + 5y = 20$
29. $x - 1 = y$
30. $2x + 3y = 8$
31. $2x - 1 = y$
32. $x - 3 = y$
33. $4x - 3y = 12$
34. $3x - 2 = y$
35. $7x + 2y = 6$
36. $6x - 2y = 18$
37. $y = -4 - 4x$
38. $3x + 4y = 5$
39. $-3x = 6y - 2$
40. $y = -3 - 3x$
41. $3 = 2x - 5y$
42. $-4x = 8y - 5$
43. $x + 2y = 0$
44. $y - 3x = 0$

Write an equation for each graph.

45.

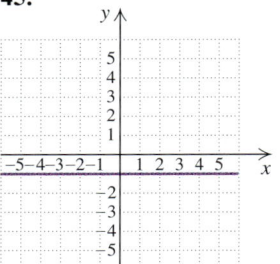

46.

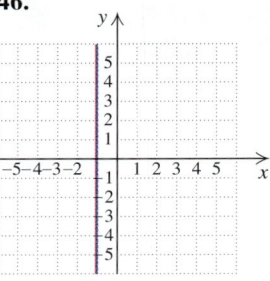

47.

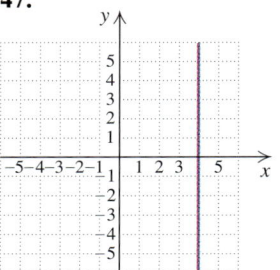

48.

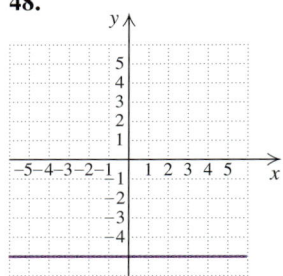

49.

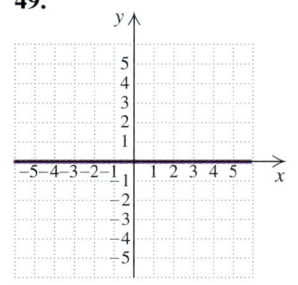

50.

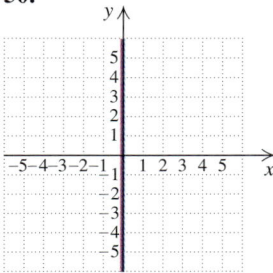

Graph.

51. $x = -3$
52. $x = -2$

53. $y = 2$
54. $y = 4$
55. $x = 7$
56. $x = 3$
57. $x = 0$
58. $y = 0$
59. $y = \frac{5}{2}$
60. $x = -\frac{3}{2}$
61. $-3y = -15$
62. $12y = 45$
63. $4x + 3 = 0$
64. $-3x + 12 = 0$
65. $18 - 3y = 0$
66. $63 + 7y = 0$

SKILL MAINTENANCE

Write the prime factorization of each number.

67. 98
68. 240
69. 275

Simplify.

70. $\frac{36}{90}$
71. $\frac{12}{84}$
72. $\frac{125}{75}$

SYNTHESIS

73. ◆ If the graph of $Ax + By = C$ is a horizontal line, what can you conclude about A? Why?
74. ◆ Explain in your own words why the graph of $x = 7$ is a vertical line.
75. ◆ Examine Example 7 in Section 3.2. Why might a business want to determine the horizontal, or t-intercept, of the graph?
76. ◆ Can *every* equation of the type $Ax + By = C$ be written in the form $y = mx + b$? Why or why not?

77. Write an equation for the x-axis.
78. Write an equation for the y-axis.
79. Find the coordinates of the point of intersection of the graphs of $y = x$ and $y = 6$.
80. Find the coordinates of the point of intersection of the graphs of the equations $x = -3$ and $y = x$.
81. Write an equation of the line shown in Exercise 1.
82. Write an equation of the line shown in Exercise 4.
83. Write an equation of a line parallel to the x-axis and passing through $(-3, -4)$.
84. Find the value of m such that the graph of $y = mx + 6$ has an x-intercept of $(2, 0)$.
85. Find the value of C such that the graph of $3x + C = 5y$ has an x-intercept of $(-4, 0)$.
86. Find the value of C such that the graph of $4x = C - 3y$ has a y-intercept of $(0, -8)$.
87. ◆ For A and B nonzero, the graphs of $Ax + D = C$ and $By + D = C$ will be parallel to an axis. Explain why.

In Exercises 88–93, find the intercepts of each equation algebraically. Then adjust the window and scale so that the intercepts can be checked graphically with no further window adjustments.

88. $3x + 2y = 50$
89. $2x - 7y = 80$
90. $y = 0.2x - 9$
91. $y = 1.3x - 15$
92. $25x - 20y = 1$
93. $50x + 25y = 1$

3.4 More Applications Using Graphs

Translating to Equations • Using Graphs to Solve Problems • Rates

Suppose we are asked to find a pair of numbers that add to 5. Because there is an infinite number of such pairs, a graph provides a convenient way of showing all solutions.

Translating to Equations

When a problem-solving situation involves pairs of numbers and many solutions, we can often translate it to an equation with two variables.

146 CHAPTER 3 • INTRODUCTION TO GRAPHING

EXAMPLE 1 Translate to an equation.

a) The sum of two numbers is 5.
b) Harry is two years older than Jane.
c) A taxi ride costs $2 plus $1.50 a mile.
d) Sondra earns $100 less than twice Jim's salary.

SOLUTION

a) Let x represent one number and y represent the other. The word "sum" signifies addition, so the translation follows immediately: $x + y = 5$.
b) Let $h =$ Harry's age and $j =$ Jane's age. We reword and then translate.

Rewording: Harry's age is two more than Jane's age.

Translating: $\qquad h \qquad = \qquad j + 2$

c) Let $c =$ the cost of the taxi ride and $m =$ the number of miles driven. Then we have

Rewording: The cost of the ride is $2 plus $1.50 a mile.

Translating: $\qquad c \qquad = 2 + 1.5m$

d) Let $s =$ Sondra's salary and $j =$ Jim's salary. Then we have

Rewording: Sondra's salary is $100 less than twice Jim's salary.

Translating: $\qquad s \qquad = \qquad 2j - 100$

Once a problem-solving situation has been translated to an equation with two variables, a graph can provide a visual representation of the situation.

Using Graphs to Solve Problems

When a problem is translated into an equation that is graphed, the graph offers a quick and useful way of approximating values that could be calculated more precisely (but also more slowly) from the equation.

EXAMPLE 2 *Truck Rentals.* Ridem Trucks charges $49.95 per day plus 35¢ per mile for the rental of an 18-ft truck. To help customers predict the cost of a rental, the firm draws a graph in which mileage is measured on the horizontal axis and cost on the vertical axis. Using such a graph, estimate the distance that an 18-ft truck can be driven on a budget of $125.

SOLUTION

1. Familiarize. In Section 2.5, another truck rental problem was solved, so we are already familiar with this situation. We let $m =$ the number of miles driven and $c =$ the cost in dollars. We will graph m on the horizontal axis and c on the vertical axis.

2. **Translate.** Since the cost of a rental is $49.95 plus 35¢ for each mile, and since m miles are to be driven, we have the translation

   $c = 49.95 + 0.35m$.

3. **Carry out.** We make a table of values using some convenient choices for m.

   When $m = 0$, $\quad c = 49.95 + 0.35(0) = 49.95$.
   When $m = 100$, $\quad c = 49.95 + 0.35(100) = 84.95$.
   When $m = 300$, $\quad c = 49.95 + 0.35(300) = 154.95$.

Mileage	Cost
0	$ 49.95
100	$ 84.95
300	$154.95

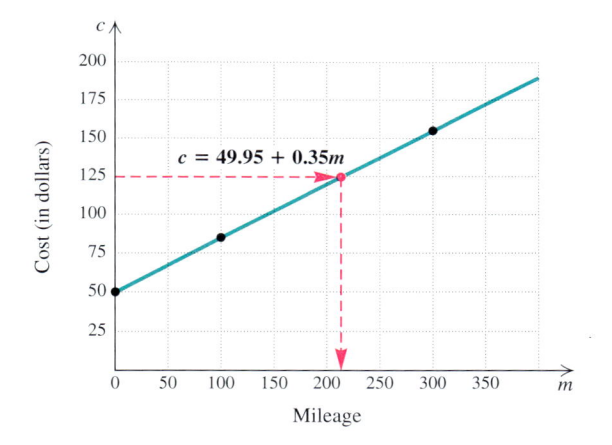

   Being careful to label the axes correctly, we draw the graph by plotting the points listed in the table and then drawing a line through them.

   To estimate how far the truck can be driven on a budget of $125, we locate $125 on the vertical axis, move horizontally to the graphed line, and then downward to the horizontal axis. It appears that the truck can be driven about 210 mi on a budget of $125.

4. **Check.** If the truck is driven 210 mi, the cost will be

   $c = 49.95 + 0.35(210)$
   $ = 49.95 + 73.50$
   $ = 123.45$.

   Since this is close to $125, our estimate of 210 is fairly accurate.

5. **State.** On a budget of $125, the truck can be driven about 210 mi.

TECHNOLOGY CONNECTION 3.4

There are at least two ways in which we can determine the coordinates of points on a graph drawn by a grapher. One approach utilizes a TRACE key. When the TRACE feature is activated, a cursor appears on the line (or curve) that has been graphed and the coordinates at that point are displayed. The left and right arrow keys allow the cursor to be moved along the graph.

Consider the problem described in Example 2. Because most graphers use x and y as variables, we graph $y = 49.95 + 0.35x$. We set the window at [0, 400, 0, 225], with Xscl = 50 and Yscl = 25.

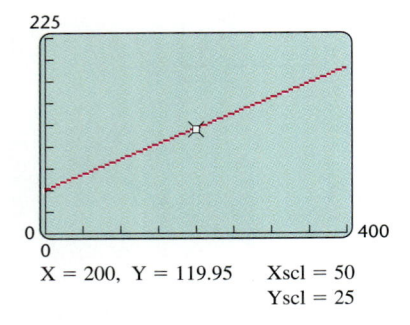

```
225
        [graph of line]
             ✳
0              400
0
X = 200, Y = 119.95    Xscl = 50
                       Yscl = 25
```

The cursor on the line indicates that TRACE has been activated. By moving the cursor, we can identify many ordered pairs at the touch of a button.

1. Use the TRACE key to find five different ordered pairs that are solutions of $y = 49.95 + 0.35x$.

2. Use the TRACE and ZOOM features to determine the value of y when x is 100.

In Exercise 2 above, it becomes clear that determining the coordinates of specific pairs can be a challenge using TRACE. The second approach to finding pairs, the TABLE feature, is available on many graphers and avoids this difficulty. To use TABLE, we first press 2nd TblSet and set TblMin to 0 and ΔTbl to 10. This means that the X values of the table will begin at 0 and increase incrementally by 10. By setting INDPNT and DEPEND to AUTO, we obtain the following when we press 2nd TABLE:

X	Y
0	49.95
10	53.45
20	56.95
30	60.45
40	63.95
50	67.45
60	70.95

X = 0

The arrow keys allow us to scroll up and down in the table.

3. Use a table to find the value of y when x is 100. Then find the value of y when x is 300.
4. Adjust the TABLE settings to INDPNT: ASK. How does the table change? Enter a number of your choice and see what happens. Use this setting to find the value of y when x is 253.

EXAMPLE 3 *Depreciation of an Office Machine.* In Example 7 of Section 3.2, we found that the value of a particular color copier could be represented by the following graph. Use the graph to estimate how long it will take for the value of the copier to drop from $2500 to $1000.

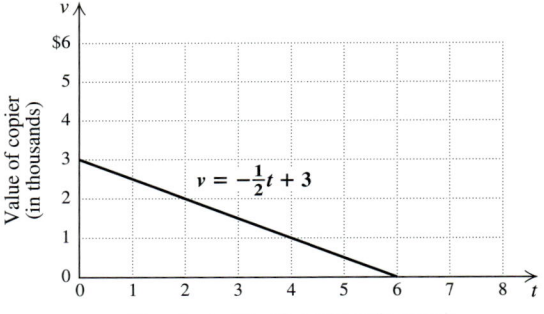

[graph: Value of copier (in thousands) vs. Time from date of purchase (in years); line $v = -\frac{1}{2}t + 3$]

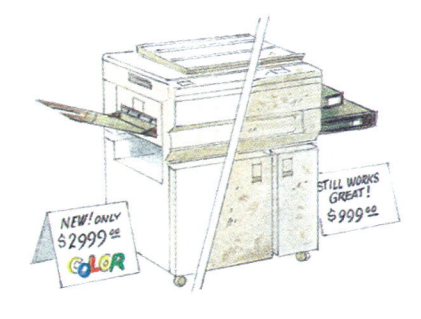

SOLUTION The graph lists values on the vertical axis, so we locate $2500 there and estimate the value with which it is paired on the horizontal (time) axis. We do this by tracing a path from the vertical axis to the line and then from the line down to the horizontal axis. It appears that the value of the copier is $2500 when the machine is about 1 yr old.

A similar procedure is followed to determine the age of the copier when its value drops to $1000. It appears that the copier is about 4 yr old when its value is $1000. The drop in value from $2500 to $1000 occurred over a period of $4 - 1$, or 3, yr.

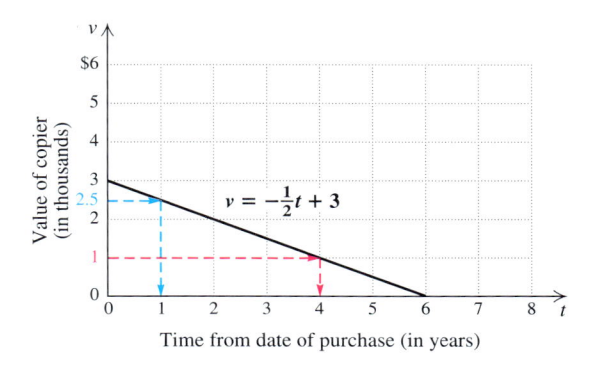

$$v = -\frac{1}{2}t + 3$$

Rates

Graphs are especially helpful in representing situations that involve *rates*.

RATE

A *rate* is a ratio that indicates how two quantities change with respect to each other.

Rates occur often in everyday life:

A town that grows by 3400 residents over a period of 2 yr has a *growth rate* of $\frac{3400}{2}$, or 1700, residents per year.

A person running 150 m in 20 sec is moving at a *rate* of $\frac{150}{20}$, or 7.5, m/sec (meters per second).

A class of 25 students pays a total of $93.75 to visit a museum. The *rate* is $\frac{\$93.75}{25}$, or $3.75, per student.

EXAMPLE 4 Determine the rate at which the cost of the truck rental in Example 2 is changing with respect to the number of miles driven.

SOLUTION The ordered pairs listed in the table below can be used to determine how the cost changes with respect to the number of miles driven.

Mileage	Cost
0	$ 49.95
100	$ 84.95
300	$154.95

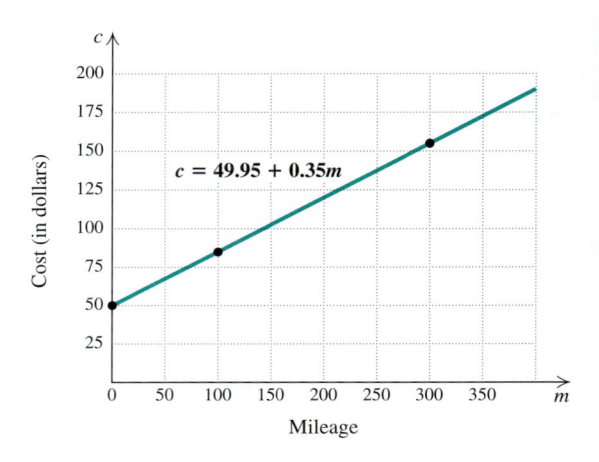

$c = 49.95 + 0.35m$

Note that as the mileage changes from 100 to 300, the cost changes from $84.95 to $154.95. We see that

a cost change of $154.95 − $84.95, or $70,

corresponds to

a mileage change of 300 − 100, or 200 mi.

The cost of the rental is changing at the rate of $70/200 mi, or 35¢ per mile. This checks with the rate given in the problem.

If a graph is a straight line, its rate of change is the same for the entire graph. Thus, in Example 4, it did not matter which two points were used to find the rate. The student should check this, using the change in cost that results from driving the first 100 mi.

In Example 4, the graph climbs upward, increasing, as we move from left to right, and the rate of change is a positive number. In the next example, the graph decreases, sloping downward, and the rate of change is negative.

EXAMPLE 5 Determine the rate at which the value of the copier in Example 3 is changing with respect to time.

SOLUTION In Example 3, we found that the value of the copier drops 2500 − 1000, or 1500, dollars in 3 yr. Thus the value is *dropping* at a rate of

$$\frac{1500}{3}, \text{ or } 500,$$

dollars per year.

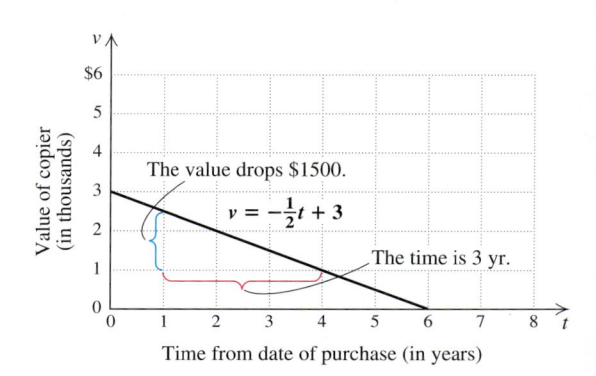

The value drops $1500.

$v = -\frac{1}{2}t + 3$

The time is 3 yr.

3.4 MORE APPLICATIONS USING GRAPHS

To reflect the fact that the value is *de*creasing, we can say that *the rate of change* is

-500 dollars per year.

Perhaps the most commonly used rate is *speed*, or rate of travel.

EXAMPLE 6   *Train Schedules and Speeds.*   The schedule below was used in France during the late 1800s. Cities, along with their distances from Dijon, are listed vertically and the hours of the day are listed horizontally. Note that each hour is divided into six 10-min intervals. Each slanted line corresponds to a different train trip. Short, horizontal, segments indicate stops for loading or unloading. At 4:30 P.M. each day, a train left Tonnerre, bound for Montereau.

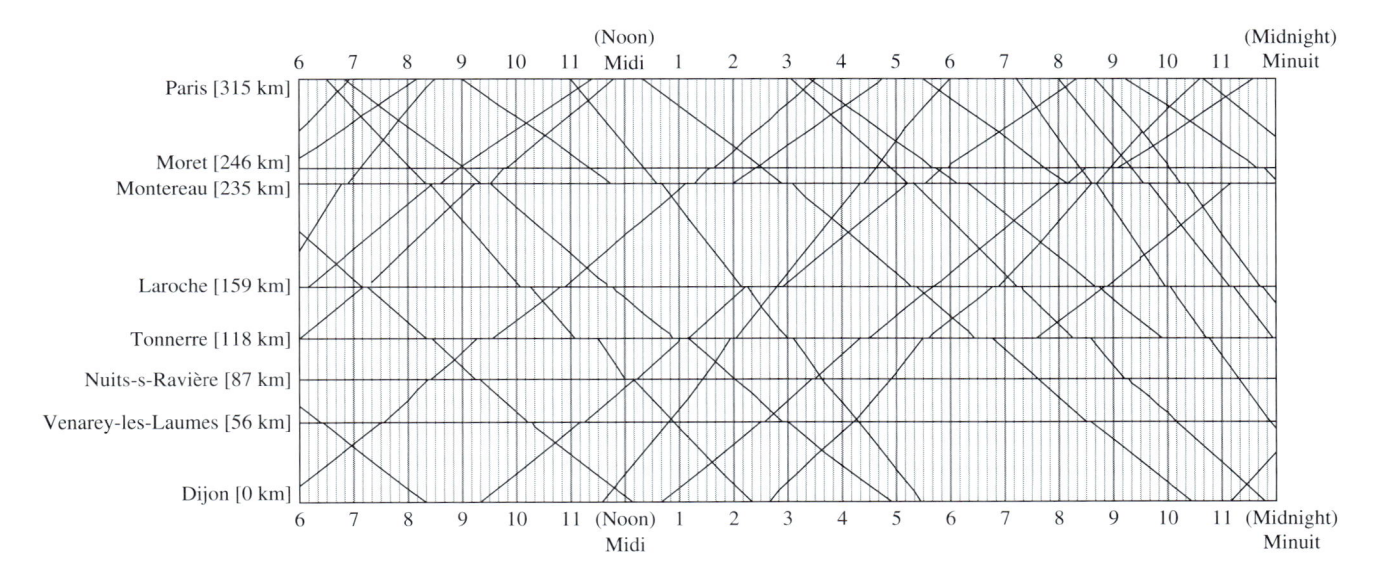

a) How far is it from Tonnerre to Montereau?
b) How much time did the 4:30 P.M. ride from Tonnerre to Montereau take?
c) Approximately how fast did the 4:30 P.M. train from Tonnerre to Montereau travel?

SOLUTION

a) To determine the distance between Tonnerre and Montereau, we look at the vertical axis:

Montereau is   235 km from Dijon

and

Tonnerre is   118 km from Dijon,

so

Montereau is $235 - 118 = 117$ km from Tonnerre.

b) To determine how much time the ride took, we locate 4:30 P.M. on the horizontal axis and, from there, move upward to the horizontal line that starts at Tonnerre on the vertical axis. From there, we follow the line representing the train's progress upward and to the right until it intersects the horizontal line that represents Montereau.

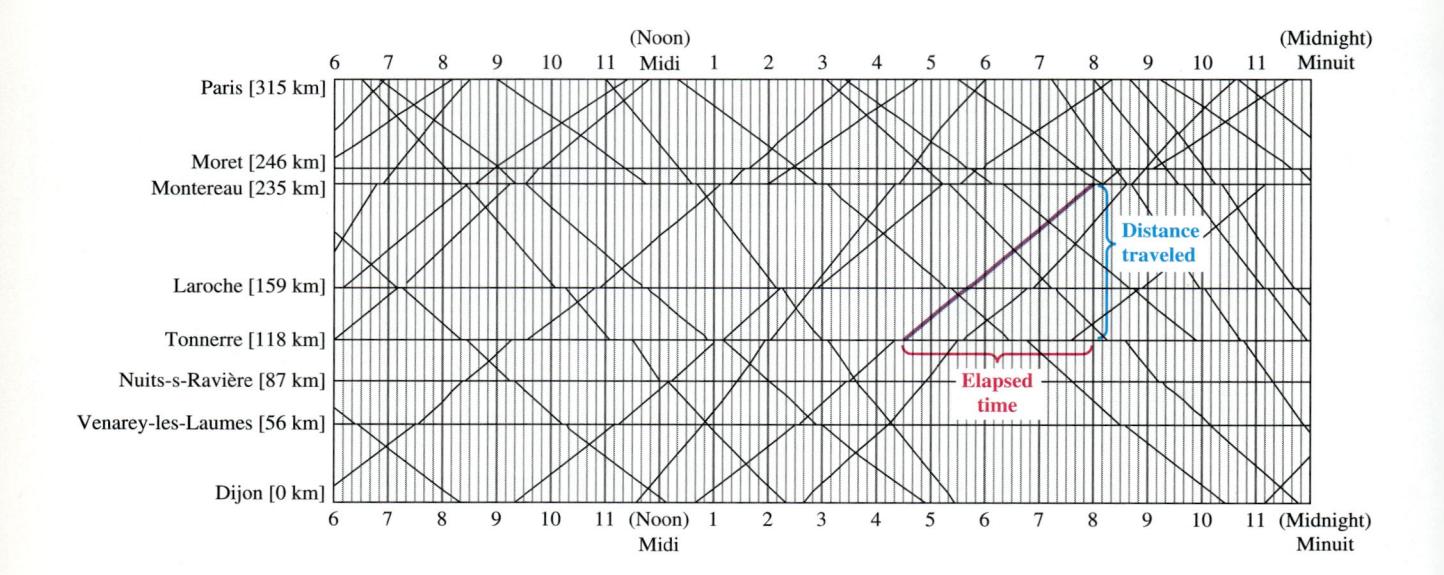

We see that the train that left Tonnerre at 4:30 P.M. reached Montereau at 8:00 P.M. Thus the ride took 3.5 hr.

c) To find the average speed of the train, or its *rate* of travel, we divide the total distance traveled by the time required for the trip. Using parts (a) and (b) above, we have

$$\text{Average speed} = \frac{\text{total distance}}{\text{total time}}$$
$$= \frac{117 \text{ km}}{3.5 \text{ hr}}$$
$$\approx 33 \text{ km/hr}.$$

Note in Example 6 that the 4:30 P.M. train from Tonnerre to Montereau was significantly slower than the 2:00 P.M. train between the same cities (the 2:00 P.M. train needed only 2 hr 20 min to travel the same route). This can be detected visually by observing that the line representing the 2:00 P.M. train is steeper than the line representing the 4:30 P.M. train. The 2:00 P.M. train had an average speed of

$$\frac{117 \text{ km}}{2\frac{1}{3} \text{ hr}} \approx 50 \text{ km/hr}.$$

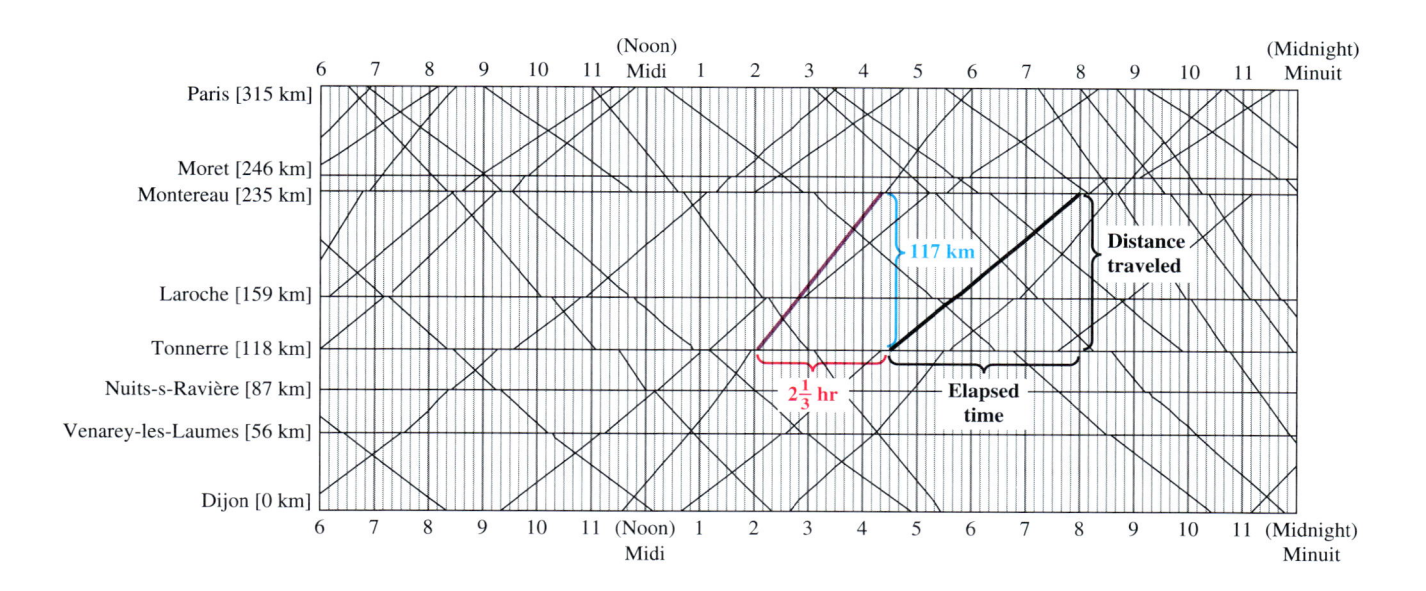

Rates play an important role in the natural and social sciences as well as in other mathematics courses. In this section, our study of rates has just begun. In Section 3.5, we will see how the notion of rate can be related to the graphs of lines in general.

EXERCISE SET 3.4

Translate each sentence to an equation containing two variables. Be sure to state what each variable represents.

1. The sum of two numbers is 19.
2. The sum of two numbers is 39.
3. A number plus twice another number is 65.
4. The sum of a number and twice another number is 93.
5. Justine paid twice as much as the necklaces were worth.
6. Evan made a $5.25 profit on each flower pot sold.
7. Jason travels 220 m during each minute he is out jogging.
8. Eva makes half as many careless mistakes as her brother does.
9. Frank is 3 less than half Cecilia's age.
10. Lois earns $200 less than 3 times Roberta's weekly salary.
11. The cost of the bike rental is $8 plus $2 per hour.
12. The photocopies cost $3 plus 5¢ per page.
13. The lake depth of 94 ft was dropping at a rate of 3 ft per week.
14. The politician's savings of $75,000 were dropping at a rate of $8000 per week.

Solve.

15. *Van rentals.* Rent King charges $59.95 plus 45¢ per mile for the rental of its 20-ft van. Draw a graph in which mileage is measured on the horizontal axis and cost on the vertical axis. Then use the graph to estimate how far a van can be driven on a budget of $150.
16. *Truck rentals.* Riverside Trucks charges $39.95 plus 55¢ per mile for the rental of a 20-ft truck. Draw a graph in which mileage is measured on the horizontal axis and cost on the vertical axis. Then use the graph to estimate how far a truck can be driven on a budget of $120.

17. *Hair growth.* After Tina gets a "buzz cut," the length L of her hair, in inches, is given by
$$L = \frac{1}{2}t + 1,$$
where t is the number of months after she got the haircut. Using the horizontal axis for t, graph the equation. Then use the graph to estimate how long it will take for Tina's hair to be $2\frac{3}{4}$ in. long.

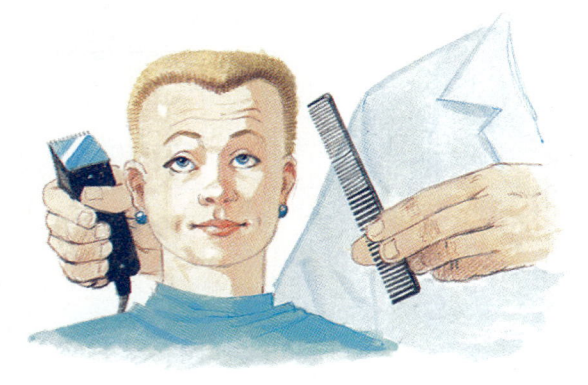

18. *Allocating resources.* Servemaster food services has $240 to spend on turkey at $4.00 per pound and/or roast beef at $6.00 per pound. If t pounds of turkey and r pounds of roast beef are bought, the equation
$$4t + 6r = 240$$
must be satisfied. Graph the equation, using r as the first coordinate. Then use the graph to estimate how much roast beef was bought if Servemaster bought 23 lb of turkey.

19. *Wages and commissions.* Each salesperson at Big Shot Appliances is paid a weekly salary of $200 plus 8% of that person's sales. Draw a graph in which sales are measured on the horizontal axis and wages on the vertical axis. Then use the graph to estimate a salesperson's sales when a week's pay is $790.

20. *Wages and commissions.* Each salesperson at the Shoe Box is paid a weekly salary of $200 plus 4% of that person's sales. Draw a graph in which sales are measured on the horizontal axis and wages on the vertical axis. Then use the graph to estimate a salesperson's sales when a week's pay is $375.

21. *Real-estate depreciation.* Because of wear and tear, rental property can be depreciated each year that it is in service. The depreciated value for some real estate is found by subtracting $\frac{1}{18}$ of the original value for each year that the property is rented. Draw a graph that can be used to find the depreciated value of a house that was valued at $150,000 when it was first rented. Let the horizontal axis represent time. Then use the graph to estimate how long it takes the house to depreciate in value from $125,000 to $75,000.

22. *Food preparation.* Harriet's Catering believes that parties for more than 10 should include a 3-lb wheel of cheese and an additional $\frac{2}{9}$ lb for each person in excess of 10. Draw a graph that can be used to predict how much cheese should be purchased for parties of 10 or more. Let the horizontal axis represent the number of people and the vertical axis the number of pounds. Then use the graph to estimate how large a party can be accommodated with 8 lb of cheese.

23. *Parking fees.* Karla's Parking charges $3.00 to park plus 50¢ for each 15-min unit of time. Draw a linear graph that can be used to estimate the cost of parking at Karla's.* Let the horizontal axis represent time and the vertical axis cost. Then use the graph to estimate how long someone was parked if charged $7.50.

24. *Cost of a road call.* Dave's Foreign Auto Village charges $35 for a road call plus $10 for each 15-min unit of time. Draw a linear graph that can be used to estimate the cost of a service call.* Let the horizontal axis represent time and the vertical axis cost. Then use the graph to estimate how long a road call lasted if the charge for the road call was $105.

25. *Copying costs.* A local Mailboxes Etc.® store charges $2.25 for binding plus 5¢ per page for each spiralbound copy of a town report. Draw a graph that can be used to determine the cost of one copy of the report. Let the horizontal axis represent the number of pages and the vertical axis the price. Then use the graph to estimate the length of a report that costs $5.20 per copy.

26. *Cost of a FedEx delivery.* In 1997, for Priority delivery of packages weighing from 10 to 50 lb, FedEx charged $46.25 for the first 10 lb plus $1.25 for each additional pound. Draw a graph that can be used to estimate the cost of FedEx Priority deliveries

---

*More precise, nonlinear models of Exercises 23 and 24 appear in Exercises 55 and 56, respectively.

that weigh from 10 to 50 lb. Let the horizontal axis represent the number of pounds being shipped and the vertical axis the cost, in dollars. Then use the graph to estimate how much weight was added to a package if the delivery charge increased from $60 to $65.

27. *Aviation.* Captain Hsu is landing a 747 from its cruising altitude of 32,200 ft. The jet descends to Denver International Airport (elevation 5200 ft) at a rate of 3000 ft/min. Draw a graph in which the altitude of the plane is measured on the vertical axis and use the graph to estimate how long the descent will take.

28. *Aviation.* Helga is landing a single-engine Tandem Taildragger from a cruising altitude of 6000 ft. Her plane is descending to Dallas–Fort Worth Airport (elevation 500 ft) at a rate of 150 ft/min. Draw a graph in which the altitude of the plane is measured on the vertical axis and use the graph to estimate how long the descent will last.

29. Use the graph in Exercise 17 to determine the rate at which Tina's hair is growing.

30. Use the graph from Exercise 16 to determine the rate at which the rental fee changes with respect to the number of miles driven.

Use the train schedule in Example 6 to answer Exercises 31–40.

31. How much time did the 1:10 P.M. train ride from Tonnerre to Laroche take? How far is it from Tonnerre to Laroche? What was the average speed of the train?

32. How much time did the 9:30 A.M. train ride from Montereau to Paris take? How far is it from Montereau to Paris? What was the average speed of the train?

33. How much time did the 11:35 A.M. train ride from Dijon to Montereau take? How far is it from Dijon to Montereau? What was the average speed of the train?

34. How much time did the 6:55 P.M. train ride from Laroche to Paris take? How far is it from Laroche to Paris? What was the average speed of the train?

35. What was the average rate of travel for the 8:25 A.M. train from Montereau to Nuits-s-Ravière? How long was its longest stop along the way?

36. What was the average rate of travel for the 12:20 P.M. train from Paris to Dijon? How many stops did it make along the way?

37. What was the quickest way to get from Paris to Tonnerre by train? How can this be determined?

38. What was the quickest way to get from Tonnerre to Paris by train? How can this be determined?

39. At what time of day did the fastest train for Dijon leave Tonnerre?

40. What was the average speed of the fastest train from Montereau to Laroche?

SKILL MAINTENANCE

41. Solve $s = vt + d$ for t.
42. Solve: $3(x - 4) + 7 = -2x + 6(x - 5)$.
43. Multiply: $2(4x + 5y - 3z)$.
44. Factor: $3x + 18y - 6z$.

SYNTHESIS

45. ◆ Write a problem for a classmate to solve. Devise the problem so that a graph is drawn to represent a situation with cost or value on the vertical axis and time on the horizontal axis. Have the problem require the classmate to use the graph to estimate the time (on the horizontal axis) that is paired with a particular monetary figure (on the vertical axis).

46. ◆ Ruby argues that the graph for Exercise 22 should consist of several points rather than a solid line. Does she have a valid argument? Why or why not?

47. ◆ Without drawing a new graph, describe how the graph in Example 2 would change if Ridem Trucks charged just $24.95 per day plus 35¢ per mile.

48. ◆ Without drawing a new graph, describe how the graph in Example 2 would change if Ridem Trucks charged $49.95 per day plus 23¢ per mile.

49. Janet lives 5 km from work. One day she ran half the distance and walked the other half. Draw a graph to represent Janet's travel. Use the horizontal axis for time and the vertical axis for distance from home.

50. *Aviation.* A Boeing 737 climbs from sea level to a cruising altitude of 34,000 ft at a rate of 6500 ft/min. After cruising for 3 min, the jet is forced to land, descending at a rate of 3500 ft/min. Draw a graph in which altitude is measured on the vertical axis and time on the horizontal axis.

51. *Wages with commissions.* Each salesperson at

156 CHAPTER 3 • INTRODUCTION TO GRAPHING

Mike's Bikes is paid $140 a week plus 13% of all sales up to $2000, and then 20% on any sales in excess of $2000. Draw a graph in which sales are measured on the horizontal axis and wages on the vertical axis. Then use the graph to estimate the wages paid when a salesperson sells $2700 in merchandise in one week.

52. *Salaries.* Peggy earns $150 less than twice Paul's weekly salary. Paul's salary is $70 more than half of Jenna's. Draw a graph in which Jenna's salary is listed on the horizontal axis and Peggy's salary on the vertical axis. Then write an equation that relates Peggy's salary, p, to Jenna's salary, j.

Use the train schedule in Example 6 to answer Exercises 53 and 54.

53. ◆ Why did travelers bound for Moret rush to catch the 3:05 P.M. train out of Paris?

54. ◆ Would it have made sense for a traveler leaving Paris for Dijon to rush for the 11:00 A.M. train out of Paris and then change trains? Why or why not?

55. (Refer to Exercise 23.) It costs as much to park at Karla's for 16 min as it does for 29 min. Thus the linear graph drawn in the solution of Exercise 23 is not a precise representation of the situation. Draw a graph with a series of "steps" that more accurately reflects the situation.

56. (Refer to Exercise 24.) A 32-min road call with Dave's costs the same as a 44-min road call. Thus the linear graph drawn in the solution of Exercise 24 is not a precise representation of the situation. Draw a graph with a series of "steps" that more accurately reflects the situation.

Solve using a grapher.

57. Weekly pay at Bikes for Hikes is $219 plus a 3.5% sales commission. If a salesperson's pay was $370.03, what did that salesperson's sales total?

58. It costs Bert's Shirts $38 plus $2.35 a shirt to print tee shirts for a day camp. Camp Weehawken paid Bert's $623.15 for shirts. How many shirts were printed?

COLLABORATIVE C•O•R•N•E•R

Focus: Graphing and rates
Time: 20–30 minutes
Group size: 2–3
Materials: Graph paper (if possible, at least 10 squares per inch)

It has been argued that the French train schedule in Example 6, from 1885, is easier to read than many of today's train schedules. Below is a portion of a recent Amtrak timetable for service between Boston and Washington, D.C.

AMTRAK® Southbound   (7 Days/week)

Train Number	Mile	171	173	177	179
Boston, MA	457	7:30 A	11:20 A	4:25 P	6:15 P
New Haven, CT	297	10:35 A	2:40 P	7:35 P	9:25 P
		10:45 A	2:50 P	7:45 P	9:35 P
New York, NY	226	12:30 P	4:36 P	9:25 P	11:15 P
		12:55 P	5:06 P	9:45 P	11:35 P
Philadelphia, PA	135	2:25 P	6:28 P	11:10 P	1:01 A
		2:35 P	6:32 P	11:13 P	1:03 A
Washington, DC	0	4:32 P	8:28 P	1:15 A	3:10 A

AMTRAK® Northbound   (7 Days/week)

Train Number	Mile	172	174	168	66
Washington, DC	0	7:15 A	10:15 A	4:10 P	10:00 P
Philadelphia, PA	135	9:11 A	12:12 P	6:00 P	12:11 A
		9:21 A	12:15 P	6:10 P	12:25 A
New York, NY	226	10:47 A	1:46 P	7:37 P	2:00 A
		11:07 A	2:06 P	8:10 P	2:30 A
New Haven, CT	297	12:53 P	3:54 P	9:45 P	4:08 A
		1:13 P	4:14 P	9:55 P	4:23 A
Boston, MA	457	4:00 P	7:00 P	12:35 A	7:49 A

ACTIVITY

1. Draw a train schedule similar to the schedule used in Example 6. Decide how many minutes each square should represent on the horizontal axis in order to fit one 24-hr day onto one page. Similarly, determine how many miles each square should represent on the vertical axis. Once the axes have been labeled, use a straightedge to draw lines for each train's travel between cities. Be sure to include horizontal segments for stops along each route.

2. Use your graph to determine the average speed of Train 177 over its entire route. Then calculate its speed by using just the numbers in the timetable. The speeds should match, or at least be close.
3. Is Train 172 faster than Train 66? Why or why not?
4. Do all the trains travel at essentially the same speed between stops? How can this be determined visually?

3.5 Slope

Rate and Slope • Horizontal and Vertical Lines • Applications

In Section 3.4, we introduced *rate* as a method of measuring how two quantities change with respect to each other. There we discussed rates involving money (cost in dollars *per* student and in cents *per* mile) and travel (speed in kilometers *per* hour). In this section, we discuss *production rate* and see how rate can be related to the slope of a line.

Rate and Slope

Suppose that a car manufacturer operates two plants: one in Michigan and one in Pennsylvania. Knowing that the Michigan plant produces 3 cars every 2 hours and the Pennsylvania plant produces 5 cars every 4 hours, we can set up tables listing the number of cars produced after various amounts of time.

| Michigan Plant ||
Hours Elapsed	Cars Produced
0	0
2	3
4	6
6	9
8	12

| Pennsylvania Plant ||
Hours Elapsed	Cars Produced
0	0
4	5
8	10
12	15
16	20

158 CHAPTER 3 • INTRODUCTION TO GRAPHING

By comparing the number of cars produced at each plant over a specified period of time, we can compare the two production rates. For example, the Michigan plant produces 3 cars every 2 hours, so its *rate* is $3 \div 2 = 1\frac{1}{2}$, or $\frac{3}{2}$ cars per hour. Since the Pennsylvania plant produces 5 cars every 4 hours, its rate is $5 \div 4 = 1\frac{1}{4}$, or $\frac{5}{4}$ cars per hour.

Let's now graph the pairs of numbers listed in the tables, using the horizontal axis for time and the vertical axis for the number of cars produced.

Michigan Plant

Points plotted: (0, 0), (2, 3), (4, 6), (6, 9), (8, 12)
Horizontal axis: Number of hours elapsed
Vertical axis: Number of cars produced

Pennsylvania Plant

Points plotted: (0, 0), (4, 5), (8, 10), (12, 15), (16, 20)
Horizontal axis: Number of hours elapsed
Vertical axis: Number of cars produced

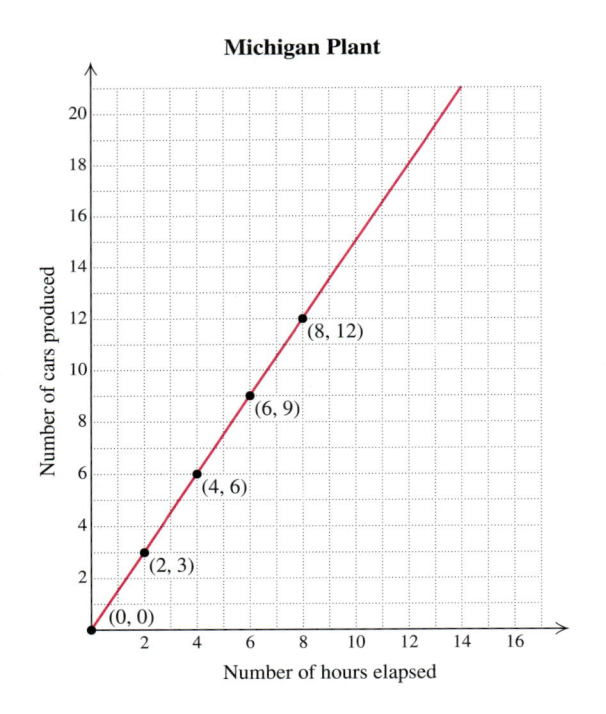

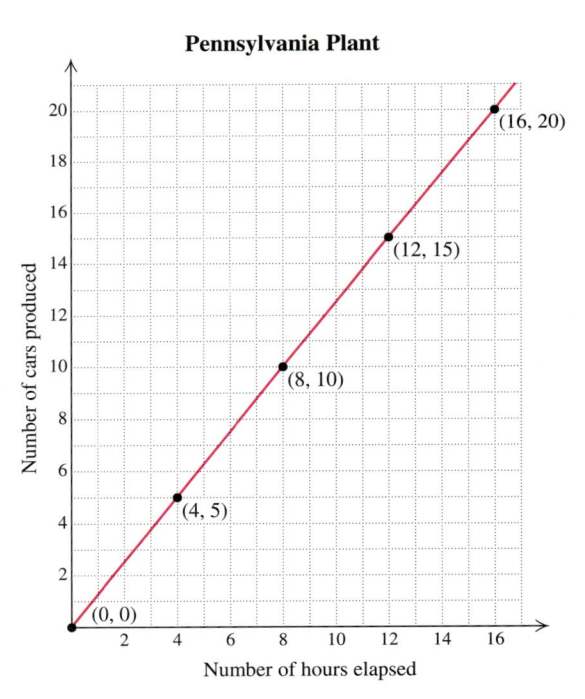

The rates $\frac{3}{2}$ and $\frac{5}{4}$ can also be found using the coordinates of any two points that are on the line. For example, we can use the points (6, 9) and (8, 12) to find the production rate for the Michigan plant. To do so, remember

Michigan Plant

[Graph showing Michigan Plant car production with points (0,0), (4,6), (6,9), (8,12). Vertical change = 6, Horizontal change = 4 between (0,0) and (4,6). Vertical change = 3, Horizontal change = 2 between (6,9) and (8,12). X-axis: Number of hours elapsed. Y-axis: Number of cars produced.]

that these coordinates tell us that after 6 hr, 9 cars have been produced, and after 8 hr, 12 cars have been produced. In the 2 hr between the 6-hr and 8-hr points, 12 − 9, or 3 cars were produced. Thus,

$$\text{Michigan production rate} = \frac{\text{change in number of cars produced}}{\text{corresponding change in time}}$$

$$= \frac{12 - 9 \text{ cars}}{8 - 6 \text{ hr}}$$

$$= \frac{3 \text{ cars}}{2 \text{ hr}} = \frac{3}{2} \text{ cars per hour.}$$

Because the line is straight, the same rate is found using *any* pair of points on the line. For instance, using (0, 0) and (4, 6), we have

$$\text{Michigan production rate} = \frac{6 - 0 \text{ cars}}{4 - 0 \text{ hr}} = \frac{6 \text{ cars}}{4 \text{ hr}} = \frac{3}{2} \text{ cars per hour.}$$

Note that the rate is always the vertical change divided by the associated horizontal change. The situation is similar to one in Section 3.4 in which we found the rate of travel.

EXAMPLE 1 Use the graph of car production at the Pennsylvania plant to find the rate of production.

Pennsylvania Plant

[Graph showing Pennsylvania Plant car production with points (0,0), (8,10), (12,15), (16,20). Vertical change = 10, Horizontal change = 8 between (0,0) and (8,10). Vertical change = 5, Horizontal change = 4 between (12,15) and (16,20). X-axis: Number of hours elapsed. Y-axis: Number of cars produced.]

SOLUTION We can use any two points on the line, such as (12, 15) and (16, 20):

$$\text{Pennsylvania production rate} = \frac{\text{change in number of cars produced}}{\text{corresponding change in time}}$$

$$= \frac{20 - 15 \text{ cars}}{16 - 12 \text{ hr}}$$

$$= \frac{5 \text{ cars}}{4 \text{ hr}}$$

$$= \frac{5}{4} \text{ cars per hour.}$$

As a check, we can use another pair of points, like (0, 0) and (8, 10):

$$\text{Pennsylvania production rate} = \frac{10 - 0 \text{ cars}}{8 - 0 \text{ hr}}$$

$$= \frac{10 \text{ cars}}{8 \text{ hr}}$$

$$= \frac{5}{4} \text{ cars per hour.}$$

When the axes of a graph are simply labeled x and y, it is useful to know the ratio of vertical change to horizontal change. This ratio gives a measure of a line's slant, or *slope,* and tells us the rate at which y is changing with respect to x.

Consider a line passing through the points (2, 3) and (6, 5), as shown below. We find the ratio of vertical change, or *rise,* to horizontal change, or *run,* as follows:

$$\text{Ratio of vertical change to horizontal change} = \frac{\text{change in } y}{\text{change in } x} = \frac{\text{rise}}{\text{run}}$$

$$= \frac{5-3}{6-2}$$

$$= \frac{2}{4}, \text{ or } \frac{1}{2}.$$

Note that these calculations can be performed without viewing a graph.

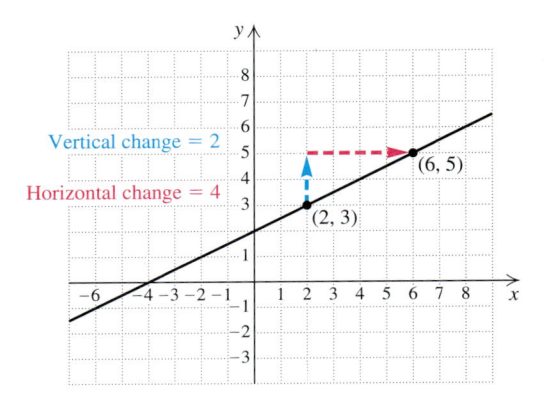

Thus the y-coordinates of points on this line increase at a rate of 2 units for every 4-unit increase in x, 1 unit for every 2-unit increase in x, or $\frac{1}{2}$ unit for every 1-unit increase in x. The slope of the line is $\frac{1}{2}$.

SLOPE

The *slope* of the line containing points (x_1, y_1) and (x_2, y_2) is given by

$$m = \frac{\text{change in } y}{\text{change in } x} = \frac{\text{rise}}{\text{run}} = \frac{y_2 - y_1}{x_2 - x_1}.$$

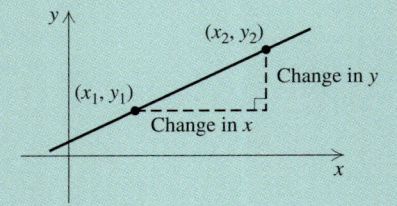

3.5 SLOPE

EXAMPLE 2   Graph the line containing the points $(-4, 3)$ and $(2, -6)$ and find the slope.

SOLUTION   The graph is shown below. From $(-4, 3)$ to $(2, -6)$, the change in y, or rise, is $-6 - 3$, or -9. The change in x, or run, is $2 - (-4)$, or 6. Thus,

$$\text{Slope} = \frac{\text{change in } y}{\text{change in } x}$$
$$= \frac{\text{rise}}{\text{run}}$$
$$= \frac{-6 - 3}{2 - (-4)}$$
$$= \frac{-9}{6}$$
$$= -\frac{9}{6}, \text{ or } -\frac{3}{2}.$$

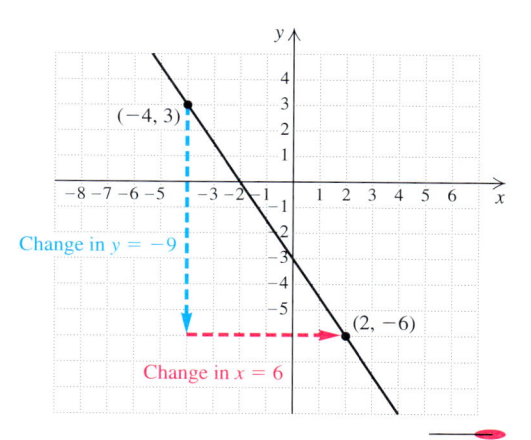

CAUTION!   When we use the formula

$$m = \frac{y_2 - y_1}{x_2 - x_1},$$

it makes no difference which of the two points is considered (x_1, y_1) or if a coordinate is negative. What matters is that we subtract the y-coordinates in the same order that we subtract the x-coordinates.

To illustrate, we reverse *both* of the subtractions in Example 2. The slope is still $-\frac{3}{2}$:

$$\text{Slope} = \frac{\text{change in } y}{\text{change in } x} = \frac{3 - (-6)}{-4 - 2} = \frac{9}{-6} = -\frac{3}{2}.$$

If a line has a positive slope, it slants up from left to right. The larger the slope, the steeper the slant. A line with negative slope slants down from left to right.

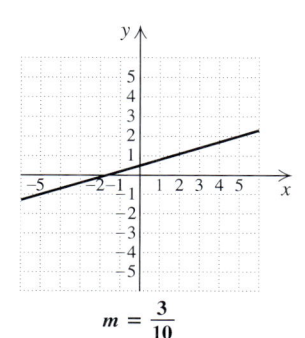

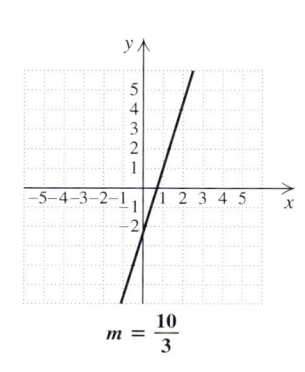

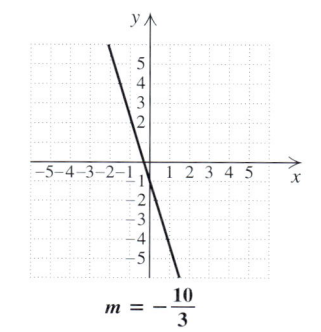

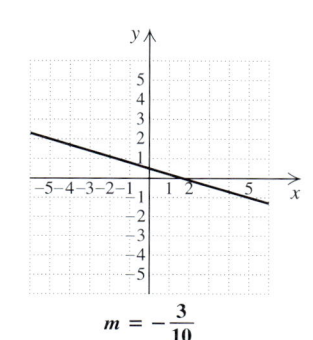

$m = \frac{3}{10}$ $m = \frac{10}{3}$ $m = -\frac{10}{3}$ $m = -\frac{3}{10}$

Horizontal and Vertical Lines

What about the slope of a horizontal or a vertical line?

EXAMPLE 3 Find the slope of the line $y = 4$.

SOLUTION Consider the points $(2, 4)$ and $(-3, 4)$, which are on the line.

The change in $y = 4 - 4$, or 0.
The change in $x = -3 - 2$, or -5.

$$m = \frac{4 - 4}{-3 - 2}$$
$$= \frac{0}{-5}$$
$$= 0$$

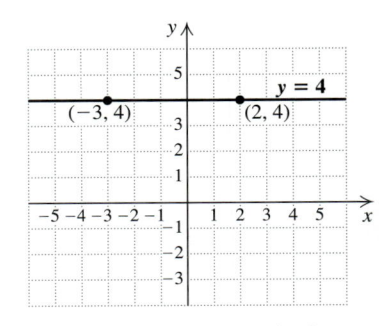

Any two points on a horizontal line have the same y-coordinate. Thus the change in y is 0, so the slope is 0.

A horizontal line has slope 0.

EXAMPLE 4 Find the slope of the line $x = -3$.

SOLUTION Consider the points $(-3, 4)$ and $(-3, -2)$, which are on the line.

The change in $y = 4 - (-2)$, or 6.
The change in $x = -3 - (-3)$, or 0.

$$m = \frac{4 - (-2)}{-3 - (-3)}$$
$$= \frac{6}{0} \quad \text{(undefined)}$$

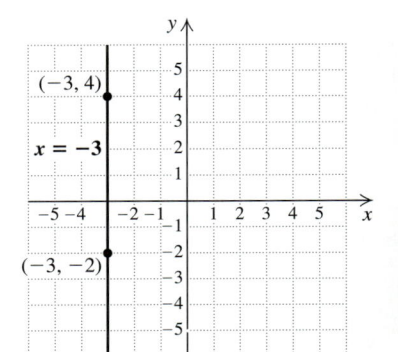

Since division by 0 is not defined, the slope of this line is not defined. The answer to a problem of this type is "The slope of this line is undefined."

The slope of a vertical line is undefined.

Applications

We have seen that slope has many real-world applications, ranging from train speed to automobile production. Some applications use slope to measure steepness. For example, numbers like 2%, 3%, and 6% are often used to represent the **grade** of a road, a measure of a road's steepness. For example, a 3% grade means that for every horizontal distance of 100 ft, the road rises or drops 3 ft. The concept of grade also occurs in skiing or snowboarding, where a 4% grade is considered very tame, but a 40% grade is considered steep.

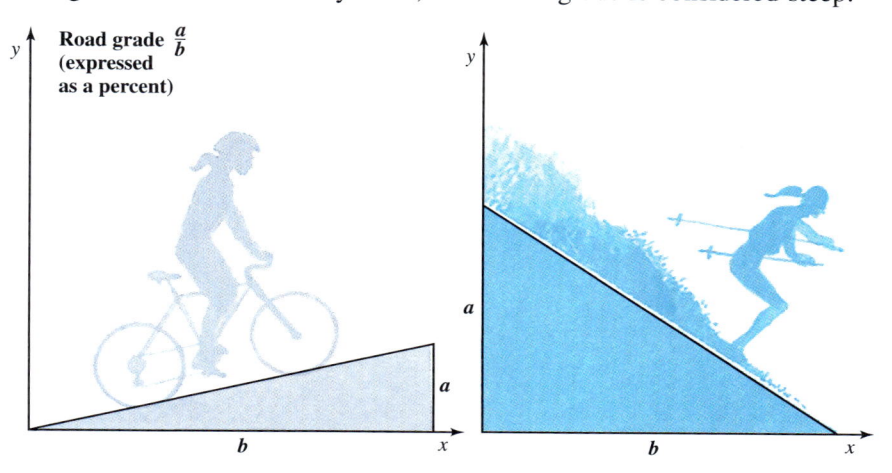

EXAMPLE 5   *Skiing.* Among the steepest skiable terrain in North America, the Headwall on Mount Washington, in New Hampshire, drops 720 ft over a horizontal distance of 900 ft. Find the grade of the Headwall.

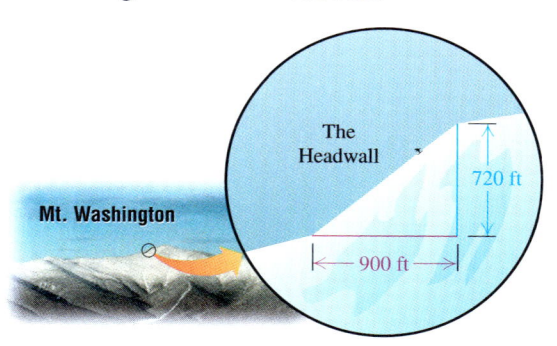

SOLUTION   The grade of the Headwall is its slope, expressed as a percent:

$$m = \frac{720}{900}$$
$$= \frac{8}{10}$$
$$= 80\%.$$

Carpenters use slope when designing stairs, ramps, or roof pitches. Another application occurs in the engineering of a dam—the force or strength of a river depends on how much the river drops over a specified distance. Ratios of change also appear in areas of research ranging from the rate at which a star is cooling (in astronomy) to the growth rate of an animal (in zoology).

EXERCISE SET 3.5

1. Find the rate of change of the U.S. population (based on information in the *Statistical Abstract of the United States*, 1995).

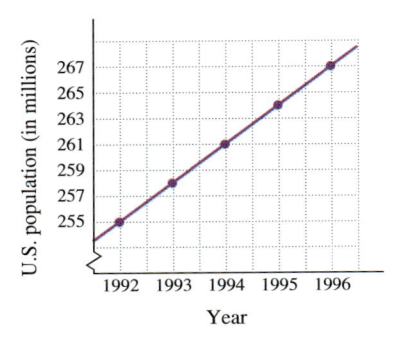

2. Find the rate at which a runner burns calories.

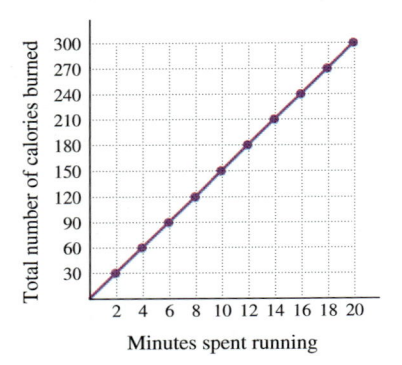

3. Find the rate of change in U.S. defense outlays (based on information in the *Statistical Abstract of the United States*, 1995).

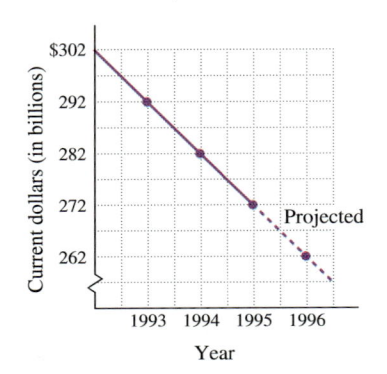

4. Find the rate of change of the tuition and fees at public two-year colleges (based on information in the *Statistical Abstract of the United States*, 1995).

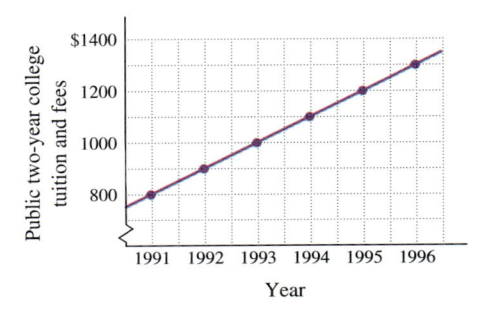

5. Find the rate of change of the tuition and fees at private four-year colleges (based on information in the *Statistical Abstract of the United States*, 1995).

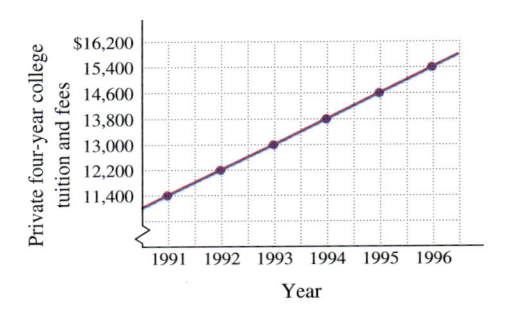

6. Find the rate of change in the number of U.S. farms (based on information in the *Statistical Abstract of the United States*, 1995).

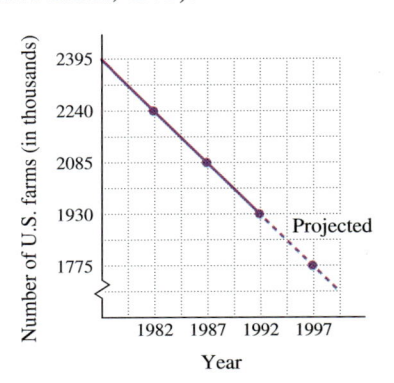

Find the slope, if it is defined, of each line.

7.

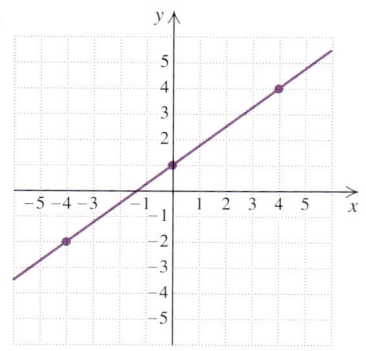

8.

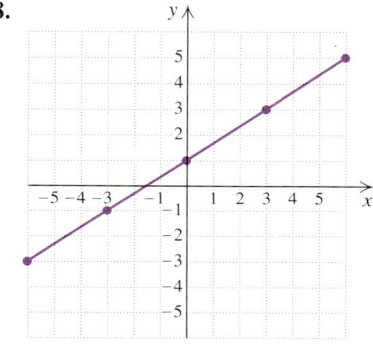

9.

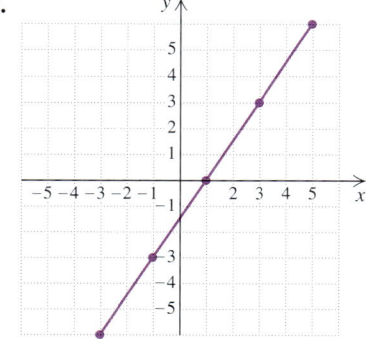

10.

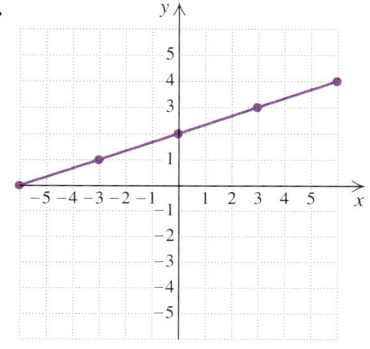

11.

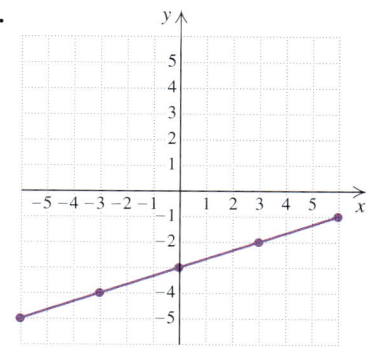

12.

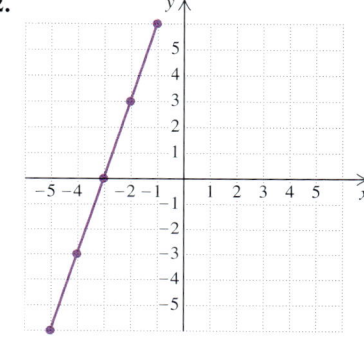

13.

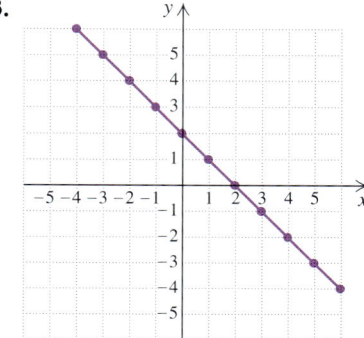

14.

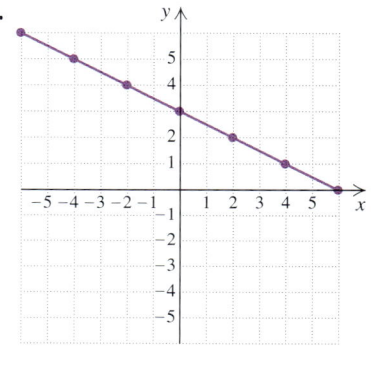

166 CHAPTER 3 • INTRODUCTION TO GRAPHING

15.

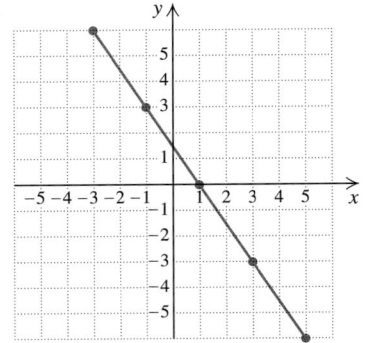

16.

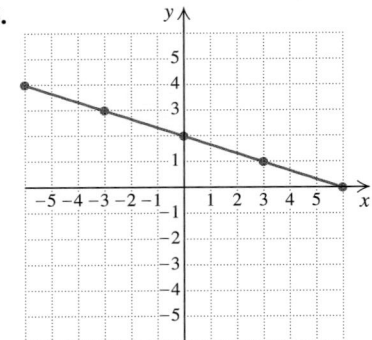

17.

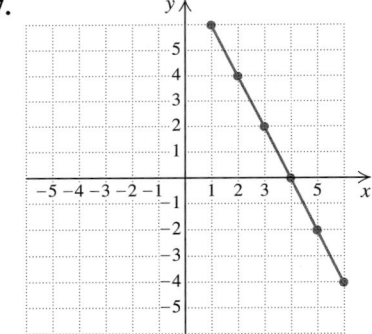

18.

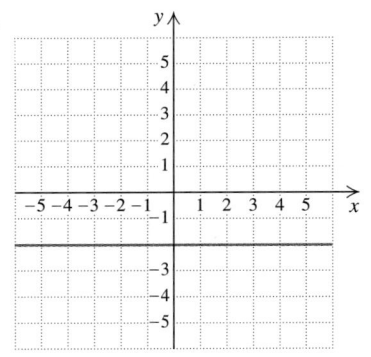

19.

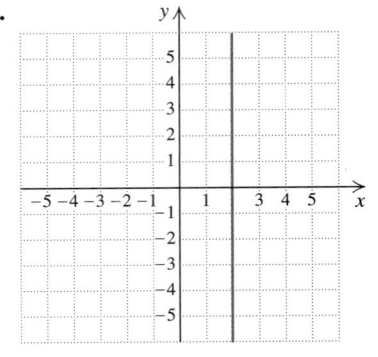

20.

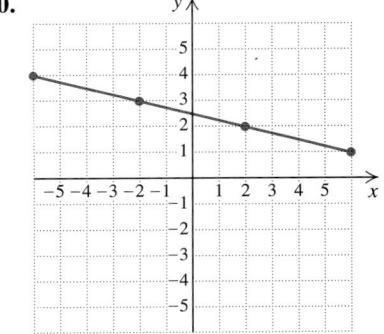

Find the slope of the line containing each given pair of points.

21. (2, 3) and (5, −1) **22.** (4, 1) and (−2, −3)
23. (−2, 4) and (3, 0) **24.** (−4, 2) and (2, −3)
25. (4, 0) and (5, 7) **26.** (3, 0) and (6, 2)
27. (0, 8) and (−3, 10) **28.** (0, 9) and (4, 7)
29. (−2, 3) and (−6, 5) **30.** (−2, 4) and (6, −7)
31. $\left(-2, \frac{1}{2}\right)$ and $\left(-5, \frac{1}{2}\right)$ **32.** (8, −3) and (10, −3)
33. (9, −4) and (9, −7) **34.** (−10, 3) and (−10, 4)

Find the slope of each line.

35. $x = -3$ **36.** $x = -4$
37. $y = 4$ **38.** $y = 17$
39. $x = 9$ **40.** $x = 6$
41. $y = -9$ **42.** $y = -4$

43. *Surveying.* Vermont Route 108 rises 106 m over a horizontal distance of 1325 m. Find the grade of the road.

44. *Navigation.* Capital Rapids drops 54 ft vertically over a horizontal distance of 1080 ft. What is the slope of the rapids?

45. *Architecture.* In order to meet federal standards, a wheelchair ramp should not rise more than 1 ft over a horizontal distance of 12 ft. Express this slope as a grade.

46. *Engineering.* At one point, the Beartooth Highway rises 315 ft over a horizontal distance of 4500 ft. Find the grade of the road.

47. *Carpentry.* Find the slope (or pitch) of the roof.

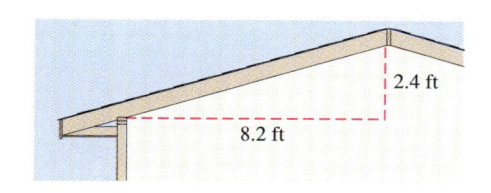

48. *Exercise.* Find the slope (or grade) of the treadmill.

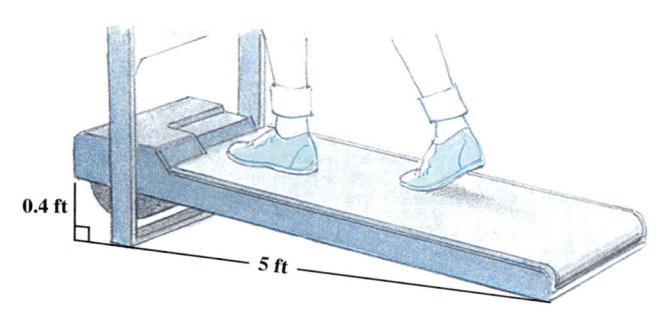

49. *Surveying.* From a base elevation of 9600 ft, Longs Peak, Colorado, rises to a summit elevation of 14,255 ft over a horizontal distance of 15,840 ft. Find the grade of Longs Peak.

50. *Construction.* Public buildings regularly include steps with 7-in. risers and 11-in. treads. Find the grade of such a stairway.

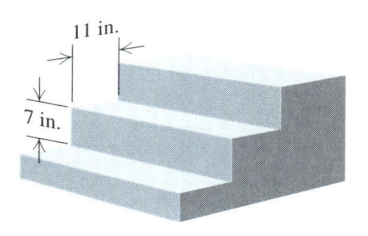

SKILL MAINTENANCE

Simplify.

51. $3^2 - 5^3$ **52.** $4(-3)^4$ **53.** $(-1)^{17}$

Solve.

54. $4x - 7 = 9x$ **55.** $3x = 1 - 5x$

56. $4 - 7x \leq -17$

SYNTHESIS

57. ◆ If one line has a slope of -3 and another has a slope of 2, which line is steeper? Why?

58. ◆ Explain why the order in which coordinates are subtracted to find slope does not matter so long as y-coordinates and x-coordinates are subtracted in the same order.

59. ◆ The graphs in Exercises 4 and 5 appear to have the same slant but different slopes. How is this possible?

60. ◆ From 1991 to 1996, prices on all items rose by about 3% per year. On the basis of this information and the information in Exercise 4, should students at public two-year colleges have complained about their increased tuition and fees? Why or why not?

61. A nonvertical line passes through $(5, -6)$. What numbers could the line have for its slope if the line never enters the first quadrant?

62. A nonvertical line passes through $(3, 4)$. What numbers could the line have for its slope if the line never enters the second quadrant?

63. *Architecture.* Architects often use the equation $x + y = 18$ to determine the height y, in inches, of the riser of a step when the tread is x inches wide. Express the slope of stairs designed with this equation in terms of x.

168 CHAPTER 3 • INTRODUCTION TO GRAPHING

64. By 3:00, Catanya and Chad had already made 46 candles. Forty minutes later, the total reached 64 candles. Find the rate at which Catanya and Chad made candles. Give your answer as a number of candles per hour.

In Exercises 65 and 66, the slope of each line is $-\frac{2}{3}$, but the numbering on one axis is missing. How many units should each tick mark on that unnumbered axis represent?

65.

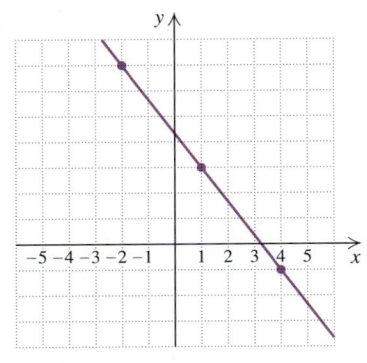

66.

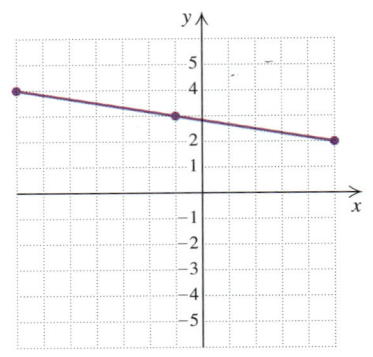

67. Marcy picks apples twice as fast as Ryan. By 4:30, Ryan had already picked 4 bushels of apples. Fifty minutes later, his total reached $5\frac{1}{2}$ bushels. Find Marcy's picking rate. Give your answer in number of bushels per hour.

68. ◆ The points $(-4, -3)$, $(1, 4)$, $(4, 2)$, and $(-1, -5)$ are vertices of a quadrilateral. Use slopes to explain why the quadrilateral is a parallelogram.

69. ◆ Can the points $(-4, 0)$, $(-1, 5)$, $(6, 2)$, and $(2, -3)$ be vertices of a parallelogram? Why or why not?

SUMMARY AND REVIEW 3

KEY TERMS

Bar graph, p. 120
Circle graph, or pie chart, p. 121
Line graph, p. 122
Axes (singular, axis), p. 123
Origin, p. 123
Coordinate, p. 123
Ordered pair, p. 123
Quadrant, p. 124
Graph, p. 130

Linear equation, p. 130
y-intercept, p. 134
x-intercept, p. 139
Rate, p. 149
Slope, p. 160
Rise, p. 160
Run, p. 160
Grade, p. 163

REVIEW EXERCISES: CHAPTER 3

IMPORTANT PROPERTIES AND FORMULAS

To Graph a Linear Equation

1. Select a value for one coordinate and calculate the corresponding value of the other coordinate. Form an ordered pair. This pair is one solution of the equation.
2. Repeat step (1) to find at least one other ordered pair.
3. Plot the ordered pairs and draw a straight line passing through the points. The line represents all solutions of the equation.

$$\text{Slope} = m = \frac{\text{change in } y}{\text{change in } x} = \frac{\text{rise}}{\text{run}} = \frac{y_2 - y_1}{x_2 - x_1}$$

Horizontal line: Slope is 0.
Vertical line: Slope is undefined.

REVIEW EXERCISES

The following circle graph shows the reasons people buy sport utility vehicles.

Sport Utility Vehicle Popularity

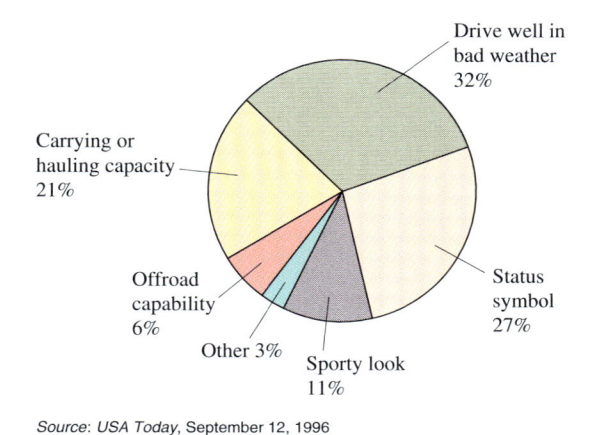

- Drive well in bad weather 32%
- Carrying or hauling capacity 21%
- Offroad capability 6%
- Other 3%
- Sporty look 11%
- Status symbol 27%

Source: USA Today, September 12, 1996

1. One year, Bartlet Motors sold 150 sport utility vehicles. Estimate the number purchased because they handle well in bad weather.

2. About 10% of the 15 million cars and light trucks sold in the United States in 1995 were sport utility vehicles. How many were purchased as a status symbol?

Plot each point.

3. (2, −3)   4. (1, 0)   5. (2, 4)

In which quadrant is each point located?

6. (5, −10)   7. (−16.5, −20.3)   8. (−14, 7)

Find the coordinates of each point in the figure.

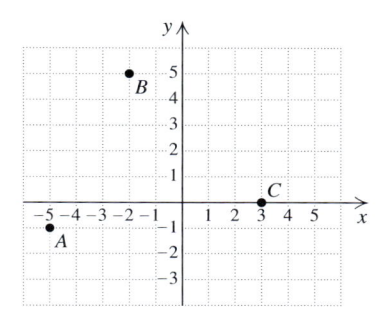

9. A   10. B   11. C

Determine whether the equation $y = 2x - 5$ has each ordered pair as a solution.

12. $(-3, 1)$ **13.** $(3, 1)$

14. Show that the ordered pairs $(0, -3)$ and $(2, 1)$ are solutions of the equation $2x - y = 3$. Then use the graph of the two points to determine another solution. Answers may vary.

Graph.

15. $y = x - 5$ **16.** $y = -\frac{1}{4}x$

17. $y = -x + 4$ **18.** $4x + y = 3$

19. $4x + 5 = 3$ **20.** $5x - 2y = 10$

Find the x- and y-intercepts. Do not graph.

21. $x - 2y = 6$ **22.** $-2x + 4y = 9$

23. Write an equation for the graph.

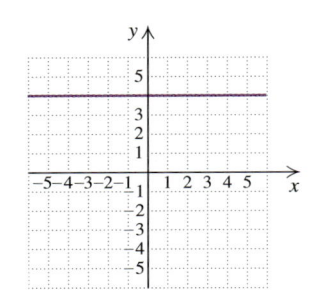

24. Translate to an equation containing two variables. State what each variable represents.

>Jan has a total of 15 wrenches and screwdrivers in her toolbox.

25. Kitchen designers recommend choosing a refrigerator on the basis of the number of residents n in the home. The appropriate size s, in cubic feet (ft³), is given by $s = \frac{3}{2}n + 13$. Graph the equation, using n as the first coordinate. Then use the graph to estimate how large a refrigerator is needed for 3 roommates sharing an apartment.

26. A family of 5 is buying a home that comes with an 18-ft³ refrigerator. Use the graph from Exercise 25 to determine whether the refrigerator will meet their needs.

27. The Amtrak Pioneer leaves Chicago at 3:05 P.M. It reaches Omaha, 501 mi away, at 12:12 A.M. It also reaches Denver, 1037 mi from Chicago, at 10:12 A.M. Determine the average speed of the train for the trip between Omaha and Denver.

Find the slope of each line.

28.

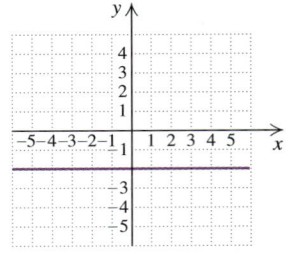

29.

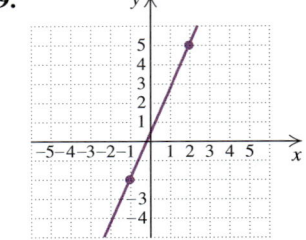

30.

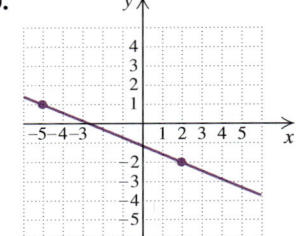

Find the slope of the line containing the given pair of points.

31. $(6, 8)$ and $(-2, -4)$

32. $(5, 1)$ and $(-1, 1)$

33. $(-3, 0)$ and $(-3, 5)$

34. $(-8.3, 4.6)$ and $(-9.9, 1.4)$

35. A road drops 369.6 ft vertically over a horizontal distance of 5280 ft. What is the grade of the road?

SKILL MAINTENANCE

36. Add and simplify: $\frac{3}{8} + \frac{5}{12}$.

37. Simplify: $\frac{26}{34}$.

38. Solve: $2(x - 3) = 10$.

39. Solve $A = \frac{m + n}{2}$ for m.

SYNTHESIS

40. ◆ Describe two ways in which a small business might make use of graphs.

41. ◆ Explain why the first coordinate of the y-intercept is always 0.

42. Find the value of m in $y = mx + 3$ such that $(-2, 5)$ is on the graph.

43. Find the value of b in $y = -5x + b$ such that $(3, 4)$ is on the graph.

44. Find the area and the perimeter of a rectangle for which $(-2, 2)$, $(7, 2)$, and $(7, -3)$ are three of the vertices.

45. Find three solutions of $y = 4 - |x|$.

CHAPTER TEST 3

The following bar graph shows the number of U.S. students who studied in various countries in 1994.

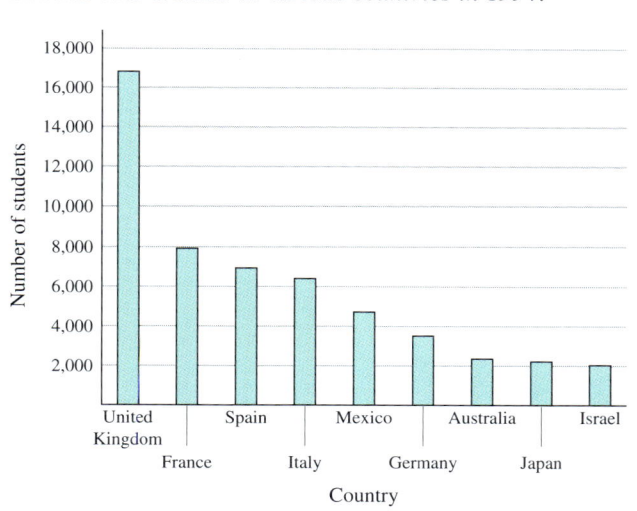

Source: USA Today, September 13, 1996

1. Spanish is spoken in both Spain and Mexico. Approximately how many students studied in these countries?

2. Approximately 76,000 U.S. students studied abroad in 1994. What percentage studied in the United Kingdom?

In which quadrant is each point located?

3. $\left(-\frac{1}{2}, 7\right)$

4. $(-5, -6)$

Find the coordinates of each point in the figure.

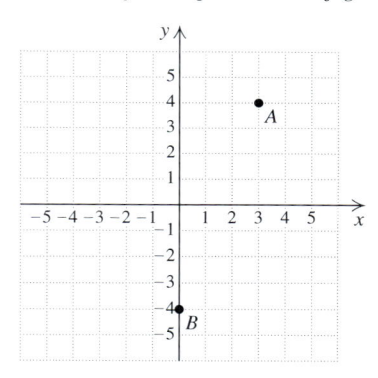

5. A

6. B

7. Show that the ordered pairs $(3, 11)$ and $(-1, 3)$ are solutions of the equation $y - 2x = 5$. Then use the graph of the two points to determine another solution. Answers may vary.

Graph.

8. $y = 2x - 1$
9. $2x - 4y = -8$
10. $y + 1 = 6$
11. $y = \frac{3}{4}x$
12. $2x - y = 3$

Find the x- and y-intercepts. Do not graph.

13. $5x - 3y = 30$

14. $x = 10 - 4y$

172 CHAPTER 3 • INTRODUCTION TO GRAPHING

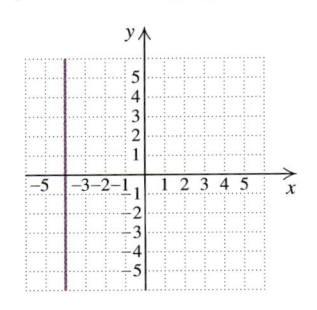

15. Write an equation for the graph.

16. Translate to an equation containing two variables. State what each variable represents.

Greta earns $150 less than twice Alice's salary.

17. The number n of secondary teachers needed each year, in thousands, can be estimated by $n = 35t + 1340$, where t is the number of years since 1994 (*Source*: *USA Today*, September 9, 1996). Graph the equation and use the graph to estimate the number of secondary teachers needed in 2002.

18. Bright Electrical Service charges $25 for a service call plus $45 per hour. Draw a graph that could be used to predict the cost of a service call. Let the horizontal axis represent time and the vertical axis cost. Then use the graph to estimate the length of a service call that cost $60.

Find the slope of the line containing each pair of points.

19. (4, 7) and (4, −1)

20. (9, 2) and (−3, −5)

SKILL MAINTENANCE

21. Divide and simplify: $\frac{3}{5} \div \frac{3}{11}$.

22. Write the prime factorization of 135.

23. Solve: $\frac{1}{3} + 2p = 4p - \frac{1}{5}$.

24. Solve $mx = b - nx$ for x.

SYNTHESIS

25. A diagonal of a square connects the points (−3, −1) and (2, 4). Find the area and the perimeter of the square.

26. Write an equation of a line parallel to the x-axis and 3 units above it.

CUMULATIVE REVIEW 1–3

1. Evaluate $\dfrac{x}{2y}$ for $x = 60$ and $y = 2$.
2. Multiply: $3(4x - 5y + 7)$.
3. Factor: $15x - 9y + 3$.
4. Find the prime factorization of 42.
5. Find decimal notation: $\dfrac{9}{20}$.
6. Find the absolute value: $|-4|$.
7. Find the opposite of $-\dfrac{1}{4}$.
8. Find the reciprocal of $-\dfrac{1}{4}$.
9. Combine like terms: $2x - 5y + (-3x) + 4y$.
10. Find decimal notation: 78.5%.

Simplify.

11. $\dfrac{3}{5} - \dfrac{5}{12}$
12. $3.4 + (-0.8)$
13. $(-2)(-1.4)(2.6)$
14. $\dfrac{3}{8} \div \left(-\dfrac{9}{10}\right)$
15. $2 - [32 \div (4 + 2^2)]$
16. $-5 + 16 \div 2 \cdot 4$
17. $y - (3y + 7)$
18. $3(x - 1) - 2[x - (2x + 7)]$

Solve.

19. $1.5 = 2.7 + x$
20. $\dfrac{2}{7}x = -6$
21. $5x - 9 = 36$
22. $\dfrac{2}{3} = \dfrac{-m}{10}$
23. $5.4 - 1.9x = 0.8x$
24. $x - \dfrac{7}{8} = \dfrac{3}{4}$
25. $2(2 - 3x) = 3(5x + 7)$
26. $\dfrac{1}{4}x - \dfrac{2}{3} = \dfrac{3}{4} + \dfrac{1}{3}x$
27. $y + 5 - 3y = 5y - 9$
28. $x - 28 < 20 - 2x$
29. $2(x + 2) \geq 5(2x + 3)$
30. Solve $A = 2\pi rh + \pi r^2$ for h.
31. In which quadrant is the point $(3, -1)$ located?
32. Graph on a number line: $-1 < x \leq 2$.

Graph.

33. $y = -2$
34. $2x + 5y = 10$
35. $y = -2x + 1$
36. $y = \dfrac{2}{3}x$

Find the coordinates of the x- and y-intercepts. Do not graph.

37. $2x - 7y = 21$
38. $y = 4x + 5$

Solve.

39. Each year 8 million Americans donate blood. This is 5% of those healthy enough to do so (*Source: Indianapolis Star*, 10/6/96). How many Americans are eligible to donate blood?

40. There are 117 million Americans with either O-positive or O-negative blood. Those with O-positive blood outnumber those with O-negative blood by 85.8 million. How many Americans have O-negative blood?

41. Tina paid $126 for a cordless drill, including a 5% sales tax. How much did the drill itself cost?

42. A 143-m wire is cut into three pieces. The second is 3 m longer than the first. The third is four fifths as long as the first. How long is each piece?

43. Cory's contract stipulates that he cannot work more than 40 hr per week. For the first 4 days of one week, he worked 7, 10, 9, and 6 hr. How many hours can he work the fifth day and not violate his contract?

44. Translate to an equation containing two variables. Be sure to state what each variable represents.

    The steak cost $5 more than 3 times the cost of the chicken.

45. The cost c of a phone line for a business, in hundreds of dollars, is given by $c = \dfrac{3}{4}t + 3$, where t is the number of months that the line has been in service. Graph the equation and use the graph to estimate the cost of a line that has been in service for 9 months.

46. Sparkle's Cleaning charges $20 a visit plus $15 an hour for commercial cleaning jobs. Draw a graph that can be used to predict the cost of a cleaning job. Let the horizontal axis represent time and the vertical axis cost. Then use the graph to estimate the length of a service call that costs $60.

47. Find the slope of the line containing the points $(-4, 1)$ and $(2, -1)$.

SYNTHESIS

48. Paula's salary at the end of a year is $26,780. This reflects a 4% salary increase that preceded a 3% cost-of-living adjustment during the year. What was her salary at the beginning of the year?

Solve.

49. $4|x| - 13 = 3$

50. $4(x + 2) = 4(x - 2) + 16$

51. $0(x + 3) + 4 = 0$

52. $\dfrac{2 + 5x}{4} = \dfrac{11}{28} + \dfrac{8x + 3}{7}$

53. $5(7 + x) = (x + 7)5$

54. Solve $p = \dfrac{2}{m + Q}$ for Q.

CHAPTER 4

Polynomials

4.1 Exponents and Their Properties
4.2 Polynomials
4.3 Addition and Subtraction of Polynomials
4.4 Multiplication of Polynomials
4.5 Special Products
4.6 Polynomials in Several Variables
4.7 Division of Polynomials
4.8 Negative Exponents and Scientific Notation
SUMMARY AND REVIEW
TEST

AN APPLICATION

For jumps that exceed 13 sec, the polynomial $173t - 369$ can be used to approximate the distance, in feet, that a skydiver has fallen in t seconds. Approximately how far has a skydiver fallen 20 sec after jumping from a plane?

THIS PROBLEM APPEARS AS EXERCISE 64 IN SECTION 4.2.

More information on skydiving is available at
http://hepg.awl.com/be/elem_5

Every time I log a jump, I use math to calculate my free-fall time using the altitudes at which I exited the plane and opened my parachute and my average free-fall speed.

PAULA PHILBROOK
Skydiving Instructor
Pepperell, Massachusetts

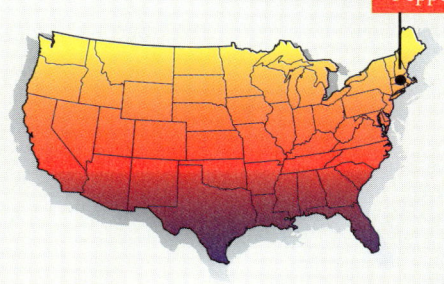

CHAPTER 4 • POLYNOMIALS

Algebraic expressions like $16t^2$, $5a^2 - 45$, and $3x^2 - 7x + 5$ are called polynomials. Polynomials occur frequently in applications and appear in most branches of mathematics. Thus learning to add, subtract, multiply, and divide polynomials is an important part of most courses in elementary algebra and is the focus of this chapter.

4.1 Exponents and Their Properties

Multiplying Powers with Like Bases • Dividing Powers with Like Bases • Zero as an Exponent • Raising a Power to a Power • Raising a Product or a Quotient to a Power

In Section 4.2, we begin our study of polynomials. Before doing so, however, we must develop some rules for manipulating exponents.

Multiplying Powers with Like Bases

We know that an expression like a^3 means $a \cdot a \cdot a$ and that a^1 means a. Now consider multiplying powers with like bases:

$a^3 \cdot a^2 = (a \cdot a \cdot a)(a \cdot a)$ There are three factors in a^3; two factors in a^2.
$ = a \cdot a \cdot a \cdot a \cdot a$ Using an associative law
$ = a^5$.

Note that the exponent in a^5 is the sum of the exponents in $a^3 \cdot a^2$. That is, $3 + 2 = 5$. Similarly,

$b^4 \cdot b^3 = (b \cdot b \cdot b \cdot b)(b \cdot b \cdot b)$
$ = b^7,$ where $4 + 3 = 7$.

Adding the exponents gives the correct result.

THE PRODUCT RULE

For any number a and any positive integers m and n,

$a^m \cdot a^n = a^{m+n}$.

(To multiply with exponential notation, if the bases are the same, keep the base and add the exponents.)

4.1 EXPONENTS AND THEIR PROPERTIES   177

EXAMPLE 1   Multiply and simplify each of the following. (Here "simplify" means express the product as one base to a power whenever possible.)

a) $x^2 \cdot x^9$ **b)** $8^4 \cdot 8^3$ **c)** $(r+s)(r+s)^6$
d) $m^5 m^{10} m^3$ **e)** $(a^3 b^2)(a^3 b^5)$

SOLUTION

a) $x^2 \cdot x^9 = x^{2+9}$ Adding exponents: $a^m \cdot a^n = a^{m+n}$
             $= x^{11}$

b) $8^4 \cdot 8^3 = 8^{4+3}$
           $= 8^7$

c) $(r+s)(r+s)^6 = (r+s)^1(r+s)^6 = (r+s)^{1+6}$
                     $= (r+s)^7$ Note: $(r+s)^7 \neq r^7 + s^7$.

d) $m^5 m^{10} m^3 = m^{5+10+3}$ Adding exponents
               $= m^{18}$

e) $(a^3 b^2)(a^3 b^5) = a^3 b^2 a^3 b^5$ Using an associative law
                    $= a^3 a^3 b^2 b^5$ Using a commutative law
                    $= a^6 b^7$ Adding exponents

Dividing Powers with Like Bases

The following suggests a rule for dividing powers with like bases, such as a^5/a^2:

$$\frac{a^5}{a^2} = \frac{a \cdot a \cdot a \cdot a \cdot a}{a \cdot a} = \frac{a \cdot a \cdot a \cdot a \cdot a}{1 \cdot a \cdot a} = \frac{a \cdot a \cdot a}{1} \cdot \frac{a \cdot a}{a \cdot a} = \frac{a \cdot a \cdot a}{1} \cdot 1$$
$$= a \cdot a \cdot a = a^3.$$

Note that the exponent in a^3 is the difference of the exponents in a^5/a^2. Similarly,

$$\frac{x^4}{x^3} = \frac{x \cdot x \cdot x \cdot x}{x \cdot x \cdot x} = \frac{x}{1} \cdot \frac{x \cdot x \cdot x}{x \cdot x \cdot x} = \frac{x}{1} \cdot 1 = x^1, \quad \text{or } x.$$

Subtracting the exponents gives the correct result.

THE QUOTIENT RULE

For any nonzero number a and any positive integers m and n for which $m > n$,

$$\frac{a^m}{a^n} = a^{m-n}.$$

(To divide with exponential notation, if the bases are the same, keep the base and subtract the exponent of the denominator from the exponent of the numerator.)

178 CHAPTER 4 • POLYNOMIALS

EXAMPLE 2   Divide and simplify. (Here "simplify" means express the quotient as one base to a power whenever possible.)

a) $\dfrac{x^8}{x^2}$ 　　 b) $\dfrac{6^5}{6^3}$ 　　 c) $\dfrac{(5a)^{12}}{(5a)^4}$ 　　 d) $\dfrac{p^5 q^7}{p^2 q^5}$

SOLUTION

a) $\dfrac{x^8}{x^2} = x^{8-2}$ 　　 Subtracting exponents: $\dfrac{a^m}{a^n} = a^{m-n}$

$\phantom{\dfrac{x^8}{x^2}} = x^6$

b) $\dfrac{6^5}{6^3} = 6^{5-3}$

$\phantom{\dfrac{6^5}{6^3}} = 6^2$

c) $\dfrac{(5a)^{12}}{(5a)^4} = (5a)^{12-4} = (5a)^8$

d) $\dfrac{p^5 q^7}{p^2 q^5} = \dfrac{p^5}{p^2} \cdot \dfrac{q^7}{q^5} = p^{5-2} \cdot q^{7-5} = p^3 q^2$

Zero as an Exponent

The quotient rule can be used to help determine what 0 should mean when it appears as an exponent. Consider a^4/a^4, where a is nonzero. Since the numerator and the denominator are the same,

$\dfrac{a^4}{a^4} = 1.$

On the other hand, using the quotient rule would give us

$\dfrac{a^4}{a^4} = a^{4-4} = a^0.$ 　　 Subtracting exponents

Since $a^0 = a^4/a^4 = 1$, this suggests that $a^0 = 1$ for any nonzero value of a.

THE EXPONENT ZERO

For any real number a, $a \neq 0$,

$a^0 = 1.$

(Any nonzero number raised to the 0 power is 1.)

EXAMPLE 3   Simplify: (a) 1957^0; (b) $(-7)^0$; (c) $(-1)7^0$; (d) $(3x)^0$.

SOLUTION

a) $1957^0 = 1$ 　　 Any nonzero number raised to the 0 power is 1.

b) $(-7)^0 = 1$ 　　 Any nonzero number raised to the 0 power is 1.

c) We have

$(-1)7^0 = (-1)1 = -1.$

Since multiplying by -1 is the same as finding the opposite, the expression $(-1)7^0$ could have been written as -7^0. Note that although $-7^0 = -1$, part (b) shows that $(-7)^0 = 1$.

d) The parentheses indicate that the base is $3x$. Thus,

$(3x)^0 = 1$ for any $x \neq 0.$

To see why 0^0 is not defined, note that $0^0 = 0^{1-1} = 0^1/0^1 = 0/0$. As we saw in Section 1.7, $0/0$ cannot be determined. Thus, 0^0 is not defined either. For this text, we will assume that expressions like a^m do not represent 0^0.

Raising a Power to a Power

Consider an expression like $(5^2)^4$.

$(5^2)^4 = (5^2)(5^2)(5^2)(5^2)$ There are four factors of 5^2.
$= (5 \cdot 5)(5 \cdot 5)(5 \cdot 5)(5 \cdot 5)$
$= 5 \cdot 5 \cdot 5 \cdot 5 \cdot 5 \cdot 5 \cdot 5 \cdot 5$ Using an associative law
$= 5^8.$

Note that in this case we could have multiplied the exponents:

$(5^2)^4 = 5^{2 \cdot 4} = 5^8.$

Likewise, $(y^7)^3 = (y^7)(y^7)(y^7) = y^{21}$. Once again, we get the same result if we multiply the exponents:

$(y^7)^3 = y^{7 \cdot 3} = y^{21}.$

THE POWER RULE

For any number a and any whole numbers m and n,

$(a^m)^n = a^{mn}.$

(To raise a power to a power, multiply the exponents and leave the base unchanged.)

EXAMPLE 4

Simplify: **(a)** $(m^2)^5$; **(b)** $(3^5)^4$.

SOLUTION

a) $(m^2)^5 = m^{2 \cdot 5}$ Multiplying exponents: $(a^m)^n = a^{mn}$
$= m^{10}$

b) $(3^5)^4 = 3^{5 \cdot 4}$
$= 3^{20}$

Raising a Product or a Quotient to a Power

When an expression inside parentheses is raised to a power, the inside expression is the base. Let us compare $2a^3$ and $(2a)^3$:

$2a^3 = 2 \cdot a \cdot a \cdot a;$ The base is a.

$(2a)^3 = (2a)(2a)(2a)$ The base is $2a$.

$= (2 \cdot 2 \cdot 2)(a \cdot a \cdot a)$ Using an associative and a commutative law

$= 2^3 a^3$

$= 8a^3.$

We see that $2a^3$ and $(2a)^3$ are *not* equivalent. Note too that $(2a)^3$ can be simplified by cubing each factor. This leads to the following rule for raising a product to a power.

RAISING A PRODUCT TO A POWER

For any numbers a and b and any whole number n,

$(ab)^n = a^n b^n.$

(To raise a product to the nth power, raise each factor to the nth power.)

EXAMPLE 5

Simplify: **(a)** $(4a)^3$; **(b)** $(5x^4)^2$; **(c)** $(-4a^5 b^3)^3$.

SOLUTION

a) $(4a)^3 = 4^3 a^3 = 64a^3$ Raising each factor to the third power and simplifying

b) $(5x^4)^2 = 5^2(x^4)^2$ Raising each factor to the second power

$= 25x^8$ Simplifying 5^2; using the power rule

c) $(-4a^5 b^3)^3 = (-4)^3 (a^5)^3 (b^3)^3$ Cubing each factor

$= -64 a^{15} b^9$ A negative number raised to an odd power is negative; using the power rule.

CAUTION! The rule $(ab)^n = a^n b^n$ applies only to products raised to a power, not to sums or differences. For example, $(3 + 4)^2 \neq 3^2 + 4^2$ since $7^2 \neq 9 + 16$.

There is a similar rule for raising a quotient to a power.

4.1 EXPONENTS AND THEIR PROPERTIES

RAISING A QUOTIENT TO A POWER

For any numbers a and b, $b \neq 0$, and any whole number n,
$$\left(\frac{a}{b}\right)^n = \frac{a^n}{b^n}.$$

(To raise a quotient to a power, raise the numerator to the power and divide by the denominator to the power.)

EXAMPLE 6 Simplify: **(a)** $\left(\frac{x}{5}\right)^2$; **(b)** $\left(\frac{5}{a^4}\right)^3$; **(c)** $\left(\frac{3a^4}{b^3}\right)^2$.

SOLUTION

a) $\left(\frac{x}{5}\right)^2 = \frac{x^2}{5^2} = \frac{x^2}{25}$ Squaring the numerator and the denominator

b) $\left(\frac{5}{a^4}\right)^3 = \frac{5^3}{(a^4)^3}$
$= \frac{125}{a^{4 \cdot 3}} = \frac{125}{a^{12}}$

c) $\left(\frac{3a^4}{b^3}\right)^2 = \frac{(3a^4)^2}{(b^3)^2}$
$= \frac{3^2(a^4)^2}{b^{3 \cdot 2}} = \frac{9a^8}{b^6}$

In the following summary of definitions and rules, we assume that no denominators are 0 and that 0^0 is not considered.

DEFINITIONS AND PROPERTIES OF EXPONENTS

For any whole numbers m and n,

1 as an exponent:	$a^1 = a$
0 as an exponent:	$a^0 = 1$
The Product Rule:	$a^m \cdot a^n = a^{m+n}$
The Quotient Rule:	$\dfrac{a^m}{a^n} = a^{m-n}$
The Power Rule:	$(a^m)^n = a^{mn}$
Raising a product to a power:	$(ab)^n = a^n b^n$
Raising a quotient to a power:	$\left(\dfrac{a}{b}\right)^n = \dfrac{a^n}{b^n}$

EXERCISE SET 4.1

Simplify. (Express each product or quotient as one base to a power whenever possible.)

1. $m^4 \cdot m^6$
2. $3^5 \cdot 3^2$
3. $8^5 \cdot 8^9$
4. $n^3 \cdot n^{20}$
5. $x^4 \cdot x^3$
6. $y^7 \cdot y^9$
7. $5^7 \cdot 5^0$
8. $t^0 \cdot t^{16}$
9. $(3y)^4(3y)^8$
10. $(2t)^8(2t)^{17}$
11. $(5t)(5t)^6$
12. $(8x)^0(8x)^1$
13. $(a^2b^7)(a^3b^2)$
14. $(m-3)^4(m-3)^5$
15. $(x+1)^5(x+1)^7$
16. $(a^8b^3)(a^4b)$
17. $r^3 \cdot r^7 \cdot r^2$
18. $s^4 \cdot s^5 \cdot s^2$
19. $(xy^4)(xy)^3$
20. $(a^3b)(ab)^4$
21. $\dfrac{7^5}{7^2}$
22. $\dfrac{4^7}{4^3}$
23. $\dfrac{x^{15}}{x^3}$
24. $\dfrac{a^{10}}{a^2}$
25. $\dfrac{y^9}{y^5}$
26. $\dfrac{x^{12}}{x^{11}}$
27. $\dfrac{(5a)^7}{(5a)^6}$
28. $\dfrac{(3m)^9}{(3m)^8}$
29. $\dfrac{(x+y)^8}{(x+y)^3}$
30. $\dfrac{(a-b)^{13}}{(a-b)^4}$
31. $\dfrac{18m^5}{6m^2}$
32. $\dfrac{30n^7}{6n^3}$
33. $\dfrac{a^9b^7}{a^2b}$
34. $\dfrac{r^{10}s^7}{r^3s^0}$
35. $\dfrac{m^9n^8}{m^0n^4}$
36. $\dfrac{a^{10}b^{12}}{a^2b^3}$

Simplify.

37. x^0 when $x = 13$
38. y^0 when $y = 38$
39. $5x^0$ when $x = -4$
40. $7m^0$ when $m = 1.7$
41. n^0, for any $n \neq 0$
42. t^0, for any $t \neq 0$
43. $9^1 - 9^0$
44. $7^0 - 7^1$

Simplify. Answers should not contain parentheses.

45. $(x^3)^4$
46. $(a^4)^6$
47. $(5^8)^2$
48. $(2^5)^3$
49. $(m^7)^5$
50. $(n^9)^2$
51. $(a^{25})^3$
52. $(a^3)^{25}$
53. $(7x)^2$
54. $(5a)^2$
55. $(-2a)^3$
56. $(-3x)^3$
57. $(4m^3)^2$
58. $(5n^4)^2$
59. $(a^2b)^7$
60. $(xy^4)^9$
61. $(a^3b^2)^5$
62. $(m^4n^5)^6$
63. $(-5x^4y^5)^2$
64. $(-3a^5b^7)^4$
65. $\left(\dfrac{a}{4}\right)^3$
66. $\left(\dfrac{3}{x}\right)^4$
67. $\left(\dfrac{7}{5a}\right)^2$
68. $\left(\dfrac{4x}{3}\right)^3$
69. $\left(\dfrac{a^4}{b^3}\right)^5$
70. $\left(\dfrac{x^5}{y^2}\right)^7$
71. $\left(\dfrac{y^3}{2}\right)^2$
72. $\left(\dfrac{a^5}{3}\right)^3$
73. $\left(\dfrac{x^2y}{z^3}\right)^4$
74. $\left(\dfrac{x^3}{y^2z}\right)^5$
75. $\left(\dfrac{a^3}{-2b^5}\right)^4$
76. $\left(\dfrac{x^5}{-3y^3}\right)^4$
77. $\left(\dfrac{2a^2}{3b^4}\right)^3$
78. $\left(\dfrac{3x^5}{4y^3}\right)^2$
79. $\left(\dfrac{4x^3y^5}{3z^7}\right)^2$
80. $\left(\dfrac{5a^7}{2b^5c}\right)^3$

SKILL MAINTENANCE

81. Factor: $3s + 3t + 24$.
82. Factor: $-7x - 14$.
83. Combine like terms: $9x + 2y - 4x - 2y$.
84. 24 is what percent of 64?
85. Graph: $y = x - 5$.
86. Graph: $2x + y = 8$.

SYNTHESIS

87. ◆ Under what conditions does a^n represent a negative number? Why?
88. ◆ Using the quotient rule, explain why 9^0 is 1.
89. ◆ Suppose that the width of a square is 3 times the width of a second square. How do the areas of the squares compare? Why?

90. ◆ Explain in your own words when exponents should be added and when they should be multiplied.

Simplify.

91. $(y^{2x})(y^{3x})$
92. $a^{5k} \div a^{3k}$
93. $\dfrac{a^{6t}(a^{7t})}{a^{9t}}$
94. $\dfrac{\left(\frac{1}{2}\right)^4}{\left(\frac{1}{2}\right)^5}$

Replace ▨ with $>$, $<$, or $=$ to write a true sentence.

95. 3^5 ▨ 3^4
96. 4^2 ▨ 4^3
97. 4^3 ▨ 5^3
98. 4^3 ▨ 3^4
99. 9^7 ▨ 3^{13}
100. 25^8 ▨ 125^5

Find a value of the variable that shows that the two expressions are not *equivalent. Answers may vary.*

101. $(a + 5)^2$; $a^2 + 5^2$ **102.** $3x^2$; $(3x)^2$

103. $\dfrac{a+7}{7}$; a **104.** $\dfrac{t^6}{t^2}$; t^3

105. Solve for x:
$$\dfrac{t^{38}}{t^x} = t^x.$$

Interest compounded annually. If a principal P is invested at interest rate r, compounded annually, in t years it will grow to an amount A given by
$$A = P(1 + r)^t.$$

106. Suppose that $10,400 is invested at 8.5% compounded annually. How much is in the account at the end of 5 years?

107. Suppose that $20,800 is invested at 4.5%, compounded annually. How much is in the account at the end of 6 years?

108. Suppose that the width of a cube is twice the width of a second cube. How do the volumes of the cubes compare? Why?

COLLABORATIVE C·O·R·N·E·R

Focus: Estimation and properties of exponents

Time: 20–30 minutes

Group size: 2

According to legend, the inventor of chess was offered compensation for his game by an enthusiastic, chess-playing king. The inventor slyly asked that the king place *one* grain of gold on a corner square of his chessboard, *two* grains on the next square, *four* grains on the next, *eight* grains on the next, and so on, doubling the number of grains in each square until each of the 64 squares of the chessboard had been accounted for.

ACTIVITY

1. Express, as a power of 2, the number of grains of gold on each of the first 7 squares and also on the last (64th) square. Each group member should then estimate the volume of gold that would be used for the 64th square. Would it fill a shoebox? a refrigerator? a garage? something smaller? bigger?
2. When 32 grains of gold are placed side by side, they form a line that is about 1 cm long. Determine the number of grains of gold in one cubic centimeter. How can this be represented as a power of 2?
3. Determine how many cubic centimeters are in one cubic meter. Approximate this number with a power of 2, using the fact that $2^{10} \approx 10^3$.
4. Use the results of parts (2) and (3) to estimate the number of grains of gold in one cubic meter.
5. Use the results of parts (1) and (4) to determine the volume—in cubic meters—of gold that the king would have needed for the 64th square of the chessboard. (Use powers of 2.)
6. Las Vegas, Nevada, a city known for goldseekers, has about 128, or 2^7, square kilometers of land that has not yet been built upon. Using the calculations from part (5), how deep a layer of gold could you spread on this land? Which estimate from part (1) comes closest to this volume?
7. Gold recently sold for $360 per ounce, which is the same as 75.6¢ per grain. Use this to estimate the value of the gold needed for the 64th square. (Can you see why, according to legend, the king had the inventor beheaded?)

184 CHAPTER 4 • POLYNOMIALS

4.2 Polynomials

Terms • Types of Polynomials • Degree and Coefficients • Combining Like Terms • Evaluating Polynomials and Applications

We now examine an important algebraic expression known as a *polynomial*. Certain polynomials have appeared earlier in this text so you already have some experience working with them.

Terms

At this point, we have seen a variety of algebraic expressions like

$$4x^9, \quad 2l + 2w, \quad rt, \quad 5x^2 - 3x + 2, \quad \text{and} \quad -17.$$

Of these expressions, $4x^9$, rt, and -17 are examples of *terms*. A **term** can be a number (say, -3 or 28), a variable (say, x or a), or a product of numbers and/or variables, which may be raised to powers (say, $2x^3$ or $7a^2b$).*

Types of Polynomials

If a term is a number, a variable, or a product of numbers and variables raised to whole-number powers, it is a **monomial** (pronounced mä-nō′mē-əl). Thus the terms $2x^7$, -56, and ab^2c are monomials, but the expressions d/r, $2/x$, and $a^5/(bc^2)$ are not. Other examples of monomials are

$$9, \quad z, \quad -4a^{17}, \quad \text{and} \quad \tfrac{2}{5}x^2y^3.$$

Algebraic expressions like the following are **polynomials**:

$$\tfrac{3}{4}y^5, \quad -2, \quad 5y + 3, \quad 3x^2 + 2x - 5, \quad -7a^3 + 4ab, \quad 6x, \quad 37p^4, \quad x, \quad 0.$$

POLYNOMIAL

A *polynomial* is a monomial or a sum of monomials.

The following algebraic expressions are *not* polynomials:

$$(1) \; \frac{x+3}{x-4}, \quad (2) \; 5x^3 - 2x^2 + \frac{1}{x}, \quad (3) \; \frac{1}{x^3 - 2}.$$

Expressions (1) and (3) are not polynomials because they represent quotients, not sums. Expression (2) is not a polynomial because $1/x$ is not a monomial.

---

*Later in this text, expressions like $5x^{3/2}$ and $2a^{-7}b$ will be discussed. Such expressions are also considered terms.

When a polynomial is written as a sum of monomials, each monomial is called a *term of the polynomial*.

EXAMPLE 1 Identify the terms of the polynomial $3t^4 - 5t^6 - 4t + 2$.

SOLUTION The terms are $3t^4$, $-5t^6$, $-4t$, and 2. We can see this by rewriting all subtractions as additions of opposites:

$$3t^4 - 5t^6 - 4t + 2 = 3t^4 + (-5t^6) + (-4t) + 2.$$

These are the terms of the polynomial.

A polynomial that is composed of two terms is called a **binomial**, whereas those composed of three terms are called **trinomials**. Polynomials with four or more terms have no special name.

Monomials	Binomials	Trinomials	No Special Name
$4x^2$	$2x + 4$	$3t^3 + 4t + 7$	$4x^3 - 5x^2 + xy - 8$
9	$3a^5 + 6bc$	$6x^7 - 8z^2 + 4$	$z^5 + 2z^4 - z^3 + 7z + 3$
$-7a^{19}b^5$	$-9x^7 - 6$	$4x^2 - 6x - \frac{1}{2}$	$4x^6 - 3x^5 + x^4 - x^3 + 2x - 1$

Degree and Coefficients

When a term includes just one variable, the **degree** of that term is the exponent of the variable, or 0 if the term is a constant.*

EXAMPLE 2 Determine the degree of each term: **(a)** $8x^4$; **(b)** $3x$; **(c)** 7.

SOLUTION

a) The degree of $8x^4$ is 4.
b) The degree of $3x$ is 1.   Recall that $x = x^1$.
c) The degree of 7 is 0.   Think of 7 as $7x^0$. Recall that $x^0 = 1$.

The number 8 in the term $8x^4$ is said to be the **coefficient** of that term. The coefficient of $3x$ is 3, and the coefficient for the term 7 is simply 7.

EXAMPLE 3 Identify the coefficient of each term in the polynomial

$$4x^3 - 7x^2y + x - 8.$$

SOLUTION

The coefficient of $4x^3$ is 4.
The coefficient of $-7x^2y$ is -7.
The coefficient of the third term is 1, since $x = 1x$.
The coefficient of -8 is simply -8.

---

*A more detailed definition for terms containing several variables is given in Section 4.6.

The **leading term** of a polynomial is the term of highest degree. Its coefficient is called the **leading coefficient** and its degree is considered the **degree of the polynomial.** To illustrate how this terminology is used, consider the polynomial

$$3x^2 - 8x^3 + 5x^4 + 7x - 6.$$

The *terms* are $\quad\quad\quad\quad 3x^2,\ -8x^3,\ 5x^4,\ 7x,\ \text{and}\ -6.$
The *coefficients* are $\quad\quad 3,\ -8,\ 5,\ 7,\ \text{and}\ -6.$
The *degree of each term* is $\ 2,\ 3,\ 4,\ 1,\ \text{and}\ 0.$
The *leading term* is $5x^4$ and the *leading coefficient* is 5.
The *degree of the polynomial* is 4.

Combining Like Terms

Recall from Section 1.8 that *like,* or *similar, terms* are either constant terms or terms containing the same variable(s) raised to the same power(s). To simplify certain polynomials, we can often *combine,* or *collect,* like terms.

EXAMPLE 4 Identify the like terms in $4x^3 + 5x - 7x^2 + 2x^3 + x^2$.

SOLUTION

Like terms: $\quad 4x^3\ \text{and}\ 2x^3 \quad\quad$ Same variable and exponent
Like terms: $\quad -7x^2\ \text{and}\ \ x^2 \quad\quad$ Same variable and exponent

EXAMPLE 5 Combine like terms.

a) $2x^3 - 6x^3$
b) $5x^2 + 7 + 2x^4 + 4x^2 - 11 - 2x^4$
c) $7a^3 - 5a^2 + 9a^3 + a^2$
d) $\frac{2}{3}x^4 - x^3 - \frac{1}{6}x^4 + \frac{2}{5}x^3 - \frac{3}{10}x^3$

SOLUTION

a) $2x^3 - 6x^3 = (2 - 6)x^3 \quad\quad$ Using the distributive law
$ = -4x^3$

b) $5x^2 + 7 + 2x^4 + 4x^2 - 11 - 2x^4$
$= (5 + 4)x^2 + (2 - 2)x^4 + (7 - 11)$ ⎱ These steps are often
$= 9x^2 + 0x^4 + (-4)$ ⎰ done mentally.
$= 9x^2 - 4$

c) $7a^3 - 5a^2 + 9a^3 + a^2 = 7a^3 - 5a^2 + 9a^3 + 1a^2 \quad$ When a variable to a power appears without a coefficient, we can write in 1.
$ = 16a^3 - 4a^2$

d) $\frac{2}{3}x^4 - x^3 - \frac{1}{6}x^4 + \frac{2}{5}x^3 - \frac{3}{10}x^3 = \left(\frac{2}{3} - \frac{1}{6}\right)x^4 + \left(-1 + \frac{2}{5} - \frac{3}{10}\right)x^3$
$= \left(\frac{4}{6} - \frac{1}{6}\right)x^4 + \left(-\frac{10}{10} + \frac{4}{10} - \frac{3}{10}\right)x^3$
$= \frac{3}{6}x^4 - \frac{9}{10}x^3$
$= \frac{1}{2}x^4 - \frac{9}{10}x^3$

Note in Examples 5(b), (c), and (d) that the solutions are written so that the term of highest degree appears first, followed by the term of next highest degree, and so on. This is known as **descending order** and is the form in which answers will normally appear.

Evaluating Polynomials and Applications

When each variable in a polynomial is replaced with a number, the polynomial then represents a number, or *value*, that can be calculated using the rules for the order of operations.

EXAMPLE 6 Evaluate $-x^2 + 3x + 9$ for $x = -2$.

SOLUTION For $x = -2$, we have

$-x^2 + 3x + 9 = -(-2)^2 + 3(-2) + 9$ The negative sign in front of x^2 remains.
$= -4 + (-6) + 9$
$= -10 + 9 = -1.$

EXAMPLE 7 *Games in a Sports League.* In a sports league of n teams in which each team plays every other team twice, the total number of games to be played is given by the polynomial

$n^2 - n.$

A women's softball league has 10 teams. How many games are played if each team plays every other team twice?

SOLUTION We evaluate the polynomial for $n = 10$:

$n^2 - n = 10^2 - 10$
$= 100 - 10$
$= 90.$

The league plays 90 games.

EXAMPLE 8 *Medical Dosage.* The concentration, in parts per million, of a certain antibiotic in the bloodstream after t hours is given by the polynomial

$-0.05t^2 + 2t + 2.$

Find the concentration after 2 hr.

SOLUTION To find the concentration after 2 hr, we evaluate the polynomial for $t = 2$:

$-0.05t^2 + 2t + 2 = -0.05(2)^2 + 2(2) + 2$
$= -0.05(4) + 2(2) + 2$
$= -0.2 + 4 + 2$
$= 5.8.$

The concentration after 2 hr is 5.8 parts per million.

188 CHAPTER 4 • POLYNOMIALS

TECHNOLOGY CONNECTION 4.2

One way to evaluate $-x^2 + 3x + 9$ for $x = -2$ (see Example 6) is to graph $y_1 = -x^2 + 3x + 9$. On many graphers, a CALC key accesses a menu in which the option VALUE appears. This allows us to evaluate $-x^2 + 3x + 9$ for any x-value we choose.

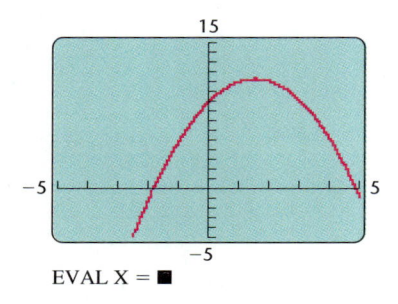

After we enter the x-value in which we are interested (in this case, -2), the value of the polynomial appears as y, and the cursor immediately appears at $(-2, -1)$.

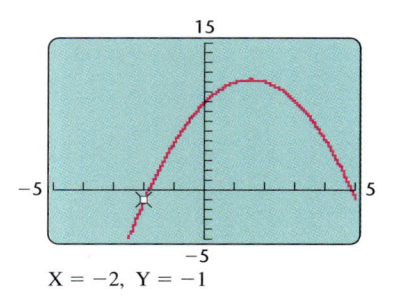

EVAL X = ■

X = −2, Y = −1

1. Use the approach above to find the value of $-x^2 + 3x + 9$ for $x = 4$.

Sometimes, when a graph has been provided, the value of a polynomial for a particular replacement can be estimated visually.

EXAMPLE 9

Medical Dosage. In the following graph, the polynomial from Example 8 has been graphed by evaluating it for several choices of t. Use the graph to estimate the concentration c of antibiotic in the bloodstream after 14 hr.

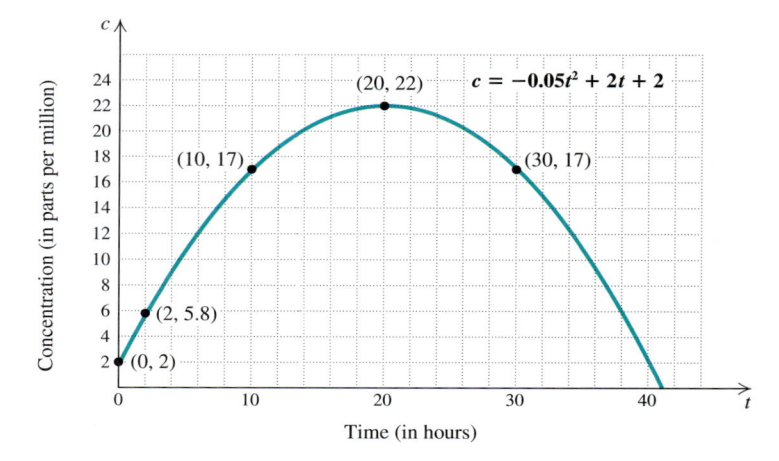

$c = -0.05t^2 + 2t + 2$

Points shown: (0, 2), (2, 5.8), (10, 17), (20, 22), (30, 17)

Concentration (in parts per million) vs. Time (in hours)

SOLUTION To estimate the concentration after 14 hr, we locate 14 on the horizontal axis. From there, we move vertically to the graph of the equation and then horizontally to the c-axis (see the top of the following page). This locates a c-value of about 20.

After 14 hr, the concentration of antibiotic in the bloodstream is about 20 parts per million. (For $t = 14$, the value of $-0.05t^2 + 2t + 2$ is approximately 20.)

EXERCISE SET 4.2

Identify the terms of each polynomial.
1. $3x^4 - 7x^3 + x - 5$
2. $5a^3 + 4a^2 - 9a - 7$
3. $-t^4 + 2t^3 - 5t^2 + 3$
4. $n^5 - 4n^3 + 2n - 8$

Determine the coefficient and the degree of each term in each polynomial.
5. $7x^3 - 5x$
6. $9a^3 - 4a^2$
7. $9t^2 - 3t + 4$
8. $7x^4 + 5x - 3$
9. $5a^4 + 9a + a^3$
10. $6t^5 - 3t^2 - t$
11. $x^4 - x^3 + 4x - 3$
12. $3a^4 - a^3 + a - 9$

For each of the following polynomials, (a) list the degree of each term; (b) determine the leading term and the leading coefficient; and (c) determine the degree of the polynomial.
13. $4a^3 + 7a^5 + a^2$
14. $5x - 9x^2 + 3x^6$
15. $2t + 3 + 4t^2$
16. $3a^2 - 7 + 2a^5$
17. $-5x^4 + x^2 - x + 3$
18. $-7x^3 + 6x^2 - 3x - 4$
19. $9a - a^4 + 3 + 2a^3$
20. $-x + 2x^5 - x^3 + 2$

21. Complete the following table for the polynomial $3x^2 + 8x^5 - 4x^3 + 6 - \frac{1}{2}x^4$.

Term	Coefficient	Degree of the Term	Degree of the Polynomial
		5	
$-\frac{1}{2}x^4$			
	-4		
		2	
	6		

22. Complete the following table for the polynomial
$-7x^4 + 6x^3 - 3x^2 + 8x - 2$.

Term	Coefficient	Degree of the Term	Degree of the Polynomial
	-7		
$6x^3$			
		2	
		1	
	-2		

Classify each polynomial as a monomial, binomial, trinomial, or none of these.

23. $x^2 - 10x + 25$ **24.** $-6x^4$
25. $x^3 - 7x^2 + 2x - 4$ **26.** $x^2 - 9$
27. $4x^2 - 25$ **28.** $4x^2 + 12x + 9$
29. $40x$
30. $2x^4 - 7x^3 + x^2 + x - 6$

Combine like terms. Write all answers in descending order.

31. $3x^2 + 5x + 4x^2$
32. $5a + 7a^2 + 3a$
33. $3a^4 - 2a + 2a + a^4$
34. $6b^5 + 3b^2 - 2b^5 - 3b^2$
35. $2x^2 - 6x + 3x + 4x^2$
36. $\frac{1}{4}x^5 - 5 + \frac{1}{2}x^5 - 2x - 37$
37. $\frac{1}{3}x^3 + 2x - \frac{1}{6}x^3 + 4 - 16$
38. $6x^2 + 2x^4 - 2x^2 - x^4 - 4x^2$
39. $8x^2 + 2x^3 - 3x^3 - 4x^2 - 4x^2$
40. $\frac{1}{4}x^3 - x^2 - \frac{1}{6}x^2 + \frac{3}{8}x^3 + \frac{5}{16}x^3$
41. $\frac{1}{5}x^4 + \frac{1}{5} - 2x^2 + \frac{1}{10} - \frac{3}{15}x^4 + 2x^2 - \frac{3}{10}$
42. $3x^4 - 5x^6 - 2x^4 + 6x^6$
43. $-1 + 5x^3 - 3 - 7x^3 + x^4 + 5$
44. $-2x + 4x^3 - 7x + 9x^3 + 8$
45. $2x - \frac{5}{6} + 4x^3 + x + \frac{1}{3} - 2x$
46. $5a^2 - \frac{2}{3} + a^3 - 9a^2 - \frac{4}{3}a^3 + 1$

Evaluate each polynomial for $x = 3$.

47. $-7x + 5$ **48.** $-5x + 9$
49. $2x^2 - 3x + 7$ **50.** $4x^2 - 6x + 9$

Evaluate each polynomial for $x = -2$.

51. $5x + 7$ **52.** $7 - 3x$
53. $x^2 - 3x + 1$ **54.** $5x - 9 + x^2$
55. $-3x^3 + 7x^2 - 4x - 5$
56. $-2x^3 - 4x^2 + 3x + 1$

Memorizing words. *Participants in a psychology experiment were able to memorize an average of M words in t minutes, where $M = -0.001t^3 + 0.1t^2$. Use the following graph for Exercises 57–62.*

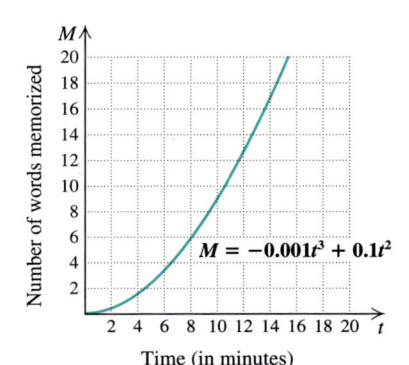

57. Estimate the number of words memorized after 10 min.
58. Estimate the number of words memorized after 14 min.
59. Find the approximate value of $-0.001t^3 + 0.1t^2$ for $t = 8$.
60. Find the approximate value of $-0.001t^3 + 0.1t^2$ for $t = 12$.
61. Estimate the value of $-0.001t^3 + 0.1t^2$ when t is 13.
62. Estimate the value of $-0.001t^3 + 0.1t^2$ when t is 7.
63. *Skydiving.* During the first 13 sec of a jump, the number of feet that a skydiver falls in t seconds is approximated by the polynomial
$11.12t^2$.

Approximately how far has a skydiver fallen 10 sec after jumping from a plane?

64. *Skydiving.* For jumps that exceed 13 sec, the polynomial $173t - 369$ can be used to approximate the distance, in feet, that a skydiver has fallen in t seconds. Approximately how far has a skydiver fallen 20 sec after jumping from a plane?

Daily accidents. The average number of accidents per day involving drivers of age r can be approximated by the polynomial

$$0.4r^2 - 40r + 1039.$$

65. Evaluate the polynomial for $r = 18$ to find the daily number of accidents involving 18-year-old drivers.
66. Evaluate the polynomial for $r = 20$ to find the daily number of accidents involving 20-year-old drivers.

Total revenue. Hadley Electronics is marketing a new kind of stereo. Total revenue is the total amount of money taken in. The firm determines that when it sells x stereos, it will take in

$$280x - 0.4x^2 \text{ dollars.}$$

67. What is the total revenue from the sale of 75 stereos?
68. What is the total revenue from the sale of 100 stereos?

Total cost. Hadley Electronics determines that the total cost of producing x stereos is given by

$$5000 + 0.6x^2 \text{ dollars.}$$

69. What is the total cost of producing 500 stereos?
70. What is the total cost of producing 650 stereos?

Circumference. The circumference of a circle of radius r is given by the polynomial $2\pi r$, where π is an irrational number. For an approximation of π, use 3.14.

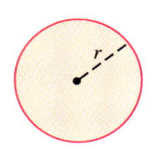

71. Find the circumference of a circle with radius 10 cm.
72. Find the circumference of a circle with radius 5 ft.

Area of a circle. The area of a circle of radius r is given by the polynomial πr^2.

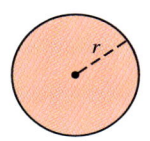

73. Find the area of a circle with radius 5 m.
74. Find the area of a circle with radius 10 in.

SKILL MAINTENANCE

75. Solve: $3(x + 2) = 5x - 9$.
76. The sum of the page numbers on the facing pages of a book is 549. What are the page numbers?
77. A family spent $2011 to drive a car one year, during which the car was driven 14,800 mi. The family spent $972 for insurance and $114 for registration and oil. The only other cost was for gasoline. How much did gasoline cost per mile?
78. Solve $cx = ab - r$ for b.

SYNTHESIS

79. ◆ Is every monomial a term and is every term a monomial? Why or why not?
80. ◆ Explain why an understanding of the rules for order of operations is essential when evaluating polynomials.
81. ◆ Is it better to evaluate a polynomial before or after like terms have been combined? Why?
82. ◆ Suppose that the coefficients of a polynomial are all integers and the polynomial is evaluated for some integer. Must the value of the polynomial then also be an integer? Why or why not?
83. Construct a polynomial in x (meaning that x is the variable) of degree 5 with four terms and coefficients that are integers.
84. Construct a trinomial in y of degree 4 with coefficients that are rational numbers.
85. What is the degree of $(5m^5)^2$?
86. Construct three like terms of degree 4.

Simplify.

87. $\frac{9}{2}x^8 + \frac{1}{9}x^2 + \frac{1}{2}x^9 + \frac{9}{2}x + \frac{9}{2}x^9 + \frac{8}{9}x^2 + \frac{1}{2}x - \frac{1}{2}x^8$
88. $(3x^2)^3 + 4x^2 \cdot 4x^4 - x^4(2x)^2 + ((2x)^2)^3 - 100x^2(x^2)^2$
89. ▦ Evaluate $s^2 - 50s + 675$ and $-s^2 + 50s - 675$ for $s = 18$, $s = 25$, and $s = 32$.
90. ▦ *Daily accidents.* The average number of accidents per day involving drivers of age r can be approximated by the polynomial

$$0.4r^2 - 40r + 1039.$$

For what age is the number of daily accidents smallest?

192 CHAPTER 4 • POLYNOMIALS

In Exercises 91 and 92, complete the table for the given choices of t. Then plot the points and connect them with a smooth curve representing the graph of the polynomial.

91.

t	$-t^2 + 10t - 18$
3	
4	
5	
6	
7	

92.

t	$-t^2 + 6t - 4$
1	
2	
3	
4	
5	

d	$-0.0064d^2 + 0.8d + 2$
0	
30	
60	
90	
120	

93. *Path of the Olympic arrow.* The Olympic flame at the 1992 Summer Olympics was lit by a flaming arrow. As the arrow moved d meters horizontally from the archer, its height, in meters, was approximated by the polynomial

$$-0.0064d^2 + 0.8d + 2.$$

Complete the table for the choices of d given. Then plot the points and draw a graph representing the path of the arrow.

94. A polynomial in x has degree 3. The coefficient of x^2 is 3 less than the coefficient of x^3. The coefficient of x is 3 times the coefficient of x^2. The remaining constant is 2 more than the coefficient of x^3. The sum of the coefficients is -4. Find the polynomial.

4.3 Addition and Subtraction of Polynomials

Addition of Polynomials • Opposites of Polynomials •
Subtraction of Polynomials • Problem Solving

Addition of Polynomials

To add two polynomials, we write a plus sign between them and combine like terms.

EXAMPLE 1 Add.

a) $(-3x^3 + 2x - 4) + (4x^3 + 3x^2 + 2)$

b) $\left(\frac{2}{3}x^4 + 3x^2 - 2x + \frac{1}{2}\right) + \left(-\frac{1}{3}x^4 + 5x^3 - 3x^2 + 3x - \frac{1}{2}\right)$

4.3 ADDITION AND SUBTRACTION OF POLYNOMIALS

SOLUTION

a) $(-3x^3 + 2x - 4) + (4x^3 + 3x^2 + 2)$
$= (-3 + 4)x^3 + 3x^2 + 2x + (-4 + 2)$ Combining like terms; using the distributive law
$= x^3 + 3x^2 + 2x - 2$

b) $\left(\frac{2}{3}x^4 + 3x^2 - 2x + \frac{1}{2}\right) + \left(-\frac{1}{3}x^4 + 5x^3 - 3x^2 + 3x - \frac{1}{2}\right)$
$= \left(\frac{2}{3} - \frac{1}{3}\right)x^4 + 5x^3 + (3 - 3)x^2 + (-2 + 3)x + \left(\frac{1}{2} - \frac{1}{2}\right)$ Combining like terms
$= \frac{1}{3}x^4 + 5x^3 + x$

After some practice, polynomial addition is often performed mentally.

EXAMPLE 2   Add: $(3x^2 - 2x + 2) + (5x^3 - 2x^2 + 3x - 4)$.

SOLUTION

$(3x^2 - 2x + 2) + (5x^3 - 2x^2 + 3x - 4)$
$= 5x^3 + (3 - 2)x^2 + (-2 + 3)x + (2 - 4)$ You might do this step mentally.
$= 5x^3 + x^2 + x - 2$ Then you would write only this.

We can also add polynomials by writing like terms in columns. Sometimes this makes like terms easier to see.

EXAMPLE 3   Add: $9x^5 - 2x^3 + 6x^2 + 3$ and $5x^4 - 7x^2 + 6$ and $3x^6 - 5x^5 + x^2 + 5$.

SOLUTION   We arrange the polynomials with like terms in columns.

$$
\begin{array}{r}
9x^5 \phantom{{}+{}} - 2x^3 + 6x^2 + 3 \\
5x^4 \phantom{{}+ 2x^3} - 7x^2 + 6 \\
3x^6 - 5x^5 \phantom{{}+ 5x^4 - 2x^3} + 1x^2 + 5 \\
\hline
3x^6 + 4x^5 + 5x^4 - 2x^3 \phantom{{}+ 1x^2} + 14
\end{array}
$$

We leave spaces for missing terms.
Writing x^2 as $1x^2$
Adding

The answer is $3x^6 + 4x^5 + 5x^4 - 2x^3 + 14$.

Opposites of Polynomials

In Section 1.8, we used the property of -1 to show that the opposite of a sum is the sum of the opposites. This idea can be extended to polynomials with any number of terms.

THE OPPOSITE OF A POLYNOMIAL

To find an equivalent polynomial for the *opposite*, or *additive inverse*, of a polynomial, replace each term with its opposite—that is, *change the sign of every term*. This is the same as multiplying the polynomial by -1.

EXAMPLE 4 Find two equivalent expressions for the opposite of $4x^5 - 7x^3 - 8x + \frac{5}{6}$.

SOLUTION

i) $-\left(4x^5 - 7x^3 - 8x + \frac{5}{6}\right)$

ii) $-4x^5 + 7x^3 + 8x - \frac{5}{6}$ Changing the sign of every term

Thus, $-\left(4x^5 - 7x^3 - 8x + \frac{5}{6}\right)$ is equivalent to $-4x^5 + 7x^3 + 8x - \frac{5}{6}$. Both expressions represent the opposite of $4x^5 - 7x^3 - 8x + \frac{5}{6}$.

EXAMPLE 5 Simplify: $-\left(-7x^4 - \frac{5}{9}x^3 + 8x^2 - x + 67\right)$.

SOLUTION

$$-\left(-7x^4 - \frac{5}{9}x^3 + 8x^2 - x + 67\right) = 7x^4 + \frac{5}{9}x^3 - 8x^2 + x - 67$$

Subtraction of Polynomials

We can now subtract one polynomial from another by adding the opposite of the polynomial being subtracted.

EXAMPLE 6 Subtract.

a) $(9x^5 + x^3 - 2x^2 + 4) - (-2x^5 + x^4 - 4x^3 - 3x^2)$
b) $(7x^5 + x^3 - 9x) - (3x^5 - 4x^3 + 5)$

SOLUTION

a) $(9x^5 + x^3 - 2x^2 + 4) - (-2x^5 + x^4 - 4x^3 - 3x^2)$
   $= 9x^5 + x^3 - 2x^2 + 4 + 2x^5 - x^4 + 4x^3 + 3x^2$ Adding the opposite
   $= 11x^5 - x^4 + 5x^3 + x^2 + 4$ Combining like terms

b) $(7x^5 + x^3 - 9x) - (3x^5 - 4x^3 + 5)$
   $= 7x^5 + x^3 - 9x + (-3x^5) + 4x^3 - 5$ Adding the opposite
   $= 7x^5 + x^3 - 9x - 3x^5 + 4x^3 - 5$ Try to go directly to this step.
   $= 4x^5 + 5x^3 - 9x - 5$ Combining like terms

To use columns to subtract, we replace coefficients by their opposites and then add.

EXAMPLE 7 Write in columns and subtract: $(5x^2 - 3x + 6) - (9x^2 - 5x - 3)$.

SOLUTION

i) $\quad 5x^2 - 3x + 6$
   $\underline{-(9x^2 - 5x - 3)}$ Writing similar terms in columns

ii) $\quad 5x^2 - 3x + 6$
    $\underline{-9x^2 + 5x + 3}$ Changing signs and removing parentheses

iii) $\quad 5x^2 - 3x + 6$
     $\underline{-9x^2 + 5x + 3}$
     $\quad -4x^2 + 2x + 9$ Adding

TECHNOLOGY CONNECTION 4.3

To check polynomial addition or subtraction, we can let y_1 = the expression before the addition or subtraction has been performed and y_2 = the simplified sum or difference. If the addition or subtraction is correct, y_1 will equal y_2 and $y_2 - y_1$ will be 0. We enter $y_2 - y_1$ as y_3, using the Y-VARS key. Below is a check of Example 6(b) in which

$$y_1 = (7x^5 + x^3 - 9x) - (3x^5 - 4x^3 + 5),$$
$$y_2 = 4x^5 + 5x^3 - 9x - 5,$$

and

$$y_3 = y_2 - y_1.$$

We graph only y_3. Note that since $y_3 = 0$ for all x, the graph coincides with the x-axis. The TRACE or TABLE features (if available) can confirm that y_3 is always 0.

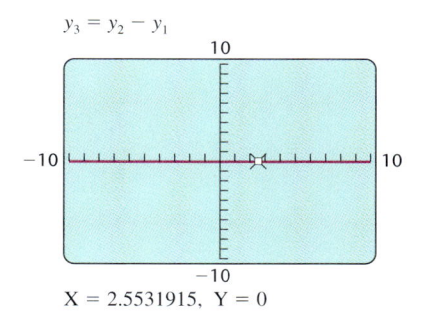

$y_3 = y_2 - y_1$

X = 2.5531915, Y = 0

1. Use a grapher to check Examples 1, 2, and 6(a).

If you can do so without error, you can arrange the polynomials in columns and write just the answer.

EXAMPLE 8 Write in columns and subtract: $(x^3 + x^2 + 2x - 12) - (-2x^3 + x^2 - 3x)$.

SOLUTION We have

$$\begin{array}{r} x^3 + x^2 + 2x - 12 \\ -(-2x^3 + x^2 - 3x) \\ \hline 3x^3 + 5x - 12. \end{array}$$

Leaving space for the missing term

Problem Solving

EXAMPLE 9 Find a polynomial for the sum of the areas of rectangles A, B, C, and D.

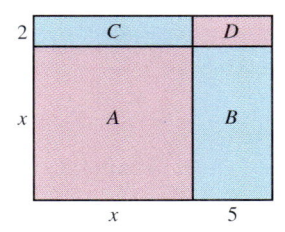

SOLUTION

1. **Familiarize.** Recall that the area of a rectangle is the product of its length and width.
2. **Translate.** We translate the problem to mathematical language. The sum

196 CHAPTER 4 • POLYNOMIALS

of the areas is a sum of products. We find each product and then add:

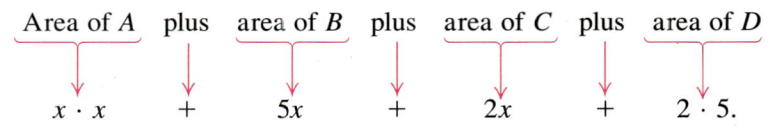

Area of A plus area of B plus area of C plus area of D
$x \cdot x \quad + \quad 5x \quad + \quad 2x \quad + \quad 2 \cdot 5.$

3. **Carry out.** We combine like terms:

$$x^2 + 5x + 2x + 10 = x^2 + 7x + 10.$$

4. **Check.** A partial check is to replace x with a number, say 3. Then we evaluate $x^2 + 7x + 10$ and compare that result with the total area for that choice of x:

$$3^2 + 7 \cdot 3 + 10 = 9 + 21 + 10 = 40.$$

When we substitute 3 for x and calculate the areas of A, B, C, and D all at once, regarding the figure as one large rectangle, we should also get 40:

Total area $= (x + 5)(x + 2) = (3 + 5)(3 + 2) = 8 \cdot 5 = 40.$

Our check is only partial, since it is also possible for an incorrect answer to equal 40 when evaluated for $x = 3$. This would be very unlikely, especially if a second choice of x, say $x = 5$, also checks. We leave that check to the student.

5. **State.** A polynomial for the sum of the areas is $x^2 + 7x + 10$.

EXAMPLE 10 A 4-ft by 4-ft sandbox is placed on a square lawn x ft on a side. Find a polynomial for the remaining area.

SOLUTION

1. **Familiarize.** We draw a picture of the situation as follows.

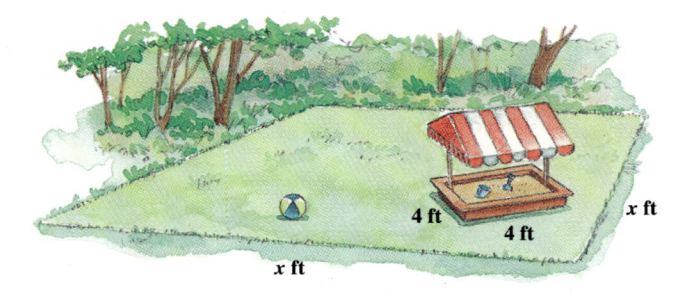

2. **Translate.** We reword the problem and translate as follows.

Rewording: Area of lawn $-$ area of sandbox $=$ area left over

Translating: $x \cdot x \quad - \quad 4 \cdot 4 \quad =$ Area left over

3. **Carry out.** We carry out the manipulation by multiplying:

$x^2 - 16 =$ Area left over.

4. **Check.** We leave the check to the student.
5. **State.** The remaining area in the yard is $(x^2 - 16)$ ft^2.

EXERCISE SET 4.3

Add.
1. $(5x + 3) + (-7x + 1)$
2. $(4x + 1) + (-9x + 5)$
3. $(-6x + 2) + (x^2 + x - 3)$
4. $(x^2 - 5x + 4) + (8x - 9)$
5. $(3x^2 - 5x + 10) + (2x^2 + 8x - 40)$
6. $(6x^4 + 3x^3 - 1) + (4x^2 - 3x + 3)$
7. $(1.2x^3 + 4.5x^2 - 3.8x) + (-3.4x^3 - 4.7x^2 + 23)$
8. $(0.5x^4 - 0.6x^2 + 0.7) + (2.3x^4 + 1.8x - 3.9)$
9. $(3 + 5x + 7x^2 + 8x^3) + (8 - 3x + 9x^2 - 8x^3)$
10. $(2x^4 - 5x - 4x^2 + 6) + (7x^2 - 3x^3 - 2 + 8x)$
11. $(9x^8 - 7x^4 + 2x^2 + 5) + (8x^7 + 4x^4 - 2x)$
12. $(4x^5 - 6x^3 - 9x + 1) + (6x^3 + 9x^2 + 9x)$
13. $\left(\frac{1}{4}x^4 + \frac{2}{3}x^3 + \frac{5}{8}x^2 + 7\right) + \left(-\frac{3}{4}x^4 + \frac{3}{8}x^2 - 7\right)$
14. $\left(\frac{1}{3}x^9 + \frac{1}{5}x^5 - \frac{1}{2}x^2 + 7\right) + \left(-\frac{1}{5}x^9 + \frac{1}{4}x^4 - \frac{3}{5}x^5\right)$
15. $(0.02x^5 - 0.2x^3 + x + 0.08) +$
    $(-0.01x^5 + x^4 - 0.8x - 0.02)$
16. $(0.03x^6 + 0.05x^3 + 0.22x + 0.05) +$
    $\left(\frac{7}{100}x^6 - \frac{3}{100}x^3 + 0.5\right)$
17. $-3x^4 + 6x^2 + 2x - 1$
    $\underline{- 3x^2 + 2x + 1}$
18. $-4x^3 + 8x^2 + 3x - 2$
    $\underline{- 4x^2 + 3x + 2}$
19. $0.15x^4 + 0.10x^3 - 0.9x^2$
    $- 0.01x^3 + 0.01x^2 + x$
    $1.25x^4 + 0.11x^2 + 0.01$
    $0.27x^3 + 0.99$
    $\underline{-0.35x^4 + 15x^2 - 0.03}$
20. $0.05x^4 + 0.12x^3 - 0.5x^2$
    $- 0.02x^3 + 0.02x^2 + 2x$
    $1.5x^4 + 0.01x^2 + 0.15$
    $0.25x^3 + 0.85$
    $\underline{-0.25x^4 + 10x^2 - 0.04}$

Find two equivalent expressions for the opposite of each polynomial.
21. $-x^2 + 9x - 4$
22. $-4x^3 - 5x^2 + 2x$
23. $12x^4 - 3x^3 + 3$
24. $4x^3 - 6x^2 - 8x + 1$

Simplify.
25. $-(8x - 9)$
26. $-(-6x + 5)$
27. $-(4x^2 - 3x + 2)$
28. $-(-6a^3 + 2a^2 - 9a + 1)$
29. $-\left(-4x^4 + 6x^2 + \frac{3}{4}x - 8\right)$
30. $-(-5x^4 + 4x^3 - x^2 + 0.9)$

Subtract.
31. $(7x + 4) - (-2x + 1)$
32. $(5x + 6) - (-2x + 4)$
33. $(-6x + 2) - (x^2 + x - 3)$
34. $(x^2 - 5x + 4) - (8x - 9)$
35. $(6x^4 + 3x^3 - 1) - (4x^2 - 3x + 3)$
36. $(-4x^2 + 2x) - (3x^3 - 5x^2 + 3)$
37. $(1.2x^3 + 4.5x^2 - 3.8x) - (-3.4x^3 - 4.7x^2 + 23)$
38. $(0.5x^4 - 0.6x^2 + 0.7) - (2.3x^4 + 1.8x - 3.9)$
39. $(7x^3 - 2x^2 + 6) - (7x^2 + 2x - 4)$
40. $(6x^5 - 3x^4 + x + 1) - (8x^5 + 3x^4 - 1)$
41. $(6x^2 + 2x) - (-3x^2 - 7x + 8)$
42. $7x^3 - (-3x^2 - 2x + 1)$
43. $\left(\frac{5}{8}x^3 - \frac{1}{4}x - \frac{1}{3}\right) - \left(-\frac{1}{8}x^3 + \frac{1}{4}x - \frac{1}{3}\right)$
44. $\left(\frac{1}{5}x^3 + 2x^2 - 0.1\right) - \left(-\frac{2}{5}x^3 + 2x^2 + 0.01\right)$
45. $(0.08x^3 - 0.02x^2 + 0.01x) - (0.02x^3 + 0.03x^2 - 1)$
46. $(0.8x^4 + 0.2x - 1) - \left(\frac{7}{10}x^4 + \frac{1}{5}x - 0.1\right)$
47. $x^2 + 5x + 6$
    $\underline{-(x^2 + 2x)}$
48. $x^3 + 1$
    $\underline{-(x^3 + x^2)}$
49. $5x^4 + 6x^3 - 9x^2$
    $\underline{-(-6x^4 - 6x^3 + 8x + 9)}$

198 CHAPTER 4 • POLYNOMIALS

50. $\quad 5x^4 \quad\quad + 6x^2 - 3x + 6$
$\quad\quad -(\quad\quad 6x^3 + 7x^2 - 8x - 9)$

51. $\quad\quad 3x^4 + 6x^2 + 8x - 1$
$\quad -(4x^5 - 6x^4 \quad\quad - 8x - 7)$

52. $\quad 6x^5 \quad\quad + 3x^2 - 7x + 2$
$\quad -(10x^5 + 6x^3 - 5x^2 - 2x + 4)$

53. Solve.
  a) Find a polynomial for the sum of the areas of the rectangles shown in the figure.
  b) Find the sum of the areas when $x = 5$ and $x = 7$.

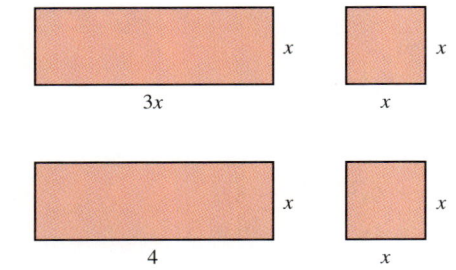

54. Solve.
  a) Find a polynomial for the sum of the areas of the circles shown in the figure.
  b) Find the sum of the areas when $r = 5$ and $r = 11.3$.

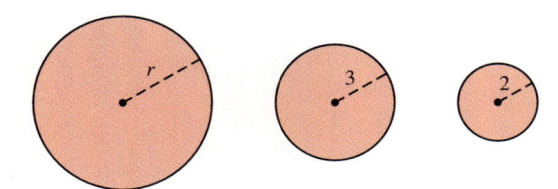

Find a polynomial for the perimeter of each figure in Exercises 55 and 56.

55.

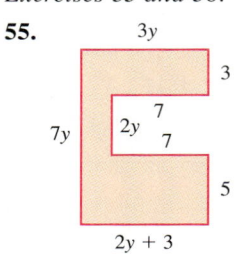

56.

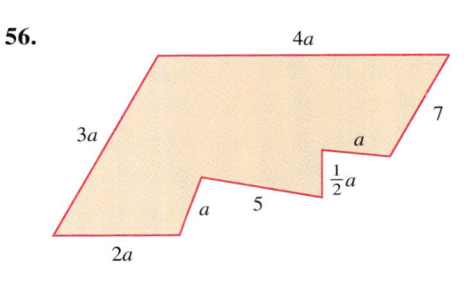

Find two algebraic expressions for the area of each figure. First, regard the figure as one large rectangle, and then regard the figure as a sum of four smaller rectangles.

57.

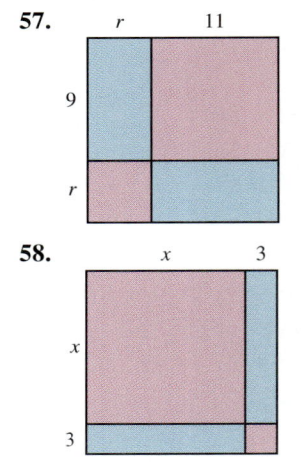

58.

Find a polynomial for the shaded area of each figure.

59.

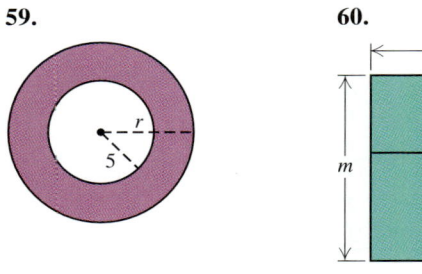

60.

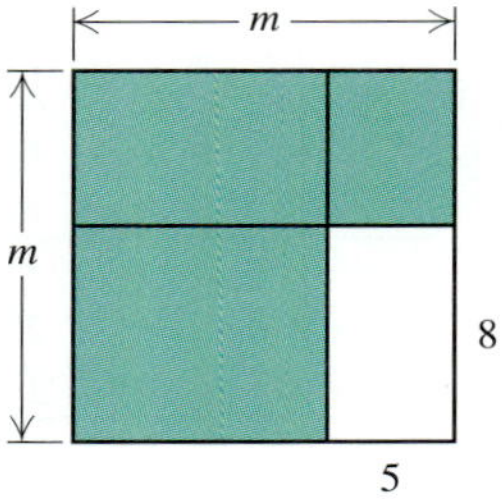

61.

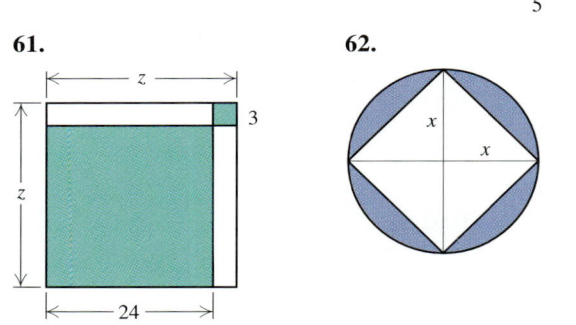

62.

63. Find $(y - 2)^2$ using the four parts of this square.

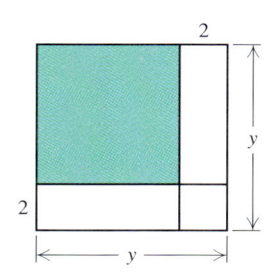

64. Find $(10 - 2x)^2$ using the nine parts of this square.

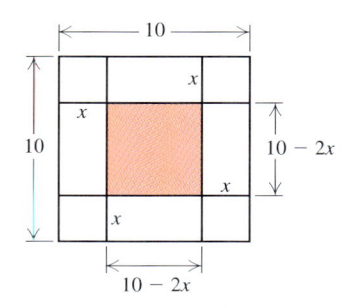

SKILL MAINTENANCE

Solve.

65. $1.5x - 2.7x = 23 - 5.6x$
66. $3x - 3 = -4x + 4$
67. $8(x - 2) = 16$
68. $4(x - 5) = 7(x + 8)$
69. $3x - 7 \leq 5x + 13$
70. $2(x - 4) > 5(x - 3) + 7$

SYNTHESIS

71. ◆ Under what conditions will the sum of two binomials be a monomial?
72. ◆ Which, if any, of the commutative, associative, and distributive laws are needed for adding polynomials? Why?
73. ◆ Is the sum of two binomials ever a trinomial? Why or why not?
74. ◆ What advice would you offer to a student who is successful at adding, but not subtracting, polynomials?

Simplify.

75. $(5a^2 - 8a) + (7a^2 - 9a - 13) - (7a - 5)$
76. $(3x^2 - 4x + 6) - (-2x^2 + 4) + (-5x - 3)$
77. $(-8y^2 - 4) - (3y + 6) - (2y^2 - y)$
78. $(5x^3 - 4x^2 + 6) - (2x^3 + x^2 - x) + (x^3 - x)$
79. $(-y^4 - 7y^3 + y^2) + (-2y^4 + 5y - 2) - (-6y^3 + y^2)$
80. $(-4 + x^2 + 2x^3) - (-6 - x + 3x^3) - (-x^2 - 5x^3)$
81. 🧮 $(345.099x^3 - 6.178x) - (94.508x^3 - 8.99x)$

Find a polynomial for the surface area of the right rectangular solid.

82.

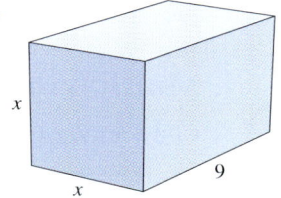

83.

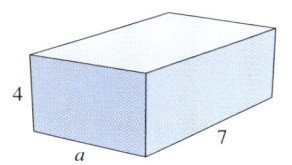

84. *Total profit.* Hadley Electronics is marketing a new kind of stereo. Total revenue is the total amount of money taken in. The firm determines that when it sells x stereos, its total revenue is given by

$$R = 280x - 0.4x^2.$$

Total cost is the total cost of producing x stereos. Hadley Electronics determines that the total cost of producing x stereos is given by

$$C = 5000 + 0.6x^2.$$

The total profit is

(Total Revenue) $-$ (Total Cost) $= R - C$.

a) Find a polynomial for total profit.
b) What is the total profit on the production and sale of 75 stereos?
c) What is the total profit on the production and sale of 100 stereos?

85. ◆ Does replacing each occurrence of the variable x in $5x^3 - 3x^2 + 2x$ with its opposite result in the opposite of the polynomial? Why or why not?

4.4 Multiplication of Polynomials

Multiplying Monomials • Multiplying a Monomial and a Polynomial • Multiplying Any Two Polynomials • Checking by Evaluating

We now multiply polynomials using techniques based largely on the distributive, associative, and commutative laws and the rules for exponents.

Multiplying Monomials

Consider $(3x)(4x)$. We multiply as follows:

$$(3x)(4x) = 3 \cdot x \cdot 4 \cdot x \qquad \text{Using an associative law}$$
$$= 3 \cdot 4 \cdot x \cdot x \qquad \text{Using a commutative law}$$
$$= (3 \cdot 4) \cdot x \cdot x \qquad \text{Using an associative law}$$
$$= 12x^2.$$

TO MULTIPLY MONOMIALS:

To find an equivalent expression for the product of two monomials, multiply the coefficients and then multiply the variables using the product rule for exponents.

EXAMPLE 1 Multiply: **(a)** $(5x)(6x)$; **(b)** $(3a)(-a)$; **(c)** $(-7x^5)(4x^3)$.

SOLUTION

a) $(5x)(6x) = (5 \cdot 6)(x \cdot x)$ Multiplying the coefficients; multiplying the variables
$\qquad\quad\;\; = 30x^2$ Simplifying

b) $(3a)(-a) = (3a)(-1a)$ Writing $-a$ as $-1a$ can ease calculations.
$\qquad\quad\;\;\, = (3)(-1)(a \cdot a)$ Using an associative and a commutative law
$\qquad\quad\;\;\, = -3a^2$

c) $(-7x^5)(4x^3) = (-7 \cdot 4)(x^5 \cdot x^3)$
$\qquad\qquad\quad\; = -28x^{5+3}$ $\Big\}$ Using the product rule for exponents
$\qquad\qquad\quad\; = -28x^8$

After some practice, you can try writing only the answer.

Multiplying a Monomial and a Polynomial

To find an equivalent expression for the product of a monomial, such as $2x$, and a polynomial, such as $5x + 3$, we use the distributive law.

EXAMPLE 2 Multiply: **(a)** x and $x + 3$; **(b)** $5x(2x^2 - 3x + 4)$.

SOLUTION

a) $x(x + 3) = x \cdot x + x \cdot 3$ ⟵ Using the distributive law
$ = x^2 + 3x$

b) $5x(2x^2 - 3x + 4) = (5x)(2x^2) - (5x)(3x) + (5x)(4)$ ⟵ Using the distributive law
$ = 10x^3 - 15x^2 + 20x$ ⟵ Performing the three multiplications

The product in Example 2(a) can be visualized as the area of a rectangle with width x and length $x + 3$.

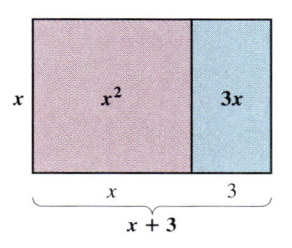

Note that the total area can be expressed as $x(x + 3)$ or, by adding the two smaller areas, $x^2 + 3x$.

THE PRODUCT OF A MONOMIAL AND A POLYNOMIAL

> To multiply a monomial and a polynomial, multiply each term of the polynomial by the monomial.

Try to do this mentally, when possible.

EXAMPLE 3 Multiply: $2x^2(x^3 - 7x^2 + 10x - 4)$.

SOLUTION

Think: $2x^2 \cdot x^3 - 2x^2 \cdot 7x^2 + 2x^2 \cdot 10x - 2x^2 \cdot 4$

$2x^2(x^3 - 7x^2 + 10x - 4) = 2x^5 - 14x^4 + 20x^3 - 8x^2$

Multiplying Any Two Polynomials

Before considering the product of *any* two polynomials, let's look at the product of two binomials.

To find an equivalent expression for the product of two binomials, we again begin by using the distributive law. This time, however, it is a *binomial* rather than a monomial that is being distributed.

EXAMPLE 4

Multiply each of the following.

a) $x + 5$ and $x + 4$
b) $4x - 3$ and $x - 2$

SOLUTION

a) $(x + 5)(x + 4) = (x + 5)x + (x + 5)4$ Using the distributive law

$= x(x + 5) + 4(x + 5)$ Using the commutative law for multiplication

$= x \cdot x + x \cdot 5 + 4 \cdot x + 4 \cdot 5$ Using the distributive law (twice)

$= x^2 + 5x + 4x + 20$ Multiplying the monomials

$= x^2 + 9x + 20$ Combining like terms

b) $(4x - 3)(x - 2) = (4x - 3)x - (4x - 3)2$ Using the distributive law

$= x(4x - 3) - 2(4x - 3)$ Using the commutative law for multiplication. This step is often omitted.

$= x \cdot 4x - x \cdot 3 - 2 \cdot 4x - 2(-3)$ Using the distributive law (twice)

$= 4x^2 - 3x - 8x + 6$ Multiplying the monomials

$= 4x^2 - 11x + 6$ Combining like terms

To visualize the product in Example 4(a), consider a rectangle of length $x + 5$ and width $x + 4$.

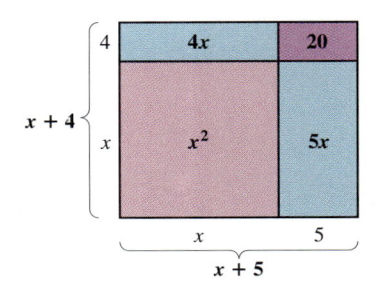

The total area can be expressed as $(x + 5)(x + 4)$ or, by adding the four smaller areas, $x^2 + 5x + 4x + 20$.

Let us consider the product of a binomial and a trinomial. Again we make repeated use of the distributive law.

EXAMPLE 5

Multiply: $(x^2 + 2x - 3)(x + 4)$.

SOLUTION

$(x^2 + 2x - 3)(x + 4)$

$= (x^2 + 2x - 3)x + (x^2 + 2x - 3)4$ Using the distributive law

$= x(x^2 + 2x - 3) + 4(x^2 + 2x - 3)$ Using the commutative law

$= x \cdot x^2 + x \cdot 2x - x \cdot 3 + 4 \cdot x^2 + 4 \cdot 2x - 4 \cdot 3$ Using the distributive law (twice)

$= x^3 + 2x^2 - 3x + 4x^2 + 8x - 12$ Multiplying the monomials

$= x^3 + 6x^2 + 5x - 12$ Combining like terms

Perhaps you have discovered the following in the preceding examples.

THE PRODUCT OF TWO POLYNOMIALS

To multiply two polynomials P and Q, select one of the polynomials, say P. Then multiply each term of P by every term of Q and combine like terms.

To use columns for long multiplications, multiply each term in the top row by every term in the bottom row. We write like terms in columns, and then add the results. Such multiplication is like multiplying with whole numbers:

```
      3 2 1              300 + 20 + 1
   ×    1 2           ×         10 + 2
   ─────────          ──────────────────
      6 4 2              600 + 40 + 2      Multiplying the top row by 2
    3 2 1             3000 + 200 + 10      Multiplying the bottom row by 10
   ─────────          ──────────────────
    3 8 5 2           3000 + 800 + 50 + 2   Adding
```

EXAMPLE 6

Multiply: $(4x^3 - 2x^2 + 3x)(x^2 + 2x)$.

SOLUTION

```
         4x³ − 2x² + 3x
               x² + 2x
         ─────────────────
         8x⁴ − 4x³ + 6x²              Multiplying the top row by 2x
   4x⁵ − 2x⁴ + 3x³                    Multiplying the top row by x²
   ───────────────────────
   4x⁵ + 6x⁴ − x³ + 6x²               Combining like terms
    ↑    ↑    ↑    ↑
    └────┴────┴────┴──── Line up like terms in columns.
```

If a term is missing, it helps to leave space for it so that like terms will be aligned as we multiply.

EXAMPLE 7 Multiply: $(-2x^2 - 3)(5x^3 - 3x + 4)$.

SOLUTION

$$
\begin{array}{r}
5x^3 - 3x + 4 \\
- 2x^2 - 3 \\
\hline
- 15x^3 + 9x - 12 \\
-10x^5 + 6x^3 - 8x^2 \\
\hline
-10x^5 - 9x^3 - 8x^2 + 9x - 12
\end{array}
$$

Multiplying by -3
Multiplying by $-2x^2$
Combining like terms

With practice some steps can be skipped. Sometimes we multiply horizontally, while still aligning like terms.

EXAMPLE 8 Multiply: $(2x^3 + 3x^2 - 4x + 6)(3x + 5)$.

SOLUTION

Multiplying by $3x$

$$(2x^3 + 3x^2 - 4x + 6)(3x + 5) = 6x^4 + 9x^3 - 12x^2 + 18x$$
$$+ 10x^3 + 15x^2 - 20x + 30$$

Multiplying by 5

$$= 6x^4 + 19x^3 + 3x^2 - 2x + 30$$

TECHNOLOGY CONNECTION 4.4

The TABLE feature, available on many graphers, was discussed with regard to graphing in Section 3.4. Tables can also be used to check that polynomials have been correctly multiplied. To illustrate, we can check Example 8 by entering $y_1 = (2x^3 + 3x^2 - 4x + 6)(3x + 5)$ and $y_2 = 6x^4 + 19x^3 + 3x^2 - 2x + 30$.

When TABLE is then pressed, we are shown two columns of values—one for y_1 and one for y_2. If our multiplication was correct, both columns of values will match, regardless of how far we scroll up or down.

X	Y1	Y2
-3	36	36
-2	-10	-10
-1	22	22
0	30	30
1	56	56
2	286	286
3	1050	1050

X = -3

1. Form a Table and scroll up and down to check Example 7.

Checking by Evaluating

How can we be certain that our multiplication (or addition or subtraction) of polynomials is correct? One check is to simply review our calculations. A different type of check, used in Example 9 of Section 4.3, makes use of the fact that equivalent expressions have the same value when evaluated for the same replacement. Thus a quick, partial, check of Example 8 can be made by selecting a convenient replacement for x (say, 1) and comparing the values of the expressions $(2x^3 + 3x^2 - 4x + 6)(3x + 5)$ and $6x^4 + 19x^3 + 3x^2 - 2x + 30$:

$(2x^3 + 3x^2 - 4x + 6)(3x + 5)$
$= (2 \cdot 1^3 + 3 \cdot 1^2 - 4 \cdot 1 + 6)(3 \cdot 1 + 5)$
$= (2 + 3 - 4 + 6)(3 + 5)$
$= 7 \cdot 8 = 56;$

$6x^4 + 19x^3 + 3x^2 - 2x + 30$
$= 6 \cdot 1^4 + 19 \cdot 1^3 + 3 \cdot 1^2 - 2 \cdot 1 + 30$
$= 6 + 19 + 3 - 2 + 30$
$= 28 - 2 + 30 = 56.$

Since the value of both expressions is 56, the multiplication in Example 8 is very likely correct.

EXERCISE SET 4.4

Multiply.
1. $(3x^4)8$
2. $(4x^3)7$
3. $(-x^2)(-x)$
4. $(-x^3)(x^4)$
5. $(-x^5)(x^3)$
6. $(-x^6)(-x^2)$
7. $(7t^5)(4t^3)$
8. $(10a^2)(3a^2)$
9. $(-0.1x^6)(0.2x^4)$
10. $(0.3x^3)(-0.4x^6)$
11. $\left(-\frac{1}{5}x^3\right)\left(-\frac{1}{3}x\right)$
12. $\left(-\frac{1}{4}x^4\right)\left(\frac{1}{5}x^8\right)$
13. $19t^2 \cdot 0$
14. $(-5n^3)(-1)$
15. $(3x^2)(-4x^3)(2x^6)$
16. $(-2y^5)(10y^4)(-3y^3)$
17. $3x(-x + 5)$
18. $2x(4x - 6)$
19. $4x(x + 1)$
20. $3x(x + 2)$
21. $(x + 7)5x$
22. $(x - 6)3x$
23. $x^2(x^3 + 1)$
24. $-2x^3(x^2 - 1)$
25. $3x(2x^2 - 6x + 1)$
26. $-4x(2x^3 - 6x^2 - 5x + 1)$
27. $4x^2(3x + 6)$
28. $5x^2(-2x + 1)$
29. $-6x^2(x^2 + x)$
30. $-4x^2(x^2 - x)$
31. $\frac{2}{3}a^4\left(6a^5 - 12a^3 - \frac{5}{8}\right)$
32. $\frac{3}{4}t^5\left(8t^6 - 12t^4 + \frac{12}{7}\right)$
33. $(x + 6)(x + 3)$
34. $(x + 5)(x + 2)$
35. $(x + 5)(x - 2)$
36. $(x + 6)(x - 2)$
37. $(x - 4)(x - 3)$
38. $(x - 7)(x - 3)$
39. $(x + 3)(x - 3)$
40. $(x + 6)(x - 6)$
41. $(5 - x)(5 - 2x)$
42. $(3 + x)(6 + 2x)$
43. $\left(t + \frac{3}{2}\right)\left(t + \frac{4}{3}\right)$
44. $\left(a - \frac{2}{5}\right)\left(a + \frac{5}{2}\right)$
45. $\left(\frac{1}{4}a + 2\right)\left(\frac{3}{4}a - 1\right)$
46. $\left(\frac{2}{5}t - 1\right)\left(\frac{3}{5}t + 1\right)$

Draw and label rectangles similar to those following Examples 2 and 4 to illustrate each product.
47. $x(x + 5)$
48. $x(x + 2)$
49. $(x + 1)(x + 2)$
50. $(x + 3)(x + 1)$
51. $(x + 5)(x + 3)$
52. $(x + 4)(x + 6)$
53. $(3x + 2)(3x + 2)$
54. $(5x + 3)(5x + 3)$

Multiply and check.
55. $(x^2 - x + 3)(x + 1)$
56. $(x^2 + x - 2)(x - 1)$
57. $(2a + 5)(a^2 - 3a + 2)$
58. $(3t + 4)(t^2 - 5t + 1)$
59. $(y^2 - 3)(2y^3 + y + 1)$
60. $(a^2 + 2)(5a^3 - 3a - 1)$
61. $(5x^3 - 7x^2 + 1)(x - 3x^2)$
62. $(4x^3 - 5x - 3)(1 + 2x^2)$
63. $(x^2 - 3x + 2)(x^2 + x + 1)$
64. $(x^2 + 5x - 1)(x^2 - x + 3)$
65. $(2x^2 + 3x - 4)(2x^2 + x - 2)$
66. $(2x^2 - x - 3)(2x^2 - 5x - 2)$
67. $(x + 1)(x^3 + 7x^2 + 5x + 4)$
68. $(x + 2)(x^3 + 5x^2 + 9x + 3)$
69. $\left(x - \frac{1}{2}\right)\left(2x^3 - 4x^2 + 3x - \frac{2}{5}\right)$
70. $\left(x + \frac{1}{3}\right)\left(6x^3 - 12x^2 - 5x + \frac{1}{2}\right)$

SKILL MAINTENANCE
71. Simplify: $10 - 2 + (-6)^2 \div 3 \cdot 2$.
72. Graph: $y = \frac{1}{2}x - 3$.
73. Factor: $15x - 18y + 12$.
74. Solve: $4(x - 3) = 5(2 - 3x) + 1$.

SYNTHESIS
75. ◆ Under what conditions will the product of two binomials be a trinomial?
76. ◆ The polynomials
$$(a + b + c + d) \text{ and } (r + s + m + p)$$
are multiplied. Without performing the multiplication, determine how many terms the product will contain. Provide a justification for your answer.
77. ◆ Is it possible to understand polynomial

206 CHAPTER 4 • POLYNOMIALS

multiplication without first understanding the distributive law? Why or why not?

78. ◆ How can the following figure be used to show that $(x + 3)^2 \neq x^2 + 9$?

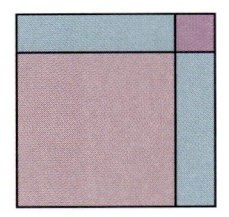

Find a polynomial for the shaded area of each figure.

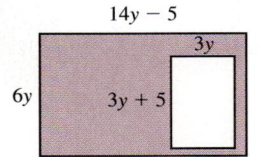

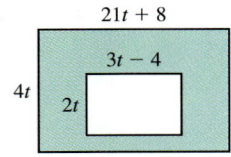

79. $14y - 5$, $3y$, $6y$, $3y + 5$

80. $21t + 8$, $3t - 4$, $4t$, $2t$

For each figure, determine what the missing number must be in order for the figure to have the given area.

81. Area is $x^2 + 7x + 10$

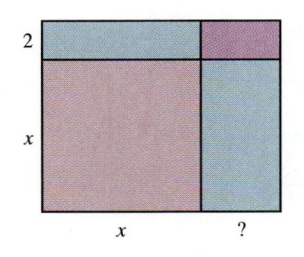

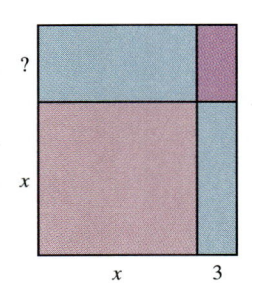

82. Area is $x^2 + 8x + 15$

83. A box with a square bottom is to be made from a 12-in.-square piece of cardboard. Squares with side x are cut out of the corners and the sides are folded up. Find the polynomials for the volume and the outside surface area of the box.

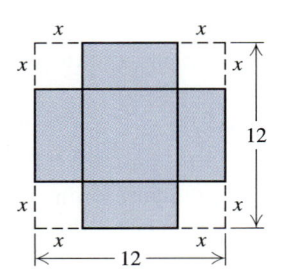

84. An open wooden box is a cube with side x cm. The box, including its bottom, is made of wood that is 1 cm thick. Find a polynomial for the interior volume of the cube.

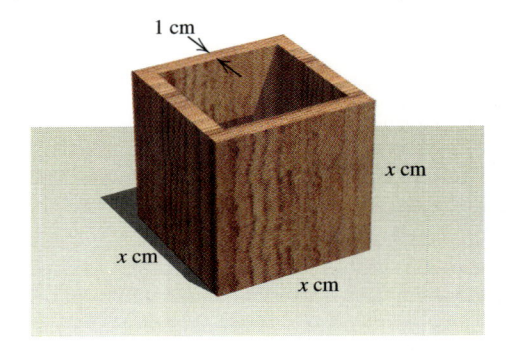

85. A side of a cube is $(x + 2)$ cm long. Find a polynomial for the volume of the cube.

86. A rectangular garden is twice as long as it is wide and is surrounded by a sidewalk that is 4 ft wide (see the figure below). The area of the sidewalk is 256 ft². Find the dimensions of the garden.

Compute and simplify.

87. $(x + 3)(x + 6) + (x + 3)(x + 6)$
88. $(x - 2)(x - 7) + (x - 2)(x - 7)$
89. $(x + 5)^2 - (x - 3)^2$
90. $(x - 6)^2 + (4 - x)^2$
91. Use a grapher to check your answers to Exercises 21, 41, and 61. Use graphs, tables, or both, as directed by your instructor.

COLLABORATIVE CORNER

Focus: Multiplication, addition, and evaluation of polynomials

Time: 10–20 minutes

Group size: 3

Best Built, Inc., is designing a cardboard carton for Sweet Dreams Confections. The length of the box must be twice the width, and the height must be 10 in. less than the length.

ACTIVITY

1. The cost of each carton depends on the surface area. To help determine area, each group member should select a different dimension (*l*, *w*, or *h*) and, using no other variables, write a polynomial for the surface area of the carton.

2. Determine, as a group, which of the polynomials from part (1) would be the easiest to use for each situation below. In each case, the group member who wrote the expression should perform the calculation.

   a) Sweet Dreams needs to ship chocolate bunnies that are 15 in. tall and must be stored upright. Find the surface area of the smallest carton that satisfies the specifications above.

   b) Sweet Dreams needs to ship licorice ropes that can be up to 48 in. long. Find the surface area of the smallest carton that satisfies the specifications preceding part (1). Assume that the licorice is not coiled or curved when shipped.

   c) Inside each carton, Sweet Dreams plans to stack and ship heart-shaped candy boxes that are 10 in. wide and 10 in. long. Find the surface area of the smallest carton that satisfies the specifications preceding part (1).

4.5 Special Products

Products of Two Binomials • Multiplying Sums and Differences of Two Terms • Squaring Binomials • Multiplications of Various Types

Certain products of two binomials occur so often that it is helpful to be able to compute them quickly. In this section, we develop methods for computing "special" products more quickly than we were able to in Section 4.4.

Products of Two Binomials

To multiply two binomials, we can select either binomial and multiply each term of that binomial by every term of the other. Then we combine like terms. Consider the product $(x + 5)(x + 4)$:

$$(x + 5)(x + 4) = x \cdot x + x \cdot 4 + 5 \cdot x + 5 \cdot 4$$
$$= x^2 + 4x + 5x + 20$$
$$= x^2 + 9x + 20.$$

Note that the product $x \cdot x$ is found by multiplying the *First* terms of each binomial, $x \cdot 4$ is found by multiplying the *Outside* terms of the two binomials, $5 \cdot x$ is the product of the *Inside* terms of the two binomials, and $5 \cdot 4$ is the product of the *Last* terms of each binomial:

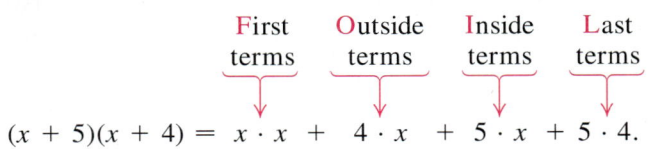

$$\underbrace{\text{First terms}}_{} \quad \underbrace{\text{Outside terms}}_{} \quad \underbrace{\text{Inside terms}}_{} \quad \underbrace{\text{Last terms}}_{}$$

$$(x + 5)(x + 4) = x \cdot x + 4 \cdot x + 5 \cdot x + 5 \cdot 4.$$

To remember this method of multiplying, use the initials **FOIL**.

THE FOIL METHOD

To multiply two binomials, $A + B$ and $C + D$, multiply the First terms AC, the Outside terms AD, the Inside terms BC, and then the Last terms BD. Then combine like terms, if possible.

$$(A + B)(C + D) = AC + AD + BC + BD$$

1. Multiply First terms: AC.
2. Multiply Outside terms: AD.
3. Multiply Inside terms: BC.
4. Multiply Last terms: BD.

↓

FOIL

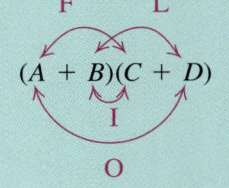

EXAMPLE 1 Multiply: $(x + 8)(x^2 + 5)$.

SOLUTION

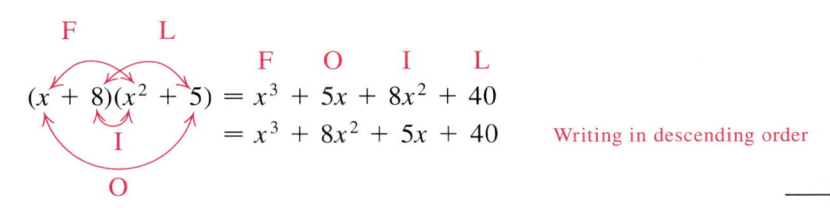

$$(x + 8)(x^2 + 5) = x^3 + 5x + 8x^2 + 40$$
$$= x^3 + 8x^2 + 5x + 40 \qquad \text{Writing in descending order}$$

After multiplying, remember to combine any like terms.

EXAMPLE 2

Multiply.

a) $(x + 7)(x + 4)$
b) $(y + 3)(y - 2)$
c) $(4t^3 + 5t)(3t^2 - 2)$
d) $(3 - 4x)(7 - 5x^3)$

SOLUTION

a) $(x + 7)(x + 4) = x^2 + 4x + 7x + 28$ Using FOIL
$= x^2 + 11x + 28$ Combining like terms

b) $(y + 3)(y - 2) = y^2 - 2y + 3y - 6$
$= y^2 + y - 6$

c) $(4t^3 + 5t)(3t^2 - 2) = 12t^5 - 8t^3 + 15t^3 - 10t$ Remember to add exponents when multiplying terms with the same base.

$= 12t^5 + 7t^3 - 10t$

d) $(3 - 4x)(7 - 5x^3) = 21 - 15x^3 - 28x + 20x^4$
$= 21 - 28x - 15x^3 + 20x^4$ Because the original binomials are in *ascending* order, we write the answer that way.

Multiplying Sums and Differences of Two Terms

Consider the product of the sum and difference of the same two terms, such as

$(x + 2)(x - 2)$.

Since this is the product of two binomials, we can use FOIL. However, products of this type occur so often that a faster method has been developed. To see where such a method comes from, look for a pattern in the following:

a) $(x + 2)(x - 2) = x^2 - 2x + 2x - 4$
$= x^2 - 4;$

b) $(3a - 5)(3a + 5) = 9a^2 + 15a - 15a - 25$
$= 9a^2 - 25;$

c) $\left(x^3 + \frac{2}{7}\right)\left(x^3 - \frac{2}{7}\right) = x^6 - \frac{2}{7}x^3 + \frac{2}{7}x^3 - \frac{4}{49}$
$= x^6 - \frac{4}{49}.$

Perhaps you discovered in each case that when we multiply the two binomials, the "outer" and "inner" products add to 0 and "drop out."

THE PRODUCT OF A SUM AND DIFFERENCE

The product of the sum and difference of the same two terms is the square of the first term minus the square of the second term:

$(A + B)(A - B) = \underbrace{A^2 - B^2}.$

This is called a *difference of squares*.

CHAPTER 4 • POLYNOMIALS

EXAMPLE 3

Multiply.

a) $(x + 4)(x - 4)$
b) $(5 + 2w)(5 - 2w)$
c) $(3a^4 - 5)(3a^4 + 5)$

SOLUTION

$(A + B)(A - B) = A^2 - B^2$ Saying the words can help.

a) $(x + 4)(x - 4) = x^2 - 4^2$ "The square of the first term, x^2, minus the square of the second, 4^2."

$= x^2 - 16$ Simplifying

b) $(5 + 2w)(5 - 2w) = 5^2 - (2w)^2$

$= 25 - 4w^2$ Squaring both 5, 2, and w

c) $(3a^4 - 5)(3a^4 + 5) = (3a^4)^2 - 5^2$

$= 9a^8 - 25$ Using the rules for exponents. Remember to multiply exponents when raising a power to a power.

Squaring Binomials

Consider the square of a binomial, such as $(x + 3)^2$. This can be expressed as $(x + 3)(x + 3)$. Since this is the product of two binomials, we can use FOIL. But again, this product occurs so often that a faster method has been developed. Look for a pattern in the following:

a) $(x + 3)^2 = (x + 3)(x + 3)$
$= x^2 + 3x + 3x + 9$
$= x^2 + 6x + 9;$

b) $(5 - 3p)^2 = (5 - 3p)(5 - 3p)$
$= 25 - 15p - 15p + 9p^2$
$= 25 - 30p + 9p^2;$

c) $(a^3 - 7)^2 = (a^3 - 7)(a^3 - 7)$
$= a^6 - 7a^3 - 7a^3 + 49$
$= a^6 - 14a^3 + 49.$

Perhaps you noticed that in each product the "outer" and "inner" products are identical. The other two terms, the "first" and "last" products, are squares.

THE SQUARE OF A BINOMIAL

The square of a binomial is the square of the first term, plus twice the product of the two terms, plus the square of the last term:

$(A + B)^2 = A^2 + 2AB + B^2;$
$(A - B)^2 = A^2 - 2AB + B^2.$

EXAMPLE 4 Multiply: **(a)** $(x + 7)^2$; **(b)** $(t - 5)^2$; **(c)** $(3a + 0.4)^2$; **(d)** $(5x - 3x^4)^2$.

SOLUTION

$$(A + B)^2 = A^2 + 2 \cdot A \cdot B + B^2$$

Saying the words can help.

a) $(x + 7)^2 = x^2 + 2 \cdot x \cdot 7 + 7^2$ "The square of the first term, x^2, plus twice the product of the terms, $2 \cdot 7x$, plus the square of the second term, 7^2."

$ = x^2 + 14x + 49$

b) $(t - 5)^2 = t^2 - 2 \cdot t \cdot 5 + 5^2$
$ = t^2 - 10t + 25$

c) $(3a + 0.4)^2 = (3a)^2 + 2 \cdot 3a \cdot 0.4 + 0.4^2$
$ = 9a^2 + 2.4a + 0.16$

d) $(5x - 3x^4)^2 = (5x)^2 - 2 \cdot 5x \cdot 3x^4 + (3x^4)^2$
$ = 25x^2 - 30x^5 + 9x^8$ Using the rules for exponents

CAUTION! Remember that the square of a sum is *not* the sum of the squares:

The term $2AB$ is missing.

$$(A + B)^2 \neq A^2 + B^2.$$

To confirm this inequality, note that

$$(20 + 5)^2 = 25^2 = 625,$$

whereas

$$20^2 + 5^2 = 400 + 25 = 425, \quad \text{and } 425 \neq 625.$$

Geometrically, $(A + B)^2$ can be viewed as shown. The area of the large square is

$$(A + B)(A + B) = (A + B)^2.$$

This is equal to the sum of the areas of the smaller rectangles:

$$A^2 + AB + AB + B^2 = A^2 + 2AB + B^2.$$

Thus,

$$(A + B)^2 = A^2 + 2AB + B^2.$$

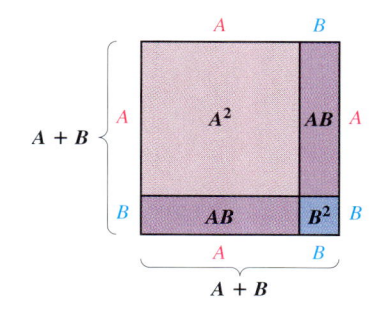

Multiplications of Various Types

Recognizing patterns often helps when new problems are encountered. To simplify a new multiplication problem, always examine what type of product it is so that the best method for finding that product can be used. To do this, ask yourself questions similar to the following.

MULTIPLYING TWO POLYNOMIALS

1. Is the multiplication the product of a monomial and a polynomial? If so, multiply each term of the polynomial by the monomial.
2. Is the multiplication the product of two binomials? If so:
   a) Is it the product of the sum and difference of the *same* two terms? If so, use the pattern
   $$(A + B)(A - B) = A^2 - B^2.$$
   b) Is the product the square of a binomial? If so, use the pattern
   $$(A + B)(A + B) = (A + B)^2 = A^2 + 2AB + B^2,$$
   or
   $$(A - B)(A - B) = (A - B)^2 = A^2 - 2AB + B^2.$$
   c) Is it the product of two binomials other than those above? If so, use FOIL.
3. Is the multiplication the product of two polynomials other than those above? If so, multiply each term of one by every term of the other. Use columns if you wish.

EXAMPLE 5

Multiply.

a) $(x + 3)(x - 3)$ b) $(t + 7)(t - 5)$ c) $(x + 7)(x + 7)$
d) $2x^3(9x^2 + x - 7)$ e) $(p + 3)(p^2 + 2p - 1)$ f) $\left(3x + \tfrac{1}{4}\right)^2$

SOLUTION

a) $(x + 3)(x - 3) = x^2 - 9$ The product of the sum and difference of the same two terms

b) $(t + 7)(t - 5) = t^2 - 5t + 7t - 35$ Using FOIL
$= t^2 + 2t - 35$

c) $(x + 7)(x + 7) = x^2 + 14x + 49$ The product is the square of a binomial.

d) $2x^3(9x^2 + x - 7) = 18x^5 + 2x^4 - 14x^3$ Multiplying each term of the trinomial by the monomial

e)
$$\begin{array}{r} p^2 + 2p - 1 \\ p + 3 \\ \hline 3p^2 + 6p - 3 \\ p^3 + 2p^2 - p \\ \hline p^3 + 5p^2 + 5p - 3 \end{array}$$
Using columns to multiply a binomial and a trinomial

Multiplying by 3
Multiplying by p

f) $\left(3x + \tfrac{1}{4}\right)^2 = 9x^2 + 2(3x)\left(\tfrac{1}{4}\right) + \tfrac{1}{16}$ Squaring a binomial
$= 9x^2 + \tfrac{3}{2}x + \tfrac{1}{16}$

EXERCISE SET 4.5

Multiply.

1. $(x + 5)(x^2 + 1)$
2. $(x^2 - 3)(x - 1)$
3. $(x^3 + 6)(x + 2)$
4. $(x^4 + 2)(x + 12)$
5. $(y + 2)(y - 3)$
6. $(a + 2)(a + 2)$
7. $(3x + 2)(3x + 3)$
8. $(4x + 1)(2x + 2)$
9. $(5x - 6)(x + 2)$
10. $(t - 9)(t + 9)$
11. $(3t - 1)(3t + 1)$
12. $(2m + 3)(2m + 3)$
13. $(2x - 7)(x - 1)$
14. $(2x - 1)(3x + 1)$
15. $\left(p - \frac{1}{4}\right)\left(p + \frac{1}{4}\right)$
16. $\left(q + \frac{3}{4}\right)\left(q + \frac{3}{4}\right)$
17. $(x - 0.1)(x + 0.1)$
18. $(x + 0.3)(x - 0.4)$
19. $(2x^2 + 6)(x + 1)$
20. $(2x^2 + 3)(2x - 1)$
21. $(-2x + 1)(x + 6)$
22. $(3x + 4)(2x - 4)$
23. $(a + 7)(a + 7)$
24. $(2y + 7)(2y + 7)$
25. $(1 + 3t)(1 - 5t)$
26. $(-3x - 2)(x + 1)$
27. $(x^2 + 3)(x^3 - 1)$
28. $(x^4 - 3)(2x + 1)$
29. $(3x^2 - 2)(x^4 - 2)$
30. $(x^{10} + 3)(x^{10} - 3)$
31. $(3x^5 + 2)(2x^2 + 6)$
32. $(1 - 2x)(1 + 3x^2)$
33. $(8x^3 + 5)(x^2 + 2)$
34. $(4 - 2x)(5 - 2x^2)$
35. $(4x^2 + 3)(x - 3)$
36. $(7x - 2)(2x - 7)$

Multiply. Try to recognize what type of product each multiplication is before multiplying.

37. $(x + 8)(x - 8)$
38. $(x + 1)(x - 1)$
39. $(2x + 1)(2x - 1)$
40. $(x^2 + 1)(x^2 - 1)$
41. $(5m - 2)(5m + 2)$
42. $(3x^4 + 2)(3x^4 - 2)$
43. $(2x^2 + 3)(2x^2 - 3)$
44. $(6x^5 - 5)(6x^5 + 5)$
45. $(3x^4 - 1)(3x^4 + 1)$
46. $(t^2 - 0.2)(t^2 + 0.2)$
47. $(x^6 - x^2)(x^6 + x^2)$
48. $(2x^3 - 0.3)(2x^3 + 0.3)$
49. $(x^4 + 3x)(x^4 - 3x)$
50. $\left(\frac{3}{4} + 2x^3\right)\left(\frac{3}{4} - 2x^3\right)$
51. $(2y^8 + 3)(2y^8 - 3)$
52. $\left(m - \frac{2}{3}\right)\left(m + \frac{2}{3}\right)$
53. $(x + 2)^2$
54. $(2x - 1)^2$
55. $(3x^2 + 1)^2$
56. $\left(3x + \frac{3}{4}\right)^2$
57. $\left(a - \frac{2}{5}\right)^2$
58. $\left(2a - \frac{1}{5}\right)^2$
59. $(x^2 + 3)^2$
60. $(8x - x^2)^2$
61. $(2 - 3x^4)^2$
62. $(6x^3 - 2)^2$
63. $(5 + 6t^2)^2$
64. $(3p^2 - p)^2$
65. $(7x - 0.3)^2$
66. $(4a - 0.6)^2$
67. $5a^3(2a^2 - 1)$
68. $(a - 3)(a^2 + 2a - 4)$
69. $(x^2 - 5)(x^2 + x - 1)$
70. $9x^4(3x^2 - x)$
71. $(3 - 2x^3)^2$
72. $(x - 4x^3)^2$
73. $4x(x^2 + 6x - 3)$
74. $8x(-x^5 + 6x^2 + 9)$
75. $\left(2x^2 - \frac{1}{2}\right)\left(2x^2 - \frac{1}{2}\right)$
76. $(-x^2 + 1)^2$
77. $(-1 + 3p)(1 + 3p)$
78. $(-3q + 2)(3q + 2)$
79. $3t^2(5t^3 - t^2 + t)$
80. $-5x^3(x^2 + 8x - 9)$
81. $(6x^4 - 3)^2$
82. $(8a + 5)^2$
83. $(3x + 2)(4x^2 + 5)$
84. $(2x^2 - 7)(3x^2 + 9)$
85. $(8 - 6x^4)^2$
86. $\left(\frac{1}{3}t^2 + 5\right)\left(\frac{2}{3}t^2 - 1\right)$
87. $(a + 1)(a^2 - a + 1)$
88. $(x - 5)(x^2 + 5x + 25)$

Find the total area of all shaded rectangles.

89.

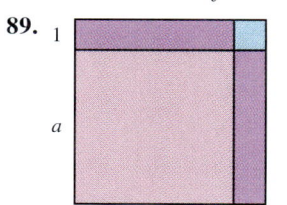

90.

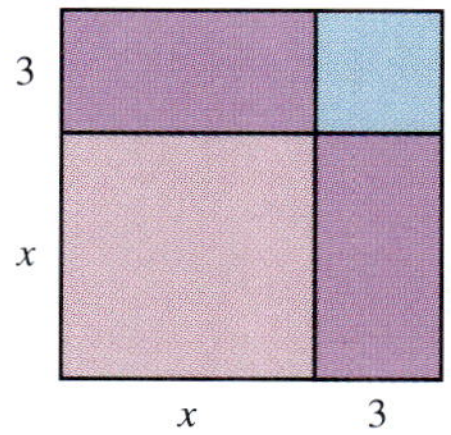

91.

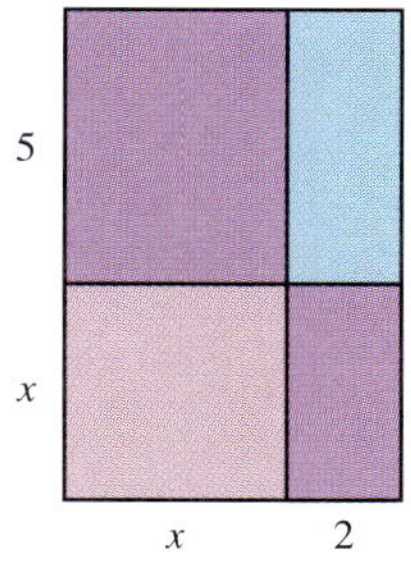

92.

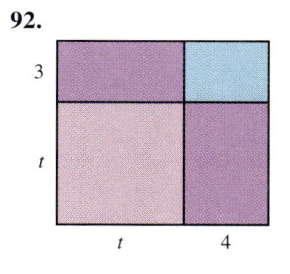

93.

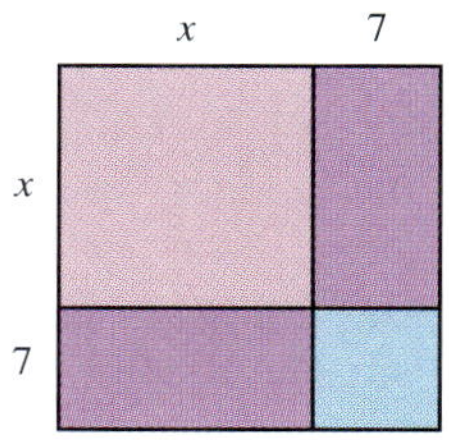

94.

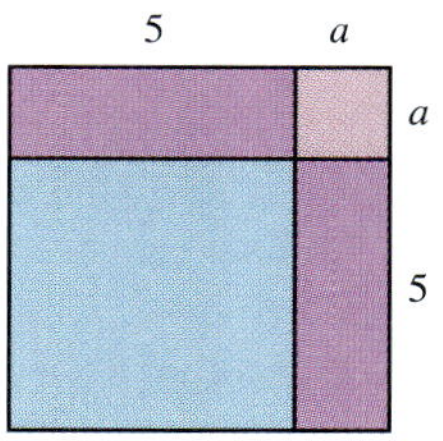

95. [figure: rectangle with sides $t+4$ and $9+t$]

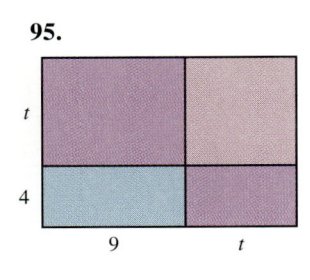

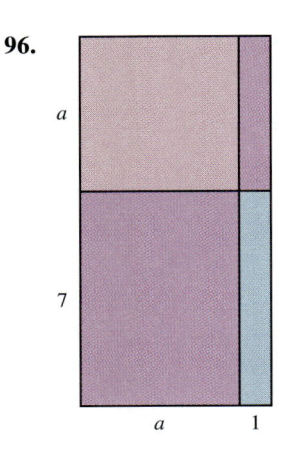

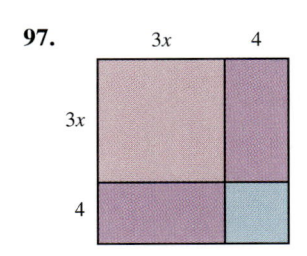

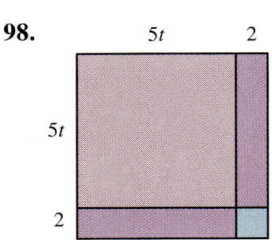

96. [figure: rectangle with sides $a+1$ and $7+a$... wait, labeled a, 7, a, 1]

97. [figure: square with sides $3x+4$]

98. [figure: square with sides $5t+2$]

Draw and label rectangles similar to those in Exercises 89–98 to illustrate each of the following.

99. $(x + 6)^2$
100. $(x + 8)^2$
101. $(t + 9)^2$
102. $(a + 12)^2$
103. $(4a + 1)^2$
104. $(2t + 3)^2$

SKILL MAINTENANCE

105. *Energy use.* In an apartment, lamps, an air conditioner, and a television set are all operating at the same time. The lamps take 10 times as many watts as the television set, and the air conditioner takes 40 times as many watts as the television set. The total wattage used in the apartment is 2550 watts. How many watts are used by each appliance?

106. Solve: $3x - 8x = 4(7 - 8x)$.

107. Solve $ab - c = ad$ for a.

108. In what quadrant is the point $(2, -5)$ located?

SYNTHESIS

109. ◆ Under what conditions is the product of two binomials a binomial?

110. ◆ Todd feels that since he can find the product of any two binomials using FOIL, he needn't study the other special products. What advice would you give him?

111. ◆ Anais claims that by writing $19 \cdot 21$ as $(20 - 1)(20 + 1)$, she can find the product mentally. How is this possible?

112. ◆ The product $(A + B)^2$ can be regarded as the sum of the areas of four regions (as shown following Example 4). How might one visually represent $(A + B)^3$? Why?

Multiply.

113. $5x(3x - 1)(2x + 3)$
114. $[(2x - 3)(2x + 3)](4x^2 + 9)$
115. $[(a - 5)(a + 5)]^2$
116. $(a - 3)^2(a + 3)^2$
[*Hint*: Examine Exercise 115.]
117. $(3t^4 - 2)^2(3t^4 + 2)^2$
[*Hint*: Examine Exercise 115.]
118. ▦ $(32.41x + 5.37)^2$

Calculate as the difference of squares.

119. 18×22 [*Hint*: $(20 - 2)(20 + 2)$.]
120. 93×107

Solve.

121. $(x + 2)(x - 5) = (x + 1)(x - 3)$
122. $(2x + 5)(x - 4) = (x + 5)(2x - 4)$

The height of a box is 1 more than its length l, and the length is 1 more than its width w. Find a polynomial for the volume V in terms of the following.

123. The length l
124. The width w

Find two expressions for the total shaded area in each figure.

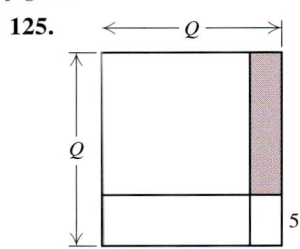

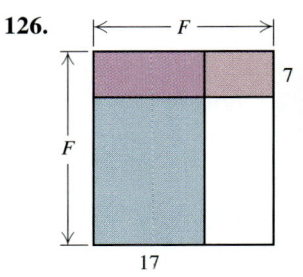

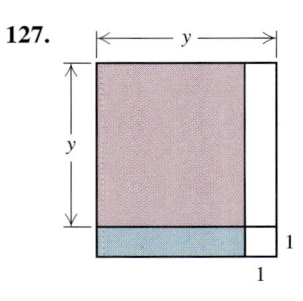

125. [figure: square of side Q, with subregion 14 by 5]

126. [figure: square of side F, with corner 17 by 7]

127. [figure: square of side y, with 1 by 1 corner]

128. Find three consecutive integers for which the sum of the squares is 65 more than 3 times the square of the smallest integer.

129. A polynomial for the shaded area in this rectangle is $(A + B)(A - B)$.

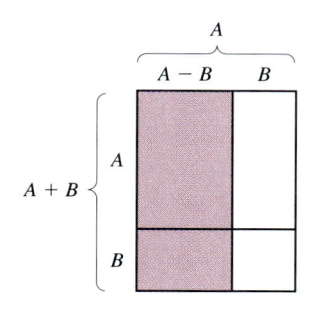

a) Find a polynomial for the area of the entire figure.
b) Find a polynomial for the sum of the areas of the two small unshaded rectangles.
c) Find a polynomial for the area in part (a) minus the area in part (b).
d) Find a polynomial for the area of the shaded region and compare this with the polynomial found in part (c).

130. Use a grapher to check your answers to Exercises 17, 49, and 87.

4.6 Polynomials in Several Variables

Evaluating Polynomials • Like Terms and Degree • Addition and Subtraction • Multiplication

Thus far, the polynomials that we have studied have had only one variable. Polynomials such as

$$5x + x^2y - 3y + 7, \quad 9ab^2c - 2a^3b^2 + 8a^2b^3 + 15, \quad \text{and} \quad 4m^2 - 9n^2$$

contain two or more variables. In this section, we will add, subtract, multiply, and evaluate such **polynomials in several variables.**

Evaluating Polynomials

To evaluate a polynomial in two or more variables, we substitute numbers for the variables. Then we compute, using the rules for order of operations.

EXAMPLE 1 Evaluate the polynomial $4 + 3x + xy^2 + 8x^3y^3$ for $x = -2$ and $y = 5$.

SOLUTION We replace x with -2 and y with 5:

$$4 + 3x + xy^2 + 8x^3y^3 = 4 + 3(-2) + (-2) \cdot 5^2 + 8(-2)^3 \cdot 5^3$$
$$= 4 - 6 - 50 - 8000 = -8052.$$

EXAMPLE 2

Surface Area of a Right Circular Cylinder. The surface area of a right circular cylinder is given by the polynomial

$$2\pi rh + 2\pi r^2,$$

where h is the height and r is the radius of the base. A 12-oz beverage can has a height of 4.7 in. and a radius of 1.2 in. Approximate its surface area.

SOLUTION We evaluate the polynomial for $h = 4.7$ and $r = 1.2$. If 3.14 is used to approximate π, we have

$$\begin{aligned}2\pi rh + 2\pi r^2 &\approx 2(3.14)(1.2)(4.7) + 2(3.14)(1.2)^2 \\ &\approx 2(3.14)(1.2)(4.7) + 2(3.14)(1.44) \\ &\approx 35.4192 + 9.0432 = 44.4624.\end{aligned}$$

If the π key of a calculator is used, we have

$$\begin{aligned}2\pi rh + 2\pi r^2 &\approx 2(3.141592654)(1.2)(4.7) + 2(3.141592654)(1.2)^2 \\ &\approx 44.48495198.\end{aligned}$$

The surface area is about 44.5 in^2 (square inches).

Like Terms and Degree

To determine the **degree of a term** in a polynomial of several variables, we add the exponents of the variables in that term. Thus, in Example 1,

the degree of the term $8x^3y^3$ is $3 + 3$, or 6,

and

the degree of the term xy^2, or x^1y^2, is $1 + 2$, or 3.

As we learned in Section 4.2, the degree of a polynomial is the degree of the term of highest degree.

EXAMPLE 3

Identify the coefficient and the degree of each term and the degree of the polynomial

$$9x^2y^3 - 14xy^2z^3 + xy + 4y + 5x^2 + 7.$$

SOLUTION

Term	Coefficient	Degree	Degree of the Polynomial
$9x^2y^3$	9	5	
$-14xy^2z^3$	-14	6	6
xy	1	2	
$4y$	4	1	
$5x^2$	5	2	
7	7	0	

Note in Example 3 that although both xy and $5x^2$ have degree 2, they are *not* like terms. *Like,* or *similar, terms* either have exactly the same variables with exactly the same exponents or are constants. For example,

$8a^4b^7$ and $-3a^4b^7$ are like terms

and

17 and 3 are like terms,

but

$5x^2y$ and $9xy^2$ are *not* like terms.

As always, combining like terms is based on the distributive law.

EXAMPLE 4 Combine like terms.

a) $9x^2y + 3xy^2 - 5x^2y - xy^2$
b) $7ab - 5ab^2 + 3ab^2 + 6a^3 + 9ab - 11a^3 + b - 1$

SOLUTION
a) $9x^2y + 3xy^2 - 5x^2y - xy^2 = (9 - 5)x^2y + (3 - 1)xy^2$
$= 4x^2y + 2xy^2$ Try to go directly to this step.

b) $7ab - 5ab^2 + 3ab^2 + 6a^3 + 9ab - 11a^3 + b - 1$
$= -2ab^2 + 16ab - 5a^3 + b - 1$

Addition and Subtraction

The procedure used for adding polynomials in one variable is used to add polynomials in several variables.

EXAMPLE 5 Add.

a) $(-5x^3 + 3y - 5y^2) + (8x^3 + 4x^2 + 7y^2)$
b) $(5ab^2 - 4a^2b + 5a^3 + 2) + (3ab^2 - 2a^2b + 3a^3b - 5)$

SOLUTION
a) $(-5x^3 + 3y - 5y^2) + (8x^3 + 4x^2 + 7y^2)$
$= (-5 + 8)x^3 + 4x^2 + 3y + (-5 + 7)y^2$ Try to do this step mentally.
$= 3x^3 + 4x^2 + 3y + 2y^2$

b) $(5ab^2 - 4a^2b + 5a^3 + 2) + (3ab^2 - 2a^2b + 3a^3b - 5)$
$= 8ab^2 - 6a^2b + 5a^3 + 3a^3b - 3$

When subtracting a polynomial, remember to find the opposite of each term in that polynomial and then add.

EXAMPLE 6 Subtract: $(4x^2y + x^3y^2 + 3x^2y^3 + 6y) - (4x^2y - 6x^3y^2 + x^2y^2 - 5y)$.

SOLUTION
$$(4x^2y + x^3y^2 + 3x^2y^3 + 6y) - (4x^2y - 6x^3y^2 + x^2y^2 - 5y)$$
$$= 4x^2y + x^3y^2 + 3x^2y^3 + 6y - 4x^2y + 6x^3y^2 - x^2y^2 + 5y$$
$$= 7x^3y^2 + 3x^2y^3 - x^2y^2 + 11y \quad \text{Combining like terms}$$

Multiplication

To multiply polynomials in several variables, multiply each term of one polynomial by every term of the other, just as we did in Sections 4.4 and 4.5.

EXAMPLE 7 Multiply: $(3x^2y - 2xy + 3y)(xy + 2y)$.

SOLUTION
$$\begin{array}{r} 3x^2y - 2xy + 3y \\ xy + 2y \\ \hline 6x^2y^2 - 4xy^2 + 6y^2 \\ 3x^3y^2 - 2x^2y^2 + 3xy^2 \\ \hline 3x^3y^2 + 4x^2y^2 - xy^2 + 6y^2 \end{array}$$

Multiplying by $2y$
Multiplying by xy
Adding

The special products that we have studied can be used to speed up our multiplication of polynomials in several variables.

EXAMPLE 8 Multiply.

a) $(p + 5q)(2p - 3q)$
b) $(3x + 2y)^2$
c) $(a^3 - 7a^2b)^2$
d) $(3x^2y + 2y)(3x^2y - 2y)$
e) $(-2x^3y^2 + 5t)(2x^3y^2 + 5t)$
f) $(2x + 3 - 2y)(2x + 3 + 2y)$

SOLUTION

$$\text{F}\text{O}\text{I}\text{L}$$
a) $(p + 5q)(2p - 3q) = 2p^2 - 3pq + 10pq - 15q^2$
$ = 2p^2 + 7pq - 15q^2$ Combining like terms

$$(A + B)^2 = A^2 + 2 \cdot A \cdot B + B^2$$
b) $(3x + 2y)^2 = (3x)^2 + 2(3x)(2y) + (2y)^2$ Squaring a binomial
$ = 9x^2 + 12xy + 4y^2$

$$(A - B)^2 = A^2 - 2 \cdot A \cdot B + B^2$$
c) $(a^3 - 7a^2b)^2 = (a^3)^2 - 2(a^3)(7a^2b) + (7a^2b)^2$ Squaring a binomial
$ = a^6 - 14a^5b + 49a^4b^2$ Using the rules for exponents

$$(A + B)(A - B) = A^2 - B^2$$
d) $(3x^2y + 2y)(3x^2y - 2y) = (3x^2y)^2 - (2y)^2 = 9x^4y^2 - 4y^2$

e) $(-2x^3y^2 + 5t)(2x^3y^2 + 5t) = (5t - 2x^3y^2)(5t + 2x^3y^2)$ Using the commutative law for addition twice

$= (5t)^2 - (2x^3y^2)^2$ Multiplying the sum and the difference of the same two terms

$= 25t^2 - 4x^6y^4$

$(\;A\;-\;B\;)(\;A\;+\;B\;) = A^2 - B^2$
$\;\downarrow\;\;\;\;\;\downarrow\;\;\;\;\;\downarrow\;\;\;\;\;\downarrow\;\;\;\;\;\downarrow\;\;\;\;\;\downarrow$

f) $(\;2x + 3\;-\;2y)(\;2x + 3\;+\;2y) = (\;2x + 3\;)^2 - (2y)^2$ Multiplying a sum and a difference

$= 4x^2 + 12x + 9 - 4y^2$ Squaring a binomial

Note that in Example 8 we recognized patterns that might have eluded some students, particularly in parts (e) and (f). In part (e), we could have used FOIL, and in part (f), we could have used long multiplication, but doing so would have been slower. By carefully inspecting a problem before "jumping in," we can often save ourselves considerable work.

EXERCISE SET 4.6

Evaluate each polynomial for $x = 3$ and $y = -2$.

1. $x^2 - 3y^2 + 2xy$ **2.** $x^2 + 5y^2 - 4xy$

Evaluate each polynomial for $x = 2$, $y = -3$, and $z = -1$.

3. $xyz^2 - z$ **4.** $xy - xz + yz$

Lung capacity. The polynomial
  $0.041h - 0.018A - 2.69$
can be used to estimate the lung capacity, in liters, of a female with height h, in centimeters, and age A, in years.

5. Find the lung capacity of a 50-yr-old woman who is 160 cm tall.

6. Find the lung capacity of a 20-yr-old woman who is 165 cm tall.

Altitude of a launched object. *The altitude of an object, in meters, is given by the polynomial*
  $h + vt - 4.9t^2$,
where h is the height, in meters, at which the launch occurs, v is the initial upward speed (or velocity), in meters per second, and t is the number of seconds for which the object is airborne.

7. A model rocket is launched from atop the Leaning Tower of Pisa, 50 m above the ground. If the initial upward speed is 40 meters per second (m/s), how high above the ground will the rocket be 2 sec after having been launched?

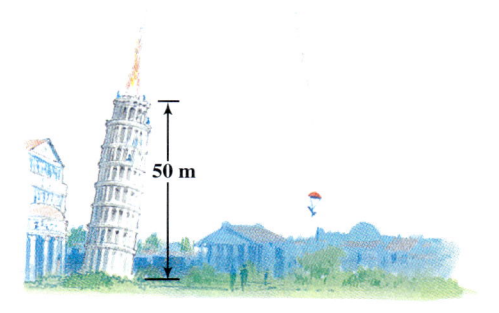

8. A golf ball is thrown upward with an initial speed of 30 m/s by a golfer atop the Washington Monument, 160 m above the ground. How high above the ground will the ball be after 3 sec?

Surface area of a silo. A silo is a structure that is shaped like a right circular cylinder with a half sphere on top. The surface area of a silo of height h and radius r (including the area of the base) is given by the polynomial $2\pi rh + \pi r^2$.

9. A $1\frac{1}{2}$-oz bottle of roll-on deodorant has a height of 4 in. and a radius of $\frac{3}{4}$ in. Find the surface area of the bottle if the bottle is shaped like a silo. Use 3.14 for π.

10. A container of tennis balls is silo-shaped, with a height of $7\frac{1}{2}$ in. and a radius of $1\frac{1}{4}$ in. Find the surface area of the container. Use 3.14 for π.

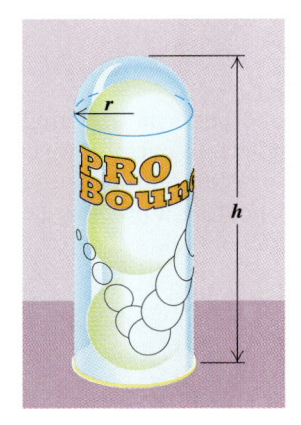

Identify the coefficient and the degree of each term of each polynomial. Then find the degree of each polynomial.

11. $x^3y - 2xy + 3x^2 - 5$
12. $5y^3 - y^2 + 15y + 1$
13. $17x^2y^3 - 3x^3yz - 7$
14. $6 - xy + 8x^2y^2 - y^5$

Combine like terms.

15. $5a + b - 4a - 3b$
16. $y^2 - 1 + y - 6 - y^2$
17. $3x^2y - 2xy^2 + x^2$
18. $m^3 + 2m^2n - 3m^2 + 3mn^2$
19. $2u^2v - 3uv^2 + 6u^2v - 2uv^2$
20. $3x^2 + 6xy + 3y^2 - 5x^2 - 10xy - 5y^2$
21. $8uv + 3av + 14au + 7av$
22. $3x^2y - 2z^2y + 3xy^2 + 5z^2y$

Add or subtract, as indicated.

23. $(2x^2 - xy + y^2) + (-x^2 - 3xy + 2y^2)$
24. $(r^3 + 3rs - 5s^2) - (5r^3 + rs + 4s^2)$
25. $(7a^4 - 5ab + 6ab^2) - (9a^4 + 3ab - ab^2)$
26. $(r - 2s + 3) + (2r + 3s - 7)$
27. $(b^3a^2 - 2b^2a^3 + 3ba + 4) + (b^2a^3 - 4b^3a^2 + 2ba - 1)$
28. $(2x^2 - 3xy + y^2) + (-4x^2 - 6xy - y^2) + (x^2 + xy - y^2)$
29. $(x^3 - y^3) - (-2x^3 + x^2y - xy^2 + 2y^3)$
30. $(xy - ab) - (xy - 3ab)$
31. $(3y^4x^2 + 2y^3x - 3y) - (2y^4x^2 + 2y^3x - 4y - 2x)$
32. $(5a^2b + 7ab) + (9a^2b - 5ab) + (a^2b - 6ab)$
33. Subtract $7x + 3y$ from the sum of $4x + 5y$ and $-5x + 6y$.
34. Subtract $5a + 2b$ from the sum of $2a + b$ and $3a - 4b$.

Multiply.

35. $(3z - u)(2z + 3u)$
36. $(a^2b - 2)(a^2b - 5)$
37. $(xy + 7)(xy - 4)$
38. $(a^3 + bc)(a^3 - bc)$
39. $(m^2 + n^2 - mn)(m^2 + mn + n^2)$
40. $(y^4x + y^2 + 1)(y^2 + 1)$
41. $(a - b)(a^2 + ab + b^2)$
42. $(3xy - 1)(4xy + 2)$
43. $(m^3n + 8)(m^3n - 6)$
44. $(3 - c^2d^2)(4 + c^2d^2)$
45. $(6x - 2y)(5x - 3y)$
46. $(m^2 - n^2)(m + n)$
47. $(pq + 0.2)(0.4pq - 0.1)$
48. $(xy + x^5y^5)(x^4y^4 - xy)$
49. $(x + h)^2$
50. $(3a + 2b)^2$
51. $(r^3t^2 - 4)^2$
52. $(3a^2b - b^2)^2$
53. $(c^2 - d)(c^2 + d)$
54. $(p^3 - 5q)(p^3 + 5q)$
55. $(ab + cd^2)(ab - cd^2)$
56. $(xy + pq)(xy - pq)$
57. $(x + y - 3)(x + y + 3)$
58. $[x + y + z][x - (y + z)]$
59. $[a + b + c][a - (b + c)]$
60. $(a + b + c)(a - b - c)$

Find the total area of each shaded area.

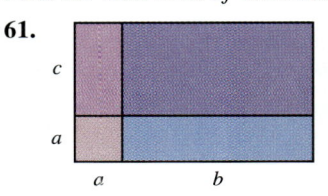

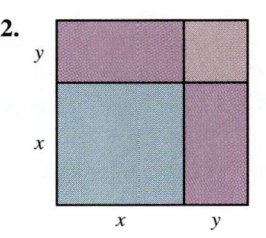

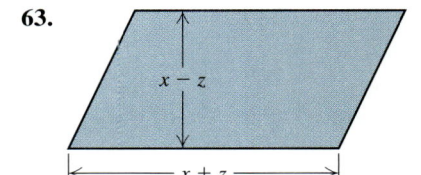

61.

62.

63.

64.

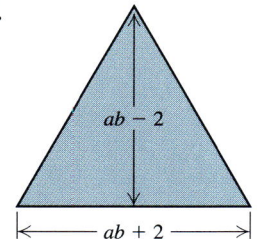

[Triangle with height $ab - 2$ and base $ab + 2$]

65.

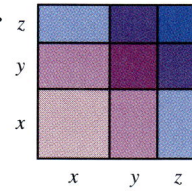

[Grid with columns labeled x, y, z and rows labeled z, y, x]

66.

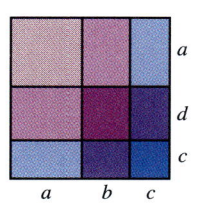

[Grid with columns labeled a, b, c and rows labeled a, d, c]

67.

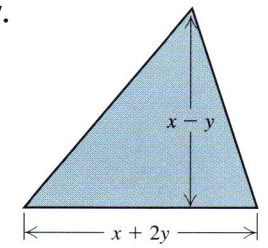

[Triangle with height $x - y$ and base $x + 2y$]

68.

[Parallelogram with height $m + n$ and base $m - n$]

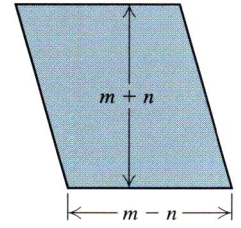

Draw and label rectangles similar to those in Exercises 61, 62, 65, and 66 to illustrate each product.

69. $(r + s)(u + v)$ **70.** $(m + r)(n + v)$

71. $(a + b + c)(a + d + f)$ **72.** $(r + s + t)^2$

SKILL MAINTENANCE

The graph at right shows the prices paid for a ton of white office paper and for a ton of newsprint by recyclers in the Boston market. (Source: Burlington Free Press, 7/13/92, Burlington, VT. Reprinted with permission.)

73. How much was being paid for white office paper in June 1994?

74. At what date did recyclers switch from paying for newsprint to charging for accepting it?

75. When did the value of newsprint peak?

76. When did the price that was paid for white office paper drop to $20 per ton?

77. During what six-month period did the price paid for newsprint increase the most?

78. During what six-month period did the price paid for white office paper decrease the most?

SYNTHESIS

79. ◆ Can the sum of two trinomials in several variables be a binomial in one variable? Why or why not?

80. ◆ Is it possible for a polynomial in 4 variables to have a degree less than 4? Why or why not?

81. ◆ A fourth-degree polynomial is multiplied by a third-degree polynomial. What is the degree of the product? Explain your reasoning.

82. ◆ Can the sum of two trinomials in several variables be a trinomial in one variable? Why or why not?

Find a polynomial for the shaded area. (Leave results in terms of π where appropriate.)

83.

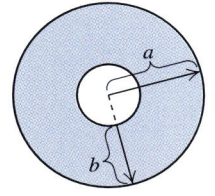

[Circle with outer radius b and inner radius a]

84.

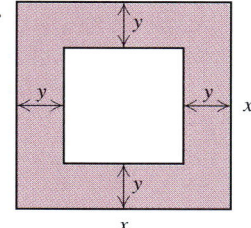

[Square with side x containing a smaller square; distance y between them on each side]

85.

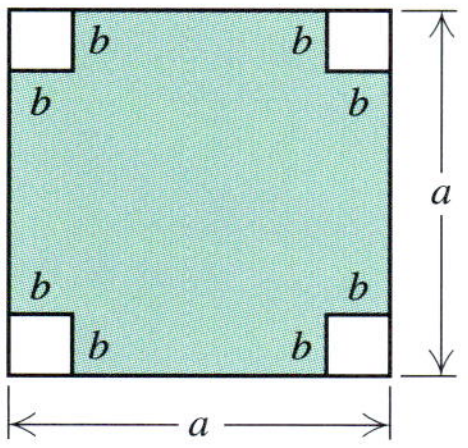

[Square with side a having smaller squares of side b cut from each corner]

86.

[Rounded rectangle shape with middle rectangle of height y and semicircular ends of radius x]

[Graph showing prices paid per ton in the Boston market from Jun 1991 to Jun 1996, with curves for White office paper and Newsprint, y-axis from -$60 to $140]

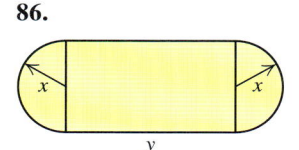

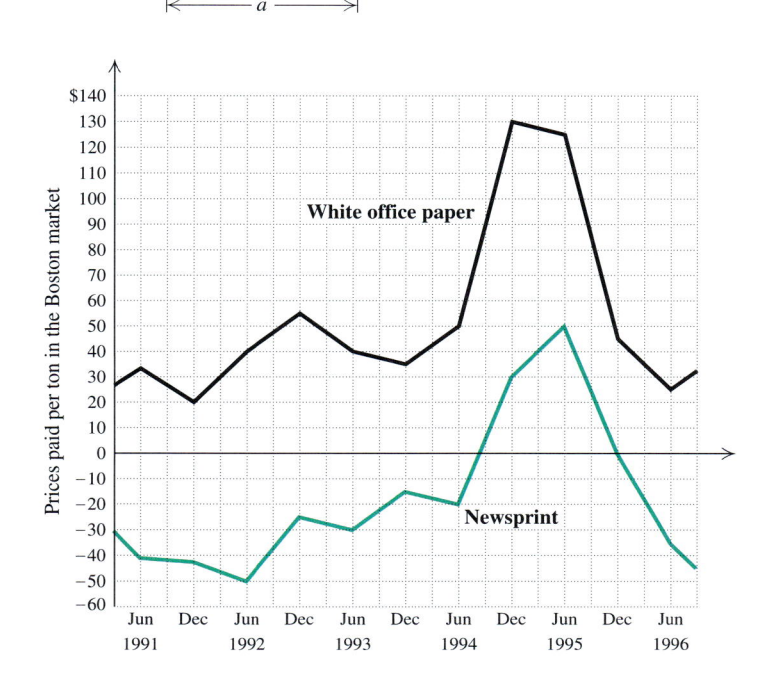

Find a polynomial for the surface area of each solid object shown. (Leave results in terms of π.)

87.

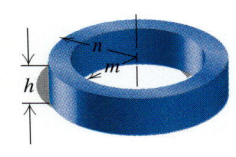

88.

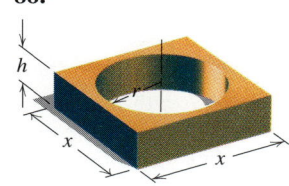

89. *Interest compounded annually.* An amount of money P that is invested at the yearly interest rate r grows to the amount

$$P(1 + r)^t$$

after t years. Find a polynomial that can be used to determine the amount to which P will grow after 2 yr.

90. *Yearly depreciation.* An investment P that drops in value at the yearly rate r drops in value to

$$P(1 - r)^t$$

after t years. Find a polynomial that can be used to determine the value to which P has dropped after 2 yr.

91. ◆ The observatory at Danville University is shaped like a silo that is 40 ft high and 30 ft wide (see Exercise 9). The Heavenly Bodies Astronomy Club is to paint the exterior of the observatory using paint that covers 250 ft² per gallon. How many gallons should they purchase? Explain your reasoning.

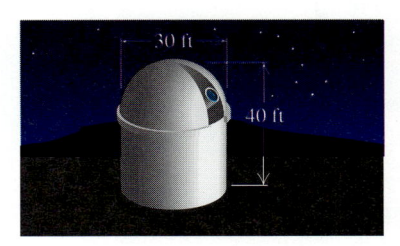

COLLABORATIVE CORNER

Focus: Evaluating polynomials in several variables

Time: 15–25 minutes

Group size: 3

Materials: A coin for each person

As a team nears the end of its schedule in first place, fans begin to discuss the team's "magic number." A team's magic number is the combined number of wins by that team and losses by the second-place team that guarantee the leading team a first-place finish. For example, if the Cubs' magic number is 3 over the Reds, any combination of Cubs wins and Reds losses that totals 3 will guarantee a first-place finish for the Cubs, regardless of how subsequent games are decided. A team's magic number is computed using the polynomial

$$G - P - L + 1,$$

where G is the length of the season, in games, P is the number of games that the leading team has played, and L is the total number of games that the second-place team has lost minus the total number of games that the leading team has lost.

ACTIVITY

1. The standings below are from a fictitious baseball league. Together, the group should calculate the Jaguars' magic number with respect to the Catamounts as well as the Jaguars' magic number with respect to the Wildcats. (Assume that the schedule is 162 games long.)

	W	L
Jaguars	90	62
Catamounts	88	64
Wildcats	87	64

2. Each group member should play the role of one of the teams. To simulate each team's remaining games, coin tosses will be performed. If a group member correctly predicts the side (heads or tails) that comes up, the coin toss represents a win for that team. Should the other side appear, the toss represents a

loss. Assume that all remaining games are against other (unlisted) teams in the league. Each group member should perform four coin tosses and then update the standings.
3. Recalculate the two magic numbers, using the updated standings from part (2).
4. Slowly—one coin toss at a time—play out the remainder of the season. Record all wins and losses, update the standings, and recalculate the magic numbers each time all three group members have completed a round of coin tosses.
5. Examine the work in part (4) and explain why a magic number of 0 indicates that a team has been eliminated from contention.

4.7 Division of Polynomials

Dividing by a Monomial • Dividing by a Binomial

In this section, we consider the division of polynomials. You will see that polynomial division is similar to division in arithmetic.

Dividing by a Monomial

We first consider division by a monomial. When dividing a monomial by a monomial, we use the quotient rule of Section 4.1 to subtract exponents when bases are the same. For example,

$$\frac{15x^{10}}{3x^4} = 5x^{10-4}$$
$$= 5x^6$$

CAUTION! The coefficients are divided but the exponents are subtracted.

and

$$\frac{42a^2b^5}{-3ab^2} = \frac{42}{-3}a^{2-1}b^{5-2}$$
$$= -14ab^3.$$

To divide a polynomial by a monomial, we note that since

$$\frac{A}{C} + \frac{B}{C} = \frac{A+B}{C},$$

it follows that

$$\frac{A+B}{C} = \frac{A}{C} + \frac{B}{C}.$$

This is actually how we perform divisions like $86 \div 2$: Although we might simply write

$$\frac{86}{2} = 43,$$

we are really saying

$$\frac{80+6}{2} = \frac{80}{2} + \frac{6}{2} = 40 + 3.$$

Similarly, to divide a polynomial by a monomial, we divide each term by the monomial.

EXAMPLE 1 Divide $9x^8 + 12x^6$ by $3x^2$.

SOLUTION This is equivalent to

$$\frac{9x^8}{3x^2} + \frac{12x^6}{3x^2}. \quad \text{To see this, add and get the original expression.}$$

We can now perform the separate divisions:

$$\frac{9x^8}{3x^2} + \frac{12x^6}{3x^2} = \frac{9}{3}x^{8-2} + \frac{12}{3}x^{6-2} \quad \text{Dividing coefficients and subtracting exponents}$$

$$= 3x^6 + 4x^4.$$

To check, we multiply the quotient by $3x^2$:

$$(3x^6 + 4x^4)3x^2 = 9x^8 + 12x^6. \quad \text{The answer checks.}$$

EXAMPLE 2 Divide and check.

a) $(x^3 + 10x^2 - 8x) \div 2x$
b) $(10a^5b^4 - 2a^3b^2 + 6a^2b) \div 2a^2b$

SOLUTION

a) $$\frac{x^3 + 10x^2 - 8x}{2x} = \frac{x^3}{2x} + \frac{10x^2}{2x} - \frac{8x}{2x}$$

$$= \frac{1}{2}x^{3-1} + \frac{10}{2}x^{2-1} - \frac{8}{2}x^{1-1} \quad \text{Dividing coefficients and subtracting exponents}$$

$$= \frac{1}{2}x^2 + 5x - 4$$

Check: We check by multiplying the quotient by $2x$:

$$\begin{array}{r} \frac{1}{2}x^2 + 5x - 4 \\ \underline{ 2x} \\ x^3 + 10x^2 - 8x \end{array} \quad \begin{array}{l} \text{Multiplying} \\ \\ \text{The answer checks.} \end{array}$$

b) $$\frac{10a^5b^4 - 2a^3b^2 + 6a^2b}{2a^2b} = \frac{10a^5b^4}{2a^2b} - \frac{2a^3b^2}{2a^2b} + \frac{6a^2b}{2a^2b}$$

$$= \frac{10}{2}a^{5-2}b^{4-1} - \frac{2}{2}a^{3-2}b^{2-1} + \frac{6}{2}$$

$$= 5a^3b^3 - ab + 3$$

Check:
$$5a^3b^3 - ab + 3$$
$$\underline{2a^2b}$$ Multiplying
$$10a^5b^4 - 2a^3b^2 + 6a^2b$$ The answer checks.

Dividing by a Binomial

For divisors with more than one term, we use long division, much as we do in arithmetic. Polynomials are written in descending order and any missing terms are written in using 0 for the coefficients.

EXAMPLE 3   Divide $x^2 + 5x + 6$ by $x + 2$.

SOLUTION   We have

$$\begin{array}{r} x \\ x+2\overline{)x^2 + 5x + 6} \\ \underline{x^2 + 2x} \\ 3x \end{array}$$

Divide the first term, x^2, by the first term in the divisor: $x^2/x = x$. Ignore the term 2 for the moment.

Multiply x above by the divisor, $x + 2$.

Subtract: $(x^2 + 5x) - (x^2 + 2x) = x^2 + 5x - x^2 - 2x = 3x$.

Now we "bring down" the next term—in this case, 6—and repeat the procedure:

$$\begin{array}{r} x + 3 \\ x+2\overline{)x^2 + 5x + 6} \\ \underline{x^2 + 2x} \\ 3x + 6 \\ \underline{3x + 6} \\ 0 \end{array}$$

Consider the "remainder" $3x + 6$. Divide its first term by the first term of the divisor: $3x/x = 3$.

The 6 has been "brought down."

Multiply 3 by the divisor, $x + 2$.

Subtract: $(3x + 6) - (3x + 6) = 0$.

The quotient is $x + 3$. The remainder is 0, expressed as R 0. A remainder of 0 is generally not listed in an answer.

Check: To check, we multiply the quotient by the divisor and add the remainder, if any, to see if we get the dividend:

$$\underbrace{(x+2)}_{\text{Divisor}} \underbrace{(x+3)}_{\text{Quotient}} + \underbrace{0}_{\text{Remainder}} = \underbrace{x^2 + 5x + 6}_{\text{Dividend}}.$$ The division checks.

EXAMPLE 4   Divide: $(2x^2 + 5x - 1) \div (2x - 1)$.

SOLUTION   We have

$$\begin{array}{r} x \\ 2x-1\overline{)2x^2 + 5x - 1} \\ \underline{2x^2 - x} \\ 6x \end{array}$$

Divide the first term by the first term: $2x^2/(2x) = x$.

Multiply x above by the divisor, $2x - 1$.

$(2x^2 + 5x) - (2x^2 - x) = 2x^2 + 5x - 2x^2 + x = 6x$.

Now, we bring down the next term of the dividend, -1.

$$
\begin{array}{r}
x + 3 \\
2x - 1 \overline{\smash{\big)}\, 2x^2 + 5x - 1} \\
\underline{2x^2 - x} \\
6x - 1 \\
\underline{6x - 3} \\
2
\end{array}
$$

- $x+3$ ← Divide the first term of $6x - 1$ by the first term of the divisor: $6x/(2x) = 3$.
- $6x - 1$ ← The -1 has been "brought down."
- $6x - 3$ ← Multiply 3 by the divisor, $2x - 1$.
- 2 ← Subtract: $(6x - 1) - (6x - 3) = 6x - 1 - 6x + 3 = 2$.

The answer is $x + 3$ with R 2, or

Quotient $\underbrace{x + 3}$ + $\dfrac{\overset{\text{Remainder}}{2}}{\underset{\text{Divisor}}{2x - 1}}$

(This is the way answers will be given at the back of the book.)

Check: In arithmetic, to check that $9 \div 4 = 2\frac{1}{4}$, we can multiply and add: $4 \cdot 2 + 1 = 9$. A similar procedure, used to check that

$$(2x^2 + 5x - 1) \div (2x - 1) = x + 3 + \frac{2}{2x - 1},$$

is to multiply the quotient by the divisor and add the remainder:

$$(2x - 1)(x + 3) + 2 = 2x^2 + 5x - 3 + 2$$
$$= 2x^2 + 5x - 1. \quad \text{Our answer checks.}$$

Our division procedure ends when the degree of the remainder is less than the degree of the divisor. Check that this was the case in Example 4.

EXAMPLE 5 Divide each of the following.

a) $(x^3 + 1) \div (x + 1)$
b) $(x^4 - 3x^2 + 2x - 3) \div (x^2 - 5)$

SOLUTION

a)
$$
\begin{array}{r}
x^2 - x + 1 \\
x + 1 \overline{\smash{\big)}\, x^3 + 0x^2 + 0x + 1} \\
\underline{x^3 + x^2} \\
-x^2 + 0x \\
\underline{-x^2 - x} \\
x + 1 \\
\underline{x + 1} \\
0
\end{array}
$$

- ← Fill in the missing terms.
- ← Subtracting $x^3 + x^2$ from $x^3 + 0x^2$ and bringing down the $0x$
- ← Subtracting $-x^2 - x$ from $-x^2 + 0x$ and bringing down the 1

The answer is $x^2 - x + 1$.

Check: $(x + 1)(x^2 - x + 1) = x^3 - x^2 + x + x^2 - x + 1$
$= x^3 + 1.$

b) $\phantom{x^2-5\overline{)}}\!\!\begin{array}{r}x^2+2\end{array}$

$x^2-5\overline{\smash{)}x^4+0x^3-3x^2+2x-3}$ Writing in the missing term

$\phantom{x^2-5\overline{)}}\underline{x^4-5x^2}$

$\phantom{x^2-5\overline{)}x^4+0x^3-}2x^2+2x-3$ Subtracting x^4-5x^2 from x^4-3x^2 and bringing down $2x-3$

$\phantom{x^2-5\overline{)}x^4+0x^3-}\underline{2x^2-10}$

$\phantom{x^2-5\overline{)}x^4+0x^3-2x^2+}2x+7$ ← Subtracting $2x^2-10$ from $2x^2+2x-3$

Since the remainder, $2x + 7$, is of lower degree than the divisor, the division process stops. The answer is $x^2 + 2$, with R $2x + 7$, or

$$x^2 + 2 + \frac{2x+7}{x^2-5}.$$

Check: $(x^2 - 5)(x^2 + 2) + 2x + 7$
$= x^4 + 2x^2 - 5x^2 - 10 + 2x + 7$
$= x^4 - 3x^2 + 2x - 3.$

EXERCISE SET 4.7

Divide and check.

1. $\dfrac{32x^5 - 16x}{8}$

2. $\dfrac{12a^4 - 3a^2}{6}$

3. $\dfrac{u - 2u^2 + u^7}{u}$

4. $\dfrac{50x^5 - 7x^4 + x^2}{x}$

5. $(15t^3 - 24t^2 + 6t) \div (3t)$
6. $(25t^3 - 15t^2 + 30t) \div (5t)$
7. $(35x^6 - 20x^4 - 5x^2) \div (-5x^2)$
8. $(16x^6 + 32x^5 - 8x^2) \div (-8x^2)$
9. $(24x^5 - 40x^4 + 6x^3) \div (4x^3)$
10. $(18x^6 - 27x^5 - 3x^3) \div (9x^3)$

11. $\dfrac{8x^2 - 10x + 1}{2}$

12. $\dfrac{6x^2 + 3x - 2}{3}$

13. $\dfrac{2x^3 + 6x^2 + 4x}{2x}$

14. $\dfrac{2x^4 - 3x^3 + 5x^2}{x^2}$

15. $\dfrac{9r^2s^2 + 3r^2s - 6rs^2}{-3rs}$

16. $\dfrac{4x^4y - 8x^6y^2 + 12x^8y^6}{4x^4y}$

17. $(x^2 + 4x - 12) \div (x - 2)$
18. $(x^2 - 6x + 8) \div (x - 4)$
19. $(x^2 - 10x - 25) \div (x - 5)$
20. $(x^2 + 8x - 16) \div (x + 4)$
21. $(2x^2 + 11x - 5) \div (x + 6)$
22. $(3x^2 - 2x - 13) \div (x - 2)$

23. $\dfrac{x^2 - 9}{x + 3}$

24. $\dfrac{x^2 - 25}{x + 5}$

25. $(3x^2 + 11x - 4) \div (3x - 1)$
26. $(10x^2 + 13x - 3) \div (5x - 1)$

27. $\dfrac{8x^3 - 22x^2 - 5x + 12}{4x + 3}$

28. $\dfrac{2x^3 - 9x^2 + 11x - 3}{2x - 3}$

29. $(x^4 - 2x^2 + 4x - 5) \div (x^2 - 3)$
30. $(x^4 + 4x^2 + 3x - 6) \div (x^2 + 5)$
31. $(t^3 - t^2 + t - 1) \div (t + 1)$

228 CHAPTER 4 • POLYNOMIALS

32. $(t^3 - t^2 + t - 1) \div (t - 1)$
33. $(6x^4 - 3x^2 + x - 4) \div (2x^2 + 1)$
34. $(4x^4 - 4x^2 - x - 3) \div (2x^2 - 3)$

SKILL MAINTENANCE

35. The perimeter of a rectangle is 640 ft. The length is 15 ft greater than the width. Find the area of the rectangle.
36. Solve: $3(2x - 1) = 7x - 5$.
37. Solve: $2x > 12 + 7x$.
38. Plot the points $(4, -1)$, $(0, 5)$, $(-2, 3)$, and $(-3, 0)$.
39. In which quadrant are both coordinates negative?
40. Graph: $3x - 2y = 12$.

SYNTHESIS

41. ◆ Explain how the equation
$$(2x + 3)(3x - 1) = 6x^2 + 7x - 3$$
can be used to devise two division problems.

42. ◆ On an assignment, a student *incorrectly* writes
$$\frac{12x^3 - 6x}{3x} = 4x^2 - 6x.$$
What mistake do you think the student is making and how might you convince the student that a mistake has been made?

43. ◆ Can the quotient of two binomials be a trinomial? Why or why not?
44. ◆ How is the distributive law used when dividing a polynomial by a polynomial?

Divide.

45. $(45x^{8k} + 30x^{6k} - 60x^{4k}) \div (3x^{2k})$
46. $(10a^{9k} - 32a^{6k} + 28a^{3k}) \div (2a^{3k})$
47. $(y^4 + a^2) \div (y + a)$
48. $(5a^3 + 8a^2 - 23a - 1) \div (5a^2 - 7a - 2)$
49. $(15y^3 - 30y + 7 - 19y^2) \div (3y^2 - 2 - 5y)$
50. Divide the sum of $4x^5 - 14x^3 - x^2 + 3$ and $2x^5 + 3x^4 + x^3 - 3x^2 + 5x$ by $3x^3 - 2x - 1$.
51. Divide $5x^7 - 3x^4 + 2x^2 - 10x + 2$ by the sum of $(x - 3)^2$ and $5x - 8$.
52. Divide $6a^{3h} + 13a^{2h} - 4a^h - 15$ by $2a^h + 3$.

If the remainder is 0 when one polynomial is divided by another, the divisor is a factor of the dividend. Find the value(s) of c for which $x - 1$ is a factor of each polynomial.

53. $x^2 - 4x + c$
54. $2x^2 - 3cx - 8$
55. $c^2x^2 + 2cx + 1$

4.8 Negative Exponents and Scientific Notation

Negative Integers as Exponents • Scientific Notation •
Multiplying and Dividing Using Scientific Notation •
Problem Solving with Scientific Notation

We now attach a meaning to negative exponents. Once we understand both positive and negative exponents, we can study a method of writing numbers known as *scientific notation*.

Negative Integers as Exponents

If we define negative exponents carefully, the rules that apply to whole-number exponents will hold for all integer exponents.

Consider $5^3/5^7$ and first simplify by removing a factor equal to 1:

$$\frac{5^3}{5^7} = \frac{5 \cdot 5 \cdot 5}{5 \cdot 5 \cdot 5 \cdot 5 \cdot 5 \cdot 5 \cdot 5}$$

$$= \frac{5 \cdot 5 \cdot 5}{5 \cdot 5 \cdot 5} \cdot \frac{1}{5 \cdot 5 \cdot 5 \cdot 5}$$

$$= \frac{1}{5^4}.$$

Next, suppose we were to subtract exponents, as in Section 4.1:

$$\frac{5^3}{5^7} = 5^{3-7} = 5^{-4}.$$

The two expressions for $5^3/5^7$ suggest that

$$5^{-4} = \frac{1}{5^4}.$$

This leads to our definition of negative exponents.

NEGATIVE EXPONENTS

For any real number a that is nonzero and any integer n,

$$a^{-n} = \frac{1}{a^n}.$$

(The numbers a^{-n} and a^n are reciprocals.)

EXAMPLE 1 Express using positive exponents. Then simplify, if possible.

a) m^{-3} b) 4^{-2} c) $(-3)^{-2}$ d) ab^{-1}

SOLUTION

a) $m^{-3} = \dfrac{1}{m^3}$ m^{-3} is the reciprocal of m^3.

b) $4^{-2} = \dfrac{1}{4^2} = \dfrac{1}{16}$ 4^{-2} is the reciprocal of 4^2. Note that $4^{-2} \neq 4(-2)$.

c) $(-3)^{-2} = \dfrac{1}{(-3)^2} = \dfrac{1}{(-3)(-3)} = \dfrac{1}{9}$ $\begin{cases}(-3)^{-2} \text{ is the reciprocal of } (-3)^2.\\ \text{Note that } (-3)^{-2} \neq -\dfrac{1}{3^2}.\end{cases}$

d) $ab^{-1} = a\left(\dfrac{1}{b^1}\right) = a\left(\dfrac{1}{b}\right) = \dfrac{a}{b}$ b^{-1} is the reciprocal of b^1.

Note in Examples 1(b) and 1(c) that a negative exponent does not, in itself, indicate that an expression represents a negative number.

The following is another way to understand why negative exponents are defined as they are.

| On this side, we divide by 5 at each step. | $125 = 5^3$
$25 = 5^2$
$5 = 5^1$
$1 = 5^0$
$\dfrac{1}{5} = 5^?$
$\dfrac{1}{25} = 5^?$ | On this side, the exponents decrease by 1. |

To continue the pattern, it follows that

$$\dfrac{1}{5} = \dfrac{1}{5^1} = 5^{-1}, \qquad \dfrac{1}{25} = \dfrac{1}{5^2} = 5^{-2}, \quad \text{and, in general,} \quad \dfrac{1}{a^n} = a^{-n}.$$

EXAMPLE 2 Express $\dfrac{1}{x^7}$ using negative exponents.

SOLUTION We know that $\dfrac{1}{a^n} = a^{-n}$. Thus,

$$\dfrac{1}{x^7} = x^{-7}.$$

The rules for powers still hold when exponents are negative.

EXAMPLE 3 Simplify.

a) $a^5 \cdot a^{-2}$ b) $\dfrac{x}{x^7}$ c) $\dfrac{b^{-4}}{b^{-5}}$

d) $(y^{-5})^{-7}$ e) $(5x^2y^{-3})^4$ f) $1/x^{-3}$

SOLUTION

a) $a^5 \cdot a^{-2} = a^{5+(-2)}$ Adding exponents
$\phantom{a^5 \cdot a^{-2}} = a^3$

b) $\dfrac{x}{x^7} = x^{1-7}$ Subtracting exponents

$\phantom{\dfrac{x}{x^7}} = x^{-6}$, or $\dfrac{1}{x^6}$

c) $\dfrac{b^{-4}}{b^{-5}} = b^{-4-(-5)} = b^1 = b$ We subtract exponents even if the exponent in the denominator is negative.

d) $(y^{-5})^{-7} = y^{(-5)(-7)}$ Multiplying exponents
$\phantom{(y^{-5})^{-7}} = y^{35}$

e) $(5x^2y^{-3})^4 = 5^4(x^2)^4(y^{-3})^4$ Raising each factor to the fourth power
$\phantom{(5x^2y^{-3})^4} = 625x^8y^{-12}$, or $\dfrac{625x^8}{y^{12}}$

f) Since $\dfrac{1}{a^n} = a^{-n}$, we have

$$\dfrac{1}{x^{-3}} = x^{-(-3)} = x^3.$$

Some manipulations with negative exponents can be shortened when certain patterns are discovered. For example, since $m^{-5} = 1/m^5$ and $1/x^{-3} = x^3$, we have

$$\boxed{\dfrac{m^{-5}}{x^{-3}} = m^{-5} \cdot \dfrac{1}{x^{-3}} = \dfrac{1}{m^5} \cdot x^3 = \dfrac{x^3}{m^5}.}$$

Note how the signs of the exponents change.

EXAMPLE 4 Simplify: $\left(\dfrac{y^3}{5}\right)^{-2}$.

SOLUTION

$\left(\dfrac{y^3}{5}\right)^{-2} = \dfrac{(y^3)^{-2}}{5^{-2}}$ Raising a quotient to a power

$= \dfrac{y^{-6}}{5^{-2}}$ Using the power rule

$= \dfrac{5^2}{y^6}$ Rewriting with positive exponents

$= \dfrac{25}{y^6}$

DEFINITIONS AND PROPERTIES OF EXPONENTS

The following summary assumes that no denominators are 0 and that 0^0 is not considered. For any integers m and n,

1 as an exponent:	$a^1 = a$
0 as an exponent:	$a^0 = 1$
Negative exponents:	$a^{-n} = \dfrac{1}{a^n}$
The Product Rule:	$a^m \cdot a^n = a^{m+n}$
The Quotient Rule:	$\dfrac{a^m}{a^n} = a^{m-n}$
The Power Rule:	$(a^m)^n = a^{mn}$
Raising a product to a power:	$(ab)^n = a^n b^n$
Raising a quotient to a power:	$\left(\dfrac{a}{b}\right)^n = \dfrac{a^n}{b^n}$

Scientific Notation

When we are working with the very large or very small numbers that frequently occur in science, **scientific notation** provides a useful way of writing numbers. The following are examples of scientific notation.

The distance from the earth to the sun:

9.3×10^7 mi $= 93{,}000{,}000$ mi

The mass of a hydrogen atom:

1.7×10^{-24} g $= 0.00000000000000000000000017$ g

SCIENTIFIC NOTATION

Scientific notation for a number is an expression of the type

$N \times 10^n,$

where N is at least 1 but less than 10 ($1 \leq N < 10$), N is expressed in decimal notation, and n is an integer.

Converting from scientific to decimal notation involves multiplying by a power of 10. Consider the following.

Scientific Notation $N \times 10^n$	*Multiplication*	*Decimal Notation*
4.52×10^2	4.52×100	$452.$
4.52×10^1	4.52×10	45.2
4.52×10^0	4.52×1	4.52
4.52×10^{-1}	4.52×0.1	0.452
4.52×10^{-2}	4.52×0.01	0.0452

Note that when n, the power of 10, is positive, the decimal point moves right n places in decimal notation. When n is negative, the decimal point moves left n places. We generally try to perform this multiplication mentally.

EXAMPLE 5 Convert to decimal notation: **(a)** 7.893×10^5; **(b)** 4.7×10^{-8}.

SOLUTION

a) Since the exponent is positive, the decimal point moves to the right:

7.89300.
5 places

$7.893 \times 10^5 = 789{,}300$ The decimal point moves 5 places to the right.

b) Since the exponent is negative, the decimal point moves to the left:

0.00000004.7
8 places

$4.7 \times 10^{-8} = 0.000000047$ The decimal point moves 8 places to the left.

To convert from decimal to scientific notation, the above procedure is reversed.

EXAMPLE 6 Write in scientific notation: **(a)** 7800; **(b)** 0.0549.

SOLUTION

a) We must have $7800 = N \times 10^n$, where $1 \leq N < 10$. Because multiplication by 10^n moves only the decimal point, we must have $N = 7.8$:

$$7800 = 7.8 \times 10^n.$$

Multiplying 7.8 by 10^3 moves the decimal point 3 places to the right. Thus, n is 3 and

$$7800 = 7.8 \times 10^3.$$

b) In scientific notation, 0.0549 is written as 5.49×10^n. Multiplying 5.49 by 10^{-2} moves the decimal point 2 places to the left. Thus, n is -2 and

$$0.0549 = 5.49 \times 10^{-2}.$$

Conversions to and from scientific notation are often made mentally. Remember that positive exponents are used when representing large numbers and negative exponents are used when representing numbers between 0 and 1.

Multiplying and Dividing Using Scientific Notation

Products and quotients of numbers written in scientific notation are found using the rules for exponents.

EXAMPLE 7 Simplify.

a) $(1.8 \times 10^9) \cdot (2.3 \times 10^{-4})$ **b)** $(3.41 \times 10^5) \div (1.1 \times 10^{-3})$

SOLUTION

a) $(1.8 \times 10^9) \cdot (2.3 \times 10^{-4})$
$= 1.8 \times 2.3 \times 10^9 \times 10^{-4}$ Using the associative and commutative laws
$= 4.14 \times 10^{9+(-4)}$ Adding exponents
$= 4.14 \times 10^5$

b) $(3.41 \times 10^5) \div (1.1 \times 10^{-3})$
$= \dfrac{3.41 \times 10^5}{1.1 \times 10^{-3}}$
$= \dfrac{3.41}{1.1} \times \dfrac{10^5}{10^{-3}}$
$= 3.1 \times 10^{5-(-3)}$ Subtracting exponents
$= 3.1 \times 10^8$

TECHNOLOGY CONNECTION 4.8

A key labeled EXP or EE is often used to enter scientific notation into a calculator. Sometimes this is a secondary function, meaning that another key— often labeled SHIFT or 2nd —must be pressed first.

To check Example 8(b) on page 234, we press

7.2 EXP 7 +/− ÷ 8.0 EXP 6

When we then press = or ENTER, the result 9 E⁻14 or 9.⁻14 appears. This must be interpreted as 9.0×10^{-14}. As always, keystrokes may vary depending on the calculator.

Calculate.
1. $(3.8 \times 10^9) \cdot (4.5 \times 10^7)$
2. $(2.9 \times 10^{-8}) \div (5.4 \times 10^6)$
3. $(9.2 \times 10^7) \div (2.5 \times 10^{-9})$

When a problem is stated using scientific notation, it is customary to use scientific notation for the answer.

EXAMPLE 8 Simplify.

a) $(3.1 \times 10^5) \cdot (4.5 \times 10^{-3})$ b) $(7.2 \times 10^{-7}) \div (8.0 \times 10^6)$

SOLUTION

a) We have

$$(3.1 \times 10^5) \cdot (4.5 \times 10^{-3}) = 3.1 \times 4.5 \times 10^5 \times 10^{-3}$$
$$= 13.95 \times 10^2.$$

Our answer is not yet in scientific notation because 13.95 is not between 1 and 10. We convert to scientific notation as follows:

$$13.95 \times 10^2 = 1.395 \times 10^1 \times 10^2 \quad \text{Substituting } 1.395 \times 10^1 \text{ for } 13.95$$
$$= 1.395 \times 10^3. \quad \text{Adding exponents}$$

b) $(7.2 \times 10^{-7}) \div (8.0 \times 10^6) = \dfrac{7.2 \times 10^{-7}}{8.0 \times 10^6} = \dfrac{7.2}{8.0} \times \dfrac{10^{-7}}{10^6}$

$$= 0.9 \times 10^{-13}$$
$$= 9.0 \times 10^{-1} \times 10^{-13} \quad \text{Substituting } 9.0 \times 10^{-1} \text{ for } 0.9$$
$$= 9.0 \times 10^{-14} \quad \text{Adding exponents}$$

Problem Solving with Scientific Notation

EXAMPLE 9 Light traveling at a rate of 300,000 kilometers per second (km/s) takes 499 sec to reach the earth from the sun. Find the distance, expressed in scientific notation, from the sun to the earth.

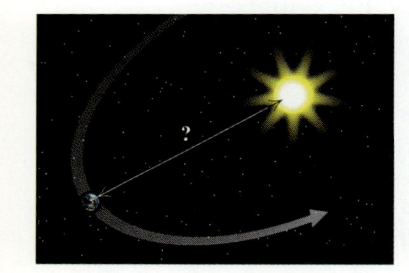

SOLUTION

1. **Familiarize.** The time t that it takes for light to reach the earth from the sun is 4.99×10^2 sec (s). The speed r is 3.0×10^5 km/s. Recall that distance can be expressed in terms of speed and time:

   $$d = rt$$

   Distance = Speed × Time.

2. **Translate.** We translate the problem to mathematical language by substituting 3.0×10^5 for r and 4.99×10^2 for t:

   $$d = rt$$
   $$= (3.0 \times 10^5)(4.99 \times 10^2).$$

3. **Carry out.** We carry out the computation and express the result using scientific notation for the answer:

   $$d = (3.0 \times 10^5)(4.99 \times 10^2)$$
   $$= 14.97 \times 10^7$$
   $$= 1.497 \times 10^8 \text{ km.} \quad \Big\} \text{ Converting to scientific notation}$$

4.8 NEGATIVE EXPONENTS AND SCIENTIFIC NOTATION

4. **Check.** We can check by reviewing our computations. Note too that our answer is a very large number, as expected.
5. **State.** The distance from the sun to the earth is 1.497×10^8 km.

EXERCISE SET 4.8

Express using positive exponents. Then, if possible, simplify.

1. 6^{-2}
2. 2^{-4}
3. 10^{-4}
4. 5^{-3}
5. $(-2)^{-6}$
6. $(-3)^{-4}$
7. a^{-5}
8. x^{-2}
9. $\dfrac{1}{y^{-4}}$
10. $\dfrac{1}{t^{-7}}$
11. $\dfrac{1}{z^{-9}}$
12. $\dfrac{1}{h^{-8}}$
13. 7^{-1}
14. $\left(\dfrac{2}{3}\right)^{-1}$
15. $\left(\dfrac{1}{4}\right)^{-2}$
16. $\left(\dfrac{4}{5}\right)^{-2}$

Express using negative exponents.

17. $\dfrac{1}{4^3}$
18. $\dfrac{1}{5^2}$
19. $\dfrac{1}{t^6}$
20. $\dfrac{1}{y^2}$
21. $\dfrac{1}{a^4}$
22. $\dfrac{1}{t^5}$
23. $\dfrac{1}{p^8}$
24. $\dfrac{1}{m^{12}}$
25. $\dfrac{1}{5}$
26. $\dfrac{1}{8}$
27. $\dfrac{1}{t}$
28. $\dfrac{1}{m}$

Simplify.

29. $2^{-5} \cdot 2^8$
30. $5^{-8} \cdot 5^9$
31. $x^{-2} \cdot x$
32. $x \cdot x^{-1}$
33. $x^{-7} \cdot x^{-6}$
34. $y^{-5} \cdot y^{-8}$
35. $\dfrac{m^6}{m^{12}}$
36. $\dfrac{p^4}{p^5}$
37. $\dfrac{(8x)^6}{(8x)^{10}}$
38. $\dfrac{(9t)^4}{(9t)^{11}}$
39. $\dfrac{18^9}{18^9}$
40. $\dfrac{(6y)^7}{(6y)^7}$
41. $(a^{-5}b^{-7})(a^{-3}b^{-6})$
42. $(x^{-2}y^{-7})(x^{-3}y^{-2})$
43. $\dfrac{x^7}{x^{-2}}$
44. $\dfrac{t^8}{t^{-3}}$
45. $\dfrac{z^{-6}}{z^{-2}}$
46. $\dfrac{y^{-7}}{y^{-3}}$
47. $\dfrac{a^{-6}}{a^{-10}}$
48. $\dfrac{y^{-4}}{y^{-9}}$
49. $\dfrac{x}{x^{-1}}$
50. $\dfrac{x^6}{x}$
51. $(a^{-3})^5$
52. $(x^{-5})^6$
53. $(a^{-5})^{-6}$
54. $(x^{-3})^{-4}$
55. $(n^{-2})^8$
56. $(m^{-3})^7$
57. $(mn)^{-5}$
58. $(ab)^{-3}$
59. $(4xy)^{-2}$
60. $(5ab)^{-2}$
61. $(3a^{-4})^4$
62. $(6x^{-5})^2$
63. $(t^5x^3)^{-4}$
64. $(x^4y^5)^{-3}$
65. $(x^{-2}y^{-7})^{-5}$
66. $(x^{-6}y^{-2})^{-4}$
67. $(x^3y^{-4}z^{-5})(x^{-4}y^{-2}z^9)$
68. $(a^{-5}b^7c^{-2})(a^{-3}b^{-2}c^6)$
69. $(m^{-4}n^7p^3)(m^9n^{-2}p^{-10})$
70. $(t^{-9}p^{10}m^8)(t^{-5}p^{-7}m^{-2})$
71. $\left(\dfrac{y^2}{2}\right)^{-2}$
72. $\left(\dfrac{a^4}{3}\right)^{-2}$
73. $\left(\dfrac{3}{a^2}\right)^4$
74. $\left(\dfrac{7}{x^7}\right)^2$
75. $\left(\dfrac{x^2y}{z^4}\right)^3$
76. $\left(\dfrac{m}{n^4p}\right)^3$
77. $\left(\dfrac{a^2b}{cd^3}\right)^{-5}$
78. $\left(\dfrac{2a^2}{3b^4}\right)^{-3}$

Convert to decimal notation.

79. 9.12×10^4
80. 8.92×10^2
81. 6.92×10^{-3}
82. 7.26×10^{-4}

83. 2.04×10^8
84. 1.35×10^7
85. 8.764×10^{-10}
86. 9.043×10^{-3}
87. 10^7
88. 10^4
89. 10^{-4}
90. 10^{-7}

Convert to scientific notation.
91. 370,000
92. 71,500
93. 0.00583
94. 0.0814
95. 78,000,000,000
96. 3,700,000,000,000
97. 907,000,000,000,000,000
98. 168,000,000,000,000
99. 0.00000486
100. 0.000000000275
101. 0.000000018
102. 0.00000000002
103. 10,000,000
104. 100,000,000,000

Multiply or divide, and write scientific notation for the result.
105. $(4 \times 10^7)(2 \times 10^5)$
106. $(1.9 \times 10^8)(3.4 \times 10^{-3})$
107. $(3.8 \times 10^9)(6.5 \times 10^{-2})$
108. $(7.1 \times 10^{-7})(8.6 \times 10^{-5})$
109. $(8.7 \times 10^{-12})(4.5 \times 10^{-5})$
110. $(4.7 \times 10^5)(6.2 \times 10^{-12})$
111. $\dfrac{8.5 \times 10^8}{3.4 \times 10^{-5}}$
112. $\dfrac{5.6 \times 10^{-2}}{2.5 \times 10^5}$
113. $(3.0 \times 10^6) \div (6.0 \times 10^9)$
114. $(1.5 \times 10^{-3}) \div (1.6 \times 10^{-6})$
115. $\dfrac{7.5 \times 10^{-9}}{2.5 \times 10^{12}}$
116. $\dfrac{4.0 \times 10^{-3}}{8.0 \times 10^{20}}$

Write scientific notation for each answer.

117. *Orange juice consumption.* Americans drink 3 million gal of orange juice each day. How much orange juice do Americans consume in 1 yr?

118. *River discharge.* The average discharge at the mouth of the Amazon River is 4,200,000 cubic feet per second. How much water is discharged from the Amazon River in 1 hr? in 1 yr?

119. *Water contamination.* In the United States, 200 million gal of used motor oil is improperly disposed of each year. One gallon of used oil can contaminate one million gallons of drinking water (*Source*: *The Macmillan Visual Almanac* 1996). How many gallons of drinking water can 200 million gallons of oil contaminate?

120. *Flight delays.* In 1993, the 22 busiest airports in the United States experienced over 20,000 hr of flight delays (*Source*: *The Macmillan Visual Almanac* 1996). Each hour cost airlines $1.6 million. How much money did airlines lose to delays at these airports in 1993?

121. *Computers.* A gigabyte is a measure of a computer's storage capacity. One gigabyte holds about one billion bytes of information. If a firm's computer network contains 2500 gigabytes of memory, how many bytes are in the network?

122. *Two-person households.* In 1993, there were about 31.2 million two-person households in the United States. The average income of these households was about $42,400 (*Source*: *Statistical Abstract of the United States,* 1995). Find the total income generated by two-person households in 1993.

SKILL MAINTENANCE

123. Simplify: $\frac{3}{4} - 5\left(-\frac{1}{2}\right)^2 + \frac{1}{3}$.

Combine like terms.
124. $-9a + 17a$
125. $-12x + (-5x)$
126. Plot the points $(-4, 1)$, $(-3, -2)$, $(5, 2)$, and $(-1, 4)$.
127. In which two quadrants is the first coordinate positive?
128. Solve $cx - bt = r$ for t.

SYNTHESIS

129. ◆ Under what conditions will x^{-n} represent a negative number?

130. ◆ Explain in your own words how the quotient rule helped us to determine what negative exponents would mean.

Carry out the indicated operations. Write scientific notation for the result.

131. $\dfrac{(2.5 \times 10^{-8})(6.1 \times 10^{-11})}{1.28 \times 10^{-3}}$

132. $\dfrac{7.4 \times 10^{29}}{(5.4 \times 10^{-6})(2.8 \times 10^{8})}$

133. $\dfrac{5.8 \times 10^{17}}{(4.0 \times 10^{-13})(2.3 \times 10^{4})}$

134. $\dfrac{(7.8 \times 10^{7})(8.4 \times 10^{23})}{2.1 \times 10^{-12}}$

135. Simplify:
$$\dfrac{4.2 \times 10^{8}[(2.5 \times 10^{-5}) \div (5.0 \times 10^{-9})]}{3.0 \times 10^{27}}.$$

136. Find the reciprocal. Express in scientific notation.
   a) 6.25×10^{-3}
   b) 4.0×10^{10}

137. Write $4^{-3} \cdot 8 \cdot 16$ as a power of 2.
138. Write $2^{8} \cdot 16^{-3} \cdot 64$ as a power of 4.

Simplify.

139. $(5^{-12})^{2} \cdot 5^{25}$

140. $49^{18} \cdot 7^{-35}$

141. $9^{23} \cdot 27^{-6}$

142. $\left(\dfrac{1}{a}\right)^{-n}$

Determine whether each of the following is true for all pairs of integers m and n and all positive numbers x and y.

143. $x^{m} \cdot y^{n} = (xy)^{mn}$

144. $x^{m} \cdot y^{m} = (xy)^{2m}$

145. $(x - y)^{m} = x^{m} - y^{m}$

COLLABORATIVE C•O•R•N•E•R

Focus: Scientific notation and estimation
Time: 10–20 minutes
Group size: 2
Materials: Calculators and possibly a watermelon to share after the activity

As its name suggests, a watermelon contains a lot of water: 90% by weight. In the following activity, each group member will determine the number of water molecules in a 10-lb watermelon. One member will work without a calculator on parts (1), (2), and (3) while the other person will use a calculator on parts (4), (5), and (6).

ACTIVITY

1. How many pounds of water are in a 10-lb watermelon (see above)? About how many *kilograms* of water? (1 kg ≈ 2 lb) About how many grams of water? (1 kg = 1000 g)
2. Water (H_2O) is a *molecule* composed of two hydrogen atoms and one oxygen atom. Estimate the number of molecules per gram of water if 6×10^{23} molecules weighs 18 g.
3. Using the results from parts (1) and (2), estimate the number of molecules in a 10-lb watermelon. Express your answer in scientific notation.
4. Repeat part (1), but use the calculation 1 kg ≈ 2.2 lb.
5. Repeat part (2), assuming that 6.02×10^{23} molecules weighs 18 g.
6. Repeat part (3), but use the results from parts (4) and (5).
7. Compare your answers for parts (1)–(6). Which method *seems* more reliable—estimation or the use of a calculator? Do you believe that the result found with a calculator in part (6) is *really* more accurate? Why or why not?

SUMMARY AND REVIEW 4

KEY TERMS

Polynomial, p. 184
Term, p. 184
Monomial, p. 184
Binomial, p. 185
Trinomial, p. 185
Degree of a term, pp. 185, 216
Coefficient, p. 185
Leading term, p. 186
Leading coefficient, p. 186
Degree of a polynomial, p. 186
Descending order, p. 187
Opposite of a polynomial, p. 193
FOIL, p. 208
Difference of squares, p. 209
Polynomial in several variables, p. 215
Like terms, p. 217
Scientific notation, p. 232

IMPORTANT PROPERTIES AND FORMULAS

Definitions and Properties of Exponents

Assuming that no denominator is 0 and that 0^0 is not considered, for any integers m and n,

1 as an exponent:	$a^1 = a$
0 as an exponent:	$a^0 = 1$
Negative exponents:	$a^{-n} = \dfrac{1}{a^n}$
The Product Rule:	$a^m \cdot a^n = a^{m+n}$
The Quotient Rule:	$\dfrac{a^m}{a^n} = a^{m-n}$
The Power Rule:	$(a^m)^n = a^{mn}$
Raising a product to a power:	$(ab)^n = a^n b^n$
Raising a quotient to a power:	$\left(\dfrac{a}{b}\right)^n = \dfrac{a^n}{b^n}$

Special Products of Polynomials

$(A + B)(A - B) = A^2 - B^2$
$(A + B)(A + B) = A^2 + 2AB + B^2$
$(A - B)(A - B) = A^2 - 2AB + B^2$

Scientific notation: $N \times 10^n$, where $1 \leq N < 10$ and n is an integer

REVIEW EXERCISES

Simplify.
1. $y^7 \cdot y^3 \cdot y$
2. $(3x)^5 \cdot (3x)^9$
3. $t^8 \cdot t^0$
4. $\dfrac{4^5}{4^2}$
5. $\dfrac{(a+b)^4}{(a+b)^4}$
6. $\left(\dfrac{3t^4}{2s^3}\right)^2$
7. $(-2xy^2)^3$
8. $(2x^3)(-3x)^2$
9. $(a^2b)(ab)^5$

Identify the terms of each polynomial.
10. $3x^2 + 6x + \tfrac{1}{2}$
11. $-4y^5 + 7y^2 - 3y - 2$

List the coefficients of the terms in each polynomial.
12. $6x^2 + x + 5$
13. $4x^3 + 6x^2 - 5x + \tfrac{5}{3}$

For each polynomial, (a) list the degree of each term; (b) determine the leading term and the leading coefficient; and (c) determine the degree of the polynomial.
14. $4t^2 + 6 + 15t^5$
15. $-2x^5 + x^4 - 3x^2 + x$

Classify each polynomial as a monomial, a binomial, a trinomial, or none of these.
16. $4x^3 - 1$
17. $4 - 9t^3 - 7t^4 + 10t^2$
18. $7y^2$

Combine like terms and write in descending order.
19. $5x - x^2 + 4x$
20. $\tfrac{3}{4}x^3 + 4x^2 - x^3 + 7$
21. $-2x^4 + 16 + 2x^4 + 9 - 3x^5$
22. $3x^2 - 2x + 3 - 5x^2 - 1 - x$
23. $-x + \tfrac{1}{2} + 14x^4 - 7x^2 - 1 - 4x^4$

Evaluate each polynomial for $x = -1$.
24. $7x - 10$
25. $x^2 - 3x + 6$

Add or subtract.
26. $(3x^4 - x^3 + x - 4) + (x^5 + 7x^3 - 3x - 5)$
27. $(3x^4 - 5x^3 + 3x^2) + (4x^5 + 4x^3) + (-5x^5 - 5x^2)$
28. $(5x^2 - 4x + 1) - (3x^2 + 7)$

29. $(3x^5 - 4x^4 + 2x^2 + 3) - (2x^5 - 4x^4 + 3x^3 + 4x^2 - 5)$

30. $\begin{array}{r} -\tfrac{3}{4}x^4 + \tfrac{1}{2}x^3 \phantom{- x^2 - \tfrac{7}{4}x} + \tfrac{7}{8} \\ -\tfrac{1}{4}x^3 - x^2 - \tfrac{7}{4}x \phantom{+\tfrac{7}{8}} \\ +\tfrac{3}{2}x^4 \phantom{+\tfrac{1}{2}x^3} + \tfrac{2}{3}x^2 \phantom{-\tfrac{7}{4}x} - \tfrac{1}{2} \\ \hline \end{array}$

31. $\begin{array}{r} 2x^5 - x^3 + x + 3 \\ -(3x^5 - x^4 + 4x^3 + 2x^2 - x + 3) \\ \hline \end{array}$

32. The length of a rectangle is 3 m greater than its width.

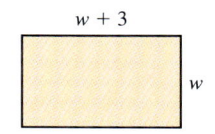

[rectangle with width w and length $w + 3$]

a) Find a polynomial for the perimeter.
b) Find a polynomial for the area.

Multiply.
33. $3x(-4x^2)$
34. $(7x + 1)^2$
35. $\left(x + \tfrac{2}{3}\right)\left(x + \tfrac{1}{2}\right)$
36. $(m+5)(m-5)$
37. $(4x^2 - 5x + 1)(3x - 2)$
38. $(x - 9)^2$
39. $\tfrac{2}{5}a^2\left(5a^3 - 10a + \tfrac{10}{3}\right)$
40. $(x+4)(x-7)$
41. $(x - 0.3)(x - 0.75)$
42. $(x^4 - 2x + 3)(x^3 + x - 1)$
43. $(3y^2 - 2y)^2$
44. $(2t^2 + 3)(t^2 - 7)$
45. $(2a^2 + a)(2a^2 + a)$
46. $(3x^2 + 4)(3x^2 - 4)$
47. $(2 - x)(2 + x)$
48. $(13x - 3)(x - 13)$
49. Evaluate $2 - 5xy + y^2 - 4xy^3 + x^6$ for $x = -1$ and $y = 2$.

Identify the coefficient and the degree of each term of each polynomial. Then find the degree of each polynomial.
50. $x^5y - 7xy + 9x^2 - 8$
51. $x^2y^5z^9 - y^{40} + x^{13}z^{10}$

Combine like terms.

52. $y + w - 2y + 8w - 5$

53. $6m^3 + 3m^2n + 4mn^2 + m^2n - 5mn^2$

Add or subtract.

54. $(5x^2 - 7xy + y^2) + (-6x^2 - 3xy - y^2)$

55. $(6x^3y^2 - 4x^2y - 6x) - (-5x^3y^2 + 4x^2y + 6x^2 - 6)$

Multiply.

56. $(p - q)(p^2 + pq + q^2)$ **57.** $\left(3a^4 - \frac{1}{3}b^3\right)^2$

58. Find a polynomial for the shaded area.

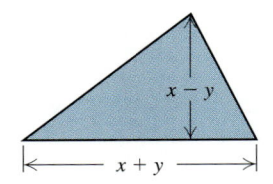

Divide.

59. $(10x^3 - x^2 + 6x) \div 2x$

60. $(6x^3 - 5x^2 - 13x + 13) \div (2x + 3)$

61. $\dfrac{t^4 + t^3 + 2t^2 - t - 3}{t + 1}$

62. Express using a positive exponent: y^{-4}.

63. Express using a negative exponent: $\dfrac{1}{t^5}$.

Simplify.

64. $7^2 \cdot 7^{-4}$ **65.** $\dfrac{a^{-5}b}{a^8b^8}$ **66.** $(x^3)^{-4}$

67. $(2x^{-3}y)^{-2}$ **68.** $\left(\dfrac{2x}{y}\right)^{-3}$

69. Convert to decimal notation: 8.3×10^6.

70. Convert to scientific notation: 0.0000328.

Multiply or divide and write scientific notation for the result.

71. $(3.8 \times 10^4)(5.5 \times 10^{-1})$

72. $\dfrac{1.28 \times 10^{-8}}{2.5 \times 10^{-4}}$

73. In 1995, 1.4 million people in the United States bought their first home.* If the average price paid was $140,000, find the total amount that new homeowners paid for houses in 1995. Write scientific notation for the answer.

SKILL MAINTENANCE

74. Find the perimeter of the figure.

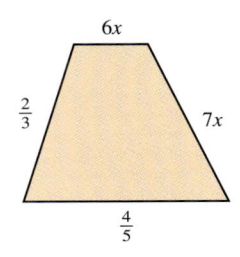

75. The perimeter of a rectangle is 540 m. The width is 19 m less than the length. Find the width and the length.

76. Solve: $3 - 2x \leq 7$.

77. In which quadrant is the point $(3, -1)$ located?

SYNTHESIS

78. ◆ Explain why $5x^3$ and $(5x)^3$ are not equivalent expressions.

79. ◆ If two polynomials of degree n are added, is the sum also of degree n? Why or why not?

80. If a and b are positive, how many terms are there in each of the following?
   a) $(x - a)(x - b) + (x - a)(x - b)$
   b) $(x + a)(x - b) + (x - a)(x + b)$

81. Combine like terms:
$-3x^5 \cdot 3x^3 - x^6(2x)^2 + (3x^4)^2 + (2x^2)^4 - 40x^2(x^3)^2$.

82. A polynomial has degree 4. The x^2-term is missing. The coefficient of x^4 is 2 times the coefficient of x^3. The coefficient of x is 3 less than the coefficient of x^4. The remaining coefficient is 7 less than the coefficient of x. The sum of the coefficients is 15. Find the polynomial.

83. Multiply: $[(x - 4) - x^3][(x + 4) + 4x^3]$.

84. Solve: $(x - 7)(x + 10) = (x - 4)(x - 6)$.

---

*Perspective, Better Homes and Gardens® Real Estate News (source quoted: *Today's Realtor*) September 1996.

CHAPTER TEST 4

Simplify.

1. $x^6 \cdot x^2 \cdot x$
2. $(x + 3)^5(x + 3)^6$
3. $\dfrac{3^5}{3^2}$
4. $\dfrac{(2x)^5}{(2x)^5}$
5. $(x^3)^2$
6. $(-3y^2)^3$
7. $(3x^2)(-2x^5)^3$
8. $(a^3b^2)(ab)^3$

9. Classify the polynomial as a monomial, a binomial, a trinomial, or none of these:

    $3 - x^2 + 2x$.

10. Identify the coefficient of each term of the polynomial:

    $\tfrac{1}{3}x^5 - x + 7$.

11. Determine the degree of each term, the leading term and the leading coefficient, and the degree of the polynomial:

    $2t^3 - t + 7t^5 + 4$.

12. Evaluate the polynomial $x^2 + 5x - 1$ for $x = -2$.

Combine like terms and write in descending order.

13. $4a^2 - 6 + a^2$
14. $y^2 - 3y - y + \tfrac{3}{4}y^2$
15. $3 - x^2 + 2x^3 + 5x^2 - 6x - 2x + x^5$

Add or subtract.

16. $(3x^5 + 5x^3 - 5x^2 - 3) + (x^5 + x^4 - 3x^3 - 3x^2 + 2x - 4)$
17. $\left(x^4 + \tfrac{2}{3}x + 5\right) + \left(4x^4 + 5x^2 + \tfrac{1}{3}x\right)$
18. $(2x^4 + x^3 - 8x^2 - 6x - 3) - (6x^4 - 8x^2 + 2x)$
19. $(x^3 - 0.4x^2 - 12) - (x^5 - 0.3x^3 + 0.4x^2 - 9)$

Multiply.

20. $-3x^2(4x^2 - 3x - 5)$
21. $\left(x - \tfrac{1}{3}\right)^2$
22. $(3x + 10)(3x - 10)$
23. $(3b + 5)(b - 3)$
24. $(x^6 - 4)(x^8 + 4)$
25. $(8 - y)(6 + 5y)$
26. $(2x + 1)(3x^2 - 5x - 3)$
27. $(8a + 3)(8a + 3)$

28. Combine like terms:

    $x^3y - y^3 + xy^3 + 8 - 6x^3y - x^2y^2 + 11$.

29. Subtract:

    $(8a^2b^2 - ab + b^3) - (-6ab^2 - 7ab - ab^3 + 5b^3)$.

30. Multiply: $(3x^5 - 4y^5)(3x^5 + 4y^5)$.

Divide.

31. $(12x^4 + 9x^3 - 15x^2) \div 3x^2$
32. $(6x^3 - 8x^2 - 14x + 13) \div (3x + 2)$
33. Express using a positive exponent: 5^{-3}.
34. Express using a negative exponent: $\dfrac{1}{y^8}$.

Simplify.

35. $6^{-2} \cdot 6^{-3}$
36. $\dfrac{x^3y^2}{x^8y^{-3}}$
37. $(2a^3b^{-1})^{-4}$
38. $\left(\dfrac{ab}{c}\right)^{-3}$

39. Convert to scientific notation: 3,900,000,000.
40. Convert to decimal notation: 5×10^{-8}.

Multiply or divide and write scientific notation for the result.

41. $\dfrac{5.6 \times 10^6}{3.2 \times 10^{-11}}$
42. $(2.4 \times 10^5)(5.4 \times 10^{16})$

43. A CD-rom can contain about 600 million pieces of information. How many sound files, each needing 40,000 pieces of information, can a CD-rom hold? Write scientific notation for the answer.

SKILL MAINTENANCE

44. Solve: $7x - 4x - 2 > 37$.
45. Plot the point $(-1, 5)$.
46. Add: $\tfrac{2}{5} + \left(-\tfrac{3}{4}\right)$.
47. The first angle of a triangle is twice as large as the second. The measure of the third angle is 20° greater than that of the first. How large are the angles?

SYNTHESIS

48. The height of a box is 1 less than its length, and the length is 2 more than its width. Express the volume in terms of the length.
49. Solve: $x^2 + (x - 7)(x + 4) = 2(x - 6)^2$.

CHAPTER 5

Polynomials and Factoring

5.1 Introduction to Factoring
5.2 Factoring Trinomials of the Type $x^2 + bx + c$
5.3 Factoring Trinomials of the Type $ax^2 + bx + c$
5.4 Factoring Perfect-Square Trinomials and Differences of Squares
5.5 Factoring: A General Strategy
5.6 Solving Quadratic Equations by Factoring
5.7 Solving Applications
SUMMARY AND REVIEW
TEST

AN APPLICATION

The mainsail of a Lightning sailboat is a right triangle in which the hypotenuse is called the leech. If a 24-ft-tall mainsail has a leech length of 26 ft and if Dacron® sailcloth costs $10 per square foot, find the cost of a new mainsail.

THIS PROBLEM APPEARS AS EXERCISE 51 IN SECTION 5.7.

More information on sailboat design is available at
http://hepg.awl.com/be/elem_5

*Y*ou need a solid grounding in geometry and algebra to tackle the basics of sail design. Without it, you'll never leave the dock.

BRIAN F. MEDEIROS
Sail Consultant
Jamestown, Rhode Island

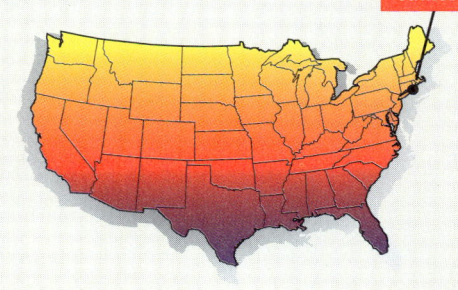

CHAPTER 5 • POLYNOMIALS AND FACTORING

In Chapter 1, we learned that factoring *is multiplying reversed. To factor a polynomial is to find an equivalent expression that is a product. Factoring polynomials requires a solid command of the multiplication methods studied in Chapter 4.*

One important reason for studying factoring is that it can be used to solve certain types of equations that arise in applications. Factoring is also very important elsewhere in algebra, as we will see in later chapters of this text.

5.1 Introduction to Factoring

Factoring Monomials • Factoring When Terms Have a Common Factor • Factoring by Grouping • Checking by Evaluating

Just as a number like 15 can be factored as $3 \cdot 5$, a polynomial like $x^2 + 7x$ can be factored as $x(x + 7)$. In both cases, we ask ourselves, "What was multiplied to obtain the given result?" The situation is much like a popular television game show in which an "answer" is given and participants must find a "question" to which the answer corresponds.

FACTORING

To *factor* a polynomial is to find an equivalent expression that is a product.

Factoring Monomials

To factor a monomial, we find two monomials whose product is equivalent to the original monomial. For example, $20x^2$ can be factored as $2 \cdot 10x^2$, $4x \cdot 5x$, or $10x \cdot 2x$ as well as several other ways.

EXAMPLE 1 Find three factorizations of $15x^3$.

SOLUTION

a) $15x^3 = (3 \cdot 5)(x \cdot x^2)$ Thinking of how 15 and x^3 factor
$ = (3x)(5x^2)$ The factors here are $3x$ and $5x^2$.

b) $15x^3 = (3 \cdot 5)(x^2 \cdot x)$
$ = (3x^2)(5x)$ The factors here are $3x^2$ and $5x$.

c) $15x^3 = ((-5)(-3))x^3$
$ = (-5)(-3x^3)$ The factors here are -5 and $-3x^3$.

Recall from Section 1.2 that the word "factor" can be a verb or a noun, depending on the context in which it appears.

Factoring When Terms Have a Common Factor

To multiply a polynomial of two or more terms by a monomial, we multiply each term by the monomial, using the distributive law $a(b + c) = ab + ac$. To factor, we do the reverse. We express a polynomial as a product, using the same law, read from right to left: $ab + ac = a(b + c)$. Consider the following:

Multiply
$3x(x^2 + 2x - 4)$
$= 3x \cdot x^2 + 3x \cdot 2x - 3x \cdot 4$
$= 3x^3 + 6x^2 - 12x;$

Factor
$3x^3 + 6x^2 - 12x$
$= 3x \cdot x^2 + 3x \cdot 2x - 3x \cdot 4$
$= 3x(x^2 + 2x - 4).$

In the factorization on the right, note that since $3x$ appears as a factor of $3x^3$, $6x^2$, and $-12x$, it is a *common factor* for all the terms of the trinomial $3x^3 + 6x^2 - 12x$.

To factor a polynomial with two or more terms, always try to first find a factor common to all terms. In some cases, there may not be a common factor (other than 1). If a common factor *does* exist, we generally use the common factor with the largest possible coefficient and the largest possible exponent. Such a factor is called the *largest*, or *greatest, common factor*.

EXAMPLE 2   Factor: $5x^2 + 15$.

SOLUTION   We have

$5x^2 + 15 = 5 \cdot x^2 + 5 \cdot 3$ Factoring each term
$= 5(x^2 + 3).$ Factoring out the common factor, 5

To check, we multiply: $5(x^2 + 3) = 5 \cdot x^2 + 5 \cdot 3 = 5x^2 + 15$.

CAUTION! $5 \cdot x^2 + 5 \cdot 3$ is a factorization of the *terms* of $5x^2 + 15$, but not of the polynomial itself. The factorization of $5x^2 + 15$ is $5(x^2 + 3)$.

When asked to factor a polynomial in which all terms contain the same letter raised to various powers, we factor out the largest power possible.

EXAMPLE 3   Factor: $24x^3 + 30x^2$.

SOLUTION   The largest factor common to 24 and 30 is 6. The largest power of x common to x^3 and x^2 is x^2. (To see this, think of x^3 as $x^2 \cdot x$.) Thus the largest common factor of $24x^3$ and $30x^2$ is $6x^2$. We factor as follows:

$24x^3 + 30x^2 = 6x^2 \cdot 4x + 6x^2 \cdot 5$ Factoring each term
$= 6x^2(4x + 5).$ Factoring out $6x^2$

Suppose in Example 3 that you did not recognize the *largest* common factor, and removed only part of it, as follows:

$$24x^3 + 30x^2 = 2x^2 \cdot 12x + 2x^2 \cdot 15$$
$$= 2x^2(12x + 15). \quad \text{12x + 15 still has a common factor.}$$

Note that $12x + 15$ still has a common factor, 3. To find the largest common factor, continue factoring out common factors, as follows, until no more exist:

$$= 2x^2[3(4x + 5)] \quad \text{Factoring } 12x + 15$$
$$= 6x^2(4x + 5). \quad \text{Using an associative law}$$

EXAMPLE 4

Factor: $15x^5 - 12x^4 + 27x^3 - 3x^2$.

SOLUTION We have

$$15x^5 - 12x^4 + 27x^3 - 3x^2$$
$$= 3x^2 \cdot 5x^3 - 3x^2 \cdot 4x^2 + 3x^2 \cdot 9x - 3x^2 \cdot 1 \quad \text{Try to do this mentally.}$$
$$= 3x^2(5x^3 - 4x^2 + 9x - 1). \quad \text{Factoring out } 3x^2$$

CAUTION! Don't forget the term -1.

Since $5x^3 - 4x^2 + 9x - 1$ has no common factor, we are finished, except for a check:

$$3x^2(5x^3 - 4x^2 + 9x - 1) = 15x^5 - 12x^4 + 27x^3 - 3x^2. \quad \text{Our factorization checks.}$$

If you spot the largest common factor without writing out a factorization of each term, you can write the answer in one step.

EXAMPLE 5

Factor: **(a)** $8m^3 - 16m$; **(b)** $14p^2y^3 - 8py^2 + 2py$; **(c)** $\frac{4}{5}x^2 + \frac{1}{5}x + \frac{2}{5}$.

SOLUTION

a) $8m^3 - 16m = 8m(m^2 - 2)$
b) $14p^2y^3 - 8py^2 + 2py = 2py(7py^2 - 4y + 1)$
c) $\frac{4}{5}x^2 + \frac{1}{5}x + \frac{2}{5} = \frac{1}{5}(4x^2 + x + 2)$

Determine the largest common factor by inspection; then carefully fill in the parentheses.

The checks are left to the student.

TIPS FOR FACTORING

1. Whenever factoring, try to factor out the largest common factor.
2. Factoring can always be checked by multiplying. Multiplication should yield the original polynomial.

Factoring by Grouping

Sometimes algebraic expressions contain a common factor that is a polynomial with two or more terms.

EXAMPLE 6

Factor: $x^2(x + 1) + 2(x + 1)$.

SOLUTION The binomial $x + 1$ is a factor of both $x^2(x + 1)$ and $2(x + 1)$. Thus, $x + 1$ is a common factor:

$x^2(x + 1) + 2(x + 1) = (x + 1)x^2 + (x + 1)2$ Using a commutative law twice

$= (x + 1)(x^2 + 2)$.   Factoring out the common factor, $x + 1$

The factorization is $(x + 1)(x^2 + 2)$.

In Example 6, the common binomial factor was obvious. How do we find such a factor in a polynomial like $5x^3 - x^2 + 15x - 3$? Although there is no factor, other than 1, common to all terms, $5x^3 - x^2$ and $15x - 3$ can be grouped and factored separately:

$5x^3 - x^2 = x^2(5x - 1)$ and $15x - 3 = 3(5x - 1)$.

Note that $5x^3 - x^2$ and $15x - 3$ share a common factor of $5x - 1$. This means that the original polynomial, $5x^3 - x^2 + 15x - 3$, can be factored:

$5x^3 - x^2 + 15x - 3 = (5x^3 - x^2) + (15x - 3)$ Using an associative law. This is generally done mentally.

$= x^2(5x - 1) + 3(5x - 1)$ Factoring each binomial

$= (5x - 1)(x^2 + 3)$.   Factoring out the common factor, $5x - 1$

If a polynomial can be split into two groups of terms and both groups share a common factor, then the original polynomial can be factored. This method, known as **factoring by grouping,** can be tried on any polynomial with four terms.

EXAMPLE 7

Factor by grouping.

a) $8x^4 + 6x - 28x^3 - 21$
b) $2x^3 + 8x^2 + x + 4$

SOLUTION

a) $8x^4 + 6x - 28x^3 - 21 = 2x(4x^3 + 3) - 7(4x^3 + 3)$ Factoring two binomials. Using -7 gives a common binomial factor.

$= (4x^3 + 3)(2x - 7)$ Factoring out the common factor, $4x^3 + 3$

Check: $(4x^3 + 3)(2x - 7) = 8x^4 - 28x^3 + 6x - 21$
$= 8x^4 + 6x - 28x^3 - 21$ This is the original polynomial.

b) $2x^3 + 8x^2 + x + 4 = 2x^2(x + 4) + 1(x + 4)$ Factoring $2x^3 + 8x^2$ to find a common binomial factor. Writing the 1 helps with the next step.

$= (x + 4)(2x^2 + 1)$ Factoring out the common factor, $x + 4$. The 1 is essential in the factor $2x^2 + 1$.

The check is left to the student.

Although factoring by grouping can be useful, many polynomials, like $x^3 + x^2 + 2x - 2$, cannot be factored, no matter how we group:

$x^3 + x^2 + 2x - 2 = x^2(x + 1) + 2(x - 1);$ There is no common factor.

$x^3 + x^2 + 2x - 2 = x^3 + 2x + x^2 - 2$
$= x(x^2 + 2) + x^2 - 2;$ There is no common factor.

$x^3 + x^2 + 2x - 2 = x^3 - 2 + x^2 + 2x$
$= x^3 - 2 + x(x + 2).$ There is no common factor.

Checking by Evaluating

We have seen that one way to check a factorization is to multiply. A second type of check, discussed toward the end of Section 4.4, uses the fact that equivalent expressions have the same value when evaluated for the same replacement. Thus a quick, partial check of Example 7(b) can be made by using a convenient replacement for x (say, 1) and evaluating both $2x^3 + 8x^2 + x + 4$ and $(x + 4)(2x^2 + 1)$:

$2 \cdot 1^3 + 8 \cdot 1^2 + 1 + 4$
$= 2 + 8 + 1 + 4$
$= 15;$

$(1 + 4)(2 \cdot 1^2 + 1)$
$= 5 \cdot 3$
$= 15.$

Since the value of both expressions is the same, the factorization is probably correct.

Keep in mind that it is possible, by chance, for two expressions that are not equivalent to share the same value when evaluated. Because of this, evaluating provides only a partial check. Consult with your instructor before making extensive use of this type of check.

TECHNOLOGY CONNECTION 5.1

We saw in Technology Connection 4.4 that, on many graphers, a Table of values can be used to check that two expressions are equal. Thus to check Example 7(b), we let $y_1 = 2x^3 + 8x^2 + x + 4$ and $y_2 = (x + 4)(2x^2 + 1)$. Although any table settings can be used, we use TblMin = 0 and △Tbl = 1:

X	Y₁	Y₂
0	4	4
1	15	15
2	54	54
3	133	133
4	264	264
5	459	459
6	730	730

X = 0

No matter how far up or down we scroll, $y_1 = y_2$. Thus Example 7(b) is correct.

1. Use a Table to check Example 7(a).

EXERCISE SET 5.1

Find three factorizations for each monomial.
1. $10x^3$
2. $6x^3$
3. $-6x^5$
4. $-15x^6$
5. $26x^5$
6. $25x^4$

Factor. Remember to use the largest common factor and to check by multiplying.
7. $x^2 - 6x$
8. $x^2 + 8x$
9. $8x^2 + 4x$
10. $5x^2 - 10x$
11. $x^3 + 6x^2$
12. $4x^4 + x^2$
13. $8x^4 - 24x^2$
14. $5x^5 + 10x^3$
15. $2x^2 + 2x - 8$
16. $6x^2 + 3x - 15$
17. $5x^6 - 10x^3 + 8x^2$
18. $10x^5 + 6x^4 - 3x^3$
19. $2x^8 + 4x^6 - 8x^4 + 10x^2$
20. $5x^4 - 15x^3 - 25x - 10$
21. $x^5y^5 + x^4y^3 + x^3y^3 - x^2y^2$
22. $x^9y^6 - x^7y^5 + x^4y^4 + x^3y^3$
23. $\frac{5}{3}x^6 + \frac{4}{3}x^5 + \frac{1}{3}x^4 + \frac{1}{3}x^3$
24. $\frac{5}{7}x^7 + \frac{3}{7}x^5 - \frac{6}{7}x^3 - \frac{1}{7}x$

Factor.
25. $y(y + 3) + 7(y + 3)$
26. $b(b - 5) + 3(b - 5)$
27. $x^2(x + 3) - 7(x + 3)$
28. $3z^2(2z + 9) + (2z + 9)$
29. $y^2(y + 8) + (y + 8)$
30. $x^2(x - 7) - 3(x - 7)$

Factor by grouping, if possible, and check.
31. $x^3 + 3x^2 + 4x + 12$
32. $6z^3 + 3z^2 + 2z + 1$
33. $2x^3 + 6x^2 + 3x + 9$
34. $3x^3 + 2x^2 + 3x + 2$
35. $9x^3 - 12x^2 + 3x - 4$
36. $10x^3 - 25x^2 + 4x - 10$
37. $5x^3 - 2x^2 + 5x - 2$
38. $18x^3 - 21x^2 + 30x - 35$
39. $x^3 + 8x^2 - 3x - 24$
40. $2x^3 + 12x^2 - 5x - 30$
41. $w^3 - 7w^2 + 4w - 28$
42. $y^3 + 8y^2 - 2y - 16$
43. $x^3 - x^2 - 2x + 5$
44. $p^3 + p^2 - 3p + 10$
45. $2x^3 - 8x^2 - 9x + 36$
46. $20g^3 - 4g^2 - 25g + 5$

SKILL MAINTENANCE

47. Graph: $y = x - 6$.
48. Solve: $4x - 8x + 16 \geq 6(x - 2)$.
49. Subtract: $-13 - (-25)$.
50. Solve $A = \dfrac{p + q}{2}$ for p.

Multiply.
51. $(y + 5)(y + 7)$
52. $(y + 7)^2$
53. $(y + 7)(y - 7)$
54. $(y - 7)^2$

SYNTHESIS

55. ◆ Write a two-sentence paragraph in which the word "factor" is used at least once as a noun and once as a verb.
56. ◆ Josh says that for Exercises 1–46 there is no need to print answers in the back of the text. Is he correct in saying this? Why or why not?
57. ◆ Explain how someone could construct a polynomial with four terms that can be factored by grouping.
58. ◆ In answering a factoring problem, Taylor says the largest common factor is $-5x^2$ and Natasha says the largest common factor is $5x^2$. Can they both be correct? Why or why not?

Factor, if possible.
59. $4x^5 + 6x^3 + 6x^2 + 9$
60. $x^6 + x^4 + x^2 + 1$
61. $x^{12} + x^7 + x^5 + 1$
62. $x^3 + x^2 - 2x + 2$
63. $5x^5 - 5x^4 + x^3 - x^2 + 3x - 3$
64. $ax^2 + 2ax + 3a + x^2 + 2x + 3$

65. ◆ Explain how to construct a polynomial of degree 9 for which $5x^3y^2$ is the largest common factor.
66. ◆ Marlene recognizes that evaluating provides only a partial check of her factoring. Because of this, she often performs a second check with a different replacement value. Is this a good idea? Why or why not?

5.2 Factoring Trinomials of the Type $x^2 + bx + c$

Constant Term Positive • Constant Term Negative

We now learn how to factor trinomials like

$x^2 + 5x + 4$ or $x^2 + 3x - 10$,

for which no common factor exists and the leading coefficient is 1. As preparation for the factoring that follows, compare the following multiplications:

$$\begin{array}{c} \quad\quad\quad\quad\ \ \text{F}\quad\ \ \text{O}\quad\ \ \text{I}\quad\ \ \text{L} \\ \quad\quad\quad\quad\ \ \downarrow\quad\ \ \downarrow\quad\ \ \downarrow\quad\ \ \downarrow \\ (x+2)(x+5) = x^2 + 5x + 2x + 2\cdot 5 \\ = x^2 + \ \ 7x\ \ + 10; \end{array}$$

$$(x-2)(x-5) = x^2 - 5x - 2x + (-2)(-5)$$
$$= x^2 - \ \ 7x\ \ + \ \ 10;$$

$$(x+3)(x-7) = x^2 - 7x + 3x + 3(-7)$$
$$= x^2 - \ \ 4x\ \ - \ \ 21;$$

$$(x-3)(x+7) = x^2 + 7x - 3x + (-3)7$$
$$= x^2 + \ \ 4x\ \ - \ \ 21.$$

Note that for all four products:

- The product of the two binomials is a trinomial.
- The coefficient of x in the trinomial is the sum of the constant terms in the binomials.
- The constant term in the trinomial is the product of the constant terms in the binomials.

These observations lead to a method for factoring certain trinomials. The first type we consider has a positive constant term, just as in the first two multiplications above.

Constant Term Positive

To factor a polynomial like $x^2 + 7x + 10$, we think of FOIL in reverse. The x^2 resulted from x times x, which suggests that the first term of each binomial factor is x. Next, we look for numbers p and q such that

$x^2 + 7x + 10 = (x + p)(x + q)$.

To get the middle term and the last term of the trinomial, we need two numbers p and q whose product is 10 and whose sum is 7. Those numbers are 2 and 5. Thus the factorization is

$(x + 2)(x + 5)$.   The check appears as the first equation in this section.

5.2 FACTORING TRINOMIALS OF THE TYPE $x^2 + bx + c$

EXAMPLE 1 Factor: $x^2 + 5x + 6$.

SOLUTION Think of FOIL in reverse. The first term of each factor is x:

$(x + \ )(x + \ ).$

To complete the factorization, we need a constant term for each of these binomial factors. The constants must have a product of 6 and a sum of 5. We list some pairs of numbers that multiply to 6.

Pairs of Factors of 6	Sums of Factors
1, 6	7
2, 3	5 ← The numbers we seek are 2 and 3.
−1, −6	−7
−2, −3	−5

Since $2 \cdot 3 = 6$ and $2 + 3 = 5$, the factorization of $x^2 + 5x + 6$ is $(x + 2)(x + 3)$. To check, we simply multiply the two binomials.

Check: $(x + 2)(x + 3) = x^2 + 3x + 2x + 6$
$= x^2 + 5x + 6.$ The product is the original polynomial.

Note that since 5 and 6 are both positive, when factoring $x^2 + 5x + 6$ we need not consider negative factors of 6. Note too that changing the signs of the factors changes only the sign of the sum.

At the beginning of this section, we considered the multiplication $(x - 2)(x - 5)$. For this product, the resulting trinomial, $x^2 - 7x + 10$, has a positive constant term but a negative coefficient of x. This is because the *product* of two negative numbers is always positive, whereas the *sum* of two negative numbers is always negative.

TO FACTOR $x^2 + bx + c$ WHEN c IS POSITIVE:

When the constant term of a trinomial is positive, look for two numbers with the same sign. The sign is that of the middle term:

$x^2 - 7x + 10 = (x - 2)(x - 5);$

$x^2 + 7x + 10 = (x + 2)(x + 5).$

EXAMPLE 2 Factor: $y^2 - 8y + 12$.

SOLUTION Since the constant term is positive and the coefficient of the middle term is negative, we look for a factorization of 12 in which both factors are negative. Their sum must be −8.

Pairs of Factors of 12	Sums of Factors
−1, −12	−13
−2, −6	−8
−3, −4	−7

We need a sum of −8.
The numbers we need are −2 and −6.

The factorization of $y^2 - 8y + 12$ is $(y - 2)(y - 6)$. The check is left to the student.

Constant Term Negative

As we saw in two of the multiplications earlier in this section, the product of two binomials can have a negative constant term:

$$(x + 3)(x - 7) = x^2 - 4x - 21$$

and

$$(x - 3)(x + 7) = x^2 + 4x - 21.$$

Note that when the signs of the constants in the binomials are reversed, only the sign of the middle term in the product changes.

EXAMPLE 3   Factor: $x^2 - 8x - 20$.

SOLUTION   The constant term, −20, must be expressed as the product of a negative number and a positive number. Since the sum of these two numbers must be negative (specifically, −8), the negative number must have the greater absolute value.

Pairs of Factors of −20	Sums of Factors
1, −20	−19
2, −10	−8
4, −5	−1
5, −4	1
10, −2	8
20, −1	19

The numbers we need are 2 and −10.

Since the positive factor in each of these pairs has the larger absolute value, the sums are all positive. For this problem, we can disregard these pairs. Note that changing the signs of the factors changes the sign of the sum.

The numbers that we are looking for are 2 and −10. Therefore, the factorization is $(x + 2)(x - 10)$.

Check: $(x + 2)(x - 10) = x^2 - 10x + 2x - 20$
$= x^2 - 8x - 20.$

5.2 FACTORING TRINOMIALS OF THE TYPE $x^2 + bx + c$ 253

TO FACTOR $x^2 + bx + c$ WHEN c IS NEGATIVE:

When the constant term of a trinomial is negative, look for two numbers whose product is negative. One must be positive and the other negative:

$$x^2 - 4x - 21 = (x + 3)(x - 7);$$

$$x^2 + 4x - 21 = (x - 3)(x + 7).$$

Select the two numbers so that the number with the larger absolute value has the same sign as b, the coefficient of the middle term.

EXAMPLE 4   Factor: $t^2 - 24 + 5t$.

SOLUTION   It helps to first write the trinomial in descending order: $t^2 + 5t - 24$. The factorization of the constant term, -24, must have one factor positive and one factor negative. The sum must be 5, so the positive factor must have the larger absolute value. Thus we consider only pairs of factors in which the positive factor has the larger absolute value.

Pairs of Factors of -24	Sums of Factors
$-1, 24$	23
$-2, 12$	10
$-3, 8$	5 ← The numbers we need are -3 and 8.
$-4, 6$	2

The factorization is $(t - 3)(t + 8)$. The check is left to the student.

Polynomials in two or more variables, such as $a^2 + 4ab - 21b^2$, are factored in a similar manner.

EXAMPLE 5   Factor: $a^2 + 4ab - 21b^2$.

SOLUTION   It may help to write the trinomial in the equivalent form

$$a^2 + 4ba - 21b^2.$$

This way we think of $-21b^2$ as the "constant" term and $4b$ as the "coefficient" of the middle term. Then we try to express $-21b^2$ as a product of two factors whose sum is $4b$. Those factors are $-3b$ and $7b$. Thus the factorization is

$$(a - 3b)(a + 7b).$$

Check: $(a - 3b)(a + 7b) = a^2 + 7ab - 3ba - 21b^2$
$ = a^2 + 4ab - 21b^2.$

EXAMPLE 6 Factor: $x^2 - x + 5$.

SOLUTION Since 5 has very few factors, we can easily check all possibilities.

Pairs of Factors of 5	Sums of Factors
5, 1	6
−5, −1	−6

Since there are no factors whose sum is −1, the polynomial is *not* factorable into binomials.

A polynomial like $x^2 - x + 5$ that cannot be factored further is said to be **prime**.

Often factoring requires two or more steps. In general, when told to factor, we should *factor completely*. This means that the final factorization should not contain any factors that can be factored further.

EXAMPLE 7 Factor: $2x^3 - 20x^2 + 50x$.

SOLUTION *Always* look first for a common factor. This time there is one, $2x$, which we factor out first:

$$2x^3 - 20x^2 + 50x = 2x(x^2 - 10x + 25).$$

Now consider $x^2 - 10x + 25$. Since the constant term is positive and the coefficient of the middle term is negative, we look for a factorization of 25 in which both factors are negative. Their sum must be −10.

Pairs of Factors of 25	Sums of Factors
−25, −1	−26
−5, −5	−10 ← The numbers we need are −5 and −5.

The factorization of

$$x^2 - 10x + 25$$

is

$$(x - 5)(x - 5), \quad \text{or} \quad (x - 5)^2.$$

Remember, this is not the entire factorization of $2x^3 - 20x^2 + 50x$.

Thus,

$$2x^3 - 20x^2 + 50x = 2x(x^2 - 10x + 25)$$
$$= 2x(x - 5)(x - 5), \quad \text{or} \quad 2x(x - 5)^2.$$

Check: $2x(x - 5)(x - 5) = 2x[x^2 - 10x + 25]$ Multiplying binomials
$$= 2x^3 - 20x^2 + 50x.$$ Using the distributive law

TO FACTOR $x^2 + bx + c$:

1. Find a pair of factors that have c as their product and b as their sum.

   a) If c is positive, its factors will have the same sign as b.

   b) If c is negative, one factor will be positive and the other will be negative. Select the factors such that the factor with the larger absolute value is the factor with the same sign as b.

2. Check by multiplying.

EXERCISE SET 5.2

Factor completely. Remember that you can check by multiplying. If a polynomial is prime, state this.

1. $x^2 + 6x + 8$
2. $x^2 + 7x + 6$
3. $x^2 + 9x + 8$
4. $x^2 + 7x + 12$
5. $y^2 + 11y + 28$
6. $x^2 - 6x + 9$
7. $a^2 + 11a + 30$
8. $x^2 + 9x + 14$
9. $x^2 - 5x + 4$
10. $b^2 + 5b + 4$
11. $z^2 - 8z + 7$
12. $a^2 - 4a - 12$
13. $x^2 - 8x + 15$
14. $d^2 - 7d + 10$
15. $x^2 - 2x - 15$
16. $y^2 - 11y + 10$
17. $x^2 + 2x - 15$
18. $x^2 + x - 42$
19. $3y^2 - 9y - 84$
20. $2x^2 - 14x - 36$
21. $x^3 - x^2 - 42x$
22. $x^3 - 6x^2 - 16x$
23. $x^2 - 7x - 60$
24. $y^2 - 4y - 45$
25. $x^2 - 72 + 6x$
26. $-2x - 99 + x^2$
27. $5b^2 + 25b - 120$
28. $c^4 + c^3 - 56c^2$
29. $x^5 + x^4 - 2x^3$
30. $2a^2 + 4a - 70$
31. $x^2 + 2x + 3$
32. $x^2 + x + 1$
33. $11 - 3w + w^2$
34. $7 - 2p + p^2$
35. $x^2 + 20x + 99$
36. $x^2 + 20x + 100$
37. $2x^3 - 40x^2 + 192x$
38. $3x^3 - 63x^2 - 300x$
39. $4x^2 + 40x + 100$
40. $x^2 - 21x - 72$
41. $y^2 - 21y + 108$
42. $x^2 - 25x + 144$
43. $a^6 + 9a^5 - 90a^4$
44. $a^4 + a^3 - 132a^2$
45. $x^2 - \frac{2}{5}x + \frac{1}{25}$
46. $t^2 + \frac{2}{3}t + \frac{1}{9}$
47. $112 + 9y - y^2$
48. $108 - 3x - x^2$
49. $t^2 - 0.3t - 0.10$
50. $y^2 - 0.2y - 0.08$
51. $p^2 + 3pq - 10q^2$
52. $a^2 - 2ab - 3b^2$
53. $m^2 + 5mn + 5n^2$
54. $x^2 - 11xy + 24y^2$
55. $s^2 - 2st - 15t^2$
56. $b^2 + 8bc - 20c^2$
57. $6a^{10} - 30a^9 - 84a^8$
58. $7x^9 - 28x^8 - 35x^7$

SKILL MAINTENANCE

Solve.

59. $3x - 8 = 0$
60. $2x + 7 = 0$

Multiply.

61. $(x + 6)(3x + 4)$
62. $(7w + 6)^2$

63. In a recent year, 29,090 people were arrested for counterfeiting. This figure was down 1.2% from the year before. How many people were arrested the year before?

64. The first angle of a triangle is 4 times as large as the second. The measure of the third angle is 30° greater than that of the second. How large are the angles?

SYNTHESIS

65. ◆ When searching for a factorization, why do we list pairs of numbers with the correct *product* instead of pairs of numbers with the correct *sum*?

256 CHAPTER 5 • POLYNOMIALS AND FACTORING

66. ◆ Why are we urged to always factor out common factors first?
67. ◆ Without multiplying $(x - 17)(x - 18)$, explain why it cannot possibly be a factorization of $x^2 + 35x + 306$.
68. ◆ Marge factors $x^3 - 8x^2 + 15x$ as $(x^2 - 5x)(x - 3)$. Is she wrong? Why or why not? What advice would you offer?
69. Find all integers m for which $y^2 + my + 50$ can be factored.
70. Find all integers b for which $a^2 + ba - 50$ can be factored.

Factor completely.

71. $x^2 + \frac{1}{4}x - \frac{1}{8}$
72. $x^2 + \frac{1}{2}x - \frac{3}{16}$
73. $\frac{1}{3}a^3 - \frac{1}{3}a^2 - 2a$
74. $a^7 - \frac{25}{7}a^5 - \frac{30}{7}a^6$
75. $x^{2m} + 11x^m + 28$
76. $t^{2n} - 7t^n + 10$
77. $(x + 1)a^2 + (x + 1)3a + (x + 1)2$
78. $ax^2 - 5x^2 + 8ax - 40x - (a - 5)9$
(*Hint:* See Exercise 77.)

Find a polynomial in factored form for the shaded area in each figure. (Leave answers in terms of π.)

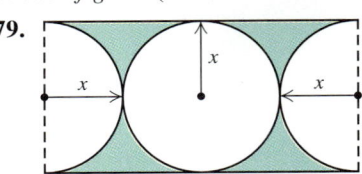

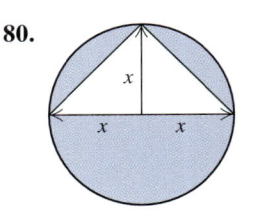

79.

80.

COLLABORATIVE CORNER

Focus: Visualizing factoring*
Time: 20–30 minutes
Group size: 2
Materials: Graph paper and scissors

The product $(x + 2)(x + 3)$ can be regarded as the area of a rectangle with width $x + 2$ and length $x + 3$. Similarly, factoring a polynomial like $x^2 + 5x + 6$ can be thought of as determining the length and the width of a rectangle that has area $x^2 + 5x + 6$. This is the approach used in parts (1) and (2) below.

ACTIVITY

1. a) To factor $x^2 + 11x + 10$ geometrically, the group must first determine areas representing x^2, $11x$, and 10. To represent x^2, each student should select a value for x—say, 4—and use the squares of the graph paper for guidance in cutting out a square that is x units on each side.
   b) Each group member should then represent $11x$ by cutting out 11 strips that are each 1 unit wide and x units long. Use the squares of the graph paper for guidance. Be sure to use the same x-value that was chosen in part (a).
   c) There are two rectangles with whole-number dimensions and an area of 10: One is 2 units by 5 units and the other is 1 unit by 10 units. Each group member should cut out one of these and then attempt to use it and all of the above shapes to form one large rectangle. Only one of the group members will be successful.
   d) From the large rectangle formed in part (c), use the length and the width to determine the factorization of $x^2 + 11x + 10$.
2. a) To factor $x^2 + 8x + 12$, the group can reuse the above shapes for x^2 and x. Represent $x^2 + 8x$ in a variety of configurations using the square for x^2 and eight strips for $8x$.

---

*Instructors may wish to make use of Cuisinaire® Algebra Tiles for this activity. Please contact your local Addison Wesley Longman sales consultant for more information.

b) Cut out all possible rectangles that have an area of 12 and whole numbers for the length and the width.
c) Use one of the rectangles from part (b), along with the shapes for $x^2 + 8x$, to form one large rectangle.
d) From the large rectangle formed in part (c), determine the factorization of $x^2 + 8x + 12$.

3. Where do the dimensions of the rectangle representing 10 show up in the factorization of $x^2 + 11x + 10$? Where do the dimensions of the rectangle representing 12 show up in the factorization of $x^2 + 8x + 12$?

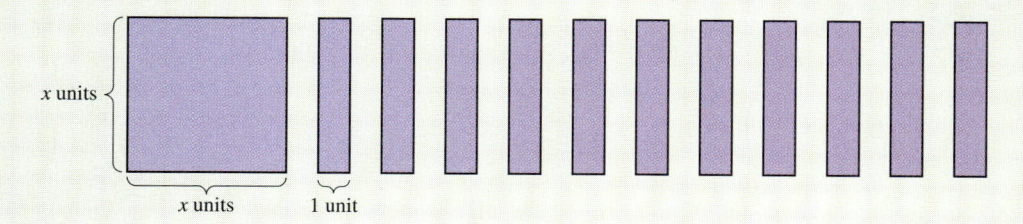

5.3 Factoring Trinomials of the Type $ax^2 + bx + c$

Factoring with FOIL • The Grouping Method

In Section 5.2, we learned a FOIL-based method for factoring trinomials of the type $x^2 + bx + c$. Now we learn to factor trinomials in which the leading, or x^2, coefficient is not 1. First we will study another FOIL-based method and then we will consider an alternative method that involves factoring by grouping. Use the method that you prefer or the approach selected by your instructor.

Factoring with FOIL

We want to factor trinomials of the type $ax^2 + bx + c$. Consider the following multiplication:

$$\begin{array}{c} \text{F} \text{O} \text{I} \text{L} \\ (2x+5)(3x+4) = 6x^2 + 8x + 15x + 20 \\ = 6x^2 + 23x + 20 \end{array}$$

To factor $6x^2 + 23x + 20$, we reverse the multiplication above and look for two binomials whose product is this trinomial. The product of the First terms must be $6x^2$. The product of the Outside terms plus the product of the Inside terms must be $23x$. The product of the Last terms must be 20. We see from above that the desired factorization is

$(2x + 5)(3x + 4)$.

258 CHAPTER 5 • POLYNOMIALS AND FACTORING

How can such a factorization be found without first seeing the corresponding multiplication? One approach relies heavily on trial and error (guessing and checking) and FOIL.

TO FACTOR $ax^2 + bx + c$ **USING FOIL:**

1. Factor out the largest common factor, if one exists.
2. Find two **F**irst terms whose product is ax^2:

   $(\boxed{}x +)(\boxed{}x +) = ax^2 + bx + c.$

   ————FOIL

3. Find two **L**ast terms whose product is c:

   $(x + \boxed{})(x + \boxed{}) = ax^2 + bx + c.$

   ————FOIL

4. Repeat steps (2) and (3) until a combination is found for which the sum of the **O**uter and **I**nner products is bx:

   $(\boxed{}x + \boxed{})(\boxed{}x + \boxed{}) = ax^2 + bx + c.$

   I
   O
   FOIL

EXAMPLE 1

Factor: $3x^2 - 10x - 8$.

SOLUTION

1. First, check for a common factor. In this case, there is none (other than 1 or -1).
2. Find two **F**irst terms whose product is $3x^2$.

   The only possibilities for the **F**irst terms are $3x$ and x, so any factorization must be of the form

   $(3x +)(x +).$

3. Find two **L**ast terms whose product is -8.

   Possible factorizations of -8 are

   $(-8) \cdot 1, \quad 8 \cdot (-1), \quad (-2) \cdot 4, \quad \text{and} \quad 2 \cdot (-4).$

   Since the **F**irst terms are not identical, we must also consider

   $1 \cdot (-8), \quad (-1) \cdot 8, \quad 4 \cdot (-2), \quad \text{and} \quad (-4) \cdot 2.$

4. Inspect the **O**uter and **I**nner products resulting from steps (2) and (3). Look for a combination in which the sum of the products is the middle term, $-10x$.

5.3 FACTORING TRINOMIALS OF THE TYPE $ax^2 + bx + c$

Trial	Product	
$(3x - 8)(x + 1)$	$3x^2 + 3x - 8x - 8$ $= 3x^2 - 5x - 8$	Wrong middle term
$(3x + 8)(x - 1)$	$3x^2 - 3x + 8x - 8$ $= 3x^2 + 5x - 8$	Wrong middle term
$(3x - 2)(x + 4)$	$3x^2 + 12x - 2x - 8$ $= 3x^2 + 10x - 8$	Wrong middle term
$(3x + 2)(x - 4)$	$3x^2 - 12x + 2x - 8$ $= 3x^2 - 10x - 8$	Correct middle term!
$(3x + 1)(x - 8)$	$3x^2 - 24x + x - 8$ $= 3x^2 - 23x - 8$	Wrong middle term
$(3x - 1)(x + 8)$	$3x^2 + 24x - x - 8$ $= 3x^2 + 23x - 8$	Wrong middle term
$(3x + 4)(x - 2)$	$3x^2 - 6x + 4x - 8$ $= 3x^2 - 2x - 8$	Wrong middle term
$(3x - 4)(x + 2)$	$3x^2 + 6x - 4x - 8$ $= 3x^2 + 2x - 8$	Wrong middle term

The correct factorization is $(3x + 2)(x - 4)$.

Two observations can be made from Example 1. First, we listed all possible trials even though we could have stopped after finding the correct factorization. We did this to show that each trial differs only in the middle term of the product. Second, note that as in Section 5.2, only the sign of the middle term changes when the signs in the binomials are reversed.

EXAMPLE 2

Factor: $24x^2 - 76x + 40$.

SOLUTION

1. First we factor out the largest common factor, 4:

   $4(6x^2 - 19x + 10)$.

   Now we factor the trinomial $6x^2 - 19x + 10$.

2. Since $6x^2$ can be factored as $3x \cdot 2x$ or $6x \cdot x$, we have two possibilities:

   $(3x +\ \ )(2x +\ \ )$ or $(6x +\ \ )(x +\ \ )$.

3. There are four pairs of factors of 10 and each can be listed two ways:

   10, 1 $-10, -1$ 5, 2 $-5, -2$

   and

   1, 10 $-1, -10$ 2, 5 $-2, -5$.

4. The two possibilities from step (2) and the eight possibilities from step (3) give $2 \cdot 8$, or 16 possibilities for factorizations. We look for **O**uter and **In**ner products resulting from steps (2) and (3) for which the sum is the middle term, $-19x$. Since the sign of the middle term is negative, but the sign of the last term, 10, is positive, the two factors of 10 must both be

negative. This means only four pairings from step (3) need be considered. We first try these factors with $(3x +)(2x +)$. If none gives the correct factorization of $6x^2 - 19x + 10$, then we will consider $(6x +)(x +)$.

Trial	Product	
$(3x - 10)(2x - 1)$	$6x^2 - 3x - 20x + 10$ $= 6x^2 - 23x + 10$	Wrong middle term
$(3x - 1)(2x - 10)$	$6x^2 - 30x - 2x + 10$ $= 6x^2 - 32x + 10$	Wrong middle term
$(3x - 5)(2x - 2)$	$6x^2 - 6x - 10x + 10$ $= 6x^2 - 16x + 10$	Wrong middle term
$(3x - 2)(2x - 5)$	$6x^2 - 15x - 4x + 10$ $= 6x^2 - 19x + 10$	Correct middle term!

Since we have a correct factorization, we don't need to consider

$(6x +)(x +)$.

Look again at the possibility $(3x - 5)(2x - 2)$. Without multiplying, we can reject such a possibility. To see why, note that

$$(3x - 5)(2x - 2) = 2(3x - 5)(x - 1).$$

The expression $2x - 2$ has a common factor, 2. But we removed the *largest* common factor in the first step. If $2x - 2$ were one of the factors, then 2 would have to be a common factor in addition to the original 4. Thus, $(2x - 2)$ cannot be part of the factorization of the original trinomial. Similar reasoning can be used to reject $(3x - 1)(2x - 10)$ as a possible factorization.

Given that the largest common factor is factored out at the outset, we need not consider factorizations in which a factor has a common factor.

The factorization of $6x^2 - 19x + 10$ is $(3x - 2)(2x - 5)$, but do not forget the common factor! We must include it in order to factor the original trinomial:

$$24x^2 - 76x + 40 = 4(6x^2 - 19x + 10)$$
$$= 4(3x - 2)(2x - 5).$$

EXAMPLE 3

Factor: $10x^2 + 37x + 7$.

SOLUTION

1. There is no common factor (other than 1 or -1).
2. Because $10x^2$ factors as $10x \cdot x$ or $5x \cdot 2x$, we have these possibilities:

   $(10x +)(x +)$ or $(5x +)(2x +)$.

3. There are two pairs of factors of 7 and each can be listed two ways:

   1, 7 $-1, -7$

   and

   7, 1 $-7, -1$.

4. From steps (2) and (3), we see that there are 8 possibilities for factorizations. Look for **O**uter and **I**nner products for which the sum is the middle term. Because all coefficients in $10x^2 + 37x + 7$ are positive, we need consider only positive factors of 7. The possibilities are

$(10x + 1)(x + 7) = 10x^2 + 71x + 7,$
$(10x + 7)(x + 1) = 10x^2 + 17x + 7,$
$(5x + 7)(2x + 1) = 10x^2 + 19x + 7,$
$(5x + 1)(2x + 7) = 10x^2 + 37x + 7.$

The factorization is $(5x + 1)(2x + 7)$.

TIPS FOR FACTORING $ax^2 + bx + c$

1. Always factor out the largest common factor, if one exists.
2. Once the largest common factor has been factored out of the original trinomial, no binomial factor can contain a common factor (other than 1 or −1).
3. If c is positive, then the signs in both binomial factors must match the sign of b. (This assumes $a > 0$.)
4. Reversing the signs in the binomials reverses the sign of the middle term of their product.
5. Organize your work so that you can keep track of which possibilities have or have not been checked.
6. Always check by multiplying.

Factoring trinomials of the type $ax^2 + bx + c$ involves trial and error. The more you practice, the more you will find that certain trials can either be skipped or performed mentally.

EXAMPLE 4   Factor: $6p^2 - 13pq - 28q^2$.

SOLUTION   Since no common factor exists, we examine the first term, $6p^2$. There are two possibilities:

$(2p + \quad)(3p + \quad)$ or $(6p + \quad)(p + \quad)$.

The last term, $-28q^2$, has the following pairs of factors:

$\quad 28q, -q \qquad 14q, -2q \qquad 7q, -4q$

and

$\quad -28q, \ q \qquad -14q, \ 2q \qquad -7q, \ 4q,$

as well as each of the pairings reversed.

Some trials, like $(2p + 28q)(3p - q)$ and $(2p + 14q)(3p - 2q)$, cannot be correct because $(2p + 28q)$ and $(2p + 14q)$ contain a common factor, 2. We try $(2p + 7q)(3p - 4q)$:

$(2p + 7q)(3p - 4q) = 6p^2 - 8pq + 21pq - 28q^2$
$\qquad\qquad\qquad\quad = 6p^2 + 13pq - 28q^2.$

Our trial is incorrect, but only because of the sign of the middle term. To correctly factor $6p^2 - 13pq - 28q^2$, we simply change the signs in the binomials:

$$(2p - 7q)(3p + 4q).$$

Check: $(2p - 7q)(3p + 4q) = 6p^2 - 13pq - 28q^2.$

The Grouping Method

Another method of factoring trinomials of the type $ax^2 + bx + c$ is known as the *grouping method*. The grouping method relies on finding two numbers, p and q, for which $p + q = b$ and $ax^2 + px + qx + c$ can be factored by grouping. To develop this method, consider the following*:

$$\begin{aligned}(2x + 5)(3x + 4) &= 2x \cdot 3x + 2x \cdot 4 + 5 \cdot 3x + 5 \cdot 4 \quad \text{Using FOIL}\\ &= 2 \cdot 3 \cdot x^2 + 2 \cdot 4x + 5 \cdot 3x + 5 \cdot 4\\ &= 2 \cdot 3 \cdot x^2 + (2 \cdot 4 + 5 \cdot 3)x + 5 \cdot 4\\ &\quad\quad\; a \quad\quad\quad\quad\quad\; b \quad\quad\quad\;\; c\\ &= 6x^2 + 23x + 20.\end{aligned}$$

Note that reversing these steps shows that $6x^2 + 23x + 20$ can be rewritten as $6x^2 + 8x + 15x + 20$ and then factored by grouping. Note that the number b (in this case, $2 \cdot 4 + 5 \cdot 3$, or 23) is written as a sum of two numbers whose product is ac (in this case, $2 \cdot 3 \cdot 5 \cdot 4$, or $6 \cdot 20$).

TO FACTOR $ax^2 + bx + c$, USING THE GROUPING METHOD:

1. Factor out the largest common factor, if one exists.
2. Multiply the leading coefficient a and the constant c.
3. Find a pair of factors of ac whose sum is b.
4. Rewrite the middle term, bx, as a sum or difference using the factors found in step (3).
5. Then factor by grouping.

EXAMPLE 5

Factor: $3x^2 - 10x - 8$.

SOLUTION

1. First note that there is no common factor (other than 1 or -1).
2. We multiply the leading coefficient, 3, and the constant, -8:

   $3(-8) = -24.$

3. We then look for a factorization of -24 in which the sum of the factors is the coefficient of the middle term, -10.

---

*This discussion was inspired by a lecture given by Irene Doo at Austin Community College.

Pairs of Factors of −24	Sums of Factors
1, −24	−23
−1, 24	23
2, −12	−10 ←
−2, 12	10
3, −8	−5
−3, 8	5
4, −6	−2
−4, 6	2

2 + (−12) = −10

We normally stop listing pairs of factors once we have found the one we are after.

4. Next, we express the middle term as a sum or difference using the factors found in step (3):

   $-10x = 2x - 12x.$

5. We now factor by grouping as follows:

   $3x^2 - 10x - 8 = 3x^2 + 2x - 12x - 8$ Substituting $2x - 12x$ for $-10x$

   $= x(3x + 2) - 4(3x + 2)$ Factoring by grouping; see Section 5.1

   $= (3x + 2)(x - 4).$ Factoring out the common factor, $3x + 2$

   Check: $(3x + 2)(x - 4) = 3x^2 - 10x - 8.$

EXAMPLE 6   Factor: $8x^3 + 22x^2 - 6x$.

SOLUTION

1. We factor out the largest common factor, $2x$:

   $8x^3 + 22x^2 - 6x = 2x(4x^2 + 11x - 3).$

2. To factor $4x^2 + 11x - 3$ by grouping, we multiply the leading coefficient, 4, and the constant term, -3:

   $4(-3) = -12.$

3. We next look for factors of -12 that add to 11.

Pairs of Factors of −12	Sums of Factors
1, −12	−11
−1, 12	11 ←
.	.
.	.
.	.

Since $-1 + 12 = 11$, there is no need to list other pairs of factors.

4. We then rewrite the middle term, $11x$:

   $11x = -1x + 12x,$ or $11x = 12x - 1x.$

5. Next, we factor by grouping:

$$4x^2 + 11x - 3 = 4x^2 + 12x - 1x - 3 \quad \text{Rewriting the middle term}$$
$$= 4x(x + 3) - 1(x + 3) \quad \text{Factoring by grouping. Removing } -1 \text{ reveals the common factor, } x + 3.$$
$$= (x + 3)(4x - 1). \quad \text{Factoring out the common factor}$$

The factorization of $4x^2 + 11x - 3$ is $(x + 3)(4x - 1)$. But don't forget the common factor, $2x$, when giving the factorization of the original trinomial:

$$8x^3 + 22x^2 - 6x = 2x(x + 3)(4x - 1).$$

EXERCISE SET 5.3

Factor completely. If a polynomial is prime, state this.

1. $3x^2 + x - 4$
2. $2x^2 + 7x - 4$
3. $5x^2 - x - 18$
4. $3x^2 - 4x - 15$
5. $6x^2 - 13x + 6$
6. $6x^2 - 23x + 7$
7. $3x^2 + 4x + 1$
8. $7x^2 + 15x + 2$
9. $4x^2 + 4x - 15$
10. $9x^2 + 6x - 8$
11. $15x^2 - 19x - 10$
12. $3x^2 - 5x - 2$
13. $9x^2 + 18x - 16$
14. $2x^2 - x - 1$
15. $2x^2 - 5x + 2$
16. $18x^2 - 3x - 10$
17. $12x^2 - 31x + 20$
18. $15x^2 + 19x + 6$
19. $28x^2 + 38x - 6$
20. $35x^2 + 34x + 8$
21. $9x^2 + 18x + 8$
22. $4 - 13x + 6x^2$
23. $49 + 42x + 9x^2$
24. $25x^2 + 40x + 16$
25. $24x^2 + 47x - 2$
26. $16a^2 + 78a + 27$
27. $35x^2 - 57x - 44$
28. $18t^2 + 24t - 10$
29. $2x^2 - 6x - 19$
30. $2x^2 - x - 15$
31. $12x^2 + 28x - 24$
32. $6x^2 + 33x + 15$
33. $30x^2 - 24x - 54$
34. $20x^2 - 25x + 5$
35. $6x^2 + 33x + 15$
36. $12x^2 + 28x - 24$
37. $-9 + 18x^2 + 21x$
38. $4x + 1 + 3x^2$

Factor. Use factoring by grouping even though it would seem reasonable to first combine like terms.

39. $y^2 + 4y - 2y - 8$
40. $x^2 + 5x + 2x + 10$
41. $x^2 - 4x - x + 4$
42. $a^2 + 5a - 2a - 10$
43. $6x^2 + 4x + 9x + 6$
44. $3x^2 - 2x + 3x - 2$
45. $3x^2 - 4x - 12x + 16$
46. $24 - 18y - 20y + 15y^2$
47. $35x^2 - 40x + 21x - 24$
48. $8x^2 - 6x - 28x + 21$
49. $4x^2 + 6x - 6x - 9$
50. $2x^4 - 6x^2 - 5x^2 + 15$

Factor completely. If a polynomial is prime, state this.

51. $9x^2 - 42x + 49$
52. $25t^2 + 80t + 64$
53. $18t^2 + 3t - 10$
54. $15x^2 - 25x - 10$
55. $14x^2 - 35x + 14$
56. $2x^2 + 6x - 14$
57. $6x^3 - 4x^2 - 10x$
58. $18x^3 + 21x^2 - 9x$
59. $47 - 42y + 9y^2$
60. $89x + 64 + 25x^2$
61. $144x^5 - 168x^4 + 48x^3$
62. $168x^3 + 45x^2 + 3x$
63. $70a^4 - 68a^3 + 16a^2$
64. $14t^4 - 19t^3 - 3t^2$
65. $12m^2 + mn - 20n^2$
66. $6a^2 - ab - 15b^2$
67. $3p^2 - 16pq - 12q^2$
68. $9a^2 + 18ab + 8b^2$
69. $10s^2 + 4st - 6t^2$
70. $35p^2 + 34pq + 8q^2$
71. $30a^2 + 87ab + 30b^2$
72. $18x^2 - 6xy - 24y^2$
73. $15a^2 - 5ab - 20b^2$
74. $24a^2 - 34ab + 12b^2$
75. $9x^2y^2 + 18xy^2 - 16$
76. $4x^2y + 10xy + 2y$
77. $18x^2y^2 - 3xy - 10$
78. $9a^2b^2 - 15ab - 2$
79. $35x^2 + 34x^3 + 8x^4$
80. $19x^3 - 3x^2 + 14x^4$
81. $18a + 8 + 9a^2$
82. $40a + 16 + 25a^2$

SKILL MAINTENANCE

83. The earth is a sphere (or ball) that is about 40,000 km in circumference. Find the radius of the earth, in kilometers and in miles. Use 3.14 for π. (*Hint*: 1 km ≈ 0.62 mi.)

84. The second angle of a triangle is 10° less than twice the first. The third angle is 15° more than 4 times the first. Find the measure of the second angle.

85. Graph: $y = \frac{2}{5}x - 1$.

86. Divide: $\dfrac{y^{12}}{y^4}$.

Multiply.

87. $(3x - 5)(3x + 5)$

88. $(4a - 3)^2$

SYNTHESIS

89. ◆ A student presents the following work:
$$4x^2 + 28x + 48 = (2x + 6)(2x + 8)$$
$$= 2(x + 3)(x + 4).$$
Is this correct? Explain.

90. ◆ Of the six tips listed after Example 3, which one do you find the most useful? Why?

91. ◆ Asked to factor $2x^2 - 18x + 36$, a student *incorrectly* answers
$$2x^2 - 18x + 36 = 2(x^2 + 9x + 18)$$
$$= 2(x + 3)(x + 6).$$
If this were a 10-point quiz question, how many points would you take off? Why?

92. ◆ Explain how the answer to Exercise 21 can be used to answer Exercise 68.

Factor.

93. $16x^{10} - 8x^5 + 1$

94. $9x^{10} + 12x^5 + 4$

95. $20x^{2n} + 16x^n + 3$

96. $-15x^{2m} + 26x^m - 8$

97. $3x^{6a} - 2x^{3a} - 1$

98. $x^{2n+1} - 2x^{n+1} + x$

99. $3(a + 1)^{n+1}(a + 3)^2 - 5(a + 1)^n(a + 3)^3$

100. $7(t - 3)^{2n} + 5(t - 3)^n - 2$

5.4 Factoring Perfect-Square Trinomials and Differences of Squares

Recognizing Perfect-Square Trinomials • Factoring Perfect-Square Trinomials • Recognizing Differences of Squares • Factoring Differences of Squares • Factoring Completely

In Chapter 4, we studied some shortcuts for finding certain products of binomials. Reversing these procedures provides shortcuts for factoring certain polynomials.

Recognizing Perfect-Square Trinomials

Some trinomials are squares of binomials. For example, $x^2 + 10x + 25$ is the square of the binomial $x + 5$. To see this, we can calculate $(x + 5)^2$. It is $x^2 + 2 \cdot x \cdot 5 + 5^2$, or $x^2 + 10x + 25$. A trinomial that is the square of a binomial is called a **perfect-square trinomial.**

In Chapter 4, we considered squaring binomials as a special-product rule:

$(A + B)^2 = A^2 + 2AB + B^2;$
$(A - B)^2 = A^2 - 2AB + B^2.$

Written from right to left, these equations can be used to factor perfect-square trinomials. Note that in order for a trinomial to be the square of a binomial, it must have the following:

1. Two terms, A^2 and B^2, must be squares, such as

    $4, \quad x^2, \quad 81m^2, \quad 16t^2.$

2. There must be no minus sign before A^2 or B^2.
3. The remaining term is either $2 \cdot A \cdot B$ or $-2 \cdot A \cdot B$, where A and B are the square roots of A^2 and B^2.

EXAMPLE 1 Determine whether each of the following is a perfect-square trinomial.

a) $x^2 + 6x + 9$ b) $t^2 - 8t - 9$ c) $16x^2 + 49 - 56x$

SOLUTION

a) To see if $x^2 + 6x + 9$ is a perfect-square trinomial, note that:

1. Two terms, x^2 and 9, are squares.
2. There is no minus sign before x^2 or 9.
3. The remaining term, $6x$, is $2 \cdot x \cdot 3$, where x and 3 are the square roots of x^2 and 9.

Thus, $x^2 + 6x + 9$ is a perfect-square trinomial.

b) To see if $t^2 - 8t - 9$ is a perfect-square trinomial, note that:

1. Two terms, t^2 and 9, are squares. But:
2. Since 9 is being subtracted, $t^2 - 8t - 9$ is *not* a perfect-square trinomial.

c) To see if $16x^2 + 49 - 56x$ is a perfect-square trinomial, it helps to first write it in descending order:

$16x^2 - 56x + 49.$

Next, note that:

1. Two terms, $16x^2$ and 49, are squares.
2. There is no minus sign before $16x^2$ or 49.
3. Twice the product of the square roots, $2 \cdot 4x \cdot 7$, is $56x$, the opposite of the remaining term, $-56x$.

Thus, $16x^2 + 49 - 56x$ *is* a perfect-square trinomial.

Factoring Perfect-Square Trinomials

Either of the factoring methods from Section 5.3 can be used to factor perfect-square trinomials, but a faster method is to recognize the following patterns.

FACTORING A PERFECT-SQUARE TRINOMIAL

$$A^2 + 2AB + B^2 = (A + B)^2; \qquad A^2 - 2AB + B^2 = (A - B)^2$$

The factorization uses the square roots of the squared terms and the sign of the remaining term.

EXAMPLE 2 Factor: (a) $x^2 + 6x + 9$; (b) $x^2 + 49 - 14x$; (c) $16x^2 - 40x + 25$.

SOLUTION

a) $x^2 + 6x + 9 = x^2 + 2 \cdot x \cdot 3 + 3^2 = (x + 3)^2$ The sign of the middle term is positive.
$$A^2 + 2 \; A \; B + B^2 = (A + B)^2$$

b) $x^2 + 49 - 14x = x^2 - 14x + 49$ Using a commutative law to write descending order

$\qquad\qquad\qquad\quad = x^2 - 2 \cdot x \cdot 7 + 7^2$ The sign of the middle term is negative.

$\qquad\qquad\qquad\quad = (x - 7)^2$ Factoring the perfect-square trinomial

c) $16x^2 - 40x + 25 = (4x)^2 - 2 \cdot 4x \cdot 5 + 5^2 = (4x - 5)^2$
$$A^2 - 2 \; A \; B + B^2 = (A - B)^2$$

With practice, it is possible to spot perfect-square trinomials as they occur and factor them quickly.

EXAMPLE 3 Factor: (a) $4p^2 - 12pq + 9q^2$; (b) $75m^3 + 60m^2 + 12m$.

SOLUTION

a) $4p^2 - 12pq + 9q^2 = (2p)^2 - 2(2p)(3q) + (3q)^2$ Recognizing the perfect-square trinomial

$\qquad\qquad\qquad\qquad\quad = (2p - 3q)^2$ The sign of the middle term is negative.

Check: $(2p - 3q)(2p - 3q) = 4p^2 - 12pq + 9q^2$.

b) *Always* look first for a common factor. This time there is one, $3m$:

$75m^3 + 60m^2 + 12m = 3m[25m^2 + 20m + 4]$ Factoring out the largest common factor

$\qquad\qquad\qquad\qquad\; = 3m[(5m)^2 + 2(5m)(2) + 2^2]$ Recognizing the perfect-square trinomial. Try to do this mentally.

$\qquad\qquad\qquad\qquad\; = 3m(5m + 2)^2.$

Check: $3m(5m + 2)^2 = 3m(5m + 2)(5m + 2)$
$\qquad\qquad\qquad\quad\; = 3m(25m^2 + 20m + 4)$
$\qquad\qquad\qquad\quad\; = 75m^3 + 60m^2 + 12m.$

Recognizing Differences of Squares

Some binomials represent the difference of two squares. For example, the binomial $16x^2 - 9$ is a difference of two expressions, $16x^2$ and 9, that are squares. To see this, note that $16x^2 = (4x)^2$ and $9 = 3^2$.

Any expression, like $16x^2 - 9$, that can be written in the form $A^2 - B^2$ is called a **difference of squares.** Note that for a binomial to be a difference of squares, it must have the following:

1. There must be two expressions, both squares, such as

    $25x^2$, 9, $4x^2y^2$, 1, x^6, $49y^8$.

2. The terms in the binomial must have different signs.

Note that in order for a term to be a square, its coefficient must be a perfect square and the power(s) of the variable(s) must be even.

EXAMPLE 4 Determine whether each of the following is a difference of squares.

a) $9x^2 - 64$ **b)** $25 - t^3$ **c)** $-4x^{10} + 16$

SOLUTION

a) To see if $9x^2 - 64$ is a difference of squares, note that:

1. The first expression is a square: $9x^2 = (3x)^2$.
   The second expression is a square: $64 = 8^2$.
2. The terms have different signs.

Thus, $9x^2 - 64$ is a difference of squares, $(3x)^2 - 8^2$.

b) To see if $25 - t^3$ is a difference of squares, note that:

1. The expression t^3 is not a square.

Thus, $25 - t^3$ is not a difference of squares.

c) To see if $-4x^{10} + 16$ is a difference of squares, note that:

1. The expressions $4x^{10}$ and 16 are squares: $4x^{10} = (2x^5)^2$ and $16 = 4^2$.
2. The terms have different signs.

Thus, $-4x^{10} + 16$ is a difference of squares, $4^2 - (2x^5)^2$.

Factoring Differences of Squares

To factor a difference of squares, we reverse a pattern from Section 4.5:

FACTORING A DIFFERENCE OF SQUARES

$$A^2 - B^2 = (A + B)(A - B).$$

To factor a difference of squares $A^2 - B^2$, we first determine A and B. Then we use A and B to form the two factors $A + B$ and $A - B$.

5.4 FACTORING PERFECT-SQUARE TRINOMIALS AND DIFFERENCES OF SQUARES   269

EXAMPLE 5   Factor: **(a)** $x^2 - 4$; **(b)** $m^2 - 9p^2$; **(c)** $9 - 16t^{10}$; **(d)** $50x^2 - 8x^8$.

SOLUTION

a) $x^2 - 4 = x^2 - 2^2 = (x + 2)(x - 2)$
$\ \ A^2 - B^2 = (A + B)(A - B)$

b) $m^2 - 9p^2 = m^2 - (3p)^2 = (m + 3p)(m - 3p)$
$\ A^2 - B^2 = (A + B)(A - B)$

c) $9 - 16t^{10} = 3^2 - (4t^5)^2$ Using the rules for powers
$\phantom{9 - 16t^{10} = }A^2 - B^2$

$\phantom{9 - 16t^{10}}\ = (3 + 4t^5)(3 - 4t^5)$ Try to go directly to this step.
$\phantom{9 - 16t^{10} = }\ (A + B)(A - B)$

d) *Always* check first for a common factor. This time there is one, $2x^2$:

$50x^2 - 8x^8 = 2x^2(25 - 4x^6)$ Factoring out the common factor
$\ = 2x^2[5^2 - (2x^3)^2]$ Recognizing $A^2 - B^2$. Try to do this mentally.
$\ = 2x^2(5 + 2x^3)(5 - 2x^3).$ Factoring the difference of squares

Check: $2x^2(5 + 2x^3)(5 - 2x^3) = 2x^2(25 - 4x^6)$
$= 50x^2 - 8x^8.$

CAUTION! Note carefully in these examples that a difference of squares is *not* the square of the difference; that is,

$A^2 - B^2 \neq (A - B)^2.$

For example,

$8^2 - 3^2 = 64 - 9 = 55,$

but

$(8 - 3)^2 = 5^2 = 25.$

Factoring Completely

When you are factoring, if a factor with more than one term can itself be factored, be sure to factor it. When no factor can be factored further, you have factored completely. Always factor completely when asked to factor.

EXAMPLE 6   Factor: $p^4 - 16$.

SOLUTION

$$p^4 - 16 = (p^2)^2 - 4^2 \quad \text{Recognizing } A^2 - B^2$$
$$= (p^2 + 4)(p^2 - 4) \quad \text{Factoring a difference of squares}$$
$$= (p^2 + 4)(p + 2)(p - 2) \quad \text{Factoring further. The factor } p^2 - 4 \text{ is itself a difference of squares.}$$

Check: $(p^2 + 4)(p + 2)(p - 2) = (p^2 + 4)(p^2 - 4)$
$$= p^4 - 16.$$

Note in Example 6 that the factor $p^2 + 4$ is a *sum* of squares that cannot be factored further.

CAUTION! Apart from possibly removing a common factor, you cannot factor a sum of squares. In particular,

$$A^2 + B^2 \neq (A + B)^2.$$

Consider $25x^2 + 100$. Here a sum of squares has a common factor, 25. Factoring, we get $25(x^2 + 4)$, where $x^2 + 4$ is prime.

As you proceed through the exercises, these suggestions may prove helpful.

TIPS FOR FACTORING

1. Always look first for a common factor! If there is one, factor it out.
2. Be alert for perfect-square trinomials and differences of squares. Once recognized, they can be factored without trial and error.
3. Always factor completely.
4. Check by multiplying.

EXERCISE SET 5.4

Determine whether each of the following is a perfect-square trinomial.

1. $x^2 - 18x + 81$
2. $x^2 - 16x + 64$
3. $x^2 + 16x - 64$
4. $x^2 - 14x - 49$
5. $x^2 - 3x + 9$
6. $x^2 + 2x + 4$
7. $9x^2 - 36x + 24$
8. $36x^2 - 24x + 16$

Factor completely. Remember to look first for a common factor and to check by multiplying.

9. $x^2 - 16x + 64$
10. $x^2 - 14x + 49$
11. $x^2 + 14x + 49$
12. $x^2 + 16x + 64$
13. $x^2 - 2x + 1$
14. $5x^2 - 10x + 5$
15. $4 + 4x + x^2$
16. $4 + x^2 - 4x$
17. $18x^2 - 12x + 2$
18. $25x^2 + 10x + 1$
19. $49 + 56y + 16y^2$
20. $120m + 75 + 48m^2$
21. $x^5 - 18x^4 + 81x^3$
22. $2x^2 - 40x + 200$
23. $2x^3 - 4x^2 + 2x$
24. $x^3 + 24x^2 + 144x$
25. $20x^2 + 100x + 125$
26. $12x^2 + 36x + 27$
27. $49 - 42x + 9x^2$
28. $64 - 112x + 49x^2$

29. $5y^2 + 10y + 5$
30. $2a^2 + 28a + 98$
31. $2 + 20x + 50x^2$
32. $7 - 14a + 7a^2$
33. $4p^2 + 12pq + 9q^2$
34. $25m^2 + 20mn + 4n^2$
35. $a^2 - 14ab + 49b^2$
36. $x^2 - 6xy + 9y^2$
37. $64m^2 + 16mn + n^2$
38. $81p^2 - 18pq + q^2$
39. $16s^2 - 40st + 25t^2$
40. $36a^2 + 96ab + 64b^2$

Determine whether each of the following is a difference of squares.

41. $x^2 - 100$
42. $x^2 - 36$
43. $x^2 + 36$
44. $x^2 + 4$
45. $x^2 - 35$
46. $x^2 - 50y^2$
47. $16x^2 - 25y^2$
48. $-1 + 36x^2$

Factor completely. Remember to look first for a common factor.

49. $y^2 - 4$
50. $x^2 - 36$
51. $p^2 - 9$
52. $q^2 - 1$
53. $-49 + t^2$
54. $-64 + m^2$
55. $a^2 - b^2$
56. $p^2 - q^2$
57. $m^2n^2 - 49$
58. $25 - a^2b^2$
59. $200 - 2t^2$
60. $81 - w^2$
61. $16a^2 - 9$
62. $25x^2 - 4$
63. $4x^2 - 25y^2$
64. $9a^2 - 16b^2$
65. $8x^2 - 98$
66. $24x^2 - 54$
67. $36x - 49x^3$
68. $16x - 81x^3$
69. $49a^4 - 81$
70. $25a^4 - 9$
71. $5x^4 - 5$
72. $x^4 - 16$
73. $4x^4 - 64$
74. $5x^4 - 80$
75. $1 - y^8$
76. $x^8 - 1$
77. $3x^3 - 24x^2 + 48x$
78. $2a^4 - 36a^3 + 162a^2$
79. $x^{12} - 16$
80. $x^8 - 81$
81. $y^2 - \frac{1}{16}$
82. $x^2 - \frac{1}{25}$
83. $a^8 - 2a^7 + a^6$
84. $x^8 - 8x^7 + 16x^6$
85. $16 - m^4n^4$
86. $1 - a^4b^4$

SKILL MAINTENANCE

87. Bonnie is taking an astronomy course. To get an A, she must average at least 90 after four exams. Bonnie scored 96, 98, and 89 on the first three tests. Determine (in terms of an inequality) what scores on the last test will earn her an A.

88. About 5 L of oxygen can be dissolved in 100 L of water at 0°C. This is 1.6 times the amount that can be dissolved in the same volume of water at 20°C. How much oxygen can be dissolved at 20°C?

Simplify.

89. $(x^3y^5)(x^9y^7)$
90. $(5a^2b^3)^2$

Graph.

91. $y = \frac{3}{2}x - 3$
92. $3x - 5y = 30$

SYNTHESIS

93. ◆ Is it possible to thoroughly understand Examples 1–6 without having a solid understanding of the material in Section 4.5? Why or why not?

94. ◆ Leon concludes that since $x^2 - 9 = (x - 3)(x + 3)$, it must follow that $x^2 + 9 = (x + 3)(x - 3)$. What mistake(s) is he making?

95. ◆ Explain in your own words how to determine whether a polynomial is a perfect-square trinomial.

96. ◆ Explain in your own words how to determine whether a polynomial is a difference of squares.

Factor completely. If a polynomial is prime, state this.

97. $x^2 - 2.25$
98. $81x^2 + 216$
99. $x^2 - 5x + 25$
100. $x^8 - 2^8$
101. $3x^2 - \frac{1}{3}$
102. $18x^3 - \frac{8}{25}x$
103. $0.49p - p^3$
104. $0.64x^2 - 1.21$
105. $(x + 3)^4 - 81$
106. $(y - 5)^4 - z^8$
107. $x^2 - \left(\frac{1}{x}\right)^2$
108. $a^{2n} - 49b^{2n}$
109. $81 - b^{4k}$
110. $x^4 - 8x^2 - 9$
111. $9b^{2n} + 12b^n + 4$
112. $16x^4 - 96x^2 + 144$
113. $(y + 3)^2 + 2(y + 3) + 1$
114. $49(x + 1)^2 - 42(x + 1) + 9$
115. $27x^3 - 63x^2 - 147x + 343$
116. Subtract $(x^2 + 1)^2$ from $x^2(x + 1)^2$ and factor the result.

Factor by grouping. Look for a grouping of three terms that is a perfect-square trinomial.

117. $a^2 + 2a + 1 - 9$
118. $y^2 + 6y + 9 - x^2 - 8x - 16$

Find c such that each polynomial is the square of a binomial.

119. $cy^2 + 6y + 1$
120. $cy^2 - 24y + 9$

121. Find the value of a if $x^2 + a^2x + a^2$ factors into $(x + a)^2$.

122. Show that the difference of the squares of two consecutive integers is the sum of the integers. (*Hint*: Use x for the smaller number.)

COLLABORATIVE CORNER

Focus: Perfect-square trinomials and differences of squares

Time: 15–25 minutes

Group size: 2

Materials: Graph paper and scissors

The factorizations

$$a^2 + 2ab + b^2 = (a + b)^2$$

and

$$a^2 - b^2 = (a + b)(a - b)$$

can be visualized using rectangular areas.

ACTIVITY

Trace or duplicate the following figures on graph paper. Then label each area and cut out each shape.*

1. One group member should assemble the four shapes to show, using areas, that $a^2 + 2ab + b^2 = (a + b)^2$. The other group member should then explain how this can be used to factor $x^2 + 6x + 9$.

2. Working together, use the shapes already made to help trace and cut out the following shapes:
    i) A rectangle with area $(a - b)b$.
    ii) A square with a notched corner; the area should be $a^2 - b^2$.
    iii) A rectangle with area $(a - b)a$.

3. Working together, and using any or all of the seven shapes from parts (1) and (2) above, use areas to show that $a^2 - b^2 = (a - b)(a + b)$. [*Hint*: Note that $(a - b)(a + b) = (a - b)a + (a - b)b$.] Then explain how this procedure can be used to factor $x^2 - 25$.

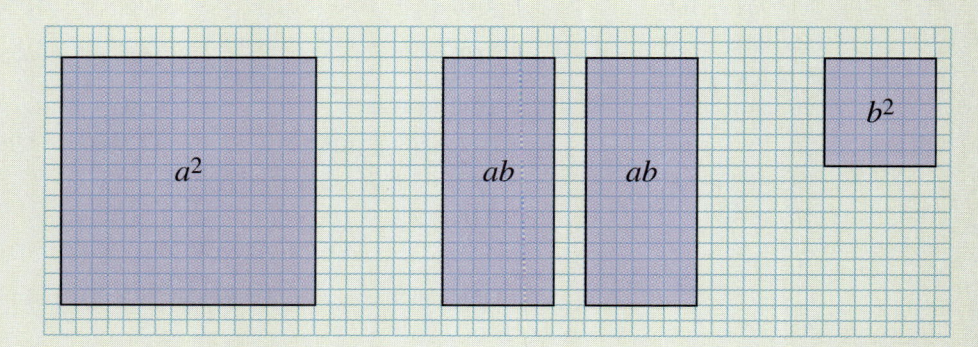

---

*Instructors may wish to make use of Cuisinaire® Algebra Tiles for this activity. Please contact your local Addison Wesley Longman sales consultant for more information.

5.5 Factoring: A General Strategy

Choosing the Right Method

Choosing the Right Method

We now combine all of our factoring techniques and consider a general strategy for factoring polynomials. Here we will encounter polynomials of various types, in random order, so you will have to determine which method to use.

TO FACTOR A POLYNOMIAL:

A. Always look first for a common factor. If there is one, factor out the largest common factor. Be sure to include it in your final answer.

B. Then look at the number of terms.

Two terms: If you have a difference of squares, factor accordingly. Do not try to factor a sum of squares: $A^2 + B^2$.

Three terms: Determine whether the trinomial is a perfect square. If so, factor accordingly. If not, try trial and error, using the standard method or grouping.

Four terms: Try factoring by grouping.

C. Always *factor completely*. If a factor with more than one term can itself be factored, be sure to factor it.

D. Check by multiplying.

EXAMPLE 1 Factor: $5t^4 - 80$.

SOLUTION

A. We look for a common factor:

$5t^4 - 80 = 5(t^4 - 16)$.   5 is the largest common factor.

B. The factor $t^4 - 16$ has only two terms. It is a difference of squares: $(t^2)^2 - 4^2$. We factor it, being careful to rewrite the common factor:

$5t^4 - 80 = 5(t^2 + 4)(t^2 - 4)$. $t^4 - 16 = (t^2 + 4)(t^2 - 4)$

C. Since one of the factors is again a difference of squares, we factor it:

$5t^4 - 80 = 5(t^2 + 4)(t - 2)(t + 2)$. $t^2 - 4 = (t - 2)(t + 2)$
↑
This is a sum of squares. It cannot be factored!

We have factored completely because no factor can be factored further.

274 CHAPTER 5 • POLYNOMIALS AND FACTORING

D. Check: $5(t^2 + 4)(t - 2)(t + 2) = 5(t^2 + 4)(t^2 - 4)$
$= 5(t^4 - 16) = 5t^4 - 80.$

EXAMPLE 2

Factor: $2x^3 + 10x^2 + x + 5$.

SOLUTION

A. We look for a common factor. There is none.
B. Because there are four terms, we try factoring by grouping:

$2x^3 + 10x^2 + x + 5$
$= (2x^3 + 10x^2) + (x + 5)$ Separating into two binomials
$= 2x^2(x + 5) + 1(x + 5)$ Factoring out the largest common factor from each binomial. The 1 serves as an aid.
$= (x + 5)(2x^2 + 1)$ Factoring out the common factor, $x + 5$

C. No factor can be factored further, so we have factored completely.
D. Check: $(x + 5)(2x^2 + 1) = 2x^3 + x + 10x^2 + 5$
$= 2x^3 + 10x^2 + x + 5.$

EXAMPLE 3

Factor: $x^5 - 2x^4 - 35x^3$.

SOLUTION

A. We look first for a common factor. This time there is one, x^3:

$x^5 - 2x^4 - 35x^3 = x^3(x^2 - 2x - 35).$

B. The factor $x^2 - 2x - 35$ has three terms, but it is not a perfect-square trinomial. We factor it using trial and error:

$x^5 - 2x^4 - 35x^3 = x^3(x^2 - 2x - 35)$
$= x^3(x - 7)(x + 5).$ Don't forget to rewrite the common factor.

C. No factor with more than one term can be factored further, so we have factored completely.
D. Check: $x^3(x - 7)(x + 5) = x^3(x^2 - 2x - 35)$
$= x^5 - 2x^4 - 35x^3.$

EXAMPLE 4

Factor: $x^2 - 20x + 100$.

SOLUTION

A. We look first for a common factor. There is none.
B. There are three terms. This polynomial is a perfect-square trinomial, so we factor it accordingly:

$x^2 - 20x + 100 = x^2 - 2 \cdot x \cdot 10 + 10^2$ Try to do this step mentally.
$= (x - 10)^2.$

C. No factor can be factored further, so we have factored completely.
D. Check: $(x - 10)(x - 10) = x^2 - 20x + 100.$

EXAMPLE 5

Factor: $6x^2y^4 - 21x^3y^5 + 3x^2y^6$.

SOLUTION

A. We first factor out the largest common factor, $3x^2y^4$:

$$6x^2y^4 - 21x^3y^5 + 3x^2y^6 = 3x^2y^4(2 - 7xy + y^2).$$

B. There are three terms in $2 - 7xy + y^2$. Since only y^2 is a square, we do not have a perfect-square trinomial. Can $2 - 7xy + y^2$ be factored by trial and error? A key to the answer is that x appears only in $-7xy$. If $2 - 7xy + y^2$ could be factored into a form like $(1 - y)(2 - y)$, there would be no x in the middle term. Thus, $2 - 7xy + y^2$ cannot be factored.

C. Have we factored completely? Yes, because no factor with more than one term can be factored further.

D. *Check:* $3x^2y^4(2 - 7xy + y^2) = 6x^2y^4 - 21x^3y^5 + 3x^2y^6$.

EXAMPLE 6

Factor: $px + py + qx + qy$.

SOLUTION

A. We look first for a common factor. There is none.

B. There are four terms. We try factoring by grouping:

$$px + py + qx + qy = p(x + y) + q(x + y)$$
$$= (x + y)(p + q).$$

C. Since no factor can be factored further, we have factored completely.

D. *Check:* $(x + y)(p + q) = xp + xq + yp + yq$
$$= px + py + qx + qy.$$

EXAMPLE 7

Factor: $25x^2 + 20xy + 4y^2$.

SOLUTION

A. We look first for a common factor. There is none.

B. There are three terms. We determine whether the trinomial is a square. The first term and the last terms are squares:

$$25x^2 = (5x)^2 \quad \text{and} \quad 4y^2 = (2y)^2.$$

Since twice the product of $5x$ and $2y$ is the other term,

$$2 \cdot 5x \cdot 2y = 20xy,$$

the trinomial is a perfect square.

We factor by writing a binomial squared. The binomial is found by writing the square roots of the square terms and the sign of the middle term:

$$25x^2 + 20xy + 4y^2 = (5x + 2y)^2.$$

C. No factor can be factored further, so we have factored completely.

D. *Check:* $(5x + 2y)(5x + 2y) = 25x^2 + 20xy + 4y^2$.

EXAMPLE 8

Factor: $p^2q^2 + 7pq + 12$.

SOLUTION

A. We look first for a common factor. There is none.
B. There are three terms. Since only one term is a square, we do not have a perfect-square trinomial. We use trial and error, thinking of the product pq as a single variable. The binomials will then be of the form

$$(pq + \quad)(pq + \quad).$$

We factor the last term, 12. All the signs are positive, so we consider only positive factors. Possibilities are 1, 12 and 2, 6 and 3, 4. The pair 3, 4 gives a sum of 7 for the coefficient of the middle term. Thus,

$$p^2q^2 + 7pq + 12 = (pq + 3)(pq + 4).$$

C. No factor can be factored further, so we have factored completely.
D. The check is left to the student.

EXAMPLE 9

Factor: $a^4 - 16b^4$.

SOLUTION

A. We look first for a common factor. There is none.
B. There are two terms. Since $a^4 = (a^2)^2$ and $16b^4 = (4b^2)^2$, we see that we do have a difference of squares. Thus,

$$a^4 - 16b^4 = (a^2 + 4b^2)(a^2 - 4b^2).$$

C. The factor $(a^2 - 4b^2)$ is itself a difference of squares. Thus,

$$a^4 - 16b^4 = (a^2 + 4b^2)(a + 2b)(a - 2b). \quad \text{Factoring } a^2 - 4b^2$$

D. **Check:** $(a^2 + 4b^2)(a + 2b)(a - 2b) = (a^2 + 4b^2)(a^2 - 4b^2)$
$= a^4 - 16b^4.$

EXERCISE SET 5.5

Factor completely. If a polynomial is prime, state this.

1. $5x^2 - 45$
2. $10a^2 - 640$
3. $a^2 + 25 + 10a$
4. $y^2 + 49 - 14y$
5. $2x^2 - 11x + 12$
6. $8y^2 + 18y - 5$
7. $x^3 - 24x^2 + 144x$
8. $x^3 - 18x^2 + 81x$
9. $x^3 + 3x^2 - 4x - 12$
10. $x^3 - 5x^2 - 25x + 125$
11. $24x^2 - 54$
12. $8x^2 - 98$
13. $20x^3 - 4x^2 - 72x$
14. $9x^3 + 12x^2 - 45x$
15. $x^2 + 4$
16. $t^2 + 25$
17. $a^4 + 8a^2 + 8a^3 + 64a$
18. $t^4 + 7t^2 - 3t^3 - 21t$
19. $x^5 - 14x^4 + 49x^3$
20. $2x^6 + 8x^5 + 8x^4$
21. $20 - 6x - 2x^2$
22. $45 - 3x - 6x^2$
23. $x^2 + 3x + 1$
24. $x^2 + 5x + 2$
25. $4x^4 - 64$
26. $5x^5 - 80x$
27. $t^8 - 1$
28. $1 - n^8$
29. $x^5 - 4x^4 + 3x^3$
30. $x^6 - 2x^5 + 7x^4$
31. $x^2 - y^2$
32. $p^2q^2 - r^2$
33. $12n^2 + 24n^3$
34. $ax^2 + ay^2$

35. $ab^2 - a^2b$
36. $36mn - 9m^2n^2$
37. $2\pi rh + 2\pi r^2$
38. $10p^4q^4 + 35p^3q^3 + 10p^2q^2$
39. $(a + b)(x - 3) + (a + b)(x + 4)$
40. $5c(a^3 + b) - (a^3 + b)$
41. $(x - 5)(x + 2) - y(x + 2)$
42. $n^2 + 2n + np + 2p$
43. $x^2 + x + xy + y$
44. $2x^2 - 4x + xz - 2z$
45. $a^2 - 3a + ay - 3y$
46. $x^2 + y^2 - 2xy$
47. $6y^2 - 3y + 2py - p$
48. $9c^2 + 6cd + d^2$
49. $4b^2 + a^2 - 4ab$
50. $7p^4 - 7q^4$
51. $16x^2 + 24xy + 9y^2$
52. $25z^2 + 10zy + y^2$
53. $4x^2y^2 + 12xyz + 9z^2$
54. $a^5 + 4a^4b - 5a^3b^2$
55. $a^4b^4 - 16$
56. $a^2 - ab - 2b^2$
57. $4p^2q + pq^2 + 4p^3$
58. $2mn - 360n^2 + m^2$
59. $3b^2 - 17ab - 6a^2$
60. $m^2n^2 - 4mn - 32$
61. $15 + x^2y^2 + 8xy$
62. $a^5b^2 + 3a^4b - 10a^3$
63. $p^2q^2 + 7pq + 6$
64. $112xy + 49x^2 + 64y^2$
65. $4ab^5 - 32b^4 + a^2b^6$
66. $2s^6t^2 + 10s^3t^3 + 12t^4$
67. $x^6 + x^5y - 2x^4y^2$
68. $a^2 + 2a^2bc + a^2b^2c^2$
69. $36a^2 - 15a + \frac{25}{16}$
70. $\frac{1}{81}x^2 - \frac{8}{27}x + \frac{16}{9}$
71. $\frac{1}{4}a^2 + \frac{1}{3}ab + \frac{1}{9}b^2$
72. $1 - 16x^{12}y^{12}$
73. $b^4a - 81a^5$
74. $0.01x^2 - 0.1xy + 0.25y^2$
75. $w^3 - 7w^2 - 4w + 28$
76. $y^3 + 8y^2 - y - 8$

SKILL MAINTENANCE

77. Show that the pairs $(-1, 11)$, $(0, 7)$, and $(3, -5)$ are solutions of $y = -4x + 7$.
78. Solve: $2x - 7 = 0$.
79. Solve: $3x + 4 = 0$.
80. Graph: $y = -\frac{1}{2}x + 4$.
81. Solve $A = aX + bX - 7$ for X.
82. Solve: $4(x - 9) - 2(x + 7) < 14$.

SYNTHESIS

83. ◆ Is it possible to express a third-degree polynomial as a product of four binomials? Why or why not?
84. ◆ Describe in your own words a strategy for factoring polynomials.
85. ◆ Kelly factored $16 - 8x + x^2$ as $(x - 4)^2$, while Tony factored it as $(4 - x)^2$. Evaluate each expression for several values of x. Then explain why $(x - 4)^2$ and $(4 - x)^2$ are equivalent.
86. ◆ There are third-degree polynomials in x that we are not yet able to factor, despite the fact that they are not prime. Explain how such a polynomial could be created.

Factor.
87. $-(x^5 + 7x^3 - 18x)$
88. $18 + a^3 - 9a - 2a^2$
89. $3a^4 - 15a^2 + 12$
90. $x^4 - 7x^2 - 18$
91. $x^3 - x^2 - 4x + 4$
92. $y^2(y + 1) - 4y(y + 1) - 21(y + 1)$
93. $y^2(y - 1) - 2y(y - 1) + (y - 1)$
94. $6(x - 1)^2 + 7y(x - 1) - 3y^2$
95. $(y + 4)^2 + 2x(y + 4) + x^2$
96. $2(a + 3)^2 - (a + 3)(b - 2) - (b - 2)^2$
97. Factor $x^{2k} - 2^{2k}$ when $k = 4$.

Check that each factorization is most likely correct by evaluating with the values given.

98. $6x^2 - xy - 15y^2 = (2x + 3y)(3x - 5y)$; $x = 2, y = -1$
99. $6x^2y^2 - 23xy + 20 = (2xy - 5)(3xy - 4)$; $x = -1, y = 1$

5.6 Solving Quadratic Equations by Factoring

The Principle of Zero Products • Factoring to Solve Equations • Graphing and Quadratic Equations

In this section, we will see how factoring can be used to solve equations like

$x^2 - 8x = -16$ and $x^2 + x - 156 = 0$. Second-degree equations of this type are said to be **quadratic**.

QUADRATIC EQUATION

A *quadratic equation* is an equation equivalent to one of the form
$$ax^2 + bx + c = 0, \quad \text{where } a \neq 0.$$

In order to solve quadratic equations, we need to develop a new principle.

The Principle of Zero Products

Suppose we are told that the product of two numbers is 10. On the basis of this information, it is impossible to know the value of either number—the product could be $2 \cdot 5$, $10 \cdot 1$, $20 \cdot \frac{1}{2}$, and so on. However, if we are told that the product of two numbers is 0, we know that at least one of the two numbers being multiplied must itself be 0. For example, if $(x + 3)(x - 2) = 0$, we can conclude that either $x + 3$ is 0 or $x - 2$ is 0.

THE PRINCIPLE OF ZERO PRODUCTS

An equation $AB = 0$ is true if and only if $A = 0$ or $B = 0$, or both. (A product is 0 if and only if at least one factor is 0.)

EXAMPLE 1   Solve: $(x + 3)(x - 2) = 0$.

SOLUTION   We are told that the product of $x + 3$ and $x - 2$ is 0. In order for a product to be 0, at least one of the factors must be 0. We reason that either

$x + 3 = 0 \quad or \quad x - 2 = 0.$ Using the principle of zero products

We solve each equation:

$$x + 3 = 0 \quad or \quad x - 2 = 0$$
$$x = -3 \quad or \quad x = 2.$$

Each of the numbers -3 and 2 is a solution of the original equation, as shown in the following checks.

Check: For -3:

$$\frac{(x + 3)(x - 2) = 0}{(-3 + 3)(-3 - 2) \: ? \: 0}$$
$$0(-5) \bigg|$$
$$0 \: \bigg| \: 0 \quad \text{TRUE}$$

For 2:

$$\frac{(x + 3)(x - 2) = 0}{(2 + 3)(2 - 2) \: ? \: 0}$$
$$5(0) \bigg|$$
$$0 \: \bigg| \: 0 \quad \text{TRUE}$$

5.6 SOLVING QUADRATIC EQUATIONS BY FACTORING

The equation $(x + 3)(x - 2) = 0$ has two solutions, -3 and 2.

When we are using the principle of zero products, the word "or" is meant to emphasize that any one of the factors could be the one that represents 0.

EXAMPLE 2   Solve: $(5x + 1)(x - 7) = 0$.

SOLUTION   We have

$(5x + 1)(x - 7) = 0$
$5x + 1 = 0$ or $x - 7 = 0$ Using the principle of zero products
$5x = -1$ or $x = 7$ Solving the two equations separately
$x = -\frac{1}{5}$ or $x = 7$.

Check: For $-\frac{1}{5}$:

$$\frac{(5x + 1)(x - 7) = 0}{\left(5\left(-\frac{1}{5}\right) + 1\right)\left(-\frac{1}{5} - 7\right) \;?\; 0}$$
$(-1 + 1)\left(-7\frac{1}{5}\right)$
$0\left(-7\frac{1}{5}\right)$
$0 \;|\; 0$ TRUE

For 7:
$$\frac{(5x + 1)(x - 7) = 0}{(5(7) + 1)(7 - 7) \;?\; 0}$$
$(35 + 1)0$
$36 \cdot 0$
$0 \;|\; 0$ TRUE

The solutions are $-\frac{1}{5}$ and 7.

The principle of zero products can be used whenever a product equals 0—even if a factor has only one term.

EXAMPLE 3   Solve: $x(2x - 9) = 0$.

SOLUTION   We have

$x(2x - 9) = 0$
$x = 0$ or $2x - 9 = 0$ Using the principle of zero products
$x = 0$ or $2x = 9$
$x = 0$ or $x = \frac{9}{2}$.

The solutions are 0 and $\frac{9}{2}$. The check is left to the student.

Factoring to Solve Equations

By factoring and using the principle of zero products, we can now solve a variety of quadratic equations.

EXAMPLE 4   Solve: $x^2 + 5x + 6 = 0$.

SOLUTION   This equation differs from those solved in Chapter 2. There are no like terms to combine, and there is a squared term. We first factor the polynomial. Then we use the principle of zero products:

$$x^2 + 5x + 6 = 0$$
$$(x + 2)(x + 3) = 0 \qquad \text{Factoring}$$
$$x + 2 = 0 \quad \text{or} \quad x + 3 = 0 \qquad \text{Using the principle of zero products}$$
$$x = -2 \quad \text{or} \quad x = -3.$$

Check: For -2:

$$\frac{x^2 + 5x + 6 = 0}{(-2)^2 + 5(-2) + 6 \;?\; 0}$$
$$4 - 10 + 6$$
$$-6 + 6$$
$$0 \quad | \quad 0 \quad \text{TRUE}$$

For -3:

$$\frac{x^2 + 5x + 6 = 0}{(-3)^2 + 5(-3) + 6 \;?\; 0}$$
$$9 - 15 + 6$$
$$-6 + 6$$
$$0 \quad | \quad 0 \quad \text{TRUE}$$

The solutions are -2 and -3.

CAUTION! We *must* have 0 on one side before using the principle of zero products. Get all nonzero terms on one side and 0 on the other.

EXAMPLE 5 Solve: **(a)** $x^2 - 8x = -16$; **(b)** $x^2 + 5x = 0$; **(c)** $4x^2 = 25$.

SOLUTION

a) We first add 16 to get 0 on one side:

$$x^2 - 8x = -16$$
$$x^2 - 8x + 16 = 0 \qquad \text{Adding 16 on both sides to get 0 on one side}$$
$$(x - 4)(x - 4) = 0 \qquad \text{Factoring}$$
$$x - 4 = 0 \quad \text{or} \quad x - 4 = 0 \qquad \text{Using the principle of zero products}$$
$$x = 4 \quad \text{or} \quad x = 4.$$

There is only one solution, 4. The check is left to the student.

b) $x^2 + 5x = 0$
$$x(x + 5) = 0 \qquad \text{Factoring out a common factor}$$
$$x = 0 \quad \text{or} \quad x + 5 = 0 \qquad \text{Using the principle of zero products}$$
$$x = 0 \quad \text{or} \quad x = -5$$

The solutions are 0 and -5. The check is left to the student.

c)
$$4x^2 = 25$$
$$4x^2 - 25 = 0 \qquad \text{Subtracting 25 on both sides to get 0 on one side}$$
$$(2x - 5)(2x + 5) = 0 \qquad \text{Factoring a difference of squares}$$
$$2x - 5 = 0 \quad \text{or} \quad 2x + 5 = 0$$
$$2x = 5 \quad \text{or} \quad 2x = -5$$
$$x = \tfrac{5}{2} \quad \text{or} \quad x = -\tfrac{5}{2}$$

The solutions are $\tfrac{5}{2}$ and $-\tfrac{5}{2}$. The check is left to the student.

EXAMPLE 6 Solve: $(x + 3)(2x - 1) = 9$.

SOLUTION Be careful with an equation like this! Since we need 0 on one side, we multiply out the product on the left and then subtract 9 from both sides:

$(x + 3)(2x - 1) = 9$
$2x^2 + 5x - 3 = 9$ Multiplying on the left
$2x^2 + 5x - 3 - 9 = 9 - 9$ Subtracting 9 on both sides to get 0 on one side
$2x^2 + 5x - 12 = 0$
$(2x - 3)(x + 4) = 0$ Factoring
$2x - 3 = 0$ or $x + 4 = 0$ Using the principle of zero products
$2x = 3$ or $x = -4$
$x = \frac{3}{2}$ or $x = -4$.

Check: For $\frac{3}{2}$:

$$\frac{(x + 3)(2x - 1) = 9}{(\frac{3}{2} + 3)(2 \cdot \frac{3}{2} - 1) \: ? \: 9}$$
$$(\tfrac{9}{2})(2)$$
$$9 \: | \: 9 \quad \text{TRUE}$$

For -4:

$$\frac{(x + 3)(2x - 1) = 9}{(-4 + 3)(2(-4) - 1) \: ? \: 9}$$
$$(-1)(-9)$$
$$9 \: | \: 9 \quad \text{TRUE}$$

The solutions are $\frac{3}{2}$ and -4.

Graphing and Quadratic Equations

In Chapter 3, we graphed linear equations of the form $Ax + By = C$ and $y = mx + b$. Recall that to find the x-intercept, we replaced y with 0 and solved for x. This same procedure can be used to find the x-intercepts when an equation of the form $y = ax^2 + bx + c$ ($a \neq 0$) is graphed. Equations like this are graphed in Chapter 10. Their graphs are shaped like the following curves:

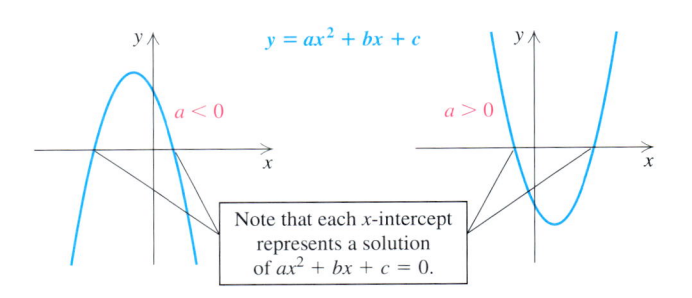

$y = ax^2 + bx + c$

$a < 0$ $a > 0$

Note that each x-intercept represents a solution of $ax^2 + bx + c = 0$.

282 CHAPTER 5 • POLYNOMIALS AND FACTORING

EXAMPLE 7 Find the *x*-intercepts for the graph of the equation shown. (The axes are intentionally not scaled.)

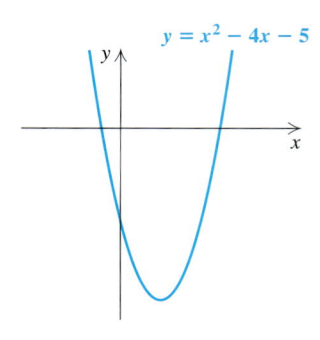

$y = x^2 - 4x - 5$

SOLUTION To find the *x*-intercepts, we let $y = 0$ and solve for *x*:

$0 = x^2 - 4x - 5$ Substituting 0 for *y*
$0 = (x - 5)(x + 1)$ Factoring
$x - 5 = 0$ or $x + 1 = 0$ Using the principle of zero products
$x = 5$ or $x = -1$. Solving for *x*

The *x*-intercepts are (5, 0) and (−1, 0).

TECHNOLOGY CONNECTION 5.6

A grapher allows us to solve quadratic equations by locating any *x*-intercepts that might exist. This technique is especially useful when the equation cannot be solved by factoring. As an example, in the following figure, there appears to be an *x*-intercept between −2 and −1 and another one between 4 and 5.

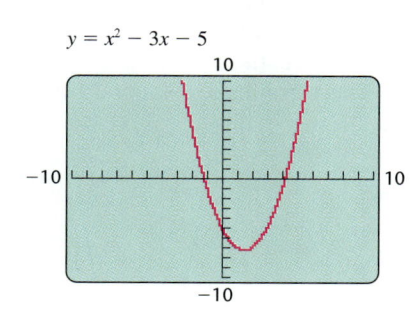

$y = x^2 - 3x - 5$

To more precisely determine the intercepts, we could use ZOOM and TRACE several times. A more direct approach, however, utilizes the ROOT option of the CALC menu. This option requires the user to enter an *x*-value slightly less than the *x*-intercept as a LOWER BOUND. An *x*-value slightly more than the *x*-intercept is then entered as an UPPER BOUND. Finally, a GUESS value, between the lower and upper bounds, is entered and the *x*-intercept, or *root*, or *zero*, is displayed.

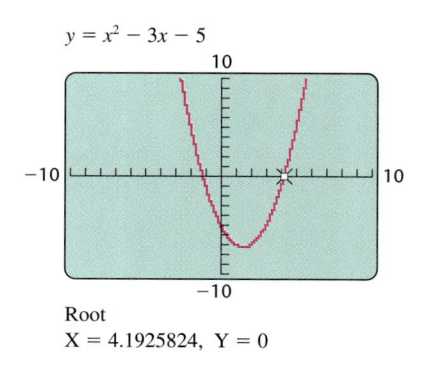

$y = x^2 - 3x - 5$

Root
X = 4.1925824, Y = 0

Use a grapher to find the solutions, if they exist, accurate to two decimal places.

1. $x^2 + 4x - 3 = 0$
2. $x^2 - 5x - 2 = 0$
3. $x^2 + 13.54x + 40.95 = 0$
4. $x^2 - 4.43x + 6.32 = 0$

EXERCISE SET 5.6

Solve using the principle of zero products.

1. $(x + 7)(x + 6) = 0$
2. $(x + 1)(x + 2) = 0$
3. $(x - 3)(x + 5) = 0$
4. $(x + 9)(x - 3) = 0$
5. $(2x - 9)(x + 4) = 0$
6. $(3x - 5)(x + 1) = 0$
7. $(10x - 7)(4x + 9) = 0$
8. $(2x - 7)(3x + 4) = 0$
9. $x(x + 6) = 0$
10. $t(t + 9) = 0$
11. $\left(\frac{2}{3}x - \frac{12}{11}\right)\left(\frac{7}{4}x - \frac{1}{12}\right) = 0$
12. $\left(\frac{1}{9} - 3x\right)\left(\frac{1}{5} + 2x\right) = 0$
13. $7x(2x + 9) = 0$
14. $12x(4x + 7) = 0$
15. $(20 - 0.4x)(7 - 0.1x) = 0$
16. $(1 - 0.05x)(1 - 0.3x) = 0$
17. $(x + 3)(2x - 5)(x - 6) = 0$
18. $(x - 4)(x + 9)(3x - 1) = 0$

Solve by factoring and using the principle of zero products.

19. $x^2 - 7x + 6 = 0$
20. $x^2 - 6x + 5 = 0$
21. $x^2 - 4x - 21 = 0$
22. $x^2 - 7x - 18 = 0$
23. $x^2 + 9x + 14 = 0$
24. $x^2 + 8x + 15 = 0$
25. $x^2 + 3x = 0$
26. $x^2 + 8x = 0$
27. $x^2 - 9x = 0$
28. $x^2 + 4x = 0$
29. $100 = x^2$
30. $x^2 = 16$
31. $4x^2 - 9 = 0$
32. $9x^2 - 4 = 0$
33. $0 = 25 + x^2 + 10x$
34. $0 = 6x + x^2 + 9$
35. $1 + x^2 = 2x$
36. $x^2 + 16 = 8x$
37. $9x^2 = 4x$
38. $3x^2 = 7x$
39. $3x^2 - 7x = 20$
40. $6x^2 - 4x = 10$
41. $2y^2 + 12y = -10$
42. $12y^2 - 5y = 2$
43. $(x - 5)(x + 1) = 16$
44. $(x + 1)(x - 7) = -15$
45. $y(3y + 1) = 2$
46. $t(t - 5) = 14$
47. $81x^2 - 5 = 20$
48. $36m^2 - 9 = 40$
49. $(x - 1)(5x + 4) = 2$
50. $(x + 3)(3x + 5) = 7$
51. $x^2 - 2x = 18 + 5x$
52. $3x^2 - 2x = 9 - 8x$
53. $(6a + 1)(a + 1) = 21$
54. $(2t + 1)(4t - 1) = 14$

55. Use the following graph to solve $x^2 + x - 6 = 0$.

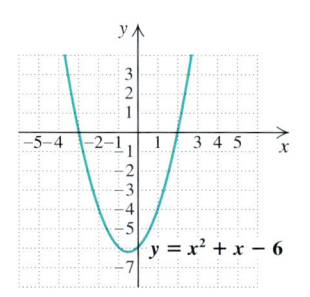

56. Use the following graph to solve $x^2 - 3x - 4 = 0$.

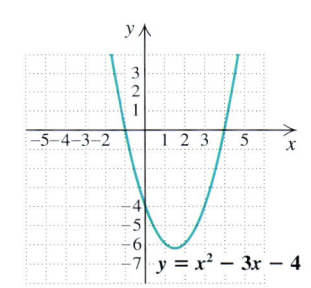

Find the x-intercepts for the graph of each equation.

57. $y = x^2 + 3x - 4$ 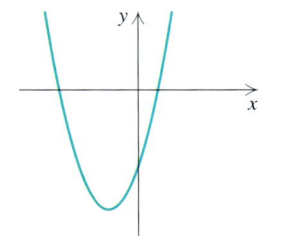
58. $y = x^2 - x - 6$ 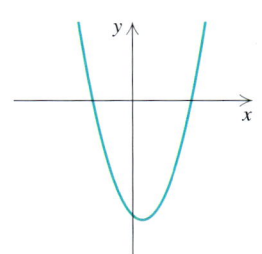
59. $y = x^2 - 2x - 15$ 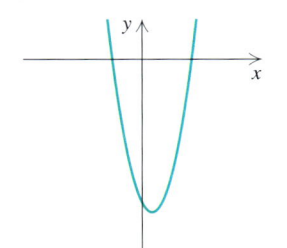
60. $y = x^2 + 2x - 8$

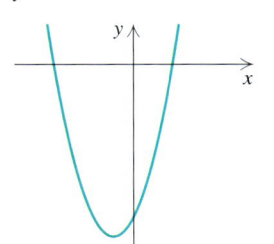

284 CHAPTER 5 • POLYNOMIALS AND FACTORING

61. $y = 2x^2 + x - 10$ **62.** $y = 2x^2 + 3x - 9$

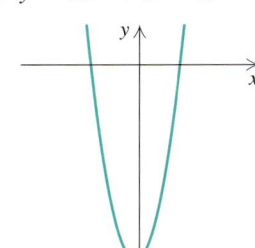

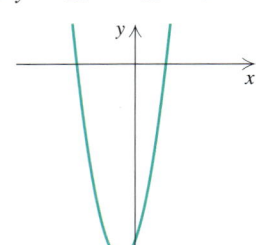

73. Find an equation with integer coefficients that has the given numbers as solutions. For example, 3 and -2 are solutions to $x^2 - x - 6 = 0$.
   a) $-3, 4$ b) $-3, -4$ c) $\frac{1}{2}, \frac{1}{2}$
   d) $5, -5$ e) $0, 0.1, \frac{1}{4}$

Solve.

74. $16(x - 1) = x(x + 8)$ **75.** $a(9 + a) = 4(2a + 5)$

76. $(t - 5)^2 = 2(5 - t)$ **77.** $x^2 - \frac{9}{25} = 0$

78. $x^2 - \frac{25}{36} = 0$ **79.** $\frac{5}{16}x^2 = 5$

80. $\frac{27}{25}x^2 = \frac{1}{3}$

81. For each equation on the left, find an equivalent equation on the right.
   a) $x^2 + 10x - 2 = 0$ $4x^2 + 8x + 36 = 0$
   b) $(x - 6)(x + 3) = 0$ $(2x + 8)(2x - 5) = 0$
   c) $5x^2 - 5 = 0$ $9x^2 - 12x + 24 = 0$
   d) $(2x - 5)(x + 4) = 0$ $(x + 1)(5x - 5) = 0$
   e) $x^2 + 2x + 9 = 0$ $x^2 - 3x - 18 = 0$
   f) $3x^2 - 4x + 8 = 0$ $2x^2 + 20x - 4 = 0$

82. ◆ Explain how to construct an equation that has seven solutions.

83. ◆ Explain how the graph in Exercise 57 can be used to visualize the solutions of
$$x^2 + 3x - 4 = -6.$$

Use a grapher to find the solutions of each equation, accurate to two decimal places.

84. $x^2 + 1.80x - 5.69 = 0$

85. $x^2 - 9.10x + 15.77 = 0$

86. $-x^2 + 0.63x + 0.22 = 0$

87. $x^2 + 13.74x + 42.00 = 0$

88. $6.4x^2 - 8.45x - 94.06 = 0$

89. $-0.25x^2 - 2.50x - 5.48 = 0$

SKILL MAINTENANCE

Translate to an algebraic expression.

63. The square of the sum of a and b
64. The sum of the squares of a and b
65. The sum of two consecutive integers

Translate to an inequality.

66. 5 more than twice a number is less than 19.
67. 7 less than half of a number exceeds 24.
68. 3 less than a number is at least 34.

SYNTHESIS

69. ◆ What is the difference between a quadratic polynomial and a quadratic equation?

70. ◆ What is wrong with solving $x^2 = 3x$ by dividing both sides of the equation by x?

71. ◆ The equation $x^2 + 1$ has no real-number solutions. What implications does this have for the graph of $y = x^2 + 1$?

72. ◆ When the principle of zero products is used to solve a quadratic equation, will there always be two solutions? Why or why not?

5.7 Solving Applications

Applications • The Pythagorean Theorem

Applications

We can use the five-step problem-solving process and our new methods for solving quadratic equations to solve problems.

5.7 SOLVING APPLICATIONS

EXAMPLE 1

Page Numbers. The product of the page numbers on two consecutive pages of a book is 156. Find the page numbers.

SOLUTION

1. **Familiarize.** Recall that consecutive page numbers are one apart, like 49 and 50. Let x = the first page number; then $x + 1$ = the next page number.

2. **Translate.** We reword the problem before translating:

   Rewording: The first page number times the next page number is 156.

   Translating: $x \cdot (x + 1) = 156$

3. **Carry out.** We solve the equation as follows:

   $x(x + 1) = 156$
   $x^2 + x = 156$ Multiplying
   $x^2 + x - 156 = 0$ Subtracting 156 to get 0 on one side
   $(x - 12)(x + 13) = 0$ Factoring
   $x - 12 = 0$ or $x + 13 = 0$ Using the principle of zero products
   $x = 12$ or $x = -13$.

4. **Check.** The solutions of the equation are 12 and -13. Since page numbers cannot be negative, -13 is not a solution. On the other hand, if x is 12, then $x + 1$ is 13 and $12 \cdot 13 = 156$. Thus, 12 checks.

5. **State.** The page numbers are 12 and 13.

EXAMPLE 2

Dimensions of a Garden. A vacant square-shaped lot is being turned into a community garden. Because a path 2 m wide is needed at one end, only 48 m^2 of the lot will be garden space. Find the dimensions of the lot.

SOLUTION

1. **Familiarize.** We first make a drawing. Recall that the area of any rectangle, including a square, is length · width. We let x = the length (or width) of the square lot. Note that the path is a rectangle 2 m wide and x m long.

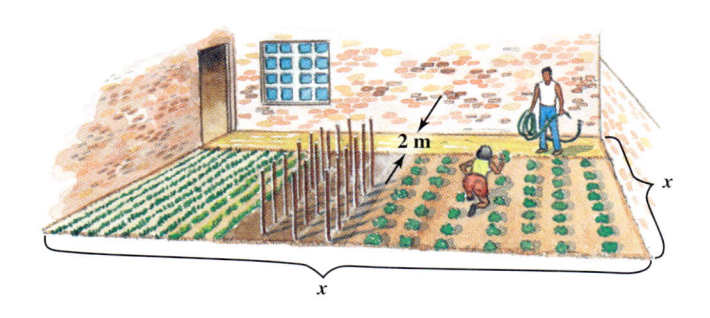

2. **Translate.** It helps to reword this problem before translating.

286 CHAPTER 5 • POLYNOMIALS AND FACTORING

Rewording: The area of the lot minus the area of the path is 48 m².

Translating: $\quad x^2 \quad - \quad 2 \cdot x \quad = \quad 48$

3. **Carry out.** We solve the equation as follows:

$$x^2 - 2x = 48$$
$$x^2 - 2x - 48 = 0 \quad \text{Subtracting 48 to get 0 on one side}$$
$$(x - 8)(x + 6) = 0 \quad \text{Factoring}$$
$$x - 8 = 0 \quad \text{or} \quad x + 6 = 0 \quad \text{Using the principle of zero products}$$
$$x = 8 \quad \text{or} \quad x = -6.$$

4. **Check.** Since measurements cannot be negative, we disregard -6 as a solution of the original problem. If $x = 8$, the area of the lot is $8 \cdot 8 = 64$ m² and the area of the path is $2 \cdot 8 = 16$ m², leaving $64 - 16 = 48$ m² for the garden. Thus, 8 checks. Another approach to this problem is to express the area of the garden as $x(x - 2)$ and set this equal to 48. This provides a second check, since $8(8 - 2) = 8 \cdot 6 = 48$.

5. **State.** The lot is 8 m long and 8 m wide.

EXAMPLE 3   *Dimensions of a Sail.* The mainsail of Brenda's Lightning-styled sailboat has an area of 125 ft². If the sail is 15 ft taller than it is wide, find the height and the width of the sail.

SOLUTION

1. **Familiarize.** We first make a drawing. The formula for the area of a triangle is Area $= \frac{1}{2} \cdot$ (base) $\cdot$ (height). We let $b =$ the width, in feet, of the triangle's base and $b + 15 =$ the height, in feet.

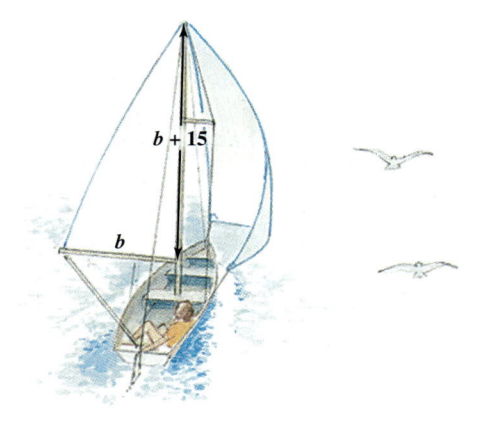

2. **Translate.** We reword the problem and translate:

Rewording: $\frac{1}{2}$ times the base times the base plus 15 is 125.

Translating: $\quad \frac{1}{2} \quad \cdot \quad b \quad \cdot \quad (b + 15) \quad = 125$

3. Carry out. We solve the equation as follows:

$$\tfrac{1}{2} \cdot b \cdot (b+15) = 125$$
$$\tfrac{1}{2}(b^2 + 15b) = 125 \quad \text{Multiplying}$$
$$b^2 + 15b = 250 \quad \text{Multiplying by 2 to clear fractions}$$
$$b^2 + 15b - 250 = 0 \quad \text{Subtracting 250 to get 0 on one side}$$
$$(b+25)(b-10) = 0 \quad \text{Factoring}$$
$$b + 25 = 0 \quad \text{or} \quad b - 10 = 0 \quad \text{Using the principle of zero products}$$
$$b = -25 \quad \text{or} \quad b = 10.$$

4. Check. The base of a triangle cannot have a negative length, so -25 cannot be a solution. Suppose the base is 10 ft. Then the height is 15 ft more than the base, so the height is 25 ft and the area is $\tfrac{1}{2}(10)(25)$, or 125 ft². These numbers check in the original problem.

5. State. Brenda's mainsail is 25 ft tall and 10 ft wide.

EXAMPLE 4

Games in a League's Schedule. In a sports league of n teams in which each team plays every other team twice, the total number N of games to be played is given by

$$n^2 - n = N.$$

The Colchester Youth Soccer League plays a total of 240 games, with every team playing each of the others twice. How many teams are in the league?

SOLUTION

1. **Familiarize.** To familiarize yourself with this equation, reread Example 7 in Section 4.2, where we first considered it.

2. **Translate.** We are trying to find the number of teams n in a league in which 240 games are played and all teams play each other twice. The number 240 can be substituted into the formula above to create an equation in one variable:

$$n^2 - n = 240. \quad \text{Substituting 240 for } N$$

3. **Carry out.** We solve the equation as follows:

$$n^2 - n = 240$$
$$n^2 - n - 240 = 0 \quad \text{Subtracting 240 to get 0 on one side}$$
$$(n-16)(n+15) = 0 \quad \text{Factoring}$$
$$n - 16 = 0 \quad \text{or} \quad n + 15 = 0 \quad \text{Using the principle of zero products}$$
$$n = 16 \quad \text{or} \quad n = -15.$$

4. **Check.** The solutions of the equation are 16 and -15. Since the number of teams must be positive, -15 cannot be a solution. However, 16 checks, since $16^2 - 16 = 256 - 16 = 240$.

5. **State.** There are 16 teams in the league.

The Pythagorean Theorem

The following problems involve the Pythagorean theorem, which relates the lengths of the sides of a *right* triangle. A triangle is a **right triangle** if it has a 90°, or *right,* angle. The side opposite the 90° angle is called the **hypotenuse**. The other sides are called **legs**.

THE PYTHAGOREAN THEOREM

The sum of the squares of the legs of a right triangle is equal to the square of the hypotenuse:

$$a^2 + b^2 = c^2.$$

EXAMPLE 5

Right Triangle Geometry. One leg of a right triangle is 7 m longer than the other. The length of the hypotenuse is 13 m. Find the lengths of the legs.

SOLUTION

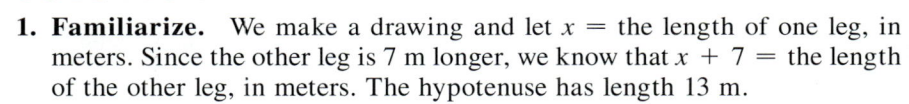

1. **Familiarize.** We make a drawing and let $x = $ the length of one leg, in meters. Since the other leg is 7 m longer, we know that $x + 7 = $ the length of the other leg, in meters. The hypotenuse has length 13 m.

2. **Translate.** Applying the Pythagorean theorem, we obtain the following translation:

   $a^2 + b^2 = c^2$
   $x^2 + (x + 7)^2 = 13^2.$ Substituting

3. **Carry out.** We solve the equation as follows:

$x^2 + (x^2 + 14x + 49) = 169$	Squaring the binomial and 13
$2x^2 + 14x + 49 = 169$	Combining like terms
$2x^2 + 14x - 120 = 0$	Subtracting 169 to get 0 on one side
$2(x^2 + 7x - 60) = 0$	Factoring out a common factor
$2(x + 12)(x - 5) = 0$	Factoring
$x + 12 = 0$ or $x - 5 = 0$	Using the principle of zero products
$x = -12$ or $x = 5.$	

4. **Check.** The integer -12 cannot be a length of a side because it is negative. When $x = 5$, $x + 7 = 12$, and $5^2 + 12^2 = 13^2$. So 5 checks.

5. **State.** The lengths of the legs are 5 m and 12 m.

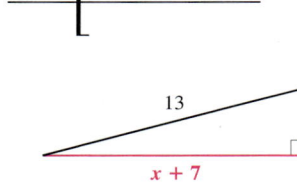

EXAMPLE 6

Roadway Design. Elliott Street is 24 ft wide when it ends at Main Street in Brattleboro, Vermont. A 40-ft–long diagonal crosswalk allows pedestrians to cross Main Street to or from either corner of Elliott Street (see the figure). Determine the width of Main Street.

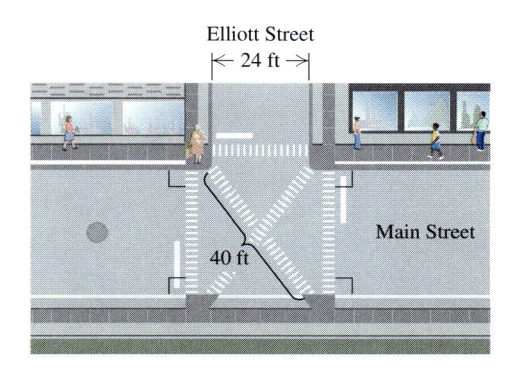

SOLUTION

1. **Familiarize.** A drawing has already been provided, but we can redraw and label the relevant part.

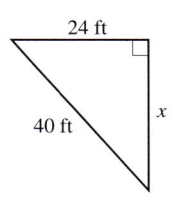

   Note that the two streets intersect at a right angle. We let $x =$ the width of Main Street, in feet.

2. **Translate.** Since a right triangle is formed, we can use the Pythagorean theorem:

   $a^2 + b^2 = c^2$
   $x^2 + 24^2 = 40^2$.   Substituting

3. **Carry out.** We solve the equation as follows:

   $\quad\quad x^2 + 576 = 1600$ Squaring 24 and 40
   $\quad\quad x^2 - 1024 = 0$ Subtracting 1600 on both sides
   $(x - 32)(x + 32) = 0$ Note that $1024 = 32^2$. A calculator might be helpful here.
   $x - 32 = 0 \quad$ or $\quad x + 32 = 0$ Using the principle of zero products
   $\quad\quad x = 32 \quad$ or $\quad\quad x = -32$.

4. **Check.** Since the width of a street must be positive, -32 is not a solution. If the width is 32 ft, we have $32^2 + 24^2 = 1024 + 576 = 1600$, which is 40^2. Thus, 32 checks.

5. **State.** The width of Main Street is 32 ft.

EXERCISE SET 5.7

Solve. Be sure to label all variables.

1. A number is 2 less than its square. Find all such numbers.
2. A number is 6 less than its square. Find all such numbers.
3. Eight times a number is 15 more than the square of that number. Find all such numbers.
4. Six times a number is 8 more than the square of that number. Find all such numbers.
5. *Page numbers.* The product of the page numbers on two facing pages of a book is 110. Find the page numbers.
6. *Page numbers.* The product of the page numbers on two facing pages of a book is 210. Find the page numbers.
7. The product of two consecutive even integers is 224. Find the integers.
8. The product of two consecutive odd integers is 255. Find the integers.
9. *Calculator design.* A rectangular calculator is 5 cm longer than it is wide. The area of the rectangle is 84 cm². Find the length and the width.

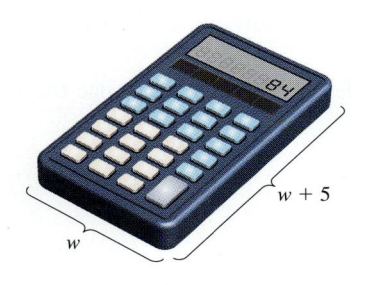

10. *Area of a garden.* The length of a rectangular garden is 4 m greater than the width. The area of the rectangle is 96 m². Find the length and the width.

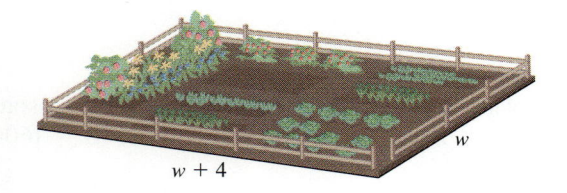

11. *Dimensions of a triangle.* The height of a triangle is 3 cm less than the length of the base. If the area of the triangle is 35 cm², find the height and the length of the base.

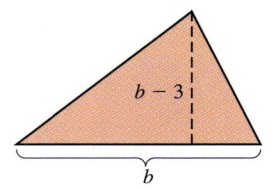

12. *Dimensions of a triangle.* A triangle is 10 cm wider than it is tall. The area is 28 cm². Find the height and the base.

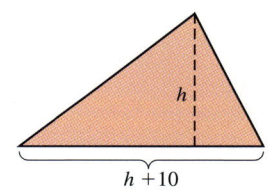

13. *Dimensions of a sail.* The height of the jib sail on a Lightning sailboat is 5 ft greater than the length of its "foot." If the area of the sail is 42 ft², find the length of the foot and the height of the sail.

14. *Road design.* A triangular traffic island has a base half as long as its height. Find the base and the height if the island has an area of 64 m².

15. *Physical education.* An outdoor-education ropes course includes a cable that slopes downward from a height of 37 ft to a resting place 30 ft above the ground. The trees that the cable connects are 24 ft apart. How long is the cable?

16. *Aviation.* Engine failure forced Geraldine to pilot her Cessna 150 to an emergency landing. To land, Geraldine's plane glided 17,000 ft over a 15,000-ft stretch of deserted highway. From what altitude did the descent begin?

Games in a league. Use the formula from Example 4, $n^2 - n = N$, for Exercises 17–20. Assume that in each league teams play each other twice.

17. A women's volleyball league has 20 teams. What is the total number of games to be played?

18. A chess league has 14 teams. What is the total number of games to be played?

19. A women's softball league plays a total of 132 games. How many teams are in the league?

20. A basketball league plays a total of 90 games. How many teams are in the league?

21. *Reach of a ladder.* Jason has a 26-ft ladder leaning against his house. If the bottom of the ladder is 10 ft from the base of the house, how high does the ladder reach?

22. *Construction.* The diagonal braces in a lookout tower are 15 ft long and span a distance of 12 ft. How high does each brace reach vertically?

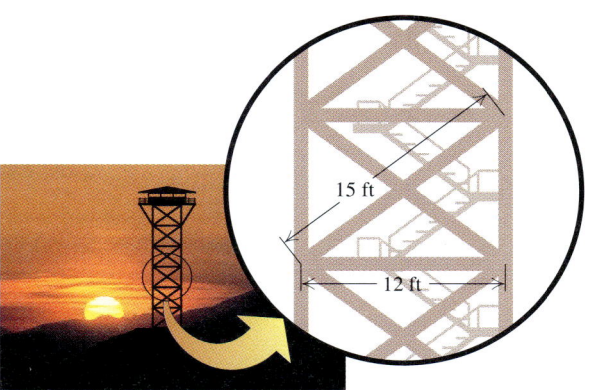

Number of handshakes. The number of possible handshakes N within a group of n people is given by $N = \frac{1}{2}(n^2 - n)$.

23. At a meeting, there are 12 people. How many handshakes are possible?

24. At a party, there are 30 people. How many handshakes are possible?

25. *High-fives.* After winning the championship, all members of a team exchanged "high-fives." Altogether, there were 300 high-fives. How many people were on the team?

26. *Toasting.* During a toast at a party, there were 190 "clicks" of glasses. How many people took part in the toast?

27. *Guy wire.* The guy wire on a TV antenna is 1 m longer than the height of the antenna. If the guy wire is anchored 3 m from the foot of the antenna, how tall is the antenna?

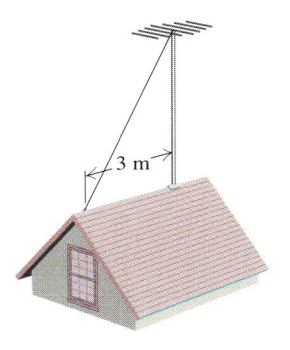

28. *Architecture.* An architect has allocated a rectangular space of 264 ft² for a square dining room

and a 10-ft wide kitchen. Find the dimensions of each room.

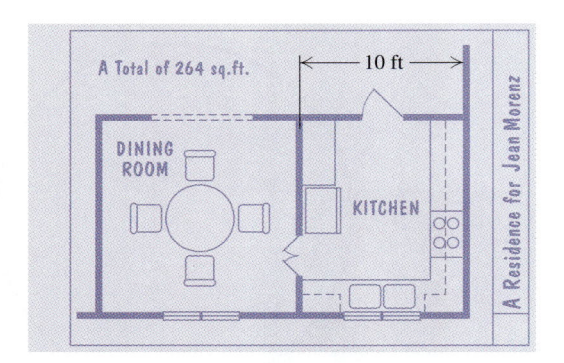

Height of a rocket. For Exercises 29–32, assume that a water rocket is launched upward with an initial velocity of 48 ft/sec. Its height h, in feet, after t seconds, is given by $h = 48t - 16t^2$.

29. Determine the height of the rocket $\frac{1}{2}$ sec after it has been launched.
30. Determine the height 1.5 sec after the rocket has been launched.
31. When will the rocket be exactly 32 ft above the ground?
32. When will the rocket crash into the ground?

SKILL MAINTENANCE

Graph.

33. $y = -\frac{2}{3}x + 1$
34. $y = \frac{3}{5}x - 1$

Simplify.

35. $7x^0$ when $x = -4$
36. $\frac{m^7 n^9}{mn^3}$
37. $-\frac{2}{3} + \frac{3}{4} \cdot \frac{1}{2}$
38. $\frac{5}{8} - \frac{7}{3} \cdot \frac{9}{8}$

SYNTHESIS

39. ◆ An archaeologist has a 3-ft, a 4-ft, and a 5-ft measuring stick. Explain how she could draw a 4-ft by 3-ft rectangle on a piece of land being excavated.
40. ◆ Write a problem in which a quadratic equation must be solved in order to solve the problem.
41. ◆ Write a problem for a classmate to solve such that only one of two solutions of a quadratic equation can be used as an answer.
42. ◆ Can we solve any problem that translates to a quadratic equation? Why or why not?

43. Solve for *x*.

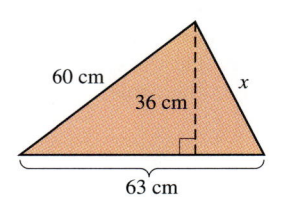

44. *Pool sidewalk.* A cement walk of uniform width is built around a 20-ft by 40-ft rectangular pool. The total area of the pool and the walk is 1500 ft². Find the width of the walk.

45. *Roofing.* Determine the area of the roof of the house shown in the following figure.

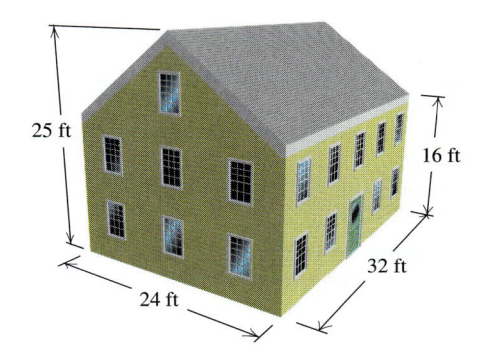

46. The ones digit of a number less than 100 is 4 greater than the tens digit. The sum of the number and the product of the digits is 58. Find the number.

47. *Dimensions of a closed box.* The total surface area of a closed box is 350 m². The box is 9 m high and has a square base and lid. Find the length of a side of the base.

48. *Dimensions of an open box.* A rectangular piece of cardboard is twice as long as it is wide. A 4-cm square is cut out of each corner, and the sides are turned up to make a box with an open top. The

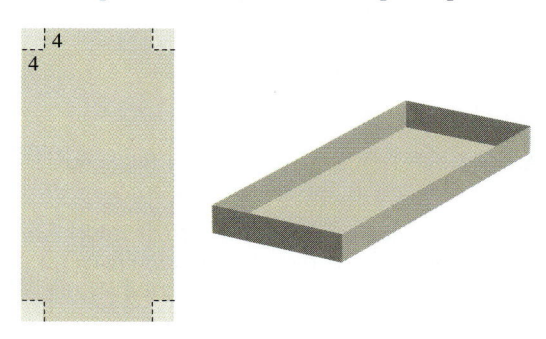

volume of the box is 616 cm³. Find the original dimensions of the cardboard.

49. *Rain-gutter design.* An open rectangular gutter is made by turning up the sides of a piece of metal 20 in. wide. The area of the cross-section of the gutter is 50 in². Find the depth of the gutter.

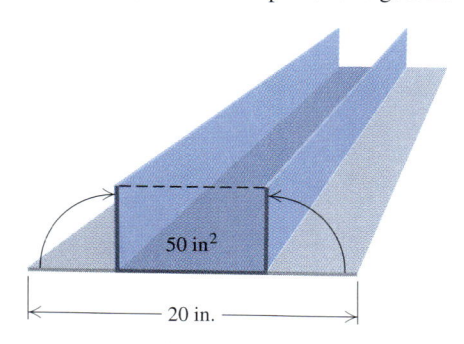

50. The length of each side of a square is increased by 5 cm to form a new square. The area of the new square is $2\frac{1}{4}$ times the area of the original square. Find the area of each square.

51. *Sailing.* The mainsail of a Lightning sailboat is a right triangle in which the hypotenuse is called the leech. If a 24-ft tall mainsail has a leech length of 26 ft and if Dacron® sailcloth costs $10 per square foot, find the cost of a new mainsail.

52. *Television screens.* All television sets have screens with the same ratio of height to width. Chris just bought a new television that has a 27-in. screen, measured diagonally. If Chris's old set measured $13\frac{1}{2}$ in. diagonally, how many times larger is his new screen?

COLLABORATIVE CORNER

Focus: The Pythagorean theorem
Time: 10–20 minutes
Group size: 2

Believe it or not, our twentieth president, James Garfield, is credited with one of the many proofs of the Pythagorean theorem. His proof uses the figure shown. Note that the outer boundary of this figure is a trapezoid.

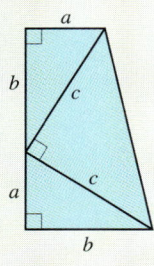

ACTIVITY

1. Each group member should find the area of the trapezoid at left using a different approach:
   a) One student should use the formula for the area of a trapezoid (see Section 6.8 and/or Exercise Set 2.3).
   b) The other student should find the sum of the areas of the three triangles.

2. Use the results of parts (a) and (b) above to form an equation. Then use the equation to prove the Pythagorean theorem.

SUMMARY AND REVIEW 5

KEY TERMS

Factor, p. 244
Common factor, p. 245
Factoring by grouping, p. 247
Prime polynomial, p. 254
Factor completely, p. 254
Perfect-square trinomial, p. 265

Difference of squares, p. 268
Quadratic equation, p. 278
Right triangle, p. 288
Hypotenuse, p. 288
Leg, p. 288

IMPORTANT PROPERTIES AND FORMULAS

Factoring Formulas

$A^2 + 2AB + B^2 = (A + B)^2$;
$A^2 - 2AB + B^2 = (A - B)^2$;
$A^2 - B^2 = (A + B)(A - B)$

To factor a polynomial:

A. Look first for a common factor. If there is one, factor out the largest common factor. Be sure to include it in your final answer.

B. Look at the number of terms.

Two terms: If you have a difference of squares, factor accordingly.

Three terms: If you have a perfect-square trinomial, factor accordingly. If not, try trial and error, using the standard method or grouping.

Four terms: Try factoring by grouping.

C. Always factor completely.
D. Check by multiplying.

The Principle of Zero Products: $AB = 0$ is true if and only if $A = 0$ or $B = 0$, or both.

The Pythagorean theorem: $a^2 + b^2 = c^2$

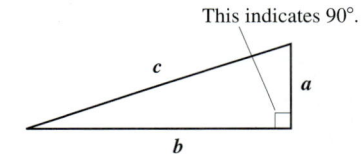

This indicates 90°.

REVIEW EXERCISES

Find three factorizations of each monomial.
1. $36x^3$
2. $-20x^5$

Factor completely. If a polynomial is prime, state this.
3. $2x^4 + 6x^3$
4. $x^2 - 3x$
5. $9x^2 - 4$
6. $x^2 + 4x - 12$
7. $x^2 + 14x + 49$
8. $6x^3 + 12x^2 + 3x$
9. $6x^3 + 9x^2 + 2x + 3$
10. $6x^2 - 5x + 1$
11. $x^4 - 81$
12. $9x^3 + 12x^2 - 45x$
13. $2x^2 - 50$
14. $x^4 + 4x^3 - 2x - 8$
15. $16x^4y^4 - 1$
16. $8x^6 - 32x^5 + 4x^4$
17. $75 + 12x^2 + 60x$
18. $x^2 + 9$
19. $x^3 - x^2 - 30x$
20. $4x^2 - 25$
21. $9x^2 + 25 - 30x$
22. $6x^2 - 28x - 48$
23. $x^2 - 6x + 9$
24. $2x^2 - 7x - 4$
25. $18x^2 - 12x + 2$
26. $3x^2 - 27$
27. $15 - 8x + x^2$
28. $25x^2 - 20x + 4$
29. $x^2y^2 + xy - 12$
30. $12a^2 + 84ab + 147b^2$
31. $m^2 + 5m + mt + 5t$
32. $32x^4 - 128y^4z^4$

Solve.
33. $(x - 1)(x + 3) = 0$
34. $x^2 + 2x - 35 = 0$
35. $9x^2 = 1$
36. $3x^2 + 2 = 5x$
37. $2x^2 + 5x = 12$
38. $(x + 1)(x - 2) = 4$

39. The square of a number is 12 more than the number. Find all such numbers.
40. Find the x-intercepts for the graph of
$$y = 2x^2 - 3x - 5.$$ 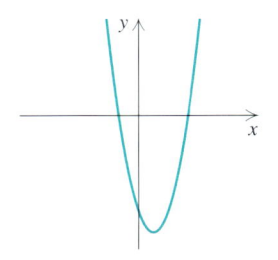
41. A triangular sign is as wide as it is tall. Its area is 800 cm². Find the height and the base. 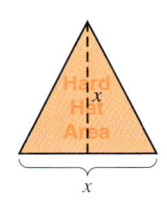
42. A guy wire from a ham radio antenna is 26 m long. It reaches from the top of the antenna to a point on the ground 10 m from the base of the antenna. How tall is the antenna?

SKILL MAINTENANCE

43. Subtract: $\frac{2}{5} - \left(-\frac{1}{10}\right)$.
44. Graph: $y = -\frac{3}{4}x$.
45. Divide and simplify: $\frac{m^3n^{10}}{mn^5}$.
46. Translate to an inequality: 2 less than half a number is no less than 10.

SYNTHESIS

47. ◆ On a quiz, Edith writes the factorization of $4x^2 - 100$ as $(2x - 10)(2x + 10)$. If this were a 10-point question, how many points would you give Edith? Why?
48. ◆ How do the equations solved in this chapter differ from those solved in previous chapters?

Solve.
49. The pages of a book measure 15 cm by 20 cm. Margins of equal width surround the printing on each page and constitute one half of the area of the page. Find the width of the margins.

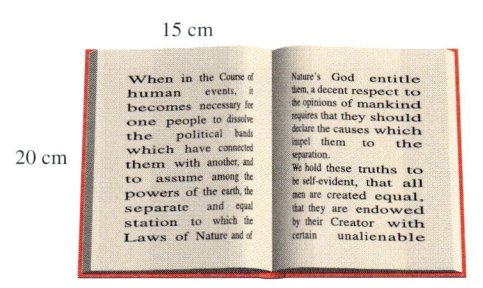

50. The cube of a number is the same as twice the square of the number. Find the number.
51. The length of a rectangle is 2 times its width. When the length is increased by 20 cm and the width is decreased by 1 cm, the area is 160 cm². Find the original length and width.

Solve.
52. $x^2 + 25 = 0$
53. $(x - 2)2x^2 + x(x - 2) - (x - 2)15 = 0$

CHAPTER TEST 5

1. Find three factorizations of $8x^4$.

Factor completely.

2. $x^2 - 7x + 10$
3. $x^2 + 25 - 10x$
4. $6y^2 - 8y^3 + 4y^4$
5. $x^3 + x^2 + 2x + 2$
6. $x^2 - 5x$
7. $x^3 + 2x^2 - 3x$
8. $28x - 48 + 10x^2$
9. $4x^2 - 9$
10. $x^2 - x - 12$
11. $6m^3 + 9m^2 + 3m$
12. $3w^2 - 75$
13. $60x + 45x^2 + 20$
14. $3x^4 - 48$
15. $49x^2 - 84x + 36$
16. $5x^2 - 26x + 5$
17. $x^4 + 2x^3 - 3x - 6$
18. $80 - 5x^4$
19. $4x^2 - 4x - 15$
20. $6t^3 + 9t^2 - 15t$
21. $3m^2 - 9mn - 30n^2$

Solve.

22. $x^2 - x - 20 = 0$
23. $2x^2 + 7x = 15$
24. $x(x - 3) = 28$

25. The length of a rectangle is 2 m more than the width. The area of the rectangle is 48 m². Find the length and the width.

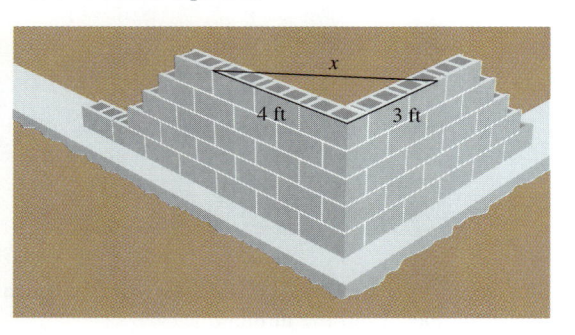

26. A mason wants to be sure she has a right corner in the foundation she is laying. She marks a point 3 ft from the corner along one wall and another point 4 ft from the corner along the other wall. If the corner is a right angle, what should the distance be between the two marked points?

SKILL MAINTENANCE

27. Simplify: $-2.8 - 3.5 + 4.2 - (-1.7)$.
28. The width of a rectangle must be 8 cm. For what lengths will the area be less than 104 cm²? Use an inequality.
29. Graph: $y = \frac{3}{4}x + 1$.
30. Simplify: $(-7a^3b^5)^2$.

SYNTHESIS

31. The length of a rectangle is 5 times its width. When the length is decreased by 3 and the width is increased by 2, the area of the new rectangle is 60. Find the original length and width.
32. Factor: $(a + 3)^2 - 2(a + 3) - 35$.

CHAPTER 6

Rational Expressions and Equations

- 6.1 Rational Expressions
- 6.2 Multiplication and Division
- 6.3 Addition, Subtraction, and Least Common Denominators
- 6.4 Addition and Subtraction with Unlike Denominators
- 6.5 Complex Rational Expressions
- 6.6 Solving Rational Equations
- 6.7 Applications Using Rational Equations and Proportions
- 6.8 Formulas
  SUMMARY AND REVIEW
  TEST

AN APPLICATION

To determine the size of a park's moose population, naturalists catch 69 moose, tag them, and then set them free. Months later, 40 moose are caught, of which 15 have tags. Estimate the size of the moose population.

THIS PROBLEM APPEARS AS EXERCISE 43 IN SECTION 6.7.

More information on moose is available at
http://hepg.awl.com/be/elem_5

Managing wildlife populations involves using statistics to examine trends in population size and to study correlations between those trends and the numbers of visitors to the park.

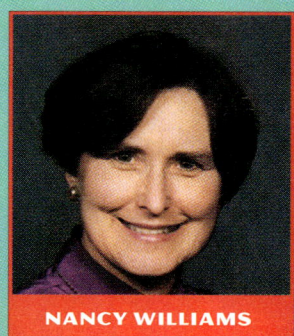

NANCY WILLIAMS
Naturalist
Gustavus, Alaska

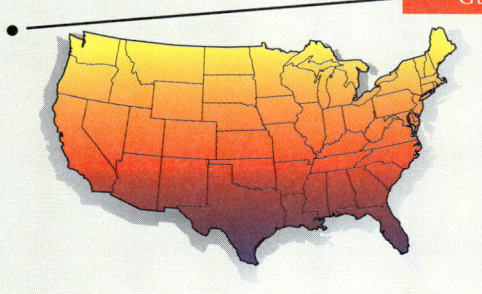

Just as fractions are needed to solve certain arithmetic problems, rational expressions similar to those in the following pages are needed to solve algebra problems. We now learn how to simplify, as well as add, subtract, multiply, and divide, rational expressions. These skills are important for solving problems like the one on the preceding page.

6.1 Rational Expressions

Simplifying Rational Expressions • Factors That Are Opposites

Whereas a rational number is a quotient of two integers, a **rational expression** is a quotient of two polynomials. The following are rational expressions:

$$\frac{7}{3}, \quad \frac{5}{x+6}, \quad \frac{t^2 - 5t + 6}{4t^2 - 7}.$$

Rational expressions are examples of what are sometimes called *algebraic fractions*. They are also examples of *fractional expressions*.

Because rational expressions indicate division, we must be careful to avoid denominators that are 0. When a variable is replaced with a number that produces a denominator of 0, the rational expression is undefined. For example, in the expression

$$\frac{x+3}{x-7},$$

when x is replaced with 7, the denominator is 0, and the expression is undefined:

$$\frac{x+3}{x-7} = \frac{7+3}{7-7} = \frac{10}{0}. \leftarrow \text{Undefined}$$

When x is replaced with a number other than 7—say, 6—the expression *is* defined because the denominator is nonzero:

$$\frac{x+3}{x-7} = \frac{6+3}{6-7} = \frac{9}{-1} = -9.$$

EXAMPLE 1 Find all numbers for which the rational expression

$$\frac{x+4}{x^2 - 3x - 10}$$

is undefined.

SOLUTION To determine which numbers make the rational expression undefined, we set the denominator equal to 0 and solve:

$$x^2 - 3x - 10 = 0$$
$$(x - 5)(x + 2) = 0 \qquad \text{Factoring}$$
$$x - 5 = 0 \quad \text{or} \quad x + 2 = 0 \qquad \text{Using the principle of zero products}$$
$$x = 5 \quad \text{or} \quad x = -2.$$

Check:
For $x = 5$:

$$\frac{x + 4}{x^2 - 3x - 10} = \frac{5 + 4}{5^2 - 3 \cdot 5 - 10}$$
$$= \frac{9}{25 - 15 - 10} = \frac{9}{0},$$

which is undefined.

For $x = -2$:

$$\frac{x + 4}{x^2 - 3x - 10} = \frac{-2 + 4}{(-2)^2 - 3(-2) - 10}$$
$$= \frac{2}{4 + 6 - 10} = \frac{2}{0},$$

which is undefined.

Thus, $\dfrac{x + 4}{x^2 - 3x - 10}$ is undefined for $x = 5$ and $x = -2$.

TECHNOLOGY CONNECTION 6.1

To check Example 1 with a grapher, we can let $y_1 = x^2 - 3x - 10$ and $y_2 = (x + 4)/(x^2 - 3x - 10)$ or $(x + 4)/y_1$ and use the TABLE feature. Since $x^2 - 3x - 10$ is 0 for $x = -2$, we cannot evaluate y_2 for $x = -2$.

TBL MIN = −2   ΔTBL = 1

X	Y1	Y2
−2	0	ERROR
−1	−6	−.5
0	−10	−.4
1	−12	−.4167
2	−12	−.5
3	−10	−.7
4	−6	−1.333

X = −2

When the ZDECIMAL option of the ZOOM feature is used, we use TRACE to see that the graph of y_2 does not include the x-value −2.

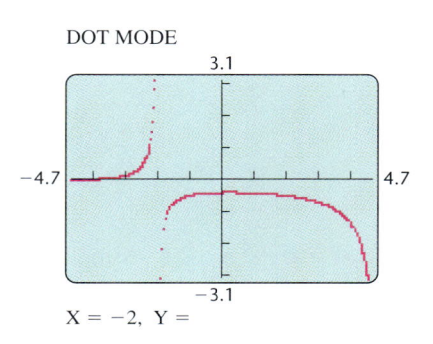

DOT MODE

X = −2, Y =

1. Use a TABLE to show that y_2 above is undefined for $x = 5$.
2. Use TRACE to show that y_2 is undefined for $x = 5$.

Simplifying Rational Expressions

Simplifying rational expressions is similar to simplifying the fractional expressions studied in Section 1.3. We saw, for example, that an expression like $\frac{15}{40}$ could be simplified as follows:

$$\frac{15}{40} = \frac{3 \cdot 5}{8 \cdot 5}$$ Factoring the numerator and the denominator. Note the common factor of 5.

$$= \frac{3}{8} \cdot \frac{5}{5}$$ Rewriting as a product of two fractions; $\frac{5}{5} = 1$

$$= \frac{3}{8} \cdot 1$$

$$= \frac{3}{8}.$$

Using the identity property of 1. We call this "removing a factor equal to 1."

The same steps are followed when simplifying rational expressions: We factor and remove a factor equal to 1, using the fact that

$$\frac{ab}{cb} = \frac{a}{c} \cdot \frac{b}{b}.$$

EXAMPLE 2   Simplify: $\frac{8x^2}{24x}$.

SOLUTION

$$\frac{8x^2}{24x} = \frac{8 \cdot x \cdot x}{3 \cdot 8 \cdot x}$$ Factoring the numerator and the denominator. Note the common factor of $8 \cdot x$.

$$= \frac{x}{3} \cdot \frac{8x}{8x}$$ Rewriting as a product of two rational expressions

$$= \frac{x}{3} \cdot 1$$ $\frac{8x}{8x} = 1$

$$= \frac{x}{3}$$ Removing the factor 1

When two or more terms appear in a numerator or a denominator, we factor as we did in Chapter 5. Then we try to remove a factor equal to 1.

EXAMPLE 3   Simplify.

a) $\dfrac{5a + 15}{10}$ b) $\dfrac{6x + 12}{7x + 14}$

c) $\dfrac{6a^2 + 4a}{2a^2 + 2a}$ d) $\dfrac{x^2 + 3x + 2}{x^2 - 1}$

SOLUTION

a) $\dfrac{5a + 15}{10} = \dfrac{5(a + 3)}{5 \cdot 2}$ Factoring the numerator and the denominator

$= \dfrac{5}{5} \cdot \dfrac{a + 3}{2}$ Rewriting as a product of two rational expressions

$= 1 \cdot \dfrac{a + 3}{2}$

$= \dfrac{a + 3}{2}$ Removing a factor equal to 1: $\dfrac{5}{5} = 1$

b) $\dfrac{6x + 12}{7x + 14} = \dfrac{6(x + 2)}{7(x + 2)}$ Factoring the numerator and the denominator

$= \dfrac{6}{7} \cdot \dfrac{x + 2}{x + 2}$ Rewriting as a product of two rational expressions

$= \dfrac{6}{7} \cdot 1$

$= \dfrac{6}{7}$ Removing a factor equal to 1: $\dfrac{x + 2}{x + 2} = 1$

c) $\dfrac{6a^2 + 4a}{2a^2 + 2a} = \dfrac{2a(3a + 2)}{2a(a + 1)}$ Factoring the numerator and the denominator

$= \dfrac{2a}{2a} \cdot \dfrac{3a + 2}{a + 1}$ Rewriting as a product of two rational expressions

$= 1 \cdot \dfrac{3a + 2}{a + 1}$ $\dfrac{2a}{2a} = 1$

$= \dfrac{3a + 2}{a + 1}$ Removing the factor 1. Note in this step that you *cannot* remove the remaining a's because they are not factors of the entire numerator and the entire denominator.

d) $\dfrac{x^2 + 3x + 2}{x^2 - 1} = \dfrac{(x + 2)(x + 1)}{(x + 1)(x - 1)}$ Factoring

$= \dfrac{x + 1}{x + 1} \cdot \dfrac{x + 2}{x - 1}$ Rewriting as a product of two rational expressions

$= 1 \cdot \dfrac{x + 2}{x - 1}$

$= \dfrac{x + 2}{x - 1}$

Canceling is a shortcut that can be used—and easily *misused*—when working with rational expressions. As we stated in Section 1.3, canceling must be done with care and understanding. Essentially, canceling streamlines the steps in which a rational expression is simplified by removing a factor

302   CHAPTER 6 • RATIONAL EXPRESSIONS AND EQUATIONS

equal to 1. Example 3(d) could have been streamlined as follows:

$$\frac{x^2 + 3x + 2}{x^2 - 1} = \frac{(x + 2)(x + 1)}{(x + 1)(x - 1)}$$ When a factor equal to 1 is noted, it is "canceled": $\frac{x + 1}{x + 1} = 1$.

$$= \frac{x + 2}{x - 1}.$$ Simplifying

CAUTION!   Canceling is often used incorrectly. The following cancellations are *incorrect*:

$$\frac{x + 2}{x + 3}; \quad \frac{a^2 - 5}{5}; \quad \frac{6x^2 + 5x + 1}{4x^2 - 3x}.$$

Wrong!   Wrong!   Wrong!

None of the above cancellations removes a factor equal to 1. Factors are parts of products. For example, in $x \cdot 2$, x and 2 are factors, but in $x + 2$, x and 2 are *not* factors. If you can't factor, you can't cancel!

EXAMPLE 4   Simplify: $\dfrac{3x^2 - 2x - 1}{x^2 - 3x + 2}$.

SOLUTION   We factor the numerator and the denominator and look for common factors:

$$\frac{3x^2 - 2x - 1}{x^2 - 3x + 2} = \frac{(3x + 1)(x - 1)}{(x - 2)(x - 1)}$$ Try to visualize this as $\dfrac{3x + 1}{x - 2} \cdot \dfrac{x - 1}{x - 1}$.

$$= \frac{3x + 1}{x - 2}.$$ Removing a factor equal to 1: $\dfrac{x - 1}{x - 1} = 1$

When a rational expression is simplified, the result is an equivalent expression. Example 3(a) says that

$$\frac{5a + 15}{10} \quad \text{is equivalent to} \quad \frac{a + 3}{2}.$$

This result can be partially checked using a value of a. For instance, if $a = 2$, then

$$\frac{5a + 15}{10} = \frac{5 \cdot 2 + 15}{10} = \frac{25}{10} = \frac{5}{2}$$

and

> To see why this check is not foolproof, see Exercise 59.

$$\frac{a + 3}{2} = \frac{2 + 3}{2} = \frac{5}{2}.$$

If evaluating both expressions yields differing results, we know that a mistake has been made. For instance, if $(5a + 15)/10$ is incorrectly simplified as

$(a + 15)/2$ and we evaluate using $a = 2$, we have

$$\frac{5a + 15}{10} = \frac{5 \cdot 2 + 15}{10} = \frac{5}{2}$$

and

$$\frac{a + 15}{2} = \frac{2 + 15}{2} = \frac{17}{2},$$

which proves that a mistake has been made.

Factors That Are Opposites

Consider

$$\frac{x - 4}{8 - 2x}, \quad \text{or, equivalently,} \quad \frac{x - 4}{2(4 - x)}.$$

At first glance, the numerator and the denominator do not appear to have any common factors. But $x - 4$ and $4 - x$ are opposites, or additive inverses, of each other. Thus we can find a common factor by factoring out -1 in one expression.

EXAMPLE 5 Simplify: $\dfrac{x - 4}{8 - 2x}$.

SOLUTION We have

$$\frac{x - 4}{8 - 2x} = \frac{x - 4}{2(4 - x)} \qquad \text{Factoring}$$

$$= \frac{x - 4}{2(-1)(x - 4)} \qquad \text{Note that } 4 - x = -(x - 4).$$

$$= \frac{x - 4}{-2(x - 4)} \qquad \text{Try to go straight to this step.}$$

$$= \frac{1}{-2} \cdot \frac{x - 4}{x - 4} \qquad \text{Rewriting as a product. It is important to write the 1 in the numerator.}$$

$$= -\frac{1}{2}.$$

As a partial check, note that for any choice of x other than 4, the value of the rational expression is $-\frac{1}{2}$. For instance, if $x = 6$, then

$$\frac{x - 4}{8 - 2x} = \frac{6 - 4}{8 - 2 \cdot 6}$$

$$= \frac{2}{-4}$$

$$= -\frac{1}{2}.$$

EXERCISE SET 6.1

List all numbers for which each rational expression is undefined.

1. $\dfrac{17}{-3x}$
2. $\dfrac{14}{-5y}$
3. $\dfrac{t-4}{t+6}$
4. $\dfrac{a-8}{a+7}$
5. $\dfrac{9}{2a-10}$
6. $\dfrac{x^2-9}{4x-12}$
7. $\dfrac{x^2+11}{x^2-3x-28}$
8. $\dfrac{p^2-9}{p^2-7p+10}$
9. $\dfrac{m^3-2m}{m^2-25}$
10. $\dfrac{7-3x+x^2}{49-x^2}$

Simplify by removing a factor equal to 1. Show all steps.

11. $\dfrac{50a^2b}{40ab^3}$
12. $\dfrac{45x^3y^2}{9x^5y}$
13. $\dfrac{35x^2y}{14x^3y^5}$
14. $\dfrac{12a^5b^6}{18a^3b}$
15. $\dfrac{9x+15}{6x+10}$
16. $\dfrac{14x-7}{10x-5}$
17. $\dfrac{a^2-9}{a^2+4a+3}$
18. $\dfrac{a^2+5a+6}{a^2-9}$

Simplify, if possible. Then check by evaluating.

19. $\dfrac{36x^6}{24x^9}$
20. $\dfrac{76a^5}{24a^3}$
21. $\dfrac{-2y+6}{-4y}$
22. $\dfrac{4x-12}{4x}$
23. $\dfrac{6a^2-3a}{7a^2-7a}$
24. $\dfrac{3m^2+3m}{6m^2+9m}$
25. $\dfrac{t^2-25}{t^2+t-20}$
26. $\dfrac{a^2-4}{a^2+5a+6}$
27. $\dfrac{3a^2-9a-12}{6a^2+30a+24}$
28. $\dfrac{2t^2+6t+4}{4t^2-12t-16}$
29. $\dfrac{x^2+8x+16}{x^2-16}$
30. $\dfrac{x^2-25}{x^2-10x+25}$
31. $\dfrac{t^2-1}{t+1}$
32. $\dfrac{a^2-1}{a-1}$
33. $\dfrac{y^2+4}{y+2}$
34. $\dfrac{x^2+1}{x+1}$
35. $\dfrac{5x^2-20}{10x^2-40}$
36. $\dfrac{6x^2-54}{4x^2-36}$
37. $\dfrac{5y+5}{y^2+7y+6}$
38. $\dfrac{6t+12}{t^2-t-6}$
39. $\dfrac{y^2-3y-18}{y^2-2y-15}$
40. $\dfrac{a^2-10a+21}{a^2-11a+28}$
41. $\dfrac{(a-3)^2}{a^2-9}$
42. $\dfrac{t^2-4}{(t+2)^2}$
43. $\dfrac{x-8}{8-x}$
44. $\dfrac{6-x}{x-6}$
45. $\dfrac{q-p}{-p+q}$
46. $\dfrac{a-b}{b-a}$
47. $\dfrac{5a-15}{3-a}$
48. $\dfrac{6t-12}{2-t}$
49. $\dfrac{3x^2-3y^2}{2y^2-2x^2}$
50. $\dfrac{7a^2-7}{1-a}$

SKILL MAINTENANCE

Factor.

51. x^2+8x+7
52. $x^2-9x+14$

Find the intercepts. Then graph.

53. $5x+2y=20$
54. $2x-4y=8$

Simplify.

55. $\dfrac{2}{3}-\left(\dfrac{3}{4}\right)^2$
56. $\dfrac{7}{9}-\dfrac{2}{3}\cdot\dfrac{6}{7}$

SYNTHESIS

57. ◆ Explain how someone could form a rational expression that is undefined for $x=-3$ and $x=4$.

58. ◆ How is canceling related to the identity property of 1?

59. ◆ Xavier *incorrectly* simplifies
$$\dfrac{x^2+x-2}{x^2+3x+2} \text{ as } \dfrac{x-1}{x+2}.$$
He then checks his simplification by evaluating both expressions for $x=1$. Use this situation to explain why evaluating is not a foolproof check.

60. ◆ How could you convince someone that $a - b$ and $b - a$ are opposites of each other?

Simplify.

61. $\dfrac{16y^4 - x^4}{(x^2 + 4y^2)(x - 2y)}$

62. $\dfrac{(a - b)^2}{b^2 - a^2}$

63. $\dfrac{(t^4 - 1)(t^2 - 9)(t - 9)^2}{(t^4 - 81)(t^2 + 1)(t + 1)^2}$

64. $\dfrac{(t + 2)^3(t^2 + 2t + 1)(t + 1)}{(t + 1)^3(t^2 + 4t + 4)(t + 2)}$

65. $\dfrac{(x^2 - y^2)(x^2 - 2xy + y^2)}{(x + y)^2(x^2 - 4xy - 5y^2)}$

66. $\dfrac{(x - 1)(x^4 - 1)(x^2 - 1)}{(x^2 + 1)(x - 1)^2(x^4 - 2x^2 + 1)}$

67. $\dfrac{x^5 - 2x^3 + 4x^2 - 8}{x^7 + 2x^4 - 4x^3 - 8}$

68. $\dfrac{10t^4 - 8t^3 + 15t - 12}{8 - 10t + 12t^2 - 15t^3}$

69. ◆ Select any number x, multiply by 2, add 5, multiply by 5, subtract 25, and divide by 10. What do you get? Explain how this procedure can be used for a number trick.

6.2 Multiplication and Division

Multiplication • Division

Multiplication and division of rational expressions is similar to multiplication and division with fractional notation.

Multiplication

Recall that to multiply fractions, we simply multiply their numerators and multiply their denominators. Rational expressions are multiplied similarly.

THE PRODUCT OF TWO RATIONAL EXPRESSIONS

To multiply rational expressions, multiply numerators and multiply denominators:

$$\dfrac{A}{B} \cdot \dfrac{C}{D} = \dfrac{AC}{BD}.$$

Then factor and simplify the result if possible.

For example,

$$\dfrac{3}{5} \cdot \dfrac{8}{11} = \dfrac{24}{55} \quad \text{and} \quad \dfrac{x}{3} \cdot \dfrac{x + 2}{y} = \dfrac{x(x + 2)}{3y}.$$

Fraction bars are grouping symbols, so parentheses are needed when writing some products. Because we generally simplify, we often leave parentheses in the product. There is no need to multiply further.

306 CHAPTER 6 • RATIONAL EXPRESSIONS AND EQUATIONS

EXAMPLE 1 Multiply and simplify.

a) $\dfrac{5a^3}{4} \cdot \dfrac{2}{5a}$ b) $\dfrac{x^2 + 6x + 9}{x^2 - 4} \cdot \dfrac{x - 2}{x + 3}$

c) $\dfrac{x^2 + x - 2}{15} \cdot \dfrac{5}{2x^2 - 3x + 1}$

SOLUTION

a) $\dfrac{5a^3}{4} \cdot \dfrac{2}{5a} = \dfrac{5a^3(2)}{4(5a)}$ Forming the product of the numerators and the product of the denominators

$= \dfrac{2 \cdot 5 \cdot a \cdot a \cdot a}{2 \cdot 2 \cdot 5 \cdot a}$ Factoring the numerator and the denominator

$= \dfrac{\cancel{2} \cdot \cancel{5} \cdot \cancel{a} \cdot a \cdot a}{\cancel{2} \cdot 2 \cdot \cancel{5} \cdot \cancel{a}}$ Removing a factor equal to 1: $\dfrac{2 \cdot 5 \cdot a}{2 \cdot 5 \cdot a} = 1$

$= \dfrac{a^2}{2}$

b) $\dfrac{x^2 + 6x + 9}{x^2 - 4} \cdot \dfrac{x - 2}{x + 3} = \dfrac{(x^2 + 6x + 9)(x - 2)}{(x^2 - 4)(x + 3)}$ Multiplying the numerators and the denominators

$= \dfrac{(x + 3)(x + 3)(x - 2)}{(x + 2)(x - 2)(x + 3)}$ Factoring the numerator and the denominator

$= \dfrac{\cancel{(x + 3)}(x + 3)\cancel{(x - 2)}}{(x + 2)\cancel{(x - 2)}\cancel{(x + 3)}}$ Removing a factor equal to 1: $\dfrac{(x + 3)(x - 2)}{(x + 3)(x - 2)} = 1$

$= \dfrac{x + 3}{x + 2}$

c) $\dfrac{x^2 + x - 2}{15} \cdot \dfrac{5}{2x^2 - 3x + 1} = \dfrac{(x^2 + x - 2)5}{15(2x^2 - 3x + 1)}$ Multiplying the numerators and the denominators

$= \dfrac{(x + 2)(x - 1)5}{5(3)(x - 1)(2x - 1)}$ Factoring the numerator and the denominator. Try to go directly to this step.

$= \dfrac{(x + 2)\cancel{(x - 1)}\cancel{5}}{\cancel{5}(3)\cancel{(x - 1)}(2x - 1)}$ Removing a factor equal to 1: $\dfrac{5(x - 1)}{5(x - 1)} = 1$

$= \underbrace{\dfrac{x + 2}{3(2x - 1)}}$

You need not carry out this multiplication.

Division

As with fractions, reciprocals of rational expressions are found by interchanging the numerator and the denominator. For example,

the reciprocal of $\dfrac{2}{7}$ is $\dfrac{7}{2}$, and the reciprocal of $\dfrac{3x}{x + 5}$ is $\dfrac{x + 5}{3x}$.

6.2 MULTIPLICATION AND DIVISION 307

THE QUOTIENT OF TWO RATIONAL EXPRESSIONS

To divide by a rational expression, multiply by its reciprocal:

$$\frac{A}{B} \div \frac{C}{D} = \frac{A}{B} \cdot \frac{D}{C} = \frac{AD}{BC}.$$

Then factor and simplify if possible.

EXAMPLE 2 Divide: (a) $\dfrac{x}{5} \div \dfrac{7}{y}$; (b) $(x + 1) \div \dfrac{x - 1}{x + 3}$.

SOLUTION

a) $\dfrac{x}{5} \div \dfrac{7}{y} = \dfrac{x}{5} \cdot \dfrac{y}{7}$ Multiplying by the reciprocal of the divisor

$= \dfrac{xy}{35}$ Multiplying rational expressions

b) $(x + 1) \div \dfrac{x - 1}{x + 3} = \dfrac{x + 1}{1} \cdot \dfrac{x + 3}{x - 1}$ Multiplying by the reciprocal of the divisor. Writing $x + 1$ as $\dfrac{x + 1}{1}$ can be helpful.

$= \dfrac{(x + 1)(x + 3)}{x - 1}$

As usual, we should simplify when possible.

EXAMPLE 3 Divide and simplify.

a) $\dfrac{x + 1}{x^2 - 1} \div \dfrac{x + 1}{x^2 - 2x + 1}$

b) $\dfrac{a^2 + 3a + 2}{a^2 + 4} \div (5a^2 + 10a)$

c) $\dfrac{x^2 - 2x - 3}{x^2 - 4} \div \dfrac{x + 1}{x + 5}$

SOLUTION

a) $\dfrac{x + 1}{x^2 - 1} \div \dfrac{x + 1}{x^2 - 2x + 1} = \dfrac{x + 1}{x^2 - 1} \cdot \dfrac{x^2 - 2x + 1}{x + 1}$ Multiplying by the reciprocal

$= \dfrac{(x + 1)(x - 1)(x - 1)}{(x + 1)(x - 1)(x + 1)}$ Multiplying rational expressions and factoring numerators and denominators

$= \dfrac{(x + 1)(x - 1)(x - 1)}{(x + 1)(x - 1)(x + 1)}$ Removing a factor equal to 1: $\dfrac{(x + 1)(x - 1)}{(x + 1)(x - 1)} = 1$

$= \dfrac{x - 1}{x + 1}$

b) $\dfrac{a^2 + 3a + 2}{a^2 + 4} \div (5a^2 + 10a) = \dfrac{a^2 + 3a + 2}{a^2 + 4} \cdot \dfrac{1}{5a^2 + 10a}$ Multiplying by the reciprocal

$= \dfrac{(a + 2)(a + 1)}{(a^2 + 4)5a(a + 2)}$ Multiplying rational expressions and factoring

$= \dfrac{(\cancel{a + 2})(a + 1)}{(a^2 + 4)5a(\cancel{a + 2})}$ $\Bigg\}$ Removing a factor equal to 1: $\dfrac{a + 2}{a + 2} = 1$

$= \dfrac{a + 1}{(a^2 + 4)5a}$

c) $\dfrac{x^2 - 2x - 3}{x^2 - 4} \div \dfrac{x + 1}{x + 5} = \dfrac{x^2 - 2x - 3}{x^2 - 4} \cdot \dfrac{x + 5}{x + 1}$ Multiplying by the reciprocal

$= \dfrac{(x - 3)(x + 1)(x + 5)}{(x - 2)(x + 2)(x + 1)}$ Multiplying rational expressions and factoring

$= \dfrac{(x - 3)(\cancel{x + 1})(x + 5)}{(x - 2)(x + 2)(\cancel{x + 1})}$ $\Bigg\}$ Removing a factor equal to 1: $\dfrac{x + 1}{x + 1} = 1$

$= \dfrac{(x - 3)(x + 5)}{(x - 2)(x + 2)}$

EXERCISE SET 6.2

Multiply. Leave parentheses in each product.

1. $\dfrac{7x}{5} \cdot \dfrac{x - 3}{2x + 1}$

2. $\dfrac{3x}{4} \cdot \dfrac{5x + 2}{x - 1}$

3. $\dfrac{a - 4}{a + 6} \cdot \dfrac{a + 2}{a + 6}$

4. $\dfrac{a + 3}{a + 6} \cdot \dfrac{a + 3}{a - 1}$

5. $\dfrac{2x + 3}{4} \cdot \dfrac{x + 1}{x - 5}$

6. $\dfrac{-5}{3x - 4} \cdot \dfrac{-6}{5x + 6}$

7. $\dfrac{a - 5}{a^2 + 1} \cdot \dfrac{a + 2}{a^2 - 1}$

8. $\dfrac{t + 3}{t^2 - 2} \cdot \dfrac{t + 3}{t^2 - 2}$

9. $\dfrac{x + 1}{2 + x} \cdot \dfrac{x - 1}{x + 1}$

10. $\dfrac{m^2 + 5}{m + 8} \cdot \dfrac{m^2 - 4}{m^2 - 4}$

17. $\dfrac{a^2 - 25}{a^2 - 4a + 3} \cdot \dfrac{2a - 5}{2a + 5}$

18. $\dfrac{x + 3}{x^2 + 9} \cdot \dfrac{x^2 + 5x + 4}{x + 9}$

19. $\dfrac{a^2 - 9}{a^2} \cdot \dfrac{a^2 - 3a}{a^2 + a - 12}$

20. $\dfrac{x^2 + 10x - 11}{x^2 - 1} \cdot \dfrac{x + 1}{x + 11}$

21. $\dfrac{4a^2}{3a^2 - 12a + 12} \cdot \dfrac{3a - 6}{2a}$

22. $\dfrac{5v + 5}{v - 2} \cdot \dfrac{v^2 - 4v + 4}{v^2 - 1}$

23. $\dfrac{t^2 + 2t - 3}{t^2 + 4t - 5} \cdot \dfrac{t^2 - 3t - 10}{t^2 + 5t + 6}$

24. $\dfrac{x^2 + 5x + 4}{x^2 - 6x + 8} \cdot \dfrac{x^2 + 5x - 14}{x^2 + 8x + 7}$

Multiply and, if possible, simplify.

11. $\dfrac{5a^4}{8a} \cdot \dfrac{2}{a}$

12. $\dfrac{10}{t^7} \cdot \dfrac{3t^2}{25t}$

13. $\dfrac{3c}{d^2} \cdot \dfrac{4d}{6c^3}$

14. $\dfrac{3x^2y}{2} \cdot \dfrac{4}{xy^3}$

15. $\dfrac{x^2 - 3x - 10}{(x - 2)^2} \cdot \dfrac{x - 2}{x - 5}$

16. $\dfrac{t^2}{t^2 - 4} \cdot \dfrac{t^2 - 5t + 6}{t^2 - 3t}$

25. $\dfrac{5a^2 - 180}{10a^2 - 10} \cdot \dfrac{20a + 20}{2a - 12}$

26. $\dfrac{2t^2 - 98}{4t^2 - 4} \cdot \dfrac{8t + 8}{16t - 112}$

27. $\dfrac{x^2-1}{x^2-9} \cdot \dfrac{(x-3)^4}{(x+1)^2}$

28. $\dfrac{(x+2)^5}{(x-1)^3} \cdot \dfrac{x^2-1}{x^2+5x+6}$

29. $\dfrac{(t-2)^3}{(t-1)^3} \cdot \dfrac{t^2-2t+1}{t^2-4t+4}$

30. $\dfrac{(y+4)^3}{(y+2)^3} \cdot \dfrac{y^2+4y+4}{y^2+8y+16}$

Find the reciprocal of each expression.

31. $\dfrac{x}{9}$

32. $\dfrac{3-x}{x^2+4}$

33. $a^3 - 8a$

34. $\dfrac{7}{a^2-b^2}$

35. $\dfrac{x^2+2x-5}{x^2-4x+7}$

36. $\dfrac{x^2-3xy+y^2}{x^2+7xy-y^2}$

Divide and, if possible, simplify.

37. $\dfrac{5}{9} \div \dfrac{2}{7}$

38. $\dfrac{3}{8} \div \dfrac{5}{2}$

39. $\dfrac{x}{3} \div \dfrac{x}{12}$

40. $\dfrac{x}{4} \div \dfrac{5}{x}$

41. $\dfrac{x^5}{y^2} \div \dfrac{x^2}{y}$

42. $\dfrac{a^5}{b^4} \div \dfrac{a^3}{b}$

43. $\dfrac{a+2}{a-3} \div \dfrac{a-1}{a+3}$

44. $\dfrac{y+2}{4} \div \dfrac{y}{2}$

45. $\dfrac{x^2-1}{x} \div \dfrac{x+1}{x-1}$

46. $\dfrac{4y-8}{y+2} \div \dfrac{y-2}{y^2-4}$

47. $\dfrac{x+1}{6} \div \dfrac{x+1}{3}$

48. $\dfrac{a}{a-b} \div \dfrac{b}{a-b}$

49. $(y^2-9) \div \dfrac{y^2-2y-3}{y^2+1}$

50. $(x^2-5x-6) \div \dfrac{x^2-1}{x+6}$

51. $\dfrac{5x-5}{16} \div \dfrac{x-1}{6}$

52. $\dfrac{-4+2x}{8} \div \dfrac{x-2}{2}$

53. $\dfrac{-6+3x}{5} \div \dfrac{4x-8}{25}$

54. $\dfrac{-12+4x}{4} \div \dfrac{-6+2x}{6}$

55. $\dfrac{a+2}{a-1} \div \dfrac{3a+6}{a-5}$

56. $\dfrac{t-3}{t+2} \div \dfrac{4t-12}{t+1}$

57. $(x-5) \div \dfrac{2x^2-11x+5}{4x^2-1}$

58. $(a+7) \div \dfrac{3a^2+14a-49}{a^2+8a+7}$

59. $\dfrac{x^2-4}{x} \div \dfrac{x-2}{x+2}$

60. $\dfrac{x+y}{x-y} \div \dfrac{x^2+y}{x^2-y^2}$

61. $\dfrac{x^2+7x+12}{x^2-x-20} \div \dfrac{x^2+6x+9}{x^2-10x+25}$

62. $\dfrac{a^2+2a-3}{a^2+3a} \div \dfrac{a+1}{a}$

63. $\dfrac{t^2+4}{t^2-6t+9} \div \dfrac{t^4+16}{t^2+6t+9}$

64. $\dfrac{x^2-4}{x^2+2x+1} \div \dfrac{x^4+1}{x-1}$

65. $\dfrac{c^2+10c+21}{c^2-2c-15} \div (c^2+2c-35)$

66. $\dfrac{1-z}{1+2z-z^2} \div (1-z)$

67. $\dfrac{(t+5)^3}{(t-5)^3} \div \dfrac{(t+5)^2}{(t-5)^2}$

68. $\dfrac{(y-3)^3}{(y+3)^3} \div \dfrac{(y-3)^2}{(y+3)^2}$

SKILL MAINTENANCE

69. Sixteen more than the square of a number is 8 times the number. Find the number.

Subtract.

70. $(6x^2+7) - (4x^2-9)$

71. $(8x^3-3x^2+7) - (8x^2+3x-5)$

72. $(0.08y^3-0.04y^2+0.01y) - (0.02y^3+0.05y^2+1)$

Simplify.

73. $\dfrac{2}{5} - \left(\dfrac{3}{2}\right)^2$

74. $\dfrac{5}{9} + \dfrac{2}{3} \cdot \dfrac{4}{5}$

SYNTHESIS

75. ◆ Is the reciprocal of a product the product of the two reciprocals? Why or why not?

76. ◆ Why is it important to insert parentheses when multiplying rational expressions in which the numerators and the denominators contain more than one term?

77. ◆ Explain why the quotient
$$\dfrac{x+3}{x-5} \div \dfrac{x-7}{x+1}$$
is undefined for $x = 5$, $x = -1$, and $x = 7$.

78. ◆ A student claims to be able to divide, but not multiply, rational expressions. Why is this claim difficult to believe?

Simplify.

79. $\dfrac{2a^2 - 5ab}{c - 3d} \div (4a^2 - 25b^2)$

80. $(x - 2a) \div \dfrac{a^2x^2 - 4a^4}{a^2x + 2a^3}$

81. $\dfrac{3a^2 - 5ab - 12b^2}{3ab + 4b^2} \div (3b^2 - ab)^2$

82. $\dfrac{3x^2 - 2xy - y^2}{x^2 - y^2} \div (3x^2 + 4xy + y^2)^2$

83. $\dfrac{y^2 - 4xy}{y - x} \div \dfrac{16x^2y^2 - y^4}{4x^2 - 3xy - y^2} \div \dfrac{4}{x^3y^3}$

84. $\dfrac{z^2 - 8z + 16}{z^2 + 8z + 16} \div \dfrac{(z - 4)^5}{(z + 4)^5} \div \dfrac{3z + 12}{z^2 - 16}$

85. $\dfrac{x^2 - x + xy - y}{x^2 + 6x - 7} \div \dfrac{x^2 + 2xy + y^2}{4x + 4y}$

86. $\dfrac{3x + 3y + 3}{9x} \div \dfrac{x^2 + 2xy + y^2 - 1}{x^4 + x^2}$

87. $\dfrac{t^4 - 1}{t^4 - 81} \cdot \dfrac{t^2 - 9}{t^2 + 1} \div \dfrac{(t + 1)^2}{(t - 9)^2}$

88. $\dfrac{(t + 2)^3}{(t + 1)^3} \div \dfrac{t^2 + 4t + 4}{t^2 + 2t + 1} \cdot \dfrac{t + 1}{t + 2}$

89. $\left(\dfrac{y^2 + 5y + 6}{y^2} \cdot \dfrac{3y^3 + 6y^2}{y^2 - y - 12}\right) \div \dfrac{y^2 - y}{y^2 - 2y - 8}$

90. $\dfrac{a^4 - 81b^4}{a^2c - 6abc + 9b^2c} \cdot \dfrac{a + 3b}{a^2 + 9b^2} \div \dfrac{a^2 + 6ab + 9b^2}{(a - 3b)^2}$

6.3 Addition, Subtraction, and Least Common Denominators

Addition When Denominators Are the Same • Subtraction When Denominators Are the Same • Least Common Multiples and Denominators

Addition When Denominators Are the Same

Recall that to add fractions having the same denominator, like $\frac{2}{7}$ and $\frac{3}{7}$, we add the numerators: $\frac{2}{7} + \frac{3}{7} = \frac{5}{7}$. The same procedure is used when rational expressions share a common denominator.

THE SUM OF TWO RATIONAL EXPRESSIONS

To add when the denominators are the same, add the numerators and keep the common denominator:

$$\dfrac{A}{B} + \dfrac{C}{B} = \dfrac{A + C}{B}.$$

Whenever possible, we simplify the final result.

6.3 ADDITION, SUBTRACTION, AND LEAST COMMON DENOMINATORS 311

EXAMPLE 1 Add.

a) $\dfrac{4}{a} + \dfrac{3+a}{a}$ b) $\dfrac{3x}{x-5} + \dfrac{2x+1}{x-5}$

c) $\dfrac{2x^2 + 3x - 7}{2x+1} + \dfrac{x^2 + x - 8}{2x+1}$ d) $\dfrac{x-5}{x^2-9} + \dfrac{2}{x^2-9}$

SOLUTION

a) $\dfrac{4}{a} + \dfrac{3+a}{a} = \dfrac{7+a}{a}$ When the denominators are alike, add the numerators.

b) $\dfrac{3x}{x-5} + \dfrac{2x+1}{x-5} = \dfrac{5x+1}{x-5}$ Adding the numerators

c) $\dfrac{2x^2 + 3x - 7}{2x+1} + \dfrac{x^2 + x - 8}{2x+1} = \dfrac{(2x^2 + 3x - 7) + (x^2 + x - 8)}{2x+1}$

$= \dfrac{3x^2 + 4x - 15}{2x+1}$ Combining like terms in the numerator

d) $\dfrac{x-5}{x^2-9} + \dfrac{2}{x^2-9} = \dfrac{x-3}{x^2-9}$ Combining like terms in the numerator: $x - 5 + 2 = x - 3$

$= \dfrac{x-3}{(x-3)(x+3)}$ Factoring

$= \dfrac{1 \cdot (x-3)}{(x-3)(x+3)}$ Removing a factor equal to 1: $\dfrac{x-3}{x-3} = 1$

$= \dfrac{1}{x+3}$

Subtraction When Denominators Are the Same

When two fractions have the same denominator, we subtract by subtracting one numerator from the other—for example, $\frac{5}{7} - \frac{2}{7} = \frac{3}{7}$. The same procedure is used with rational expressions.

THE DIFFERENCE OF TWO RATIONAL EXPRESSIONS

To subtract when the denominators are the same, subtract the second numerator from the first and keep the common denominator:

$$\dfrac{A}{B} - \dfrac{C}{B} = \dfrac{A-C}{B}.$$

CAUTION! Keep in mind that a fraction bar is a grouping symbol. When a numerator is being subtracted, remember to subtract *every* term in that expression.

312 CHAPTER 6 • RATIONAL EXPRESSIONS AND EQUATIONS

EXAMPLE 2   Subtract: **(a)** $\dfrac{3x}{x+2} - \dfrac{x-5}{x+2}$;   **(b)** $\dfrac{x^2}{x-4} - \dfrac{x+12}{x-4}$.

SOLUTION

a) $\dfrac{3x}{x+2} - \dfrac{x-5}{x+2} = \dfrac{3x - (x-5)}{x+2}$ The parentheses are needed to make sure that we subtract both terms.

$= \dfrac{3x - x + 5}{x+2}$ Removing the parentheses and changing signs

$= \dfrac{2x+5}{x+2}$ Combining like terms

b) $\dfrac{x^2}{x-4} - \dfrac{x+12}{x-4} = \dfrac{x^2 - (x+12)}{x-4}$ Remember the parentheses!

$= \dfrac{x^2 - x - 12}{x-4}$ Removing parentheses

$= \dfrac{(x-4)(x+3)}{x-4}$ Factoring, in hopes of simplifying

$= \dfrac{(x-4)(x+3)}{x-4}$ Removing a factor equal to 1: $\dfrac{x-4}{x-4} = 1$

$= x + 3$

Least Common Multiples and Denominators

Thus far, every pair of rational expressions that we have added or subtracted shared a common denominator. To add or subtract rational expressions that lack a common denominator, we must first find equivalent rational expressions that *do* have a common denominator.

In algebra, we find a common denominator much as we do in arithmetic. Recall that to add $\frac{5}{12}$ and $\frac{7}{30}$, we first identify the smallest number that contains both 12 and 30 as factors. Such a number, the **least common multiple (LCM)** of the denominators, is then used as the **least common denominator (LCD)**.

Let's find the LCM of 12 and 30 using a method that can also be used with polynomials. We begin by writing the prime factorization of 12:

$12 = 2 \cdot 2 \cdot 3.$

Next, we write the prime factorization of 30:

$30 = 2 \cdot 3 \cdot 5.$

The LCM must include the factors of each number, so it must include each prime factor the greatest number of times that it appears in either of the factorizations. To find the LCM for 12 and 30, we select one factorization, say

$2 \cdot 2 \cdot 3,$

and note that because it lacks a factor of 5, it does not contain the entire fac-

6.3 ADDITION, SUBTRACTION, AND LEAST COMMON DENOMINATORS   313

torization of 30. If we multiply 2 · 2 · 3 by 5, every prime factor occurs just often enough to contain both 12 and 30 as factors.

$$\text{LCM} = 2 \cdot 2 \cdot 3 \cdot 5$$

12 is a factor of the LCM.
30 is a factor of the LCM.

Note that each prime factor—2, 3, and 5— is used the greatest number of times that it appears in either of the individual factorizations. The factor 2 occurs twice and the factors 3 and 5 once each.

TO FIND THE LEAST COMMON DENOMINATOR (LCD):

1. Write the prime factorization of each denominator.
2. Select one of the factorizations and inspect it to see if it contains the other.

   a) If it does, it represents the LCM of the denominators.
   b) If it does not, multiply that factorization by any factors of the other denominator that it lacks. The final product is the LCM of the denominators.

The LCD is the LCM of the denominators. It should contain each factor the greatest number of times that it occurs in any one factorization.

Let's finish adding $\frac{5}{12}$ and $\frac{7}{30}$:

$$\frac{5}{12} + \frac{7}{30} = \frac{5}{2 \cdot 2 \cdot 3} + \frac{7}{2 \cdot 3 \cdot 5}.$$

The least common denominator, or LCD, is 2 · 2 · 3 · 5. To get the LCD in the first denominator, we need a 5. To get the LCD in the second denominator, we need another 2. We get these numbers by multiplying by different forms of 1:

$$\frac{5}{12} + \frac{7}{30} = \frac{5}{2 \cdot 2 \cdot 3} \cdot \frac{5}{5} + \frac{7}{2 \cdot 3 \cdot 5} \cdot \frac{2}{2} \qquad \frac{5}{5} = 1 \text{ and } \frac{2}{2} = 1$$

$$= \frac{25}{2 \cdot 2 \cdot 3 \cdot 5} + \frac{14}{2 \cdot 3 \cdot 5 \cdot 2} \qquad \text{The denominators are now the LCD.}$$

$$= \frac{39}{2 \cdot 2 \cdot 3 \cdot 5} \qquad \text{Adding the numerators and keeping the LCD}$$

$$= \frac{3 \cdot 13}{2 \cdot 2 \cdot 3 \cdot 5}$$

$$= \frac{13}{20}. \qquad \text{Simplifying by removing a factor equal to 1: } \frac{3}{3} = 1$$

Expressions like $\dfrac{7}{24x}$ and $\dfrac{5}{36x^2}$ are added in much the same manner.

EXAMPLE 3 Find the LCD of $\dfrac{7}{24x}$ and $\dfrac{5}{36x^2}$.

SOLUTION

1. We begin by writing the prime factorizations of $24x$ and $36x^2$:

   $24x = 2 \cdot 2 \cdot 2 \cdot 3 \cdot x,$
   $36x^2 = 2 \cdot 2 \cdot 3 \cdot 3 \cdot x \cdot x.$

2. Note that the factorization of $36x^2$ contains the entire factorization of $24x$ except for a third factor of 2. To find the smallest product that contains both $24x$ and $36x^2$ as factors, we need to multiply $2 \cdot 2 \cdot 3 \cdot 3 \cdot x \cdot x$ by a third factor of 2:

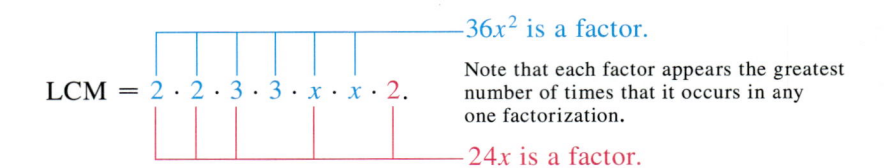

   $\text{LCM} = 2 \cdot 2 \cdot 3 \cdot 3 \cdot x \cdot x \cdot 2.$

   $36x^2$ is a factor.
   Note that each factor appears the greatest number of times that it occurs in any one factorization.
   $24x$ is a factor.

The LCM is thus $2^3 \cdot 3^2 \cdot x^2$, or $72x^2$, so the LCD is $72x^2$.

Now let's finish adding $\dfrac{7}{24x}$ and $\dfrac{5}{36x^2}$:

$$\dfrac{7}{24x} + \dfrac{5}{36x^2} = \dfrac{7}{2 \cdot 2 \cdot 2 \cdot 3 \cdot x} + \dfrac{5}{2 \cdot 2 \cdot 3 \cdot 3 \cdot x \cdot x}.$$

In Example 3, we found that the LCD is $2 \cdot 2 \cdot 2 \cdot 3 \cdot 3 \cdot x \cdot x$. To obtain equivalent expressions with this LCD, we multiply each expression by 1, using the missing factors of the LCD to write 1:

$$\dfrac{7}{24x} + \dfrac{5}{36x^2} = \dfrac{7}{2 \cdot 2 \cdot 2 \cdot 3 \cdot x} \cdot \dfrac{3 \cdot x}{3 \cdot x} + \dfrac{5}{2 \cdot 2 \cdot 3 \cdot 3 \cdot x \cdot x} \cdot \dfrac{2}{2}$$

The LCD requires additional factors of 3 and x.   The LCD requires another factor of 2.

$$= \dfrac{21x}{2 \cdot 2 \cdot 2 \cdot 3 \cdot x \cdot 3 \cdot x} + \dfrac{10}{2 \cdot 2 \cdot 3 \cdot 3 \cdot x \cdot x \cdot 2}$$

Both denominators are now the LCD.

$$= \dfrac{21x + 10}{72x^2}.$$

You now have the "big" picture of why LCMs are needed when adding rational expressions. For the remainder of this section, we will practice finding LCMs and rewriting rational expressions so that they have the LCD as the denominator. In Section 6.4, we will return to the addition and subtraction of rational expressions.

6.3 ADDITION, SUBTRACTION, AND LEAST COMMON DENOMINATORS

EXAMPLE 4 For each pair of polynomials, find the least common multiple.

a) $15a$ and $35b$
b) $21x^3y^6$ and $7x^5y^2$
c) $x^2 + 5x - 6$ and $x^2 - 1$

SOLUTION

a) We write the prime factorizations and then construct the LCM:

$15a = 3 \cdot 5 \cdot a$
$35b = 5 \cdot 7 \cdot b$

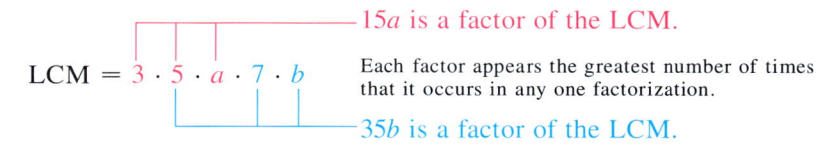

$\text{LCM} = 3 \cdot 5 \cdot a \cdot 7 \cdot b$

— $15a$ is a factor of the LCM.
Each factor appears the greatest number of times that it occurs in any one factorization.
— $35b$ is a factor of the LCM.

The LCM is $3 \cdot 5 \cdot a \cdot 7 \cdot b$, or $105ab$.

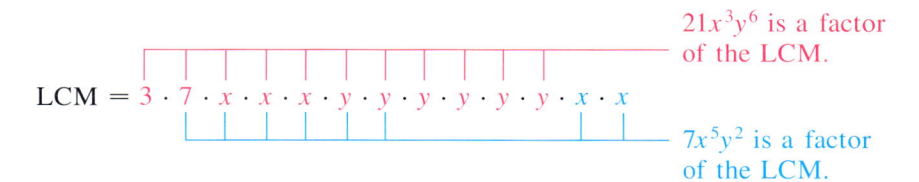

b) $21x^3y^6 = 3 \cdot 7 \cdot x \cdot x \cdot x \cdot y \cdot y \cdot y \cdot y \cdot y \cdot y$
$7x^5y^2 = 7 \cdot x \cdot x \cdot x \cdot x \cdot x \cdot y \cdot y$

Try to visualize the factors of x and y mentally.

$\text{LCM} = 3 \cdot 7 \cdot x \cdot x \cdot x \cdot y \cdot y \cdot y \cdot y \cdot y \cdot y \cdot x \cdot x$

— $21x^3y^6$ is a factor of the LCM.
— $7x^5y^2$ is a factor of the LCM.

Note that we used the highest power of each factor in $21x^3y^6$ and $7x^5y^2$. The LCM is $21x^5y^6$.

c) $x^2 + 5x - 6 = (x - 1)(x + 6)$
$x^2 - 1 = (x - 1)(x + 1)$

$\text{LCM} = (x - 1)(x + 6)(x + 1)$

— $x^2 + 5x - 6$ is a factor of the LCM.
— $x^2 - 1$ is a factor of the LCM.

The LCM is $(x - 1)(x + 6)(x + 1)$. There is no need to multiply this out.

The above procedure can be used to find the LCM of three polynomials as well. We factor each polynomial and then construct the LCM using each factor the greatest number of times that it appears in any one factorization.

EXAMPLE 5 For each group of polynomials, find the LCM.

a) $12x$, $16y$, and $8xyz$
b) $x^2 + 4$, $x + 1$, and 5

SOLUTION

a) $12x = 2 \cdot 2 \cdot 3 \cdot x$
$16y = 2 \cdot 2 \cdot 2 \cdot 2 \cdot y$
$8xyz = 2 \cdot 2 \cdot 2 \cdot x \cdot y \cdot z$

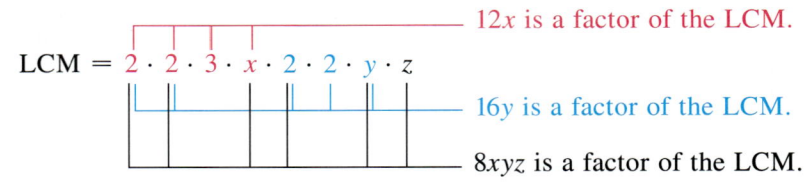

LCM = $2 \cdot 2 \cdot 3 \cdot x \cdot 2 \cdot 2 \cdot y \cdot z$

— $12x$ is a factor of the LCM.
— $16y$ is a factor of the LCM.
— $8xyz$ is a factor of the LCM.

The LCM is $2^4 \cdot 3 \cdot xyz$, or $48xyz$.

b) Since $x^2 + 4$, $x + 1$, and 5 are not factorable, the LCM is their product: $5(x^2 + 4)(x + 1)$.

Remember that to add or subtract expressions with different denominators, it is important to be able to write equivalent expressions that have the LCD.

EXAMPLE 6 Find equivalent expressions that have the LCD:

$$\frac{x+3}{x^2+5x-6}, \quad \frac{x+7}{x^2-1}.$$

SOLUTION From Example 4(c), we know that the LCD is $(x + 6)(x - 1)(x + 1)$. Since $x^2 + 5x - 6 = (x + 6)(x - 1)$, the factor of the LCD that is missing from the first denominator is $x + 1$. We multiply by 1 using $(x + 1)/(x + 1)$:

$$\frac{x+3}{x^2+5x-6} = \frac{x+3}{(x+6)(x-1)} \cdot \frac{x+1}{x+1}$$

$$= \frac{(x+3)(x+1)}{(x+6)(x-1)(x+1)}.$$

For the second expression, we have $x^2 - 1 = (x + 1)(x - 1)$. The factor of the LCD that is missing is $x + 6$. We multiply by 1 using $(x + 6)/(x + 6)$:

$$\frac{x+7}{x^2-1} = \frac{x+7}{(x+1)(x-1)} \cdot \frac{x+6}{x+6}$$

$$= \frac{(x+7)(x+6)}{(x+1)(x-1)(x+6)}.$$

We leave the results in factored form. In Section 6.4, we will carry out the actual addition and subtraction of such rational expressions.

EXERCISE SET 6.3

Perform the indicated operation. Simplify, if possible.

1. $\dfrac{4}{x} + \dfrac{9}{x}$

2. $\dfrac{4}{a^2} + \dfrac{9}{a^2}$

3. $\dfrac{x}{15} + \dfrac{2x+1}{15}$

4. $\dfrac{a}{7} + \dfrac{3a-4}{7}$

5. $\dfrac{5}{a+3} + \dfrac{1}{a+3}$

6. $\dfrac{5}{x+2} + \dfrac{8}{x+2}$

7. $\dfrac{9}{a+2} - \dfrac{3}{a+2}$

8. $\dfrac{8}{x+7} - \dfrac{2}{x+7}$

9. $\dfrac{3y+9}{2y} - \dfrac{y+1}{2y}$

10. $\dfrac{5+3t}{4t} - \dfrac{2t+1}{4t}$

11. $\dfrac{9x+8}{x+1} + \dfrac{2x+3}{x+1}$

12. $\dfrac{3a+13}{a+4} + \dfrac{2a+7}{a+4}$

13. $\dfrac{9x+8}{x+1} - \dfrac{2x+3}{x+1}$

14. $\dfrac{3a+13}{a+4} - \dfrac{2a+7}{a+4}$

15. $\dfrac{a^2}{a-4} + \dfrac{a-20}{a-4}$

16. $\dfrac{x^2}{x+5} + \dfrac{7x+10}{x+5}$

17. $\dfrac{x^2}{x-2} - \dfrac{6x-8}{x-2}$

18. $\dfrac{a^2}{a+3} - \dfrac{2a+15}{a+3}$

19. $\dfrac{t^2+4t}{t-1} + \dfrac{2t-7}{t-1}$

20. $\dfrac{y^2+6y}{y+2} + \dfrac{2y+12}{y+2}$

21. $\dfrac{x+1}{x^2+5x+6} + \dfrac{2}{x^2+5x+6}$

22. $\dfrac{-7}{x^2-4x+3} + \dfrac{x+4}{x^2-4x+3}$

23. $\dfrac{a^2+5}{a^2+5a-6} - \dfrac{6}{a^2+5a-6}$

24. $\dfrac{a^2-1}{a^2-7a+12} - \dfrac{8}{a^2-7a+12}$

25. $\dfrac{t^2-3t}{t^2+6t+9} + \dfrac{2t-12}{t^2+6t+9}$

26. $\dfrac{y^2-7y}{y^2+8y+16} + \dfrac{6y-20}{y^2+8y+16}$

27. $\dfrac{2x^2+x}{x^2-8x+12} - \dfrac{x^2-2x+10}{x^2-8x+12}$

28. $\dfrac{2x^2+3}{x^2-6x+5} - \dfrac{x^2-5x+9}{x^2-6x+5}$

29. $\dfrac{7-2x}{x^2-6x+8} + \dfrac{3-3x}{x^2-6x+8}$

30. $\dfrac{3-2t}{t^2-5t+4} + \dfrac{2-3t}{t^2-5t+4}$

31. $\dfrac{x-7}{x^2+3x-4} - \dfrac{2x-3}{x^2+3x-4}$

32. $\dfrac{5-3x}{x^2-2x+1} - \dfrac{x+1}{x^2-2x+1}$

Find the LCM.

33. 15, 27

34. 10, 15

35. 8, 9

36. 12, 15

37. 6, 9, 21

38. 8, 36, 40

Find the LCM.

39. $6x^2$, $12x^3$

40. $2a^2b$, $8ab^2$

41. $15a^4b^7$, $10a^2b^8$

42. $6a^2b^7$, $9a^5b^2$

43. $2(y-3)$, $6(y-3)$

44. $4(x-1)$, $8(x-1)$

45. x^2-4, x^2+5x+6

46. x^2+3x+2, x^2-4

47. t^3+4t^2+4t, t^2-4t

48. y^3-y^2, y^4-y^2

49. $10x^2y$, $6y^2z$, $5xz^3$

50. $8x^3z$, $12xy^2$, $4y^5z^2$

51. $a+1$, $(a-1)^2$, a^2-1

52. x^2-y^2, $2x+2y$, $x^2+2xy+y^2$

53. m^2-5m+6, m^2-4m+4

54. $2x^2+5x+2$, $2x^2-x-1$

55. $10v^2+30v$, $5v^2+35v+60$

56. $12a^2+24a$, $4a^2+20a+24$

57. $9x^3-9x^2-18x$, $6x^5-24x^4+24x^3$

58. x^5-4x^3, x^3+4x^2+4x

Find equivalent expressions that have the LCD.

59. $\dfrac{13}{6x^5}$, $\dfrac{y}{12x^3}$

60. $\dfrac{3}{10a^3}$, $\dfrac{b}{5a^6}$

61. $\dfrac{3}{2a^2b}$, $\dfrac{5}{8ab^2}$

62. $\dfrac{7}{3x^4y^2}$, $\dfrac{4}{9xy^3}$

63. $\dfrac{x+1}{x^2-4}$, $\dfrac{x-2}{x^2+5x+6}$

64. $\dfrac{x-4}{x^2-9}$, $\dfrac{x+2}{x^2+11x+24}$

SKILL MAINTENANCE

Factor.

65. $x^2 - 19x + 60$
66. $x^2 + 9x - 36$

Find a polynomial that can represent the shaded area of each figure.

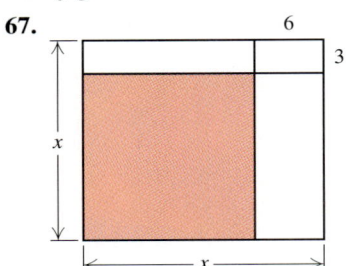

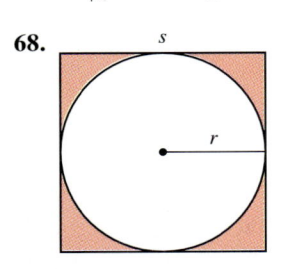

67.

68.

Simplify.

69. $-\frac{7}{24} + \frac{5}{18}$
70. $\frac{2}{15} - \frac{7}{20}$

SYNTHESIS

71. ◆ Explain why the product of two numbers is not always their LCM.

72. ◆ Is every LCM an LCD? Why or why not?

73. ◆ Is it possible to add or subtract rational expressions without knowing how to factor? Why or why not?

74. ◆ If the LCM of a binomial and a trinomial is the trinomial, what relationship exists between the two expressions?

Perform the indicated operations and, if possible, simplify.

75. $\frac{x+y}{x^2-y^2} + \frac{x-y}{x^2-y^2} - \frac{2x}{x^2-y^2}$

76. $\frac{6x-1}{x-1} + \frac{3(2x+5)}{x-1} + \frac{3(2x-3)}{x-1}$

77. $\frac{2x+11}{x-3} \cdot \frac{3}{x+4} + \frac{-1}{4+x} \cdot \frac{6x+3}{x-3}$

78. $\frac{x^2}{3x^2-5x-2} - \frac{2x}{3x+1} \cdot \frac{1}{x-2}$

Find the LCM.

79. 72, 90, 96

80. $8x^2 - 8$, $6x^2 - 12x + 6$, $10 - 10x$

81. Beginning at 5:00 A.M., a hotel shuttle bus leaves Salton Airport every 25 min, and the downtown shuttle bus leaves the airport every 35 min. What time will it be when both shuttles again leave at the same time?

82. Kim and Jed leave the starting point of a fitness loop at the same time. Kim jogs a lap in 6 min and Jed jogs one in 8 min. Assuming they continue to run at the same pace, when will they next meet at the starting place?

83. ◆ Explain how evaluating can be used to perform a partial check on the result of Example 1(d):

$$\frac{x-5}{x^2-9} + \frac{2}{x^2-9} = \frac{1}{x+3}.$$

84. ◆ On page 313, the second step in finding an LCD is to select one of the factorizations of the denominators. Does it matter which one is selected? Why or why not?

COLLABORATIVE CORNER

Focus: Least common multiples and proportions
Time: 20–30 minutes
Group size: 2

Travel between different countries usually necessitates an exchange of currencies. Recently, one Canadian dollar was worth 72 cents in U.S. funds. Use this exchange rate for the activity that follows.

ACTIVITY

1. Within each group of two students, one student should play the role of a U.S. citizen planning a visit to Canada. The other student should play the role of a Canadian citizen planning a visit to the United States.
    a) Determine how much Canadian money the U.S. citizen would receive in exchange for $72 of U.S. funds. Remember: The U.S. dollar is worth more ($1.00 vs. 72¢) than the Canadian dollar. Then determine how much U.S. money the Canadian would receive in exchange for $72 of Canadian funds. (*Hint:* Use a proportion.)
    b) Determine how much U.S. money the Canadian citizen would receive in exchange for $100. Then determine how much the U.S. citizen would receive in exchange for $100 of U.S. funds.

2. The answers to parts (a) and (b) above should indicate that change (coins) is needed to exchange $72 of Canadian money for U.S. funds, or to exchange $100 of U.S. funds for Canadian money.
    a) What is the smallest amount of Canadian dollars that can be exchanged for U.S. dollars without requiring coins?
    b) What is the smallest amount of U.S. dollars that can be exchanged for Canadian dollars without requiring coins? (*Hint:* See part a.)
    c) Use your results from parts (a) and (b) of this question to find two other amounts of U.S. currency and two other amounts of Canadian currency that can be exchanged without requiring coins. Answers may vary.
    d) Use your results from parts (a) and (b) of this question to find the smallest number x for which neither conversion—from x Canadian dollars to U.S. funds or from x U.S. dollars to Canadian funds—will require coins.

3. At one time in 1997, one New Zealand dollar was worth about 76 cents in U.S. funds. Find the smallest number x for which neither conversion—from x New Zealand dollars to U.S. funds or from x U.S. dollars to New Zealand funds—will require coins. (*Hint:* See part d above.)

6.4 Addition and Subtraction with Unlike Denominators

Adding and Subtracting with LCDs • When Factors Are Opposites

Adding and Subtracting with LCDs

We now know how to rewrite two rational expressions in an equivalent form that uses the LCD. Once rational expressions share a common denominator,

CHAPTER 6 • RATIONAL EXPRESSIONS AND EQUATIONS

they can be added or subtracted just as they were at the beginning of Section 6.3.

TO ADD OR SUBTRACT RATIONAL EXPRESSIONS HAVING DIFFERENT DENOMINATORS:

1. Find the LCD.
2. Multiply each rational expression by a form of 1 made up of the factors of the LCD missing from that expression's denominator.
3. Add or subtract the numerators, as indicated. Write the sum or difference over the LCD.
4. Simplify, if possible.

EXAMPLE 1

Add: $\dfrac{5x^2}{8} + \dfrac{7x}{12}$.

SOLUTION

1. First, we find the LCD:

$$\left.\begin{array}{l} 8 = 2 \cdot 2 \cdot 2 \\ 12 = 2 \cdot 2 \cdot 3 \end{array}\right\} \quad \text{LCD} = 2 \cdot 2 \cdot 2 \cdot 3, \text{ or } 24.$$

2. The denominator 8 needs to be multiplied by 3 in order to obtain the LCD. The denominator 12 needs to be multiplied by 2 in order to obtain the LCD. Thus we multiply by $\frac{3}{3}$ and $\frac{2}{2}$ to get the LCD:

$$\dfrac{5x^2}{8} + \dfrac{7x}{12} = \dfrac{5x^2}{2 \cdot 2 \cdot 2} + \dfrac{7x}{2 \cdot 2 \cdot 3}$$

$$= \dfrac{5x^2}{2 \cdot 2 \cdot 2} \cdot \dfrac{3}{3} + \dfrac{7x}{2 \cdot 2 \cdot 3} \cdot \dfrac{2}{2} \quad \text{Multiplying each expression by a form of 1 to get the LCD}$$

$$= \dfrac{15x^2}{24} + \dfrac{14x}{24}.$$

3. Next, we add the numerators:

$$\dfrac{15x^2}{24} + \dfrac{14x}{24} = \dfrac{15x^2 + 14x}{24}.$$

4. Although $15x^2 + 14x$ can be factored, $\dfrac{15x^2 + 14x}{24}$ cannot be simplified, so we are done.

Subtraction is performed in much the same way.

EXAMPLE 2

Subtract: $\dfrac{7}{8x} - \dfrac{5}{12x^2}$.

SOLUTION We follow, but do not list, the four steps shown above. First, we find the LCD:

$$\left.\begin{array}{l}8x = 2 \cdot 2 \cdot 2 \cdot x \\ 12x^2 = 2 \cdot 2 \cdot 3 \cdot x \cdot x\end{array}\right\} \quad \text{LCD} = 2 \cdot 2 \cdot 3 \cdot x \cdot x \cdot 2, \text{ or } 24x^2.$$

The denominator $8x$ must be multiplied by $3x$ in order to obtain the LCD. The denominator $12x^2$ must be multiplied by 2 in order to obtain the LCD. Thus we multiply by $\frac{3x}{3x}$ and $\frac{2}{2}$ to get the LCD. Then we subtract and, if possible, simplify.

$$\frac{7}{8x} - \frac{5}{12x^2} = \frac{7}{8x} \cdot \frac{3 \cdot x}{3 \cdot x} - \frac{5}{12x^2} \cdot \frac{2}{2}$$

$$= \frac{21x}{24x^2} - \frac{10}{24x^2} \quad \leftarrow \text{ **CAUTION!** Do not simplify *these* rational expressions or you will lose the LCD.}$$

$$= \frac{21x - 10}{24x^2} \quad \text{This cannot be simplified, so we are done.}$$

When denominators contain polynomials with two or more terms, the same steps are used.

EXAMPLE 3 Add:

$$\frac{2a}{a^2 - 1} + \frac{1}{a^2 + a}.$$

SOLUTION First, we find the LCD:

$$\left.\begin{array}{l}a^2 - 1 = (a - 1)(a + 1) \\ a^2 + a = a(a + 1)\end{array}\right\} \quad \text{LCD} = (a - 1)(a + 1)a.$$

We multiply by a form of 1 to get the LCD in each expression:

$$\frac{2a}{a^2 - 1} + \frac{1}{a^2 + a} = \frac{2a}{(a - 1)(a + 1)} \cdot \frac{a}{a} + \frac{1}{a(a + 1)} \cdot \frac{a - 1}{a - 1} \quad \text{Multiplying by } \frac{a}{a} \text{ and } \frac{a-1}{a-1} \text{ to get the LCD}$$

$$= \frac{2a^2}{(a - 1)(a + 1)a} + \frac{a - 1}{(a - 1)(a + 1)a}.$$

$$= \frac{2a^2 + a - 1}{(a - 1)(a + 1)a} \quad \text{Adding numerators}$$

$$= \frac{(2a - 1)(a + 1)}{(a - 1)(a + 1)a} \quad \text{Simplifying by factoring and removing a factor equal to 1:}$$

$$= \frac{2a - 1}{(a - 1)a} \quad \frac{a+1}{a+1} = 1$$

CHAPTER 6 • RATIONAL EXPRESSIONS AND EQUATIONS

EXAMPLE 4 Perform the indicated operations.

a) $\dfrac{x+4}{x-2} - \dfrac{x-7}{x+5}$ \qquad b) $\dfrac{x}{x^2+11x+30} + \dfrac{-5}{x^2+9x+20}$

c) $\dfrac{x}{x^2+5x+6} - \dfrac{2}{x^2+3x+2}$

SOLUTION

a) First, we find the LCD. It is just the product of the denominators:

$$\text{LCD} = (x-2)(x+5).$$

We multiply by a form of 1 to get the LCD in each expression. Then we subtract and try to simplify.

$\dfrac{x+4}{x-2} - \dfrac{x-7}{x+5} = \dfrac{x+4}{x-2} \cdot \dfrac{x+5}{x+5} - \dfrac{x-7}{x+5} \cdot \dfrac{x-2}{x-2}$

$= \dfrac{x^2+9x+20}{(x-2)(x+5)} - \dfrac{x^2-9x+14}{(x-2)(x+5)}$ \qquad Multiplying out numerators (but not denominators)

$= \dfrac{x^2+9x+20 - (x^2-9x+14)}{(x-2)(x+5)}$ \qquad When subtracting a numerator with more than one term, parentheses are important.

$= \dfrac{x^2+9x+20 - x^2+9x-14}{(x-2)(x+5)}$ \qquad Removing parentheses and subtracting every term

$= \dfrac{18x+6}{(x-2)(x+5)}$

Although $18x+6$ can be factored as $6(3x+1)$, doing so will not enable us to simplify our result.

b) $\dfrac{x}{x^2+11x+30} + \dfrac{-5}{x^2+9x+20}$

$= \dfrac{x}{(x+5)(x+6)} + \dfrac{-5}{(x+5)(x+4)}$ \qquad Factoring the denominators in order to find the LCD. The LCD is $(x+5)(x+6)(x+4)$.

$= \dfrac{x}{(x+5)(x+6)} \cdot \dfrac{x+4}{x+4} + \dfrac{-5}{(x+5)(x+4)} \cdot \dfrac{x+6}{x+6}$ \qquad Multiplying to get the LCD

$= \dfrac{x^2+4x}{(x+5)(x+6)(x+4)} + \dfrac{-5x-30}{(x+5)(x+6)(x+4)}$ \qquad Multiplying in each numerator

$= \dfrac{x^2+4x-5x-30}{(x+5)(x+6)(x+4)}$ \qquad Adding numerators

$= \dfrac{x^2-x-30}{(x+5)(x+6)(x+4)}$ \qquad Combining like terms in the numerator

$= \dfrac{(x+5)(x-6)}{(x+5)(x+6)(x+4)}$ \qquad Always simplify the result, if possible, by removing a factor equal to 1. (Here $(x+5)/(x+5) = 1$.)

$= \dfrac{x-6}{(x+6)(x+4)}$

c) $\dfrac{x}{x^2 + 5x + 6} - \dfrac{2}{x^2 + 3x + 2}$

$= \dfrac{x}{(x + 2)(x + 3)} - \dfrac{2}{(x + 2)(x + 1)}$ Factoring denominators. The LCD is $(x + 2)(x + 3)(x + 1)$.

$= \dfrac{x}{(x + 2)(x + 3)} \cdot \dfrac{x + 1}{x + 1} - \dfrac{2}{(x + 2)(x + 1)} \cdot \dfrac{x + 3}{x + 3}$

$= \dfrac{x^2 + x}{(x + 2)(x + 3)(x + 1)} - \dfrac{2x + 6}{(x + 2)(x + 3)(x + 1)}$

$= \dfrac{x^2 + x - (2x + 6)}{(x + 2)(x + 3)(x + 1)}$ Don't forget the parentheses!

$= \dfrac{x^2 + x - 2x - 6}{(x + 2)(x + 3)(x + 1)}$ Remember to subtract each term in $2x + 6$.

$= \dfrac{x^2 - x - 6}{(x + 2)(x + 3)(x + 1)}$ Combining like terms in the numerator

$= \dfrac{(x + 2)(x - 3)}{(x + 2)(x + 3)(x + 1)}$ Factoring and simplifying; $\dfrac{x + 2}{x + 2} = 1$

$= \dfrac{x - 3}{(x + 3)(x + 1)}$

When Factors Are Opposites

Recall from Section 6.1 that expressions of the form $a - b$ and $b - a$ are opposites of each other. When either of these binomials is multiplied by -1, the result is the other binomial:

$-1(a - b) = -a + b = b + (-a) = b - a;$
$-1(b - a) = -b + a = a + (-b) = a - b.$

Multiplication by -1 reverses the order in which subtraction occurs.

EXAMPLE 5   Add: $\dfrac{x}{x - 5} + \dfrac{7}{5 - x}$.

SOLUTION   Since the denominators are opposites of each other, we can find a common denominator by multiplying either rational expression by $-1/-1$. Because polynomials are most often written in descending order, we choose to reverse the subtraction in the second denominator:

$\dfrac{x}{x - 5} + \dfrac{7}{5 - x} = \dfrac{x}{x - 5} + \dfrac{7}{5 - x} \cdot \dfrac{-1}{-1}$ Writing 1 as $-1/-1$ and multiplying to obtain a common denominator

$= \dfrac{x}{x - 5} + \dfrac{-7}{-5 + x}$

$= \dfrac{x}{x - 5} + \dfrac{-7}{x - 5}$ Note that $-5 + x = x + (-5) = x - 5$.

$= \dfrac{x - 7}{x - 5}.$

324 CHAPTER 6 • RATIONAL EXPRESSIONS AND EQUATIONS

Sometimes, after factoring to find the LCD, we find a factor in one denominator that is the opposite of a factor in the other denominator. When this happens, multiplication by $-1/-1$ can again be helpful.

EXAMPLE 6 Perform the indicated operations and simplify.

a) $\dfrac{x}{x^2 - 25} + \dfrac{3}{5 - x}$

b) $\dfrac{x + 9}{x^2 - 4} + \dfrac{6 - x}{4 - x^2} - \dfrac{1 + x}{x^2 - 4}$

SOLUTION

a) $\dfrac{x}{x^2 - 25} + \dfrac{3}{5 - x} = \dfrac{x}{(x - 5)(x + 5)} + \dfrac{3}{5 - x}$ Factoring

$= \dfrac{x}{(x - 5)(x + 5)} + \dfrac{3}{5 - x} \cdot \dfrac{-1}{-1}$ Multiplying by $-1/-1$ changes $5 - x$ to $x - 5$.

$= \dfrac{x}{(x - 5)(x + 5)} + \dfrac{-3}{x - 5}$ Note that $(5 - x)(-1) = x - 5$.

$= \dfrac{x}{(x - 5)(x + 5)} + \dfrac{-3}{(x - 5)} \cdot \dfrac{x + 5}{x + 5}$ The LCD is $(x - 5)(x + 5)$.

$= \dfrac{x}{(x - 5)(x + 5)} + \dfrac{-3x - 15}{(x - 5)(x + 5)}$

$= \dfrac{-2x - 15}{(x - 5)(x + 5)}$

b) Since $4 - x^2$ is the opposite of $x^2 - 4$, multiplying the second rational expression by $-1/-1$ will lead to a common denominator:

$\dfrac{x + 9}{x^2 - 4} + \dfrac{6 - x}{4 - x^2} - \dfrac{1 + x}{x^2 - 4} = \dfrac{x + 9}{x^2 - 4} + \dfrac{6 - x}{4 - x^2} \cdot \dfrac{-1}{-1} - \dfrac{1 + x}{x^2 - 4}$

$= \dfrac{x + 9}{x^2 - 4} + \dfrac{x - 6}{x^2 - 4} - \dfrac{1 + x}{x^2 - 4}$

$= \dfrac{x + 9 + x - 6 - 1 - x}{x^2 - 4}$ Adding and subtracting numerators

$= \dfrac{x + 2}{x^2 - 4}$

$= \dfrac{(x + 2) \cdot 1}{(x + 2)(x - 2)}$ Simplifying

$= \dfrac{1}{x - 2}.$

EXERCISE SET 6.4

Perform the indicated operation. Simplify, if possible.

1. $\dfrac{3}{x} + \dfrac{4}{x^2}$
2. $\dfrac{5}{x} + \dfrac{6}{x^2}$
3. $\dfrac{5}{6r} - \dfrac{5}{8r}$
4. $\dfrac{2}{9t} - \dfrac{11}{6t}$
5. $\dfrac{7}{xy^2} + \dfrac{3}{x^2y}$
6. $\dfrac{2}{c^2d} + \dfrac{7}{cd^3}$
7. $\dfrac{8}{9t^3} - \dfrac{5}{6t^2}$
8. $\dfrac{-2}{3xy^2} - \dfrac{6}{x^2y^3}$
9. $\dfrac{x+5}{8} + \dfrac{x-3}{12}$
10. $\dfrac{x-4}{9} + \dfrac{x+5}{6}$
11. $\dfrac{a+2}{2} - \dfrac{a-4}{4}$
12. $\dfrac{x-2}{6} - \dfrac{x+1}{3}$
13. $\dfrac{2a-1}{3a^2} + \dfrac{5a+1}{9a}$
14. $\dfrac{a+4}{16a} + \dfrac{3a+4}{4a^2}$
15. $\dfrac{x-1}{4x} - \dfrac{2x+3}{x}$
16. $\dfrac{4z-9}{3z} - \dfrac{3z-8}{4z}$
17. $\dfrac{2c-d}{c^2d} + \dfrac{c+d}{cd^2}$
18. $\dfrac{x+y}{xy^2} + \dfrac{3x+y}{x^2y}$
19. $\dfrac{5x+3y}{2x^2y} - \dfrac{3x+4y}{xy^2}$
20. $\dfrac{4x+2t}{3xt^2} - \dfrac{5x-3t}{x^2t}$
21. $\dfrac{2}{x-1} + \dfrac{2}{x+1}$
22. $\dfrac{3}{x-2} + \dfrac{3}{x+2}$
23. $\dfrac{2z}{z-1} - \dfrac{3z}{z+1}$
24. $\dfrac{5}{x+5} - \dfrac{3}{x-5}$
25. $\dfrac{2}{x+5} + \dfrac{3}{4x}$
26. $\dfrac{3}{x+1} + \dfrac{2}{3x}$
27. $\dfrac{8}{x^2-4} - \dfrac{3}{x+2}$
28. $\dfrac{3}{2t^2-2t} - \dfrac{5}{2t-2}$
29. $\dfrac{4x}{x^2-25} + \dfrac{x}{x+5}$
30. $\dfrac{2x}{x^2-16} + \dfrac{x}{x-4}$
31. $\dfrac{t}{t-3} - \dfrac{5}{4t-12}$
32. $\dfrac{6}{z+4} - \dfrac{2}{3z+12}$
33. $\dfrac{2}{x+3} + \dfrac{4}{(x+3)^2}$
34. $\dfrac{3}{x-1} + \dfrac{2}{(x-1)^2}$
35. $\dfrac{2}{5x^2+5x} - \dfrac{4}{3x+3}$
36. $\dfrac{2t}{t^2-9} - \dfrac{3}{t-3}$
37. $\dfrac{3a}{4a-20} + \dfrac{9a}{6a-30}$
38. $\dfrac{4a}{5a-10} + \dfrac{3a}{10a-20}$
39. $\dfrac{t}{y-t} - \dfrac{y}{y+t}$
40. $\dfrac{a}{x+a} - \dfrac{a}{x-a}$
41. $\dfrac{x}{x-5} + \dfrac{x-5}{x}$
42. $\dfrac{x+4}{x} + \dfrac{x}{x+4}$
43. $\dfrac{x}{x^2+5x+6} - \dfrac{2}{x^2+3x+2}$
44. $\dfrac{x}{x^2+9x+20} - \dfrac{4}{x^2+7x+12}$
45. $\dfrac{x}{x^2+2x+1} + \dfrac{1}{x^2+5x+4}$
46. $\dfrac{7}{a^2+a-2} + \dfrac{5}{a^2-4a+3}$
47. $\dfrac{x}{x^2+15x+56} - \dfrac{6}{x^2+13x+42}$
48. $\dfrac{-5}{x^2+17x+16} - \dfrac{3}{x^2+9x+8}$
49. $\dfrac{3}{x^2-9} + \dfrac{2}{x^2-x-6}$
50. $\dfrac{3z}{z^2-4z+4} + \dfrac{10}{z^2+z-6}$
51. $\dfrac{5}{x-1} - \dfrac{6}{1-x}$
52. $\dfrac{x}{4} - \dfrac{3x-5}{-4}$
53. $\dfrac{t^2}{t-2} + \dfrac{4}{2-t}$
54. $\dfrac{y^2}{y-3} + \dfrac{9}{3-y}$
55. $\dfrac{a-3}{a^2-25} + \dfrac{a-3}{25-a^2}$
56. $\dfrac{b-7}{b^2-16} + \dfrac{7-b}{16-b^2}$
57. $\dfrac{4-p}{25-p^2} + \dfrac{p+1}{p-5}$
58. $\dfrac{y+2}{y-7} + \dfrac{3-y}{49-y^2}$
59. $\dfrac{5x}{x^2-9} - \dfrac{4}{3-x}$
60. $\dfrac{8x}{16-x^2} - \dfrac{5}{x-4}$
61. $\dfrac{3x+2}{3x+6} + \dfrac{x}{4-x^2}$
62. $\dfrac{a}{a^2-1} + \dfrac{2a}{a-a^2}$
63. $\dfrac{4-a^2}{a^2-9} - \dfrac{a-2}{3-a}$
64. $\dfrac{4x}{x^2-y^2} - \dfrac{6}{y-x}$

Simplify.

65. $\dfrac{4y}{y^2 - 1} - \dfrac{2}{y} - \dfrac{2}{y + 1}$

66. $\dfrac{x + 6}{4 - x^2} - \dfrac{x + 3}{x + 2} + \dfrac{x - 3}{2 - x}$

67. $\dfrac{2z}{1 - 2z} + \dfrac{3z}{2z + 1} - \dfrac{3}{4z^2 - 1}$

68. $\dfrac{1}{x + y} + \dfrac{1}{x - y} - \dfrac{2x}{x^2 - y^2}$

69. $\dfrac{2r}{r^2 - s^2} + \dfrac{1}{r + s} - \dfrac{1}{r - s}$

70. $\dfrac{3}{2c - 1} - \dfrac{1}{c + 2} - \dfrac{5}{2c^2 + 3c - 2}$

SKILL MAINTENANCE

Graph.

71. $y = \tfrac{1}{2}x - 5$

72. $y = -\tfrac{1}{2}x - 5$

73. $y = 3$

74. $x = -5$

Solve.

75. $3x - 7 = 5x + 9$

76. $2a + 8 = 13 - 4a$

77. $x^2 - 8x + 15 = 0$

78. $x^2 - 7x - 18 = 0$

SYNTHESIS

79. ◆ Are parentheses as important for adding rational expressions as they are for subtracting rational expressions? Why or why not?

80. ◆ How could you convince someone that

$$\dfrac{1}{3 - x} \text{ and } \dfrac{1}{x - 3}$$

are opposites of each other?

81. ◆ Why is it best to use the *least* common denominator—rather than just *any* common denominator—when adding or subtracting rational expressions?

82. ◆ Describe in your own words a procedure that can be used to add any two rational expressions.

Write expressions for the perimeter and the area of each rectangle.

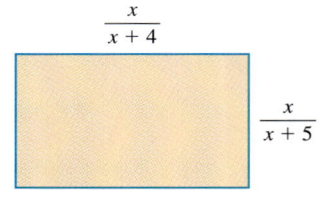

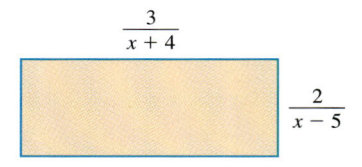

83. Rectangle with dimensions $x + 4$ (top) and $x + 5$ (right, labeled x).

84. Rectangle with dimensions $\dfrac{3}{x + 4}$ (top) and $\dfrac{2}{x - 5}$ (right).

Perform the indicated operations.

85. $\dfrac{2x + 11}{x - 3} \cdot \dfrac{3}{x + 4} + \dfrac{2x + 1}{4 + x} \cdot \dfrac{3}{3 - x}$

86. $\dfrac{x^2}{3x^2 - 5x - 2} - \dfrac{2x}{3x + 1} \cdot \dfrac{1}{x - 2}$

87. $\dfrac{x}{x^4 - y^4} - \left(\dfrac{1}{x + y}\right)^2$

88. $\dfrac{1}{ay - 3a + 2xy - 6x} - \dfrac{xy + ay}{a^2 - 4x^2} \left(\dfrac{1}{y - 3}\right)^2$

89. $\dfrac{2x^2 + 5x - 3}{2x^2 - 9x + 9} + \dfrac{x + 1}{3 - 2x} + \dfrac{4x^2 + 8x + 3}{x - 3} \cdot \dfrac{x + 3}{9 - 4x^2}$

90. $\left(\dfrac{a}{a - b} + \dfrac{b}{a + b}\right)\left(\dfrac{1}{3a + b} + \dfrac{2a + 6b}{9a^2 - b^2}\right)$

91. Express

$$\dfrac{a - 3b}{a - b}$$

as a sum of two rational expressions with denominators that are opposites of each other. Answers may vary.

6.5 Complex Rational Expressions

Multiplying by the LCD • Using Division to Simplify

A **complex rational expression,** or **complex fractional expression,** is a rational expression that has one or more rational expressions within its numerator or denominator. Here are some examples:

$$\frac{1 + \frac{2}{x}}{3}, \quad \frac{x+y}{\frac{2}{x+1}}, \quad \frac{\frac{1}{3} + \frac{1}{5}}{\frac{2}{x} - \frac{x}{y}}.$$

These are rational expressions within the complex rational expression.

We will consider two methods for simplifying complex rational expressions. Each method offers certain advantages.

Multiplying by the LCD (Method 1)

Our first method for simplifying complex rational expressions relies on multiplying by an expression equal to 1.

TO SIMPLIFY A COMPLEX RATIONAL EXPRESSION BY MULTIPLYING BY THE LCD:

1. Find the LCD of all rational expressions *within* the complex rational expression.
2. Multiply the complex rational expression by a form of 1, using the LCD to write 1.
3. Distribute and simplify. No fractional expressions should remain within the complex rational expression.
4. Factor and simplify, if possible.

EXAMPLE 1 Simplify: $\dfrac{\frac{1}{2} + \frac{3}{4}}{\frac{5}{6} - \frac{3}{8}}$.

SOLUTION

1. The denominators *within* the complex rational expression, 2, 4, 6, and 8, have 24 as the LCD.

2. We multiply by a form of 1, using the LCD:

$$\frac{\frac{1}{2}+\frac{3}{4}}{\frac{5}{6}-\frac{3}{8}} = \frac{\frac{1}{2}+\frac{3}{4}}{\frac{5}{6}-\frac{3}{8}} \cdot \frac{24}{24} \qquad \text{Multiplying by a form of 1, using the LCD: } \frac{24}{24} = 1$$

3. Using the distributive law, we perform the multiplication:

$$\frac{\frac{1}{2}+\frac{3}{4}}{\frac{5}{6}-\frac{3}{8}} \cdot \frac{24}{24} = \frac{\left(\frac{1}{2}+\frac{3}{4}\right)24}{\left(\frac{5}{6}-\frac{3}{8}\right)24} \qquad \begin{array}{l}\leftarrow \text{Multiplying the numerator by 24} \\ \text{Don't forget the parentheses!} \\ \leftarrow \text{Multiplying the denominator by 24}\end{array}$$

$$= \frac{\frac{1}{2}(24) + \frac{3}{4}(24)}{\frac{5}{6}(24) - \frac{3}{8}(24)} \qquad \text{Using the distributive law}$$

$$= \frac{12 + 18}{20 - 9}, \text{ or } \frac{30}{11}. \qquad \text{Simplifying}$$

4. The result, $\frac{30}{11}$, cannot be factored or simplified, so we are done.

Multiplying in this manner has the effect of clearing fractions in both the top and bottom of the complex rational expression. In Example 2, we follow, but do not list, the same four steps.

EXAMPLE 2 Simplify.

a) $\dfrac{\dfrac{3}{x}+\dfrac{1}{2x}}{\dfrac{1}{3x}-\dfrac{3}{4x}}$

b) $\dfrac{1-\dfrac{1}{x}}{1-\dfrac{1}{x^2}}$

SOLUTION

a) The denominators within the complex expression are x, $2x$, $3x$, and $4x$, so the LCD is $12x$. We multiply by 1 using $(12x)/(12x)$:

$$\frac{\frac{3}{x}+\frac{1}{2x}}{\frac{1}{3x}-\frac{3}{4x}} = \frac{\frac{3}{x}+\frac{1}{2x}}{\frac{1}{3x}-\frac{3}{4x}} \cdot \frac{12x}{12x} = \frac{\frac{3}{x}(12x)+\frac{1}{2x}(12x)}{\frac{1}{3x}(12x)-\frac{3}{4x}(12x)}. \qquad \text{Using the distributive law}$$

When the multiplications by $12x$ are performed, all fractions in the numerator and the denominator of the complex rational expression are cleared, as shown at the top of the following page.

$$\frac{\dfrac{3}{x}(12x) + \dfrac{1}{2x}(12x)}{\dfrac{1}{3x}(12x) - \dfrac{3}{4x}(12x)} = \frac{36+6}{4-9} = -\frac{42}{5}.$$

b) $\dfrac{1-\dfrac{1}{x}}{1-\dfrac{1}{x^2}} = \dfrac{1-\dfrac{1}{x}}{1-\dfrac{1}{x^2}} \cdot \dfrac{x^2}{x^2}$ The LCD is x^2 so we multiply by 1 using x^2/x^2.

$= \dfrac{1 \cdot x^2 - \dfrac{1}{x} \cdot x^2}{1 \cdot x^2 - \dfrac{1}{x^2} \cdot x^2}$ Using the distributive law

$= \dfrac{x^2 - x}{x^2 - 1}$ All fractions have been cleared within the complex rational expression.

$= \dfrac{x(x-1)}{(x+1)(x-1)}$ Factoring and simplifying: $\dfrac{x-1}{x-1} = 1$

$= \dfrac{x}{x+1}$

Using Division to Simplify (Method 2)

A second method for simplifying complex rational expressions involves rewriting the expression as a quotient of two rational expressions.

TO SIMPLIFY A COMPLEX RATIONAL EXPRESSION BY DIVIDING:

1. Add or subtract, as needed, to get a single rational expression in the numerator.
2. Add or subtract, as needed, to get a single rational expression in the denominator.
3. Divide the numerator by the denominator (invert and multiply).
4. If possible, simplify by removing a factor equal to 1.

The key here is to express a complex rational expression as one rational expression divided by another. We can then proceed as in Section 6.2.

EXAMPLE 3 Simplify.

a) $\dfrac{\dfrac{x}{x-3}}{\dfrac{4}{5x-15}}$ b) $\dfrac{\dfrac{5}{2a} + \dfrac{1}{a}}{\dfrac{1}{4a} - \dfrac{5}{6}}$ c) $\dfrac{\dfrac{x^2}{y} - \dfrac{5}{x}}{xy}$

330 CHAPTER 6 • RATIONAL EXPRESSIONS AND EQUATIONS

SOLUTION

a) When the numerator and the denominator are already single rational expressions, we can start by dividing (step 3 on the preceding page):

$$\dfrac{\dfrac{x}{x-3}}{\dfrac{4}{5x-15}} = \dfrac{x}{x-3} \div \dfrac{4}{5x-15} \qquad \text{Rewriting with a division symbol}$$

$$= \dfrac{x}{x-3} \cdot \dfrac{5x-15}{4} \qquad \text{Multiplying by the reciprocal of the divisor (inverting and multiplying)}$$

$$= \dfrac{x}{\cancel{x-3}} \cdot \dfrac{5(\cancel{x-3})}{4} \qquad \text{Factoring and removing a factor equal to 1: } \dfrac{x-3}{x-3} = 1$$

$$= \dfrac{5x}{4}.$$

b) $\dfrac{\dfrac{5}{2a} + \dfrac{1}{a}}{\dfrac{1}{4a} - \dfrac{5}{6}} = \dfrac{\dfrac{5}{2a} + \dfrac{1}{a} \cdot \dfrac{2}{2}}{\dfrac{1}{4a} \cdot \dfrac{3}{3} - \dfrac{5}{6} \cdot \dfrac{2a}{2a}}$ ⟵ Multiplying by 1 to get the LCD, $2a$, for the numerator

⟵ Multiplying by 1 to get the LCD, $12a$, for the denominator

$$= \dfrac{\dfrac{5}{2a} + \dfrac{2}{2a}}{\dfrac{3}{12a} - \dfrac{10a}{12a}} = \dfrac{\dfrac{7}{2a}}{\dfrac{3-10a}{12a}} \qquad \text{Adding in the numerator; subtracting in the denominator}$$

$$= \dfrac{7}{2a} \div \dfrac{3-10a}{12a} \qquad \text{Rewriting with a division symbol. This is often done mentally.}$$

$$= \dfrac{7}{2a} \cdot \dfrac{12a}{3-10a} \qquad \text{Multiplying by the reciprocal of the divisor (inverting and multiplying)}$$

$$= \dfrac{7}{\cancel{2a}} \cdot \dfrac{\cancel{2a} \cdot 6}{3-10a} \qquad \text{Removing a factor equal to 1: } \dfrac{2a}{2a} = 1$$

$$= \dfrac{42}{3-10a}$$

c) $\dfrac{\dfrac{x^2}{y} - \dfrac{5}{x}}{xy} = \dfrac{\dfrac{x^2}{y} \cdot \dfrac{x}{x} - \dfrac{5}{x} \cdot \dfrac{y}{y}}{xy}$ ⟵ Multiplying by 1 to get the LCD, xy, for the numerator

$$= \dfrac{\dfrac{x^3}{xy} - \dfrac{5y}{xy}}{xy} = \dfrac{\dfrac{x^3-5y}{xy}}{xy} \qquad \text{⟵ Subtracting in the numerator}$$

$$= \dfrac{x^3-5y}{xy} \div (xy) \qquad \text{Rewriting with a division symbol}$$

$$= \dfrac{x^3-5y}{xy} \cdot \dfrac{1}{xy} \qquad \text{Multiplying by the reciprocal of the divisor (inverting and multiplying)}$$

$$= \dfrac{x^3-5y}{x^2y^2}$$

EXERCISE SET 6.5

Simplify. Use either method or the method specified by your instructor.

1. $\dfrac{3 + \dfrac{1}{2}}{9 - \dfrac{1}{4}}$

2. $\dfrac{1 - \dfrac{3}{4}}{1 + \dfrac{9}{16}}$

3. $\dfrac{1 + \dfrac{1}{3}}{5 - \dfrac{5}{27}}$

4. $\dfrac{1 + \dfrac{1}{5}}{1 - \dfrac{3}{5}}$

5. $\dfrac{\dfrac{s}{3} + s}{\dfrac{3}{s} + s}$

6. $\dfrac{\dfrac{1}{x} - 5}{\dfrac{1}{x} + 3}$

7. $\dfrac{\dfrac{3}{x}}{\dfrac{2}{x} + \dfrac{1}{x^2}}$

8. $\dfrac{\dfrac{4}{x} - \dfrac{1}{x^2}}{\dfrac{2}{x^2}}$

9. $\dfrac{\dfrac{2a - 5}{3a}}{\dfrac{a - 1}{6a}}$

10. $\dfrac{\dfrac{a + 4}{a^2}}{\dfrac{a - 2}{3a}}$

11. $\dfrac{\dfrac{x}{4} - \dfrac{4}{x}}{\dfrac{1}{4} + \dfrac{1}{x}}$

12. $\dfrac{\dfrac{3}{x} + \dfrac{3}{8}}{\dfrac{x}{8} - \dfrac{3}{x}}$

13. $\dfrac{\dfrac{1}{3} + \dfrac{1}{x}}{\dfrac{3 + x}{3}}$

14. $\dfrac{\dfrac{1}{5} - \dfrac{1}{a}}{\dfrac{5 - a}{5}}$

15. $\dfrac{\dfrac{1}{t^2} + 1}{\dfrac{1}{t} - 1}$

16. $\dfrac{2 + \dfrac{1}{x}}{2 - \dfrac{1}{x^2}}$

17. $\dfrac{\dfrac{x^2}{x^2 - y^2}}{\dfrac{x}{x + y}}$

18. $\dfrac{\dfrac{a^2}{a - 3}}{\dfrac{2a}{a^2 - 9}}$

19. $\dfrac{\dfrac{3}{a} - \dfrac{4}{a^2}}{\dfrac{2}{a^3} + \dfrac{3}{a}}$

20. $\dfrac{\dfrac{5}{x^3} + \dfrac{1}{x^2}}{\dfrac{2}{x} - \dfrac{3}{x^2}}$

21. $\dfrac{\dfrac{2}{7a^4} - \dfrac{1}{14a}}{\dfrac{3}{5a^2} + \dfrac{2}{15a}}$

22. $\dfrac{\dfrac{5}{4x^3} - \dfrac{3}{8x}}{\dfrac{3}{2x} + \dfrac{3}{4x^3}}$

23. $\dfrac{\dfrac{x}{5y^3} - \dfrac{3}{10y}}{\dfrac{x}{10y} + \dfrac{3}{y^4}}$

24. $\dfrac{\dfrac{a}{6b^3} + \dfrac{4}{9b^2}}{\dfrac{5}{6b} - \dfrac{1}{9b^3}}$

25. $\dfrac{\dfrac{5}{ab^4} + \dfrac{2}{a^3b}}{\dfrac{5}{a^3b} - \dfrac{3}{ab}}$

26. $\dfrac{\dfrac{2}{x^2y} + \dfrac{3}{xy^2}}{\dfrac{2}{xy^3} + \dfrac{1}{x^2y}}$

27. $\dfrac{2 - \dfrac{3}{x^2}}{2 + \dfrac{3}{x^4}}$

28. $\dfrac{3 - \dfrac{2}{a^4}}{2 + \dfrac{3}{a^3}}$

29. $\dfrac{t - \dfrac{2}{t}}{t + \dfrac{5}{t}}$

30. $\dfrac{x + \dfrac{3}{x}}{x - \dfrac{2}{x}}$

31. $\dfrac{7 - \dfrac{5}{ab^3}}{\dfrac{4 + a}{a^2b}}$

32. $\dfrac{5 + \dfrac{3}{x^2y}}{\dfrac{3 + x}{x^3y}}$

33. $\dfrac{\dfrac{a - 7}{a^3}}{\dfrac{3}{a^2} + \dfrac{2}{a}}$

34. $\dfrac{\dfrac{x + 5}{x^2}}{\dfrac{2}{x} - \dfrac{3}{x^2}}$

35. $\dfrac{1 + \dfrac{a}{5b^2}}{\dfrac{a}{10b} - 1}$

36. $\dfrac{x - 3 + \dfrac{2}{x}}{x - 4 + \dfrac{3}{x}}$

37. $\dfrac{x - 2 + \dfrac{x}{3}}{x + 7 - \dfrac{4}{5x}}$

38. $\dfrac{a + 5 - \dfrac{3}{a}}{a - 3 + \dfrac{5}{a}}$

SKILL MAINTENANCE

Solve.

39. $3x - 5 + 2(4x - 1) = 12x - 3$

40. $(x - 1)7 - (x + 1)9 = 4(x + 2)$

41. Subtract:
$(5x^4 - 6x^3 + 23x^2 - 79x + 24)$
$- (-18x^4 - 56x^3 + 84x - 17)$.

42. The length of a rectangle is 3 yd greater than the width. The area of the rectangle is 10 yd^2. Find the perimeter.

SYNTHESIS

43. ◆ Which of the two methods presented would you use to simplify Exercise 18? Why?

44. ◆ Which of the two methods presented would you use to simplify Exercise 26? Why?

45. ◆ Why is the distributive law especially important when using Method 1 of this section?

46. ◆ Why is factoring an important skill when simplifying complex rational expressions?

47. Find the simplified form for the reciprocal of
$$\dfrac{2}{x - 1} - \dfrac{1}{3x - 2}.$$

Simplify.

48. $\dfrac{\dfrac{a}{b} - \dfrac{c}{d}}{\dfrac{b}{a} - \dfrac{d}{c}}$

49. $\dfrac{\dfrac{a}{b} + \dfrac{c}{d}}{\dfrac{b}{a} + \dfrac{d}{c}}$

50. $\left[\dfrac{\dfrac{x+1}{x-1} + 1}{\dfrac{x+1}{x-1} - 1}\right]^5$

51. $1 + \dfrac{1}{1 + \dfrac{1}{1 + \dfrac{1}{x}}}$

52. $\dfrac{\dfrac{z}{1 - \dfrac{z}{2+2z}} - 2z}{\dfrac{2z}{5z-2} - 3}$

53. ◆ Explain how evaluating can be used as a partial check for Exercise 33.

54. Use a grapher to check Example 2.

6.6 Solving Rational Equations

Solving a New Type of Equation • A Visual Interpretation

In Chapters 1 and 2, we first distinguished between *equivalent expressions*, like $x + x$ and $2x$, and *equivalent equations*, like $3(x + 2) = 18$ and $3x + 6 = 18$. Recall that equivalent equations have the same solution set. In Sections 6.1–6.5, we saw how to write equivalent expressions but nowhere did we solve an equation. We now return to solving equations by examining a type of equation that we could not have solved prior to this chapter.

Solving a New Type of Equation

A **rational**, or **fractional**, **equation** is an equation containing one or more rational expressions, often with the variable in a denominator. Here are some examples:

$$\dfrac{2}{3} + \dfrac{5}{6} = \dfrac{x}{9}, \quad x + \dfrac{6}{x} = -5, \quad \dfrac{x^2}{x-1} = \dfrac{1}{x-1}.$$

TO SOLVE A RATIONAL EQUATION:

1. List any restrictions that exist. No solution can make a denominator equal 0.
2. Clear the equation of fractions by multiplying both sides by the LCD of all rational expressions in the equation.
3. Solve the resulting equation using the addition principle, the multiplication principle, and the principle of zero products, as needed.
4. Check the possible solution(s) in the original equation.

6.6 SOLVING RATIONAL EQUATIONS

As we proceed, you will see why the fourth step, checking, is so important.

In the examples that follow, we *do not* use the LCD to add rational expressions. Instead, we use the LCD as a multiplier that will clear fractions.

EXAMPLE 1 Solve: $\dfrac{x}{6} - \dfrac{x}{8} = \dfrac{1}{12}$.

SOLUTION The LCD is 24. We multiply both sides by 24:

$$24\left(\dfrac{x}{6} - \dfrac{x}{8}\right) = 24 \cdot \dfrac{1}{12}$$ Using the multiplication principle to multiply both sides by the LCD. Parentheses are important!

$$24 \cdot \dfrac{x}{6} - 24 \cdot \dfrac{x}{8} = 24 \cdot \dfrac{1}{12}$$ Using the distributive law

Be sure to multiply *each* term by the LCD.

$$4x - 3x = 2$$ Simplifying. Note that $\dfrac{24x}{6} = 4x$, $\dfrac{24x}{8} = 3x$, and $\dfrac{24}{12} = 2$. All fractions have been cleared.

$$x = 2.$$

Check:
$$\dfrac{x}{6} - \dfrac{x}{8} = \dfrac{1}{12}$$
$$\dfrac{2}{6} - \dfrac{2}{8} \;?\; \dfrac{1}{12}$$
$$\dfrac{1}{3} - \dfrac{1}{4}$$
$$\dfrac{4}{12} - \dfrac{3}{12}$$
$$\dfrac{1}{12} \;\bigg|\; \dfrac{1}{12} \quad \text{TRUE}$$

This checks, so the solution is 2.

Up to now, the multiplication principle has been used only to multiply both sides of an equation by a nonzero constant. Because rational equations often contain variables in a denominator, clearing fractions may now require us to multiply both sides of an equation by a variable expression. Since a variable expression could represent 0, multiplying both sides of an equation by a variable expression does not always produce an equivalent equation. Thus checking in the original equation is very important.

EXAMPLE 2 Solve.

a) $\dfrac{2}{3x} + \dfrac{1}{x} = 10$

b) $x + \dfrac{6}{x} = -5$

c) $1 + \dfrac{3x}{x+2} = \dfrac{-6}{x+2}$

d) $\dfrac{3}{x-5} + \dfrac{1}{x+5} = \dfrac{2}{x^2 - 25}$

SOLUTION

a) Note that x cannot be 0. The LCD is $3x$, so we multiply both sides by $3x$:

$$\frac{2}{3x} + \frac{1}{x} = 10 \qquad \text{The LCD is } 3x;\ x \neq 0.$$

$$3x\left(\frac{2}{3x} + \frac{1}{x}\right) = 3x \cdot 10 \qquad \text{Using the multiplication principle to multiply both sides by the LCD. Don't forget the parentheses!}$$

$$3x \cdot \frac{2}{3x} + 3x \cdot \frac{1}{x} = 3x \cdot 10 \qquad \text{Multiplying to remove parentheses}$$

$$2 + 3 = 30x \qquad \text{Simplifying}$$

$$5 = 30x$$

$$\frac{5}{30} = x, \quad \text{so } x = \frac{1}{6}. \qquad \text{Since } \frac{1}{6} \neq 0, \text{ this } \textit{should} \text{ check.}$$

Check:

$$\frac{\dfrac{2}{3x} + \dfrac{1}{x} = 10}{\dfrac{2}{3 \cdot \frac{1}{6}} + \dfrac{1}{\frac{1}{6}} \ ?\ 10}$$

$$\dfrac{2}{\frac{1}{2}} + \dfrac{1}{\frac{1}{6}}$$

$$4 + 6$$

$$10 \ \bigg|\ 10 \quad \text{TRUE}$$

The solution is $\frac{1}{6}$.

b)

$$x + \frac{6}{x} = -5 \qquad \text{The LCD is } x. \text{ Again, we must have } x \neq 0.$$

$$x\left(x + \frac{6}{x}\right) = -5x \qquad \text{Multiplying on both sides by } x$$

$$x \cdot x + x \cdot \frac{6}{x} = -5x \qquad \text{Note that each term on the left is now multiplied by } x.$$

$$x^2 + 6 = -5x \qquad \text{Simplifying. Note that we have a quadratic equation.}$$

$$x^2 + 5x + 6 = 0 \qquad \text{Using the addition principle to add } 5x \text{ on both sides}$$

$$(x + 3)(x + 2) = 0 \qquad \text{Factoring}$$

$$x + 3 = 0 \quad \text{or} \quad x + 2 = 0 \qquad \text{Using the principle of zero products}$$

$$x = -3 \quad \text{or} \quad x = -2 \qquad \text{Since neither solution is 0, they should both check.}$$

Check: For -3:

$$\frac{x + \dfrac{6}{x} = -5}{-3 + \dfrac{6}{-3} \ ?\ -5}$$

$$-3 - 2$$

$$-5 \ \bigg|\ -5 \quad \text{TRUE}$$

For -2:

$$\frac{x + \dfrac{6}{x} = -5}{-2 + \dfrac{6}{-2} \ ?\ -5}$$

$$-2 - 3$$

$$-5 \ \bigg|\ -5 \quad \text{TRUE}$$

Both of these check, so there are two solutions, -3 and -2.

c)
$$1 + \frac{3x}{x+2} = \frac{-6}{x+2}$$
The LCD is $x+2$. To avoid dividing by 0, we must have $x \neq -2$.

$$(x+2)\left(1 + \frac{3x}{x+2}\right) = (x+2)\frac{-6}{x+2}$$
Multiplying on both sides by $x+2$

$$(x+2) \cdot 1 + (x+2)\frac{3x}{x+2} = (x+2)\frac{-6}{x+2}$$
Using the distributive law

$$x + 2 + 3x = -6$$ Simplifying
$$4x + 2 = -6$$
$$4x = -8$$
$$x = -2$$ Above, we stated that $x \neq -2$.

Check:
$$\frac{1 + \dfrac{3x}{x+2} = \dfrac{-6}{x+2}}{\begin{array}{c|c} 1 + \dfrac{3(-2)}{-2+2} & \dfrac{-6}{-2+2} \\ 1 + \dfrac{-6}{0} & \dfrac{-6}{0} \end{array}} \quad ? \quad$$

$\dfrac{-6}{0}$ is undefined.

We see that -2 is *not* a solution of the original equation because it results in division by 0. The equation has no solution.

d)
$$\frac{3}{x-5} + \frac{1}{x+5} = \frac{2}{x^2-25}$$
Note that $x \neq 5$ and $x \neq -5$; the LCD is $(x-5)(x+5)$.

$$(x-5)(x+5)\left(\frac{3}{x-5} + \frac{1}{x+5}\right) = \frac{2}{(x-5)(x+5)}(x-5)(x+5)$$

$$\frac{(x-5)(x+5) \cdot 3}{x-5} + \frac{(x-5)(x+5)}{x+5} = \frac{2(x-5)(x+5)}{(x-5)(x+5)}$$
Using the distributive law

$$(x+5)3 + (x-5) = 2$$
Removing factors equal to 1: $\dfrac{x-5}{x-5} = 1$, $\dfrac{x+5}{x+5} = 1$, and $\dfrac{(x-5)(x+5)}{(x-5)(x+5)} = 1$

$$3x + 15 + x - 5 = 2$$ Using the distributive law
$$4x + 10 = 2$$
$$4x = -8$$
$$x = -2 \quad$$ $-2 \neq 5$ and $-2 \neq -5$, so -2 *should* check.

We leave it to the student to check that the number -2 is the solution.

A Visual Interpretation

It is possible to solve a rational equation by graphing. The procedure consists of graphing each side of the equation and then determining the first coordi-

TECHNOLOGY CONNECTION 6.6

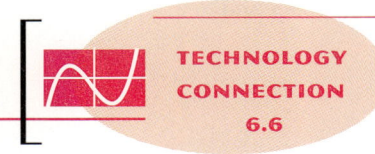

A grapher can be used to check that Example 2(b),

$$x + \frac{6}{x} = -5,$$

has two solutions. To do so, we graph

$$y_1 = x + \frac{6}{x} \quad \text{and} \quad y_2 = -5$$

on the same set of axes.

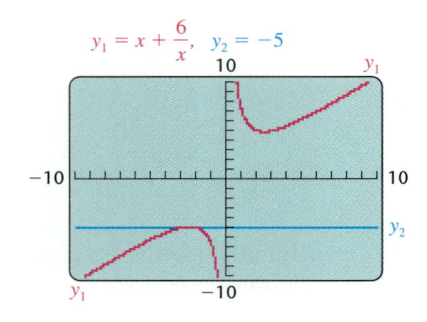

Next, we use TRACE and ZOOM or the INTERSECT option of the CALC menu to confirm that the points of intersection occur when $x = -3$ and $x = -2$.

nate(s) of any point(s) of intersection. (With the advent of the graphing calculator, producing such graphs requires little work.) For example, the equation

$$\frac{x}{4} + \frac{x}{2} = 6$$

can be solved by graphing the equations

$$y = \frac{x}{4} + \frac{x}{2} \quad \text{and} \quad y = 6$$

on the same set of axes.

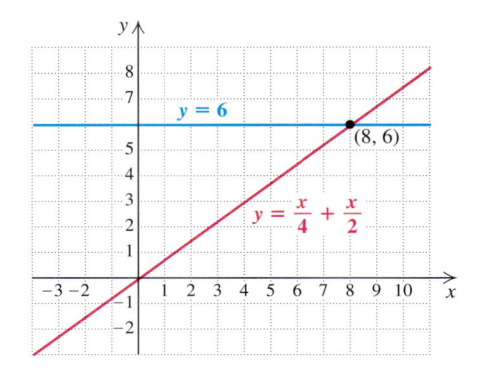

As we can see in the graph above, when $x = 8$, the value of $x/4 + x/2$ is 6. Thus, 8 is the solution of $x/4 + x/2 = 6$. We can check by substitution:

$$\frac{x}{4} + \frac{x}{2} = \frac{8}{4} + \frac{8}{2} = 2 + 4 = 6.$$

EXERCISE SET 6.6

Solve.

1. $\dfrac{4}{5} - \dfrac{2}{3} = \dfrac{x}{9}$

2. $\dfrac{3}{8} - \dfrac{4}{5} = \dfrac{x}{20}$

3. $\dfrac{3}{5} + \dfrac{1}{8} = \dfrac{1}{x}$

4. $\dfrac{2}{3} + \dfrac{5}{6} = \dfrac{1}{x}$

5. $\dfrac{1}{8} + \dfrac{1}{10} = \dfrac{1}{t}$

6. $\dfrac{1}{6} + \dfrac{1}{8} = \dfrac{1}{t}$

7. $x + \dfrac{3}{x} = -4$

8. $x + \dfrac{4}{x} = -5$

9. $\dfrac{x}{7} - \dfrac{7}{x} = 0$

10. $\dfrac{x}{6} - \dfrac{6}{x} = 0$

11. $\dfrac{4}{x} = \dfrac{5}{x} - \dfrac{1}{2}$

12. $\dfrac{5}{x} = \dfrac{6}{x} - \dfrac{1}{3}$

13. $\dfrac{3}{4x} + \dfrac{5}{x} = 1$

14. $\dfrac{5}{3x} + \dfrac{3}{x} = 1$

15. $\dfrac{a - 2}{a + 3} = \dfrac{3}{8}$

16. $\dfrac{x - 7}{x + 2} = \dfrac{1}{4}$

17. $\dfrac{5}{x - 1} = \dfrac{3}{x + 2}$

18. $\dfrac{2}{x + 1} = \dfrac{1}{x - 2}$

19. $\dfrac{x}{8} - \dfrac{x}{12} = \dfrac{1}{8}$

20. $\dfrac{x}{6} - \dfrac{x}{10} = \dfrac{1}{6}$

21. $\dfrac{x + 2}{5} - 1 = \dfrac{x - 2}{4}$

22. $\dfrac{x + 1}{3} - 1 = \dfrac{x - 1}{2}$

23. $\dfrac{x-7}{x-9} = \dfrac{2}{x-9}$

24. $\dfrac{x-1}{x-5} = \dfrac{4}{x-5}$

25. $\dfrac{3}{x+4} = \dfrac{5}{x}$

26. $\dfrac{2}{x+3} = \dfrac{7}{x}$

27. $\dfrac{2b-3}{3b+2} = \dfrac{2b+1}{3b-2}$

28. $\dfrac{x-2}{x-3} = \dfrac{x-1}{x+1}$

29. $\dfrac{4}{x-3} + \dfrac{2x}{x^2-9} = \dfrac{1}{x+3}$

30. $\dfrac{x}{x+4} - \dfrac{4}{x-4} = \dfrac{x^2+16}{x^2-16}$

31. $\dfrac{5}{y-3} - \dfrac{30}{y^2-9} = 1$

32. $\dfrac{1}{x+3} + \dfrac{1}{x-3} = \dfrac{1}{x^2-9}$

33. $\dfrac{4}{8-a} = \dfrac{4-a}{a-8}$

34. $\dfrac{t+10}{7-t} = \dfrac{3}{t-7}$

SKILL MAINTENANCE

Simplify.

35. $\left(\dfrac{3}{5}\right)^{-2}$

36. $(-4)^{-3}$

37. $(a^2 b^5)^{-3}$

38. $(x^{-2} y^{-3})^{-4}$

39. $\dfrac{5}{x^2-9} - \dfrac{2}{x^2-5x+6}$

40. $\dfrac{2}{3x-12} + \dfrac{5}{x^2-16}$

SYNTHESIS

41. ◆ Explain the difference between adding rational expressions and solving rational equations.

42. ◆ Without multiplying by the LCD and solving, explain why the rational equation

$$\dfrac{x}{x+2} = \dfrac{-2}{x+2}$$

cannot have a solution. (*Hint*: Examine both numerators and denominators carefully.)

43. ◆ Why is it especially important to check the possible solutions of a rational equation?

44. ◆ How can a graph be used to determine how many solutions an equation has?

Solve.

45. $x + 1 + \dfrac{x-1}{x-3} = \dfrac{2}{x-3}$

46. $\dfrac{4}{y-2} - \dfrac{2y-3}{y^2-4} = \dfrac{5}{y+2}$

47. $\dfrac{x}{x^2+3x-4} + \dfrac{x+1}{x^2+6x+8} = \dfrac{2x}{x^2+x-2}$

48. $\dfrac{12-6x}{x^2-4} = \dfrac{3x}{x+2} - \dfrac{2x-3}{x-2}$

49. $\dfrac{x^2}{x^2-4} = \dfrac{x}{x+2} - \dfrac{2x}{2-x}$

50. $2 - \dfrac{a-2}{a+3} = \dfrac{a^2-4}{a+3}$

51. $\dfrac{1}{x-1} + x + 1 = \dfrac{5x-4}{x-1}$

52. $\dfrac{3a-5}{a^2+4a+3} + \dfrac{2a+2}{a+3} = \dfrac{a-3}{a+1}$

53. 📊 Use a grapher to check the solutions to Examples 1 and 2(c).

54. 📊 Use a grapher to check your answers to Exercises 9, 25, and 45.

6.7 Applications Using Rational Equations and Proportions

Problem Solving • Problems Involving Work • Problems Involving Motion • Problems Involving Proportions

In many areas of study, applications involving rates, proportions, or reciprocals translate to rational equations. By using the five steps for problem solving and the lessons of Section 6.6, we can now solve such problems.

Problem Solving

EXAMPLE 1 A number, plus three times its reciprocal, is -4. Find the number.

SOLUTION

1. **Familiarize.** Let's try to guess the number. Try 2: $2 + 3 \cdot \frac{1}{2} = \frac{7}{2}$. Although $\frac{7}{2} \neq -4$, the guess helps us to better understand how the problem can be translated. We let $x =$ the number.

2. **Translate.** From the *Familiarize* step, we can translate directly:

   A number, plus three times its reciprocal, is -4.

   $$x + 3 \cdot \frac{1}{x} = -4$$

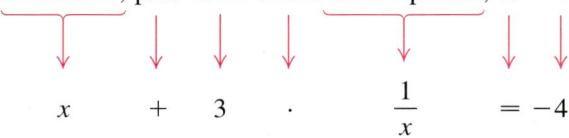

3. **Carry out.** We solve the equation:

$$x + 3 \cdot \frac{1}{x} = -4$$

$$x\left(x + \frac{3}{x}\right) = x(-4) \quad \text{Multiplying on both sides by the LCD, } x$$

$$x \cdot x + x \cdot \frac{3}{x} = -4x \quad \text{Using the distributive law}$$

$$x^2 + 3 = -4x \quad \text{Simplifying}$$

$$x^2 + 4x + 3 = 0$$

$$(x + 3)(x + 1) = 0 \quad \text{Using the principle of zero products}$$

$$x + 3 = 0 \quad \text{or} \quad x + 1 = 0$$

$$x = -3 \quad \text{or} \quad x = -1.$$

4. **Check.** Three times the reciprocal of -3 is $3 \cdot \frac{1}{-3}$, or -1. Since $-3 + (-1) = -4$, the number -3 is a solution.
   Three times the reciprocal of -1 is $3 \cdot \frac{1}{-1}$, or -3. Since $-1 + (-3) = -4$, the number -1 is also a solution.

5. **State.** The solutions are -3 and -1.

Problems Involving Work

EXAMPLE 2 *Sorting Recyclables.* Cecilia and Aaron work as volunteers at a town's recycling depot. Cecilia can sort a day's accumulation of recyclables in 4 hr, while Aaron requires 6 hr to do the same job. How long would it take them, working together, to sort the recyclables?

SOLUTION

1. **Familiarize.** We familiarize ourselves with the problem by exploring two common, but *incorrect,* ways of translating the problem to mathematical language.

6.7 APPLICATIONS USING RATIONAL EQUATIONS AND PROPORTIONS   **339**

a) One common incorrect approach is to simply add the two times:

$$4 \text{ hr} + 6 \text{ hr} = 10 \text{ hr}.$$

Let's think about this. If Cecilia can do the sorting *alone* in 4 hr, then Cecilia and Aaron *together* should take *less* than 4 hr. Thus we reject 10 hr as a solution and reason that the answer must be less than 4 hr.

b) Another incorrect way to translate the problem is to assume that Cecilia does half the sorting and Aaron does the other half. Then

Cecilia sorts $\frac{1}{2}$ of the accumulation in $\frac{1}{2}(4 \text{ hr})$, or 2 hr, and

Aaron sorts $\frac{1}{2}$ of the accumulation in $\frac{1}{2}(6 \text{ hr})$, or 3 hr.

This would waste time since Cecilia would finish 1 hr earlier than Aaron. In reality, Cecilia would help Aaron after completing her half, so that Aaron would actually sort less than half of the accumulation. This tells us that the entire job will take them between 2 hr and 3 hr.

A correct approach is to consider how much of the sorting is finished in 1 hr, 2 hr, 3 hr, and so on. It takes Cecilia 4 hr to sort the recyclables alone. Thus, in 1 hr, she can do $\frac{1}{4}$ of the job. It takes Aaron 6 hr to do the sorting alone. Thus, in 1 hr, he can do $\frac{1}{6}$ of the job. Working together, they can complete

$$\frac{1}{4} + \frac{1}{6}, \quad \text{or} \quad \frac{5}{12} \text{ of the sorting in 1 hr.}$$

In 2 hr, Cecilia can do $\frac{1}{4} \cdot 2$ of the sorting and Aaron can do $\frac{1}{6} \cdot 2$ of the sorting. Working together, they can complete

$$\frac{1}{4} \cdot 2 + \frac{1}{6} \cdot 2, \quad \text{or} \quad \frac{5}{6} \text{ of the sorting in 2 hr.}$$

Continuing this reasoning, we can form a table.

Time	Fraction of the Sorting Completed		
	Cecilia	**Aaron**	**Together**
1 hr	$\frac{1}{4}$	$\frac{1}{6}$	$\frac{1}{4} + \frac{1}{6}$, or $\frac{5}{12}$
2 hr	$\frac{1}{4} \cdot 2$	$\frac{1}{6} \cdot 2$	$\frac{1}{4} \cdot 2 + \frac{1}{6} \cdot 2$, or $\frac{5}{6}$
3 hr	$\frac{1}{4} \cdot 3$	$\frac{1}{6} \cdot 3$	$\frac{1}{4} \cdot 3 + \frac{1}{6} \cdot 3$, or $1\frac{1}{4}$
t hr	$\frac{1}{4} \cdot t$	$\frac{1}{6} \cdot t$	$\frac{1}{4} \cdot t + \frac{1}{6} \cdot t$

From the table, we see that if they work 3 hr, the fraction of the sorting that they complete is $1\frac{1}{4}$, which is more of the job than needs to be done. We need to find a number t for which the fraction of the sorting that is completed in t hours is 1; that is, the job is just completed—not more and not less.

2. **Translate.** From the table, we see that the time we want is some number t for which

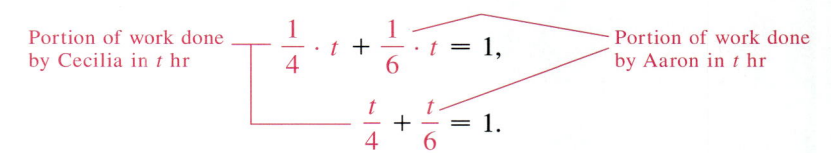

$$\underbrace{\frac{1}{4} \cdot t}_{\text{Portion of work done by Cecilia in } t \text{ hr}} + \underbrace{\frac{1}{6} \cdot t}_{\text{Portion of work done by Aaron in } t \text{ hr}} = 1,$$

or

$$\frac{t}{4} + \frac{t}{6} = 1.$$

3. **Carry out.** We solve the equation:

$$\frac{t}{4} + \frac{t}{6} = 1$$

$$12\left(\frac{t}{4} + \frac{t}{6}\right) = 12 \cdot 1 \qquad \text{The LCD is } 2 \cdot 2 \cdot 3, \text{ or } 12.$$

$$12 \cdot \frac{t}{4} + 12 \cdot \frac{t}{6} = 12 \qquad \text{Using the distributive law}$$

$$3t + 2t = 12 \qquad \text{Simplifying}$$

$$5t = 12$$

$$t = \frac{12}{5}, \text{ or } 2\frac{2}{5} \text{ hr}.$$

4. **Check.** The check can be done following the pattern used in the table of the *Familiarize* step above:

$$\frac{1}{4} \cdot \frac{12}{5} + \frac{1}{6} \cdot \frac{12}{5} = \frac{3}{5} + \frac{2}{5} = \frac{5}{5} = 1.$$

A second, partial, check is that (as we predicted in step 1) the answer is between 2 hr and 3 hr.

5. **State.** Together, it takes Cecilia and Aaron $2\frac{2}{5}$ hr to complete the sorting.

THE WORK PRINCIPLE

Suppose that $a =$ the time it takes A to complete a task, $b =$ the time it takes B to complete the same task, and $t =$ the time it takes them to complete the task working together. Then

$$\frac{1}{a} \cdot t + \frac{1}{b} \cdot t = 1, \text{ or } \frac{t}{a} + \frac{t}{b} = 1.$$

Problems Involving Motion

Problems that deal with distance, speed (or rate), and time are called **motion problems.** Translation of these problems involves the distance formula, $d = r \cdot t$, and/or the equivalent formulas $r = d/t$ and $t = d/r$.

6.7 APPLICATIONS USING RATIONAL EQUATIONS AND PROPORTIONS

EXAMPLE 3

Driving Speed. Claude drives 20 mph faster than his mother, Mae. In the same time that Claude travels 180 mi, his mother travels 120 mi. Find their speeds.

SOLUTION

1. **Familiarize.** Suppose that Mae drives 30 mph. Claude would then be driving at 30 + 20, or 50 mph. Thus, if r is the speed of Mae's car, in miles per hour, then the speed of Claude's car is $r + 20$.

   If Mae drove 30 mph, she would drive 120 mi in 120/30, or 4 hr. At 50 mph, Claude would drive 180 mi at 180/50, or $3\frac{3}{5}$ hr. Because we know that both drivers spend the same amount of time traveling, and because 4 hr $\neq 3\frac{3}{5}$ hr, we see that our guess of 30 mph is incorrect. We let $t =$ the time, in hours, that is spent traveling and create a table.

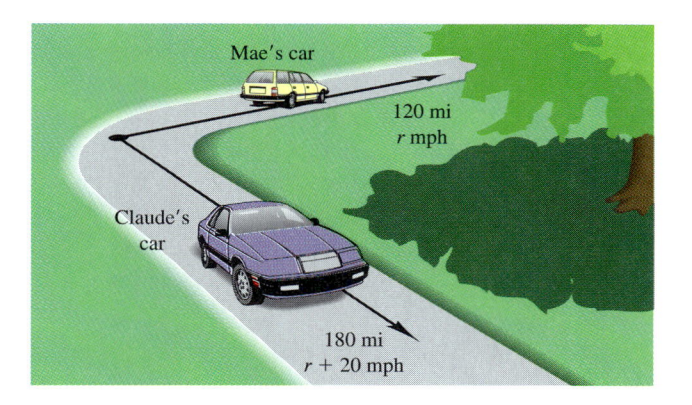

$$d = r \cdot t$$

	Distance	Speed	Time
Mae's Car	120	r	t
Claude's Car	180	$r + 20$	t

2. **Translate.** Examine how we checked our guess. We found, and then compared, the two driving times. The times were found by dividing the distances, 120 mi and 180 mi, by the rates, 30 mph and 50 mph, respectively. Thus the t's in the table above can be replaced, using the formula $t = d/r$. This yields a table that uses only one variable.

	Distance	Speed	Time
Mae's Car	120	r	120/r
Claude's Car	180	$r + 20$	180/($r + 20$)

← The times must be the same.

Since the times must be the same for both cars, we have the equation

$$\frac{120}{r} = \frac{180}{r + 20}.$$

3. **Carry out.** To solve the equation, we first multiply both sides by the LCD, $r(r + 20)$:

$$r(r + 20) \cdot \frac{120}{r} = r(r + 20) \cdot \frac{180}{r + 20} \quad \text{Multiplying on both sides by the LCD, } r(r + 20)$$

$$120(r + 20) = 180r \quad \text{Simplifying}$$

$$120r + 2400 = 180r \quad \text{Removing parentheses}$$

$$2400 = 60r \quad \text{Subtracting } 120r \text{ on both sides}$$

$$40 = r. \quad \text{Dividing on both sides by 60}$$

We now have a possible solution. The speed of Mae's car is 40 mph, and the speed of Claude's car is $40 + 20$, or 60 mph.

4. **Check.** We first reread the problem to see what we were to find. Note that if Claude drives 60 mph and Mae drives 40 mph, Claude is indeed going 20 mph faster than his mother. If Claude travels 180 mi at 60 mph, he drives for 180/60, or 3 hr. If Mae travels 120 mi at 40 mph, she drives for 120/40, or 3 hr. Since the times are the same, the speeds check.

5. **State.** Mae is driving at 40 mph, while Claude is driving at 60 mph.

Problems Involving Proportions

A **ratio** of two quantities is their quotient. For example, 37% is the ratio of 37 to 100, or $\frac{37}{100}$. A **proportion** is an equation stating that two ratios are equal.

PROPORTION

An equality of ratios, $A/B = C/D$, is called a *proportion*. The numbers within a proportion are said to be *proportional* to each other.

Proportions can be used to solve applied problems in which one ratio can be expressed in more than one way.

EXAMPLE 4

Mileage. A Ford Taurus travels 135 mi on 6 gal of gas. Find the amount of gas required for a 360-mi trip.

SOLUTION By assuming that the car always burns gas at the same rate, we can form a proportion in which the ratio of miles to gallons is expressed in two ways:

$$\begin{array}{c}\text{Miles} \rightarrow\\ \text{Gas} \rightarrow\end{array} \frac{135}{6} = \frac{360}{x} \begin{array}{c}\leftarrow \text{Miles}\\ \leftarrow \text{Gas}\end{array}.$$

6.7 APPLICATIONS USING RATIONAL EQUATIONS AND PROPORTIONS 343

To solve for x, we multiply both sides of the equation by the LCD, $6x$:

$$6x \cdot \frac{135}{6} = 6x \cdot \frac{360}{x}$$

$$6 \cdot \frac{135x}{6} = x \cdot \frac{6 \cdot 360}{x} \qquad \text{Removing factors equal to 1: } \frac{6}{6} = 1 \text{ and } \frac{x}{x} = 1$$

$$135x = 6 \cdot 360$$

$$x = \frac{6 \cdot 360}{135} \qquad \text{Dividing on both sides by 135}$$

$$x = 16. \qquad \text{Simplifying}$$

The trip will require 16 gal of gas.

Proportions arise in geometry when we are studying *similar triangles*. If two triangles are **similar**, then their corresponding angles have the same measure and their corresponding sides are proportional. To illustrate, if triangle ABC is similar to triangle RST, then angles A and R have the same measure, angles B and S have the same measure, angles C and T have the same measure, and

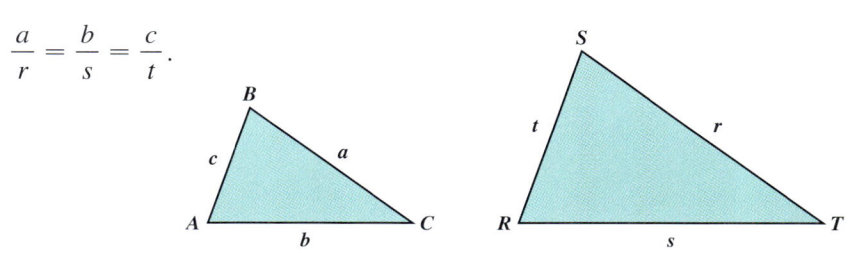

$$\frac{a}{r} = \frac{b}{s} = \frac{c}{t}.$$

EXAMPLE 5 *Similar Triangles.* Triangles ABC and XYZ are similar. Solve for z if $x = 10$, $a = 8$, and $c = 5$.

SOLUTION We make a sketch, write a proportion, and then solve. Note that side a is always opposite angle A, side x is always opposite angle X, and so on.

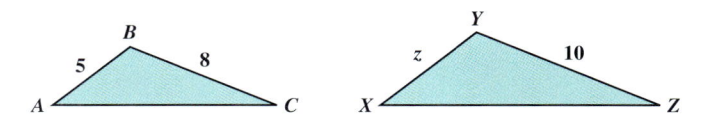

We have

$$\frac{z}{5} = \frac{10}{8} \qquad \text{The proportions } \frac{5}{z} = \frac{8}{10}, \frac{5}{8} = \frac{z}{10}, \text{ or } \frac{8}{5} = \frac{10}{z} \text{ could also be used.}$$

$$z = \frac{10}{8} \cdot 5 \qquad \text{Multiplying on both sides by 5}$$

$$z = \frac{50}{8}, \text{ or } 6.25.$$

EXAMPLE 6

Estimating Populations. To determine the number of fish in a lake, a park ranger catches 225 fish, tags them, and throws them back into the lake. Later, 108 fish are caught, and 15 are found to be tagged. Estimate how many fish are in the lake.

SOLUTION

1. **Familiarize.** If we knew that the 225 tagged fish constituted, say, 10% of the fish population, we could easily calculate the total fish population from the proportion

$$\frac{225}{F} = \frac{10}{100},$$

where F is the fish population. Unfortunately, we are *not* told the percentage of fish that were tagged. We must reread the problem, looking for numbers that could be used to approximate the percentage of the total fish population that was tagged.

2. **Translate.** Since 15 of 108 fish that were later caught had tags, we can use the ratio 15/108 to estimate the percentage of the total population that was originally tagged. Then we can translate to a proportion:

Fish tagged originally $\rightarrow \dfrac{225}{F} = \dfrac{15}{108}$ $\leftarrow$ Tagged fish caught later
Fish in lake $\rightarrow$ $\qquad\quad\;\;$ $\leftarrow$ Fish caught later

3. **Carry out.** To solve the proportion, we multiply by the LCD, 108F:

$$108F \cdot \frac{225}{F} = 108F \cdot \frac{15}{108} \qquad \text{Multiplying on both sides by } 108F$$

$$108 \cdot 225 = F \cdot 15$$

$$\frac{108 \cdot 225}{15} = F \quad \text{or} \quad F = 1620.$$

4. **Check.** We leave the check to the student.

5. **State.** There are about 1620 fish in the lake.

EXERCISE SET 6.7

Solve.

1. A number, minus 5 times its reciprocal, is 4. Find the number.

2. A number, minus 4 times its reciprocal, is 3. Find the number.

3. The sum of a number and its reciprocal is 2. Find the number.

4. The sum of a number and 5 times its reciprocal is 6. Find the number.

5. *Carpentry.* By checking work records, a carpenter finds that Juanita can build a small shed in 12 hr. Antoine can do the same job in 16 hr. How long would it take if they worked together?

6. *Construction.* It takes Oscar 5 hr to put up paneling in a room. Erika takes 4 hr to do the same

job. How long would it take them, working together, to cover the room with paneling?

7. *Shoveling.* Vern can shovel the snow from his driveway in 45 min. Nina can do the same job in 60 min. How long would it take Nina and Vern to shovel the driveway if they worked together?

8. *Raking.* Zoë can rake her yard in 4 hr. Steffi does the same job in 3 hr. How long would it take the two of them, working together, to rake the yard?

9. *Plumbing.* By checking work records, a plumber finds that Raul can plumb a house in 48 hr. Mira can do the same job in 36 hr. How long would it take if they worked together?

10. *Water intake.* A water tank can be filled in 12 hr by pipe A alone and in 9 hr by pipe B alone. How long would it take to fill the tank if both pipes were working?

11. *Harvesting.* Bobbi can pick a quart of raspberries in 20 min. Blanche can pick a quart in 25 min. How long would it take if Bobbi and Blanche worked together?

12. *Masonry.* By checking work records, a contractor finds that it takes Red Bryck 6 hr to construct a wall of a certain size. It takes Lotta Mudd 8 hr to construct the same wall. How long would it take if they worked together?

13. *Fax machines.* The Brother MFC4500® can fax a year-end report in 10 min while the Xerox 850® can fax the same report in 8 min. How long would it take the two machines, working together, to fax the report? (Assume that the recipient has at least two machines for incoming faxes.)

14. *Mowing.* Zack mows the backyard in 40 min, while Angela can mow the same yard in 50 min. How long would it take them, working together with two mowers, to mow the yard?

15. *Speed of travel.* A police cruiser is moving 40 mph faster than a freight train. In the time that it takes the train to travel 300 mi, the police car travels 700 mi. Find their speeds.

    Complete the tables as part of the familiarization. Do not use t's in the second table.

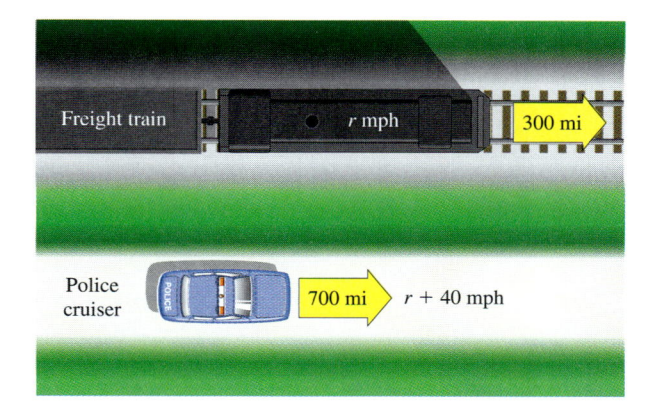

$$d = r \cdot t$$

	Distance	Speed	Time
Train	300	r	
Police Car	700		t

	Distance	Speed	Time
Train	300	r	$\dfrac{300}{r}$
Police Car	700		

16. *Driving speed.* Hillary's Lexus travels 30 mph faster than Bill's Harley. In the same time that Bill travels 75 mi, Hillary travels 120 mi. Find their speeds.

17. *Train speeds.* The speed of a freight train is 14 km/h slower than the speed of a passenger train.

The freight train travels 330 km in the same time that it takes the passenger train to travel 400 km. Find the speed of each train.

Complete the tables as part of the familiarization. Do not use t's in the second table.

| d | = | r | · | t |

	Distance	Speed	Time
Freight	330		t
Passenger	400	r	

	Distance	Speed	Time
Freight	330		
Passenger	400	r	

18. *Train speed.* The speed of a freight train is 15 km/h slower than the speed of a passenger train. The freight train travels 390 km in the same time that it takes the passenger train to travel 480 km. Find the speed of each train.

19. *Bicycle speed.* Joni bicycles 5 km/h faster than Ted. In the time that it takes Ted to bicycle 42 km, Joni can bicycle 57 km. How fast does each bicyclist travel?

20. *Snowmobile speed.* Ellie's snowmobile travels 5 km/h faster than Mark's. In the time that it takes Mark to travel 42 km, Ellie can travel 48 km. Find the speed of each snowmobile.

21. *Tractor speed.* Manley's tractor is just as fast as Caledonia's. It takes Manley 1 hr more than it takes Caledonia to drive to town. If Manley is 20 mi from town and Caledonia is 15 mi from town, how long does it take Caledonia to drive to town?

22. *Boat speed.* Tory and Emilio's motorboats both travel at the same speed. Tory pilots her boat 40 km before docking. Emilio continues for another 2 hr, traveling a total of 100 km before docking. How long did it take Tory to navigate the 40 km?

Find the ratio of the following. Simplify, if possible.

23. 750 students, 50 faculty
24. 800 mi, 50 gal

Solve.

25. *Speed of sound.* Sound travels 5440 ft in 5 sec. What is its speed in feet per second?

26. *Speed of a snake.* A black racer snake travels 4.6 km in 2 hr. What is its speed in kilometers per hour?

Geometry. *When three parallel lines are crossed by two or more lines (transversals), the lengths of corresponding segments are proportional (see the following figure).*

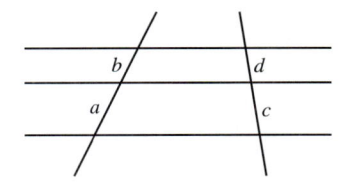

27. If a is 8 cm when b is 5 cm, find d when c is 6 cm.
28. If d is 7 cm when c is 10 cm, find a when b is 9 cm.
29. If c is 2 m longer than b and d is 2 m shorter than b, find all four lengths when a is 15 m.
30. If d is 2 m shorter than b and c is 3 m longer than b, find b, c, and d when a is 18 m.
31. *Coffee harvest.* The coffee beans from 14 trees are needed to produce 7.7 kg of coffee. (This is the average amount that each person in the United States consumes each year.) How many trees are needed to produce 320 kg of coffee?
32. *Walking speed.* Wanda walked 234 km in 14 days. At this rate, how far would she walk in 42 days?
33. *Baking.* In a potato bread recipe, the ratio of milk to flour is $\frac{3}{13}$. If 5 cups of milk are used, how many cups of flour are used?
34. *Hemoglobin.* A normal 10-cc specimen of human blood contains 1.2 g of hemoglobin. How much hemoglobin would 16 cc of the same blood contain?

Similar triangles. *For each pair of similar triangles, find the value of the indicated letter.*

35. *b*

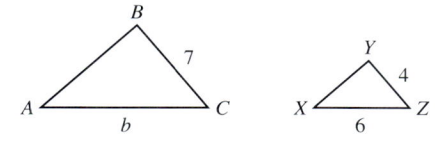

36. *a*

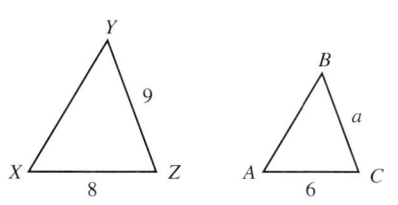

37. *f*

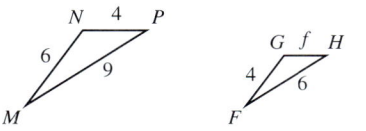

38. *r*

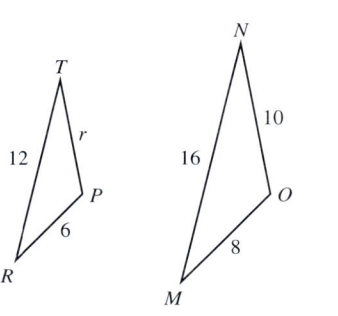

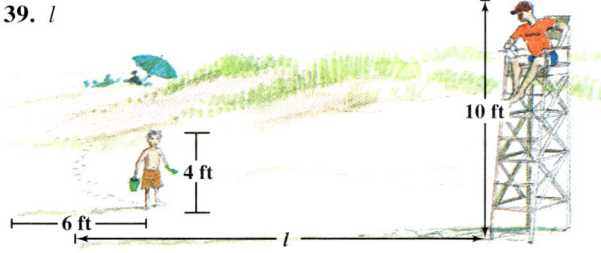

39. *l*

40. *h*

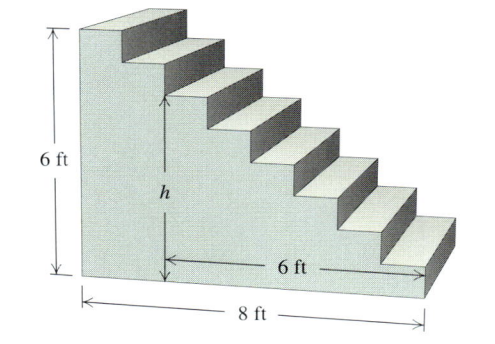

Solve.

41. *Fish population.* To determine the number of trout in a lake, a naturalist catches 112 trout, tags them, and throws them back into the lake. Later, 82 trout are caught; 32 of them have tags. Estimate the number of trout in the lake.

42. *Deer population.* To determine the number of deer in a game preserve, a game warden catches 318 deer, tags them, and lets them loose. Later, 168 deer are caught; 56 of them have tags. Estimate the number of deer in the preserve.

43. *Moose population.* To determine the size of a park's moose population, naturalists catch 69 moose, tag them, and then set them free. Months later, 40 moose are caught, of which 15 have tags. Estimate the size of the moose population.

44. *Light bulbs.* A sample of 184 light bulbs contained 6 defective bulbs. How many defective bulbs would you expect in a sample of 1288 bulbs?

45. *Miles driven.* Emmanuel is allowed to drive his leased car for 45,000 mi in 4 yr without penalty. In the first $1\frac{1}{2}$ yr, Emmanuel has driven 16,000 mi. At this rate will he exceed the mileage allowed for 4 yr?

46. *Firecrackers.* A sample of 144 firecrackers contained 9 "duds." How many duds would you expect in a sample of 320 firecrackers?

47. *Deer population.* To determine the number of deer in a park, a naturalist catches, tags, and then releases 25 deer. Later, 36 deer are caught; 4 of them have tags. Estimate the deer population of the park.

48. *Weight on the moon.* The ratio of the weight of an object on the moon to the weight of that object on Earth is 0.16 to 1.
    a) How much would a 12-ton rocket weigh on the moon?
    b) How much would a 180-lb astronaut weigh on the moon?

49. *Weight on Mars.* The ratio of the weight of an object on Mars to the weight of that object on Earth is 0.4 to 1.
    a) How much would a 12-ton rocket weigh on Mars?
    b) How much would a 120-lb astronaut weigh on Mars?

50. Simplest fractional notation for a rational number is $\frac{9}{17}$. Find an equivalent ratio where the sum of the numerator and the denominator is 104.

SKILL MAINTENANCE

Subtract.

51. $(x + 2) - (x + 1)$

52. $(x^2 + x) - (x + 1)$
53. $(4y^3 - 5y^2 + 7y - 24) - (-9y^3 + 9y^2 - 5y + 49)$
54. The perimeter of a rectangle is 642 ft. The length is 15 ft greater than the width. Find the area of the rectangle.
55. Solve $ar + st = n$ for t.
56. Solve $uv - av = m$ for v.

SYNTHESIS

57. ◈ Is it correct to assume that two workers will complete a task twice as quickly as one person working alone? Why or why not?
58. ◈ Write a problem similar to Example 2 for a classmate to solve. Design the problem so that the translation step is
$$\frac{t}{7} + \frac{t}{5} = 1.$$
59. ◈ Write a problem similar to Example 3 for a classmate to solve. Design the problem so that the translation step is
$$\frac{30}{r + 4} = \frac{18}{r}.$$
60. ◈ If two triangles are exactly the same shape and size, are they similar? Why or why not?
61. *Sewing a quilt.* Ann and Betty work together and sew a quilt in 4 hr. Working alone, Betty would need 6 hr more than Ann to sew a quilt. How long would it take each of them working alone?
62. *Boating.* The speed of a boat in still water is 10 mph. It travels 24 mi upstream and 24 mi downstream in a total time of 5 hr. What is the speed of the current?
63. How soon after 5 o'clock will the hands on a clock first be together?
64. Given that
$$\frac{A}{B} = \frac{C}{D},$$
write three other proportions using A, B, C, and D.
65. *Programming.* Rosina, Ng, and Oscar can write a computer program in 3 days. Rosina can write the program in 8 days and Ng can do it in 10 days. How many days will it take Oscar to write the program?
66. *Waxing.* Together, Michelle, Sal, and Kristen can wax a car in 1 hr 20 min. To complete the job alone, Michelle needs twice the time that Sal needs and 2 hr more than Kristen. How long would it take each to wax the car working alone?
67. *Wiring.* Janet can wire a house in 28 hr. Linus can wire a house in 34 hr. How long will it take Janet and Linus, working together, to wire *two* houses?
68. *Distances.* The shadow from a 40-ft cliff just reaches across a water-filled quarry at the same time that a 6-ft tall diver casts a 10-ft shadow. How wide is the quarry?

69. *Elections.* Melanie beat her opponent for the presidency of the student senate by a 3-to-2 ratio. If 450 votes were cast, how many votes did Melanie receive?
70. *Commuting.* To reach an appointment 50 mi away, Dr. Wright allowed 1 hr. After driving 30 mi, she realized that her speed would have to be increased 15 mph for the remainder of the trip. What was her speed for the first 30 mi?
71. ◈ Are the equations
$$\frac{A + B}{B} = \frac{C + D}{D} \quad \text{and} \quad \frac{A}{B} = \frac{C}{D}$$
equivalent? Why or why not?
72. ◈ If two triangles are similar, are their areas and perimeters proportional? Why or why not?

COLLABORATIVE CORNER

Focus: Modeling, estimation, and work problems

Time: 20 minutes

Group size: 3

Materials: A watch or clock to measure time in seconds, 3 pieces of paper with 5-in. by 5-in. squares already drawn.

Many tasks can be done by two people working together. If both people work at the same rate, each does half the task, and the project is completed in half the time. If, however, the work rates differ, the faster worker performs more of the task than the slower worker.

ACTIVITY

1. Each group should begin with 3 pieces of paper, each with a 5-in. by 5-in. square already drawn.
2. The project is to color in a square. To do this, one group member will work as quickly as possible, another group member will work slowly and carefully, and the other group member will serve as timekeeper. Beginning at opposite corners of the same square, the two workers should color in the square while the third group member records their time. The time required should be recorded in the appropriate box of the table below.
3. Next, one of the workers should be timed coloring in another square, working alone and at the same rate as in part (2). Record that time in the table. The group should then estimate how long it will take the second worker, working at the same rate as in part (2), to color in the remaining square.
4. The second worker, working at the same rate as in part (2), should be timed coloring in the third square. Record this time and compare it with the estimate from part (3). How far off was the group's estimate?
5. Let t_1, t_2, and t_3 represent the times required for the first worker, the second worker, and the two workers together, respectively, to complete a task. Then develop a model that can be used to find t_2 when t_1 and t_3 are known.
6. Compare the actual experimental time from part (4) with the time predicted by the model in part (5). List reasons that might account for any discrepancy.

Time Required Working Together	Time Required for One of the Workers, Working Alone	Estimated Time for the Other Worker, Working Alone	Actual Time Required for the Other Worker, Working Alone

6.8 Formulas

Formulas from Applications • Tips for Solving Formulas

Formulas arise frequently in the natural and social sciences, business, and engineering. When a formula takes the form of a rational equation, we can use our equation-solving techniques to solve for any specified letter.

EXAMPLE 1 *Intelligence Quotient.* The formula $Q = \dfrac{100m}{c}$ is used to determine the intelligence quotient, Q, of a person of mental age m and chronological age c. Solve for c.

SOLUTION We have

$$Q = \frac{100m}{c}$$

$$c \cdot Q = c \cdot \frac{100m}{c} \qquad \text{Multiplying on both sides by } c$$

$$cQ = 100m \qquad \text{Removing a factor equal to 1: } \frac{c}{c} = 1$$

$$c = \frac{100m}{Q}. \qquad \text{Dividing on both sides by } Q$$

This formula can be used to determine a person's chronological, or actual, age from his or her mental age and intelligence quotient.

The procedure we follow is similar to the one used in Section 2.3.

TO SOLVE A FORMULA FOR A GIVEN LETTER:

1. Identify the letter for which you are solving and multiply on both sides to clear fractions or decimals, if necessary. Multiply, as needed, to remove parentheses.
2. Make sure that all terms with the letter for which you are solving are on one side of the equation and all other terms are on the other side. Use the addition principle as needed.
3. Factor out the letter for which you are solving if it appears in more than one term.
4. Multiply or divide to solve for the letter in question.

EXAMPLE 2 *A Work Formula.* The formula $t/a + t/b = 1$ was used in Section 6.7. Solve this formula for t.

SOLUTION

$$\frac{t}{a} + \frac{t}{b} = 1$$

$$ab\left(\frac{t}{a} + \frac{t}{b}\right) = ab \cdot 1 \quad \text{Multiplying by the LCD, } ab, \text{ to clear fractions}$$

$$\left.\begin{array}{c}\dfrac{abt}{a} + \dfrac{abt}{b} = ab \\ bt + at = ab\end{array}\right\} \quad \text{Multiplying to remove parentheses and removing factors equal to 1: } \dfrac{a}{a} = 1 \text{ and } \dfrac{b}{b} = 1$$

$$(b + a)t = ab \quad \text{Factoring out } t, \text{ the letter for which we are solving}$$

$$t = \frac{ab}{b + a} \quad \text{Dividing on both sides by } b + a$$

The answer to Example 2 can be used to find solutions to problems such as Example 2 in Section 6.7:

$$t = \frac{4 \cdot 6}{6 + 4} = \frac{24}{10} = 2\tfrac{2}{5}. \quad \text{This matches the solution found on p. 340.}$$

EXAMPLE 3   *The Area of a Trapezoid.*   The area A of a trapezoid is half the product of the height h and the sum of the lengths a and b of the parallel sides:

$$A = \tfrac{1}{2}(a + b)h.$$

Solve for b.

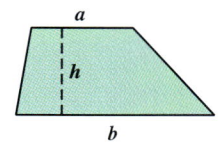

SOLUTION

To isolate b on the right-hand side of $A = \tfrac{1}{2}(a + b)h$, we can first isolate $a + b$:

$$A = \frac{1}{2}(a + b)h$$

$$A = \frac{h}{2}(a + b) \quad \tfrac{1}{2} \cdot h = \tfrac{h}{2}$$

$$\frac{2}{h}A = a + b \quad \text{Multiplying on both sides by } \frac{2}{h} \text{ to isolate } a + b.$$

$$\frac{2}{h}A - a = b. \quad \text{Using the addition principle: adding } -a \text{ on both sides}$$

In Examples 1 and 2, the letter for which we solved finished on the left side of the equation. In Example 3, the letter finished on the right. The location of the letter is unimportant, since all equations are reversible.

Recall from Section 2.3 that the variable to be solved for must be alone on one side of the equation, with *no* occurrence of that variable on the other side.

EXAMPLE 4 *Young's Rule.* Young's rule for determining the size of a particular child's medicine dosage c is

$$c = \frac{a}{a+12} \cdot d,$$

where a is the child's age and d is the usual adult's dosage.* Solve for a.

SOLUTION We have

$$c = \frac{a}{a+12} \cdot d$$

$(a+12)c = (a+12)\dfrac{a}{a+12} \cdot d$ Multiplying by $a+12$ to clear the fraction

$ac + 12c = ad$ Simplifying

> **CAUTION!** If we next divide by d, we will not isolate a since a would still appear on both sides of the equation.

$12c = ad - ac$ Subtracting ac to isolate all a-terms on one side

$12c = a(d-c)$ Factoring out a since it is in more than one term

$\dfrac{12c}{d-c} = a.$ Dividing by $d-c$

EXERCISE SET 6.8

Solve.

1. $S = 2\pi rh$, for h
2. $A = P(1 + rt)$, for t
   (An interest formula)
3. $A = \frac{1}{2}bh$, for h
   (The area of a triangle) 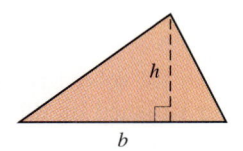
4. $s = \frac{1}{2}gt^2$, for g
   (A physics formula for distance)

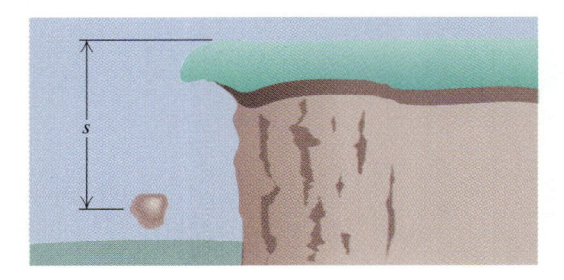

---

*Olsen, June Looby, Leon J. Ablon, and Anthony Patrick Giangrasso, *Medical Dosage Calculations*, 6th ed. (Redwood City, CA: Addison-Wesley), p. A-31.

5. $\dfrac{1}{180} = \dfrac{n-2}{s}$, for n **6.** $S = \dfrac{n}{2}(a+l)$, for a

7. $V = \frac{1}{3}k(B + b - 2n)$, for b

8. $A = P + Prt$, for P
(An interest problem)

9. $rl - rS = L$, for r

10. $T = mg - mf$, for m

11. $A = \frac{1}{2}h(b_1 + b_2)$, for h

12. $S = 2\pi r(r + h)$, for h
(The surface area of a right circular cylinder)

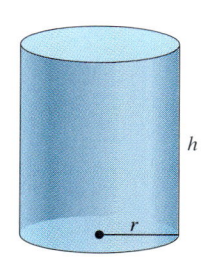

13. $ab = ac + d$, for a **14.** $mn + p = np$, for n

15. $\dfrac{m}{r} = s$, for r **16.** $\dfrac{V}{m} = d$, for m

17. $a + b = \dfrac{c}{d}$, for d **18.** $\dfrac{m}{n} = p - q$, for n

19. $I = \dfrac{V_1 - V_2}{R}$, for R
(An electricity formula)

20. $\dfrac{M - g}{t} = r + s$, for t **21.** $\dfrac{1}{p} + \dfrac{1}{q} = \dfrac{1}{f}$, for f
(An optics formula)

22. $\dfrac{1}{R} = \dfrac{1}{r_1} + \dfrac{1}{r_2}$, for R
(An electricity formula)

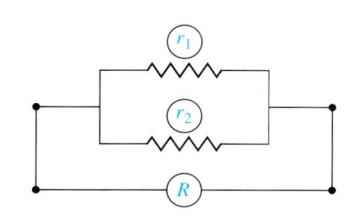

23. $a = \dfrac{v_2 - v_1}{t}$, for t **24.** $r = \dfrac{v^2 pL}{a}$, for p
(A physics formula)

25. $P = 2(l + w)$, for l **26.** $\dfrac{a}{c} = n + bn$, for n

27. $ab - ac = \dfrac{Q}{M}$, for a **28.** $S = \dfrac{a + 2b}{3b}$, for b

29. $C = \dfrac{Ka - b}{a}$, for a

30. $C = \frac{5}{9}(F - 32)$, for F
(A temperature conversion formula)

31. $V = \frac{4}{3}\pi r^3$, for r^3
(The volume of a sphere)

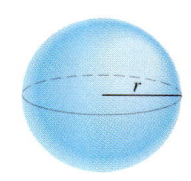

32. $f = \dfrac{gm - t}{m}$, for g
(A physics formula)

33. $S = \dfrac{rl - a}{r - l}$, for r **34.** $f = \dfrac{gm - t}{m}$, for m

SKILL MAINTENANCE

35. Graph: $y = \frac{4}{5}x + 1$.

36. Graph: $y = -\frac{2}{3}x + 1$.

37. Solve: $-\frac{3}{5}x = \frac{9}{20}$.

38. Find the intercepts and graph: $3x + 4y = 24$.

39. Factor: $x^2 - 13x - 30$.

40. Subtract: $(5x^3 - 7x^2 + 9) - (8x^3 - 2x^2 + 4)$.

SYNTHESIS

41. ◆ Is it easier to solve
$$\dfrac{1}{25} + \dfrac{1}{23} = \dfrac{1}{x} \text{ for } x,$$
or to solve
$$\dfrac{1}{p} + \dfrac{1}{q} = \dfrac{1}{f} \text{ for } f?$$
Explain why.

42. ◆ Explain why someone might want to solve
$A = \frac{1}{2}bh$ for b.
(See Exercise 3.)

43. ◆ As a step in solving a formula for a certain variable, a student takes the reciprocal on both sides of the equation. Is a mistake being made? Why or why not?

44. ◆ Describe a situation in which the result of Example 2,
$$t = \dfrac{ab}{b + a},$$
would be especially useful.

Solve.

45. $\dfrac{n_1}{p_1} + \dfrac{n_2}{p_2} = \dfrac{n_2 - n_1}{R}$, for n_2

46. $u = -F\left(E - \dfrac{P}{T}\right)$, for T

47. The formula
$$N = \dfrac{(b + d)f_1 - v}{(b - v)f_2}$$
is used when monitoring the water in fisheries. Solve for v.

48. The formula
$$C = \tfrac{5}{9}(F - 32)$$
is used to convert the Fahrenheit temperature F to the Celsius temperature C. At what temperature are the Fahrenheit and Celsius readings the same?

COLLABORATIVE CORNER

Focus: Developing a formula

Time: 15–20 minutes

Group size: 3

Materials: Each student should bring a Nutrition Facts panel from a food that has some fat content.

The nutrition label found on most food products contains a great deal of information. Unfortunately, the percentage of a food's calories that come from carbohydrate, protein, or fat is *not* listed on these labels. In the activity that follows, you will develop formulas for each of these percentages.

ACTIVITY

1. Each student should bring to class a Nutrition Facts panel from a food that he or she enjoys. The food should contain at least a small amount of fat.
2. Each group should form a table with headings as shown below. Then fill in the table with the per serving information from the Nutrition Facts panels.
3. Use the information in the table to estimate how many calories are contained in one gram of fat. Decide, as a group, the whole number to which this should be rounded.
4. The "calories not from fat" come from carbohydrates and proteins that contain nearly equal numbers of calories per gram. Use the information above to find a whole-number estimate for the number of calories per gram of carbohydrate or protein.
5. Suppose that a food serving contains c grams of carbohydrate, f grams of fat, and p grams of protein. Each group member should use c, f, and p, along with information in parts (3) and (4), to develop a formula for approximating one of the following:

   a) The percentage of that food's calories coming from carbohydrate (call it P_c);
   b) The percentage of that food's calories coming from fat (call it P_f);
   c) The percentage of that food's calories coming from protein (call it P_p).

6. Group members should check each other's work (does $P_c + P_f + P_p = 100\%$?) and then compare formulas with other groups. Why might answers vary?

Total Fat (in grams)	Total Carbohydrate (in grams)	Total Protein (in grams)	Calories from Fat	Calories Not from Fat

SUMMARY AND REVIEW 6

KEY TERMS

Rational expression, p. 298
Least Common Multiple, LCM, p. 312
Least Common Denominator, LCD, p. 312
Complex rational expression, p. 327
Rational equation, p. 332

Motion problem, p. 340
Ratio, p. 342
Proportion, p. 342
Similar triangles, p. 343

IMPORTANT PROPERTIES AND FORMULAS

To add, subtract, multiply, and divide rational expressions:

$$\frac{A}{B} \cdot \frac{C}{D} = \frac{AC}{BD}; \quad \frac{A}{B} \div \frac{C}{D} = \frac{A}{B} \cdot \frac{D}{C} = \frac{AD}{BC};$$

$$\frac{A}{B} + \frac{C}{B} = \frac{A+C}{B}; \quad \frac{A}{B} - \frac{C}{B} = \frac{A-C}{B}.$$

To find the least common denominator (LCD):

1. Write the prime factorization of each denominator.
2. Select one of the factorizations and inspect it to see if it contains the other.
   a) If it does, it represents the LCM of the denominators.
   b) If it does not, multiply that factorization by any factors of the other denominator that it lacks. The final product is the LCM of the denominators.

The LCD is the LCM of the denominators. It contains each factor the greatest number of times that it occurs in any one factorization.

To add or subtract rational expressions that have different denominators:

1. Find the LCD.
2. Multiply each rational expression by a form of 1 made up of the factors of the LCD missing from that expression's denominator.
3. Add or subtract the numerators, as indicated. Write the sum or difference over the LCD.
4. Simplify, if possible.

To simplify a complex rational expression by multiplying by the LCD:

1. Find the LCD of all rational expressions *within* the complex rational expression.
2. Multiply the complex rational expression by a form of 1, using the LCD to write 1.
3. Distribute and simplify. No fractional expressions should remain within the complex rational expression.
4. Factor and simplify, if possible.

(continued)

To simplify a complex rational expression by dividing:

1. Add or subtract, as needed, to get a single rational expression in the numerator.
2. Add or subtract, as needed, to get a single rational expression in the denominator.
3. Divide the numerator by the denominator (invert and multiply).
4. If possible, simplify by removing a factor equal to 1.

To solve a rational equation:

1. List any restrictions that exist. No solution can make a denominator equal 0.
2. Clear the equation of fractions by multiplying both sides by the LCD of all rational expressions in the equation.
3. Solve the resulting equation using the addition principle, the multiplication principle, and the principle of zero products, as needed.
4. Check the possible solution(s) in the original equation.

The Work Principle

Suppose that a = the time it takes A to complete a task, b = the time it takes B to complete the same task, and t = the time it takes them to complete the task working together. Then

$$\frac{1}{a} \cdot t + \frac{1}{b} \cdot t = 1, \quad \text{or} \quad \frac{t}{a} + \frac{t}{b} = 1.$$

To solve a formula for a given letter:

1. Identify the letter for which you are solving and multiply on both sides to clear fractions or decimals, if necessary. Multiply, as needed, to remove parentheses.
2. Make sure that all terms with the letter for which you are solving are on one side of the equation and all other terms are on the other side. Use the addition principle as needed.
3. Factor out the letter for which you are solving if it appears in more than one term.
4. Multiply or divide to solve for the letter in question.

REVIEW EXERCISES

List all numbers for which each expression is undefined.

1. $\dfrac{35}{-x^2}$

2. $\dfrac{4}{x-6}$

3. $\dfrac{x+5}{x^2-36}$

4. $\dfrac{x^2-3x+2}{x^2+x-30}$

5. $\dfrac{-4}{(x+2)^2}$

Simplify.

6. $\dfrac{4x^2-8x}{4x^2+4x}$

7. $\dfrac{14x^2-x-3}{2x^2-7x+3}$

8. $\dfrac{(y-5)^2}{y^2-25}$

9. $\dfrac{5x^2-20y^2}{2y-x}$

Multiply or divide and simplify, if possible.

10. $\dfrac{a^2-36}{10a} \cdot \dfrac{2a}{a+6}$

11. $\dfrac{6t-6}{2t^2+t-1} \cdot \dfrac{t^2-1}{t^2-2t+1}$

12. $\dfrac{10-5t}{3} \div \dfrac{t-2}{12t}$

13. $\dfrac{4x^4}{x^2-1} \div \dfrac{2x^3}{x^2-2x+1}$

14. $\dfrac{x^2+1}{x-2} \cdot \dfrac{2x+1}{x+1}$

15. $(t^2+3t-4) \div \dfrac{t^2-1}{t+4}$

Find the LCM.
16. $8a^2b^7$, $6a^5b^3$
17. $x^2 - x$, $x^5 - x^3$, x^4
18. $y^2 - y - 2$, $y^2 - 4$

Add or subtract and simplify, if possible.
19. $\dfrac{x + 8}{x + 7} + \dfrac{10 - 4x}{x + 7}$
20. $\dfrac{3}{3x - 9} + \dfrac{x - 2}{3 - x}$
21. $\dfrac{6x - 3}{x^2 - x - 12} - \dfrac{2x - 15}{x^2 - x - 12}$
22. $\dfrac{3x - 1}{2x} - \dfrac{x - 3}{x}$
23. $\dfrac{x + 3}{x - 2} - \dfrac{x}{2 - x}$
24. $\dfrac{2a}{a + 1} - \dfrac{4a}{1 - a^2}$
25. $\dfrac{d^2}{d - c} + \dfrac{c^2}{c - d}$
26. $\dfrac{1}{x^2 - 25} - \dfrac{x - 5}{x^2 - 4x - 5}$
27. $\dfrac{3x}{x + 2} - \dfrac{x}{x - 2} + \dfrac{8}{x^2 - 4}$
28. $\dfrac{2}{5x} + \dfrac{3}{2x + 4}$

Simplify.
29. $\dfrac{\dfrac{1}{z} + 1}{\dfrac{1}{z^2} - 1}$
30. $\dfrac{2 + \dfrac{1}{xy^2}}{\dfrac{1 + x}{x^4 y}}$
31. $\dfrac{\dfrac{c}{d} - \dfrac{d}{c}}{\dfrac{1}{c} + \dfrac{1}{d}}$

Solve.
32. $\dfrac{3}{y} - \dfrac{1}{4} = \dfrac{1}{y}$
33. $\dfrac{5}{x + 3} = \dfrac{3}{x + 2}$
34. $\dfrac{15}{x} - \dfrac{15}{x + 2} = 2$

35. Rhetta can polish a helicopter rotor blade in 9 hr. Jason can do the same job in 12 hr. How long would it take if they worked together?

36. The distance by highway between Richmond and Waterbury is 70 km, and the distance by rail is 60 km. A car and a train leave Richmond at the same time and arrive in Waterbury at the same time, the car having traveled 15 km/h faster than the train. Find the speed of the car and the speed of the train.

37. The reciprocal of 1 more than a number is twice the reciprocal of the number itself. What is the number?

38. A sample of 25 doorknobs contained 2 defective doorknobs. How many defective doorknobs would you expect among 1000 doorknobs?

39. Triangles ABC and XYZ are similar. Find the value of x.

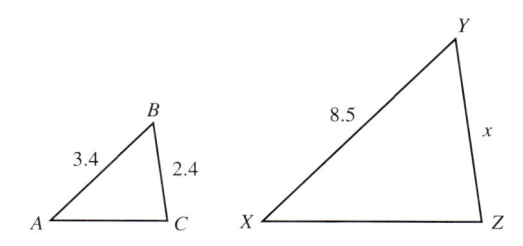

Solve.
40. $\dfrac{1}{r} + \dfrac{1}{s} = \dfrac{1}{t}$, for s
41. $F = \dfrac{9C + 160}{5}$, for C

SKILL MAINTENANCE

42. Solve: $-3 + x = 8$.
43. Find the intercepts and graph: $2x - y = 6$.
44. Factor: $x^2 + 8x - 48$.
45. Subtract:
$(5x^3 - 4x^2 + 3x - 4) - (7x^3 - 7x^2 - 9x + 14)$.

SYNTHESIS

46. ◆ Why is factoring an important skill to master before beginning a study of rational equations?
47. ◆ A student insists on finding a common denominator by always multiplying the denominators of the expressions being added. How could this approach be improved?

Simplify.
48. $\dfrac{2a^2 + 5a - 3}{a^2} \cdot \dfrac{5a^3 + 30a^2}{2a^2 + 7a - 4} \div \dfrac{a^2 + 6a}{a^2 + 7a + 12}$
49. $\dfrac{12a}{(a - b)(b - c)} - \dfrac{2a}{(b - a)(c - b)}$

CHAPTER TEST 6

List all numbers for which each expression is undefined.

1. $\dfrac{8-x}{3x}$
2. $\dfrac{5}{x+8}$
3. $\dfrac{x-7}{x^2-49}$
4. $\dfrac{x^2+x-30}{x^2-3x+2}$

5. Simplify: $\dfrac{6x^2+17x+7}{2x^2+7x+3}$.

Multiply or divide and simplify, if possible.

6. $\dfrac{a^2-25}{9a} \cdot \dfrac{6a}{5-a}$
7. $\dfrac{25y^2-1}{9y^2-6y} \div \dfrac{5y^2+9y-2}{3y^2+y-2}$
8. $\dfrac{4x^2-1}{x^2-2x+1} \div \dfrac{x-2}{x^2+1}$
9. $(x^2+6x+9) \cdot \dfrac{(x-3)^2}{x^2-9}$

10. Find the LCM:
$y^2-9, \; y^2+10y+21, \; y^2+4y-21$.

Add or subtract. Simplify, if possible.

11. $\dfrac{16+x}{x^3} + \dfrac{7-4x}{x^3}$
12. $\dfrac{5-t}{t^2+1} - \dfrac{t-3}{t^2+1}$
13. $\dfrac{x-4}{x-3} + \dfrac{x-1}{3-x}$
14. $\dfrac{x-4}{x-3} - \dfrac{x-1}{3-x}$
15. $\dfrac{5}{t-1} + \dfrac{3}{t}$
16. $\dfrac{1}{x^2-16} - \dfrac{x+4}{x^2-3x-4}$
17. $\dfrac{1}{x-1} + \dfrac{4}{x^2-1} - \dfrac{2}{x^2-2x+1}$

Simplify.

18. $\dfrac{9-\dfrac{1}{y^2}}{3-\dfrac{1}{y}}$
19. $\dfrac{\dfrac{3}{a^2b}-\dfrac{2}{ab^3}}{\dfrac{1}{ab}+\dfrac{2}{a^4b}}$

Solve.

20. $\dfrac{7}{y} - \dfrac{1}{3} = \dfrac{1}{4}$
21. $\dfrac{15}{x} - \dfrac{15}{x-2} = -2$

22. Kopy Kwik has 2 copiers. One can copy a year-end report in 20 min. The other can copy the same document in 30 min. How long would it take both machines, working together, to copy the report?

23. A recipe for pizza crust calls for $3\tfrac{1}{2}$ cups of whole wheat flour and $1\tfrac{1}{4}$ cups of warm water. If 6 cups of whole wheat flour are used, how much water should be used?

24. One car travels 20 km/h faster than another. In the same time that one goes 225 km, the other goes 325 km. Find the speed of each car.

25. Solve $d = rt + wt$ for t.

SKILL MAINTENANCE

26. Solve: $-3y = \dfrac{9}{7}$.
27. Find the intercepts and graph: $2x + 5y = 20$.
28. Factor: $x^2 - 4x - 45$.
29. Subtract:
$(5x^2 - 19x + 34) - (-8x^2 + 10x - 42)$.

SYNTHESIS

30. Reggie and Rema work together to mulch the flower beds around an office complex in $2\tfrac{6}{7}$ hr. Working alone, it would take Reggie 6 hr more than it would take Rema. How long would it take each of them to complete the landscaping working alone?

31. Simplify: $1 + \dfrac{1}{1+\dfrac{1}{1+\dfrac{1}{a}}}$.

CUMULATIVE REVIEW 1–6

1. Use the commutative law of addition to write an expression equivalent to $a + 2b$.
2. Write a true sentence using either $<$ or $>$:
   -3.1 ___ -3.15.
3. Evaluate $(y - 1)^2$ for $y = -6$.
4. Simplify: $-4[2(x - 3) - 1]$.

Simplify.

5. $-\frac{1}{2} + \frac{3}{8} + (-6) + \frac{3}{4}$
6. $-\frac{72}{108} \div \left(-\frac{2}{3}\right)$
7. $-6.262 \div 1.01$
8. $4 \div (-2) \cdot 2 + 3 \cdot 4$

Solve.

9. $3(x - 2) = 24$
10. $49 = x^2$
11. $-4x = -18$
12. $5x + 7 = -3x - 9$
13. $4(y - 5) = -2(y + 2)$
14. $x^2 + 11x + 10 = 0$
15. $\frac{1}{3}x - \frac{2}{9} = \frac{2}{3} + \frac{4}{9}x$
16. $\frac{4}{x} + x = 5$
17. $3 - y \geq 2y + 5$
18. $\frac{2}{x - 3} = \frac{5}{3x + 1}$
19. $2x^2 + 7x = 4$
20. $4(x + 7) < 5(x - 3)$
21. $\frac{x^2}{x + 2} = \frac{4}{x + 2}$
22. $(2x + 7)(x - 5) = 0$
23. $\frac{2}{x^2 - 9} + \frac{5}{x - 3} = \frac{3}{x + 3}$

Solve each formula.

24. $A = \frac{4b}{t}$, for t
25. $\frac{1}{t} = \frac{1}{m} - \frac{1}{n}$, for n
26. $r = \frac{a - b}{c}$, for c

Combine like terms.

27. $x + 2y - 2z + \frac{1}{2}x - z$
28. $2x^3 - 7 + \frac{3}{7}x^2 - 6x^3 - \frac{4}{7}x^2 + 5$

Graph.

29. $y = 1 - \frac{1}{2}x$
30. $x = -3$
31. $x - 6y = 6$

32. Find the slope of the line containing the points $(1, 5)$ and $(2, 3)$.

Simplify.

33. $\frac{x^{-5}}{x^{-3}}$
34. $y^2 \cdot y^{-10}$
35. $-(2a^2b^7)^2$
36. Subtract:
    $(-8y^2 - y + 2) - (y^3 - 6y^2 + y - 5)$.

Multiply.

37. $4(3x + 4y + z)$
38. $(2x^2 - 1)(x^3 + x - 3)$
39. $(6x - 5y)^2$
40. $(x + 3)(2x - 7)$
41. $(2x^3 + 1)(2x^3 - 1)$

Factor.

42. $6x - 2x^2 - 24x^4$
43. $16x^2 - 81$
44. $x^2 - 10x + 24$
45. $8x^2 + 10x + 3$
46. $6x^2 - 28x + 16$
47. $2x^2 - 18$
48. $16x^2 + 40x + 25$
49. $3x^2 + 10x - 8$
50. $x^4 + 2x^3 - 3x - 6$

Simplify.

51. $\frac{y^2 - 36}{2y + 8} \cdot \frac{y + 4}{y + 6}$
52. $\frac{x^2 - 1}{x^2 - x - 2} \div \frac{x - 1}{x - 2}$
53. $\frac{5ab}{a^2 - b^2} + \frac{a + b}{a - b}$
54. $\frac{x + 2}{4 - x} - \frac{x + 3}{x - 4}$
55. $\dfrac{1 + \dfrac{2}{x}}{1 - \dfrac{4}{x^2}}$
56. $\dfrac{\dfrac{1}{t} + 2t}{t - \dfrac{2}{t^2}}$

Divide.

57. $\dfrac{15x^4 - 12x^3 + 6x^2 + 2x + 18}{3x^2}$
58. $(15x^4 - 12x^3 + 6x^2 + 2x + 18) \div (x + 3)$

Solve.

59. Linnae has $36 budgeted for stationery. Engraved stationery costs $20 for the first 25 sheets and $0.08 for each additional sheet. How many sheets of stationery can Linnae order and still stay within her budget?

60. The price of a box of cereal increased 15% to $4.14. What was the price of the cereal before the increase?

61. If the sides of a square are increased by 2 ft, the area of the original square plus the area of the enlarged square is 452 ft². Find the length of a side of the original square.

62. The sum of two consecutive even integers is −554. Find the integers.

63. It takes Vikki 75 min to shovel 4 in. of snow from her driveway. It takes Jerri 50 min to do the same job. How long would it take if they worked together?

64. Millie's car travels 10 km/h faster than Harley's. In the same time that Harley drives 120 km, Millie travels 150 km. Find the speed of each car.

65. A 78-in. board is to be cut into two pieces. One piece must be twice as long as the other. How long should the shorter piece be?

SYNTHESIS

66. Simplify: $(x + 7)(x - 4) - (x + 8)(x - 5)$.

67. Solve: $\frac{1}{3}|n| + 8 = 56$.

68. Multiply: $[4y^3 - (y^2 - 3)][4y^3 + (y^2 - 3)]$.

69. Factor: $2a^{32} - 13{,}122b^{40}$.

70. Solve: $x(x^2 + 3x - 28) - 12(x^2 + 3x - 28) = 0$.

71. Simplify: $-\left|0.875 - \left(-\frac{1}{8}\right) - 8\right|$.

CHAPTER 7

More with Graphing

7.1 Slope–Intercept Form
7.2 Point–Slope Form
7.3 Linear Inequalities in Two Variables
7.4 Direct and Inverse Variation
SUMMARY AND REVIEW
TEST

AN APPLICATION

The formula

$$T = -\frac{3}{4}a + 165$$

can be used to determine the target heart rate of a person, a years old, participating in aerobic exercise. Graph the equation and interpret the significance of its slope.

THIS PROBLEM APPEARS AS EXERCISE 80 IN SECTION 7.1.

More information on aerobic activity is available at
http://hepg.awl.com/be/elem_5

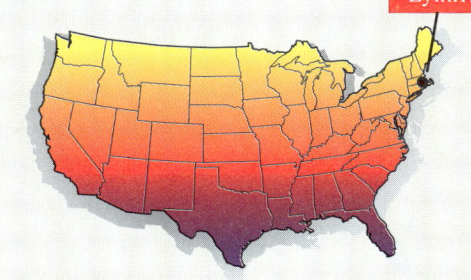

As an athletic trainer for a professional sports team, I use math daily to calculate players' target heart rates. It's crucial in setting up appropriate training protocols to ensure optimal performance.

DON DELNIGRO
Athletic Trainer
Lynnfield, Massachusetts

CHAPTER 7 • MORE WITH GRAPHING

The basics of graphing were introduced in Chapter 3. There we learned to graph equations by using tables or intercepts. We also saw how the notion of rate and slope can be related to a graph. In this chapter, we will see how slope can be related to the equation of a line. This will lead to faster and easier ways of graphing certain equations. Greater skill with graphing will prove useful later in the chapter when we consider linear inequalities and variation.

7.1 Slope–Intercept Form

Using the *y*-intercept and the Slope to Graph a Line • Equations in Slope–Intercept Form • Graphing and Slope–Intercept Form

If we know the slope of a line and the point at which the *y*-axis is crossed, it is possible to draw a graph of the line. In this section, we will learn how to find a line's slope and *y*-intercept from its equation. We can then graph certain equations quite easily.

Using the *y*-intercept and the Slope to Graph a Line

Let's modify the car production situation that first appeared in Section 3.5. Suppose that as a new workshift begins, 4 cars have already been produced. At the Michigan plant, 3 cars were being produced every 2 hours, a rate of $\frac{3}{2}$ cars per hour. If this rate remains the same regardless of how many cars have already been produced, the table and graph shown here can be made.

| Michigan Plant ||
Hours Elapsed	Cars Produced
0	4
2	7
4	10
6	13
8	16

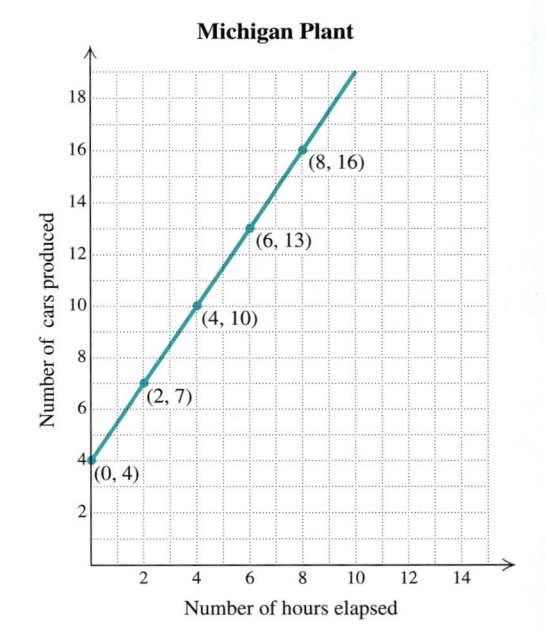

To confirm that the production rate is still $\frac{3}{2}$, we calculate the slope. Recall that

$$\text{Slope} = \frac{\text{change in } y}{\text{change in } x} = \frac{\text{rise}}{\text{run}} = \frac{y_2 - y_1}{x_2 - x_1},$$

where (x_1, y_1) and (x_2, y_2) are any two points on the graphed line. Here we select (0, 4) and (2, 7):

$$\text{Slope} = \frac{\text{change in } y}{\text{change in } x} = \frac{7 - 4}{2 - 0} = \frac{3}{2}.$$

Knowing that the slope is $\frac{3}{2}$, we could have drawn the graph by plotting (0, 4) and from there moving *up* 3 units and *to the right* 2 units. This would have located the point (2, 7). Using (0, 4) and (2, 7), we then could have drawn the line. This is the method used in the following example.

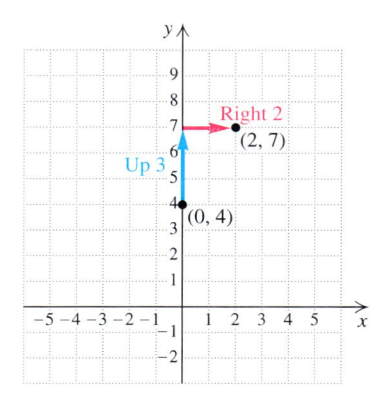

EXAMPLE 1   Draw a line that has slope $\frac{1}{4}$ and y-intercept (0, 2).

SOLUTION   We plot (0, 2) and from there move *up* 1 unit and *to the right* 4 units. This locates the point (4, 3). We plot (4, 3) and draw a line passing through (0, 2) and (4, 3), as shown on the right below.

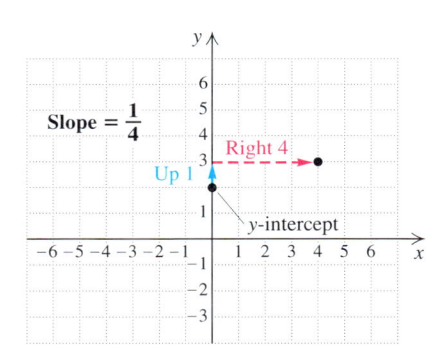

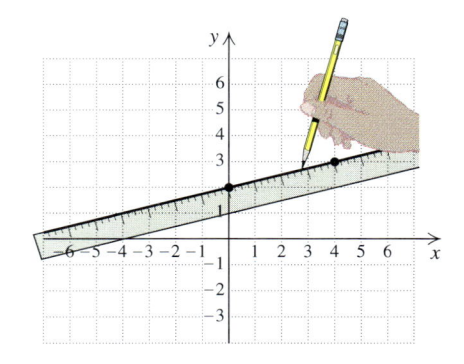

Equations in Slope–Intercept Form

It is not difficult to find a line's slope and y-intercept from its equation. Recall from Section 3.3 that to find the y-intercept of an equation's graph, we replace x with 0 and solve the resulting equation for y. For example, to find the y-intercept of the graph of $y = 2x + 3$, we replace x with 0 and solve as follows:

$$y = 2 \cdot 0 + 3 = 0 + 3 = 3.$$

364 CHAPTER 7 • MORE WITH GRAPHING

The y-intercept of the graph of $y = 2x + 3$ is $(0, 3)$. It can be similarly shown that the graph of $y = mx + b$ has the y-intercept $(0, b)$.

To calculate the slope of the graph of $y = 2x + 3$, we need two ordered pairs that are solutions of the equation. The y-intercept $(0, 3)$ is one pair; a second pair, $(1, 5)$, can be found easily by substituting 1 for x. We then have

$$\text{Slope} = \frac{\text{change in } y}{\text{change in } x} = \frac{5 - 3}{1 - 0} = \frac{2}{1} = 2.$$

Note that the slope, 2, is also the x-coefficient in $y = 2x + 3$. It can be similarly shown that the graph of any equation of the form $y = mx + b$ has slope m (see Exercise 77).

THE SLOPE–INTERCEPT EQUATION

The equation $y = mx + b$ is called the *slope–intercept equation.* The equation represents a line of slope m with y-intercept $(0, b)$.

The equation of any nonvertical line can be written in this form.

EXAMPLE 2 Find the slope and the y-intercept of each line.

a) $y = \frac{4}{5}x - 8$ b) $2x + y = 5$ c) $3x + 4y = 7$

SOLUTION

a) We rewrite $y = \frac{4}{5}x - 8$ as $y = \frac{4}{5}x + (-8)$. Now we simply read the slope and the y-intercept from the equation:

$$y = \tfrac{4}{5}x + (-8).$$

The slope is $\frac{4}{5}$. The y-intercept is $(0, -8)$.

b) We first solve for y to find an equivalent equation in the form $y = mx + b$:

$$2x + y = 5$$
$$y = -2x + 5. \quad \text{Adding } -2x \text{ on both sides}$$

The slope is -2. The y-intercept is $(0, 5)$.

c) We rewrite the equation in the form $y = mx + b$:

$$3x + 4y = 7$$
$$4y = -3x + 7 \quad \text{Adding } -3x \text{ on both sides}$$
$$y = \tfrac{1}{4}(-3x + 7) \quad \text{Multiplying by } \tfrac{1}{4} \text{ on both sides}$$
$$y = -\tfrac{3}{4}x + \tfrac{7}{4}. \quad \text{Using the distributive law}$$

The slope is $-\frac{3}{4}$. The y-intercept is $\left(0, \frac{7}{4}\right)$.

EXAMPLE 3 A line has slope $-\frac{12}{5}$ and y-intercept $(0, 11)$. Find an equation of the line.

SOLUTION We use the slope–intercept equation, substituting $-\frac{12}{5}$ for m and 11 for b:
$$y = mx + b = -\frac{12}{5}x + 11.$$
The desired equation is $y = -\frac{12}{5}x + 11$.

EXAMPLE 4 Determine an equation for the graph of car production shown at the beginning of this section.

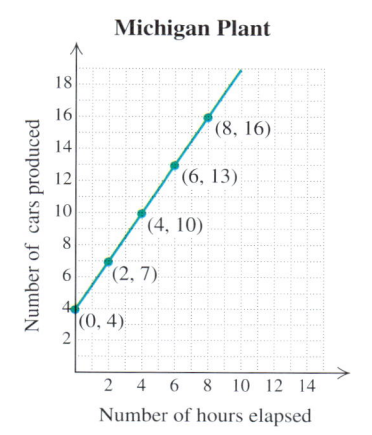

Michigan Plant

Number of cars produced vs. Number of hours elapsed, with points $(0, 4)$, $(2, 7)$, $(4, 10)$, $(6, 13)$, $(8, 16)$.

SOLUTION To write an equation for a line, we can use slope–intercept form, provided the slope and the y-intercept are known. Using the coordinates of two points, we already found that the slope, or rate of production, is $\frac{3}{2}$. Since $(0, 4)$ is given, we know the y-intercept as well. The desired equation is
$$y = \frac{3}{2}x + 4,$$
where y is the number of cars produced after x hours.

Graphing and Slope–Intercept Form

Our work in Examples 1 and 2 can be combined.

EXAMPLE 5 Graph: **(a)** $y = \frac{2}{5}x + 4$; **(b)** $2x + 3y = 3$.

SOLUTION

a) From the equation $y = \frac{2}{5}x + 4$, we see that the slope of the graph is $\frac{2}{5}$ and the y-intercept is $(0, 4)$. We plot $(0, 4)$ and then consider the slope, $\frac{2}{5}$. Starting at $(0, 4)$, we plot a second point by moving *up* 2 units (since the numerator is *positive* and corresponds to the change in y) and *to the right* 5 units (since the denominator is *positive* and corresponds to the change in x). We reach a new point, $(5, 6)$.

We can also rewrite the slope as $\frac{-2}{-5}$. We again start at the y-intercept, $(0, 4)$, but move *down* 2 units (since the numerator is *negative* and corresponds to the change in *y*) and *to the left* 5 units (since the denominator is *negative* and corresponds to the change in *x*). We reach another point, $(-5, 2)$. Once two or three points have been plotted, the line representing all solutions of $y = \frac{2}{5}x + 4$ can be drawn.

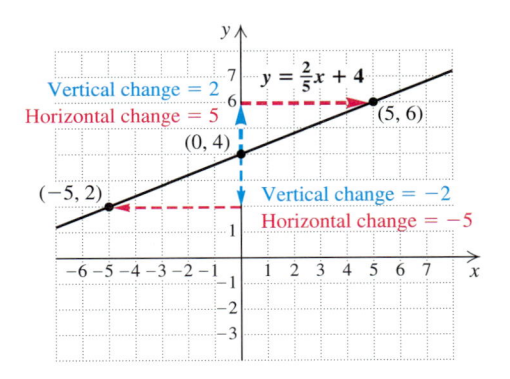

b) To graph $2x + 3y = 3$, we first rewrite it in slope–intercept form:

$2x + 3y = 3$

$\quad 3y = -2x + 3$ Adding $-2x$ on both sides

$\quad\ \ y = \frac{1}{3}(-2x + 3)$ Multiplying by $\frac{1}{3}$ on both sides

$\quad\ \ y = -\frac{2}{3}x + 1.$ Using the distributive law

To graph $y = -\frac{2}{3}x + 1$, we first plot the *y*-intercept, $(0, 1)$. We can think of the slope as $\frac{-2}{3}$. Starting at $(0, 1)$ and using the slope, we find a second point by moving *down* 2 units (since the numerator is *negative*) and *to the right* 3 units (since the denominator is *positive*). We plot the new point, $(3, -1)$. In a similar manner, we can move from the point $(3, -1)$ to locate a third point, $(6, -3)$. The line can then be drawn.

If the slope is thought of as $\frac{2}{-3}$, we can again start at $(0, 1)$, but this time move *up* 2 units (since the numerator is *positive*) and *to the left* 3 units (since the denominator is *negative*). We get another point on the graph, $(-3, 3)$.

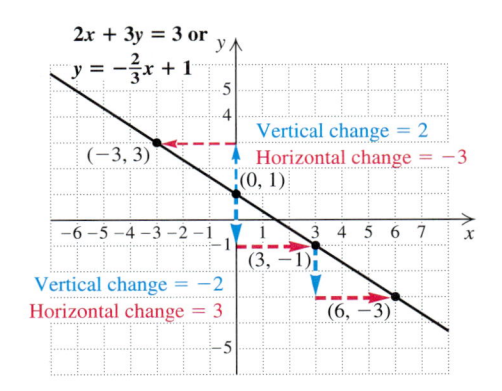

Slope–intercept form allows us to quickly determine the slope of a line by simply inspecting its equation. This can be especially helpful when attempting to decide whether two lines are parallel.

EXAMPLE 6 Determine whether the graphs of $y = -3x + 4$ and $6x + 2y = -10$ are parallel.

SOLUTION Recall that *parallel* lines extend indefinitely without intersecting. Thus, when two lines have the same slope but different y-intercepts, they are parallel.

One of the two equations given,

$$y = -3x + 4,$$

represents a line with slope -3 and y-intercept $(0, 4)$. To find the slope of the other line, we need to rewrite

$$6x + 2y = -10$$

in slope–intercept form:

$$6x + 2y = -10$$
$$2y = -6x - 10$$
$$y = -3x - 5. \quad \text{The slope is } -3 \text{ and the } y\text{-intercept is } (0, -5).$$

Since both lines have slope -3 but different y-intercepts, the graphs are parallel. There is no need for us to actually graph either equation.

TECHNOLOGY CONNECTION 7.1

Using a standard $[-10, 10, -10, 10]$ window, graph the equations $y_1 = \frac{2}{3}x + 1$, $y_2 = \frac{3}{8}x + 1$, $y_3 = \frac{2}{3}x + 5$, and $y_4 = \frac{3}{8}x + 5$. If you can, use your grapher in the MODE that will graph equations *simultaneously*. Once the four lines have been drawn, try to decide which equation corresponds to each line. After matching equations with lines, you can check your matches by using TRACE and the up and down arrow keys to move from one line to the next. The number of the equation will appear in a corner of the screen.

1. Graph the equations $y_1 = -\frac{3}{4}x - 2$, $y_2 = -\frac{1}{5}x - 2$, $y_3 = -\frac{3}{4}x - 5$, and $y_4 = -\frac{1}{5}x - 5$ using the SIMULTANEOUS mode. Then match each line with the corresponding equation. Check using TRACE.

EXERCISE SET 7.1

Draw a line that has the given slope and y-intercept.

1. Slope $\frac{3}{5}$; y-intercept $(0, 2)$
2. Slope $\frac{2}{5}$; y-intercept $(0, -1)$
3. Slope $\frac{5}{3}$; y-intercept $(0, -3)$
4. Slope $\frac{5}{2}$; y-intercept $(0, 1)$
5. Slope $-\frac{3}{4}$; y-intercept $(0, 5)$
6. Slope $-\frac{4}{5}$; y-intercept $(0, 6)$
7. Slope 2; y-intercept $(0, -4)$
8. Slope -2; y-intercept $(0, -3)$
9. Slope -3; y-intercept $(0, 2)$
10. Slope 3; y-intercept $(0, 4)$

Find the slope and the y-intercept of each line.

11. $y = \frac{3}{7}x + 6$
12. $y = -\frac{3}{8}x + 7$
13. $y = -\frac{5}{6}x + 2$
14. $y = \frac{7}{2}x + 4$
15. $y = \frac{9}{4}x - 7$
16. $y = \frac{2}{9}x - 1$
17. $y = -\frac{2}{5}x$
18. $y = \frac{4}{3}x$
19. $-2x + y = 4$
20. $-5x + y = 5$
21. $3x - 4y = 12$
22. $3x - 2y = 18$
23. $x - 5y = -8$
24. $x - 6y = 9$
25. $y = 4$
26. $y - 3 = 5$

Find the slope–intercept equation for the line with the indicated slope and y-intercept.

27. Slope 3; y-intercept (0, 7)
28. Slope -4; y-intercept $(0, -2)$
29. Slope $\frac{7}{8}$; y-intercept $(0, -1)$
30. Slope $\frac{5}{7}$; y-intercept $(0, 4)$
31. Slope $-\frac{5}{3}$; y-intercept $(0, -8)$
32. Slope $\frac{3}{4}$; y-intercept $(0, 23)$
33. Slope -2; y-intercept $(0, 3)$
34. Slope 7; y-intercept $(0, -6)$

Graph.

35. $y = \frac{3}{5}x + 2$
36. $y = -\frac{3}{5}x - 1$
37. $y = -\frac{3}{5}x + 1$
38. $y = \frac{3}{5}x - 2$
39. $y = \frac{5}{3}x + 3$
40. $y = \frac{5}{3}x - 2$
41. $y = -\frac{3}{2}x - 2$
42. $y = -\frac{4}{3}x + 3$
43. $2x + y = 1$
44. $3x + y = 2$
45. $3x - y = 4$
46. $2x - y = 5$
47. $2x + 3y = 9$
48. $4x + 5y = 15$
49. $x - 4y = 12$
50. $x + 5y = 20$

Solve.

51. *Cost of water.* Freda and Phil are keeping track of their water bills. One month, while they were away on vacation, they used no water and were billed $9. The next month they used 70,000 gal and their bill was for $16. Let y represent the size of a monthly bill and x represent the amount of water used (in 10,000-gal units). Find and graph an equation of the form $y = mx + b$. Then determine the rate that they pay in dollars per 10,000 gallons.

52. *Cost of cable TV.* Allegra and Larry are keeping track of how much their cable TV service costs. When service began, they paid $50 for installation. After one month, their total costs had risen to $70, and after two months they had paid a total of $90 for their cable service. Let y represent the amount paid for x months of service. Find and graph an equation of the form $y = mx + b$. Then determine their monthly rate.

53. *Refrigerator size.* Kitchen designers recommend that a refrigerator be selected on the basis of the number of people in the household. For 1–2 people, a 16-ft^3 model is suggested. For each additional person, an additional 1.5 ft^3 is recommended. If x is the number of residents over 2, find the slope–intercept equation for the recommended size of a refrigerator.

54. *Internet service.* In December of 1996, America Online introduced a plan in which a customer's monthly fee is $4.95 plus $2.50 for each hour of use beyond the first three. If x is the number of hours beyond three, find the slope–intercept equation for the monthly bill.

Determine whether each pair of equations represents parallel lines.

55. $y = \frac{2}{3}x + 7$,
    $y = \frac{2}{3}x - 5$

56. $y = -\frac{5}{4}x + 1$,
    $y = \frac{5}{4}x + 3$

57. $y = 2x - 5$,
    $4x + 2y = 9$

58. $y = -3x + 1$,
    $6x + 2y = 8$

59. $3x + 4y = 8$,
    $7 - 12y = 9x$

60. $3x = 5y - 2$,
    $10y = 4 - 6x$

SKILL MAINTENANCE

61. Solve: $2x^2 + 6x = 0$.
62. Factor: $x^3 + 5x^2 - 14x$.
63. The product of two consecutive odd integers is 195. Find the integers.
64. Eleven less than the square of a number is 10 times the number. Find the number.
65. Solve: $3x - 4(9 - x) = 17$.
66. Solve: $2(5 + 2y) + 4y = 13$.

SYNTHESIS

67. ◆ Can horizontal lines be graphed using the method of Example 5? Why or why not?
68. ◆ Can vertical lines be graphed using the method of Example 5? Why or why not?
69. ◆ Explain how it is possible for an incorrect graph to be drawn, even after plotting three points that line up.

70. ◆ What would you prefer, and why: graphing an equation of the form $y = mx + b$ or graphing an equation of the form $Ax + By = C$?

Two lines are perpendicular *if either the product of their slopes is* -1, *or one line is vertical and the other horizontal. For Exercises 71–76, determine whether each pair of equations represents perpendicular lines.*

71. $3y = 5x - 3$,
    $3x + 5y = 10$

72. $y + 3x = 10$,
    $2x - 6y = 18$

73. $3x + 5y = 10$,
    $15x + 9y = 18$

74. $10 - 4y = 7x$,
    $7y + 21 = 4x$

75. $x = 5$,
    $y = \frac{1}{2}$

76. $y = -2x$,
    $x = \frac{1}{2}$

77. Show that the slope of the line given by $y = mx + b$ is m. (*Hint*: Substitute both 0 and 1 for x to find two pairs of coordinates. Then use the formula Slope = change in y/change in x.)

78. Find an equation of the line with the same slope as the line given by $3x - 2y = 8$ and the same y-intercept as the line given by $2y + 3x = -4$.

79. Find an equation of the line perpendicular to the line given by $2x + 5y = 6$ (see Exercises 71–76) that passes through (2, 6). (*Hint*: Draw a graph.)

80. ◆ *Aerobic exercise.* The formula $T = -\frac{3}{4}a + 165$ can be used to determine the *target heart rate*, in beats per minute, of a person, a years old, participating in aerobic exercise. Graph the equation and interpret the significance of its slope.

COLLABORATIVE C•O•R•N•E•R

Focus: Slope–intercept form

Time: 15–20 minutes

Group size: 3

Materials: Graph paper and straightedges

It is important not only to be able to graph equations written in slope–intercept form, but to be able to match a linear graph with an appropriate equation.

ACTIVITY

1. Each group member should select a different one of the following sets of equations:

   A. $y = \frac{4}{3}x - 5$,
   $y = \frac{4}{3}x + 2$,
   $y = \frac{1}{2}x - 5$,
   $y = \frac{1}{2}x + 2$;

   B. $y = \frac{2}{5}x + 1$,
   $y = \frac{3}{4}x + 1$,
   $y = \frac{2}{5}x - 1$,
   $y = \frac{3}{4}x - 1$;

   C. $y = \frac{3}{5}x + 2$,
   $y = \frac{3}{5}x - 2$,
   $y = \frac{4}{3}x + 2$,
   $y = \frac{4}{3}x - 2$.

2. Working independently, each group member should graph the four equations he or she has selected. Do not label the graphs with their corresponding equations, but instead list the four equations (in any random sequence) across the top of the graph paper.

3. After all group members have completed part (2), the sheets should be passed, clockwise, to the person on the left. This person should then attempt to match each of the four equations listed at the top of the graph paper with the appropriate graph below. If no graph appears to be appropriate, discuss the relevant equation with the group member who drew the graphs. If necessary, turn to the third group member for guidance in identifying any incorrect graphs.

4. Once all four equations and graphs have been matched, share your answers with the rest of the group. Make sure everyone agrees on all of the matches.

7.2 Point–Slope Form

Writing Equations in Point–Slope Form • Graphing and Point–Slope Form

We now learn how to write an equation of a line using the line's slope and any one point through which the line passes.

Writing Equations in Point–Slope Form

Consider a line with slope 2 passing through the point (4, 1), as shown in the figure. In order for a point (x, y) to be on the line, the coordinates x and y must be solutions of the slope equation

$$\frac{y - 1}{x - 4} = 2.$$

Multiplying on both sides by $x - 4$, we have

$$y - 1 = 2(x - 4).$$

This is considered **point–slope form** for the line shown at right.

To generalize, a line with slope m passing through the point (x_1, y_1) will include a point (x, y) if x and y are solutions of the slope equation

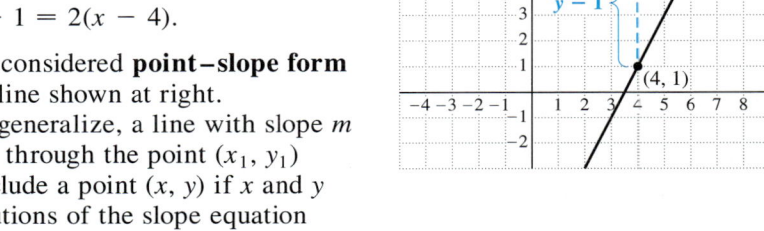

$$\frac{y - y_1}{x - x_1} = m.$$

Multiplying on both sides by $x - x_1$ gives us the *point–slope equation*

$$y - y_1 = m(x - x_1).$$

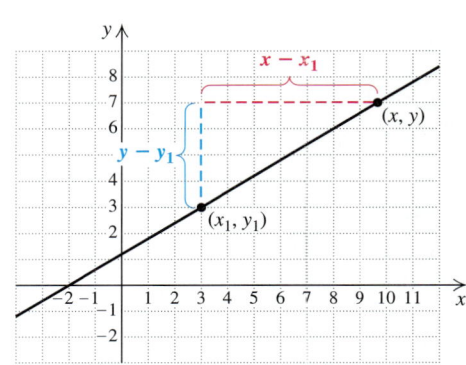

THE POINT–SLOPE EQUATION

The equation $y - y_1 = m(x - x_1)$ is called the *point–slope equation* for the line with slope m that contains the point (x_1, y_1).

Point–slope form is especially useful in more advanced mathematics courses, where problems similar to the following often arise.

EXAMPLE 1   Find a point–slope equation for the line with slope $\frac{1}{5}$ that contains the point $(-2, -3)$.

SOLUTION   We substitute $\frac{1}{5}$ for m, -2 for x_1, and -3 for y_1:

$$y - y_1 = m(x - x_1) \quad \text{Using the point–slope equation}$$
$$y - (-3) = \tfrac{1}{5}(x - (-2)). \quad \text{Substituting}$$

EXAMPLE 2   Find a slope–intercept equation for the line with slope 3 that contains the point $(1, 7)$.

SOLUTION   There are two parts to this solution. First we write an equation in point–slope form:

$$y - y_1 = m(x - x_1)$$
$$y - 7 = 3(x - 1). \quad \text{Substituting}$$

Next, we find an equivalent equation of the form $y = mx + b$:

$$y - 7 = 3(x - 1)$$
$$y - 7 = 3x - 3 \quad \text{Using the distributive law}$$
$$y = 3x + 4. \quad \text{Adding 7 on both sides to get slope–intercept form}$$

EXAMPLE 3   A line passes through the points $(3, -5)$ and $(-4, 9)$. Find an equation for the line **(a)** in point–slope form and **(b)** in slope–intercept form.

SOLUTION

a) To find a point–slope equation, we first compute the slope:

$$m = \frac{9 - (-5)}{-4 - 3} = \frac{14}{-7} = -2.$$

Next, we use the point–slope equation and substitute, using -2 for m and either $(3, -5)$ or $(-4, 9)$ for (x_1, y_1):

$$y - y_1 = m(x - x_1)$$
$$y - (-5) = -2(x - 3). \quad \text{Using } (3, -5) \text{ for } (x_1, y_1)$$

b) To write an equation in slope–intercept form, we use the result of part (a) above:

$$y - (-5) = -2(x - 3)$$
$$y + 5 = -2x + 6 \quad \text{Simplifying the subtraction and using the distributive law}$$
$$y = -2x + 1. \quad \text{Subtracting 5 on both sides to get slope–intercept form}$$

Had we used $(-4, 9)$ as (x_1, y_1) in part (a), we would have obtained the same slope–intercept equation:

$y - 9 = -2(x - (-4))$ Using $(-4, 9)$ for (x_1, y_1)
$y - 9 = -2(x + 4)$
$y - 9 = -2x - 8$
$y = -2x + 1.$ There is only one slope–intercept equation for any given linear equation.

Graphing and Point–Slope Form

We can graph equations written in point–slope form by reading the slope and a point on the graph from the equation.

EXAMPLE 4   Graph: $y - 2 = 3(x - 4)$.

SOLUTION   Since $y - 2 = 3(x - 4)$ is in point–slope form, we know that the line has slope 3, or $\frac{3}{1}$, and passes through the point $(4, 2)$. We plot $(4, 2)$ and then find a second point by moving *up* 3 units and *to the right* 1 unit. The line can then be drawn, as shown below.

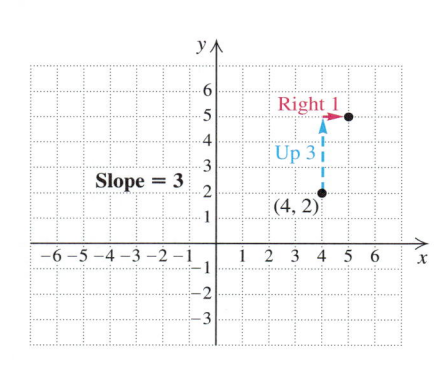

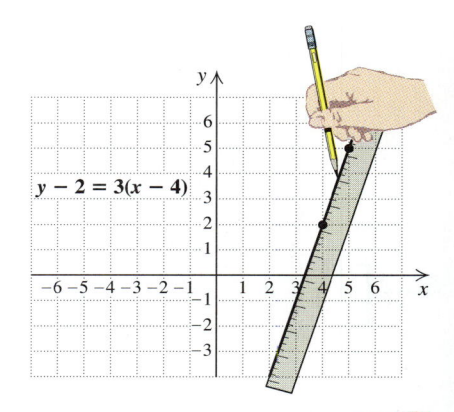

EXERCISE SET 7.2

Find a point–slope equation for the line containing the given point and having the given slope.

1. $(3, 7), m = 5$
2. $(-3, 0), m = -2$
3. $(2, 4), m = \frac{3}{4}$
4. $(\frac{1}{2}, 2), m = -1$
5. $(2, -6), m = 1$
6. $(4, -2), m = 6$
7. $(-4, 0), m = -3$
8. $(0, 3), m = -3$
9. $(5, 6), m = \frac{2}{3}$
10. $(2, 7), m = \frac{5}{6}$

Find the slope–intercept equation for the line containing the given point and having the given slope.

11. $(1, 4), m = 2$
12. $(1, 5), m = 4$
13. $(3, 5), m = -1$
14. $(2, -3), m = 1$
15. $(-2, 3), m = \frac{1}{2}$
16. $(6, -4), m = -\frac{1}{2}$
17. $(-6, -5), m = -\frac{1}{3}$
18. $(-5, 7), m = \frac{1}{5}$
19. $(4, -3), m = \frac{5}{4}$
20. $(-3, 8), m = \frac{4}{3}$

Find the slope–intercept equation for the line containing the given pair of points. (*Hint*: First use point–slope form.)

21. $(-6, 1)$ and $(2, 3)$
22. $(12, 16)$ and $(1, 5)$
23. $(0, 4)$ and $(4, 2)$
24. $(0, 0)$ and $(4, 2)$
25. $(3, 2)$ and $(1, 5)$
26. $(-4, 1)$ and $(-1, 4)$
27. $(5, 0)$ and $(0, -2)$
28. $(-2, -2)$ and $(1, 3)$
29. $(-2, -4)$ and $(2, -1)$
30. $(-3, 5)$ and $(-1, -3)$

Graph.

31. $y - 4 = \frac{1}{2}(x - 3)$
32. $y - 2 = \frac{1}{3}(x - 5)$
33. $y - 2 = -\frac{1}{2}(x - 5)$
34. $y - 1 = -\frac{1}{4}(x - 3)$
35. $y + 1 = \frac{1}{2}(x - 3)$
36. $y - 2 = \frac{1}{3}(x + 5)$
37. $y + 2 = 3(x + 1)$
38. $y + 4 = 2(x + 1)$
39. $y - 4 = -2(x + 1)$
40. $y + 3 = -1(x - 4)$
41. $y + 3 = -(x + 2)$
42. $y + 4 = 2(x + 2)$

SKILL MAINTENANCE

Graph.

43. $5x - 2y = 5$
44. $3x + 4y = 16$
45. $y = -\frac{4}{3}x - 6$
46. $y = \frac{2}{3}x - 5$

Solve.

47. $3x - 5 \leq 7x + 2$
48. $2(x - 4) > 9 - 5x$

SYNTHESIS

49. ◆ Can equations for horizontal or vertical lines be written in point–slope form? Why or why not?
50. ◆ Describe in your own words a procedure that can be used to write a slope–intercept equation for any nonvertical line passing through two given points.
51. ◆ The graph of $(y - 1)/(x - 4) = 2$ includes all the points found in the graph of $y - 1 = 2(x - 4)$ except $(4, 1)$. Why is this?
52. ◆ Any nonvertical line has many equations in point–slope form, but only one in slope–intercept form. Why is this?
53. Find an equation of the line that contains the point $(-5, 2)$ and that has the same slope as the line $3x - y + 4 = 0$.
54. Find an equation of the line that has the same y-intercept as the line $x - 3y = 6$ and contains the point $(5, -1)$.
55. Find the slope–intercept equation of the line that contains the point $(-1, 5)$ and is parallel to the line passing through $(2, 7)$ and $(-1, -3)$.
56. Find an equation of the line that has x-intercept $(-2, 0)$ and is parallel to $4x - 8y = 12$.
57. ◆ Why is slope–intercept form more useful than point–slope form when using a grapher? How can point–slope form be modified so that it is more easily used with graphers?

7.3 Linear Inequalities in Two Variables

Graphing Linear Inequalities • Linear Inequalities in One Variable

Just as the solutions of linear equations like $5x + 4y = 13$ or $y = \frac{1}{2}x + 1$ can be graphed, so too can the solutions of *linear inequalities* like $5x + 4y < 13$ or $y > \frac{1}{2}x + 1$ be represented graphically.

Graphing Linear Inequalities

In Section 2.6, we found that solutions of inequalities like $5x + 9 \leq 4x + 3$ can be represented by a shaded portion of a number line. When a solution included an endpoint, we drew a solid dot, and when the endpoint was excluded, we drew an open dot. To graph inequalities like $y > \frac{1}{2}x + 1$ or $2x + 3y \leq 6$,

374 CHAPTER 7 • MORE WITH GRAPHING

we will shade a region of a plane. That region will be either above or below the graph of a "boundary line" (in this case, $y = \frac{1}{2}x + 1$ or $2x + 3y = 6$). If the symbol is $\leq$ or $\geq$, we will draw the boundary line solid, since it is part of the solution. When the boundary is excluded—that is, if $<$ or $>$ is used— we will draw a dashed line.

EXAMPLE 1 Graph: $y > \frac{1}{2}x + 1$.

SOLUTION We begin by graphing the boundary line $y = \frac{1}{2}x + 1$. The slope is $\frac{1}{2}$ and the y-intercept is $(0, 1)$. This line is drawn dashed since the symbol $>$ is used.

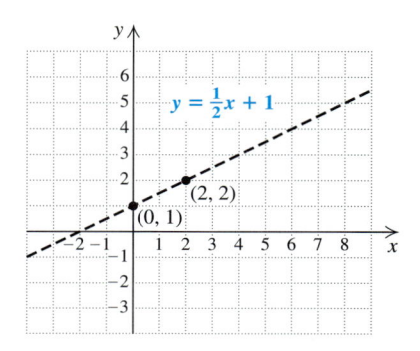

Note that the plane is now split into two regions. If we consider the coordinates of a few points above the line, we will find that all are solutions of $y > \frac{1}{2}x + 1$.

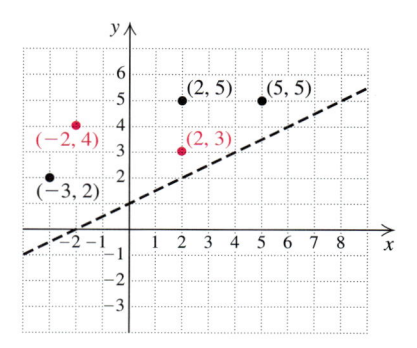

Here is a check for the points $(2, 3)$ and $(-2, 4)$:

$$\begin{array}{c|c}
y > \frac{1}{2}x + 1 \\
\hline
3 \; ? \; \frac{1}{2} \cdot 2 + 1 \\
 \; | \; 1 + 1 \\
3 \; | \; 2 \quad \text{TRUE}
\end{array} \qquad \begin{array}{c|c}
y > \frac{1}{2}x + 1 \\
\hline
4 \; ? \; \frac{1}{2}(-2) + 1 \\
 \; | \; -1 + 1 \\
4 \; | \; 0 \quad \text{TRUE}
\end{array}$$

7.3 LINEAR INEQUALITIES IN TWO VARIABLES

The student can check that *any* point on the same side of the dashed line as (2, 3) or (−2, 4) is a solution. If one point in a region solves an inequality, then *all* points in that region are solutions. The graph of

$$y > \tfrac{1}{2}x + 1$$

is shown below. Note that the solution set consists of all points in the shaded region. Furthermore, note that for any inequality of the form $y > mx + b$ or $y \geq mx + b$, we shade the region *above* the boundary line.

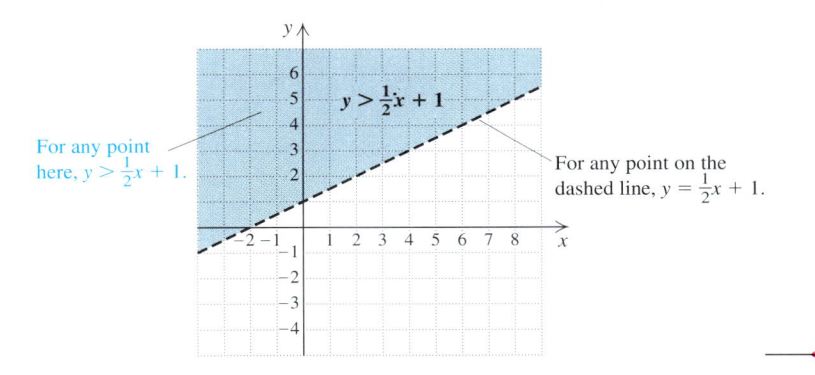

For any point here, $y > \tfrac{1}{2}x + 1$.

For any point on the dashed line, $y = \tfrac{1}{2}x + 1$.

EXAMPLE 2   Graph: $2x + 3y \leq 6$.

SOLUTION   First, we graph $2x + 3y = 6$. This can be done either by using the intercepts, (0, 2) and (3, 0), or by finding slope–intercept form, $y = -\tfrac{2}{3}x + 2$. Since the inequality contains the symbol $\leq$, we draw a solid boundary line to indicate that any pair on the line is a solution. The graph of this inequality includes the line and either the region above it or the region below it. By using a "test point" that is clearly above or below the line, we can determine which region to shade. The origin, (0, 0), is often a convenient test point:

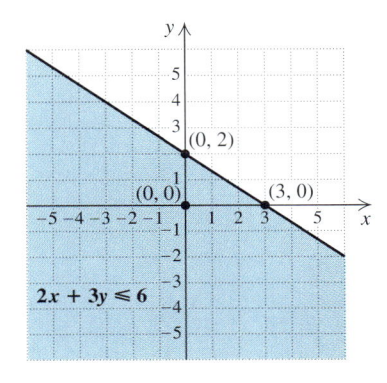

$$\begin{array}{c|c} 2x + 3y \leq 6 \\ \hline 2 \cdot 0 + 3 \cdot 0 \; ? \; 6 \\ 0 \; | \; 6 \quad \text{TRUE} \end{array}$$

The point (0, 0) is a solution and it appears in the region below the boundary line. Thus this region, along with the line itself, represents the solution.

The original inequality is equivalent to $y \leq -\tfrac{2}{3}x + 2$. Note that for any inequality of the form $y \leq mx + b$ or $y < mx + b$, we shade the region *below* the boundary line.

TO GRAPH A LINEAR INEQUALITY:

1. Draw the boundary line by replacing the inequality symbol with an equals sign and graphing the resulting equation. If the inequality symbol is $<$ or $>$, the line is dashed. If the symbol is $\leq$ or $\geq$, the line is solid.
2. Shade the region on one side of the boundary line. To determine which side, select a point not on the line as a test point. If that point's coordinates are a solution of the inequality, shade the region containing the point. If not, shade the other region.

Linear Inequalities in One Variable

EXAMPLE 3 Graph $y \leq -2$ on a plane.

SOLUTION We graph $y = -2$ as a solid line to indicate that all points on the line are solutions. Again, we select (0, 0) as a test point. It may help to write $y \leq -2$ as $y \leq 0 \cdot x - 2$:

$$\frac{y \leq 0 \cdot x - 2}{0 \ ? \ 0 \cdot 0 - 2}$$
$$0 \ | \ -2 \quad \text{FALSE}$$

Since (0, 0) is *not* a solution, we do not shade the region in which it appears. Instead, we shade below the boundary line as shown. Note that the solution consists of all ordered pairs whose y-coordinates are less than or equal to -2.

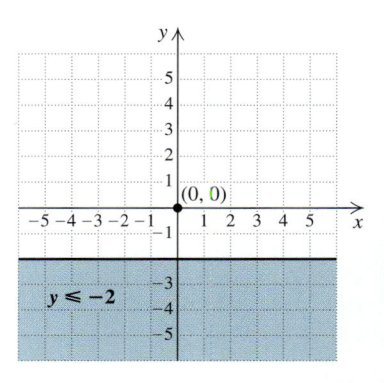

EXAMPLE 4 Graph $x < 3$ on a plane.

SOLUTION We graph the equation $x = 3$ using a dashed line. To determine which region to shade, we again use the test point, (0, 0). It may help to write $x < 3$ as $x + 0 \cdot y < 3$:

$$\frac{x + 0 \cdot y < 3}{0 + 0 \cdot 0 \ ? \ 3}$$
$$0 \ | \ 3 \quad \text{TRUE}$$

Since (0, 0) is a solution, we shade left of the boundary line, as shown. The solution consists of all ordered pairs with first coordinates less than 3.

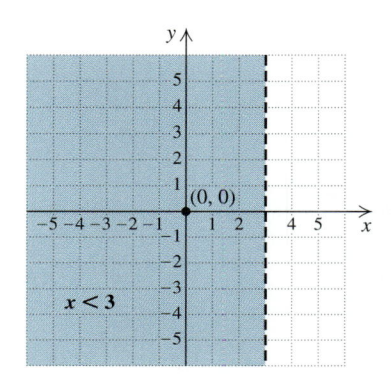

EXERCISE SET 7.3

1. Determine whether $(-3, -5)$ is a solution of $x + 3y < -18$.
2. Determine whether $(5, -3)$ is a solution of $-2x + 4y \leq -2$.
3. Determine whether $\left(-1, \frac{1}{2}\right)$ is a solution of $6y - 5x \geq 7$.
4. Determine whether $(-6, 5)$ is a solution of $x + 0 \cdot y < 3$.

Graph on a plane.

5. $y \leq x + 3$
6. $y \leq x - 5$
7. $y < x - 1$
8. $y < x + 4$
9. $y \geq x - 2$
10. $y \geq x - 1$
11. $y \leq 2x - 1$
12. $y \leq 3x + 2$
13. $x - y \leq 3$
14. $x + y \leq 4$
15. $x + y > 7$
16. $x - y > -2$
17. $y \geq 1 - 2x$
18. $y - x < 0$
19. $y - 3x > 0$
20. $x \leq 0$
21. $x \geq 3$
22. $x > -4$
23. $y \leq 3$
24. $y > -1$
25. $y \geq -5$
26. $y < 0$
27. $x < 4$
28. $x - 3y < 6$
29. $x - y < -10$
30. $y - 2x \leq -1$
31. $2x + 3y \leq 12$
32. $5x + 4y \geq 20$

SKILL MAINTENANCE

Multiply.

33. $(2x^2 - x - 1)(x + 3)$
34. $(3x - 5)(3x + 5)$
35. Factor: $3a^3 + 18a^2 - 4a - 24$.
36. Simplify: $\dfrac{x^2 - 9x + 14}{x^2 + 3x - 10}$.

Solve.

37. $2(y - 5) - 3y = 8$
38. $4x - 3(8 - x) = 19$

SYNTHESIS

39. ◆ Examine the solution of Example 2. Why is the point $(4.5, -1)$ *not* a good choice for a test point?
40. ◆ Why is $(0, 0)$ such a "convenient" test point to use?
41. ◆ Why is the graph of any inequality of the form $y > mx + b$ always shaded *above* the line $y = mx + b$?
42. ◆ Describe a procedure that could be used to graph any inequality of the form $Ax + By < C$.
43. *Elevators.* Many elevators have a capacity of 1 metric ton (1000 kg). Suppose c children, each weighing 35 kg, and a adults, each weighing 75 kg, are on an elevator. Find and graph an inequality that asserts that the elevator is overloaded.
44. *Hockey wins and losses.* A hockey team figures that it needs at least 60 points for the season in order to make the playoffs. A win w is worth 2 points and a tie t is worth 1 point. Find and graph an inequality that describes the situation.
45. *Architecture.* Most architects agree that the sum of a step's riser r and tread t, in inches, should not be less than 17 in. Find and graph an inequality that describes the situation.

Find an inequality for each graph shown.

46.

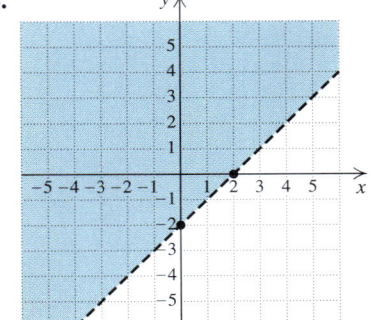

47.

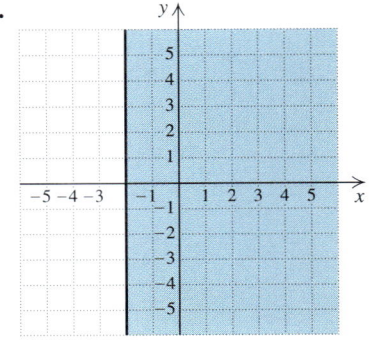

Graph on a plane. (Hint: Use several test points.)

48. $xy \leq 0$
49. $xy \geq 0$

COLLABORATIVE C·O·R·N·E·R

Focus: Linear inequalities
Time: 20–30 minutes
Group size: 3–5
Materials: Graph paper

The game of Battleship® or "find the hidden point" was explained in the Collaborative Corner on page 128. In that game, one group member secretly chose a point with integer coordinates. Only integers from −10 to 10 could be used, and the other group members were allowed a total of ten "yes/no" questions to determine the coordinates of the hidden point.

ACTIVITY

1. Repeat the aforementioned activity with the following modification: Once the point's coordinates have been secretly written down, each question must be either:
   - phrased using a linear inequality (for example, "Are the point's coordinates a solution of $y > x$?"), or
   - an actual guess at the coordinates (for example, "Are the coordinates of the hidden point $(-3, 7)$?").
2. Be sure to use a graph to keep track of what points are still under consideration after each question has been answered.
3. Repeat the activity, giving each group member a chance to choose the hidden point.

7.4 Direct and Inverse Variation

Equations of Direct Variation • Problem Solving with Direct Variation • Equations of Inverse Variation • Problem Solving with Inverse Variation

Many problems lead to equations of the form $y = kx$ or $y = k/x$, for some constant k. Such equations are called *equations of variation*.

Equations of Direct Variation

A bicycle tour is traveling at a speed of 15 km/h. In 1 hr, it goes 15 km. In 2 hr, it goes 30 km. In 3 hr, it goes 45 km, and so on. In the graph at the top of the following page, we will use the number of hours as the first coordinate and the number of kilometers traveled as the second coordinate: (1, 15), (2, 30), (3, 45), (4, 60), and so on. Note that the second coordinate is always 15 times the first.

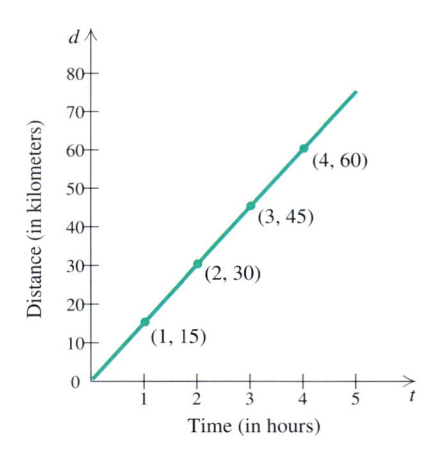

7.4 DIRECT AND INVERSE VARIATION

In this example, distance is a constant multiple of time, so we say that there is **direct variation** and that distance **varies directly** as time. The **equation of variation** is $d = 15t$.

DIRECT VARIATION

When a situation translates to an equation described by $y = kx$, where k is a constant, $y = kx$ is called an *equation of direct variation*, and k is called the *variation constant*. We say that y *varies directly* as x.

The terminologies

"y varies as x,"

"y is directly proportional to x," and

"y is proportional to x"

also imply direct variation and are used in many situations. The constant k is called the **constant of proportionality.** It can be found if one pair of values of x and y is known. Once k is known, other pairs can be determined.

EXAMPLE 1 If y varies directly as x and $y = 2$ when $x = 5$, find the equation of variation.

SOLUTION We substitute to find k:

$$y = kx$$
$$2 = k \cdot 5$$
$$\tfrac{2}{5} = k, \quad \text{or} \quad k = 0.4.$$

Thus the equation of variation is $y = 0.4x$. A visualization of the situation is shown here.

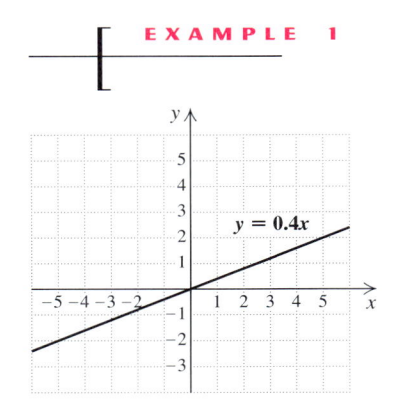

A visualization of Example 1

Note from the graph at left that when y varies directly as x, the constant of proportionality is also the slope of the associated graph—the rate at which y changes with respect to x.

EXAMPLE 2 Find an equation in which s varies directly as t and $s = 10$ when $t = 15$. Then find the value of s when $t = 32$.

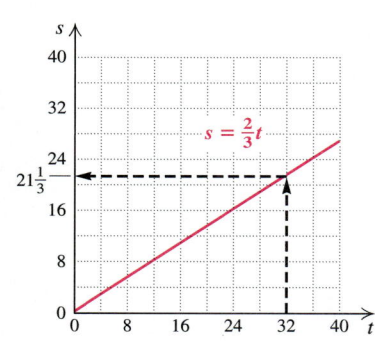

SOLUTION We have

$s = kt$ We know that s varies directly as t.
$10 = k \cdot 15$ Substituting 10 for s and 15 for t
$\frac{10}{15} = k$, or $k = \frac{2}{3}$.

Thus the equation of variation is $s = \frac{2}{3}t$. When $t = 32$, we have

$s = \frac{2}{3}t$
$s = \frac{2}{3} \cdot 32$ Substituting 32 for t in the equation of variation
$s = \frac{64}{3}$, or $21\frac{1}{3}$.

The value of s is $21\frac{1}{3}$ when $t = 32$.

A visualization of Example 2

Problem Solving with Direct Variation

In applications, it is often necessary to find an equation of variation and then use it to find other values, much as we did in Example 2.

EXAMPLE 3 The karat rating R of a gold object varies directly as the actual percentage P of gold in the object. A 14-karat gold ring is 58.25% gold. What is the percentage of gold in a 24-karat gold ring?

SOLUTION

1., 2. **Familiarize** and **Translate.** The problem states that we have direct variation between R and P. Thus an equation $R = kP$ applies.

3. **Carry out.** We find an equation of variation:

$R = kP$
$14 = k(0.5825)$ Substituting 14 for R and 58.25%, or 0.5825, for P
$\dfrac{14}{0.5825} = k$
$24.03 \approx k.$ Dividing and rounding to the nearest hundredth

The equation of variation is $R = 24.03P$. When $R = 24$, we have

$R = 24.03P$
$24 = 24.03P$ Substituting 24 for R
$\dfrac{24}{24.03} = P$ Solving for P
$0.999 \approx P$
$99.9\% \approx P.$

4. **Check.** The check might be done by repeating the computations. You might also note that as the karat rating increased from 14 to 24, the percentage increased from 58.25% to 99.9%. The ratios 14/0.5825 and 24/0.999 are both about 24.03.

5. **State.** A 24-karat gold ring is 99.9% gold.

Equations of Inverse Variation

A car is traveling a distance of 20 mi. At a speed of 5 mph, the trip will take 4 hr. At 20 mph, it will take 1 hr. At 40 mph, it will take $\frac{1}{2}$ hr, and so on. This determines a set of pairs of numbers:

$(5, 4),$ $(20, 1),$ $\left(40, \frac{1}{2}\right),$ and so on.

Note that the product of speed and time for each of these pairs is 20. Note too that as the speed *increases,* the time *decreases.*

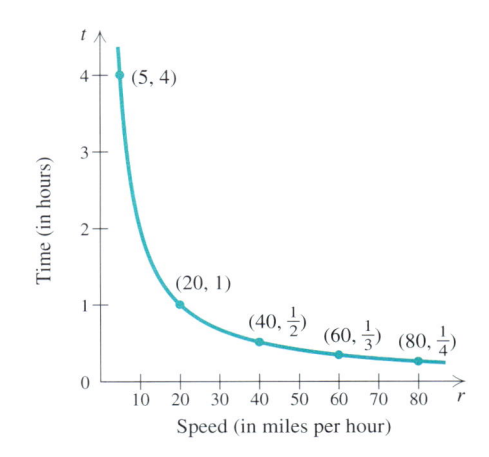

In this case, the product of speed and time is constant so we say that there is **inverse variation** and that time **varies inversely** as speed. The equation of variation is

$$rt = 20 \text{ (a constant)}, \quad \text{or} \quad t = \frac{20}{r}.$$

INVERSE VARIATION

> When a situation translates to an equation described by $y = k/x$, where k is a constant, then $y = k/x$ is called an *equation of inverse variation*. We say that y *varies inversely* as x.

The terminology

"y is inversely proportional to x"

also implies inverse variation and is used in some situations. The constant k is again called the *constant of proportionality.*

EXAMPLE 4   If y varies inversely as x and $y = 145$ when $x = 0.8$, find the equation of variation.

SOLUTION We substitute to find k:

$$y = \frac{k}{x}$$

$$145 = \frac{k}{0.8} \quad \text{We are told } y = 145 \text{ when } x = 0.8.$$

$$(0.8)145 = k$$

$$116 = k.$$

The equation of variation is $y = \dfrac{116}{x}$.

Problem Solving with Inverse Variation

Often in applications, we must decide what kind of variation, if any, applies.

EXAMPLE 5   It takes 4 hr for 20 people to raise a barn. How long would it take 25 people to complete the job?

SOLUTION

1. **Familiarize.** Think about the situation. What kind of variation applies? It seems reasonable that the greater the number of people working on a job, the less time it will take. Thus we assume that inverse variation applies. We let $T =$ the time to complete the job, in hours, and $N =$ the number of people working.

2. **Translate.** Since inverse variation applies, we have

$$T = \frac{k}{N}.$$

3. **Carry out.** We find an equation of variation:

$$T = \frac{k}{N}$$

$$4 = \frac{k}{20} \quad \text{Substituting 4 for } T \text{ and 20 for } N$$

$$20 \cdot 4 = k$$

$$80 = k.$$

The equation of variation is $T = \dfrac{80}{N}$. When $N = 25$, we have

$$T = \frac{80}{25} \quad \text{Substituting 25 for } N$$

$$T = 3.2.$$

4. **Check.** A check might be done by repeating the computations or by noting that $(3.2)(25)$ and $(4)(20)$ are both 80. Also, as the number of people increases, the time needed to complete the job decreases, as expected.

5. **State.** It should take 3.2 hr for 25 people to raise a barn.

EXERCISE SET 7.4

For each of the following, find an equation of variation in which y varies directly as x and the following are true.

1. $y = 28$, when $x = 7$
2. $y = 30$, when $x = 8$
3. $y = 0.7$, when $x = 0.4$
4. $y = 0.8$, when $x = 0.5$
5. $y = 400$, when $x = 125$
6. $y = 630$, when $x = 175$
7. $y = 200$, when $x = 300$
8. $y = 500$, when $x = 60$

For each of the following, find an equation of variation in which y varies inversely as x and the following are true.

9. $y = 45$, when $x = 2$
10. $y = 8$, when $x = 10$
11. $y = 7$, when $x = 10$
12. $y = 0.125$, when $x = 8$
13. $y = 6.25$, when $x = 0.16$
14. $y = 42$, when $x = 25$
15. $y = 42$, when $x = 50$
16. $y = 0.2$, when $x = 0.3$

Solve.

17. *Wages.* A person's paycheck P varies directly as the number of hours worked H. For 15 hr of work, the pay is $78.75. Find the pay for 35 hr of work.

18. *Manufacturing.* The number of bolts B that a machine can make varies directly as the time T that it operates. It can make 6578 bolts in 2 hr. How many can it make in 5 hr?

19. *Turkey servings.* The number of servings S of meat that can be obtained from a turkey varies directly as its weight W. From a turkey weighing 14 kg, one can get 40 servings of meat. How many servings can be obtained from an 8-kg turkey?

20. *Gas volume.* The volume V of a gas varies inversely as the pressure P on it. The volume of a gas is 200 cm³ (cubic centimeters) under a pressure of 32 kg/cm². What will be its volume under a pressure of 20 kg/cm²?

21. *Electrical current.* The current I in an electrical conductor varies inversely as the resistance R of the conductor. The current is 2 amperes when the resistance is 960 ohms. What is the current when the resistance is 540 ohms?

22. *Lunar weight.* The weight M of an object on the moon varies directly as its weight E on Earth. A person who weighs 171.6 lb on Earth weighs 28.6 lb on the moon. How much would a person who weighs 110 lb on Earth weigh on the moon?

23. *Weight on Mars.* The weight M of an object on Mars varies directly as its weight E on Earth. A person who weighs 209 lb on Earth weighs 79.42 lb on Mars. How much would a person who weighs 176 lb on Earth weigh on Mars?

24. *Musical tones.* The pitch P of a musical tone varies inversely as its wavelength W. One tone has a pitch of 660 vibrations per second and a wavelength of 1.6 ft. Find the wavelength of another tone that has a pitch of 440 vibrations per second.

25. *Pumping time.* The time t required to empty a tank varies inversely as the rate r of pumping. A pump can empty a tank in 90 min at the rate of 1200 L/min. How long will it take the pump to empty the tank at the rate of 2000 L/min?

26. *Cost of television.* The cost c of operating a television varies directly as the number of hours n that it is in operation. It costs $14.00 to operate a standard-size color television continuously for 30 days. At this rate, how much would it cost to operate the television for 1 day? for 1 hr?

27. *Answering questions.* The number of minutes m that a student should allow for each question on a quiz is inversely proportional to the number of questions n on the quiz. If a 16-question quiz means that students have 2.5 min per question, how many questions would appear on a quiz in which students have 4 min per question?

28. *Chartering a boat.* The cost per person c of a chartered fishing boat is inversely proportional to the number of people n who are chartering the boat. If it costs $17.50 per person when 9 people charter a boat, how many people would be going fishing if the cost were $31.50 per person?

SKILL MAINTENANCE

29. Solve $2a - 3b = 4$ for b.
30. Solve $4m + 2n = p$ for m.

Subtract.

31. $(5x + 9y) - (2x - 3y)$

32. $(4a - 5b) - (7a - 6b)$

Graph.

33. $3x - 5y = 6$ **34.** $2x - 7y = 14$

SYNTHESIS

State whether each situation represents direct variation, inverse variation, or neither. Give reasons for your answers.

35. ◆ The cost of mailing a letter in the United States and the distance that it travels

36. ◆ A runner's speed in a race and the time it takes to run the race

37. ◆ The weight of a turkey and the cooking time

38. ◆ The number of plays it takes to go 80 yd for a touchdown and the average gain per play

Write an equation of variation to describe each situation. If possible, give a value for k.

39. *Geometry.* The perimeter P of an equilateral octagon varies directly as the length S of a side.

40. *Geometry.* The circumference C of a circle varies directly as the radius r.

41. *Peanut sales.* The number of bags of peanuts B sold at the circus varies directly as the number of people N in attendance.

42. *Geometry.* The area of a circle varies directly as the square of the length of the radius.

43. *Ecology.* In a stream, the amount of salt S carried varies directly as the sixth power of the speed of the stream v.

44. *Acoustics.* The square of the pitch P of a vibrating string varies directly as the tension t on the string.

45. *Lighting.* The intensity of illumination I from a light source varies inversely as the square of the distance d from the source.

46. *Density.* The density D of a given mass varies inversely as its volume V.

47. *Geometry.* The volume V of a sphere varies directly as the cube of the radius r.

48. *Wind energy.* The power P in a windmill varies directly as the cube of the wind speed v.

49. ◆ If a varies directly as b and b varies directly as c, does it follow that a varies directly as c? Why or why not?

SUMMARY AND REVIEW 7

KEY TERMS

Slope–intercept equation, p. 364
Parallel lines, p. 367
Point–slope equation, p. 370
Linear inequality, p. 373

Direct variation, p. 379
Constant of proportionality, p. 379
Inverse variation, p. 381

IMPORTANT PROPERTIES AND FORMULAS

Slope–intercept equation: $y = mx + b$
Parallel lines: Slopes equal
Point–slope equation: $y - y_1 = m(x - x_1)$

To graph a linear inequality:

1. Draw the boundary line by replacing the inequality symbol with an equals sign and graphing the resulting equation. If the inequality symbol is $<$ or $>$, the line is dashed. If the symbol is $\leq$ or $\geq$, the line is solid.
2. Shade the region on one side of the boundary line. To determine which side, select a point not on the line as a test point. If that point's coordinates are a solution of the inequality, shade the region containing the point. If not, shade the other region.

Direct variation: $y = kx$, where k is a constant
Inverse variation: $y = k/x$, where k is a constant

REVIEW EXERCISES

Find the slope and the y-intercept of each line.

1. $y = -9x + 46$
2. $x + y = 9$
3. $2x - 6y = 4$
4. $y = 8$

Find the slope–intercept equation for each line with the indicated slope and y-intercept.

5. Slope 2; y-intercept $(0, -4)$
6. Slope $-\frac{3}{2}$; y-intercept $(0, 1)$

Graph.

7. $y = -\frac{3}{4}x - 2$
8. $y + \frac{1}{2}x = 2$
9. $y - 2 = 3(x - 6)$
10. $y + 1 = \frac{2}{5}(x - 3)$
11. $4y = 3x - 8$
12. $y - 4 = -\frac{5}{3}(x + 2)$

Determine whether each pair of equations represents parallel lines.

13. $4x + y = 6$,
    $4x = 8 - y$
14. $3x - y = 6$,
    $3x + y = 8$

15. City Gas charges a monthly service fee of $12.00 plus $0.50 for every hundred cubic feet of gas used. If x is the number of hundred cubic feet of gas, find the slope–intercept form of the equation for a monthly bill.

Find a point–slope equation for each line containing the given point and having the given slope.

16. $(1, 2)$, $m = 3$
17. $(-2, -5)$, $m = \frac{2}{3}$

Find the slope–intercept equation for each line containing the given pair of points.

18. $(5, 7)$ and $(-1, 1)$
19. $(2, 0)$ and $(-4, -3)$

Graph on a plane.

20. $x \leq y$
21. $x - 2y \geq 4$
22. $y > -2$
23. $y \geq \frac{2}{3}x - 5$
24. $2x + y < 1$
25. $x < 4$

26. If y varies inversely as x and $y = 81$ when $x = 3$, find the equation of variation.

386 CHAPTER 7 • MORE WITH GRAPHING

27. The number of sandwiches S that can be made at a buffet varies directly as the number of pounds of cold cuts C in the buffet. From 6 lb of cold cuts, 25 sandwiches can be made. How many pounds of cold cuts are needed for 40 sandwiches?

SKILL MAINTENANCE

28. Multiply: $(\frac{1}{2}y + 3)^2$.
29. Factor: $x^3 - x^2 + 2x - 2$.

Subtract.

30. $(5x^3 + 2x + 7) - (x^2 - 4x + 2)$
31. $(9a + 14b) - (7a - 14b)$

SYNTHESIS

32. ◆ Which is easier: to convert an equation from point–slope form to slope–intercept form or to convert from slope–intercept form to point–slope form? Why?
33. ◆ Why is the boundary line part of the graph when a linear inequality contains the symbol ≤, but not when it contains the symbol <?
34. Find an equation of the line for which the second coordinate is the opposite of the first coordinate.
35. Find the slope and the intercepts of a line whose equation is
$$\frac{x}{a} + \frac{y}{c} = 1, \text{ where } a \neq 0 \text{ and } c \neq 0.$$
36. Find an equation of the line having the same y-intercept as the line $2x - y = 3$ and the same slope as the line $2x + y = 3$.
37. Find an equation of the line that contains the point $(-1, 2)$ and is perpendicular to the line $2x + 3y = 1$. (See Exercises 71–76 in Section 7.1.)

CHAPTER TEST 7

Find the slope and the y-intercept of each line.

1. $y = -\frac{1}{5}x - 7$
2. $y = -3$
3. $y = 2x + \frac{1}{4}$
4. $-4x + 3y = -6$

Find the slope–intercept equation for each line with the indicated slope and y-intercept.

5. Slope $\frac{1}{2}$; y-intercept $(0, -7)$
6. Slope -4; y-intercept $(0, 3)$

Find a point–slope equation for each line containing the given point and having the given slope.

7. $(3, 5), m = 1$
8. $(-2, 0), m = -\frac{1}{2}$

Find the slope–intercept equation for each line containing the given pair of points.

9. $(1, 1)$ and $(2, -2)$
10. $(4, -1)$ and $(-4, -3)$

Graph.

11. $2x + y = 5$
12. $y = \frac{2}{3}x - 6$
13. $y + 4 = -\frac{1}{2}(x - 1)$
14. $3y = 2x - 9$

15. Determine whether the following pair of equations represents parallel lines:
$2x + y = 8,$
$2x + y = 4.$

Graph on a plane.

16. $y > x - 1$
17. $2x - y \leq 4$
18. $y < -2$

19. If y varies directly as x and $y = 9$ when $x = 2$, find the equation of variation.
20. It takes 45 min for 2 people to shovel a driveway. How long would it take 5 people to shovel the same driveway?

SKILL MAINTENANCE

Multiply.

21. $(-3y^4)(y^3 - 3y + 7)$
22. $(x + 0.1)(x - 0.1)$
23. Solve $3x - 5y = 2z$ for x.
24. Subtract: $(27x - 34y) - (19x - 34y)$.

SYNTHESIS

25. Find the slope–intercept equation for the line that contains the point $(-4, 1)$ and has the same slope as the line $2x - 3y = -6$.
26. Find an equation for the steepest line that can pass through $(-2, -5)$ without touching the second quadrant.

CHAPTER 8

Systems of Equations and Problem Solving

8.1 Systems of Equations and Graphing
8.2 Systems of Equations and Substitution
8.3 Systems of Equations and Elimination
8.4 More Applications Using Systems
8.5 Systems of Linear Inequalities
SUMMARY AND REVIEW
TEST

AN APPLICATION

The octane rating of a gasoline is a measure of the amount of isooctane in the gas. How much 87-octane gas and 93-octane gas should be blended in order to make 12 gal of 91-octane gas?

THIS PROBLEM APPEARS AS EXERCISE 23 IN SECTION 8.4.

More information on gas station memorabilia is available at
http://hepg.awl.com/be/elem_5

The fundamental design of a gas dispenser is based on the equation Volume × Unit price = Total sale. More complex equations determine the ratio used to blend gas of different octane ratings.

DEBORAH JOINES
Engineer
Salisbury, Maryland

388 CHAPTER 8 • SYSTEMS OF EQUATIONS AND PROBLEM SOLVING

In fields such as business, engineering, sociology, science, and psychology, problems often are most easily solved using a system of equations. *In this chapter, we study three methods for solving systems of two equations in which two variables appear. These methods are then used in a variety of applications and as an aid in solving systems of inequalities.*

8.1 Systems of Equations and Graphing

Solutions of Systems • Graphing Systems of Equations

In Section 2.5, the following problem appeared as Example 4:

A rectangular community garden is to be enclosed with 92 m of fencing. In order to allow for compost storage, the garden must be 4 m longer than it is wide. Determine the dimensions of the garden.

The problem describes two separate relationships between the unknown quantities. Each relationship can be translated to an equation in two variables. Because solving the problem means finding a solution that makes *both* equations true, the two equations are considered a *system of equations*:

$2l + 2w = 92$,   The perimeter is 92.
$l = w + 4$.   The length is 4 more than the width.

Since it is often easier to translate a problem into a system of equations than into one equation in one variable (as we did in Chapter 2), it is important to know how to solve a system when one arises.

Solutions of Systems

SOLUTION OF A SYSTEM

A *solution* of a system of two equations is an ordered pair that makes both equations true.

EXAMPLE 1   Consider the system

$2l + 2w = 92$,
$l = w + 4$.

Determine if each pair is a solution of the system: **(a)** (25, 21); **(b)** (16, 12).

SOLUTION

a) We check by substituting (alphabetically) 25 for l and 21 for w:

$$\begin{array}{c|c} 2l + 2w = 92 & l = w + 4 \\ \hline 2 \cdot 25 + 2 \cdot 21 \;?\; 92 & 25 \;?\; 21 + 4 \\ 50 + 42 & 25 \;|\; 25 \quad \text{TRUE} \\ 92 \;|\; 92 \quad \text{TRUE} & \end{array}$$

Since (25, 21) checks in *both* equations, it is a solution of the system.

b) We substitute 16 for l and 12 for w:

$$\begin{array}{c|c} 2l + 2w = 92 & l = w + 4 \\ \hline 2 \cdot 16 + 2 \cdot 12 \;?\; 92 & 16 \;?\; 12 + 4 \\ 32 + 24 & 16 \;|\; 16 \quad \text{TRUE} \\ 56 \;|\; 92 \quad \text{FALSE} & \end{array}$$

Since (16, 12) is not a solution of $2l + 2w = 92$, it is *not* a solution of the system.

How is a solution of a system found? One method involves graphing.

Graphing Systems of Equations

Recall that a graph of an equation is a set of points representing its solution set. Each point on the graph corresponds to an ordered pair that is a solution of the equation. By graphing two equations on the same set of axes, we can identify a solution of both equations by looking for a point of intersection.

EXAMPLE 2 Solve this system of equations by graphing:

$$x + y = 7,$$
$$y = 3x - 1.$$

SOLUTION We graph the equations using any method studied earlier. The equation $x + y = 7$ can be graphed easily using the intercepts, (0, 7) and (7, 0). The equation $y = 3x - 1$ is in slope–intercept form, so we graph the line by plotting its y-intercept, $(0, -1)$, and "counting off" a slope of 3.

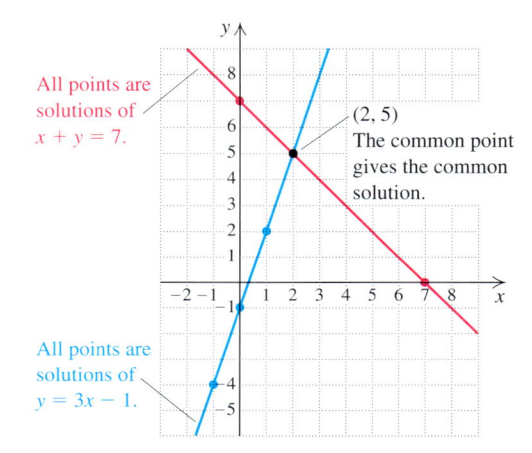

The "apparent" solution of the system, (2, 5), should be checked in both equations:

Check:

$$\begin{array}{c|c} x + y = 7 \\ \hline 2 + 5 \: ? \: 7 \\ 7 \: | \: 7 \quad \text{TRUE} \end{array} \qquad \begin{array}{c|c} y = 3x - 1 \\ \hline 5 \: ? \: 3 \cdot 2 - 1 \\ 5 \: | \: 5 \quad \text{TRUE} \end{array}$$

Since it checks in both equations, (2, 5) is a solution of the system.

A system of equations that has at least one solution, like the systems in Examples 1 and 2, is said to be **consistent**. A system for which there is no solution is said to be **inconsistent**.

EXAMPLE 3

Solve this system of equations by graphing:

$y = 3x + 4,$
$y = 3x - 3.$

SOLUTION Both equations are in slope–intercept form so it is easy to see that both lines have the same slope, 3. The y-intercepts differ so the lines are parallel, as shown in the figure at right.

Because the lines are parallel, there is no point of intersection. Thus the system is inconsistent and has no solution.

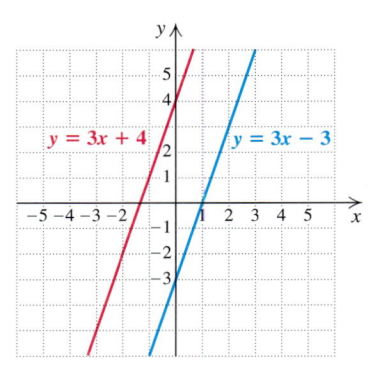

TECHNOLOGY CONNECTION 8.1

Graphers can be used to solve systems of equations, provided each equation has been solved for y. Thus, to solve Example 2 with a grapher, we must rewrite $x + y = 7$ as $y = -x + 7$ to form the equivalent system

$y = -x + 7,$
$y = 3x - 1.$

We then enter the two equations as y_1 and y_2. After the equations have been graphed, we use the INTERSECT option of the CALC menu to identify and display the point of intersection. If the INTERSECT option is unavailable, ZOOM and TRACE can be used.

1. Use a grapher to check Example 2.
2. Graph the system

   $y = 0.23x + 1.49,$
   $y = 0.23x + 3.49.$

   Show that the vertical separation between the two lines is uniform by using the TRACE feature or a TABLE to show that for any given x-value, y_1 and y_2 differ by exactly 2. Does this system have a solution? Why or why not?

8.1 SYSTEMS OF EQUATIONS AND GRAPHING

Sometimes both equations in a system have the same graph.

EXAMPLE 4 Solve this system of equations by graphing:

$$2x + 3y = 6,$$
$$-8x - 12y = -24.$$

SOLUTION Graphing the equations, we see that they both represent the same line. This can also be seen by solving each equation for y, obtaining the equivalent slope–intercept form, $y = -\frac{2}{3}x + 2$. Because the equations are equivalent, any solution of one equation is a solution of the other equation as well. We show four such solutions.

We check one solution, $(0, 2)$, in each of the original equations.

$$\frac{2x + 3y = 6}{2(0) + 3(2) \; ? \; 6}$$
$$0 + 6$$
$$6 \; | \; 6 \quad \text{TRUE}$$

$$\frac{-8x - 12y = -24}{-8(0) - 12(2) \; ? \; -24}$$
$$0 - 24$$
$$-24 \; | \; -24 \quad \text{TRUE}$$

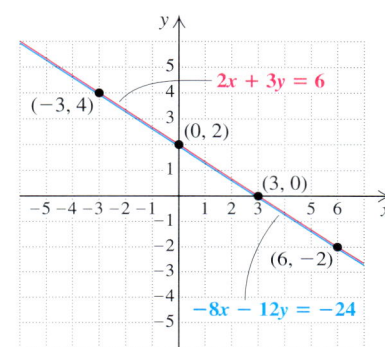

On your own, check that $(3, 0)$ is also a solution of the system. If two points are solutions, then all points on the line containing them are solutions. The lines coincide, so there is an infinite number of solutions. Since a solution exists, the system is consistent.

The equations in Example 4 are equivalent. When equivalent equations appear in a system, we call the equations **dependent**. Thus the equations in Example 4 are dependent, but those in Examples 2 and 3 are **independent**.

When a system of two linear equations is graphed, one of the following must occur:

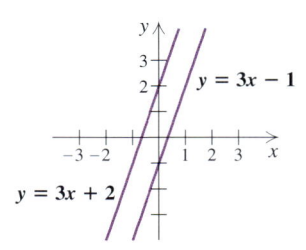

Graphs are parallel.
The system is *inconsistent* because there is no solution. The equations are *independent* since the graphs differ.

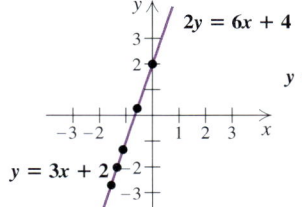

Equations have the same graph.
The system is *consistent* and has an infinite number of solutions. The equations are *dependent* since the graphs are the same.

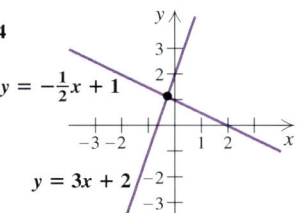

Graphs intersect.
The system is *consistent* and has one solution. Since the graphs differ, the equations are *independent*.

Although graphing lets us "see" the solution of a system, it does not always allow us to find a precise solution. For example, the solution of the system

$$3x + 7y = 5,$$
$$6x - 7y = 1$$

is $\left(\frac{2}{3}, \frac{3}{7}\right)$, but finding that precise solution from a graph—*even with a computer or graphing calculator*—can be difficult. Fortunately, systems like this can be solved precisely with methods discussed in Sections 8.2 and 8.3.

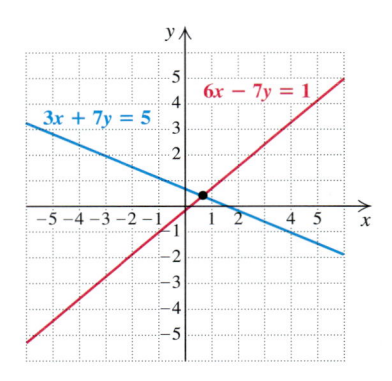

EXERCISE SET 8.1

Determine whether each ordered pair is a solution of the system of equations. Use alphabetical order of the variables.

1. $(3, 2)$; $2x + 3y = 12$,
   $x - 4y = -5$

2. $(1, 5)$; $5x - 2y = -5$,
   $3x - 7y = -32$

3. $(3, 2)$; $3b - 2a = 0$,
   $b + 2a = 15$

4. $(2, -2)$; $b + 2a = 2$,
   $b - a = -4$

5. $(15, 20)$; $3x - 2y = 5$,
   $6x - 5y = -10$

6. $(-1, -3)$; $3r + s = -6$,
   $2r = 1 + s$

Solve each system of equations by graphing. If there is no solution or an infinite number of solutions, state this.

7. $x + y = 6$,
   $x - y = 2$

8. $x - y = 1$,
   $x + y = 3$

9. $y = -2x + 5$,
   $x - y = 1$

10. $y = 2x - 5$,
    $x + y = 4$

11. $3x + y = 4$,
    $x - y = 4$

12. $4x + 2y = 6$,
    $3x - y = 2$

13. $4x - 5y = 20$,
    $8x - 10y = 12$

14. $6x + 12 = 2y$,
    $6 - y = -3x$

15. $x = 2$,
    $y = -1$

16. $y = -3x + 2$,
    $y = 3x$

17. $x + 3y = 6$,
    $4x + 12y = 24$

18. $x = -4$,
    $y = 2$

19. $3y - 9 = 6x,$
$y = x$

20. $2x = 3y - 6,$
$x = 3y$

21. $y = \frac{1}{5}x + 4,$
$x + y = -2$

22. $y + 4x = -1,$
$8x + 2y = 6$

23. $2x + y = -1,$
$y = 3x - 6$

24. $2x - y = 2,$
$y = \frac{1}{2}x + 1$

25. $2x - 3y = 5,$
$x - 2y = 6$

26. $3x + 4y = 8,$
$x + 2y = 10$

27. $3x + 2y = 1,$
$2x + 5y = -14$

28. $4x + 2y = -2,$
$5x - y = -20$

29. $x = 3y,$
$y = 2$

30. $x = \frac{1}{2}y,$
$x = 3$

SKILL MAINTENANCE

Simplify.

31. $\dfrac{x+2}{x-4} - \dfrac{x+1}{x+4}$

32. $\dfrac{2x^2 - x - 15}{x^2 - 9}$

Classify each polynomial as a monomial, binomial, trinomial, or none of these.

33. $5x^2 - 3x + 7$

34. $4x^3 - 2x^2$

Solve.

35. $2(7 - y) + 3y = 12$

36. $4x - 5(9 - 2x) = 7$

SYNTHESIS

37. ◆ Why is slope–intercept form especially useful when solving systems of equations by graphing?

38. ◆ Is it possible for a system of two linear equations to have exactly two solutions? Why or why not?

39. ◆ Suppose that the equations in a system of two linear equations are dependent. Does it follow that the system is consistent? Why or why not?

40. ◆ It can be shown that if $3x - 1 = 9 - 2x$, then $x = 2$. How can this information be used to determine where the graphs of $y = 3x - 1$ and $y = 9 - 2x$ intersect?

41. Which of the systems in Exercises 7–30 contain dependent equations?

42. Which of the systems in Exercises 7–30 are consistent?

43. Which of the systems in Exercises 7–30 are inconsistent?

44. Which of the systems in Exercises 7–30 contain independent equations?

45. Write a system of equations that has $(2, -4)$ as the solution. Answers may vary.

46. Write an equation that can be paired with $4x + 3y = 6$ to form a system that has $(3, -2)$ as the solution. Answers may vary.

47. The solution of the following system is $(2, -3)$. Find A and B.
$Ax - 3y = 13,$
$x - By = 8$

48. Solve by graphing:
$4x - 8y = -7,$
$2x + 3y = 7.$
(*Hint*: Try different scales on your graph.)

49. *Reducing utility costs.* Elizabeth owns a three-family house with a 20-yr-old electric water heater that uses $100 of electricity per month. A new gas water heater will cost $25 per month to operate, but costs $250 to purchase and $150 to install.

 a) Create cost equations for both water heaters, taking into consideration the purchase and installation of the new heater.

 b) Graph both cost equations on the same set of axes.

 c) After how many months of use with the gas water heater will Elizabeth break even?

50. *Copying costs.* Shelby occasionally goes to Mailboxes Etc.® with small copying jobs. He can purchase a "copy card" for $18 that will entitle him to 300 copies, or he can simply pay 8¢ per page.

 a) Create cost equations for each method of paying for a number (up to 300) of copies.

 b) Graph both cost equations on the same set of axes.

 c) Use the graph to determine how many copies Shelby must make if the card is to be more economical.

51. 📈 Use a grapher to solve the system
$y = 1.2x - 32.7,$
$y = -0.7x + 46.15.$

COLLABORATIVE CORNER

Focus: Modeling, systems of equations, and graphing
Time: 20–30 minutes
Group size: 3
Materials: Graph paper and straightedges

From the minute a new car is driven out of the dealership, its value drops with the passing of time. The N.A.D.A.® Official Used Car Guide (often called the "blue book" despite its orange cover) is a monthly listing of the trade-in and retail values of used cars. The data below are taken from the New England edition of the N.A.D.A. guides for April and December of 1996.

Car	Trade-in Value in April 1996	Trade-in Value in December 1996
1995 Honda Accord 4-door sedan DX	$10,775	$10,125
1995 Ford Taurus 4-door sedan LX	$12,125	$10,600
1995 Chevrolet Camaro 2-door with T-Top	$11,450	$10,400

ACTIVITY

1. Each group member should select a different one of the cars listed in the table as his or her own. Then, sharing one graph, each student should draw a line representing the trade-in value of his or her car. Let the horizontal axis represent the time, in months, since April 1996, and let the vertical axis represent the trade-in value of each car. Decide as a group how many months or dollars each square should represent. Make the drawing as neat as possible. (Assume that the values are dropping linearly.)
2. Using the graph from part (1), the "owner" of the Accord should determine when the Accord and the Taurus had, or will have, the same trade-in value. Similarly, the Taurus owner should determine when the Taurus and the Camaro had, or will have, the same trade-in value. Finally, the Camaro owner should determine when the Camaro and the Accord had, or will have, the same trade-in value.
3. At what *rate* is each car depreciating and how are the different rates illustrated in the graph of part (1)?
4. Each student should find an equation for the line that he or she drew. Then, sharing all three equations, the group should check the three answers from part (2).
5. If one of the three cars had to be sold in December 1996, which one would your group have sold and why? Compare answers with other groups.

8.2 Systems of Equations and Substitution

The Substitution Method • Solving for the Variable First • Problem Solving

Near the end of Section 8.1, we mentioned that graphing can be an imprecise method for solving systems. In this section and the next, we develop methods of finding exact solutions using algebra.

The Substitution Method

One nongraphical method for solving systems is known as the **substitution method.** It uses algebra and is thus considered an *algebraic* method.

EXAMPLE 1 Solve the system

$$x + y = 7, \quad (1)$$
$$y = 3x - 1. \quad (2)$$

SOLUTION The second equation says that y and $3x - 1$ represent the same value. Thus, in the first equation, we can substitute $3x - 1$ for y:

$x + y = 7,$ Equation (1)
$x + 3x - 1 = 7.$ Substituting $3x - 1$ for y

The equation $x + 3x - 1 = 7$ has only one variable, for which we now solve:

$4x - 1 = 7$ Combining like terms
$4x = 8$ Adding 1 on both sides
$x = 2.$ Dividing by 4 on both sides

We have found the x-value of the solution. To find the y-value, we return to the original pair of equations. Substituting into either equation will give us the y-value. We choose equation (1):

$x + y = 7$ Equation (1)
$2 + y = 7$ Substituting 2 for x
$y = 5.$ Subtracting 2 on both sides

The ordered pair $(2, 5)$ appears to be a solution. We check:

$$\begin{array}{c|c} x + y = 7 & y = 3x - 1 \\ \hline 2 + 5 \; ? \; 7 & 5 \; ? \; 3 \cdot 2 - 1 \\ 7 \; | \; 7 \quad \text{TRUE} & 5 \; | \; 5 \quad \text{TRUE} \end{array}$$

Since $(2, 5)$ checks, it is the solution. For this particular system, we can also check by examining Example 2 in Section 8.1. There we found the same solution by graphing.

> **CAUTION!** A solution of a system of equations in two variables is a *pair* of numbers. Once you have solved for one variable, don't forget the other.

EXAMPLE 2   Solve:
$$x = 13 - 3y, \quad (1)$$
$$5 - x = y. \quad (2)$$

SOLUTION   We substitute $13 - 3y$ for x in the second equation:

$$5 - x = y, \qquad \text{Equation (2)}$$
$$5 - (13 - 3y) = y. \qquad \text{Substituting } 13 - 3y \text{ for } x. \text{ The parentheses are important.}$$

Now we solve for y:

$$5 - 13 + 3y = y \qquad \text{Removing parentheses and changing signs}$$
$$-8 + 3y = y$$
$$-8 = -2y \qquad \text{Solving for } y$$
$$4 = y.$$

Next, we substitute 4 for y in equation (1) of the original system:

$$x = 13 - 3y \qquad \text{Equation (1)}$$
$$x = 13 - 3 \cdot 4 \qquad \text{Substituting 4 for } y$$
$$x = 1. \qquad \text{Simplifying}$$

The pair $(1, 4)$ is the solution. A graph is shown in the margin as a check.

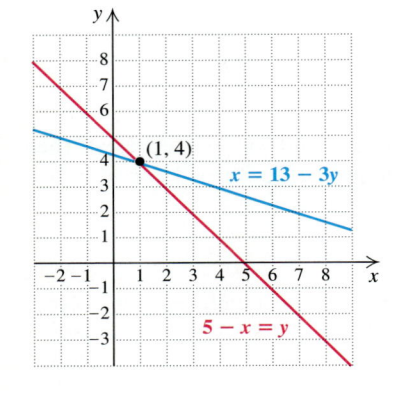

Solving for the Variable First

Sometimes neither equation has a variable alone on one side. In that case, we solve one equation for one of the variables and then proceed as before.

EXAMPLE 3   Solve:
$$x - 2y = 6, \quad (1)$$
$$3x + 2y = 4. \quad (2)$$

SOLUTION   We can solve either equation for one variable. Since the coefficient of x is 1 in equation (1), it is easier to solve that equation for x:

$$x - 2y = 6 \qquad \text{Equation (1)}$$
$$x = 6 + 2y. \qquad \text{Adding } 2y \qquad (3)$$

8.2 SYSTEMS OF EQUATIONS AND SUBSTITUTION

TECHNOLOGY CONNECTION 8.2

To check Example 3 with a grapher, we must first solve each equation for y. When we do so, equation (1) becomes $y = (6 - x)/(-2)$ and equation (2) becomes $y = (4 - 3x)/2$.

1. Use the INTERSECT option of the CALC menu or use ZOOM and TRACE to determine the solution of the system.

We substitute $6 + 2y$ for x in equation (2) of the original pair and solve for y:

$$3x + 2y = 4 \quad \text{Equation (2)}$$
$$3(6 + 2y) + 2y = 4 \quad \text{Substituting } 6 + 2y \text{ for } x$$

Remember to use parentheses when you substitute.

$$18 + 6y + 2y = 4 \quad \text{Using the distributive law}$$
$$18 + 8y = 4 \quad \text{Combining like terms}$$
$$8y = -14 \quad \text{Subtracting 18}$$
$$y = \frac{-14}{8} = -\frac{7}{4}. \quad \text{Dividing by 8}$$

To find x, we can substitute $-\frac{7}{4}$ for y in equation (1), (2), or (3). Because it is generally easier to use an equation that has already been solved for a specific variable, we decide to use equation (3):

$$x = 6 + 2y = 6 + 2\left(-\frac{7}{4}\right) = 6 - \frac{7}{2}, \text{ or } \frac{5}{2}.$$

We check the ordered pair $\left(\frac{5}{2}, -\frac{7}{4}\right)$.

Check:

$$\begin{array}{c|c}
x - 2y = 6 & 3x + 2y = 4 \\
\hline
\frac{5}{2} - 2\left(-\frac{7}{4}\right) \; ? \; 6 & 3 \cdot \frac{5}{2} + 2\left(-\frac{7}{4}\right) \; ? \; 4 \\
\frac{5}{2} + \frac{7}{2} & \frac{15}{2} - \frac{7}{2} \\
\frac{12}{2} & \frac{8}{2} \\
6 \;\big|\; 6 \quad \text{TRUE} & 4 \;\big|\; 4 \quad \text{TRUE}
\end{array}$$

Since $\left(\frac{5}{2}, -\frac{7}{4}\right)$ checks, it is the solution.

Some systems have no solution and some have an infinite number of solutions.

EXAMPLE 4 Solve each system.

a) $y = 3x + 4$, (1)
   $y = 3x - 3$ (2)

b) $2y = 6x + 4$, (1)
   $y = 3x + 2$ (2)

SOLUTION

a) We solved this system graphically in Example 3 of Section 8.1. The lines are parallel and the system has no solution. Let's see what happens if we try to solve this system by substituting $3x - 3$ for y in the first equation:

$$y = 3x + 4 \quad \text{Equation (1)}$$
$$3x - 3 = 3x + 4 \quad \text{Substituting } 3x - 3 \text{ for } y$$
$$-3 = 4. \quad \text{Subtracting } 3x \text{ on both sides}$$

When we subtract $3x$ on both sides, we obtain a *false* equation. In such a case, when solving algebraically leads to a false equation, we state that the system has no solution and thus is inconsistent.

b) The graph of this system is shown on page 391. The equations have the same graph so the system has an infinite number of solutions. Solving by substitution, we would replace y in the first equation with $3x + 2$:

$$2y = 6x + 4 \qquad \text{Equation (1)}$$
$$2(3x + 2) = 6x + 4 \qquad \text{Substituting } 3x + 2 \text{ for } y$$
$$6x + 4 = 6x + 4.$$

Since this last equation is true for *any* choice of x, the system has an infinite number of solutions. Whenever the algebraic solution of a system of two equations leads to an equation that is always true, we state that the equations in the system are dependent.

Problem Solving

Now let's use the substitution method in problem solving.

EXAMPLE 5

Supplementary Angles. Two angles are supplementary. One angle is 30° more than 2 times the other. Find the angles.

SOLUTION

1. **Familiarize.** Recall that two angles are supplementary if their sum is 180°. We could try to guess a solution, but instead we make a sketch and translate. Let x and y represent the two angles.

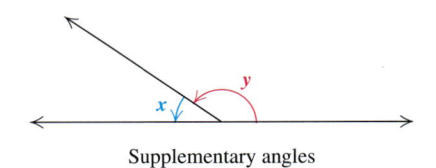

Supplementary angles

2. **Translate.** Since we are told that the angles are supplementary, one equation is

$$x + y = 180. \qquad (1)$$

The second sentence can be translated as follows:

One angle is 30° more than two times the other.
$$y = 2x + 30 \qquad (2)$$

We now have a system of two equations in two unknowns:

$$x + y = 180, \qquad (1)$$
$$y = 2x + 30. \qquad (2)$$

3. **Carry out.** We solve the system using substitution:

$$x + (2x + 30) = 180 \quad \text{Substituting } 2x + 30 \text{ for } y \text{ in equation (1)}$$
$$3x + 30 = 180$$
$$3x = 150 \quad \text{Subtracting 30}$$
$$x = 50. \quad \text{Dividing by 3}$$

Substituting 50 for x in equation (1) then gives us

$$x + y = 180 \quad \text{Equation (1)}$$
$$50 + y = 180 \quad \text{Substituting 50 for } x$$
$$y = 130.$$

4. **Check.** If one angle is 50° and the other is 130°, then the sum of the angles is 180°. Thus the angles are supplementary. If 30° is added to twice the smaller angle, we have $2 \cdot 50° + 30°$, or 130°, which is the other angle. The pair of angles checks.

5. **State.** One angle is 50° and the other is 130°.

EXERCISE SET 8.2

Solve each system using the substitution method. If a system has no solution or an infinite number of solutions, state this.

1. $x + y = -2,$
   $y = x - 6$
2. $x + y = 10,$
   $x = y + 8$
3. $x = y + 1,$
   $x + 2y = 4$
4. $y = x - 3,$
   $3x + y = 5$
5. $y = 2x - 5,$
   $3y - x = 5$
6. $y = 2x + 1,$
   $x + y = 4$
7. $4x + y = 2,$
   $y = -2x$
8. $r = -3s,$
   $r + 4s = 10$
9. $2x + 3y = 8,$
   $x = y - 6$
10. $x = y - 8,$
    $3x + 2y = 1$
11. $x = 2y + 1,$
    $3x - 6y = 2$
12. $y = 3x - 1,$
    $6x - 2y = 2$
13. $s + t = -4,$
    $s - t = 2$
14. $x - y = 6,$
    $x + y = -2$
15. $x - y = 5,$
    $x + 2y = 7$
16. $y - 2x = -6,$
    $2y - x = 5$
17. $x - 2y = 7,$
    $3x - 21 = 6y$
18. $x - 4y = 3,$
    $2x - 6 = 8y$
19. $y = -2x + 3,$
    $2y = -4x + 6$
20. $y = 2x + 5,$
    $y = 2x - 5$
21. $x + 2y = 10,$
    $3x + 4y = 8$
22. $2x + 3y = -2,$
    $2x - y = 9$
23. $3a + 2b = 2,$
    $-2a + b = 8$
24. $x - y = -3,$
    $2x + 3y = -6$
25. $y - 2x = 0,$
    $3x + 7y = 17$
26. $r - 2s = 0,$
    $4r - 3s = 15$
27. $8x + 2y = 6,$
    $4x = 3 - y$
28. $x - 3y = 7,$
    $-4x + 12y = 28$
29. $x - 3y = -1,$
    $5y - 2x = 4$
30. $x - 2y = 5,$
    $2y - 3x = 1$
31. $5x - y = 0,$
    $5x - y = -2$
32. $2x = y - 3,$
    $2x = y + 5$

Solve.

33. The sum of two numbers is 49. One number is 3 more than the other. Find the numbers.
34. The sum of two numbers is 56. One number is 2 more than the other. Find the numbers.
35. Find two numbers for which the sum is 58 and the difference is 14.

36. Find two numbers for which the sum is 66 and the difference is 12.
37. The difference between two numbers is 16. Three times the larger number is 7 times the smaller. What are the numbers?
38. The difference between two numbers is 18. Twice the smaller number plus 3 times the larger is 74. What are the numbers?
39. *Supplementary angles.* Two angles are supplementary. One angle is 30° less than twice the other. Find the angles.
40. *Supplementary angles.* Two angles are supplementary. One angle is 8° less than three times the other. Find the angles.
41. *Complementary angles.* Two angles are complementary. Their difference is 34°. Find the angles. (*Complementary angles* are angles for which the sum is 90°.)

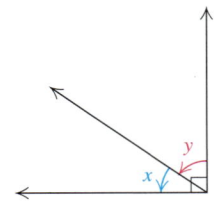

Complementary angles

42. *Complementary angles.* Two angles are complementary. One angle is 42° more than one-half the other. Find the angles.
43. *Perimeter of a rectangle.* The state of Colorado is a rectangle with a perimeter of 1300 mi. The length is 110 mi more than the width. Find the length and the width.

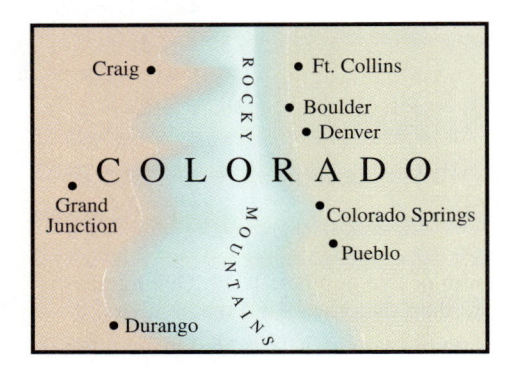

44. *Perimeter of a rectangle.* The state of Wyoming is a rectangle with a perimeter of 1280 mi. The width is 90 mi less than the length. Find the length and the width. 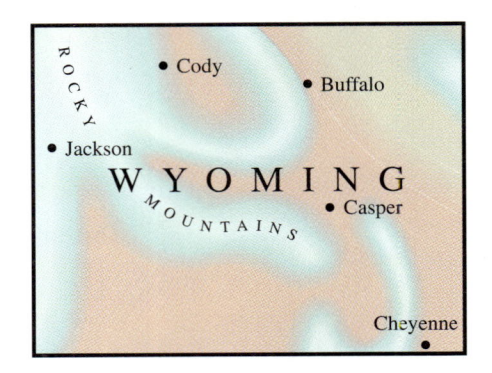
45. *Racquetball.* A regulation racquetball court should have a perimeter of 120 ft, with a length that is twice the width. Find the length and the width of a court.

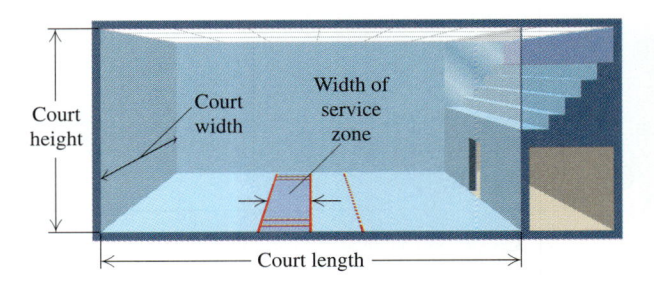

46. *Racquetball.* The height of the front wall of a standard racquetball court is 4 times the width of the service zone (see the figure). Together, these measurements total 25 ft. Find the height and the width.
47. *Soccer.* The perimeter of a soccer field is 340 yd. The length exceeds the width by 50 yd. Find the length and the width.

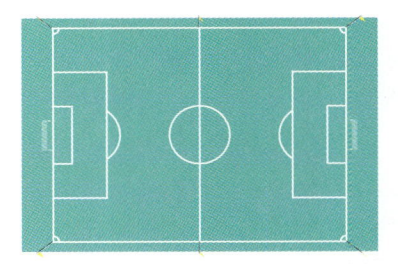

48. *Football.* The perimeter of a football field (including the end zones) is $346\frac{2}{3}$ yd. The length is

$66\frac{2}{3}$ yd longer than the width. Find the length and the width.

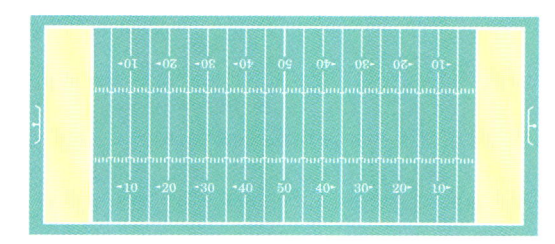

SKILL MAINTENANCE

Factor completely. If a polynomial is prime, state this.

49. $6x^2 - 13x + 6$
50. $4p^2 - p - 3$
51. $4x^2 + 3x + 2$
52. $9a^2 - 25$

Simplify.

53. $2(5x - 3y) - 5(2x + y)$
54. $4(2x + 3y) + 3(5x - 4y)$

SYNTHESIS

55. ◆ Joel solves every system of two equations (in x and y) by first solving for y in the first equation and then substituting into the second equation. Is he using the best approach? Why or why not?

56. ◆ Describe two advantages of the substitution method over the graphing method for solving systems of equations.

57. ◆ Under what circumstances can a system of equations be solved more easily by graphing than by substitution?

58. ◆ Janine can tell by inspection that the system
$$x = 2y - 1,$$
$$x = 2y + 3$$
has no solution. How can she tell?

Solve by the substitution method.

59. $y - 2.35x = -5.97,$
    $2.14y - x = 4.88$

60. $\frac{1}{4}(a - b) = 2,$
    $\frac{1}{6}(a + b) = 1$

61. $\frac{x}{4} + \frac{3y}{4} = 1,$
    $\frac{x}{5} - \frac{y}{2} = 3$

62. $0.4x + 0.7y = 0.1,$
    $0.5x - 0.1y = 1.1$

Exercises 63 and 64 contain systems of three equations in three variables. A solution is an ordered triple of the form (x, y, z). Use the substitution method to solve.

63. $x + y + z = 4,$
    $x - 2y - z = 1,$
    $y = -1$

64. $x + y + z = 180,$
    $x = z - 70,$
    $2y - z = 0$

65. ◆ Solve Example 3 by first solving for $2y$ in equation (1) and then substituting for $2y$ in equation (2). Is this method easier than the procedure used in Example 3? Why or why not?

66. Write a system of two linear equations that can be solved more quickly—but still precisely—by a grapher than by substitution. Time yourself using both methods to solve the system.

8.3 Systems of Equations and Elimination

Solving by the Elimination Method • Comparing the Three Methods • Problem Solving

We have seen that graphing is not always a precise method of solving a system of equations, especially when fractional solutions are involved. The substitution method, considered in Section 8.2, is precise but sometimes difficult to use. For example, to solve the system

$$2x + 3y = 13, \quad (1)$$
$$4x - 3y = 17 \quad (2)$$

by substitution, we would need to first solve for a variable in one of the equations. Were we to solve equation (1) for y, we would find (after several steps) that $y = \frac{13}{3} - \frac{2}{3}x$. We could then use the expression $\frac{13}{3} - \frac{2}{3}x$ in equation (2) as a replacement for y:

$$4x - 3\left(\frac{13}{3} - \frac{2}{3}x\right) = 17.$$

As you can see, although substitution *could* be used to solve this system, doing so is not easy. Fortunately, another method, *elimination*, can be used to solve systems and, on problems like this, is simpler to use.

Solving by the Elimination Method

The **elimination method** for solving systems of equations makes use of the addition principle. To see how it works, we use it to solve the system discussed above.

EXAMPLE 1 Solve the system

$$2x + 3y = 13, \quad (1)$$
$$4x - 3y = 17. \quad (2)$$

SOLUTION According to equation (2), $4x - 3y$ and 17 are the same number. Thus we can add $4x - 3y$ to the left side of equation (1) and 17 to the right side:

$$2x + 3y = 13 \quad (1)$$
$$\underline{4x - 3y = 17} \quad (2)$$
$$6x + 0y = 30. \qquad \text{Adding. Note that } y \text{ has been "eliminated."}$$

The resulting equation has just one variable:

$$6x = 30.$$

We now solve for x and obtain $x = 5$.

Next, we substitute 5 for x in either of the original equations:

$$2x + 3y = 13 \qquad \text{Equation (1)}$$
$$2 \cdot 5 + 3y = 13 \qquad \text{Substituting 5 for } x$$
$$10 + 3y = 13$$
$$3y = 3$$
$$y = 1. \qquad \text{Solving for } y$$

We check the ordered pair (5, 1).

Check:

$$\begin{array}{c|c} 2x + 3y = 13 \\ \hline 2(5) + 3(1) \ ? \ 13 \\ 10 + 3 \\ 13 \ | \ 13 \ \text{TRUE} \end{array} \qquad \begin{array}{c|c} 4x - 3y = 17 \\ \hline 4(5) - 3(1) \ ? \ 17 \\ 20 - 3 \\ 17 \ | \ 17 \ \text{TRUE} \end{array}$$

Since (5, 1) checks in both equations, it is the solution.

The system in Example 1 is easier to solve by elimination than by substitution because the term $-3y$ in equation (2) is the opposite of the term $3y$ in equation (1). When a system has no pair of terms that are opposites, we need to multiply one or both of the equations by appropriate numbers to create a pair of terms that are opposites.

EXAMPLE 2 Solve:

$$2x + 3y = 8, \quad (1)$$
$$x + 3y = 7. \quad (2)$$

SOLUTION For these equations, addition will not eliminate a variable. However, if the $3y$ were $-3y$ in one equation, we could eliminate y. We multiply on both sides of equation (2) by -1 to find an equivalent equation that contains $-3y$, and then add:

$$\begin{aligned} 2x + 3y &= 8 & &\text{Equation (1)} \\ -x - 3y &= -7 & &\text{Multiplying on both sides of equation (2) by } -1 \\ \hline x &= 1. & &\text{Adding} \end{aligned}$$

Next, we substitute 1 for x in either of the original equations:

$$\begin{aligned} x + 3y &= 7 & &\text{Equation (2)} \\ 1 + 3y &= 7 & &\text{Substituting 1 for } x \\ 3y &= 6 \\ y &= 2. \end{aligned} \Bigg\} \text{Solving for } y$$

We can check the ordered pair (1, 2).

Check:

$$\begin{array}{c|c} 2x + 3y = 8 \\ \hline 2 \cdot 1 + 3 \cdot 2 \ ? \ 8 \\ 2 + 6 \\ 8 \ | \ 8 \ \text{TRUE} \end{array} \qquad \begin{array}{c|c} x + 3y = 7 \\ \hline 1 + 3 \cdot 2 \ ? \ 7 \\ 1 + 6 \\ 7 \ | \ 7 \ \text{TRUE} \end{array}$$

Since (1, 2) checks in both equations, it is the solution.

In Example 2, we used the multiplication principle, multiplying by -1. We often need to multiply by a number other than -1.

EXAMPLE 3 Solve.

$$3x + 6y = -6, \quad (1)$$
$$5x - 2y = 14. \quad (2)$$

404   CHAPTER 8   •   SYSTEMS OF EQUATIONS AND PROBLEM SOLVING

SOLUTION   No terms in $3x + 6y = -6$ and $5x - 2y = 14$ are opposites, but if equation (2) is multiplied by 3 (or if equation (1) is multiplied by $\frac{1}{3}$), the coefficients of y will be opposites:

$$\begin{aligned} 3x + 6y &= -6 &&\text{Equation (1)} \\ \underline{15x - 6y = 42} &&&\text{Multiplying on both sides of equation (2) by 3} \\ 18x &= 36 &&\text{Adding} \\ x &= 2. &&\text{Solving for } x \end{aligned}$$

We then go back to equation (1) and substitute 2 for x:

$$\begin{aligned} 3 \cdot 2 + 6y &= -6 &&\text{Substituting 2 for } x \text{ in equation (1)} \\ 6 + 6y &= -6 \\ 6y &= -12 \\ y &= -2. \end{aligned}\Bigg\} \text{Solving for } y$$

We leave it to the student to confirm that $(2, -2)$ checks and is the solution. The graph in the margin also serves as a check.

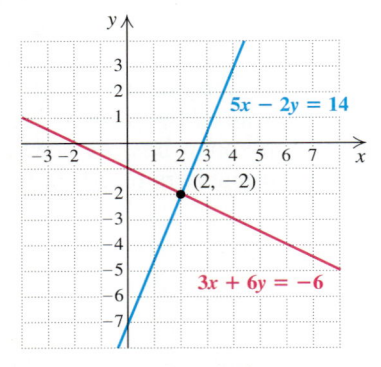

A visualization of Example 3

EXAMPLE 4   Solve.

$$3y + 1 + 2x = 0,$$
$$5x = 7 - 4y.$$

SOLUTION   It is often helpful to write both equations in the form $Ax + By = C$ before attempting to eliminate a variable:

$$\begin{aligned} 2x + 3y &= -1, &&\text{Subtracting 1 on both sides and rearranging the terms of the first equation} \\ 5x + 4y &= 7. &&\text{Adding } 4y \text{ on both sides of the second equation} \end{aligned}$$

Since neither coefficient of x is a multiple of the other and neither coefficient of y is a multiple of the other, we use the multiplication principle with *both* equations:

$$\begin{aligned} 2x + 3y &= -1, &&(1) \\ 5x + 4y &= 7. &&(2) \end{aligned} \quad \text{Note that the LCM of 2 and 5 is 10.}$$

We can eliminate the x-term by multiplying on both sides of equation (1) by 5 and on both sides of equation (2) by -2:

$$\begin{aligned} 10x + 15y &= -5 &&\text{Multiplying on both sides of equation (1) by 5} \\ \underline{-10x - 8y = -14} &&&\text{Multiplying on both sides of equation (2) by } -2 \\ 7y &= -19 &&\text{Adding} \\ y &= \tfrac{-19}{7} = -\tfrac{19}{7}. &&\text{Dividing by 7} \end{aligned}$$

We substitute $-\frac{19}{7}$ for y in equation (1):

$$\begin{aligned} 2x + 3y &= -1 &&\text{Equation (1)} \\ 2x + 3\left(-\tfrac{19}{7}\right) &= -1 &&\text{Substituting } -\tfrac{19}{7} \text{ for } y \\ 2x - \tfrac{57}{7} &= -1 \\ 2x &= -1 + \tfrac{57}{7} &&\text{Adding } \tfrac{57}{7} \text{ on both sides} \end{aligned}$$

and

$$2x = -\frac{7}{7} + \frac{57}{7} = \frac{50}{7}$$
$$x = \frac{50}{7} \cdot \frac{1}{2} = \frac{25}{7}. \quad \text{Solving for } x$$

We check the ordered pair $\left(\frac{25}{7}, -\frac{19}{7}\right)$.

Check:

$$\begin{array}{c|c}
3y + 1 + 2x = 0 \\ \hline
3\left(-\frac{19}{7}\right) + 1 + 2 \cdot \frac{25}{7} \; ? \; 0 \\
-\frac{57}{7} + \frac{7}{7} + \frac{50}{7} \\
0 \;\bigg|\; 0 \quad \text{TRUE}
\end{array}
\qquad
\begin{array}{c|c}
5x = 7 - 4y \\ \hline
5 \cdot \frac{25}{7} \; ? \; 7 - 4\left(-\frac{19}{7}\right) \\
\frac{125}{7} \;\bigg|\; \frac{49}{7} + \frac{76}{7} \\
\frac{125}{7} \;\bigg|\; \frac{125}{7} \quad \text{TRUE}
\end{array}$$

The solution is $\left(\frac{25}{7}, -\frac{19}{7}\right)$.

Next, we consider a system with no solution and see what happens when we apply the elimination method.

EXAMPLE 5 Solve:

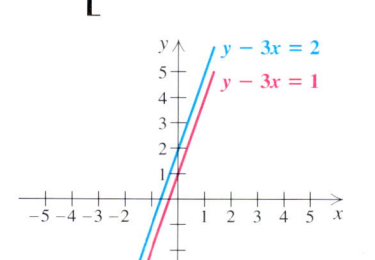

$$y - 3x = 2, \quad (1)$$
$$y - 3x = 1. \quad (2)$$

SOLUTION To eliminate y, we multiply on both sides of equation (2) by -1. Then we add:

$$\begin{array}{rl}
y - 3x = & 2 \\
-y + 3x = & -1 \quad \text{Multiplying on both sides of equation (2) by } -1 \\ \hline
0 = & 1. \quad \text{Adding}
\end{array}$$

Note that in eliminating y, we eliminated x as well. The resulting equation, $0 = 1$, is false for any pair (x, y), so there is *no solution*.

A visualization of Example 5

Sometimes there is an infinite number of solutions. Consider a system that we graphed in Example 4 of Section 8.1.

EXAMPLE 6 Solve:

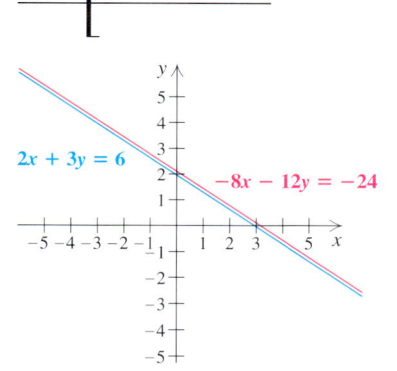

$$2x + 3y = 6, \quad (1)$$
$$-8x - 12y = -24. \quad (2)$$

SOLUTION To eliminate x, we multiply on both sides of equation (1) by 4 and then add the two equations:

$$\begin{array}{rl}
8x + 12y = & 24 \quad \text{Multiplying on both sides of equation (1) by 4} \\
-8x - 12y = & -24 \\ \hline
0 = & 0. \quad \text{Adding}
\end{array}$$

Again, we have eliminated *both* variables. The resulting equation, $0 = 0$, is always true, indicating that the equations are dependent. Such a system has an infinite number of solutions.

A visualization of Example 6

CHAPTER 8 • SYSTEMS OF EQUATIONS AND PROBLEM SOLVING

When decimals or fractions appear, we can first multiply to clear them. Then we proceed as before.

EXAMPLE 7 Solve:

$$\frac{1}{2}x + \frac{3}{4}y = 2, \quad (1)$$
$$x + 3y = 7. \quad (2)$$

SOLUTION The number 4 is the LCD for equation (1). Thus we multiply on both sides of equation (1) by 4 to clear fractions:

$$4\left(\frac{1}{2}x + \frac{3}{4}y\right) = 4 \cdot 2$$
$$4 \cdot \frac{1}{2}x + 4 \cdot \frac{3}{4}y = 8 \quad \text{Using the distributive law}$$
$$2x + 3y = 8.$$

The resulting system is

$$2x + 3y = 8, \quad \text{This equation is equivalent to equation (1).}$$
$$x + 3y = 7.$$

As we saw in Example 2, the solution of this system is (1, 2).

Comparing the Three Methods

The following summary compares the graphical, substitution, and elimination methods for solving systems of equations.

Method	Strengths	Weaknesses
Graphical	Solutions are displayed visually. Works with any system that can be graphed.	Inexact when solutions involve numbers that are not integers or are very large and off the graph.
Substitution	Always yields exact solutions. Easy to use when a variable is alone on one side of an equation.	Introduces extensive computations with fractions when solving more complicated systems. Solutions are not graphically displayed.
Elimination	Always yields exact solutions. Easy to use when fractions or decimals appear in the system.	Solutions are not graphically displayed.

To determine the best method to use for solving a particular system, consider the strengths and weaknesses listed above. As you gain experience with these methods, it will become easier to choose the best method for the system you are solving.

Problem Solving

We now use the elimination method to solve a problem.

EXAMPLE 8

Car Rental Rates. At one time, Budget Rent-A-Car rented compact cars at a daily rate of $43.95 plus 40 cents per mile. Thrifty Rent-A-Car rented compact cars at a daily rate of $42.95 plus 42 cents per mile. For what mileage was the cost the same?

SOLUTION

1. **Familiarize.** To become familiar with the problem, we make a guess. Suppose a person rented a compact car from each rental agency and drove it 100 mi. The cost at Budget would have been

   $43.95 + $0.40(100) = $43.95 + $40.00, or $83.95.

   The cost at Thrifty would have been

   $42.95 + $0.42(100) = $42.95 + $42.00, or $84.95.

   Note that we converted all monetary units to dollars. Because $83.95 ≠ $84.95, our guess is incorrect.

   To use algebra to solve the problem, we let m = the number of miles driven and c = the total cost of the car rental.

2. **Translate.** We translate the first statement, using $0.40 for 40 cents. It helps to reword the problem before translating.

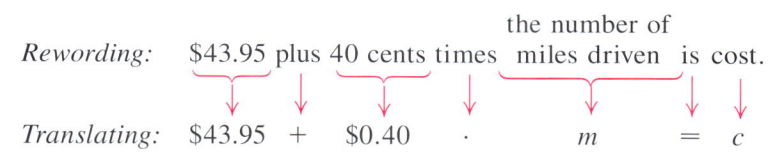

   Rewording: $43.95 plus 40 cents times the number of miles driven is cost.

   Translating: $43.95 + $0.40 · m = c

   We translate the second statement, but again it helps to reword it first.

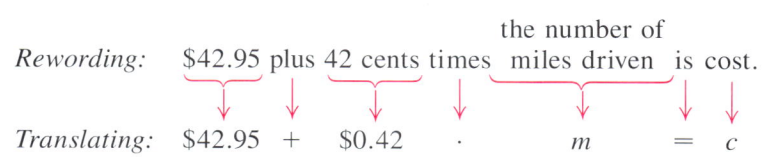

   Rewording: $42.95 plus 42 cents times the number of miles driven is cost.

   Translating: $42.95 + $0.42 · m = c

   We have now translated the problem to a system of equations:

   $43.95 + 0.40m = c$,
   $42.95 + 0.42m = c$.

3. **Carry out.** To solve the system, we clear each equation of decimals by multiplying on both sides by 100. Then we multiply the second equation by -1 and add to eliminate c:

   $$\begin{array}{r} 4395 + 40m = 100c \\ -4295 - 42m = -100c \\ \hline 100 - 2m = 0 \\ 100 = 2m \\ 50 = m. \end{array}$$

   Although we are not asked to solve for the total cost c, we will do so as part of the check.

4. Check. For 50 mi, the cost of the Budget car would have been
$$43.95 + 0.40(50), \text{ or } 43.95 + 20, \text{ or } \$63.95,$$
and the cost of the Thrifty car would have been
$$42.95 + 0.42(50), \text{ or } 42.95 + 21, \text{ or } \$63.95.$$
Thus the costs were the same when the mileage was 50.

5. State. For cars driven 50 mi, the costs were the same.

EXERCISE SET 8.3

Solve using the elimination method. If a system has no solution or an infinite number of solutions, state this.

1. $x + y = 12,$
   $x - y = 6$

2. $x - y = 7,$
   $x + y = 3$

3. $x + y = 6,$
   $-x + 2y = 15$

4. $x + y = 6,$
   $-x + 3y = -2$

5. $3x - y = 9,$
   $2x + y = 6$

6. $4x - y = 1,$
   $3x + y = 13$

7. $2a + 3b = 7,$
   $-2a + b = 5$

8. $7c + 5d = 18,$
   $c - 5d = -2$

9. $8x - 5y = -9,$
   $3x + 5y = -2$

10. $3a - 3b = -15,$
    $-3a - 3b = -3$

11. $7a - 6b = 8,$
    $-7a + 6b = -8$

12. $2x + 3y = 4,$
    $-2x - 3y = -4$

13. $-x - y = 8,$
    $2x - y = -1$

14. $x + y = -7,$
    $3x + y = -9$

15. $x + 3y = 19,$
    $x - y = -1$

16. $3x - y = 8,$
    $x + 2y = 5$

17. $x + y = 5,$
    $5x - 3y = 17$

18. $x - y = 7,$
    $4x - 5y = 25$

19. $2w - 3z = -1,$
    $3w + 4z = 24$

20. $7p + 5q = 2,$
    $8p - 9q = 17$

21. $2a + 3b = -1,$
    $3a + 5b = -2$

22. $3x - 4y = 16,$
    $5x + 6y = 14$

23. $x = 3y,$
    $5x + 14 = y$

24. $5a = 2b,$
    $2a + 11 = 3b$

25. $4x - 10y = 13,$
    $-2x + 5y = 8$

26. $2p + 5q = 9,$
    $3p - 2q = 4$

27. $3x = 8y + 11,$
    $x + 6y - 8 = 0$

28. $m = 32 + n,$
    $3m = 8n + 6$

29. $3x + 5y = 4,$
    $-2x + 3y = 10$

30. $2x + y = 13,$
    $4x + 2y = 23$

31. $0.06x + 0.05y = 0.07,$
    $0.4x - 0.3y = 1.1$

32. $x - \frac{3}{2}y = 13,$
    $\frac{3}{2}x - y = 17$

33. $x + \frac{9}{2}y = \frac{15}{4},$
    $\frac{9}{10}x - y = \frac{9}{20}$

34. $1.8x - 2y = 0.9,$
    $0.04x + 0.18y = 0.15$

Solve.

35. *Car rentals.* At one time, Avis Rent-A-Car rented an intermediate-sized car at a daily rate of $53.95 plus 30¢ per mile. Another company rented an intermediate-sized car for $54.95 plus 20¢ per mile. For what mileage was the cost the same?

36. *Car rentals.* Budget rented a basic car at a daily rate of $45.95 plus 40¢ per mile. Another company rented a basic car for $46.95 plus 20¢ per mile. For what mileage was the cost the same?

37. *Complementary angles.* Two angles are complementary. One angle is 12° more than twice the other. Find the angles.

38. *Complementary angles.* Two angles are complementary. Their difference is 26°. Find the angles.

39. *Supplementary angles.* Two angles are supplementary. One is 5° more than 4 times the other. Find the angles.

40. *Supplementary angles.* Two angles are supplementary. One is 45° less than 2 times the other. Find the angles.

41. *Farming.* Sleek Meadows Horse Farm plants 31 acres of hay and oats. The owners know that their needs are best met if they plant 9 acres more of hay than oats. How many acres of each should they plant?

42. *Planting grapes.* South Wind Vineyards uses 820 acres to plant Chardonnay and Riesling grapes. The vintner knows the profits will be greatest by planting 140 acres more of Chardonnay than Riesling. How many acres of each grape should be planted?

43. *Framing.* Angel has 12 ft of molding from which he needs to make a rectangular frame. Because of the dimensions of the picture being framed, the frame must be twice as long as it is wide. What should the dimensions of the frame be?

44. *Gardening.* Patrice has 30 yd of fencing for a rectangular garden. If the garden's length is to be $1\frac{1}{2}$ times its width, what should the garden's dimensions be?

SKILL MAINTENANCE

Simplify.

45. $\dfrac{(a^2 b^{-3})^4}{a^5 b^{-6}}$

46. $\left(\dfrac{3a^2}{2b^3}\right)^2$

Factor.

47. $4x^2 + 20x + 25$

48. $9a^2 - 24a + 16$

Simplify.

49. $3.7(5) + 2.9(4)$

50. $0.3(8) + 0.5(9)$

SYNTHESIS

51. ◆ Describe a method that could be used for writing inconsistent systems of equations.

52. ◆ Describe a method that could be used for writing systems that contain dependent equations.

53. ◆ If a system has an infinite number of solutions, does it follow that *all* ordered pairs are solutions? Why or why not?

54. ◆ Explain how the multiplication and addition principles are used in this section. Then count the number of times that these principles were used in Example 4.

Solve using either substitution or elimination.

55. $y = 3x + 4,$
    $3 + y = 2(y - x)$

56. $x + y = 7,$
    $3(y - x) = 9$

57. $0.05x + y = 4,$
    $\dfrac{x}{2} + \dfrac{y}{3} = 1\dfrac{1}{3}$

58. $2(5a - 5b) = 10,$
    $-5(2a + 6b) = 10$

Solve for x and y.

59. $y = ax + b,$
    $y = x + c$

60. $ax + by + c = 0,$
    $ax + cy + b = 0$

61. *Caged rabbits and pheasants.* Several ancient Chinese books included problems that can be solved by translating to systems of equations. *Arithmetical Rules in Nine Sections* is a book of 246 problems compiled by a Chinese mathematician, Chang Tsang, who died in 152 B.C. One of the problems is: Suppose there are a number of rabbits and pheasants confined in a cage. In all, there are 35 heads and 94 feet. How many rabbits and how many pheasants are there? Solve the problem.

62. *Age.* Patrick's age is 20% of his mother's age. Twenty years from now, Patrick's age will be 52% of his mother's age. How old are Patrick and his mother now?

63. *Age.* If 5 is added to a man's age and the total is divided by 5, the result will be his daughter's age. Five years ago, the man's age was 8 times his daughter's age. Find their present ages.

64. *Dimensions of a triangle.* When the base of a triangle is increased by 1 ft and the height is increased by 2 ft, the height changes from being two thirds of the base to being four fifths of the base. Find the original dimensions of the triangle.

8.4 More Applications Using Systems

Total Value Problems • Mixture Problems

The five steps for problem solving and our methods for solving systems of equations can be used in a variety of applications.

Total Value Problems

EXAMPLE 1

Basketball Scores. In a recent NBA game, the Miami Heat scored 79 points on a combination of 35 two- and three-point baskets. How many two-point shots were made and how many three-pointers were made?

SOLUTION

1. **Familiarize.** Suppose that of the 35 baskets, 20 were two-pointers and 15 were three-pointers. These 35 baskets would then amount to a total of $20 \cdot 2 + 15 \cdot 3 = 40 + 45 = 85$ points. Although our guess is incorrect, checking the guess has familiarized us with the problem. We let w = the number of two-pointers made and r = the number of three-pointers made.

2. **Translate.** Since a total of 35 baskets was made, we must have

    $w + r = 35.$

    To find a second equation, we reword some information and focus on the points scored, just as when we checked our guess above.

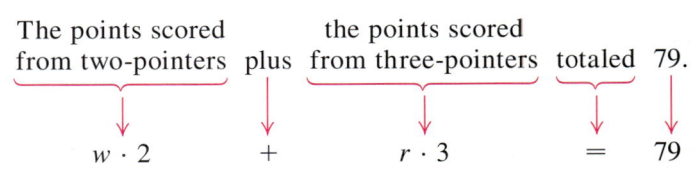

    Rewording: The points scored from two-pointers plus the points scored from three-pointers totaled 79.

    Translating: $w \cdot 2 \quad + \quad r \cdot 3 \quad = \quad 79$

    The problem has been translated to the following system of equations:

    $w + r = 35,$ (1)
    $2w + 3r = 79.$ (2)

3. **Carry out.** For purposes of review, we solve by substitution. First we solve equation (1) for w:

    $w + r = 35$ Equation (1)
    $w = 35 - r.$ Solving for w (3)

Next, we replace w in equation (2) with $35 - r$:

$2w + 3r = 79$ Equation (2)
$2(35 - r) + 3r = 79$ Substituting $35 - r$ for w
$70 - 2r + 3r = 79$ Using the distributive law
$70 + r = 79$
$r = 9.$ } Solving for r

We find w by substituting 9 for r in equation (3):

$w = 35 - r = 35 - 9 = 26.$

4. **Check.** If the Heat made 26 two-pointers and 9 three-pointers, they would have made 35 shots, for a total of $26 \cdot 2 + 9 \cdot 3 = 52 + 27 = 79$ points. The numbers check.

5. **State.** The Miami Heat made 26 two-point baskets and 9 three-point baskets.

EXAMPLE 2   *Paid Admissions.* The manager of a movie theater noted that 411 people attended a movie but forgot to note the numbers of adults and children. Admission was $7.00 each for adults and $3.75 each for children. Receipts totaled $2678.75. How many adults and how many children attended?

SOLUTION

1. **Familiarize.** One way to become familiar with a problem is to compare it to a similar problem. Here, instead of counting two-pointers and three-pointers as in Example 1, we count adults and children. Instead of point values, we use ticket prices. We let $a =$ the number of adults in attendance and $c =$ the number of children in attendance.

2. **Translate.** Since a total of 411 people attended the movie, we have

$a + c = 411.$

To find a second equation, we reword some information and focus on the amount of money taken in:

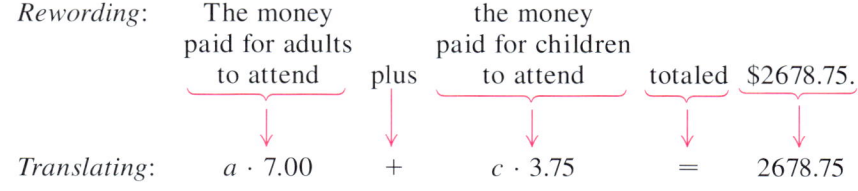

Rewording: The money paid for adults to attend plus the money paid for children to attend totaled $2678.75.

Translating: $a \cdot 7.00 + c \cdot 3.75 = 2678.75$

Presenting the information in a table can be helpful.

	Adults	Children	Total	
Admission	$7.00	$3.75		
Number Attending	a	c	411	→ $a + c = 411$
Money Taken In	7.00a	3.75c	2678.75	→ $7.00a + 3.75c = 2678.75$

We have translated to a system of equations:

$$a + c = 411, \quad (1)$$
$$7a + 3.75c = 2678.75. \quad (2)$$

3. **Carry out.** To solve the system, we first multiply equation (2) by 100 to clear decimals. The resulting system is then

$$a + c = 411, \quad (1)$$
$$700a + 375c = 267{,}875. \quad (3)$$

We solve by elimination:

$$-375a - 375c = -154{,}125 \quad \text{Multiplying on both sides of equation (1) by } -375$$
$$\underline{700a + 375c = 267{,}875} \quad \text{Equation (3)}$$
$$325a = 113{,}750 \quad \text{Adding}$$
$$a = \frac{113{,}750}{325} \quad \text{Dividing on both sides by 325}$$
$$a = 350.$$

We go back to equation (1) and substitute 350 for a:

$$a + c = 411$$
$$350 + c = 411$$
$$c = 61.$$

4. **Check.** If $c = 61$ and $a = 350$, a total of 411 people attended. The amount paid was 350($7.00), or $2450, for adults, and 61($3.75), or $228.75, for children. The total receipts were then $2450 + $228.75, or $2678.75, as stated in the problem. The numbers check.

5. **State.** The movie was attended by 350 adults and 61 children.

Mixture Problems

EXAMPLE 3

Blending Coffees. The Java Joint wants to mix Kenyan beans that sell for $8.25 per pound with Venezuelan beans that sell for $9.50 per pound to form a 50-lb batch of Morning Blend that sells for $9.00 per pound. How many pounds of Kenyan beans and how many pounds of Venezuelan beans should go into the blend?

SOLUTION

1. **Familiarize.** This problem seems similar to Example 2. Instead of adults and children, we have pounds of Kenyan coffee and pounds of Venezuelan coffee. Instead of two different prices of admission, we have two different prices per pound. Finally, instead of having the total receipts, we know the weight and price per pound of the batch of Morning Blend that is to be made. Note that we can easily find the value of the batch of Morning Blend by multiplying 50 lb times $9.00 per pound. We let k = the number of pounds of Kenyan coffee used and v = the number of pounds of Venezuelan coffee used.

2. **Translate.** Since a 50-lb batch is being made, we must have

$$k + v = 50.$$

To find a second equation, we consider the total value of the 50-lb batch. That value must be the same as the value of the Kenyan beans and the value of the Venezuelan beans that go into the blend:

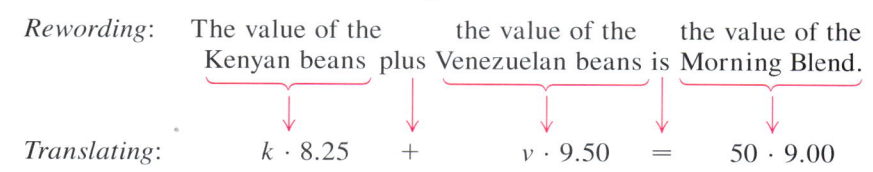

Rewording: The value of the Kenyan beans plus the value of the Venezuelan beans is the value of the Morning Blend.

Translating: $k \cdot 8.25 \;+\; v \cdot 9.50 \;=\; 50 \cdot 9.00$

This information can be presented in a table.

	Kenyan	**Venezuelan**	**Morning Blend**	
Price per Pound	$8.25	$9.50	$9.00	
Number of Pounds	k	v	50	⟶ $k + v = 50$
Value of Beans	$8.25k$	$9.50v$	$50 \cdot 9$, or 450	⟶ $8.25k + 9.50v = 450$

We have translated to a system of equations:

$$k + v = 50, \quad (1)$$
$$8.25k + 9.50v = 450. \quad (2)$$

3. **Carry out.** When equation (1) is solved for k, we have $k = 50 - v$. We then substitute $50 - v$ for k in equation (2):

$$8.25(50 - v) + 9.50v = 450 \quad \text{Solving by substitution}$$
$$412.50 - 8.25v + 9.50v = 450 \quad \text{Using the distributive law}$$
$$1.25v = 37.50 \quad \text{Combining like terms; subtracting 412.50 on both sides}$$
$$v = 30. \quad \text{Dividing on both sides by 1.25}$$

If $v = 30$, we see from equation (1) that $k = 20$.

4. **Check.** If 20 lb of Kenyan beans and 30 lb of Venezuelan beans are mixed, a 50-lb blend will result. The value of 20 lb of Kenyan beans is 20($8.25), or $165. The value of 30 lb of Venezuelan beans is 30($9.50), or $285, so the value of the blend is $165 + $285 = $450. A 50-lb blend priced at $9.00 a pound is also worth $450, so our answer checks.

5. **State.** The Morning Blend should be made by combining 20 lb of Kenyan beans with 30 lb of Venezuelan beans.

EXAMPLE 4   *Paint Colors.* At a local "paint swap," Gayle found large supplies of Skylite Pink (12.5% red pigment) and MacIntosh Red (20% red pigment). How many gallons of each color should be mixed in order to create a 10-gal batch of Summer Rose (17% red pigment)?

SOLUTION

1. **Familiarize.** This problem is similar to Example 3. Instead of mixing two types of coffee and keeping an eye on the price of the mixture, we are mixing two types of paint and keeping an eye on the amount of pigment in the mixture.

   To visualize this problem, think of the pigment as a solid that, given time, would settle to the bottom of each can. We let $p =$ the number of gallons of Skylite Pink needed and $m =$ the number of gallons of MacIntosh Red needed.

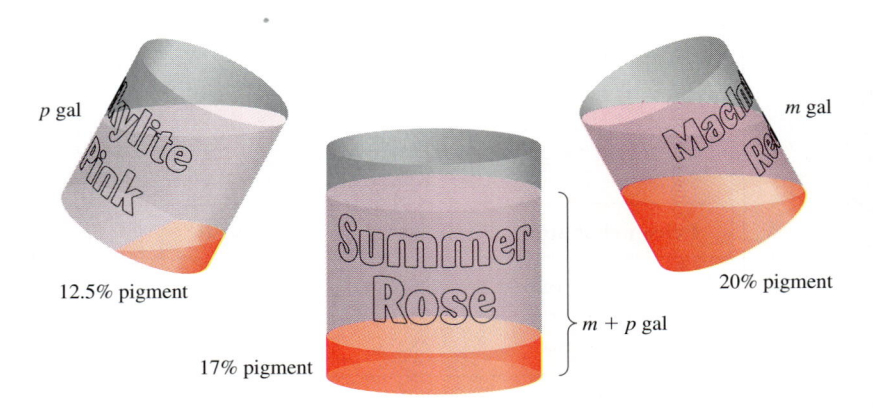

   Suppose that 2 gal of Skylite Pink and 8 gal of MacIntosh Red are mixed. The Skylite Pink would contribute 12.5% of 2 gal, or 0.25 gal of pigment, and the MacIntosh Red would contribute 20% of 8 gal, or 1.6 gal of pigment. Thus the 10-gal mixture would contain $0.25 + 1.6 = 1.85$ gal of pigment. Since Gayle wants the 10 gal of Summer Rose to be 17% pigment, and since 17% of 10 gal is 1.7 gal, our guess is incorrect.

   The given and unknown information can be arranged in a table.

	Skylight Pink	**MacIntosh Red**	**Summer Rose**
Amount of Paint (in gallons)	p	m	10
Percent Pigment	12.5%	20%	17%
Amount of Pigment (in gallons)	$0.125p$	$0.2m$	0.17×10, or 1.7

2. **Translate.** A system of two equations can be formed by reading across the first and third rows of the table. Since Gayle needs 10 gal of mixture, we must have

   $p + m = 10.$ ⟵ Total amount of paint

   Since the pigment in the Summer Rose paint comes from the pigment in both the Skylite Pink and the MacIntosh Red paint, we have

$0.125p + 0.2m = 1.7.$ ← Total amount of pigment

We have translated to a system of equations:

$$p + m = 10, \quad (1)$$
$$0.125p + 0.2m = 1.7. \quad (2)$$

3. **Carry out.** We note that if we multiply on both sides of equation (2) by -5, we can eliminate m (other approaches will also work):

$$p + m = 10$$
$$\underline{-0.625p - m = -8.5} \quad \text{We observed that } (-5)(0.2m) = -m.$$
$$0.375p = 1.5$$
$$p = 4. \quad \text{Dividing on both sides by } 0.375$$

If $p = 4$, we see from equation (1) that $m = 6$.

4. **Check.** Clearly, 4 gal of Skylite Pink and 6 gal of MacIntosh Red do combine to make a 10-gal mixture. To see if the mixture is the right color, Summer Rose, we calculate the amount of pigment in the mixture: $0.125 \cdot 4 + 0.2 \cdot 6 = 0.5 + 1.2 = 1.7$. Since 1.7 is 17% of 10, the mixture is the correct color.

5. **State.** Gayle needs 4 gal of Skylite Pink and 6 gal of MacIntosh Red in order to make 10 gal of Summer Rose.

Re-examine Examples 1–4, looking for similarities. Examples 3 and 4 are often called *mixture problems,* but they have much in common with Examples 1 and 2.

PROBLEM-SOLVING TIP

When solving a problem, see if it is patterned or modeled after a problem that you have already solved.

EXERCISE SET 8.4

Solve. Use the five steps for problem solving.

1. *Selling vehicles.* A firm sells cars and trucks. There is room on its lot for 260 vehicles. They know that profits are greatest if there are 90 more cars than trucks on the lot. How many of each vehicle should the firm have on the lot for the greatest profit?

2. *Driving and hiking.* The Derricks drove and hiked 45 km to reach a campsite. They drove 23 km more than they walked. How far did they walk?

3. *Food prices.* Rita's charges $1.99 for a slice of pizza and a soda and $5.48 for three slices of pizza and two sodas. Determine the cost of one soda and the cost of one slice of pizza.

4. *Food prices.* Janey's Quick Stop is running a promotion in which a burger and two pieces of chicken cost $2.39 and a burger and one piece of chicken cost $1.69. Determine the cost of one burger.

5. *Household waste.* The Hendersons generate two and a half times as much trash as their neighbors, the Savickis. Together, the two households produce 14 bags of trash each month. How much trash does each household produce?

6. *Insulation.* The Mazzas' attic required three and a half times as much insulation as did the Kranepools'. Together, the two attics required 36 rolls of insulation. How much insulation did each attic require?

7. *Basketball scoring.* Shaquille O'Neil recently scored 36 points on 22 shots. O'Neil shot only two-pointers and foul shots (each foul shot is worth one point). How many foul shots did he make?

8. *Basketball scoring.* Wilt Chamberlain once scored 100 points in an NBA game. Chamberlain took only two-point shots and foul shots (see Exercise 7) and made a total of 64 shots. How many shots of each type did he make?

9. *Returnable bottles.* The Dixville Cub Scout troop collected 436 returnable bottles and cans, some worth 5 cents each and the rest worth 10 cents each. If the total value of the cans and bottles was $26.60, how many 5-cent bottles or cans and how many 10-cent bottles or cans were collected?

10. *Ice cream cones.* A busload of campers stopped at a dairy stand for ice cream. They ordered 75 cones, some soft-serve at $1.25 and the rest hard-pack at $1.50. If the total bill was $104.25, how many of each type of cone were ordered?

11. *Zoo admissions.* The Bronx Zoo charges $3.00 for adults and half price for seniors and children. One Wednesday, a total of $967.50 was collected from 435 admissions. How many full-price and how many half-price admissions were there?

12. *Paid admissions.* The Colchester High School production of *South Pacific* attracted an audience of 490 people. Students paid $4 per ticket and everyone else paid $7.00. If receipts totaled $2905, how many tickets of each type were sold?

13. *Paid admissions.* There were 203 tickets sold for a volleyball game. For activity-card holders the price was $1.25, and for noncard holders the price was $2. The total amount of money collected was $310. How many of each type of ticket were sold?

14. *Paid admissions.* There were 200 tickets sold for a women's basketball game. Tickets for students were $2 each and for adults were $3 each. The total amount collected was $530. How many of each type of ticket were sold?

15. *Coffee blends.* Cafe Europa mixes Brazilian coffee worth $19 per kilogram with Turkish coffee worth $22 per kilogram. The mixture should be worth $20 per kilogram. How much of each type of coffee should be used to make a 300-kg mixture?

16. *Seed mix.* Sunflower seed is worth $1.00 per pound and rolled oats are worth $1.35 per pound. How much of each would you use to make 50 lb of a mixture worth $1.14 per pound?

17. *Mixed nuts.* A grocer wishes to mix peanuts worth $2.52 per pound with Brazil nuts worth $3.80 per pound to make 480 lb of a mixture worth $3.44 per pound. How much of each should be used?

18. *Mixed nuts.* The Nuthouse has 10 kg of mixed cashews and pecans worth $8.40 per kilogram. Cashews alone sell for $8.00 per kilogram, and pecans sell for $9.00 per kilogram. How many kilograms of each are in the mixture?

19. *Acid mixtures.* Jerome's experiment requires him to mix a 50%-acid solution with an 80%-acid solution to create 200 mL of a 68%-acid solution. How much 50%-acid solution and how much 80%-acid solution should he use? Complete the following table as part of the *Familiarization* step.

Type of Solution	50%-Acid	80%-Acid	68%-Acid Mix
Amount of Solution	x	y	
Percent Acid	50%		68%
Amount of Acid in Solution		$0.8y$	

20. *Production.* Clear Shine window cleaner is 12% alcohol and Sunstream window cleaner is 30%

alcohol. How much of each should be used to make 90 oz of a cleaner that is 20% alcohol?

21. *Horticulture.* A solution containing 28% fungicide is to be mixed with a solution containing 40% fungicide to make 300 L of a solution containing 36% fungicide. How much of each solution should be used?

22. *Chemistry.* A chemist has one solution that is 80% base and another solution that is 30% base. What is needed is 200 L of a solution that is 62% base. The chemist will prepare it by mixing the two solutions on hand. How much of each should be used?

23. *Octane ratings.* The octane rating of a gasoline is a measure of the amount of isooctane in the gas. How much 87-octane gas and 93-octane gas should be blended in order to make 12 gal of 91-octane gas?

24. *Test scores.* Amy is taking a test in which items of type A are worth 10 points and items of type B are worth 15 points. She answers 16 questions and scores 180 points. How many questions of each type did Amy answer correctly?

25. *Metallurgy.* Francine stocks two alloys that are different purities of gold. The first is three-fourths pure gold and the second is five-twelfths pure gold. How many ounces of each should she melt and mix in order to obtain a 60-oz mixture that is two-thirds pure gold?

26. *Coin value.* A collection of dimes and quarters is worth $15.25. There are 103 coins in all. How many of each are there?

27. *Coin value.* A collection of quarters and nickels is worth $1.25. There are 13 coins in all. How many of each are there?

28. *Printing.* Using some pages that hold 1300 words per page and others that hold 1850 words per page, a typesetter is able to completely fill 12 pages with an 18,350-word document. How many pages of each kind were used?

29. *Catering.* Sandy's Catering needs to provide 10 lb of mixed nuts for a wedding reception. Peanuts cost $2.50 per pound and fancy nuts cost $7 per pound. If $40 has been allocated for nuts, how many pounds of each type should be mixed?

30. *Painting.* Campus Painters has two kinds of paint. If 9 gal of the inexpensive paint is mixed with 7 gal of the expensive paint, the mixture will be worth $19.70 per gallon. If 3 gal of the inexpensive paint is mixed with 5 gal of the expensive paint, the mixture will be worth $19.825 per gallon. What is the price per gallon of each type of paint?

SKILL MAINTENANCE

Factor.

31. $25x^2 - 81$
32. $36 - a^2$
33. $4x^2 + 100$

Solve.

34. $\dfrac{x^2}{x+4} = \dfrac{16}{x+4}$
35. $x^2 - 10x + 25 = 0$
36. $x^2 = 100$

SYNTHESIS

37. ◆ Write a problem for a classmate to solve by translating to a system of two equations in two unknowns.

38. ◆ What characteristics do Examples 1–4 share when they are being translated to systems of equations?

39. ◆ Which of the five problem-solving steps have you found the most challenging? Why?

40. ◆ Why might fractional answers be acceptable on problems like Examples 3 and 4, but not on problems like Examples 1 and 2?

41. *Automobile maintenance.* An automobile radiator contains 16 L of antifreeze and water. This mixture is 30% antifreeze. How much should be drained and replaced with pure antifreeze so that the mixture will be 50% antifreeze?

42. *Chemistry.* A tank contains 8000 L of a solution that is 40% acid. How much water should be added in order to make a solution that is 30% acid?

43. *Octane rating.* Many cars need gasoline with an octane rating of at least 87. After mistakenly putting 5 gal of 85-octane gas in her empty gas tank, Kim plans to add 91-octane gas until the mixture's octane rating is 87. How much 91-octane gas should she add?

44. *Dairy farming.* Farmer Benz has 100 L of milk that is 4.6% butterfat. How much skim milk (no butterfat) should be added to make milk that is 3.2% butterfat?

45. *Payroll.* Ace Engineering pays a total of $325 an hour when employing some workers at $20 an hour and others at $25 an hour. When the number of $20 workers is increased by 50% and the number of $25 workers is decreased by 20%, the cost per hour is $400. How many workers were originally employed at each rate?

46. *Investing.* Eduardo invested $54,000, part of it at 6% and the rest at 6.5%. The total yield after one year is $3385. How much was invested at each rate?

47. *Investing.* One year Shannon made $288 from two investments: $1100 was invested at one yearly rate and $1800 at a rate that was 1.5% higher. Find the two rates of interest.

48. A two-digit number is 6 times the sum of its digits. The tens digit is 1 more than the ones digit. Find the number.

49. The sum of the digits of a two-digit number is 12. When the digits are reversed, the number is decreased by 18. Find the original number.

50. *Sporting-goods prices.* Together, a bat, ball, and glove cost $99.00. The bat costs $9.95 more than the ball, and the glove costs $65.45 more than the bat. How much does each cost?

51. *Literature.* In Lewis Carroll's *Through the Looking Glass,* Tweedledum says to Tweedledee, "The sum of your weight and twice mine is 361 pounds." Then Tweedledee says to Tweedledum, "Contrariwise, the sum of your weight and twice mine is 362 pounds." Find the weights of Tweedledum and Tweedledee.

COLLABORATIVE CORNER

Focus: Mixture problems

Time: 20–30 minutes

Group size: 3

Materials: Calculators are optional, as instructors see fit.

Sunoco® gasoline stations pride themselves in offering customers the "custom blending pump." While most competitors offer just three octane levels—87, 93, and a blend that is 89—Sunoco customers can select an octane level of 86, 87, 89, 92, or 94. The Sunoco supplier brings 86- and 94-octane gasoline to each station and a computerized pump blends the two to create the blend that a customer desires.

ACTIVITY

1. Assume that your group's gas station has a generous supply of 86-octane and 94-octane gasoline. Each group member should select a different one of the custom blends: 87, 89, or 92.

2. *Following an agreed-upon series of steps,* group members should determine how many gallons of 86-octane gas and how many of 94-octane should be mixed in order to form 100 gal of each of the selected blends. Use substitution when solving the systems of equations that arise, agree on what each variable represents, solve for the same variable first each time, and so on. Check that all work is done correctly and consistently.

3. Suppose that instead of an 87-, 89-, or 92-octane blend, a blend of octane level C is needed, where $86 \leq C \leq 94$. Reuse the steps from part (2) to find a formula for the number of gallons of 86-octane and 94-octane gas required for 100 gal of a blend that has an octane level of C. How might a pump manufacturer use such a formula?

4. The pumps in use at most gas stations blend mixtures in $\frac{1}{10}$-gal "batches." How can the results of parts (2) and (3) be adapted so that $\frac{1}{10}$-gal, not 100-gal, blends are formulated?

8.5 Systems of Linear Inequalities

Graphing Systems of Inequalities • Locating Solution Sets

Linear inequalities in two variables were first graphed in Section 7.3. We now consider *systems of linear inequalities* in two variables, such as

$x + y \leq 3,$
$x - y < 3.$

When systems of equations were solved graphically in Section 8.1, we searched for any points common to both lines. To solve a system of inequalities graphically, we again look for points common to both graphs. This is accomplished by graphing each inequality and determining where the graphs overlap.

EXAMPLE 1 Graph the solutions of the system

$x + y \leq 3,$
$x - y < 3.$

SOLUTION To graph $x + y \leq 3$, we draw the graph of $x + y = 3$ using a solid line (see the graph on the left, below). Next, since (0, 0) is a solution of $x + y \leq 3$, we shade (in red) all points on that side of the line. The arrows near the ends of the line also indicate the region that contains solutions.

Next, we superimpose the graph of $x - y < 3$, using a dashed line for $x - y = 3$ and again using (0, 0) as a test point. Since (0, 0) is a solution, we shade (in blue) the region on the same side of the dashed line as (0, 0).

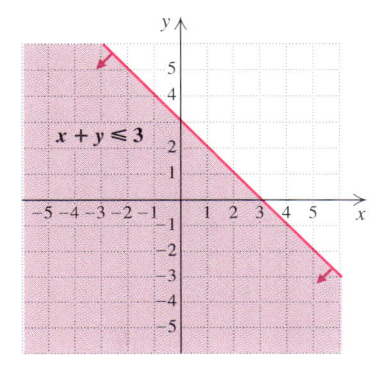

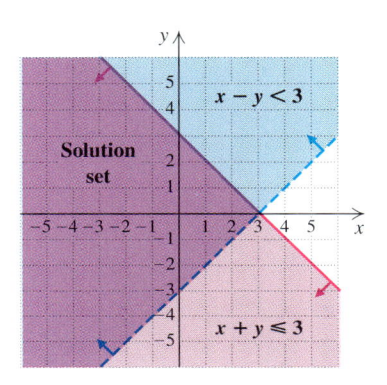

The solution set of the system is the region shaded purple along with the purple portion of the line $x + y = 3$.

EXAMPLE 2

Graph the solutions of the system

$$x \geq 3,$$
$$x - 3y < 6.$$

SOLUTION We draw the graphs of $x \geq 3$ and $x - 3y < 6$ using the same set of axes. The solution set is the purple region along with the purple portion of the solid line.

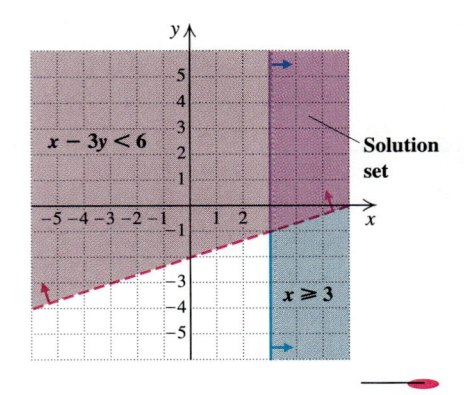

EXAMPLE 3

Graph the solutions of the system

$$x - 2y < 0,$$
$$-2x + y > 2.$$

SOLUTION We graph $x - 2y < 0$ using red and $-2x + y > 2$ using blue. The region that is purple is the solution set of the system since those points solve both inequalities.

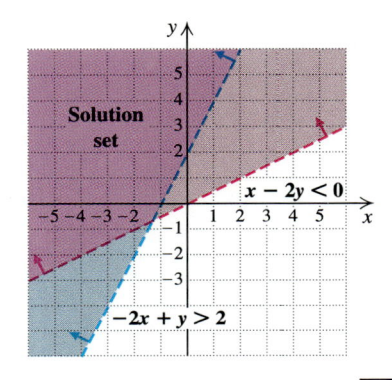

EXERCISE SET 8.5

Graph the solutions of each system.

1. $x + y \leq 2,$
   $x - y \leq 7$

2. $x + y \leq 3,$
   $x - y \leq 4$

3. $y - 2x > 1,$
   $y - 2x < 3$

4. $x + y < 2,$
   $x + y > 0$

5. $y \geq -3,$
   $x > 2 + y$

6. $x > 3,$
   $x + y \leq 4$

7. $y > 3x - 2,$
   $y < -x + 4$

8. $y \geq x,$
   $y \leq 1 - x$

9. $x \leq 4,$
   $y \leq 5$

10. $x \geq -2,$
    $y \geq -3$

11. $x \leq 0,$
    $y \leq 0$

12. $x \geq 0,$
    $y \geq 0$

13. $2x - 3y \geq 9,$
    $2y + x > 6$

14. $3x - 2y \leq 8,$
    $2x + y > 6$

15. $y > 5x + 2,$
    $y \leq 1 - x$

16. $y > 4,$
    $2y + x \leq 4$

17. $x + y \leq 5,$
    $x \geq 0,$
    $y \geq 0,$
    $y \leq 3$

18. $x + 2y \leq 8,$
    $x \leq 6,$
    $x \geq 0,$
    $y \geq 0$

19. $y - x \geq 1,$
    $y - x \leq 3,$
    $x \leq 5,$
    $x \geq 2$

20. $x - 2y \leq 0,$
    $y - 2x \leq 2,$
    $x \leq 2,$
    $y \leq 2$

21. $y \leq x,$
    $x \geq -2,$
    $x \leq -y$

22. $y > 0,$
    $2y + x \geq -6,$
    $x + 2 \leq 2y$

SKILL MAINTENANCE

Subtract.

23. $\dfrac{3}{x^2 - 4} - \dfrac{2}{3x + 6}$

24. $\dfrac{5}{4a^3 - 12a^2} - \dfrac{2}{a^2 - 4a + 3}$

Evaluate each polynomial for $x = -3$.

25. $x^2 - 5x + 2$
26. $4x^3 - 5x^2$

Factor.

27. $x^2 - 10x + 25$
28. $3a^2 - 18a + 27$

SYNTHESIS

29. ◆ Could (0, 0) have been used as a test point in Example 3? Why or why not?

30. ◆ Does the solution set of Example 1 include any points that are on the solid line? If so, which points and why? If not, why not?

31. ◆ Under what condition(s) will a system of two linear inequalities have no solution?

32. ◆ If shadings are used for each inequality in a system, will the darkest region always represent the solution set? Why or why not?

Graph the solutions of each system.

33. $5a + 3b \geq 30,$
    $2a + 3b \geq 21,$
    $3a + 6b \geq 36,$
    $a \geq 0,$
    $b \geq 0$

34. $2u + v \geq 8,$
    $4u + 3v \geq 22,$
    $2u + 5v \geq 18,$
    $u \geq 0,$
    $v \geq 0$

SUMMARY AND REVIEW 8

KEY TERMS

System of equations, p. 388
Consistent, p. 390
Inconsistent, p. 390
Dependent, p. 391
Independent, p. 391
Substitution method, p. 395
Elimination method, p. 402
System of linear inequalities, p. 419

IMPORTANT PROPERTIES AND FORMULAS

When graphing a system of two linear equations, one of the following must occur:

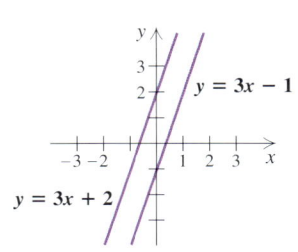

Graphs are parallel.
The system is *inconsistent* because there is no solution. The equations are *independent* since the graphs differ.

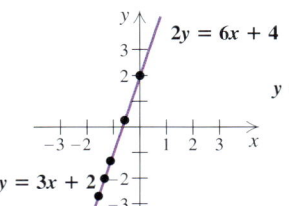

Equations have the same graph.
The system is *consistent* and has an infinite number of solutions. The equations are *dependent* since the graphs are the same.

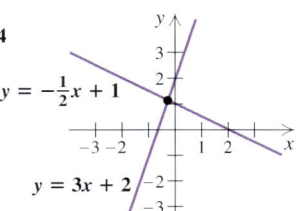

Graphs intersect.
The system is *consistent* and has one solution. Since the graphs differ, the equations are *independent*.

A Comparison of Methods for Solving Systems of Linear Equations

Method	Strengths	Weaknesses
Graphical	Solutions are displayed visually. Works with any system that can be graphed.	Inexact when solutions involve numbers that are not integers or are very large and off the graph.
Substitution	Always yields exact solutions. Easy to use when a variable is alone on one side of an equation.	Introduces extensive computations with fractions when solving more complicated systems. Solutions are not graphically displayed.
Elimination	Always yields exact solutions. Easy to use when fractions or decimals appear in the system.	Solutions are not graphically displayed.

Problem-Solving Tip

When solving a problem, see if it is patterned or modeled after a problem that you have already solved.

REVIEW EXERCISES

Determine whether each ordered pair is a solution of the system of equations.

1. $(-3, 2)$; $x + 2y = 1$,
$x - 3y = -9$
2. $(4, -1)$; $3x - y = 13$,
$2x + y = -9$

Solve by graphing. If there is no solution or an infinite number of solutions, state this.

3. $2x - y = 4$,
$3x - y = 5$
4. $x - y = 8$,
$x + y = 4$
5. $3x - 4y = 8$,
$4y - 3x = 6$
6. $2x + y = 3$,
$4x + 2y = 6$

Solve using the substitution method. If there is no solution or an infinite number of solutions, state this.

7. $y = 4 - x$,
$3x + 4y = 21$
8. $x + 2y = 6$,
$2x + y = 8$
9. $x + y = 4$,
$y = 2 - x$
10. $x + y = 6$,
$y = 3 - 2x$
11. $3x - y = 7$,
$2x + 3y = 23$
12. $3x - y = 5$,
$6x = 2y + 10$

Solve using the elimination method. If there is no solution or an infinite number of solutions, state this.

13. $x + 2y = 9$,
$3x - 2y = -3$
14. $x - y = 8$,
$2x + y = 7$
15. $x - \frac{1}{3}y = -\frac{13}{3}$,
$3x - y = -13$
16. $2x + 3y = 8$,
$5x + 2y = -2$
17. $5x - 2y = 11$,
$3x - 7y = -5$
18. $-x - y = -5$,
$2x - y = 4$
19. $4x - 6y = 9$,
$-2x + 3y = 6$
20. $2x + 6y = 4$,
$7x + 10y = -8$

Solve.

21. The sum of two numbers is 27. One half of the first number plus one third of the second number is 11. Find the numbers.

22. The perimeter of a rectangle is 96 cm. The length is 27 cm more than the width. Find the length and the width.

23. In a recent basketball game, the Hoopsters made 48 shots for a total of 81 points. If only two-pointers and foul shots (each foul shot is worth one point) were taken, how many foul shots were made?

24. There were 825 people at an orchestra's family-series performance. Adults' tickets were $15.00 each and children's tickets were $10.00 each. The total amount collected was $10,300.00. Find the number of adults' tickets and the number of children's tickets sold.

25. Café Rich instant flavored coffee gets 40% of its calories from fat. Café Light coffee gets 25% of its calories from fat. How much of each brand of coffee should be mixed in order to make 200 g of instant coffee with 30% of its calories from fat?

Graph the solutions of each system.

26. $x \geq 1$,
    $y \leq -1$

27. $x - y > 2$,
    $x + y < 1$

SKILL MAINTENANCE

Simplify.

28. $\dfrac{a^2 - 4}{2a^2 - 3a - 2}$

29. $\dfrac{9a^2b^4}{6ab^{-3}}$

30. Multiply: $3a^2b(5a^4 - 2ab)$.

31. Factor: $9x^3 - 36x^2 + 36x$.

SYNTHESIS

32. ◈ Explain why any solution of a system of equations is a point of intersection of the graphs of each equation in the system.

33. ◈ Monroe sketches the boundary lines of a system of two linear inequalities and notes that the lines are parallel. Since there is no point of intersection, he concludes that the solution set is empty. What is wrong with this conclusion?

34. The solution of the following system is (6, 2). Find C and D.
    $2x - Dy = 6$,
    $Cx + 4y = 14$

35. Solve using the substitution method:
    $x - y + 2z = -3$,
    $2x + y - 3z = 11$,
    $z = -2$.

36. Solve:
    $3(x - y) = 4 + x$,
    $x = 5y + 2$.

37. For a two-digit number, the sum of the ones digit and the tens digit is 6. When the digits are reversed, the new number is 18 more than the original number. Find the original number.

38. A stable boy agreed to work for one year. At the end of that time, he was to receive $240 and one horse. After 7 months, the boy quit the job, but still received the horse and $100. What was the value of the boy's yearly salary?

CHAPTER TEST 8

1. Determine whether $(-1, -2)$ is a solution of the following system of equations:
   $3x - 4y = 5$,
   $2x + 3y = -8$.

Solve by graphing. If there is no solution or an infinite number of solutions, state this.

2. $2x + y = 5$,
   $4x - y = 1$

3. $2y - x = 7$,
   $2x - 4y = 4$

Solve using the substitution method. If there is no solution or an infinite number of solutions, state this.

4. $2x - 22 = 3y$,
   $y = 6 - x$

5. $x + y = 2$,
   $x + 2y = 5$

6. $x = 5y - 10$,
   $10y = 2x + 20$

Solve using the elimination method. If there is no solution or an infinite number of solutions, state this.

7. $x - y = 6,$
   $3x + y = -2$

8. $\frac{1}{2}x - \frac{1}{3}y = 8,$
   $\frac{2}{3}x + \frac{1}{2}y = 5$

9. $4x + 5y = 5,$
   $6x + 7y = 7$

10. $2x + 3y = 13,$
    $3x - 5y = 10$

Solve.

11. A chemist has one solution that is 25% acid and another solution that is 40% acid. How much of each is needed to make 60 L of a solution that is 30% acid?

12. Two angles are complementary. One angle is 18° less than twice the other. Find the angles.

13. Kate bought 18 sheets of plywood for $750. Oak plywood cost $45 for a sheet and pine cost $25. How many sheets of each kind of plywood did she buy?

Graph the solutions of each system.

14. $y \geq x + 1,$
    $y > 2x$

15. $x + y \leq 3,$
    $x \geq 0,$
    $y \geq 0$

SKILL MAINTENANCE

16. Subtract: $\dfrac{1}{x^2 - 16} - \dfrac{x - 4}{x^2 - 3x - 4}$.

17. Evaluate $-3x^4 + 4x^3 - 5x + 2$ for $x = -1$.

Solve.

18. $2(y - 7) + 3 = 4(y + 1)$

19. $4x^2 = 36$

SYNTHESIS

20. You are in line at a ticket window. There are two more people ahead of you in line than there are behind you. In the entire line, there are three times as many people as there are behind you. How many are in the line?

21. Graph on a plane: $|x| \leq 4$.

22. Find the numbers C and D such that $(-2, 3)$ is a solution of the system
    $Cx - 4y = 7,$
    $3x + Dy = 8.$

CHAPTER 9

Radical Expressions and Equations

- **9.1** Introduction to Square Roots and Radical Expressions
- **9.2** Multiplying and Simplifying Radical Expressions
- **9.3** Quotients Involving Square Roots
- **9.4** More Operations with Radicals
- **9.5** Radical Equations
- **9.6** Applications Using Right Triangles
- **9.7** Higher Roots and Rational Exponents
  SUMMARY AND REVIEW
  TEST

AN APPLICATION

A pilot is instructed to descend from 30,000 ft to 20,000 ft over a horizontal distance of 50,000 ft. What distance will the plane travel during this descent?

THIS PROBLEM APPEARS AS EXERCISE 27 IN SECTION 9.6.

More information on aviation is available at
http://hepg.awl.com/be/elem_5

Without math, you can't answer the basic questions of flying: How long will it take me to get to my destination? How much fuel do I need to get where I'm going?

TRACY L. WILLIAMS
Director of Airports
San Diego, California

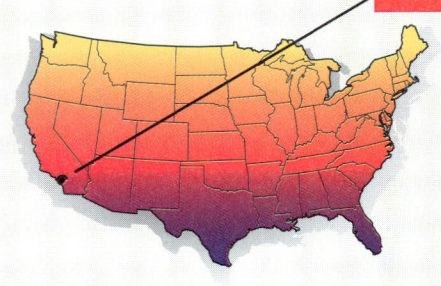

CHAPTER 9 • RADICAL EXPRESSIONS AND EQUATIONS

Many of us already have some familiarity with the notion of square roots. For example, 3 is a square root of 9 because $3^2 = 9$. In this chapter, we learn how to manipulate square roots of polynomials and rational expressions. Later in this chapter, these radical expressions will appear in equations and in problem-solving situations.

9.1 Introduction to Square Roots and Radical Expressions

Square Roots • Radicands and Radical Expressions • Irrational Numbers • Square Roots and Absolute Value • Problem Solving

We begin our study of square roots by examining square roots of numbers, square roots of variable expressions, and an application involving a formula.

Square Roots

Often in this text we have found the result of squaring a number. When the process is reversed, we say that we are looking for a number's *square root*.

SQUARE ROOT

The number c is a *square root* of a if $c^2 = a$.

Every positive number has two square roots. For example, the square roots of 25 are 5 and -5 because $5^2 = 25$ and $(-5)^2 = 25$.

EXAMPLE 1 Find the square roots of each number: **(a)** 81; **(b)** 100.

SOLUTION

a) The square roots of 81 are 9 and -9. To check, note that $9^2 = 81$ and $(-9)^2 = (-9)(-9) = 81$.
b) The square roots of 100 are 10 and -10, since $10^2 = 100$ and $(-10)^2 = 100$.

The positive square root of a number is also called the **principal square root**. A **radical sign**, $\sqrt{}$, is generally used when finding square roots and indicates the principal root. Thus, $\sqrt{25} = 5$ and $\sqrt{25} \neq -5$.

EXAMPLE 2 Find each of the following: (a) $\sqrt{225}$; (b) $-\sqrt{64}$.

SOLUTION

a) The principal square root of 225 is its positive square root, so $\sqrt{225} = 15$.
b) The symbol $-\sqrt{64}$ represents the opposite of $\sqrt{64}$. Since $\sqrt{64} = 8$, we have $-\sqrt{64} = -8$.

Radicands and Radical Expressions

A **radical expression** is an algebraic expression that contains at least one radical sign. Here are some examples:

$$\sqrt{14}, \quad 7 + \sqrt{2x}, \quad \sqrt{x^2 + 4}, \quad \sqrt{\frac{x^2 - 5}{2}}.$$

The expression under the radical is called the **radicand.**

EXAMPLE 3 Identify the radicand in each expression: (a) $\sqrt{x}$; (b) $\sqrt{y^2 - 5}$.

SOLUTION

a) In $\sqrt{x}$, the radicand is x.
b) In $\sqrt{y^2 - 5}$, the radicand is $y^2 - 5$.

The square of any nonzero number is always positive. For example, $8^2 = 64$ and $(-11)^2 = 121$. No real number, squared, is equal to a negative number. Thus the following expressions are not real numbers:

$$\sqrt{-100}, \quad \sqrt{-49}, \quad -\sqrt{-3}.$$

Numbers like $\sqrt{-100}$, $\sqrt{-49}$, and $-\sqrt{-3}$ are discussed in Chapter 10.

Irrational Numbers

In Section 1.4, we learned that numbers like $\sqrt{2}$ cannot be written as a ratio of two integers. These numbers are real but not rational. We call numbers like $\sqrt{2}$ *irrational*. The square root of any whole number that is not a perfect square is irrational.

EXAMPLE 4 Classify each of the following numbers as rational or irrational.

a) $\sqrt{3}$ b) $\sqrt{25}$ c) $\sqrt{35}$ d) $-\sqrt{9}$

SOLUTION

a) $\sqrt{3}$ is irrational, since 3 is not a perfect square.
b) $\sqrt{25}$ is rational, since 25 is a perfect square: $\sqrt{25} = 5$.
c) $\sqrt{35}$ is irrational, since 35 is not a perfect square.
d) $-\sqrt{9}$ is rational, since 9 is a perfect square: $-\sqrt{9} = -3$.

Often, when square roots are irrational, we need decimal approximations. Such approximations can be found using a calculator with a square root key $\boxed{\sqrt{}}$. They can also be found using Table 2 at the back of the book.

TECHNOLOGY CONNECTION

9.1

Graphing equations that contain radical expressions often involves approximating irrational numbers. Also, since the square root of a negative number is not real, such graphs may not exist for all choices of x. For example, the graph of $y = \sqrt{x - 1}$ does not exist for x-values that are less than 1. Similarly, the graph of $y = \sqrt{2 - x}$ does not exist for x-values that exceed 2.

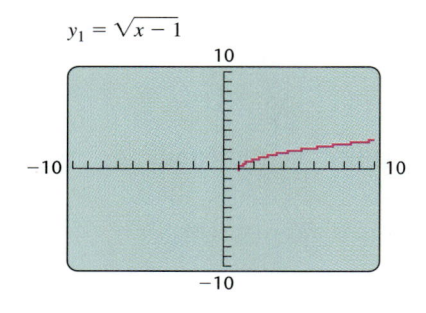

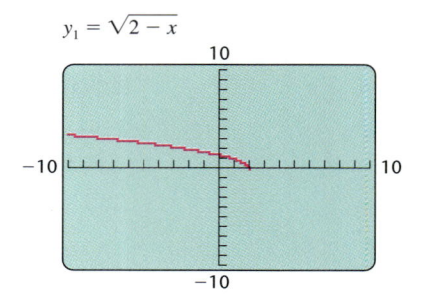

$y_1 = \sqrt{x - 1}$

$y_1 = \sqrt{2 - x}$

Graph.

1. $y = \sqrt{x}$
2. $y = \sqrt{2x}$
3. $y = \sqrt{x^2}$
4. $y = \sqrt{(2x)^2}$
5. $y = \sqrt{x + 4}$
6. $y = \sqrt{6 - x}$

EXAMPLE 5 Use a calculator or Table 2 to approximate $\sqrt{10}$. Round to three decimal places.

SOLUTION Calculators vary in their methods of operation. In most cases, however, we simply enter the number and then press $\boxed{\sqrt{}}$:

$\sqrt{10} \approx 3.162277660.$ Using a calculator with a 10-digit readout

The actual decimal representation of an irrational number would be nonrepeating and nonending. Rounding to the third decimal place, we have $\sqrt{10} \approx 3.162$.

Square Roots and Absolute Value

Note that $\sqrt{(-5)^2} = \sqrt{25} = 5$ and $\sqrt{5^2} = \sqrt{25} = 5$, so it appears that squaring a number and then taking its square root is the same as taking the absolute value of the number: $|-5| = 5$ and $|5| = 5$. In short, the principal square root of the square of A is the absolute value of A:

For any real number A, $\sqrt{A^2} = |A|$.

EXAMPLE 6 Simplify $\sqrt{(3x)^2}$ given that x can represent any real number.

SOLUTION If x represents a negative number, then $3x$ is negative. Since the principal square root is always positive, to write $\sqrt{(3x)^2} = 3x$ would be incorrect. Instead, we write

$\sqrt{(3x)^2} = |3x|.$

Fortunately, in many uses of radicals, it can be assumed that radicands never represent the square of a negative number. When this assumption is made, the need for absolute-value symbols disappears:

For $A \geq 0$, $\sqrt{A^2} = A$.

EXAMPLE 7 Simplify each expression. Assume that all variables represent nonnegative numbers.

a) $\sqrt{(3x)^2}$
b) $\sqrt{a^2b^2}$

SOLUTION

a) $\sqrt{(3x)^2} = 3x$ Since $3x$ is assumed to be nonnegative, $|3x| = 3x$.
b) $\sqrt{a^2b^2} = \sqrt{(ab)^2} = ab$ Since ab is assumed to be nonnegative, $|ab| = ab$.

To reduce the need for absolute-value symbols, we will assume in Sections 9.2–9.6 that all variables represent nonnegative numbers.

Problem Solving

Radical expressions often appear in applications.

EXAMPLE 8 *Parking-Lot Arrival Spaces.* The attendants at a parking lot use spaces to leave cars before they are taken to long-term parking stalls. The required number N of such spaces is approximated by the formula

$$N = 2.5\sqrt{A},$$

where A is the average number of arrivals in peak hours. Find the number of spaces needed when an average of 43 cars arrive during peak hours.

SOLUTION We substitute 43 into the formula. We use a calculator or Table 2 to find an approximation:

$$N = 2.5\sqrt{43} \approx 2.5(6.557) \approx 16.393 \approx 17.$$

Note that we round *up* to 17 spaces because rounding down would create some overcrowding. Thus, for an average of 43 arrivals, 17 spaces are needed.

Calculator note. Generally, when using a calculator for a calculation like that in Example 8, we round at the *end* of our calculations. Using a calculator, we might find

$$N = 2.5\sqrt{43} \approx 2.5(6.557438524) = 16.39359631 \approx 16.394.$$

Note that this gives a variance in the third decimal place. When using a calculator for approximation, be aware of possible variations in answers. You may get answers that differ from those given at the back of the text. Answers to the exercises have been found by rounding at the end of the calculations.

EXERCISE SET 9.1

Find the square roots.
1. 9
2. 4
3. 16
4. 1
5. 49
6. 121
7. 169
8. 144

Simplify.
9. $\sqrt{4}$
10. $\sqrt{9}$
11. $-\sqrt{1}$
12. $-\sqrt{25}$
13. $\sqrt{0}$
14. $-\sqrt{81}$
15. $-\sqrt{121}$
16. $\sqrt{361}$
17. $\sqrt{400}$
18. $\sqrt{441}$
19. $\sqrt{169}$
20. $\sqrt{144}$
21. $-\sqrt{625}$
22. $-\sqrt{900}$

Identify the radicand.
23. $\sqrt{a-4}$
24. $\sqrt{t+3}$
25. $5\sqrt{t^2+1}$
26. $8\sqrt{x^2+5}$
27. $x^2y\sqrt{\dfrac{3}{x+2}}$
28. $ab^2\sqrt{\dfrac{a}{a-b}}$

Classify each square root as rational or irrational.
29. $\sqrt{16}$
30. $\sqrt{6}$
31. $\sqrt{8}$
32. $\sqrt{10}$
33. $\sqrt{32}$
34. $\sqrt{100}$
35. $\sqrt{98}$
36. $\sqrt{75}$
37. $-\sqrt{4}$
38. $-\sqrt{1}$
39. $-\sqrt{12}$
40. $-\sqrt{14}$

Use a calculator or Table 2 to approximate each square root. Round to three decimal places.
41. $\sqrt{5}$
42. $\sqrt{6}$
43. $\sqrt{17}$
44. $\sqrt{19}$
45. $\sqrt{93}$
46. $\sqrt{43}$

Simplify. Remember that we have assumed that all variables represent nonnegative numbers.
47. $\sqrt{t^2}$
48. $\sqrt{x^2}$
49. $\sqrt{25x^2}$
50. $\sqrt{9a^2}$
51. $\sqrt{(7a)^2}$
52. $\sqrt{(4x)^2}$
53. $\sqrt{(17x)^2}$
54. $\sqrt{(8ab)^2}$

Parking spaces. *Solve. Use the formula* $N = 2.5\sqrt{A}$ *of Example 8.*

55. Find the number of spaces needed when the average number of arrivals is (a) 36; (b) 29.

56. Find the number of spaces needed when the average number of arrivals is (a) 49; (b) 53.

Hang time. *An athlete's hang time (time airborne for a jump), T, in seconds, is given by* $T = 0.144\sqrt{V}$,* *where V is the athlete's vertical leap, in inches.*

57. Anfernee Hardaway of the Orlando Magic can jump 36 in. vertically. Find his hang time.

*Based on an article by Peter Brancazio, "The Mechanics of a Slam Dunk," *Popular Mechanics*, November 1991. Courtesy of Peter Brancazio, Brooklyn College.

58. Dikembe Mutombo of the Atlanta Hawks can jump 25 in. vertically. Find his hang time.

SKILL MAINTENANCE

Multiply.

59. $5x^3 \cdot 2x^6$

60. $(3a^3b^2)(5ab^7)$

61. Find the slope of the line containing the points $(-3, 4)$ and $(5, -6)$.

62. Find the slope of the line $-3x + 5y = 15$.

63. Find an equation of the line containing the point $(-3, 4)$ with slope 2.

64. Find an equation of the line containing the points $(-3, 4)$ and $(5, -6)$.

SYNTHESIS

65. ◆ Explain in your own words why $\sqrt{A^2} \ne A$ when A is negative.

66. ◆ Which is the more exact way to write the square root of 12: 3.464101615 or $\sqrt{12}$? Why?

67. ◆ What is the difference between saying "*the* square root of 10" and saying "*a* square root of 10"?

68. ◆ One number has only one square root. What is the number and why is it unique in this regard?

Simplify.

69. $\sqrt{\sqrt{16}}$

70. $\sqrt{3^2 + 4^2}$

71. Between what two consecutive integers is $-\sqrt{33}$?

72. Find a number that is the square of an integer and the cube of a different integer.

Solve. If no solution exists, state this.

73. $\sqrt{x^2} = 8$

74. $\sqrt{y^2} = -5$

75. $-\sqrt{x^2} = -3$

76. $t^2 = 49$

Simplify. Assume that all variables represent positive numbers.

77. $\sqrt{(9a^3b^4)^2}$

78. $(\sqrt{3a})^2$

79. $\sqrt{\dfrac{144x^8}{36y^6}}$

80. $\sqrt{\dfrac{y^{12}}{8100}}$

81. $\sqrt{\dfrac{400}{m^{16}}}$

82. $\sqrt{\dfrac{p^2}{3600}}$

83. Use the graph of $y = \sqrt{x}$, shown below, to find rational approximations for each of the following numbers: **(a)** $\sqrt{3}$; **(b)** $\sqrt{5}$; **(c)** $\sqrt{7}$. Answers may vary.

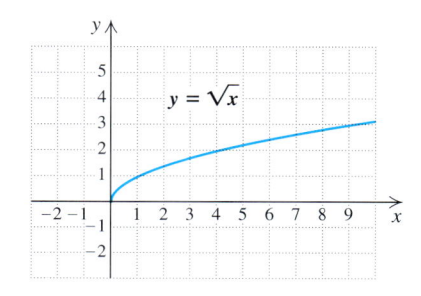

Speed of sound. The speed V of sound traveling through air, in feet per second, is given by

$$V = \dfrac{1087\sqrt{273 + t}}{16.52},$$

where t is the temperature, in degrees Celsius. Using a calculator, find the speed of sound through air at each of the following temperatures. Round to the nearest tenth.

84. 28°C

85. 5°C

86. -10°C

87. 100°C

88. Use a grapher to draw the graphs of $y_1 = \sqrt{x - 2}$, $y_2 = \sqrt{x + 7}$, $y_3 = 5 + \sqrt{x}$, and $y_4 = -4 + \sqrt{x}$. If possible, graph all four equations using the SIMULTANEOUS mode and a $[-10, 10, -10, 10]$ window. Then determine which equation corresponds to each curve.

COLLABORATIVE CORNER

Focus: Square roots and modeling
Time: 25–35 minutes
Group size: 3
Materials: Rulers, clocks or watches (to measure seconds), pendulums (see below), calculators

A pendulum is simply a string, a rope, or a chain with a weight of some sort attached at one end. When the unweighted end is held, a pendulum can swing freely from side to side. A shoe hanging from a shoelace, a yo-yo, a pendant hanging from a chain, a fishing weight hanging from a fish line, or a hairbrush tied to a length of dental floss are all examples of a pendulum. In this activity, each group will develop a mathematical model (formula) that relates a pendulum's length L to the time T that it takes for one complete swing back and forth (one "cycle").

ACTIVITY

1. Using a pendulum of some type (see above), one group member should hold the string or chain so that its length is 1 ft. A second group member should lift the weight to one side and then release (do not throw) the weight. The third group member should determine the time, in seconds, for one complete swing (cycle), by timing *five* cycles and dividing by five. Repeat this procedure for each pendulum length listed, so that the table below is completed.

L (in feet)	T (in seconds)
1	
1.5	
2	
2.5	
3	
3.5	

2. Examine the table your group has created. Can you find one number, a, such that $T \approx aL$ for all pairs of values on the chart?
3. To see if a better model can be found, add a third column to the chart and fill in $\sqrt{L}$ for each value of L listed. Can you find one number, b, such that $T \approx b\sqrt{L}$? Does this appear to be a more accurate model than $T = aL$?
4. Use the model for part (3) to predict T when L is 4 ft. Then check your prediction by measuring T as you did in part (1) above. Was your prediction "acceptable"? Compare your results with those of other groups.
5. In Section 9.3, we use a formula equivalent to

$$T = \frac{2\pi}{\sqrt{32}} \cdot \sqrt{L}.$$

How does your value of b from part (3) compare with $2\pi/\sqrt{32}$?

9.2 Multiplying and Simplifying Radical Expressions

Multiplying • Simplifying and Factoring • Simplifying Square Roots of Powers • Multiplying and Simplifying

We now learn to multiply and simplify radical expressions.

Multiplying

To see how to multiply with radical notation, consider the following:

$\sqrt{9} \cdot \sqrt{4} = 3 \cdot 2 = 6$;   This is a product of square roots.
$\sqrt{9 \cdot 4} = \sqrt{36} = 6$.   This is the square root of a product.

Note that $\sqrt{9} \cdot \sqrt{4} = \sqrt{9 \cdot 4}$.

THE PRODUCT RULE FOR SQUARE ROOTS

For any real numbers $\sqrt{A}$ and $\sqrt{B}$,
$$\sqrt{A} \cdot \sqrt{B} = \sqrt{A \cdot B}.$$

(To multiply square roots, multiply the radicands and take the square root.)

EXAMPLE 1 Multiply: (a) $\sqrt{5}\sqrt{7}$; (b) $\sqrt{6}\sqrt{6}$; (c) $\sqrt{\frac{2}{3}}\sqrt{\frac{7}{5}}$; (d) $\sqrt{2x}\sqrt{3y}$.

SOLUTION

a) $\sqrt{5}\sqrt{7} = \sqrt{5 \cdot 7} = \sqrt{35}$
b) $\sqrt{6}\sqrt{6} = \sqrt{6 \cdot 6} = \sqrt{36} = 6$ Try to do this one directly: $\sqrt{6}\sqrt{6} = 6$.
c) $\sqrt{\frac{2}{3}}\sqrt{\frac{7}{5}} = \sqrt{\frac{2}{3} \cdot \frac{7}{5}} = \sqrt{\frac{14}{15}}$
d) $\sqrt{2x}\sqrt{3y} = \sqrt{6xy}$

Simplifying and Factoring

To factor a square root, we can use the product rule in reverse. That is,
$$\sqrt{AB} = \sqrt{A}\sqrt{B}.$$

This property is especially useful when a radicand that is not a perfect square contains a perfect-square factor. For instance, the radicand 48 in $\sqrt{48}$

434 CHAPTER 9 • RADICAL EXPRESSIONS AND EQUATIONS

is not a perfect square, but one of its factors, 16, *is* a perfect square. Thus,

$$\sqrt{48} = \sqrt{16 \cdot 3}$$
$$= \sqrt{16} \cdot \sqrt{3}$$
$$= 4\sqrt{3}.$$

It is not always obvious that a radicand contains a factor that is a perfect square. In such a case, writing a prime factorization of the radicand is helpful. For example, if we did not immediately see that 50 contains a perfect square as a factor, we could write

$$\sqrt{50} = \sqrt{2 \cdot 5 \cdot 5}\qquad \text{Factoring into prime factors}$$
$$= \sqrt{2} \cdot \sqrt{5 \cdot 5}\qquad \text{Grouping pairs of like factors;}$$
$$\qquad\qquad\qquad\text{5 · 5 is a perfect square.}$$
$$= \sqrt{2} \cdot 5.$$

To avoid any uncertainty as to what is under the radical sign, it is customary to write the radical factor last. Thus, $\sqrt{50} = 5\sqrt{2}$.

A radical expression, like $\sqrt{26}$, in which the radicand has no perfect-square factors, is considered to be in simplest form.

SIMPLIFIED FORM OF A SQUARE ROOT

A radical expression for a square root is simplified when its radicand has no factor other than 1 that is a perfect square.

EXAMPLE 2   Simplify by factoring (remember that all variables are assumed to represent nonnegative numbers).

a) $\sqrt{18}$ \qquad\qquad\qquad b) $\sqrt{48t}$
c) $\sqrt{a^2 b}$ \qquad\qquad\qquad d) $\sqrt{98t^2}$

SOLUTION

a) $\sqrt{18} = \sqrt{9 \cdot 2}$ \qquad Identifying a perfect-square factor and factoring the radicand. The factor 9 is a perfect square.
$\qquad\quad = \sqrt{9}\sqrt{2}$ \qquad Factoring into a product of radicals
$\qquad\quad = 3\sqrt{2}$ \qquad The radicand has no factors that are perfect squares.

b) $\sqrt{48t} = \sqrt{16 \cdot 3t}$ \qquad Identifying a perfect-square factor and factoring the radicand. The factor 16 is a perfect square.
$\qquad\quad = \sqrt{16}\sqrt{3t}$ \qquad Factoring into a product of radicals
$\qquad\quad = 4\sqrt{3t}$

c) $\sqrt{a^2 b} = \sqrt{a^2}\sqrt{b}$ \qquad Identifying a perfect-square factor and factoring into a product of radicals
$\qquad\quad = a\sqrt{b}$ \qquad No absolute-value signs are necessary since a is assumed to be nonnegative.

d) $\sqrt{98t^2} = \sqrt{2 \cdot 7 \cdot 7 \cdot t \cdot t}$ Writing the prime factorization
$= \sqrt{7^2} \cdot \sqrt{t^2} \cdot \sqrt{2}$ Factoring into a product of radicals. 7^2 and t^2 are perfect squares.
$= 7t\sqrt{2}$ Taking square roots. No absolute-value signs are necessary since t is assumed to be nonnegative.

Simplifying Square Roots of Powers

To take the square root of an even power such as x^{10}, note that $x^{10} = (x^5)^2$. Then

$$\sqrt{x^{10}} = \sqrt{(x^5)^2} = x^5.$$

The exponent of the square root is half the exponent of the radicand. That is,

$$\sqrt{x^{10}} = x^5. \quad \leftarrow \tfrac{1}{2}(10) = 5$$

EXAMPLE 3 Simplify: (a) $\sqrt{x^6}$; (b) $\sqrt{x^8}$; (c) $\sqrt{t^{22}}$.

SOLUTION

a) $\sqrt{x^6} = \sqrt{(x^3)^2} = x^3$ Half of 6 is 3.
b) $\sqrt{x^8} = \sqrt{(x^4)^2} = x^4$ Half of 8 is 4.
c) $\sqrt{t^{22}} = \sqrt{(t^{11})^2} = t^{11}$ Half of 22 is 11.

If a radicand is an odd power, we can simplify by factoring, as in the following example.

EXAMPLE 4 Simplify: (a) $\sqrt{x^9}$; (b) $\sqrt{32x^{15}}$.

SOLUTION

a) $\sqrt{x^9} = \sqrt{x^8 \cdot x}$ x^8 is the largest perfect-square factor of x^9.
$= \sqrt{x^8}\sqrt{x}$
$= x^4\sqrt{x}$

> **CAUTION!** The square root of x^9 is not x^3.

b) $\sqrt{32x^{15}} = \sqrt{16x^{14} \cdot 2x}$ 16 is the largest perfect-square factor of 32; x^{14} is the largest perfect-square factor of x^{15}.
$= \sqrt{16}\sqrt{x^{14}}\sqrt{2x}$
$= 4x^7\sqrt{2x}$ Simplifying. Since $2x$ has no perfect-square factor, we are done.

Multiplying and Simplifying

Sometimes we can simplify after multiplying. To do so, we again try to identify any perfect-square factors of the radicand.

CHAPTER 9 • RADICAL EXPRESSIONS AND EQUATIONS

EXAMPLE 5 Multiply and, if possible, simplify.

a) $\sqrt{2}\sqrt{14}$ b) $\sqrt{5x}\sqrt{3x}$ c) $\sqrt{2x^8}\sqrt{9x^3}$

SOLUTION

a) $\sqrt{2}\sqrt{14} = \sqrt{2 \cdot 14}$ Multiplying
$\phantom{\sqrt{2}\sqrt{14}} = \sqrt{2 \cdot 2 \cdot 7}$ Writing the prime factorization
$\phantom{\sqrt{2}\sqrt{14}} = \sqrt{2^2}\sqrt{7}$ Note that $2 \cdot 2$, or 4, is a perfect-square factor.
$\phantom{\sqrt{2}\sqrt{14}} = 2\sqrt{7}$ Simplifying

b) $\sqrt{5x}\sqrt{3x} = \sqrt{15x^2}$ Multiplying
$\phantom{\sqrt{5x}\sqrt{3x}} = \sqrt{x^2}\sqrt{15}$
$\phantom{\sqrt{5x}\sqrt{3x}} = x\sqrt{15}$ Simplifying

c) To simplify $\sqrt{2x^8}\sqrt{9x^3}$, note that x^8 and 9 are perfect-square factors. This allows us to simplify *before* multiplying:

$\sqrt{2x^8}\sqrt{9x^3} = \sqrt{2} \cdot \sqrt{(x^4)^2} \cdot \sqrt{9} \cdot \sqrt{x^3}$ $(x^4)^2$ and 9 are perfect squares.
$\phantom{\sqrt{2x^8}\sqrt{9x^3}} = \sqrt{2} \cdot x^4 \cdot 3 \cdot \sqrt{x^3}$ Simplifying
$\phantom{\sqrt{2x^8}\sqrt{9x^3}} = 3x^4\sqrt{2}\sqrt{x^3}.$ Using a commutative law

The result, $3x^4\sqrt{2}\sqrt{x^3}$, or $3x^4\sqrt{2x^3}$, can be simplified further:

$3x^4\sqrt{2x^3} = 3x^4\sqrt{x^2 \cdot 2x}$ x^3 has a perfect-square factor, x^2.
$\phantom{3x^4\sqrt{2x^3}} = 3x^4\sqrt{x^2}\sqrt{2x}$
$\phantom{3x^4\sqrt{2x^3}} = 3x^4 x\sqrt{2x}$
$\phantom{3x^4\sqrt{2x^3}} = 3x^5\sqrt{2x}.$

EXERCISE SET 9.2

Multiply.

1. $\sqrt{5}\sqrt{7}$
2. $\sqrt{3}\sqrt{5}$
3. $\sqrt{4}\sqrt{3}$
4. $\sqrt{2}\sqrt{9}$
5. $\sqrt{\frac{2}{5}}\sqrt{\frac{3}{4}}$
6. $\sqrt{\frac{3}{8}}\sqrt{\frac{1}{5}}$
7. $\sqrt{8}\sqrt{8}$
8. $\sqrt{18}\sqrt{18}$
9. $\sqrt{25}\sqrt{3}$
10. $\sqrt{36}\sqrt{2}$
11. $\sqrt{2}\sqrt{x}$
12. $\sqrt{3}\sqrt{a}$
13. $\sqrt{7}\sqrt{3x}$
14. $\sqrt{5}\sqrt{4x}$
15. $\sqrt{x}\sqrt{7y}$
16. $\sqrt{5m}\sqrt{2n}$
17. $\sqrt{3a}\sqrt{2c}$
18. $\sqrt{3x}\sqrt{yz}$

Simplify by factoring.

19. $\sqrt{28}$
20. $\sqrt{12}$
21. $\sqrt{8}$
22. $\sqrt{45}$
23. $\sqrt{500}$
24. $\sqrt{200}$
25. $\sqrt{9x}$
26. $\sqrt{4y}$
27. $\sqrt{75a}$
28. $\sqrt{40m}$
29. $\sqrt{16a}$
30. $\sqrt{49b}$
31. $\sqrt{64y^2}$
32. $\sqrt{9x^2}$
33. $\sqrt{13x^2}$
34. $\sqrt{29t^2}$
35. $\sqrt{8t^2}$
36. $\sqrt{125a^2}$
37. $\sqrt{80}$
38. $\sqrt{98}$
39. $\sqrt{288y}$
40. $\sqrt{363p}$
41. $\sqrt{x^{20}}$
42. $\sqrt{x^{30}}$
43. $\sqrt{x^{12}}$
44. $\sqrt{x^{16}}$
45. $\sqrt{x^5}$
46. $\sqrt{x^3}$
47. $\sqrt{t^{19}}$
48. $\sqrt{p^{17}}$

49. $\sqrt{36m^3}$
50. $\sqrt{250y^3}$
51. $\sqrt{8a^5}$
52. $\sqrt{12b^7}$
53. $\sqrt{104p^{17}}$
54. $\sqrt{90m^{23}}$

Multiply and, if possible, simplify.

55. $\sqrt{15} \cdot \sqrt{5}$
56. $\sqrt{3}\sqrt{6}$
57. $\sqrt{3} \cdot \sqrt{27}$
58. $\sqrt{14} \cdot \sqrt{21}$
59. $\sqrt{3x}\sqrt{12y}$
60. $\sqrt{5x}\sqrt{20y}$
61. $\sqrt{10}\sqrt{10}$
62. $\sqrt{11}\sqrt{11x}$
63. $\sqrt{5b}\sqrt{15b}$
64. $\sqrt{6a}\sqrt{18a}$
65. $\sqrt{7x} \cdot \sqrt{7x}$
66. $\sqrt{3a}\sqrt{3a}$
67. $\sqrt{ab}\sqrt{ac}$
68. $\sqrt{xy}\sqrt{xz}$
69. $\sqrt{2x}\sqrt{4x^5}$
70. $\sqrt{15m^6}\sqrt{5m^2}$
71. $\sqrt{x^2y^3}\sqrt{xy^4}$
72. $\sqrt{x^3y^2}\sqrt{xy}$
73. $\sqrt{50ab}\sqrt{10a^2b^4}$
74. $\sqrt{10xy^2}\sqrt{5x^2y^3}$

Speed of a skidding car. *The formula*
$$r = 2\sqrt{5L}$$
can be used to approximate the speed r, in miles per hour, of a car that has left a skid mark L feet long.

75. What was the speed of a car that left skid marks of 20 ft? of 150 ft?
76. What was the speed of a car that left skid marks of 30 ft? of 70 ft?

SKILL MAINTENANCE

77. *Driving distances.* A car leaves Hereford traveling north at a speed of 56 km/h. Another car leaves Hereford one hour later, traveling north at 84 km/h. How far from Hereford will the second car overtake the first?
78. Multiply: $(a - 5b)(a + 5b)$.

Simplify.

79. $\dfrac{15a^3b^7}{5ab^2}$
80. $\dfrac{12x^8y^6}{3x^2y^2}$

SYNTHESIS

81. ◆ Are the rules for manipulating exponents important when simplifying radical expressions? Why or why not?
82. ◆ Explain why $\sqrt{16x^4} = 4x^2$, but $\sqrt{4x^{16}} \neq 2x^4$.
83. ◆ What is wrong with the following?
$$\sqrt{x^2 - 25} = \sqrt{x^2} - \sqrt{25} = x - 5$$
84. ◆ Simplify $\sqrt{49}$, $\sqrt{490}$, $\sqrt{4900}$, $\sqrt{49{,}000}$, and $\sqrt{490{,}000}$; then describe the pattern you see.

Simplify.

85. $\sqrt{0.01}$
86. $\sqrt{0.25}$
87. $\sqrt{0.0625}$
88. $\sqrt{0.000001}$

Use the proper symbol ($>$, $<$, or $=$) between each pair of values to make a true sentence. Do not use a calculator.

89. 15 $4\sqrt{14}$
90. $15\sqrt{2}$ $\sqrt{450}$
91. 16 $\sqrt{15}\sqrt{17}$
92. $3\sqrt{11}$ $7\sqrt{2}$
93. $5\sqrt{7}$ $4\sqrt{11}$
94. 8 $\sqrt{15} + \sqrt{17}$

Multiply and then simplify by factoring.

95. $\sqrt{27(x+1)}\sqrt{12y(x+1)^2}$
96. $\sqrt{18(x-2)}\sqrt{20(x-2)^3}$
97. $\sqrt{x}\sqrt{2x}\sqrt{10x^5}$
98. $\sqrt{2^{109}}\sqrt{x^{306}}\sqrt{x^{11}}$

Simplify.

99. $\sqrt{x^{8n}}$
100. $\sqrt{0.04x^{4n}}$
101. Simplify $\sqrt{y^n}$, where n is an odd whole number greater than or equal to 3.

9.3 Quotients Involving Square Roots

Dividing Radical Expressions • Rationalizing Denominators

In this section, we divide radical expressions and simplify quotients containing radicals.

Dividing Radical Expressions

To see how to divide with radical notation, consider the following:

$$\frac{\sqrt{25}}{\sqrt{16}} = \frac{5}{4} \text{ since } \sqrt{25} = 5 \text{ and } \sqrt{16} = 4;$$

$$\sqrt{\frac{25}{16}} = \frac{5}{4} \text{ since } \frac{5}{4} \cdot \frac{5}{4} = \frac{25}{16}.$$

We see that $\dfrac{\sqrt{25}}{\sqrt{16}} = \sqrt{\dfrac{25}{16}}$.

THE QUOTIENT RULE FOR RADICALS

For any real numbers $\sqrt{A}$ and $\sqrt{B}$, $B \neq 0$,

$$\frac{\sqrt{A}}{\sqrt{B}} = \sqrt{\frac{A}{B}}.$$

(To divide two square roots, divide the radicands and take the square root.)

EXAMPLE 1 Divide and simplify: (a) $\dfrac{\sqrt{27}}{\sqrt{3}}$; (b) $\dfrac{\sqrt{8a^7}}{\sqrt{2a}}$.

SOLUTION

a) $\dfrac{\sqrt{27}}{\sqrt{3}} = \sqrt{\dfrac{27}{3}}$ We can now simplify the radicand: $\dfrac{27}{3} = 9$.

$\phantom{\dfrac{\sqrt{27}}{\sqrt{3}}} = \sqrt{9} = 3$

b) $\dfrac{\sqrt{8a^7}}{\sqrt{2a}} = \sqrt{\dfrac{8a^7}{2a}}$ We do this because $\dfrac{8a^7}{2a}$ can be simplified.

$\phantom{\dfrac{\sqrt{8a^7}}{\sqrt{2a}}} = \sqrt{4a^6} = 2a^3$

9.3 QUOTIENTS INVOLVING SQUARE ROOTS

The quotient rule for radicals can also be read from right to left:

$$\sqrt{\frac{A}{B}} = \frac{\sqrt{A}}{\sqrt{B}}.$$

We generally select the form that makes simplification easier.

EXAMPLE 2 Simplify by taking square roots in the numerator and the denominator separately.

a) $\sqrt{\dfrac{25}{9}}$ b) $\sqrt{\dfrac{1}{16}}$ c) $\sqrt{\dfrac{49}{t^2}}$

SOLUTION

a) $\sqrt{\dfrac{25}{9}} = \dfrac{\sqrt{25}}{\sqrt{9}} = \dfrac{5}{3}$ Taking the square root of the numerator and the square root of the denominator. This is sometimes done mentally, in one step.

b) $\sqrt{\dfrac{1}{16}} = \dfrac{\sqrt{1}}{\sqrt{16}} = \dfrac{1}{4}$ Taking the square root of the numerator and the square root of the denominator

c) $\sqrt{\dfrac{49}{t^2}} = \dfrac{\sqrt{49}}{\sqrt{t^2}} = \dfrac{7}{t}$ We have assumed $t > 0$.

Sometimes a rational expression can be simplified to one that has a perfect-square numerator and a perfect-square denominator.

EXAMPLE 3 Simplify: (a) $\sqrt{\dfrac{18}{50}}$; (b) $\sqrt{\dfrac{48x^3}{3x^7}}$.

SOLUTION

a) $\sqrt{\dfrac{18}{50}} = \sqrt{\dfrac{9 \cdot 2}{25 \cdot 2}}$

$= \sqrt{\dfrac{9 \cdot \cancel{2}}{25 \cdot \cancel{2}}}$ Removing a factor equal to 1: $\dfrac{2}{2} = 1$

$= \dfrac{\sqrt{9}}{\sqrt{25}} = \dfrac{3}{5}$

b) $\sqrt{\dfrac{48x^3}{3x^7}} = \sqrt{\dfrac{16 \cdot \cancel{3x^3}}{x^4 \cdot \cancel{3x^3}}}$ Removing a factor equal to 1: $\dfrac{3x^3}{3x^3} = 1$

$= \dfrac{\sqrt{16}}{\sqrt{x^4}} = \dfrac{4}{x^2}$

Rationalizing Denominators

A procedure for finding an equivalent expression without a radical in the denominator is sometimes useful. This makes long division involving decimal

440 CHAPTER 9 • RADICAL EXPRESSIONS AND EQUATIONS

approximations easier to perform. The procedure, called **rationalizing the denominator**, relies on multiplying by a carefully selected form of 1 and the fact that (for $x \geq 0$) $\sqrt{x} \cdot \sqrt{x} = x$.

EXAMPLE 4 Rationalize each denominator: (a) $\dfrac{8}{\sqrt{7}}$; (b) $\sqrt{\dfrac{5}{a}}$.

SOLUTION

a) By writing 1 as $\sqrt{7}/\sqrt{7}$, we can find an expression that is equivalent to $8/\sqrt{7}$ without a radical in the denominator:

$$\dfrac{8}{\sqrt{7}} = \dfrac{8}{\sqrt{7}} \cdot \dfrac{\sqrt{7}}{\sqrt{7}} \quad \text{Multiplying by 1, using the denominator, } \sqrt{7}, \text{ to write 1}$$

$$= \dfrac{8\sqrt{7}}{7}. \quad \sqrt{7} \cdot \sqrt{7} = 7$$

b) $\sqrt{\dfrac{5}{a}} = \dfrac{\sqrt{5}}{\sqrt{a}}$ The square root of a quotient is the quotient of the square roots.

$$= \dfrac{\sqrt{5}}{\sqrt{a}} \cdot \dfrac{\sqrt{a}}{\sqrt{a}} \quad \text{Multiplying by 1, using the denominator, } \sqrt{a}, \text{ to write 1}$$

$$= \dfrac{\sqrt{5a}}{a} \quad \sqrt{a} \cdot \sqrt{a} = a; \text{ we assume } a \geq 0.$$

It is usually easiest to rationalize a denominator after the expression has been simplified.

EXAMPLE 5 Rationalize each denominator.

a) $\sqrt{\dfrac{5}{18}}$ b) $\dfrac{\sqrt{7a}}{\sqrt{20}}$

SOLUTION

a) $\sqrt{\dfrac{5}{18}} = \dfrac{\sqrt{5}}{\sqrt{18}}$ The square root of a quotient is the quotient of the square roots.

$$= \dfrac{\sqrt{5}}{\sqrt{9}\sqrt{2}}$$

Simplifying the denominator. Note that 9 is a perfect square.

$$= \dfrac{\sqrt{5}}{3\sqrt{2}}$$

$$= \dfrac{\sqrt{5}}{3\sqrt{2}} \cdot \dfrac{\sqrt{2}}{\sqrt{2}} \quad \text{Multiplying by 1, using } \sqrt{2} \text{ to write 1}$$

$$= \dfrac{\sqrt{10}}{3 \cdot 2} = \dfrac{\sqrt{10}}{6}$$

b) $\dfrac{\sqrt{7a}}{\sqrt{20}} = \dfrac{\sqrt{7a}}{\sqrt{4}\sqrt{5}}$ Simplifying the denominator. Note that 4 is a perfect square.

$= \dfrac{\sqrt{7a}}{2\sqrt{5}}$

$= \dfrac{\sqrt{7a}}{2\sqrt{5}} \cdot \dfrac{\sqrt{5}}{\sqrt{5}}$ Multiplying by 1

$= \dfrac{\sqrt{35a}}{2 \cdot 5}$

$= \dfrac{\sqrt{35a}}{10}$

CAUTION! Our solutions in Example 5 cannot be simplified any further. A common mistake is to remove a factor of 1 that does not exist. For example, $\dfrac{\sqrt{10}}{6}$ *cannot* be simplified to $\dfrac{\sqrt{5}}{3}$ because $\sqrt{10}$ and 6 do not share a common factor.

EXERCISE SET 9.3

Simplify.

1. $\dfrac{\sqrt{12}}{\sqrt{3}}$
2. $\dfrac{\sqrt{20}}{\sqrt{5}}$
3. $\dfrac{\sqrt{60}}{\sqrt{15}}$
4. $\dfrac{\sqrt{72}}{\sqrt{2}}$
5. $\dfrac{\sqrt{75}}{\sqrt{15}}$
6. $\dfrac{\sqrt{18}}{\sqrt{3}}$
7. $\dfrac{\sqrt{5}}{\sqrt{80}}$
8. $\dfrac{\sqrt{3}}{\sqrt{48}}$
9. $\dfrac{\sqrt{12}}{\sqrt{75}}$
10. $\dfrac{\sqrt{18}}{\sqrt{32}}$
11. $\dfrac{\sqrt{8x}}{\sqrt{2x}}$
12. $\dfrac{\sqrt{18b}}{\sqrt{2b}}$
13. $\dfrac{\sqrt{63y^3}}{\sqrt{7y}}$
14. $\dfrac{\sqrt{48x^3}}{\sqrt{3x}}$
15. $\dfrac{\sqrt{27x^5}}{\sqrt{3x}}$
16. $\dfrac{\sqrt{20a^8}}{\sqrt{5a^2}}$
17. $\dfrac{\sqrt{75x}}{\sqrt{25x^9}}$
18. $\dfrac{\sqrt{6x^9}}{\sqrt{2x^3}}$
19. $\sqrt{\dfrac{36}{25}}$
20. $\sqrt{\dfrac{9}{49}}$
21. $\sqrt{\dfrac{49}{16}}$
22. $\sqrt{\dfrac{1}{4}}$
23. $-\sqrt{\dfrac{1}{81}}$
24. $-\sqrt{\dfrac{25}{49}}$
25. $\sqrt{\dfrac{64}{144}}$
26. $\sqrt{\dfrac{81}{121}}$
27. $\sqrt{\dfrac{36}{a^2}}$
28. $\sqrt{\dfrac{25}{x^2}}$
29. $\sqrt{\dfrac{9a^2}{625}}$
30. $\sqrt{\dfrac{x^2y^2}{144}}$

Rationalize each denominator.

31. $\sqrt{\dfrac{5}{3}}$
32. $\sqrt{\dfrac{2}{7}}$
33. $\sqrt{\dfrac{7}{20}}$
34. $\sqrt{\dfrac{1}{12}}$
35. $\sqrt{\dfrac{1}{45}}$
36. $\sqrt{\dfrac{7}{18}}$
37. $\dfrac{3}{\sqrt{5}}$
38. $\dfrac{4}{\sqrt{3}}$
39. $\sqrt{\dfrac{8}{3}}$
40. $\sqrt{\dfrac{12}{5}}$
41. $\sqrt{\dfrac{3}{x}}$
42. $\sqrt{\dfrac{2}{x}}$

43. $\dfrac{\sqrt{7}}{\sqrt{3}}$ 44. $\dfrac{\sqrt{11}}{\sqrt{7}}$ 45. $\dfrac{\sqrt{9}}{\sqrt{8}}$

46. $\dfrac{\sqrt{4}}{\sqrt{27}}$ 47. $\dfrac{\sqrt{3}}{\sqrt{14}}$ 48. $\dfrac{\sqrt{3}}{\sqrt{2}}$

49. $\dfrac{\sqrt{7}}{\sqrt{12}}$ 50. $\dfrac{\sqrt{5}}{\sqrt{18}}$ 51. $\dfrac{\sqrt{x}}{\sqrt{32}}$

52. $\dfrac{\sqrt{a}}{\sqrt{40}}$ 53. $\dfrac{4y}{\sqrt{3}}$ 54. $\dfrac{8x}{\sqrt{5}}$

55. $\dfrac{\sqrt{6a}}{\sqrt{8}}$ 56. $\dfrac{\sqrt{3x}}{\sqrt{27}}$ 57. $\dfrac{\sqrt{72}}{\sqrt{20x}}$

58. $\dfrac{\sqrt{45}}{\sqrt{8a}}$ 59. $\dfrac{\sqrt{27c}}{\sqrt{32c^3}}$ 60. $\dfrac{\sqrt{7x^3}}{\sqrt{12x}}$

Period of a swinging pendulum. The period T of a pendulum is the time it takes to move from one side to the other and back. A formula for the period is

$$T = 2\pi\sqrt{\dfrac{L}{32}},$$

where T is in seconds and L is in feet. Use 3.14 for π.

61. Find the periods of pendulums of lengths 2 ft, 8 ft, 64 ft, and 100 ft.

62. Find the period of a pendulum of length $\tfrac{2}{3}$ in.

63. The pendulum of a grandfather clock is $45/\pi^2$ ft long. How long does it take to swing from one side to the other?

64. The pendulum of a grandfather clock is $32/\pi^2$ ft long. How long does it take to swing from one side to the other?

SKILL MAINTENANCE

Solve.

65. $x = y + 2,$
    $x + y = 6$

66. $2x - 3y = 7,$
    $2x + 3y = 9$

Multiply.

67. $(3x - 7)(3x + 7)$

68. $(4a - 5b)(4a + 5b)$

Simplify.

69. $9x - 5y + 12x - 4y$

70. $17a + 9b - 3a - 15b$

SYNTHESIS

71. ◆ Is it always best to rewrite an expression of the form $\sqrt{a}/\sqrt{b}$ as $\sqrt{a/b}$ before simplifying? Why or why not?

72. ◆ When using long division, why is it easier to approximate $\sqrt{2}/2$ than it is to approximate $1/\sqrt{2}$?

73. ◆ Why is it important to know how to multiply radical expressions before learning how to divide them?

74. ◆ Describe a method that could be used to rationalize the *numerator* of a radical expression.

Rationalize each denominator and, if possible, simplify.

75. $\sqrt{\dfrac{7}{800}}$ 76. $\sqrt{\dfrac{3}{1000}}$

77. $\sqrt{\dfrac{5}{8x^7}}$ 78. $\sqrt{\dfrac{3x^2y}{a^2x^5}}$

79. $\sqrt{\dfrac{2a}{5b^3c^9}}$ 80. $\sqrt{\dfrac{1}{5zw^2}}$

Simplify.

81. $\sqrt{\dfrac{1}{x^2} - \dfrac{2}{xy} + \dfrac{1}{y^2}}$ 82. $\sqrt{2 - \dfrac{4}{z^2} + \dfrac{2}{z^4}}$

9.4 More Operations with Radicals

Adding and Subtracting Radical Expressions • More with Multiplication • More with Rationalizing Denominators

We now consider addition and subtraction of radical expressions as well as some new types of multiplication and simplification.

Adding and Subtracting Radical Expressions

The sum of a rational number and an irrational number, like $5 + \sqrt{2}$, *cannot* be simplified. However, the sum of **like radicals**—that is, radical expressions that have a common radical factor—*can* be simplified.

EXAMPLE 1 Add: $3\sqrt{5} + 4\sqrt{5}$.

SOLUTION Recall that to simplify an expression like $3x + 4x$, we use the distributive law, as follows:

$$3x + 4x = (3 + 4)x = 7x.$$

In this example, x is replaced with $\sqrt{5}$:

$3\sqrt{5} + 4\sqrt{5} = (3 + 4)\sqrt{5}$ Using the distributive law to factor out $\sqrt{5}$
$\qquad\qquad\quad = 7\sqrt{5}.$ $3\sqrt{5}$ and $4\sqrt{5}$ are like radicals.

To simplify in this manner, the radical factors must be the same.

EXAMPLE 2 Simplify.

a) $9\sqrt{17} - 3\sqrt{17}$
b) $7\sqrt{x} + \sqrt{x}$
c) $5\sqrt{2} - \sqrt{18}$
d) $\sqrt{5} + \sqrt{20} + \sqrt{7}$

SOLUTION

a) $9\sqrt{17} - 3\sqrt{17} = (9 - 3)\sqrt{17}$ Using the distributive law. Try to do this mentally.
$\qquad\qquad\qquad\quad = 6\sqrt{17}$

b) $7\sqrt{x} + \sqrt{x} = (7 + 1)\sqrt{x}$ Using the distributive law. Try to do this mentally.
$\qquad\qquad\quad = 8\sqrt{x}$

c) $5\sqrt{2} - \sqrt{18} = 5\sqrt{2} - \sqrt{9 \cdot 2}$
$\qquad\qquad\quad = 5\sqrt{2} - \sqrt{9}\sqrt{2}$ Simplifying $\sqrt{18}$
$\qquad\qquad\quad = 5\sqrt{2} - 3\sqrt{2}$ We now have like radicals.
$\qquad\qquad\quad = 2\sqrt{2}$ Using the distributive law mentally: $5\sqrt{2} - 3\sqrt{2} = (5 - 3)\sqrt{2}$

d) $\sqrt{5} + \sqrt{20} + \sqrt{7} = \sqrt{5} + \sqrt{4}\sqrt{5} + \sqrt{7}$ Simplifying $\sqrt{20}$
$= \sqrt{5} + 2\sqrt{5} + \sqrt{7}$ We now have like radicals.
$= 3\sqrt{5} + \sqrt{7}$ Adding like radicals; $3\sqrt{5} + \sqrt{7}$ cannot be simplified.

> **CAUTION!** It is *not true* that the sum of two square roots is the square root of the sum: $\sqrt{A} + \sqrt{B} \neq \sqrt{A + B}$. For example, $\sqrt{9} + \sqrt{16} \neq \sqrt{9 + 16}$ since $3 + 4 \neq 5$.

More with Multiplication

Radical expressions with more than one term are multiplied in much the same way that polynomials with more than one term are multiplied.

EXAMPLE 3 Multiply.

a) $\sqrt{2}(\sqrt{3} + \sqrt{5})$
b) $(4 + \sqrt{7})(2 + \sqrt{7})$
c) $(2 - \sqrt{5})(2 + \sqrt{5})$
d) $(2 + \sqrt{3})(5 - 4\sqrt{3})$

SOLUTION

a) $\sqrt{2}(\sqrt{3} + \sqrt{5}) = \sqrt{2}\sqrt{3} + \sqrt{2}\sqrt{5}$ Using the distributive law
$= \sqrt{6} + \sqrt{10}$ Using the product rule for radicals

b) $(4 + \sqrt{7})(2 + \sqrt{7}) = 4 \cdot 2 + 4 \cdot \sqrt{7} + \sqrt{7} \cdot 2 + \sqrt{7} \cdot \sqrt{7}$ Using FOIL
$= 8 + 4\sqrt{7} + 2\sqrt{7} + 7$
$= 15 + 6\sqrt{7}$ Combining like terms

c) Note that $(2 - \sqrt{5})(2 + \sqrt{5})$ is of the form $(A - B)(A + B)$:
$(2 - \sqrt{5})(2 + \sqrt{5}) = 2 \cdot 2 + 2 \cdot \sqrt{5} - \sqrt{5} \cdot 2 - \sqrt{5} \cdot \sqrt{5}$
$= 4 + 2\sqrt{5} - 2\sqrt{5} - 5$ As expected, $2\sqrt{5}$ and $-2\sqrt{5}$ are opposites.
$= 4 - 5$
$= -1$

d) $(2 + \sqrt{3})(5 - 4\sqrt{3}) = 2 \cdot 5 - 2 \cdot 4\sqrt{3} + \sqrt{3} \cdot 5 - \sqrt{3} \cdot 4\sqrt{3}$
$= 10 - 8\sqrt{3} + 5\sqrt{3} - 4 \cdot 3$ $2 \cdot 5 = 10, 2 \cdot 4 = 8,$ and $\sqrt{3} \cdot \sqrt{3} = 3$
$= 10 - 3\sqrt{3} - 12$ Adding like radicals
$= -2 - 3\sqrt{3}$

More with Rationalizing Denominators

Note in Example 3(c) that the result has no radicals. This will happen whenever expressions like $\sqrt{a} + \sqrt{b}$ and $\sqrt{a} - \sqrt{b}$ are multiplied:

$(\sqrt{a} + \sqrt{b})(\sqrt{a} - \sqrt{b}) = (\sqrt{a})^2 - (\sqrt{b})^2 = a - b.$

9.4 MORE OPERATIONS WITH RADICALS **445**

Expressions such as $\sqrt{3} - \sqrt{5}$ and $\sqrt{3} + \sqrt{5}$ are said to be **conjugates** of each other. So too are expressions like $2 + \sqrt{7}$ and $2 - \sqrt{7}$. Once the conjugate of a denominator has been found, it can be used to rationalize the denominator.

EXAMPLE 4   Rationalize each denominator and, if possible, simplify.

a) $\dfrac{3}{2 + \sqrt{5}}$

b) $\dfrac{2}{\sqrt{7} - \sqrt{3}}$

SOLUTION

a) We multiply by a form of 1, using the conjugate of $2 + \sqrt{5}$, which is $2 - \sqrt{5}$, as the numerator and the denominator:

$\dfrac{3}{2 + \sqrt{5}} = \dfrac{3}{2 + \sqrt{5}} \cdot \dfrac{2 - \sqrt{5}}{2 - \sqrt{5}}$ Multiplying by 1

$= \dfrac{3(2 - \sqrt{5})}{(2 + \sqrt{5})(2 - \sqrt{5})}$

$= \dfrac{3(2 - \sqrt{5})}{2^2 - (\sqrt{5})^2}$ Using $(A + B)(A - B) = A^2 - B^2$

$= \dfrac{3(2 - \sqrt{5})}{-1}$ Simplifying the denominator. See Example 3(c).

$= \dfrac{6 - 3\sqrt{5}}{-1}$

$= -6 + 3\sqrt{5}.$ Dividing *both* terms in the numerator by -1

b) $\dfrac{2}{\sqrt{7} - \sqrt{3}} = \dfrac{2}{\sqrt{7} - \sqrt{3}} \cdot \dfrac{\sqrt{7} + \sqrt{3}}{\sqrt{7} + \sqrt{3}}$ Multiplying by 1, using $\sqrt{7} + \sqrt{3}$, the conjugate of $\sqrt{7} - \sqrt{3}$

$= \dfrac{2(\sqrt{7} + \sqrt{3})}{(\sqrt{7} - \sqrt{3})(\sqrt{7} + \sqrt{3})}$

$= \dfrac{2(\sqrt{7} + \sqrt{3})}{(\sqrt{7})^2 - (\sqrt{3})^2}$ Using $(A - B)(A + B) = A^2 - B^2$

$= \dfrac{2(\sqrt{7} + \sqrt{3})}{7 - 3}$ The denominator is free of radicals.

$= \dfrac{2(\sqrt{7} + \sqrt{3})}{4}$ Since 2 is a common factor, we simplify.

$= \dfrac{2(\sqrt{7} + \sqrt{3})}{2 \cdot 2}$ Factoring and removing a factor equal to 1: $\dfrac{2}{2} = 1$

$= \dfrac{\sqrt{7} + \sqrt{3}}{2}$

EXERCISE SET 9.4

Add or subtract. Simplify by combining like radical terms, if possible.

1. $7\sqrt{2} + 4\sqrt{2}$
2. $4\sqrt{3} + 3\sqrt{3}$
3. $9\sqrt{5} - 6\sqrt{5}$
4. $8\sqrt{2} - 5\sqrt{2}$
5. $6\sqrt{x} + 7\sqrt{x}$
6. $9\sqrt{y} + 3\sqrt{y}$
7. $9\sqrt{x} - 11\sqrt{x}$
8. $6\sqrt{a} - 14\sqrt{a}$
9. $5\sqrt{2a} + 3\sqrt{2a}$
10. $7\sqrt{6x} + 2\sqrt{6x}$
11. $9\sqrt{10y} - \sqrt{10y}$
12. $12\sqrt{14y} - \sqrt{14y}$
13. $5\sqrt{7} + 2\sqrt{7} + 4\sqrt{7}$
14. $3\sqrt{5} + 7\sqrt{5} + 5\sqrt{5}$
15. $7\sqrt{2} - 9\sqrt{2} + 4\sqrt{2}$
16. $5\sqrt{6} - 7\sqrt{6} + 2\sqrt{6}$
17. $5\sqrt{3} + \sqrt{8}$
18. $2\sqrt{5} + \sqrt{45}$
19. $\sqrt{x} - \sqrt{9x}$
20. $\sqrt{25a} - \sqrt{a}$
21. $5\sqrt{8} + 15\sqrt{2}$
22. $3\sqrt{12} + 2\sqrt{300}$
23. $\sqrt{27} - 2\sqrt{3}$
24. $7\sqrt{50} - 3\sqrt{2}$
25. $\sqrt{72} + \sqrt{98}$
26. $\sqrt{45} + \sqrt{80}$
27. $4\sqrt{12} + \sqrt{27} - \sqrt{8}$
28. $9\sqrt{8} - \sqrt{72} + \sqrt{98}$
29. $5\sqrt{18} - 2\sqrt{32} - \sqrt{50}$
30. $\sqrt{18} - 3\sqrt{8} + \sqrt{75}$
31. $\sqrt{9x} + \sqrt{49x} - 9\sqrt{x}$
32. $\sqrt{16a} - 4\sqrt{a} + \sqrt{25a}$

Multiply.

33. $\sqrt{2}(\sqrt{5} + \sqrt{7})$
34. $\sqrt{5}(\sqrt{2} + \sqrt{11})$
35. $\sqrt{5}(\sqrt{6} - \sqrt{10})$
36. $\sqrt{6}(\sqrt{15} - \sqrt{7})$
37. $(4 + \sqrt{2})(5 + \sqrt{2})$
38. $(5 + \sqrt{11})(3 + \sqrt{11})$
39. $(\sqrt{6} - 2)(\sqrt{6} - 5)$
40. $(\sqrt{10} + 4)(\sqrt{10} - 7)$
41. $(\sqrt{5} + 7)(\sqrt{5} - 7)$
42. $(1 + \sqrt{5})(1 - \sqrt{5})$
43. $(\sqrt{6} - \sqrt{3})(\sqrt{6} + \sqrt{3})$
44. $(\sqrt{2} + \sqrt{6})(\sqrt{2} - \sqrt{6})$
45. $(5 + 3\sqrt{2})(1 - \sqrt{2})$
46. $(8 - \sqrt{7})(3 + 2\sqrt{7})$
47. $(7 + \sqrt{3})^2$
48. $(2 + \sqrt{5})^2$
49. $(1 - 2\sqrt{3})^2$
50. $(6 - 3\sqrt{5})^2$
51. $(\sqrt{x} - \sqrt{10})^2$
52. $(\sqrt{a} - \sqrt{6})^2$

Rationalize each denominator and, if possible, simplify.

53. $\dfrac{9}{5 + \sqrt{2}}$
54. $\dfrac{2}{3 + \sqrt{5}}$
55. $\dfrac{6}{2 - \sqrt{7}}$
56. $\dfrac{3}{7 - \sqrt{2}}$
57. $\dfrac{2}{\sqrt{7} + 3}$
58. $\dfrac{6}{\sqrt{10} + 5}$
59. $\dfrac{\sqrt{6}}{\sqrt{6} - 5}$
60. $\dfrac{\sqrt{10}}{\sqrt{10} - 7}$
61. $\dfrac{\sqrt{5}}{\sqrt{5} - \sqrt{3}}$
62. $\dfrac{\sqrt{7}}{\sqrt{7} - \sqrt{5}}$
63. $\dfrac{\sqrt{3}}{\sqrt{5} + \sqrt{3}}$
64. $\dfrac{\sqrt{6}}{\sqrt{7} - \sqrt{6}}$
65. $\dfrac{2}{\sqrt{7} - \sqrt{2}}$
66. $\dfrac{6}{\sqrt{5} - \sqrt{3}}$
67. $\dfrac{\sqrt{6} + \sqrt{5}}{\sqrt{6} - \sqrt{5}}$
68. $\dfrac{\sqrt{10} - \sqrt{7}}{\sqrt{10} + \sqrt{7}}$

SKILL MAINTENANCE

Solve.

69. $3x + 5 + 2(x - 3) = 4 - 6x$
70. $3(x - 4) - 2 = 8(2x + 3)$
71. $x^2 - 5x = 6$
72. $x^2 + 10 = 7x$

73. *Mixing juice.* Jolly Juice is 3% real fruit juice, and Real Squeeze is 6% real fruit juice. How many liters of each should be used in order to make an 8-L mixture that is 5.4% real fruit juice?

74. *Commuting costs.* Thelma says it costs at least $7.50 every time she drives to work: $3 in tolls and $1.50 per hour for parking. For how long does Thelma park?

SYNTHESIS

75. ◆ Describe a method that could be used to rationalize a numerator that is the sum of two radical expressions.

76. ◆ Explain why it is important for the signs within each pair of conjugates to differ.
77. ◆ Why must you know how to add and subtract radical expressions before you can rationalize denominators with two terms?
78. ◆ Is it possible to square the sum of two radical expressions without adding like radicals? Why or why not?

Add or subtract.

79. $7\sqrt{\dfrac{1}{2}} + \dfrac{5}{2}\sqrt{18} + \sqrt{98}$

80. $\sqrt{a^3 b^6} - b\sqrt{a^5} + a\sqrt{a^7}$

81. $\sqrt{\dfrac{9}{x}} + \dfrac{\sqrt{x}}{2x}$

82. $\sqrt{ab^6} + b\sqrt{a^3} + a\sqrt{a}$

83. $x\sqrt{2y} - \sqrt{8x^2 y} + \dfrac{x}{3}\sqrt{18y}$

84. $7x\sqrt{12xy^2} - 9y\sqrt{27x^3} + 5\sqrt{300x^3 y^2}$

85. Can you find any pairs of nonnegative numbers a and b for which $\sqrt{a} + \sqrt{b} = \sqrt{a+b}$? If so, name them.

86. Three students were asked to simplify $\sqrt{10} + \sqrt{50}$. Their answers were $\sqrt{10}(1 + \sqrt{5})$, $\sqrt{10} + 5\sqrt{2}$, and $\sqrt{2}(5 + \sqrt{5})$. Which answer(s), if any, is correct?

9.5 Radical Equations

Solving Radical Equations • Problem Solving and Applications

An equation in which a variable appears in a radicand is called a **radical equation**. The following are examples:

$$\sqrt{2x} - 4 = 7, \quad 2\sqrt{x+2} = \sqrt{x+10}, \quad \text{and} \quad 3 + \sqrt{27 - 3x} = x.$$

We now learn to solve such equations and use them in problem solving.

Solving Radical Equations

An equation with a square root can be rewritten without the radical by using *the principle of squaring*.

THE PRINCIPLE OF SQUARING

If $a = b$, then $a^2 = b^2$.

If, before using the principle of squaring, we isolate a radical on one side of an equation, that radical will be eliminated after both sides have been squared.

EXAMPLE 1   Solve.

a) $\sqrt{x} + 3 = 7$

b) $3\sqrt{x} = \sqrt{x + 32}$

SOLUTION

a) $\sqrt{x} + 3 = 7$

$\sqrt{x} = 4$ Subtracting 3 to get the radical alone on one side

$(\sqrt{x})^2 = 4^2$ Squaring both sides (using the principle of squaring)

$x = 16$

Check:
$$\begin{array}{c|c} \sqrt{x} + 3 = 7 \\ \hline \sqrt{16} + 3 \; ? \; 7 \\ 4 + 3 \\ 7 & 7 \quad \text{TRUE} \end{array}$$

The solution is 16.

b) $3\sqrt{x} = \sqrt{x + 32}$

$(3\sqrt{x})^2 = (\sqrt{x + 32})^2$ Squaring both sides (using the principle of squaring)

$3^2(\sqrt{x})^2 = x + 32$ Squaring the product on the left; simplifying on the right

$9x = x + 32$ Simplifying on the left

$\left. \begin{array}{l} 8x = 32 \\ x = 4. \end{array} \right\}$ Solving for x

Check:
$$\begin{array}{c|c} 3\sqrt{x} = \sqrt{x + 32} \\ \hline 3\sqrt{4} \; ? \; \sqrt{4 + 32} \\ 3 \cdot 2 & \sqrt{36} \\ 6 & 6 \quad \text{TRUE} \end{array}$$

The number 4 checks. The solution is 4.

The principle of squaring states that whenever $a = b$ is true, $a^2 = b^2$ is also true. It does *not* say that whenever $a^2 = b^2$ is true, $a = b$ is also true. For example, if a is replaced with -5 and b with 5, the equation $a^2 = b^2$ is true (since $(-5)^2 = 5^2$), but the equation $a = b$ is false (since $-5 \neq 5$). Thus, although the principle of squaring enables us to find any solutions that a radical equation might have, it can also lead us to numbers that are not solutions of the original equation.

CAUTION! When the principle of squaring is used to solve an equation, all possible solutions *must* be checked in the original equation!

EXAMPLE 2   Solve: $\sqrt{2x} = -5$.

SOLUTION   We have

$\sqrt{2x} = -5$

$(\sqrt{2x})^2 = (-5)^2$ Squaring both sides (using the principle of squaring)

$\left. \begin{array}{l} 2x = 25 \\ x = \frac{25}{2}. \end{array} \right\}$ Solving for x

9.5 RADICAL EQUATIONS 449

Check:
$$\sqrt{2x} = -5$$
$$\sqrt{2 \cdot \frac{25}{2}} \;?\; -5$$
$$\sqrt{25}$$
$$5 \;|\; -5 \quad \text{FALSE}$$

There are no solutions. You might have suspected this from the start since no number has a principal square root that is negative.

In some cases, we may need to apply the principle of zero products (see Section 5.6) after squaring.

EXAMPLE 3 Solve: **(a)** $x - 5 = \sqrt{x + 7}$; **(b)** $3 + \sqrt{27 - 3x} = x$.

SOLUTION

a)
$$x - 5 = \sqrt{x + 7}$$
$$(x - 5)^2 = (\sqrt{x + 7})^2 \quad \text{Using the principle of squaring}$$
$$x^2 - 10x + 25 = x + 7 \quad \text{Squaring a binomial on the left side}$$
$$x^2 - 11x + 18 = 0 \quad \text{Adding } -x - 7 \text{ on both sides}$$
$$(x - 9)(x - 2) = 0 \quad \text{Factoring}$$
$$x - 9 = 0 \quad \text{or} \quad x - 2 = 0 \quad \text{Using the principle of zero products}$$
$$x = 9 \quad \text{or} \quad x = 2$$

Check: For 9:
$$x - 5 = \sqrt{x + 7}$$
$$9 - 5 \;?\; \sqrt{9 + 7}$$
$$4 \;|\; 4 \quad \text{TRUE}$$

For 2:
$$x - 5 = \sqrt{x + 7}$$
$$2 - 5 \;?\; \sqrt{2 + 7}$$
$$-3 \;|\; 3 \quad \text{FALSE}$$

The number 9 checks, but 2 does not. Thus the solution is 9.

TECHNOLOGY CONNECTION 9.5

Solutions of radical equations can be visualized easily with a grapher. To "see" that 9 is the solution of Example 3(a), we let $y_1 = x - 5$ (the left side of $x - 5 = \sqrt{x + 7}$) and $y_2 = \sqrt{x + 7}$ (the right side of $x - 5 = \sqrt{x + 7}$). Using the INTERSECT option of CALC, we see that (9, 4) is on both curves. This indicates that when $x = 9$, y_1 and y_2 are both 4. No intersection occurs at $x = 2$, which confirms our check in Example 3(a).

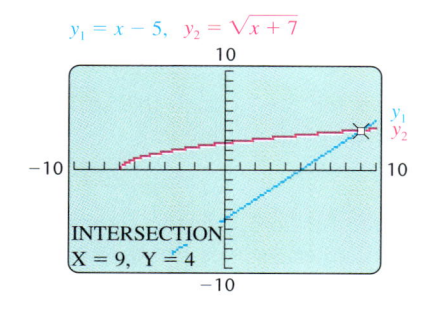

Note that, in the absence of the INTERSECT option, ZOOM and TRACE can be used.

1. Use a grapher to visualize the solution of Example 1(b).
2. Use a grapher to visualize the solution of Example 3(b).

$y_1 = x - 5, \; y_2 = \sqrt{x + 7}$

INTERSECTION
X = 9, Y = 4

b)
$$3 + \sqrt{27 - 3x} = x$$
$$\sqrt{27 - 3x} = x - 3 \quad \text{Subtracting 3 to isolate the radical}$$
$$(\sqrt{27 - 3x})^2 = (x - 3)^2 \quad \text{Using the principle of squaring}$$
$$27 - 3x = x^2 - 6x + 9$$
$$0 = x^2 - 3x - 18 \quad \text{Adding } 3x - 27 \text{ on both sides}$$
$$0 = (x - 6)(x + 3) \quad \text{Factoring}$$
$$x - 6 = 0 \quad \text{or} \quad x + 3 = 0 \quad \text{Using the principle of zero products}$$
$$x = 6 \quad \text{or} \quad x = -3$$

Check: For 6:

$$\begin{array}{c|c} 3 + \sqrt{27 - 3x} = x \\ \hline 3 + \sqrt{27 - 3 \cdot 6} \;?\; 6 \\ 3 + \sqrt{9} \\ 3 + 3 \\ 6 \;\big|\; 6 \quad \text{TRUE} \end{array}$$

For −3:

$$\begin{array}{c|c} 3 + \sqrt{27 - 3x} = x \\ \hline 3 + \sqrt{27 - 3 \cdot (-3)} \;?\; -3 \\ 3 + \sqrt{27 + 9} \\ 3 + \sqrt{36} \\ 3 + 6 \\ 9 \;\big|\; -3 \\ \quad \text{FALSE} \end{array}$$

The number 6 checks, but −3 does not. The solution is 6.

Problem Solving and Applications

Many applications involve the use of radicals. For example, there is a formula to determine how far you can see from a given height. At a height of h meters, you can see V kilometers to the horizon, where

$$V = 3.5\sqrt{h}.$$

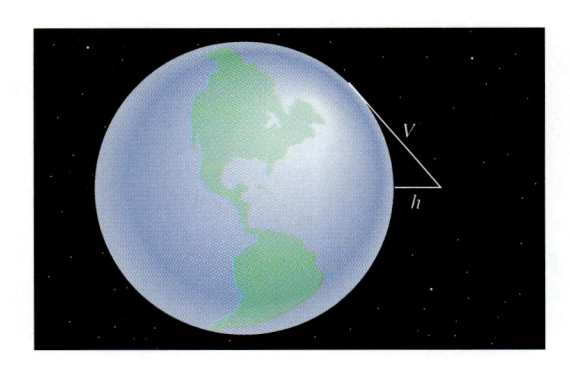

EXAMPLE 4   *Sighting to the Horizon.*   Elaine can see 50.4 km to the horizon from the top of Sunset Rock. What is the altitude of Elaine's eyes?

SOLUTION

1. Familiarize. A sketch can be helpful here (see the figure at the top of the following page). The altitude of Elaine's eyes, in meters, is labeled h.

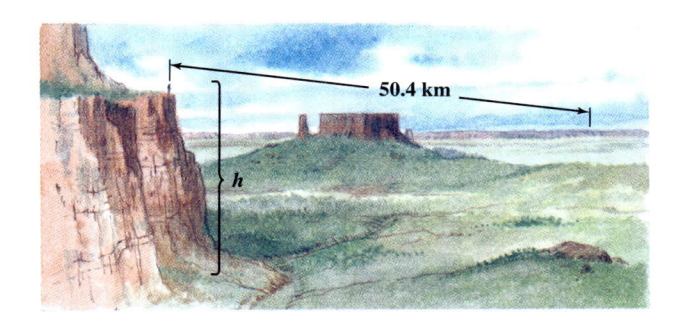

2. Translate. We substitute 50.4 for V in the equation $V = 3.5\sqrt{h}$:

$$50.4 = 3.5\sqrt{h}.$$

3. Carry out. We solve the equation for h:

$$50.4 = 3.5\sqrt{h}$$
$$\frac{50.4}{3.5} = \sqrt{h}$$
$$14.4 = \sqrt{h}$$
$$(14.4)^2 = (\sqrt{h})^2 \quad \text{Using the principle of squaring}$$
$$207.36 = h.$$

4. Check. We leave the check to the student.

5. State. The altitude of Elaine's eyes is about 207 m.

EXERCISE SET 9.5

Solve.

1. $\sqrt{x} = 8$
2. $\sqrt{x} = 7$
3. $\sqrt{x + 3} = 4$
4. $\sqrt{x + 4} = 11$
5. $\sqrt{2x + 4} = 9$
6. $\sqrt{2x + 1} = 13$
7. $3 + \sqrt{x - 1} = 5$
8. $4 + \sqrt{y - 3} = 11$
9. $6 - 2\sqrt{3n} = 0$
10. $8 - 4\sqrt{5n} = 0$
11. $\sqrt{4x + 7} = \sqrt{2x + 13}$
12. $\sqrt{4x - 5} = \sqrt{x + 9}$
13. $\sqrt{x} = -2$
14. $\sqrt{x} = -9$
15. $\sqrt{2y + 6} = \sqrt{2y - 5}$
16. $\sqrt{3x - 5} = \sqrt{2x + 1}$
17. $x - 7 = \sqrt{x - 5}$
18. $\sqrt{3x - 2} = x - 4$
19. $\sqrt{x + 18} = x - 2$
20. $x - 9 = \sqrt{x - 3}$
21. $x - 5 = \sqrt{15 - 3x}$
22. $x - 1 = 6\sqrt{x - 9}$
23. $\sqrt{5x + 21} = x + 3$
24. $\sqrt{22 - x} = x - 2$
25. $x + 4 = 4\sqrt{x + 1}$
26. $1 + 2\sqrt{x - 1} = x$
27. $\sqrt{x^2 + 6} - x + 3 = 0$
28. $\sqrt{x^2 + 5} - x + 2 = 0$
29. $\sqrt{(p + 6)(p + 1)} - 2 = p + 1$
30. $\sqrt{(4x + 5)(x + 4)} = 2x + 5$
31. $\sqrt{x - 2} = \sqrt{5 - 2x}$
32. $\sqrt{7 - 3x} = \sqrt{x - 2}$
33. $x - 1 = \sqrt{(x + 1)(x - 2)}$
34. $x = 1 + \sqrt{1 - x}$

452 CHAPTER 9 • RADICAL EXPRESSIONS AND EQUATIONS

Sighting to the horizon. Solve. Use $V = 3.5\sqrt{h}$ from Example 4.

35. A steeplejack can see 21 km to the horizon from the top of a building. What is the altitude of the steeplejack's eyes?

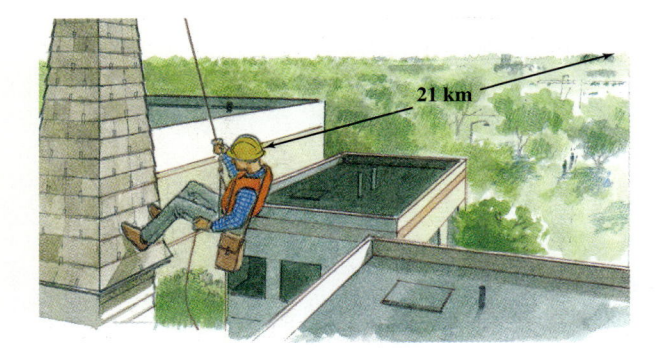

36. A scout can see 84 km to the horizon from atop a firetower. What is the altitude of the scout's eyes?
37. Amelia can see 378 km to the horizon from an airplane window. How high is the airplane?
38. Ahab can see 99.4 km to the horizon from the top of a mast. How high is the mast?

Period of a swinging pendulum. *The formula $T = 2\pi\sqrt{L/32}$ can be used to find the period T, in seconds, of a pendulum of length L, in feet.*

39. What is the length of a pendulum that has a period of 1.6 sec? Use 3.14 for π.
40. What is the length of a pendulum that has a period of 3 sec? Use 3.14 for π.

Speed of a skidding car. *The formula $r = 2\sqrt{5L}$ can be used to approximate the speed r, in miles per hour, of a car that has left a skid mark of length L, in feet.*

41. How far will a car skid at 40 mph? at 80 mph?
42. How far will a car skid at 48 mph? at 60 mph?
43. Find a number such that the opposite of 3 times its square root is -33.
44. Find a number such that 1 less than the square root of twice the number is 7.

SKILL MAINTENANCE

Simplify.

45. $(-2)^5$
46. $(-5)^3$

Multiply and simplify.

47. $\dfrac{7x^9}{27} \cdot \dfrac{9}{7x^3}$
48. $\dfrac{3}{x^2 - 9} \cdot \dfrac{x^2 - 6x + 9}{12}$

Add or subtract as indicated. Simplify, if possible.

49. $\dfrac{x}{x+5} + \dfrac{x^2 - 20}{x+5}$
50. $\dfrac{9}{a-1} - \dfrac{4}{1-a}$

SYNTHESIS

51. ◆ Review Example 2 and those exercises to which there is no solution. How could someone easily write a radical equation to which there is no solution?
52. ◆ Explain what would have happened in Example 1(a) if we had not isolated the radical before squaring. Could we still have solved the equation? Why or why not?
53. ◆ Explain in your own words why possible solutions of radical equations must be checked.
54. ◆ Do you believe that the principle of squaring can be extended to powers other than 2? That is, if $a = b$, does it follow that $a^n = b^n$ for any integer n? Why or why not?

Sometimes the principle of squaring must be used more than once in order to solve an equation. Solve Exercises 55–62 by using the principle of squaring as often as necessary.

55. $5 - \sqrt{x} = \sqrt{x - 5}$
56. $1 + \sqrt{x} = \sqrt{x + 9}$
57. $\sqrt{3x + 1} = 1 - \sqrt{x + 4}$
58. $\sqrt{y + 8} - \sqrt{y} = 2$
59. $1 + \sqrt{19 - x} = 3 + \sqrt{4 - x}$
60. $\sqrt{y + 1} - \sqrt{2y - 5} = \sqrt{y - 2}$
61. $2\sqrt{x - 1} - \sqrt{3x - 5} = \sqrt{x - 9}$
62. $x + (2 - x)\sqrt{x} = 0$
63. *Changing elevations.* A mountain climber pauses to rest and view the horizon. Using the formula

$V = 3.5\sqrt{h}$, the climber computes the distance to the horizon and then climbs another 100 m. At this higher elevation, the horizon is 20 km farther than before. At what height was the climber when the first computation was made? (*Hint*: Use a system of equations.)

64. Solve $A = \sqrt{1 + \sqrt{a/b}}$ for b.

Graph. Use a calculator or Table 2 to find y-values.

65. $y = \sqrt{x}$
66. $y = \sqrt{x - 2}$
67. $y = \sqrt{x - 3}$
68. $y = \sqrt{x + 1}$

69. Graph $y = x - 7$ and $y = \sqrt{x - 5}$ using the same set of axes. Determine where the graphs intersect in order to estimate a solution of $x - 7 = \sqrt{x - 5}$.

70. Graph $y = 1 + \sqrt{x}$ and $y = \sqrt{x + 9}$ using the same set of axes. Determine where the graphs intersect in order to estimate a solution of $1 + \sqrt{x} = \sqrt{x + 9}$.

Use a grapher to solve Exercises 71 and 72. Round answers to the nearest hundredth when appropriate.

71. $\sqrt{x + 3} = 2x - 1$
72. $-\sqrt{x + 3} = 2x - 1$

COLLABORATIVE C•O•R•N•E•R

Focus: Radicals, equations, graphing, and estimating
Time: 20–30 minutes
Group size: 2
Materials: Graph paper; calculators are optional.

When the wind is blowing, the temperature reading on a thermometer is not an accurate indicator of how cold it *feels* outdoors. Meteorologists use the *wind chill temperature* to describe how cold it feels for a given wind speed and thermometer reading.

ACTIVITY

1. Below is a list of some wind chill temperatures for two different thermometer readings. Each group member should choose one chart and plot the points on a graph. On the basis of the pattern graphed, each student should estimate the wind chill temperature for a 25-mph wind at that same thermometer reading.

THERMOMETER READING: 30°F

Wind Speed	Wind Chill Temperature
5 mph	27°F
10 mph	16°F
15 mph	9°F
20 mph	4°F

THERMOMETER READING: 20°F

Wind Speed	Wind Chill Temperature
5 mph	16°F
10 mph	3°F
15 mph	−5°F
20 mph	−10°F

2. A formula for finding the wind chill temperature, T_w, is

$$T_w = 91.4 - \frac{(10.45 + 6.68\sqrt{v} - 0.447v)(457 - 5T)}{110},$$

where v is the wind speed, in miles per hour, and T is the thermometer reading, in degrees Fahrenheit. Using the formula, each student should compute the wind chill temperature that was being estimated in part (1). How far off were the estimates?

3. Each group member should privately choose a thermometer reading and, using the formula, calculate the wind chill temperatures at that thermometer reading for at least two different wind speeds. The results should be listed in a table similar to those given above, but the thermometer reading should *not* be included. Exchange the charts when finished.

4. Using the information given, each student should determine the thermometer reading corresponding to the wind chill temperatures calculated. Check with your partner when you are both finished.

9.6 Applications Using Right Triangles

Right Triangles • Problem Solving

Radicals frequently occur in problem-solving situations in which the Pythagorean theorem is used. In Section 5.7, when we first used the Pythagorean theorem, we had not yet studied square roots. We now know that if $x^2 = n$, then x is a square root of n.

Right Triangles

For convenience, we restate the Pythagorean theorem.

THE PYTHAGOREAN THEOREM*

In any right triangle, if a and b are the lengths of the legs and c is the length of the hypotenuse, then

$$a^2 + b^2 = c^2.$$

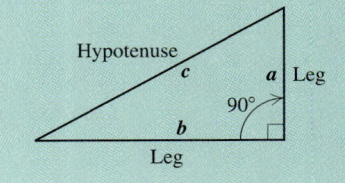

When the Pythagorean theorem is used to find a length, we need not concern ourselves with negative square roots, since length cannot be negative.

EXAMPLE 1 Find the length of the hypotenuse of the triangle shown. Give an exact answer and an approximation to three decimal places.

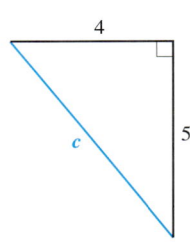

---

*The converse of the Pythagorean theorem also holds. That is, if a, b, and c are the lengths of the sides of a triangle and $a^2 + b^2 = c^2$, then the triangle is a right triangle.

SOLUTION   We have

$$a^2 + b^2 = c^2$$
$$4^2 + 5^2 = c^2 \quad \text{Substituting the lengths of the legs}$$
$$16 + 25 = c^2$$
$$41 = c^2.$$

We now use the fact that if $x^2 = n$, then $x = \sqrt{n}$ or $x = -\sqrt{n}$. In this case, since c is a length, it follows that c is the positive square root of 41:

$$c = \sqrt{41} \quad \text{This is an exact answer.}$$
$$c \approx 6.403. \quad \text{Using a calculator or Table 2 for an approximation}$$

EXAMPLE 2   Find the length of the indicated leg in each triangle. In each case, give an exact answer and an approximation to three decimal places.

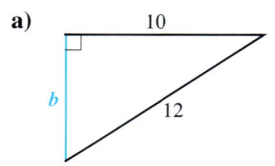

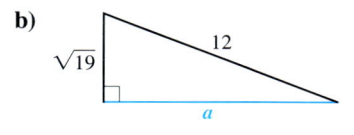

a) [triangle with legs 10 and b, hypotenuse 12]

b) [triangle with leg $\sqrt{19}$, leg a, hypotenuse 12]

SOLUTION

a) $10^2 + b^2 = 12^2$ Substituting in the Pythagorean theorem
   $100 + b^2 = 144$
   $b^2 = 44$ Subtracting 100 on both sides
   $b = \sqrt{44}$ The exact answer is $\sqrt{44}$. Since the length is positive, only the principal square root is used.
   $b \approx 6.633$ Approximating $\sqrt{44}$ with a calculator or Table 2

b) $a^2 + (\sqrt{19})^2 = 12^2$ Substituting
   $a^2 + 19 = 144$
   $a^2 = 125$ Subtracting 19 on both sides
   $a = \sqrt{125}$ The exact answer is $\sqrt{125}$.
   $a \approx 11.180$ Using a calculator

Problem Solving

The five-step process and the Pythagorean theorem can be used for problem solving.

EXAMPLE 3   *Reach of a Ladder.*   A 32-ft ladder is leaning against a house. The bottom of the ladder is 7 ft from the house. How high is the top of the ladder? Give an exact answer and an approximation to three decimal places.

SOLUTION

1. **Familiarize.**   First we make a sketch (see the following page). In it, there is a right triangle. We label the unknown height h.

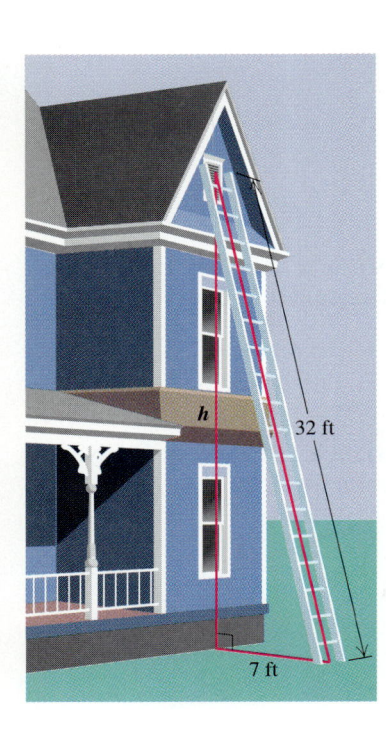

2. Translate. We use the Pythagorean theorem, substituting 7 for a, h for b, and 32 for c:

$$7^2 + h^2 = 32^2.$$

3. Carry out. We solve the equation:

$$7^2 + h^2 = 32^2$$
$$49 + h^2 = 1024$$
$$h^2 = 975$$
$$h = \sqrt{975} \quad \text{This answer is exact.}$$
$$h \approx 31.225. \quad \text{Approximating with a calculator}$$

4. Check. We check by substituting 7, $\sqrt{975}$, and 32:

$$\begin{array}{c|c} a^2 + b^2 = c^2 \\ \hline 7^2 + (\sqrt{975})^2 \; ? \; 32^2 \\ 49 + 975 & 1024 \\ 1024 & 1024 \quad \text{TRUE} \end{array}$$

5. State. The top of the ladder is $\sqrt{975}$, or about 31.225 ft from the ground.

EXAMPLE 4

Softball Dimensions. A softball diamond is a square 65 ft on a side. How far is it from home plate to second base? Give an exact answer and an approximation to three decimal places.

SOLUTION

1. Familiarize. We first make a drawing. Note that the first- and second-base lines, together with a line from home to second, form a right triangle. We label the unknown distance d.

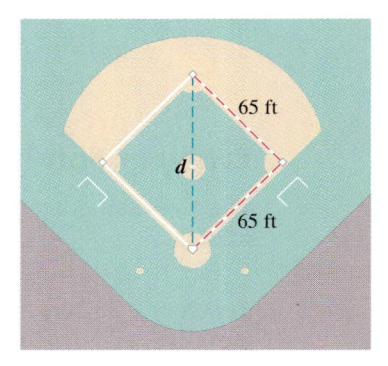

2. Translate. We substitute 65 for a, 65 for b, and d for c in the Pythagorean theorem:

$$65^2 + 65^2 = d^2.$$

3. Carry out. We solve the equation:

$$65^2 + 65^2 = d^2$$
$$4225 + 4225 = d^2$$
$$8450 = d^2$$
$$\sqrt{8450} = d \quad \text{This is exact.}$$
$$91.924 \approx d. \quad \text{This is approximate.}$$

4. Check. We check by substituting 65, 65, and $\sqrt{8450}$:

$$\frac{a^2 + b^2 = c^2}{65^2 + 65^2 \; ? \; (\sqrt{8450})^2}$$
$$\begin{array}{c|c} 4225 + 4225 & 8450 \\ 8450 & 8450 \end{array} \quad \text{TRUE}$$

5. State. From home plate to second base is $\sqrt{8450}$, or about 91.924 ft.

EXERCISE SET 9.6

Find the length of the third side of each triangle. Give an exact answer and, where appropriate, an approximation to three decimal places.

1. 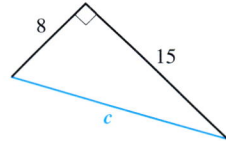
2. 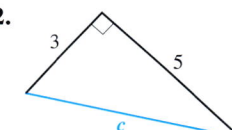
3. 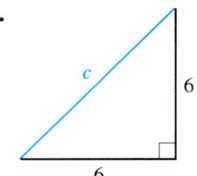
4. 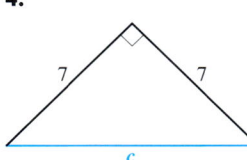
5. 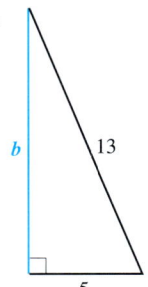
6. 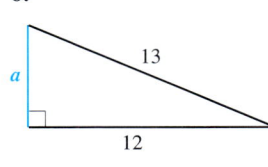
7. 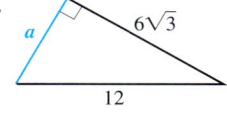
8.

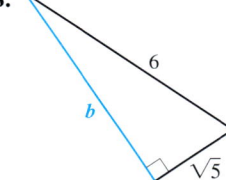

In a right triangle, find the length of the side not given. Give an exact answer and, where appropriate, an approximation to three decimal places. Keep in mind that a and b are the lengths of the legs and c is the length of the hypotenuse.

9. $a = 10, b = 24$
10. $a = 5, b = 12$
11. $a = 9, c = 15$
12. $a = 18, c = 30$
13. $b = 1, c = \sqrt{5}$
14. $b = 1, c = \sqrt{2}$
15. $a = 1, c = \sqrt{3}$
16. $a = \sqrt{3}, b = \sqrt{5}$
17. $c = 10, b = 5\sqrt{3}$
18. $a = 5, b = 5$

Solve. Don't forget to make drawings. Give an exact answer and an approximation to three decimal places.

19. **Home maintenance.** A 14-m ladder is leaning against a building. The bottom of the ladder is 7 m from the building. How high is the top of the ladder?

20. *Masonry.* Find the length of a diagonal of a square tile that has sides 4 cm long.

21. *Plumbing.* A new water pipe is being prepared so that it will run diagonally under a kitchen floor. If the kitchen is 8 ft wide and 12 ft long, how long should the pipe be?

22. *Guy wires.* How long must a guy wire be to reach from the top of a 13-m telephone pole to a point on the ground 9 m from the foot of the pole?

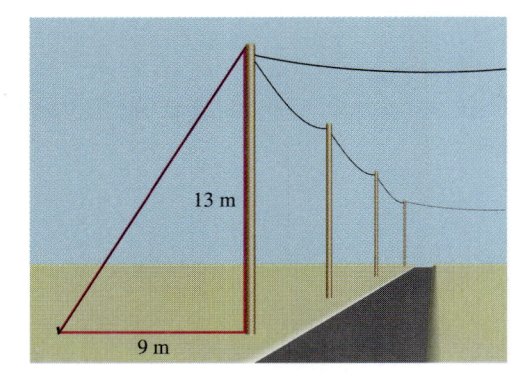

23. *Soccer fields.* The smallest regulation soccer field is 50 yd wide and 100 yd long. Find the length of a diagonal of such a field.

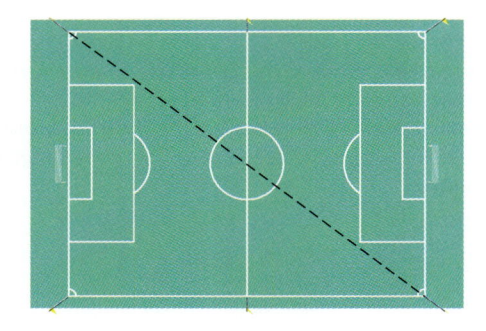

24. *Soccer fields.* The largest regulation soccer field is 100 yd wide and 130 yd long. Find the length of a diagonal of such a field.

25. *Cordless telephones.* Becky's new cordless telephone has clear reception up to 300 ft from its base. Her phone is located near a window in her apartment, 180 ft above ground level. How far into her backyard can Becky use her phone?

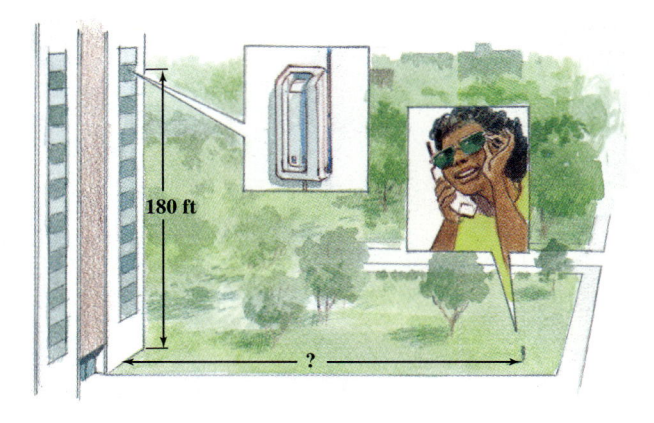

26. *Baseball.* A baseball diamond is a square 90 ft on a side. How far is it from first base to third base?

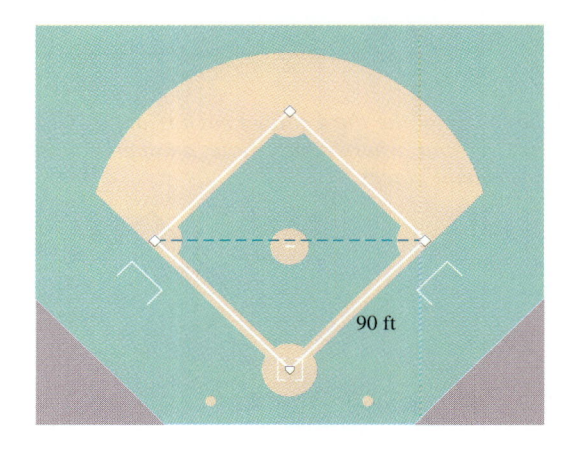

27. *Aviation.* A pilot is instructed to descend from 30,000 ft to 20,000 ft over a horizontal distance of 50,000 ft. What distance will the plane travel during this descent?

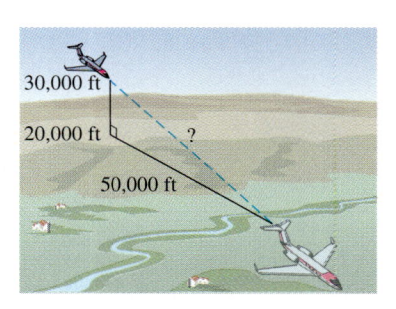

28. *Surveying.* A surveyor had poles located at points *P*, *Q*, and *R*. The distances that she was able to measure are marked in the figure. What is the approximate length of the lake?

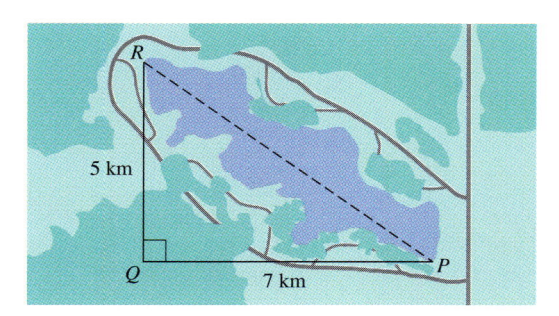

SKILL MAINTENANCE

29. Find the slope of the line $4 - x = 3y$.

30. Find the slope of the line containing the points $(8, -3)$ and $(0, -8)$.

Solve.

31. $-\frac{3}{5}x < 15$

32. $-2x + 6 \geq 7x - 3$

Simplify.

33. $(3x)^4$

34. $\left(-\frac{2}{5}\right)^3$

SYNTHESIS

35. ◆ Can the length of a triangle's hypotenuse ever equal the combined lengths of the two legs? Why or why not?

36. ◆ In an *equilateral triangle*, all sides have the same length. Can a right triangle be equilateral? Why or why not?

37. ◆ In an *isosceles triangle*, two sides have the same length. Can a right triangle be isosceles? Why or why not?

38. ◆ Can a carpenter use a 28-ft ladder to repair clapboard that is 28 ft above ground level? Why or why not?

39. *Cordless telephones.* In 1996, Sanyo unveiled a cordless phone with a range of one quarter mile. Vance has a corner office in the Empire State Building, 900 ft above street level. Can Vance locate the Sanyo's phone base in his office and use the handset at a restaurant at street level on the opposite corner? Show your work.

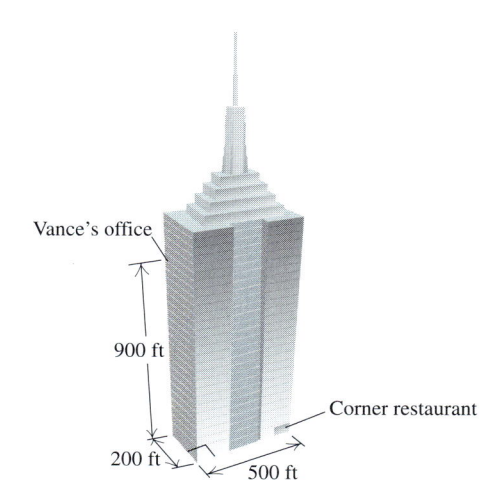

40. The diagonal of a square has a length of $8\sqrt{2}$ ft. Find the length of a side of the square.

41. Find the length of a side of a square that has an area of 7 m².

42. A right triangle has sides with lengths that are consecutive integers. Find the lengths of the sides.

43. Find the length of the diagonal of a cube with sides of length *s*.

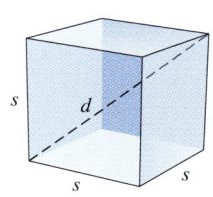

44. Figure *ABCD* is a square. Find the length of a diagonal, *AC*.

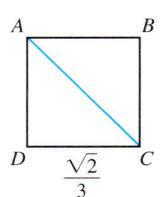

45. Express the height h of an equilateral triangle in terms of the length of a side a.

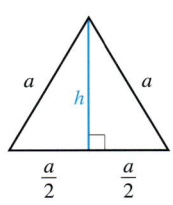

46. *Racquetball.* A racquetball court is 20 ft by 20 ft by 40 ft. What is the longest straight-line distance that can be measured in this racquetball court?

47. *Distance driven.* Two cars leave a service station at the same time. One car travels east at a speed of 50 mph, and the other travels south at a speed of 60 mph. After one half hour, how far apart are they?

Find x in each of the following.

48.

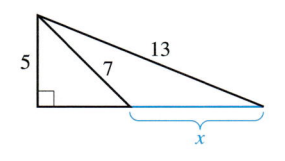

49.

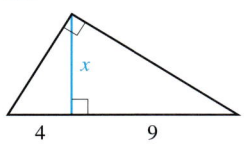

50. The area of square $PQRS$ is 100 ft^2, and A, B, C, and D are midpoints of the sides on which they lie. Find the area of square $ABCD$.

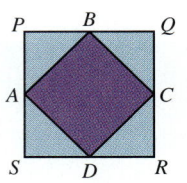

51. *Ranching.* If 2 mi of fencing encloses a square plot of land with an area of 160 acres, how large a square, in acres, will 4 mi of fencing enclose?

COLLABORATIVE C•O•R•N•E•R

Focus: Pythagorean theorem
Time: 15–25 minutes
Group size: 2–4
Materials: Tape measure and chalk; string and scissors are optional.

We mentioned in the footnote on page 454 that the converse of the Pythagorean theorem is also true: If a, b, and c are the lengths of the sides of a triangle and $a^2 + b^2 = c^2$, then the triangle is a right triangle. This provides carpenters, masons, archaeologists, and others with a handy way of locating a line that forms a 90° angle with another line.

ACTIVITY

1. We have seen in examples and exercises that $3^2 + 4^2 = 5^2$ and $12^2 + 5^2 = 13^2$. Thus we know that any triangle with sides of 3 m, 4 m, and 5 m will be a right triangle, as will any triangle with sides of 12 ft, 5 ft, and 13 ft. Each group member should check that $k \cdot 3$, $k \cdot 4$, and $k \cdot 5$, or $12 \cdot k$, $5 \cdot k$, and

$13 \cdot k$ will satisfy $a^2 + b^2 = c^2$ for any choice of k. Each such three-number solution is called a *Pythagorean triple*.

2. Suppose that, in the process of building a deck, your group needs to position an 8-ft piece of lumber so that it forms a 90° angle with a wall. Use a tape measure, chalk, Pythagorean triples, and—if desired—string and scissors to construct a right angle at a specified point at the base of some wall in or near your classroom. (*Hint*: If the point on the wall is 4 ft from a second point on the wall, there is only one point that is precisely 3 ft from the first point and 5 ft from the second point.)

3. To check that the angle formed in part (2) is truly 90°, repeat the procedure, at the same point on the wall, with a different Pythagorean triple.

9.7 Higher Roots and Rational Exponents

Higher Roots • Products and Quotients Involving Higher Roots • Rational Exponents • Calculators

In this section, we study *higher* roots, such as cube roots or fourth roots, and exponents that are not integers.

Higher Roots

Recall that c is a square root of a if $c^2 = a$. A similar definition exists for *cube roots*.

CUBE ROOT

The number c is the *cube root* of a if $c^3 = a$.

The symbolism $\sqrt[3]{a}$ is used to represent the cube root of a. In the radical $\sqrt[3]{a}$, the number 3 is called the **index**, and a is called the **radicand**.

462 CHAPTER 9 • RADICAL EXPRESSIONS AND EQUATIONS

EXAMPLE 1 Find the cube root of each number: (a) 8; (b) −125.

SOLUTION

a) The cube root of 8 is the number whose cube is 8. Since $2^3 = 2 \cdot 2 \cdot 2 = 8$, the cube root of 8 is 2: $\sqrt[3]{8} = 2$.

b) The cube root of −125 is the number whose cube is −125. Since $(-5)^3 = (-5)(-5)(-5) = -125$, the cube root of −125 is −5: $\sqrt[3]{-125} = -5$.

Positive numbers always have *two* nth roots when n is even but when we refer to *the* nth root of a positive number a, denoted $\sqrt[n]{a}$, we mean the *positive* nth root. Thus, although −3 and 3 are both fourth roots of 81 (since $(-3)^4 = 3^4 = 81$), 3 is considered *the* fourth root of 81. In symbols,

$\sqrt[4]{81} = 3$.

nTH ROOT

The number c is an *nth root* of a if $c^n = a$.

If n is odd, then there is exactly one nth root and $\sqrt[n]{a}$ represents that root.

If n is even, then $\sqrt[n]{a}$ represents the nonnegative nth root.

Even roots of negative numbers are not real numbers.

EXAMPLE 2 Find each root: (a) $\sqrt[4]{16}$; (b) $\sqrt[5]{-32}$; (c) $\sqrt[4]{-16}$; (d) $-\sqrt[3]{64}$.

SOLUTION

a) $\sqrt[4]{16} = 2$ Since $2^4 = 2 \cdot 2 \cdot 2 \cdot 2 = 16$

b) $\sqrt[5]{-32} = -2$ Since $(-2)^5 = (-2)(-2)(-2)(-2)(-2) = -32$

c) $\sqrt[4]{-16}$ is not a real number, because it is an even root of a negative number.

d) $-\sqrt[3]{64} = -(\sqrt[3]{64})$ This is the opposite of $\sqrt[3]{64}$.
   $= -4$ $4^3 = 4 \cdot 4 \cdot 4 = 64$

Some roots occur so frequently that you may want to memorize them.

Square Roots		Cube Roots	Fourth Roots	Fifth Roots
$\sqrt{1} = 1$	$\sqrt{4} = 2$	$\sqrt[3]{1} = 1$	$\sqrt[4]{1} = 1$	$\sqrt[5]{1} = 1$
$\sqrt{9} = 3$	$\sqrt{16} = 4$	$\sqrt[3]{8} = 2$	$\sqrt[4]{16} = 2$	$\sqrt[5]{32} = 2$
$\sqrt{25} = 5$	$\sqrt{36} = 6$	$\sqrt[3]{27} = 3$	$\sqrt[4]{81} = 3$	$\sqrt[5]{243} = 3$
$\sqrt{49} = 7$	$\sqrt{64} = 8$	$\sqrt[3]{64} = 4$	$\sqrt[4]{256} = 4$	
$\sqrt{81} = 9$	$\sqrt{100} = 10$	$\sqrt[3]{125} = 5$	$\sqrt[4]{625} = 5$	
$\sqrt{121} = 11$	$\sqrt{144} = 12$	$\sqrt[3]{216} = 6$		

Products and Quotients Involving Higher Roots

The rules for working with products and quotients of square roots can be extended to products and quotients of nth roots. Prime factorizations can again be used when no simplification is readily apparent.

THE PRODUCT AND QUOTIENT RULES

$$\sqrt[n]{AB} = \sqrt[n]{A} \, \sqrt[n]{B}, \qquad \sqrt[n]{\frac{A}{B}} = \frac{\sqrt[n]{A}}{\sqrt[n]{B}}$$

EXAMPLE 3 Simplify: (a) $\sqrt[3]{40}$; (b) $\sqrt[3]{\dfrac{125}{27}}$; (c) $\sqrt[4]{1250}$; (d) $\sqrt[5]{\dfrac{2}{243}}$.

SOLUTION

a) $\sqrt[3]{40} = \sqrt[3]{8 \cdot 5}$ Note that $40 = 2 \cdot 2 \cdot 2 \cdot 5 = 8 \cdot 5$ and 8 is a perfect cube.
$\phantom{\sqrt[3]{40}} = \sqrt[3]{8} \cdot \sqrt[3]{5}$
$\phantom{\sqrt[3]{40}} = 2\sqrt[3]{5}$

b) $\sqrt[3]{\dfrac{125}{27}} = \dfrac{\sqrt[3]{125}}{\sqrt[3]{27}} = \dfrac{5}{3}$ $125 = 5 \cdot 5 \cdot 5$ and $27 = 3 \cdot 3 \cdot 3$, so 125 and 27 are perfect cubes. See the chart on p. 462.

c) $\sqrt[4]{1250} = \sqrt[4]{625 \cdot 2}$ Note that $1250 = 2 \cdot 5 \cdot 5 \cdot 5 \cdot 5 = 2 \cdot 625$ and 625 is a perfect fourth power.
$\phantom{\sqrt[4]{1250}} = \sqrt[4]{625} \cdot \sqrt[4]{2}$
$\phantom{\sqrt[4]{1250}} = 5\sqrt[4]{2}$

d) $\sqrt[5]{\dfrac{2}{243}} = \dfrac{\sqrt[5]{2}}{\sqrt[5]{243}}$
$\phantom{\sqrt[5]{\dfrac{2}{243}}} = \dfrac{\sqrt[5]{2}}{3}$ $243 = 3 \cdot 3 \cdot 3 \cdot 3 \cdot 3$, so 243 is a perfect fifth power. See the chart on p. 462.

Rational Exponents

Expressions containing rational exponents, like $8^{1/3}$, $4^{5/2}$, and $81^{-3/4}$, are defined in a manner that ensures that the laws of exponents still hold. For example, if the product rule, $a^m \cdot a^n = a^{m+n}$, is to hold, then

$$a^{1/2} \cdot a^{1/2} = a^{1/2 + 1/2}$$
$$= a^1 = a.$$

This says that the square of $a^{1/2}$ is a, which suggests that $a^{1/2}$ is a square root of a:

$a^{1/2}$ means $\sqrt{a}.$

This idea is generalized as follows.

THE EXPONENT $1/n$

$a^{1/n}$ means $\sqrt[n]{a}$. If a is negative, then the index n must be odd. (Note that $a^{1/2}$ is written $\sqrt{a}$.)

EXAMPLE 4 Simplify: (a) $8^{1/3}$; (b) $100^{1/2}$; (c) $81^{1/4}$; (d) $(-243)^{1/5}$.

SOLUTION

a) $8^{1/3} = \sqrt[3]{8} = 2$

b) $100^{1/2} = \sqrt{100} = 10$

c) $81^{1/4} = \sqrt[4]{81} = 3$

d) $(-243)^{1/5} = \sqrt[5]{-243} = -3$

If we still wish to multiply exponents when raising a power to a power, we must have $a^{2/3} = (a^{1/3})^2$ and $a^{2/3} = (a^2)^{1/3}$. This suggests that $a^{2/3} = (\sqrt[3]{a})^2$ and $a^{2/3} = \sqrt[3]{a^2}$.

POSITIVE RATIONAL EXPONENTS

For any natural numbers m and n ($n \neq 1$) and any real number a for which $\sqrt[n]{a}$ exists,

$a^{m/n}$ means $(\sqrt[n]{a})^m$, or equivalently, $a^{m/n}$ means $\sqrt[n]{a^m}$.

Usually simplifications are most easily performed using $(\sqrt[n]{a})^m$.

EXAMPLE 5 Simplify: (a) $27^{2/3}$; (b) $8^{5/3}$; (c) $81^{3/4}$.

SOLUTION

a) $27^{2/3} = (27^{1/3})^2 = (\sqrt[3]{27})^2 = 3^2 = 9$

b) $8^{5/3} = (8^{1/3})^5 = (\sqrt[3]{8})^5 = 2^5 = 32$

c) $81^{3/4} = (81^{1/4})^3 = (\sqrt[4]{81})^3 = 3^3 = 27$

Negative rational exponents are defined in much the same way that negative integer exponents are.

NEGATIVE RATIONAL EXPONENTS

For any rational number m/n and any nonzero real number a for which $a^{m/n}$ exists,

$$a^{-m/n} = \frac{1}{a^{m/n}}.$$

EXAMPLE 6

Simplify: **(a)** $16^{-1/2}$; **(b)** $27^{-1/3}$; **(c)** $32^{-2/5}$; **(d)** $64^{-3/2}$.

SOLUTION

a) $16^{-1/2} = \dfrac{1}{16^{1/2}} = \dfrac{1}{\sqrt{16}} = \dfrac{1}{4}$

b) $27^{-1/3} = \dfrac{1}{27^{1/3}} = \dfrac{1}{\sqrt[3]{27}} = \dfrac{1}{3}$

c) $32^{-2/5} = \dfrac{1}{32^{2/5}} = \dfrac{1}{(32^{1/5})^2} = \dfrac{1}{(\sqrt[5]{32})^2} = \dfrac{1}{2^2} = \dfrac{1}{4}$

d) $64^{-3/2} = \dfrac{1}{64^{3/2}} = \dfrac{1}{(\sqrt{64})^3} = \dfrac{1}{8^3} = \dfrac{1}{512}$

CAUTION! A negative exponent does not indicate that the expression in which it appears is negative.

Calculators

A calculator with a key for finding powers can be used to approximate numbers like $\sqrt[5]{8}$. Generally such keys are labeled $\boxed{x^y}$, $\boxed{a^x}$, or $\boxed{\wedge}$. We find approximations by entering the radicand, pressing the power key, entering the exponent, and then pressing $\boxed{=}$ or $\boxed{\text{ENTER}}$. Thus, $\sqrt[5]{8}$ can be approximated by entering 8, pressing the power key, entering 0.2 or (1 ÷ 5), and pressing $\boxed{=}$ to get $\sqrt[5]{8} \approx 1.515716567$. Consult an owner's manual or your instructor if your calculator works differently.

EXERCISE SET 9.7

Simplify. If an expression does not represent a real number, state this.

1. $\sqrt[3]{-64}$
2. $\sqrt[3]{-8}$
3. $\sqrt[3]{1000}$
4. $\sqrt[3]{-27}$
5. $-\sqrt[3]{125}$
6. $-\sqrt[3]{8}$
7. $\sqrt[3]{216}$
8. $\sqrt[3]{-343}$
9. $\sqrt[4]{625}$
10. $\sqrt[4]{81}$
11. $\sqrt[5]{0}$
12. $\sqrt[5]{1}$
13. $\sqrt[5]{-1}$
14. $\sqrt[5]{-243}$
15. $\sqrt[4]{-81}$
16. $\sqrt[4]{-1}$
17. $\sqrt[4]{10{,}000}$
18. $\sqrt[5]{100{,}000}$
19. $\sqrt[3]{5^3}$
20. $\sqrt[4]{7^4}$
21. $\sqrt[6]{64}$
22. $\sqrt[6]{1}$
23. $\sqrt[9]{a^9}$
24. $\sqrt[3]{n^3}$
25. $\sqrt[3]{32}$
26. $\sqrt[3]{54}$
27. $\sqrt[4]{48}$
28. $\sqrt[5]{160}$
29. $\sqrt[3]{\dfrac{64}{125}}$
30. $\sqrt[3]{\dfrac{125}{27}}$
31. $\sqrt[5]{\dfrac{32}{243}}$
32. $\sqrt[4]{\dfrac{625}{256}}$
33. $\sqrt[3]{\dfrac{17}{8}}$
34. $\sqrt[5]{\dfrac{11}{32}}$
35. $\sqrt[4]{\dfrac{13}{81}}$
36. $\sqrt[3]{\dfrac{10}{27}}$

Simplify.

37. $25^{1/2}$
38. $9^{1/2}$
39. $1000^{1/3}$
40. $125^{1/3}$
41. $32^{1/5}$
42. $16^{1/4}$

43. $16^{3/4}$
44. $8^{4/3}$
45. $4^{5/2}$
46. $9^{3/2}$
47. $64^{2/3}$
48. $32^{2/5}$
49. $8^{5/3}$
50. $16^{5/4}$
51. $25^{5/2}$
52. $4^{-1/2}$
53. $36^{-1/2}$
54. $32^{-1/5}$
55. $256^{-1/4}$
56. $100^{-3/2}$
57. $81^{-3/4}$
58. $16^{-3/4}$
59. $81^{-5/4}$
60. $32^{-2/5}$
61. $8^{-2/3}$
62. $625^{-3/4}$

SKILL MAINTENANCE

Factor.

63. $x^2 - 10x + 25$
64. $a^2 - 12a + 36$
65. $t^2 + 18t + 81$

66. Find an equation of variation in which y varies directly as x, and $y = 2.5$ when $x = 4$.

Perform the indicated operation and, if possible, simplify.

67. $\dfrac{a+5}{a^2-25} - \dfrac{3a+15}{a^2-25}$

68. $\dfrac{x}{x^2-9} \cdot \dfrac{x^2-4x+3}{3x^4}$

SYNTHESIS

69. ◆ Explain in your own words why $\sqrt[n]{a}$ is negative when n is odd and a is negative.

70. ◆ Expressions of the form $a^{m/n}$ can be rewritten as $(\sqrt[n]{a})^m$ or $\sqrt[n]{a^m}$. Which radical expression would you use when simplifying $25^{3/2}$ and why?

71. ◆ If $a > b$, does it follow that $a^{1/n} > b^{1/n}$? Why or why not?

72. ◆ Under what condition(s) will $a^{-3/5}$ be negative?

Using a calculator, approximate each of the following to three decimal places.

73. $10^{4/5}$
74. $24^{1/4}$
75. $10^{3/2}$
76. $36^{5/8}$

Simplify.

77. $(x^{2/3})^{7/3}$
78. $a^{1/4} a^{3/2}$
79. $\dfrac{p^{4/5}}{p^{2/3}}$
80. $m^{-2/3} m^{3/4} m^{1/2}$

Graph.

81. $y = \sqrt[3]{x}$
82. $y = \sqrt[4]{x}$

83. Use a grapher to draw the graphs of $y_1 = x^{2/3}$, $y_2 = x^1$, $y_3 = x^{5/4}$, and $y_4 = x^{3/2}$. Use the window $[-1, 17, -1, 32]$ and the SIMULTANEOUS mode. Then determine which curve corresponds to each equation.

SUMMARY AND REVIEW 9

KEY TERMS

Square root, p. 426
Principal square root, p. 426
Radical sign, p. 426
Radical expression, p. 427
Radicand, p. 427
Irrational number, p. 427
Rationalize the denominator, p. 440
Like radicals, p. 443
Conjugates, p. 445
Radical equation, p. 447
Principle of squaring, p. 447
Pythagorean theorem, p. 454
Higher root, p. 461
Cube root, p. 461
Index, p. 461
nth root, p. 462
Rational exponent, p. 463

IMPORTANT PROPERTIES AND FORMULAS

For any real number A, $\sqrt{A^2} = |A|$.

For $A \geq 0$, $\sqrt{A^2} = A$.

The Product Rule: $\quad \sqrt{A}\sqrt{B} = \sqrt{A \cdot B}; \quad \sqrt[n]{AB} = \sqrt[n]{A}\sqrt[n]{B}$

The Quotient Rule: $\quad \dfrac{\sqrt{A}}{\sqrt{B}} = \sqrt{\dfrac{A}{B}}; \quad \sqrt[n]{\dfrac{A}{B}} = \dfrac{\sqrt[n]{A}}{\sqrt[n]{B}}$

Simplified Form of a Square Root

A radical expression for a square root is simplified when its radicand has no factor other than 1 that is a perfect square.

The principle of squaring: If $a = b$, then $a^2 = b^2$.

The Pythagorean theorem: $a^2 + b^2 = c^2$, where a and b are the lengths of the legs of a right triangle and c is the length of the hypotenuse.

The exponent $1/n$: $\quad a^{1/n} = \sqrt[n]{a}$

Rational exponents: $\quad a^{m/n} = (\sqrt[n]{a})^m = \sqrt[n]{a^m}; \quad a^{-m/n} = \dfrac{1}{a^{m/n}}$

REVIEW EXERCISES

Find the square roots.

1. 25
2. 196
3. 900
4. 225

Simplify.

5. $-\sqrt{144}$
6. $\sqrt{81}$
7. $\sqrt{49}$
8. $-\sqrt{169}$

Identify each radicand.

9. $5x\sqrt{2x^2y}$
10. $a\sqrt{\dfrac{a}{b}}$

Determine whether each square root is rational or irrational.

11. $-\sqrt{36}$
12. $-\sqrt{12}$
13. $\sqrt{18}$
14. $\sqrt{25}$

Use a calculator or Table 2 to approximate each square root. Round to three decimal places.

15. $\sqrt{3}$
16. $\sqrt{99}$

Simplify. Remember for Exercises 17–37 that we assume that all variables represent nonnegative numbers.

17. $\sqrt{p^2}$
18. $\sqrt{(7x)^2}$
19. $\sqrt{16m^2}$
20. $\sqrt{(ac)^2}$

Simplify by factoring.

21. $\sqrt{48}$
22. $\sqrt{98t^2}$
23. $\sqrt{32p}$
24. $\sqrt{x^6}$
25. $\sqrt{12a^{13}}$
26. $\sqrt{36m^{15}}$

468 CHAPTER 9 • RADICAL EXPRESSIONS AND EQUATIONS

Multiply and, if possible, simplify.

27. $\sqrt{3}\ \sqrt{11}$
28. $\sqrt{6}\ \sqrt{10}$
29. $\sqrt{5x}\ \sqrt{7t}$
30. $\sqrt{3a}\ \sqrt{8a}$
31. $\sqrt{5x}\ \sqrt{10xy^2}$
32. $\sqrt{20a^3b}\ \sqrt{5a^2b^2}$

Simplify.

33. $\dfrac{\sqrt{35}}{\sqrt{45}}$
34. $\dfrac{\sqrt{30y^9}}{\sqrt{54y}}$
35. $\sqrt{\dfrac{25}{64}}$
36. $\sqrt{\dfrac{20}{45}}$
37. $\sqrt{\dfrac{49}{t^2}}$
38. $10\sqrt{5} + 3\sqrt{5}$
39. $\sqrt{80} - \sqrt{45}$
40. $2\sqrt{x} - \sqrt{25x}$
41. $(2 + \sqrt{3})^2$
42. $(2 + \sqrt{3})(2 - \sqrt{3})$
43. $(1 + 2\sqrt{7})(3 - \sqrt{7})$

Rationalize the denominator.

44. $\sqrt{\dfrac{1}{2}}$
45. $\sqrt{\dfrac{5}{8}}$
46. $\sqrt{\dfrac{5}{y}}$
47. $\dfrac{2}{\sqrt{3}}$
48. $\dfrac{4}{2 + \sqrt{3}}$
49. $\dfrac{1 + \sqrt{5}}{2 - \sqrt{5}}$

Solve.

50. $\sqrt{x - 3} = 7$
51. $\sqrt{5x + 3} = \sqrt{2x - 1}$
52. $\sqrt{x + 5} = x - 1$
53. $1 + x = \sqrt{1 + 5x}$

54. The formula $r = 2\sqrt{5L}$ can be used to approximate the speed r, in miles per hour, of a car that has left a skid mark of length L, in feet. How far will a car skid at a speed of 90 mph?

In a right triangle, find the length of the side not given. Give an exact answer and, where appropriate, an approximation to three decimal places. Keep in mind that a and b are the lengths of the legs and c is the length of the hypotenuse.

55. $a = 15,\ c = 25$
56. $a = 1,\ b = \sqrt{2}$

57. One wire steadying a radio tower stretches from a point 100 ft high on the tower to a point on the ground 25 ft from the base of the tower. How long is the wire?

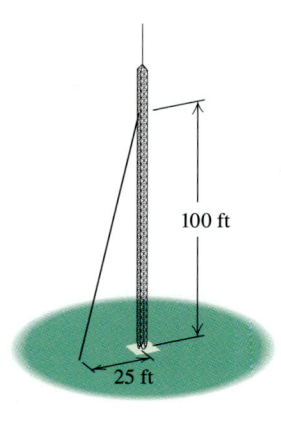

Simplify. If an expression does not represent a real number, state this.

58. $\sqrt[5]{32}$
59. $\sqrt[4]{-16}$
60. $\sqrt[3]{-27}$
61. $\sqrt[4]{32}$

Simplify.

62. $100^{1/2}$
63. $9^{-1/2}$
64. $16^{3/2}$
65. $81^{-3/4}$

SKILL MAINTENANCE

Solve.

66. $x + 2y = 7,$
    $x + y = 10$
67. $-2x > 17$

68. Find the slope and the y-intercept of the graph of $2x - 5y = 14$.

69. Multiply and, if possible, simplify:
$$\dfrac{x^2 + x}{x^2 - 4} \cdot \dfrac{x^2 + 4x + 4}{x^2}.$$

SYNTHESIS

70. ◆ Jesse automatically eliminates any possible solutions of a radical equation that are negative. What mistake is he making?

71. ◆ Why should you simplify each term in a radical expression before attempting to collect like radical terms?

72. Simplify: $\sqrt{\sqrt{\sqrt{256}}}$.

73. Solve: $\sqrt{x^2} = -10$.

74. Use square roots to factor $x^2 - 5$.

75. Solve $A = \sqrt{a^2 + b^2}$ for b.

CHAPTER TEST 9

1. Find the square roots of 81.

Simplify.
2. $\sqrt{64}$
3. $-\sqrt{25}$
4. Identify the radicand in $3\sqrt{x+4}$.

Determine whether each square root is rational or irrational.
5. $\sqrt{10}$
6. $\sqrt{16}$

Approximate using a calculator or Table 2. Round to three decimal places.
7. $\sqrt{87}$
8. $\sqrt{7}$

Simplify. Remember: Assume $a, y \geq 0$.
9. $\sqrt{a^2}$
10. $\sqrt{36y^2}$

Simplify by factoring.
11. $\sqrt{24}$
12. $\sqrt{27x^6}$
13. $\sqrt{4t^5}$

Perform the indicated operation and, if possible, simplify.
14. $\sqrt{5}\sqrt{6}$
15. $\sqrt{5}\sqrt{10}$
16. $\sqrt{7x}\sqrt{2y}$
17. $\sqrt{2t}\sqrt{8t}$
18. $\sqrt{3ab}\sqrt{6ab^3}$
19. $\dfrac{\sqrt{18}}{\sqrt{32}}$
20. $\dfrac{\sqrt{35x}}{\sqrt{80xy^2}}$
21. $\sqrt{\dfrac{27}{12}}$
22. $\sqrt{\dfrac{144}{a^2}}$
23. $3\sqrt{18} - 5\sqrt{18}$
24. $\sqrt{27} + 2\sqrt{12}$
25. $(4-\sqrt{5})^2$
26. $(4-\sqrt{5})(4+\sqrt{5})$

Rationalize each denominator.
27. $\sqrt{\dfrac{2}{5}}$
28. $\dfrac{2x}{\sqrt{y}}$
29. $\dfrac{10}{4-\sqrt{5}}$

30. The legs of a right triangle are 8 cm and 4 cm long. Find the length of the hypotenuse. Give an exact answer and an approximation to three decimal places.

Solve.
31. $\sqrt{3x} + 2 = 14$
32. $\sqrt{6x+13} = x+3$

33. Valerie calculates that she can see 247.49 km to the horizon from an airplane window. How high is the airplane? Use the formula $V = 3.5\sqrt{h}$, where h is the altitude, in meters, and V is the distance to the horizon, in kilometers.

Simplify. If an expression does not represent a real number, state this.
34. $\sqrt[4]{16}$
35. $-\sqrt[6]{1}$
36. $\sqrt[3]{-64}$
37. $\sqrt[4]{-81}$
38. $9^{1/2}$
39. $27^{-1/3}$
40. $100^{3/2}$
41. $16^{-5/4}$

SKILL MAINTENANCE

42. Simplify: $(-3)^3$.
43. Find the slope–intercept equation for a line with slope $-\frac{1}{2}$ and y-intercept $(0, -1)$.
44. Divide and, if possible, simplify:
$$\dfrac{a^2-1}{a} \div \dfrac{a-1}{a+1}.$$
45. Add and, if possible, simplify:
$$\dfrac{2x+1}{3x-5} + \dfrac{x-3}{5-3x}.$$

SYNTHESIS

46. Solve: $\sqrt{1-x} + 1 = \sqrt{6-x}$.
47. Simplify: $\sqrt{y^{16n}}$.

CHAPTER 10

Quadratic Equations

10.1 Solving Quadratic Equations: The Principle of Square Roots
10.2 Solving Quadratic Equations: Completing the Square
10.3 The Quadratic Formula and Applications
10.4 Complex Numbers as Solutions of Quadratic Equations
10.5 Graphs of Quadratic Equations
10.6 Functions
SUMMARY AND REVIEW
TEST

AN APPLICATION

Stuntman Dar Robinson once fell 700 ft from the top of the CN Tower in Toronto before opening a parachute. How long did the free-fall portion of his jump last?

THIS PROBLEM APPEARS AS EXERCISE 45 IN SECTION 10.3.

More information on stunts is available at
http://hepg.awl.com/be/elem_5

Before I leap from a moving train or fall from a building, I carefully plan every detail using math from algebra through trigonometry. Doing so can mean the difference between life and death.

LOREN JANES
Stunt Director
Canyon Country, California

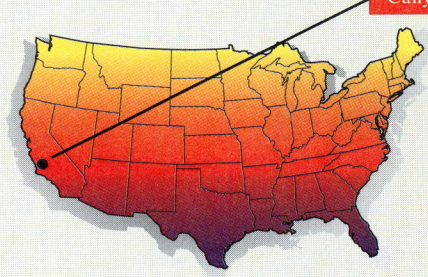

10.1 Solving Quadratic Equations: The Principle of Square Roots

> Quadratic equations first appeared in Section 5.6. At that time, we used the principle of zero products because all of the equations could be solved by factoring. In this chapter, we will learn methods for solving any quadratic equation. These methods are then used in applications and in graphing.

The Principle of Square Roots • Solving Quadratic Equations of the Type $(x + k)^2 = p$

The following are examples of quadratic equations:

$$x^2 - 7x + 9 = 0, \quad 5t^2 - 4t = 8, \quad 6y^2 = -9y, \quad m^2 = 49.$$

We saw in Chapter 5 that one way to solve an equation like $m^2 = 49$ is to subtract 49 on both sides, factor, and then use the principle of zero products:

$$m^2 - 49 = 0$$
$$(m + 7)(m - 7) = 0$$
$$m + 7 = 0 \quad \text{or} \quad m - 7 = 0$$
$$m = -7 \quad \text{or} \quad m = 7.$$

This approach relies on our ability to factor. By using the *principle of square roots,* we can develop a method for solving equations like $m^2 = 49$ that allows us to solve equations when factoring is impractical.

The Principle of Square Roots

Another way to solve $m^2 = 49$ relies on the definition of square root: If $c^2 = a$, then c is a square root of a. Thus if $m^2 = 49$, then m is a square root of 49, namely, -7 or 7. This approach was used to solve right triangles in Section 9.6, but there only positive square roots appeared, since length is never negative.

THE PRINCIPLE OF SQUARE ROOTS

For any nonnegative real number k, if $x^2 = k$, then $x = \sqrt{k}$ or $x = -\sqrt{k}$.

10.1 SOLVING QUADRATIC EQUATIONS: THE PRINCIPLE OF SQUARE ROOTS

EXAMPLE 1   Solve: $x^2 = 16$.

SOLUTION   We use the principle of square roots:

$$x^2 = 16$$
$$x = \sqrt{16} \quad \text{or} \quad x = -\sqrt{16} \quad \text{Using the principle of square roots}$$
$$x = 4 \quad \text{or} \quad x = -4. \quad \text{Simplifying}$$

We check mentally that $4^2 = 16$ and $(-4)^2 = 16$. The solutions are 4 and -4.

Unlike the principle of zero products, the principle of square roots can be used to solve quadratic equations that have irrational solutions.

EXAMPLE 2   Solve: **(a)** $x^2 = 17$; **(b)** $5x^2 = 15$; **(c)** $-3x^2 + 7 = 0$.

SOLUTION

a) $\qquad x^2 = 17$
$\qquad x = \sqrt{17} \quad \text{or} \quad x = -\sqrt{17} \qquad$ Using the principle of square roots

Check: For $\sqrt{17}$:

$$\begin{array}{c|c} x^2 = 17 \\ \hline (\sqrt{17})^2 \;?\; 17 \\ 17 \;|\; 17 \quad \text{TRUE} \end{array}$$

For $-\sqrt{17}$:

$$\begin{array}{c|c} x^2 = 17 \\ \hline (-\sqrt{17})^2 \;?\; 17 \\ 17 \;|\; 17 \quad \text{TRUE} \end{array}$$

The solutions are $\sqrt{17}$ and $-\sqrt{17}$.

b) $\qquad 5x^2 = 15$
$\qquad\quad x^2 = 3 \qquad$ Solving for x^2
$\qquad\quad x = \sqrt{3} \quad \text{or} \quad x = -\sqrt{3} \qquad$ Using the principle of square roots

We leave the check to the student. The solutions are $\sqrt{3}$ and $-\sqrt{3}$.

c) $\qquad -3x^2 + 7 = 0$
$\qquad\quad -3x^2 = -7 \qquad$ Subtracting 7
$\qquad\quad x^2 = \dfrac{-7}{-3} \qquad$ Dividing on both sides by -3
$\qquad\quad x^2 = \dfrac{7}{3}$
$\qquad\quad x = \sqrt{\dfrac{7}{3}} \quad \text{or} \quad x = -\sqrt{\dfrac{7}{3}} \qquad$ Using the principle of square roots

$\Big($If you wish to rationalize denominators, these answers can also be written $\dfrac{\sqrt{21}}{3}$ and $-\dfrac{\sqrt{21}}{3}.\Big)$

TECHNOLOGY CONNECTION

10.1

We can visualize Example 2(a) on a grapher by letting $y_1 = x^2$ and $y_2 = 17$ and finding the x-coordinate at each point of intersection. To do this, we use INTERSECT or TRACE and ZOOM. Below, we show the intersection when $x \approx -\sqrt{17}$.

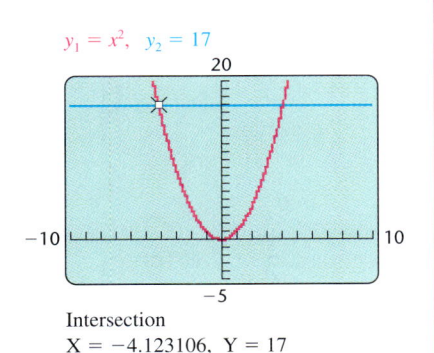

$y_1 = x^2, \; y_2 = 17$

Intersection
X = −4.123106, Y = 17

You can confirm that $(-4.123106)^2 \approx 17$. Thus the visualization serves as a check of the algebraic approach.

1. Use a grapher to visualize and check the solutions of Examples 2(b) and 2(c).

Check: For $\sqrt{\frac{7}{3}}$:

$$-3x^2 + 7 = 0$$
$$-3\left(\sqrt{\frac{7}{3}}\right)^2 + 7 \;?\; 0$$
$$-3 \cdot \frac{7}{3} + 7$$
$$-7 + 7$$
$$0 \;|\; 0 \quad \text{TRUE}$$

For $-\sqrt{\frac{7}{3}}$:

$$-3x^2 + 7 = 0$$
$$-3\left(-\sqrt{\frac{7}{3}}\right)^2 + 7 \;?\; 0$$
$$-3 \cdot \frac{7}{3} + 7$$
$$-7 + 7$$
$$0 \;|\; 0 \quad \text{TRUE}$$

The solutions of $-3x^2 + 7 = 0$ are $\sqrt{\frac{7}{3}}$ and $-\sqrt{\frac{7}{3}}$.

Solving Quadratic Equations of the Type $(x + k)^2 = p$

Equations like $(x - 5)^2 = 9$ or $(x + 2)^2 = 7$ are of the form $(x + k)^2 = p$. The principle of square roots can be used to solve such equations.

EXAMPLE 3

Solve: **(a)** $(x - 5)^2 = 9$; **(b)** $(x + 2)^2 = 7$.

SOLUTION

a) $(x - 5)^2 = 9$
$x - 5 = 3 \quad \text{or} \quad x - 5 = -3$ Using the principle of square roots; $\sqrt{9} = 3$ and $-\sqrt{9} = -3$
$x = 8 \quad \text{or} \quad x = 2$ Adding 5 on both sides

The solutions are 8 and 2. We leave the check to the student.

b) $(x + 2)^2 = 7$
$x + 2 = \sqrt{7} \quad \text{or} \quad x + 2 = -\sqrt{7}$ Using the principle of square roots
$x = -2 + \sqrt{7} \quad \text{or} \quad x = -2 - \sqrt{7}$ Adding -2 on both sides

The solutions are $-2 + \sqrt{7}$ and $-2 - \sqrt{7}$, or simply $-2 \pm \sqrt{7}$ (read "-2 plus or minus $\sqrt{7}$"). We leave the check to the student.

In Example 3, the left sides of the equations are squares of binomials. Sometimes factoring can be used to express an equation in that form.

EXAMPLE 4

Solve by factoring and using the principle of square roots.

a) $x^2 + 8x + 16 = 49$ **b)** $x^2 + 6x + 9 = 10$

SOLUTION

a) $x^2 + 8x + 16 = 49$ The left side is a perfect-square trinomial.
$(x + 4)^2 = 49$ Factoring
$x + 4 = 7 \quad \text{or} \quad x + 4 = -7$ Using the principle of square roots
$x = 3 \quad \text{or} \quad x = -11$

The solutions are 3 and -11.

b) $x^2 + 6x + 9 = 10$ The left side is a perfect-square trinomial.
$(x + 3)^2 = 10$ Factoring
$x + 3 = \sqrt{10}$ or $x + 3 = -\sqrt{10}$ Using the principle of square roots
$x = -3 + \sqrt{10}$ or $x = -3 - \sqrt{10}$

The solutions are $-3 + \sqrt{10}$ and $-3 - \sqrt{10}$, or simply $-3 \pm \sqrt{10}$.

EXERCISE SET 10.1

Solve. Use the principle of square roots.

1. $x^2 = 100$
2. $x^2 = 25$
3. $a^2 = 36$
4. $a^2 = 49$
5. $t^2 = 23$
6. $n^2 = 13$
7. $3x^2 = 30$
8. $5x^2 = 70$
9. $9t^2 = 72$
10. $7t^2 = 140$
11. $4 - 9x^2 = 0$
12. $25 - 4a^2 = 0$
13. $49y^2 - 5 = 15$
14. $4y^2 - 3 = 9$
15. $5x^2 - 120 = 0$
16. $25x^2 - 35 = 0$
17. $(x + 1)^2 = 25$
18. $(x - 2)^2 = 49$
19. $(x - 4)^2 = 81$
20. $(x + 3)^2 = 36$
21. $(m + 3)^2 = 6$
22. $(m - 4)^2 = 21$
23. $(a - 13)^2 = 64$
24. $(a + 13)^2 = 8$
25. $(x - 5)^2 = 14$
26. $(x - 7)^2 = 12$
27. $(t + 2)^2 = 25$
28. $(x + 9)^2 = 34$
29. $\left(y - \frac{3}{4}\right)^2 = \frac{17}{16}$
30. $\left(x + \frac{3}{2}\right)^2 = \frac{7}{2}$
31. $x^2 + 10x + 25 = 100$
32. $x^2 - 6x + 9 = 64$
33. $p^2 - 8p + 16 = 1$
34. $y^2 + 14y + 49 = 4$
35. $t^2 + 6t + 9 = 13$
36. $m^2 - 2m + 1 = 5$
37. $x^2 - 12x + 36 = 18$
38. $x^2 + 4x + 4 = 12$

SKILL MAINTENANCE

Determine whether each pair of equations represents parallel lines.

39. $y + 5 = 2x$,
    $y - 2x = 7$
40. $y - 4x = 6$,
    $x + 4y = 8$
41. $3x - 2y = 9$,
    $4y - 6x = 8$
42. $2x - 3y = 12$,
    $3y - 2x = 24$

Multiply.

43. $\left(x - \frac{3}{8}\right)^2$
44. $\left(t + \frac{5}{4}\right)^2$

SYNTHESIS

45. ◆ If a quadratic equation can be solved by factoring, what type of number(s) will be solutions? Why?
46. ◆ Explain why 9 is not *the* solution of $x^2 = 81$.
47. ◆ Write a quadratic equation that is most easily solved using the principle of square roots. Explain why the principle of zero products would not work as easily on the equation you wrote.
48. ◆ Tina finds the solution of a quadratic equation to be $3 \pm \sqrt{14}$ and states that there is only one solution. What mistake is she likely making?

Factor the left side of each equation. Then solve.

49. $x^2 + \frac{7}{3}x + \frac{49}{36} = \frac{7}{36}$
50. $x^2 - 5x + \frac{25}{4} = \frac{13}{4}$
51. $m^2 - \frac{3}{2}m + \frac{9}{16} = \frac{17}{16}$
52. $t^2 + 3t + \frac{9}{4} = \frac{49}{4}$
53. 📱 $x^2 + 2.5x + 1.5625 = 9.61$
54. 📱 $a^2 - 3.8a + 3.61 = 27.04$

Use the graph of $y = (x + 3)^2$ to solve each equation.

55. $(x + 3)^2 = 4$
56. $(x + 3)^2 = 9$

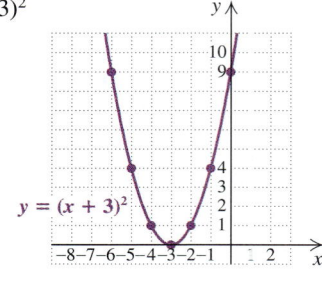

57. *Gravitational force.* Newton's law of gravitation states that the gravitational force f between objects of mass M and m, at a distance d from each other, is given by
$$f = \frac{kMm}{d^2},$$
where k is a constant. Solve for d.

COLLABORATIVE CORNER

Focus: Principle of square roots; formulas
Time: 15–25 minutes
Group size: 3
Materials: Calculators are optional.

Frankie, Johnnie & Luigi, Too! has the best pizza in the area, according to a poll of Stanford University students. Pizzas there have 12-in., 14-in., or 16-in. diameters.

ACTIVITY

1. Suppose that, as part of a promotion, Frankie, Johnnie & Luigi, Too! decides to offer a "mini" pie that has half the area of a 12-in. pie, a "personal" pie that has half the area of a 14-in. pie, and a "junior" pie that has half the area of a 16-in. pie. Each group member should calculate the radius, and then the diameter, of a different one of the "new" pies.
2. After checking each other's work, group members should develop a formula that could be used to determine the diameter of a circle that has half the area of a circle of known diameter. That is, if a circle has diameter d, what is the diameter of a circle with half the area? Try to follow the steps used in part (1) above.
3. If the diameter of one pizza is twice the diameter of another pizza, should the first pizza cost twice as much? Why or why not? (*Optional*: At your favorite pizzeria, determine which size pizza offers the best value.)

10.2 Solving Quadratic Equations: Completing the Square

Completing the Square • Solving by Completing the Square

In Section 10.1, we solved equations like $(x - 5)^2 = 7$ using the principle of square roots. Equations like $x^2 + 8x + 16 = 12$ were also solved using the principle of square roots because the expression on the left side is a perfect-square trinomial. We now learn to solve equations like $x^2 - 10x = 4$, in which the left side is not (yet) a perfect-square trinomial. The new procedure involves *completing the square* and enables us to solve any quadratic equation.

Completing the Square

Recall that

$$(x + 3)^2 = (x + 3)(x + 3)$$
$$= x^2 + 3x + 3x + 9$$
$$= x^2 + 6x + 9 \quad \text{This is a perfect-square trinomial.}$$

and, in general,

$$(x + a)^2 = x^2 + 2ax + a^2. \quad \text{This is also a perfect-square trinomial.}$$

In $x^2 + 6x + 9$, note that 9 is the square of half of the coefficient of x: $\frac{1}{2} \cdot 6 = 3$, and $3^2 = 9$. Similarly, in $x^2 + 2ax + a^2$, note that a^2 is also the square of half of the coefficient of x: $\frac{1}{2} \cdot 2a = a$.

Consider the following quadratic equation:

$$x^2 + 10x = 4.$$

We need to find a number that can be added to both sides and that will make the left side a perfect-square trinomial. Such a number is described in the discussion above as the square of half of the coefficient of x: $\frac{1}{2} \cdot 10 = 5$, and $5^2 = 25$. Thus we add 25 on both sides:

$$x^2 + 10x = 4 \quad \textit{Think}: \text{Half of 10 is 5; } 5^2 = 25.$$
$$x^2 + 10x + 25 = 4 + 25 \quad \text{Adding 25 on both sides}$$
$$(x + 5)^2 = 29. \quad \text{Factoring the perfect-square trinomial}$$

By adding 25 to $x^2 + 10x$, we have *completed the square.* The resulting equation contains the square of a binomial on one side. Solutions can then be found using the principle of square roots, as in Section 10.1:

$$(x + 5)^2 = 29$$
$$x + 5 = \sqrt{29} \quad \text{or} \quad x + 5 = -\sqrt{29} \quad \text{Using the principle of square roots}$$
$$x = -5 + \sqrt{29} \quad \text{or} \quad x = -5 - \sqrt{29}.$$

The solutions are $-5 \pm \sqrt{29}$.

COMPLETING THE SQUARE

To *complete the square* for an expression like $x^2 + bx$, add half of the coefficient of x, squared. That is, add $(b/2)^2$.

A visual interpretation of completing the square is sometimes helpful. Consider the figures shown on the following page.

478 CHAPTER 10 • QUADRATIC EQUATIONS

In all figures, the sum of the red and purple areas is $x^2 + 10x$. However, by splitting the purple area in half, we can "complete" a square by adding the blue area. The blue area is $5 \cdot 5$, or 25 square units.

EXAMPLE 1 Complete the square: **(a)** $x^2 - 12x$; **(b)** $x^2 + 5x$.

SOLUTION

a) To complete the square for $x^2 - 12x$, note that the coefficient of x is -12. Half of -12 is -6 and $(-6)^2$ is 36. Thus, $x^2 - 12x$ becomes a perfect-square trinomial when 36 is added:

$x^2 - 12x + 36$ is the square of $x - 6$.

That is,

$x^2 - 12x + 36 = (x - 6)^2$.

b) To complete the square for $x^2 + 5x$, we take half of the coefficient of x and square it:

$\left(\frac{5}{2}\right)^2 = \frac{25}{4}$.   Half of 5 is $\frac{5}{2}$; $\left(\frac{5}{2}\right)^2 = \frac{5}{2} \cdot \frac{5}{2} = \frac{25}{4}$.

The trinomial $x^2 + 5x + \frac{25}{4}$ is the square of $x + \frac{5}{2}$. That is,

$x^2 + 5x + \frac{25}{4} = \left(x + \frac{5}{2}\right)^2$.

Solving by Completing the Square

The concept of completing the square can now be used to solve equations like $x^2 + 10x = 4$ much as we did on page 477, prior to Example 1.

EXAMPLE 2 Solve by completing the square.

a) $x^2 + 6x = -8$

b) $x^2 - 10x + 14 = 0$

SOLUTION

a) To solve $x^2 + 6x = -8$, we take half of 6 and square it, to get 9. Then we add 9 on both sides of the equation. This makes the left side the square of

10.2 SOLVING QUADRATIC EQUATIONS: COMPLETING THE SQUARE

a binomial:

$$x^2 + 6x + 9 = -8 + 9 \quad \text{Adding 9 on both sides to complete the square}$$
$$(x + 3)^2 = 1 \quad \text{Factoring}$$
$$x + 3 = 1 \quad \text{or} \quad x + 3 = -1 \quad \text{Using the principle of square roots}$$
$$x = -2 \quad \text{or} \quad x = -4.$$

The solutions are -2 and -4.

b) We have

$$x^2 - 10x + 14 = 0$$
$$x^2 - 10x = -14 \quad \text{Subtracting 14 on both sides}$$
$$x^2 - 10x + 25 = -14 + 25 \quad \text{Adding 25 on both sides to complete the square: } (-10/2)^2 = 25$$
$$(x - 5)^2 = 11 \quad \text{Factoring}$$
$$x - 5 = \sqrt{11} \quad \text{or} \quad x - 5 = -\sqrt{11} \quad \text{Using the principle of square roots}$$
$$x = 5 + \sqrt{11} \quad \text{or} \quad x = 5 - \sqrt{11}.$$

The solutions are $5 + \sqrt{11}$ and $5 - \sqrt{11}$, or simply $5 \pm \sqrt{11}$.

To complete the square, the coefficient of x^2 must be 1. When the x^2-coefficient is not 1, we can multiply or divide on both sides to find an equivalent equation with an x^2-coefficient of 1.

EXAMPLE 3 Solve by completing the square.

a) $3x^2 + 24x = 3$ **b)** $2x^2 - 3x - 1 = 0$

SOLUTION

a)
$$3x^2 + 24x = 3$$
$$\left.\begin{array}{r}\tfrac{1}{3}(3x^2 + 24x) = \tfrac{1}{3} \cdot 3 \\ x^2 + 8x = 1\end{array}\right\} \quad \text{We multiply by } \tfrac{1}{3} \text{ (or divide by 3) to ensure an } x^2\text{-coefficient of 1.}$$
$$x^2 + 8x + 16 = 1 + 16 \quad \text{Adding 16 on both sides to complete the square: } \left(\tfrac{8}{2}\right)^2 = 16$$
$$(x + 4)^2 = 17 \quad \text{Factoring}$$
$$x + 4 = \sqrt{17} \quad \text{or} \quad x + 4 = -\sqrt{17}$$
$$x = -4 + \sqrt{17} \quad \text{or} \quad x = -4 - \sqrt{17}$$

The solutions are $-4 \pm \sqrt{17}$.

b)
$$2x^2 - 3x - 1 = 0$$
$$\tfrac{1}{2}(2x^2 - 3x - 1) = \tfrac{1}{2} \cdot 0 \quad \text{Multiplying by } \tfrac{1}{2} \text{ to make the } x^2\text{-coefficient 1}$$
$$x^2 - \tfrac{3}{2}x - \tfrac{1}{2} = 0$$
$$x^2 - \tfrac{3}{2}x = \tfrac{1}{2} \quad \text{Adding } \tfrac{1}{2} \text{ on both sides}$$

and

$$x^2 - \frac{3}{2}x + \frac{9}{16} = \frac{1}{2} + \frac{9}{16}$$ Adding $\frac{9}{16}$ on both sides: $\left[\frac{1}{2}\left(-\frac{3}{2}\right)\right]^2 = \left[-\frac{3}{4}\right]^2 = \frac{9}{16}$. This completes the square on the left side.

$$\left(x - \frac{3}{4}\right)^2 = \frac{8}{16} + \frac{9}{16}$$ Factoring and finding a common denominator

$$\left(x - \frac{3}{4}\right)^2 = \frac{17}{16}$$

$$x - \frac{3}{4} = \frac{\sqrt{17}}{4} \quad \text{or} \quad x - \frac{3}{4} = -\frac{\sqrt{17}}{4}$$ Using the principle of square roots

$$x = \frac{3}{4} + \frac{\sqrt{17}}{4} \quad \text{or} \quad x = \frac{3}{4} - \frac{\sqrt{17}}{4}$$

The solutions of $2x^2 - 3x - 1 = 0$ are $\dfrac{3 \pm \sqrt{17}}{4}$.

The steps in Example 3 can be used to solve any quadratic equation.

TO SOLVE $ax^2 + bx + c = 0$ BY COMPLETING THE SQUARE:

1. If $a \neq 1$, multiply by $1/a$ or divide by a on both sides so that the x^2-coefficient is 1.
2. When the x^2-coefficient is 1, rewrite the equation in the form
$$x^2 + bx = -c, \quad \text{or, if step (1) has been applied,}$$
$$x^2 + \frac{b}{a}x = -\frac{c}{a}.$$
3. Take half of the x-coefficient and square it. Add the result on both sides of the equation.
4. Express the left side as the square of a binomial. (Factor.)
5. Use the principle of square roots and complete the solution.

EXERCISE SET 10.2

Complete the square.

1. $x^2 + 8x$
2. $x^2 + 4x$
3. $x^2 - 14x$
4. $x^2 - 12x$
5. $x^2 - 3x$
6. $x^2 - x$
7. $t^2 + t$
8. $y^2 - 9y$
9. $x^2 + \frac{5}{4}x$
10. $x^2 + \frac{4}{3}x$
11. $m^2 - \frac{9}{2}m$
12. $r^2 - \frac{2}{5}r$

Solve by completing the square.

13. $x^2 - 8x + 15 = 0$
14. $x^2 - 6x - 7 = 0$
15. $x^2 + 22x + 21 = 0$
16. $x^2 + 14x + 40 = 0$
17. $3x^2 - 6x - 15 = 0$
18. $3x^2 - 12x - 33 = 0$
19. $x^2 - 22x + 102 = 0$
20. $x^2 - 18x + 74 = 0$
21. $x^2 + 10x - 4 = 0$
22. $x^2 - 7x - 2 = 0$

10.2 SOLVING QUADRATIC EQUATIONS: COMPLETING THE SQUARE

23. $x^2 + 5x - 2 = 0$
24. $2x^2 + 6x - 56 = 0$
25. $x^2 + \frac{3}{2}x - 2 = 0$
26. $x^2 - \frac{3}{2}x - 2 = 0$
27. $2x^2 + 3x - 16 = 0$
28. $2x^2 - 3x - 8 = 0$
29. $3x^2 + 4x - 1 = 0$
30. $3x^2 - 4x - 3 = 0$
31. $2x^2 = 9 + 5x$
32. $2x^2 = 5 + 9x$
33. $4x^2 + 12x = 7$
34. $6x^2 + 11x = 10$

SKILL MAINTENANCE

Solve each system using the substitution method.

35. $y - x = 5,$
    $y + 2x = 7$
36. $2x + 3y = 8,$
    $x = y - 6$

Graph.

37. $y = \frac{3}{5}x - 1$
38. $y - 4 = \frac{2}{3}(x + 1)$

Simplify.

39. $\sqrt{84}$
40. $\sqrt{90}$

SYNTHESIS

41. ◆ Is the addition principle used every time we complete the square? Why or why not?
42. ◆ Explain in your own words how completing the square enables us to solve equations we could not otherwise have solved.
43. ◆ Sal states that "since solving a quadratic equation by completing the square relies on the principle of square roots, the solutions are always opposites of each other." Is Sal correct? Why or why not?
44. ◆ When completing the square, what determines if the number being added is a whole number or a fraction?

Find b such that each trinomial is a square.

45. $x^2 + bx + 49$
46. $x^2 + bx + 36$
47. $x^2 + bx + 50$
48. $x^2 + bx + 45$
49. $x^2 - bx + 48$
50. $4x^2 + bx + 16$

📈 *Solve each of the following by letting y_1 represent the left side of each equation, letting y_2 represent the right side, and graphing y_1 and y_2 on the same set of axes.* INTERSECT *or* TRACE *and* ZOOM *can then be used to determine the x-coordinate at any point of intersection. Find solutions accurate to two decimal places.*

51. $(x + 4)^2 = 13$
52. $(x - 6)^2 = 2$
53. $x^2 + 5x - 2 = 0$ (Exercise 23)
54. $x^2 - 7x - 2 = 0$ (Exercise 22)
55. $2x^2 = 9 + 5x$ (Exercise 31)
56. $2x^2 = 5 + 9x$ (Exercise 32)

COLLABORATIVE C•O•R•N•E•R

Focus: Visualizing completion of the square

Time: 15–25 minutes

Group size: 2

Materials: Rulers and graph paper may be helpful.

It is not difficult to draw a visual representation of completing the square. To do so, we use areas and the fact that the area of any rectangle is given by multiplying the length and the width. For example, the following sequence of figures can be drawn to explain why 9 completes the square for $x^2 + 6x$. *(continued)*

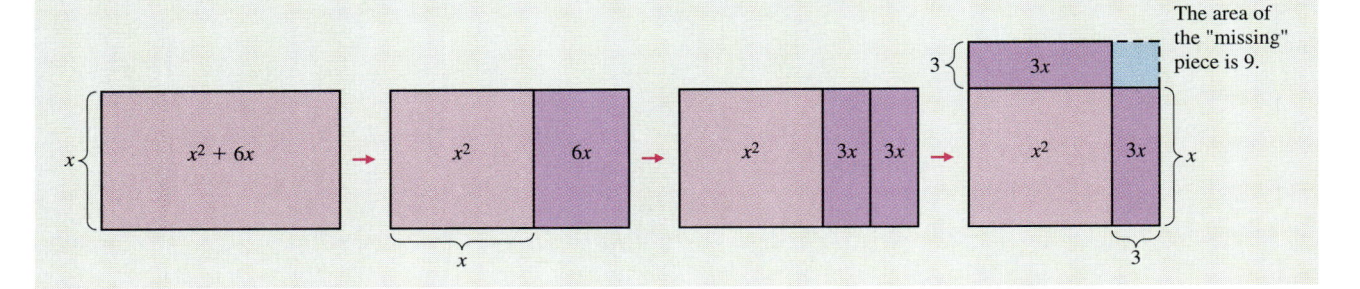

ACTIVITY

1. Draw a sequence of four figures, similar to those shown on the preceding page, to complete the square for $x^2 + 8x$. Group members should take turns, so that each person draws and labels two of the figures.
2. Repeat part (1) to complete the square for $x^2 + 14x$. The person who drew the first drawing in part (1) should take the second turn this time.
3. Keep in mind that when we complete the square, we are not forming an equivalent expression. For this reason, completing the square is generally performed by using the addition principle to form an equivalent *equation*. Use the work in parts (1) and (2) to solve the equations $x^2 + 8x = 9$ and $x^2 + 14x = 15$.
4. Each equation in part (3) has two solutions. Can both be represented geometrically? Why or why not?

10.3 The Quadratic Formula and Applications

The Quadratic Formula • Problem Solving

We now derive the *quadratic formula*. This formula enables us to solve quadratic equations more quickly than the method of completing the square.

The Quadratic Formula

When mathematicians use a procedure repeatedly, they often try to find a formula for the procedure. The quadratic formula condenses the many steps used to solve a quadratic equation by completing the square.

Consider a quadratic equation in *standard form*, $ax^2 + bx + c = 0$, with $a > 0$. Our plan is to solve this equation for x by completing the square. As the steps are performed, compare them with those in Example 3(b) on pages 479 and 480.

$$ax^2 + bx + c = 0$$

$$\frac{1}{a}(ax^2 + bx + c) = \frac{1}{a} \cdot 0 \qquad \text{Multiplying by } \frac{1}{a} \text{ to make the } x^2\text{-coefficient 1}$$

$$x^2 + \frac{b}{a}x + \frac{c}{a} = 0$$

$$x^2 + \frac{b}{a}x = -\frac{c}{a} \qquad \text{Adding } -\frac{c}{a} \text{ on both sides}$$

Half of b/a is $b/(2a)$, and the square of $b/(2a)$ is $b^2/(4a^2)$. We add $b^2/(4a^2)$ on both sides to complete the square:

$$x^2 + \frac{b}{a}x + \frac{b^2}{4a^2} = -\frac{c}{a} + \frac{b^2}{4a^2}. \qquad \text{Adding } \frac{b^2}{4a^2} \text{ on both sides}$$

Then

$$\left(x + \frac{b}{2a}\right)^2 = -\frac{4ac}{4a^2} + \frac{b^2}{4a^2} \quad \text{Factoring and finding a common denominator}$$

$$\left(x + \frac{b}{2a}\right)^2 = \frac{b^2 - 4ac}{4a^2}$$

$$x + \frac{b}{2a} = \sqrt{\frac{b^2 - 4ac}{4a^2}} \quad \text{or} \quad x + \frac{b}{2a} = -\sqrt{\frac{b^2 - 4ac}{4a^2}}. \quad \text{Using the principle of square roots}$$

Since we assumed $a > 0$, $\sqrt{4a^2} = 2a$. Thus we can simplify as follows:

$$x + \frac{b}{2a} = \frac{\sqrt{b^2 - 4ac}}{2a} \quad \text{or} \quad x + \frac{b}{2a} = -\frac{\sqrt{b^2 - 4ac}}{2a}.$$

Thus,

$$x = -\frac{b}{2a} + \frac{\sqrt{b^2 - 4ac}}{2a} \quad \text{or} \quad x = -\frac{b}{2a} - \frac{\sqrt{b^2 - 4ac}}{2a},$$

so $\quad x = -\dfrac{b}{2a} \pm \dfrac{\sqrt{b^2 - 4ac}}{2a},$

or $\quad x = \dfrac{-b \pm \sqrt{b^2 - 4ac}}{2a}.$ $\quad$ Unless $b^2 - 4ac$ is 0, this represents two solutions.

This last equation is the result we sought. It is so useful that it is worth memorizing.

THE QUADRATIC FORMULA

The solutions of $ax^2 + bx + c = 0$ are given by

$$x = \frac{-b \pm \sqrt{b^2 - 4ac}}{2a}.$$

The quadratic formula also holds when $a < 0$. A similar proof would show this, but we will not consider it here.

EXAMPLE 1   Solve using the quadratic formula.

a) $4x^2 + 5x - 6 = 0$
b) $x^2 = 4x + 7$
c) $x^2 + x = -1$

SOLUTION

a) We identify a, b, and c and substitute into the quadratic formula:

$4x^2 + 5x - 6 = 0.$
$\ \ \uparrow \quad\ \ \uparrow \quad\ \uparrow$
$\ \ a \quad\ \ \ b \quad\ \ c$

484 CHAPTER 10 • QUADRATIC EQUATIONS

$$x = \frac{-b \pm \sqrt{b^2 - 4ac}}{2a}$$

$$x = \frac{-5 \pm \sqrt{5^2 - 4 \cdot 4(-6)}}{2 \cdot 4} \qquad \text{Substituting for } a, b, \text{ and } c$$

Be sure to write the fraction bar all the way across.

$$x = \frac{-5 \pm \sqrt{25 - (-96)}}{8}$$

$$x = \frac{-5 \pm \sqrt{121}}{8}$$

$$x = \frac{-5 \pm 11}{8}$$

$$x = \frac{-5 + 11}{8} \quad \text{or} \quad x = \frac{-5 - 11}{8}$$

$$x = \frac{6}{8} \quad \text{or} \quad x = \frac{-16}{8}$$

$$x = \frac{3}{4} \quad \text{or} \quad x = -2.$$

The solutions are $\frac{3}{4}$ and -2.

b) We rewrite $x^2 = 4x + 7$ in standard form, identify a, b, and c, and solve using the quadratic formula:

$$1x^2 - 4x - 7 = 0 \qquad \text{Subtracting } 4x + 7 \text{ on both sides}$$
$\uparrow \uparrow \uparrow$
$a b c$

$$x = \frac{-(-4) \pm \sqrt{(-4)^2 - 4(1)(-7)}}{2 \cdot 1} \qquad \text{Substituting into the quadratic formula}$$

$$x = \frac{4 \pm \sqrt{16 + 28}}{2} = \frac{4 \pm \sqrt{44}}{2}.$$

Since $\sqrt{44}$ can be simplified, we have

$$x = \frac{4 \pm \sqrt{4}\sqrt{11}}{2} = \frac{4 \pm 2\sqrt{11}}{2}.$$

Finally, since 2 is a common factor of 4 and $2\sqrt{11}$, we can simplify the fraction by removing a factor equal to 1:

$$x = \frac{2(2 \pm \sqrt{11})}{2 \cdot 1} \qquad \text{Factoring}$$

$$x = \frac{2}{2} \cdot \frac{2 \pm \sqrt{11}}{1}. \qquad \text{Removing a factor equal to 1: } \frac{2}{2} = 1$$

The solutions are $2 + \sqrt{11}$ and $2 - \sqrt{11}$, or $2 \pm \sqrt{11}$.

10.3 THE QUADRATIC FORMULA AND APPLICATIONS   485

c) We rewrite $x^2 + x = -1$ in standard form and use the quadratic formula:

$1x^2 + 1x + 1 = 0$ Adding 1 on both sides
 ↑   ↑   ↑
 a b c

$$x = \frac{-1 \pm \sqrt{1^2 - 4 \cdot 1 \cdot 1}}{2 \cdot 1}$$ Substituting into the quadratic formula

$$x = \frac{-1 \pm \sqrt{1 - 4}}{2}$$

$$x = \frac{-1 \pm \sqrt{-3}}{2}.$$

Since the radicand, -3, is negative, there are no real-number solutions and we state this as the answer. In Section 10.4, we will study a number system in which solutions of this equation can be found.

The following summary compares the three main methods for solving quadratic equations.

Method	Advantages	Disadvantages
The quadratic formula	Can be used to solve *any* quadratic equation.	Can be slower than factoring or the principle of square roots.
The principle of square roots	Fastest way to solve equations of the form $ax^2 = p$, or $(x + k)^2 = p$. Can be used to solve *any* quadratic equation.	Can be slow when completing the square is required.
Factoring	Can be very fast.	Can be used only on certain equations. Many equations can be difficult or impossible to solve by factoring.

> **TECHNOLOGY CONNECTION**
> **10.3**
>
> To see that no real solutions exist for Example 1(c), we let $y_1 = x^2 + x$ and $y_2 = -1$.
>
> $y_1 = x^2 + x,\ y_2 = -1$
>
> The absence of any point of intersection supports the conclusion that no real-number solution exists.
>
> **1.** What happens when the INTERSECT feature is used with the graph above?
> **2.** How can the graph of $y = x^2 + x + 1$ be used to provide still another check of Example 1(c)?

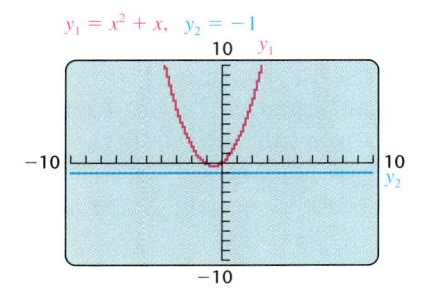

Problem Solving

EXAMPLE 2 *Diagonals in a Polygon.* The number of diagonals d of a polygon of n sides is given by the formula

$$d = \frac{n^2 - 3n}{2}.$$

If a polygon has 27 diagonals, how many sides does it have?

SOLUTION

1. **Familiarize.** A sketch can help us to become familiar with the problem. We draw a hexagon (6 sides) and count the diagonals. As the formula predicts, for $n = 6$, there are

$$\frac{6^2 - 3 \cdot 6}{2} = \frac{36 - 18}{2}$$

$$= \frac{18}{2} = 9 \text{ diagonals.}$$

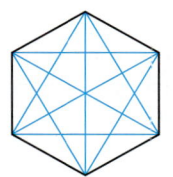

We might suspect that tripling the number of diagonals requires tripling the number of sides. Using the above formula, you can confirm that this is *not* the case. Rather than continue guessing, we proceed to a translation.

2. **Translate.** Since the number of diagonals is 27, we substitute 27 for d:

$$27 = \frac{n^2 - 3n}{2}.$$

This gives us a translation.

3. **Carry out.** We solve the equation for n, first reversing the equation for convenience:

$$\frac{n^2 - 3n}{2} = 27$$

$n^2 - 3n = 54$ Multiplying by 2 on both sides to clear fractions

$n^2 - 3n - 54 = 0$ Subtracting 54 on both sides

$(n - 9)(n + 6) = 0$ Factoring. There is no need for the quadratic formula here.

$n - 9 = 0$ or $n + 6 = 0$

$n = 9$ or $n = -6$.

4. **Check.** Since the number of sides cannot be negative, -6 cannot be a solution. We leave it to the student to show by substitution that 9 checks.

5. **State.** The polygon has 9 sides (it is a nonagon).

EXAMPLE 3

Free-falling Objects. The World Trade Center in New York City is 1368 ft tall. How many seconds will it take an object to fall from the top? Round to the nearest hundredth.

SOLUTION

1. **Familiarize.** If we did not know anything about this problem, we might consider looking up a formula in a mathematics or physics book. A formula that fits this situation is

$$s = 16t^2,$$

where s is the distance, in feet, traveled by a body falling freely from rest in t seconds. This formula is actually an approximation because it does not account for air resistance. In this problem, we know the distance s to be

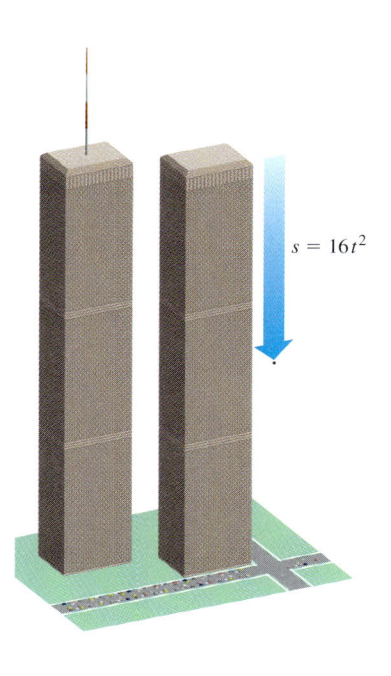

1368. We want to determine the time t for the object to reach the ground. If we check a couple of guesses, we can see that the time t must be between 5 and 10 sec.

2. **Translate.** The distance is 1368 ft and we need to solve for t. We substitute 1368 for s in the formula above to get the following translation:

$$1368 = 16t^2.$$

3. **Carry out.** Because there is no t-term, we can use the principle of square roots to solve:

$$1368 = 16t^2$$

$$\frac{1368}{16} = t^2 \qquad \text{Solving for } t^2$$

$$\sqrt{\frac{1368}{16}} = t \quad or \quad -\sqrt{\frac{1368}{16}} = t \qquad \text{Using the principle of square roots}$$

$$\frac{\sqrt{1368}}{4} = t \quad or \quad \frac{-\sqrt{1368}}{4} = t$$

$$9.25 \approx t \quad or \quad -9.25 \approx t. \qquad \text{Using a calculator or Table 2 and rounding to the nearest hundredth}$$

4. **Check.** The number -9.25 cannot be a solution because time cannot be negative in this situation. We substitute 9.25 in the original equation:

$$s = 16(9.25)^2 = 16(85.5625) = 1369.$$

This is close. Remember that we approximated a solution. As we expected in step (1), the solution is between 5 and 10 sec.

5. **State.** It takes about 9.25 sec for the object to fall to the ground from the top of the World Trade Center.

EXAMPLE 4 *Right Triangles.* The hypotenuse of a right triangle is 6 m long. One leg is 1 m longer than the other. Find the lengths of the legs. Round to the nearest hundredth.

SOLUTION

1. **Familiarize.** We first make a drawing and label it. We let s = the length of one leg. Then $s + 1$ = the length of the other leg.

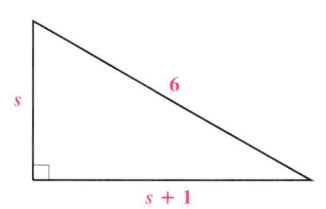

Note that if $s = 3$, then $s + 1 = 4$ and $3^2 + 4^2 = 25 \neq 6^2$. Thus, because of the Pythagorean theorem, we see that $s \neq 3$. Another guess, $s = 4$, is too big since $4^2 + (4 + 1)^2 = 41 \neq 6^2$. Although we have not guessed the solution, we expect s to be between 3 and 4.

2. **Translate.** To translate, we use the Pythagorean theorem:
$$s^2 + (s + 1)^2 = 6^2.$$

3. **Carry out.** We solve the equation:
$$s^2 + (s + 1)^2 = 6^2$$
$$s^2 + s^2 + 2s + 1 = 36$$
$$2s^2 + 2s - 35 = 0 \quad \text{This cannot be factored so we use the quadratic formula.}$$
$$\uparrow \quad \uparrow \quad \uparrow$$
$$a \quad b \quad c$$

$$s = \frac{-2 \pm \sqrt{2^2 - 4 \cdot 2(-35)}}{2 \cdot 2} \qquad \text{Remember: } s = \frac{-b \pm \sqrt{b^2 - 4ac}}{2a}.$$

$$s = \frac{-2 \pm \sqrt{4 + 280}}{4} = \frac{-2 \pm \sqrt{284}}{4}$$

$$s \approx 3.71 \quad \text{or} \quad s \approx -4.71. \qquad \text{Using a calculator or Table 2 and rounding to the nearest hundredth}$$

4. **Check.** Length cannot be negative, so -4.71 does not check. Note that if the smaller leg is 3.71, the other leg is 4.71. Then
$$(3.71)^2 + (4.71)^2 = 13.7641 + 22.1841 = 35.9482$$
and since $35.9482 \approx 6^2$, our approximation checks. Also, note that the value of s, 3.71, is between 3 and 4, as predicted in step (1).

5. **State.** One leg is about 3.71 m long; the other is about 4.71 m long.

EXERCISE SET 10.3

Solve. Try factoring first. If factoring is not possible or is difficult, use the quadratic formula. If no real-number solutions exist, state this.

1. $x^2 - 7x = 18$
2. $x^2 + 4x = 21$
3. $x^2 = 6x - 9$
4. $x^2 = 8x - 16$
5. $3y^2 + 7y + 4 = 0$
6. $3y^2 + 2y - 8 = 0$
7. $4x^2 - 12x = 7$
8. $4x^2 + 4x = 15$
9. $x^2 - 64 = 0$
10. $x^2 - 4 = 0$
11. $x^2 + 4x - 7 = 0$
12. $x^2 + 2x - 2 = 0$
13. $y^2 - 10y + 22 = 0$
14. $y^2 + 6y - 1 = 0$
15. $x^2 + 2x + 1 = 7$
16. $x^2 - 4x + 4 = 5$
17. $3x^2 + 4x - 2 = 0$
18. $3x^2 - 8x + 2 = 0$
19. $2x^2 - 5x = 1$
20. $2x^2 + 2x = 3$
21. $4y^2 + 4y - 1 = 0$
22. $4y^2 - 4y - 1 = 0$
23. $2t^2 - 3t + 2 = 0$
24. $4y^2 + 2y + 3 = 0$
25. $3x^2 = 5x + 4$
26. $2x^2 + 3x = 1$
27. $2y^2 - 6y = 10$
28. $5m^2 = 3 + 11m$

29. $10x^2 - 15x = 0$
30. $7x^2 + 2 = 6x$
31. $5t^2 - 7t = -4$
32. $15t^2 + 10t = 0$
33. $9y^2 = 162$
34. $5t^2 = 100$

Solve using the quadratic formula. Use a calculator or Table 2 to approximate the solutions to the nearest thousandth.

35. $x^2 - 4x - 7 = 0$
36. $x^2 + 2x - 2 = 0$
37. $y^2 - 6y - 1 = 0$
38. $y^2 + 10y + 22 = 0$
39. $4x^2 + 4x = 1$
40. $4x^2 = 4x + 1$

Solve. If an irrational answer occurs, round to the nearest hundredth.

41. A polygon has 35 diagonals. How many sides does it have? See Example 2.

42. A polygon has 20 diagonals. How many sides does it have? See Example 2.

43. *Free-fall time.* Library Square Tower, in Los Angeles, is 1012 ft tall. How long would it take an object to fall from the top?

44. *Free-fall time.* The height of the Amoco building in Chicago is 1136 ft. How long would it take an object to fall from the top?

45. *Free-fall record.* Stuntman Dar Robinson once fell 700 ft from the top of the CN Tower in Toronto before opening a parachute (*Source*: *The Guinness Book of Records* 1996). How long did the free fall portion of his jump last?

46. *Free-fall record.* The world record for free-fall to the ground without a parachute by a woman is 175 ft and is held by Kitty O'Neill. Approximately how long did the fall take?

47. *Right triangles.* The hypotenuse of a right triangle is 25 ft long. One leg is 17 ft longer than the other. Find the lengths of the legs.

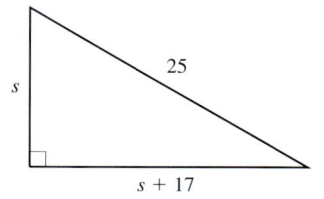

48. *Right triangles.* The hypotenuse of a right triangle is 26 yd long. One leg is 14 yd longer than the other. Find the lengths of the legs.

49. *Area of a rectangle.* The length of a rectangle is 4 cm greater than the width. The area is 60 cm². Find the length and the width.

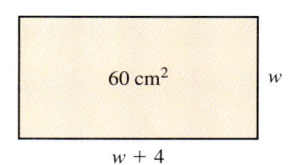

50. *Area of a rectangle.* The length of a rectangle is 3 m greater than the width. The area is 70 m². Find the length and the width.

51. *Plumbing.* A water pipe runs diagonally under a rectangular yard that is 5 m longer than it is wide. If the pipe is 25 m long, determine the dimensions of the yard.

52. *Guy wires.* A 26-ft long guy wire is anchored 10 ft from the base of a telephone pole. How far up the pole does the wire reach?

53. *Right triangles.* The area of a right triangle is 13 m². One leg is 2.5 m longer than the other. Find the lengths of the legs.

54. *Right triangles.* The area of a right triangle is 15.5 cm². One leg is 1.2 cm longer than the other. Find the lengths of the legs.

55. *Area of a rectangle.* The length of a rectangle is 2 in. greater than the width. The area is 20 in². Find the length and the width.

56. *Area of a rectangle.* The length of a rectangle is 3 ft greater than the width. The area is 15 ft². Find the length and the width.

490 CHAPTER 10 • QUADRATIC EQUATIONS

57. *Area of a rectangle.* The length of a rectangle is twice the width. The area is 16 m². Find the length and the width.

58. *Area of a rectangle.* The length of a rectangle is twice the width. The area is 20 cm². Find the length and the width.

Investments. *The formula $A = P(1 + r)^t$ is used to find the value A to which P dollars grows when invested for t years at an annual interest rate r. In Exercises 59–62, find the interest rate for the given information.*

59. $1000 grows to $1440 in 2 years

60. $2560 grows to $3610 in 2 years

61. $6250 grows to $6760 in 2 years

62. $6250 grows to $7290 in 2 years

63. *Gardening.* Laura has enough mulch to cover 250 ft² of garden space. How wide is the largest circular flower garden that Laura can cover with mulch?

64. *Environmental science.* A circular oil slick is 20,000 m² in area. How wide is the oil slick?

SKILL MAINTENANCE

Solve.

65. $5(2x - 3) + 4x = 9 - 6x$

66. $\dfrac{3}{x - 4} = \dfrac{2}{x + 4}$

Simplify.

67. $\sqrt{40} - 2\sqrt{10} + \sqrt{90}$ **68.** $\sqrt{9000x^{10}}$

69. $(-1)^7$ **70.** $(-1)^{64}$

SYNTHESIS

71. Under what condition(s) is the quadratic formula *not* the easiest way to solve a quadratic equation?

72. A student claims to be able to solve any quadratic equation by completing the square. The same student claims to be incapable of understanding why the quadratic formula works. Does this strike you as odd? Why or why not?

73. Write a problem for a classmate to solve. Devise the problem so that the quadratic formula is used to solve it and the solution is an irrational number.

74. Is it possible for a quadratic equation to have one solution that is rational and one that is irrational? Why or why not?

Solve.

75. $x(3x + 7) - 3x = 0$ **76.** $5x + x(x - 7) = 0$

77. $x(5x - 7) = 1$ **78.** $3 - x(x - 3) = 4$

79. $x^2 + (x + 2)^2 = 7$ **80.** $(y + 4)(y + 3) = 15$

81. $\dfrac{x^2}{x + 5} - \dfrac{7}{x + 5} = 0$ **82.** $\dfrac{x^2}{x + 3} - \dfrac{5}{x + 3} = 0$

83. $\dfrac{1}{x} + \dfrac{1}{x + 6} = \dfrac{1}{5}$ **84.** $\dfrac{1}{x} + \dfrac{1}{x + 1} = \dfrac{1}{3}$

85. Find *r* in this figure. Round to the nearest hundredth.

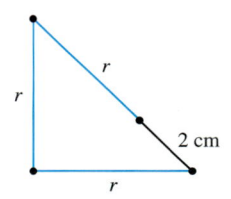

86. *Area of a square.* Find the area of a square for which the diagonal is one unit longer than the length of the sides.

87. *Flagpoles.* A 20-ft flagpole is struck by lightning and, while not completely broken, falls over and touches the ground 10 ft from the bottom of the pole. How high up did the pole break?

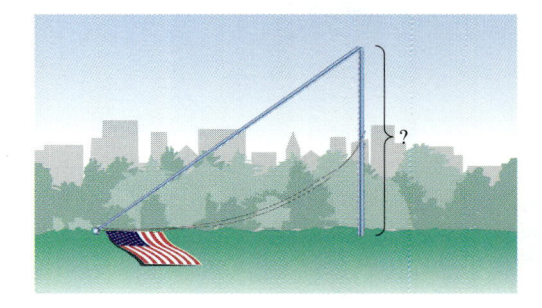

88. *Dimensions of a square.* How long is the side of a square whose diagonal is 3 cm longer than a side?

89. 🖩 *Investments.* $8000 is invested at interest rate r. In 2 yr, it grows to $8904.20. What is the interest rate?

90. 🖩 *Investments.* In 2 yr, you want to have $3000. How much do you need to invest now if you can get an interest rate of 5.75% compounded annually?

91. *Enlarged strike zone.* In baseball, a batter's strike zone is a rectangular area about 15 in. wide and 40 in. high. Many batters subconsciously enlarge this area by 40% when fearful that if they don't swing, the umpire will call the pitch a strike. Assuming that the strike zone is enlarged by an invisible band of uniform width around the actual zone, find the dimensions of the enlarged strike zone.

92. 📈 Use a grapher to approximate to the nearest thousandth the solutions of Exercises 35–40. Compare your answers with those found using a calculator or Table 2.

10.4 Complex Numbers as Solutions of Quadratic Equations

The Complex-Number System • Solutions of Equations

The Complex-Number System

Because negative numbers do not have square roots that are real numbers, mathematicians have devised a larger set of numbers known as *complex numbers*. In the complex-number system, the number i is used to represent the square root of -1.

THE NUMBER i

We define the number $i = \sqrt{-1}$. That is, $i = \sqrt{-1}$ and $i^2 = -1$.

EXAMPLE 1 Express in terms of i.

a) $\sqrt{-3}$ b) $\sqrt{-25}$
c) $-\sqrt{-10}$ d) $\sqrt{-24}$

SOLUTION

a) $\sqrt{-3} = \sqrt{-1 \cdot 3} = \sqrt{-1} \cdot \sqrt{3} = i\sqrt{3}$, or $\sqrt{3}i$ ← *i is not under the radical.*

b) $\sqrt{-25} = \sqrt{-1 \cdot 25} = \sqrt{-1} \cdot \sqrt{25} = i \cdot 5 = 5i$

c) $-\sqrt{-10} = -\sqrt{-1 \cdot 10} = -\sqrt{-1} \cdot \sqrt{10} = -i\sqrt{10}$, or $-\sqrt{10}i$

d) $\sqrt{-24} = \sqrt{4(-1)6} = \sqrt{4}\sqrt{-1}\sqrt{6} = 2i\sqrt{6}$, or $2\sqrt{6}i$

IMAGINARY NUMBERS

An *imaginary* number is a number that can be written in the form $a + bi$, where a and b are real numbers and $b \neq 0$.

The following are examples of imaginary numbers:

$$3 + 8i, \quad \sqrt{7} - 2i, \quad 4 + \sqrt{6}i, \quad \text{and} \quad 4i \text{ (here } a = 0).^{\dagger}$$

The imaginary numbers together with the real numbers form the set of **complex numbers.**

COMPLEX NUMBERS

A *complex number* is any number that can be written as $a + bi$, where a and b are real numbers. (Note that a and b both can be 0.)

It may help to remember that every real number is a complex number ($a + bi$ with $b = 0$), but not every complex number is real ($a + bi$ with $b \neq 0$). For example, numbers like $2 + 3i$ and $-7i$ are complex but not real.

Solutions of Equations

As we saw in Example 1(c) of Section 10.3, not all quadratic equations have real-number solutions. All quadratic equations *do* have complex-number solutions. These solutions are usually written in the form $a + bi$ unless a or b is zero.

EXAMPLE 2 Solve: **(a)** $x^2 + 3x + 4 = 0$; **(b)** $x^2 + 2 = 2x$.

SOLUTION

a) We use the quadratic formula:

$$1x^2 + 3x + 4 = 0$$

$$x = \frac{-3 \pm \sqrt{3^2 - 4 \cdot 1 \cdot 4}}{2 \cdot 1}.$$

Remember: $x = \dfrac{-b \pm \sqrt{b^2 - 4ac}}{2a}$.

---

*The name "imaginary" should not lead you to believe that these numbers are not useful. Imaginary numbers have important applications in engineering and the physical sciences.

†Numbers like $4i$ are often called *pure imaginary* since they are of the form $0 + bi$.

Then

$$x = \frac{-3 \pm \sqrt{-7}}{2} \quad \text{Simplifying}$$

$$x = \frac{-3 \pm \sqrt{-1}\sqrt{7}}{2}$$

$$x = \frac{-3 \pm i\sqrt{7}}{2}$$

$$x = -\frac{3}{2} \pm \frac{\sqrt{7}}{2}i. \quad \text{Writing in the form } a + bi$$

The solutions are $-\frac{3}{2} + \frac{\sqrt{7}}{2}i$ and $-\frac{3}{2} - \frac{\sqrt{7}}{2}i$.

b) We have

$$x^2 + 2 = 2x$$
$$1x^2 - 2x + 2 = 0 \quad \text{Rewriting in standard form}$$

$$x = \frac{-(-2) \pm \sqrt{(-2)^2 - 4 \cdot 1 \cdot 2}}{2 \cdot 1} \quad \text{Remember: } x = \frac{-b \pm \sqrt{b^2 - 4ac}}{2a}.$$

$$x = \frac{2 \pm \sqrt{-4}}{2} \quad \text{Simplifying}$$

$$x = \frac{2 \pm \sqrt{-1}\sqrt{4}}{2}$$

$$x = \frac{2 \pm i2}{2} \quad \sqrt{-1} = i \text{ and } \sqrt{4} = 2$$

$$x = \frac{2}{2} \pm \frac{2i}{2} \quad \text{Rewriting in the form } a + bi$$

$$x = 1 \pm i. \quad \text{Simplifying}$$

The solutions are $1 + i$ and $1 - i$.

EXERCISE SET 10.4

Express in terms of i.
1. $\sqrt{-1}$
2. $\sqrt{-36}$
3. $\sqrt{-9}$
4. $\sqrt{-81}$
5. $\sqrt{-50}$
6. $\sqrt{-44}$
7. $-\sqrt{-20}$
8. $-\sqrt{-45}$
9. $-\sqrt{-18}$
10. $-\sqrt{-28}$
11. $4 + \sqrt{-49}$
12. $7 + \sqrt{-4}$
13. $7 + \sqrt{-16}$
14. $-8 - \sqrt{-36}$
15. $3 - \sqrt{-98}$
16. $-2 + \sqrt{-125}$

Solve.
17. $x^2 + 9 = 0$
18. $x^2 + 4 = 0$
19. $x^2 = -28$
20. $x^2 = -48$

494 CHAPTER 10 • QUADRATIC EQUATIONS

21. $x^2 - 4x + 6 = 0$
22. $x^2 + 4x + 5 = 0$
23. $(x - 3)^2 = -25$
24. $(x + 1)^2 = -16$
25. $x^2 + 2x + 2 = 0$
26. $x^2 + 5 = 2x$
27. $x^2 + 7 = 4x$
28. $x^2 + 7 + 4x = 0$
29. $2t^2 + 6t + 5 = 0$
30. $4y^2 + 3y + 2 = 0$
31. $1 + 2m + 3m^2 = 0$
32. $4p^2 + 3 = 6p$

SKILL MAINTENANCE

Graph.

33. $y = \frac{3}{5}x - 2$
34. $2x - 3y = 10$
35. $y = -4$
36. $x = 2$

Simplify.

37. $-1 - 4^2$
38. $-1(-4)^2$

SYNTHESIS

39. ◆ Under what condition(s) will an equation of the form $x^2 = c$ have imaginary-number solutions?

40. ◆ When using the quadratic formula, why is it only necessary to examine $b^2 - 4ac$ to determine if the solutions are imaginary?

41. ◆ Is it possible for a quadratic equation to have one imaginary-number solution and one real-number solution? Why or why not?

42. ◆ Can imaginary-number solutions of a quadratic equation be found using the method of completing the square? Why or why not?

Solve.

43. $(x + 1)^2 + (x + 3)^2 = 0$
44. $(p + 5)^2 + (p + 1)^2 = 0$
45. $\dfrac{2x - 1}{5} - \dfrac{2}{x} = \dfrac{x}{2}$
46. $\dfrac{1}{a - 1} - \dfrac{2}{a - 1} = 3a$

47. Use a grapher to confirm that there are no real-number solutions of Examples 2(a) and 2(b).

10.5 Graphs of Quadratic Equations

Graphing Equations of the Form $y = ax^2$ •
Graphing Equations of the Form $y = ax^2 + bx + c$

In this section, we will graph equations of the form

$$y = ax^2 + bx + c, \quad a \neq 0.$$

The polynomial on the right side of the equation is of second degree, or **quadratic** in x. Examples of the types of equations we will graph are

$$y = x^2, \quad y = x^2 + 2x - 3, \quad \text{and} \quad y = -5x^2 + 4.$$

Graphs of quadratic equations of the type $y = ax^2 + bx + c$ (where $a \neq 0$) are always cupped upward or downward. These graphs are symmetric with respect to an **axis of symmetry,** as shown in the figures at the top of the following page. If the graph of a quadratic equation is folded along its axis of symmetry, the two halves match exactly.

The point at which the graph of a quadratic equation crosses its axis of symmetry is called the **vertex** (plural, vertices). The y-coordinate of the vertex is the graph's largest value of y (if the curve opens downward) or smallest value of y (if the curve opens upward). Graphs of quadratic equations are called **parabolas.**

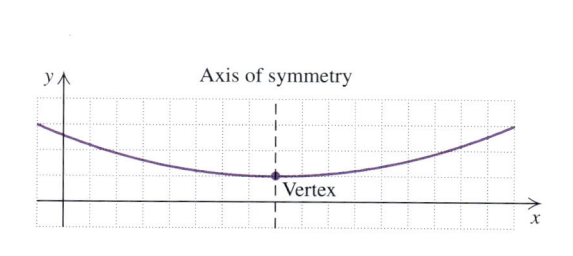

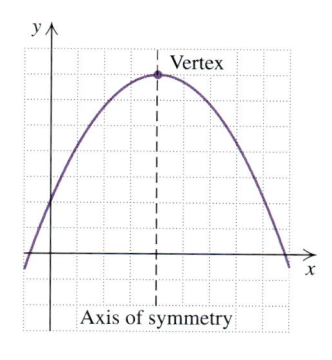

Graphing Equations of the Form $y = ax^2$

The simplest parabolas to sketch are given by equations of the form $y = ax^2$.

EXAMPLE 1   Graph: $y = x^2$.

SOLUTION   We choose numbers for x and find the corresponding values for y:

If $x = -2$, then $y = (-2)^2 = 4$.   We get the pair $(-2, 4)$.
If $x = -1$, then $y = (-1)^2 = 1$.   We get the pair $(-1, 1)$.
If $x = 0$, then $y = 0^2 = 0$.   We get the pair $(0, 0)$.
If $x = 1$, then $y = 1^2 = 1$.   We get the pair $(1, 1)$.
If $x = 2$, then $y = 2^2 = 4$.   We get the pair $(2, 4)$.

The following table lists these solutions of the equation $y = x^2$. After several ordered pairs are found, we plot them and connect them with a smooth curve.

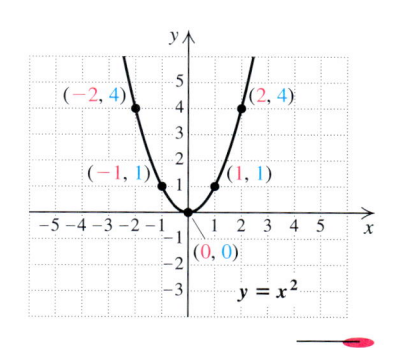

x	$y = x^2$	(x, y)
-2	4	$(-2, 4)$
-1	1	$(-1, 1)$
0	0	$(0, 0)$
1	1	$(1, 1)$
2	4	$(2, 4)$

In Example 1, the vertex is $(0, 0)$ and the axis of symmetry is the y-axis. This will be the case for any parabola having an equation of the form $y = ax^2$.

EXAMPLE 2   Graph: $y = -\frac{1}{2}x^2$.

SOLUTION   We select numbers for x, find the corresponding y-values, plot the resulting ordered pairs, and connect them with a smooth curve.

496 CHAPTER 10 • QUADRATIC EQUATIONS

x	$y = -\frac{1}{2}x^2$	(x, y)
-4	-8	$(-4, -8)$
-2	-2	$(-2, -2)$
0	0	$(0, 0)$
2	-2	$(2, -2)$
4	-8	$(4, -8)$

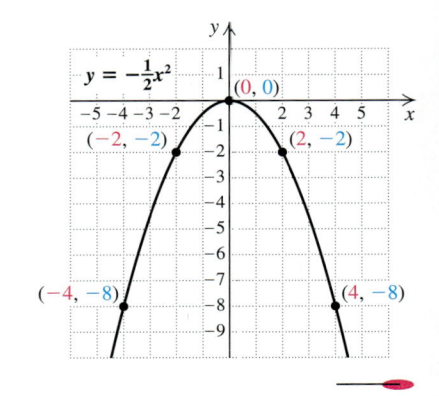

Graphing Equations of the Form $y = ax^2 + bx + c$

Recall from our earlier work with lines that it is often very useful to know a line's x- and y-intercepts. Intercepts are also useful when graphing parabolas. To find the intercepts of a parabola, we use the same approach we used with lines.

THE INTERCEPTS OF A PARABOLA

To find the y-intercept of the graph of $y = ax^2 + bx + c$, replace x with 0 and solve for y.

To find any x-intercept(s) of the graph of $y = ax^2 + bx + c$, replace y with 0 and solve for x.

EXAMPLE 3 Find all y- and x-intercepts of the parabola given by $y = 2x^2 - x - 28$.

SOLUTION To find the y-intercept, we replace x with 0 and solve for y:

$$y = 2 \cdot 0^2 - 0 - 28$$
$$y = 0 - 0 - 28$$
$$y = -28.$$

When x is 0, we have $y = -28$. Thus the y-intercept is $(0, -28)$.

To find the x-intercept(s), we replace y with 0 and solve for x:

$$0 = 2x^2 - x - 28.$$

The quadratic formula could be used, but factoring is faster:

$$0 = (2x + 7)(x - 4) \quad \text{Factoring}$$

$2x + 7 = 0 \quad \text{or} \quad x - 4 = 0$
$\quad 2x = -7 \quad \text{or} \quad \quad x = 4$
$\quad \quad x = -\frac{7}{2} \quad \text{or} \quad \quad x = 4.$

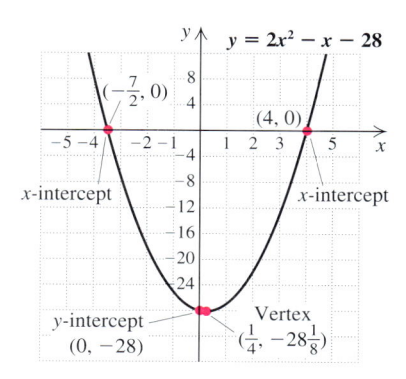

A visualization of Example 3

The x-intercepts are $(4, 0)$ and $\left(-\frac{7}{2}, 0\right)$. The y-intercept is $(0, -28)$.

Although we were not asked to graph the equation in Example 3, we did so to show that the x-coordinate of the vertex, $\frac{1}{4}$, is exactly midway between the x-intercepts. The quadratic formula, $x = \dfrac{-b \pm \sqrt{b^2 - 4ac}}{2a}$, can be used to locate x-intercepts when an equation of the form $y = ax^2 + bx + c$ is graphed. If one x-intercept is determined by $\dfrac{-b - \sqrt{b^2 - 4ac}}{2a}$ and the other by $\dfrac{-b + \sqrt{b^2 - 4ac}}{2a}$, then the average of these two values can be used to find the x-coordinate of the vertex. In Exercise 41, you are asked to show that this x-value is $\dfrac{-b}{2a}$. A more complicated approach can be used to show that the x-value of the vertex is $\dfrac{-b}{2a}$ even when no x-intercepts exist.

THE VERTEX OF A PARABOLA

For a parabola given by the quadratic equation $y = ax^2 + bx + c$:

1. The x-coordinate of the vertex is $-\dfrac{b}{2a}$.

2. The y-coordinate of the vertex is found by substituting $-\dfrac{b}{2a}$ for x and solving for y.

EXAMPLE 4   Graph: $y = x^2 + 2x - 3$.

SOLUTION   Our plan is to plot the vertex and some points on both sides of the vertex, and then draw a parabola passing through these points.

498 CHAPTER 10 • QUADRATIC EQUATIONS

To locate the vertex, we use $-b/(2a)$ to find its x-coordinate:

$$x\text{-coordinate of the vertex} = -\frac{b}{2a} = -\frac{2}{2 \cdot (1)}$$
$$= -1.$$

We substitute -1 for x to find the y-coordinate of the vertex:

$$y\text{-coordinate of the vertex} = (-1)^2 + 2(-1) - 3 = 1 - 2 - 3$$
$$= -4.$$

The vertex is $(-1, -4)$. The axis of symmetry is $x = -1$.

We choose some x-values on both sides of the vertex and graph the parabola.

x	y $y = x^2 + 2x - 3$	(x, y)	
-4	5	$(-4, 5)$	
-3	0	$(-3, 0)$	← x-intercept
-2	-3	$(-2, -3)$	
-1	-4	$(-1, -4)$	← Vertex
0	-3	$(0, -3)$	← y-intercept
1	0	$(1, 0)$	← x-intercept
2	5	$(2, 5)$	

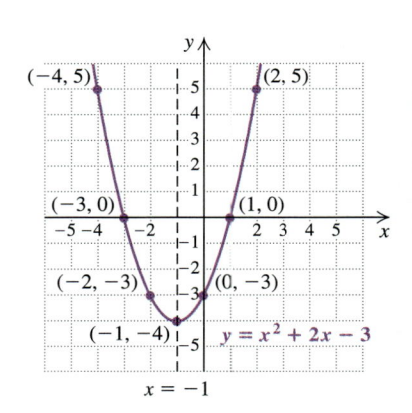

One tip for graphing quadratic equations involves the coefficient of x^2. Note that a in $y = ax^2 + bx + c$ tells us whether the graph opens upward or downward. When a is positive, the graph opens upward, as in Examples 1, 3, and 4; when a is negative, the graph opens downward, as in Example 2.

EXAMPLE 5   Graph: $y = -2x^2 + 4x + 1$.

SOLUTION   Since the coefficient of x^2 is negative, we know that the graph opens downward. To locate the vertex, we first find its x-coordinate:

$$x\text{-coordinate of the vertex} = -\frac{b}{2a} = -\frac{4}{2 \cdot (-2)}$$
$$= 1.$$

We substitute 1 for x to find the y-coordinate of the vertex:

$$y\text{-coordinate of the vertex} = -2 \cdot 1^2 + 4 \cdot 1 + 1 = -2 + 4 + 1$$
$$= 3.$$

The vertex is $(1, 3)$. The axis of symmetry is $x = 1$.

We choose some x-values on both sides of the vertex, calculate their corresponding y-values, and graph the parabola.

x	y $y = -2x^2 + 4x + 1$	(x, y)
3	−5	(3, −5)
2	1	(2, 1)
1	3	(1, 3) ← This is the vertex.
0	1	(0, 1)
−1	−5	(−1, −5)

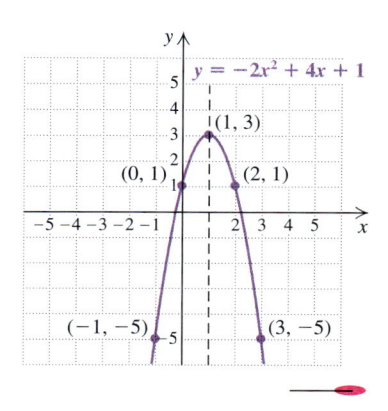

A second tip for graphing quadratic equations can cut our calculation time in half. In Examples 1–5, you may have noted that any x-value to the left of the vertex is paired with the same y-value as an x-value the same distance to the right of the vertex. Thus, since the vertex for Example 5 is (1, 3), we know that the x-values 2 and 0 will both be paired with the same y-value. Being aware of such symmetries, you can plot *two* points after calculating just *one* and also check your work as you proceed.

GUIDELINES FOR GRAPHING QUADRATIC EQUATIONS

1. Graphs of quadratic equations, $y = ax^2 + bx + c$, are parabolas. They are cupped upward if $a > 0$ and downward if $a < 0$.
2. Use the formula $x = -b/(2a)$ to find the x-coordinate of the vertex. After calculating the y-coordinate, plot the vertex and some points on either side of it.
3. After a point has been graphed, a second point with the same y-coordinate can be found on the opposite side of the axis of symmetry.
4. Graph the y-intercept and, if requested, any x-intercepts.

EXERCISE SET 10.5

Graph each quadratic equation, labeling the vertex and the y-intercept.

1. $y = x^2 - 2$
2. $y = 2x^2$
3. $y = -1 \cdot x^2$
4. $y = x^2 + 1$
5. $y = -x^2 + 2x$
6. $y = x^2 + x - 6$
7. $y = 3x^2 - 12x + 11$
8. $y = x^2 + 2x + 1$
9. $y = x^2 - 2x - 3$
10. $y = 2x^2 - 12x + 13$
11. $y = -2x^2 - 4x + 1$
12. $y = -3x^2 - 2x + 8$
13. $y = \frac{1}{4}x^2$
14. $y = -\frac{1}{3}x^2$
15. $y = -\frac{1}{2}x^2 + 5$
16. $y = \frac{1}{2}x^2 - 7$
17. $y = x^2 - 3x$
18. $y = x^2 + 4x$

Graph each equation, labeling the vertex, the y-intercept, and any x-intercepts. If an x-intercept is irrational, round to three decimal places.

19. $y = x^2 - x - 6$
20. $y = x^2 - 2x - 8$
21. $y = 2x^2 - 5x$
22. $y = 2x^2 + 7x$
23. $y = -x^2 - x + 12$
24. $y = -x^2 + 3x + 10$
25. $y = -3x^2 + 6x - 1$
26. $y = -3x^2 - 12x - 11$
27. $y = x^2 - 2x + 3$
28. $y = -x^2 + 2x - 3$
29. $y = 1 - 4x - 2x^2$
30. $y = 3 - 4x - 2x^2$

SKILL MAINTENANCE

31. *Construction.* A 24-ft pipe is leaning against a building. The bottom of the pipe is 12 ft from the building. How high is the top of the pipe?
32. *Court dimensions.* The maximum length of a regulation basketball court is 94 ft and the maximum width is 50 ft. Find the length of a diagonal for such a court.
33. Write a slope–intercept equation for the line containing the points $(-2, 7)$ and $(4, -3)$.
34. Find the slope of the line given by the equation $8 - 6x = 2y$.

Evaluate.

35. $5a^3 - 2a$, for $a = -1$
36. $3t^4 + 5t^2$, for $t = -2$

SYNTHESIS

37. ◆ What effect does $|a|$ have on the graph of $y = ax^2 + bx + c$?
38. ◆ Suppose that both x-intercepts of a parabola are known. What is the easiest way to find the coordinates of the vertex?
39. ◆ Does the graph of every equation of the type $y = ax^2 + bx + c$ have a y-intercept? Why or why not? What if $a = 0$?
40. ◆ Describe a method that could be used to find an equation for a parabola that has x-intercepts at r_1 and r_2.
41. Show that the average of
$$\frac{-b - \sqrt{b^2 - 4ac}}{2a} \quad \text{and} \quad \frac{-b + \sqrt{b^2 - 4ac}}{2a}$$
is $-\dfrac{b}{2a}$.

42. *Height of a projectile.* The height H, in feet, of a projectile with an initial velocity of 96 ft/sec is given by the equation
$$H = -16t^2 + 96t,$$
where t is the number of seconds from launch. Use the graph of this equation, shown below, or any equation-solving technique to answer the following.

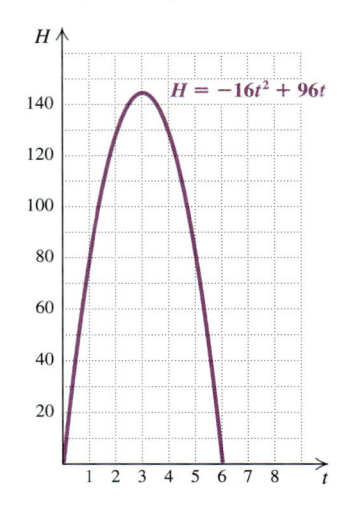

a) How many seconds after launch is the projectile 128 ft above ground?
b) When does the projectile reach its maximum height?
c) How many seconds after launch does the projectile return to the ground?

43. *Stopping distance.* In how many feet can a car stop if it is traveling at a speed of r mph? One estimate, developed in Britain, is as follows. The distance d, in feet, is given by

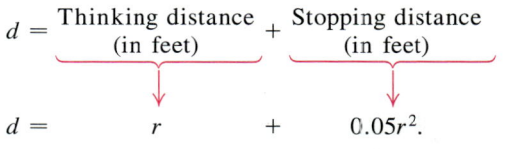

$$d = \underbrace{\text{Thinking distance}}_{\text{(in feet)}} + \underbrace{\text{Stopping distance}}_{\text{(in feet)}}$$

$$d = \quad r \quad + \quad 0.05r^2.$$

a) How many feet would it take to stop a car traveling 25 mph? 40 mph? 55 mph? 65 mph? 75 mph? 100 mph?
b) Graph the equation, assuming $r \geq 0$.

44. On one set of axes, graph $y = x^2$, $y = (x - 3)^2$, and $y = (x + 1)^2$. Compare the three graphs. How can the graph of $y = (x - h)^2$ be obtained from the graph of $y = x^2$?

45. *Seller's supply.* As the price of a product increases, the seller is willing to sell, or *supply*, more of the

product. Suppose that the supply for a certain product is given by
$$S = p^2 + p + 10,$$
where p is the price in dollars and S is the number sold, in thousands, at that price. Graph the equation for values of p such that $0 \leq p \leq 6$.

46. *Consumer's demand.* As the price of a product increases, consumers purchase, or *demand*, less of the product. Suppose that the demand for a certain product is given by
$$D = (p - 6)^2,$$
where p is the price in dollars and D is the number sold, in thousands, at that price. Graph the equation for values of p such that $0 \leq p \leq 6$.

47. *Equilibrium point.* The price p at which the consumer and the seller agree determines the *equilibrium point.* Find p such that
$$D = S$$
for the demand and supply curves in Exercises 45 and 46. How many units of the product will be sold at that price?

48. On one set of axes, graph $y = x^2$, $y = x^2 - 5$, and $y = x^2 + 2$. Compare the graphs. How can the graph of $y = x^2 + k$ be obtained from the graph of $y = x^2$?

49. Use a grapher to draw the graph of $y = x^2 - 5$ and then, using the graph, estimate $\sqrt{5}$ to four decimal places.

COLLABORATIVE C•O•R•N•E•R

Focus: Graphs of quadratic equations

Time: 30 minutes

Group size: 2–4

Materials: Graph paper, calculators

The population of downtown Indianapolis started to rise in the 1990s after having been on the decline for two decades. This rise was due at least in part to renewed development in the downtown area. The following chart lists the population for several years.

1970	1980	1990	1996
22,834	13,070	12,179	13,000

ACTIVITY

1. Each group member should graph the four data points. Then, using the graph, each should estimate the population in 2000 as well as the year in which the population would again exceed 22,000.

2. What would be the shape of a curve drawn through the points? Could such a curve be described by a linear equation?

3. Using a process called *regression*, a computer or graphing calculator will give a quadratic equation that closely fits the given data points. If we use $x = 0$ for 1970, that equation is

$$y = 34.045x^2 - 1242.727x + 22,659.584.$$

   a) Use this equation to estimate the population in 2000. Compare this value with your estimates from part (1) above.

   b) Use the quadratic formula and this equation to predict the year in which the population will next reach 22,000. Compare with the previous estimates.

4. Graph the regression equation on the graph from part (1) above. Does the curve actually pass through any of the points? How well does it follow the pattern established by the points?

5. Discuss how such equations and estimates could be used by city planners and developers in decision-making.

10.6 Functions

Identifying Functions • Functions Written as Formulas • Function Notation • Graphs of Functions • Recognizing Graphs of Functions

Functions are enormously important in modern mathematics. The more mathematics you study, the more you will use functions.

Identifying Functions

Functions appear regularly in magazines and newspapers although they are usually not referred to as such. Consider the following table.

Year	Global Average Temperature (in degrees Celsius)
1986	12.17
1987	12.33
1988	12.35
1989	12.25
1990	12.47
1991	12.41
1992	12.25
1993	12.29
1994	12.38

Note that to each year there corresponds *exactly one* temperature. A correspondence of this sort is called a **function**.

FUNCTION

A *function* is a correspondence (or rule) that assigns to each member of some set (called the *domain*) exactly one member of a set (called the *range*).

Sometimes the members of the domain are called **inputs**, and the members of the range **outputs**.

Domain (Set of Inputs)	Range (Set of Outputs)
1986	12.17
1987	12.33
1988	12.35
1989	12.25
1990	12.47
1991	12.41
1992	
1993	12.29
1994	12.38

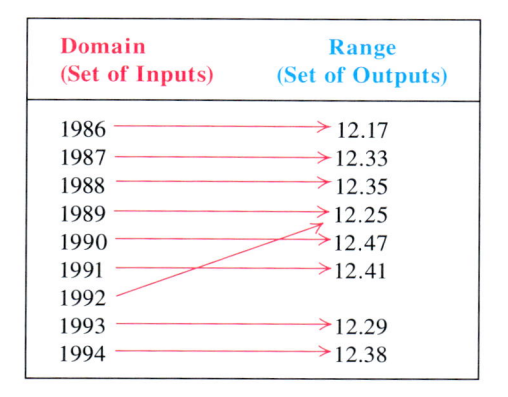

Note that each input has exactly one output, even though one of the outputs, 12.25, is used twice.

EXAMPLE 1 Determine whether or not each of the following correspondences is a function.

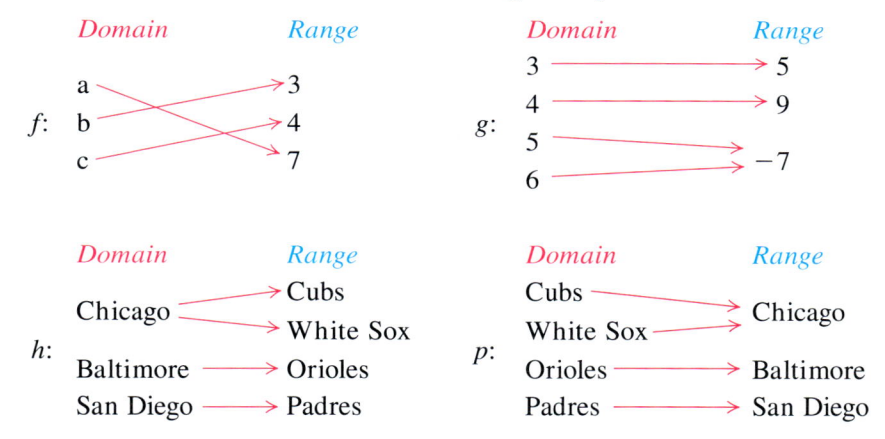

f:
Domain — Range
a, b, c → 3, 4, 7

g:
Domain — Range
3, 4, 5, 6 → 5, 9, −7

h:
Domain — Range
Chicago → Cubs
Chicago → White Sox
Baltimore → Orioles
San Diego → Padres

p:
Domain — Range
Cubs → Chicago
White Sox → Chicago
Orioles → Baltimore
Padres → San Diego

SOLUTION Correspondence f is a function because each member of the domain is matched to only one member of the range.

Correspondence g is also a function because each member of the domain is matched to only one member of the range.

Correspondence h is *not* a function because one member of the domain, Chicago, is matched to more than one member of the range.

Correspondence p is a function because each member of the domain is paired with only one member of the range.

Functions Written as Formulas

Many functions are described by formulas. Equations like $y = x + 3$ and $y = 4x^2$ are examples of such formulas. Outputs are found by substituting members of the domain for x.

EXAMPLE 2

Thunderstorm Distance. During a thunderstorm, it is possible to calculate how far away lightning is by using the formula

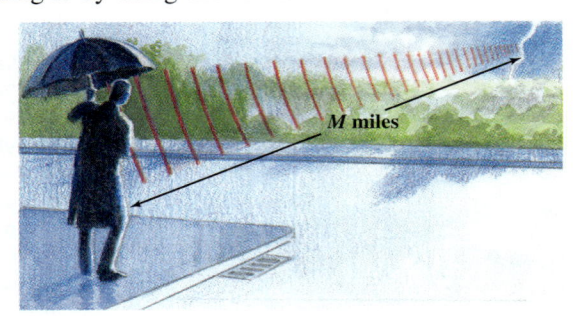

$$M = \tfrac{1}{5}t.$$

Here M is the distance, in miles, that a storm is from an observer when the sound of thunder arrives t seconds after the lightning has been sighted.

Complete the following table for this function.

t (in seconds)	0	1	2	3	4	5	6	10
M (in miles)	0	$\tfrac{1}{5}$						

SOLUTION To complete the table, we substitute values of t and compute M.

For $t = 2$, $M = \tfrac{1}{5} \cdot 2 = \tfrac{2}{5}$; For $t = 3$, $M = \tfrac{1}{5} \cdot 3 = \tfrac{3}{5}$;

For $t = 4$, $M = \tfrac{1}{5} \cdot 4 = \tfrac{4}{5}$; For $t = 5$, $M = \tfrac{1}{5} \cdot 5 = 1$;

For $t = 6$, $M = \tfrac{1}{5} \cdot 6 = \tfrac{6}{5}$, or $1\tfrac{1}{5}$; For $t = 10$, $M = \tfrac{1}{5} \cdot 10 = 2$

Function Notation

In Example 2, it was somewhat time-consuming to repeatedly write "For $t = \underline{}$, $M = \tfrac{1}{5} \cdot \underline{}$." Function notation clearly and concisely presents inputs and outputs together. The notation $M(t)$, read "M of t," denotes the output that is paired with the input t by the function M. Thus, for Example 2,

$$M(2) = \tfrac{1}{5} \cdot 2 = \tfrac{2}{5}, \quad M(3) = \tfrac{1}{5} \cdot 3 = \tfrac{3}{5}, \quad \text{and, in general,} \quad M(t) = \tfrac{1}{5} \cdot t.$$

The notation $M(4) = \tfrac{4}{5}$ makes clear that when 4 is the input, $\tfrac{4}{5}$ is the output.

> **CAUTION!** $M(4)$ *does not* mean M times 4 and should not be read that way.

Equations for nonvertical lines can be written in function notation. For example, $f(x) = x + 2$, read "f of x equals x plus 2," can be used instead of $y = x + 2$ when we are discussing functions, although both equations describe the same correspondence.

EXAMPLE 3 For the function given by $f(x) = x + 2$, find each of the following.

a) $f(8)$ **b)** $f(-3)$ **c)** $f(0)$

SOLUTION

a) $f(8) = 8 + 2$, or 10 *$f(8)$ is read "f of 8"; $f(8)$ does not mean "f times 8"!*
b) $f(-3) = -3 + 2$, or -1 *$f(-3)$ is the output corresponding to the input -3.*
c) $f(0) = 0 + 2$, or 2

It is sometimes helpful to think of a function as a machine that gives an output for each input that enters the machine. The following diagram is one way in which the function given by $g(t) = 2t^2 + t$ can be illustrated.

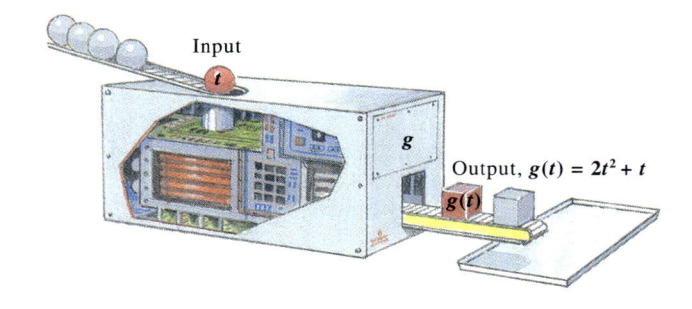

EXAMPLE 4 For the function $g(t) = 2t^2 + t$, find each of the following.

a) $g(3)$ **b)** $g(0)$ **c)** $g(-2)$

SOLUTION

a) $g(3) = 2 \cdot 3^2 + 3$ *Using 3 for each occurrence of t*
$= 2 \cdot 9 + 3$
$= 21$

b) $g(0) = 2 \cdot 0^2 + 0$ *Using 0 for each occurrence of t*
$= 0$

c) $g(-2) = 2(-2)^2 + (-2)$ *Using -2 for each occurrence of t*
$= 2 \cdot 4 - 2$
$= 6$

Outputs are also called **function values.** In Example 4, $g(-2) = 6$. We can say that the "function value is 6 at -2," or "when x is -2, the value of the function is 6." Most often we say "g of -2 is 6."

TECHNOLOGY CONNECTION

10.6

There are several ways in which to check Example 4 with a grapher. Once $y_1 = 2x^2 + x$ has been entered, we can use TRACE, CALC and VALUE, or the TABLE feature to find the function value for any selected input. Another way to check is to use the Y-VARS key to access the function y_1. Parentheses can then be used to complete the writing of $y_1(3)$. When ENTER is pressed, we see the output 21. Use this approach to find $y_1(0)$ and $y_1(-2)$.

Graphs of Functions

To graph a function, we find ordered pairs (x, y) or $(x, f(x))$, plot them, and connect the points. Note that y and $f(x)$ are often used interchangeably when working with functions and their graphs.

506 CHAPTER 10 • QUADRATIC EQUATIONS

EXAMPLE 5 Graph: $f(x) = x + 2$.

SOLUTION A list of some function values is shown in this table. We plot the points and connect them. The graph is a straight line.

x	$f(x)$
-4	-2
-3	-1
-2	0
-1	1
0	2
1	3
2	4
3	5
4	6

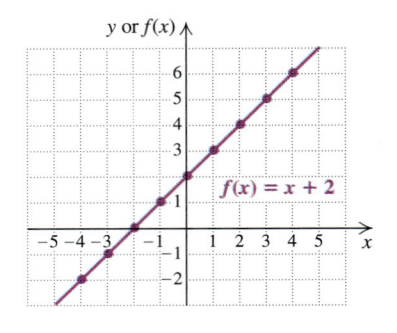

EXAMPLE 6 Graph: $g(x) = 4 - x^2$.

SOLUTION Recall from Section 10.5 that the graph is a parabola. We calculate some function values and draw the curve.

$$g(0) = 4 - 0^2 = 4 - 0 = 4,$$
$$g(-1) = 4 - (-1)^2 = 4 - 1 = 3,$$
$$g(2) = 4 - (2)^2 = 4 - 4 = 0,$$
$$g(-3) = 4 - (-3)^2 = 4 - 9 = -5$$

x	$g(x)$
-3	-5
-2	0
-1	3
0	4
1	3
2	0
3	-5

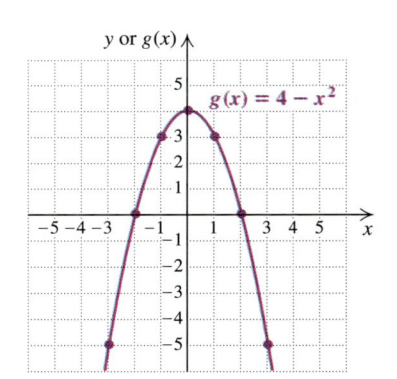

EXAMPLE 7 Graph: $h(x) = |x|$.

SOLUTION A list of some function values is shown in the following table. We plot the points and connect them. The graph is V-shaped, rising on either side of the vertical axis.

x	$h(x)$
-3	3
-2	2
-1	1
0	0
1	1
2	2
3	3

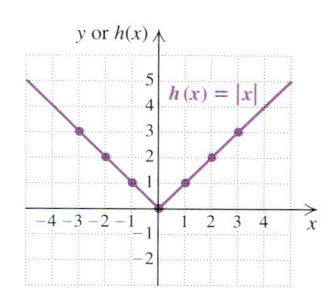

Recognizing Graphs of Functions

Consider the function f described by $f(x) = x^2 - 5$. Its graph is shown at left. It is also the graph of the equation $y = x^2 - 5$.

To find a function value, like $f(3)$, from a graph, we locate the input on the horizontal axis, move vertically to the graph of the function, and then horizontally to find the output on the vertical axis, where members of the range can be found.

Recall that when one member of the domain is paired with two or more different members of the range, the correspondence is not a function. Thus, when a graph contains two or more different points with the same first coordinate, the graph cannot represent a function. Points sharing a common first coordinate are vertically above and below each other, as shown in the following figure.

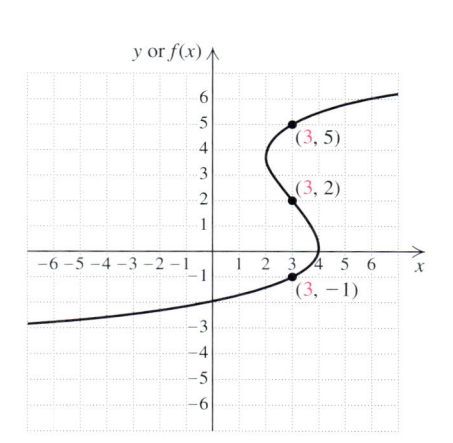

Since 3 is paired with more than one member of the range, the graph does not represent a function.

This observation leads to the *vertical-line test*.

THE VERTICAL-LINE TEST

A graph represents a function if it is impossible to draw a vertical line that intersects the graph more than once.

CHAPTER 10 • QUADRATIC EQUATIONS

EXAMPLE 8 Determine whether each of the following is the graph of a function.

a)

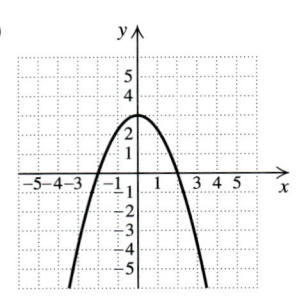

b)

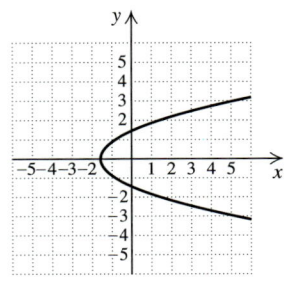

c)

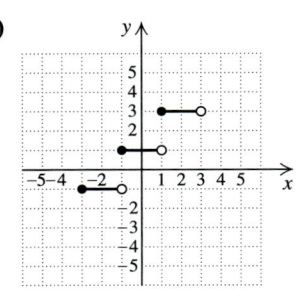

d)

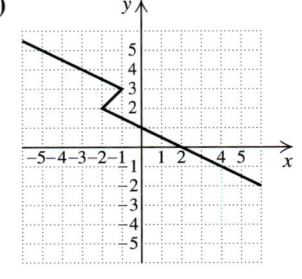

SOLUTION

a) The graph is that of a function because no vertical line can cross the graph at more than one point. This can be confirmed with a ruler or straightedge.

b) The graph is *not* that of a function because a vertical line—say, $x = 1$—crosses the graph at more than one point.

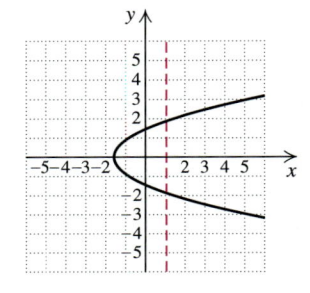

c) The graph is that of a function because no vertical line can cross the graph at more than one point. Note that the open dots indicate the absence of a point.

d) The graph is *not* that of a function because a vertical line—say, $x = -2$—crosses the graph at more than one point.

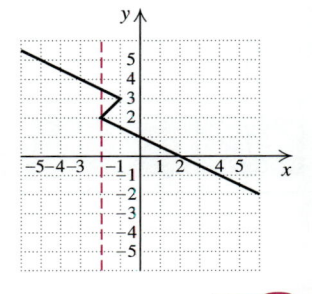

EXERCISE SET 10.6

Determine whether each correspondence is a function.

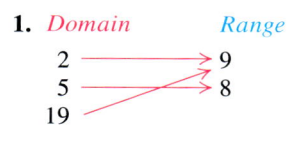

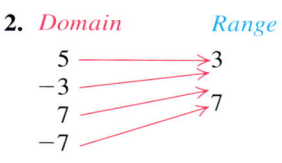

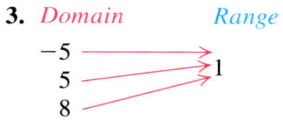

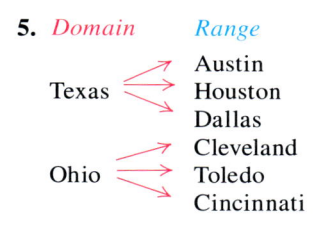

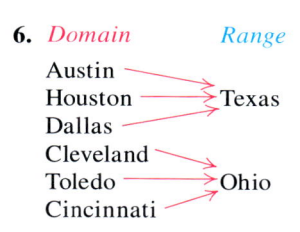

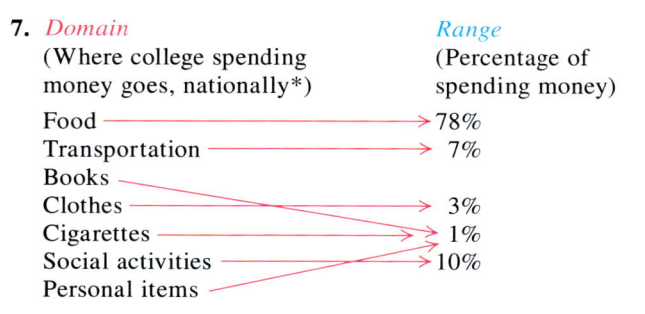

1. Domain → Range
2 → 9
5 → 8
19

2. Domain → Range
5 → 3
−3
7 → 7
−7

3. Domain → Range
−5
5 → 1
8

4. Domain → Range
6 → −6
7 → −7
3 → −3

5. Domain → Range
Texas → Austin, Houston, Dallas
Ohio → Cleveland, Toledo, Cincinnati

6. Domain → Range
Austin, Houston, Dallas → Texas
Cleveland, Toledo, Cincinnati → Ohio

7. Domain (Where college spending money goes, nationally*) → Range (Percentage of spending money)
Food → 78%
Transportation → 7%
Books
Clothes → 3%
Cigarettes → 1%
Social activities → 10%
Personal items

8. Domain (Brand of single-serving pizza) → Range (Number of calories)
Old Chicago Pizza-lite → 324
Weight Watchers Cheese
Banquet Zap Cheese → 310
Lean Cuisine Cheese
Pizza Hut Supreme Personal Pan → 647
Celeste Suprema Pizza-For-One → 678

---

*Due to rounding, total exceeds 100%. Source: USA Today, September 23, 1996.

Find the indicated outputs.

9. $f(4), f(7),$ and $f(-3)$

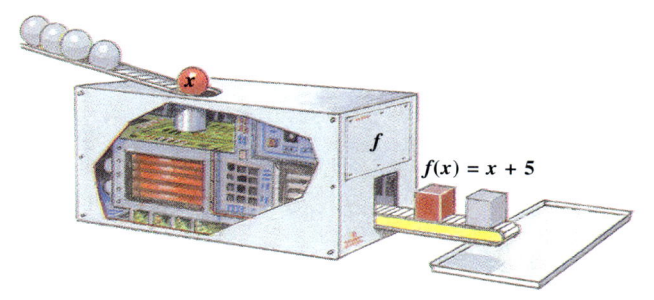

$f(x) = x + 5$

10. $g(0), g(6),$ and $g(13)$

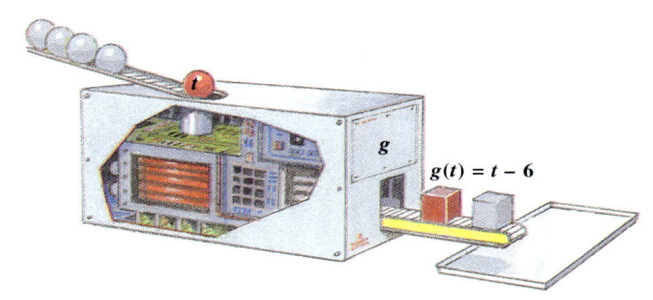

$g(t) = t - 6$

11. $h(-7), h(5),$ and $h(14)$

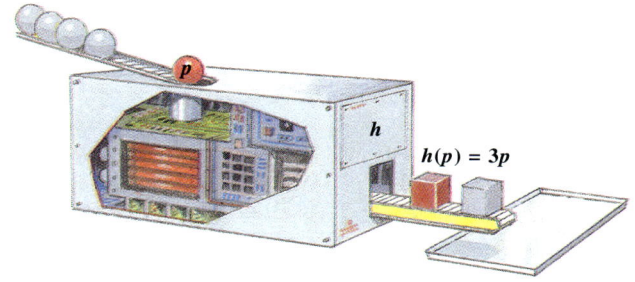

$h(p) = 3p$

12. $f(6), f\left(-\frac{1}{2}\right),$ and $f(20)$

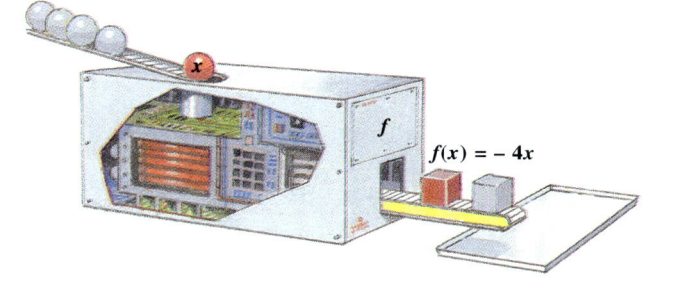

$f(x) = -4x$

510 CHAPTER 10 • QUADRATIC EQUATIONS

13. $g(s) = 3s + 4$; find $g(1)$, $g(-7)$, and $g(6.7)$.
14. $h(x) = 19$; find $h(4)$, $h(-16)$, and $h(12.5)$.
15. $F(x) = 2x^2 - 3x$; find $F(0)$, $F(-1)$, and $F(2)$.
16. $P(x) = 3x^2 - 2x$; find $P(0)$, $P(-2)$, and $P(3)$.
17. $f(t) = |t| + 1$; find $f(-5)$, $f(0)$, and $f\left(-\frac{9}{4}\right)$.
18. $f(x) = |x| - 2$; find $f(-3)$, $f(93)$, and $f(-100)$.
19. $g(t) = t^3 + 3$; find $g(1)$, $g(-5)$, and $g(0)$.
20. $h(x) = x^4 - 3$; find $h(0)$, $h(-1)$, and $h(3)$.

21. *Predicting heights.* An anthropologist can estimate the height of a male or a female, given the lengths of certain bones. A *humerus* is the bone from the elbow to the shoulder. The height, in centimeters, of a female with a humerus of x centimeters is given by
$$F(x) = 2.75x + 71.48.$$

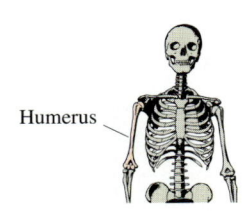

Humerus

If a humerus is known to be from a female, how tall was the female if the bone is (a) 32 cm long? (b) 30 cm long?

22. When a humerus (see Exercise 21) is from a male, the function given by $M(x) = 2.89x + 70.64$ can be used to find the male's height, in centimeters. If a humerus is known to be from a male, how tall was the male if the bone is (a) 30 cm long? (b) 35 cm long?

23. *Pressure at sea depth.* The function given by $P(d) = 1 + (d/33)$ gives the pressure, in *atmospheres* (atm), at a depth of d feet, in the sea. Note that $P(0) = 1$ atm, $P(33) = 2$ atm, and so on. Find the pressure at 20 ft, 30 ft, and 100 ft.

24. *Temperature as a function of depth.* The function given by $T(d) = 10d + 20$ gives the temperature, in degrees Celsius, inside the earth as a function of the depth d, in kilometers. Find the temperature at 5 km, 20 km, and 1000 km.

25. *Melting snow.* The function given by $W(d) = 0.112d$ approximates the amount, in centimeters, of water that results from d centimeters of snow melting. Find the amount of water that results from snow melting from depths of 16 cm, 25 cm, and 100 cm.

26. *Temperature conversions.* The function given by $C(F) = \frac{5}{9}(F - 32)$ determines the Celsius temperature that corresponds to F degrees Fahrenheit. Find the Celsius temperature that corresponds to 62°F, 77°F, and 23°F.

Graph each function.

27. $f(x) = 3x - 1$
28. $g(x) = 2x + 5$
29. $g(x) = -2x + 3$
30. $f(x) = -\frac{1}{2}x + 2$
31. $f(x) = \frac{1}{2}x + 1$
32. $f(x) = -\frac{3}{4}x - 2$
33. $g(x) = 2|x|$
34. $h(x) = -|x|$
35. $g(x) = x^2$
36. $f(x) = x^2 - 1$
37. $f(x) = x^2 - x - 2$
38. $g(x) = x^2 + 6x + 5$

Determine whether each graph is that of a function.

39.

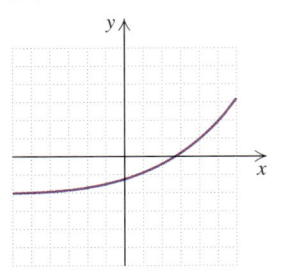

40.

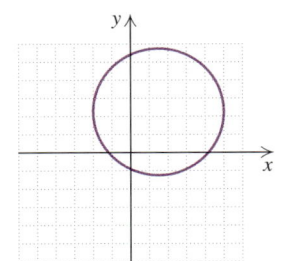

41.

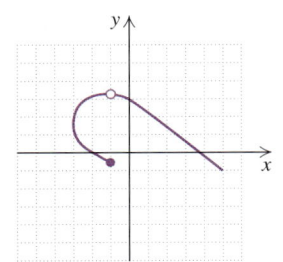

42.

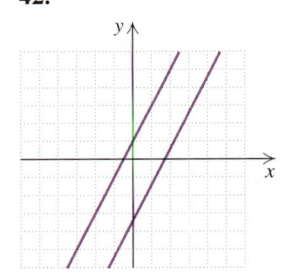

43.

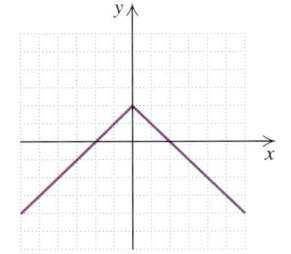

44.

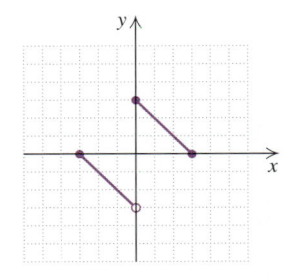

SKILL MAINTENANCE

Determine whether each pair of equations represents parallel lines.

45. $y = \frac{3}{4}x - 7$,
    $3x + 4y = 7$
46. $y = \frac{3}{5}x$,
    $y = -\frac{5}{3}$

Solve each system using the substitution method.

47. $2x - y = 6$,
$4x - 2y = 5$

48. $x - 3y = 2$,
$3x - 9y = 6$

SYNTHESIS

49. ◆ Is it possible for a function to have more numbers as outputs than as inputs? Why or why not?

50. ◆ Is it possible for a function to have more numbers as inputs than as outputs? Why or why not?

51. ◆ Explain in your own words how the vertical-line test works.

52. ◆ Look up the word "function" in a dictionary. Explain how that definition might be related to the mathematical one given in this section.

Graph.

53. $g(x) = x^3$

54. $f(x) = 2 + \sqrt{x}$

55. $f(x) = |x| + x$

56. $g(x) = |x| - x$

57. Sketch a graph that is not that of a function.

58. If $f(-1) = -7$ and $f(3) = 8$, find a linear equation for $f(x)$.

59. If $g(0) = -4$, $g(-2) = 0$, and $g(2) = 0$, find a quadratic equation for $g(x)$.

Find the range of each function for the given domain.

60. $f(x) = 3x + 5$, when the domain is the set of whole numbers less than 4

61. $g(t) = t^2 - 5$, when the domain is the set of integers between -4 and 2

62. $h(x) = |x| - x$, when the domain is the set of integers between -2 and 20

63. $f(m) = m^3 + 1$, when the domain is the set of integers between -3 and 3

64. Use a grapher to check your answers to Exercises 53–56 and 58–63.

SUMMARY AND REVIEW 10

KEY TERMS

Quadratic equation, p. 472
Principle of square roots, p. 472
Completing the square, p. 477
Quadratic formula, p. 482
Standard form, p. 482
Imaginary number, p. 492

Complex number, p. 492
Axis of symmetry, p. 494
Vertex (plural, vertices), p. 494
Parabola, p. 494
Function, p. 502

Domain, p. 502
Range, p. 502
Input, p. 502
Output, p. 502
Function value, p. 505
Vertical-line test, p. 507

IMPORTANT PROPERTIES AND FORMULAS

Principle of square roots:

If $x^2 = k$, then $x = \sqrt{k}$ or $x = -\sqrt{k}$.

Quadratic formula:

$$x = \frac{-b \pm \sqrt{b^2 - 4ac}}{2a}$$

To Solve a Quadratic Equation

1. If it is in the form $ax^2 = p$ or $(x + k)^2 = p$, use the principle of square roots.
2. When it is not in the form of (1), write it in standard form, $ax^2 + bx + c = 0$.

(continued)

3. Try to factor and use the principle of zero products.
4. If it is not possible to factor or if factoring seems difficult, use the quadratic formula.

The solutions of a quadratic equation can always be found using the quadratic formula, but not always by factoring. When the radicand, $b^2 - 4ac$, is nonnegative, the equation has real-number solutions. When $b^2 - 4ac$ is negative, the equation has no real-number solutions.

The Intercepts of a Parabola

To find the y-intercept of the graph of $y = ax^2 + bx + c$, replace x with 0 and solve for y.
To find any x-intercept(s) of the graph of $y = ax^2 + bx + c$, replace y with 0 and solve for x.

The Vertex of a Parabola

For a parabola given by the quadratic equation $y = ax^2 + bx + c$:

1. The x-coordinate of the vertex is $-\dfrac{b}{2a}$.

2. The y-coordinate of the vertex is found by substituting $-\dfrac{b}{2a}$ for x and solving for y.

Guidelines for Graphing Quadratic Equations

1. Graphs of quadratic equations, $y = ax^2 + bx + c$, are parabolas. They are cupped upward if $a > 0$ and downward if $a < 0$.
2. Use the formula $x = -b/(2a)$ to find the x-coordinate of the vertex. Calculate the y-coordinate and plot the vertex and some points on either side of it.
3. After a point has been graphed, a second point with the same y-coordinate can be found on the opposite side of the axis of symmetry.
4. Graph the y-intercept and, if requested, any x-intercepts.

The Vertical-Line Test

A graph represents a function if it is impossible to draw a vertical line that intersects the graph more than once.

REVIEW EXERCISES

Solve by completing the square.

1. $3x^2 + 2x - 5 = 0$
2. $x^2 = 3x - 1$

Solve.

3. $8x^2 = 24$
4. $5x^2 - 8x + 3 = 0$
5. $x^2 - 2x - 10 = 0$
6. $x^2 + 64 = 0$
7. $3y^2 + 5y = 2$
8. $(x + 8)^2 = 13$
9. $(x - 3)^2 = -4$
10. $9x^2 - 6x - 9 = 0$
11. $9x^2 = 0$
12. $x^2 + 6x = 9$
13. $x^2 + x + 1 = 0$
14. $1 + 4x^2 = 8x$
15. $x^2 + 1 = 2x$
16. $40 = 5y^2$
17. $3m = 4 + 5m^2$
18. $6x^2 + 11x = 35$

Approximate the solutions to the nearest thousandth.

19. $x^2 = 3x - 1$
20. $4y^2 + 8y + 1 = 0$
21. The hypotenuse of a right triangle is 5 m long. One leg is 3 m longer than the other. Find the lengths of the legs. Round to the nearest tenth.
22. $1000 is invested at interest rate r, compounded annually. In 2 years, it grows to $1102.50. What is the interest rate?
23. The length of a rectangle is 3 m greater than the width. The area is 108 m². Find the length and the width.
24. The height of the Lake Point Towers in Chicago is 645 ft. How long would it take an object to fall from the top?

Express in terms of i.

25. $2 - \sqrt{-64}$ **26.** $\sqrt{-24}$

Graph. Label the vertex, the y-intercept, and any x-intercepts.

27. $y = 2 - x^2$ **28.** $y = x^2 - 4x - 1$
29. $y = 3x^2 - 11x$ **30.** $y = x^2 - 2x + 1$

31. If $g(x) = 2x - 5$, find $g(2)$, $g(-1)$, and $g(3.5)$.

32. If $h(x) = |x| - 1$, find $h(1)$, $h(-1)$, and $h(-20)$.

33. Each day a moderately active person needs about 15 calories per pound of body weight. The function given by $C(p) = 15p$ approximates the number of calories that are needed to maintain body weight p, in pounds. How many calories are needed to maintain a body weight of 180 lb?

Graph.

34. $g(x) = x + 7$ **35.** $f(x) = x^2 - 3$
36. $h(x) = 3|x|$

Determine whether each graph is that of a function.

37.

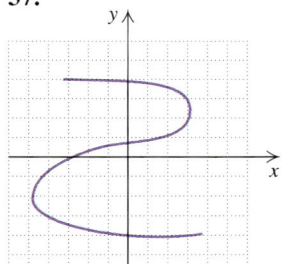

38.

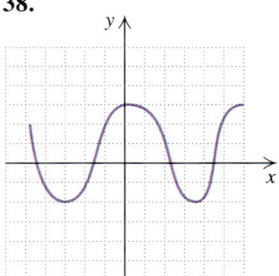

SKILL MAINTENANCE

39. Determine whether the pair of equations represents parallel lines:
$$y = 2x - 5,$$
$$2x - y = 3.$$

Simplify.

40. $\sqrt{84p^6}$ **41.** $(3 + \sqrt{2})(3 - \sqrt{2})$

42. Write a slope–intercept equation for the line containing the points $(-2, 5)$ and $(4, -3)$.

SYNTHESIS

43. ◆ Explain why it is helpful to know the general shape of a graph before connecting plotted points.

44. ◆ Explain how the radicand in the quadratic formula, $b^2 - 4ac$, can indicate how many x-intercepts the graph of a quadratic equation has.

45. Two consecutive integers have squares that differ by 63. Find the integers.

46. Find b such that the trinomial $x^2 + bx + 49$ is a square.

47. Solve: $x - 4\sqrt{x} - 5 = 0$.

48. A square with sides of length s has the same area as a circle with radius 5 in. Find s.

CHAPTER TEST 10

Solve.

1. $9x^2 = 45$ **2.** $3x^2 - 7x = 0$
3. $x^2 = x + 3$ **4.** $3y^2 + 4y = 15$
5. $48 = t^2 + 2t$ **6.** $x^2 = -49$
7. $(x - 2)^2 = 5$ **8.** $x^2 - 4x = -4$
9. $x^2 - 4x = -8$ **10.** $m^2 - 3m = 7$
11. $10 = 4x + x^2$ **12.** $3x^2 - 7x + 1 = 0$

13. Solve by completing the square:
$$x^2 - 4x - 10 = 0.$$

14. Approximate the solutions to the nearest thousandth:
$$x^2 - 3x - 8 = 0.$$

15. The width of a rectangle is 4 m less than the length. The area is 16.25 m². Find the length and the width.

16. A polygon has 44 diagonals. How many sides does it have? (*Hint*: $d = (n^2 - 3n)/2$.)

Express in terms of i.

17. $\sqrt{-49}$
18. $3 - \sqrt{-32}$

Graph. Label the vertex, the y-intercept, and any x-intercepts.

19. $y = -x^2 + x - 5$ **20.** $y = x^2 + 2x - 15$

21. If $f(x) = \frac{1}{2}x + 1$, find $f(0)$, $f(1)$, and $f(2)$.

22. If $g(t) = -2|t| + 3$, find $g(-1)$, $g(0)$, and $g(3)$.

23. The world record for the 10,000-m run has been decreasing steadily since 1940. The function given by $R(t) = 30.18 - 0.06t$ estimates the record, in minutes, t years since 1940. Predict what the record will be in 2010.

Graph.

24. $h(x) = x - 4$ **25.** $g(x) = x^2 - 4$

Determine whether each graph is that of a function.

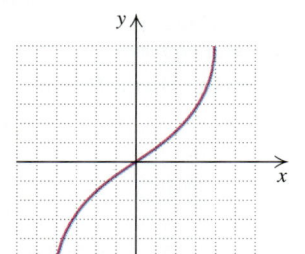

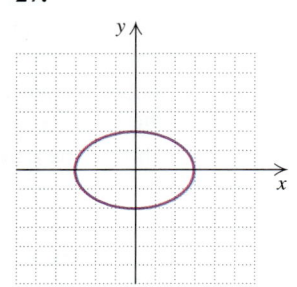

26.   **27.**

SKILL MAINTENANCE

28. Write a slope–intercept equation for the line of slope -3 passing through the point $(-5, 4)$.

29. Solve:
$$x + 2y = 10,$$
$$3x - y = 2.$$

30. Simplify: $2\sqrt{3} - \sqrt{27}$.

31. Graph: $y = -\frac{1}{3}x + 2$.

SYNTHESIS

32. Find the side of a square whose diagonal is 5 ft longer than a side.

33. Solve this system for x. Use the substitution method.
$$x - y = 2,$$
$$xy = 4$$

CUMULATIVE REVIEW
1–10

1. Write exponential notation for $x \cdot x \cdot x$.
2. Evaluate $(x - 3)^2 + 5$ for $x = 10$.
3. Use the commutative and associative laws to write an expression equivalent to $6(xy)$.
4. Find the LCM of 15 and 48.
5. Find the absolute value: $|-7|$.

Compute and simplify.

6. $-6 + 12 + (-4) + 7$
7. $2.8 - (-12.2)$
8. $-\frac{3}{8} \div \frac{5}{2}$
9. $13 \cdot 6 \div 3 \cdot 2 \div 13$
10. Remove parentheses and simplify:
$$4m + 9 - (6m + 13).$$

Solve.

11. $-5x = 45$
12. $-5x > 45$
13. $3(y - 1) - 2(y + 2) = 0$
14. $x^2 - 8x + 15 = 0$
15. $y - x = 1,$
    $y = 3 - x$
16. $\dfrac{x}{x + 1} = \dfrac{3}{2x} + 1$
17. $4x - 3y = 3,$
    $3x - 2y = 4$
18. $x^2 - x - 6 = 0$
19. $x^2 + 3x = 5$
20. $3 - x = \sqrt{x^2 - 3}$
21. $5 - 9x \leq 19 + 5x$
22. $-\frac{7}{8}x + 7 = \frac{3}{8}x - 3$
23. $0.6x - 1.8 = 1.2x$
24. $x + y = 15,$
    $x - y = 15$
25. $x^2 + 2x + 5 = 0$
26. $3y^2 = 30$
27. $(x - 3)^2 = 6$
28. $\dfrac{6x - 2}{2x - 1} = \dfrac{9x}{3x + 1}$
29. $12x^2 + x = 20$
30. $\dfrac{2x}{x + 3} + \dfrac{6}{x} + 7 = \dfrac{18}{x^2 + 3x}$
31. $\sqrt{x + 9} = \sqrt{2x - 3}$

Solve the formula for the given letter.

32. $A = \dfrac{4s + 3}{t}$, for t
33. $\dfrac{1}{t} = \dfrac{1}{m} - \dfrac{1}{n}$, for m

Simplify. Write scientific notation for the result.

34. $(2.1 \times 10^7)(1.3 \times 10^{-12})$
35. $\dfrac{5.2 \times 10^{-1}}{2.6 \times 10^{-15}}$

Simplify.

36. $x^{-6} \cdot x^2$
37. $\dfrac{y^3}{y^{-4}}$
38. $(2y^6)^2$

39. Combine like terms and arrange in descending order:
$$2x - 3 + 5x^3 - 2x^3 + 7x^3 + x.$$

Perform the indicated operation and simplify.

40. $(4x^3 + 3x^2 - 5) + (3x^3 - 5x^2 + 4x - 12)$
41. $(6x^2 - 4x + 1) - (-2x^2 + 7)$
42. $-2y^2(4y^2 - 3y + 1)$
43. $(2t - 3)(3t^2 - 4t + 2)$
44. $\left(t - \frac{1}{4}\right)\left(t + \frac{1}{4}\right)$
45. $(3m - 2)^2$
46. $(15x^2y^3 + 10xy^2 + 5) - (5xy^2 - x^2y^2 - 2)$
47. $(x^2 - 0.2y)(x^2 + 0.2y)$
48. $(3p + 4q^2)^2$
49. $\dfrac{4}{2x - 6} \cdot \dfrac{x - 3}{x + 3}$
50. $\dfrac{3a^4}{a^2 - 1} \div \dfrac{2a^3}{a^2 - 2a + 1}$
51. $\dfrac{3}{3x - 1} + \dfrac{4}{5x}$
52. $\dfrac{2}{x^2 - 16} - \dfrac{x - 3}{x^2 - 9x + 20}$
53. $(x^3 + 7x^2 - 2x + 3) \div (x - 2)$

Factor.

54. $49x^2 - 4$
55. $x^3 - 8x^2 - 5x + 40$
56. $16x^3 - x$
57. $m^2 - 8m + 16$
58. $15x^2 + 14x - 8$
59. $18x^5 + 4x^4 - 10x$

516   CUMULATIVE REVIEW: CHAPTERS 1–10

60. $6p^2 - 5p - 6$
61. $2ac - 6ab - 3db + dc$
62. $9x^2 + 30xy + 25y^2$
63. $3t^4 + 6t^2 - 72$
64. $49a^2b^2 - 9$

Simplify.

65. $\dfrac{\dfrac{3}{x} + \dfrac{1}{2x}}{\dfrac{1}{3x} - \dfrac{3}{4x}}$

66. $\sqrt{49}$

67. $-\sqrt[4]{625}$

68. $\sqrt{64x^2}$ (Assume $x \geq 0$.)

69. Multiply: $\sqrt{a+b}\,\sqrt{a-b}$.
70. Multiply and simplify: $\sqrt{32ab}\,\sqrt{6a^4b^2}$ ($a, b \geq 0$).

Simplify.

71. $8^{-2/3}$
72. $\sqrt{243x^3y^2}$ ($x, y \geq 0$)
73. $\sqrt{\dfrac{100}{81}}$
74. $(2\sqrt{3} + \sqrt{5})(2\sqrt{3} - \sqrt{5})$
75. $4\sqrt{12} + 2\sqrt{48}$

76. Divide and simplify: $\dfrac{\sqrt{72}}{\sqrt{45}}$.

77. The hypotenuse of a right triangle is 41 cm long and the length of a leg is 9 cm. Find the length of the other leg.

Graph on a plane.

78. $y = \frac{1}{3}x - 2$
79. $2x + 3y = -6$
80. $y = -3$
81. $4x - 3y > 12$
82. $y = x^2 + 2x + 1$
83. $x \geq -3$

84. Solve $9x^2 - 12x - 2 = 0$ by completing the square.
85. Approximate the solutions of $4x^2 = 4x + 1$ to the nearest thousandth.
86. Graph $y = x^2 + 2x - 5$. Label the vertex, the y-intercept, and the x-intercepts.

Solve.

87. In 1993, 50% of the 96.4 million U.S. households had answering machines (*Source*: U.S. Census Bureau; U.S. Energy Information Administration. *The MacMillan Visual Almanac*, 1996). How many households had answering machines?

88. Find all numbers such that the sum of the number and 10 is less than 3 times the number.
89. The product of two consecutive even integers is 224. Find the integers.
90. The length of a rectangle is 7 m more than the width. The length of a diagonal is 13 m. Find the length.
91. Three-fifths of the automobiles entering Dalton each morning will fill the city parking lots. There are 3654 such parking spaces. How many cars enter Dalton each morning?
92. A candy shop mixes nuts worth $1.10 per pound with another variety worth $0.80 per pound in order to make 42 lb of a mixture worth $0.90 per pound. How many pounds of each kind of nut should be used?
93. Kathy and Ken work in a textbook warehouse. Kathy can fill an order of 100 books in 40 min. Ken can fill the same order in 60 min. How long would it take them, working together, to fill the order?
94. The cost c of cable television, in dollars, is given by $c = 20t + 25$, where t is the number of months that a household has had cable service. Graph the equation and use the graph to estimate the cost of a line that has been in service for 10 months.
95. For the function f described by
$$f(x) = 2x^2 + 7x - 4,$$
find $f(0)$, $f(-4)$, and $f\left(\frac{1}{2}\right)$.
96. Determine whether the graphs of the following equations are parallel:
$$y - x = 4,$$
$$3y = 5 + 3x.$$
97. Graph the following system of inequalities:
$$x + y \leq 6,$$
$$x + y \geq 2,$$
$$x \leq 3,$$
$$x \geq 1.$$
98. Find the slope and the y-intercept:
$$-6x + 3y = -24.$$
99. Find the slope of the line containing the points $(-5, -6)$ and $(-4, 9)$.
100. Find a point–slope equation for the line containing $(1, -3)$ and having slope $m = -\frac{1}{2}$.
101. Simplify: $-\sqrt{-25}$.

SYNTHESIS

102. Find b such that the trinomial $x^2 - bx + 225$ is a square.

103. Find x.

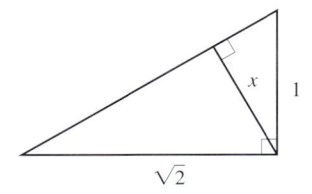

Determine whether each pair of expressions is equivalent.

104. $x^2 - 9$, $(x - 3)(x + 3)$

105. $\dfrac{x + 3}{3}$, x

106. $(x + 5)^2$, $x^2 + 25$

107. $\sqrt{x^2 + 16}$, $x + 4$

108. $\sqrt{x^2}$, $|x|$

Appendixes

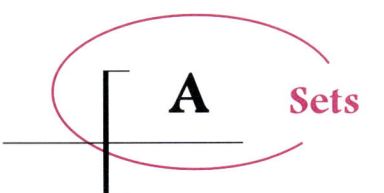

A Sets

Naming Sets • Membership • Subsets • Intersections • Unions

The notion of a "set" is used frequently in more advanced mathematics. We provide a basic introduction to sets in this appendix.

Naming Sets

To name the set of whole numbers less than 6, we can use *roster notation,* as follows:

$\{0, 1, 2, 3, 4, 5\}$.

The set of real numbers x such that x is less than 6 cannot be named by listing all its members because there is an infinite number of them. We name such a set using *set-builder notation,* as follows:

$\{x | x < 6\}$.

This is read

"The set of all x such that x is less than 6."

See Section 2.6 for more on this notation.

Membership

The symbol $\in$ means *is a member of* or *belongs to,* or *is an element of.* Thus,

$x \in A$

means

x is a member of A

or x belongs to A

or x is an element of A.

EXAMPLE 1

Classify each of the following as true or false.

a) $1 \in \{1, 2, 3\}$
b) $1 \in \{2, 3\}$
c) $4 \in \{x \mid x \text{ is an even whole number}\}$
d) $5 \in \{x \mid x \text{ is an even whole number}\}$

SOLUTION

a) Since 1 is listed as a member of the set, $1 \in \{1, 2, 3\}$ is true.
b) Since 1 is *not* a member of $\{2, 3\}$, the statement $1 \in \{2, 3\}$ is false.
c) Since 4 is an even whole number, $4 \in \{x \mid x \text{ is an even whole number}\}$ is true.
d) Since 5 is *not* even, $5 \in \{x \mid x \text{ is an even whole number}\}$ is false.

Set membership can be illustrated with a diagram, as shown below.

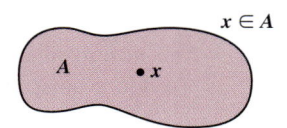

Subsets

If every element of A is an element of B, then A is a *subset* of B. This is denoted $A \subseteq B$.

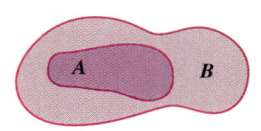

The set of whole numbers is a subset of the set of integers. The set of rational numbers is a subset of the set of real numbers.

EXAMPLE 2

Classify each of the following as true or false.

a) $\{1, 2\} \subseteq \{1, 2, 3, 4\}$
b) $\{p, q, r, w\} \subseteq \{a, p, r, z\}$
c) $\{x \mid x < 6\} \subseteq \{x \mid x \leq 11\}$

SOLUTION

a) Since every element of $\{1, 2\}$ is in the set $\{1, 2, 3, 4\}$, it follows that $\{1, 2\} \subseteq \{1, 2, 3, 4\}$ is true.
b) Since $q \in \{p, q, r, w\}$, but $q \notin \{a, p, r, z\}$, it follows that $\{p, q, r, w\} \subseteq \{a, p, r, z\}$ is false.
c) Since every number that is less than 6 is also less than 11, the statement $\{x \mid x < 6\} \subseteq \{x \mid x \leq 11\}$ is true.

Intersections

The *intersection* of sets A and B, denoted A ∩ B, is the set of members common to both sets.

EXAMPLE 3   Find each intersection.

a) $\{0, 1, 3, 5, 25\} \cap \{2, 3, 4, 5, 6, 7, 9\}$
b) $\{a, p, q, w\} \cap \{p, q, t\}$

SOLUTION
a) $\{0, 1, 3, 5, 25\} \cap \{2, 3, 4, 5, 6, 7, 9\} = \{3, 5\}$
b) $\{a, p, q, w\} \cap \{p, q, t\} = \{p, q\}$

Set intersection can be illustrated with a diagram, as shown below.

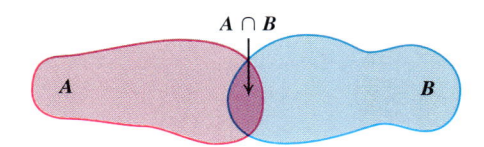

A ∩ B

The set without members is known as the *empty set*, and is written ∅, and sometimes { }. Each of the following is a description of the empty set:

The set of all six-eyed algebra teachers;
$\{2, 3\} \cap \{5, 6, 7\}$;
$\{x \mid x \text{ is an even natural number}\} \cap \{x \mid x \text{ is an odd natural number}\}$.

Unions

Two sets A and B can be combined to form a set that contains the members of A as well as those of B. The new set is called the *union* of A and B, denoted A ∪ B.

EXAMPLE 4   Find each union.

a) $\{0, 5, 7, 13, 27\} \cup \{0, 2, 3, 4, 5\}$ b) $\{a, c, e, g\} \cup \{b, d, f\}$

SOLUTION
a) $\{0, 5, 7, 13, 27\} \cup \{0, 2, 3, 4, 5\} = \{0, 2, 3, 4, 5, 7, 13, 27\}$
Note that the 0 and the 5 are *not* listed twice in the solution.
b) $\{a, c, e, g\} \cup \{b, d, f\} = \{a, b, c, d, e, f, g\}$

Set union can be illustrated with a diagram, as shown below.

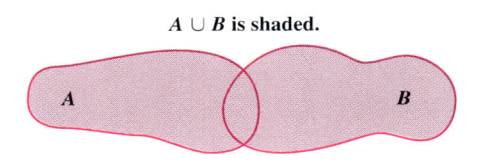

A ∪ B is shaded.

EXERCISE SET A

Name each set using the roster method.

1. The set of whole numbers 3 through 8
2. The set of whole numbers 101 through 107
3. The set of odd numbers between 40 and 50
4. The set of multiples of 5 between 10 and 40
5. $\{x|$ the square of x is $9\}$
6. $\{x|\ x$ is the cube of $0.2\}$

Classify each statement as true or false.

7. $2 \in \{x|\ x$ is an odd number$\}$
8. $7 \in \{x|\ x$ is an odd number$\}$
9. Bruce Springsteen $\in$ The set of all rock stars
10. Apple $\in$ The set of all fruit
11. $-3 \in \{-4, -3, 0, 1\}$
12. $0 \in \{-4, -3, 0, 1\}$
13. $\frac{2}{3} \in \{x|\ x$ is a rational number$\}$
14. Heads $\in$ The set of outcomes of flipping a penny
15. $\{4, 5, 8\} \subseteq \{1, 3, 4, 5, 6, 7, 8, 9\}$
16. The set of vowels $\subseteq$ The set of consonants
17. $\{-1, -2, -3, -4, -5\} \subseteq \{-1, 2, 3, 4, 5\}$
18. The set of integers $\subseteq$ The set of rational numbers

Find each intersection.

19. $\{a, b, c, d, e\} \cap \{c, d, e, f, g\}$
20. $\{a, e, i, o, u\} \cap \{q, u, i, c, k\}$
21. $\{1, 2, 5, 10\} \cap \{0, 1, 7, 10\}$
22. $\{0, 1, 7, 10\} \cap \{0, 1, 2, 5\}$
23. $\{1, 2, 5, 10\} \cap \{3, 4, 7, 8\}$
24. $\{a, e, i, o, u\} \cap \{m, n, f, g, h\}$

Find each union.

25. $\{a, e, i, o, u\} \cup \{q, u, i, c, k\}$
26. $\{a, b, c, d, e\} \cup \{c, d, e, f, g\}$
27. $\{0, 1, 7, 10\} \cup \{0, 1, 2, 5\}$
28. $\{1, 2, 5, 10\} \cup \{0, 1, 7, 10\}$
29. $\{a, e, i, o, u\} \cup \{m, n, f, g, h\}$
30. $\{1, 2, 5, 10\} \cup \{a, b\}$

SYNTHESIS

31. ◆ What advantage(s) does set-builder notation have over roster notation?
32. ◆ What advantage(s) does roster notation have over set-builder notation?
33. Find the union of the set of integers and the set of whole numbers.
34. Find the intersection of the set of odd integers and the set of even integers.
35. Find the union of the set of rational numbers and the set of irrational numbers.
36. Find the intersection of the set of even integers and the set of positive rational numbers.
37. Find the intersection of the set of rational numbers and the set of irrational numbers.
38. Find the union of the set of negative integers, the set of positive integers, and the set containing 0.
39. For a set A, find each of the following.
    a) $A \cup \emptyset$
    b) $A \cup A$
    c) $A \cap A$
    d) $A \cap \emptyset$
40. A set is *closed* under an operation if, when the operation is performed on its members, the result is in the set. For example, the set of real numbers is closed under the operation of addition since the sum of any two real numbers is a real number.
    a) Is the set of even numbers closed under addition?
    b) Is the set of odd numbers closed under addition?
    c) Is the set $\{0, 1\}$ closed under addition?
    d) Is the set $\{0, 1\}$ closed under multiplication?
    e) Is the set of real numbers closed under multiplication?
    f) Is the set of integers closed under division?
41. Experiment with sets of various types and determine whether the following distributive law for sets is true:
    $$A \cap (B \cup C) = (A \cap B) \cup (A \cap C).$$

B Factoring Sums or Differences of Cubes

Factoring a Sum of Two Cubes • Factoring a Difference of Two Cubes

Although a sum of two squares cannot be factored unless a common factor exists, a sum of two cubes can always be factored. A difference of two cubes can also be factored.

Consider the following products:

$$(A + B)(A^2 - AB + B^2) = A(A^2 - AB + B^2) + B(A^2 - AB + B^2)$$
$$= A^3 - A^2B + AB^2 + A^2B - AB^2 + B^3$$
$$= A^3 + B^3$$

and

$$(A - B)(A^2 + AB + B^2) = A(A^2 + AB + B^2) - B(A^2 + AB + B^2)$$
$$= A^3 + A^2B + AB^2 - A^2B - AB^2 - B^3$$
$$= A^3 - B^3.$$

These equations show how we can factor a sum or a difference of two cubes.

TO FACTOR A SUM OR DIFFERENCE OF CUBES:

$$A^3 + B^3 = (A + B)(A^2 - AB + B^2),$$
$$A^3 - B^3 = (A - B)(A^2 + AB + B^2)$$

This table of cubes will help in the examples that follow.

N	0.2	0.1	0	1	2	3	4	5	6	7	8
N^3	0.008	0.001	0	1	8	27	64	125	216	343	512

EXAMPLE 1 Factor: $x^3 - 8$.

SOLUTION We have

$$x^3 - 8 = x^3 - 2^3 = (x - 2)(x^2 + x \cdot 2 + 2^2).$$
$$A^3 - B^3 = (A - B)(A^2 + A\,B + B^2)$$

This tells us that $x^3 - 8 = (x - 2)(x^2 + 2x + 4)$. Note that we cannot factor $x^2 + 2x + 4$. (It is not a perfect-square trinomial nor can it be factored by trial and error or grouping.)

EXAMPLE 2   Factor: $x^3 + 125$.

SOLUTION   We have

$$x^3 + 125 = x^3 + 5^3 = (x + 5)(x^2 - x \cdot 5 + 5^2).$$
$$\uparrow\uparrow\uparrow\uparrow\uparrow\uparrow\uparrow\uparrow$$
$$A^3 + B^3 = (A + B)(A^2 - AB + B^2)$$

Thus, $x^3 + 125 = (x + 5)(x^2 - 5x + 25)$.

EXAMPLE 3   Factor: $16a^7b + 54ab^7$.

SOLUTION   We first look for a common factor:

$$\begin{aligned}16a^7b + 54ab^7 &= 2ab[8a^6 + 27b^6] \\ &= 2ab[(2a^2)^3 + (3b^2)^3] \quad \text{This is of the form } A^3 + B^3, \\ & \text{where } A = 2a^2 \text{ and } B = 3b^2. \\ &= 2ab[(2a^2 + 3b^2)(4a^4 - 6a^2b^2 + 9b^4)].\end{aligned}$$

EXAMPLE 4   Factor: $y^3 - 0.001$.

SOLUTION   Since $0.001 = (0.1)^3$, we have a difference of cubes:

$$y^3 - 0.001 = (y - 0.1)(y^2 + 0.1y + 0.01).$$

Remember the following about factoring sums or differences of squares and cubes:

Difference of cubes: $A^3 - B^3 = (A - B)(A^2 + AB + B^2),$
Sum of cubes: $A^3 + B^3 = (A + B)(A^2 - AB + B^2),$
Difference of squares: $A^2 - B^2 = (A + B)(A - B),$
Sum of squares: $A^2 + B^2$ **cannot be factored unless a common factor exists.**

EXERCISE SET B

Factor completely.

1. $t^3 + 27$
2. $p^3 + 8$
3. $a^3 - 1$
4. $w^3 - 64$
5. $z^3 + 125$
6. $x^3 + 1$
7. $8a^3 - 1$
8. $27x^3 - 1$
9. $y^3 - 27$
10. $p^3 - 8$
11. $64 + 125x^3$
12. $8 + 27b^3$
13. $125p^3 - 1$
14. $64w^3 - 1$
15. $27m^3 + 64$
16. $8t^3 + 27$
17. $p^3 - q^3$
18. $a^3 + b^3$
19. $x^3 + \frac{1}{8}$
20. $y^3 + \frac{1}{27}$
21. $2y^3 - 128$
22. $3z^3 - 3$
23. $24a^3 + 3$
24. $54x^3 + 2$
25. $rs^3 + 64r$
26. $ab^3 + 125a$
27. $5x^3 - 40z^3$
28. $2y^3 - 54z^3$
29. $x^3 + 0.001$
30. $y^3 + 0.125$

SYNTHESIS

31. ◆ Dino incorrectly believes that
$$a^3 - b^3 = (a - b)(a^2 + b^2).$$
How could you convince him that he is wrong?

32. ◆ If $x^3 + c$ is prime, what can you conclude about c? Why?

Factor. Assume that variables in exponents represent natural numbers.

33. $125c^6 - 8d^6$
34. $64x^6 - 8t^6$
35. $3x^{3a} + 24y^{3b}$
36. $\frac{8}{27}x^3 + \frac{1}{64}y^3$
37. $\frac{1}{24}x^3y^3 + \frac{1}{3}z^3$
38. $\frac{1}{16}x^{3a} + \frac{1}{2}y^{6a}z^{9b}$

Tables

1 Fractional and Decimal Equivalents
2 Squares and Square Roots
3 Geometric Formulas

TABLE 1 Fractional and Decimal Equivalents

Fractional Notation	$\frac{1}{10}$	$\frac{1}{8}$	$\frac{1}{6}$	$\frac{1}{5}$	$\frac{1}{4}$	$\frac{3}{10}$	$\frac{1}{3}$	$\frac{3}{8}$	$\frac{2}{5}$	$\frac{1}{2}$
Decimal Notation	0.1	0.125	$0.16\overline{6}$	0.2	0.25	0.3	$0.33\overline{3}$	0.375	0.4	0.5
Percent Notation	10%	12.5% or $12\frac{1}{2}\%$	$16.6\overline{6}\%$ or $16\frac{2}{3}\%$	20%	25%	30%	$33.3\overline{3}\%$ or $33\frac{1}{3}\%$	37.5% or $37\frac{1}{2}\%$	40%	50%
Fractional Notation	$\frac{3}{5}$	$\frac{5}{8}$	$\frac{2}{3}$	$\frac{7}{10}$	$\frac{3}{4}$	$\frac{4}{5}$	$\frac{5}{6}$	$\frac{7}{8}$	$\frac{9}{10}$	$\frac{1}{1}$
Decimal Notation	0.6	0.625	$0.666\overline{6}$	0.7	0.75	0.8	$0.83\overline{3}$	0.875	0.9	1
Percent Notation	60%	62.5% or $62\frac{1}{2}\%$	$66.6\overline{6}\%$ or $66\frac{2}{3}\%$	70%	75%	80%	$83.3\overline{3}\%$ or $83\frac{1}{3}\%$	87.5% or $87\frac{1}{2}\%$	90%	100%

TABLE 2 Squares and Square Roots

N	$\sqrt{N}$	N^2	N	$\sqrt{N}$	N^2	N	$\sqrt{N}$	N^2	N	$\sqrt{N}$	N^2
1	1	1	26	5.099	676	51	7.141	2601	76	8.718	5776
2	1.414	4	27	5.196	729	52	7.211	2704	77	8.775	5929
3	1.732	9	28	5.292	784	53	7.280	2809	78	8.832	6084
4	2	16	29	5.385	841	54	7.348	2916	79	8.888	6241
5	2.236	25	30	5.477	900	55	7.416	3025	80	8.944	6400
6	2.449	36	31	5.568	961	56	7.483	3136	81	9	6561
7	2.646	49	32	5.657	1024	57	7.550	3249	82	9.055	6724
8	2.828	64	33	5.745	1089	58	7.616	3364	83	9.110	6889
9	3	81	34	5.831	1156	59	7.681	3481	84	9.165	7056
10	3.162	100	35	5.916	1225	60	7.746	3600	85	9.220	7225
11	3.317	121	36	6	1296	61	7.810	3721	86	9.274	7396
12	3.464	144	37	6.083	1369	62	7.874	3844	87	9.327	7569
13	3.606	169	38	6.164	1444	63	7.937	3969	88	9.381	7744
14	3.742	196	39	6.245	1521	64	8	4096	89	9.434	7921
15	3.873	225	40	6.325	1600	65	8.062	4225	90	9.487	8100
16	4	256	41	6.403	1681	66	8.124	4356	91	9.539	8281
17	4.123	289	42	6.481	1764	67	8.185	4489	92	9.592	8464
18	4.243	324	43	6.557	1849	68	8.246	4624	93	9.644	8649
19	4.359	361	44	6.633	1936	69	8.307	4761	94	9.695	8836
20	4.472	400	45	6.708	2025	70	8.367	4900	95	9.747	9025
21	4.583	441	46	6.782	2116	71	8.426	5041	96	9.798	9216
22	4.690	484	47	6.856	2209	72	8.485	5184	97	9.849	9409
23	4.796	529	48	6.928	2304	73	8.544	5329	98	9.899	9604
24	4.899	576	49	7	2401	74	8.602	5476	99	9.950	9801
25	5	625	50	7.071	2500	75	8.660	5625	100	10	10,000

TABLE 3 Geometric Formulas

PLANE GEOMETRY

Rectangle
Area: $A = lw$
Perimeter: $P = 2l + 2w$

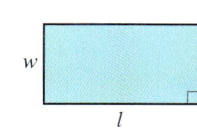

Square
Area: $A = s^2$
Perimeter: $P = 4s$

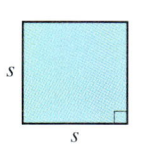

Triangle
Area: $A = \frac{1}{2}bh$

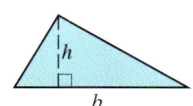

Triangle
Sum of Angle Measures:
$A + B + C = 180°$

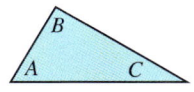

Right Triangle
Pythagorean Theorem (Equation):
$a^2 + b^2 = c^2$

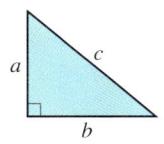

Parallelogram
Area: $A = bh$

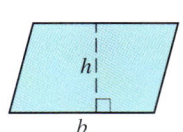

Trapezoid
Area: $A = \frac{1}{2}h(b_1 + b_2)$

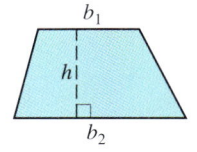

Circle
Area: $A = \pi r^2$
Circumference:
$C = \pi d = 2\pi r$
($\frac{22}{7}$ and 3.14 are different approximations for π)

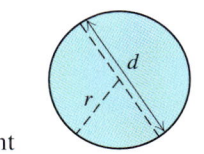

SOLID GEOMETRY

Rectangular Solid
Volume: $V = lwh$

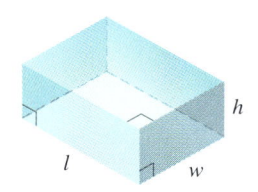

Cube
Volume: $V = s^3$

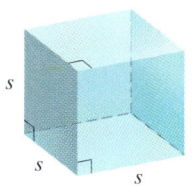

Right Circular Cylinder
Volume: $V = \pi r^2 h$
Total Surface Area:
$S = 2\pi rh + 2\pi r^2$

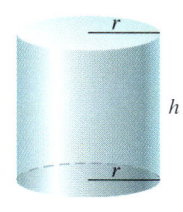

Right Circular Cone
Volume: $V = \frac{1}{3}\pi r^2 h$
Total Surface Area:
$S = \pi r^2 + \pi rs$
Slant Height:
$s = \sqrt{r^2 + h^2}$

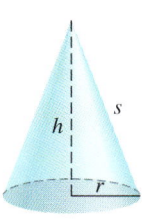

Sphere
Volume: $V = \frac{4}{3}\pi r^3$
Surface Area: $S = 4\pi r^2$

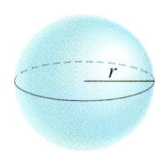

Answers

CHAPTER 1

EXERCISE SET 1.1, PP. 7–9

1. 45 **3.** 8 **5.** 8 **7.** 2 **9.** 5 **11.** 3 **13.** 165 mi
15. 100.1 sq cm **17.** 15 sq cm **19.** $7x$ or $x7$
21. $b + 6$, or $6 + b$ **23.** $c - 9$ **25.** $6 + q$, or $q + 6$
27. Let n represent "a number"; $n + 9$, or $9 + n$
29. $y - x$ **31.** $x \div w$, or $\dfrac{x}{w}$ **33.** $n - m$
35. Let a and b represent the numbers; $a + b$, or $b + a$
37. $9 \cdot 2m$ **39.** Let y represent "some number"; $\dfrac{1}{4}y$, or $\dfrac{y}{4}$
41. Let x represent "some number"; 64% of x, or $0.64x$
43. $\$50 - x$ **45.** Yes **47.** No **49.** Yes **51.** Yes
53. Let x represent the unknown number; $73 + x = 201$
55. Let n represent the number; $42n = 2352$
57. Let s represent the number of squares your opponent wins; $s + 35 = 64$
59. Let w = the length of the average commute in the West, in minutes; $w = 24.5 - 1.8$ **61.** ◆ **63.** ◆
65. Let n represent "a number"; $3n - 5$
67. $l + w + l + w$, or $2l + 2w$ **69.** $(a + 2) + 7$, or $a + 9$ **71.** 158.75 sq cm **73.** 5 **75.** 9 **77.** d

EXERCISE SET 1.2, PP. 14–15

1. $x + 3$ **3.** $bc + a$ **5.** $3y + 9x$ **7.** $2(3 + a)$
9. $a \cdot 8$ **11.** ts **13.** $5 + ba$ **15.** $(a + 3)2$
17. $a + (3 + b)$ **19.** $(r + t) + 9$ **21.** $ab + (c + d)$
23. $3(xy)$ **25.** $(4x)y$ **27.** $(3 \cdot 2)(a + b)$
29. $(6 + t) + r$, $(6 + r) + t$; answers may vary
31. $(ba)5$, $a(5b)$; answers may vary
33. $(7 + x) + 2 = (x + 7) + 2$ Commutative law
$ = x + (7 + 2)$ Associative law
$ = x + 9$ Simplifying
35. $(m3)7 = m(3 \cdot 7)$ Associative law
$ = (3 \cdot 7)m$ Commutative law
$ = 21m$ Simplifying
37. $3a + 12$ **39.** $6 + 6x$ **41.** $3x + 3$ **43.** $4 + 4y$
45. $18x + 54$ **47.** $5r + 10 + 15t$ **49.** $2a + 2b$
51. $5x + 5y + 10$ **53.** $7(x + z)$ **55.** $5(1 + y)$
57. $3(6x + y)$ **59.** $5(x + 2 + 3y)$ **61.** $3(4x + 3)$
63. $3(3x + y)$ **65.** $2(a + 8b + 32)$

67. $11(x + 4y + 11)$ **69.** $t - 9$ **71.** ◆ **73.** ◆
75. Yes; distributive law
77. Yes; distributive law and commutative law of multiplication
79. No; for example, let $x = 1$ and $y = 2$. Then $30 \cdot 2 + 1 \cdot 15 = 60 + 15 = 75$, but $5[2(1 + 3 \cdot 2)] = 5[2(7)] = 5 \cdot 14 = 70$. **81.** ◆

EXERCISE SET 1.3, PP. 22–23

1. $2 \cdot 15$, $3 \cdot 10$; 1, 2, 3, 5, 6, 10, 15, 30
3. $3 \cdot 14$, $6 \cdot 7$; 1, 2, 3, 6, 7, 14, 21, 42 **5.** $2 \cdot 11$
7. $2 \cdot 3 \cdot 5$ **9.** $2 \cdot 5 \cdot 5$ **11.** $3 \cdot 3 \cdot 3$ **13.** $2 \cdot 3 \cdot 3$
15. $2 \cdot 2 \cdot 2 \cdot 5$ **17.** Prime **19.** $2 \cdot 3 \cdot 5 \cdot 7$
21. $5 \cdot 23$ **23.** $\dfrac{5}{7}$ **25.** $\dfrac{7}{2}$ **27.** $\dfrac{1}{8}$ **29.** 8 **31.** $\dfrac{1}{4}$
33. 5 **35.** $\dfrac{15}{16}$ **37.** $\dfrac{60}{41}$ **39.** $\dfrac{15}{7}$ **41.** $\dfrac{3}{10}$ **43.** $\dfrac{51}{8}$
45. $\dfrac{1}{2}$ **47.** $\dfrac{7}{6}$ **49.** $\dfrac{3b}{7a}$ **51.** $\dfrac{7}{a}$ **53.** $\dfrac{5}{6}$ **55.** 1
57. $\dfrac{5}{18}$ **59.** $\dfrac{5}{9}$ **61.** $\dfrac{35}{18}$ **63.** $\dfrac{10}{3}$ **65.** 14 **67.** $\dfrac{5}{36}$
69. 60 **71.** $\dfrac{x}{8}$ **73.** $\dfrac{5b}{3a}$ **75.** $\dfrac{x-2}{6}$
77. $5(3 + x)$; answers may vary **79.** ◆ **81.** ◆
83. $\dfrac{4}{3}$ **85.** $\dfrac{2}{5}$ **87.** $\dfrac{6}{25}$ **89.** 24 in. **91.** $\dfrac{28}{45}$ m²
93. $\dfrac{20}{9}$ m **95.** ◆

EXERCISE SET 1.4, PP. 30–31

1. $-3, 7$ **3.** $-508, 186.84$ **5.** $-1286, 29{,}029$
7. $750, -125$ **9.** $20, -150, 300$
11.

$$\xleftarrow{}\underset{-5\ -4\ -3\ -2\ -1\ \ 0\ \ 1\ \ 2\ \ 3\ \ 4\ \ 5}{\stackrel{\stackrel{\tfrac{10}{3}}{\bullet}}{\rule{7cm}{0.4pt}}}\xrightarrow{}$$

13.

$$\xleftarrow{}\underset{-5\ -4\ -3\ -2\ -1\ \ 0\ \ 1\ \ 2\ \ 3\ \ 4\ \ 5}{\stackrel{\stackrel{-4.3}{\bullet}}{\rule{7cm}{0.4pt}}}\xrightarrow{}$$

15.

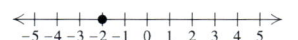

$$\xleftarrow{}\underset{-5\ -4\ -3\ -2\ -1\ \ 0\ \ 1\ \ 2\ \ 3\ \ 4\ \ 5}{\rule{7cm}{0.4pt}}\xrightarrow{}$$

17. 0.875 **19.** -0.75 **21.** $1.\overline{16}$ **23.** $0.\overline{6}$ **25.** -0.5

27. 0.1 **29.** > **31.** < **33.** < **35.** < **37.** >
39. < **41.** < **43.** $a < -2$ **45.** $y \geq -10$ **47.** True
49. False **51.** True **53.** 4 **55.** 17 **57.** 5.6
59. 329 **61.** $\frac{9}{7}$ **63.** 0 **65.** 8 **67.** $-23, -4.7, 0, \frac{5}{9}$,
8.31, 62 **69.** $-23, 0, 62$ **71.** $-23, -4.7, 0, \frac{5}{9}, \pi, \sqrt{17}$,
8.31, 62 **73.** $\frac{3}{5}$ **75.** $5 + ab$; answers may vary
77. ◆ **79.** ◆ **81.** $-17, -12, 5, 13$
83. $-\frac{4}{3}, \frac{4}{9}, \frac{4}{8}, \frac{4}{6}, \frac{4}{5}, \frac{4}{3}, \frac{4}{2}$ **85.** > **87.** = **89.** <
91. $7, -7$ **93.** $-4, -3, 3, 4$ **95.** $\frac{3}{3}$ **97.** ◆

EXERCISE SET 1.5, PP. 36–37
1. -4 **3.** 4 **5.** 0 **7.** -8 **9.** -15 **11.** -8
13. 0 **15.** -41 **17.** 0 **19.** 1 **21.** -2 **23.** 11
25. -33 **27.** 0 **29.** 18 **31.** -32 **33.** 0 **35.** 20
37. -1.7 **39.** -9.1 **41.** $-\frac{2}{3}$ **43.** $-\frac{10}{9}$ **45.** $-\frac{1}{6}$
47. $-\frac{23}{24}$ **49.** -3 **51.** 50 **53.** $69
55. $183,415 profit **57.** 13,796 ft above sea level
59. $0 **61.** $13a$ **63.** $9x$ **65.** $11x$ **67.** $-2m$
69. $-7a$ **71.** $1 - 2x$ **73.** $12x + 17$ **75.** $18n + 16$
77. $21z + 7y + 14$ **79.** ◆ **81.** ◆ **83.** $65\frac{1}{4}$
85. $-5y$ **87.** $-7m$ **89.** $\frac{7}{2}x$

EXERCISE SET 1.6, PP. 42–43
1. -39 **3.** 9 **5.** 3.14 **7.** -23 **9.** $\frac{14}{3}$ **11.** -0.101
13. 72 **15.** $-\frac{2}{5}$ **17.** 1 **19.** -7 **21.** -9 **23.** -7
25. 2 **27.** 0 **29.** -4 **31.** -7 **33.** -5 **35.** 0
37. 0 **39.** 12 **41.** 11 **43.** -14 **45.** 5 **47.** -1
49. 15 **51.** -5 **53.** -3 **55.** -21 **57.** 5 **59.** -8
61. 10 **63.** -23 **65.** -68 **67.** -73 **69.** 116
71. 0 **73.** $-\frac{1}{4}$ **75.** $\frac{2}{15}$ **77.** $-\frac{17}{12}$ **79.** -2.8
81. -0.91 **83.** $-\frac{1}{2}$ **85.** $\frac{6}{7}$ **87.** $3.8 - (-5.2); 9$
89. $114 - (-79); 193$ **91.** -58 **93.** 34
95. Negative one point eight minus two point seven; -4.5
97. Negative two hundred fifty minus negative four hundred twenty-five; 175 **99.** 41 **101.** -62
103. -139 **105.** 6 **107.** $-7x, -4y$
109. $-5, 3m, -6mn$ **111.** $-3x$ **113.** $-5a + 4$
115. $-7n - 9$ **117.** $-6x + 5$ **119.** $-7 - 8t$
121. $8 - 12x$ **123.** $15x + 66$ **125.** 100°F
127. 14,494 ft above sea level **129.** 1767 m
131. 432 ft^2 **133.** ◆ **135.** ◆
137. False. For example, let $m = -3$ and $n = -5$. Then $-3 > -5$, but $-3 + (-5) = -8 \not> 0$.
139. False. For example, let $m = 2$ and $n = -2$. Then 2 and -2 are opposites, but $2 - (-2) = 4 \neq 0$. **141.** ◆

EXERCISE SET 1.7, PP. 49–50
1. -27 **3.** -56 **5.** -24 **7.** -72 **9.** 16 **11.** 45
13. -120 **15.** -238 **17.** 1200 **19.** 98 **21.** -78

23. 21.7 **25.** $-\frac{2}{5}$ **27.** $\frac{1}{12}$ **29.** -17.01 **31.** $-\frac{5}{12}$
33. 252 **35.** -90 **37.** $\frac{1}{28}$ **39.** 150 **41.** 0
43. -720 **45.** $-30,240$ **47.** -6 **49.** -4 **51.** -2
53. 4 **55.** -8 **57.** 2 **59.** -12 **61.** -8
63. Undefined **65.** $-\frac{88}{9}$ **67.** 0 **69.** Indeterminate
71. $-\frac{8}{3}, \frac{8}{-3}$ **73.** $-\frac{29}{35}, \frac{-29}{35}$ **75.** $\frac{-7}{3}, \frac{7}{-3}$
77. $-\frac{x}{2}, \frac{x}{-2}$ **79.** $-\frac{5}{4}$ **81.** $\frac{13}{47}$ **83.** $-\frac{1}{10}$ **85.** $\frac{1}{4.3}$
87. $-\frac{4}{9}$ **89.** $-\frac{1}{1}$, or -1 **91.** $\frac{21}{20}$ **93.** $\frac{12}{55}$ **95.** -1
97. $\frac{18}{7}$ **99.** $-\frac{8}{11}$ **101.** $-\frac{7}{4}$ **103.** -12 **105.** -3
107. $\frac{2}{3}$ **109.** 7 **111.** $-\frac{7}{9}$ **113.** $-\frac{7}{10}$ **115.** $-\frac{7}{6}$
117. $\frac{6}{7}$ **119.** $-\frac{14}{15}$ **121.** $\frac{22}{39}$ **123.** ◆ **125.** ◆
127. For 2 and 3, the reciprocal of the sum is $1/(2 + 3)$, or $1/5$. But $1/5 \neq 1/2 + 1/3$. **129.** Negative
131. Negative **133.** Negative **135.** (a) m and n have different signs; (b) either m or n is zero; (c) m and n have the same sign **137.** ◆

EXERCISE SET 1.8, PP. 56–57
1. 17^3 **3.** x^7 **5.** $(6y)^4$ **7.** 81 **9.** 9 **11.** -1
13. 64 **15.** 625 **17.** 7 **19.** $16x^4$ **21.** $-343x^3$
23. 26 **25.** 86 **27.** 7 **29.** 5 **31.** 12 **33.** 250
35. -7 **37.** -7 **39.** -4 **41.** -85 **43.** 14
45. 1880 **47.** 16 **49.** 1 **51.** -26 **53.** 37 **55.** -6
57. 4 **59.** 36 **61.** -6 **63.** -3 **65.** 45 **67.** 9
69. $\frac{55}{4}$ **71.** 6 **73.** -17 **75.** $-9x - 1$ **77.** $-7 + 2x$
79. $-4a + 3b - 7c$ **81.** $-3x^2 - 5x + 1$ **83.** $3x - 7$
85. $-3a + 9$ **87.** $5x - 6$ **89.** $-3t - 11r$
91. $9y - 25z$ **93.** $x^2 + 2$ **95.** $-t^3 - 2t$
97. $37a^2 - 23ab + 35b^2$ **99.** $-22t^3 - t^2 + 9t$
101. $-8x + 34$ **103.** $4a^3 + 2a - 10$ **105.** $10x - 24$
107. $2x + 9$ **109.** ◆ **111.** ◆
113. $-5t - 6r + 21$ **115.** $-2x - f$ **117.** ◆
119. True **121.** True **123.** False **125.** True

REVIEW EXERCISES: CHAPTER 1, PP. 59–60
1. [1.1] 15 **2.** [1.1] 6 **3.** [1.1] 5 **4.** [1.1] 4
5. [1.8] -15 **6.** [1.8] -5 **7.** [1.1] $z - 8$ **8.** [1.1] xz
9. [1.1] Let m and n represent the numbers; $mn + 1$, or $1 + mn$ **10.** [1.1] No
11. [1.1] Let c represent the number of countries that participated in the 1988 Olympics; $c + 37 = 197$
12. [1.2] $x \cdot 2 + y$ **13.** [1.2] $2x + (y + z)$
14. [1.2] $4(yx), (4y)x, y(4x)$ **15.** [1.2] $18x + 30y$

16. [1.2] $40x + 24y + 16$ **17.** [1.2] $3(7x + 5y)$
18. [1.2] $7(5x + 2 + y)$ **19.** [1.3] $2 \cdot 2 \cdot 13$ **20.** [1.3] $\frac{5}{12}$
21. [1.3] $\frac{9}{4}$ **22.** [1.3] $\frac{31}{36}$ **23.** [1.3] $\frac{3}{16}$ **24.** [1.3] $\frac{3}{5}$
25. [1.3] $\frac{72}{25}$ **26.** [1.4] $-45, 72$
27. [1.4]
$\xleftarrow{\hspace{1cm}\bullet\hspace{2cm}}_{-5\,-4\,-3\,-2\,-1\ 0\ 1\ 2\ 3\ 4\ 5}^{\ \ \ \ \ \ \ \ \ \ \ \ \frac{-1}{3}}$
28. [1.4] $x > -3$
29. [1.4] True **30.** [1.4] False **31.** [1.4] -0.875
32. [1.4] 1 **33.** [1.6] -5 **34.** [1.5] -3 **35.** [1.5] $-\frac{7}{12}$
36. [1.5] -4 **37.** [1.5] -5 **38.** [1.6] 4 **39.** [1.6] $-\frac{7}{5}$
40. [1.6] -7.9 **41.** [1.7] 54 **42.** [1.7] -9.18
43. [1.7] $-\frac{2}{7}$ **44.** [1.7] -210 **45.** [1.7] -7
46. [1.7] -3 **47.** [1.7] $\frac{3}{4}$ **48.** [1.8] 92 **49.** [1.8] 62
50. [1.8] 48 **51.** [1.8] 168 **52.** [1.8] $\frac{21}{8}$ **53.** [1.8] $\frac{103}{17}$
54. [1.5] $7a - 3b$ **55.** [1.6] $-2x + 5y$ **56.** [1.6] 7
57. [1.7] $-\frac{1}{7}$ **58.** [1.8] $(2x)^4$ **59.** [1.8] $-27y^3$
60. [1.8] $-3a + 9$ **61.** [1.8] $-2b + 21$
62. [1.8] $-3x + 9$ **63.** [1.8] $12y - 34$
64. [1.8] $5x + 24$ **65.** [1.8] $-15x + 25$
66. [1.1] ◆ The value of a constant never varies. A variable can represent a variety of numbers.
67. [1.2] ◆ A term is one of the parts of an expression that is separated from the other parts by plus signs. A factor is part of a product.
68. [1.2], [1.5], [1.8] ◆ The distributive law is used in factoring algebraic expressions, multiplying algebraic expressions, collecting like terms, finding the opposite of a sum, and subtracting algebraic expressions.
69. [1.8] ◆ A negative quantity raised to an even power is positive; a negative quantity raised to an odd power is negative. **70.** [1.8] 25,281 **71.** [1.4] (a) $\frac{3}{11}$; (b) $\frac{10}{11}$
72. [1.8] $-\frac{5}{8}$ **73.** [1.8] -2.1

TEST: CHAPTER 1, P. 61

1. [1.1] 6 **2.** [1.1] Let x represent the number; $x - 9$
3. [1.1] 240 ft^2 **4.** [1.2] $q + 3p$ **5.** [1.2] $(x \cdot 4) \cdot y$
6. [1.1] Yes **7.** [1.1] Let p represent the maximum production capability; $p - 282 = 2518$ **8.** [1.2] $18 - 3x$
9. [1.2] $-5y + 5$ **10.** [1.2] $11(1 - 4x)$
11. [1.2] $7(x + 3 + 2y)$ **12.** [1.3] $2 \cdot 2 \cdot 3 \cdot 5 \cdot 5$
13. [1.3] $\frac{2}{7}$ **14.** [1.4] $<$ **15.** [1.4] $>$ **16.** [1.4] $\frac{9}{4}$
17. [1.4] 2.7 **18.** [1.6] $-\frac{2}{3}$ **19.** [1.7] $-\frac{7}{4}$ **20.** [1.6] 8
21. [1.4] $-2 \geq x$ **22.** [1.6] 7.8 **23.** [1.5] -8
24. [1.5] $\frac{7}{40}$ **25.** [1.6] 10 **26.** [1.6] -2.5 **27.** [1.6] $\frac{7}{8}$
28. [1.7] -48 **29.** [1.7] $\frac{3}{16}$ **30.** [1.7] -9 **31.** [1.7] $\frac{3}{4}$
32. [1.7] -9.728 **33.** [1.8] -173 **34.** [1.6] 12
35. [1.8] -4 **36.** [1.8] 448 **37.** [1.6] $22y + 21a$
38. [1.8] $16x^4$ **39.** [1.8] $2x + 7$ **40.** [1.8] $9a - 12b - 7$
41. [1.8] $68y - 8$ **42.** [1.1] 15 **43.** [1.3] $\frac{23}{70}$
44. [1.8] 15 **45.** [1.8] $4a$

CHAPTER 2

EXERCISE SET 2.1, PP. 69–70

1. 13 **3.** -20 **5.** -14 **7.** -13 **9.** 15 **11.** -14
13. -6 **15.** 20 **17.** -6 **19.** $\frac{7}{3}$ **21.** $-\frac{13}{10}$ **23.** $\frac{41}{24}$
25. $-\frac{1}{20}$ **27.** 1.5 **29.** -5 **31.** 16 **33.** 9
35. 12 **37.** -23 **39.** 8 **41.** -7 **43.** -6 **45.** 6
47. 48 **49.** 36 **51.** -21 **53.** $-\frac{8}{9}$ **55.** $\frac{3}{2}$ **57.** $\frac{9}{2}$
59. -5.9 **61.** -2.5 **63.** -12 **65.** $-\frac{1}{2}$ **67.** -15
69. 24 **71.** $7x$ **73.** $x - 4$ **75.** ◆ **77.** ◆
79. 342.246 **81.** All real numbers **83.** 12, -12
85. All real numbers **87.** No solution **89.** 9.4
91. 6 **93.** 8 **95.** 11,074 **97.** ◆

EXERCISE SET 2.2, PP. 76–77

1. 9 **3.** 8 **5.** 7 **7.** 14 **9.** -8 **11.** -5 **13.** -7
15. 15 **17.** 3 **19.** -12 **21.** 6 **23.** 8 **25.** 1
27. -20 **29.** 6 **31.** 7 **33.** 3 **35.** 2 **37.** 10
39. 4 **41.** 0 **43.** $-\frac{2}{5}$ **45.** $\frac{64}{3}$ **47.** $\frac{2}{5}$ **49.** 3
51. -4 **53.** 0.8 **55.** $-\frac{40}{37}$ **57.** 2 **59.** 2
61. 6 **63.** 8 **65.** 2 **67.** -4 **69.** -8 **71.** 2
73. 6 **75.** $\frac{17}{6}$ **77.** $-\frac{15}{4}$ **79.** $\frac{11}{18}$ **81.** $-\frac{51}{31}$ **83.** 2
85. -6.5 **87.** $<$ **89.** ◆ **91.** ◆ **93.** 1.2497
95. -4 **97.** All real numbers **99.** $\frac{2}{3}$ **101.** 0
103. $\frac{52}{45}$

EXERCISE SET 2.3, PP. 81–83

1. 57,000 Btu's **3.** 54 in^2 **5.** 3450 watts **7.** 255 mg
9. $b = \dfrac{A}{h}$ **11.** $r = \dfrac{d}{t}$ **13.** $P = \dfrac{I}{rt}$ **15.** $m = 65 - H$
17. $l = \dfrac{P - 2w}{2}$ **19.** $\pi = \dfrac{A}{r^2}$ **21.** $h = \dfrac{2A}{b}$
23. $m = \dfrac{E}{c^2}$ **25.** $d = 2Q - c$ **27.** $b = 3A - a - c$
29. $A = Ms$ **31.** $y = \dfrac{C - Ax}{B}$ **33.** $S = \dfrac{360A}{\pi r^2}$
35. $h = \dfrac{2A}{a + b}$ **37.** -42 **39.** -29 **41.** ◆
43. ◆ **45.** $K = 917 + 6\left(\dfrac{w}{2.2046} + \dfrac{h}{0.3937} - a\right)$
47. $a = \dfrac{wd}{c}$ **49.** $T = t - \dfrac{h}{100}, 0 \leq h \leq 12{,}000$
51. $c = \dfrac{d}{a - b}$ **53.** $a = \dfrac{c}{3 + b + d}$

EXERCISE SET 2.4, PP. 88–89

1. 0.32 3. 0.07 5. 0.241 7. 0.0046 9. 454%
11. 99.8% 13. 200% 15. 134% 17. 0.68%
19. 37.5% 21. 28% 23. $66.\overline{6}$%, or $66\frac{2}{3}$% 25. 52%
27. 24% 29. 60 31. 2.5 33. 273 35. 125%
37. 0.8 39. 5% 41. 285 43. About 92.4 million
45. About 60% 47. 7410 49. 7% 51. 52%
53. $940 55. $138.95 57. $14.40 per hour
59. About 165 calories 61. 0.68 63. −13.2
65. ◆ 67. ◆ 69. About 77% 71. About 35%
73. 20%

EXERCISE SET 2.5, PP. 98–100

1. 11 3. 11 5. $90 7. 15, 16, 17 9. 24, 26
11. 63, 65 13. $699\frac{1}{3}$ mi
15. Bathrooms: 11\frac{2}{3}$ billion; kitchens: 23\frac{1}{3}$ billion
17. First: 22.5°; second: 90°; third: 67.5° 19. 70°
21. 142 and 143 23. 63 mm, 65 mm, 67 mm
25. Width: 165 ft, length: 265 ft; area: 43,725 ft^2
27. Width: 275 mi, length: 365 mi 29. $6600
31. 450.5 mi 33. 65°, 25° 35. 2020
37. $5(a + 2b − 9)$ 39. $13x − 12$ 41. ◆ 43. ◆
45. 37 47. If Ed intends to drive fewer than 465 mi, the Reston rental is less expensive. If he drives more than 465 mi, the Long Haul rental is less expensive.
49. $4x + 7 = 1863 − 1776$; 20 yr 51. 87°, 89°, 91°, 93°
53. 120 55. $600 57. 0.726175 in. 59. 10 61. ◆

EXERCISE SET 2.6, PP. 107–108

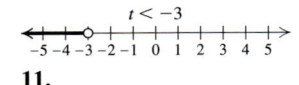

1. (a) Yes; (b) yes; (c) no; (d) no; (e) yes
3. (a) No; (b) no; (c) yes; (d) yes; (e) no
5. $x \geq 6$
7. $t < -3$
9. $m > -4$
11. $-3 < x \leq 5$
13. $0 < x < 3$
15. $\{x \mid x > -1\}$
17. $\{x \mid x \leq 2\}$ 19. $\{x \mid x < -2\}$ 21. $\{x \mid x \geq 0\}$
23. $\{y \mid y > 6\}$, 
25. $\{x \mid x \leq -18\}$, 
27. $\{x \mid x < 10\}$,
29. $\{x \mid x \geq 8\}$, 
31. $\{y \mid y > -5\}$, 
33. $\{x \mid x \leq 5\}$, 
35. $\{x \mid x \geq 5\}$ 37. $\{y \mid y \leq \frac{1}{2}\}$ 39. $\{x \mid x > \frac{5}{8}\}$
41. $\{x \mid x < 0\}$
43. $\{x \mid x < 7\}$,
45. $\{y \mid y \leq 9\}$, 
47. $\{x \mid x > -\frac{13}{7}\}$, 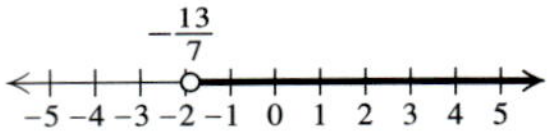
49. $\{x \mid x > -3\}$, 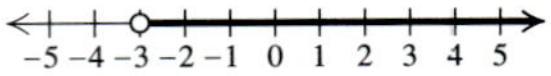
51. $\{y \mid y \geq -\frac{2}{7}\}$ 53. $\{x \mid x > \frac{17}{5}\}$ 55. $\{y \mid y \geq -\frac{1}{12}\}$
57. $\{x \mid x > \frac{4}{5}\}$ 59. $\{x \mid x < 9\}$ 61. $\{y \mid y \geq 4\}$
63. $\{x \mid x \leq 6\}$ 65. $\{x \mid x < -3\}$ 67. $\{y \mid y < -4\}$
69. $\{x \mid x > -4\}$ 71. $\{y \mid y < -\frac{10}{3}\}$ 73. $\{x \mid x > -10\}$
75. $\{y \mid y < 2\}$ 77. $\{y \mid y \geq 3\}$ 79. $\{x \mid x > -4\}$
81. $\{x \mid x > -4\}$ 83. $\{n \mid n \geq 70\}$ 85. $\{x \mid x \leq 15\}$
87. $\{y \mid y \leq -7\}$ 89. $\{y \mid y < 6\}$ 91. $\{t \mid t \leq -4\}$
93. $\{r \mid r > -3\}$ 95. $\{x \mid x \geq 8\}$ 97. $\{x \mid x < \frac{11}{18}\}$
99. 140 101. $-2x − 23$ 103. ◆ 105. ◆
107. $\{t \mid t > -\frac{27}{19}\}$ 109. $\{x \mid x \leq -4a\}$
111. $\{x \mid x > \frac{y-b}{a}\}$ 113. (a) No; (b) yes; (c) no; (d) yes; (e) no; (f) yes

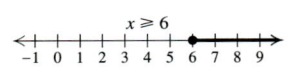

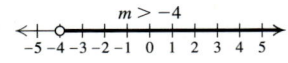

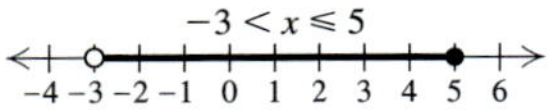

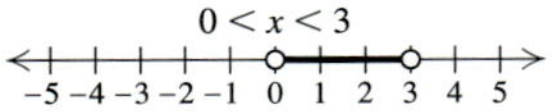

EXERCISE SET 2.7, PP. 111–114

1. Let n represent the number; $n < 9$ 3. Let b represent the weight of the bag; $b \geq 2$ 5. Let s represent the average speed; $90 < s < 110$ 7. Let a represent the number of people attending; $a \leq 1{,}200{,}000$
9. Let c represent the cost, per gallon, of gasoline; $c \geq \$1.20$ 11. Scores of 84 and higher
13. Mileages less than or equal to 525.8 mi
15. 5 min or more 17. 4 servings or more 19. 437.5 ft or less 21. Less than 21.5 cm 23. The blue-book value is at least $10,625. 25. Years after 1934 27. When the puppy is more than 18 weeks old 29. Temperatures greater than 37°C 31. 21 or more 33. Widths greater than 14 cm 35. Lengths greater than 6 cm
37. George: more than 12 hr; Joan: more than 15 hr
39. At most 9 lb 41. There are at least $5\frac{1}{3}$ g of fat.
43. 9 cm or more 45. 2001 and beyond 47. −160
49. $4x^2 − 4x − 7$ 51. $−3ab + 9b − 12a$ 53. ◆
55. ◆ 57. 47 and 49 59. Between −15°C and $-9\frac{4}{9}$°C 61. Less than $20,000 63. The fat content is at least 7.5 g. 65. ◆

REVIEW EXERCISES: CHAPTER 2, PP. 116–117

1. [2.1] −22 2. [2.1] 7 3. [2.1] −192 4. [2.1] 1
5. [2.1] $-\frac{7}{3}$ 6. [2.1] 1.11 7. [2.1] $\frac{1}{2}$ 8. [2.1] $-\frac{15}{64}$
9. [2.2] $\frac{38}{5}$ 10. [2.1] −8 11. [2.2] −5 12. [2.2] $-\frac{1}{3}$

13. [2.2] 4   14. [2.2] 3   15. [2.2] 4   16. [2.2] 16
17. [2.2] 6   18. [2.2] −3   19. [2.2] 12   20. [2.2] 4
21. [2.3] $d = \dfrac{C}{\pi}$ 22. [2.3] $B = \dfrac{3V}{h}$
23. [2.3] $a = 2A - b$ 24. [2.4] 0.001   25. [2.4] 44%
26. [2.4] 20%   27. [2.4] 360   28. [2.6] Yes
29. [2.6] No   30. [2.6] Yes
31. [2.6]   32. [2.6]

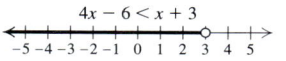

$4x - 6 < x + 3$
-5-4-3-2-1 0 1 2 3 4 5

$-2 < x \leq 5$
-5-4-3-2-1 0 1 2 3 4 5

33. [2.6]

$y > 0$
-5-4-3-2-1 0 1 2 3 4 5

34. [2.6] $\{y \mid y \geq -\tfrac{1}{2}\}$ 35. [2.6] $\{x \mid x \geq 7\}$
36. [2.6] $\{y \mid y > 2\}$ 37. [2.6] $\{y \mid y \leq -4\}$
38. [2.6] $\{x \mid x < -11\}$ 39. [2.6] $\{y \mid y > -7\}$
40. [2.6] $\{x \mid x > -6\}$ 41. [2.6] $\{x \mid x > -\tfrac{9}{11}\}$
42. [2.6] $\{y \mid y \leq 7\}$ 43. [2.6] $\{x \mid x \geq -\tfrac{1}{12}\}$
44. [2.5] $317   45. [2.5] 20   46. [2.5] 3 m, 5 m
47. [2.5] 5880   48. [2.5] 57, 59
49. [2.5] Width: 11 cm; length: 17 cm   50. [2.5] $160
51. [2.5] $17.5 billion   52. [2.5] 35°, 85°, 60°
53. [2.7] At most $35   54. [2.7] Widths greater than 17 cm   55. [1.1] 5   56. [1.2] $12t + 8 + 4s$
57. [1.7] −2.3   58. [1.8] $-43x + 8y$
59. [2.1], [2.6] ◆ Multiplying on both sides of an equation by *any* nonzero number results in an equivalent equation. When multiplying on both sides of an inequality, the sign of the number being multiplied by must be considered. If the number is positive, the direction of the inequality symbol remains unchanged; if the number is negative, the direction of the inequality symbol must be reversed to produce an equivalent inequality.
60. [2.1], [2.6] ◆ The solutions of an equation can usually each be checked. The solutions of an inequality are normally too numerous to check. Checking a few numbers from the solution set found cannot guarantee that the answer is correct, although if any number does not check, the answer found is incorrect.   61. [2.5] Amazon: 6437 km; Nile: 6671 km   62. [2.5] $14,150
63. [1.4], [2.2] 23, −23   64. [1.4], [2.1] 20, −20
65. [2.3] $a = \dfrac{y - 3}{2 - b}$

TEST: CHAPTER 2, PP. 117–118
1. [2.1] 8   2. [2.1] 26   3. [2.1] −6   4. [2.1] 49
5. [2.2] −12   6. [2.2] 2   7. [2.1] −8   8. [2.1] $-\tfrac{7}{20}$
9. [2.2] 7   10. [2.2] −5   11. [2.2] $\tfrac{23}{3}$
12. [2.6] $\{x \mid x \leq -4\}$ 13. [2.6] $\{x \mid x > -13\}$
14. [2.6] $\{x \mid x < \tfrac{21}{8}\}$ 15. [2.6] $\{y \mid y \leq -13\}$

16. [2.6] $\{y \mid y \leq -8\}$ 17. [2.6] $\{x \mid x \leq -\tfrac{1}{20}\}$
18. [2.6] $\{x \mid x < -6\}$ 19. [2.6] $\{x \mid x \leq -1\}$
20. [2.3] $r = \dfrac{A}{2\pi h}$ 21. [2.3] $l = \dfrac{2w - P}{-2}$ 22. [2.4] 5
23. [2.4] 5.4%   24. [2.4] 21   25. [2.4] 44%
26. [2.6]   27. [2.6]

$y < 9$
-10-8-6-4-2 0 2 4 6 8 10

$-2 \leq x \leq 2$
-5-4-3-2-1 0 1 2 3 4 5

28. [2.5] Width: 7 cm; length: 11 cm   29. [2.5] 30 mi
30. [2.5] 81 mm, 83 mm, 85 mm   31. [2.5] $2700
32. [2.7] All numbers greater than 6
33. [2.7] Lengths of at least 174 yd   34. [1.1] $x - 10$
35. [1.2] $3(a + 8b + 4)$ 36. [1.7] $\tfrac{3}{10}$ 37. [1.8] 5
38. [2.3] $d = \dfrac{ca - 1}{c}$ 39. [1.4], [2.2] 15, −15
40. [2.5] 60

CHAPTER 3

EXERCISE SET 3.1, PP. 125–128
1. 5   3. The person weighs at least 120 lb.   5. $1010.88
7. $8170.75   9. 19.2 million tons   11. 0.07 lb   13. 4%
15. 1982   17. About $123 billion (using 4.5%)
19. $1016 billion (using 5%)
21.   23.

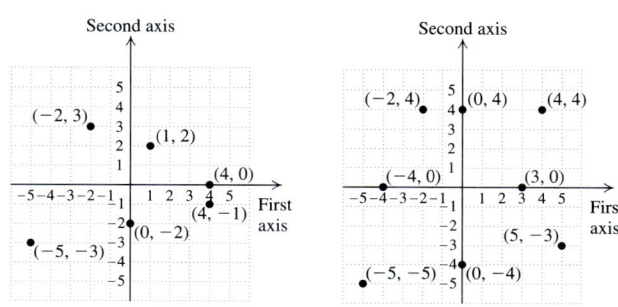

[Graph 21: points plotted at (−2, 3), (1, 2), (4, 0), (4, −1), (0, −2), (−5, −3)]

[Graph 23: points plotted at (−2, 4), (0, 4), (4, 4), (−4, 0), (3, 0), (5, −3), (−5, −5), (0, −4)]

25. A: (−4, 5); B: (−3, −3); C: (0, 4); D: (3, 4); E: (3, −4)
27. A: (4, 1); B: (0, −5); C: (−4, 0); D: (−3, −2); E: (3, 0)
29. II   31. IV   33. III   35. I
37. Negative, negative
39.

[Graph: Percentage of female first-year students from 1970 to 1990, curve rising from about 45 to about 55]

A-6 ANSWERS

41.

[Graph: Red meat consumption (in pounds) vs Year, showing values around 125-126 from 1975-1985, declining to about 115 by 1995]

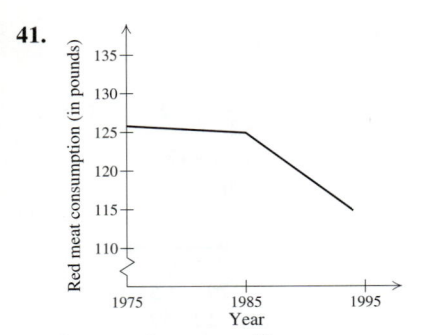

43. $\frac{8}{9}$ **45.** $\frac{2}{3}$ **47.** $-\frac{13}{35}$ **49.** ◆ **51.** ◆
53. II or III **55.** II or IV **57.** $(-1, -5)$
59. **61.** 26 units
63. Latitude 32.5° North, longitude 64.5° West
65. ◆

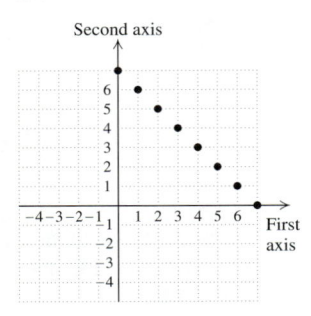

[Graph with Second axis and First axis, showing plotted points]

7.
$$\begin{array}{c|c}
y = x - 5 \\ \hline
2 \; ? \; 7 - 5 \\
2 \; | \; 2 \quad \text{TRUE}
\end{array}
\qquad
\begin{array}{c|c}
y = x - 5 \\ \hline
-4 \; ? \; 1 - 5 \\
-4 \; | \; -4 \quad \text{TRUE}
\end{array}$$
$(0, -5)$; answers may vary

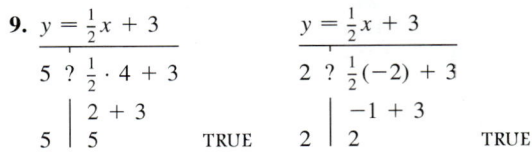

9.
$$\begin{array}{c|c}
y = \tfrac{1}{2}x + 3 \\ \hline
5 \; ? \; \tfrac{1}{2} \cdot 4 + 3 \\
\; | \; 2 + 3 \\
5 \; | \; 5 \quad \text{TRUE}
\end{array}
\qquad
\begin{array}{c|c}
y = \tfrac{1}{2}x + 3 \\ \hline
2 \; ? \; \tfrac{1}{2}(-2) + 3 \\
\; | \; -1 + 3 \\
2 \; | \; 2 \quad \text{TRUE}
\end{array}$$
$(0, 3)$; answers may vary

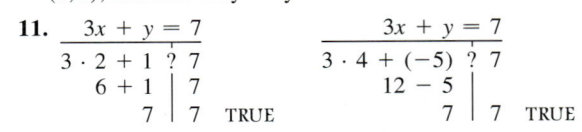

11.
$$\begin{array}{c|c}
3x + y = 7 \\ \hline
3 \cdot 2 + 1 \; ? \; 7 \\
6 + 1 \; | \; 7 \\
7 \; | \; 7 \quad \text{TRUE}
\end{array}
\qquad
\begin{array}{c|c}
3x + y = 7 \\ \hline
3 \cdot 4 + (-5) \; ? \; 7 \\
12 - 5 \; | \; \\
7 \; | \; 7 \quad \text{TRUE}
\end{array}$$
$(1, 4)$; answers may vary

13.
$$\begin{array}{c|c}
4x - 2y = 10 \\ \hline
4 \cdot 0 - 2(-5) \; ? \; 10 \\
0 + 10 \; | \; \\
10 \; | \; 10 \quad \text{TRUE}
\end{array}
\qquad
\begin{array}{c|c}
4x - 2y = 10 \\ \hline
4 \cdot 4 - 2 \cdot 3 \; ? \; 10 \\
16 - 6 \; | \; \\
10 \; | \; 10 \quad \text{TRUE}
\end{array}$$
$(2, -1)$; answers may vary

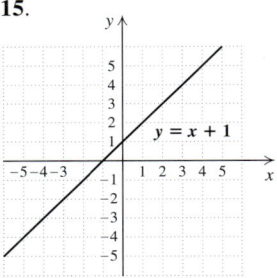

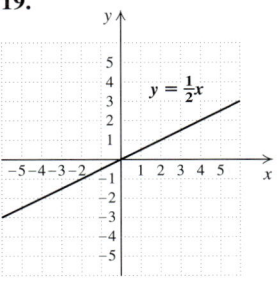

15. [Graph of $y = x + 1$] **17.** [Graph of $y = x$]

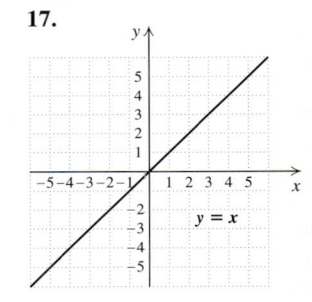

19. [Graph of $y = \tfrac{1}{2}x$] **21.** [Graph of $y = x - 3$]

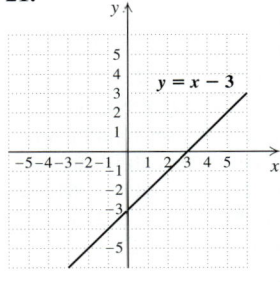

23. [Graph of $y = 3x - 2$] **25.** [Graph of $y = \tfrac{1}{2}x + 1$]

TECHNOLOGY CONNECTION 3.2, P. 135

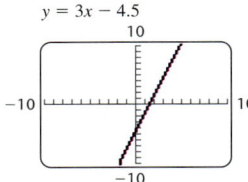

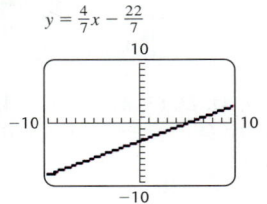

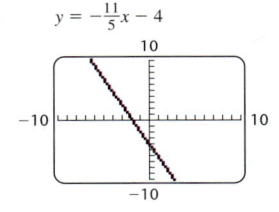

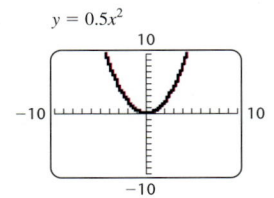

1. $y = -5x + 6.5$ [calculator graph]
2. $y = 3x - 4.5$ [calculator graph]
3. $y = \tfrac{4}{7}x - \tfrac{22}{7}$ [calculator graph]
4. $y = -\tfrac{11}{5}x - 4$ [calculator graph]
5. $y = 0.5x^2$ [calculator graph]
6. $y = 8 - x^2$ [calculator graph]

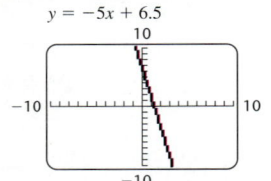

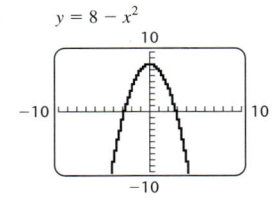

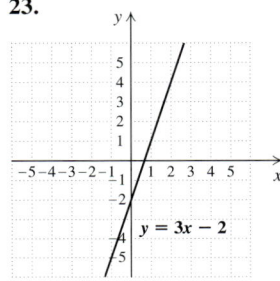

EXERCISE SET 3.2, PP. 135–138
1. No **3.** No **5.** Yes

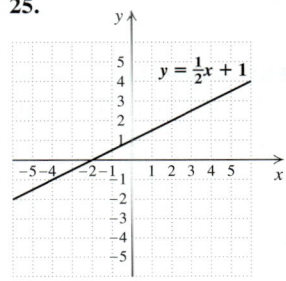

27. [graph of $x+y=-5$]

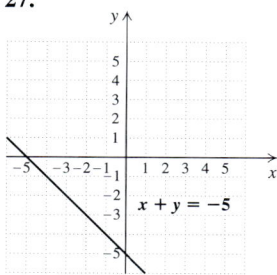

29. [graph of $y=\frac{5}{3}x-2$]

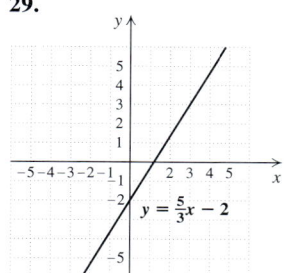

31. [graph of $x+2y=8$]

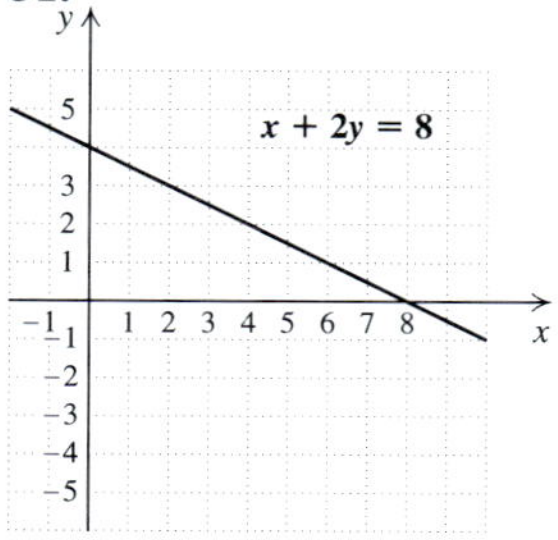

33. [graph of $y=\frac{3}{2}x+1$]

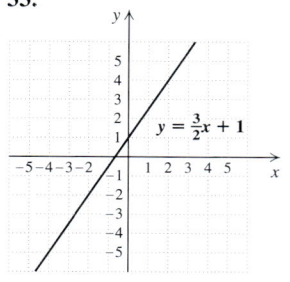

35. [graph of $8x-4y=12$]

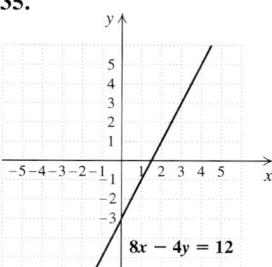

37. [graph of $8y+2x=-4$]

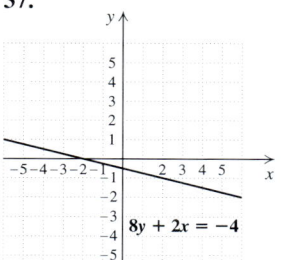

39. [graph: Value of software (in hundreds) vs. Time from date of purchase (in years)]; $300

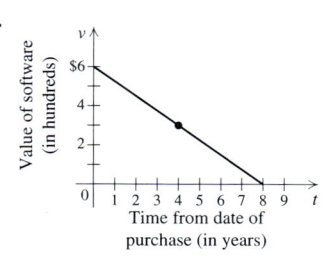

41. [graph: Risk of death of exsmoker over nonsmoker vs. Number of years since smoking stopped]; $12\frac{1}{2}$ times

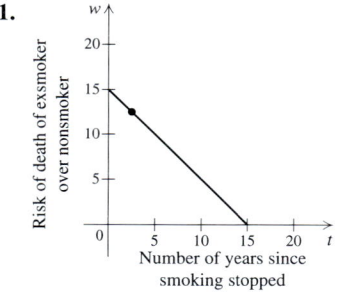

43. [graph: Tuition and fees (in hundreds) vs. Number of credits]; $1500

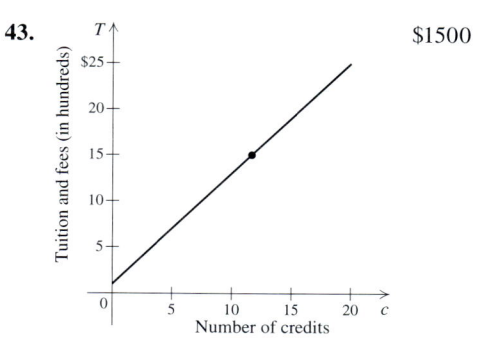

45. [graph: Average tea consumption (in gallons) vs. Number of years since 1991]; 7.6 gal

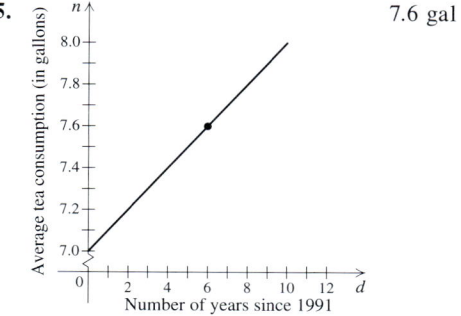

47. -9 **49.** $\dfrac{16}{5}$ **51.** $y=\dfrac{C-Ax}{B}$ **53.** ◆

55. ◆

57. (0, 7), (1, 6), (2, 5), (3, 4), (4, 3), (5, 2), (6, 1), (7, 0); [graph of $x+y=7$]

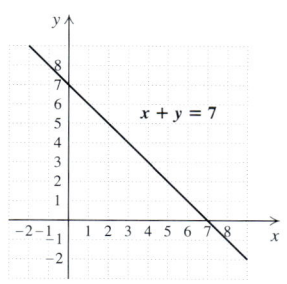

59. ◆ **61.** $y=-x+2$, or $x+y=2$ **63.** $y=3x$

65. [graph of $25d+5l=225$]

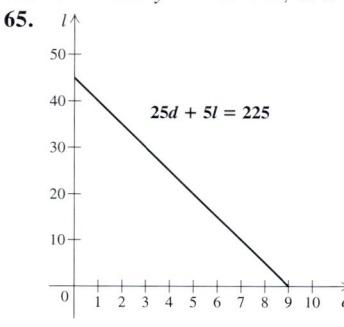

Answers may vary. 1 dinner, 40 lunches; 5 dinners, 20 lunches; 8 dinners, 5 lunches

A-8 ANSWERS

67. [graph of $y = \frac{1}{2}x^3$]

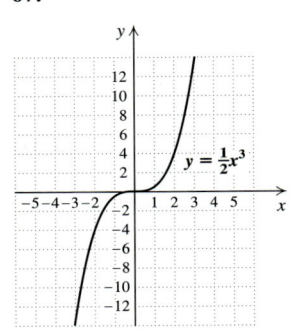

69. [graph of $y = |x| - 3$]

9. (a) $\left(0, \frac{10}{3}\right)$; (b) $\left(-\frac{5}{2}, 0\right)$ **11.** (a) $\left(0, -\frac{1}{3}\right)$; (b) $\left(\frac{1}{2}, 0\right)$

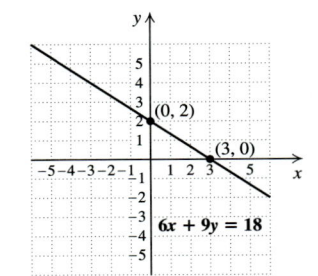

13. [graph of $2x + y = 6$ with (0, 6) and (3, 0)]

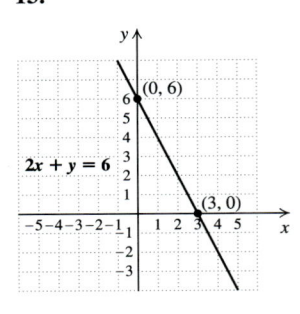

15. [graph of $6x + 9y = 18$ with (0, 2) and (3, 0)]

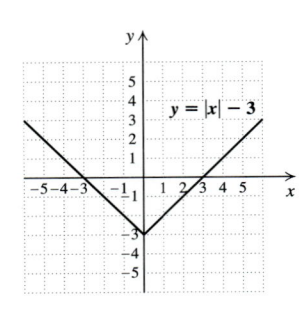

71. $y = -2.8x + 3.5$ [calculator graph]

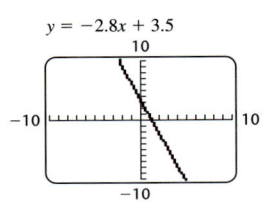

73. $y = 2.8x - 3.5$ [calculator graph]

17. [graph of $-x + 3y = 9$ with (0, 3) and (−9, 0)]

19. [graph of $2x - y = 8$ with (4, 0) and (0, −8)]

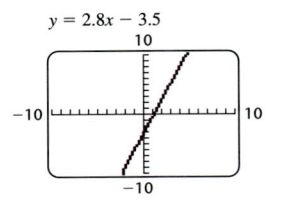

75. $y = x^2 + 4x + 1$ [calculator graph]

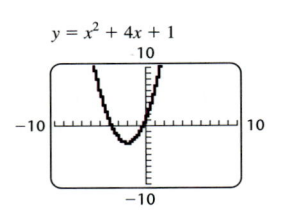

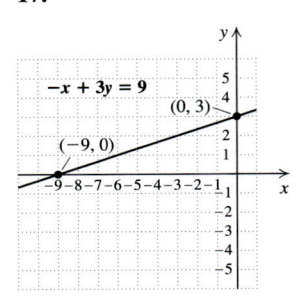

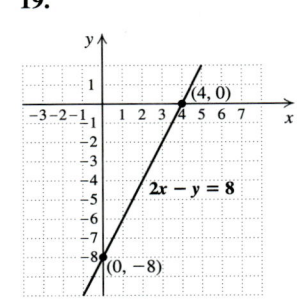

77. ◆

TECHNOLOGY CONNECTION 3.3, P. 141

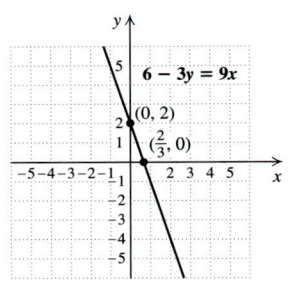

1. $y = -0.72x - 15$ [calculator graph, Xscl = 5, Yscl = 5]

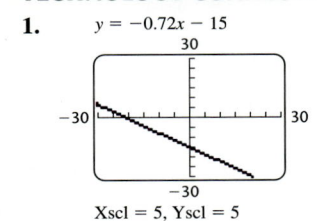

2. $y - 2.13x = 27$, or $y = 2.13x + 27$ [calculator graph, Xscl = 5, Yscl = 5]

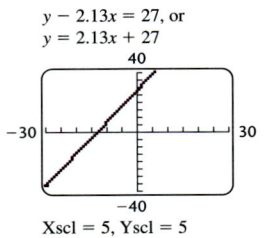

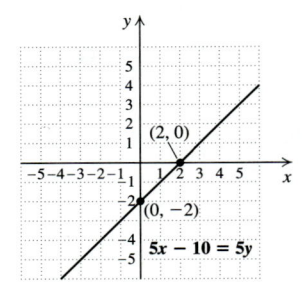

21. [graph of $6 - 3y = 9x$ with (0, 2) and $\left(\frac{2}{3}, 0\right)$]

23. [graph of $5x - 10 = 5y$ with (2, 0) and (0, −2)]

3. $5x + 6y = 84$, or $y = -\frac{5}{6}x + 14$ [calculator graph, Xscl = 5, Yscl = 5]

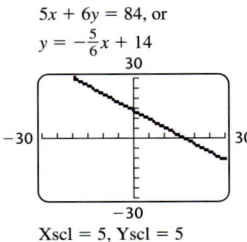

4. $2x - 7y = 150$, or $y = \frac{2}{7}x - \frac{150}{7}$ [calculator graph, Xscl = 10, Yscl = 5]

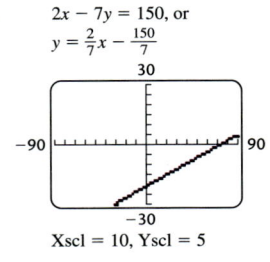

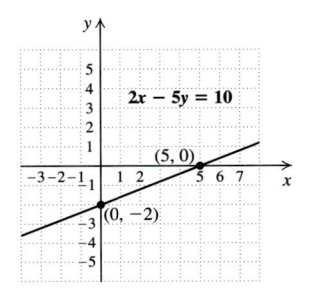

25. [graph of $2x - 5y = 10$ with (5, 0) and (0, −2)]

27. [graph of $2x + 6y = 12$ with (0, 2) and (6, 0)]

EXERCISE SET 3.3, PP. 144–145

1. (a) (0, 5); (b) (2, 0)   **3.** (a) (0, −4); (b) (3, 0)
5. (a) (0, 3); (b) (5, 0)   **7.** (a) (0, −14); (b) (4, 0)

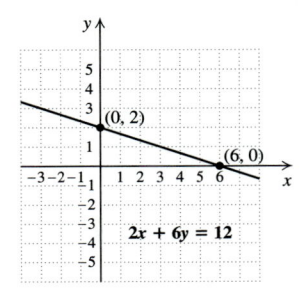

29. [graph: $x - 1 = y$, points (1, 0), (0, −1)]

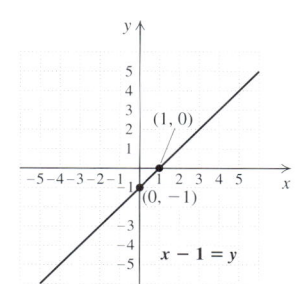

31. [graph: $2x - 1 = y$, points $(\frac{1}{2}, 0)$, (0, −1)]

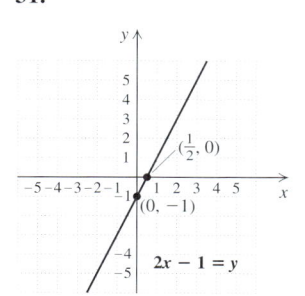

33. [graph: $4x - 3y = 12$, points (3, 0), (0, −4)]

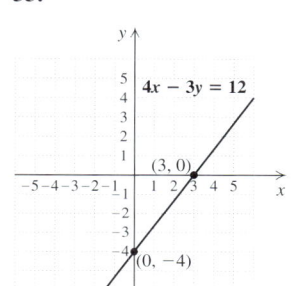

35. [graph: $7x + 2y = 6$, points (0, 3), $(\frac{6}{7}, 0)$]

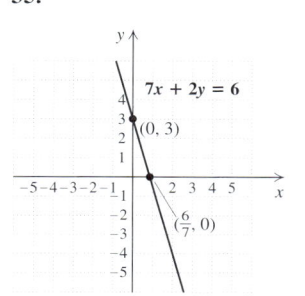

37. [graph: $y = -4 - 4x$, points (−1, 0), (0, −4)]

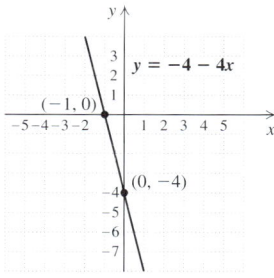

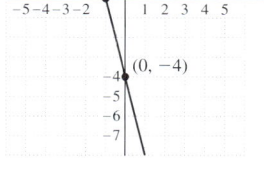

39. [graph: $-3x = 6y - 2$, points $(0, \frac{1}{3})$, $(\frac{2}{3}, 0)$]

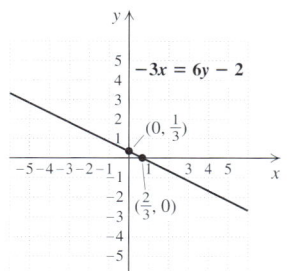

41. [graph: $3 = 2x - 5y$, points $(\frac{3}{2}, 0)$, $(0, -\frac{3}{5})$]

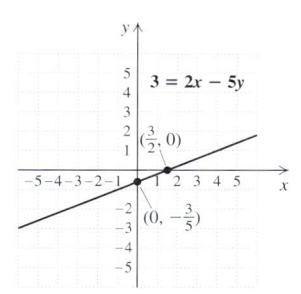

43. [graph: $x + 2y = 0$, point (0, 0)]

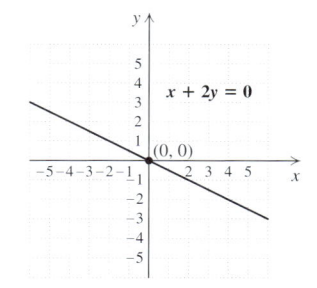

45. $y = -1$ **47.** $x = 4$ **49.** $y = 0$

51. [graph: $x = -3$]

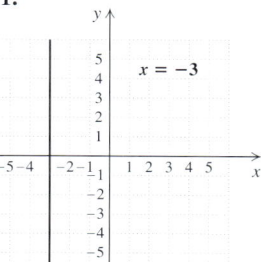

53. [graph: $y = 2$]

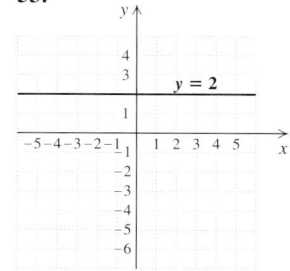

55. [graph: $x = 7$]

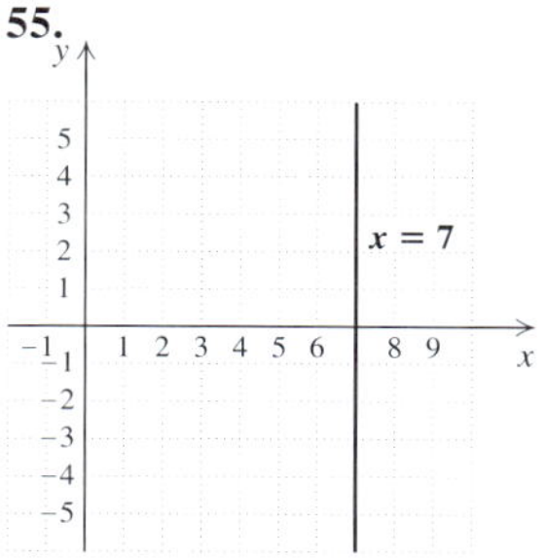

57. [graph: $x = 0$]

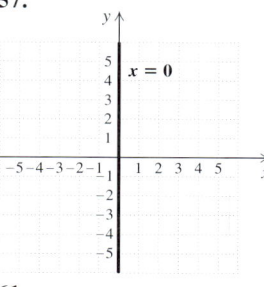

59. [graph: $y = \frac{5}{2}$]

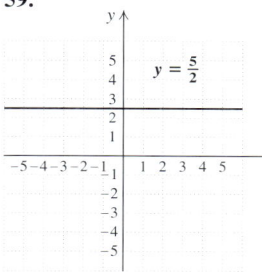

61. [graph: $-3y = -15$]

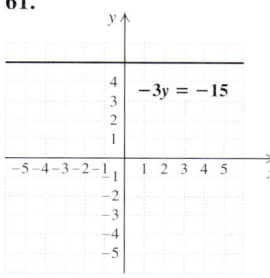

63. [graph: $4x + 3 = 0$]

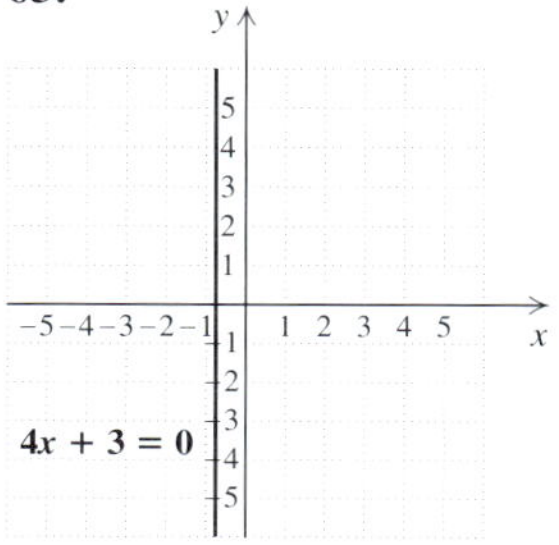

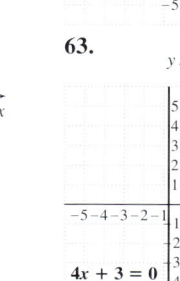

65. [graph: $18 - 3y = 0$]

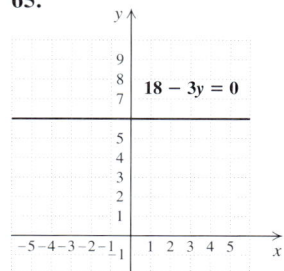

67. $2 \cdot 7 \cdot 7$ **69.** $5 \cdot 5 \cdot 11$ **71.** $\frac{1}{7}$ **73.** ◆   **75.** ◆
77. $y = 0$ **79.** (6, 6)   **81.** $5x + 2y = 10$ **83.** $y = -4$
85. 12   **87.** ◆   **89.** $(0, -11.\overline{428571})$; (40, 0)
91. $(0, -15)$; $(11.\overline{538461}, 0)$ **93.** (0, 0.04); (0.02, 0)

TECHNOLOGY CONNECTION 3.4, P. 148

1. Answers may vary. (12.765957, 54.418085), (51.06383, 67.82234), (85.106383, 79.737234), (123.40426, 93.141489), (157.44681, 105.05638)   **2.** 84.95   **3.** 84.95; 154.95
4. The table becomes blank; 138.5

EXERCISE SET 3.4, PP. 153–156

1. Let x represent one number and y the other; $x + y = 19$
3. Let x represent one number and y the other; $x + 2y = 65$
5. Let a = the amount Justine paid and n = the amount the necklaces were worth; $a = 2n$ **7.** Let d = the distance Jason travels, in meters, and t = the number of minutes he jogs; $d = 220t$ **9.** Let f = Frank's age and c = Cecilia's age; $f = \frac{1}{2}c - 3$ **11.** Let c = the cost of the bike rental and t = the number of hours the bike is rented; $c = 8 + 2t$ **13.** Let d = the depth of the lake, in feet, and w = the number of weeks; $d = 94 - 3w$

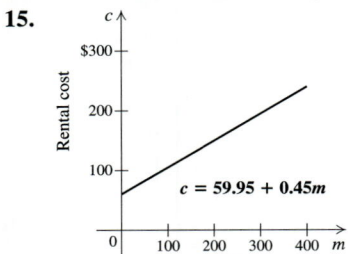

15. About 200 miles

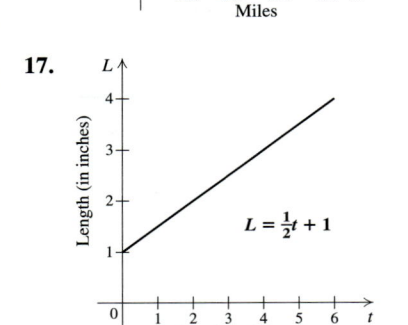

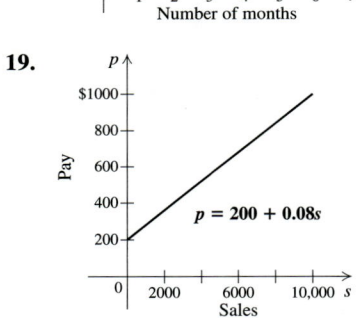

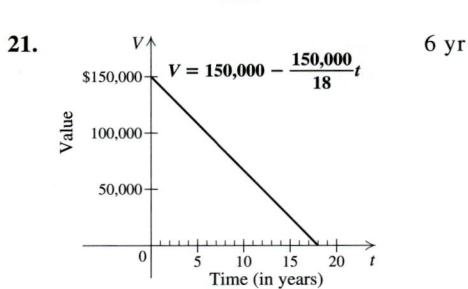

17. $3\frac{1}{2}$ months

19. About $7400

21. 6 yr

23. 135 min, or 2 hr 15 min

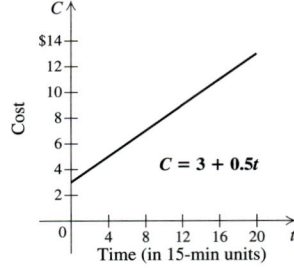

25. About 60 pages

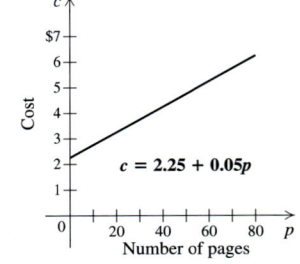

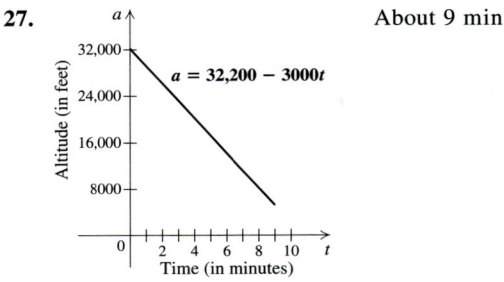

27. About 9 min

29. $\frac{1}{2}$ inch per month   **31.** 1 hr, 5 min; 41 km; about 38 km/h   **33.** 4 hr, 45 min; 235 km; about 49 km/h   **35.** About 41 km/h; 25 min   **37.** The 7:15 P.M. train. The line representing that train is the steepest.   **39.** 3:05 P.M.
41. $t = \dfrac{s - d}{v}$ **43.** $8x + 10y - 6z$ **45.** ◆   **47.** ◆
49. Answers may vary.

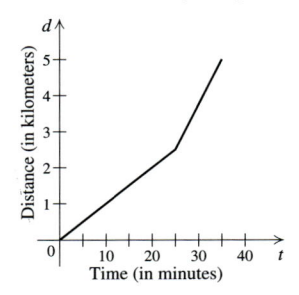

51.

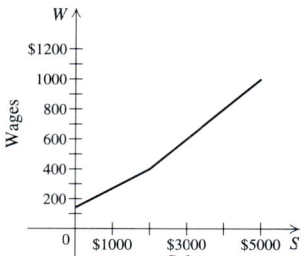

[Graph: Wages vs Sales, line from ~$150 at $0 rising to ~$1000 at $5000]

About $550

53. ◆

55.

[Step graph: Cost vs Time (in 15-min units)]

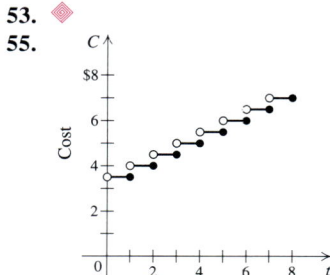

57. $4315.14

EXERCISE SET 3.5, PP. 164–168

1. 3 million people per year **3.** Going down at $10,000,000,000 per year **5.** $800 per year **7.** $\frac{3}{4}$ **9.** $\frac{3}{2}$
11. $\frac{1}{3}$ **13.** -1 **15.** $-\frac{3}{2}$ **17.** -2 **19.** Undefined
21. $-\frac{4}{3}$ **23.** $-\frac{4}{5}$ **25.** 7 **27.** $-\frac{2}{3}$ **29.** $-\frac{1}{2}$ **31.** 0
33. Undefined **35.** Undefined **37.** 0 **39.** Undefined
41. 0 **43.** 8% **45.** $8.\overline{3}\%$ **47.** $\frac{12}{41}$, or about 29%
49. About 29% **51.** -116 **53.** -1 **55.** $\frac{1}{8}$ **57.** ◆
59. ◆ **61.** $\{m \mid -\frac{6}{5} \leq m \leq 0\}$ **63.** $\frac{18-x}{x}$ **65.** $\frac{1}{2}$
67. 3.6 bushels per hour **69.** ◆

REVIEW EXERCISES, CHAPTER 3, PP. 169–171

1. [3.1] 48 **2.** [3.1] 405,000
3.–5. [3.1]

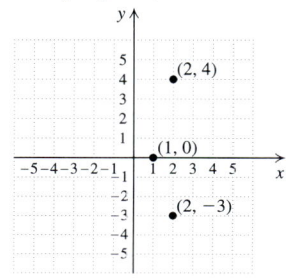

[Graph showing points (2, 4), (1, 0), (2, −3)]

6. [3.1] IV **7.** [3.1] III **8.** [3.1] II **9.** [3.1] $(-5, -1)$
10. [3.1] $(-2, 5)$ **11.** [3.1] $(3, 0)$ **12.** [3.2] No
13. [3.2] Yes

14. [3.2]
$$\begin{array}{c|c} 2x - y = 3 \\ \hline 2 \cdot 0 - (-3) \,?\, 3 \\ 0 + 3 \\ 3 \mid 3 \text{ TRUE} \end{array} \qquad \begin{array}{c|c} 2x - y = 3 \\ \hline 2 \cdot 2 - 1 \,?\, 3 \\ 4 - 1 \\ 3 \mid 3 \text{ TRUE} \end{array}$$

$(-1, -5)$; answers may vary

15. [3.2] [Graph: $y = x - 5$]

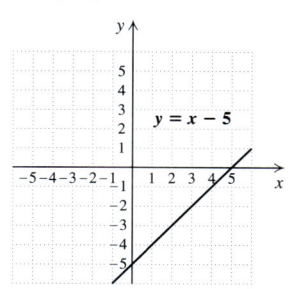

16. [3.2] [Graph: $y = -\frac{1}{4}x$]

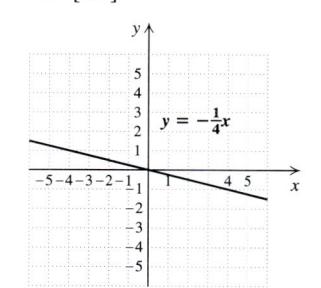

17. [3.2] [Graph: $y = -x + 4$]

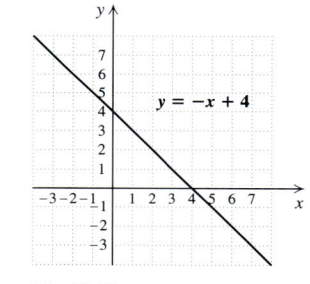

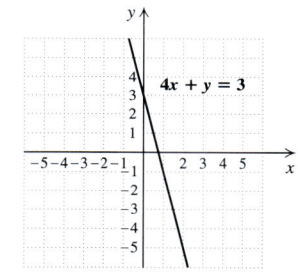

18. [3.2] [Graph: $4x + y = 3$]

19. [3.3] [Graph: $4x + 5 = 3$]

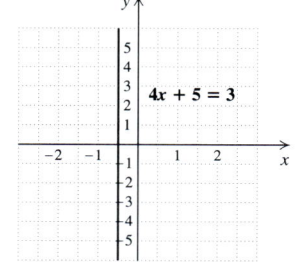

20. [3.3] [Graph: $5x - 2y = 10$]

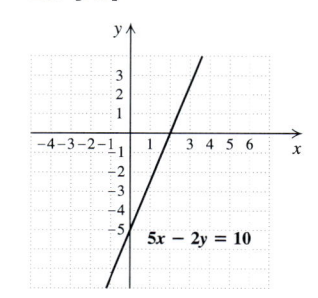

21. [3.3] x-intercept: $(6, 0)$; y-intercept: $(0, -3)$
22. [3.3] x-intercept: $\left(-\frac{9}{2}, 0\right)$; y-intercept: $\left(0, \frac{9}{4}\right)$
23. [3.3] $y = 4$ **24.** [3.4] Let w represent the number of wrenches and s the number of screwdrivers; $w + s = 15$
25. [3.4]

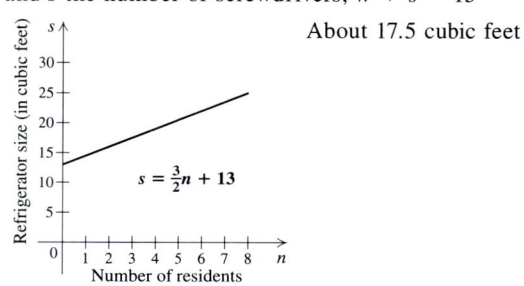

[Graph: Refrigerator size (in cubic feet) vs Number of residents, $s = \frac{3}{2}n + 13$]

About 17.5 cubic feet

A-12 ANSWERS

26. [3.4] No **27.** [3.4] 53.6 mph **28.** [3.5] 0
29. [3.5] $\frac{7}{3}$ **30.** [3.5] $-\frac{3}{7}$ **31.** [3.5] $\frac{3}{2}$ **32.** [3.5] 0
33. [3.5] Undefined **34.** [3.5] 2 **35.** [3.5] 7%
36. [1.3] $\frac{19}{24}$ **37.** [1.3] $\frac{13}{17}$ **38.** [2.2] 8
39. [2.3] $m = 2A - n$
40. [3.4] ◆ A business might use a graph to quickly look up prices (as in the rental truck example) or to plot how total sales change from year to year. Many other applications exist.
41. [3.2] ◆ The y-intercept is the point at which the graph crosses the y-axis. Since a point on the y-axis is neither left nor right of the origin, the first or x-coordinate of the point is 0.
42. [3.2] -1 **43.** [3.2] 19
44. [3.1] Area: 45 sq units; perimeter: 28 units
45. [3.2] (0, 4), (1, 3), (-1, 3); answers may vary

TEST: CHAPTER 3, PP. 171–172

1. [3.1] About 12,000 **2.** [3.1] About 22% **3.** [3.1] II
4. [3.1] III **5.** [3.1] (3, 4) **6.** [3.1] (0, -4)
7. [3.2]

$$\begin{array}{c|c} y - 2x = 5 \\ \hline 11 - 2 \cdot 3 \;?\; 5 \\ 11 - 6 \\ 5 \;|\; 5 \quad \text{TRUE} \end{array} \qquad \begin{array}{c|c} y - 2x = 5 \\ \hline 3 - 2(-1) \;?\; 5 \\ 3 + 2 \\ 5 \;|\; 5 \quad \text{TRUE} \end{array}$$

(0, 5); answers may vary

8. [3.2] **9.** [3.3]

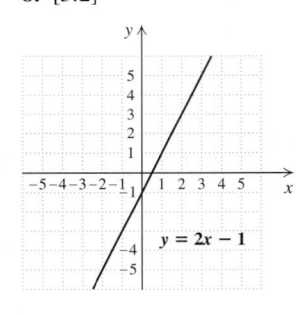

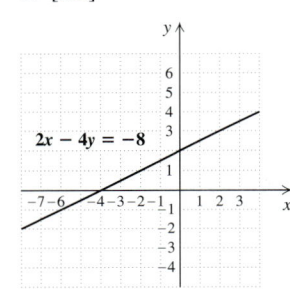

10. [3.3] 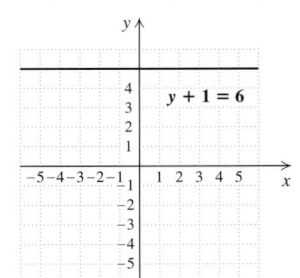 **11.** [3.2]

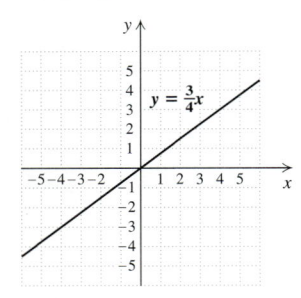

12. [3.2]

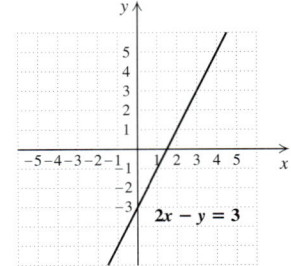

13. [3.3] x-intercept: (6, 0); y-intercept: (0, -10)
14. [3.3] x-intercept: (10, 0); y-intercept: $\left(0, \frac{5}{2}\right)$
15. [3.3] $x = -4$ **16.** [3.4] Let g = Greta's earnings and a = Alice's salary; $g = 2a - 150$
17. [3.4]

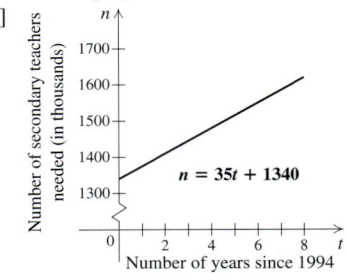

About 1600 thousand, or 1,600,000
18. [3.4] About $\frac{3}{4}$ hr

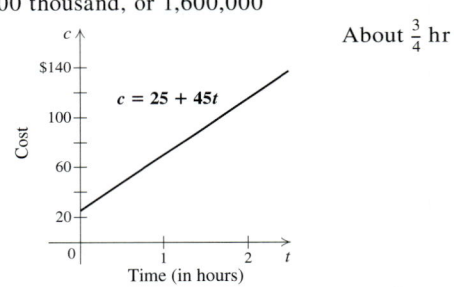

19. [3.5] Undefined **20.** [3.5] $\frac{7}{12}$ **21.** [1.3] $\frac{11}{5}$
22. [1.3] $3 \cdot 3 \cdot 3 \cdot 5$ **23.** [2.2] $\frac{4}{15}$
24. [2.3] $x = \dfrac{b}{m + n}$ **25.** [3.1] Area: 25 square units; perimeter: 20 units **26.** [3.3] $y = 3$

CUMULATIVE REVIEW: 1–3, PP. 173–174

1. [1.1] 15 **2.** [1.2] $12x - 15y + 21$
3. [1.2] $3(5x - 3y + 1)$ **4.** [1.3] $2 \cdot 3 \cdot 7$
5. [1.4] 0.45 **6.** [1.4] 4 **7.** [1.6] $\frac{1}{4}$ **8.** [1.7] -4
9. [1.6] $-x - y$ **10.** [2.4] 0.785 **11.** [1.3] $\frac{11}{60}$
12. [1.5] 2.6 **13.** [1.7] 7.28 **14.** [1.7] $-\frac{5}{12}$
15. [1.8] -2 **16.** [1.8] 27 **17.** [1.8] $-2y - 7$
18. [1.8] $5x + 11$ **19.** [2.1] -1.2 **20.** [2.1] -21

21. [2.2] 9   **22.** [2.1] $-\frac{20}{3}$ **23.** [2.2] 2   **24.** [2.1] $\frac{13}{8}$
25. [2.2] $-\frac{17}{21}$ **26.** [2.2] -17 **27.** [2.2] 2
28. [2.6] $\{x \mid x < 16\}$ **29.** [2.6] $\{x \mid x \leq -\frac{11}{8}\}$
30. [2.3] $h = \dfrac{A - \pi r^2}{2\pi r}$ **31.** [3.1] IV
32. [2.6] $-1 < x \leq 2$ (number line from -5 to 5)

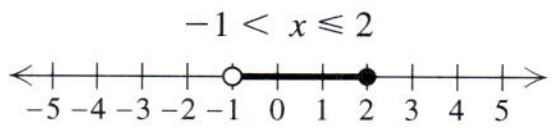

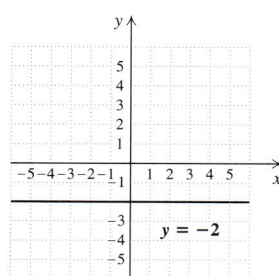

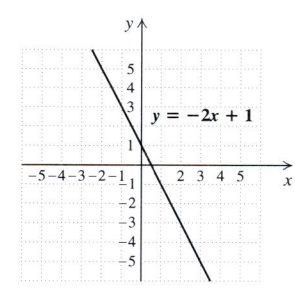

33. [3.3] (graph of $y = -2$)   **34.** [3.3] (graph of $2x + 5y = 10$)

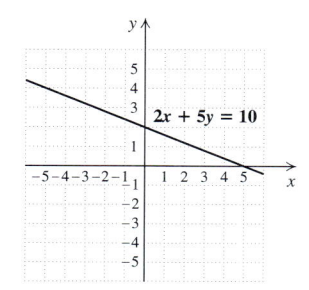

35. [3.2] (graph of $y = -2x + 1$)   **36.** [3.2] (graph of $y = \frac{2}{3}x$)

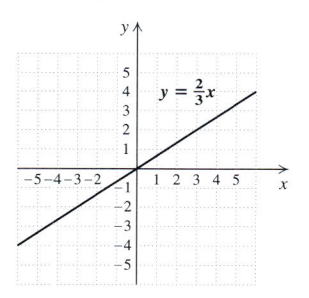

37. [3.3] x-intercept: $(10.5, 0)$; y-intercept: $(0, -3)$
38. [3.3] x-intercept: $\left(-\frac{5}{4}, 0\right)$; y-intercept: $(0, 5)$
39. [2.4] 160 million   **40.** [2.5] 15.6 million
41. [2.4] $120   **42.** [2.5] 50 m, 53 m, 40 m
43. [2.7] 8 hr or less   **44.** [3.4] Let $s = $ the cost of steak, in dollars, and $c = $ the cost of chicken, in dollars; $s = 5 + 3c$

45. [3.4] (graph of $c = \frac{3}{4}t + 3$, Cost in hundreds vs Time in months) About $1000

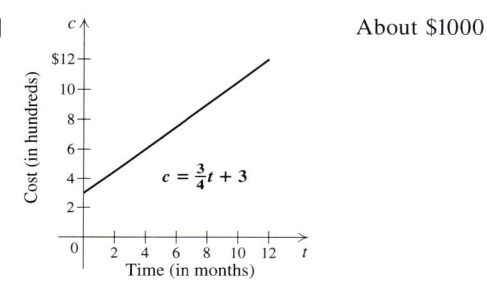

46. [3.4] (graph of $c = 20 + 15t$, Cost vs Time in hours) About 3 hr

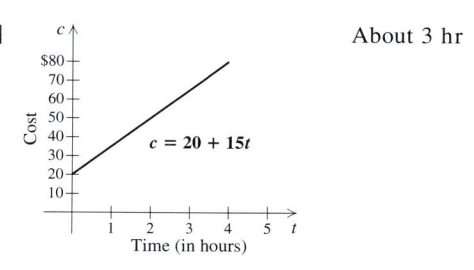

47. [3.5] $-\frac{1}{3}$ **48.** [2.4], [2.5] $25,000
49. [1.4], [2.2] $4, -4$ **50.** [2.2] All real numbers
51. [2.2] No solution   **52.** [2.2] 3
53. [2.2] All real numbers   **54.** [2.3] $Q = \dfrac{2 - pm}{p}$

CHAPTER 4

EXERCISE SET 4.1, PP. 182–183
1. m^{10} **3.** 8^{14} **5.** x^7 **7.** 5^7 **9.** $(3y)^{12}$ **11.** $(5t)^7$
13. a^5b^9 **15.** $(x+1)^{12}$ **17.** r^{12} **19.** x^4y^7 **21.** 7^3
23. x^{12} **25.** y^4 **27.** $5a$ **29.** $(x+y)^5$ **31.** $3m^3$
33. a^7b^6 **35.** m^9n^4 **37.** 1   **39.** 5   **41.** 1   **43.** 8
45. x^{12} **47.** 5^{16} **49.** m^{35} **51.** a^{75} **53.** $49x^2$
55. $-8a^3$ **57.** $16m^6$ **59.** $a^{14}b^7$ **61.** $a^{15}b^{10}$
63. $25x^8y^{10}$ **65.** $\dfrac{a^3}{64}$ **67.** $\dfrac{49}{25a^2}$ **69.** $\dfrac{a^{20}}{b^{15}}$ **71.** $\dfrac{y^6}{4}$
73. $\dfrac{x^8y^4}{z^{12}}$ **75.** $\dfrac{a^{12}}{16b^{20}}$ **77.** $\dfrac{8a^6}{27b^{12}}$ **79.** $\dfrac{16x^6y^{10}}{9z^{14}}$
81. $3(s + t + 8)$ **83.** $5x$

85. (graph of $y = x - 5$)

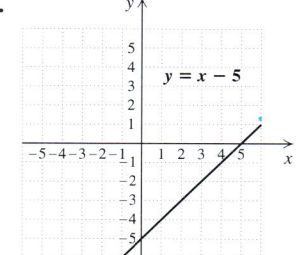

87. ◆   **89.** ◆   **91.** y^{5x} **93.** a^{4t} **95.** $>$
97. $<$ **99.** $>$
101. Let $a = 1$; then $(a + 5)^2 = 36$, but $a^2 + 5^2 = 26$.
103. Let $a = 0$; then $\dfrac{a + 7}{7} = 1$, but $a = 0$.   **105.** 19
107. $27,087.01

TECHNOLOGY CONNECTION 4.2, P. 188
1. 5

EXERCISE SET 4.2, PP. 189–192

1. $3x^4, -7x^3, x, -5$ 3. $-t^4, 2t^3, -5t^2, 3$
5. Coefficients: 7, −5; degrees: 3, 1
7. Coefficients: 9, −3, 4; degrees: 2, 1, 0
9. Coefficients: 5, 9, 1; degrees: 4, 1, 3
11. Coefficients: 1, −1, 4, −3; degrees: 4, 3, 1, 0
13. (a) 3, 5, 2; (b) $7a^5$, 7; (c) 5
15. (a) 1, 0, 2; (b) $4t^2$, 4; (c) 2
17. (a) 4, 2, 1, 0; (b) $-5x^4$, −5; (c) 4
19. (a) 1, 4, 0, 3; (b) $-a^4$, −1; (c) 4
21.

Term	Coefficient	Degree of Term	Degree of Polynomial
$8x^5$	8	5	5
$-\frac{1}{2}x^4$	$-\frac{1}{2}$	4	
$-4x^3$	−4	3	
$3x^2$	3	2	
6	6	0	

23. Trinomial 25. None of these 27. Binomial
29. Monomial 31. $7x^2 + 5x$ 33. $4a^4$ 35. $6x^2 - 3x$
37. $\frac{1}{6}x^3 + 2x - 12$ 39. $-x^3$ 41. 0
43. $x^4 - 2x^3 + 1$ 45. $4x^3 + x - \frac{1}{2}$ 47. −16 49. 16
51. −3 53. 11 55. 55 57. 9 59. 6 61. 15
63. 1112 ft 65. Approximately 449 67. $18,750
69. $155,000 71. 62.8 cm 73. 78.5 m² 75. $\frac{15}{2}$
77. $0.0625, or 6.25¢ 79. ◆ 81. ◆
83. $-6x^5 + 14x^4 - x^2 + 11$; answers may vary
85. 10 87. $5x^9 + 4x^8 + x^2 + 5x$
89. 99, −99; 50, −50; 99, −99
91.

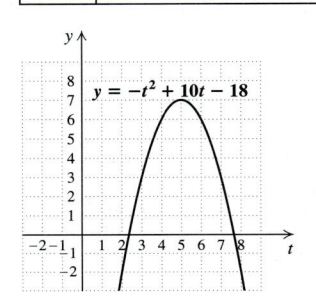

t	$-t^2 + 10t - 18$
3	3
4	6
5	7
6	6
7	3

93.

d	$-0.0064d^2 + 0.8d + 2$
0	2
30	20.24
60	26.96
90	22.16
120	5.84

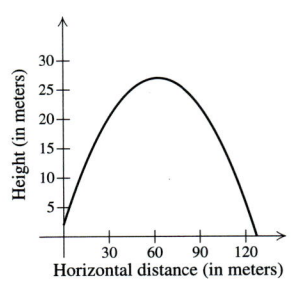

TECHNOLOGY CONNECTION 4.3, P. 195

1. In each case, let $y_1 =$ the expression before the addition or subtraction has been performed, $y_2 =$ the simplified sum or difference, and $y_3 = y_2 - y_1$; and note that the graph of y_3 coincides with the x-axis. That is, $y_3 = 0$.

EXERCISE SET 4.3, PP. 197–199

1. $-2x + 4$ 3. $x^2 - 5x - 1$ 5. $5x^2 + 3x - 30$
7. $-2.2x^3 - 0.2x^2 - 3.8x + 23$ 9. $11 + 2x + 16x^2$
11. $9x^8 + 8x^7 - 3x^4 + 2x^2 - 2x + 5$
13. $-\frac{1}{2}x^4 + \frac{2}{3}x^3 + x^2$
15. $0.01x^5 + x^4 - 0.2x^3 + 0.2x + 0.06$
17. $-3x^4 + 3x^2 + 4x$
19. $1.05x^4 + 0.36x^3 + 14.22x^2 + x + 0.97$
21. $-(-x^2 + 9x - 4), x^2 - 9x + 4$
23. $-(12x^4 - 3x^3 + 3), -12x^4 + 3x^3 - 3$ 25. $-8x + 9$
27. $-4x^2 + 3x - 2$ 29. $4x^4 - 6x^2 - \frac{3}{4}x + 8$
31. $9x + 3$ 33. $-x^2 - 7x + 5$
35. $6x^4 + 3x^3 - 4x^2 + 3x - 4$
37. $4.6x^3 + 9.2x^2 - 3.8x - 23$
39. $7x^3 - 9x^2 - 2x + 10$ 41. $9x^2 + 9x - 8$
43. $\frac{3}{4}x^3 - \frac{1}{2}x$ 45. $0.06x^3 - 0.05x^2 + 0.01x + 1$
47. $3x + 6$ 49. $11x^4 + 12x^3 - 9x^2 - 8x - 9$
51. $-4x^5 + 9x^4 + 6x^2 + 16x + 6$
53. (a) $5x^2 + 4x$; (b) 145, 273 55. $14y + 25$
57. $(r + 9)(r + 11); 9r + 99 + r^2 + 11r$
59. $\pi r^2 - 25\pi$ 61. $27z - 144$ 63. $y^2 - 4y + 4$
65. $\frac{115}{22}$ 67. 4 69. $\{x \mid x \geq -10\}$ 71. ◆
73. ◆ 75. $12a^2 - 24a - 8$ 77. $-10y^2 - 2y - 10$
79. $-3y^4 - y^3 + 5y - 2$ 81. $250.591x^3 + 2.812x$
83. $22a + 56$ 85. ◆

CHAPTER 4 — A-15

TECHNOLOGY CONNECTION 4.4, P. 204

1. Let $y_1 = (-2x^2 - 3)(5x^3 - 3x + 4)$ and $y_2 = -10x^5 - 9x^3 - 8x^2 + 9x - 12$. With the table set in AUTO mode, note that the values in the Y1- and Y2-columns match, regardless of how far we scroll up or down.

EXERCISE SET 4.4, PP. 205–206

1. $24x^4$ 3. x^3 5. $-x^8$ 7. $28t^8$ 9. $-0.02x^{10}$
11. $\frac{1}{15}x^4$ 13. 0 15. $-24x^{11}$ 17. $-3x^2 + 15x$
19. $4x^2 + 4x$ 21. $5x^2 + 35x$ 23. $x^5 + x^2$
25. $6x^3 - 18x^2 + 3x$ 27. $12x^3 + 24x^2$
29. $-6x^4 - 6x^3$ 31. $4a^9 - 8a^7 - \frac{5}{12}a^4$
33. $x^2 + 9x + 18$ 35. $x^2 + 3x - 10$
37. $x^2 - 7x + 12$ 39. $x^2 - 9$ 41. $25 - 15x + 2x^2$
43. $t^2 + \frac{17}{6}t + 2$ 45. $\frac{3}{16}a^2 + \frac{5}{4}a - 2$

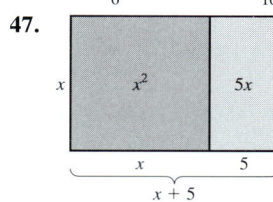

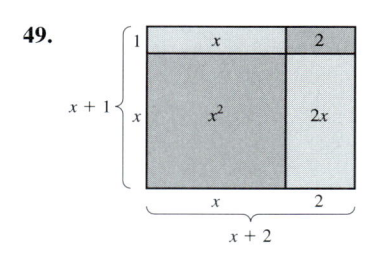

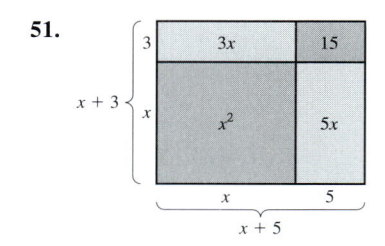

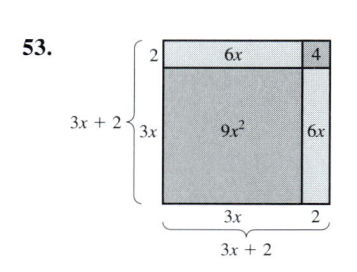

47.

49.

51.

53.

55. $x^3 + 2x + 3$ 57. $2a^3 - a^2 - 11a + 10$
59. $2y^5 - 5y^3 + y^2 - 3y - 3$
61. $-15x^5 + 26x^4 - 7x^3 - 3x^2 + x$
63. $x^4 - 2x^3 - x + 2$ 65. $4x^4 + 8x^3 - 9x^2 - 10x + 8$
67. $x^4 + 8x^3 + 12x^2 + 9x + 4$
69. $2x^4 - 5x^3 + 5x^2 - \frac{19}{10}x + \frac{1}{5}$ 71. 32
73. $3(5x - 6y + 4)$ 75. ◆ 77. ◆
79. $75y^2 - 45y$ 81. 5 83. $V = 4x^3 - 48x^2 + 144x$; $S = -4x^2 + 144$ 85. $x^3 + 6x^2 + 12x + 8$ cm^3
87. $2x^2 + 18x + 36$ 89. $16x + 16$ 91. 📺

EXERCISE SET 4.5, PP. 213–215

1. $x^3 + 5x^2 + x + 5$ 3. $x^4 + 2x^3 + 6x + 12$
5. $y^2 - y - 6$ 7. $9x^2 + 15x + 6$ 9. $5x^2 + 4x - 12$
11. $9t^2 - 1$ 13. $2x^2 - 9x + 7$ 15. $p^2 - \frac{1}{16}$
17. $x^2 - 0.01$ 19. $2x^3 + 2x^2 + 6x + 6$
21. $-2x^2 - 11x + 6$ 23. $a^2 + 14a + 49$
25. $1 - 2t - 15t^2$ 27. $x^5 + 3x^3 - x^2 - 3$
29. $3x^6 - 2x^4 - 6x^2 + 4$ 31. $6x^7 + 18x^5 + 4x^2 + 12$
33. $8x^5 + 16x^3 + 5x^2 + 10$ 35. $4x^3 - 12x^2 + 3x - 9$
37. $x^2 - 64$ 39. $4x^2 - 1$ 41. $25m^2 - 4$
43. $4x^4 - 9$ 45. $9x^8 - 1$ 47. $x^{12} - x^4$
49. $x^8 - 9x^2$ 51. $4y^{16} - 9$ 53. $x^2 + 4x + 4$
55. $9x^4 + 6x^2 + 1$ 57. $a^2 - \frac{4}{5}a + \frac{4}{25}$
59. $x^4 + 6x^2 + 9$ 61. $4 - 12x^4 + 9x^8$
63. $25 + 60t^2 + 36t^4$ 65. $49x^2 - 4.2x + 0.09$
67. $10a^5 - 5a^3$ 69. $x^4 + x^3 - 6x^2 - 5x + 5$
71. $9 - 12x^3 + 4x^6$ 73. $4x^3 + 24x^2 - 12x$
75. $4x^4 - 2x^2 + \frac{1}{4}$ 77. $-1 + 9p^2$
79. $15t^5 - 3t^4 + 3t^3$ 81. $36x^8 - 36x^4 + 9$
83. $12x^3 + 8x^2 + 15x + 10$ 85. $64 - 96x^4 + 36x^8$
87. $a^3 + 1$ 89. $a^2 + 2a + 1$ 91. $x^2 + 7x + 10$
93. $x^2 + 14x + 49$ 95. $t^2 + 13t + 36$
97. $9x^2 + 24x + 16$

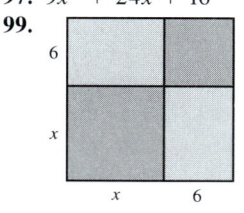

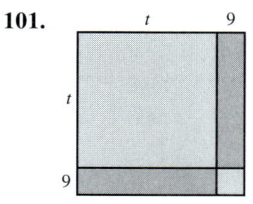

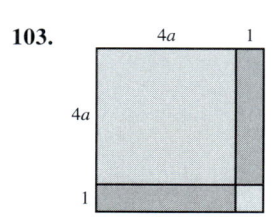

99.

101.

103.

105. Television: 50 watts; lamps: 500 watts; air conditioner: 2000 watts

107. $a = \dfrac{c}{b - d}$ **109.** ◆   **111.** ◆
113. $30x^3 + 35x^2 - 15x$ **115.** $a^4 - 50a^2 + 625$
117. $81t^{16} - 72t^8 + 16$ **119.** $400 - 4 = 396$ **121.** -7
123. $l^3 - l$ **125.** $Q(Q - 14) - 5(Q - 14)$,
$(Q - 5)(Q - 14)$; other equivalent expressions are possible.
127. $(y + 1)(y - 1)$, $y(y + 1) - y - 1$; other equivalent expressions are possible.
129. (a) $A^2 + AB$; (b) $AB + B^2$; (c) $A^2 - B^2$; (d) $A^2 - B^2$

EXERCISE SET 4.6, PP. 219–222

1. -15 **3.** -5 **5.** 2.97 L   **7.** 110.4 m
9. 20.60625 in^2
11. Coefficients: 1, -2, 3, -5; degrees: 4, 2, 2, 0; 4
13. Coefficients: 17, -3, -7; degrees: 5, 5, 0; 5
15. $a - 2b$ **17.** $3x^2y - 2xy^2 + x^2$ **19.** $8u^2v - 5uv^2$
21. $8uv + 10av + 14au$ **23.** $x^2 - 4xy + 3y^2$
25. $-2a^4 - 8ab + 7ab^2$ **27.** $-3b^3a^2 - b^2a^3 + 5ba + 3$
29. $3x^3 - x^2y + xy^2 - 3y^3$ **31.** $y^4x^2 + y + 2x$
33. $-8x + 8y$ **35.** $6z^2 + 7zu - 3u^2$
37. $x^2y^2 + 3xy - 28$ **39.** $m^4 + m^2n^2 + n^4$
41. $a^3 - b^3$ **43.** $m^6n^2 + 2m^3n - 48$
45. $30x^2 - 28xy + 6y^2$ **47.** $0.4p^2q^2 - 0.02pq - 0.02$
49. $x^2 + 2xh + h^2$ **51.** $r^6t^4 - 8r^3t^2 + 16$ **53.** $c^4 - d^2$
55. $a^2b^2 - c^2d^4$ **57.** $x^2 + 2xy + y^2 - 9$
59. $a^2 - b^2 - 2bc - c^2$ **61.** $a^2 + ac + ab + bc$
63. $x^2 - z^2$ **65.** $x^2 + 2xy + y^2 + 2xz + 2yz + z^2$
67. $\frac{1}{2}x^2 + \frac{1}{2}xy - y^2$
69. [figure]   **71.** [figure]
73. $50 per ton   **75.** June 1995
77. June to December 1994   **79.** ◆   **81.** ◆
83. $2\pi ab - \pi b^2$ **85.** $a^2 - 4b^2$
87. $2\pi nh + 2\pi mh + 2\pi n^2 - 2\pi m^2$ **89.** $P + 2Pr + Pr^2$
91. ◆

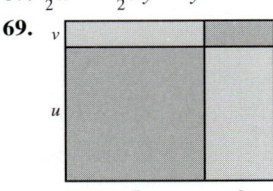

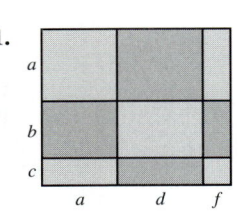

EXERCISE SET 4.7, PP. 227–228

1. $4x^5 - 2x$ **3.** $1 - 2u + u^6$ **5.** $5t^2 - 8t + 2$
7. $-7x^4 + 4x^2 + 1$ **9.** $6x^2 - 10x + \dfrac{3}{2}$
11. $4x^2 - 5x + \dfrac{1}{2}$ **13.** $x^2 + 3x + 2$
15. $-3rs - r + 2s$ **17.** $x + 6$ **19.** $x - 5 + \dfrac{-50}{x - 5}$
21. $2x - 1 + \dfrac{1}{x + 6}$ **23.** $x - 3$ **25.** $x + 4$

27. $2x^2 - 7x + 4$ **29.** $x^2 + 1 + \dfrac{4x - 2}{x^2 - 3}$
31. $t^2 - 2t + 3 + \dfrac{-4}{t + 1}$ **33.** $3x^2 - 3 + \dfrac{x - 1}{2x^2 + 1}$
35. 25,543.75 ft^2 **37.** $\left\{x \mid x < -\dfrac{12}{5}\right\}$ **39.** III
41. ◆   **43.** ◆   **45.** $15x^{6k} + 10x^{4k} - 20x^{2k}$
47. $y^3 - ay^2 + a^2y - a^3 + \dfrac{a^2 + a^4}{y + a}$
49. $5y + 2 + \dfrac{-10y + 11}{3y^2 - 5y - 2}$
51. $5x^5 + 5x^4 - 8x^2 - 8x + 2$ **53.** 3   **55.** -1

TECHNOLOGY CONNECTION 4.8, P. 233

1. 1.71×10^{17} **2.** $5.\overline{370} \times 10^{-15}$ **3.** 3.68×10^{16}

EXERCISE SET 4.8, PP. 235–237

1. $\dfrac{1}{6^2} = \dfrac{1}{36}$ **3.** $\dfrac{1}{10^4} = \dfrac{1}{10{,}000}$ **5.** $\dfrac{1}{(-2)^6} = \dfrac{1}{64}$ **7.** $\dfrac{1}{a^5}$
9. y^4 **11.** z^9 **13.** $\dfrac{1}{7}$ **15.** 16   **17.** 4^{-3} **19.** t^{-6}
21. a^{-4} **23.** p^{-8} **25.** 5^{-1} **27.** t^{-1} **29.** 2^3, or 8
31. x^{-1}, or $\dfrac{1}{x}$ **33.** x^{-13}, or $\dfrac{1}{x^{13}}$ **35.** m^{-6}, or $\dfrac{1}{m^6}$
37. $(8x)^{-4}$, or $\dfrac{1}{(8x)^4}$ **39.** 1   **41.** $a^{-8}b^{-13}$, or $\dfrac{1}{a^8b^{13}}$
43. x^9 **45.** z^{-4}, or $\dfrac{1}{z^4}$ **47.** a^4 **49.** x^2
51. a^{-15}, or $\dfrac{1}{a^{15}}$ **53.** a^{30} **55.** n^{-16}, or $\dfrac{1}{n^{16}}$
57. $m^{-5}n^{-5}$, or $\dfrac{1}{m^5n^5}$ **59.** $4^{-2}x^{-2}y^{-2}$, or $\dfrac{1}{16x^2y^2}$
61. $81a^{-16}$, or $\dfrac{81}{a^{16}}$ **63.** $t^{-20}x^{-12}$, or $\dfrac{1}{t^{20}x^{12}}$ **65.** $x^{10}y^{35}$
67. $x^{-1}y^{-6}z^4$, or $\dfrac{z^4}{xy^6}$ **69.** $m^5n^5p^{-7}$, or $\dfrac{m^5n^5}{p^7}$
71. $\dfrac{y^{-4}}{2^{-2}}$, or $\dfrac{4}{y^4}$ **73.** $\dfrac{81}{a^8}$ **75.** $\dfrac{x^6y^3}{z^{12}}$
77. $\dfrac{a^{-10}b^{-5}}{c^{-5}d^{-15}}$, or $\dfrac{c^5d^{15}}{a^{10}b^5}$ **79.** 91,200   **81.** 0.00692
83. 204,000,000   **85.** 0.0000000008764   **87.** 10,000,000
89. 0.0001   **91.** 3.7×10^5 **93.** 5.83×10^{-3}
95. 7.8×10^{10} **97.** 9.07×10^{17} **99.** 4.86×10^{-6}
101. 1.8×10^{-8} **103.** 10^7 **105.** 8×10^{12}
107. 2.47×10^8 **109.** 3.915×10^{-16} **111.** 2.5×10^{13}
113. 5×10^{-4} **115.** 3×10^{-21} **117.** 1.095×10^9 gal
119. 2×10^{14} gal   **121.** 2.5×10^{12} **123.** $-\dfrac{1}{6}$
125. $-17x$ **127.** I and IV   **129.** ◆
131. $1.19140625 \times 10^{-15}$ **133.** $6.304347826 \times 10^{25}$
135. 7×10^{-16} **137.** 2^1 **139.** 5   **141.** 3^{28}
143. False   **145.** False

REVIEW EXERCISES: CHAPTER 4, PP. 239–240

1. [4.1] y^{11} 2. [4.1] $(3x)^{14}$ 3. [4.1] t^8 4. [4.1] 4^3
5. [4.1] 1   6. [4.1] $\dfrac{9t^8}{4s^6}$ 7. [4.1] $-8x^3y^6$ 8. [4.1] $18x^5$
9. [4.1] a^7b^6 10. [4.2] $3x^2, 6x, \dfrac{1}{2}$
11. [4.2] $-4y^5, 7y^2, -3y, -2$ 12. [4.2] 6, 1, 5
13. [4.2] $4, 6, -5, \dfrac{5}{3}$
14. [4.2] (a) 2, 0, 5; (b) $15t^5$, 15; (c) 5
15. [4.2] (a) 5, 4, 2, 1; (b) $-2x^5$, -2; (c) 5
16. [4.2] Binomial   17. [4.2] None of these
18. [4.2] Monomial   19. [4.2] $-x^2 + 9x$
20. [4.2] $-\dfrac{1}{4}x^3 + 4x^2 + 7$ 21. [4.2] $-3x^5 + 25$
22. [4.2] $-2x^2 - 3x + 2$ 23. [4.2] $10x^4 - 7x^2 - x - \dfrac{1}{2}$
24. [4.2] -17 25. [4.2] 10
26. [4.3] $x^5 + 3x^4 + 6x^3 - 2x - 9$
27. [4.3] $-x^5 + 3x^4 - x^3 - 2x^2$ 28. [4.3] $2x^2 - 4x - 6$
29. [4.3] $x^5 - 3x^3 - 2x^2 + 8$
30. [4.3] $\dfrac{3}{4}x^4 + \dfrac{1}{4}x^3 - \dfrac{1}{3}x^2 - \dfrac{7}{4}x + \dfrac{3}{8}$
31. [4.3] $-x^5 + x^4 - 5x^3 - 2x^2 + 2x$
32. (a) [4.3] $4w + 6$; (b) [4.4] $w^2 + 3w$ 33. [4.4] $-12x^3$
34. [4.5] $49x^2 + 14x + 1$ 35. [4.5] $x^2 + \dfrac{7}{6}x + \dfrac{1}{3}$
36. [4.5] $m^2 - 25$ 37. [4.4] $12x^3 - 23x^2 + 13x - 2$
38. [4.5] $x^2 - 18x + 81$ 39. [4.4] $2a^5 - 4a^3 + \dfrac{4}{3}a^2$
40. [4.5] $x^2 - 3x - 28$ 41. [4.5] $x^2 - 1.05x + 0.225$
42. [4.4] $x^7 + x^5 - 3x^4 + 3x^3 - 2x^2 + 5x - 3$
43. [4.5] $9y^4 - 12y^3 + 4y^2$ 44. [4.5] $2t^4 - 11t^2 - 21$
45. [4.5] $4a^4 + 4a^3 + a^2$ 46. [4.5] $9x^4 - 16$
47. [4.5] $4 - x^2$ 48. [4.5] $13x^2 - 172x + 39$
49. [4.6] 49
50. [4.6] Coefficients: 1, -7, 9, -8; degrees: 6, 2, 2, 0; 6
51. [4.6] Coefficients: 1, -1, 1; degrees: 16, 40, 23; 40
52. [4.6] $9w - y - 5$ 53. [4.6] $6m^3 + 4m^2n - mn^2$
54. [4.6] $-x^2 - 10xy$
55. [4.6] $11x^3y^2 - 8x^2y - 6x^2 - 6x + 6$
56. [4.6] $p^3 - q^3$ 57. [4.6] $9a^8 - 2a^4b^3 + \dfrac{1}{9}b^6$
58. [4.6] $\dfrac{1}{2}x^2 - \dfrac{1}{2}y^2$ 59. [4.7] $5x^2 - \dfrac{1}{2}x + 3$
60. [4.7] $3x^2 - 7x + 4 + \dfrac{1}{2x+3}$ 61. [4.7] $t^3 + 2t - 3$
62. [4.8] $\dfrac{1}{y^4}$ 63. [4.8] t^{-5} 64. [4.8] $\dfrac{1}{7^2}$, or $\dfrac{1}{49}$
65. [4.8] $\dfrac{1}{a^{13}b^7}$ 66. [4.8] $\dfrac{1}{x^{12}}$ 67. [4.8] $\dfrac{x^6}{4y^2}$
68. [4.8] $\dfrac{y^3}{8x^3}$ 69. [4.8] 8,300,000
70. [4.8] 3.28×10^{-5} 71. [4.8] 2.09×10^4
72. [4.8] 5.12×10^{-5} 73. [4.8] $\$1.96 \times 10^{11}$
74. [1.5] $13x + \dfrac{22}{15}$ 75. [2.5] $w = 125.5$ m, $l = 144.5$ m
76. [2.6] $\{x \mid x \geq -2\}$ 77. [3.1] IV
78. [4.1] ◆ In the expression $5x^3$, the exponent refers only to the x. In the expression $(5x)^3$, the entire expression within the parentheses is cubed.
79. [4.3] ◆ The sum of two polynomials of degree n will also have degree n, since only the coefficients are added and the variables remain unchanged. An exception to this occurs when the leading terms of the two polynomials are opposites. The sum of those terms is then zero and the sum of the polynomials will have a degree less than n.
80. [4.2], [4.5] (a) 3; (b) 2   81. [4.1], [4.2] $-28x^8$
82. [4.2] $8x^4 + 4x^3 + 5x - 2$
83. [4.5] $-4x^6 + 3x^4 - 20x^3 + x^2 - 16$
84. [2.2], [4.5] $\dfrac{94}{13}$

TEST: CHAPTER 4, P. 241

1. [4.1] x^9 2. [4.1] $(x+3)^{11}$ 3. [4.1] 3^3 4. [4.1] 1
5. [4.1] x^6 6. [4.1] $-27y^6$ 7. [4.1] $-24x^{17}$
8. [4.1] a^6b^5 9. [4.2] Trinomial   10. [4.2] $\dfrac{1}{3}, -1, 7$
11. [4.2] Degrees of terms: 3, 1, 5, 0; leading term: $7t^5$; leading coefficient: 7; degree of polynomial: 5
12. [4.2] -7 13. [4.2] $5a^2 - 6$ 14. [4.2] $\dfrac{7}{4}y^2 - 4y$
15. [4.2] $x^5 + 2x^3 + 4x^2 - 8x + 3$
16. [4.3] $4x^5 + x^4 + 2x^3 - 8x^2 + 2x - 7$
17. [4.3] $5x^4 + 5x^2 + x + 5$
18. [4.3] $-4x^4 + x^3 - 8x - 3$
19. [4.3] $-x^5 + 1.3x^3 - 0.8x^2 - 3$
20. [4.4] $-12x^4 + 9x^3 + 15x^2$ 21. [4.5] $x^2 - \dfrac{2}{3}x + \dfrac{1}{9}$
22. [4.5] $9x^2 - 100$ 23. [4.5] $3b^2 - 4b - 15$
24. [4.5] $x^{14} - 4x^8 + 4x^6 - 16$
25. [4.5] $48 + 34y - 5y^2$ 26. [4.4] $6x^3 - 7x^2 - 11x - 3$
27. [4.5] $64a^2 + 48a + 9$
28. [4.6] $-5x^3y - x^2y^2 + xy^3 - y^3 + 19$
29. [4.6] $8a^2b^2 + 6ab - 4b^3 + 6ab^2 + ab^3$
30. [4.6] $9x^{10} - 16y^{10}$ 31. [4.7] $4x^2 + 3x - 5$
32. [4.7] $2x^2 - 4x - 2 + \dfrac{17}{3x+2}$ 33. [4.8] $\dfrac{1}{5^3}$
34. [4.8] y^{-8} 35. [4.8] $\dfrac{1}{6^5}$ 36. [4.8] $\dfrac{y^5}{x^5}$
37. [4.8] $\dfrac{b^4}{16a^{12}}$ 38. [4.8] $\dfrac{c^3}{a^3b^3}$ 39. [4.8] 3.9×10^9
40. [4.8] 0.00000005   41. [4.8] 1.75×10^{17}
42. [4.8] 1.296×10^{22} 43. [4.8] 1.5×10^4
44. [2.6] $\{x \mid x > 13\}$
45. [3.1]

46. [1.5] $-\dfrac{7}{20}$ 47. [2.5] 64°, 32°, 84°

48. [4.4], [4.5] $V = l(l-2)(l-1) = l^3 - 3l^2 + 2l$
49. [2.2], [4.5] $\frac{100}{21}$

CHAPTER 5

TECHNOLOGY CONNECTION 5.1, P. 248

1. Let $y_1 = 8x^4 + 6x - 28x^3 - 21$ and $y_2 = (4x^3 + 3)(2x - 7)$. Note that the Y1- and Y2-columns of the table match regardless of how far we scroll up or down.

EXERCISE SET 5.1, P. 249

1. Answers may vary. $(10x)(x^2)$, $(5x^2)(2x)$, $(-2)(-5x^3)$
3. Answers may vary. $(-3x^2)(2x^3)$, $(-x)(6x^4)$, $(2x^2)(-3x^3)$
5. Answers may vary. $(2x)(13x^4)$, $(13x^5)(2)$, $(-x^3)(-26x^2)$
7. $x(x-6)$ **9.** $4x(2x+1)$ **11.** $x^2(x+6)$
13. $8x^2(x^2-3)$ **15.** $2(x^2+x-4)$
17. $x^2(5x^4-10x+8)$ **19.** $2x^2(x^6+2x^4-4x^2+5)$
21. $x^2y^2(x^3y^3+x^2y+xy-1)$
23. $\frac{1}{3}x^3(5x^3+4x^2+x+1)$ **25.** $(y+3)(y+7)$
27. $(x+3)(x^2-7)$ **29.** $(y+8)(y^2+1)$
31. $(x+3)(x^2+4)$ **33.** $(x+3)(2x^2+3)$
35. $(3x-4)(3x^2+1)$ **37.** $(5x-2)(x^2+1)$
39. $(x+8)(x^2-3)$ **41.** $(w-7)(w^2+4)$
43. Not factorable by grouping   **45.** $(x-4)(2x^2-9)$
47.

$y = x - 6$

49. 12   **51.** $y^2 + 12y + 35$ **53.** $y^2 - 49$ **55.** ◆
57. ◆   **59.** $(2x^2+3)(2x^3+3)$ **61.** $(x^5+1)(x^7+1)$
63. $(x-1)(5x^4+x^2+3)$ **65.** ◆

EXERCISE SET 5.2, PP. 255–256

1. $(x+2)(x+4)$ **3.** $(x+1)(x+8)$
5. $(y+4)(y+7)$ **7.** $(a+5)(a+6)$
9. $(x-1)(x-4)$ **11.** $(z-1)(z-7)$
13. $(x-3)(x-5)$ **15.** $(x+3)(x-5)$
17. $(x-3)(x+5)$ **19.** $3(y+4)(y-7)$
21. $x(x+6)(x-7)$ **23.** $(x+5)(x-12)$
25. $(x-6)(x+12)$ **27.** $5(b-3)(b+8)$
29. $x^3(x+2)(x-1)$ **31.** Prime   **33.** Prime
35. $(x+9)(x+11)$ **37.** $2x(x-8)(x-12)$
39. $4(x+5)^2$ **41.** $(y-9)(y-12)$
43. $a^4(a-6)(a+15)$ **45.** $\left(x-\frac{1}{5}\right)^2$

47. $(16-y)(7+y)$ **49.** $(t-0.5)(t+0.2)$
51. $(p+5q)(p-2q)$ **53.** Prime   **55.** $(s-5t)(s+3t)$
57. $6a^8(a-7)(a+2)$ **59.** $\frac{8}{3}$ **61.** $3x^2 + 22x + 24$
63. 29,443   **65.** ◆   **67.** ◆
69. 15, −15, 27, −27, 51, −51   **71.** $\left(x+\frac{1}{2}\right)\left(x-\frac{1}{4}\right)$
73. $\frac{1}{3}a(a-3)(a+2)$ **75.** $(x^m+4)(x^m+7)$
77. $(x+1)(a+2)(a+1)$ **79.** $2x^2(4-\pi)$

EXERCISE SET 5.3, PP. 264–265

1. $(3x+4)(x-1)$ **3.** $(5x+9)(x-2)$
5. $(2x-3)(3x-2)$ **7.** $(3x+1)(x+1)$
9. $(2x+5)(2x-3)$ **11.** $(5x+2)(3x-5)$
13. $(3x+8)(3x-2)$ **15.** $(2x-1)(x-2)$
17. $(3x-4)(4x-5)$ **19.** $2(7x-1)(2x+3)$
21. $(3x+4)(3x+2)$ **23.** $(7+3x)^2$
25. $(x+2)(24x-1)$ **27.** $(7x+4)(5x-11)$
29. Prime   **31.** $4(3x-2)(x+3)$ **33.** $6(5x-9)(x+1)$
35. $3(2x+1)(x+5)$ **37.** $3(2x+3)(3x-1)$
39. $(y+4)(y-2)$ **41.** $(x-4)(x-1)$
43. $(3x+2)(2x+3)$ **45.** $(3x-4)(x-4)$
47. $(7x-8)(5x+3)$ **49.** $(2x+3)(2x-3)$
51. $(3x-7)^2$ **53.** $(3t-2)(6t+5)$
55. $7(2x-1)(x-2)$ **57.** $2x(3x-5)(x+1)$
59. Prime   **61.** $24x^3(3x-2)(2x-1)$
63. $2a^2(7a-4)(5a-2)$ **65.** $(4m-5n)(3m+4n)$
67. $(3p+2q)(p-6q)$ **69.** $2(5s-3t)(s+t)$
71. $3(2a+5b)(5a+2b)$ **73.** $5(3a-4b)(a+b)$
75. Prime   **77.** $(6xy-5)(3xy+2)$
79. $x^2(4x+7)(2x+5)$ **81.** $(3a+4)(3a+2)$
83. 6369 km, 3949 mi
85.

$y = \frac{2}{5}x - 1$

87. $9x^2 - 25$ **89.** ◆   **91.** ◆   **93.** $(4x^5-1)^2$
95. $(10x^n+3)(2x^n+1)$ **97.** $(x^{3a}-1)(3x^{3a}+1)$
99. $-2(a+1)^n(a+3)^2(a+6)$

EXERCISE SET 5.4, PP. 270–271

1. Yes   **3.** No   **5.** No   **7.** No   **9.** $(x-8)^2$
11. $(x+7)^2$ **13.** $(x-1)^2$ **15.** $(2+x)^2$
17. $2(3x-1)^2$ **19.** $(7+4y)^2$ **21.** $x^3(x-9)^2$
23. $2x(x-1)^2$ **25.** $5(2x+5)^2$ **27.** $(7-3x)^2$
29. $5(y+1)^2$ **31.** $2(1+5x)^2$ **33.** $(2p+3q)^2$
35. $(a-7b)^2$ **37.** $(8m+n)^2$ **39.** $(4s-5t)^2$ **41.** Yes

43. No **45.** No **47.** Yes **49.** $(y + 2)(y - 2)$
51. $(p + 3)(p - 3)$ **53.** $(t + 7)(t - 7)$
55. $(a + b)(a - b)$ **57.** $(mn + 7)(mn - 7)$
59. $2(10 + t)(10 - t)$ **61.** $(4a + 3)(4a - 3)$
63. $(2x + 5y)(2x - 5y)$ **65.** $2(2x + 7)(2x - 7)$
67. $x(6 + 7x)(6 - 7x)$ **69.** $(7a^2 + 9)(7a^2 - 9)$
71. $5(x^2 + 1)(x + 1)(x - 1)$
73. $4(x^2 + 4)(x + 2)(x - 2)$
75. $(1 + y^4)(1 + y^2)(1 + y)(1 - y)$ **77.** $3x(x - 4)^2$
79. $(x^6 + 4)(x^3 + 2)(x^3 - 2)$ **81.** $\left(y + \frac{1}{4}\right)\left(y - \frac{1}{4}\right)$
83. $a^6(a - 1)^2$ **85.** $(4 + m^2n^2)(2 + mn)(2 - mn)$
87. $s \geq 77$ **89.** $x^{12}y^{12}$

91.

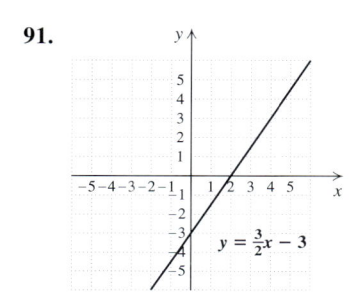

$y = \frac{3}{2}x - 3$

93. ◆ **95.** ◆ **97.** $(x + 1.5)(x - 1.5)$ **99.** Prime
101. $3\left(x + \frac{1}{3}\right)\left(x - \frac{1}{3}\right)$ **103.** $p(0.7 + p)(0.7 - p)$
105. $x(x + 6)(x^2 + 6x + 18)$ **107.** $\left(x + \frac{1}{x}\right)\left(x - \frac{1}{x}\right)$
109. $(9 + b^{2k})(3 + b^k)(3 - b^k)$ **111.** $(3b^n + 2)^2$
113. $(y + 4)^2$ **115.** $(3x + 7)(3x - 7)^2$
117. $(a + 4)(a - 2)$ **119.** 9 **121.** 0, 2

EXERCISE SET 5.5, PP. 276–277

1. $5(x + 3)(x - 3)$ **3.** $(a + 5)^2$ **5.** $(2x - 3)(x - 4)$
7. $x(x - 12)^2$ **9.** $(x + 3)(x + 2)(x - 2)$
11. $6(2x + 3)(2x - 3)$ **13.** $4x(x - 2)(5x + 9)$
15. Prime **17.** $a(a^2 + 8)(a + 8)$ **19.** $x^3(x - 7)^2$
21. $-2(x - 2)(x + 5)$ **23.** Prime
25. $4(x^2 + 4)(x + 2)(x - 2)$
27. $(t^4 + 1)(t^2 + 1)(t + 1)(t - 1)$ **29.** $x^3(x - 3)(x - 1)$
31. $(x + y)(x - y)$ **33.** $12n^2(1 + 2n)$ **35.** $ab(b - a)$
37. $2\pi r(h + r)$ **39.** $(a + b)(2x + 1)$
41. $(x + 2)(x - 5 - y)$ **43.** $(x + 1)(x + y)$
45. $(a - 3)(a + y)$ **47.** $(2y - 1)(3y + p)$
49. $(a - 2b)^2$, or $(2b - a)^2$ **51.** $(4x + 3y)^2$
53. $(2xy + 3z)^2$ **55.** $(a^2b^2 + 4)(ab + 2)(ab - 2)$
57. $p(2p + q)^2$ **59.** $(3b + a)(b - 6a)$
61. $(xy + 5)(xy + 3)$ **63.** $(pq + 6)(pq + 1)$
65. $b^4(ab + 8)(ab - 4)$ **67.** $x^4(x + 2y)(x - y)$
69. $\left(6a - \frac{5}{4}\right)^2$ **71.** $\left(\frac{1}{2}a + \frac{1}{3}b\right)^2$
73. $a(b^2 + 9a^2)(b + 3a)(b - 3a)$

75. $(w - 7)(w + 2)(w - 2)$
77.

$y = -4x + 7$		$y = -4x + 7$	
11 ? $-4(-1) + 7$		7 ? $-4 \cdot 0 + 7$	
$\mid$ $4 + 7$		$\mid$ $0 + 7$	
11 $\mid$ 11	TRUE	7 $\mid$ 7	TRUE

$y = -4x + 7$	
-5 ? $-4 \cdot 3 + 7$	
$\mid$ $-12 + 7$	
-5 $\mid$ -5	TRUE

79. $-\frac{4}{3}$ **81.** $X = \frac{A + 7}{a + b}$ **83.** ◆ **85.** ◆
87. $-x(x^2 + 9)(x^2 - 2)$
89. $3(a + 1)(a - 1)(a + 2)(a - 2)$
91. $(x - 1)(x + 2)(x - 2)$ **93.** $(y - 1)^3$
95. $[(y + 4) + x]^2$ **97.** $(x^4 + 16)(x^2 + 4)(x + 2)(x - 2)$
99. $49 = 49$; probably correct

TECHNOLOGY CONNECTION 5.6, P. 282

1. $-4.65, 0.65$ **2.** $-0.37, 5.37$ **3.** $-8.98, -4.56$
4. No solution

EXERCISE SET 5.6, PP. 283–284

1. $-7, -6$ **3.** $3, -5$ **5.** $\frac{9}{2}, -4$ **7.** $\frac{7}{10}, -\frac{9}{4}$ **9.** $0, -6$
11. $\frac{18}{11}, \frac{1}{21}$ **13.** $0, -\frac{9}{2}$ **15.** $50, 70$ **17.** $-3, \frac{5}{2}, 6$
19. $1, 6$ **21.** $-3, 7$ **23.** $-7, -2$ **25.** $0, -3$
27. $0, 9$ **29.** $-10, 10$ **31.** $-\frac{3}{2}, \frac{3}{2}$ **33.** -5 **35.** 1
37. $0, \frac{4}{9}$ **39.** $-\frac{5}{3}, 4$ **41.** $-5, -1$ **43.** $-3, 7$
45. $-1, \frac{2}{3}$ **47.** $-\frac{5}{9}, \frac{5}{9}$ **49.** $-1, \frac{6}{5}$ **51.** $-2, 9$
53. $-\frac{5}{2}, \frac{4}{3}$ **55.** $-3, 2$ **57.** $(-4, 0), (1, 0)$
59. $(-3, 0), (5, 0)$ **61.** $\left(-\frac{5}{2}, 0\right), (2, 0)$ **63.** $(a + b)^2$
65. Let x represent the first integer; $x + (x + 1)$
67. Let n represent the number; $\frac{1}{2}n - 7 > 24$ **69.** ◆
71. ◆ **73.** (a) $x^2 - x - 12 = 0$; (b) $x^2 + 7x + 12 = 0$;
(c) $4x^2 - 4x + 1 = 0$; (d) $x^2 - 25 = 0$;
(e) $40x^3 - 14x^2 + x = 0$ **75.** $-5, 4$ **77.** $-\frac{3}{5}, \frac{3}{5}$
79. $-4, 4$ **81.** (a) $2x^2 + 20x - 4 = 0$;
(b) $x^2 - 3x - 18 = 0$; (c) $(x + 1)(5x - 5) = 0$;
(d) $(2x + 8)(2x - 5) = 0$; (e) $4x^2 + 8x + 36 = 0$;
(f) $9x^2 - 12x + 24 = 0$ **83.** ◆ **85.** $2.33, 6.77$
87. $-9.15, -4.59$ **89.** $-6.75, -3.25$

EXERCISE SET 5.7, PP. 290–293

1. $-1, 2$ **3.** $3, 5$ **5.** 10 and 11 **7.** 14 and 16, -16 and -14 **9.** Length: 12 cm; width: 7 cm
11. Height: 7 cm; base: 10 cm **13.** Foot: 7 ft; height: 12 ft **15.** 25 ft **17.** 380 **19.** 12 **21.** 24 ft
23. 66 **25.** 25 **27.** 4 m **29.** 20 ft **31.** At 1 sec and at 2 sec after it has been launched

33.

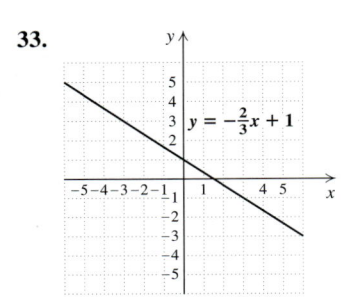

[Graph of $y = -\frac{2}{3}x + 1$]

35. 7   **37.** $-\frac{7}{24}$ **39.** ◆   **41.** ◆   **43.** 39 cm
45. 960 ft^2 **47.** 7 m   **49.** 5 in.   **51.** $1200

REVIEW EXERCISES: CHAPTER 5, P. 295

1. [5.1] Answers may vary. $(12x)(3x^2)$, $(-9x^2)(-4x)$, $(6x)(6x^2)$ **2.** [5.1] Answers may vary. $(-4x^3)(5x^2)$, $(2x^4)(-10x)$, $(-5x)(4x^4)$ **3.** [5.1] $2x^3(x+3)$
4. [5.1] $x(x-3)$ **5.** [5.4] $(3x+2)(3x-2)$
6. [5.2] $(x+6)(x-2)$ **7.** [5.4] $(x+7)^2$
8. [5.1] $3x(2x^2+4x+1)$ **9.** [5.1] $(2x+3)(3x^2+1)$
10. [5.3] $(3x-1)(2x-1)$
11. [5.4] $(x^2+9)(x+3)(x-3)$
12. [5.3] $3x(3x-5)(x+3)$ **13.** [5.4] $2(x+5)(x-5)$
14. [5.1] $(x^3-2)(x+4)$
15. [5.4] $(4x^2y^2+1)(2xy+1)(2xy-1)$
16. [5.1] $4x^4(2x^2-8x+1)$ **17.** [5.4] $3(2x+5)^2$
18. [5.4] Prime   **19.** [5.2] $x(x-6)(x+5)$
20. [5.4] $(2x+5)(2x-5)$ **21.** [5.4] $(3x-5)^2$
22. [5.3] $2(3x+4)(x-6)$ **23.** [5.4] $(x-3)^2$
24. [5.3] $(2x+1)(x-4)$ **25.** [5.4] $2(3x-1)^2$
26. [5.4] $3(x+3)(x-3)$ **27.** [5.2] $(x-5)(x-3)$
28. [5.4] $(5x-2)^2$ **29.** [5.2] $(xy+4)(xy-3)$
30. [5.4] $3(2a+7b)^2$ **31.** [5.1] $(m+t)(m+5)$
32. [5.4] $32(x^2-2y^2z^2)(x^2+2y^2z^2)$ **33.** [5.6] 1, -3
34. [5.6] $-7, 5$ **35.** [5.6] $-\frac{1}{3}, \frac{1}{3}$ **36.** [5.6] $\frac{2}{3}, 1$
37. [5.6] $-4, \frac{3}{2}$ **38.** [5.6] $-2, 3$ **39.** [5.7] $-3, 4$
40. [5.6] $(-1, 0), (\frac{5}{2}, 0)$ **41.** [5.7] 40 cm   **42.** [5.7] 24 m
43. [1.6] $\frac{1}{2}$
44. [3.2]

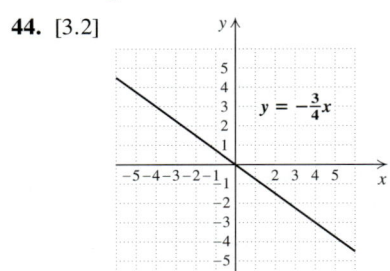

[Graph of $y = -\frac{3}{4}x$]

45. [4.1] m^2n^5 **46.** [2.7] Let x represent the number; $\frac{1}{2}x - 2 \geq 10$ **47.** [5.4] ◆ Answers may vary. Because Edith did not first factor out the largest common factor, 4, her factorization will not be "complete" until she removes a common factor of 2 from each binomial. Awarding 3 to 7 points would seem reasonable.   **48.** [5.6] ◆ The equations solved in this chapter have an x^2-term (are quadratic), whereas those solved previously have no x^2-term (are linear). The principle of zero products is used to solve quadratic equations and is not used to solve linear equations.   **49.** [5.7] $2\frac{1}{2}$ cm   **50.** [5.7] 0, 2
51. [5.7] $l = 12$ cm, $w = 6$ cm   **52.** [5.6] No real solution
53. [5.6] $-3, 2, \frac{5}{2}$

TEST: CHAPTER 5, P. 296

1. [5.1] Answers may vary. $(2x^2)(4x^2)$, $(8x)(x^3)$, $(-4x^3)(-2x)$ **2.** [5.2] $(x-5)(x-2)$
3. [5.4] $(x-5)^2$ **4.** [5.1] $2y^2(3-4y+2y^2)$
5. [5.1] $(x^2+2)(x+1)$ **6.** [5.1] $x(x-5)$
7. [5.2] $x(x+3)(x-1)$ **8.** [5.3] $2(5x-6)(x+4)$
9. [5.4] $(2x+3)(2x-3)$ **10.** [5.2] $(x-4)(x+3)$
11. [5.3] $3m(2m+1)(m+1)$ **12.** [5.4] $3(w+5)(w-5)$
13. [5.4] $5(3x+2)^2$ **14.** [5.4] $3(x^2+4)(x+2)(x-2)$
15. [5.4] $(7x-6)^2$ **16.** [5.3] $(5x-1)(x-5)$
17. [5.1] $(x^3-3)(x+2)$
18. [5.4] $5(4+x^2)(2+x)(2-x)$
19. [5.3] $(2x-5)(2x+3)$ **20.** [5.3] $3t(2t+5)(t-1)$
21. [5.2] $3(m+2n)(m-5n)$ **22.** [5.6] $5, -4$
23. [5.6] $-5, \frac{3}{2}$ **24.** [5.6] $-4, 7$ **25.** [5.7] Length: 8 m; width: 6 m   **26.** [5.7] 5 ft   **27.** [1.6] -0.4
28. [2.7] $\{l \mid l < 13 \text{ cm}\}$
29. [3.2]

[Graph of $y = \frac{3}{4}x + 1$]

30. [4.1] $49a^6b^{10}$ **31.** [5.7] $l = 15, w = 3$
32. [5.2] $(a-4)(a+8)$

CHAPTER 6

TECHNOLOGY CONNECTION 6.1, P. 299

1.

X	Y1	Y2
3	−10	−.7
4	−6	−1.333
5	0	ERROR
6	8	1.25
7	18	.61111
8	30	.4
9	44	.29545

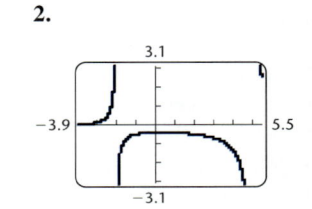

X = 5

2.

[Graph with window -3.9 to 5.5 and -3.1 to 3.1]

EXERCISE SET 6.1, PP. 304–305

1. 0 **3.** −6 **5.** 5 **7.** −4, 7 **9.** −5, 5 **11.** $\dfrac{5a}{4b^2}$
13. $\dfrac{5}{2xy^4}$ **15.** $\dfrac{3}{2}$ **17.** $\dfrac{a-3}{a+1}$ **19.** $\dfrac{3}{2x^3}$ **21.** $\dfrac{y-3}{2y}$
23. $\dfrac{3(2a-1)}{7(a-1)}$ **25.** $\dfrac{t-5}{t-4}$ **27.** $\dfrac{a-4}{2(a+4)}$ **29.** $\dfrac{x+4}{x-4}$
31. $t-1$ **33.** $\dfrac{y^2+4}{y+2}$ **35.** $\dfrac{1}{2}$ **37.** $\dfrac{5}{y+6}$
39. $\dfrac{y-6}{y-5}$ **41.** $\dfrac{a-3}{a+3}$ **43.** −1 **45.** 1 **47.** −5
49. $-\dfrac{3}{2}$ **51.** $(x+1)(x+7)$
53.

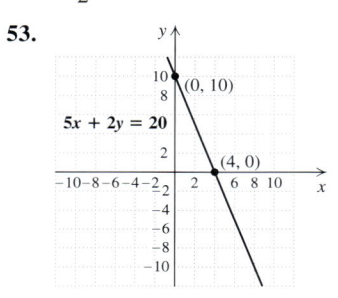

55. $\dfrac{5}{48}$ **57.** ◆ **59.** ◆ **61.** $-x-2y$
63. $\dfrac{(t-1)(t-9)^2}{(t^2+9)(t+1)}$ **65.** $\dfrac{(x-y)^3}{(x+y)^2(x-5y)}$
67. $\dfrac{x^3+4}{(x^2+2)(x^3+2)}$ **69.** ◆

EXERCISE SET 6.2, PP. 308–310

1. $\dfrac{7x(x-3)}{5(2x+1)}$ **3.** $\dfrac{(a-4)(a+2)}{(a+6)(a+6)}$ **5.** $\dfrac{(2x+3)(x+1)}{4(x-5)}$
7. $\dfrac{(a-5)(a+2)}{(a^2+1)(a^2-1)}$ **9.** $\dfrac{(x+1)(x-1)}{(2+x)(x+1)}$ **11.** $\dfrac{5a^2}{4}$
13. $\dfrac{2}{dc^2}$ **15.** $\dfrac{x+2}{x-2}$ **17.** $\dfrac{(a+5)(a-5)(2a-5)}{(a-3)(a-1)(2a+5)}$
19. $\dfrac{(a+3)(a-3)}{a(a+4)}$ **21.** $\dfrac{2a}{a-2}$ **23.** $\dfrac{t-5}{t+5}$
25. $\dfrac{5(a+6)}{a-1}$ **27.** $\dfrac{(x-1)(x-3)^3}{(x+3)(x+1)}$ **29.** $\dfrac{t-2}{t-1}$ **31.** $\dfrac{9}{x}$
33. $\dfrac{1}{a^3-8a}$ **35.** $\dfrac{x^2-4x+7}{x^2+2x-5}$ **37.** $\dfrac{35}{18}$ **39.** 4
41. $\dfrac{x^3}{y}$ **43.** $\dfrac{(a+2)(a+3)}{(a-3)(a-1)}$ **45.** $\dfrac{(x-1)^2}{x}$ **47.** $\dfrac{1}{2}$
49. $\dfrac{(y+3)(y^2+1)}{y+1}$ **51.** $\dfrac{15}{8}$ **53.** $\dfrac{15}{4}$ **55.** $\dfrac{a-5}{3(a-1)}$
57. $2x+1$ **59.** $\dfrac{(x+2)^2}{x}$ **61.** $\dfrac{x-5}{x+3}$

63. $\dfrac{(t^2+4)(t+3)^2}{(t-3)^2(t^4+16)}$ **65.** $\dfrac{1}{(c-5)^2}$ **67.** $\dfrac{t+5}{t-5}$ **69.** 4
71. $8x^3-11x^2-3x+12$ **73.** $-\dfrac{37}{20}$ **75.** ◆
77. ◆ **79.** $\dfrac{a}{(c-3d)(2a+5b)}$ **81.** $\dfrac{1}{b^3(a-3b)}$
83. $\dfrac{x^3y^2}{4}$ **85.** $\dfrac{4}{x+7}$ **87.** $\dfrac{(t-1)(t-9)(t-9)}{(t^2+9)(t+1)}$
89. $\dfrac{3(y+2)^3}{y(y-1)}$

EXERCISE SET 6.3, PP. 317–318

1. $\dfrac{13}{x}$ **3.** $\dfrac{3x+1}{15}$ **5.** $\dfrac{6}{a+3}$ **7.** $\dfrac{6}{a+2}$ **9.** $\dfrac{y+4}{y}$
11. 11 **13.** $\dfrac{7x+5}{x+1}$ **15.** $a+5$ **17.** $x-4$
19. $t+7$ **21.** $\dfrac{1}{x+2}$ **23.** $\dfrac{a+1}{a+6}$ **25.** $\dfrac{t-4}{t+3}$
27. $\dfrac{x+5}{x-6}$ **29.** $\dfrac{-5}{x-4}$ **31.** $\dfrac{-1}{x-1}$ **33.** 135 **35.** 72
37. 126 **39.** $12x^3$ **41.** $30a^4b^8$ **43.** $6(y-3)$
45. $(x+2)(x-2)(x+3)$ **47.** $t(t+2)^2(t-4)$
49. $30x^2y^2z^3$ **51.** $(a-1)^2(a+1)$ **53.** $(m-2)^2(m-3)$
55. $10v(v+3)(v+4)$ **57.** $18x^3(x-2)^2(x+1)$
59. $\dfrac{26}{12x^5}, \dfrac{x^2y}{12x^5}$ **61.** $\dfrac{12b}{8a^2b^2}, \dfrac{5a}{8a^2b^2}$
63. $\dfrac{(x+3)(x+1)}{(x+3)(x+2)(x-2)}, \dfrac{(x-2)^2}{(x+3)(x+2)(x-2)}$
65. $(x-4)(x-15)$ **67.** $x^2-9x+18$ **69.** $-\dfrac{1}{72}$
71. ◆ **73.** ◆ **75.** 0 **77.** $\dfrac{30}{(x-3)(x+4)}$
79. 1440 **81.** 7:55 A.M. **83.** ◆

EXERCISE SET 6.4, PP. 325–326

1. $\dfrac{3x+4}{x^2}$ **3.** $\dfrac{5}{24r}$ **5.** $\dfrac{7x+3y}{x^2y^2}$ **7.** $\dfrac{16-15t}{18t^3}$
9. $\dfrac{5x+9}{24}$ **11.** $\dfrac{a+8}{4}$ **13.** $\dfrac{5a^2+7a-3}{9a^2}$
15. $\dfrac{-7x-13}{4x}$ **17.** $\dfrac{c^2+3cd-d^2}{c^2d^2}$
19. $\dfrac{3y^2-3xy-6x^2}{2x^2y^2}$ **21.** $\dfrac{4x}{(x-1)(x+1)}$
23. $\dfrac{-z^2+5z}{(z-1)(z+1)}$ **25.** $\dfrac{11x+15}{4x(x+5)}$ **27.** $\dfrac{14-3x}{(x+2)(x-2)}$
29. $\dfrac{x^2-x}{(x+5)(x-5)}$ **31.** $\dfrac{4t-5}{4(t-3)}$ **33.** $\dfrac{2x+10}{(x+3)^2}$
35. $\dfrac{6-20x}{15x(x+1)}$ **37.** $\dfrac{9a}{4(a-5)}$ **39.** $\dfrac{t^2+2ty-y^2}{(y-t)(y+t)}$

A-22 ANSWERS

41. $\dfrac{2x^2 - 10x + 25}{x(x - 5)}$ **43.** $\dfrac{x - 3}{(x + 1)(x + 3)}$
45. $\dfrac{x^2 + 5x + 1}{(x + 1)^2(x + 4)}$ **47.** $\dfrac{x^2 - 48}{(x + 7)(x + 8)(x + 6)}$
49. $\dfrac{5x + 12}{(x + 3)(x - 3)(x + 2)}$ **51.** $\dfrac{11}{x - 1}$ **53.** $t + 2$
55. 0   **57.** $\dfrac{p^2 + 7p + 1}{(p + 5)(p - 5)}$ **59.** $\dfrac{9x + 12}{(x + 3)(x - 3)}$
61. $\dfrac{-3x^2 + 7x + 4}{3(x + 2)(2 - x)}$ **63.** $\dfrac{a - 2}{(a + 3)(a - 3)}$ **65.** $\dfrac{2}{y(y - 1)}$
67. $\dfrac{z - 3}{2z - 1}$ **69.** $\dfrac{2}{r + s}$
71. [graph of $y = \tfrac{1}{2}x - 5$]   **73.** [graph of $y = 3$]
75. -8 **77.** 3, 5   **79.** ◆   **81.** ◆
83. Perimeter: $\dfrac{4x^2 + 18x}{(x + 4)(x + 5)}$; area: $\dfrac{x^2}{(x + 4)(x + 5)}$
85. $\dfrac{30}{(x - 3)(x + 4)}$ **87.** $\dfrac{x^2 + xy - x^3 + x^2y - xy^2 + y^3}{(x^2 + y^2)(x + y)^2(x - y)}$
89. $\dfrac{-x^2 - 3}{(2x - 3)(x - 3)}$
91. Answers may vary; $\dfrac{a}{a - b} + \dfrac{3b}{b - a}$

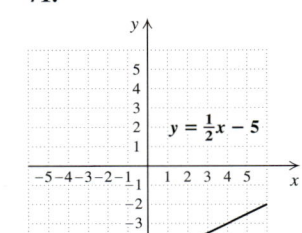

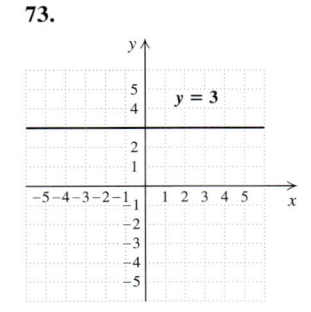

EXERCISE SET 6.5, PP. 331–332
1. $\dfrac{2}{5}$ **3.** $\dfrac{18}{65}$ **5.** $\dfrac{4s^2}{9 + 3s^2}$ **7.** $\dfrac{3x}{2x + 1}$ **9.** $\dfrac{4a - 10}{a - 1}$
11. $x - 4$ **13.** $\dfrac{1}{x}$ **15.** $\dfrac{1 + t^2}{t - t^2}$ **17.** $\dfrac{x}{x - y}$
19. $\dfrac{3a^2 - 4a}{2 + 3a^2}$ **21.** $\dfrac{60 - 15a^3}{126a^2 + 28a^3}$ **23.** $\dfrac{2xy - 3y^3}{xy^3 + 30}$
25. $\dfrac{5a^2 + 2b^3}{5b^3 - 3a^2b^3}$ **27.** $\dfrac{2x^4 - 3x^2}{2x^4 + 3}$ **29.** $\dfrac{t^2 - 2}{t^2 + 5}$
31. $\dfrac{7a^2b^3 - 5a}{4b^2 + ab^2}$ **33.** $\dfrac{a - 7}{3a + 2a^2}$ **35.** $\dfrac{10b^2 + 2a}{ab - 10b^2}$
37. $\dfrac{20x^2 - 30x}{15x^2 + 105x - 12}$ **39.** -4
41. $23x^4 + 50x^3 + 23x^2 - 163x + 41$ **43.** ◆
45. ◆   **47.** $\dfrac{(x - 1)(3x - 2)}{5x - 3}$ **49.** $\dfrac{ac}{bd}$ **51.** $\dfrac{3x + 2}{2x + 1}$
53. ◆

EXERCISE SET 6.6, PP. 336–337
1. $\dfrac{6}{5}$ **3.** $\dfrac{40}{29}$ **5.** $\dfrac{40}{9}$ **7.** $-3, -1$ **9.** $-7, 7$ **11.** 2
13. $\dfrac{23}{4}$ **15.** 5   **17.** $-\dfrac{13}{2}$ **19.** 3   **21.** -2
23. No solution   **25.** -10 **27.** $\dfrac{1}{5}$ **29.** No solution
31. 2   **33.** No solution   **35.** $\dfrac{25}{9}$ **37.** $a^{-6}b^{-15}$, or $\dfrac{1}{a^6b^{15}}$
39. $\dfrac{3x - 16}{(x + 3)(x - 3)(x - 2)}$ **41.** ◆   **43.** ◆
45. -2 **47.** $-\dfrac{1}{6}$ **49.** $-1, 0$ **51.** 4   **53.** [graph]

EXERCISE SET 6.7, PP. 344–348
1. $-1, 5$ **3.** 1   **5.** $6\tfrac{6}{7}$ hr   **7.** $25\tfrac{5}{7}$ min   **9.** $20\tfrac{4}{7}$ hr
11. $11\tfrac{1}{9}$ min   **13.** $4\tfrac{4}{9}$ min
15.

Speed	Time
r	t
$r + 40$	t

Speed	Time
r	$\dfrac{300}{r}$
$r + 40$	$\dfrac{700}{r + 40}$

Train: 30 mph; police car: 70 mph

17.

Speed	Time
$r - 14$	t
r	t

Speed	Time
$r - 14$	$\dfrac{330}{r - 14}$
r	$\dfrac{400}{r}$

Passenger: 80 km/h; freight: 66 km/h
19. Ted: 14 km/h; Joni: 19 km/h   **21.** 3 hr
23. 15 students per faculty member   **25.** 1088 ft/sec
27. $\dfrac{15}{4}$ cm   **29.** $b = 3$ m, $c = 5$ m, $d = 1$ m; $b = 10$ m, $c = 12$ m, $d = 8$ m   **31.** 582   **33.** $21\tfrac{2}{3}$ cups   **35.** 10.5
37. $\dfrac{8}{3}$ **39.** 15 ft   **41.** 287   **43.** 184   **45.** No
47. 225   **49.** (a) 4.8 tons; (b) 48 lb   **51.** 1
53. $13y^3 - 14y^2 + 12y - 73$ **55.** $t = \dfrac{n - ar}{s}$ **57.** ◆
59. ◆   **61.** Ann: 6 hr; Betty: 12 hr
63. $27\tfrac{3}{11}$ minutes after 5:00   **65.** $9\tfrac{3}{13}$ days   **67.** $30\tfrac{22}{31}$ hr
69. 270   **71.** ◆

EXERCISE SET 6.8, PP. 352–354
1. $h = \dfrac{S}{2\pi r}$ **3.** $h = \dfrac{2A}{b}$ **5.** $n = \dfrac{s}{180} + 2$, or $n = \dfrac{s + 360}{180}$ **7.** $b = \dfrac{3V - kB + 2kn}{k}$ **9.** $r = \dfrac{L}{l - S}$

11. $h = \dfrac{2A}{b_1 + b_2}$ 13. $a = \dfrac{d}{b - c}$ 15. $r = \dfrac{m}{s}$

17. $d = \dfrac{c}{a + b}$ 19. $R = \dfrac{V_1 - V_2}{I}$ 21. $f = \dfrac{pq}{q + p}$

23. $t = \dfrac{v_2 - v_1}{a}$ 25. $l = \dfrac{P - 2w}{2}$, or $l = \dfrac{P}{2} - w$

27. $a = \dfrac{Q}{M(b - c)}$ 29. $a = \dfrac{b}{K - C}$ 31. $r^3 = \dfrac{3V}{4\pi}$

33. $r = \dfrac{Sl - a}{S - l}$ 35. [graph of $y = \tfrac{4}{5}x + 1$]

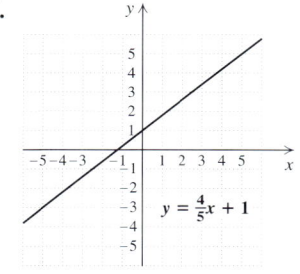

37. $-\dfrac{3}{4}$ 39. $(x + 2)(x - 15)$ 41. ◆   43. ◆

45. $n_2 = \dfrac{n_1 p_2 R + p_1 p_2 n_1}{p_1 p_2 - p_1 R}$ 47. $v = \dfrac{Nbf_2 - bf_1 - df_1}{Nf_2 - 1}$

REVIEW EXERCISES: CHAPTER 6, PP. 356–357

1. [6.1] 0   2. [6.1] 6   3. [6.1] $-6, 6$ 4. [6.1] $-6, 5$
5. [6.1] -2 6. [6.1] $\dfrac{x - 2}{x + 1}$ 7. [6.1] $\dfrac{7x + 3}{x - 3}$
8. [6.1] $\dfrac{y - 5}{y + 5}$ 9. [6.1] $-5(x + 2y)$ 10. [6.2] $\dfrac{a - 6}{5}$
11. [6.2] $\dfrac{6}{2t - 1}$ 12. [6.2] $-20t$ 13. [6.2] $\dfrac{2x^2 - 2x}{x + 1}$
14. [6.2] $\dfrac{(x^2 + 1)(2x + 1)}{(x - 2)(x + 1)}$ 15. [6.2] $\dfrac{(t + 4)^2}{t + 1}$
16. [6.3] $24a^5 b^7$ 17. [6.3] $x^4(x + 1)(x - 1)$
18. [6.3] $(y - 2)(y + 2)(y + 1)$ 19. [6.3] $\dfrac{-3x + 18}{x + 7}$
20. [6.4] -1 21. [6.3] $\dfrac{4}{x - 4}$ 22. [6.4] $\dfrac{x + 5}{2x}$
23. [6.4] $\dfrac{2x + 3}{x - 2}$ 24. [6.4] $\dfrac{2a}{a - 1}$ 25. [6.4] $d + c$
26. [6.4] $\dfrac{-x^2 + x + 26}{(x - 5)(x + 5)(x + 1)}$ 27. [6.4] $\dfrac{2(x - 2)}{x + 2}$
28. [6.4] $\dfrac{19x + 8}{10x(x + 2)}$ 29. [6.5] $\dfrac{z}{1 - z}$
30. [6.5] $\dfrac{2x^4 y^2 + x^3}{y + xy}$ 31. [6.5] $c - d$ 32. [6.6] 8
33. [6.6] $-\dfrac{1}{2}$ 34. [6.6] $3, -5$ 35. [6.7] $5\tfrac{1}{7}$ hr
36. [6.7] Car: 105 km/h; train: 90 km/h   37. [6.7] -2

38. [6.7] 80   39. [6.7] 6   40. [6.8] $s = \dfrac{rt}{r - t}$
41. [6.8] $C = \tfrac{5}{9}(F - 32)$, or $C = \tfrac{5}{9}F - \tfrac{160}{9}$ 42. [2.1] 11
43. [3.3] $(3, 0), (0, -6)$ [graph of $2x - y = 6$]

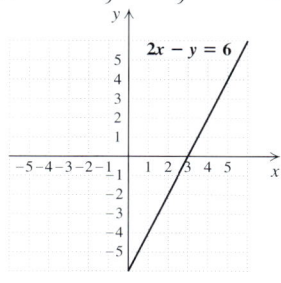

44. [5.2] $(x + 12)(x - 4)$
45. [4.3] $-2x^3 + 3x^2 + 12x - 18$
46. [6.3], [6.6] ◆ A student should master factoring before beginning a study of rational equations because it is necessary to factor when finding the LCD of the rational expressions. It may also be necessary to factor to use the principle of zero products after fractions have been cleared.
47. [6.3] ◆ Although multiplying the denominators of the expressions being added results in a common denominator, it is often not the *least* common denominator. Using a common denominator other than the LCD makes the expressions more complicated, requires additional simplifying after the addition has been performed, and leaves more room for error.
48. [6.2] $\dfrac{5(a + 3)^2}{a}$ 49. [6.4] $\dfrac{10a}{(a - b)(b - c)}$

TEST: CHAPTER 6, P. 358

1. [6.1] 0   2. [6.1] -8 3. [6.1] $-7, 7$ 4. [6.1] 1, 2
5. [6.1] $\dfrac{3x + 7}{x + 3}$ 6. [6.2] $\dfrac{-2(a + 5)}{3}$
7. [6.2] $\dfrac{(5y + 1)(y + 1)}{3y(y + 2)}$
8. [6.2] $\dfrac{(2x + 1)(2x - 1)(x^2 + 1)}{(x - 1)^2(x - 2)}$ 9. [6.2] $(x + 3)(x - 3)$
10. [6.3] $(y - 3)(y + 3)(y + 7)$ 11. [6.3] $\dfrac{23 - 3x}{x^3}$
12. [6.3] $\dfrac{8 - 2t}{t^2 + 1}$ 13. [6.4] $\dfrac{-3}{x - 3}$ 14. [6.4] $\dfrac{2x - 5}{x - 3}$
15. [6.4] $\dfrac{8t - 3}{t(t - 1)}$ 16. [6.4] $\dfrac{-x^2 - 7x - 15}{(x + 4)(x - 4)(x + 1)}$
17. [6.4] $\dfrac{x^2 + 2x - 7}{(x - 1)^2(x + 1)}$ 18. [6.5] $\dfrac{3y + 1}{y}$
19. [6.5] $\dfrac{3a^2 b^2 - 2a^3}{a^3 b^2 + 2b^2}$ 20. [6.6] 12   21. [6.6] $-3, 5$
22. [6.7] 12 min   23. [6.7] $2\tfrac{1}{7}$ cups
24. [6.7] 45 km/h, 65 km/h   25. [6.8] $t = \dfrac{d}{r + w}$

A-24 ANSWERS

26. [2.1] $-\dfrac{3}{7}$
27. [3.3] (10, 0), (0, 4)

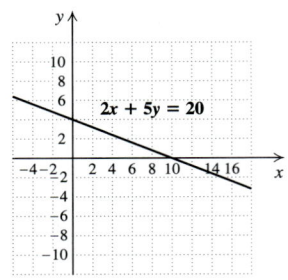

(Graph of $2x + 5y = 20$)

28. [5.2] $(x - 9)(x + 5)$ **29.** [4.3] $13x^2 - 29x + 76$
30. [6.7] Reggie: 10 hr; Rema: 4 hr **31.** [6.5] $\dfrac{3a + 2}{2a + 1}$

32. [3.5] -2 **33.** [4.8] x^{-2}, or $\dfrac{1}{x^2}$ **34.** [4.8] y^{-8}, or $\dfrac{1}{y^8}$
35. [4.1] $-4a^4b^{14}$ **36.** [4.3] $-y^3 - 2y^2 - 2y + 7$
37. [1.2] $12x + 16y + 4z$
38. [4.4] $2x^5 + x^3 - 6x^2 - x + 3$
39. [4.5] $36x^2 - 60xy + 25y^2$ **40.** [4.5] $2x^2 - x - 21$
41. [4.5] $4x^6 - 1$ **42.** [5.1] $2x(3 - x - 12x^3)$
43. [5.4] $(4x + 9)(4x - 9)$ **44.** [5.2] $(x - 6)(x - 4)$
45. [5.3] $(2x + 1)(4x + 3)$ **46.** [5.3] $2(3x - 2)(x - 4)$
47. [5.4] $2(x + 3)(x - 3)$ **48.** [5.4] $(4x + 5)^2$
49. [5.3] $(3x - 2)(x + 4)$ **50.** [5.1] $(x^3 - 3)(x + 2)$
51. [6.2] $\dfrac{y - 6}{2}$ **52.** [6.2] 1 **53.** [6.4] $\dfrac{a^2 + 7ab + b^2}{a^2 - b^2}$
54. [6.4] $\dfrac{-2x - 5}{x - 4}$ **55.** [6.5] $\dfrac{x}{x - 2}$ **56.** [6.5] $\dfrac{t + 2t^3}{t^3 - 2}$
57. [4.7] $5x^2 - 4x + 2 + \dfrac{2}{3x} + \dfrac{6}{x^2}$
58. [4.7] $15x^3 - 57x^2 + 177x - 529 + \dfrac{1605}{x + 3}$
59. [2.7] At most 225 **60.** [2.4] \$3.60 **61.** [5.7] 14 ft
62. [2.5] $-278, -276$ **63.** [6.7] 30 min
64. [6.7] Harley: 40 km/h; Millie: 50 km/h
65. [2.5] 26 in. **66.** [4.3], [4.5] 12
67. [1.4], [2.2] $-144, 144$
68. [4.5] $16y^6 - y^4 + 6y^2 - 9$
69. [5.4] $2(a^{16} + 81b^{20})(a^8 + 9b^{10})(a^4 + 3b^5)(a^4 - 3b^5)$
70. [5.6] $4, -7, 12$ **71.** [1.4], [1.6] -7

CUMULATIVE REVIEW: 1–6, PP. 359–360

1. [1.2] $2b + a$ **2.** [1.4] $-3.1 > -3.15$ **3.** [1.8] 49
4. [1.8] $-8x + 28$ **5.** [1.5] $-\dfrac{43}{8}$ **6.** [1.7] 1
7. [1.7] -6.2 **8.** [1.8] 8 **9.** [2.2] 10 **10.** [5.6] $-7, 7$
11. [2.1] $\dfrac{9}{2}$ **12.** [2.2] -2 **13.** [2.2] $\dfrac{8}{3}$
14. [5.6] $-10, -1$ **15.** [2.2] -8 **16.** [6.6] 1, 4
17. [2.6] $\{y \mid y \leq -\dfrac{2}{3}\}$ **18.** [6.6] -17 **19.** [5.6] $-4, \dfrac{1}{2}$
20. [2.6] $\{x \mid x > 43\}$ **21.** [6.6] 2 **22.** [5.6] $-\dfrac{7}{2}, 5$
23. [6.6] -13 **24.** [6.8] $t = \dfrac{4b}{A}$ **25.** [6.8] $n = \dfrac{tm}{t - m}$
26. [6.8] $c = \dfrac{a - b}{r}$ **27.** [1.6] $\dfrac{3}{2}x + 2y - 3z$
28. [4.2] $-4x^3 - \dfrac{1}{7}x^2 - 2$
29. [3.2] **30.** [3.3]

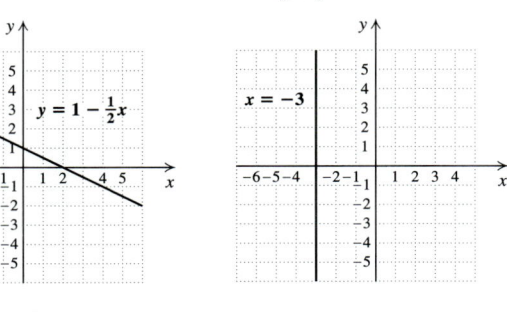

(Graph of $y = 1 - \tfrac{1}{2}x$) *(Graph of $x = -3$)*

31. [3.3]

(Graph of $x - 6y = 6$)

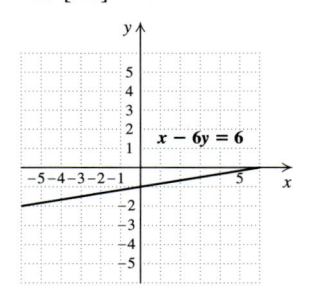

CHAPTER 7

TECHNOLOGY CONNECTION 7.1, P. 367

1. $y_1 = -\tfrac{3}{4}x - 2, \; y_2 = -\tfrac{1}{5}x - 2,$
$y_3 = -\tfrac{3}{4}x - 5, \; y_4 = -\tfrac{1}{5}x - 5$

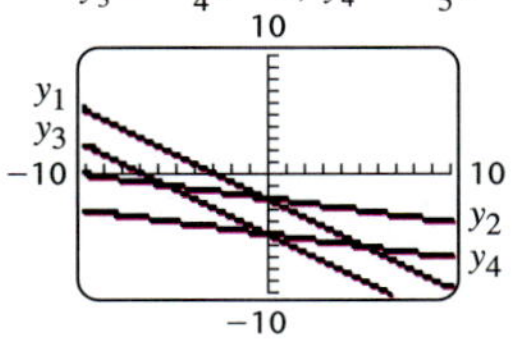

EXERCISE SET 7.1, PP. 367–369

1. **3.**

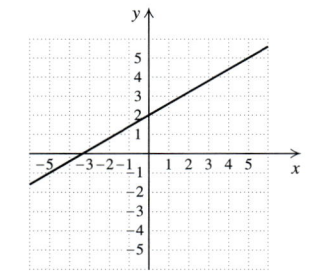

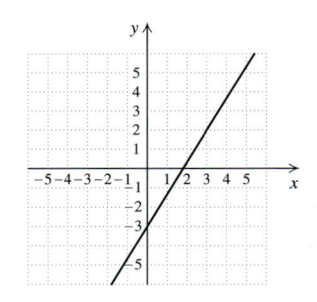

5. [graph]

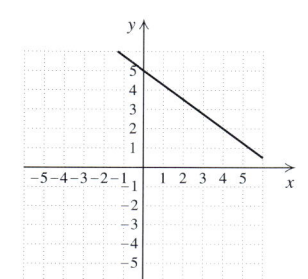

7. [graph]

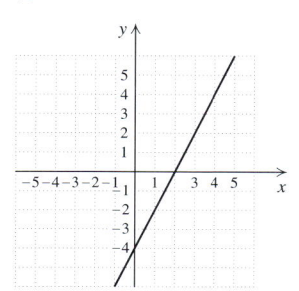

9. [graph]

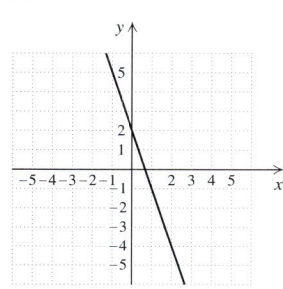

11. $\frac{3}{7}$; (0, 6)   **13.** $-\frac{5}{6}$; (0, 2)   **15.** $\frac{9}{4}$; (0, −7)
17. $-\frac{2}{5}$; (0, 0)   **19.** 2; (0, 4)   **21.** $\frac{3}{4}$; (0, −3)
23. $\frac{1}{5}$; $\left(0, \frac{8}{5}\right)$ **25.** 0; (0, 4)   **27.** $y = 3x + 7$
29. $y = \frac{7}{8}x - 1$ **31.** $y = -\frac{5}{3}x - 8$ **33.** $y = -2x + 3$

35. [graph of $y = \frac{3}{5}x + 2$]

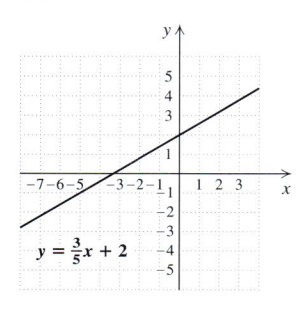

37. [graph of $y = -\frac{3}{5}x + 1$]

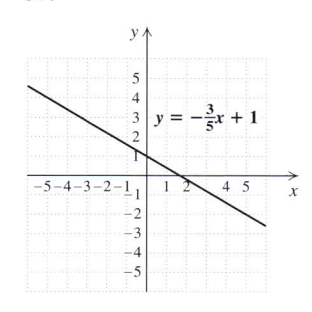

39. [graph of $y = \frac{5}{3}x + 3$]

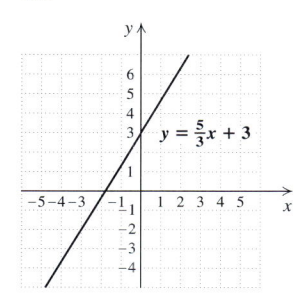

41. [graph of $y = -\frac{3}{2}x - 2$]

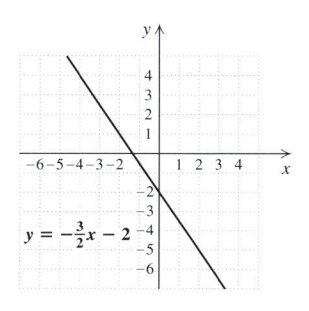

43. [graph of $2x + y = 1$]

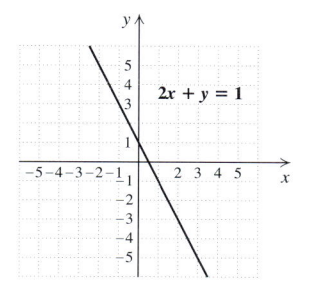

45. [graph of $3x - y = 4$]

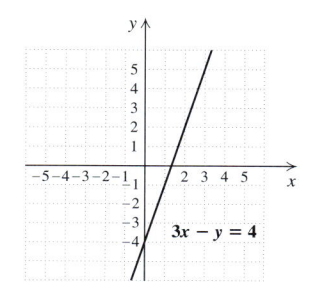

47. [graph of $2x + 3y = 9$]

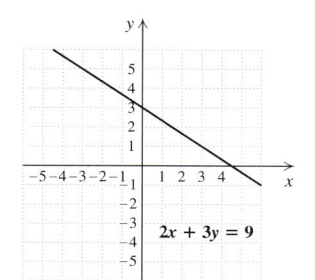

49. [graph of $x - 4y = 12$]

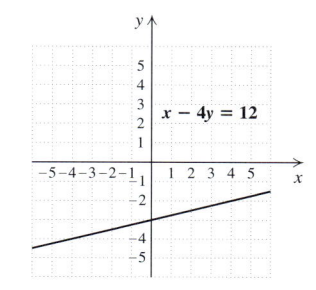

51. [graph of $y = x + 9$, Cost vs. Amount of water used (in 10,000 gallons)] $1 per 10,000 gallons

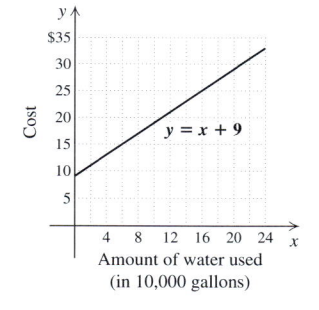

53. $y = 1.5x + 16$ **55.** Yes   **57.** No   **59.** Yes
61. −3, 0   **63.** 13 and 15, −15 and −13   **65.** $\frac{53}{7}$
67. ◆   **69.** ◆   **71.** Yes   **73.** No   **75.** Yes
77. When $x = 0$, $y = b$, so $(0, b)$ is on the line. When $x = 1$, $y = m + b$, so $(1, m + b)$ is on the line. Then
$$\text{slope} = \frac{(m + b) - b}{1 - 0} = m.$$
79. $y = \frac{5}{2}x + 1$

EXERCISE SET 7.2, PP. 372–373

1. $y - 7 = 5(x - 3)$ **3.** $y - 4 = \frac{3}{4}(x - 2)$
5. $y - (-6) = 1 \cdot (x - 2)$ **7.** $y - 0 = -3(x - (-4))$
9. $y - 6 = \frac{2}{3}(x - 5)$ **11.** $y = 2x + 2$ **13.** $y = -x + 8$
15. $y = \frac{1}{2}x + 4$ **17.** $y = -\frac{1}{3}x - 7$ **19.** $y = \frac{5}{4}x - 8$
21. $y = \frac{1}{4}x + \frac{5}{2}$ **23.** $y = -\frac{1}{2}x + 4$ **25.** $y = -\frac{3}{2}x + \frac{13}{2}$
27. $y = \frac{2}{5}x - 2$ **29.** $y = \frac{3}{4}x - \frac{5}{2}$

A-26 ANSWERS

31. [graph: $y - 4 = \frac{1}{2}(x - 3)$]

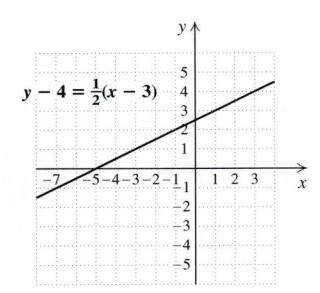

33. [graph: $y - 2 = -\frac{1}{2}(x - 5)$]

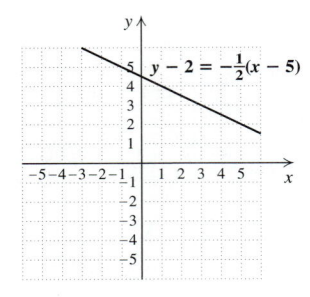

35. [graph: $y + 1 = \frac{1}{2}(x - 3)$]

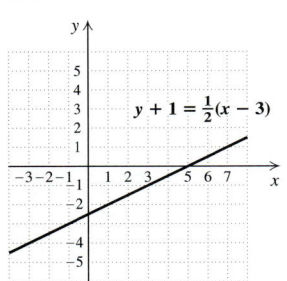

37. [graph: $y + 2 = 3(x + 1)$]

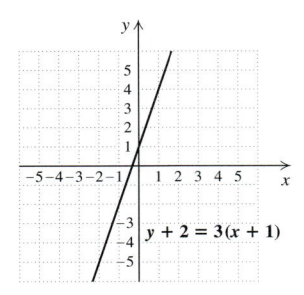

39. [graph: $y - 4 = -2(x + 1)$]

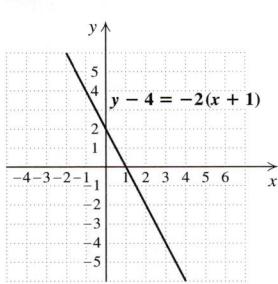

41. [graph: $y + 3 = -(x + 2)$]

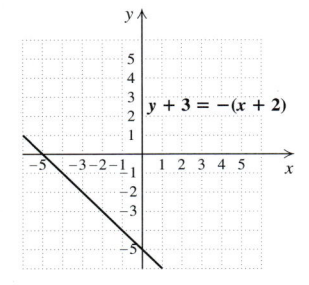

43. [graph: $5x - 2y = 5$]

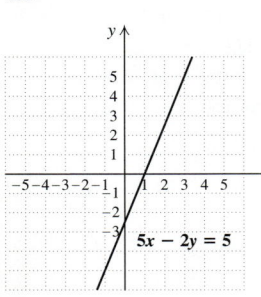

45. [graph: $y = -\frac{4}{3}x - 6$]

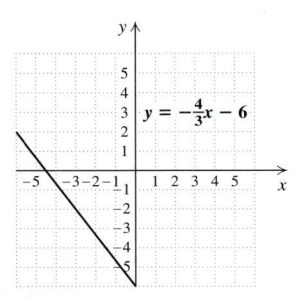

47. $\{x \mid x \geq -\frac{7}{4}\}$ **49.** ◆   **51.** ◆
53. $y = 3x + 17$ **55.** $y = \frac{10}{3}x + \frac{25}{3}$ **57.** ◆

EXERCISE SET 7.3, P. 377
1. No   **3.** Yes

5. [graph: $y \leq x + 3$]

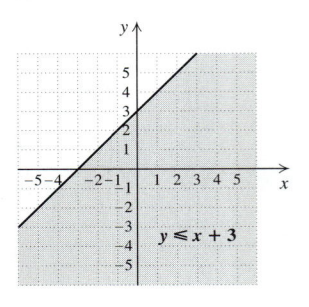

7. [graph: $y < x - 1$]

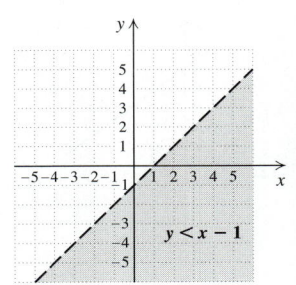

9. [graph: $y \geq x - 2$]

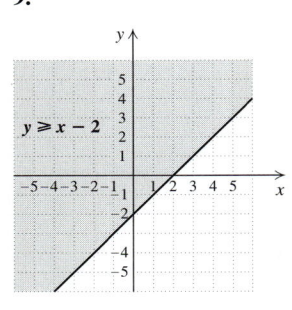

11. [graph: $y \leq 2x - 1$]

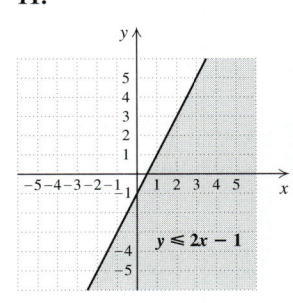

13. [graph: $x - y \leq 3$]

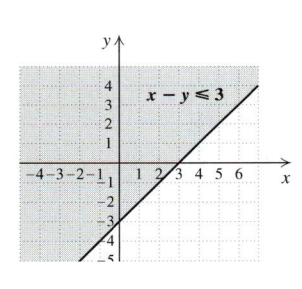

15. [graph: $x + y > 7$]

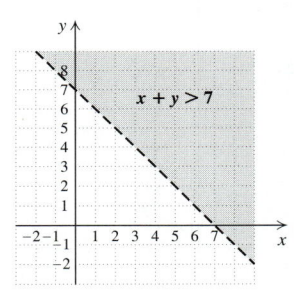

17. [graph: $x - 3y < 6$]

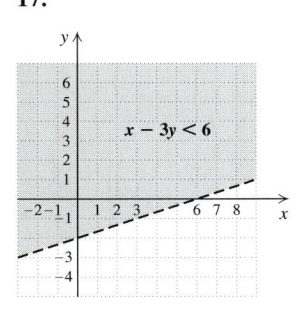

19. [graph: $2x + 3y \leq 12$]

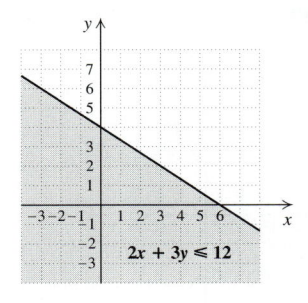

21. [graph: $y \geq 1 - 2x$]

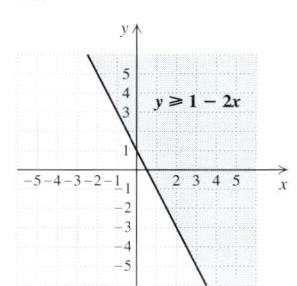

23. [graph: $y - 3x > 0$]

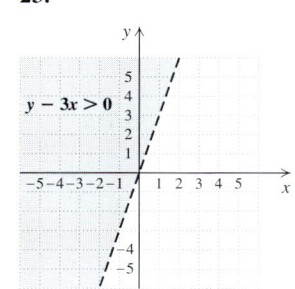

25. [graph: $x < 4$]

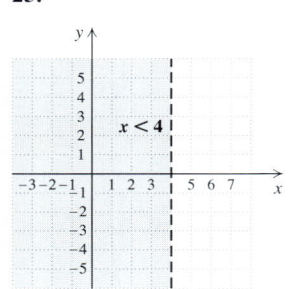

27. [graph: $x \geq 3$]

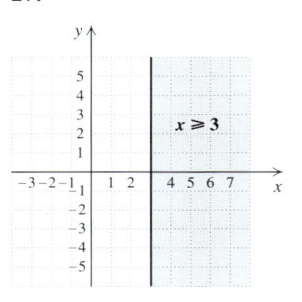

29. [graph: $y \leq 3$]

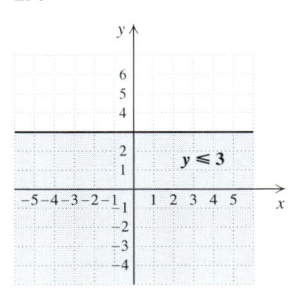

31. [graph: $y \geq -5$]

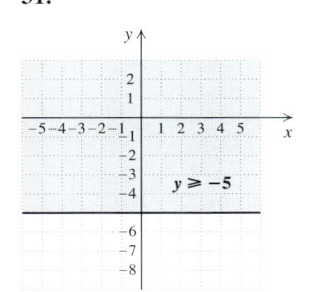

33. $2x^3 + 5x^2 - 4x - 3$ **35.** $(a + 6)(3a^2 - 4)$
37. -18 **39.** ◆   **41.** ◆
43. $35c + 75a > 1000$

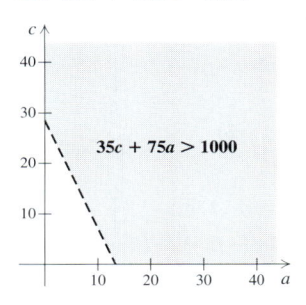

[graph: $35c + 75a > 1000$]

45. $r + t \geq 17$ [graph: $r + t \geq 17$]

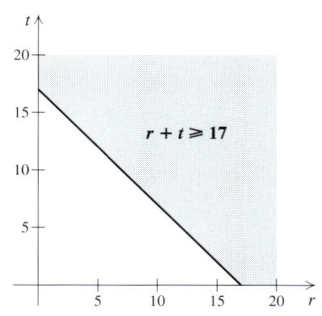

47. $x \geq -2$ **49.** [graph: $xy \geq 0$]

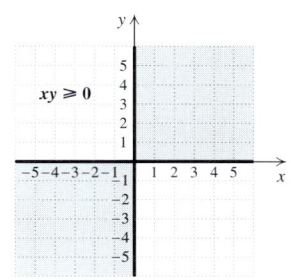

EXERCISE SET 7.4, PP. 383–384

1. $y = 4x$ **3.** $y = 1.75x$ **5.** $y = 3.2x$ **7.** $y = \dfrac{2}{3}x$
9. $y = \dfrac{90}{x}$ **11.** $y = \dfrac{70}{x}$ **13.** $y = \dfrac{1}{x}$ **15.** $y = \dfrac{2100}{x}$
17. $183.75   **19.** $22\frac{6}{7}$ **21.** $3\frac{5}{9}$ amperes   **23.** 66.88 lb
25. 54 min   **27.** 10   **29.** $b = \dfrac{2a - 4}{3}$ **31.** $3x + 12y$
33. [graph: $3x - 5y = 6$]

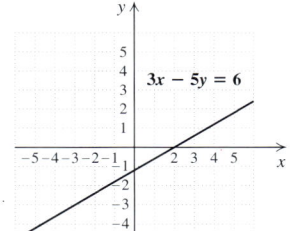

35. ◆   **37.** ◆
39. $P = kS$, $k = 8$
41. $B = kN$ **43.** $S = kv^6$ **45.** $I = \dfrac{k}{d^2}$
47. $V = kr^3$, $k = \frac{4}{3}\pi$ **49.** ◆

REVIEW EXERCISES: CHAPTER 7, PP. 385–386

1. [7.1] -9; $(0, 46)$ **2.** [7.1] -1; $(0, 9)$
3. [7.1] $\frac{1}{3}$; $\left(0, -\frac{2}{3}\right)$ **4.** [7.1] 0; $(0, 8)$
5. [7.1] $y = 2x - 4$ **6.** [7.1] $y = -\frac{3}{2}x + 1$

A-28 ANSWERS

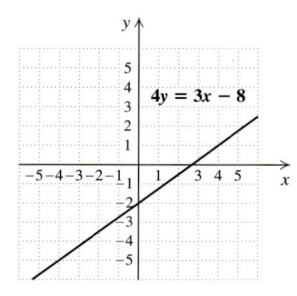

7. [7.1] [graph of $y = -\frac{3}{4}x - 2$]

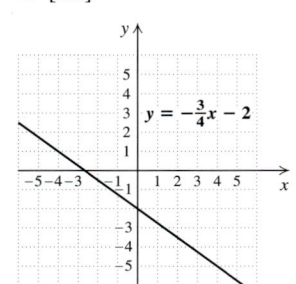

8. [7.1] [graph of $y + \frac{1}{2}x = 2$]

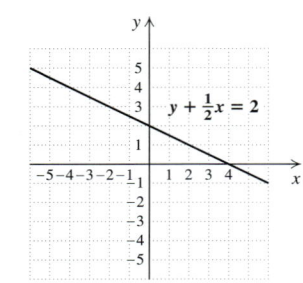

9. [7.2] [graph of $y - 2 = 3(x - 6)$]

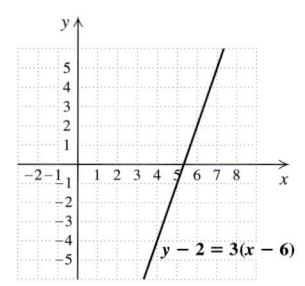

10. [7.2] [graph of $y + 1 = \frac{2}{5}(x - 3)$]

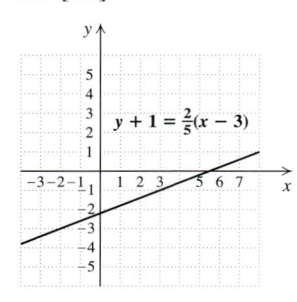

11. [7.1] [graph of $4y = 3x - 8$]

12. [7.2] [graph of $y - 4 = -\frac{5}{3}(x + 2)$]

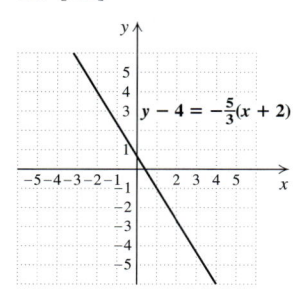

13. [7.1] Yes   **14.** [7.1] No   **15.** [7.1] $y = 0.5x + 12$
16. [7.2] $y - 2 = 3(x - 1)$
17. [7.2] $y - (-5) = \frac{2}{3}(x - (-2))$ **18.** [7.2] $y = x + 2$
19. [7.2] $y = \frac{1}{2}x - 1$
20. [7.3] [graph of $x \leq y$]

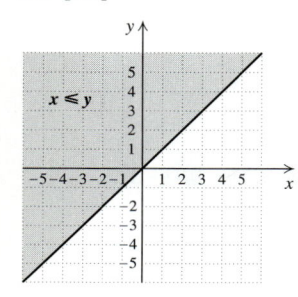

21. [7.3] [graph of $x - 2y \geq 4$]

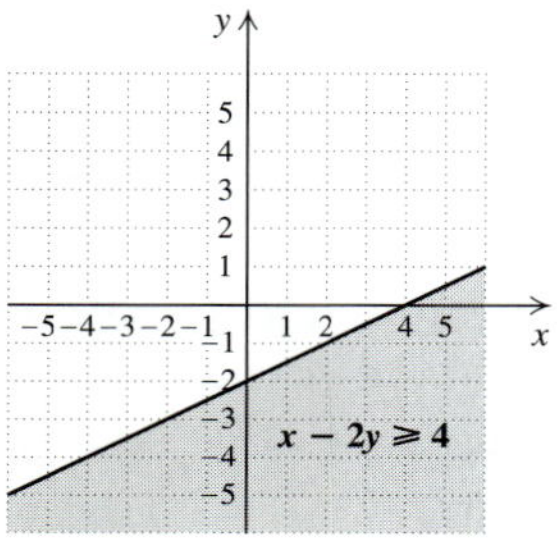

22. [7.3] [graph of $y > -2$]

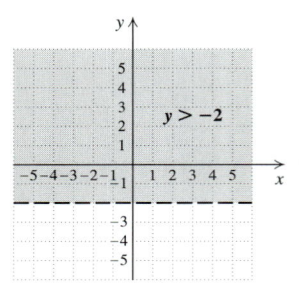

23. [7.3] [graph of $y \geq \frac{2}{3}x - 5$]

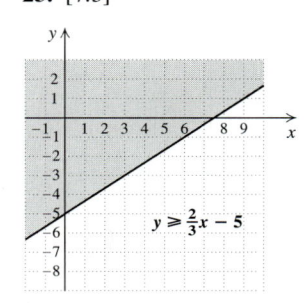

24. [7.3] [graph of $2x + y < 1$]

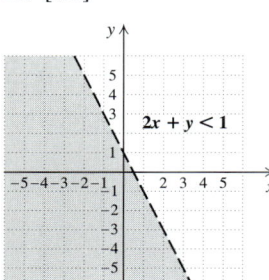

25. [7.3] [graph of $x < 4$]

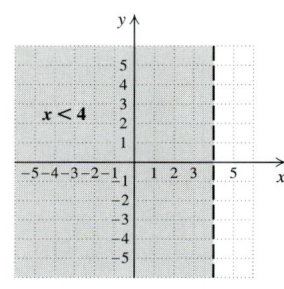

26. [7.4] $y = \dfrac{243}{x}$ **27.** [7.4] 9.6 lb
28. [4.5] $\frac{1}{4}y^2 + 3y + 9$ **29.** [5.1] $(x - 1)(x^2 + 2)$
30. [4.3] $5x^3 - x^2 + 6x + 5$ **31.** [4.3] $2a + 28b$
32. [7.1], [7.2] Answers may vary. Some will consider it easier to convert the point–slope form $y - y_1 = m(x - x_1)$ to the form $y = mx - mx_1 + y_1$, or $y = mx + b$, where $b = -mx_1 + y_1$. Others will consider it easier to convert the slope–intercept form $y = mx + b$ to the form $y - 0 = m\left(x + \dfrac{b}{m}\right)$, or $y - y_1 = m(x - x_1)$, where $y_1 = 0$ and $x_1 = -\dfrac{b}{m}$.

33. [7.3] ◆ The boundary line is part of the graph of a linear inequality $ax + by \leq c$ because the $\leq$ sign indicates that the graph of $ax + by < c$, as well as the graph of $ax + by = c$, form the solution set. The graph of $ax + by < c$ does not contain the graph of $ax + by = c$.
34. [7.1] $y = -x$ **35.** [7.1] $-\dfrac{c}{a}$; $(0, c)$, $(a, 0)$
36. [7.1] $y = -2x - 3$ **37.** [7.1], [7.2] $y = \frac{3}{2}x + \frac{7}{2}$

TEST: CHAPTER 7, P. 386

1. [7.1] $-\frac{1}{5}$; $(0, -7)$ **2.** [7.1] 0; $(0, -3)$
3. [7.1] 2, $\left(0, \frac{1}{4}\right)$ **4.** [7.1] $\frac{4}{3}$; $(0, -2)$
5. [7.1] $y = \frac{1}{2}x - 7$ **6.** [7.1] $y = -4x + 3$
7. [7.2] $y - 5 = 1(x - 3)$ **8.** [7.2] $y - 0 = -\frac{1}{2}(x - (-2))$
9. [7.2] $y = -3x + 4$ **10.** [7.2] $y = \frac{1}{4}x - 2$

11. [7.1]

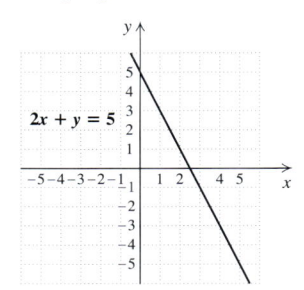

12. [7.1] $y = \frac{2}{3}x - 6$

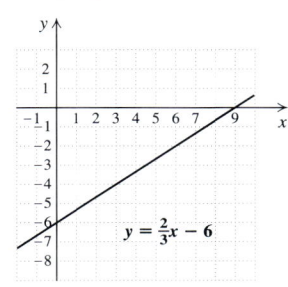

13. [7.2] $y + 4 = -\frac{1}{2}(x - 1)$

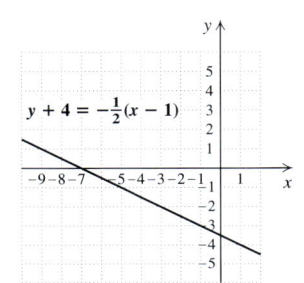

14. [7.1] $3y = 2x - 9$

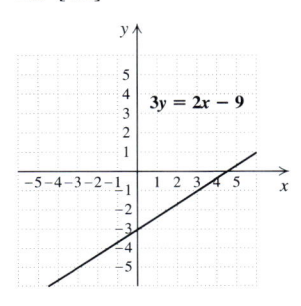

15. [7.1] Yes
16. [7.3] $y > x - 1$
17. [7.3] $2x - y \le 4$

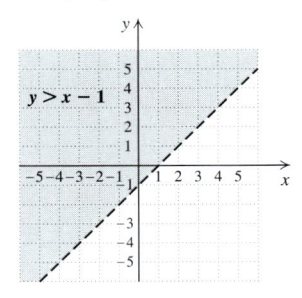

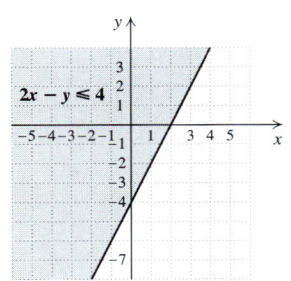

18. [7.3] $y < -2$

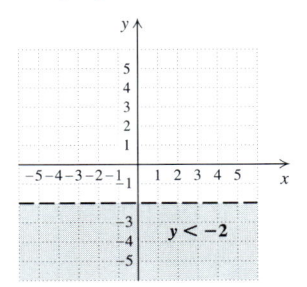

19. [7.4] $y = 4.5x$ **20.** [7.4] 18 min
21. [4.4] $-3y^7 + 9y^5 - 21y^4$ **22.** [4.5] $x^2 - 0.01$
23. [2.3] $x = \dfrac{5y + 2z}{3}$ **24.** [4.3] $8x$
25. [7.1], [7.2] $y = \frac{2}{3}x + \frac{11}{3}$
26. [7.1] $y = \frac{5}{2}x$

CHAPTER 8

TECHNOLOGY CONNECTION 8.1, P. 390
1.  ◆   **2.** No. The lines are parallel.

EXERCISE SET 8.1, PP. 392–393
1. Yes   **3.** No   **5.** Yes   **7.** (4, 2)   **9.** (2, 1)
11. (2, −2)   **13.** No solution   **15.** (2, −1)
17. Infinite number of solutions   **19.** (−3, −3)
21. (−5, 3)   **23.** (1, −3)   **25.** (−8, −7)   **27.** (3, −4)
29. (6, 2)   **31.** $\dfrac{9x + 12}{(x + 4)(x - 4)}$ **33.** Trinomial   **35.** −2
37. ◆   **39.** ◆   **41.** Exercises 14 and 17
43. Exercises 13 and 22
45. Answers may vary.
$2x - y = 8,$
$x + 3y = -10$
47. $A = 2, B = 2$
49. (a) Electric: $y = 100x$; gas: $y = 25x + 400$;
(b)

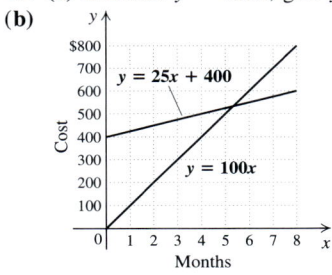

(c) $5\frac{1}{3}$ months   **51.** (41.5, 17.1)

TECHNOLOGY CONNECTION 8.2, P. 397
1. (2.5, −1.75)

EXERCISE SET 8.2, PP. 399–401
1. (2, −4)   **3.** (2, 1)   **5.** (4, 3)   **7.** (1, −2)
9. (−2, 4)   **11.** No solution   **13.** (−1, −3)
15. $\left(\frac{17}{3}, \frac{2}{3}\right)$ **17.** Infinite number of solutions
19. Infinite number of solutions   **21.** (−12, 11)
23. (−2, 4)   **25.** (1, 2)   **27.** Infinite number of solutions
29. (−7, −2)   **31.** No solution   **33.** 23 and 26
35. 36 and 22   **37.** 28 and 12   **39.** 70°, 110°
41. 62°, 28°   **43.** 380 mi, 270 mi   **45.** 40 ft, 20 ft
47. 110 yd, 60 yd   **49.** $(3x - 2)(2x - 3)$ **51.** Prime
53. $-11y$ **55.** ◆   **57.** ◆   **59.** (4.382, 4.328)
61. (10, −2)   **63.** (2, −1, 3)   **65.** ◆

EXERCISE SET 8.3, PP. 408–409
1. (9, 3)   **3.** (−1, 7)   **5.** (3, 0)   **7.** (−1, 3)

A-30 ANSWERS

9. $\left(-1, \frac{1}{5}\right)$ **11.** Infinite number of solutions
13. $(-3, -5)$ **15.** $(4, 5)$ **17.** $(4, 1)$ **19.** $(4, 3)$
21. $(1, -1)$ **23.** $(-3, -1)$ **25.** No solution   **27.** $\left(5, \frac{1}{2}\right)$
29. $(-2, 2)$ **31.** $(2, -1)$ **33.** $\left(\frac{231}{202}, \frac{117}{202}\right)$ **35.** 10 mi
37. 64°, 26°   **39.** 145°, 35°
41. Oats: 11 acres; hay: 20 acres   **43.** 2 ft by 4 ft
45. $a^3 b^{-6}$, or $\dfrac{a^3}{b^6}$ **47.** $(2x + 5)^2$ **49.** 30.1   **51.** ◆
53. ◆   **55.** $(-1, 1)$ **57.** $(0, 4)$
59. $\left(\dfrac{b - c}{1 - a}, \dfrac{b - ac}{1 - a}\right)$ **61.** 12 rabbits, 23 pheasants
63. 45, 10

EXERCISE SET 8.4, PP. 415–418

1. 175 cars, 85 trucks   **3.** Soda: $0.49; pizza: $1.50
5. Hendersons: 10 bags; Savickis: 4 bags   **7.** 8
9. 5-cent: 340; 10-cent: 96   **11.** 210 full-price;
225 half-price   **13.** 128 activity-card; 75 noncard
15. Brazilian: 200 kg; Turkish: 100 kg
17. Peanuts: 135 lb; Brazil nuts: 345 lb

19.

Type of Solution	50%-acid	80%-acid	68%-acid mix
Amount of Solution	x	y	200
Percent Acid	50%	80%	68%
Amount of Acid in Solution	$0.5x$	$0.8y$	136

80 mL of 50%; 120 mL of 80%

21. 100 L of 28%; 200 L of 40%
23. 4 gal of 87-octane; 8 gal of 93-octane
25. 45 oz of three-fourths gold; 15 oz of five-twelfths gold
27. 10 nickels; 3 quarters
29. $6\frac{2}{3}$ lb of peanuts; $3\frac{1}{3}$ lb of fancy nuts
31. $(5x + 9)(5x - 9)$ **33.** $4(x^2 + 25)$ **35.** 5
37. ◆   **39.** ◆   **41.** $4\frac{4}{7}$ L   **43.** 2.5 gal
45. 10 $20-workers; 5 $25-workers   **47.** 9%, 10.5%
49. 75   **51.** Tweedledum: 120 lb; Tweedledee: 121 lb

EXERCISE SET 8.5, PP. 420–421

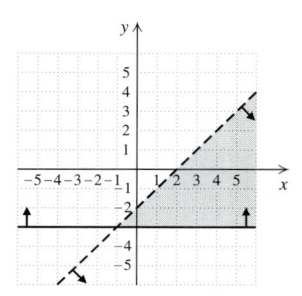

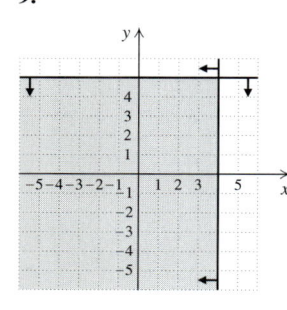

1. [graph]   **3.** [graph]

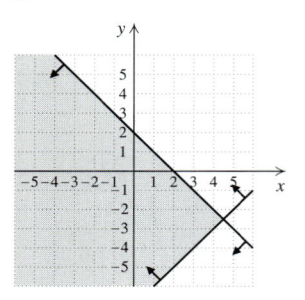

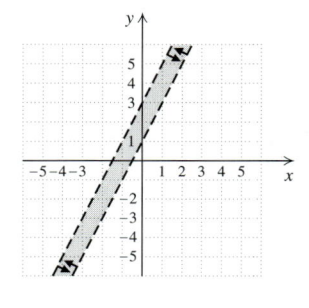

5. [graph]   **7.** [graph]

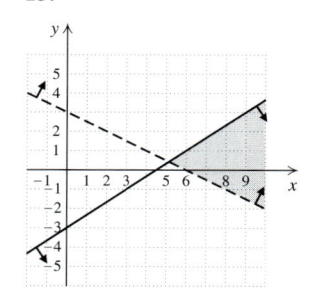

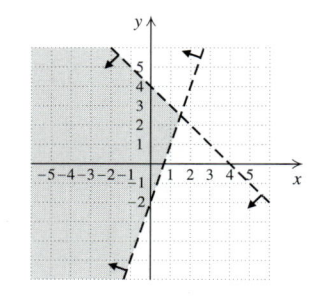

9. [graph]   **11.** [graph]

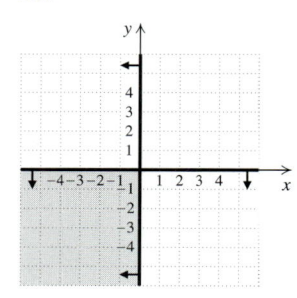

13. [graph]   **15.** [graph]

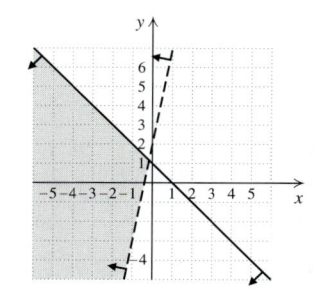

17. [graph] **19.** [graph] **26.** [8.5] [graph] **27.** [8.5] [graph]

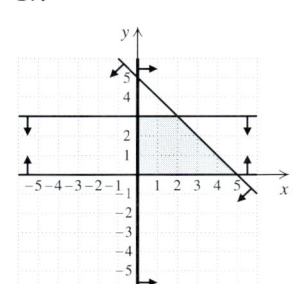

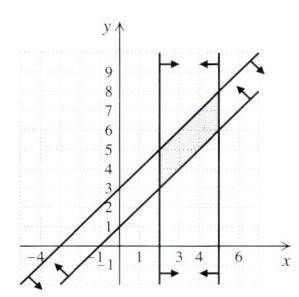

21. [graph]

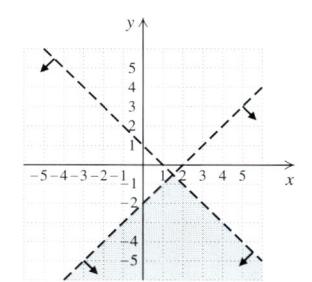

28. [6.1] $\dfrac{a+2}{2a+1}$ **29.** [4.8], [6.1] $\dfrac{3ab^7}{2}$
30. [4.4] $15a^6b - 6a^3b^2$ **31.** [5.5] $9x(x-2)^2$
32. [8.1] ◆ A solution of a system of two equations is an ordered pair that makes both equations true. The graph of an equation represents all ordered pairs that make that equation true. So in order for an ordered pair to make *both* equations true, it must be on both graphs.
33. [8.5] ◆ The solution sets of linear inequalities are regions, not lines. Thus the solution sets can intersect even if the boundary lines do not.
34. [8.1] $C = 1, D = 3$ **35.** [8.2] $(2, 1, -2)$
36. [8.2] $(2, 0)$ **37.** [8.4] 24   **38.** [8.4] $336

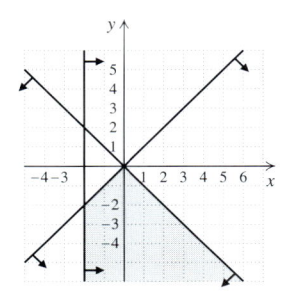

23. $\dfrac{13 - 2x}{3(x+2)(x-2)}$ **25.** 26   **27.** $(x-5)^2$ **29.** ◆
31. ◆   **33.** [graph]

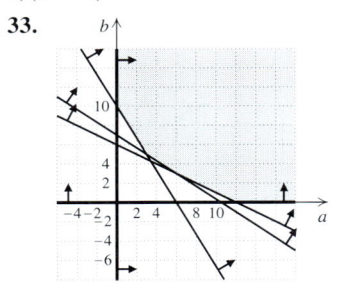

TEST: CHAPTER 8, PP. 423–424
1. [8.1] Yes   **2.** [8.1] $(1, 3)$ **3.** [8.1] No solution
4. [8.2] $(8, -2)$ **5.** [8.2] $(-1, 3)$
6. [8.2] Infinite number of solutions   **7.** [8.3] $(1, -5)$
8. [8.3] $(12, -6)$ **9.** [8.3] $(0, 1)$ **10.** [8.3] $(5, 1)$
11. [8.4] 40 L of 25%; 20 L of 40%   **12.** [8.2] 36°, 54°
13. [8.4] Oak: 15; pine: 3
14. [8.5] [graph]   **15.** [8.5] [graph]

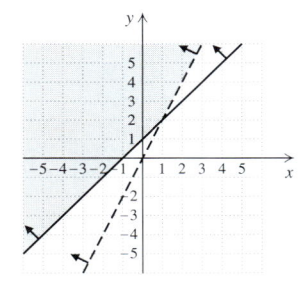

REVIEW EXERCISES: CHAPTER 8, PP. 422–423
1. [8.1] Yes   **2.** [8.1] No   **3.** [8.1] $(1, -2)$
4. [8.1] $(6, -2)$ **5.** [8.1] No solution
6. [8.1] Infinite number of solutions   **7.** [8.2] $(-5, 9)$
8. [8.2] $\left(\dfrac{10}{3}, \dfrac{4}{3}\right)$ **9.** [8.2] No solution   **10.** [8.2] $(-3, 9)$
11. [8.2] $(4, 5)$ **12.** [8.2] Infinite number of solutions
13. [8.3] $\left(\dfrac{3}{2}, \dfrac{15}{4}\right)$ **14.** [8.3] $(5, -3)$
15. [8.3] Infinite number of solutions   **16.** [8.3] $(-2, 4)$
17. [8.3] $(3, 2)$ **18.** [8.3] $(3, 2)$ **19.** [8.3] No solution
20. [8.3] $(-4, 2)$ **21.** [8.2] 12 and 15
22. [8.2] $37\tfrac{1}{2}$ cm; $10\tfrac{1}{2}$ cm   **23.** [8.4] 15
24. [8.4] 410 adults'; 415 children's
25. [8.4] Café Rich: $66\tfrac{2}{3}$ g; Café Light: $133\tfrac{1}{3}$ g

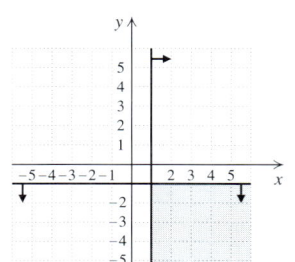

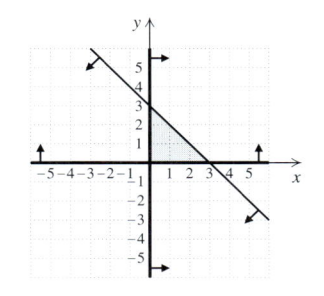

16. [6.4] $\dfrac{-x^2 + x + 17}{(x-4)(x+4)(x+1)}$ **17.** [4.2] 0
18. [2.2] $-\dfrac{15}{2}$ **19.** [5.6] $-3, 3$ **20.** [8.4] 9

21. [8.5]

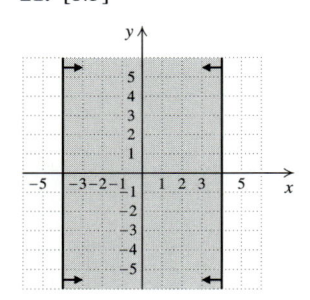

22. [8.1] $C = -\frac{19}{2}$, $D = \frac{14}{3}$

63. $y = 2x + 10$ **65.** ◆ **67.** ◆ **69.** 2
71. -6 and -5 **73.** $-8, 8$ **75.** $-3, 3$ **77.** $9a^3b^4$
79. $\frac{2x^4}{y^3}$ **81.** $\frac{20}{m^8}$ **83.** (a) 1.7; (b) 2.2; (c) 2.6. Answers may vary. **85.** 1097.1 ft/sec **87.** 1270.8 ft/sec

EXERCISE SET 9.2, PP. 436–437

1. $\sqrt{35}$ **3.** $\sqrt{12}$, or $2\sqrt{3}$ **5.** $\sqrt{\frac{3}{10}}$ **7.** 8
9. $\sqrt{75}$, or $5\sqrt{3}$ **11.** $\sqrt{2x}$ **13.** $\sqrt{21x}$ **15.** $\sqrt{7xy}$
17. $\sqrt{6ac}$ **19.** $2\sqrt{7}$ **21.** $2\sqrt{2}$ **23.** $10\sqrt{5}$ **25.** $3\sqrt{x}$
27. $5\sqrt{3a}$ **29.** $4\sqrt{a}$ **31.** $8y$ **33.** $x\sqrt{13}$ **35.** $2t\sqrt{2}$
37. $4\sqrt{5}$ **39.** $12\sqrt{2y}$ **41.** x^{10} **43.** x^6 **45.** $x^2\sqrt{x}$
47. $t^9\sqrt{t}$ **49.** $6m\sqrt{m}$ **51.** $2a^2\sqrt{2a}$ **53.** $2p^8\sqrt{26p}$
55. $5\sqrt{3}$ **57.** 9 **59.** $6\sqrt{xy}$ **61.** 10 **63.** $5b\sqrt{3}$
65. $7x$ **67.** $a\sqrt{bc}$ **69.** $2x^3\sqrt{2}$ **71.** $xy^3\sqrt{xy}$
73. $10ab^2\sqrt{5ab}$ **75.** 20 mph, 54.8 mph **77.** 168 km
79. $3a^2b^5$ **81.** ◆ **83.** ◆ **85.** 0.1 **87.** 0.25
89. $>$ **91.** $>$ **93.** $<$ **95.** $18(x + 1)\sqrt{(x + 1)y}$
97. $2x^3\sqrt{5x}$ **99.** x^{4n} **101.** $y^k\sqrt{y}$, where $k = \frac{n-1}{2}$

EXERCISE SET 9.3, PP. 441–442

1. 2 **3.** 2 **5.** $\sqrt{5}$ **7.** $\frac{1}{4}$ **9.** $\frac{2}{5}$ **11.** 2 **13.** $3y$
15. $3x^2$ **17.** $\frac{\sqrt{3}}{x^4}$ **19.** $\frac{6}{5}$ **21.** $\frac{7}{4}$ **23.** $-\frac{1}{9}$ **25.** $\frac{2}{3}$
27. $\frac{6}{a}$ **29.** $\frac{3a}{25}$ **31.** $\frac{\sqrt{15}}{3}$ **33.** $\frac{\sqrt{35}}{10}$ **35.** $\frac{\sqrt{5}}{15}$
37. $\frac{3\sqrt{5}}{5}$ **39.** $\frac{2\sqrt{6}}{3}$ **41.** $\frac{\sqrt{3x}}{x}$ **43.** $\frac{\sqrt{21}}{3}$ **45.** $\frac{3\sqrt{2}}{4}$
47. $\frac{\sqrt{42}}{14}$ **49.** $\frac{\sqrt{21}}{6}$ **51.** $\frac{\sqrt{2x}}{8}$ **53.** $\frac{4y\sqrt{3}}{3}$ **55.** $\frac{\sqrt{3a}}{2}$
57. $\frac{3\sqrt{10x}}{5x}$ **59.** $\frac{3\sqrt{6}}{8c}$
61. 1.57 sec, 3.14 sec, 8.8813 sec, 11.102 sec **63.** 1.19 sec
65. (4, 2) **67.** $9x^2 - 49$ **69.** $21x - 9y$ **71.** ◆
73. ◆ **75.** $\frac{\sqrt{14}}{40}$ **77.** $\frac{\sqrt{10x}}{4x^4}$ **79.** $\frac{\sqrt{10abc}}{5b^2c^5}$
81. $\frac{y-x}{xy}$, or $\frac{1}{x} - \frac{1}{y}$

EXERCISE SET 9.4, PP. 446–447

1. $11\sqrt{2}$ **3.** $3\sqrt{5}$ **5.** $13\sqrt{x}$ **7.** $-2\sqrt{x}$ **9.** $8\sqrt{2a}$
11. $8\sqrt{10y}$ **13.** $11\sqrt{7}$ **15.** $2\sqrt{2}$
17. $5\sqrt{3} + \sqrt{8} = 5\sqrt{3} + 2\sqrt{2}$ cannot be simplified further. **19.** $-2\sqrt{x}$ **21.** $25\sqrt{2}$ **23.** $\sqrt{3}$ **25.** $13\sqrt{2}$

CHAPTER 9

TECHNOLOGY CONNECTION 9.1, P. 429

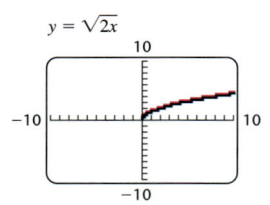

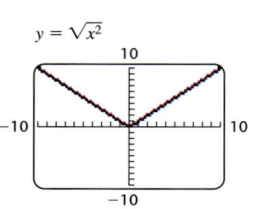

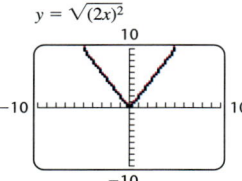

1. $y = \sqrt{x}$

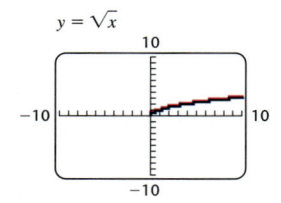

2. $y = \sqrt{2x}$

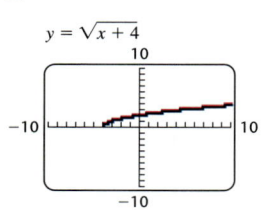

3. $y = \sqrt{x^2}$

4. $y = \sqrt{(2x)^2}$

5. $y = \sqrt{x + 4}$

6. $y = \sqrt{6 - x}$

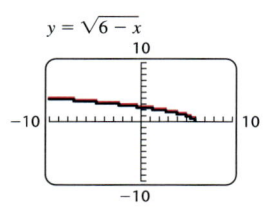

EXERCISE SET 9.1, PP. 430–431

1. 3, -3 **3.** 4, -4 **5.** 7, -7 **7.** 13, -13 **9.** 2
11. -1 **13.** 0 **15.** -11 **17.** 20 **19.** 13 **21.** -25
23. $a - 4$ **25.** $t^2 + 1$ **27.** $\frac{3}{x+2}$ **29.** Rational
31. Irrational **33.** Irrational **35.** Irrational
37. Rational **39.** Irrational **41.** 2.236 **43.** 4.123
45. 9.644 **47.** t **49.** $5x$ **51.** $7a$ **53.** $17x$
55. (a) 15; (b) 14 **57.** 0.864 sec **59.** $10x^9$ **61.** $-\frac{5}{4}$

27. $11\sqrt{3} - 2\sqrt{2}$ **29.** $2\sqrt{2}$ **31.** $\sqrt{x}$ **33.** $\sqrt{10} + \sqrt{14}$
35. $\sqrt{30} - 5\sqrt{2}$ **37.** $22 + 9\sqrt{2}$ **39.** $16 - 7\sqrt{6}$
41. -44 **43.** 3 **45.** $-1 - 2\sqrt{2}$ **47.** $52 + 14\sqrt{3}$
49. $13 - 4\sqrt{3}$ **51.** $x - 2\sqrt{10x} + 10$ **53.** $\dfrac{45 - 9\sqrt{2}}{23}$
55. $-4 - 2\sqrt{7}$ **57.** $3 - \sqrt{7}$, or $-\sqrt{7} + 3$
59. $-\dfrac{6 + 5\sqrt{6}}{19}$ **61.** $\dfrac{5 + \sqrt{15}}{2}$ **63.** $\dfrac{\sqrt{15} - 3}{2}$
65. $\dfrac{2\sqrt{7} + 2\sqrt{2}}{5}$ **67.** $11 + 2\sqrt{30}$ **69.** $\dfrac{5}{11}$ **71.** $-1, 6$
73. 1.6 L of Jolly Juice; 6.4 L of Real Squeeze **75.** ◆
77. ◆ **79.** $18\sqrt{2}$ **81.** $\dfrac{7\sqrt{x}}{2x}$ **83.** 0
85. Any pair of numbers a, b such that $a = 0$ or $b = 0$.

TECHNOLOGY CONNECTION 9.5, P. 449
1. $y_1 = 3\sqrt{x}, y_2 = \sqrt{x + 32}$; Intersection X = 4, Y = 6
2. $y_1 = 3 + \sqrt{27 - 3x}, y_2 = x$; Intersection X = 6, Y = 6

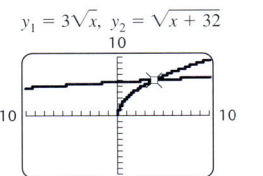

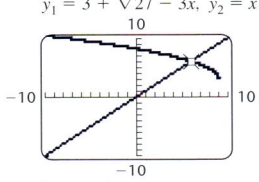

EXERCISE SET 9.5, PP. 451–453
1. 64 **3.** 13 **5.** $\dfrac{77}{2}$ **7.** 5 **9.** 3 **11.** 3
13. No solution **15.** No solution **17.** 9 **19.** 7
21. 5 **23.** 3 **25.** $0, 8$ **27.** No solution **29.** 3
31. $\dfrac{7}{3}$ **33.** 3 **35.** 36 m **37.** $11{,}664$ m
39. About 2.08 ft **41.** 80 ft, 320 ft **43.** 121 **45.** -32
47. $\dfrac{x^6}{3}$ **49.** $x - 4$ **51.** ◆ **53.** ◆ **55.** 9
57. No solution **59.** $-\dfrac{57}{16}$ **61.** 10 **63.** 34.726 m
65. [graph of $y = \sqrt{x}$]
67. [graph of $y = \sqrt{x - 3}$]

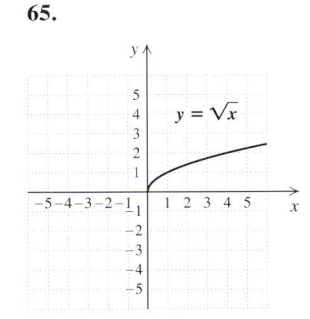

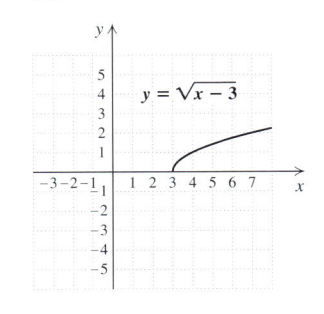

69. [graph showing $y = \sqrt{x - 5}$ and $y = x - 7$ intersecting at $(9, 2)$]; 9 **71.** 1.57

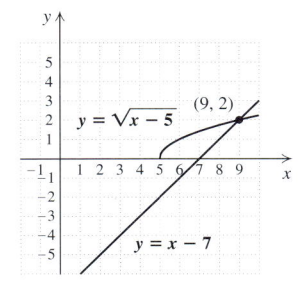

EXERCISE SET 9.6, PP. 457–460
1. 17 **3.** $\sqrt{72} \approx 8.485$ **5.** 12 **7.** 6 **9.** 26 **11.** 12
13. 2 **15.** $\sqrt{2} \approx 1.414$ **17.** 5 **19.** $\sqrt{147} \approx 12.124$ m
21. $\sqrt{208} \approx 14.422$ ft **23.** $\sqrt{12{,}500} \approx 111.803$ yd
25. 240 ft **27.** $\sqrt{2{,}600{,}000{,}000} \approx 50{,}990.195$ ft
29. $-\dfrac{1}{3}$ **31.** $\{x \mid x > -25\}$ **33.** $81x^4$ **35.** ◆
37. ◆ **39.** Yes **41.** $\sqrt{7} \approx 2.646$ m **43.** $s\sqrt{3}$
45. $h = \dfrac{a}{2}\sqrt{3}$ **47.** $\sqrt{1525} \approx 39.051$ mi **49.** 6
51. 640 acres

EXERCISE SET 9.7, PP. 465–466
1. -4 **3.** 10 **5.** -5 **7.** 6 **9.** 5 **11.** 0 **13.** -1
15. Not a real number **17.** 10 **19.** 5 **21.** 2 **23.** a
25. $2\sqrt[3]{4}$ **27.** $2\sqrt[4]{3}$ **29.** $\dfrac{4}{5}$ **31.** $\dfrac{2}{3}$ **33.** $\dfrac{\sqrt[3]{17}}{2}$
35. $\dfrac{\sqrt[4]{13}}{3}$ **37.** 5 **39.** 10 **41.** 2 **43.** 8 **45.** 32
47. 16 **49.** 32 **51.** 3125 **53.** $\dfrac{1}{6}$ **55.** $\dfrac{1}{4}$ **57.** $\dfrac{1}{27}$
59. $\dfrac{1}{243}$ **61.** $\dfrac{1}{4}$ **63.** $(x - 5)^2$ **65.** $(t + 9)^2$ **67.** $\dfrac{-2}{a - 5}$
69. ◆ **71.** ◆ **73.** 6.310 **75.** 31.623 **77.** $x^{14/9}$
79. $p^{2/15}$
81. [graph of $y = \sqrt[3]{x}$]

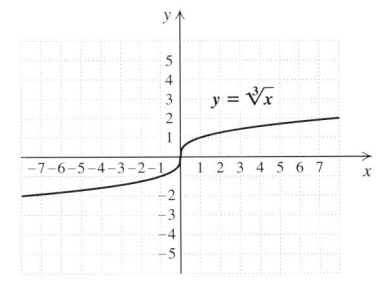

83.

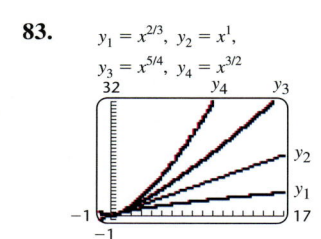

$y_1 = x^{2/3}, y_2 = x^1,$
$y_3 = x^{5/4}, y_4 = x^{3/2}$

REVIEW EXERCISES: CHAPTER 9, PP. 467–468

1. [9.1] 5, −5   **2.** [9.1] 14, −14   **3.** [9.1] 30, −30
4. [9.1] 15, −15   **5.** [9.1] −12   **6.** [9.1] 9   **7.** [9.1] 7
8. [9.1] −13   **9.** [9.1] $2x^2y$ **10.** [9.1] $\dfrac{a}{b}$
11. [9.1] Rational   **12.** [9.1] Irrational
13. [9.1] Irrational   **14.** [9.1] Rational   **15.** [9.1] 1.732
16. [9.1] 9.950   **17.** [9.1] p **18.** [9.1] $7x$ **19.** [9.1] $4m$
20. [9.1] ac **21.** [9.2] $4\sqrt{3}$ **22.** [9.2] $7t\sqrt{2}$
23. [9.2] $4\sqrt{2p}$ **24.** [9.2] x^3 **25.** [9.2] $2a^6\sqrt{3a}$
26. [9.2] $6m^7\sqrt{m}$ **27.** [9.2] $\sqrt{33}$ **28.** [9.2] $2\sqrt{15}$
29. [9.2] $\sqrt{35xt}$ **30.** [9.2] $2a\sqrt{6}$ **31.** [9.2] $5xy\sqrt{2}$
32. [9.2] $10a^2b\sqrt{ab}$ **33.** [9.3] $\dfrac{\sqrt{7}}{3}$ **34.** [9.3] $\dfrac{y^4\sqrt{5}}{3}$
35. [9.3] $\dfrac{5}{8}$ **36.** [9.3] $\dfrac{2}{3}$ **37.** [9.3] $\dfrac{7}{t}$ **38.** [9.4] $13\sqrt{5}$
39. [9.4] $\sqrt{5}$ **40.** [9.4] $-3\sqrt{x}$ **41.** [9.4] $7 + 4\sqrt{3}$
42. [9.4] 1   **43.** [9.4] $-11 + 5\sqrt{7}$ **44.** [9.3] $\dfrac{\sqrt{2}}{2}$
45. [9.3] $\dfrac{\sqrt{10}}{4}$ **46.** [9.3] $\dfrac{\sqrt{5y}}{y}$ **47.** [9.3] $\dfrac{2\sqrt{3}}{3}$
48. [9.4] $8 − 4\sqrt{3}$ **49.** [9.4] $−7 − 3\sqrt{5}$ **50.** [9.5] 52
51. [9.5] No solution   **52.** [9.5] 4   **53.** [9.5] 0, 3
54. [9.5] 405 ft   **55.** [9.6] 20   **56.** [9.6] $\sqrt{3} \approx 1.732$
57. [9.6] $\sqrt{10{,}625} \approx 103.078$ ft   **58.** [9.7] 2
59. [9.7] Not a real number   **60.** [9.7] −3
61. [9.7] $2\sqrt[4]{2}$ **62.** [9.7] 10   **63.** [9.7] $\dfrac{1}{3}$ **64.** [9.7] 64
65. [9.7] $\dfrac{1}{27}$ **66.** [8.3] (13, −3)   **67.** [2.6] $\left\{x \mid x < -\dfrac{17}{2}\right\}$
68. [7.1] $\dfrac{2}{5}$; $\left(0, -\dfrac{14}{5}\right)$ **69.** [6.2] $\dfrac{(x+1)(x+2)}{x(x-2)}$
70. [9.5] ◆ He could be mistakenly assuming that negative solutions are not possible because principal square roots are never negative. He could also be assuming that substituting a negative value for the variable in a radicand always produces a negative radicand.
71. [9.4] ◆ Some radical terms that are like terms may not appear to be so until they are in simplified form.
72. [9.1] 2   **73.** [9.1] No solution
74. [9.4] $(x + \sqrt{5})(x - \sqrt{5})$
75. [9.5] $b = \sqrt{A^2 - a^2}$ or $b = -\sqrt{A^2 - a^2}$

TEST: CHAPTER 9, P. 469

1. [9.1] 9, −9   **2.** [9.1] 8   **3.** [9.1] −5   **4.** [9.1] $x + 4$
5. [9.1] Irrational   **6.** [9.1] Rational   **7.** [9.1] 9.327
8. [9.1] 2.646   **9.** [9.1] a **10.** [9.1] $6y$ **11.** [9.2] $2\sqrt{6}$
12. [9.2] $3x^3\sqrt{3}$ **13.** [9.2] $2t^2\sqrt{t}$ **14.** [9.2] $\sqrt{30}$
15. [9.2] $5\sqrt{2}$ **16.** [9.2] $\sqrt{14xy}$ **17.** [9.2] $4t$
18. [9.2] $3ab^2\sqrt{2}$ **19.** [9.3] $\dfrac{3}{4}$ **20.** [9.3] $\dfrac{\sqrt{7}}{4y}$
21. [9.3] $\dfrac{3}{2}$ **22.** [9.3] $\dfrac{12}{a}$ **23.** [9.4] $-6\sqrt{2}$
24. [9.4] $7\sqrt{3}$ **25.** [9.4] $21 - 8\sqrt{5}$ **26.** [9.4] 11
27. [9.3] $\dfrac{\sqrt{10}}{5}$ **28.** [9.3] $\dfrac{2x\sqrt{y}}{y}$ **29.** [9.4] $\dfrac{40 + 10\sqrt{5}}{11}$
30. [9.6] $\sqrt{80} \approx 8.944$ cm   **31.** [9.5] 48
32. [9.5] −2, 2   **33.** [9.5] About 5000 m   **34.** [9.7] 2
35. [9.7] −1   **36.** [9.7] −4   **37.** [9.7] Not a real number
38. [9.7] 3   **39.** [9.7] $\dfrac{1}{3}$ **40.** [9.7] 1000   **41.** [9.7] $\dfrac{1}{32}$
42. [1.8] −27   **43.** [7.1] $y = -\dfrac{1}{2}x - 1$
44. [6.2] $\dfrac{(a+1)^2}{a}$ **45.** [6.4] $\dfrac{x+4}{3x-5}$ **46.** [9.5] −3
47. [9.2] y^{8n}

CHAPTER 10

TECHNOLOGY CONNECTION 10.1, P. 473

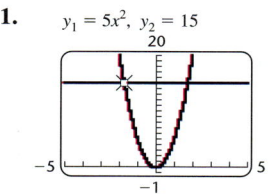

1.

$y_1 = 5x^2, y_2 = 15$
Intersection X = −1.732051, Y = 15

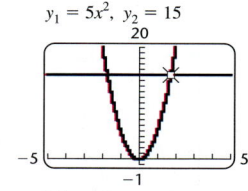

$y_1 = 5x^2, y_2 = 15$
Intersection X = 1.7320508, Y = 15

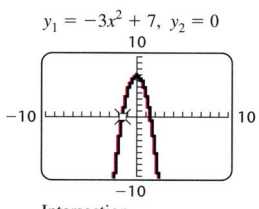

$y_1 = -3x^2 + 7, y_2 = 0$
Intersection X = −1.527525, Y = 0

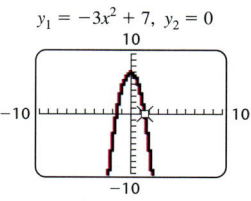

$y_1 = -3x^2 + 7, y_2 = 0$
Intersection X = 1.5275252, Y = 0

EXERCISE SET 10.1, PP. 475–476

1. 10, −10   **3.** 6, −6   **5.** $\sqrt{23}, -\sqrt{23}$ **7.** $\sqrt{10}, -\sqrt{10}$
9. $2\sqrt{2}, -2\sqrt{2}$ **11.** $\dfrac{2}{3}, -\dfrac{2}{3}$ **13.** $\dfrac{2\sqrt{5}}{7}, -\dfrac{2\sqrt{5}}{7}$
15. $2\sqrt{6}, -2\sqrt{6}$ **17.** 4, −6   **19.** 13, −5

21. $-3 \pm \sqrt{6}$ **23.** 21, 5   **25.** $5 \pm \sqrt{14}$ **27.** 3, -7
29. $\dfrac{3 \pm \sqrt{17}}{4}$ **31.** 5, -15 **33.** 5, 3   **35.** $-3 \pm \sqrt{13}$
37. $6 \pm 3\sqrt{2}$ **39.** Yes   **41.** Yes   **43.** $x^2 - \dfrac{3}{4}x + \dfrac{9}{64}$
45. ◆   **47.** ◆   **49.** $\dfrac{-7 \pm \sqrt{7}}{6}$ **51.** $\dfrac{3 \pm \sqrt{17}}{4}$
53. 1.85, -4.35 **55.** $-5, -1$ **57.** $d = \sqrt{\dfrac{kMm}{f}}$

EXERCISE SET 10.2, PP. 480–481

1. $x^2 + 8x + 16$ **3.** $x^2 - 14x + 49$ **5.** $x^2 - 3x + \dfrac{9}{4}$
7. $t^2 + t + \dfrac{1}{4}$ **9.** $x^2 + \dfrac{5}{4}x + \dfrac{25}{64}$ **11.** $m^2 - \dfrac{9}{2}m + \dfrac{81}{16}$
13. 3, 5   **15.** $-21, -1$ **17.** $1 \pm \sqrt{6}$ **19.** $11 \pm \sqrt{19}$
21. $-5 \pm \sqrt{29}$ **23.** $\dfrac{-5 \pm \sqrt{33}}{2}$ **25.** $\dfrac{-3 \pm \sqrt{41}}{4}$
27. $\dfrac{-3 \pm \sqrt{137}}{4}$ **29.** $\dfrac{-2 \pm \sqrt{7}}{3}$ **31.** $\dfrac{5 \pm \sqrt{97}}{4}$
33. $-\dfrac{7}{2}, \dfrac{1}{2}$ **35.** $\left(\dfrac{2}{3}, \dfrac{17}{3}\right)$
37. graph of $y = \dfrac{3}{5}x - 1$

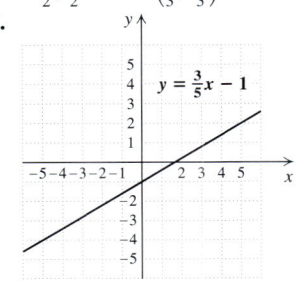

39. $2\sqrt{21}$ **41.** ◆   **43.** ◆   **45.** 14, -14
47. $\pm 10\sqrt{2}$ **49.** $\pm 8\sqrt{3}$ **51.** $-0.39, -7.61$
53. 0.37, -5.37 **55.** 3.71, -1.21

TECHNOLOGY CONNECTION 10.3, P. 485

1. An error message appears.
2. The graph has no x-intercepts, so there is no value of x for which $x^2 + x + 1 = 0$ or, equivalently, for which $x^2 + x = -1$.

EXERCISE SET 10.3, PP. 488–491

1. $-2, 9$ **3.** 3   **5.** $-\dfrac{4}{3}, -1$ **7.** $-\dfrac{1}{2}, \dfrac{7}{2}$ **9.** $-8, 8$
11. $-2 \pm \sqrt{11}$ **13.** $5 \pm \sqrt{3}$ **15.** $-1 \pm \sqrt{7}$
17. $\dfrac{-2 \pm \sqrt{10}}{3}$ **19.** $\dfrac{5 \pm \sqrt{33}}{4}$ **21.** $\dfrac{-1 \pm \sqrt{2}}{2}$
23. No real-number solutions   **25.** $\dfrac{5 \pm \sqrt{73}}{6}$
27. $\dfrac{3 \pm \sqrt{29}}{2}$ **29.** $0, \dfrac{3}{2}$ **31.** No real-number solutions

33. $\pm 3\sqrt{2}$ **35.** 5.317, -1.317 **37.** 6.162, -0.162
39. 0.207, -1.207 **41.** 10   **43.** About 7.95 sec
45. About 6.61 sec   **47.** 7 ft, 24 ft   **49.** 10 cm, 6 cm
51. 15 m by 20 m   **53.** 4 m, 6.5 m   **55.** 5.58 in., 3.58 in.
57. 5.66 m, 2.83 m   **59.** 20%   **61.** 4%   **63.** 17.84 ft
65. $\dfrac{6}{5}$ **67.** $3\sqrt{10}$ **69.** -1 **71.** ◆   **73.** ◆
75. $-\dfrac{4}{3}, 0$ **77.** $\dfrac{7 \pm \sqrt{69}}{10}$ **79.** $\dfrac{-2 \pm \sqrt{10}}{2}$ **81.** $\pm\sqrt{7}$
83. $2 \pm \sqrt{34}$ **85.** 4.83 cm   **87.** 7.5 ft from the bottom
89. 5.5%   **91.** About 19 in. by 44 in.

EXERCISE SET 10.4, PP. 493–494

1. i **3.** $3i$ **5.** $5\sqrt{2}i$ **7.** $-2\sqrt{5}i$ **9.** $-3\sqrt{2}i$
11. $4 + 7i$ **13.** $7 + 4i$ **15.** $3 - 7\sqrt{2}i$ **17.** $\pm 3i$
19. $\pm 2\sqrt{7}i$ **21.** $2 \pm \sqrt{2}i$ **23.** $3 \pm 5i$ **25.** $-1 \pm i$
27. $2 \pm \sqrt{3}i$ **29.** $-\dfrac{3}{2} \pm \dfrac{1}{2}i$ **31.** $-\dfrac{1}{3} \pm \dfrac{\sqrt{2}}{3}i$
33. graph of $y = \dfrac{3}{5}x - 2$ **35.** graph of $y = -4$

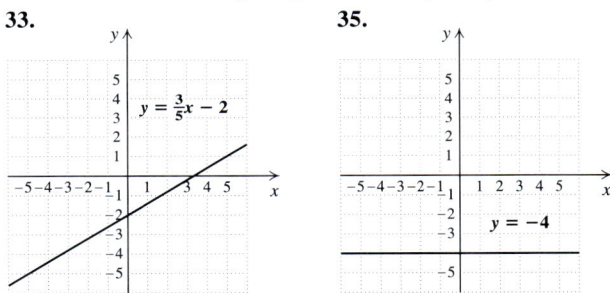

37. -17 **39.** ◆   **41.** ◆   **43.** $-2 \pm i$
45. $-1 \pm \sqrt{19}i$
47. $y = x^2 + 3x + 4$, No x-intercepts exist. $y_1 = x^2 + 2, y_2 = 2x$, No intersection exists.

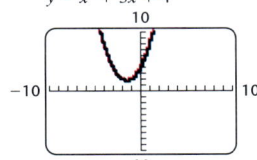

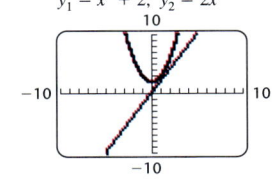

EXERCISE SET 10.5, PP. 499–501

1. $y = x^2 - 2$, $(0, -2)$
3. $y = -1 \cdot x^2$, $(0, 0)$

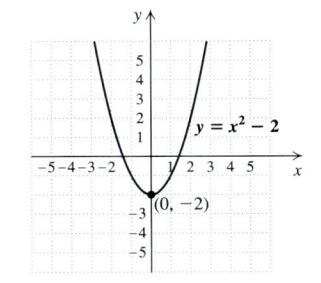

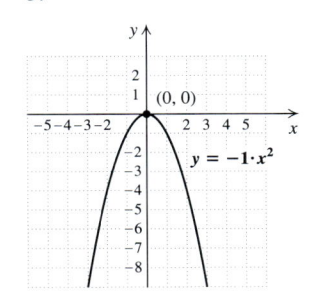

A-36 ANSWERS

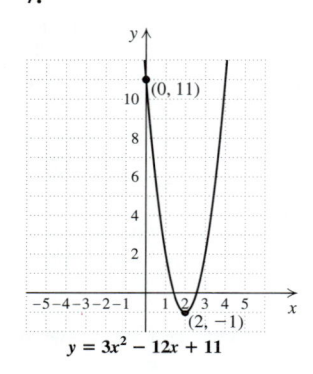

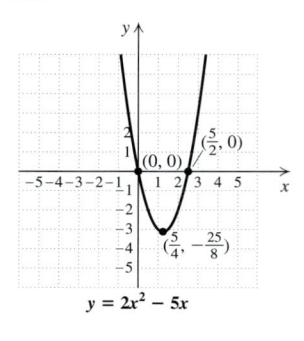

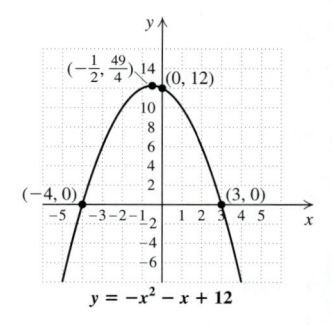

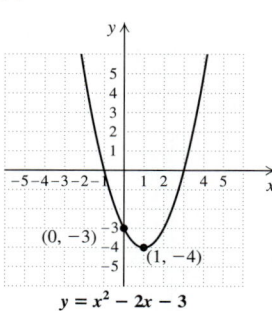

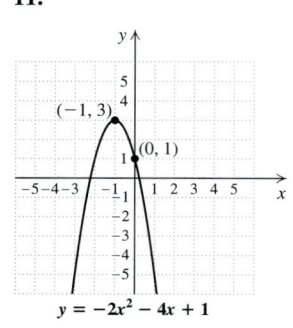

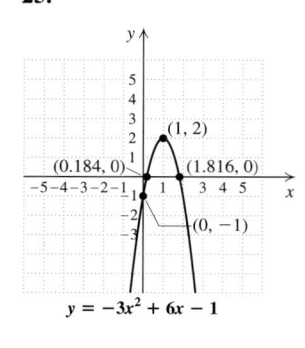

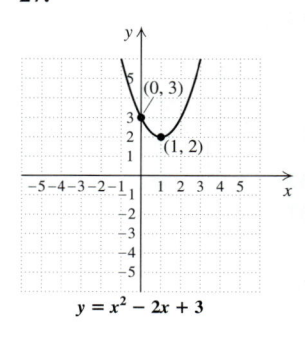

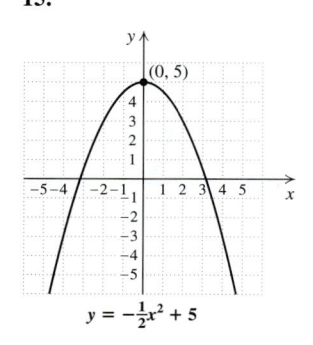

5. [graph: $y = -x^2 + 2x$, points $(0,0)$, $(1,1)$]

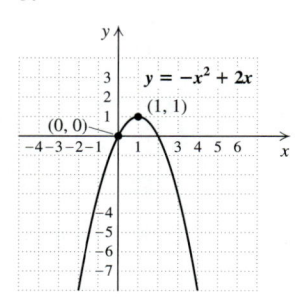

7. [graph: $y = 3x^2 - 12x + 11$, points $(0, 11)$, $(2, -1)$]

9. [graph: $y = x^2 - 2x - 3$, points $(0, -3)$, $(1, -4)$]

11. [graph: $y = -2x^2 - 4x + 1$, points $(-1, 3)$, $(0, 1)$]

13. [graph: $y = \frac{1}{4}x^2$, point $(0, 0)$]

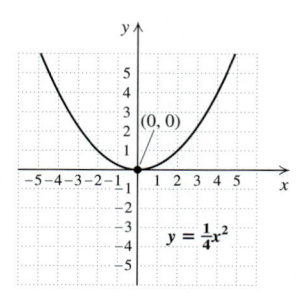

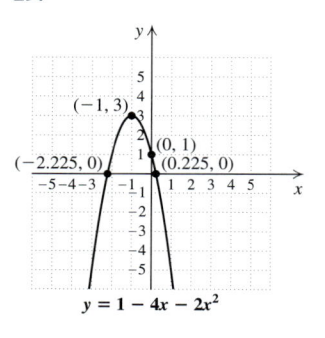

15. [graph: $y = -\frac{1}{2}x^2 + 5$, point $(0, 5)$]

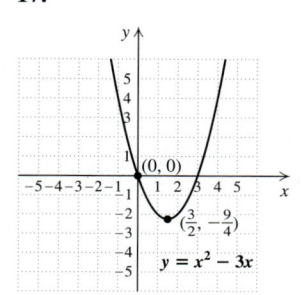

17. [graph: $y = x^2 - 3x$, points $(0, 0)$, $(\frac{3}{2}, -\frac{9}{4})$]

19. [graph: $y = x^2 - x - 6$, points $(-2, 0)$, $(3, 0)$, $(0, -6)$, $(\frac{1}{2}, -\frac{25}{4})$]

21. [graph: $y = 2x^2 - 5x$, points $(0, 0)$, $(\frac{5}{2}, 0)$, $(\frac{5}{4}, -\frac{25}{8})$]

23. [graph: $y = -x^2 - x + 12$, points $(-\frac{1}{2}, \frac{49}{4})$, $(0, 12)$, $(-4, 0)$, $(3, 0)$]

25. [graph: $y = -3x^2 + 6x - 1$, points $(0.184, 0)$, $(1, 2)$, $(1.816, 0)$, $(0, -1)$]

27. [graph: $y = x^2 - 2x + 3$, points $(0, 3)$, $(1, 2)$]

29. [graph: $y = 1 - 4x - 2x^2$, points $(-2.225, 0)$, $(-1, 3)$, $(0, 1)$, $(0.225, 0)$]

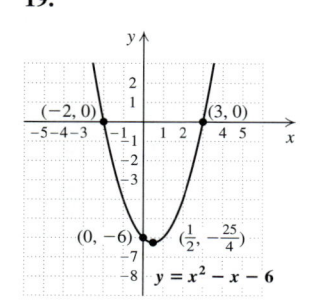

31. $\sqrt{432} \approx 20.78$ ft   **33.** $y = -\frac{5}{3}x + \frac{11}{3}$ **35.** -3

37. ◆   **39.** ◆

41. $\dfrac{\dfrac{-b - \sqrt{b^2 - 4ac}}{2a} + \dfrac{-b + \sqrt{b^2 - 4ac}}{2a}}{2} = \dfrac{\dfrac{-2b}{2a}}{2}$

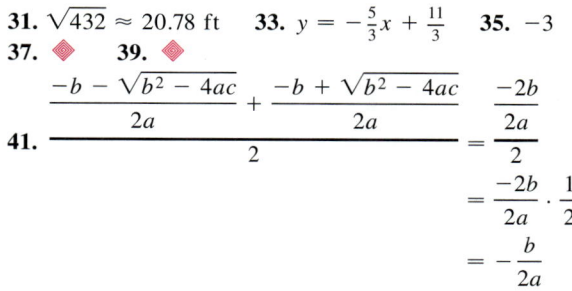

$= \dfrac{-2b}{2a} \cdot \dfrac{1}{2}$

$= -\dfrac{b}{2a}$

43. **(a)** 56.25 ft, 120 ft, 206.25 ft, 276.25 ft, 356.25 ft, 600 ft;

(b)

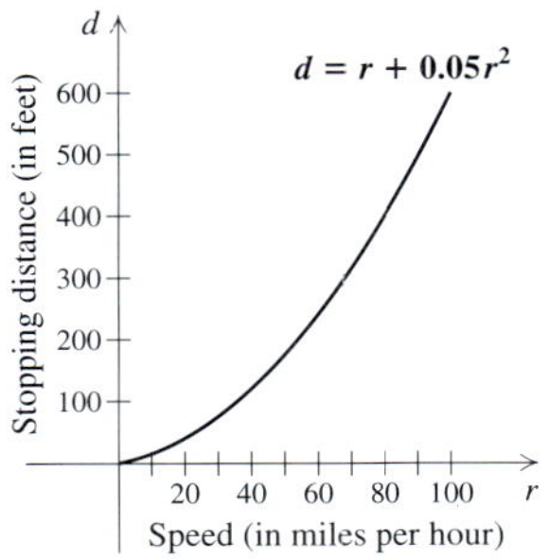

[Graph: d = r + 0.05r², Stopping distance (in feet) vs Speed (in miles per hour)]

45.

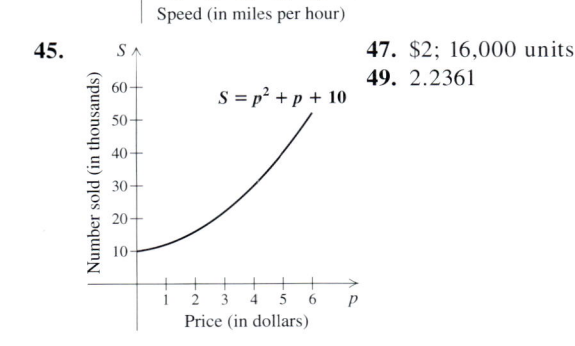

[Graph: S = p² + p + 10, Number sold (in thousands) vs Price (in dollars)]

47. $2; 16,000 units
49. 2.2361

EXERCISE SET 10.6, PP. 509–511

1. Yes **3.** Yes **5.** No **7.** Yes **9.** 9, 12, 2
11. −21, 15, 42 **13.** 7, −17, 24.1 **15.** 0, 5, 2
17. 6, 1, $\frac{13}{4}$ **19.** 4, −122, 3
21. (a) 159.48 cm; **(b)** 153.98 cm
23. $1\frac{20}{33}$ atm, $1\frac{10}{11}$ atm, $4\frac{1}{33}$ atm
25. 1.792 cm, 2.8 cm, 11.2 cm

27.

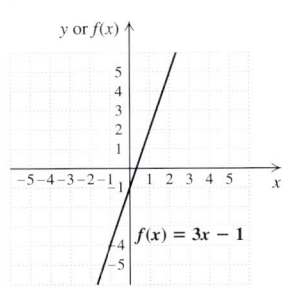

[Graph: $f(x) = 3x - 1$]

29.

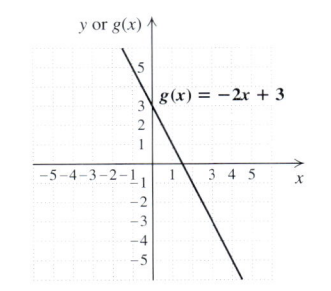

[Graph: $g(x) = -2x + 3$]

31.

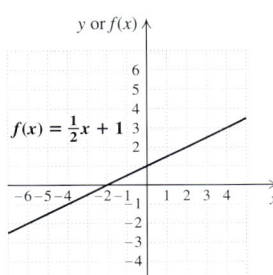

[Graph: $f(x) = \frac{1}{2}x + 1$]

33.

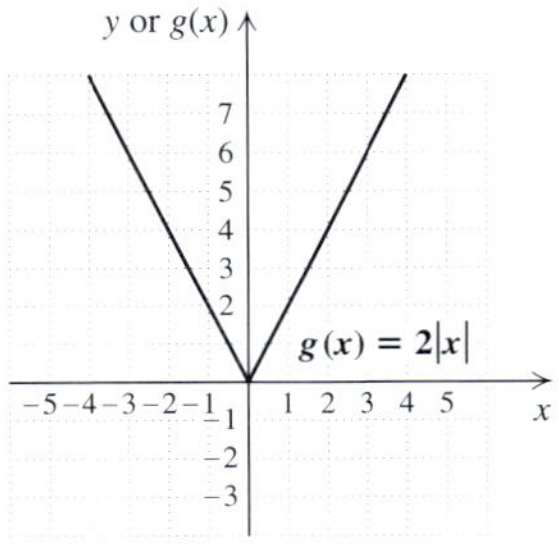

[Graph: $g(x) = 2|x|$]

35.

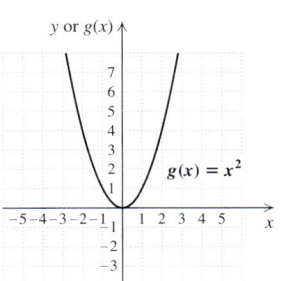

[Graph: $g(x) = x^2$]

37.

[Graph: $f(x) = x^2 - x - 2$]

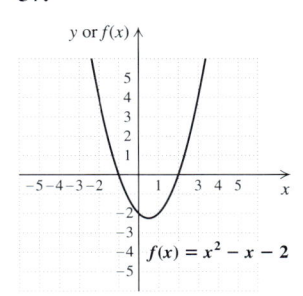

39. Yes **41.** No **43.** Yes **45.** No **47.** No solution
49. ◆ **51.** ◆
53.

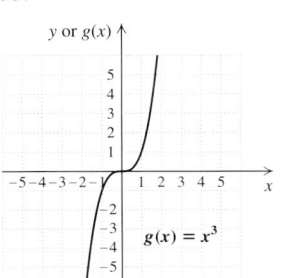

[Graph: $g(x) = x^3$]

55.

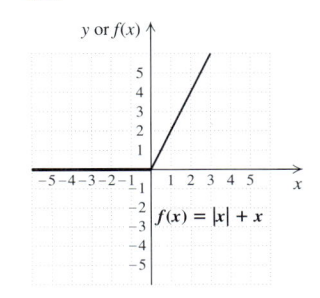

[Graph: $f(x) = |x| + x$]

57. Answers may vary.

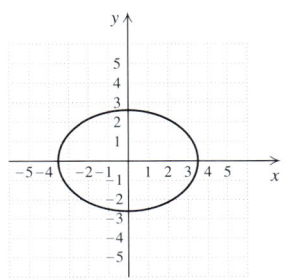

[Graph: ellipse]

59. $g(x) = x^2 - 4$ **61.** $\{-5, -4, -1, 4\}$
63. $\{-7, 0, 1, 2, 9\}$

REVIEW EXERCISES: CHAPTER 10, PP. 512–513

1. [10.2] $1, -\frac{5}{3}$ **2.** [10.2] $\frac{3 \pm \sqrt{5}}{2}$ **3.** [10.1] $\pm\sqrt{3}$
4. [10.3] $\frac{3}{5}, 1$ **5.** [10.3] $1 \pm \sqrt{11}$ **6.** [10.4] $\pm 8i$
7. [10.3] $\frac{1}{3}, -2$ **8.** [10.1] $-8 \pm \sqrt{13}$ **9.** [10.4] $3 \pm 2i$
10. [10.3] $\frac{1 \pm \sqrt{10}}{3}$ **11.** [10.1] 0 **12.** [10.3] $-3 \pm 3\sqrt{2}$
13. [10.4] $-\frac{1}{2} \pm \frac{\sqrt{3}}{2}i$ **14.** [10.3] $\frac{2 \pm \sqrt{3}}{2}$ **15.** [10.1] 1
16. [10.1] $\pm 2\sqrt{2}$ **17.** [10.4] $\frac{3}{10} \pm \frac{\sqrt{71}}{10}i$

18. [10.3] $\frac{5}{3}, -\frac{7}{2}$ **19.** [10.3] 2.618, 0.382
20. [10.3] $-0.134, -1.866$ **21.** [10.3] 1.7 m, 4.7 m
22. [10.3] 5%   **23.** [10.3] 12 m, 9 m   **24.** [10.3] 6.3 sec
25. [10.4] $2 - 8i$ **26.** [10.4] $2\sqrt{6}i$
27. [10.5]   **28.** [10.5]

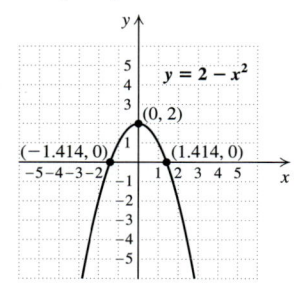

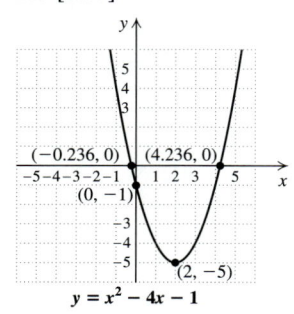

$y = 2 - x^2$; points $(-1.414, 0)$, $(1.414, 0)$, $(0, 2)$

$y = x^2 - 4x - 1$; points $(-0.236, 0)$, $(4.236, 0)$, $(0, -1)$, $(2, -5)$

29. [10.5]   **30.** [10.5]

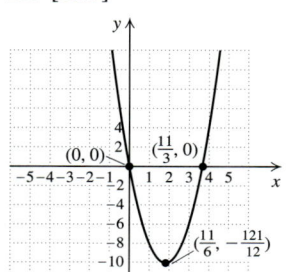

$y = 3x^2 - 11x$; points $(0, 0)$, $(\frac{11}{3}, 0)$, $(\frac{11}{6}, -\frac{121}{12})$

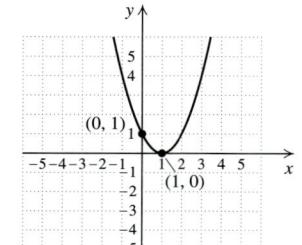

$y = x^2 - 2x + 1$; points $(0, 1)$, $(1, 0)$

31. [10.6] $-1, -7, 2$ **32.** [10.6] $0, 0, 19$
33. [10.6] 2700
34. [10.6] $g(x) = x + 7$
35. [10.6] $f(x) = x^2 - 3$
36. [10.6] $h(x) = 3|x|$

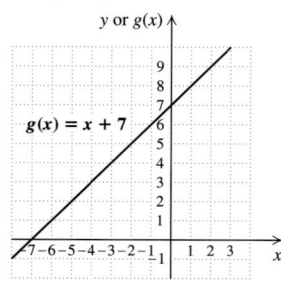

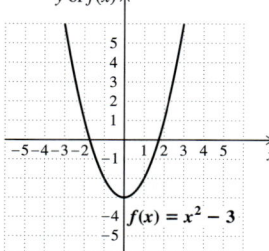

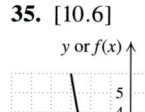

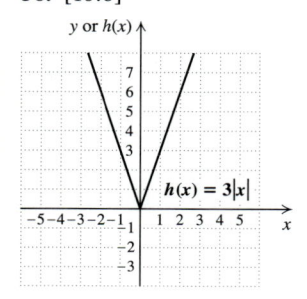

37. [10.6] No   **38.** [10.6] Yes   **39.** [7.1] Yes
40. [9.2] $2p^3\sqrt{21}$ **41.** [9.4] 7   **42.** [7.2] $y = -\frac{4}{3}x + \frac{7}{3}$
43. [10.5] ◆ The graph can be drawn more accurately and with fewer points plotted when the general shape is known.
44. [10.5] ◆ If the radicand is 0, the quadratic formula becomes $x = -b/(2a)$; thus there is only one x-intercept. If the radicand is negative, there are no real-number solutions and thus no x-intercepts. If the radicand is positive, then
$$x = \frac{-b + \sqrt{b^2 - 4ac}}{2a}$$
or
$$x = \frac{-b - \sqrt{b^2 - 4ac}}{2a}$$
so there must be two x-intercepts.
45. [10.3] 31 and 32; -32 and -31
46. [10.2] $b = 14$ or -14 **47.** [10.3] 25
48. [10.3] $s = 5\sqrt{\pi}$

TEST: CHAPTER 10, PP. 513–514

1. [10.1] $\pm\sqrt{5}$ **2.** [10.3] $0, \frac{7}{3}$ **3.** [10.3] $\frac{1 \pm \sqrt{13}}{2}$
4. [10.3] $\frac{5}{3}, -3$ **5.** [10.3] $-8, 6$ **6.** [10.4] $\pm 7i$
7. [10.1] $2 \pm \sqrt{5}$ **8.** [10.1] 2   **9.** [10.4] $2 \pm 2i$
10. [10.3] $\frac{3 \pm \sqrt{37}}{2}$ **11.** [10.3] $-2 \pm \sqrt{14}$
12. [10.3] $\frac{7 \pm \sqrt{37}}{6}$ **13.** [10.2] $2 \pm \sqrt{14}$
14. [10.3] $4.702, -1.702$ **15.** [10.3] 6.5 m, 2.5 m
16. [10.3] 11   **17.** [10.4] $7i$ **18.** [10.4] $3 - 4\sqrt{2}i$
19. [10.5]   **20.** [10.5]

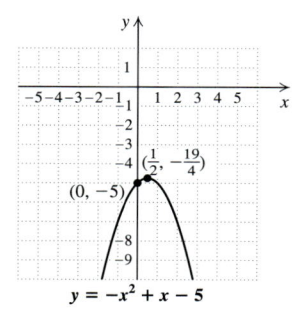

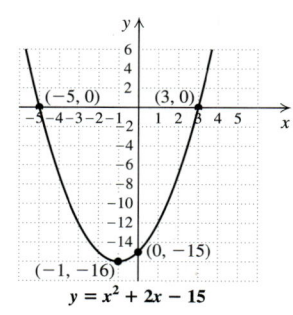

$y = -x^2 + x - 5$; points $(0, -5)$, $(\frac{1}{2}, -\frac{19}{4})$

$y = x^2 + 2x - 15$; points $(-5, 0)$, $(3, 0)$, $(0, -15)$, $(-1, -16)$

21. [10.6] $1, \frac{3}{2}, 2$ **22.** [10.6] $1, 3, -3$
23. [10.6] 25.98 min

24. [10.6]

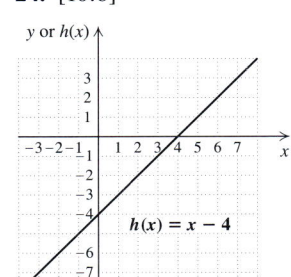

25. [10.6]

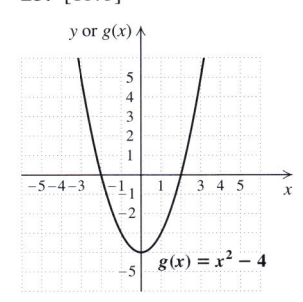

26. [10.6] Yes   **27.** [10.6] No   **28.** [7.2] $y = -3x - 11$
29. [8.2] (2, 4)   **30.** [9.4] $-\sqrt{3}$
31. [7.1]

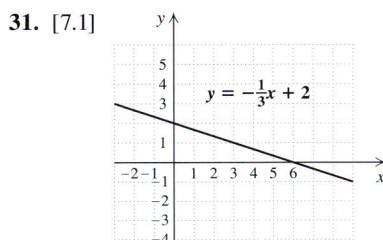

32. [10.3] $5 + 5\sqrt{2} \approx 12.071$ ft   **33.** [10.3] $1 \pm \sqrt{5}$

CUMULATIVE REVIEW: 1–10, PP. 515–517

1. [1.8] x^3 **2.** [1.8] 54   **3.** [1.2] $(6x)y, x(6y)$; there are other answers.   **4.** [6.3] 240   **5.** [1.4] 7   **6.** [1.5] 9
7. [1.6] 15   **8.** [1.7] $-\frac{3}{20}$ **9.** [1.8] 4
10. [1.8] $-2m - 4$ **11.** [2.1] -9
12. [2.6] $\{x \mid x < -9\}$ **13.** [2.2] 7   **14.** [5.6] 3, 5
15. [8.2] (1, 2)   **16.** [6.6] $-\frac{3}{5}$ **17.** [8.3] (6, 7)
18. [5.6] 3, -2 **19.** [10.3] $\dfrac{-3 \pm \sqrt{29}}{2}$ **20.** [9.5] 2
21. [2.6] $\{x \mid x \geq -1\}$ **22.** [2.2] 8   **23.** [2.2] -3
24. [8.3] (15, 0)   **25.** [10.4] $-1 \pm 2i$ **26.** [10.1] $\pm\sqrt{10}$
27. [10.1] $3 \pm \sqrt{6}$ **28.** [6.6] $\frac{2}{9}$ **29.** [5.6] $\frac{5}{4}, -\frac{4}{3}$
30. [6.6] No solution   **31.** [9.5] 12   **32.** [6.8] $t = \dfrac{4s + 3}{A}$
33. [6.8] $m = \dfrac{tn}{t + n}$ **34.** [4.8] 2.73×10^{-5}
35. [4.8] 2.0×10^{14} **36.** [4.8] x^{-4} **37.** [4.8] y^7
38. [4.1] $4y^{12}$ **39.** [4.2] $10x^3 + 3x - 3$
40. [4.3] $7x^3 - 2x^2 + 4x - 17$ **41.** [4.3] $8x^2 - 4x - 6$
42. [4.4] $-8y^4 + 6y^3 - 2y^2$
43. [4.4] $6t^3 - 17t^2 + 16t - 6$ **44.** [4.5] $t^2 - \frac{1}{16}$
45. [4.5] $9m^2 - 12m + 4$
46. [4.6] $15x^2y^3 + x^2y^2 + 5xy^2 + 7$
47. [4.6] $x^4 - 0.04y^2$ **48.** [4.6] $9p^2 + 24pq^2 + 16q^4$

49. [6.2] $\dfrac{2}{x + 3}$ **50.** [6.2] $\dfrac{3a(a - 1)}{2(a + 1)}$
51. [6.4] $\dfrac{27x - 4}{5x(3x - 1)}$ **52.** [6.4] $\dfrac{-x^2 + x + 2}{(x + 4)(x - 4)(x - 5)}$
53. [4.7] $x^2 + 9x + 16 + \dfrac{35}{x - 2}$
54. [5.4] $(7x + 2)(7x - 2)$ **55.** [5.1] $(x - 8)(x^2 - 5)$
56. [5.4] $x(4x + 1)(4x - 1)$ **57.** [5.4] $(m - 4)^2$
58. [5.3] $(5x - 2)(3x + 4)$ **59.** [5.1] $2x(9x^4 + 2x^3 - 5)$
60. [5.3] $(3p + 2)(2p - 3)$ **61.** [5.1] $(c - 3b)(2a + d)$
62. [5.4] $(3x + 5y)^2$ **63.** [5.5] $3(t^2 + 6)(t + 2)(t - 2)$
64. [5.4] $(7ab + 3)(7ab - 3)$ **65.** [6.5] $-\frac{42}{5}$ **66.** [9.1] 7
67. [9.7] -5 **68.** [9.2] $8x$ **69.** [9.2] $\sqrt{a^2 - b^2}$
70. [9.2] $8a^2b\sqrt{3ab}$ **71.** [9.7] $\frac{1}{4}$ **72.** [9.2] $9xy\sqrt{3x}$
73. [9.3] $\frac{10}{9}$ **74.** [9.4] 7   **75.** [9.4] $16\sqrt{3}$
76. [9.3] $\dfrac{2\sqrt{10}}{5}$ **77.** [9.6] 40 cm
78. [7.1]

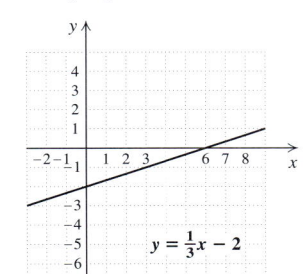

79. [3.3]

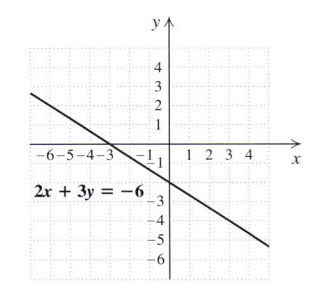

80. [3.3]

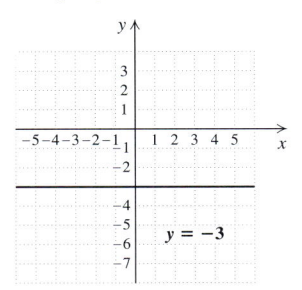

81. [7.3]

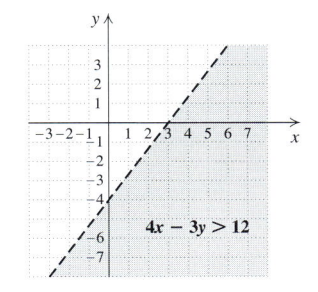

82. [10.5]

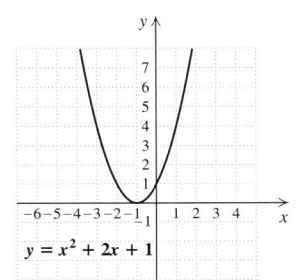

83. [7.3]

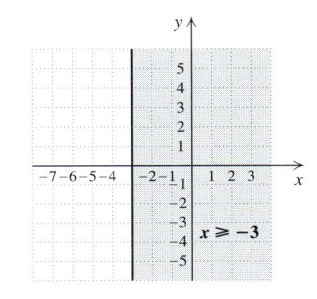

84. [10.2] $\dfrac{2 \pm \sqrt{6}}{3}$ **85.** [10.3] 1.207, −0.207
86. [10.5]

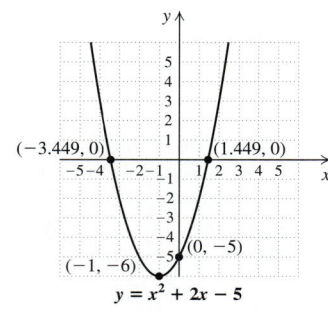

Graph of $y = x^2 + 2x - 5$ with x-intercepts $(-3.449, 0)$ and $(1.449, 0)$, vertex $(-1, -6)$, y-intercept $(0, -5)$.

87. [2.4], [2.5] 48.2 million **88.** [2.7] $\{x \mid x > 5\}$
89. [5.7] 14, 16; −16, −14 **90.** [10.3] 12 m
91. [2.5] 6090
92. [8.4] $1.10 per lb: 14 lb; $0.80 per lb: 28 lb
93. [6.7] 24 min
94. [3.4]

Graph showing Cost of cable TV $c = 20t + 25$ vs. Number of months t; point at $225.

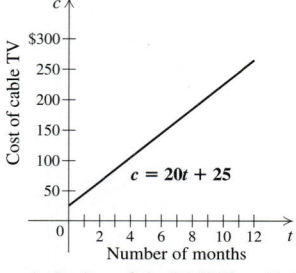

95. [10.6] −4, 0, 0   **96.** [7.1] Parallel
97. [8.5]

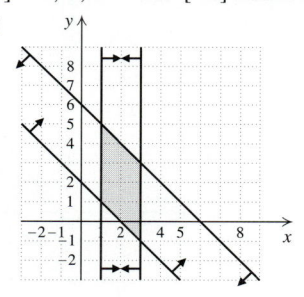

Graph of system of inequalities with shaded region.

98. [7.1] 2, (0, −8)   **99.** [3.5] 15
100. [7.2] $y + 3 = -\dfrac{1}{2}(x - 1)$ **101.** [10.4] $-5i$
102. [10.2] 30, −30   **103.** [9.6] $\dfrac{\sqrt{6}}{3}$ **104.** [4.5] Yes
105. [6.1] No   **106.** [4.5] No   **107.** [9.1] No
108. [9.1] Yes

APPENDIXES

EXERCISE SET A, P. 522

1. {3, 4, 5, 6, 7, 8}   **3.** {41, 43, 45, 47, 49}   **5.** {−3, 3}
7. False   **9.** True   **11.** True   **13.** True   **15.** True
17. False   **19.** {c, d, e}   **21.** {1, 10}   **23.** ∅
25. {a, e, i, o, u, q, c, k}   **27.** {0, 1, 2, 5, 7, 10}
29. {a, e, i, o, u, m, n, f, g, h}   **31.** ◆
33. The set of integers   **35.** The set of real numbers
37. ∅   **39.** (a) A; (b) A; (c) A; (d) ∅   **41.** True

EXERCISE SET B, P. 525

1. $(t + 3)(t^2 - 3t + 9)$ **3.** $(a - 1)(a^2 + a + 1)$
5. $(z + 5)(z^2 - 5z + 25)$ **7.** $(2a - 1)(4a^2 + 2a + 1)$
9. $(y - 3)(y^2 + 3y + 9)$ **11.** $(4 + 5x)(16 - 20x + 25x^2)$
13. $(5p - 1)(25p^2 + 5p + 1)$
15. $(3m + 4)(9m^2 - 12m + 16)$
17. $(p - q)(p^2 + pq + q^2)$ **19.** $\left(x + \dfrac{1}{2}\right)\left(x^2 - \dfrac{1}{2}x + \dfrac{1}{4}\right)$
21. $2(y - 4)(y^2 + 4y + 16)$
23. $3(2a + 1)(4a^2 - 2a + 1)$ **25.** $r(s + 4)(s^2 - 4s + 16)$
27. $5(x - 2z)(x^2 + 2xz + 4z^2)$
29. $(x + 0.1)(x^2 - 0.1x + 0.01)$ **31.** ◆
33. $(5c^2 - 2d^2)(25c^4 + 10c^2d^2 + 4d^4)$
35. $3(x^a + 2y^b)(x^{2a} - 2x^a y^b + 4y^{2b})$
37. $\dfrac{1}{3}\left(\dfrac{1}{2}xy + z\right)\left(\dfrac{1}{4}x^2y^2 - \dfrac{1}{2}xyz + z^2\right)$

Instructor's Answers

CHAPTER 1

EXERCISE SET 1.1, PP. 7–9

1. 45 **2.** $9 \cdot 7 = 63$ **3.** 8 **4.** $13 - 9 = 4$ **5.** 8
6. $\dfrac{15 + 20}{5} = \dfrac{35}{5} = 7$ **7.** 2 **8.** $\dfrac{23 - 5}{6} = \dfrac{18}{6} = 3$
9. 5 **10.** $\dfrac{54}{9} = 6$ **11.** 3 **12.** $\dfrac{5 \cdot 9}{15} = \dfrac{45}{15} = 3$
13. 165 mi **14.** $\dfrac{27{,}000}{1125} = 24$ hr **15.** 100.1 sq cm
16. $\dfrac{h}{a} = \dfrac{9}{33} = \dfrac{3}{11}$, or about 0.273 **17.** 15 sq cm
18. 5(30 sec) = 150 sec; 5(90 sec) = 450 sec; 5(2 min) = 10 min **19.** $7x$ or $x7$ **20.** $4a$
21. $b + 6$, or $6 + b$ **22.** $t + 8$, or $8 + t$ **23.** $c - 9$
24. $d - 4$ **25.** $6 + q$, or $q + 6$ **26.** $z + 11$, or $11 + z$
27. Let n represent "a number"; $n + 9$, or $9 + n$
28. $d + c$, or $c + d$ **29.** $y - x$ **30.** Let x represent "a number." Then we have $x - 2$. **31.** $x \div w$, or $\dfrac{x}{w}$
32. Let s and t represent the numbers. Then we have $s \div t$, or $\dfrac{s}{t}$. **33.** $n - m$ **34.** $q - p$
35. Let a and b represent the numbers; $a + b$, or $b + a$
36. $d + f$, or $f + d$ **37.** $9 \cdot 2m$ **38.** $t - 2r$
39. Let y represent "some number"; $\dfrac{1}{4}y$, or $\dfrac{y}{4}$
40. Let m and n represent the numbers. Then we have $\dfrac{1}{3}mn$, or $\dfrac{mn}{3}$.
41. Let x represent "some number"; 64% of x, or $0.64x$
42. Let y represent "a number." Then we have 38% of y, or $0.38y$. **43.** $\$50 - x$ **44.** $65t$ mi **45.** Yes

46. $y + 28 = 93$
$\overline{75 + 28 \; ? \; 93}$
$103 \; | \; 93 \qquad 103 = 93$ is FALSE.
75 is not a solution.
47. No **48.** $8t = 96$
$\overline{8 \cdot 12 \; ? \; 96}$
$96 \; | \; 96 \qquad 96 = 96$ is TRUE.
12 is a solution.
49. Yes **50.** $\dfrac{x}{8} = 6$
$\overline{\dfrac{52}{8} \; ? \; 6}$
$6.5 \; | \; 6 \qquad 6.5 = 6$ is FALSE.
52 is not a solution.
51. Yes **52.** $\dfrac{94}{y} = 12$
$\overline{\dfrac{94}{7} \; ? \; 12}$
$13\dfrac{3}{7} \; | \; 12 \qquad 13\dfrac{3}{7} = 12$ is FALSE.
7 is not a solution.
53. Let x represent the unknown number; $73 + x = 201$
54. Let w represent the number; $7w = 2303$ **55.** Let n represent the number; $42n = 2352$ **56.** Let x represent the number; $x + 345 = 987$ **57.** Let s represent the number of squares your opponent wins; $s + 35 = 64$
58. Let y represent the number of hours the carpenter worked; $25y = 53{,}400$ **59.** Let $w =$ the length of the average commute in the West, in minutes; $w = 24.5 - 1.8$
60. Let f represent the amount of fuel used by trucks in the United States in 1995, in billions of gallons; $f = 2(31.4)$
61. ◆ Yes; for a rectangle with length l and width w, the area A is given by $A = lw$. The area of a rectangle with length $2l$ and width w is given by $2lw$, or $2A$.

62. ◆ A *variable* is a letter that is used to stand for any number chosen from a set of numbers. An *algebraic expression* is an expression that consists of variables, constants, operation signs, and/or grouping symbols. A *variable expression* is an algebraic expression that contains a variable. An *equation* is a number sentence with the verb =. The symbol = is used to indicate that the algebraic expressions on either side of the symbol represent the same number.
63. ◆ Answers may vary. A crate of oranges contains 35 oranges. How many crates are required to hold 840 oranges?   **64.** ◆ Answers may vary. Juliet was born in 1989. How old will she be in 2003?
65. Let n represent "a number"; $3n - 5$
66. Let a and b represent the numbers. Then we have $\frac{1}{3} \cdot \frac{1}{2}ab$, or $\frac{1}{6}ab$, or $\frac{ab}{6}$.
67. $l + w + l + w$, or $2l + 2w$ **68.** $s + s + s + s$, or $4s$
69. $(a + 2) + 7$, or $a + 9$
70. Area of sign: $A = \frac{1}{2}(3)(2.5) = 3.75$ ft^2
Cost of sign: $90(3.75) = $337.50
71. 158.75 sq cm
72. $y = 8$, $x = 2y = 2 \cdot 8 = 16$;   **73.** 5
$\frac{x + y}{4} = \frac{8 + 16}{4} = \frac{24}{4} = 6$
74. $x = 9$, $y = 3x = 3 \cdot 9 = 27$;   **75.** 9
$\frac{y - x}{3} = \frac{27 - 9}{3} = \frac{18}{3} = 6$
76. The next whole number is one more than $w + 3$; $w + 3 + 1 = w + 4$ **77.** d
78. The preceding even number is 2 less than $d + 2$; $d + 2 - 2 = d$

EXERCISE SET 1.2, PP. 14–15
1. $x + 3$ **2.** $2 + a$ **3.** $bc + a$ **4.** $3y + x$
5. $3y + 9x$ **6.** $7b + 3a$ **7.** $2(3 + a)$ **8.** $9(5 + x)$
9. $a \cdot 8$ **10.** yx **11.** ts **12.** $b7$ **13.** $5 + ba$
14. $x + y3$ **15.** $(a + 3)2$ **16.** $(x + 5)9$
17. $a + (3 + b)$ **18.** $5 + (m + r)$ **19.** $(r + t) + 9$
20. $(x + 2) + y$ **21.** $ab + (c + d)$ **22.** $m + (np + r)$
23. $3(xy)$ **24.** $9(ab)$ **25.** $(4x)y$ **26.** $(9r)p$
27. $(3 \cdot 2)(a + b)$ **28.** $(5x)(2 + y)$
29. $(6 + t) + r$, $(6 + r) + t$; answers may vary
30. (a) $5 + (v + w) = (v + w) + 5 = v + (w + 5)$
 (b) $5 + (v + w) = (5 + v) + w = (v + 5) + w$
 Answers may vary.
31. $(ba)5$, $a(5b)$; answers may vary
32. (a) $x(3y) = (3y)x = 3(yx)$
 (b) $x(3y) = (x3)y = (3x)y$
 Answers may vary.

33. $(7 + x) + 2 = (x + 7) + 2$ Commutative law
$= x + (7 + 2)$ Associative law
$= x + 9$ Simplifying
34. $(2a)4 = 4(2a)$ Commutative law
$= (4 \cdot 2)a$ Associative law
$= 8a$ Simplifying
35. $(m3)7 = m(3 \cdot 7)$ Associative law
$= (3 \cdot 7)m$ Commutative law
$= 21m$ Simplifying
36. $4 + (9 + x)$
$= (4 + 9) + x$ Associative law
$= x + (4 + 9)$ Commutative law
$= x + 13$ Simplifying
37. $3a + 12$ **38.** $4x + 12$ **39.** $6 + 6x$ **40.** $6v + 24$
41. $3x + 3$ **42.** $9x + 27$ **43.** $4 + 4y$ **44.** $7s + 35$
45. $18x + 54$ **46.** $54m + 63$ **47.** $5r + 10 + 15t$
48. $20x + 32 + 12p$ **49.** $2a + 2b$ **50.** $7x + 14$
51. $5x + 5y + 10$ **52.** $12 + 6a + 6b$ **53.** $7(x + z)$
54. $5y + 5z = 5(y + z)$ **55.** $5(1 + y)$
Check: $5(y + z) = 5y + 5z$
56. $13 + 13x = 13(1 + x)$ **57.** $3(6x + y)$
Check: $13(1 + x) = 13 + 13x$
58. $5x + 20y = 5(x + 4y)$ **59.** $5(x + 2 + 3y)$
Check: $5(x + 4y) = 5x + 20y$
60. $3 + 27b + 6c = 3(1 + 9b + 2c)$
Check: $3(1 + 9b + 2c) = 3 + 27b + 6c$
61. $3(4x + 3)$
62. $6x + 6 = 6(x + 1)$ **63.** $3(3x + y)$
Check: $6(x + 1) = 6x + 6$
64. $15x + 5y = 5(3x + y)$ **65.** $2(a + 8b + 32)$
Check: $5(3x + y) = 15x + 5y$
66. $5 + 20x + 35y = 5(1 + 4x + 7y)$
Check: $5(1 + 4x + 7y) = 5 + 20x + 35y$
67. $11(x + 4y + 11)$
68. $7 + 14b + 56w = 7(1 + 2b + 8w)$
Check: $7(1 + 2b + 8w) = 7 + 14b + 56w$
69. $t - 9$ **70.** $\frac{1}{2}m$, or $\frac{m}{2}$
71. ◆ Neither subtraction nor division is associative. In general, when subtracting or dividing, the result depends on the grouping.   **72.** ◆ Neither subtraction nor division is commutative. In general, when subtracting or dividing, the result depends on the order in which the operation is performed.   **73.** ◆ No; a term is a product of factors. For example, in the expression $2x + 4y$, the terms are $2x$ and $4y$. The term $2x$ has factors 1, 2, and x while the term $4y$ has factors 1, 2, 4, and y.   **74.** ◆ No; Kara used the associative law of multiplication to get the last term of the product, $14c$:

$7(1 + a + 2c)$
$= 7 \cdot 1 + 7 \cdot a + 7(2c)$
$= 7 + 7a + (7 \cdot 2)c$ Associative law
$= 7 + 7a + 14c$

CHAPTER 1   A-43

75. Yes; distributive law
76. The expressions are not equivalent. Let $m = 1$. Then we have:
$$7 \div 3 \cdot 1 = \frac{7}{3} \cdot 1 = \frac{7}{3}, \text{ but}$$
$$1 \cdot 3 \div 7 = 3 \div 7 = \frac{3}{7}.$$

77. Yes; distributive law and commutative law of multiplication
78. The expressions are equivalent.
$yax + ax = (y + 1)ax = (1 + y)ax = ax(1 + y) = xa(1 + y)$
79. No; for example, let $x = 1$ and $y = 2$. Then $30 \cdot 2 + 1 \cdot 15 = 60 + 15 = 75$, but $5[2(1 + 3 \cdot 2)] = 5[2(7)] = 5 \cdot 14 = 70$.
80. The expressions are equivalent.
$[c(2 + 3b)]5 = 5[c(2 + 3b)] = 5c(2 + 3b) = 10c + 15bc$
81. ◆ $3(2 + x) = 3(2 + 0) = 3 \cdot 2 = 6$
$6 + x = 6 + 0 = 6$
The result indicates that $3(2 + x)$ and $6 + x$ are equivalent when $x = 0$. (By the distributive law, we know they are not equivalent for all values of x.)
82. ◆ $15x + 40 = 5(3x + 8)$
$15 \cdot 4 + 40 = 60 + 40 = 100$
$5(3 \cdot 4 + 8) = 5(12 + 8) = 5 \cdot 20 = 100$
Although the expressions $15x + 40$ and $5(3x + 8)$ are equivalent when $x = 4$, this result does not guarantee that the factorization is correct. (See Exercise 81.)

EXERCISE SET 1.3, PP. 22–23

1. $2 \cdot 15, 3 \cdot 10$; $1, 2, 3, 5, 6, 10, 15, 30$
2. Factorizations: $2 \cdot 35, 5 \cdot 14$; there are other factorizations as well.
Factors: $1, 2, 5, 7, 10, 14, 35, 70$
3. $3 \cdot 14, 6 \cdot 7$; $1, 2, 3, 6, 7, 14, 21, 42$
4. Factorizations: $2 \cdot 30, 5 \cdot 12$; there are other factorizations as well.
Factors: $1, 2, 3, 4, 5, 6, 10, 12, 15, 20, 30, 60$
5. $2 \cdot 11$ **6.** $3 \cdot 5$ **7.** $2 \cdot 3 \cdot 5$ **8.** $5 \cdot 11$ **9.** $2 \cdot 5 \cdot 5$
10. $2 \cdot 2 \cdot 5$ **11.** $3 \cdot 3 \cdot 3$ **12.** $2 \cdot 7 \cdot 7$ **13.** $2 \cdot 3 \cdot 3$
14. $2 \cdot 2 \cdot 2 \cdot 3$ **15.** $2 \cdot 2 \cdot 2 \cdot 5$ **16.** $2 \cdot 2 \cdot 2 \cdot 7$
17. Prime **18.** $2 \cdot 2 \cdot 2 \cdot 3 \cdot 5$ **19.** $2 \cdot 3 \cdot 5 \cdot 7$
20. 79 is prime. **21.** $5 \cdot 23$ **22.** $11 \cdot 13$ **23.** $\dfrac{5}{7}$
24. $\dfrac{16}{56} = \dfrac{2 \cdot \cancel{8}}{7 \cdot \cancel{8}} = \dfrac{2}{7}$ **25.** $\dfrac{7}{2}$ **26.** $\dfrac{72}{27} = \dfrac{8 \cdot \cancel{9}}{3 \cdot \cancel{9}} = \dfrac{8}{3}$
27. $\dfrac{1}{8}$ **28.** $\dfrac{12}{70} = \dfrac{2 \cdot 6}{2 \cdot 35} = \dfrac{6}{35}$ **29.** 8
30. $\dfrac{132}{11} = \dfrac{12 \cdot \cancel{11}}{1 \cdot \cancel{11}} = 12$ **31.** $\dfrac{1}{4}$ **32.** $\dfrac{17}{51} = \dfrac{1 \cdot \cancel{17}}{3 \cdot \cancel{17}} = \dfrac{1}{3}$

33. 5 **34.** $\dfrac{150}{25} = \dfrac{6 \cdot \cancel{25}}{1 \cdot \cancel{25}} = 6$ **35.** $\dfrac{15}{16}$
36. $\dfrac{42}{50} = \dfrac{\cancel{2} \cdot 21}{\cancel{2} \cdot 25} = \dfrac{21}{25}$ **37.** $\dfrac{60}{41}$ **38.** $\dfrac{75}{45} = \dfrac{5 \cdot \cancel{15}}{3 \cdot \cancel{15}} = \dfrac{5}{3}$
39. $\dfrac{15}{7}$ **40.** $\dfrac{140}{350} = \dfrac{\cancel{2} \cdot 2 \cdot \cancel{5} \cdot \cancel{7}}{\cancel{2} \cdot \cancel{5} \cdot 5 \cdot \cancel{7}} = \dfrac{2}{5}$ **41.** $\dfrac{3}{10}$
42. $\dfrac{11}{10} \cdot \dfrac{8}{5} = \dfrac{11 \cdot 8}{10 \cdot 5} = \dfrac{11 \cdot \cancel{2} \cdot 4}{\cancel{2} \cdot 5 \cdot 5} = \dfrac{44}{25}$ **43.** $\dfrac{51}{8}$
44. $\dfrac{11}{12} \cdot \dfrac{12}{11} = \dfrac{11 \cdot 12}{12 \cdot 11} = 1$ **45.** $\dfrac{1}{2}$
46. $\dfrac{1}{2} + \dfrac{1}{8} = \dfrac{4}{8} + \dfrac{1}{8} = \dfrac{5}{8}$ **47.** $\dfrac{7}{6}$
48. $\dfrac{4}{5} + \dfrac{8}{15} = \dfrac{12}{15} + \dfrac{8}{15} = \dfrac{20}{15} = \dfrac{4}{3}$ **49.** $\dfrac{3b}{7a}$
50. $\dfrac{x}{5} \cdot \dfrac{y}{z} = \dfrac{xy}{5z}$ **51.** $\dfrac{7}{a}$ **52.** $\dfrac{7}{a} - \dfrac{5}{a} = \dfrac{2}{a}$ **53.** $\dfrac{5}{6}$
54. $\dfrac{9}{8} + \dfrac{7}{12} = \dfrac{27}{24} + \dfrac{14}{24} = \dfrac{41}{24}$ **55.** 1
56. $\dfrac{12}{5} - \dfrac{2}{5} = \dfrac{10}{5} = 2$ **57.** $\dfrac{5}{18}$
58. $\dfrac{13}{15} - \dfrac{8}{45} = \dfrac{39}{45} - \dfrac{8}{45} = \dfrac{31}{45}$ **59.** $\dfrac{5}{9}$
60. $\dfrac{15}{16} - \dfrac{2}{3} = \dfrac{45}{48} - \dfrac{32}{48} = \dfrac{13}{48}$ **61.** $\dfrac{35}{18}$
62. $\dfrac{7}{5} \div \dfrac{3}{4} = \dfrac{7}{5} \cdot \dfrac{4}{3} = \dfrac{28}{15}$ **63.** $\dfrac{10}{3}$
64. $\dfrac{3}{4} \div 3 = \dfrac{3}{4} \cdot \dfrac{1}{3} = \dfrac{3 \cdot 1}{4 \cdot 3} = \dfrac{1}{4}$ **65.** 14
66. $\dfrac{1}{10} \div \dfrac{1}{5} = \dfrac{1}{10} \cdot \dfrac{5}{1} = \dfrac{5}{10} = \dfrac{1}{2}$ **67.** $\dfrac{5}{36}$
68. $\dfrac{17}{6} \div \dfrac{3}{8} = \dfrac{17}{6} \cdot \dfrac{8}{3} = \dfrac{17 \cdot \cancel{2} \cdot 4}{\cancel{2} \cdot 3 \cdot 3} = \dfrac{68}{9}$ **69.** 60
70. $78 \div \dfrac{1}{6} = \dfrac{78}{1} \cdot \dfrac{6}{1} = 468$ **71.** $\dfrac{x}{8}$
72. $\dfrac{5}{6} \div 15a = \dfrac{5}{6} \cdot \dfrac{1}{15a} = \dfrac{\cancel{5} \cdot 1}{6 \cdot \cancel{5} \cdot 3a} = \dfrac{1}{18a}$ **73.** $\dfrac{5b}{3a}$
74. $\dfrac{x}{7} \div \dfrac{4}{y} = \dfrac{x}{7} \cdot \dfrac{y}{4} = \dfrac{xy}{28}$ **75.** $\dfrac{x-2}{6}$
76. $\dfrac{9}{10} + \dfrac{x}{2} = \dfrac{9}{10} + \dfrac{5x}{10} = \dfrac{9 + 5x}{10}$
77. $5(3 + x)$; answers may vary
78. $7 + (a + b) = (a + b) + 7$; answers may vary.
79. ◆ If the fractions had the same denominator and the numerators and/or denominators were very large numbers, it would probably be easier to compute the sum of the fractions than their product. **80.** ◆ If the fractions have different denominators, it would probably be easier to compute the product of the fractions than their sum.

A-44 INSTRUCTOR'S ANSWERS

81. ◆ One factor of 6 is 3. (Factor is a noun.)
We can factor 6 as 2 · 3. (Factor is a verb.)
82. ◆ Yes; the commutative law of multiplication is true for any numbers. To illustrate that it is true for fractions, consider the following:

$$\frac{a}{b} \cdot \frac{c}{d} = \frac{ac}{bd}$$
$$= \frac{ca}{db} \quad \text{Using the commutative law in the numerator and in the denominator}$$
$$= \frac{c}{d} \cdot \frac{a}{b}$$

83. $\frac{4}{3}$ **84.** $\frac{pqrs}{qrst} = \frac{p}{t} \cdot \frac{qrs}{qrs} = \frac{p}{t}$ **85.** $\frac{2}{5}$

86. $\frac{8 \cdot 9xy}{2xy \cdot 36} = \frac{2 \cdot 4 \cdot 9 \cdot x \cdot y \cdot 1}{2 \cdot x \cdot y \cdot 4 \cdot 9} = 1$ **87.** $\frac{6}{25}$

88. $\frac{10x \cdot 12 \cdot 25y}{2 \cdot 30x \cdot 20y} = \frac{10 \cdot x \cdot 2 \cdot 2 \cdot 3 \cdot 5 \cdot 5 \cdot y}{2 \cdot 3 \cdot 10 \cdot x \cdot 2 \cdot 2 \cdot 5 \cdot y} = \frac{5}{2}$

89. 24 in.
90.

Product	56	63	36	72	140	96
Factor	7	7	2	36	14	8
Factor	8	9	18	2	10	12
Sum	15	16	20	38	24	20

Product	48	168	110	90	432	63
Factor	6	21	11	9	24	3
Factor	8	8	10	10	18	21
Sum	14	29	21	19	42	24

91. $\frac{28}{45}$ m²

92. $A = \frac{1}{2}bh = \frac{1}{2}\left(\frac{10}{7}\text{ m}\right)\left(\frac{5}{4}\text{ m}\right)$
$= \frac{1}{2}\left(\frac{10}{7}\right)\left(\frac{5}{4}\right)(\text{m})(\text{m})$
$= \frac{25}{28}$ m², or $\frac{25}{28}$ square meters

93. $\frac{20}{9}$ m

94. $P = 2l + 2w = 2\left(\frac{4}{5}\text{ m}\right) + 2\left(\frac{7}{9}\text{ m}\right)$
$= \frac{8}{5}\text{ m} + \frac{14}{9}\text{ m}$
$= \frac{142}{45}\text{ m}$

95. ◆
$x + y$
$= (x + y) \cdot 1$ Identity property of 1
$= (x + y) \cdot \frac{2}{2}$ $\left(\frac{2}{2} = 1\right)$
$= \frac{2}{2} \cdot (x + y)$ Commutative law
$= \frac{2x + 2y}{2}$ Distributive law
$= \frac{2y + 2x}{2}$ Commutative law

EXERCISE SET 1.4, PP. 30–31
1. −3, 7   **2.** 18, −2   **3.** −508, 186.84   **4.** 1200, −560
5. −1286, 29,029   **6.** Jets: −34, Strikers: 34
7. 750, −125   **8.** 27, −9.7   **9.** 20, −150, 300
10. −10, 235
11.
12. $-\frac{17}{5} = -3\frac{2}{5} = -3.4$

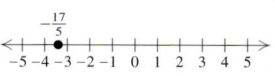

13.
14.

15.
16.

17. 0.875
18.
```
 0.1 2 5
8)1.0 0 0
  8
  2 0
  1 6
    4 0
    4 0
      0
```
$\frac{1}{8} = 0.125$, so $-\frac{1}{8} = -0.125$.

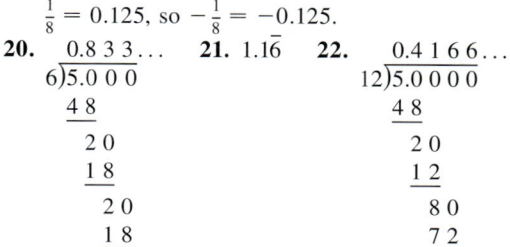

19. −0.75

20.
```
 0.8 3 3...
6)5.0 0 0
  4 8
    2 0
    1 8
      2 0
      1 8
        2
```
$\frac{5}{6} = 0.8\overline{3}$.

21. $1.1\overline{6}$

22.
```
  0.4 1 6 6...
12)5.0 0 0 0
   4 8
     2 0
     1 2
       8 0
       7 2
         8 0
         7 2
           8
```
$\frac{5}{12} = 0.41\overline{6}$.

23. $0.\overline{6}$ **24.** $\begin{array}{r} 0.25 \\ 4\overline{)1.00} \\ \underline{8} \\ 20 \\ \underline{20} \\ 0 \end{array}$ **25.** -0.5

$\frac{1}{4} = 0.25.$

26. $\begin{array}{r} 0.375 \\ 8\overline{)3.000} \\ \underline{24} \\ 60 \\ \underline{56} \\ 40 \\ \underline{40} \\ 0 \end{array}$ **27.** 0.1

$\frac{3}{8} = 0.375$, so $-\frac{3}{8} = -0.375.$

28. $\begin{array}{r} 0.35 \\ 20\overline{)7.00} \\ \underline{60} \\ 100 \\ \underline{100} \\ 0 \end{array}$ **29.** $>$ **30.** $9 > 0$

$\frac{7}{20} = 0.35$, so $-\frac{7}{20} = -0.35.$

31. $<$ **32.** $8 > -8$ **33.** $<$ **34.** $0 > -7$ **35.** $<$
36. $-4 < -3$ **37.** $>$ **38.** $-3 > -4$ **39.** $<$
40. $-10.3 > -14.5$ **41.** $<$
42. $-\frac{14}{17} \approx -0.82353$ and $-\frac{27}{35} \approx -0.77143$. Thus, $-\frac{14}{17} < -\frac{27}{35}$. **43.** $a < -2$ **44.** $9 < a$ **45.** $y \geq -10$
46. $t \leq 12$ **47.** True **48.** False **49.** False
50. True **51.** True **52.** $8 \geq 8$ is true because $8 = 8$ is true. **53.** 4 **54.** 9 **55.** 17 **56.** 3.1 **57.** 5.6
58. $\frac{2}{5}$ **59.** 329 **60.** 456 **61.** $\frac{9}{7}$ **62.** 8.02 **63.** 0
64. 1.07 **65.** 8 **66.** $|a| = |-5| = 5$
67. $-23, -4.7, 0, \frac{5}{9}, 8.31, 62$ **68.** 62 **69.** $-23, 0, 62$
70. $\pi, \sqrt{17}$ **71.** $-23, -4.7, 0, \frac{5}{9}, \pi, \sqrt{17}, 8.31, 62$
72. 0, 62 **73.** $\frac{3}{5}$ **74.** $3xy = 3 \cdot 2 \cdot 7 = 42$
75. $5 + ab$; answers may vary
76. $3x + 9 + 12y = 3(x + 3 + 4y)$
77. ◆ There are infinitely many rational numbers between 0 and 1. Consider only rational numbers of the form $\frac{1}{n}$, where n is an integer greater than 1. There are infinitely many integers greater than 1, so there are infinitely many numbers $\frac{1}{n}$, all between 0 and 1. (These numbers are a subset of the rational numbers between 0 and 1.)

78. ◆ The whole numbers are the nonnegative integers, so every nonnegative integer is a whole number.
79. ◆ The absolute value of a number is defined to be the number's distance from zero on a number line. Since distance is never negative, it is impossible for the absolute value of a number to be negative. **80.** ◆ The range of class sizes is all numbers from 10 to 35, inclusive.
81. $-17, -12, 5, 13$ **82.** $-23, -17, 0, 4$
83. $-\frac{4}{3}, \frac{4}{9}, \frac{4}{8}, \frac{4}{6}, \frac{4}{5}, \frac{4}{3}, \frac{4}{2}$
84. $-\frac{2}{3}, \frac{1}{2}, -\frac{3}{4}, -\frac{5}{6}, \frac{3}{8}, \frac{1}{6}$ can be written in decimal notation as $-0.\overline{666}, 0.5, -0.75, -0.8\overline{33}, 0.375, 0.1\overline{66}$, respectively. Listing from least to greatest, we have $-\frac{5}{6}$, $-\frac{3}{4}, -\frac{2}{3}, \frac{1}{6}, \frac{3}{8}, \frac{1}{2}$. **85.** $>$ **86.** $|4| < |-7|$ **87.** $=$
88. $|23| = |-23|$ **89.** $<$ **90.** $|-19| < |-27|$
91. $7, -7$ **92.** $-2, -1, 0, 1, 2$ **93.** $-4, -3, 3, 4$
94. $0.1\overline{1} = \frac{0.\overline{33}}{3} = \frac{\frac{1}{3}}{3} = \frac{1}{3} \cdot \frac{1}{3} = \frac{1}{9}$ **95.** $\frac{3}{3}$
96. $5.\overline{55} = 50(0.1\overline{1}) = 50 \cdot \frac{1}{9} = \frac{50}{9}$ (See Exercise 94.)
97. ◆ ⌐⌐ The number entered by hand is an approximation of $\sqrt{2}$ while the value that is squared immediately after being calculated is actually regarded by the calculator as $\sqrt{2}$.

EXERCISE SET 1.5, PP. 36–37
1. -4
2. Start at 2. Move 5 units to the left.

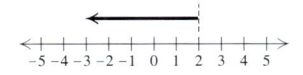

$2 + (-5) = -3$
3. 4
4. Start at 8. Move 3 units to the left.

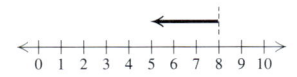

$8 + (-3) = 5$
5. 0
6. Start at 6. Move 6 units to the left.

$6 + (-6) = 0$
7. -8
8. Start at -4. Move 6 units to the left.

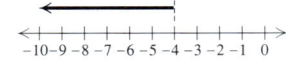

$-4 + (-6) = -10$

9. −15 **10.** −6 **11.** −8 **12.** −2 **13.** 0 **14.** 0
15. −41 **16.** −42 **17.** 0 **18.** 0 **19.** 1 **20.** 3
21. −2 **22.** −9 **23.** 11 **24.** 7 **25.** −33 **26.** 2
27. 0 **28.** −26 **29.** 18 **30.** −22 **31.** −32
32. 32 **33.** 0 **34.** 0 **35.** 20 **36.** 45 **37.** −1.7
38. −1.8 **39.** −9.1 **40.** −13.2 **41.** $-\frac{2}{3}$ **42.** $-\frac{1}{5}$
43. $-\frac{10}{9}$ **44.** $-\frac{8}{7}$ **45.** $-\frac{1}{6}$ **46.** $-\frac{3}{8}$ **47.** $-\frac{23}{24}$
48. $-\frac{3}{7} + \left(-\frac{2}{5}\right) = -\frac{15}{35} + \left(-\frac{14}{35}\right) = -\frac{29}{35}$ **49.** −3
50. $28 + (-44) + 17 + 31 + (-94) = 76 + (-138) = -62$ **51.** 50
52. $24 + 3.1 + (-44) + (-8.2) + 63 = 90.1 + (-52.2) = 37.9$ **53.** $69
54. Since $13 + 0 + (-12) + 21 + (-14) = 8$, the total gain was 8 yd. **55.** $183,415 profit
56. Since $-6 + 3 + (-14) + 4 = -13$, the total change in pressure is a 13 mb drop. **57.** 13,796 ft above sea level
58. Since $-470 + 45 + (-160) + 500 = -85$, Kyle owes the company $85. **59.** $0
60. Since $460 + (-530) + 75 + (-90) = -85$, Tony's account is $85 overdrawn. **61.** $13a$ **62.** $11x$ **63.** $9x$
64. $-5m$ **65.** $11x$ **66.** $14a$ **67.** $-2m$ **68.** $5x$
69. $-7a$ **70.** $-7n$ **71.** $1 - 2x$
72. $8a + 5 + (-a) + (-3) = 8a + (-a) + 5 + (-3) = 7a + 2$ **73.** $12x + 17$
74. $4a + 5 + 6a + 8 = 4a + 6a + 5 + 8 = 10a + 13$
75. $18n + 16$
76. $2 + 6 + 5n + 7n + 3 + 7n = 19n + 11$
77. $21z + 7y + 14$
78. $\frac{7}{2} \div \frac{3}{8} = \frac{7}{2} \cdot \frac{8}{3} = \frac{7 \cdot 8}{2 \cdot 3} = \frac{7 \cdot 2 \cdot 4}{2 \cdot 3} = \frac{7 \cdot 4}{3} = \frac{28}{3}$
79. ◆ Each nonzero integer from −50 to 50 can be added to its opposite with the sum of each pair being 0. When this is added to the remaining integer, 0, the total is 0.
80. ◆ Answers may vary. The temperature in Kansas City rose 13° F; then it dropped 7° F. After that it rose 5° F and then dropped 12° F. What was the total change in temperature? **81.** ◆ The sum will be positive when the positive number is greater than the sum of the absolute values of the negative numbers. **82.** ◆ Answers may vary. One possible explanation follows. Consider performing the addition on a number line. We start to the left of 0 and then move farther left, so the result must be a negative number. **83.** $65\frac{1}{4}$
84. Since $61 + 12 + (-17.50) = 55.50$, the original value was $55.50. **85.** $-5y$
86. $-3a + 9b + \underline{\quad} + 5a = 2a + 9b + \underline{\quad}$. This expression is equivalent to $2a - 6b$, so the missing term is the term which yields $-6b$ when added to $9b$. Since $9b + (-15b) = -6b$, the missing term is $-15b$. **87.** $-7m$
88. $\underline{\quad} + 9x + (-4y) + x = \underline{\quad} + 10x + (-4y)$. This expression is equivalent to $10x - 7y$, so the missing term is the term which yields $-7y$ when added to $-4y$. Since $-3y + (-4y) = -7y$, the missing term is $-3y$. **89.** $\frac{7}{2}x$
90. $-3 + (-3) + 2 + (-2) + 1 = -5$. Since the total is 5 under par after the five rounds and $-5 = -1 + (-1) + (-1) + (-1) + (-1)$, the golfer was 1 under par on average.

EXERCISE SET 1.6, PP. 42–43

1. −39 **2.** 17 **3.** 9 **4.** $-\frac{7}{2}$ **5.** 3.14 **6.** −48.2
7. −23 **8.** 26 **9.** $\frac{14}{3}$ **10.** $-\frac{1}{328}$ **11.** −0.101
12. 0 **13.** 72 **14.** 29 **15.** $-\frac{2}{5}$ **16.** −9.1 **17.** 1
18. 7 **19.** −7 **20.** −10 **21.** −9 **22.** −5 **23.** −7
24. −10 **25.** 2 **26.** −6 **27.** 0 **28.** 0 **29.** −4
30. −5 **31.** −7 **32.** 26 **33.** −5 **34.** 2 **35.** 0
36. 0 **37.** 0 **38.** 0 **39.** 12 **40.** 8 **41.** 11
42. −11 **43.** −14 **44.** 16 **45.** 5 **46.** −16
47. −1 **48.** −1 **49.** 15 **50.** 11 **51.** −5 **52.** −6
53. −3 **54.** −2 **55.** −21 **56.** −25 **57.** 5 **58.** 1
59. −8 **60.** −9 **61.** 10 **62.** 17 **63.** −23
64. −45 **65.** −68 **66.** −81 **67.** −73 **68.** −52
69. 116 **70.** 121 **71.** 0 **72.** 0 **73.** $-\frac{1}{4}$
74. $\frac{3}{9} - \frac{9}{9} = \frac{3}{9} + \left(-\frac{9}{9}\right) = -\frac{6}{9} = -\frac{2}{3}$ **75.** $\frac{2}{15}$
76. $\frac{5}{8} - \frac{3}{4} = \frac{5}{8} + \left(-\frac{3}{4}\right) = \frac{5}{8} + \left(-\frac{6}{8}\right) = -\frac{1}{8}$ **77.** $-\frac{17}{12}$
78. $-\frac{5}{8} - \frac{3}{4} = -\frac{5}{8} + \left(-\frac{3}{4}\right) = -\frac{5}{8} + \left(-\frac{6}{8}\right) = -\frac{11}{8}$
79. −2.8 **80.** 4.94 **81.** −0.91 **82.** −0.911
83. $-\frac{1}{2}$ **84.** $-\frac{3}{8} - \left(-\frac{1}{2}\right) = -\frac{3}{8} + \frac{1}{2} = -\frac{3}{8} + \frac{4}{8} = \frac{1}{8}$
85. $\frac{6}{7}$ **86.** 0 **87.** $3.8 - (-5.2)$; 9
88. $-2.1 - (-5.9)$; 3.8 **89.** $114 - (-79)$; 193
90. $23 - (-17)$; 40 **91.** −58 **92.** $-7 - 19 = -26$
93. 34 **94.** $-5 - (-31) = 26$
95. Negative one point eight minus two point seven; −4.5
96. $-2.7 - 5.9$ is read "negative two point seven minus five point nine;" $-2.7 - 5.9 = -8.6$
97. Negative two hundred fifty minus negative four hundred twenty-five; 175
98. $-350 - (-1000)$ is read "negative three hundred fifty minus negative one thousand;" $-350 - (-1000) = 650$
99. 41 **100.** −22 **101.** −62 **102.** 22 **103.** −139
104. 5 **105.** 6 **106.** 4 **107.** $-7x, -4y$
108. $7a, -9b$ **109.** $-5, 3m, -6mn$ **110.** $-9, -4t, 10rt$
111. $-3x$ **112.** $-11a$ **113.** $-5a + 4$ **114.** $-22x + 7$
115. $-7n - 9$ **116.** $9n - 15$ **117.** $-6x + 5$
118. $3a - 5$ **119.** $-7 - 8t$ **120.** $-2b - 12$
121. $8 - 12x$ **122.** $7x + 46$ **123.** $15x + 66$
124. $15x + 39$ **125.** 100°F **126.** $290 - $125 = $165

127. 14,494 ft above sea level
128. $29{,}028 - (-1312) = 30{,}340$ ft **129.** 1767 m
130. $-40 - (-156) = 116$ m **131.** 432 ft^2
132. $2 \cdot 2 \cdot 2 \cdot 2 \cdot 2 \cdot 3 \cdot 3 \cdot 3$
133. ◆ Doing so can simplify the computation, particularly when negative numbers are being subtracted. For example, rewriting $3 - (-2) + 7$ as $3 + 2 + 7$ eases the computation of the sum.
134. ◆ $-a + b$ is the opposite of $a + (-b)$ since $a + (-b) + (-a + b) = a + (-a) + (-b) + b = 0$.
135. ◆ Rewrite subtraction as addition of the opposite.
136. ◆ For two negative numbers a and b, $a - b$ is negative when $|a| > |b|$. **137.** False. For example, let $m = -3$ and $n = -5$. Then $-3 > -5$, but $-3 + (-5) = -8 \not> 0$. **138.** True. For example, for $m = 5$ and $n = 3$, $5 > 3$ and $5 - 3 > 0$, or $2 > 0$. For $m = -4$ and $n = -9$, $-4 > -9$ and $-4 - (-9) > 0$, or $5 > 0$. **139.** False. For example, let $m = 2$ and $n = -2$. Then 2 and -2 are opposites, but $2 - (-2) = 4 \neq 0$. **140.** True. For example, for $m = 4$ and $n = -4$, $4 = -(-4)$ and $4 + (-4) = 0$; for $m = -3$ and $n = 3$, $-3 = -3$ and $-3 + 3 = 0$. **141.** ◆ After the second "double or nothing" wager, the debt was $20, so the debt before that wager (that is, after the first "double or nothing" wager) was $10. Then the debt before the first "double or nothing" wager (that is, after the original wager) was $5. Thus, the gambler originally bet $5, this debt was doubled to $10, and that debt was doubled to $20. **142.** ◆ If n is positive and m is negative, then $-m$ is positive and $n + (-m)$ is positive.

EXERCISE SET 1.7, PP. 49–50
1. -27 **2.** -21 **3.** -56 **4.** -18 **5.** -24
6. -45 **7.** -72 **8.** -30 **9.** 16 **10.** 10
11. 45 **12.** 18 **13.** -120 **14.** 120 **15.** -238
16. 195 **17.** 1200 **18.** -1677 **19.** 98 **20.** -203.7
21. -78 **22.** -63 **23.** 21.7 **24.** 12.8 **25.** $-\frac{2}{5}$
26. $-\frac{10}{21}$ **27.** $\frac{1}{12}$ **28.** $\frac{1}{4}$ **29.** -17.01 **30.** -38.95
31. $-\frac{5}{12}$ **32.** -6 **33.** 252
34. $9 \cdot (-2) \cdot (-6) \cdot 7 = -18 \cdot (-42) = 756$ **35.** -90
36. $-4 \cdot (-3) \cdot (-5) = 12 \cdot (-5) = -60$ **37.** $\frac{1}{28}$
38. $-\frac{1}{2} \cdot \frac{3}{5} \cdot \left(-\frac{2}{7}\right) = -\frac{3}{10}\left(-\frac{2}{7}\right) = \frac{3 \cdot 2}{2 \cdot 5 \cdot 7} = \frac{3}{35}$
39. 150 **40.** $-3 \cdot (-5) \cdot (-2) \cdot (-1) = 15 \cdot 2 = 30$
41. 0 **42.** 0 **43.** -720
44. $(-7)(-8)(-9)(-10) = 56 \cdot 90 = 5040$ **45.** $-30{,}240$
46. $(-5)(-6)(-7)(-8)(-9)(-10) = 30 \cdot 56 \cdot 90 = 1680 \cdot 90 = 151{,}200$ **47.** -6 **48.** -4
49. -4 **50.** -2 **51.** -2 **52.** 8 **53.** 4 **54.** 7
55. -8 **56.** -2 **57.** 2 **58.** -25 **59.** -12

60. $\frac{64}{7}$ **61.** -8 **62.** $\frac{300}{13}$ **63.** Undefined
64. 0 **65.** $-\frac{88}{9}$ **66.** Indeterminate **67.** 0
68. Undefined **69.** Indeterminate **70.** 0
71. $-\frac{8}{3}, \frac{8}{-3}$ **72.** $\frac{12}{-7}, -\frac{12}{7}$ **73.** $-\frac{29}{35}, \frac{-29}{35}$
74. $\frac{-9}{14}, -\frac{9}{14}$ **75.** $\frac{-7}{3}, \frac{7}{-3}$ **76.** $\frac{-4}{15}, \frac{4}{-15}$
77. $-\frac{x}{2}, \frac{x}{-2}$ **78.** $\frac{-9}{a}, -\frac{9}{a}$ **79.** $-\frac{5}{4}$
80. $\frac{-9}{2}$, or $-\frac{9}{2}$ **81.** $-\frac{13}{47}$ **82.** $-\frac{12}{31}$ **83.** $-\frac{1}{10}$
84. $\frac{1}{13}$ **85.** $\frac{1}{4.3}$ **86.** $\frac{1}{-8.5}$, or $-\frac{1}{8.5}$ **87.** $-\frac{4}{9}$
88. $\frac{11}{-6}$, or $-\frac{11}{6}$ **89.** $-\frac{1}{1}$, or -1 **90.** $\frac{1}{2}$ **91.** $\frac{21}{20}$
92. $\frac{5}{18}$ **93.** $\frac{12}{55}$ **94.** $\frac{35}{12}$ **95.** -1
96. $\frac{-4}{5} + \frac{7}{-5} = \frac{-4}{5} + \frac{-7}{5} = \frac{-11}{5}$, or $-\frac{11}{5}$ **97.** $\frac{18}{7}$
98. $\left(-\frac{2}{7}\right)\left(\frac{5}{-8}\right) = \left(-\frac{2}{7}\right)\left(-\frac{5}{8}\right) = \frac{2 \cdot 5}{7 \cdot 2 \cdot 4} = \frac{5}{28}$
99. $-\frac{8}{11}$
100. $\frac{-9}{7} + \left(-\frac{4}{7}\right) = \frac{-9}{7} + \frac{-4}{7} = \frac{-13}{7}$, or $-\frac{13}{7}$
101. $-\frac{7}{4}$ **102.** $\frac{3}{4} \div \left(-\frac{2}{3}\right) = \frac{3}{4} \cdot \left(-\frac{3}{2}\right) = -\frac{9}{8}$
103. -12
104. $\frac{-5}{12} \cdot \frac{7}{15} = -\frac{5}{12} \cdot \frac{7}{15} = -\frac{5 \cdot 7}{12 \cdot 15} = -\frac{\cancel{5} \cdot 7}{12 \cdot \cancel{5} \cdot 3} = -\frac{7}{36}$ **105.** -3 **106.** $\left(-\frac{12}{5}\right) + \left(-\frac{3}{5}\right) = -\frac{15}{5} = -3$
107. $\frac{2}{3}$
108. $-\frac{5}{4} \div \left(-\frac{3}{4}\right) = -\frac{5}{4} \cdot \left(-\frac{4}{3}\right) = \frac{5 \cdot \cancel{4}}{\cancel{4} \cdot 3} = \frac{5}{3}$
109. 7 **110.** -2 **111.** $-\frac{7}{9}$
112. $\frac{-3}{7} - \frac{2}{7} = -\frac{3}{7} - \frac{2}{7} = -\frac{5}{7}$ **113.** $-\frac{7}{10}$
114. $\frac{-5}{9} + \frac{2}{-3} = \frac{-5}{9} + \frac{-2}{3} = \frac{-5}{9} + \frac{-6}{9} = \frac{-11}{9}$, or $-\frac{11}{9}$
115. $-\frac{7}{6}$
116. $\left(\frac{-3}{5}\right) \div \frac{6}{15} = -\frac{3}{5} \cdot \frac{15}{6} = -\frac{\cancel{3} \cdot 3 \cdot \cancel{5}}{\cancel{5} \cdot 2 \cdot \cancel{3}} = -\frac{3}{2}$

A-48 INSTRUCTOR'S ANSWERS

117. $\dfrac{6}{7}$ **118.** $\dfrac{4}{9} - \dfrac{1}{-9} = \dfrac{4}{9} - \left(-\dfrac{1}{9}\right) = \dfrac{4}{9} + \dfrac{1}{9} = \dfrac{5}{9}$
119. $-\dfrac{14}{15}$
120. $\dfrac{3}{-10} + \dfrac{-1}{5} = \dfrac{-3}{10} + \dfrac{-1}{5} = \dfrac{-3}{10} + \dfrac{-2}{10} = \dfrac{-5}{10} = \dfrac{-1}{2}$, or $-\dfrac{1}{2}$ **121.** $\dfrac{22}{39}$ **122.** $12x - 2y - 9$
123. ◆ Think of $3 \cdot (-5)$ as $-5 + (-5) + (-5)$. Start at -5 and move to the left 5 units to -10; then move to the left another 5 units to -15. Thus, $3(-5) = -15$.
124. ◆ You get the original number. The reciprocal of the reciprocal of a number is the original number.
125. ◆ Multiply the dividend by the reciprocal of the divisor. **126.** ◆ Yes; consider n and its opposite $-n$. The reciprocals of these numbers are $\dfrac{1}{n}$ and $\dfrac{1}{-n}$. Now $\dfrac{1}{n} + \dfrac{1}{-n} = \dfrac{1}{n} + \dfrac{-1}{n} = 0$, so the reciprocals are also opposites. **127.** For 2 and 3, the reciprocal of the sum is $1/(2 + 3)$, or $1/5$. But $1/5 \neq 1/2 + 1/3$.
128. -1 and 1 are their own reciprocals.
$[-1(-1) = 1$ and $1 \cdot 1 = 1]$

129. Negative **130.** $-n$ and $-m$ are both positive, so $\dfrac{-n}{-m}$ is positive. **131.** Negative **132.** $-m$ is positive, so $\dfrac{n}{-m}$ is negative and $-\left(\dfrac{n}{-m}\right)$ is positive. **133.** Negative
134. $-n$ and $-m$ are both positive, so $-n - m$, or $-n + (-m)$ is positive; $\dfrac{n}{m}$ is also positive, so $(-n - m)\dfrac{n}{m}$ is positive. **135.** (a) m and n have different signs; (b) either m or n is zero; (c) m and n have the same sign
136. $a(-b) + ab = a[-b + b]$ Distributive law
$= a(0)$ Law of opposites
$= 0$ Multiplicative property of 0
Therefore, $a(-b) = -(ab)$. Law of opposites
137. ◆ No; if $a > 0$ and $b < 0$, then $a > b$ but since $\dfrac{1}{a} > 0$ and $\dfrac{1}{b} < 0$, $\dfrac{1}{a} > \dfrac{1}{b}$.

EXERCISE SET 1.8, PP. 56–57
1. 17^3 **2.** 5^4 **3.** x^7 **4.** y^6 **5.** $(6y)^4$ **6.** $(5m)^5$
7. 81 **8.** 125 **9.** 9 **10.** 49 **11.** -1 **12.** -1
13. 64 **14.** 9 **15.** 625 **16.** 625 **17.** 7 **18.** -1
19. $16x^4$ **20.** $9x^2$ **21.** $-343x^3$ **22.** $625x^4$
23. 26 **24.** $9 - 4 \times 2 = 9 - 8 = 1$ **25.** 86
26. $10 \times 5 + 1 \times 1 = 50 + 1 = 51$ **27.** 7
28. $14 - 2 \times 6 + 7 = 14 - 12 + 7 = 2 + 7 = 9$

29. 5
30. $32 - 8 \div 4 - 2 = 32 - 2 - 2 = 30 - 2 = 28$
31. 12 **32.** $(2 - 5)^2 = (-3)^2 = 9$ **33.** 250
34. $3 \cdot 2^3 = 3 \cdot 8 = 24$ **35.** -7
36. $8 - (2 \cdot 3 - 9) = 8 - (6 - 9) = 8 - (-3) = 8 + 3 = 11$ **37.** -7 **38.** $(8 - 2)(3 - 9) = 6(-6) = -36$
39. -4 **40.** $32 \div (-2) \cdot (-2) = -16 \cdot (-2) = 32$
41. -85 **42.** $7 \cdot 8 - 9 \cdot 6 = 56 - 54 = 2$ **43.** 14
44. $40 - 3^2 - 2^3 = 40 - 9 - 8 = 31 - 8 = 23$
45. 1880
46. $4^3 + 10 \cdot 20 + 8^2 - 23 = 64 + 10 \cdot 20 + 64 - 23 = 64 + 200 + 64 - 23 = 264 + 64 - 23 = 328 - 23 = 305$ **47.** 16 **48.** $5^3 - 7^2 = 125 - 49 = 76$ **49.** 1
50. $\dfrac{5^2 - 3^2}{2 \cdot 6 - 4} = \dfrac{25 - 9}{12 - 4} = \dfrac{16}{8} = 2$ **51.** -26
52. $|10(-5)| + |(-1)| = |-50| - 1 = 50 - 1 = 49$
53. 37
54. $14 - 2(-6) + 7 = 14 + 12 + 7 = 26 + 7 = 33$
55. -6
56. $-32 - 8 \div 4 \cdot (-2) = -32 - 2 \cdot (-2) = -32 + 4 = -28$ **57.** 4
58. $4(-10)^3 - 5000 = 4(-1000) - 5000 = -4000 - 5000 = -9000$ **59.** 36
60. $5|8 - 3 \cdot 7| = 5|8 - 21| = 5|-13| = 5 \cdot 13 = 65$
61. -6 **62.** $7 + (-2)^3 = 7 - 8 = -1$ **63.** -3
64. $20 \div 5 \cdot 4 = 4 \cdot 4 = 16$ **65.** 45
66. $50 \div 2 \cdot 5 = 25 \cdot 5 = 125$ **67.** 9
68. $6 \cdot 2 \div 12(2)^3 = 6 \cdot 2 \div 12 \cdot 8 = 12 \div 12 \cdot 8 = 1 \cdot 8 = 8$ **69.** $\dfrac{55}{4}$
70. $-30 \div (-6)((-6) + 4)^2 = -30 \div (-6)(-2)^2 = -30 \div (-6) \cdot 4 = 5 \cdot 4 = 20$ **71.** 6
72. $(-(-3))^2 - 5(-3) = (3)^2 - 5(-3) = 9 - 5(-3) = 9 + 15 = 24$ **73.** -17
74. $\dfrac{(-2)^3 - 4(-2)}{-2(-2 - 3)} = \dfrac{-8 - 4(-2)}{-2(-5)} = \dfrac{-8 + 8}{10} = \dfrac{0}{10} = 0$
75. $-9x - 1$ **76.** $-3x - 5$ **77.** $-7 + 2x$
78. $-6x + 7$ **79.** $-4a + 3b - 7c$ **80.** $-5x + 2y + 3z$
81. $-3x^2 - 5x + 1$ **82.** $-8x^3 + 6x - 5$ **83.** $3x - 7$
84. $7y - (2y + 9) = 7y - 2y - 9 = 5y - 9$
85. $-3a + 9$
86. $11n - (3n - 7) = 11n - 3n + 7 = 8n + 7$
87. $5x - 6$
88. $3a + 2a - (4a + 7) = 3a + 2a - 4a - 7 = a - 7$
89. $-3t - 11r$
90. $4m - 9n - 3(2m - n) = 4m - 9n - 6m + 3n = -2m - 6n$ **91.** $9y - 25z$
92. $4a - b - 4(5a - 7b + 8c)$
$= 4a - b - 20a + 28b - 32c$
$= -16a + 27b - 32c$ **93.** $x^2 + 2$
94. $7x^4 + 9x - (5x^4 + 3x) = 7x^4 + 9x - 5x^4 - 3x = 2x^4 + 6x$ **95.** $-t^3 - 2t$

96. $8n^2 + n - 2(n + 3n^2) = 8n^2 + n - 2n - 6n^2 = 2n^2 - n$ **97.** $37a^2 - 23ab + 35b^2$
98. $-8a^2 + 5ab - 12b^2 - 6(2a^2 - 4ab - 10b^2)$
$= -8a^2 + 5ab - 12b^2 - 12a^2 + 24ab + 60b^2$
$= -20a^2 + 29ab + 48b^2$
99. $-22t^3 - t^2 + 9t$
100. $9t^4 + 7t - 5(9t^3 - 2t) = 9t^4 + 7t - 45t^3 + 10t = 9t^4 - 45t^3 + 17t$ **101.** $-8x + 34$
102. $5(x - 3) - 4(2x - 1) = 5x - 15 - 8x + 4 = -3x - 11$ **103.** $4a^3 + 2a - 10$
104. $2(t^2 - 2t) - 3(t^2 - 4t) + t^2$ **105.** $10x - 24$
$= 2t^2 - 4t - 3t^2 + 12t + t^2$
$= 8t$
106. $4(7x - 5) - [3(1 - 8x) + 6]$ **107.** $2x + 9$
$= 4(7x - 5) - [3 - 24x + 6]$
$= 4(7x - 5) - [9 - 24x]$
$= 28x - 20 - 9 + 24x$
$= 52x - 29$
108. Let x and y represent the numbers; $\frac{1}{2}(x + y)$.
109. ◆ No; $\frac{16}{2x} = \frac{8}{x}$, but $16/2x = 8x$.
110. ◆ Finding the opposite of a number and then squaring it is not equivalent to squaring the number and then finding the opposite of the result.
$(-x)^2 = (-1 \cdot x)^2 = (-1 \cdot x)(-1 \cdot x) =$
$(-1)(-1)(x)(x) = x^2 \neq -x^2$ for $x \neq 0$.
111. ◆ The opposite of the absolute value of a number is not equivalent to the opposite of the number. If $x < 0$ then $-|x| = -(-x) = x \neq -x$.
112. ◆ Operations should be performed in the following order.
  1. **P**arentheses: Perform all calculations within parentheses (and other grouping symbols) first.
  2. **E**xponents: Evaluate all exponential expressions.
  3. **M**ultiply and **D**ivide in order from left to right.
  4. **A**dd and **S**ubtract in order from left to right.
113. $-5t - 6r + 21$
114. $z - \{2z - [3z - (4z - 5z) - 6z] - 7z\} - 8z$
$= z - \{2z - [3z - (-z) - 6z] - 7z\} - 8z$
$= z - \{2z - [3z + z - 6z] - 7z\} - 8z$
$= z - \{2z - [-2z] - 7z\} - 8z$
$= z - \{2z + 2z - 7z\} - 8z$
$= z - \{-3z\} - 8z$
$= z + 3z - 8z$
$= -4z$
115. $-2x - f$
116. ◆ Yes; $-(ab) = -1 \cdot (ab) = -1 \cdot a \cdot b = (-a)b$ and $-(ab) = -1 \cdot (ab) = -1 \cdot a \cdot b = a \cdot (-1) \cdot b = a(-b)$.
117. ◆ Yes; $ab = 1 \cdot ab = (-1)(-1)(ab) = (-1) \cdot a \cdot (-1) \cdot b = (-a)(-b)$.
118. False; let $m = 1$ and $n = 2$. Then $-2 + 1 = -(2 - 1) = -1$, but $-(2 + 1) = -3$.

119. True **120.** True; $m - n = -n + m = -(n - m)$.
121. True **122.** False; let $m = 2$ and $n = 3$. Then $3(-3 - 2) = 3(-5) = -15$, but $-3^2 + 3 \cdot 2 = -9 + 6 = -3$. **123.** False
124. True; $-m(-n + m) = mn - m^2 = m(n - m)$.
125. True

REVIEW EXERCISES: CHAPTER 1, PP. 59–60
1. [1.1] 15 **2.** [1.1] 6 **3.** [1.1] 5 **4.** [1.1] 4
5. [1.8] -15 **6.** [1.8] -5 **7.** [1.1] $z - 8$ **8.** [1.1] xz
9. [1.1] Let m and n represent the numbers; $mn + 1$, or $1 + mn$ **10.** [1.1] No
11. [1.1] Let c represent the number of countries that participated in the 1988 Olympics; $c + 37 = 197$
12. [1.2] $x \cdot 2 + y$ **13.** [1.2] $2x + (y + z)$
14. [1.2] $4(yx)$, $(4y)x$, $y(4x)$ **15.** [1.2] $18x + 30y$
16. [1.2] $40x + 24y + 16$ **17.** [1.2] $3(7x + 5y)$
18. [1.2] $7(5x + 2 + y)$ **19.** [1.3] $2 \cdot 2 \cdot 13$ **20.** [1.3] $\frac{5}{12}$
21. [1.3] $\frac{9}{4}$ **22.** [1.3] $\frac{31}{36}$ **23.** [1.3] $\frac{3}{16}$ **24.** [1.3] $\frac{3}{5}$
25. [1.3] $\frac{72}{25}$ **26.** [1.4] $-45, 72$
27. [1.4] **28.** [1.4] $x > -3$

(number line with point at $\frac{-1}{3}$, from -5 to 5)

29. [1.4] True **30.** [1.4] False **31.** [1.4] -0.875
32. [1.4] 1 **33.** [1.6] -5 **34.** [1.5] -3 **35.** [1.5] $-\frac{7}{12}$
36. [1.5] -4 **37.** [1.5] -5 **38.** [1.6] 4 **39.** [1.6] $-\frac{7}{5}$
40. [1.6] -7.9 **41.** [1.7] 54 **42.** [1.7] -9.18
43. [1.7] $-\frac{2}{7}$ **44.** [1.7] -210 **45.** [1.7] -7
46. [1.7] -3 **47.** [1.7] $\frac{3}{4}$ **48.** [1.8] 92 **49.** [1.8] 62
50. [1.8] 48 **51.** [1.8] 168 **52.** [1.8] $\frac{21}{8}$ **53.** [1.8] $\frac{103}{17}$
54. [1.5] $7a - 3b$ **55.** [1.6] $-2x + 5y$ **56.** [1.6] 7
57. [1.7] $-\frac{1}{7}$ **58.** [1.8] $(2x)^4$ **59.** [1.8] $-27y^3$
60. [1.8] $-3a + 9$ **61.** [1.8] $-2b + 21$
62. [1.8] $-3x + 9$ **63.** [1.8] $12y - 34$
64. [1.8] $5x + 24$ **65.** [1.8] $-15x + 25$
66. [1.1] ◆ The value of a constant never varies. A variable can represent a variety of numbers.
67. [1.2] ◆ A term is one of the parts of an expression that is separated from the other parts by plus signs. A factor is part of a product. **68.** [1.2], [1.5], [1.8] ◆ The distributive law is used in factoring algebraic expressions, multiplying algebraic expressions, collecting like terms, finding the opposite of a sum, and subtracting algebraic expressions. **69.** [1.8] ◆ A negative quantity raised to an even power is positive; a negative quantity raised to an odd power is negative. **70.** [1.8] 25,281
71. [1.4] (a) $\frac{3}{11}$; (b) $\frac{10}{11}$ **72.** [1.8] $-\frac{5}{8}$ **73.** [1.8] -2.1

TEST: CHAPTER 1, P. 61

1. [1.1] 6 **2.** [1.1] Let x represent the number; $x - 9$
3. [1.1] 240 ft^2 **4.** [1.2] $q + 3p$ **5.** [1.2] $(x \cdot 4) \cdot y$
6. [1.1] Yes **7.** [1.1] Let p represent the maximum production capability; $p - 282 = 2518$ **8.** [1.2] $18 - 3x$
9. [1.2] $-5y + 5$ **10.** [1.2] $11(1 - 4x)$
11. [1.2] $7(x + 3 + 2y)$ **12.** [1.3] $2 \cdot 2 \cdot 3 \cdot 5 \cdot 5$
13. [1.3] $\frac{2}{7}$ **14.** [1.4] < **15.** [1.4] > **16.** [1.4] $\frac{9}{4}$
17. [1.4] 2.7 **18.** [1.6] $-\frac{2}{3}$ **19.** [1.7] $-\frac{7}{4}$ **20.** [1.6] 8
21. [1.4] $-2 \geq x$ **22.** [1.6] 7.8 **23.** [1.5] -8
24. [1.5] $\frac{7}{40}$ **25.** [1.6] 10 **26.** [1.6] -2.5 **27.** [1.6] $\frac{7}{8}$
28. [1.7] -48 **29.** [1.7] $\frac{3}{16}$ **30.** [1.7] -9 **31.** [1.7] $\frac{3}{4}$
32. [1.7] -9.728 **33.** [1.8] -173 **34.** [1.6] 12
35. [1.8] -4 **36.** [1.8] 448 **37.** [1.6] $22y + 21a$
38. [1.8] $16x^4$ **39.** [1.8] $2x + 7$ **40.** [1.8] $9a - 12b - 7$
41. [1.8] $68y - 8$ **42.** [1.1] 15 **43.** [1.3] $\frac{23}{70}$
44. [1.8] 15 **45.** [1.8] $4a$

CHAPTER 2

EXERCISE SET 2.1, PP. 69–70

1. 13 **2.** 3 **3.** -20 **4.** 34 **5.** -14 **6.** -21
7. -13 **8.** -31 **9.** 15 **10.** 13 **11.** -14 **12.** -11
13. -6 **14.** 18 **15.** 20 **16.** 24 **17.** -6 **18.** -15
19. $\frac{7}{3}$ **20.** $\frac{1}{4}$ **21.** $-\frac{13}{10}$
22. $x + \frac{2}{3} = -\frac{5}{6}$ **23.** $\frac{41}{24}$

$x = -\frac{5}{6} - \frac{4}{6} = -\frac{9}{6}$

$x = -\frac{3}{2}$

24. $y - \frac{3}{4} = \frac{5}{6}$ **25.** $-\frac{1}{20}$

$y = \frac{10}{12} + \frac{9}{12}$

$y = \frac{19}{12}$

26. $-\frac{1}{8} + y = -\frac{3}{4}$ **27.** 1.5 **28.** 4.7 **29.** -5

$y = -\frac{6}{8} + \frac{1}{8}$

$y = -\frac{5}{8}$

30. -10.6 **31.** 16 **32.** 13 **33.** 9 **34.** 12 **35.** 12
36. 7 **37.** -23 **38.** -100 **39.** 8 **40.** 68 **41.** -7
42. -4 **43.** -6 **44.** -7 **45.** 6 **46.** 8 **47.** 48
48. -88 **49.** 36 **50.** 20 **51.** -21 **52.** -54
53. $-\frac{8}{9}$ **54.** $-\frac{7}{9}$ **55.** $\frac{3}{2}$

56. $-\frac{2}{5}y = -\frac{4}{15}$ **57.** $\frac{9}{2}$

$-\frac{5}{2}\left(-\frac{2}{5}y\right) = -\frac{5}{2} \cdot \left(-\frac{4}{15}\right)$

$y = \frac{2}{3}$

58. $\frac{5x}{7} = -\frac{10}{14}$ **59.** -5.9

$\frac{5}{7}x = -\frac{10}{14}$

$\frac{7}{5} \cdot \frac{5}{7}x = \frac{7}{5} \cdot \left(-\frac{10}{14}\right)$

$x = -1$

60. $\frac{3}{4}x = 18$ **61.** -2.5 **62.** -5.5 **63.** -12

$x = \frac{4}{3} \cdot 18$

$x = 24$

64. $96 = -\frac{3}{4}t$ **65.** $-\frac{1}{2}$ **66.** $-\frac{14}{9}$ **67.** -15

$-\frac{4}{3} \cdot 96 = t$

$-128 = t$

68. $\frac{1}{5} + y = -\frac{3}{10}$ **69.** 24 **70.** $-\frac{19}{23}$

$y = -\frac{3}{10} - \frac{2}{10} = -\frac{5}{10}$

$y = -\frac{1}{2}$

71. $7x$ **72.** $-x + 5$ **73.** $x - 4$
74. $2 - 5(x + 5) = 2 - 5x - 25 = -5x - 23$
75. ◆ Equivalent expressions have the same value for all possible replacements for the variables. Equivalent equations have the same solution(s). **76.** ◆ The phrase means "on both the portion of the equation to the left of the equals sign and the portion to the right of the equals sign."
77. ◆ Since $a - c = b - c$ can be rewritten as $a + (-c) = b + (-c)$, it is not necessary to state a subtraction principle. **78.** ◆ For an equation $x + a = b$, add the opposite of a (or subtract a) on both sides of the equation. For an equation $ax = b$, multiply by $1/a$ (or divide by a) on both sides of the equation. **79.** 342.246
80. -8655 **81.** All real numbers **82.** For all x, $0 \cdot x = 0$. There is no solution to $0 \cdot x = 7$. **83.** 12, -12
84. $2|x| = -12$

$|x| = -6$

Absolute value cannot be negative. The equation has no solution.
85. All real numbers

86. $x + x = x$ **87.** No solution
     $2x = x$
      $x = 0$
88. $|3x| = 6$ **89.** 9.4
     $3x = -6$ or $3x = 6$
     $x = -2$ or $x = 2$
90. $x - 4 + a = a$ **91.** 6
     $x - 4 = 0$
      $x = 4$
92. $5c + cx = 7c$ **93.** 8
     $cx = 2c$
      $x = 2$
94. $|x| + 6 = 19$
     $|x| = 13$
x represents a number whose distance from 0 is 13. Thus $x = -13$ or $x = 13$.
95. 11,074
96. To "undo" the last step, divide 22.5 by 0.3.
     $22.5 \div 0.3 = 75$
Now divide 75 by 0.3.
     $75 \div 0.3 = 250$
The answer should be 250, not 22.5.
97. ◆ No; -5 is a solution of $x^2 = 25$ but not of $x = 5$.

EXERCISE SET 2.2, PP. 76–77

1. 9   **2.** $3x + 6 = 30$ **3.** 8   **4.** $6z + 3 = 57$ **5.** 7
          $3x = 24$ $6z = 54$
           $x = 8$ $z = 9$
6. $6x - 3 = 15$ **7.** 14   **8.** $5x - 7 = 48$ **9.** -8
     $6x = 18$ $5x = 55$
      $x = 3$ $x = 11$
10. $4x + 3 = -21$ **11.** -5 **12.** $-91 = 9t + 8$
      $4x = -24$ $-99 = 9t$
       $x = -6$ $-11 = t$
13. -7 **14.** $12 - 4x = 108$ **15.** 15
              $-4x = 96$
               $x = -24$
16. $-6z - 18 = -132$ **17.** 3   **18.** $4x + 5x = 45$
      $-6z = -114$ $9x = 45$
       $z = 19$ $x = 5$
19. -12 **20.** $32 - 7x = 11$ **21.** 6
               $-7x = -21$
                $x = 3$
22. $6x + 19x = 100$ **23.** 8   **24.** $-4y - 8y = 48$
      $25x = 100$ $-12y = 48$
       $x = 4$ $y = -4$
25. 1   **26.** $-10y - 3y = -39$ **27.** -20
            $-13y = -39$
             $y = 3$

28. $3.4t - 1.2t = -44$ **29.** 6   **30.** $x + \frac{1}{4}x = 10$
      $2.2t = -44$
       $t = -20$ $\frac{5}{4}x = 10$

                                                $x = \frac{4}{5} \cdot 10$
                                                $x = 8$
31. 7   **32.** $4x - 6 = 6x$ **33.** 3   **34.** $5y - 2 = 28 - y$
            $-6 = 2x$ $6y = 30$
            $-3 = x$ $y = 5$
35. 2   **36.** $5y + 3 = 2y + 15$ **37.** 10
            $3y = 12$
            $y = 4$
38. $10 - 3x = 2x - 8x + 40$ **39.** 4
      $10 - 3x = -6x + 40$
      $3x = 30$
      $x = 10$
40. $5 + 4x - 7 = 4x - 2 - x$ **41.** 0
      $4x - 2 = 3x - 2$
      $x = 0$
42. $5y - 7 + y = 7y + 21 - 5y$ **43.** $-\frac{2}{5}$
      $6y - 7 = 2y + 21$
      $4y = 28$
      $y = 7$
44. $\frac{7}{8}x - \frac{1}{4} + \frac{3}{4}x = \frac{1}{16} + x$ **45.** $\frac{64}{3}$
The least common denominator is 16.
      $14x - 4 + 12x = 1 + 16x$
      $26x - 4 = 1 + 16x$
      $10x = 5$
      $x = \frac{1}{2}$
46. $-\frac{3}{2} + x = -\frac{5}{6} - \frac{4}{3}$ **47.** $\frac{2}{5}$
The least common denominator is 6.
      $-9 + 6x = -5 - 8$
      $-9 + 6x = -13$
      $6x = -4$
      $x = -\frac{2}{3}$
48. $\frac{1}{2} + 4m = 3m - \frac{5}{2}$ **49.** 3
The least common denominator is 2.
      $1 + 8m = 6m - 5$
      $2m = -6$
      $m = -3$
50. $1 - \frac{2}{3}y = \frac{9}{5} - \frac{1}{5}y + \frac{3}{5}$ **51.** -4
The least common denominator is 15.
      $15 - 10y = 27 - 3y + 9$
      $15 - 10y = 36 - 3y$
      $-7y = 21$
      $y = -3$

A-52   INSTRUCTOR'S ANSWERS

52. $0.96y - 0.79 = 0.21y + 0.46$ **53.** 0.8
$96y - 79 = 21y + 46$
$75y = 125$
$y = \dfrac{125}{75} = \dfrac{5}{3}$

54. $1.7t + 8 - 1.62t = 0.4t - 0.32 + 8$ **55.** $-\dfrac{40}{37}$
$170t + 800 - 162t = 40t - 32 + 800$
$8t + 800 = 40t + 768$
$-32t = -32$
$t = 1$

56. $\dfrac{5}{16}y + \dfrac{3}{8}y = 2 + \dfrac{1}{4}y$ **57.** 2
The least common denominator is 16.
$5y + 6y = 32 + 4y$
$11y = 32 + 4y$
$7y = 32$
$y = \dfrac{32}{7}$

58. $5(2t - 3) = 30$ **59.** 2 **60.** $9 = 3(5x - 2)$
$10t - 15 = 30$ $9 = 15x - 6$
$10t = 45$ $15 = 15x$
$t = 4.5$, or $\dfrac{9}{2}$ $1 = x$

61. 6 **62.** $3(5 + 3m) - 8 = 88$ **63.** 8
$15 + 9m - 8 = 88$
$9m + 7 = 88$
$9m = 81$
$m = 9$

64. $6b - (3b + 8) = 16$ **65.** 2
$6b - 3b - 8 = 16$
$3b - 8 = 16$
$3b = 24$
$b = 8$

66. $5(d + 4) = 7(d - 2)$ **67.** -4
$5d + 20 = 7d - 14$
$34 = 2d$
$17 = d$

68. $8(2t + 1) = 4(7t + 7)$ **69.** -8
$16t + 8 = 28t + 28$
$-12t = 20$
$t = -\dfrac{5}{3}$

70. $5(t + 3) + 9 = 3(t - 2) + 6$ **71.** 2
$5t + 15 + 9 = 3t - 6 + 6$
$5t + 24 = 3t$
$24 = -2t$
$-12 = t$

72. $13 - (2c + 2) = 2(c + 2) + 3c$ **73.** 6
$13 - 2c - 2 = 2c + 4 + 3c$
$11 - 2c = 5c + 4$
$7 = 7c$
$1 = c$

74. $\dfrac{1}{4}(8y + 4) - 17 = -\dfrac{1}{2}(4y - 8)$ **75.** $\dfrac{17}{6}$
$2y + 1 - 17 = -2y + 4$
$2y - 16 = -2y + 4$
$4y = 20$
$y = 5$

76. $\dfrac{2}{3}(2x - 1) = 20$ **77.** $-\dfrac{15}{4}$
$2(2x - 1) = 60$
$4x - 2 = 60$
$4x = 62$
$x = \dfrac{31}{2}$

78. $\dfrac{5}{6}\left(\dfrac{3}{4}x - 2\right) + \dfrac{1}{3} = -\dfrac{2}{3}$ **79.** $\dfrac{11}{18}$
$5\left(\dfrac{3}{4}x - 2\right) + 2 = -4$
$\dfrac{15}{4}x - 10 + 2 = -4$
$\dfrac{15}{4}x = 4$
$x = \dfrac{16}{15}$

80. $\dfrac{2}{3}\left(\dfrac{7}{8} - 4x\right) - \dfrac{5}{8} = \dfrac{3}{8}$ **81.** $-\dfrac{51}{31}$
$\dfrac{7}{12} - \dfrac{8}{3}x - \dfrac{5}{8} = \dfrac{3}{8}$
$14 - 64x - 15 = 9$ Multiplying by 24
$-64x - 1 = 9$
$-64x = 10$
$x = -\dfrac{10}{64}$
$x = -\dfrac{5}{32}$

82. $0.9(2x + 8) = 20 - (x + 5)$ **83.** 2
$1.8x + 7.2 = 20 - x - 5$
$18x + 72 = 200 - 10x - 50$
$18x + 72 = 150 - 10x$
$28x = 78$
$x = \dfrac{78}{28}$
$x = \dfrac{39}{14}$

84. $0.8 - 4(b - 1) = 0.2 + 3(4 - b)$ **85.** -6.5
$0.8 - 4b + 4 = 0.2 + 12 - 3b$
$8 - 40b + 40 = 2 + 120 - 30b$
$48 - 40b = 122 - 30b$
$-74 = 10b$
$-7.4 = b$

86. $7(x - 3 - 2y)$ **87.** $<$ **88.** -14

89. ◈ We add steps to the solution to show how the commutative and associative laws are used.
$$45 - t = 13$$
$$45 - t - 45 = 13 - 45$$
$$45 + (-t) + (-45) = -32$$
$$45 + (-45) + (-t) = -32 \quad \text{Commutative law of addition}$$
$$(45 + (-45)) + (-t) = -32 \quad \text{Associative law of addition}$$
$$-t = -32$$
$$(-1)(-t) = (-1)(-32)$$
$$(-1)(-1 \cdot t) = 32$$
$$[(-1)(-1)]t = 32 \quad \text{Associative law of multiplication}$$
$$t = 32$$

90. ◈ Multiply by 100 to clear decimals. Next multiply by 12 to clear fractions. (These steps could be reversed.) Then proceed as usual. The procedure could be streamlined by multiplying by 1200 to clear decimals and fractions in one step. **91.** ◈ First multiply both sides of the equation by $\frac{1}{3}$ to "eliminate" the 3. Then proceed as shown:
$$3x + 4 = -11$$
$$\frac{1}{3}(3x + 4) = \frac{1}{3}(-11)$$
$$x + \frac{4}{3} = -\frac{11}{3}$$
$$x = -\frac{15}{3}$$
$$x = -5$$

92. ◈ **1.** Add $-b$ on both sides to get $ax = c - b$.
2. Divide by a on both sides to get $x = \frac{c-b}{a}$.

93. $1.\overline{2497}$
94. Since we are using a calculator we will not clear the decimals.
$$0.008 + 9.62x - 42.8 = 0.944x + 0.0083 - x$$
$$9.62x - 42.792 = -0.056x + 0.0083$$
$$9.676x = 42.8003$$
$$x \approx 4.4233464$$

95. -4 **96.**
$$0 = y - (-14) - (-3y)$$
$$0 = y + 14 + 3y$$
$$0 = 4y + 14$$
$$-14 = 4y$$
$$-\frac{7}{2} = y$$

97. All real numbers
98.
$$475(54x + 7856) + 9762 = 402(83x + 975)$$
$$25{,}650x + 3{,}731{,}600 + 9762 = 33{,}366x + 391{,}950$$
$$25{,}650x + 3{,}741{,}362 = 33{,}366x + 391{,}950$$
$$3{,}349{,}412 = 7716x$$
$$\frac{3{,}349{,}412}{7716} = x$$
$$\frac{837{,}353}{1929} = x$$

99. $\frac{2}{3}$ **100.**
$$x(x - 4) = 3x(x + 1) - 2(x^2 + x - 5)$$
$$x^2 - 4x = 3x^2 + 3x - 2x^2 - 2x + 10$$
$$x^2 - 4x = x^2 + x + 10$$
$$-4x = x + 10$$
$$-5x = 10$$
$$x = -2$$

101. 0 **102.**
$$-2y + 5y = 6y$$
$$3y = 6y$$
$$0 = 3y$$
$$0 = y$$
103. $\frac{52}{45}$

104.
$$\frac{5x+3}{4} + \frac{25}{12} = \frac{5+2x}{3}$$
$$3(5x+3) + 25 = 4(5+2x)$$
$$15x + 9 + 25 = 20 + 8x$$
$$15x + 34 = 20 + 8x$$
$$7x = -14$$
$$x = -2$$

EXERCISE SET 2.3, PP. 81–83

1. 57,000 Btu's **2.** $b = 50 \cdot 2500 = 125{,}000$ Btu's
3. 54 in^2 **4.** $f = \frac{21{,}345}{15} = 1423$ **5.** 3450 watts
6. $w = \frac{344}{24} = \frac{43}{3}$ m **7.** 255 mg
8. $N = 7^2 - 7 = 49 - 7 = 42$ **9.** $b = \frac{A}{h}$ **10.** $\frac{A}{b} = h$
11. $r = \frac{d}{t}$ **12.** $\frac{d}{r} = t$ **13.** $P = \frac{I}{rt}$ **14.** $\frac{I}{Pr} = t$
15. $m = 65 - H$ **16.** $d + 64 = h$ **17.** $l = \frac{P-2w}{2}$
18. $P = 2l + 2w$ **19.** $\pi = \frac{A}{r^2}$ **20.** $\frac{A}{\pi} = r^2$
$$P - 2l = 2w$$
$$\frac{P-2l}{2} = w$$
21. $h = \frac{2A}{b}$ **22.** $A = \frac{1}{2}bh$ **23.** $m = \frac{E}{c^2}$
$$2A = bh$$
$$\frac{2A}{h} = b$$
24. $\frac{E}{m} = c^2$ **25.** $d = 2Q - c$
26. $Q = \frac{p-q}{2}$ **27.** $b = 3A - a - c$
$$2Q = p - q$$
$$2Q + q = p$$
28. $A = \frac{a+b+c}{3}$ **29.** $A = Ms$
$$3A = a + b + c$$
$$3A - a - b = c$$

30. $P = \dfrac{ab}{c}$

$Pc = ab$

$\dfrac{Pc}{a} = b$

31. $y = \dfrac{C - Ax}{B}$

32. $Ax + By = C$

$Ax = C - By$

$x = \dfrac{C - By}{A}$

33. $S = \dfrac{360A}{\pi r^2}$

34. $A = P + Prt$

$A = P(1 + rt)$

$\dfrac{A}{1 + rt} = P$

35. $h = \dfrac{2A}{a + b}$

36. $R = r + \dfrac{400(W - L)}{N}$

$NR = Nr + 400(W - L)$

$NR = Nr + 400W - 400L$

$400L = Nr + 400W - NR$

$L = \dfrac{Nr + 400W - NR}{400}$

37. -42

38. $-\dfrac{2}{3} \cdot \dfrac{9}{10} = -\dfrac{2 \cdot 9}{3 \cdot 10} = -\dfrac{2 \cdot 3 \cdot 3}{3 \cdot 2 \cdot 5} = -\dfrac{3}{5}$

39. -29

40. $3|7 - (2 - 5)| = 3|7 - (-3)| = 3|7 + 3| = 3|10| = 3 \cdot 10 = 30$

41. ◆ Answers may vary. A decorator wants to have a carpet cut for a bedroom. The perimeter of the room is 54 ft and its length is 15 ft. How wide should the carpet be?

42. ◆ Answers may vary. A walker who knows how far and how long she walks each day wants to know her average speed each day.   **43.** ◆ Given the formula for converting Celsius temperature C to Fahrenheit temperature F, solve for C. This yields a formula for converting Fahrenheit temperature to Celsius temperature.   **44.** ◆ No; for a rectangle with length l and width w, $A = lw$. The area of a rectangle with length $2l$ and width $2w$ is $2l \cdot 2w = 4lw = 4A$. Thus, when a rectangle's length and width are doubled its area is quadrupled.

45. $K = 917 + 6\left(\dfrac{w}{2.2046} + \dfrac{h}{0.3937} - a\right)$

46. $2627 = 19.18(82) + 7(185) - 9.52a + 92.4$

$2627 = 1572.76 + 1295 - 9.52a + 92.4$

$2627 = 2960.16 - 9.52a$

$-333.16 = -9.52a$

$35 \approx a$

The man is about 35 years old.

47. $a = \dfrac{wd}{c}$

48. $\quad$ 8 ft = 96 in.

$700 = \dfrac{96g^2}{800}$

$560{,}000 = 96g^2$

$\dfrac{560{,}000}{96} = g^2$

$76.4 \approx g$

The girth is about 76.4 in.

49. $T = t - \dfrac{h}{100}, \ 0 \leq h \leq 12{,}000$

50. $\dfrac{y}{z} \div \dfrac{z}{t} = 1$ $\quad$ **51.** $c = \dfrac{d}{a - b}$

$\dfrac{y}{z} \cdot \dfrac{t}{z} = 1$

$\dfrac{z^2}{t} \cdot \dfrac{yt}{z^2} = \dfrac{z^2}{t} \cdot 1$

$y = \dfrac{z^2}{t}$

52. $\quad qt = r(s + t)$ **53.** $a = \dfrac{c}{3 + b + d}$

$qt = rs + rt$

$qt - rt = rs$

$t(q - r) = rs$

$t = \dfrac{rs}{q - r}$

54. ◆ The answer to Exercise 34 allows us to find the amount P that must be invested at a given rate of simple interest for a given period of time in order to return a given amount.

EXERCISE SET 2.4, PP. 88–89

1. 0.32 **2.** 0.49 **3.** 0.07 **4.** 0.913 **5.** 0.241
6. 0.02 **7.** 0.0046 **8.** 0.048 **9.** 454% **10.** 100%
11. 99.8% **12.** 73% **13.** 200% **14.** 0.57%
15. 134% **16.** 920% **17.** 0.68% **18.** 67.5%
19. 37.5% **20.** $\dfrac{3}{4} = 0.75 = 75\%$ **21.** 28%
22. $\dfrac{4}{5} = 0.8 = 80\%$ **23.** $66.\overline{6}\%$, or $66\dfrac{2}{3}\%$
24. $\dfrac{5}{6} = 0.8\overline{3} = 83.\overline{3}\%$, or $83\dfrac{1}{3}\%$ **25.** 52%
26. Solve and convert to percent notation:

$x \cdot 136 = 17$

$x = 0.125 = 12.5\%$

27. 24%
28. Solve and convert to percent notation:

$x \cdot 300 = 57$

$x = 0.19 = 19\%$

29. 60
30. Solve: $20.4 = 24\% \cdot x$ **31.** 2.5

$85 = x$

32. Solve: $7 = 175\% \cdot x$ **33.** 273

$4 = x$

34. Solve: $x = 1\% \cdot 1{,}000{,}000$
$\phantom{\text{Solve: }}x = 10{,}000$ **35.** 125%
36. Solve and convert to percent notation: **37.** 0.8
$\phantom{\text{Solve: }}x \cdot 40 = 90$
$\phantom{\text{Solve: }}x = 2.25 = 225\%$
38. Solve: $z = 40\% \cdot 2$ **39.** 5%
$\phantom{\text{Solve: }}z = 0.8$
40. Solve: $40 = 2\% \cdot x$ **41.** 285
$\phantom{\text{Solve: }}2000 = x$
42. Let $n =$ the number of women who had babies in good or excellent health.
Solve: $n = 8\% \cdot 200$
$\phantom{\text{Solve: }}n = 16$
43. About 92.4 million
44. Let $t =$ the total lottery proceeds for 1980–1994, in billions of dollars.
Solve: $47.1 = 53\% \cdot t$
$\phantom{\text{Solve: }}88.9 \approx t$
Total proceeds were approximately $\$88.9$ billion.
45. About 60%
46. Let $a =$ the percent of applicants accepted.
Solve: $600 = a \cdot 16{,}000$
$\phantom{\text{Solve: }}0.0375 = a$, or
$\phantom{\text{Solve: }}3.75\% = a$
47. 7410
48. Let $b =$ the number of bowlers you would expect to be left-handed.
Solve: $b = 17\% \cdot 160$
$\phantom{\text{Solve: }}b \approx 27$
49. 7%
50. Let $p =$ the percent that were correct.   **51.** 52%
Solve: $76 = p \cdot 88$
$\phantom{\text{Solve: }}p \approx 0.864$, or 86.4%
52. Let $c =$ the cost of the merchandise.   **53.** $\$940$
Solve: $37.80 = 105\% \cdot c$
$\phantom{\text{Solve: }}\$36 = c$
54. Let $p =$ the percentage of the meal's cost that the tip represented.
Solve: $4 = p \cdot 25$
$\phantom{\text{Solve: }}p = 0.16$, or 16%
55. $\$138.95$
56. Let $a =$ the amount the group should pay.
Solve: $157.41 = 106\% \cdot a$
$\phantom{\text{Solve: }}a = \$148.50$
57. $\$14.40$ per hour
58. Let $a =$ the amount Clara would need to earn, in dollars per hour, on her own for a comparable income.

Solve: $a = 1.2(15)$
$\phantom{\text{Solve: }}a = \$18$ per hour
59. About 165 calories
60. Let $f =$ the number of calories of fat in a serving of the leading shortbread cookie.
Solve: $35 = 60\% \cdot f$
$\phantom{\text{Solve: }}f \approx 58$ calories
61. 0.68 **62.** -90 **63.** -13.2
64. $4a - 8b - 5(5a - 4b) = 4a - 8b - 25a + 20b = -21a + 12b$ **65.** ◆ The end result is the same either way. If s is the original salary, the new salary after a 5% raise followed by an 8% raise is $1.08(1.05s)$. The new salary if the raises occur the other way around is $1.05(1.08s)$. By the commutative and associative laws of multiplication we see that these are equal. However, it would be better to receive the 8% raise first, because this increase yields a higher new salary the first year than a 5% raise.
66. ◆ No; Erin paid 75% of the original price and was offered credit for 125% of this amount, not to be used on sale items. Now 125% of 75% is 93.75%, so Erin would have a credit of 93.75% of the original price. Since this credit can be applied only to nonsale items, she has less purchasing power than if the amount she paid were refunded and she could spend it on sale items.
67. ◆ Suppose Amber invests x dollars in a CD. At 5% taxable interest, paying 28% of the amount of the interest in income tax, Amber nets $0.05x - 0.28(0.05x)$, or $0.036x$, or 3.6% of x. In a tax-free municipal bond fund, paying 4% interest, Amber will earn 4% of x in interest. Thus, the bond fund would be a better investment.   **68.** ◆ No; although over 26% of home burglaries occur between Memorial Day and Labor Day, it is possible that a larger percent occur during a different season. For example, it is possible that 30% of home burglaries occur between Labor Day and Thanksgiving.   **69.** About 77%
70. Let $p =$ the population of the community.
Solve: $0.15(0.48)p = 1332$
$\phantom{\text{Solve: }}p = 18{,}500$
71. About 35%
72. At Rollie's Music, the total cost is 107% of the price of the disk. If $R =$ the cost at Rollie's, we have:
$R = 107\%(\$11.99)$
$R = 1.07(\$11.99)$
$R = \$12.83$ Rounding
At Sound Warp, the total cost is $\$13.99 - \$2.00$ plus 7% of the original price of $\$13.99$. If $S =$ the cost at Sound Warp, we have:
$S = (\$13.99 - \$2.00) + 7\%(\$13.99)$
$S = \$11.99 + 0.07(\$13.99)$
$S = \$12.97$ Rounding
73. 20%

EXERCISE SET 2.5, PP. 98-100

1. 11

2. Let n = the number.
Solve: $10n - 2 = 78$
$n = 8$

3. 11

4. Let x = the number.
Solve: $2(x + 4) = 34$
$x = 13$

5. $90

6. Let c = the cost of the book.
Solve: $50.40 = 1.05c$
$48 = c$

7. 15, 16, 17

8. Let n = the first odd number. Then $n + 2$ = the next odd number.
Solve: $n + (n + 2) = 40$
$n = 19$
If n is 19, then $n + 2$ is 21. The numbers are 19 and 21.

9. 24, 26

10. Let n = the first even integer. Then $n + 2$ = the second even integer.
Solve: $x + (x + 2) = 106$
$x = 52$
If n is 52, then $n + 2$ is 54. The integers are 52 and 54.

11. 63, 65

12. Let d = Jenna's distance from the start of the race. Then $2d$ = her distance from the finish.
Solve: $d + 2d = 10$
$d = \frac{10}{3}$ km

13. $699\frac{1}{3}$ mi

14. Let x = the length of the first piece, in meters. Then $3x$ = the length of the second piece and $4 \cdot 3x = 12x$ = the length of the third piece.
Solve: $x + 3x + 12x = 480$
$x = 30$
If $x = 30$, then $3x = 90$ and $12x = 360$. The pieces are 30 m, 90 m, and 360 m.

15. Bathrooms: $11\frac{2}{3}$ billion; kitchens: $23\frac{1}{3}$ billion

16. Let x = the measure of the first angle. Then $3x$ = the measure of the second angle and $x + 30$ = the measure of the third angle.
Solve: $x + 3x + (x + 30) = 180$
$x = 30$
If x is 30, then $3x$ is 90 and $x + 30$ is 60, so the measures of the first, second, and third angles are 30°, 90°, and 60°, respectively.

17. First: 22.5°; second: 90°; third: 67.5°

18. Let x = the measure of the first angle. Then $3x$ = the measure of the second angle, and $x + 3x + 10 = 4x + 10$ = the measure of the third angle.
Solve: $x + 3x + (4x + 10) = 180$
$x = 21.25°$
If x is 21.25°, then the measure of the third angle is $4(21.25°) + 10° = 95°$.

19. 70°

20. Let s = the sales in the first three quarters. Then $3s$ = the sales in the last quarter.
Solve: $s + 3s = 480,000$
$s = 120,000$
If s is 120,000 then $3s$ is 360,000, so sales in the last quarter were $360,000.

21. 142 and 143

22. Let x = the first page number. Then $x + 1$ = the second page number.
Solve: $x + (x + 1) = 281$
$x = 140$
If x is 140, then $x + 1$ is 141. The page numbers are 140 and 141.

23. 63 mm, 65 mm, 67 mm

24. Let s, $s + 2$, and $s + 4$ represent the lengths of the sides.
Solve: $s + (s + 2) + (s + 4) = 396$
$s = 130$
If s is 130, then $s + 2$ is 132 and $s + 4$ is 134. The lengths of the sides are 130 mm, 132 mm, and 134 mm.

25. Width: 165 ft, length: 265 ft; area: 43,725 ft^2

26. Let w = the width of the rectangle, in feet. Then $w + 60$ = the length.
Solve: $w + w + (w + 60) + (w + 60) = 520$
$w = 100$
The width is 100 ft, the length is 160 ft, and the area is 160 ft(100 ft) = 16,000 ft^2.

27. Width: 275 mi, length: 365 mi

28. Let l = the length of the paper, in cm. Then $l - 6.3$ = the width.
Solve: $(l - 6.3) + (l - 6.3) + l + l = 99$
$l = 27.9$
If l is 27.9, then $l - 6.3$ is 21.6. The length of the paper is 27.9 cm, and the width is 21.6 cm.

29. $6600

30. Let b = the amount borrowed.
Solve: $b + 0.1b = 7194$
$b = 6540

31. 450.5 mi

32. Let m = the number of miles the tourist can travel on $90.
Solve: $43.95 + 0.10m = 90$
$m = 460.5$ mi
33. 65°, 25°
34. Let x = the measure of one angle. Then $180 - x$ = the measure of its supplement.
Solve: $x = 2(180 - x) - 45$
$x = 105$
If x is 105, then $180 - x$ is 75. The angle measures are 105° and 75°.
35. 2020   **36.** Solve: $80 = \frac{1}{4}N + 40$
$160 = N$
37. $5(a + 2b - 9)$ **38.** $3(x - 4y + 20)$ **39.** $13x - 12$
40. $2x + 9 - 3(4x - 7) = 2x + 9 - 12x + 21 = -10x + 30$
41. ◆ Most would consider the third approach the best since it is the most descriptive.   **42.** ◆ Although many of the problems in this section might be solved by guessing, using the five-step problem-solving process to solve them would give the student practice in using a technique that can be used to solve other problems whose answers are not so readily guessed.   **43.** ◆ Answers may vary. The sum of three consecutive odd integers is 375. What are the integers?   **44.** ◆ Answers may vary. Acme Rentals rents a 12-foot truck at a rate of $35 plus 20¢ per mile. Audrey has a truck-rental budget of $45 for her move to a new apartment. How many miles can she drive the rental truck without exceeding her budget?   **45.** 37
46. Let p = the highest price per mile that the person can afford.
Solve: $18.90 + 190p = 55$
$p = \$0.19$
47. If Ed intends to drive fewer than 465 mi, the Reston rental is less expensive. If he drives more than 465 mi, the Long Haul rental is less expensive.
48. Let m = the number of multiple-choice questions Pam got right. Note that she got $4 - 1$, or 3 fill-ins right.
Solve: $3 \cdot 7 + 3m = 78$
$m = 19$
49. $4x + 7 = 1863 - 1776$; 20 yr
50. Let y = the larger number. Then 25% of y, or $0.25y$ = the smaller.
Solve: $y = 0.25y + 12$
$y = 16$
The numbers are 16 and 0.25(16), or 4.
51. 87°, 89°, 91°, 93°

52. Let x = the first number. The sum of the measures of the angles of a pentagon is $(5 - 2) \cdot 180°$, or 540°.
Solve: $x + (x + 2) + (x + 4) + (x + 6) + (x + 8) = 540$
$x = 104$
The measures of the angles are 104°, 106°, 108°, 110°, and 112°.
53. 120
54. Let x = the number of additional games the Falcons will have to play. Then $\frac{x}{2}$ = the number of those games they will win, $15 + \frac{x}{2}$ = the total number of games won, and $20 + x$ = the total number of games played.
Solve: $15 + \frac{x}{2} = 0.6(20 + x)$
$x = 30$
55. $600
56. Let x = the length of the original rectangle. Then $\frac{3}{4}x$ = the width. The length and width of the enlarged rectangle are $x + 2$ and $\frac{3}{4}x + 2$, respectively.
Solve:
$\left(\frac{3}{4}x + 2\right) + \left(\frac{3}{4}x + 2\right) + (x + 2) + (x + 2) = 50$
$x = 12$
If x is 12, then $\frac{3}{4}x$ is 9. The length and width of the rectangle are 12 cm and 9 cm, respectively.
57. 0.726175 in.
58. Let s = the score on the third test.
Solve: $\frac{2 \cdot 85 + s}{3} = 82$
$s = 76$
59. 10   **60.** ◆ If the school can invest the $2000 so that it earns at least 7.5% and thus grows to at least $2150 by the end of the year, the second option should be selected. If not, the first option is preferable.   **61.** ◆ Yes; the page numbers must be consecutive integers. The only consecutive integers whose sum is 191 are 95 and 96. These cannot be the numbers of facing pages, however, because the left-hand page of a book is even-numbered.

EXERCISE SET 2.6, PP. 107–108

1. (a) Yes; **(b)** yes; **(c)** no; **(d)** no; **(e)** yes
2. (a) Yes; **(b)** no; **(c)** yes; **(d)** yes; **(e)** no
3. (a) No; **(b)** no; **(c)** yes; **(d)** yes; **(e)** no
4. (a) Yes; **(b)** yes; **(c)** yes; **(d)** no; **(e)** yes
5.   **6.**

INSTRUCTOR'S ANSWERS

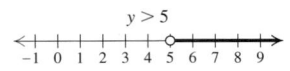

7. $t < -3$ (number line from -5 to 5, open circle at -3, shaded left)

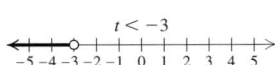

8. $y > 5$ (number line from -1 to 9, open circle at 5, shaded right)

9. $m > -4$ (number line from -5 to 5, open circle at -4, shaded right)

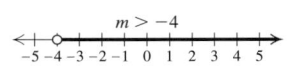

10. $p \leq 3$ (number line from -5 to 5, closed circle at 3, shaded left)

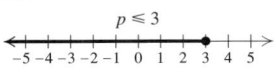

11. $-3 < x \leq 5$ (number line from -4 to 6)

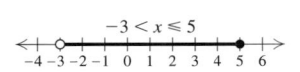

12. $-5 \leq x < 2$ (number line from -6 to 4)

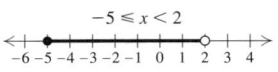

13. $0 < x < 3$ (number line from -5 to 5)

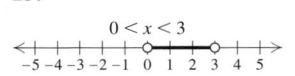

14. $-5 \leq x \leq 0$ (number line from -7 to 3)

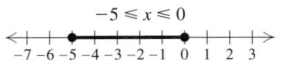

15. $\{x \mid x > -1\}$ **16.** $\{x \mid x < 3\}$ **17.** $\{x \mid x \leq 2\}$
18. $\{x \mid x \geq -2\}$ **19.** $\{x \mid x < -2\}$ **20.** $\{x \mid x > 1\}$
21. $\{x \mid x \geq 0\}$ **22.** $\{x \mid x \leq 0\}$
23. $\{y \mid y > 6\}$, (graph: open circle at 0 and 6, shaded right of 6)

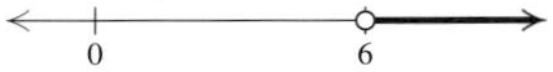

24. $y + 6 > 9$
$y > 3$
$\{y \mid y > 3\}$ (graph)

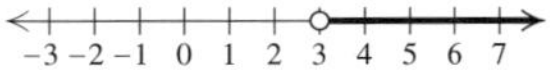

25. $\{x \mid x \leq -18\}$, (graph)

26. $x + 9 \leq -12$
$x \leq -21$
$\{x \mid x \leq -21\}$ (graph)

27. $\{x \mid x < 10\}$, (graph)

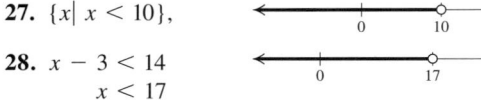

28. $x - 3 < 14$
$x < 17$
$\{x \mid x < 17\}$

29. $\{x \mid x \geq 8\}$, (graph)

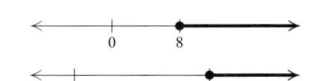

30. $x - 9 \geq 4$
$x \geq 13$
$\{x \mid x \geq 13\}$

31. $\{y \mid y > -5\}$, (graph)

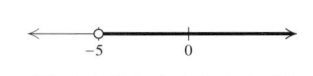

32. $y - 10 > -16$
$y > -6$
$\{y \mid y > -6\}$

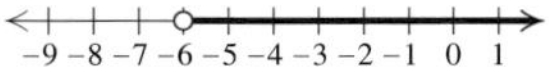

33. $\{x \mid x \leq 5\}$, (graph)

34. $2x + 4 \leq x + 1$
$x \leq -3$
$\{x \mid x \leq -3\}$

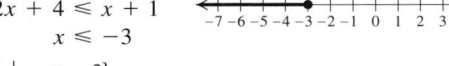

35. $\{x \mid x \geq 5\}$
36. $3x - 9 \geq 2x + 11$
$x \geq 20$
$\{x \mid x \geq 20\}$

37. $\{y \mid y \leq \frac{1}{2}\}$ **38.** $x + \frac{1}{4} \leq \frac{1}{2}$
$x \leq \frac{1}{4}$
$\{x \mid x \leq \frac{1}{4}\}$

39. $\{x \mid x > \frac{5}{8}\}$ **40.** $y - \frac{1}{3} > \frac{1}{4}$ **41.** $\{x \mid x < 0\}$
$y > \frac{7}{12}$
$\{y \mid y > \frac{7}{12}\}$

42. $-7x + 13 > 13 - 6x$
$-7x > -6x$
$0 > x$
$\{x \mid x < 0\}$

43. $\{x \mid x < 7\}$, (graph: open circles at 0 and 7)

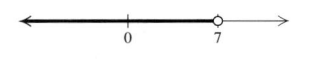

44. $8x \geq 32$
$x \geq 4$
$\{x \mid x \geq 4\}$ (graph)

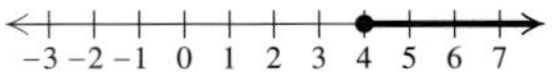

45. $\{y \mid y \leq 9\}$, (graph)

46. $10x > 240$
$x > 24$
$\{x \mid x > 24\}$ (graph)

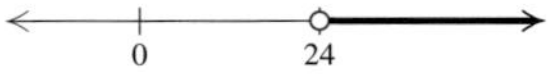

47. $\{x \mid x > -\frac{13}{7}\}$, (graph with $-\frac{13}{7}$)

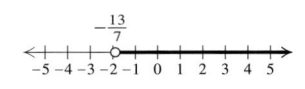

48. $8y < 17$
$y < \frac{17}{8}$
$\{y \mid y < \frac{17}{8}\}$ (graph with $\frac{17}{8}$)

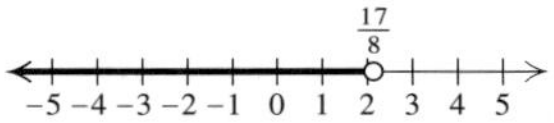

49. $\{x \mid x > -3\}$, (graph)

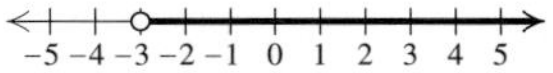

50. $-16x < -64$
$x > 4$
$\{x \mid x > 4\}$ (graph)

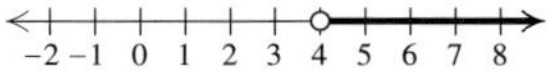

51. $\{y \mid y \geq -\frac{2}{7}\}$ **52.** $5x > -3$ **53.** $\{x \mid x > \frac{17}{5}\}$
$x > -\frac{3}{5}$
$\{x \mid x > -\frac{3}{5}\}$

54. $-3y \leq -14$ **55.** $\{y \mid y \geq -\frac{1}{12}\}$ **56.** $-2x \geq \frac{1}{5}$
$y \geq \frac{14}{3}$
$\{y \mid y \geq \frac{14}{3}\}$
$x \leq -\frac{1}{10}$
$\{x \mid x \leq -\frac{1}{10}\}$

57. $\{x \mid x > \frac{4}{5}\}$ **58.** $-\frac{5}{8} < -10y$
$-\frac{1}{10}\left(-\frac{5}{8}\right) > y$
$\frac{1}{16} > y$
$\{y \mid y < \frac{1}{16}\}$

59. $\{x|\ x < 9\}$ **60.** $5 + 4y < 37$ **61.** $\{y|\ y \geq 4\}$
$4y < 32$
$y < 8$
$\{y|\ y < 8\}$

62. $7 + 8x \geq 71$ **63.** $\{x|\ x \leq 6\}$ **64.** $5y - 9 \leq 21$
$8x \geq 64$ $5y \leq 30$
$x \geq 8$ $y \leq 6$
$\{x|\ x \geq 8\}$ $\{y|\ y \leq 6\}$

65. $\{x|\ x < -3\}$ **66.** $8y - 4 < -52$
$8y < -48$
$y < -6$
$\{y|\ y < -6\}$

67. $\{y|\ y < -4\}$ **68.** $22 < 6 - 8x$
$16 < -8x$
$-2 > x$
$\{x|\ -2 > x\}$, or $\{x|\ x < -2\}$

69. $\{x|\ x > -4\}$ **70.** $40 > 5 - 7y$
$35 > -7y$
$-5 < y$
$\{y|\ -5 < y\}$, or $\{y|\ y > -5\}$

71. $\left\{y|\ y < -\frac{10}{3}\right\}$ **72.** $8 - 2y > 14$
$-2y > 6$
$y < -3$
$\{y|\ y < -3\}$

73. $\{x|\ x > -10\}$ **74.** $-5 < 9x + 8 - 8x$
$-5 < x + 8$
$-13 < x$
$\{x|\ -13 < x\}$, or $\{x|\ x > -13\}$

75. $\{y|\ y < 2\}$ **76.** $7 - 8y > 5 - 7y$
$2 > y$
$\{y|\ 2 > y\}$, or $\{y|\ y < 2\}$

77. $\{y|\ y \geq 3\}$ **78.** $6 - 13y \leq 4 - 12y$
$2 \leq y$
$\{y|\ 2 \leq y\}$, or $\{y|\ y \geq 2\}$

79. $\{x|\ x > -4\}$ **80.** $27 - 11x > 14x - 18$
$45 > 25x$
$\frac{9}{5} > x$
$\left\{x|\ x < \frac{9}{5}\right\}$

81. $\{x|\ x > -4\}$ **82.** $0.96y - 0.79 \leq 0.21y + 0.46$
$96y - 79 \leq 21y + 46$
$75y \leq 125$
$y \leq \frac{5}{3}$
$\left\{y|\ y \leq \frac{5}{3}\right\}$

83. $\{n|\ n \geq 70\}$ **84.** $1.7t + 8 - 1.62t < 0.4t - 0.32 + 8$
$0.08t + 8 < 0.4t + 7.68$
$8t + 800 < 40t + 768$
$32 < 32t$
$1 < t$
$\{t|\ 1 < t\}$, or $\{t|\ t > 1\}$

85. $\{x|\ x \leq 15\}$ **86.** $\frac{2}{3} - \frac{x}{5} < \frac{4}{15}$
$10 - 3x < 4$
$-3x < -6$
$x > 2$
$\{x|\ x > 2\}$

87. $\{y|\ y \leq -7\}$ **88.** $\frac{3x}{5} \geq -15$
$3x \geq -75$
$x \geq -25$
$\{x|\ x \geq -25\}$

89. $\{y|\ y < 6\}$ **90.** $4(2y - 3) > 28$
$8y - 12 > 28$
$8y > 40$
$y > 5$
$\{y|\ y > 5\}$

91. $\{t|\ t \leq -4\}$ **92.** $8(2t + 1) > 4(7t + 7)$
$16t + 8 > 28t + 28$
$-12t > 20$
$t < -\frac{5}{3}$
$\left\{t|\ t < -\frac{5}{3}\right\}$

93. $\{r|\ r > -3\}$ **94.** $5(t + 3) + 9 > 3(t - 2) + 6$
$5t + 15 + 9 > 3t - 6 + 6$
$5t + 24 > 3t$
$24 > -2t$
$-12 < t$
$\{t|\ t > -12\}$

95. $\{x|\ x \geq 8\}$ **96.** $\frac{4}{5}(3x + 4) \leq 20$
$3x + 4 \leq 25$
$3x \leq 21$
$x \leq 7$
$\{x|\ x \leq 7\}$

97. $\left\{x|\ x < \frac{11}{18}\right\}$ **98.** $\frac{2}{3}\left(\frac{7}{8} - 4x\right) - \frac{5}{8} < \frac{3}{8}$
$\frac{2}{3}\left(\frac{7}{8} - 4x\right) < 1$
$\frac{7}{12} - \frac{8}{3}x < 1$
$7 - 32x < 12$
$-32x < 5$
$x > -\frac{5}{32}$
$\left\{x|\ x > -\frac{5}{32}\right\}$

99. 140
100. $10 \div 2 \cdot 5 - 3^2 + (-5)^2 = 10 \div 2 \cdot 5 - 9 + 25$
$= 5 \cdot 5 - 9 + 25$
$= 25 - 9 + 25$
$= 16 + 25$
$= 41$
101. $-2x - 23$ **102.** $9(3 + 5x) - 4(7 + 2x) = 27 + 45x - 28 - 8x = -1 + 37x$
103. ◆ The inequalities are equivalent because they have the same solution set, $\{x \mid x < 2\}$.
104. ◆ $x > -5$
$-1 \cdot x < -1(-5)$ Multiplying by -1
↑ ———The symbol has to be reversed.
$-x < 5$

Thus, all solutions of $x > -5$ are solutions of $-x < 5$.
105. ◆ For any pair of numbers, their relative position on the number line is reversed when both are multiplied by the same negative number. For example, -3 is to the left of 5 on the number line ($-3 < 5$), but 12 is to the right of -20 ($-3(-4) > 5(-4)$).   **106.** ◆ The graph of an inequality of the form $a \leq x \leq a$ consists of just one number, a.   **107.** $\{t \mid t > -\frac{27}{19}\}$
108. $27 - 4[2(4x - 3) + 7] \geq 2[4 - 2(3 - x)] - 3$
$27 - 4[8x - 6 + 7] \geq 2[4 - 6 + 2x] - 3$
$27 - 4[8x + 1] \geq 2[-2 + 2x] - 3$
$27 - 32x - 4 \geq -4 + 4x - 3$
$23 - 32x \geq -7 + 4x$
$30 \geq 36x$
$\frac{5}{6} \geq x$

$\{x \mid x \leq \frac{5}{6}\}$
109. $\{x \mid x \leq -4a\}$ **110.** $\frac{1}{2}(2x + 2b) > \frac{1}{3}(21 + 3b)$
$x + b > 7 + b$
$x > 7$
$\{x \mid x > 7\}$
111. $\{x \mid x > \frac{y-b}{a}\}$
112. $y < ax + b$ Assume $a < 0$.
$y - b < ax$
$\frac{y-b}{a} > x$ Since $a < 0$, the inequality symbol must be reversed.
$\{x \mid x < \frac{y-b}{a}\}$
113. (a) No; (b) yes; (c) no; (d) yes; (e) no; (f) yes
114.

←—+—+—○—+—+—+—+—+—+→
  −5 −4 −3 −2 −1 0 1 2 3 4 5

EXERCISE SET 2.7, PP. 111–114
1. Let n represent the number; $n < 9$ **2.** Let n represent the number; $n \geq 5$.   **3.** Let b represent the weight of the bag; $b \geq 2$ **4.** Let p represent the number of people who attended the concert; $75 < p < 100$.   **5.** Let s represent the average speed; $90 < s < 110$ **6.** Let n represent the number of people who attended the Million Man March; $n \geq 400{,}000$.   **7.** Let a represent the number of people attending; $a \leq 1{,}200{,}000$ **8.** Let a represent the amount of acid, in liters; $a \leq 40$.   **9.** Let c represent the cost, per gallon, of gasoline; $c \geq \$1.20$ **10.** Let t represent the temperature; $t \leq -2$.   **11.** Scores of 84 and higher
12. Let s represent Rod's score on the fifth quiz.
Solve: $\dfrac{73 + 75 + 89 + 91 + s}{5} \geq 85$
$s \geq 97$
13. Mileages less than or equal to 525.8 mi
14. Let m represent the number of miles driven.
Solve: $42.95 + 0.46m \leq 200$
$m \leq 341.4$ mi
15. 5 min or more
16. Let $t =$ the number of hours the car is parked. Then $2t =$ the number of half-hours it is parked.
Solve: $0.45 + 0.25(2t) \geq 2.20$
$t \geq 3.5$ hr
17. 4 servings or more
18. Let $c =$ the number of credits Millie must complete in the fourth quarter.
Solve: $\dfrac{5 + 7 + 8 + c}{4} \geq 7$
$c \geq 8$
19. 437.5 ft or less
20. Let $t =$ the number of units of time.
Solve: $50 + 15t < 70 + 10t$
$t < 4$
21. Less than 21.5 cm
22. Let $l =$ the length of the rectangle, in yd.
Solve: $16l \geq 264$.
$l \geq 16.5$ yd
23. The blue-book value is at least $10,625.
24. Let $c =$ the cost of the repair.
Solve: $c > 0.8(21{,}000)$
$c > \$16{,}800$
25. Years after 1934
26. Let $t =$ the number of years after 1920.
Solve: $-0.028t + 20.8 < 19.0$
$t > 64\dfrac{2}{7}$ year after 1920, or in years after 1984

27. When the puppy is more than 18 weeks old
28. Let w = the number of weeks after July 1.

Solve: $25 - \dfrac{2}{3}w \leq 21$

$w \geq 6$

The water level will not exceed 21 ft for dates at least 6 weeks after July 1.

29. Temperatures greater than 37°C
30. Solve: $\dfrac{9}{3} \cdot C + 32 < 88$

$C < 31.1°$

31. 21 or more
32. Let n = a number that satisfies the given condition.

Solve: $3n - 10n \geq 8n$

$n \leq 0$

33. Widths greater than 14 cm
34. Let l = the length of the rectangle.

At least 200 ft: Solve $2l + 16 \geq 200$

$l \geq 92$ ft

At most 200 ft: Solve $2l + 16 \leq 200$

$l \leq 92$ ft

35. Lengths greater than 6 cm
36. Let w = the width of the pool.

Solve: $2(2w) + 2w \leq 70$

$w \leq \dfrac{35}{3}$ ft

37. George: more than 12 hr; Joan: more than 15 hr
38. Let s = the cost of each sweater.

Solve: $21.95 + 2s \leq 120$

$s \leq \$49.02$ (Rounding down)

39. At most 9 lb
40. Let l = the length of the rectangle, in km.

Solve: $32l \geq 2048$

$l \geq 64$ km

41. There are at least $5\dfrac{1}{3}$ g of fat.
42. Let r = the amount of fat in a serving of regular peanut butter, in grams.

Solve: $12 \leq 0.75r$ (See Exercise 41.)

$r \geq 16$ g

43. 9 cm or more
44. Let b = the length of the base, in cm.

Solve: $\dfrac{1}{2} \cdot b \cdot 20 \leq 40$

$b \leq 4$ cm

45. 2001 and beyond
46. Solve: $0.027x + 0.19 \leq 6$

$x \leq 215\dfrac{5}{27}$ mi

47. -160 **48.** $4a^2 - 2$ **49.** $4x^2 - 4x - 7$
50. $3x + 2[4 - 5(2x - 1)] = 3x + 2[4 - 10x + 5]$
$= 3x + 2[9 - 10x]$
$= 3x + 18 - 20x$
$= -17x + 18$

51. $-3ab + 9b - 12a$
52. $3a^2b - 4ab + 2ab^2 - 7a^2b = -4a^2b - 4ab + 2ab^2$
53. ◆ Answers may vary. Acme rents a truck at a daily rate of $46.30 plus $0.43 per mile. The Rothmans want a one-day truck rental, but they must stay within an $85 budget. What mileages will allow them to say within their budget? Round to the nearest mile.   **54.** ◆ No; the same relationships can be expressed using the symbols $<$, $>$, $\leq$, and $\geq$. For example $a \not< b$ can also be expressed as $a \geq b$ and so forth.   **55.** ◆ Answers may vary. Todd is at least 10 years older than Frances.   **56.** ◆ If we had rounded up, the cost would have exceeded $450.   **57.** 47 and 49
58. Let s = the length of a side of the square, in cm.

Solve: $0 < s^2 \leq 64$

$-8 \leq s < 0$ or $0 < s \leq 8$

We consider only positive solutions since the length of a side cannot be negative. Thus, for positive lengths of a side no more than 8 cm the area of a square will be no more than 64 cm^2.

59. Between $-15°C$ and $-9\dfrac{4}{9}°C$
60. Let h = the number of hours the car has been parked. Then $h - 1$ = the number of hours after the first hour.

Solve: $4 + 2.50(h - 1) > 16.50$

$h > 6$ hr

61. Less than $20,000
62. Let h = the number of hours the car has been parked. Then $h - 1$ = the number of hours after the first hour.

Solve: $14 < 4 + 2.50(h - 1) < 24$

5 hr $< h <$ 9 hr

63. The fat content is at least 7.5 g.
64. ◆ Let s = Jackie's score on the tenth quiz. We determine the score required to improve her average at least 2 points. Solving $\dfrac{9 \cdot 84 + s}{10} \geq 86$, we get $s \geq 104$. Since the maximum possible score is 100, Jackie cannot improve her average two points with the next quiz.
65. ◆ Let p = the total purchases for the year. Solving $10\%p > 25$, we get $p > 250$. Thus, when a customer's purchases are more than $250 for the year, the customer saves money by purchasing a card.

REVIEW EXERCISES: CHAPTER 2, PP. 116–117

1. [2.1] -22 **2.** [2.1] 7   **3.** [2.1] -192 **4.** [2.1] 1
5. [2.1] $-\dfrac{7}{3}$ **6.** [2.1] 1.11   **7.** [2.1] $\dfrac{1}{2}$ **8.** [2.1] $-\dfrac{15}{64}$

9. [2.2] $\frac{38}{5}$ **10.** [2.1] -8 **11.** [2.2] -5 **12.** [2.2] $-\frac{1}{3}$
13. [2.2] 4 **14.** [2.2] 3 **15.** [2.2] 4 **16.** [2.2] 16
17. [2.2] 6 **18.** [2.2] -3 **19.** [2.2] 12 **20.** [2.2] 4
21. [2.3] $d = \frac{C}{\pi}$ **22.** [2.3] $B = \frac{3V}{h}$
23. [2.3] $a = 2A - b$ **24.** [2.4] 0.001 **25.** [2.4] 44%
26. [2.4] 20% **27.** [2.4] 360 **28.** [2.6] Yes
29. [2.6] No **30.** [2.6] Yes
31. [2.6] **32.** [2.6]

$4x - 6 < x + 3$
$-5\ -4\ -3\ -2\ -1\ \ 0\ \ 1\ \ 2\ \ 3\ \ 4\ \ 5$

$-2 < x \le 5$
$-5\ -4\ -3\ -2\ -1\ \ 0\ \ 1\ \ 2\ \ 3\ \ 4\ \ 5$

33. [2.6]

$y > 0$
$-5\ -4\ -3\ -2\ -1\ \ 0\ \ 1\ \ 2\ \ 3\ \ 4\ \ 5$

34. [2.6] $\{y \mid y \ge -\frac{1}{2}\}$ **35.** [2.6] $\{x \mid x \ge 7\}$
36. [2.6] $\{y \mid y > 2\}$ **37.** [2.6] $\{y \mid y \le -4\}$
38. [2.6] $\{x \mid x < -11\}$ **39.** [2.6] $\{y \mid y > -7\}$
40. [2.6] $\{x \mid x > -6\}$ **41.** [2.6] $\{x \mid x > -\frac{9}{11}\}$
42. [2.6] $\{y \mid y \le 7\}$ **43.** [2.6] $\{x \mid x \ge -\frac{1}{12}\}$
44. [2.5] $317 **45.** [2.5] 20 **46.** [2.5] 3 m, 5 m
47. [2.5] 5880 **48.** [2.5] 57, 59
49. [2.5] Width: 11 cm; length: 17 cm **50.** [2.5] $160
51. [2.5] $17.5 billion **52.** [2.5] 35°, 85°, 60°
53. [2.7] At most $35 **54.** [2.7] Widths greater than 17 cm **55.** [1.1] 5 **56.** [1.2] $12t + 8 + 4s$
57. [1.7] -2.3 **58.** [1.8] $-43x + 8y$
59. [2.1], [2.6] ◆ Multiplying on both sides of an equation by *any* nonzero number results in an equivalent equation. When multiplying on both sides of an inequality, the sign of the number being multiplied by must be considered. If the number is positive, the direction of the inequality symbol remains unchanged; if the number is negative, the direction of the inequality symbol must be reversed to produce an equivalent inequality.
60. [2.1], [2.6] ◆ The solutions of an equation can usually each be checked. The solutions of an inequality are normally too numerous to check. Checking a few numbers from the solution set found cannot guarantee that the answer is correct, although if any number does not check, the answer found is incorrect. **61.** [2.5] Amazon: 6437 km; Nile: 6671 km **62.** [2.5] $14,150
63. [1.4], [2.2] 23, -23 **64.** [1.4], [2.1] 20, -20
65. [2.3] $a = \frac{y - 3}{2 - b}$

TEST: CHAPTER 2, PP. 117–118

1. [2.1] 8 **2.** [2.1] 26 **3.** [2.1] -6 **4.** [2.1] 49
5. [2.2] -12 **6.** [2.2] 2 **7.** [2.1] -8 **8.** [2.1] $-\frac{7}{20}$
9. [2.2] 7 **10.** [2.2] -5 **11.** [2.2] $\frac{23}{3}$
12. [2.6] $\{x \mid x \le -4\}$ **13.** [2.6] $\{x \mid x > -13\}$

14. [2.6] $\{x \mid x < \frac{21}{8}\}$ **15.** [2.6] $\{y \mid y \le -13\}$
16. [2.6] $\{y \mid y \le -8\}$ **17.** [2.6] $\{x \mid x \le -\frac{1}{20}\}$
18. [2.6] $\{x \mid x < -6\}$ **19.** [2.6] $\{x \mid x \le -1\}$
20. [2.3] $r = \frac{A}{2\pi h}$ **21.** [2.3] $l = \frac{2w - P}{-2}$ **22.** [2.4] 5
23. [2.4] 5.4% **24.** [2.4] 21 **25.** [2.4] 44%
26. [2.6] **27.** [2.6]

$y < 9$
$-10\ -8\ -6\ -4\ -2\ \ 0\ \ 2\ \ 4\ \ 6\ \ 8\ \ 10$

$-2 \le x \le 2$
$-5\ -4\ -3\ -2\ -1\ \ 0\ \ 1\ \ 2\ \ 3\ \ 4\ \ 5$

28. [2.5] Width: 7 cm; length: 11 cm **29.** [2.5] 30 mi
30. [2.5] 81 mm, 83 mm, 85 mm **31.** [2.5] $2700
32. [2.7] All numbers greater than 6
33. [2.7] Lengths of at least 174 yd **34.** [1.1] $x - 10$
35. [1.2] $3(a + 8b + 4)$ **36.** [1.7] $\frac{3}{10}$ **37.** [1.8] 5
38. [2.3] $d = \frac{ca - 1}{c}$ **39.** [1.4], [2.2] 15, -15
40. [2.5] 60

CHAPTER 3

EXERCISE SET 3.1, PP. 125–128

1. 5 **2.** Approximately 3 drinks **3.** The person weighs at least 120 lb. **4.** The individual weighs more than 160 lb. **5.** $1010.88
6. Total tax: $0.16 \cdot \$26{,}000 = \$4160$
  Amount spent on community development:
  $0.09 \cdot \$4160 = \$374.40$
7. $8170.75
8. Total tax: $0.24 \cdot \$116{,}000 = \$27{,}840$
  Amount spent on law enforcement:
  $0.02 \cdot \$27{,}840 = \$556.80$
9. 19.2 million tons
10. Let $y =$ the amount of paper or cardboard, in pounds, in the solid waste generated per day by the average American in 1996.
  $y = 0.376(4.5) \approx 1.7$ lb
11. 0.07 lb
12. Amount of solid waste generated by a family of four each day: $4(4.5) = 18$ lb
  Amount that is plastic: $0.093(18) = 1.674$ lb
  Amount of plastic waste that is recycled each day:
  $0.05(1.674) \approx 0.08$ lb
13. 4% **14.** Approximately 8% **15.** 1982 **16.** 1990
17. About $123 billion (using 4.5%) **18.** From the graph we see that about 14.5% of GNP was spent on health care in 1995. Then $0.145(6977) \approx \$1012$ billion.
19. $1016 billion (using 5%)

20. From the graph we see that about 12.5% of GNP was spent on health care in 1990. Let x = the GNP in 1990, in billions of dollars.

Solve: $668 = 0.125x$
$x \approx \$5344$ billion

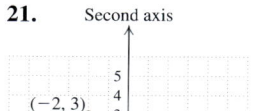

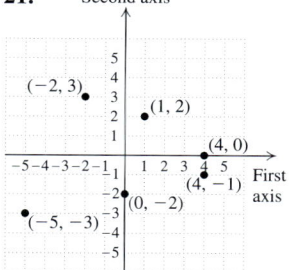

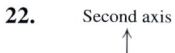

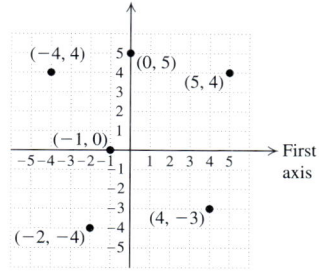

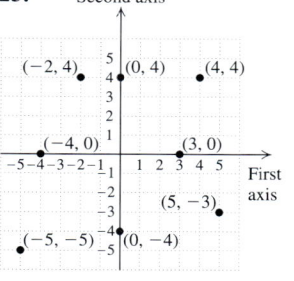

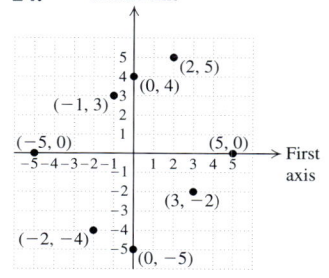

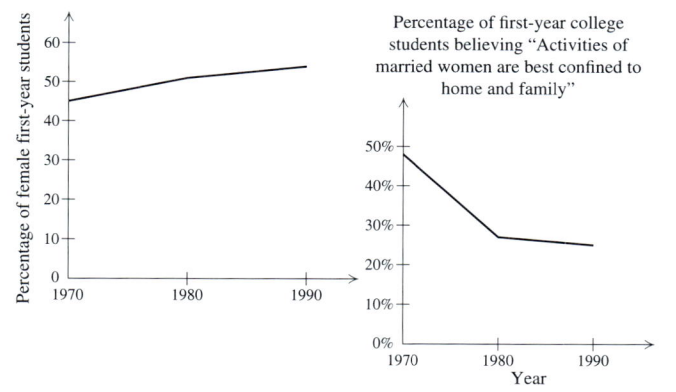

21. [Graph showing points (−2, 3), (1, 2), (4, 0), (4, −1), (0, −2), (−5, −3) on coordinate axes labeled Second axis and First axis]

22. [Graph showing points (−4, 4), (0, 5), (5, 4), (−1, 0), (−2, −4), (4, −3) on coordinate axes labeled Second axis and First axis]

23. [Graph showing points (−2, 4), (0, 4), (4, 4), (−4, 0), (3, 0), (5, −3), (−5, −5), (0, −4) on coordinate axes labeled Second axis and First axis]

24. [Graph showing points (2, 5), (0, 4), (−1, 3), (−5, 0), (5, 0), (3, −2), (−2, −4), (0, −5) on coordinate axes labeled Second axis and First axis]

25. A: $(−4, 5)$; B: $(−3, −3)$; C: $(0, 4)$; D: $(3, 4)$; E: $(3, −4)$
26. A: $(3, 3)$, B: $(0, −4)$, C: $(−5, 0)$, D: $(−1, −1)$, E: $(2, 0)$
27. A: $(4, 1)$; B: $(0, −5)$; C: $(−4, 0)$; D: $(−3, −2)$; E: $(3, 0)$
28. A: $(−5, 1)$, B: $(0, 5)$, C: $(5, 3)$, D: $(0, −1)$, E: $(2, −4)$
29. II **30.** II **31.** IV **32.** IV **33.** III **34.** III
35. I **36.** I **37.** Negative, negative **38.** Second, first
39. [Graph: Percentage of female first-year students vs Year 1970–1990, increasing from ~45 to ~55]
40. [Graph: Percentage of first-year college students believing "Activities of married women are best confined to home and family" vs Year 1970–1990, decreasing from ~50% to ~25%]

41. [Graph: Red meat consumption (in pounds) vs Year, 1975–1995, decreasing from ~130 to ~115]

42. Annual cheese consumption in the United States [Graph: Number of pounds vs Year, 1975–1995, increasing from ~15 to ~27]

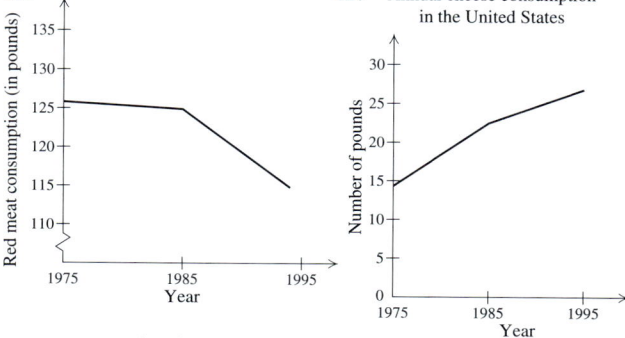

43. $\frac{8}{9}$ **44.** $\frac{5}{8} \cdot \frac{2}{15} = \frac{10}{120} = \frac{\cancel{10} \cdot 1}{\cancel{10} \cdot 12} = \frac{1}{12}$ **45.** $\frac{2}{3}$

46. $\frac{2}{3} + \frac{1}{5} = \frac{10}{15} + \frac{3}{15} = \frac{13}{15}$ **47.** $-\frac{13}{35}$ **48.** $-\frac{2}{3} \div 5 = -\frac{2}{3} \cdot \frac{1}{5} = -\frac{2}{15}$ **49.** ◆ No; a circle graph is best used to show what percent of a whole each item comprising it represents. A bar graph or line graph would be a better way to display the number of sport utility vehicles sold in each of the last five years. **50.** ◆ Pulse rate cannot continue to decrease indefinitely. After five or six months of regular exercise it reaches a level at which it remains with continued regular exercise. **51.** ◆ A line graph gives a better indication of continuous change over time.
52. ◆ The result is an arrow pointing up with the corners of its base at $(3, −2)$ and $(4, −2)$. **53.** II or III
54. I or II **55.** II or IV **56.** The coordinates must have the same sign. The point is in quadrant I or III.
57. $(−1, −5)$
58. [Three graphs showing parallelograms with vertices (−1, 2), (5, 2), (−2, −3), (4, −3); (−7, 2), (−1, 2), (−2, −3), (4, −3); and (−1, 2), (−2, −3), (3, −8), (4, −3)]

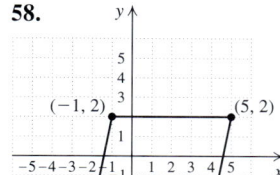

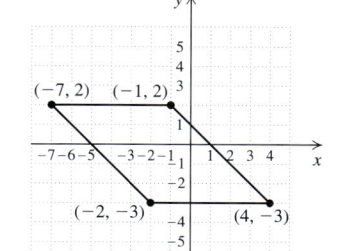

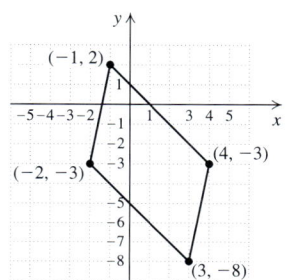

The coordinates of the fourth vertex are $(5, 2)$, $(−7, 2)$, or $(3, −8)$.

59.

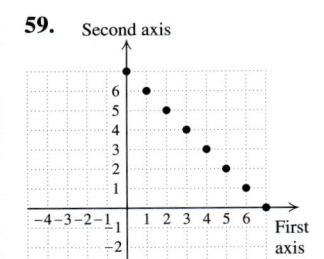

60. Answers may vary.

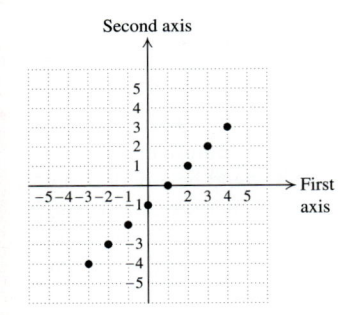

61. 26 units

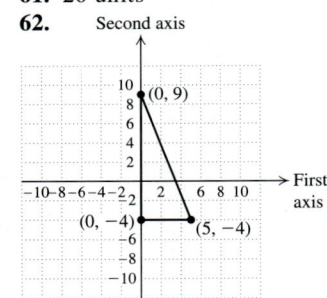

62.

The base is 5 units and the height is 13 units.
$A = \frac{1}{2}bh = \frac{1}{2} \cdot 5 \cdot 13 = \frac{65}{2}$ sq units, or $32\frac{1}{2}$ sq units
63. Latitude 32.5° North, longitude 64.5° West
64. Latitude 27° North, longitude 81° West **65.** ◆ Eight "quadrants" will exist. Think of the coordinate system being formed by the intersection of one coordinate plane with another plane perpendicular to one of its axes such that the origins of the two planes coincide. Then there are four quadrants "above" the x, y-plane and four "below" it.

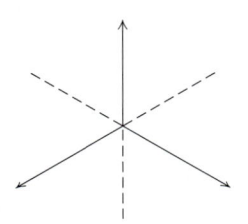

TECHNOLOGY CONNECTION 3.2, P. 135

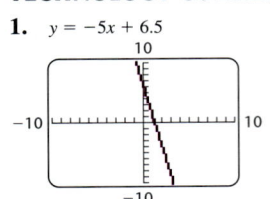

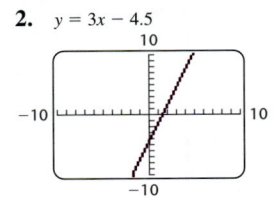

1. $y = -5x + 6.5$ **2.** $y = 3x - 4.5$

3. $y = \frac{4}{7}x - \frac{22}{7}$ **4.** $y = -\frac{11}{5}x - 4$

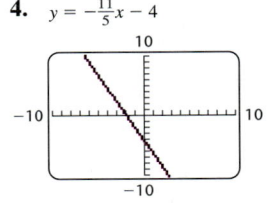

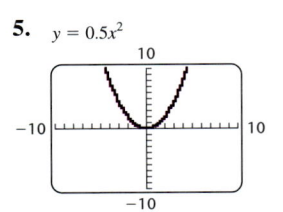

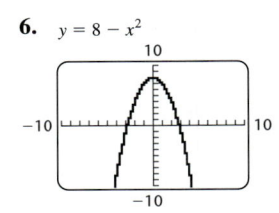

5. $y = 0.5x^2$ **6.** $y = 8 - x^2$

EXERCISE SET 3.2, PP. 135–138

1. No **2.** $y = 2x + 5$
$7 \;?\; 2 \cdot 1 + 5$
$7 \;|\; 7$ TRUE
$(1, 7)$ is a solution.

3. No **4.** $5x - 3y = 15$
$5 \cdot 0 - 3 \cdot 5 \;?\; 15$
$-15 \;|\; 15$ FALSE
$(0, 5)$ is not a solution.

5. Yes **6.** $2p - 3q = -13$
$2(-5) - 3 \cdot 1 \;?\; -13$
$-10 - 3$
$-13 \;|\; -13$ TRUE
$(-5, 1)$ is a solution.

7. $y = x - 5$ $y = x - 5$
$2 \;?\; 7 - 5$ $-4 \;?\; 1 - 5$
$2 \;|\; 2$ TRUE $-4 \;|\; -4$ TRUE
$(0, -5)$; answers may vary

8. $y = x + 3$ $y = x + 3$
$2 \;?\; -1 + 3$ $7 \;?\; 4 + 3$
$2 \;|\; 2$ TRUE $7 \;|\; 7$ TRUE
$(-1, 2)$ and $(4, 7)$ are solutions of $y = x + 3$.

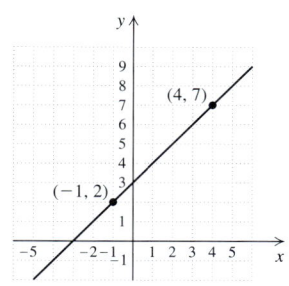

(0, 3) appears to be another solution. We check.
$$y = x + 3$$
$$3 \;?\; 0 + 3$$
$$3 \;|\; 3 \quad \text{TRUE}$$
Thus, (0, 3) is another solution. There are other correct answers, including (−5, −2), (−4, −1), (−3, 0), (−2, 1), (1, 4), (2, 5), and (3, 6).

9.
$$y = \tfrac{1}{2}x + 3 \qquad\qquad y = \tfrac{1}{2}x + 3$$
$$5 \;?\; \tfrac{1}{2}\cdot 4 + 3 \qquad\quad 2 \;?\; \tfrac{1}{2}(-2) + 3$$
$$\;2 + 3 \qquad\qquad\qquad\; -1 + 3$$
$$5 \;|\; 5 \quad \text{TRUE} \qquad 2 \;|\; 2 \quad \text{TRUE}$$
(0, 3); answers may vary

10.
$$y = \tfrac{1}{2}x - 1 \qquad\qquad y = \tfrac{1}{2}x - 1$$
$$2 \;?\; \tfrac{1}{2}\cdot 6 - 1 \qquad\quad -1 \;?\; \tfrac{1}{2}\cdot 0 - 1$$
$$\;3 - 1 \qquad\qquad\qquad -1 \;|\; -1 \quad \text{TRUE}$$
$$2 \;|\; 2 \quad \text{TRUE}$$
(6, 2) and (0, −1) are solutions of $y = \tfrac{1}{2}x - 1$.

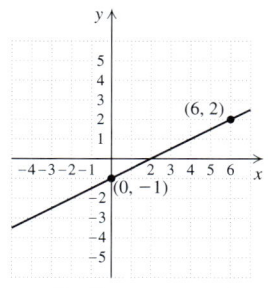

(2, 0) appears to be another solution. We check.
$$y = \tfrac{1}{2}x - 1$$
$$0 \;?\; \tfrac{1}{2}\cdot 2 - 1$$
$$\;1 - 1$$
$$0 \;|\; 0 \quad \text{TRUE}$$
Thus, (2, 0) is another solution. There are other correct answers, including (−6, −4), (−4, −3), (−2, −2), and (4, 1).

11.
$$3x + y = 7 \qquad\qquad 3x + y = 7$$
$$3\cdot 2 + 1 \;?\; 7 \qquad 3\cdot 4 + (-5) \;?\; 7$$
$$6 + 1 \;|\; 7 \qquad\qquad 12 - 5 \;|\;$$
$$\;7 \;|\; 7 \quad \text{TRUE} \qquad\;7 \;|\; 7 \quad \text{TRUE}$$
(1, 4); answers may vary

12.
$$x + 2y = 5 \qquad\qquad x + 2y = 5$$
$$-1 + 2\cdot 3 \;?\; 5 \qquad 7 + 2(-1) \;?\; 5$$
$$-1 + 6 \;|\; \qquad\qquad 7 - 2 \;|\;$$
$$\;5 \;|\; 5 \quad \text{TRUE} \qquad\;5 \;|\; 5 \quad \text{TRUE}$$
(−1, 3) and (7, −1) are solutions of $x + 2y = 5$.

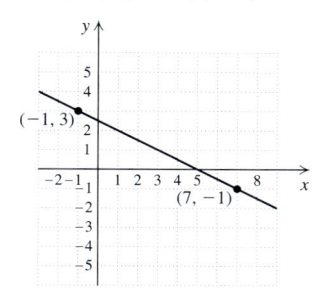

(1, 2) appears to be another solution. We check.
$$x + 2y = 5$$
$$1 + 2\cdot 2 \;?\; 5$$
$$1 + 4 \;|\;$$
$$\;5 \;|\; 5 \quad \text{TRUE}$$
Thus, (1, 2) is another solution. There are other correct answers, including (−5, 5), (−3, 4), (1, 2), (3, 1), and (5, 0).

13.
$$4x - 2y = 10 \qquad\qquad 4x - 2y = 10$$
$$4\cdot 0 - 2(-5) \;?\; 10 \qquad 4\cdot 4 - 2\cdot 3 \;?\; 10$$
$$0 + 10 \;|\; \qquad\qquad 16 - 6 \;|\;$$
$$\;10 \;|\; 10 \quad \text{TRUE} \qquad\;10 \;|\; 10 \quad \text{TRUE}$$
(2, −1); answers may vary

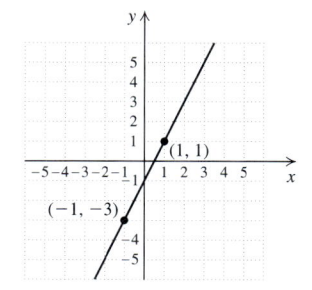

14.
$$6x - 3y = 3$$
$$6\cdot 1 - 3\cdot 1 \;?\; 3$$
$$6 - 3 \;|\;$$
$$\;3 \;|\; 3 \quad \text{TRUE}$$
$$6x - 3y = 3$$
$$6(-1) - 3(-3) \;?\; 3$$
$$-6 + 9 \;|\;$$
$$\;3 \;|\; 3 \quad \text{TRUE}$$

(1, 1) and (−1, −3) are solutions of $6x - 3y = 3$.
(0, −1) appears to be another solution. We check.
$$6x - 3y = 3$$
$$6\cdot 0 - 3(-1) \;?\; 3$$
$$\;3 \;|\; 3 \quad \text{TRUE}$$
Thus, (0, −1) is another solution. There are other correct answers, including (−2, −5), (2, 3), and (3, 5).

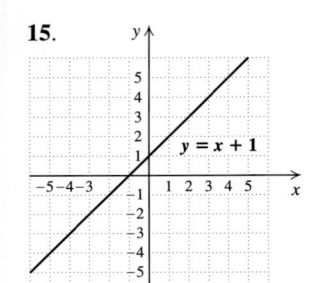

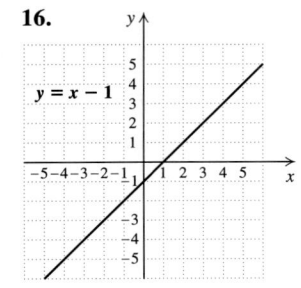

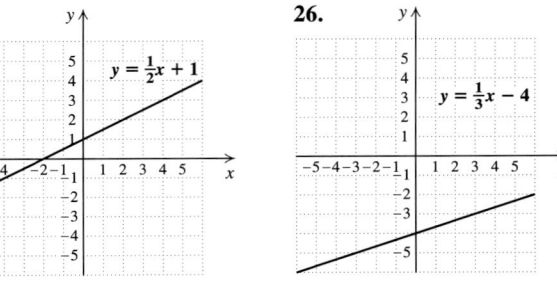

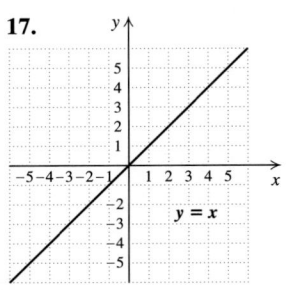

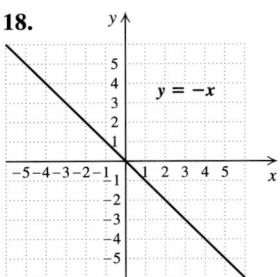

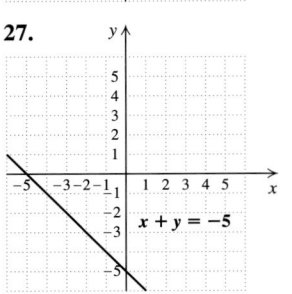

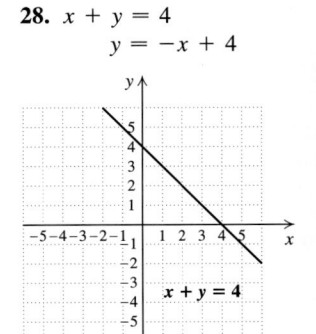

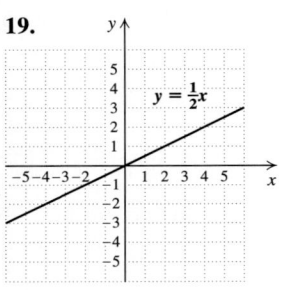

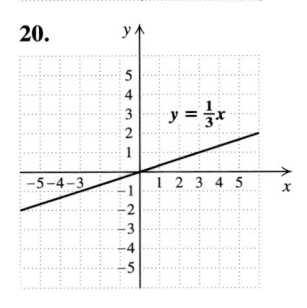

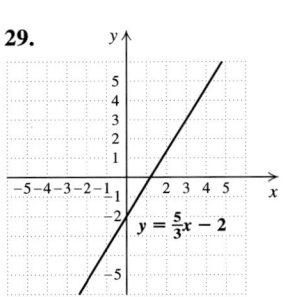

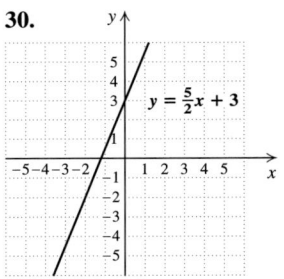

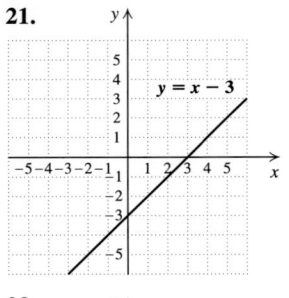

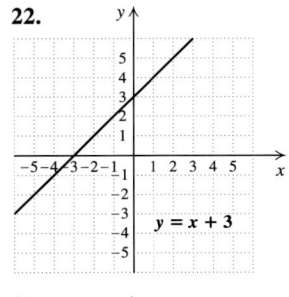

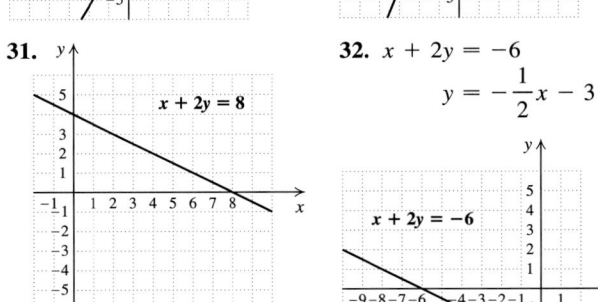

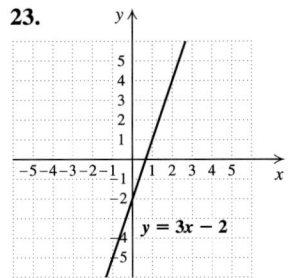

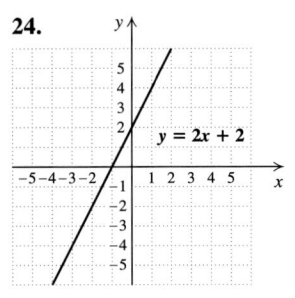

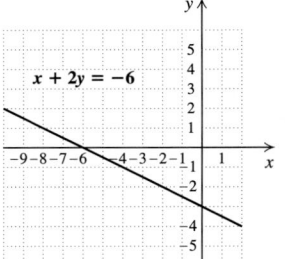

15. $y = x + 1$
16. $y = x - 1$
17. $y = x$
18. $y = -x$
19. $y = \frac{1}{2}x$
20. $y = \frac{1}{3}x$
21. $y = x - 3$
22. $y = x + 3$
23. $y = 3x - 2$
24. $y = 2x + 2$
25. $y = \frac{1}{2}x + 1$
26. $y = \frac{1}{3}x - 4$
27. $x + y = -5$
28. $x + y = 4$
    $y = -x + 4$
29. $y = \frac{5}{3}x - 2$
30. $y = \frac{5}{2}x + 3$
31. $x + 2y = 8$
32. $x + 2y = -6$
    $y = -\frac{1}{2}x - 3$

33. $y = \frac{3}{2}x + 1$

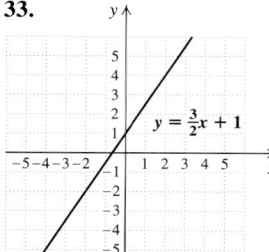

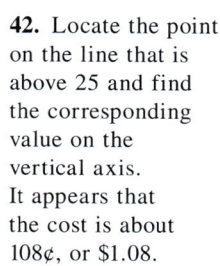

34. $y = -\frac{2}{3}x + 4$

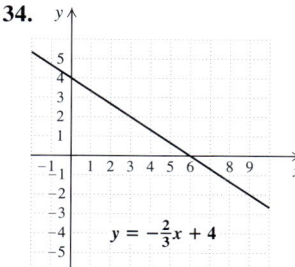

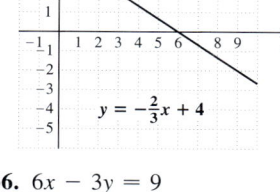

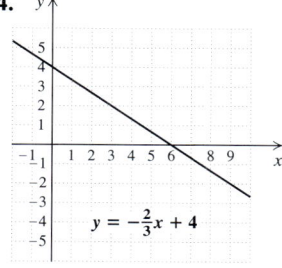

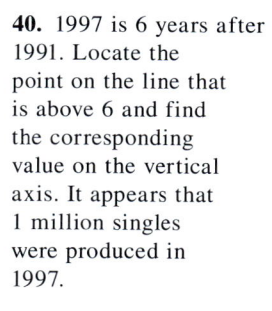

35. $8x - 4y = 12$

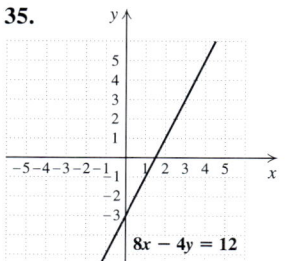

36. $6x - 3y = 9$
$y = 2x - 3$

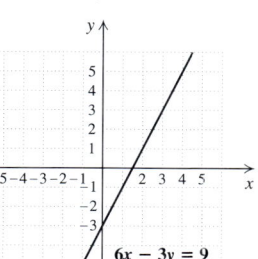

37. $8y + 2x = -4$

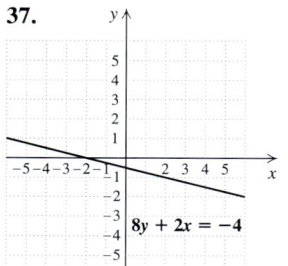

38. $6y + 2x = 8$
$y = -\frac{1}{3}x + \frac{4}{3}$

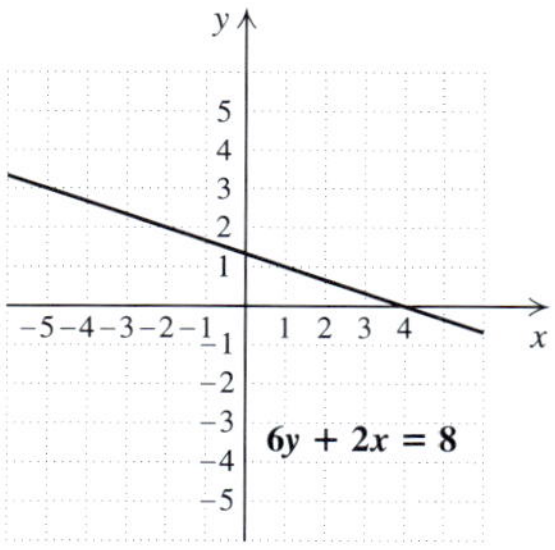

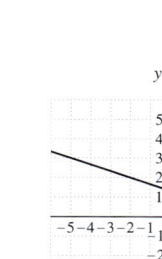

39. 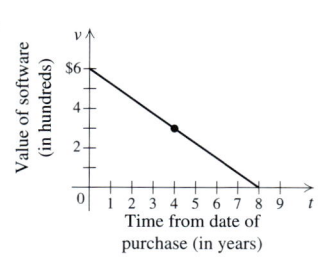 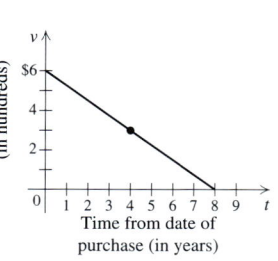 $300

40. 1997 is 6 years after 1991. Locate the point on the line that is above 6 and find the corresponding value on the vertical axis. It appears that 1 million singles were produced in 1997.

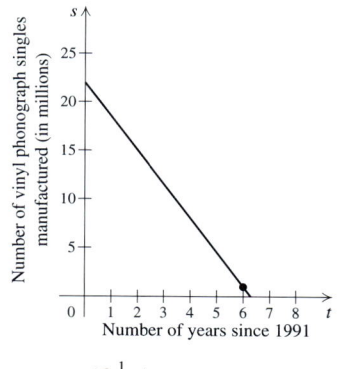

41. 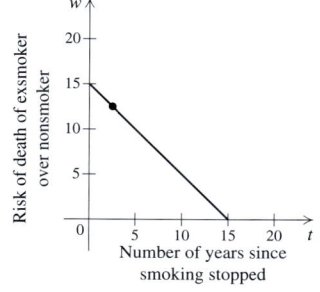 $12\frac{1}{2}$ times

42. Locate the point on the line that is above 25 and find the corresponding value on the vertical axis. It appears that the cost is about 108¢, or $1.08.

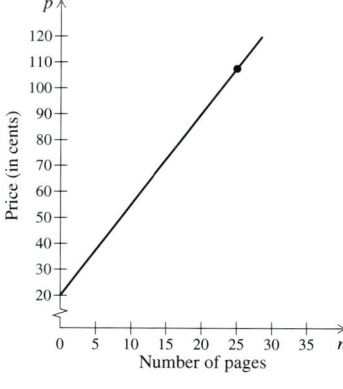

43. 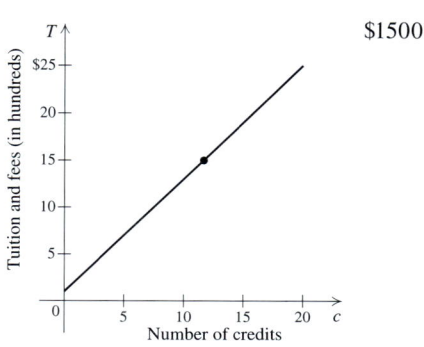 $1500

44. 2002 is 10 years after 1992. Locate the point on the line that is above 10 and find the corresponding value on the vertical axis. It appears that the cost is about $25,000.

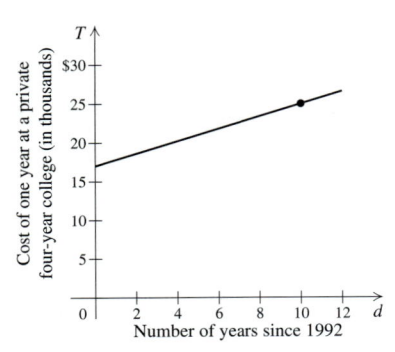

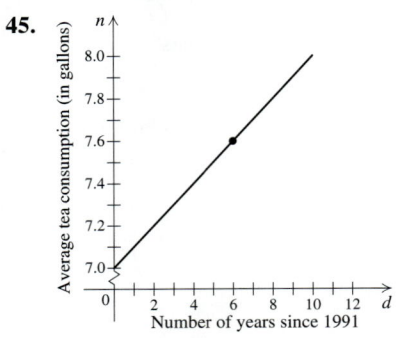

45. 7.6 gal

46. Locate the point on the line that is above 15 and find the corresponding value on the vertical axis. It appears that the temperature was about 24° F at 9:15 A.M.

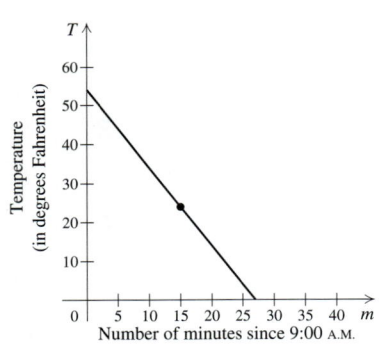

47. -9 **48.** $4(3 - 2x) = 7 - 3(5x - 1)$
$12 - 8x = 7 - 15x + 3$
$7x = -2$
$x = -\dfrac{2}{7}$
The value checks.

49. $\dfrac{16}{5}$ **50.** $pq + p = w$
$p(q + 1) = w$
$p = \dfrac{w}{q + 1}$

51. $y = \dfrac{C - Ax}{B}$ **52.** $A = \dfrac{T + Q}{2}$
$2A = T + Q$
$2A - T = Q$

53. ◆ Most would probably say that the second equation would be easier to graph because it has been solved for y. This makes it more efficient to find the y-value that corresponds to a given x-value.   **54.** ◆ Yes; an equation of the form $x = a$ has no y-intercept.   **55.** ◆ No; the three points could satisfy an equation that is different from the one being graphed.   **56.** ◆ Start on the vertical axis. Locate the point on the line that is to the right of 1.5. Then find the value on the horizontal axis that corresponds to that point.
57. (0, 7), (1, 6), (2, 5), (3, 4), (4, 3), (5, 2), (6, 1), (7, 0).

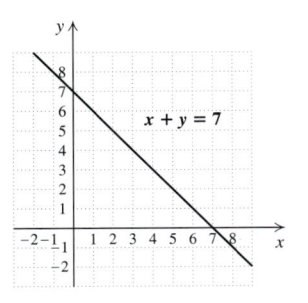

58. (0, 9), (1, 8), (2, 7), (3, 6), (4, 5), (5, 4), (6, 3), (7, 2), (8, 1), (9, 0).

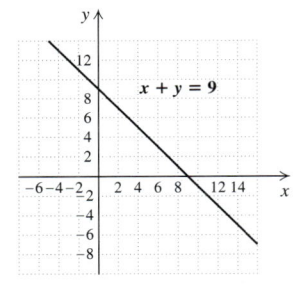

59. ◆ The graph will be a line passing through the points $(0, C)$ and $(C, 0)$.   **60.** Note that the sum of the coordinates of each point on the graph is 5. Thus, we have $x + y = 5$, or $y = -x + 5$.   **61.** $y = -x + 2$, or $x + y = 2$ **62.** Note that each y-coordinate is 2 more than the corresponding x-coordinate. Thus, we have $y = x + 2$.   **63.** $y = 3x$
64. $0.10d + 0.05n = 1.75$

From the graph we see that the following are three of the combination of dimes and nickels that total $1.75.

5 dimes, 25 nickels
10 dimes, 15 nickels
12 dimes, 11 nickels

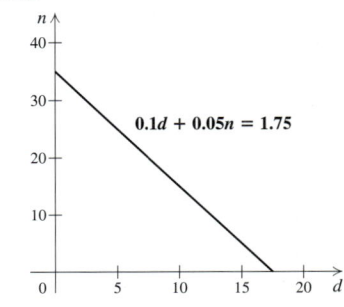

65.

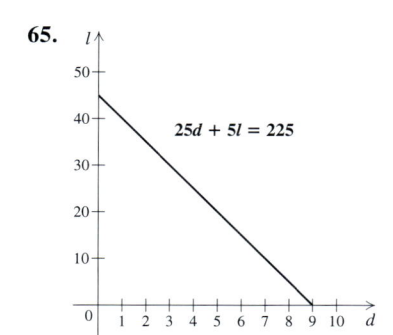

[Graph: $25d + 5l = 225$]

Answers may vary. 1 dinner, 40 lunches; 5 dinners, 20 lunches; 8 dinners, 5 lunches

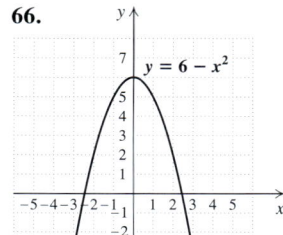

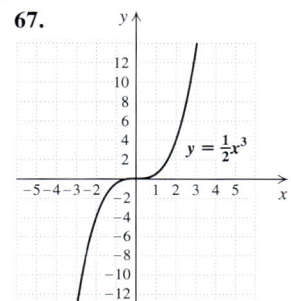

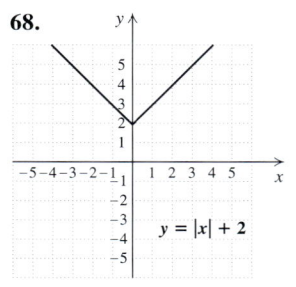

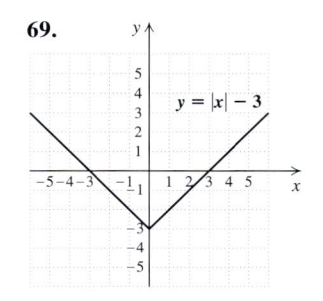

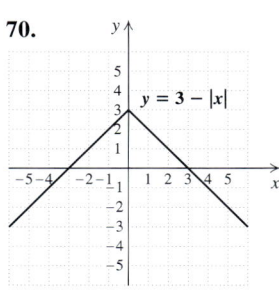

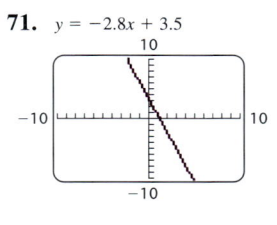

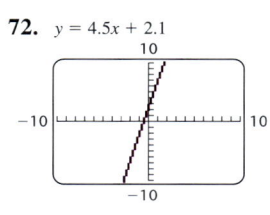

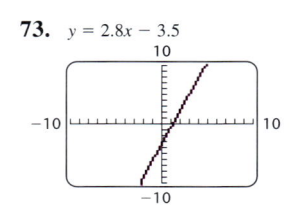

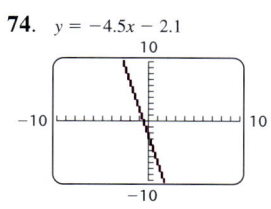

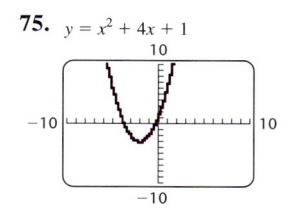

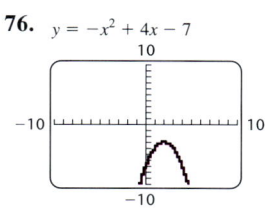

66. [Graph: $y = 6 - x^2$]

67. [Graph: $y = \frac{1}{2}x^3$]

68. [Graph: $y = |x| + 2$]

69. [Graph: $y = |x| - 3$]

70. [Graph: $y = 3 - |x|$]

71. $y = -2.8x + 3.5$ [Graph]

72. $y = 4.5x + 2.1$ [Graph]

73. $y = 2.8x - 3.5$ [Graph]

74. $y = -4.5x - 2.1$ [Graph]

75. $y = x^2 + 4x + 1$ [Graph]

76. $y = -x^2 + 4x - 7$ [Graph]

77. ◆ No; only whole number values of the variables have meaning in these applications, so only those points for which both coordinates are whole numbers are solutions of the given problems.

TECHNOLOGY CONNECTION 3.3, P. 141

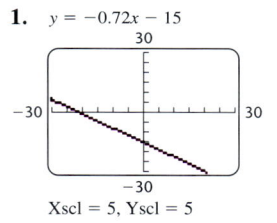

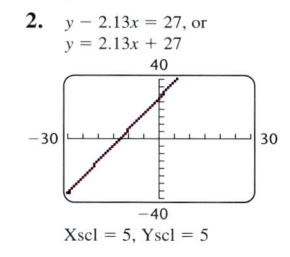

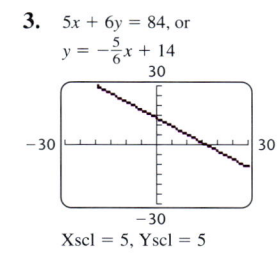

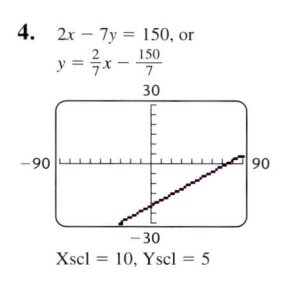

1. $y = -0.72x - 15$ [Graph] Xscl = 5, Yscl = 5

2. $y - 2.13x = 27$, or $y = 2.13x + 27$ [Graph] Xscl = 5, Yscl = 5

3. $5x + 6y = 84$, or $y = -\frac{5}{6}x + 14$ [Graph] Xscl = 5, Yscl = 5

4. $2x - 7y = 150$, or $y = \frac{2}{7}x - \frac{150}{7}$ [Graph] Xscl = 10, Yscl = 5

EXERCISE SET 3.3, PP. 144–145

1. (a) $(0, 5)$; **(b)** $(2, 0)$ **2. (a)** $(0, 3)$; **(b)** $(4, 0)$
3. (a) $(0, -4)$; **(b)** $(3, 0)$ **4. (a)** $(0, 5)$; **(b)** $(-3, 0)$
5. (a) $(0, 3)$; **(b)** $(5, 0)$
6. $5x + 2y = 20$
   (a) Solve: $2y = 20$
   $y = 10$
   The y-intercept is $(0, 10)$.
   (b) Solve: $5x = 20$
   $x = 4$
   The x-intercept is $(4, 0)$.

7. (a) $(0, -14)$; (b) $(4, 0)$

8. $3x - 4y = 24$
(a) Solve: $-4y = 24$
$y = -6$
The y-intercept is $(0, -6)$.
(b) Solve: $3x = 24$
$x = 8$
The x-intercept is $(8, 0)$.

9. (a) $\left(0, \frac{10}{3}\right)$; (b) $\left(-\frac{5}{2}, 0\right)$

10. $-2x + 3y = 7$
(a) Solve: $3y = 7$
$y = \frac{7}{3}$
The y-intercept is $\left(0, \frac{7}{3}\right)$.
(b) Solve: $-2x = 7$
$x = -\frac{7}{2}$
The x-intercept is $\left(-\frac{7}{2}, 0\right)$.

11. (a) $\left(0, -\frac{1}{3}\right)$; (b) $\left(\frac{1}{2}, 0\right)$

12. $4y - 2 = 6x$
$-6x + 4y = 2$
(a) Solve: $4y = 2$
$y = \frac{1}{2}$
The y-intercept is $\left(0, \frac{1}{2}\right)$.
(b) Solve: $-6x = 2$
$x = -\frac{1}{3}$
The x-intercept is $\left(-\frac{1}{3}, 0\right)$.

13.

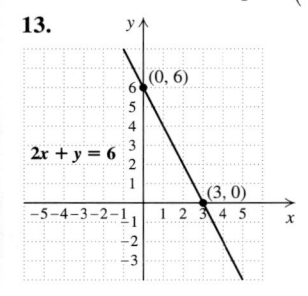

14. $3x + 2y = 12$
Find the y-intercept:
$2y = 12$ Replacing x with 0
$y = 6$
The y-intercept is $(0, 6)$.
Find the x-intercept:
$3x = 12$ Replacing y with 0
$x = 4$

The x-intercept is $(4, 0)$.
To find a third point we replace x with 2 and solve for y.
$3 \cdot 2 + 2y = 12$
$6 + 2y = 12$
$2y = 6$
$y = 3$
The point $(2, 3)$ appears to line up with the intercepts, so we draw the graph.

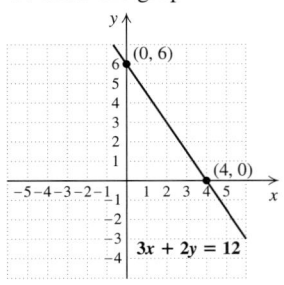

15.

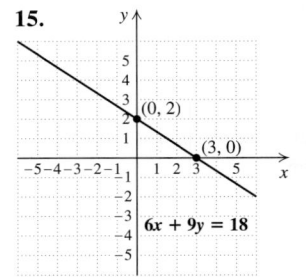

16. $x + 3y = 6$
Find the y-intercept:
$3y = 6$ Replacing x with 0
$y = 2$
The y-intercept is $(0, 2)$.
Find the x-intercept:
$x = 6$ Replacing y with 0
The x-intercept is $(6, 0)$.
To find a third point we replace x with 3 and solve for y.
$3 + 3y = 6$
$3y = 3$
$y = 1$
The point $(3, 1)$ appears to line up with the intercepts, so we draw the graph.

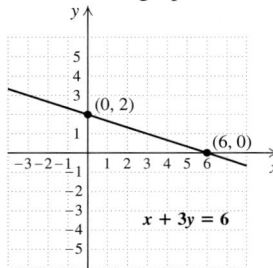

17.

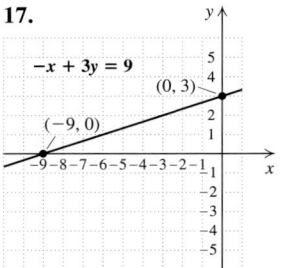

18. $-x + 2y = 4$
Find the y-intercept:
$2y = 4$ Replacing x with 0
$y = 2$
The y-intercept is $(0, 2)$.
Find the x-intercept:
$-x = 4$ Replacing y with 0
$x = -4$

The x-intercept is $(-4, 0)$.

To find a third point we replace x with 4 and solve for y.
$$-4 + 2y = 4$$
$$2y = 8$$
$$y = 4$$

The point $(4, 4)$ appears to line up with the intercepts, so we draw the graph.

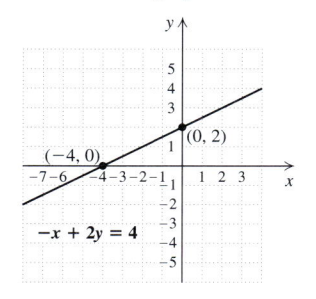

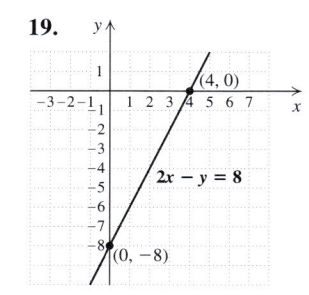

19.

20. $3x + y = 9$
Find the y-intercept:
$$y = 9 \quad \text{Replacing } x \text{ with } 0$$
The y-intercept is $(0, 9)$.

Find the x-intercept:
$$3x = 9 \quad \text{Replacing } y \text{ with } 0$$
$$x = 3$$
The x-intercept is $(3, 0)$.

To find a third point we replace x with 2 and solve for y.
$$3 \cdot 2 + y = 9$$
$$6 + y = 9$$
$$y = 3$$

The point $(2, 3)$ appears to line up with the intercepts, so we draw the graph.

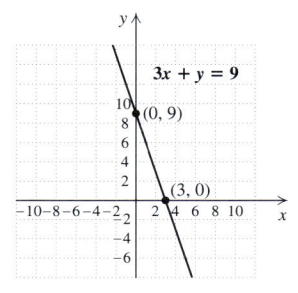

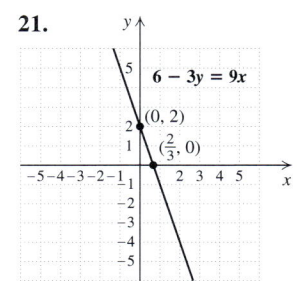

21.

22. $2y - 2 = 6x$
To find the y-intercept, let $x = 0$.
$$2y - 2 = 6 \cdot 0$$
$$2y - 2 = 0$$
$$2y = 2$$
$$y = 1$$
The y-intercept is $(0, 1)$.

Find the x-intercept:
$$-2 = 6x \quad \text{Replacing } y \text{ with } 0$$
$$-\frac{1}{3} = x$$
The x-intercept is $\left(-\frac{1}{3}, 0\right)$.

To find a third point we replace x with 1 and solve for y.
$$2y - 2 = 6 \cdot 1$$
$$2y - 2 = 6$$
$$2y = 8$$
$$y = 4$$

The point $(1, 4)$ appears to line up with the intercepts, so we draw the graph.

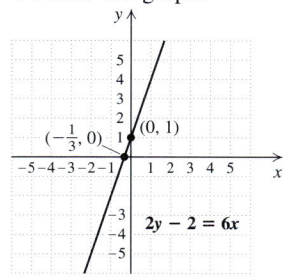

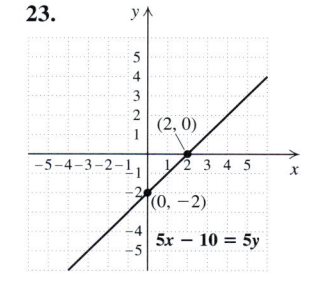

23.

24. $3x - 9 = 3y$
Find the y-intercept:
$$-9 = 3y \quad \text{Replacing } x \text{ with } 0$$
$$-3 = y$$
The y-intercept is $(0, -3)$.

To find the x-intercept, let $y = 0$.
$$3x - 9 = 3 \cdot 0$$
$$3x - 9 = 0$$
$$3x = 9$$
$$x = 3$$
The x-intercept is $(3, 0)$.

To find a third point we replace x with 1 and solve for y.
$$3 \cdot 1 - 9 = 3y$$
$$3 - 9 = 3y$$
$$-6 = 3y$$
$$-2 = y$$

The point $(1, -2)$ appears to line up with the intercepts, so we draw the graph.

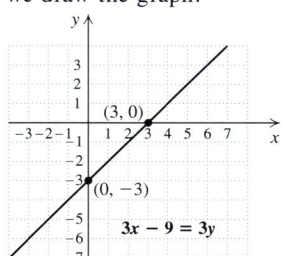

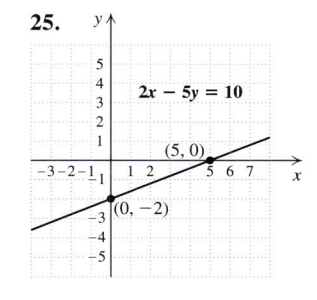

25.

26. $2x - 3y = 6$
Find the y-intercept:
$$-3y = 6 \quad \text{Replacing } x \text{ with } 0$$
$$y = -2$$
The y-intercept is $(0, -2)$.
Find the x-intercept:
$$2x = 6 \quad \text{Replacing } y \text{ with } 0$$
$$x = 3$$
The x-intercept is $(3, 0)$.
To find a third point we replace x with -3 and solve for y.
$$2(-3) - 3y = 6$$
$$-6 - 3y = 6$$
$$-3y = 12$$
$$y = -4$$
The point $(-3, -4)$ appears to line up with the intercepts, so we draw the graph.

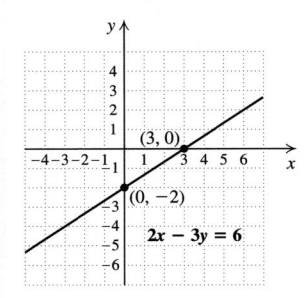

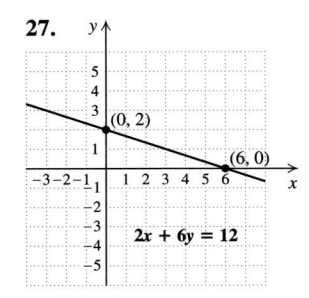

27.

28. $4x + 5y = 20$
Find the y-intercept:
$$5y = 20 \quad \text{Replacing } x \text{ with } 0$$
$$y = 4$$
The y-intercept is $(0, 4)$.
Find the x-intercept:
$$4x = 20 \quad \text{Replacing } y \text{ with } 0$$
$$x = 5$$
The x-intercept is $(5, 0)$.
To find a third point we replace x with 4 and solve for y.
$$4 \cdot 4 + 5y = 20$$
$$16 + 5y = 20$$
$$5y = 4$$
$$y = \frac{4}{5}$$

The point $\left(4, \frac{4}{5}\right)$ appears to line up with the intercepts, so we draw the graph.

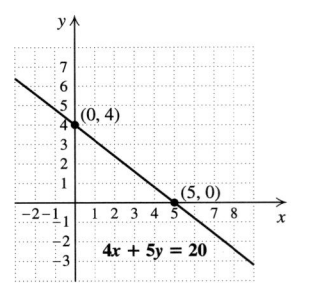

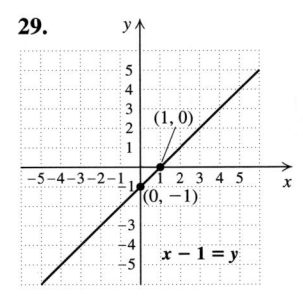

29.

30. $2x + 3y = 8$
Find the y-intercept:
$$3y = 8 \quad \text{Replacing } x \text{ with } 0$$
$$y = \frac{8}{3}$$
The y-intercept is $\left(0, \frac{8}{3}\right)$.
Find the x-intercept:
$$2x = 8 \quad \text{Replacing } y \text{ with } 0$$
$$x = 4$$
The x-intercept is $(4, 0)$.
To find a third point we replace x with 1 and solve for y.
$$2 \cdot 1 + 3y = 8$$
$$2 + 3y = 8$$
$$3y = 6$$
$$y = 2$$
The point $(1, 2)$ appears to line up with the intercepts, so we draw the graph.

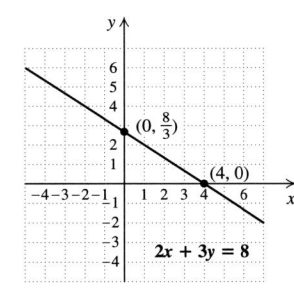

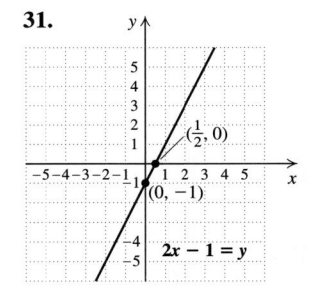

31.

32. $x - 3 = y$
Find the y-intercept:
$$-3 = y \quad \text{Replacing } x \text{ with } 0$$
The y-intercept is $(0, -3)$.
To find the x-intercept, let $y = 0$.
$$x - 3 = 0$$
$$x = 3$$
The x-intercept is $(3, 0)$.

To find a third point we replace x with -2 and solve for y.
$$-2 - 3 = y$$
$$-5 = y$$
The point $(-2, -5)$ appears to line up with the intercepts, so we draw the graph.

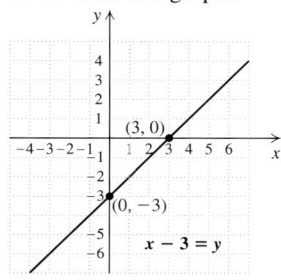

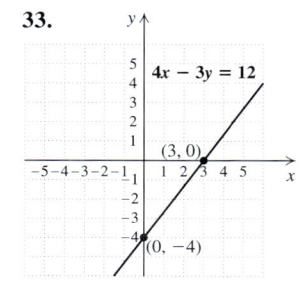

33.

The x-intercept is $(3, 0)$.
To find a third point we replace x with 1 and solve for y.
$$6 \cdot 1 - 2y = 18$$
$$6 - 2y = 18$$
$$-2y = 12$$
$$y = -6$$
The point $(1, -6)$ appears to line up with the intercepts, so we draw the graph.

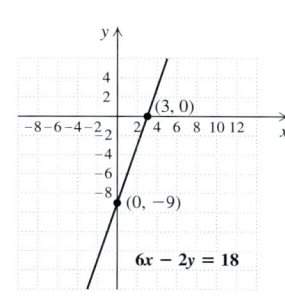

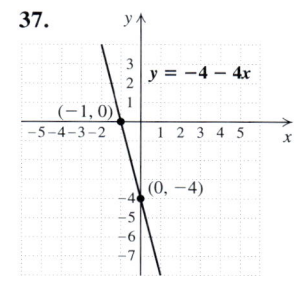

37.

34. $3x - 2 = y$
Find the y-intercept:
$$-2 = y \quad \text{Replacing } x \text{ with } 0$$
The y-intercept is $(0, -2)$.
To find the x-intercept, let $y = 0$.
$$3x - 2 = 0$$
$$3x = 2$$
$$x = \frac{2}{3}$$
The x-intercept is $\left(\frac{2}{3}, 0\right)$.
To find a third point we replace x with 2 and solve for y.
$$3 \cdot 2 - 2 = y$$
$$6 - 2 = y$$
$$4 = y$$
The point $(2, 4)$ appears to line up with the intercepts, so we draw the graph.

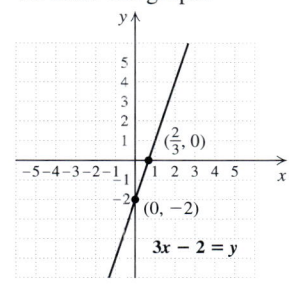

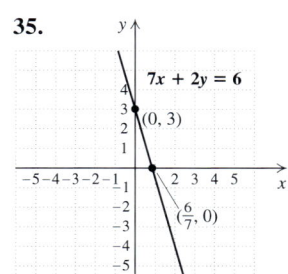

35.

38. $3x + 4y = 5$
Find the y-intercept:
$$4y = 5 \quad \text{Replacing } x \text{ with } 0$$
$$y = \frac{5}{4}$$
The y-intercept is $\left(0, \frac{5}{4}\right)$.
Find the x-intercept:
$$3x = 5 \quad \text{Replacing } y \text{ with } 0$$
$$x = \frac{5}{3}$$
The x-intercept is $\left(\frac{5}{3}, 0\right)$.
To find a third point we replace x with 3 and solve for y.
$$3 \cdot 3 + 4y = 5$$
$$9 + 4y = 5$$
$$4y = -4$$
$$y = -1$$
The point $(3, -1)$ appears to line up with the intercepts, so we draw the graph.

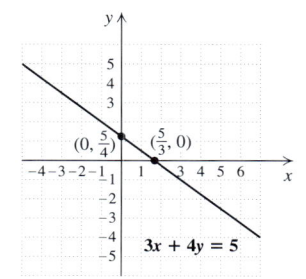

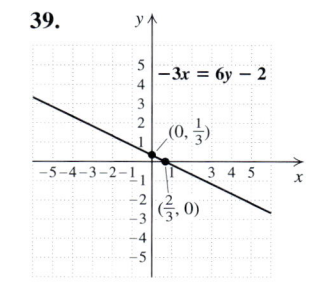

36. $6x - 2y = 18$
Find the y-intercept:
$$-2y = 18 \quad \text{Replacing } x \text{ with } 0$$
$$y = -9$$
The y-intercept is $(0, -9)$.
Find the x-intercept:
$$6x = 18 \quad \text{Replacing } y \text{ with } 0$$
$$x = 3$$

39.

40. $y = -3 - 3x$
Find the y-intercept:
$\quad y = -3 \quad$ Replacing x with 0
The y-intercept is $(0, -3)$.
To find the x-intercept, let $y = 0$.
$$0 = -3 - 3x$$
$$3x = -3$$
$$x = -1$$
The x-intercept is $(-1, 0)$.
To find a third point we replace x with -2 and solve for y.
$$y = -3 - 3 \cdot (-2)$$
$$y = -3 + 6$$
$$y = 3$$
The point $(-2, 3)$ appears to line up with the intercepts, so we draw the graph.

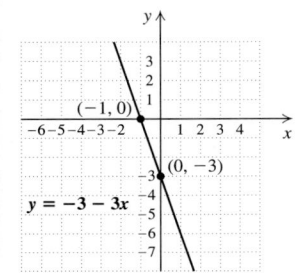

41.

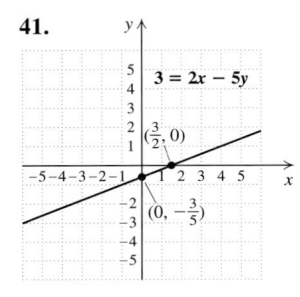

42. $-4x = 8y - 5$
To find the y-intercept, let $x = 0$.
$$-4 \cdot 0 = 8y - 5$$
$$0 = 8y - 5$$
$$5 = 8y$$
$$\frac{5}{8} = y$$
The y-intercept is $\left(0, \frac{5}{8}\right)$.
Find the x-intercept:
$\quad -4x = -5 \quad$ Replacing y with 0
$\quad x = \frac{5}{4}$
The x-intercept is $\left(\frac{5}{4}, 0\right)$.
To find a third point we replace x with -5 and solve for y.
$$-4(-5) = 8y - 5$$
$$20 = 8y - 5$$
$$25 = 8y$$
$$\frac{25}{8} = y$$

The point $\left(-5, \frac{25}{8}\right)$ appears to line up with the intercepts, so we draw the graph.

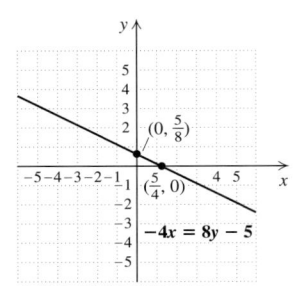

43.

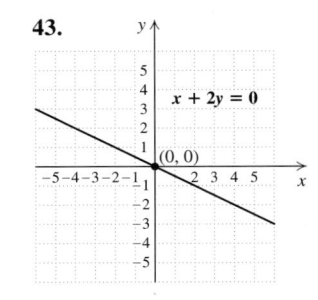

44. $y - 3x = 0$
Find the y-intercept:
$\quad y = 0 \quad$ Replacing x with 0
The y-intercept is $(0, 0)$. Note that this is also the x-intercept.
In order to graph the line, we will find a second point.
When $x = 1$, $y - 3 \cdot 1 = 0$
$$y - 3 = 0$$
$$y = 3$$
To find a third point we replace $x = -1$ and solve for y.
$$y - 3(-1) = 0$$
$$y + 3 = 0$$
$$y = -3$$
The point $(-1, -3)$ appears to line up with the other two points, so we draw the graph.

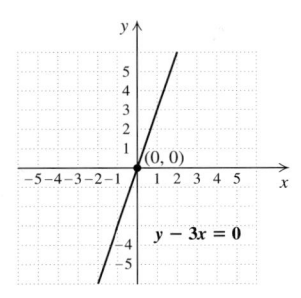

45. $y = -1$ **46.** $x = -1$ **47.** $x = 4$ **48.** $y = -5$
49. $y = 0$ **50.** $x = 0$
51. **52.**

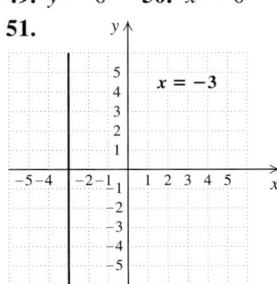

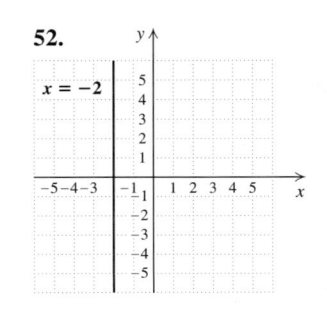

53. [graph: y = 2]

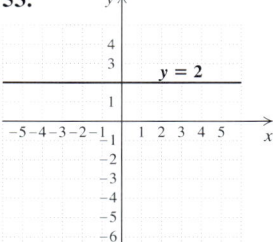

54. [graph: y = 4]

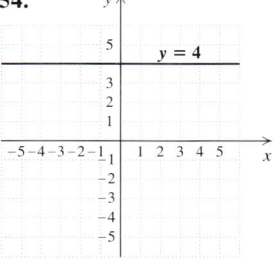

55. [graph: x = 7]

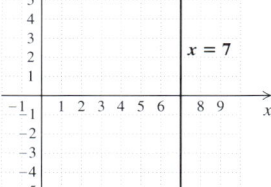

56. [graph: x = 3]

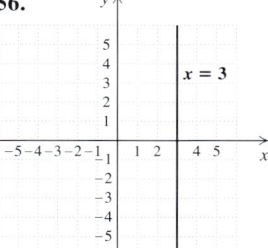

57. [graph: x = 0]

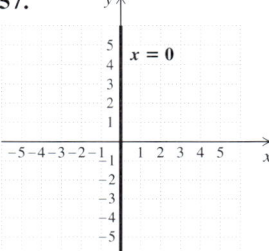

58. [graph: y = 0]

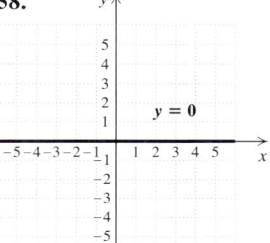

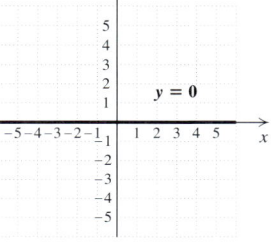

59. [graph: $y = \frac{5}{2}$]

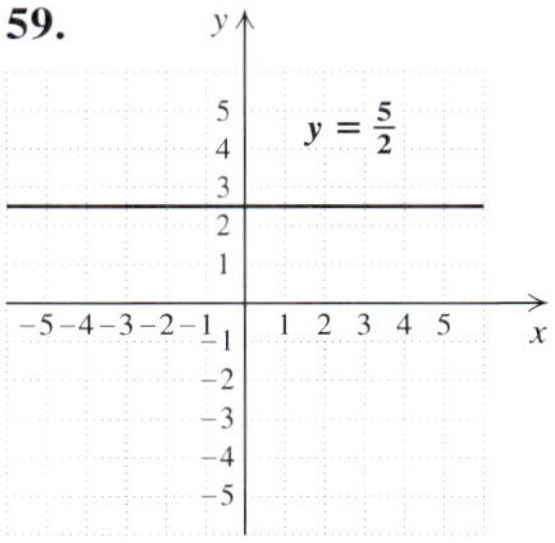

60. [graph: $x = -\frac{3}{2}$]

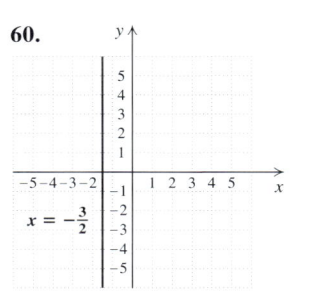

61. [graph: $-3y = -15$]

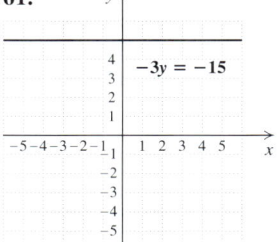

62. $12y = 45$
$y = \dfrac{45}{12} = \dfrac{15}{4}$
[graph: 12y = 45]

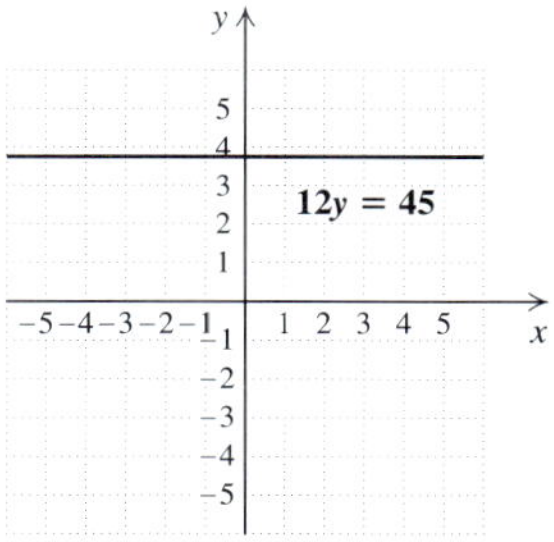

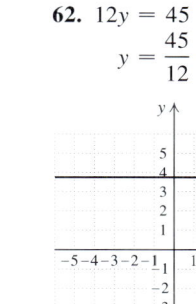

63. [graph: 4x + 3 = 0]

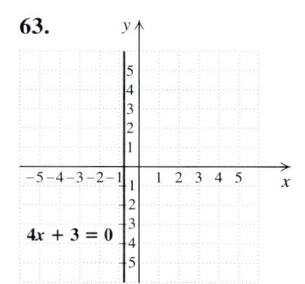

64. $-3x + 12 = 0$
$-3x = -12$
$x = 4$
[graph: −3x + 12 = 0]

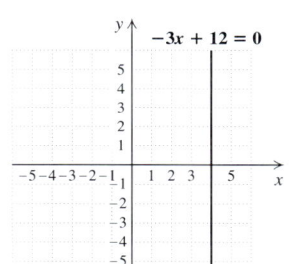

65. [graph: 18 − 3y = 0]

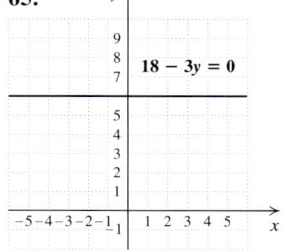

66. $63 + 7y = 0$
$7y = -63$
$y = -9$
[graph: 63 + 7y = 0]

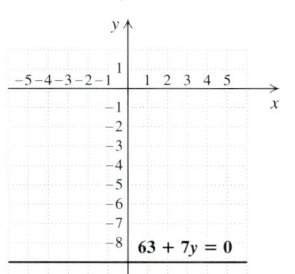

67. $2 \cdot 7 \cdot 7$ **68.** $240 = 8 \cdot 30 = 2 \cdot 4 \cdot 2 \cdot 15 = 2 \cdot 2 \cdot 2 \cdot 2 \cdot 3 \cdot 5$ **69.** $5 \cdot 5 \cdot 11$

70. $\dfrac{36}{90} = \dfrac{2 \cdot 2 \cdot 3 \cdot 3}{2 \cdot 3 \cdot 3 \cdot 5} = \dfrac{\cancel{2} \cdot 2 \cdot \cancel{3} \cdot \cancel{3}}{\cancel{2} \cdot \cancel{3} \cdot \cancel{3} \cdot 5} = \dfrac{2}{5}$

71. $\dfrac{1}{7}$ **72.** $\dfrac{125}{75} = \dfrac{5 \cdot \cancel{25}}{3 \cdot \cancel{25}} = \dfrac{5}{3}$

73. ◆ $A = 0$. If the line is horizontal, then regardless of the value of x, the value of y remains constant. Thus, Ax must be 0 and, hence, $A = 0$.

74. ◆ Any ordered pair (7, y) is a solution of $x = 7$. Thus, all points on the graph are 7 units to the right of the x-axis, so they lie on a vertical line. **75.** ◆ Knowing the coordinates of the t-intercept would tell the business in how many years from the date of purchase the copier has lost all its value. **76.** ◆ No; when $B = 0$ there is no y-term.
77. $y = 0$ **78.** $x = 0$ **79.** (6, 6)   **80.** Since the x-coordinate of the point of intersection must be -3 and y must equal x, the point of intersection is $(-3, -3)$.
81. $5x + 2y = 10$
82. The y-intercept is (0, 5), so we have $y = mx + 5$. Another point on the line is $(-3, 0)$ so we have
$$0 = m(-3) + 5$$
$$-5 = -3m$$
$$\frac{5}{3} = m$$
The equation is $y = \frac{5}{3}x + 5$, or $5x - 3y = -15$.
83. $y = -4$ **84.** Substitute 2 for x and 0 for y.
$$0 = m \cdot 2 + 6$$
$$-6 = 2m$$
$$-3 = m$$
85. 12   **86.** Substitute 0 for x and -8 for y.
$$4 \cdot 0 = C - 3(-8)$$
$$0 = C + 24$$
$$-24 = C$$
87. ◆ $Ax + D = C$
$Ax = C - D$
$x = \dfrac{C - D}{A}$, or
$x = a$ where $a = \dfrac{C - D}{A}$
and
$By + D = C$
$By = C - D$
$y = \dfrac{C - D}{B}$, or
$y = b$ where $b = \dfrac{C - D}{B}$
so the graphs will be parallel to an axis.
88. Find the y-intercept:
$2y = 50$ Covering the x-term
$y = 25$
The y-intercept is (0, 25).
Find the x-intercept:
$3x = 50$ Covering the y-term
$x = \dfrac{50}{3} = 16.\overline{6}$
The x-intercept is $\left(\dfrac{50}{3}, 0\right)$. or $(16.\overline{6}, 0)$.

89. $(0, -11.\overline{428571})$; (40, 0)
90. From the equation we see that the y-intercept is $(0, -9)$. To find the x-intercept, let $y = 0$.
$$0 = 0.2x - 9$$
$$9 = 0.2x$$
$$45 = x$$
The x-intercept is (45, 0).
91. $(0, -15)$; $(11.\overline{538461}, 0)$
92. Find the y-intercept.
$-20y = 1$ Covering the x-term
$y = -\dfrac{1}{20}$, or -0.05
The y-intercept is $\left(0, -\dfrac{1}{20}\right)$, or $(0, -0.05)$.
Find the x-intercept:
$25x = 1$ Covering the y-term
$x = \dfrac{1}{25}$, or 0.04
The x-intercept is $\left(\dfrac{1}{25}, 0\right)$, or (0.04, 0).
93. (0, 0.04); (0.02, 0)

TECHNOLOGY CONNECTION 3.4, P. 148

1. Answers may vary. (12.765957, 54.418085), (51.06383, 67.82234), (85.106383, 79.737234), (123.40426, 93.141489), (157.44681, 105.05638)   **2.** 84.95   **3.** 84.95; 154.95
4. The table becomes blank; 138.5

EXERCISE SET 3.4, PP. 153–156

1. Let x represent one number and y the other; $x + y = 19$
2. Let x and y represent the two numbers; $x + y = 39$.
3. Let x represent one number and y the other; $x + 2y = 65$
4. Let x and y represent the two numbers; $x + 2y = 93$.
5. Let a = the amount Justine paid and n = the amount the necklaces were worth; $a = 2n$ **6.** Let p = the total profit and n = the number of flower pots sold; $p = 5.25n$.
7. Let d = the distance Jason travels, in meters, and t = the number of minutes he jogs; $d = 220t$
8. Let m = the number of careless mistakes Eva makes and b = the number her brother makes; $m = \frac{1}{2}b$.
9. Let f = Frank's age and c = Cecilia's age; $f = \frac{1}{2}c - 3$ **10.** Let l = the amount Lois earns and r = Roberta's weekly salary; $l = 3r - 200$.   **11.** Let c = the cost of the bike rental and t = the number of hours the bike is rented; $c = 8 + 2t$ **12.** Let c = the total cost of the photocopies, in dollars, and p = the number of pages copied; $c = 3 + 0.05p$.   **13.** Let d = the depth of the lake, in feet, and w = the number of weeks; $d = 94 - 3w$
14. Let s = the politician's savings, in dollars, and w = the number of weeks; $s = 75{,}000 - 8000w$.

15. About 200 miles

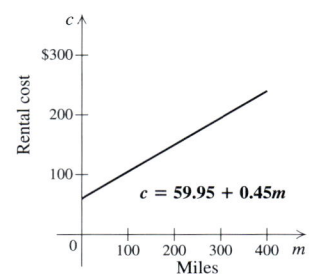

Graph: $c = 59.95 + 0.45m$, Rental cost vs. Miles

16. Let $m =$ the number of miles driven and $c =$ the cost of the rental, in dollars; $c = 39.95 + 0.55m$. We graph the equation.

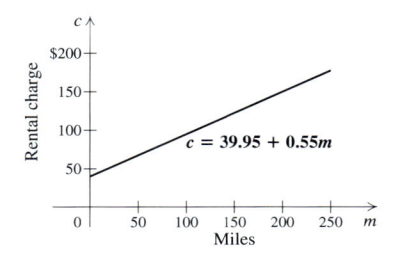

Graph: $c = 39.95 + 0.55m$, Rental charge vs. Miles

From the graph we see that the truck can be driven about 150 miles on a budget of $120.

17. $3\frac{1}{2}$ months

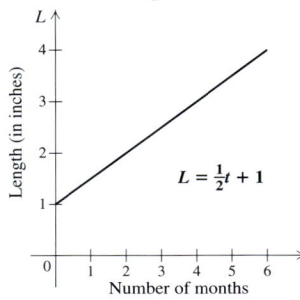

Graph: $L = \frac{1}{2}t + 1$, Length (in inches) vs. Number of months

18.

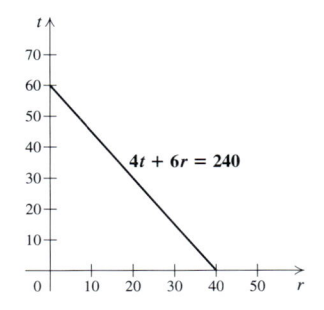

Graph: $4t + 6r = 240$

From the graph we see that when 23 pounds of turkey was bought, then about 25 pounds of roast beef was bought.

19. About $7400

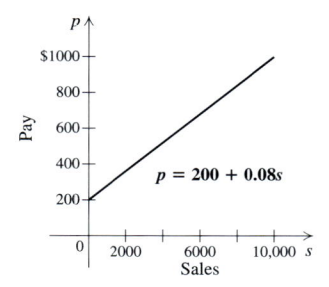

Graph: $p = 200 + 0.08s$, Pay vs. Sales

20. Let $s =$ the weekly sales, in dollars, and $w =$ the weekly wages, in dollars; $w = 200 + 0.04s$. We graph the equation.

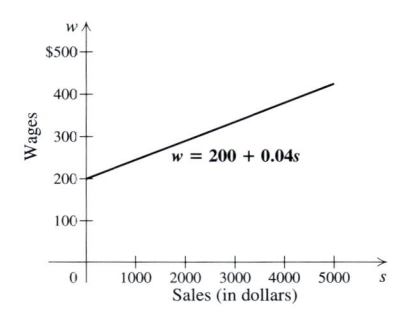

Graph: $w = 200 + 0.04s$, Wages vs. Sales (in dollars)

From the graph we see that the sales are about $4400 when a week's pay is $375.

21. 6 yr

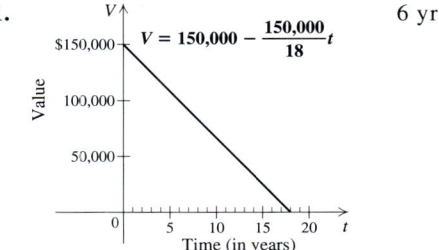

Graph: $V = 150{,}000 - \frac{150{,}000}{18}t$, Value vs. Time (in years)

22. Let $n =$ the number of people and $p =$ the number of pounds of cheese; $p = 3 + \frac{2}{9}(n - 10)$, or $p = \frac{2}{9}n + \frac{7}{9}$. We graph the equation.

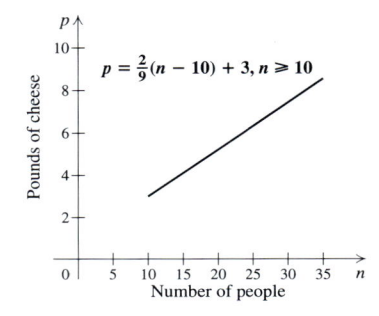

Graph: $p = \frac{2}{9}(n - 10) + 3, n \geq 10$, Pounds of cheese vs. Number of people

From the graph we see that a party of about 33 can be accommodated with 8 pounds of cheese.

23.

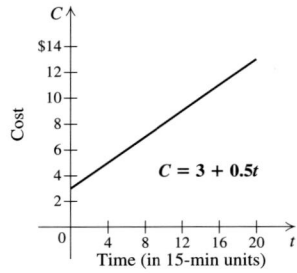

135 min, or 2 hr 15 min

Graph of $C = 3 + 0.5t$, Cost vs. Time (in 15-min units).

24. Let $t =$ the number of 15-min units of time and $c =$ the charge for the road call; $c = 35 + 10t$. We graph the equation.

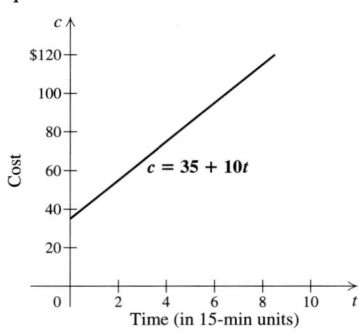

Graph of $c = 35 + 10t$, Cost vs. Time (in 15-min units).

From the graph we see that t is about 7 when c is $105, so if the charge for a road call is $105, the length of the road call is 7 15-min units of time, or $7 \cdot 15 = 105$ min, or 1 hr, 45 min.

25.

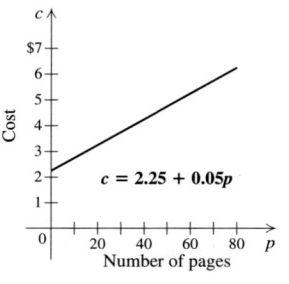

About 60 pages

Graph of $c = 2.25 + 0.05p$, Cost vs. Number of pages.

26. Let $w =$ the weight of the package, in pounds, and $c =$ the delivery cost, in dollars. Note that w is a number from 10 to 50. Then $c = 46.25 + 1.25(w - 10)$, or $c = 1.25w + 33.75$. We graph the equation.

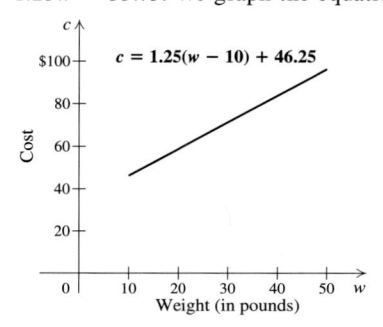

Graph of $c = 1.25(w - 10) + 46.25$, Cost vs. Weight (in pounds).

From the graph we see that a package weighing about 11 lb has a delivery charge of $60 and a package weighing about 15 lb has a delivery charge of $65. Thus, about $15 - 11$, or 4 lb, was added to the package if the delivery charge jumped from $60 to $65.

27.

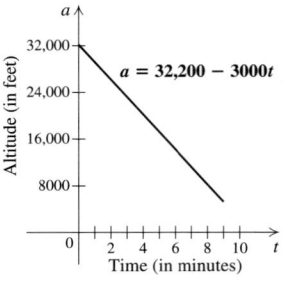

About 9 min

Graph of $a = 32{,}200 - 3000t$, Altitude (in feet) vs. Time (in minutes).

28. Let $t =$ the time of the descent, in minutes, and $a =$ the altitude, in feet; then $a = 6000 - 150t$. We graph the equation.

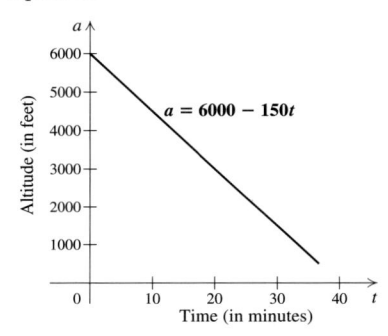

Graph of $a = 6000 - 150t$, Altitude (in feet) vs. Time (in minutes).

From the graph we see that the altitude is 500 ft about 37 minutes into the descent.

29. $\frac{1}{2}$ inch per month   **30.** The equation in Exercise 16 is $c = 39.95 + 0.55m$. Two points on the graph are $(0, 39.95)$ and $(100, 94.95)$. A cost change of $94.95 - $39.95, or $55, corresponds to a mileage change of $100 - 0$, or 100 miles. The cost of the rental is changing at the rate of $\dfrac{\$55}{100 \text{ miles}}$, or $0.55 per mile.   **31.** 1 hr, 5 min; 41 km; about 38 km/h
32. The 9:30 A.M. train from Montereau reached Paris at 11:45 A.M., so the trip took 2 hr, 15 min. The distance from Montereau to Paris is $315 - 235 = 80$ km.

$$\text{Average speed} = \frac{80 \text{ km}}{2.25 \text{ hr}} \approx 36 \text{ km/h}$$

33. 4 hr, 45 min; 235 km; about 49 km/h
34. The 6:55 P.M. train from Laroche reached Paris at 10:35 P.M., so the trip took 3 hr, 40 min. The distance from Laroche to Paris is $315 - 159 = 156$ km.

$$3 \text{ hr, } 40 \text{ min} = 3\frac{2}{3} \text{ hr} = \frac{11}{3} \text{ hr}$$

$$\text{Average speed} = \frac{156 \text{ km}}{11/3 \text{ hr}} \approx 43 \text{ km/h}$$

35. About 41 km/h; 25 min
36. The 12:20 P.M. train from Paris reached Dijon at 10:25 P.M., so the trip took 10 hr, 5 min. The distance from Paris to Dijon is 315 km.

$$10 \text{ hr, } 5 \text{ min} = 10\frac{1}{12} \text{ hr} = \frac{121}{12} \text{ hr}$$

$$\text{Average speed} = \frac{315 \text{ km}}{121/12 \text{ hr}} \approx 31 \text{ km/h}$$

There are 5 horizontal segments on the line representing the 12:20 P.M. train from Paris to Dijon. (They occur at Montereau, Laroche, Tonnerre, Nuits-s-Ravière, and Venarey-les-Laumes.) Thus, the train makes 5 stops.
37. The 7:15 P.M. train. The line representing that train is the steepest.
38. Of all the segments representing trains from Tonnerre to Paris, the one with the steepest slant represents the 2:05 P.M. train. Thus, the 2:05 P.M. train is the quickest way to travel from Tonnerre to Paris.
39. 3:05 P.M.
40. The fastest train is represented by the segment with the steepest slant. We see that the fastest train from Montereau to Laroche leaves Montereau at 8:40 P.M. and arrives in Laroche at 10:00 P.M. Thus, the trip takes 1 hr, 20 min, or $1\frac{1}{3}$ hr. The distance from Montereau to Laroche is $235 - 159 = 76$ km.

$$\text{Average speed} = \frac{76 \text{ km}}{4/3 \text{ hr}} = 57 \text{ km/h}$$

41. $t = \dfrac{s - d}{v}$
42. $3(x - 4) + 7 = -2x + 6(x - 5)$
$3x - 12 + 7 = -2x + 6x - 30$
$3x - 5 = 4x - 30$
$25 = x$
43. $8x + 10y - 6z$
44. $3x + 18y - 6z = 3(x + 6y - 2z)$
45. ◆ A telephone company provides local metered service for a monthly charge of $5 plus 12¢ per minute for each local call. Draw a graph that can be used to estimate the monthly charge. Use the horizontal axis for time and the vertical axis for cost. Then use the graph to estimate the number of minutes of local calls made when the monthly charge is $11.   **46.** ◆ Yes; the use of a solid line indicates that fractional parts of people can be considered. A collection of points gives a more realistic picture than a solid line in situations like those in Exercise 22.
47. ◆ The graph in Example 2 would be shifted down 25 units.   **48.** ◆ The graph would slant less steeply.

49. Answers may vary.

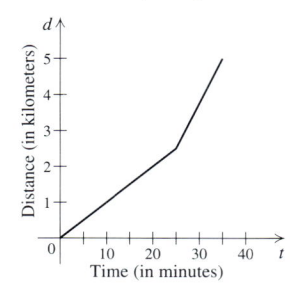

[Graph: Distance (in kilometers) vs. Time (in minutes), showing a piecewise linear increasing graph from 0 to about 40 minutes reaching about 5 km]

50. Let $t =$ flight time and $a =$ altitude. While the plane is climbing at a rate of 6500 ft/min, the equation $a = 6500t$ describes the situation. Solving $34{,}000 = 6500t$, we find that the cruising altitude of 34,000 ft is reached after about 5.23 min. Thus we graph $a = 6500t$ for $0 \le t \le 5.23$.

The plane cruises at 34,000 ft for 3 min, so we graph $a = 34{,}000$ for $5.23 < t \le 8.23$. After 8.23 min the plane descends at a rate of 3500 ft/min and lands. The equation $a = 34{,}000 - 3500(t - 8.23)$, or $a = -3500t + 62{,}805$, describes this situation. Solving $0 = -3500t + 62{,}805$, we find that the plane lands after about 17.94 min. Thus we graph $a = -3500t + 62{,}805$ for $8.23 < t \le 17.94$. The entire graph is show below.

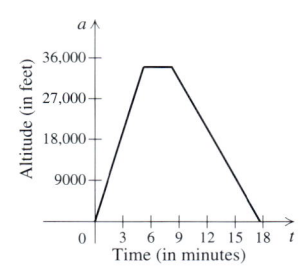

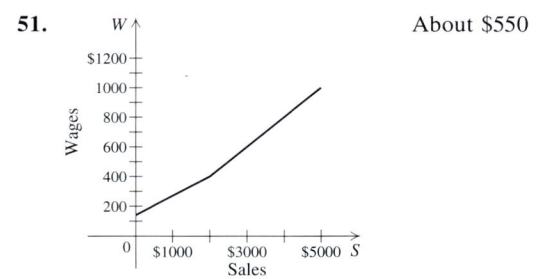

[Graph: Altitude (in feet) vs. Time (in minutes), trapezoidal shape peaking at 34,000 ft]

51.

[Graph: Wages vs. Sales, increasing piecewise linear from about $200 at $0 to about $1000 at $5000]

About $550

52. Let $s =$ Paul's weekly salary. Then we have two equations:
$$p = 2s - 150,$$
$$s = 70 + \frac{1}{2}j, \text{ or } j = 2s - 140$$

We make a table of values by choosing values for s and finding the corresponding values of j and p. Then we draw the graph by plotting the points (j, p) and drawing a line through them.

s	j	p
100	60	50
150	160	150
250	360	350

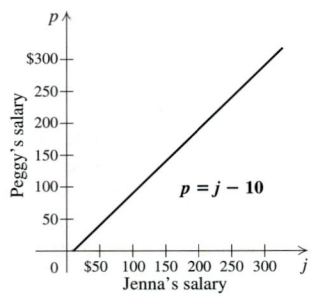

From the graph it appears that Peggy's salary is related to Jenna's by the equation $p = j - 10$. To verify this we can substitute $70 + \frac{1}{2}j$ for s in the first equation and simplify.

$p = 2s - 150$
$p = 2\left(70 + \frac{1}{2}j\right) - 150$ Substituting
$p = 140 + j - 150$
$p = j - 10$

53. ◆ The next train, leaving 20 minutes later, travels substantially slower.   **54.** ◆ Yes; the next train that leaves takes more than four hours longer to reach Dijon.
55.   **56.**

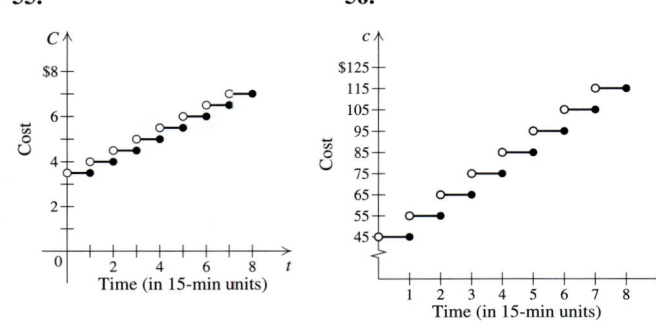

57. $4315.14   **58.** Graph $y = 38 + 2.35x$. Use Zoom and Trace to find the x-coordinate that corresponds to the y-coordinate 623.15. We find that 249 shirts were printed.

EXERCISE SET 3.5, PP. 164–168

1. 3 million people per year
2. Use (2, 30) and (6, 90).

$$\text{Rate} = \frac{90 - 30}{6 - 2}$$
$$= \frac{60}{4}$$
$$= 15 \text{ calories per minute}$$

3. Going down at $10,000,000,000 per year
4. Use (1991, 800) and (1995, 1200).   **5.** $800 per year

$$\text{Rate} = \frac{1200 - 800}{1995 - 1991}$$
$$= \frac{400}{4}$$
$$= \$100 \text{ per year}$$

6. Use (1982, 2240) and (1987, 2085).

$$\text{Rate} = \frac{2085 - 2240}{1987 - 1982}$$
$$= \frac{-155}{5}$$
$$= -31 \text{ thousand farms per year}$$

The number of U.S. farms is going down at the rate of 31 thousand farms per year.
7. $\frac{3}{4}$
8. Two points on the line are (0, 1) and (3, 3).   **9.** $\frac{3}{2}$

$$m = \frac{3 - 1}{3 - 0} = \frac{2}{3}$$

10. Two points on the line are (0, 2) and (3, 3).   **11.** $\frac{1}{3}$

$$m = \frac{3 - 2}{3 - 0} = \frac{1}{3}$$

12. Two points on the line are $(-3, 0)$ and $(-2, 3)$.

$$m = \frac{3 - 0}{-2 - (-3)} = \frac{3}{1} = 3$$

13. -1
14. Two points on the line are (0, 3) and (4, 1).   **15.** $-\frac{3}{2}$

$$m = \frac{1 - 3}{4 - 0} = \frac{-2}{4} = -\frac{1}{2}$$

16. Two points on the line are (0, 2) and (3, 1).

$$m = \frac{1 - 2}{3 - 0} = -\frac{1}{3}$$

17. -2 **18.** This is a horizontal line, so the slope is 0.
19. Undefined
20. Two points on the line are $(-2, 3)$ and $(2, 2)$.

$$m = \frac{2 - 3}{2 - (-2)} = -\frac{1}{4}$$

21. $-\frac{4}{3}$ **22.** $m = \frac{-3 - 1}{-2 - 4} = \frac{-4}{-6} = \frac{2}{3}$ **23.** $-\frac{4}{5}$

24. $m = \frac{-3 - 2}{2 - (-4)} = \frac{-5}{6} = -\frac{5}{6}$ **25.** 7

26. $m = \frac{2 - 0}{6 - 3} = \frac{2}{3}$ **27.** $-\frac{2}{3}$ **28.** $m = \frac{7 - 9}{4 - 0} = \frac{-2}{4} = -\frac{1}{2}$ **29.** $-\frac{1}{2}$ **30.** $m = \frac{-7 - 4}{6 - (-2)} = \frac{-11}{8} = -\frac{11}{8}$

31. 0 **32.** $m = \dfrac{-3 - (-3)}{10 - 8} = \dfrac{0}{2} = 0$ **33.** Undefined

34. $m = \dfrac{4 - 3}{-10 - (-10)} = \dfrac{1}{0}$, undefined **35.** Undefined

36. Vertical line; undefined **37.** 0
38. Horizontal line; $m = 0$ **39.** Undefined
40. Vertical line; undefined **41.** 0
42. Horizontal line; $m = 0$ **43.** 8%
44. $m = \dfrac{54}{1080} = 0.05$ **45.** $8.\overline{3}\%$ **46.** $m = \dfrac{315}{4500} = 0.07 = 7\%$ **47.** $\dfrac{12}{41}$, or about 29% **48.** $m = \dfrac{0.4}{5} = 0.08 = 8\%$ **49.** About 29% **50.** $m = \dfrac{7}{11} \approx 0.64 \approx 64\%$
51. -116 **52.** $4(-3)^4 = 4(81) = 324$ **53.** -1
54. $4x - 7 = 9x$ **55.** $\dfrac{1}{8}$
$-7 = 5x$
$-\dfrac{7}{5} = x$

56. $4 - 7x \leq -17$
$-7x \leq -21$
$x \geq 3$
The solution set is $\{x \mid x \geq 3\}$.
57. ◆ The line with a slope of -3 is steeper, because $|-3| > |2|$. **58.** ◆ $\dfrac{y_2 - y_1}{x_2 - x_1} = \dfrac{-(y_1 - y_2)}{-(x_1 - x_2)} = \dfrac{y_1 - y_2}{x_1 - x_2}$
59. ◆ The scales on the vertical axes are different.
60. ◆ Yes; at a rate of $100 per year, the tuition and fees are increasing at a rate greater than 3% per year.
61. $\left\{m \mid -\dfrac{6}{5} \leq m \leq 0\right\}$ **62.** If the line never enters the second quadrant and is nonvertical, then two points on the line are $(3, 4)$ and $(a, 0)$, $0 \leq a < 3$. The slope is of the form $m = \dfrac{4 - 0}{3 - a}$, or $\dfrac{4}{3 - a}$, $0 \leq a < 3$, so $m \geq \dfrac{4}{3}$. Then the numbers the line could have for its slope are $\left\{m \mid m \geq \dfrac{4}{3}\right\}$. **63.** $\dfrac{18 - x}{x}$
64. Note that 40 min $= \dfrac{2}{3}$ hr.
Rate $= \dfrac{64 - 46 \text{ candles}}{\dfrac{2}{3} \text{ hr}}$
$= \dfrac{18 \text{ candles}}{2/3 \text{ hr}}$
$= \dfrac{18}{1} \cdot \dfrac{3}{2}$ candles per hour
$= 27$ candles per hour
65. $\dfrac{1}{2}$
66. Let $t =$ the number of units each tick mark on the horizontal axis represents. Note that the graph drops 1 unit for every 6 tick marks of horizontal change. Then we have:
$\dfrac{-1}{6t} = -\dfrac{2}{3}$

$-1 = -4t$
$\dfrac{1}{4} = t$
Each tick mark on the horizontal axis represents $\dfrac{1}{4}$ unit.
67. 3.6 bushels per hour
68. ◆

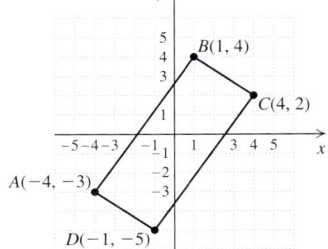

We find the slope of each side of the quadrilateral.

For side $\overline{AB}$, $m = \dfrac{4 - (-3)}{1 - (-4)} = \dfrac{7}{5}$.

For side $\overline{BC}$, $m = \dfrac{2 - 4}{4 - 1} = -\dfrac{2}{3}$.

For side $\overline{CD}$, $m = \dfrac{2 - (-5)}{4 - (-1)} = \dfrac{7}{5}$.

For side $\overline{DA}$, $m = \dfrac{-5 - (-3)}{-1 - (-4)} = -\dfrac{2}{3}$.

Since the opposite sides of the quadrilateral have the same slopes but lie on different lines, the lines on which they lie never intersect so they are parallel. Thus the quadrilateral is a parallelogram.
69. ◆

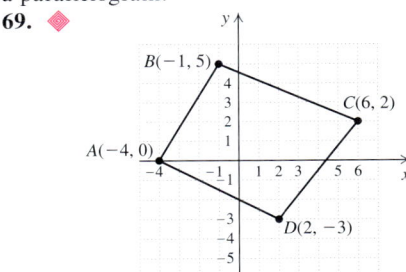

For side $\overline{AB}$, $m = \dfrac{0 - 5}{-4 - (-1)} = \dfrac{5}{3}$.

For side $\overline{CD}$, $m = \dfrac{2 - (-3)}{6 - 2} = \dfrac{5}{4}$.

Since these opposite sides have different slopes, they lie on lines that intersect. Thus they are not parallel, so the given points cannot be vertices of a parallelogram. (Sides $\overline{BC}$ and $\overline{DA}$ can also be used to show this result.)

REVIEW EXERCISES, CHAPTER 3, PP. 169–171

1. [3.1] 48 **2.** [3.1] 405,000
3.–5. [3.1]

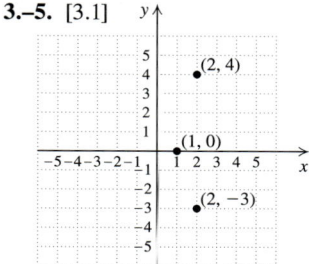

6. [3.1] IV **7.** [3.1] III **8.** [3.1] II **9.** [3.1] $(-5, -1)$
10. [3.1] $(-2, 5)$ **11.** [3.1] $(3, 0)$ **12.** [3.2] No
13. [3.2] Yes
14. [3.2]

$$\begin{array}{c|c} 2x - y = 3 \\ \hline 2 \cdot 0 - (-3) \;?\; 3 \\ 0 + 3 \\ 3 \;\big|\; 3 \quad \text{TRUE} \end{array} \qquad \begin{array}{c|c} 2x - y = 3 \\ \hline 2 \cdot 2 - 1 \;?\; 3 \\ 4 - 1 \\ 3 \;\big|\; 3 \quad \text{TRUE} \end{array}$$

$(-1, -5)$; answers may vary

15. [3.2] **16.** [3.2]

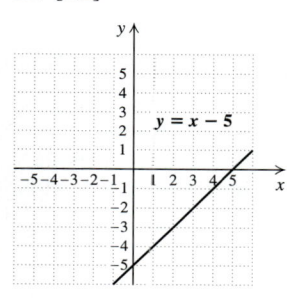

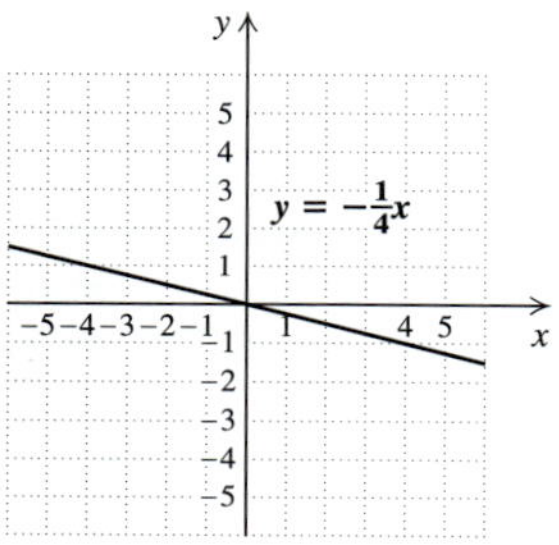

17. [3.2] 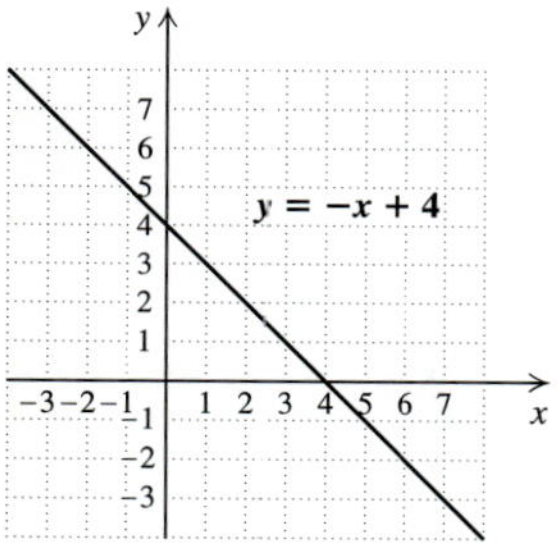 **18.** [3.2]

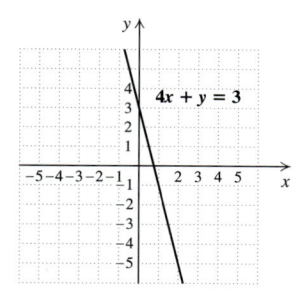

19. [3.3] 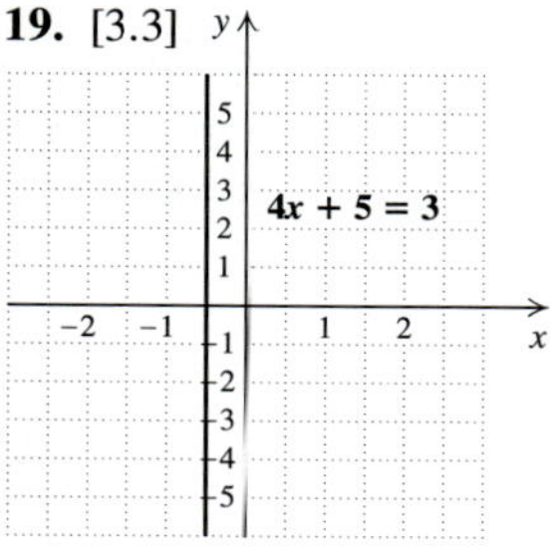 **20.** [3.3]

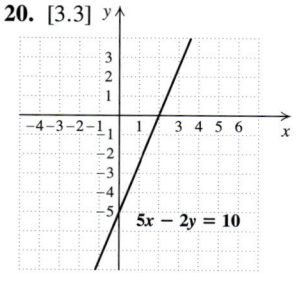

21. [3.3] x-intercept: $(6, 0)$; y-intercept: $(0, -3)$
22. [3.3] x-intercept: $\left(-\frac{9}{2}, 0\right)$; y-intercept: $\left(0, \frac{9}{4}\right)$
23. [3.3] $y = 4$ **24.** [3.4] Let w represent the number of wrenches and s the number of screwdrivers; $w + s = 15$
25. [3.4] About 17.5 cubic feet

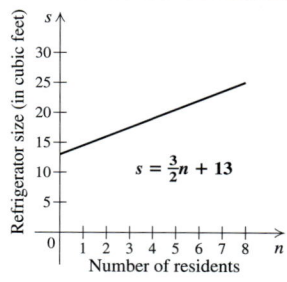

26. [3.4] No **27.** [3.4] 53.6 mph **28.** [3.5] 0
29. [3.5] $\frac{7}{3}$ **30.** [3.5] $-\frac{3}{7}$ **31.** [3.5] $\frac{3}{2}$ **32.** [3.5] 0
33. [3.5] Undefined **34.** [3.5] 2 **35.** [3.5] 7%
36. [1.3] $\frac{19}{24}$ **37.** [1.3] $\frac{13}{17}$ **38.** [2.2] 8
39. [2.3] $m = 2A - n$
40. [3.4] ◆ A business might use a graph to quickly look up prices (as in the rental truck example) or to plot how total sales change from year to year. Many other applications exist.
41. [3.2] ◆ The y-intercept is the point at which the graph crosses the y-axis. Since a point on the y-axis is neither left nor right of the origin, the first or x-coordinate of the point is 0.
42. [3.2] -1 **43.** [3.2] 19
44. [3.1] Area: 45 sq units; perimeter: 28 units
45. [3.2] $(0, 4), (1, 3), (-1, 3)$; answers may vary

TEST: CHAPTER 3, PP. 171–172

1. [3.1] About 12,000 **2.** [3.1] About 22% **3.** [3.1] II
4. [3.1] III **5.** [3.1] $(3, 4)$ **6.** [3.1] $(0, -4)$
7. [3.2]

$$\begin{array}{c|c} y - 2x = 5 \\ \hline 11 - 2 \cdot 3 \;?\; 5 \\ 11 - 6 \\ 5 \;\big|\; 5 \quad \text{TRUE} \end{array} \qquad \begin{array}{c|c} y - 2x = 5 \\ \hline 3 - 2(-1) \;?\; 5 \\ 3 + 2 \\ 5 \;\big|\; 5 \quad \text{TRUE} \end{array}$$

$(0, 5)$; answers may vary

8. [3.2] **9.** [3.3]

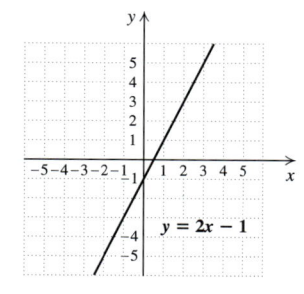

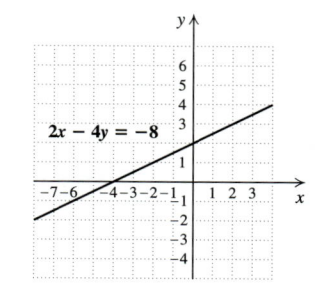

10. [3.3]

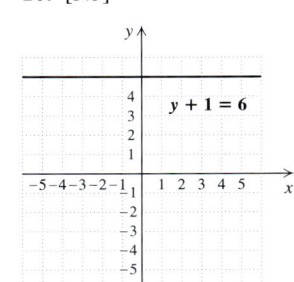

11. [3.2]

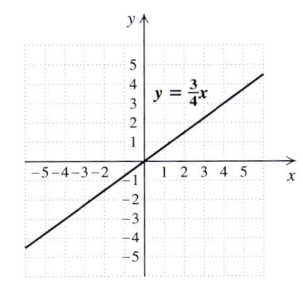

12. [3.2]

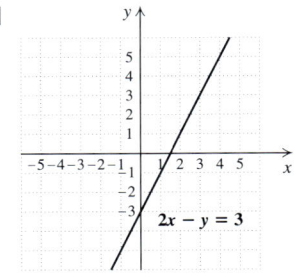

13. [3.3] x-intercept: $(6, 0)$; y-intercept: $(0, -10)$
14. [3.3] x-intercept: $(10, 0)$; y-intercept: $\left(0, \frac{5}{2}\right)$
15. [3.3] $x = -4$ **16.** [3.4] Let $g = $ Greta's earnings and $a = $ Alice's salary; $g = 2a - 150$
17. [3.4]

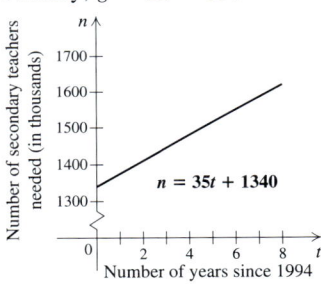

About 1600 thousand, or 1,600,000

18. [3.4]

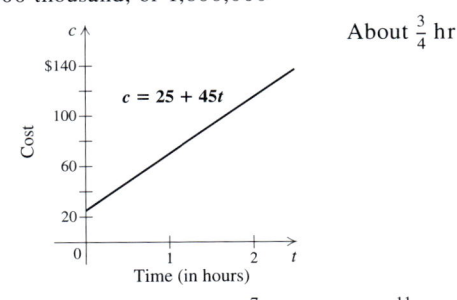

About $\frac{3}{4}$ hr

19. [3.5] Undefined   **20.** [3.5] $\frac{7}{12}$ **21.** [1.3] $\frac{11}{5}$
22. [1.3] $3 \cdot 3 \cdot 3 \cdot 5$ **23.** [2.2] $\frac{4}{15}$
24. [2.3] $x = \dfrac{b}{m + n}$ **25.** [3.1] Area: 25 square units; perimeter: 20 units   **26.** [3.3] $y = 3$

CUMULATIVE REVIEW: 1–3, PP. 173–174

1. [1.1] 15   **2.** [1.2] $12x - 15y + 21$
3. [1.2] $3(5x - 3y + 1)$ **4.** [1.3] $2 \cdot 3 \cdot 7$
5. [1.4] 0.45   **6.** [1.4] 4   **7.** [1.6] $\frac{1}{4}$ **8.** [1.7] -4
9. [1.6] $-x - y$ **10.** [2.4] 0.785   **11.** [1.3] $\frac{11}{60}$
12. [1.5] 2.6   **13.** [1.7] 7.28   **14.** [1.7] $-\frac{5}{12}$
15. [1.8] -2 **16.** [1.8] 27   **17.** [1.8] $-2y - 7$
18. [1.8] $5x + 11$ **19.** [2.1] -1.2 **20.** [2.1] -21
21. [2.2] 9   **22.** [2.1] $-\frac{20}{3}$ **23.** [2.2] 2   **24.** [2.1] $\frac{13}{8}$
25. [2.2] $-\frac{17}{21}$ **26.** [2.2] -17 **27.** [2.2] 2
28. [2.6] $\{x \mid x < 16\}$ **29.** [2.6] $\left\{x \mid x \leq -\frac{11}{8}\right\}$
30. [2.3] $h = \dfrac{A - \pi r^2}{2\pi r}$ **31.** [3.1] IV

32. [2.6] $-1 < x \leq 2$

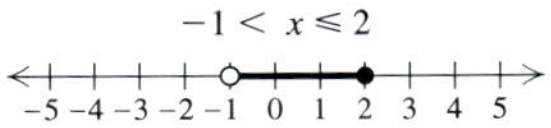

33. [3.3] 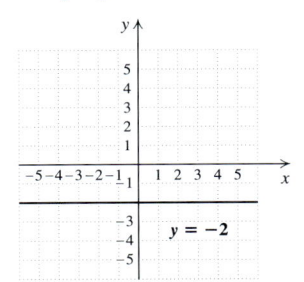   **34.** [3.3]

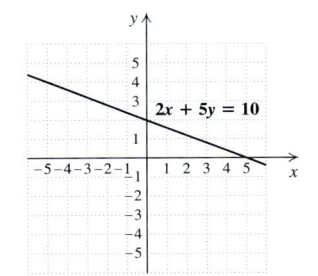

35. [3.2] 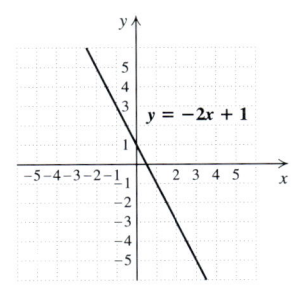   **36.** [3.2]

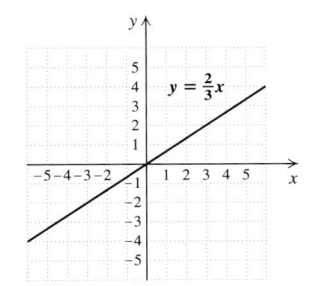

37. [3.3] x-intercept: $(10.5, 0)$; y-intercept: $(0, -3)$
38. [3.3] x-intercept: $\left(-\frac{5}{4}, 0\right)$; y-intercept: $(0, 5)$
39. [2.4] 160 million
40. [2.5] 15.6 million
41. [2.4] $120
42. [2.5] 50 m, 53 m, 40 m
43. [2.7] 8 hr or less
44. [3.4] Let $s = $ the cost of steak, in dollars, and $c = $ the cost of chicken, in dollars; $s = 5 + 3c$

45. [3.4] About $1000

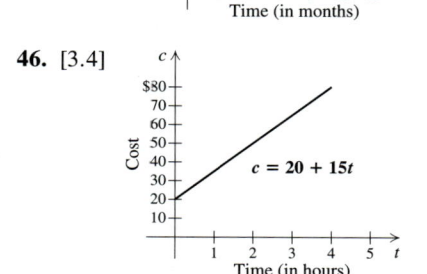

46. [3.4] About 3 hr

47. [3.5] $-\frac{1}{3}$ **48.** [2.4], [2.5] $25,000
49. [1.4], [2.2] 4, -4 **50.** [2.2] All real numbers
51. [2.2] No solution **52.** [2.2] 3
53. [2.2] All real numbers **54.** [2.3] $Q = \dfrac{2-pm}{p}$

CHAPTER 4

EXERCISE SET 4.1, PP. 182–183

1. m^{10} **2.** 3^7 **3.** 8^{14} **4.** n^{23} **5.** x^7 **6.** y^{16}
7. 5^7 **8.** t^{16} **9.** $(3y)^{12}$ **10.** $(2t)^{25}$ **11.** $(5t)^7$
12. $8x$ **13.** a^5b^9 **14.** $(m-3)^9$ **15.** $(x+1)^{12}$
16. $a^{12}b^4$ **17.** r^{12} **18.** s^{11} **19.** x^4y^7
20. $(a^3b)(ab)^4 = (a^3b)(a^4b^4) = a^7b^5$ **21.** 7^3 **22.** 4^4
23. x^{12} **24.** a^8 **25.** y^4 **26.** x **27.** $5a$ **28.** $3m$
29. $(x+y)^5$ **30.** $(a-b)^9$ **31.** $3m^3$ **32.** $5n^4$
33. a^7b^6 **34.** r^7s^7 **35.** m^9n^4 **36.** a^8b^9 **37.** 1
38. $38^0 = 1$ **39.** 5 **40.** $7(1.7)^0 = 7 \cdot 1 = 7$ **41.** 1
42. For any $t \neq 0$, $t^0 = 1$. **43.** 8 **44.** $7^0 - 7^1 = 1 - 7 = -6$ **45.** x^{12} **46.** a^{24} **47.** 5^{16} **48.** 2^{15}
49. m^{35} **50.** n^{18} **51.** a^{75} **52.** a^{75} **53.** $49x^2$
54. $25a^2$ **55.** $-8a^3$ **56.** $-27x^3$ **57.** $16m^6$
58. $25n^8$ **59.** $a^{14}b^7$ **60.** x^9y^{36} **61.** $a^{15}b^{10}$
62. $m^{24}n^{30}$ **63.** $25x^8y^{10}$ **64.** $81a^{20}b^{28}$ **65.** $\dfrac{a^3}{64}$
66. $\dfrac{81}{x^4}$ **67.** $\dfrac{49}{25a^2}$ **68.** $\dfrac{64x^3}{27}$ **69.** $\dfrac{a^{20}}{b^{15}}$ **70.** $\dfrac{x^{35}}{y^{14}}$
71. $\dfrac{y^6}{4}$ **72.** $\dfrac{a^{15}}{27}$ **73.** $\dfrac{x^8y^4}{z^{12}}$ **74.** $\dfrac{x^{15}}{y^{10}z^5}$ **75.** $\dfrac{a^{12}}{16b^{20}}$
76. $\dfrac{x^{20}}{81y^{12}}$ **77.** $\dfrac{8a^6}{27b^{12}}$ **78.** $\dfrac{9x^{10}}{16y^6}$ **79.** $\dfrac{16x^6y^{10}}{9z^{14}}$
80. $\dfrac{125a^{21}}{8b^{15}c^3}$ **81.** $3(s+t+8)$ **82.** $-7(x+2)$ **83.** $5x$

84. Solve: $24 = x \cdot 64$
$0.375 = x$
The answer is 37.5%.

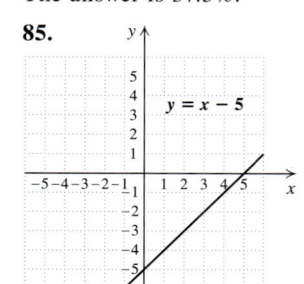

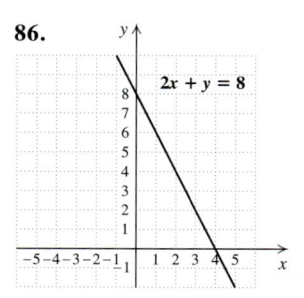

85.

86.

87. ◆ Any number raised to an even power is nonnegative. Any nonnegative number raised to an odd power is nonnegative. Any negative number raised to an odd power is negative. Thus, a must be a negative number, and n must be an odd number. **88.** ◆ $9^0 = 9^{1-1} = \dfrac{9}{9} = 1$
89. ◆ Let $s=$ the length of a side of the smaller square. Then $3s=$ the length of a side of the larger square. The area of the smaller square is s^2, and the area of the larger square is $(3s)^2$, or $9s^2$, so the area of the larger square is 9 times the area of the smaller square. **90.** ◆ Exponents are added when powers with like bases are multiplied. Exponents are multiplied when a power is raised to a power.
91. y^{5x} **92.** $a^{5k} \div a^{3k} = a^{5k-3k} = a^{2k}$ **93.** a^{4t}

94. $\dfrac{\left(\frac{1}{2}\right)^4}{\left(\frac{1}{2}\right)^5} = \dfrac{\left(\frac{1}{2}\right)^4}{\left(\frac{1}{2}\right)^4\left(\frac{1}{2}\right)} = \dfrac{\left(\frac{1}{2}\right)^{4-4}}{\frac{1}{2}} = \dfrac{\left(\frac{1}{2}\right)^0}{\frac{1}{2}} = \dfrac{1}{\frac{1}{2}} = 1 \cdot \dfrac{2}{1} = 2$

95. $>$ **96.** $4^2 < 4^3$ **97.** $<$
98. $4^3 = 64$, $3^4 = 81$, so $4^3 < 3^4$. **99.** $>$
100. $25^8 = (5^2)^8 = 5^{16}$
$125^5 = (5^3)^5 = 5^{15}$
$5^{16} > 5^{15}$, or $25^8 > 125^5$.
101. Let $a=1$; then $(a+5)^2 = 36$, but $a^2 + 5^2 = 26$.
102. Choose any number except 0. For example, let $x=1$.
$3x^2 = 3 \cdot 1^2 = 3 \cdot 1 = 3$, but
$(3x)^2 = (3 \cdot 1)^2 = 3^2 = 9$.
103. Let $a=0$; then $\dfrac{a+7}{7} = 1$, but $a=0$.
104. Choose any number except 0 or 1. For example, let $t=-1$. Then
$\dfrac{t^6}{t^2} = \dfrac{(-1)^6}{(-1)^2} = \dfrac{1}{1} = 1$, but
$t^3 = (-1)^3 = -1$.
105. 19
106. $A = \$10,400(1.085)^5 \approx \$15,638.03$

107. $27,087.01   **108.** ◆ Let s = the width of the smaller cube. Then $2s$ = the width of the larger cube. The volume of the smaller cube is s^3, and the volume of the larger cube is $(2s)^3$, or $8s^3$, so the volume of the larger cube is 8 times the volume of the smaller cube.

TECHNOLOGY CONNECTION 4.2, P. 188
1. 5

EXERCISE SET 4.2, PP. 189–192
1. $3x^4, -7x^3, x, -5$ **2.** $5a^3, 4a^2, -9a, -7$
3. $-t^4, 2t^3, -5t^2, 3$ **4.** $n^5, -4n^3, 2n, -8$
5. Coefficients: 7, −5; degrees: 3, 1
6. $9a^3 - 4a^2$

Term	Coefficient	Degree
$9a^3$	9	3
$-4a^2$	−4	2

7. Coefficients: 9, −3, 4; degrees: 2, 1, 0
8. $7x^4 + 5x - 3$

Term	Coefficient	Degree
$7x^4$	7	4
$5x$	5	1
-3	−3	0

9. Coefficients: 5, 9, 1; degrees: 4, 1, 3
10. $6t^5 - 3t^2 - t$

Term	Coefficient	Degree
$6t^5$	6	5
$-3t^2$	−3	2
$-t$	−1	1

11. Coefficients: 1, −1, 4, −3; degrees: 4, 3, 1, 0
12. $3a^4 - a^3 + a - 9$

Term	Coefficient	Degree
$3a^4$	3	4
$-a^3$	−1	3
a	1	1
-9	−9	0

13. (a) 3, 5, 2; (b) $7a^5$, 7; (c) 5
14. $5x - 9x^2 + 3x^6$
(a)

Term	$5x$	$-9x^2$	$3x^6$
Degree	1	2	6

(b) Leading term: $3x^6$; leading coefficient: 3
(c) Degree: 6

15. (a) 1, 0, 2; (b) $4t^2$, 4; (c) 2
16. $3a^2 - 7 + 2a^5$
(a)

Term	$3a^2$	-7	$2a^5$
Degree	2	0	5

(b) Leading term: $2a^5$; leading coefficient: 2
(c) Degree: 5

17. (a) 4, 2, 1, 0; (b) $-5x^4$, −5; (c) 4
18. $-7x^3 + 6x^2 - 3x - 4$
(a)

Term	$-7x^3$	$6x^2$	$-3x$	-4
Degree	3	2	1	0

(b) Leading term: $-7x^3$; leading coefficient: −7
(c) Degree: 3

19. (a) 1, 4, 0, 3; (b) $-a^4$, −1; (c) 4
20. $-x + 2x^5 - x^3 + 2$
(a)

Term	$-x$	$2x^5$	$-x^3$	2
Degree	1	5	3	0

(b) Leading term: $2x^5$; leading coefficient: 2
(c) Degree: 5

21.

Term	Coefficient	Degree of Term	Degree of Polynomial
$8x^5$	8	5	5
$-\frac{1}{2}x^4$	$-\frac{1}{2}$	4	
$-4x^3$	−4	3	
$3x^2$	3	2	
6	6	0	

22. $-7x^4 + 6x^3 - 3x^2 + 8x - 2$

Term	Coefficient	Degree of Term	Degree of Polynomial
$-7x^4$	-7	4	
$6x^3$	6	3	
$-3x^2$	-3	2	4
$8x$	8	1	
-2	-2	0	

23. Trinomial   **24.** Monomial   **25.** None of these
26. Binomial   **27.** Binomial   **28.** Trinomial
29. Monomial   **30.** None of these   **31.** $7x^2 + 5x$
32. $5a + 7a^2 + 3a = (5 + 3)a + 7a^2 = 8a + 7a^2 = 7a^2 + 8a$ **33.** $4a^4$ **34.** $6b^5 + 3b^2 - 2b^5 - 3b^2 = (6 - 2)b^5 + (3 - 3)b^2 = 4b^5$ **35.** $6x^2 - 3x$
36. $\frac{1}{4}x^5 - 5 + \frac{1}{2}x^5 - 2x - 37 = \left(\frac{1}{4} + \frac{1}{2}\right)x^5 - 2x + (-5 - 37) = \frac{3}{4}x^5 - 2x - 42$ **37.** $\frac{1}{6}x^3 + 2x - 12$
38. $6x^2 + 2x^4 - 2x^2 - x^4 - 4x^2 = 6x^2 + 2x^4 - 2x^2 - 1x^4 - 4x^2 = (6 - 2 - 4)x^2 + (2 - 1)x^4 = x^4$ **39.** $-x^3$
40. $\frac{1}{4}x^3 - x^2 - \frac{1}{6}x^2 + \frac{3}{8}x^3 + \frac{5}{16}x^3 = \left(\frac{1}{4} + \frac{3}{8} + \frac{5}{16}\right)x^3 + \left(-1 - \frac{1}{6}\right)x^2 = \frac{15}{16}x^3 - \frac{7}{6}x^2$ **41.** 0   **42.** $3x^4 - 5x^6 - 2x^4 + 6x^6 = x^4 + x^6 = x^6 + x^4$ **43.** $x^4 - 2x^3 + 1$
44. $-2x + 4x^3 - 7x + 9x^3 + 8 = -9x + 13x^3 + 8 = 13x^3 - 9x + 8$ **45.** $4x^3 + x - \frac{1}{2}$ **46.** $5a^2 - \frac{2}{3} + a^3 - 9a^2 - \frac{4}{3}a^3 + 1 = -4a^2 + \frac{1}{3} - \frac{1}{3}a^3 = -\frac{1}{3}a^3 - 4a^2 + \frac{1}{3}$
47. -16 **48.** $-5 \cdot 3 + 9 = -15 + 9 = -6$ **49.** 16
50. $4 \cdot 3^2 - 6 \cdot 3 + 9 = 36 - 18 + 9 = 27$ **51.** -3
52. $7 - 3(-2) = 7 + 6 = 13$ **53.** 11
54. $5(-2) - 9 + (-2)^2 = -10 - 9 + 4 = -15$ **55.** 55
56. $-2(-2)^3 - 4(-2)^2 + 3(-2) + 1 = -2(-8) - 4 \cdot 4 + 3(-2) + 1 = 16 - 16 - 6 + 1 = -5$ **57.** 9
58. About 17   **59.** 6   **60.** About 13   **61.** 15
62. About 5   **63.** 1112 ft   **64.** $173 - 369 = 173 \cdot 20 - 369 = 3460 - 369 = 3091$ ft   **65.** Approximately 449
66. $0.4r^2 - 40r + 1039 = 0.4(20)^2 - 40(20) + 1039 = 0.4(400) - 800 + 1039 = 160 - 800 + 1039 = 399$
67. $18,750   **68.** $280x - 0.4x^2 = 280(100) - 0.4(100)^2 = 28,000 - 4000 = $24,000$ **69.** $155,000
70. $5000 + 0.6(650)^2 = 5000 + 253,500 = $258,500$
71. 62.8 cm   **72.** $2\pi r = 2(3.14)(5) = 31.4$ ft
73. 78.5 m^2 **74.** $\pi r^2 = 3.14(10)^2 = 314$ in^2 **75.** $\frac{15}{2}$
76. Let x and $x + 1$ represent the page numbers. Solve:
$x + (x + 1) = 549$
$x = 274$, so $x + 1 = 275$ and the page numbers are 274 and 275.   **77.** $0.0625, or 6.25¢

78. $cx = ab - r$
$cx + r = ab$
$\frac{cx + r}{a} = b$

79. ◆ Every monomial is a term but not every term is a monomial. A term is a number, a variable, or a product of numbers and variables which may be raised to powers whereas a monomial is a number, a variable, or a product of numbers and variables raised to *whole number* powers. For example, the monomial $3a^2b^3$ is a term, but the term $5x^{-2}y^4$ is not a monomial.   **80.** ◆ Polynomials typically contain exponents, multiplication, addition, and subtraction.
81. ◆ It is better to evaluate a polynomial after like terms have been combined, because there are fewer steps involved.   **82.** ◆ Yes; the evaluation will yield a sum of products of integers which must be an integer.
83. $-6x^5 + 14x^4 - x^2 + 11$; answers may vary.
84. Answers may vary. Use any ay^4-term, where a is a rational number, and 2 other terms with different degrees, each less than degree 4, and rational coefficients. Three answers are $0.2y^4 - y + \frac{5}{2}$, $-\frac{8}{7}y^4 + 5.5y^3 - 2y^2$, and $2.9y^4 - 4y^2 - \frac{11}{3}$.   **85.** 10   **86.** Answers may vary. The terms must have the same variable and be of degree 4. One answer is $9y^4$, $-\frac{3}{7}y^4$, and $4.2y^4$.
87. $5x^9 + 4x^8 + x^2 + 5x$
88. $(3x^2)^3 + 4x^2 \cdot 4x^4 - x^4(2x)^2 + [(2x)^2]^3 - 100x^2(x^2)^2$
$= 27x^6 + 4x^2 \cdot 4x^4 - x^4 \cdot 4x^2 + (2x)^6 - 100x^2 \cdot x^4$
$= 27x^6 + 16x^6 - 4x^6 + 64x^6 - 100x^6$
$= 3x^6$
89. 99, -99; 50, -50; 99, -99
90. Using a calculator, evaluate $0.4r^2 - 40r + 1039$ for $r = 10, 20, 30, 40, 50, 60,$ and 70 and list the values in a table.

Age	Average Number of Accidents Per Day
r	$0.4r^2 - 40r + 1039$
10	679
20	399
30	199
40	79
50	39
60	79
70	199

The numbers in the chart increase both below and above age 50. We would assume the number of accidents is the smallest near age 50. Now we evaluate for 49 and 51.

49	39.4
50	39
51	39.4

Again the numbers increase below and above 50. We conclude that the smallest number of daily accidents occurs at age 50.

91.

t	$-t^2 + 10t - 18$
3	3
4	6
5	7
6	6
7	3

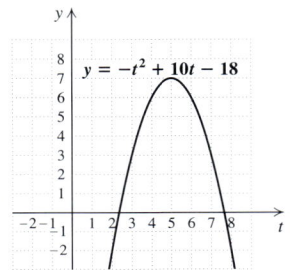

92. When $t = 1$, $-t^2 + 6t - 4 = -1^2 + 6 \cdot 1 - 4 = -1 + 6 - 4 = 1$.

When $t = 2$, $-t^2 + 6t - 4 = -2^2 + 6 \cdot 2 - 4 = -4 + 12 - 4 = 4$.

When $t = 3$, $-t^2 + 6t - 4 = -3^2 + 6 \cdot 3 - 4 = -9 + 18 - 4 = 5$.

When $t = 4$, $-t^2 + 6t - 4 = -4^2 + 6 \cdot 4 - 4 = -16 + 24 - 4 = 4$.

When $t = 5$, $-t^2 + 6t - 4 = -5^2 + 6 \cdot 5 - 4 = -25 + 30 - 4 = 1$.

We complete the table. Then we plot the points and connect them with a smooth curve.

t	$-t^2 + 6t - 4$
1	1
2	4
3	5
4	4
5	1

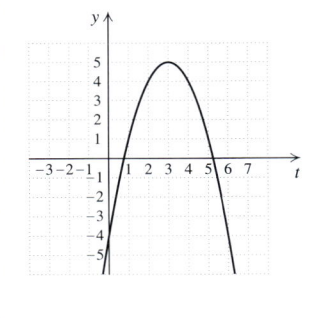

93.

d	$-0.0064d^2 + 0.8d + 2$
0	2
30	20.24
60	26.96
90	22.16
120	5.84

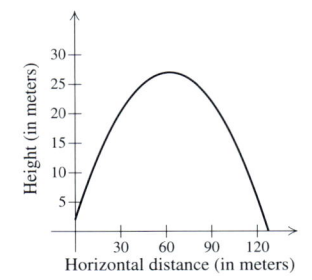

94. Let $c =$ the coefficient of x^3. Solve:

$$c + (c - 3) + 3(c - 3) + (c + 2) = -4$$
$$c + c - 3 + 3c - 9 + c + 2 = -4$$
$$6c - 10 = -4$$
$$6c = 6$$
$$c = 1$$

Coefficient of x^3, c: 1
Coefficient of x^2, $c - 3$: $1 - 3$, or -2
Coefficient of x, $3(c - 3)$: $3(1 - 3)$, or -6
Coefficient remaining (constant term), $c + 2$: $1 + 2$, or 3

The polynomial is $x^3 - 2x^2 - 6x + 3$.

TECHNOLOGY CONNECTION 4.3, P. 195

1. In each case, let $y_1 =$ the expression before the addition or subtraction has been performed, $y_2 =$ the simplified sum or difference, and $y_3 = y_2 - y_1$; and note that the graph of y_3 coincides with the x-axis. That is, $y_3 = 0$.

EXERCISE SET 4.3, PP. 197–199

1. $-2x + 4$ **2.** $-5x + 6$ **3.** $x^2 - 5x - 1$
4. $x^2 + 3x - 5$ **5.** $5x^2 + 3x - 30$ **6.** $6x^4 + 3x^3 + 4x^2 - 3x + 2$ **7.** $-2.2x^3 - 0.2x^2 - 3.8x + 23$
8. $2.8x^4 - 0.6x^2 + 1.8x - 3.2$ **9.** $11 + 2x + 16x^2$
10. $2x^4 + 3x + 3x^2 + 4 - 3x^3$, or $2x^4 - 3x^3 + 3x^2 + 3x + 4$ **11.** $9x^8 + 8x^7 - 3x^4 + 2x^2 - 2x + 5$
12. $4x^5 + 9x^2 + 1$ **13.** $-\frac{1}{2}x^4 + \frac{2}{3}x^3 + x^2$
14. $\frac{2}{15}x^9 - \frac{2}{5}x^5 + \frac{1}{4}x^4 - \frac{1}{2}x^2 + 7$
15. $0.01x^5 + x^4 - 0.2x^3 + 0.2x + 0.06$
16. $0.1x^6 + 0.02x^3 + 0.22x + 0.55$
17. $-3x^4 + 3x^2 + 4x$ **18.** $-4x^3 + 4x^2 + 6x$

19. $1.05x^4 + 0.36x^3 + 14.22x^2 + x + 0.97$
20. $1.3x^4 + 0.35x^3 + 9.53x^2 + 2x + 0.96$
21. $-(-x^2 + 9x - 4)$, $x^2 - 9x + 4$
22. $-(-4x^3 - 5x^2 + 2x)$, $4x^3 + 5x^2 - 2x$
23. $-(12x^4 - 3x^3 + 3)$, $-12x^4 + 3x^3 - 3$
24. $-(4x^3 - 6x^2 - 8x + 1)$, $-4x^3 + 6x^2 + 8x - 1$
25. $-8x + 9$ **26.** $6x - 5$ **27.** $-4x^2 + 3x - 2$
28. $6a^3 - 2a^2 + 9a - 1$ **29.** $4x^4 - 6x^2 - \frac{3}{4}x + 8$
30. $5x^4 - 4x^3 + x^2 - 0.9$ **31.** $9x + 3$
32. $(5x + 6) - (-2x + 4) = 5x + 6 + 2x - 4 = 7x + 2$
33. $-x^2 - 7x + 5$ **34.** $(x^2 - 5x + 4) - (8x - 9) =$
$x^2 - 5x + 4 - 8x + 9 = x^2 - 13x + 13$
35. $6x^4 + 3x^3 - 4x^2 + 3x - 4$ **36.** $(-4x^2 + 2x) -$
$(3x^3 - 5x^2 + 3) = -4x^2 + 2x - 3x^3 + 5x^2 - 3 =$
$-3x^3 + x^2 + 2x - 3$ **37.** $4.6x^3 + 9.2x^2 - 3.8x - 23$
38. $(0.5x^4 - 0.6x^2 + 0.7) - (2.3x^4 + 1.8x - 3.9)$
$= 0.5x^4 - 0.6x^2 + 0.7 - 2.3x^4 - 1.8x + 3.9$
$= -1.8x^4 - 0.6x^2 - 1.8x + 4.6$
39. $7x^3 - 9x^2 - 2x + 10$
40. $(6x^5 - 3x^4 + x + 1) - (8x^5 + 3x^4 - 1)$
$= 6x^5 - 3x^4 + x + 1 - 8x^5 - 3x^4 + 1$
$= -2x^5 - 6x^4 + x + 2$
41. $9x^2 + 9x - 8$ **42.** $7x^3 - (-3x^2 - 2x + 1) =$
$7x^3 + 3x^2 + 2x - 1$ **43.** $\frac{3}{4}x^3 - \frac{1}{2}x$
44. $\left(\frac{1}{5}x^3 + 2x^2 - 0.1\right) - \left(-\frac{2}{5}x^3 + 2x^2 + 0.01\right)$
$= \frac{1}{5}x^3 + 2x^2 - 0.1 + \frac{2}{5}x^3 - 2x^2 - 0.01$
$= \frac{3}{5}x^3 - 0.11$
45. $0.06x^3 - 0.05x^2 + 0.01x + 1$
46. $(0.8x^4 + 0.2x - 1) - \left(\frac{7}{10}x^4 + \frac{1}{5}x - 0.1\right)$
$= 0.8x^4 + 0.2x - 1 - \frac{7}{10}x^4 - \frac{1}{5}x + 0.1$
$= 0.1x^4 - 0.9$
47. $3x + 6$ **48.** $\begin{array}{r} x^3 + 1 \\ -(x^3 + x^2) \\ \hline x^3 + 1 \\ -x^3 - x^2 \\ \hline -x^2 + 1 \end{array}$
49. $11x^4 + 12x^3 - 9x^2 - 8x - 9$
50. $\begin{array}{r} 5x^4 + 6x^2 - 3x + 6 \\ -(6x^3 + 7x^2 - 8x - 9) \\ \hline 5x^4 + 6x^2 - 3x + 6 \\ - 6x^3 - 7x^2 + 8x + 9 \\ \hline 5x^4 - 6x^3 - x^2 + 5x + 15 \end{array}$
51. $-4x^5 + 9x^4 + 6x^2 + 16x + 6$
52. $\begin{array}{r} 6x^5 + 3x^2 - 7x + 2 \\ -(10x^5 + 6x^3 - 5x^2 - 2x + 4) \\ \hline 6x^5 + 3x^2 - 7x + 2 \\ -10x^5 - 6x^3 + 5x^2 + 2x - 4 \\ \hline -4x^5 - 6x^3 + 8x^2 - 5x - 2 \end{array}$

53. (a) $5x^2 + 4x$; (b) 145, 273
54. (a) $r^2\pi + 3^2\pi + 2^2\pi = r^2\pi + 9\pi + 4\pi = r^2\pi + 13\pi$
(b) For $r = 5$: $r^2\pi + 13\pi = 5^2\pi + 13\pi = 25\pi + 13\pi = 38\pi$
For $r = 11.3$: $r^2\pi + 13\pi = (11.3)^2\pi + 13\pi = 127.69\pi + 13\pi = 140.69\pi$
55. $14y + 25$
56. We add the lengths of the sides:
$4a + 7 + a + \frac{1}{2}a + 5 + a + 2a + 3a$
$= \left(4 + 1 + \frac{1}{2} + 1 + 2 + 3\right)a + (7 + 5)$
$= 11\frac{1}{2}a + 12$, or $\frac{23}{2}a + 12$
57. $(r + 9)(r + 11)$; $9r + 99 + r^2 + 11r$
58.

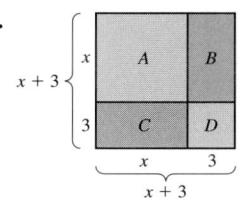

The length and width of the figure can each be expressed as $x + 3$. The area can be expressed as $(x + 3)(x + 3)$, or $(x + 3)^2$. Another way to express the area is to find an expression for the sum of the areas of the four rectangles A, B, C, and D. The area of each rectangle is the product of its length and width.

Area of A + Area of B + Area of C + Area of D
$= x \cdot x + 3 \cdot x + 3 \cdot x + 3 \cdot 3$
$= x^2 + 3x + 3x + 9$

The algebraic expressions $(x + 3)^2$ and $x^2 + 3x + 3x + 9$ represent the same area.
$(x + 3)^2 = x^2 + 3x + 3x + 9$
59. $\pi r^2 - 25\pi$
60. Area of entire square − Area not shaded = Shaded area
$m \cdot m - 8 \cdot 5 =$ Shaded area
$m^2 - 40 =$ Shaded area
61. $27z - 144$
62. The inscribed square is composed of two triangles, each with base $x + x$, or $2x$, and height x. The circle has radius x.
Shaded area = Area of circle − Area of square

Shaded area $= \pi x^2 - 2\left(\frac{1}{2} \cdot 2x \cdot x\right)$
$= \pi x^2 - 2x^2$, or $(\pi - 2)x^2$

63. $y^2 - 4y + 4$

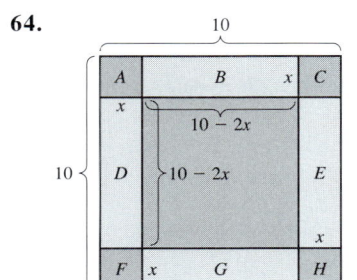

64.

$(10 - 2x)^2$ is the area of the entire square less the areas of A, B, C, D, E, F, G, and H. The areas of A, C, F, and H are each $x \cdot x$, or x^2. The areas of B, D, E, and G are each $x(10 - 2x)$, or $10x - 2x^2$. We have:
$(10 - 2x)^2 = 10^2 - 4 \cdot x^2 - 4(10x - 2x^2)$
$= 100 - 4x^2 - 40x + 8x^2$
$= 100 - 40x + 4x^2$

65. $\frac{115}{22}$ **66.** $3x - 3 = -4x + 4$ **67.** 4
$7x = 7$
$x = 1$

68. $4(x - 5) = 7(x + 8)$ **69.** $\{x \mid x \geq -10\}$
$4x - 20 = 7x + 56$
$-3x = 76$
$x = -\frac{76}{3}$

70. $2(x - 4) > 5(x - 3) + 7$
$2x - 8 > 5x - 15 + 7$
$2x - 8 > 5x - 8$
$-3x > 0$
$x < 0$
The solution set is $\{x \mid x < 0\}$.

71. ◆ The two binomials must consist of two pairs of like terms and the members of at least one pair must be opposites.

72. ◆ All three are needed. For example, consider the following.

$(3y - 2) + (-5y + 7)$
$= 3y - 5y - 2 + 7$ Using a commutative law
$= (3y - 5y) + (-2 + 7)$ Using an associative law
$= (3 - 5)y + (-2 + 7)$ Using the distributive law
$= -2y + 5$

73. ◆ Yes; consider the following.
$(x^2 + 4) + (4x - 7) = x^2 + 4x - 3$

74. ◆ Remind the student that we subtract polynomials by adding the opposite of the polynomial being subtracted.

75. $12a^2 - 24a - 8$ **76.** $5x^2 - 9x - 1$
77. $-10y^2 - 2y - 10$
78. $(5x^3 - 4x^2 + 6) - (2x^3 + x^2 - x) + (x^3 - x)$
$= 5x^3 - 4x^2 + 6 - 2x^3 - x^2 + x + x^3 - x$
$= 4x^3 - 5x^2 + 6$
79. $-3y^4 - y^3 + 5y - 2$ **80.** $2 + x + 2x^2 + 4x^3$
81. $250.591x^3 + 2.812x$ **82.** Surface area $= 2 \cdot 9 \cdot x + 2 \cdot 9 \cdot x + 2 \cdot x \cdot x = 18x + 18x + 2x^2 = 36x + 2x^2$
83. $22a + 56$
84. (a) $R - C = 280x - 0.4x^2 - (5000 + 0.6x^2)$
$= 280x - 0.4x^2 - 5000 - 0.6x^2$
$= -x^2 + 280x - 5000$
A polynomial for total profit is $-x^2 + 280x - 5000$.
(b) Evaluate the polynomial for $x = 75$:
$-x^2 + 280x - 5000$
$= -(75)^2 + 280(75) - 5000$
$= -5625 + 21,000 - 5000$
$= 10,375$

The total profit on the production and sale of 75 stereos is $10,375.
(c) Evaluate the polynomial for $x = 100$:
$-x^2 + 280x - 5000$
$= -(100)^2 + 280(100) - 5000$
$= -10,000 + 28,000 - 5000$
$= 13,000$

The total profit on the production and sale of 100 stereos is $13,000.
85. ◆ No; $5(-x)^3 - 3(-x)^2 + 2(-x) = -5x^3 - 3x^2 - 2x \neq -(5x^3 - 3x^2 + 2x)$.

TECHNOLOGY CONNECTION 4.4, P. 204

1. Let $y_1 = (-2x^2 - 3)(5x^3 - 3x + 4)$ and $y_2 = -10x^5 - 9x^3 - 8x^2 + 9x - 12$. With the table set in AUTO mode, note that the values in the Y1- and Y2-columns match, regardless of how far we scroll up or down.

EXERCISE SET 4.4, PP. 205–206

1. $24x^4$ **2.** $28x^3$ **3.** x^3 **4.** $-x^7$ **5.** $-x^8$ **6.** x^8
7. $28t^8$ **8.** $30a^4$ **9.** $-0.02x^{10}$ **10.** $-0.12x^9$
11. $\frac{1}{15}x^4$ **12.** $-\frac{1}{20}x^{12}$ **13.** 0   **14.** $5n^3$ **15.** $-24x^{11}$
16. $60y^{12}$ **17.** $-3x^2 + 15x$ **18.** $8x^2 - 12x$
19. $4x^2 + 4x$ **20.** $3x^2 + 6x$ **21.** $5x^2 + 35x$
22. $3x^2 - 18x$ **23.** $x^5 + x^2$ **24.** $-2x^5 + 2x^3$
25. $6x^3 - 18x^2 + 3x$ **26.** $-8x^4 + 24x^3 + 20x^2 - 4x$
27. $12x^3 + 24x^2$ **28.** $-10x^3 + 5x^2$ **29.** $-6x^4 - 6x^3$
30. $-4x^4 + 4x^3$ **31.** $4a^9 - 8a^7 - \frac{5}{12}a^4$
32. $6t^{11} - 9t^9 + \frac{9}{7}t^5$ **33.** $x^2 + 9x + 18$
34. $x^2 + 7x + 10$ **35.** $x^2 + 3x - 10$
36. $x^2 + 4x - 12$ **37.** $x^2 - 7x + 12$
38. $x^2 - 10x + 21$

39. $x^2 - 9$ **40.** $x^2 - 36$ **41.** $25 - 15x + 2x^2$
42. $18 + 12x + 2x^2$ **43.** $t^2 + \frac{17}{6}t + 2$
44. $a^2 + \frac{21}{10}a - 1$ **45.** $\frac{3}{16}a^2 + \frac{5}{4}a - 2$
46. $\frac{6}{25}t^2 - \frac{1}{5}t - 1$

47.

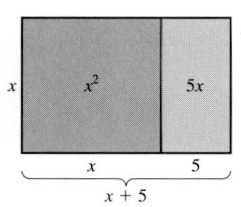

48.

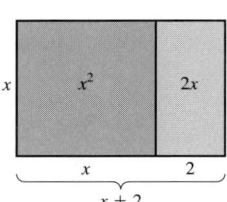

49.

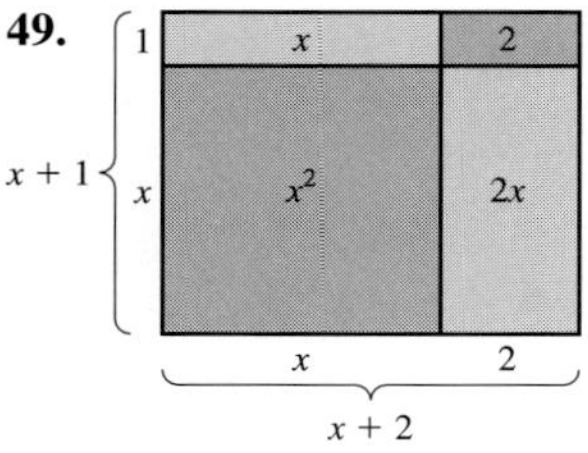

50.

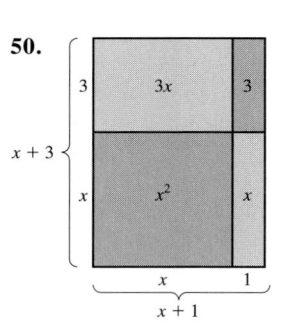

51.

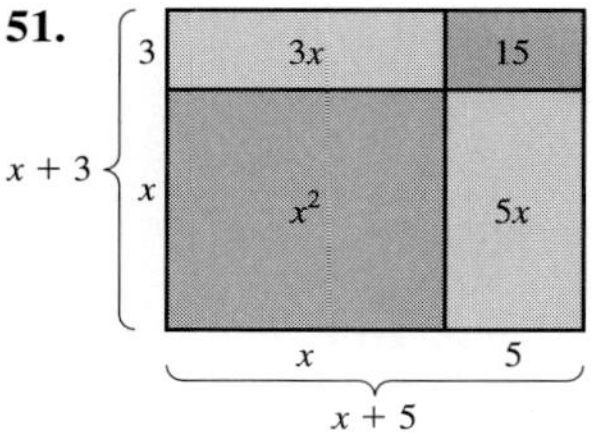

52.

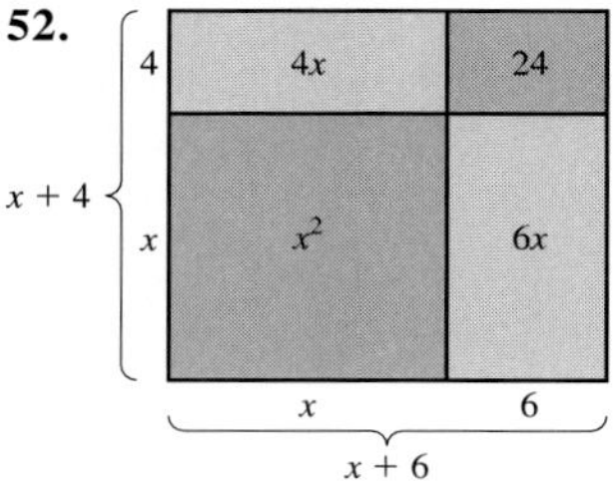

53.

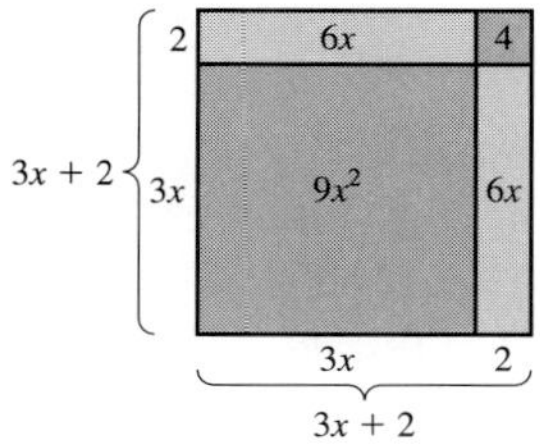

54.

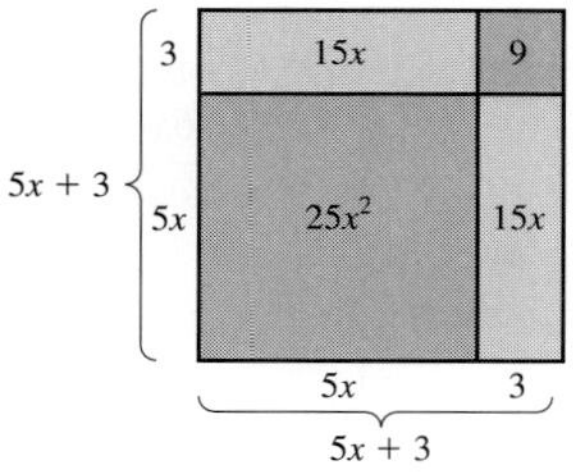

55. $x^3 + 2x + 3$
56. $\quad (x^2 + x - 2)(x - 1)$
$\quad = (x^2 + x - 2)x - (x^2 + x - 2)(1)$
$\quad = x^3 + x^2 - 2x - x^2 - x + 2$
$\quad = x^3 - 3x + 2$
57. $2a^3 - a^2 - 11a + 10$
58. $\quad (3t + 4)(t^2 - 5t + 1)$
$\quad = (3t + 4)t^2 - (3t + 4)(5t) + (3t + 4)(1)$
$\quad = 3t^3 + 4t^2 - 15t^2 - 20t + 3t + 4$
$\quad = 3t^3 - 11t^2 - 17t + 4$
59. $2y^5 - 5y^3 + y^2 - 3y - 3$
60. $\quad (a^2 + 2)(5a^3 - 3a - 1)$
$\quad = (a^2 + 2)(5a^3) - (a^2 + 2)(3a) - (a^2 + 2)(1)$
$\quad = 5a^5 + 10a^3 - 3a^3 - 6a - a^2 - 2$
$\quad = 5a^5 + 7a^3 - a^2 - 6a - 2$
61. $-15x^5 + 26x^4 - 7x^3 - 3x^2 + x$
62. $\quad (4x^3 - 5x - 3)(1 + 2x^2)$
$\quad = (4x^3 - 5x - 3)(1) + (4x^3 - 5x - 3)(2x^2)$
$\quad = 4x^3 - 5x - 3 + 8x^5 - 10x^3 - 6x^2$
$\quad = 8x^5 - 6x^3 - 6x^2 - 5x - 3$
63. $x^4 - 2x^3 - x + 2$
64.
$$\begin{array}{r} x^2 + 5x - 1 \\ x^2 - x + 3 \\ \hline 3x^2 + 15x - 3 \\ -x^3 - 5x^2 + x \\ x^4 + 5x^3 - x^2 \\ \hline x^4 + 4x^3 - 3x^2 + 16x - 3 \end{array}$$
65. $4x^4 + 8x^3 - 9x^2 - 10x + 8$
66.
$$\begin{array}{r} 2x^2 - x - 3 \\ 2x^2 - 5x - 2 \\ \hline -4x^2 + 2x + 6 \\ -10x^3 + 5x^2 + 15x \\ 4x^4 - 2x^3 - 6x^2 \\ \hline 4x^4 - 12x^3 - 5x^2 + 17x + 6 \end{array}$$
67. $x^4 + 8x^3 + 12x^2 + 9x + 4$
68. $\quad (x + 2)(x^3 + 5x^2 + 9x + 3)$
$\quad = x^4 + 5x^3 + 9x^2 + 3x$
$\qquad\qquad\quad 2x^3 + 10x^2 + 18x + 6$
$\quad = x^4 + 7x^3 + 19x^2 + 21x + 6$
69. $2x^4 - 5x^3 + 5x^2 - \frac{19}{10}x + \frac{1}{5}$
70. $\left(x + \frac{1}{3}\right)\left(6x^3 - 12x^2 - 5x + \frac{1}{2}\right)$
$\quad = 6x^4 - 12x^3 - 5x^2 + \frac{1}{2}x$
$\qquad\qquad 2x^3 - 4x^2 - \frac{5}{3}x + \frac{1}{6}$
$\quad\overline{\quad 6x^4 - 10x^3 - 9x^2 - \frac{7}{6}x + \frac{1}{6}}$
71. 32

72.

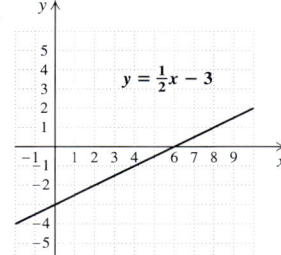

73. $3(5x - 6y + 4)$

The area of the figure is $x^2 + 3x + nx + 3n$. This is equivalent to $x^2 + 8x + 15$, so we have $3x + nx = 8x$ and $3n = 15$. Solving either equation for n, we find that the missing number is 5.

83. $V = 4x^3 - 48x^2 + 144x$; $S = -4x^2 + 144$

84.

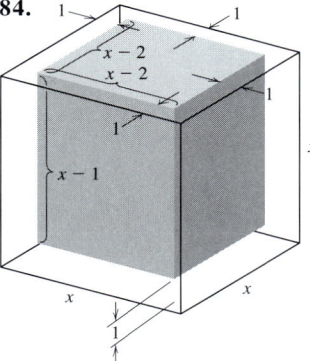

The interior dimensions of the open box are $x - 2$ cm by $x - 2$ cm by $x - 1$ cm.

$$\begin{aligned}\text{Interior volume} &= (x - 2)(x - 2)(x - 1)\\ &= (x^2 - 4x + 4)(x - 1)\\ &= x^3 - 5x^2 + 8x - 4 \text{ cm}^3\end{aligned}$$

74. $4(x - 3) = 5(2 - 3x) + 1$
$4x - 12 = 10 - 15x + 1$
$4x - 12 = 11 - 15x$
$19x = 23$
$x = \dfrac{23}{19}$

75. ◆ $(A + B)(C + D)$ will be a trinomial when there is exactly one pair of like terms among AC, AD, BC, and BD.

76. ◆ Consider $(a + b + c + d)(r + s + m + p)$. The product cannot contain like terms since each variable appears only once. Using the distributive law, we see that each of the 4 terms in the first factor produces 4 terms of the product. Thus the product will contain $4 \cdot 4$, or 16 terms.

77. ◆ No; the distributive law is the basis for polynomial multiplication.

78. ◆ Label the figure as shown.

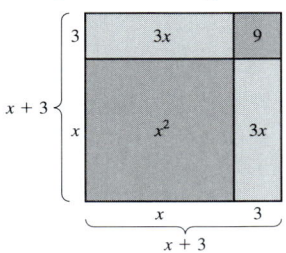

Then we see that the area of the figure is $(x + 3)^2$, or $x^2 + 3x + 3x + 9 \neq x^2 + 9$.

79. $75y^2 - 45y$

80. The shaded area is the area of the large rectangle less the area of the small rectangle:
$4t(21t + 8) - 2t(3t - 4) = 84t^2 + 32t - 6t^2 + 8t$
$= 78t^2 + 40t$

81. 5

82. Let $n = $ the missing number.

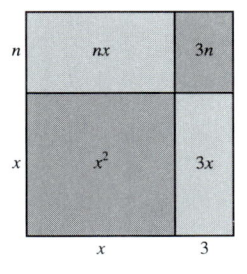

85. $x^3 + 6x^2 + 12x + 8$ cm^3

86. Let $x =$ the width of the garden. Then $2x =$ the length of the garden. The rectangular area composed of the sidewalk and garden together has dimensions $2x + 8$ by $x + 8$. Solve:

$(2x + 8)(x + 8) = 2x \cdot x + 256$
$x = 8$, so $2x = 16$.

The dimensions are 8 ft by 16 ft.

87. $2x^2 + 18x + 36$

88. $(x - 2)(x - 7) + (x - 2)(x - 7)$
$= x^2 - 2x - 7x + 14 + x^2 - 2x - 7x + 14$
$= x^2 - 9x + 14 + x^2 - 9x + 14$
$= 2x^2 - 18x + 28$

89. $16x + 16$

90. $(x - 6)^2 + (4 - x)^2$
$= (x - 6)(x - 6) + (4 - x)(4 - x)$
$= x^2 - 6x - 6x + 36 + 16 - 4x - 4x + x^2$
$= x^2 - 12x + 36 + 16 - 8x + x^2$
$= 2x^2 - 20x + 52$

91.  ⌐∿⌐

EXERCISE SET 4.5, PP. 213–215

1. $x^3 + 5x^2 + x + 5$ **2.** $x^3 - x^2 - 3x + 3$
3. $x^4 + 2x^3 + 6x + 12$ **4.** $x^5 + 12x^4 + 2x + 24$
5. $y^2 - y - 6$ **6.** $a^2 + 4a + 4$ **7.** $9x^2 + 15x + 6$
8. $8x^2 + 10x + 2$ **9.** $5x^2 + 4x - 12$ **10.** $t^2 - 81$
11. $9t^2 - 1$ **12.** $4m^2 + 12m + 9$ **13.** $2x^2 - 9x + 7$

14. $6x^2 - x - 1$ **15.** $p^2 - \frac{1}{16}$ **16.** $q^2 + \frac{3}{2}q + \frac{9}{16}$
17. $x^2 - 0.01$ **18.** $x^2 - 0.1x - 0.12$
19. $2x^3 + 2x^2 + 6x + 6$ **20.** $4x^3 - 2x^2 + 6x - 3$
21. $-2x^2 - 11x + 6$ **22.** $6x^2 - 4x - 16$
23. $a^2 + 14a + 49$ **24.** $4y^2 + 28y + 49$
25. $1 - 2t - 15t^2$ **26.** $-3x^2 - 5x - 2$
27. $x^5 + 3x^3 - x^2 - 3$ **28.** $2x^5 + x^4 - 6x - 3$
29. $3x^6 - 2x^4 - 6x^2 + 4$ **30.** $x^{20} - 9$
31. $6x^7 + 18x^5 + 4x^2 + 12$ **32.** $1 + 3x^2 - 2x - 6x^3$, or $1 - 2x + 3x^2 - 6x^3$ **33.** $8x^5 + 16x^3 + 5x^2 + 10$
34. $20 - 8x^2 - 10x + 4x^3$, or $20 - 10x - 8x^2 + 4x^3$
35. $4x^3 - 12x^2 + 3x - 9$ **36.** $14x^2 - 53x + 14$
37. $x^2 - 64$ **38.** $x^2 - 1$ **39.** $4x^2 - 1$ **40.** $x^4 - 1$
41. $25m^2 - 4$ **42.** $9x^8 - 4$ **43.** $4x^4 - 9$
44. $36x^{10} - 25$ **45.** $9x^8 - 1$ **46.** $t^4 - 0.04$
47. $x^{12} - x^4$ **48.** $4x^6 - 0.09$ **49.** $x^8 - 9x^2$
50. $\frac{9}{16} - 4x^6$ **51.** $4y^{16} - 9$ **52.** $m^2 - \frac{4}{9}$
53. $x^2 + 4x + 4$ **54.** $4x^2 - 4x + 1$
55. $9x^4 + 6x^2 + 1$ **56.** $9x^2 + \frac{9}{2}x + \frac{9}{16}$
57. $a^2 - \frac{4}{5}a + \frac{4}{25}$ **58.** $4a^2 - \frac{4}{5}a + \frac{1}{25}$
59. $x^4 + 6x^2 + 9$ **60.** $64x^2 - 16x^3 + x^4$
61. $4 - 12x^4 + 9x^8$ **62.** $36x^6 - 24x^3 + 4$
63. $25 + 60t^2 + 36t^4$ **64.** $9p^4 - 6p^3 + p^2$
65. $49x^2 - 4.2x + 0.09$ **66.** $16a^2 - 4.8a + 0.36$
67. $10a^5 - 5a^3$
68. $\quad (a - 3)(a^2 + 2a - 4)$
$\quad\quad = a^3 + 2a^2 - 4a$
$\quad\quad\quad\underline{\;\;- 3a^2 - 6a + 12}$
$\quad\quad = a^3 - a^2 - 10a + 12$
69. $x^4 + x^3 - 6x^2 - 5x + 5$ **70.** $27x^6 - 9x^5$
71. $9 - 12x^3 + 4x^6$ **72.** $x^2 - 8x^4 + 16x^6$
73. $4x^3 + 24x^2 - 12x$ **74.** $-8x^6 + 48x^3 + 72x$
75. $4x^4 - 2x^2 + \frac{1}{4}$ **76.** $x^4 - 2x^2 + 1$ **77.** $-1 + 9p^2$
78. $-9q^2 + 4$, or $4 - 9q^2$ **79.** $15t^5 - 3t^4 + 3t^3$
80. $-5x^5 - 40x^4 + 45x^3$ **81.** $36x^8 - 36x^4 + 9$
82. $64a^2 + 80a + 25$ **83.** $12x^3 + 8x^2 + 15x + 10$
84. $6x^4 - 3x^2 - 63$ **85.** $64 - 96x^4 + 36x^8$
86. $\frac{2}{9}t^4 + 3t^2 - 5$ **87.** $a^3 + 1$
88. $\quad (x - 5)(x^2 + 5x + 25)$
$\quad\quad = x^3 + 5x^2 + 25x$
$\quad\quad\quad\underline{\;\;- 5x^2 - 25x - 125}$
$\quad\quad\overline{\;\;x^3 \quad\quad\quad\quad - 125}$
89. $a^2 + 2a + 1$ **90.** $(x + 3)^2 = x^2 + 6x + 9$
91. $x^2 + 7x + 10$ **92.** $(t + 4)(t + 3) = t^2 + 7t + 12$
93. $x^2 + 14x + 49$ **94.** $(a + 5)^2 = a^2 + 10a + 25$
95. $t^2 + 13t + 36$ **96.** $(a + 7)(a + 1) = a^2 + 8a + 7$
97. $9x^2 + 24x + 16$ **98.** $(5t + 2)^2 = 25t^2 + 20t + 4$

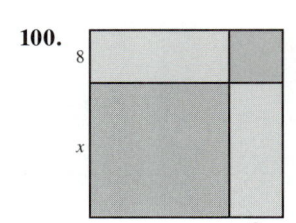

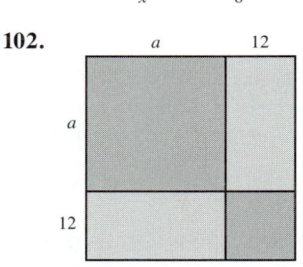

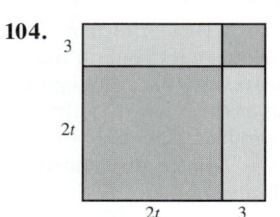

99. [diagram: square with sides $x+6$, divided at x and 6]
100. [diagram: square with sides $x+8$, divided at x and 8]
101. [diagram: square with sides $t+9$, divided at t and 9]
102. [diagram: square with sides $a+12$, divided at a and 12]
103. [diagram: square with sides $4a+1$, divided at $4a$ and 1]
104. [diagram: square with sides $2t+3$, divided at $2t$ and 3]

105. Television: 50 watts; lamps: 500 watts; air conditioner: 2000 watts

106. $3x - 8x = 4(7 - 8x)$ **107.** $a = \dfrac{c}{b - d}$ **108.** IV
$\quad\quad -5x = 28 - 32x$
$\quad\quad 27x = 28$
$\quad\quad x = \dfrac{28}{27}$

109. ◆ $(A + B)(C + D)$ is a binomial when $AD + BC = 0$. **110.** ◆ It's a good idea to study the other special products, because they allow for faster computations than the FOIL method. **111.** ◆ The computation $(20 - 1)(20 + 1) = 400 - 1 = 399$ is easily performed mentally and is equivalent to the computation $19 \cdot 21$.
112. ◆ $(A + B)^3$ can be viewed as the sum of the volumes of eight regions in a cube with side $A + B$.
113. $30x^3 + 35x^2 - 15x$
114. $\quad [(2x - 3)(2x + 3)](4x^2 + 9)$
$\quad\quad = (4x^2 - 9)(4x^2 + 9)$
$\quad\quad = 16x^4 - 81$
115. $a^4 - 50a^2 + 625$ **116.** $\quad (a - 3)^2(a + 3)^2$
$\quad\quad\quad\quad\quad\quad\quad\quad\quad\quad\quad\quad = [(a - 3)(a + 3)]^2$
$\quad\quad\quad\quad\quad\quad\quad\quad\quad\quad\quad\quad = (a^2 - 9)^2$
$\quad\quad\quad\quad\quad\quad\quad\quad\quad\quad\quad\quad = a^4 - 18a^2 + 81$
117. $81t^{16} - 72t^8 + 16$ **118.** $(32.41x + 5.37)^2 = 1050.4081x^2 + 348.0834x + 28.8369$
119. $400 - 4 = 396$ **120.** $93 \times 107 = (100 - 7)(100 + 7) = 10{,}000 - 49 = 9951$ **121.** -7

122. $(2x+5)(x-4) = (x+5)(2x-4)$
$2x^2 - 3x - 20 = 2x^2 + 6x - 20$
$-3x = 6x$
$-9x = 0$
$x = 0$

123. $l^3 - l$

124. If w = the width, then $w + 1$ = the length, and $(w+1) + 1$, or $w + 2$ = the height.
Volume = $(w+1) \cdot w \cdot (w+2) = (w^2+w)(w+2) = w^3 + 3w^2 + 2w$

125. $Q(Q-14) - 5(Q-14)$, $(Q-5)(Q-14)$; other equivalent expressions are possible.

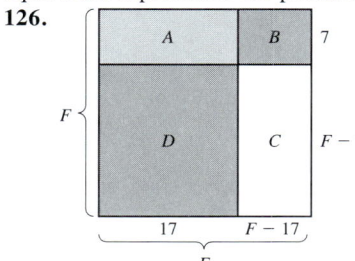

126.

The area of the entire figure is F^2. The area of the unshaded region, C, is $(F-7)(F-17)$. Then one expression for the area of the shaded region is $F^2 - (F-7)(F-17)$. To find a second expression, we add the areas of regions A, B, and D. We have:
$17 \cdot 7 + 7(F-17) + 17(F-7)$
$= 119 + 7F - 119 + 17F - 119$
$= 24F - 119$
It is possible to find other equivalent expressions also.

127. $(y+1)(y-1)$, $y(y+1) - y - 1$; other equivalent expressions are possible.

128. Solve: $x^2 + (x+1)^2 + (x+2)^2 = 3x^2 + 65$
$x = 10$
The integers are 10, 11, and 12.

129. (a) $A^2 + AB$; (b) $AB + B^2$; (c) $A^2 - B^2$; (d) $A^2 - B^2$

130.

EXERCISE SET 4.6, PP. 219–222

1. -15 **2.** $x^2 + 5y^2 - 4xy = 3^2 + 5(-2)^2 - 4 \cdot 3(-2) = 9 + 20 + 24 = 53$ **3.** -5
4. $xy - xz + yz = 2(-3) - 2(-1) + (-3)(-1) = -6 + 2 + 3 = -1$. **5.** 2.97 L
6. $0.041h - 0.018A - 2.69 = 0.041(165) - 0.018(20) - 2.69 = 3.715$ liters **7.** 110.4 m
8. $h + vt - 4.9t^2 = 160 + 30 \cdot 3 - 4.9(3)^2 = 205.9$ m
9. 20.60625 in^2 **10.** $2\pi rh + \pi r^2 \approx 2(3.14)(1.25)(7.5) + (3.14)(1.25)^2 \approx 63.78125$ in^2 **11.** Coefficients: 1, -2, 3, -5; degrees: 4, 2, 2, 0; 4 **12.** Coefficients: 5, -1, 15, 1; degrees: 3, 2, 1, 0; 3 **13.** Coefficients: 17, -3, -7; degrees: 5, 5, 0; 5 **14.** Coefficients: 6, -1, 8, -1; degrees: 0, 2, 4, 5; 5 **15.** $a - 2b$ **16.** $y - 7$

17. $3x^2y - 2xy^2 + x^2$ **18.** $m^3 + 2m^2n - 3m^2 + 3mn^2$
19. $8u^2v - 5uv^2$ **20.** $-2x^2 - 4xy - 2y^2$
21. $8uv + 10av + 14au$ **22.** $3x^2y + 3z^2y + 3xy^2$
23. $x^2 - 4xy + 3y^2$
24. $(r^3 + 3rs - 5s^2) - (5r^3 + rs + 4s^2)$
$= r^3 + 3rs - 5s^2 - 5r^3 - rs - 4s^2$
$= -4r^3 + 2rs - 9s^2$
25. $-2a^4 - 8ab + 7ab^2$ **26.** $3r + s - 4$
27. $-3b^3a^2 - b^2a^3 + 5ba + 3$ **28.** $-x^2 - 8xy - y^2$
29. $3x^3 - x^2y + xy^2 - 3y^3$
30. $(xy - ab) - (xy - 3ab)$ **31.** $y^4x^2 + y + 2x$
$= xy - ab - xy + 3ab$
$= 2ab$
32. $15a^2b - 4ab$ **33.** $-8x + 8y$
34. $(2a + b) + (3a - 4b) - (5a + 2b)$
$= 2a + b + 3a - 4b - 5a - 2b$
$= -5b$
35. $6z^2 + 7zu - 3u^2$ **36.** $a^4b^2 - 7a^2b + 10$
37. $x^2y^2 + 3xy - 28$ **38.** $a^6 - b^2c^2$
39. $m^4 + m^2n^2 + n^4$
40.
$y^4x + y^2 + 1$
$y^2 + 1$
$\overline{y^4x + y^2 + 1}$
$y^6x + y^4 + y^2$
$\overline{y^6x + y^4 + y^4x + 2y^2 + 1}$
41. $a^3 - b^3$ **42.** $12x^2y^2 + 2xy - 2$
43. $m^6n^2 + 2m^3n - 48$ **44.** $12 - c^2d^2 - c^4d^4$
45. $30x^2 - 28xy + 6y^2$ **46.** $m^3 + m^2n - mn^2 - n^3$
47. $0.4p^2q^2 - 0.02pq - 0.02$ **48.** $x^5y^5 - x^2y^2 + x^9y^9 - x^6y^6$, or $x^9y^9 - x^6y^6 + x^5y^5 - x^2y^2$
49. $x^2 + 2xh + h^2$ **50.** $9a^2 + 12ab + 4b^2$
51. $r^6t^4 - 8r^3t^2 + 16$ **52.** $9a^4b^2 - 6a^2b^3 + b^4$
53. $c^4 - d^2$ **54.** $p^6 - 25q^2$ **55.** $a^2b^2 - c^2d^4$
56. $x^2y^2 - p^2q^2$ **57.** $x^2 + 2xy + y^2 - 9$
58. $[x + y + z][x - (y + z)]$
$= [x + (y + z)][x - (y + z)]$
$= x^2 - (y + z)^2$
$= x^2 - (y^2 + 2yz + z^2)$
$= x^2 - y^2 - 2yz - z^2$
59. $a^2 - b^2 - 2bc - c^2$
60. $(a + b + c)(a - b - c)$
$= [a + (b + c)][a - (b + c)]$
$= a^2 - (b + c)^2$
$= a^2 - (b^2 + 2bc + c^2)$
$= a^2 - b^2 - 2bc - c^2$
61. $a^2 + ac + ab + bc$ **62.** The figure is a square with side $x + y$. Thus the area is $(x + y)^2 = x^2 + 2xy + y^2$.
63. $x^2 - z^2$ **64.** The figure is a triangle with base $ab + 2$ and height $ab - 2$. Its area is $\frac{1}{2}(ab+2)(ab-2) = \frac{1}{2}(a^2b^2 - 4) = \frac{1}{2}a^2b^2 - 2$.
65. $x^2 + 2xy + y^2 + 2xz + 2yz + z^2$

66. The figure is a rectangle with dimensions $a + b + c$ by $a + d + c$. Its area is

$$(a + b + c)(a + d + c)$$
$$= [(a + c) + b][(a + c) + d]$$
$$= (a + c)^2 + (a + c)d + b(a + c) + bd$$
$$= a^2 + 2ac + c^2 + ad + cd + ab + bc + bd$$

67. $\frac{1}{2}x^2 + \frac{1}{2}xy - y^2$

68. The figure is a parallelogram with base $m - n$ and height $m + n$. Its area is $(m - n)(m + n) = m^2 - n^2$.

69. We draw a rectangle with dimensions $r + s$ by $u + v$.

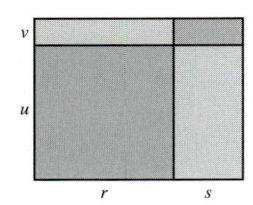

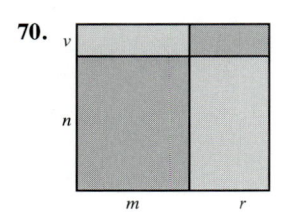

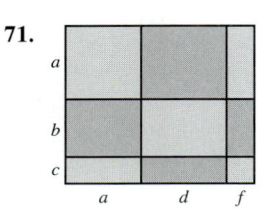

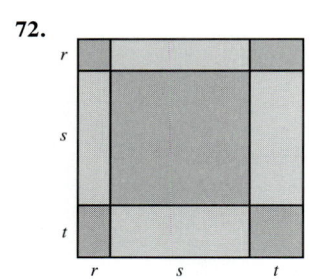

70.

71.

72.

73. $50 per ton   **74.** December 1995   **75.** June 1995
76. December 1991   **77.** June to December 1994
78. June to December 1995
79. ◆ Yes; for example, $(x^2 + xy + 1) + (3x^2 - xy + 2) = 4x^2 + 3$.
80. ◆ Yes; consider $a + b + c + d$. This is a polynomial in 4 variables but it has degree 1.
81. ◆ The degree of the product is seven. When the two leading terms are multiplied their product is the leading term of the product of the two polynomials and its degree is $4 + 3$, or 7.
82. ◆ Yes; for example, $(x^2 + x + xy) + (x^3 + 2x^2 - xy) = x^3 + 3x^2 + x$.
83. $2\pi ab - \pi b^2$

84. It is helpful to add additional labels to the figure.

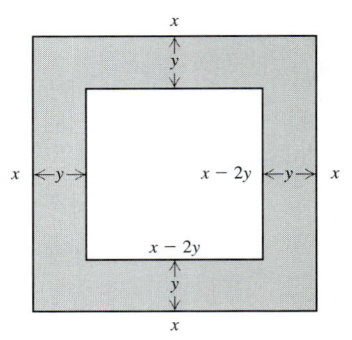

The area of the large square is $x \cdot x$, or x^2. The area of the small square is $(x - 2y)(x - 2y)$, or $(x - 2y)^2$.

$$\begin{array}{c}\text{Area of}\\ \text{shaded region}\end{array} = \begin{array}{c}\text{Area of}\\ \text{large square}\end{array} - \begin{array}{c}\text{Area of}\\ \text{small square}\end{array}$$

$$\begin{array}{c}\text{Area of}\\ \text{shaded region}\end{array} = x^2 - (x - 2y)^2$$

$$= x^2 - (x^2 - 4xy + 4y^2)$$
$$= x^2 - x^2 + 4xy - 4y^2$$
$$= 4xy - 4y^2$$

85. $a^2 - 4b^2$
86. It is helpful to add additional labels to the figure.

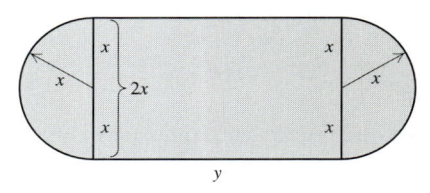

The two semicircles make a circle with radius x. The area of that circle is πx^2. The area of the rectangle is $2xy$. The sum of the two regions, $\pi x^2 + 2xy$, is the area of the shaded region.
87. $2\pi nh + 2\pi mh + 2\pi n^2 - 2\pi m^2$
88. The surface area of the solid consists of the surface area of a rectangular solid with dimensions x by x by h less the areas of 2 circles with radius r plus the lateral surface area of a right circular cylinder with radius r and height h. Thus, we have

$$2x^2 + 2xh + 2xh - 2\pi r^2 + 2\pi rh, \text{ or}$$
$$2x^2 + 4xh - 2\pi r^2 + 2\pi rh.$$

89. $P + 2Pr + Pr^2$ **90.** $P(1 - r)^2 = P(1 - 2r + r^2) = P - 2Pr + Pr^2$ **91.** ◆ The height of the observatory is 40 ft and its radius is 30/2, or 15 ft, so the surface area is $2\pi rh + \pi r^2 \approx 2(3.14)(15)(40) + (3.14)(15)^2 \approx 4474.5$ ft^2. Since 4474.5 ft^2/250 ft^2 = 17.898, 18 gallons of paint should be purchased.

EXERCISE SET 4.7, PP. 227–228

1. $4x^5 - 2x$ **2.** $2a^4 - \frac{1}{2}a^2$ **3.** $1 - 2u + u^6$
4. $50x^4 - 7x^3 + x$ **5.** $5t^2 - 8t + 2$ **6.** $5t^2 - 3t - 6$
7. $-7x^4 + 4x^2 + 1$ **8.** $-2x^4 - 4x^3 + 1$
9. $6x^2 - 10x + \frac{3}{2}$ **10.** $2x^3 - 3x^2 - \frac{1}{3}$
11. $4x^2 - 5x + \frac{1}{2}$ **12.** $2x^2 + x - \frac{2}{3}$
13. $x^2 + 3x + 2$ **14.** $2x^2 - 3x + 5$
15. $-3rs - r + 2s$ **16.** $1 - 2x^2y + 3x^4y^5$
17. $x + 6$

18.
$$\begin{array}{r} x - 2 \\ x - 4 \overline{) x^2 - 6x + 8} \\ \underline{x^2 - 4x} \\ -2x + 8 \\ \underline{-2x + 8} \\ 0 \end{array}$$

19. $x - 5 + \frac{-50}{x - 5}$

The answer is $x - 2$.

20.
$$\begin{array}{r} x + 4 \\ x + 4 \overline{) x^2 + 8x - 16} \\ \underline{x^2 + 4x} \\ 4x - 16 \\ \underline{4x + 16} \\ -32 \end{array}$$

The answer is $x + 4 + \frac{-32}{x + 4}$, or $x + 4 - \frac{32}{x + 4}$.

21. $2x - 1 + \frac{1}{x + 6}$

22.
$$\begin{array}{r} 3x + 4 \\ x - 2 \overline{) 3x^2 - 2x - 13} \\ \underline{3x^2 - 6x} \\ 4x - 13 \\ \underline{4x - 8} \\ -5 \end{array}$$

The answer is $3x + 4 + \frac{-5}{x - 2}$, or $3x + 4 - \frac{5}{x - 2}$.

23. $x - 3$

24.
$$\begin{array}{r} x - 5 \\ x + 5 \overline{) x^2 + 0x - 25} \\ \underline{x^2 + 5x} \\ -5x - 25 \\ \underline{-5x - 25} \\ 0 \end{array}$$

25. $x + 4$

The answer is $x - 5$.

26.
$$\begin{array}{r} 2x + 3 \\ 5x - 1 \overline{) 10x^2 + 13x - 3} \\ \underline{10x^2 - 2x} \\ 15x - 3 \\ \underline{15x - 3} \\ 0 \end{array}$$

The answer is $2x + 3$.

27. $2x^2 - 7x + 4$

28.
$$\begin{array}{r} x^2 - 3x + 1 \\ 2x - 3 \overline{) 2x^3 - 9x^2 + 11x - 3} \\ \underline{2x^3 - 3x^2} \\ -6x^2 + 11x \\ \underline{-6x^2 + 9x} \\ 2x - 3 \\ \underline{2x - 3} \\ 0 \end{array}$$

The answer is $x^2 - 3x + 1$.

29. $x^2 + 1 + \frac{4x - 2}{x^2 - 3}$

30.
$$\begin{array}{r} x^2 - 1 \\ x^2 + 5 \overline{) x^4 + 0x^3 + 4x^2 + 3x - 6} \\ \underline{x^4 + 5x^2} \\ -x^2 + 3x - 6 \\ \underline{-x^2 - 5} \\ 3x - 1 \end{array}$$

The answer is $x^2 - 1 + \frac{3x - 1}{x^2 + 5}$.

31. $t^2 - 2t + 3 + \frac{-4}{t + 1}$

32.
$$\begin{array}{r} t^2 + 1 \\ t - 1 \overline{) t^3 - t^2 + t - 1} \\ \underline{t^3 - t^2} \\ t - 1 \\ \underline{t - 1} \\ 0 \end{array}$$

33. $3x^2 - 3 + \frac{x - 1}{2x^2 + 1}$

The answer is $t^2 + 1$.

34.
$$\begin{array}{r} 2x^2 + 1 \\ 2x^2 - 3 \overline{) 4x^4 + 0x^3 - 4x^2 - x - 3} \\ \underline{4x^4 - 6x^2} \\ 2x^2 - x - 3 \\ \underline{2x^2 - 3} \\ -x \end{array}$$

The answer is $2x^2 + 1 + \frac{-x}{2x^2 - 3}$, or $2x^2 + 1 - \frac{x}{2x^2 - 3}$.

35. $25{,}543.75$ ft^2 **36.** $3(2x - 1) = 7x - 5$
$6x - 3 = 7x - 5$
$2 = x$

37. $\left\{x \mid x < -\frac{12}{5}\right\}$

38.

39. III

40.

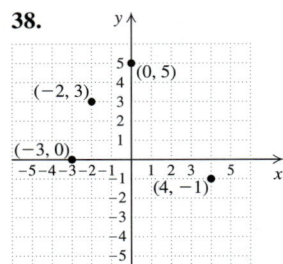

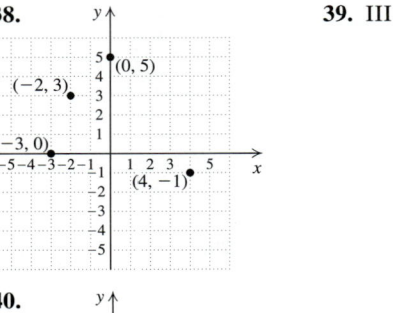

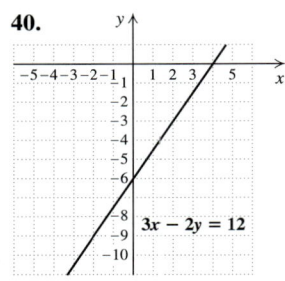

41. ◆ Let the trinomial be the dividend and let either binomial factor be the divisor. We have $(6x^2 + 7x - 3) \div (2x + 3)$ and $(6x^2 + 7x - 3) \div (3x - 1)$. **42.** ◆ The student did not divide *each* term of the polynomial by the divisor. The first term was divided by $3x$, but the second was not. Multiplying the student's "quotient" by the divisor $3x$ we get $12x^3 - 18x^2 \ne 12x^3 + 6x$. This should convince the person that a mistake has been made. **43.** ◆ Yes; for example, $(t^3 - 1) \div (t - 1) = t^2 + t + 1$. **44.** ◆ The distributive law is used in each step when a term of the quotient is multiplied by the divisor and when the subtraction is performed. The distributive law is also used when the quotient is checked.

45. $15x^{6k} + 10x^{4k} - 20x^{2k}$ **46.** $5a^{6k} - 16a^{3k} + 14$

47. $y^3 - ay^2 + a^2y - a^3 + \dfrac{a^2 + a^4}{y + a}$

48.
$$5a^2 - 7a - 2 \overline{\smash{\big)}\, 5a^3 + 8a^2 - 23a - 1} \atop \begin{array}{r} a + 3 \\ \underline{5a^3 - 7a^2 - 2a} \\ 15a^2 - 21a - 1 \\ \underline{15a^2 - 21a - 6} \\ 5 \end{array}$$

The answer is $a + 3 + \dfrac{5}{5a^2 - 7a - 2}$.

49. $5y + 2 + \dfrac{-10y + 11}{3y^2 - 5y - 2}$

50. $(4x^5 - 14x^3 - x^2 + 3) +$
$(2x^5 + 3x^4 + x^3 - 3x^2 + 5x)$
$= 6x^5 + 3x^4 - 13x^3 - 4x^2 + 5x + 3$

$$3x^3 - 2x - 1 \overline{\smash{\big)}\, 6x^5 + 3x^4 - 13x^3 - 4x^2 + 5x + 3} \atop \begin{array}{r} 2x^2 + x - 3 \\ \underline{6x^5 - 4x^3 - 2x^2} \\ 3x^4 - 9x^3 - 2x^2 + 5x \\ \underline{3x^4 - 2x^2 - x} \\ -9x^3 + 6x + 3 \\ \underline{-9x^3 + 6x + 3} \\ 0 \end{array}$$

The answer is $2x^2 + x - 3$.

51. $5x^5 + 5x^4 - 8x^2 - 8x + 2$

52.
$$2a^h + 3 \overline{\smash{\big)}\, 6a^{3h} + 13a^{2h} - 4a^h - 15} \atop \begin{array}{r} 3a^{2h} + 2a^h - 5 \\ \underline{6a^{3h} + 9a^{2h}} \\ 4a^{2h} - 4a^h \\ \underline{4a^{2h} + 6a^h} \\ -10a^h - 15 \\ \underline{-10a^h - 15} \\ 0 \end{array}$$

The answer is $3a^{2h} + 2a^h - 5$.

53. 3

54.
$$x - 1 \overline{\smash{\big)}\, 2x^2 - 3cx - 8} \atop \begin{array}{r} 2x + (-3c + 2) \\ \underline{2x^2 - 2x} \\ (-3c + 2)x - 8 \\ \underline{(-3c + 2)x - (-3c + 2)} \\ -8 + (-3c + 2) \end{array}$$

We set the remainder equal to 0:
$-8 - 3c + 2 = 0$
$-3c - 6 = 0$
$-3c = 6$
$c = -2$

Thus, c must be -2.

55. -1

TECHNOLOGY CONNECTION 4.8, P. 233

1. 1.71×10^{17} **2.** $5.\overline{370} \times 10^{-15}$ **3.** 3.68×10^{16}

EXERCISE SET 4.8, PP. 235–237

1. $\dfrac{1}{6^2} = \dfrac{1}{36}$ **2.** $2^{-4} = \dfrac{1}{2^4} = \dfrac{1}{16}$ **3.** $\dfrac{1}{10^4} = \dfrac{1}{10,000}$

4. $5^{-3} = \dfrac{1}{5^3} = \dfrac{1}{125}$ **5.** $\dfrac{1}{(-2)^6} = \dfrac{1}{64}$

6. $(-3)^{-4} = \dfrac{1}{(-3)^4} = \dfrac{1}{81}$ **7.** $\dfrac{1}{a^5}$ **8.** $\dfrac{1}{x^2}$ **9.** y^4 **10.** t^7

11. z^9 **12.** h^8 **13.** $\dfrac{1}{7}$ **14.** $\left(\dfrac{2}{3}\right)^{-1} = \dfrac{1}{\frac{2}{3}} = \dfrac{3}{2}$

15. 16 **16.** $\left(\frac{4}{5}\right)^{-2} = \frac{1}{\left(\frac{4}{5}\right)^2} = \frac{25}{16}$ **17.** 4^{-3} **18.** 5^{-2}

19. t^{-6} **20.** y^{-2} **21.** a^{-4} **22.** t^{-5} **23.** p^{-8}
24. m^{-12} **25.** 5^{-1} **26.** 8^{-1} **27.** t^{-1} **28.** m^{-1}
29. 2^3, or 8 **30.** 5 **31.** x^{-1}, or $\frac{1}{x}$ **32.** $x^0 = 1$
33. x^{-13}, or $\frac{1}{x^{13}}$ **34.** y^{-13}, or $\frac{1}{y^{13}}$ **35.** m^{-6}, or $\frac{1}{m^6}$
36. p^{-1}, or $\frac{1}{p}$ **37.** $(8x)^{-4}$, or $\frac{1}{(8x)^4}$ **38.** $(9t)^{-7}$, or $\frac{1}{(9t)^7}$
39. 1 **40.** 1 **41.** $a^{-8}b^{-13}$, or $\frac{1}{a^8 b^{13}}$
42. $x^{-5}y^{-9}$, or $\frac{1}{x^5 y^9}$ **43.** x^9 **44.** t^{11} **45.** z^{-4}, or $\frac{1}{z^4}$
46. y^{-4}, or $\frac{1}{y^4}$ **47.** a^4 **48.** y^5 **49.** x^2 **50.** x^5
51. a^{-15}, or $\frac{1}{a^{15}}$ **52.** x^{-30}, or $\frac{1}{x^{30}}$ **53.** a^{30} **54.** x^{12}
55. n^{-16}, or $\frac{1}{n^{16}}$ **56.** m^{-21}, or $\frac{1}{m^{21}}$ **57.** $m^{-5}n^{-5}$, or $\frac{1}{m^5 n^5}$
58. $(ab)^{-3} = a^{-3}b^{-3}$, or $\frac{1}{a^3 b^3}$ **59.** $4^{-2}x^{-2}y^{-2}$, or $\frac{1}{16x^2 y^2}$
60. $(5ab)^{-2} = 5^{-2}a^{-2}b^{-2}$, or $\frac{1}{5^2 a^2 b^2}$, or $\frac{1}{25a^2 b^2}$
61. $81a^{-16}$, or $\frac{81}{a^{16}}$ **62.** $(6x^{-5})^2 = 6^2 x^{-10} = 36x^{-10}$, or $\frac{36}{x^{10}}$
63. $t^{-20}x^{-12}$, or $\frac{1}{t^{20}x^{12}}$ **64.** $(x^4 y^5)^{-3} = x^{-12}y^{-15}$, or
$\frac{1}{x^{12}y^{15}}$ **65.** $x^{10}y^{35}$ **66.** $x^{24}y^8$ **67.** $x^{-1}y^{-6}z^4$, or $\frac{z^4}{xy^6}$
68. $a^{-8}b^5 c^4$, or $\frac{b^5 c^4}{a^8}$ **69.** $m^5 n^5 p^{-7}$, or $\frac{m^5 n^5}{p^7}$
70. $t^{-14}p^3 m^6$, or $\frac{p^3 m^6}{t^{14}}$ **71.** $\frac{y^{-4}}{2^{-2}}$, or $\frac{4}{y^4}$ **72.** $\left(\frac{a^4}{3}\right)^{-2} =$
$\frac{a^{-8}}{3^{-2}} = \frac{3^2}{a^8} = \frac{9}{a^8}$ **73.** $\frac{81}{a^8}$ **74.** $\frac{49}{x^{14}}$ **75.** $\frac{x^6 y^3}{z^{12}}$ **76.** $\frac{m^3}{n^{12}p^3}$
77. $\frac{a^{-10}b^{-5}}{c^{-5}d^{-15}}$, or $\frac{c^5 d^{15}}{a^{10}b^5}$ **78.** $\left(\frac{2a^2}{3b^4}\right)^{-3} = \frac{2^{-3}a^{-6}}{3^{-3}b^{-12}} =$
$\frac{3^3 b^{12}}{2^3 a^6} = \frac{27b^{12}}{8a^6}$ **79.** 91,200
80. 8.92×10^2 8.92.

       2 places

  $8.92 \times 10^2 = 892$
81. 0.00692
82. 7.26×10^{-4} .0007.26

       4 places

  $7.26 \times 10^{-4} = 0.000726$

83. 204,000,000
84. 1.35×10^7 1.3500000.

       7 places

  $1.35 \times 10^7 = 13,500,000$
85. 0.0000000008764
86. 9.043×10^{-3} .009.043

       3 places

  $9.043 \times 10^{-3} = 0.009043$
87. 10,000,000
88. $10^4 = 1 \times 10^4$ 1.0000.

       4 places

  $10^4 = 10,000$
89. 0.0001
90. $10^{-7} = 1 \times 10^{-7}$ .0000001.

       7 places

  $10^{-7} = 0.0000001$
91. 3.7×10^5
92. $71,500 = 7.15 \times 10^n$
  7.1500.

       4 places right, so n is 4
  $71,500 = 7.15 \times 10^4$
93. 5.83×10^{-3}
94. $0.0814 = 8.14 \times 10^n$
  .08.14

       2 places left, so n is -2
  $0.0814 = 8.14 \times 10^{-2}$
95. 7.8×10^{10}
96. $3,700,000,000,000 = 3.7 \times 10^n$
  3.700000000000.

       12 places right, so n is 12
  $3,700,000,000,000 = 3.7 \times 10^{12}$
97. 9.07×10^{17}
98. $168,000,000,000,000 = 1.68 \times 10^n$
  1.68000000000000.

       14 places right, so n is 14
  $168,000,000,000,000 = 1.68 \times 10^{14}$
99. 4.86×10^{-6}
100. $0.000000000275 = 2.75 \times 10^n$
  .0000000002.75

       10 places left, so n is -10
  $0.000000000275 = 2.75 \times 10^{-10}$
101. 1.8×10^{-8}
102. $0.00000000002 = 2 \times 10^n$
  0.00000000002.

       11 places left, so n is -11
  $0.00000000002 = 2 \times 10^{-11}$
103. 10^7

104. $100{,}000{,}000{,}000 = 1 \times 10^n$, or 10^n
1.00000000000.

 11 places right, so n is 11
$100{,}000{,}000{,}000 = 10^{11}$
105. 8×10^{12} **106.** $(1.9 \times 10^8)(3.4 \times 10^{-3}) = 6.46 \times 10^5$ **107.** 2.47×10^8
108. $(7.1 \times 10^{-7})(8.6 \times 10^{-5}) = 61.06 \times 10^{-12} = (6.106 \times 10) \times 10^{-12} = 6.106 \times 10^{-11}$
109. 3.915×10^{-16} **110.** $(4.7 \times 10^5)(6.2 \times 10^{-12}) = 29.14 \times 10^{-7} = (2.914 \times 10) \times 10^{-7} = 2.914 \times 10^{-6}$
111. 2.5×10^{13} **112.** $\dfrac{5.6 \times 10^{-2}}{2.5 \times 10^5} = 2.24 \times 10^{-7}$
113. 5×10^{-4} **114.** $(1.5 \times 10^{-3}) \div (1.6 \times 10^{-6}) = 0.9375 \times 10^3 = (9.375 \times 10^{-1}) \times 10^3 = 9.375 \times 10^2$
115. 3×10^{-21} **116.** $\dfrac{4.0 \times 10^{-3}}{8.0 \times 10^{20}} = 0.5 \times 10^{-23} = (5 \times 10^{-1}) \times 10^{-23} = 5 \times 10^{-24}$ **117.** 1.095×10^9 gal
118. There are 60 seconds in one minute and 60 minutes in one hour, so there are 60(60), or 3600 seconds in one hour. There are 24 hours in one day and 365 days in one year, so there are 3600(24)(365), or 31,536,000 seconds in one year.

Find the amount of water discharged in one hour:
$(4.2 \times 10^5) \times (3.6 \times 10^3) = 1.512 \times 10^{10}$ cubic feet

Find the amount of water discharged in one year:
$(4.2 \times 10^5) \times (3.1536 \times 10^7) = 1.324512 \times 10^{14}$ cubic feet

119. 2×10^{14} gal
120. Compute: $(2 \times 10^4) \times (1.6 \times 10^6) = \$3.2 \times 10^{10}$
121. 2.5×10^{12}
122. Compute: $(3.12 \times 10^7) \times (4.24 \times 10^4) = \$1.32288 \times 10^{12}$ **123.** $-\dfrac{1}{6}$ **124.** $8a$ **125.** $-17x$
126.

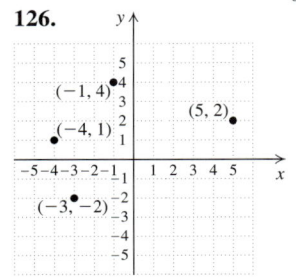

127. I and IV

128. $cx - bt = r$
$-bt = r - cx$
$t = \dfrac{r - cx}{-b}$, or $\dfrac{cx - r}{b}$

129. ◆ x^{-n} represents a negative number when x is negative and n is an odd number. **130.** ◆ See the discussion of negative integers as exponents at the beginning of Section 4.8 in the text. **131.** $1.19140625 \times 10^{-15}$

132. $\dfrac{7.4 \times 10^{29}}{(5.4 \times 10^{-6})(2.8 \times 10^8)}$
$= \dfrac{7.4}{(5.4 \cdot 2.8)} \times \dfrac{10^{29}}{(10^{-6} 10^8)}$
$\approx 0.4894179894 \times 10^{27}$
$\approx (4.894179894 \times 10^{-1}) \times 10^{27}$
$\approx 4.894179894 \times 10^{26}$

133. $6.304347826 \times 10^{25}$

134. $\dfrac{(7.8 \times 10^7)(8.4 \times 10^{23})}{2.1 \times 10^{-12}}$
$= \dfrac{(7.8 \cdot 8.4)}{2.1} \times \dfrac{(10^7 \cdot 10^{23})}{10^{-12}}$
$= 31.2 \times 10^{42}$
$= (3.12 \times 10) \times 10^{42}$
$= 3.12 \times 10^{43}$

135. 7×10^{-16}

136. (a) $\dfrac{1}{6.25 \times 10^{-3}} = 0.16 \times 10^3 = (1.6 \times 10^{-1}) \times 10^3 = 1.6 \times 10^2$

(b) $\dfrac{1}{4.0 \times 10^{10}} = 0.25 \times 10^{-10} = (2.5 \times 10^{-1}) \times 10^{-10} = 2.5 \times 10^{-11}$

137. 2^1 **138.** $2^8 \cdot 16^{-3} \cdot 64 = (2^2)^4 \cdot (4^2)^{-3} \cdot 4^3 = 4^4 \cdot 4^{-6} \cdot 4^3 = 4^1$ **139.** 5 **140.** $49^{18} \cdot 7^{-35} = (7^2)^{18} \cdot 7^{-35} = 7^{36} \cdot 7^{-35} = 7$ **141.** 3^{28}

142. $\left(\dfrac{1}{a}\right)^{-n} = \dfrac{1^{-n}}{a^{-n}} = \dfrac{a^n}{1^n} = a^n$ **143.** False

144. False; let $x = 3$, $y = 4$, and $m = 2$:
$3^2 \cdot 4^2 = 9 \cdot 16 = 144$, but
$(3 \cdot 4)^{2 \cdot 2} = 12^4 = 20{,}736$

145. False

REVIEW EXERCISES: CHAPTER 4, PP. 239–240

1. [4.1] y^{11} **2.** [4.1] $(3x)^{14}$ **3.** [4.1] t^8 **4.** [4.1] 4^3
5. [4.1] 1 **6.** [4.1] $\dfrac{9t^8}{4s^6}$ **7.** [4.1] $-8x^3 y^6$ **8.** [4.1] $18x^5$
9. [4.1] $a^7 b^6$ **10.** [4.2] $3x^2, 6x, \dfrac{1}{2}$
11. [4.2] $-4y^5, 7y^2, -3y, -2$ **12.** [4.2] 6, 1, 5
13. [4.2] 4, 6, $-5, \dfrac{5}{3}$
14. [4.2] **(a)** 2, 0, 5; **(b)** $15t^5$, 15; **(c)** 5
15. [4.2] **(a)** 5, 4, 2, 1; **(b)** $-2x^5$, -2; **(c)** 5
16. [4.2] Binomial **17.** [4.2] None of these
18. [4.2] Monomial **19.** [4.2] $-x^2 + 9x$
20. [4.2] $-\dfrac{1}{4}x^3 + 4x^2 + 7$ **21.** [4.2] $-3x^5 + 25$
22. [4.2] $-2x^2 - 3x + 2$ **23.** [4.2] $10x^4 - 7x^2 - x - \dfrac{1}{2}$
24. [4.2] -17 **25.** [4.2] 10
26. [4.3] $x^5 + 3x^4 + 6x^3 - 2x - 9$
27. [4.3] $-x^5 + 3x^4 - x^3 - 2x^2$ **28.** [4.3] $2x^2 - 4x - 6$
29. [4.3] $x^5 - 3x^3 - 2x^2 + 8$
30. [4.3] $\dfrac{3}{4}x^4 + \dfrac{1}{4}x^3 - \dfrac{1}{3}x^2 - \dfrac{7}{4}x + \dfrac{3}{8}$

31. [4.3] $-x^5 + x^4 - 5x^3 - 2x^2 + 2x$
32. (a) [4.3] $4w + 6$; (b) [4.4] $w^2 + 3w$ 33. [4.4] $-12x^3$
34. [4.5] $49x^2 + 14x + 1$ 35. [4.5] $x^2 + \frac{7}{6}x + \frac{1}{3}$
36. [4.5] $m^2 - 25$ 37. [4.4] $12x^3 - 23x^2 + 13x - 2$
38. [4.5] $x^2 - 18x + 81$ 39. [4.4] $2a^5 - 4a^3 + \frac{4}{9}a^2$
40. [4.5] $x^2 - 3x - 28$ 41. [4.5] $x^2 - 1.05x + 0.225$
42. [4.4] $x^7 + x^5 - 3x^4 + 3x^3 - 2x^2 + 5x - 3$
43. [4.5] $9y^4 - 12y^3 + 4y^2$ 44. [4.5] $2t^4 - 11t^2 - 21$
45. [4.5] $4a^4 + 4a^3 + a^2$ 46. [4.5] $9x^4 - 16$
47. [4.5] $4 - x^2$ 48. [4.5] $13x^2 - 172x + 39$
49. [4.6] 49
50. [4.6] Coefficients: 1, -7, 9, -8; degrees: 6, 2, 2, 0; 6
51. [4.6] Coefficients: 1, -1, 1; degrees: 16, 40, 23; 40
52. [4.6] $9w - y - 5$ 53. [4.6] $6m^3 + 4m^2n - mn^2$
54. [4.6] $-x^2 - 10xy$
55. [4.6] $11x^3y^2 - 8x^2y - 6x^2 - 6x + 6$
56. [4.6] $p^3 - q^3$ 57. [4.6] $9a^8 - 2a^4b^3 + \frac{1}{9}b^6$
58. [4.6] $\frac{1}{2}x^2 - \frac{1}{2}y^2$ 59. [4.7] $5x^2 - \frac{1}{2}x + 3$
60. [4.7] $3x^2 - 7x + 4 + \frac{1}{2x+3}$ 61. [4.7] $t^3 + 2t - 3$
62. [4.8] $\frac{1}{y^4}$ 63. [4.8] t^{-5} 64. [4.8] $\frac{1}{7^2}$, or $\frac{1}{49}$
65. [4.8] $\frac{1}{a^{13}b^7}$ 66. [4.8] $\frac{1}{x^{12}}$ 67. [4.8] $\frac{x^6}{4y^2}$
68. [4.8] $\frac{y^3}{8x^3}$ 69. [4.8] 8,300,000
70. [4.8] 3.28×10^{-5} 71. [4.8] 2.09×10^4
72. [4.8] 5.12×10^{-5} 73. [4.8] $\$1.96 \times 10^{11}$
74. [1.5] $13x + \frac{22}{15}$ 75. [2.5] $w = 125.5$ m, $l = 144.5$ m
76. [2.6] $\{x \mid x \geq -2\}$ 77. [3.1] IV
78. [4.1] ◆ In the expression $5x^3$, the exponent refers only to the x. In the expression $(5x)^3$, the entire expression within the parentheses is cubed.
79. [4.3] ◆ The sum of two polynomials of degree n will also have degree n, since only the coefficients are added and the variables remain unchanged. An exception to this occurs when the leading terms of the two polynomials are opposites. The sum of those terms is then zero and the sum of the polynomials will have a degree less than n.
80. [4.2], [4.5] (a) 3; (b) 2   81. [4.1], [4.2] $-28x^8$
82. [4.2] $8x^4 + 4x^3 + 5x - 2$
83. [4.5] $-4x^6 + 3x^4 - 20x^3 + x^2 - 16$
84. [2.2], [4.5] $\frac{94}{13}$

TEST: CHAPTER 4, P. 241

1. [4.1] x^9 2. [4.1] $(x + 3)^{11}$ 3. [4.1] 3^3 4. [4.1] 1
5. [4.1] x^6 6. [4.1] $-27y^6$ 7. [4.1] $-24x^{17}$
8. [4.1] a^6b^5 9. [4.2] Trinomial   10. [4.2] $\frac{1}{3}$, -1, 7

11. [4.2] Degrees of terms: 3, 1, 5, 0; leading term: $7t^5$; leading coefficient: 7; degree of polynomial: 5
12. [4.2] -7 13. [4.2] $5a^2 - 6$ 14. [4.2] $\frac{7}{4}y^2 - 4y$
15. [4.2] $x^5 + 2x^3 + 4x^2 - 8x + 3$
16. [4.3] $4x^5 + x^4 + 2x^3 - 8x^2 + 2x - 7$
17. [4.3] $5x^4 + 5x^2 + x + 5$
18. [4.3] $-4x^4 + x^3 - 8x - 3$
19. [4.3] $-x^5 + 1.3x^3 - 0.8x^2 - 3$
20. [4.4] $-12x^4 + 9x^3 + 15x^2$ 21. [4.5] $x^2 - \frac{2}{3}x + \frac{1}{9}$
22. [4.5] $9x^2 - 100$ 23. [4.5] $3b^2 - 4b - 15$
24. [4.5] $x^{14} - 4x^8 + 4x^6 - 16$
25. [4.5] $48 + 34y - 5y^2$ 26. [4.4] $6x^3 - 7x^2 - 11x - 3$
27. [4.5] $64a^2 + 48a + 9$
28. [4.6] $-5x^3y - x^2y^2 + xy^3 - y^3 + 19$
29. [4.6] $8a^2b^2 + 6ab - 4b^3 + 6ab^2 + ab^3$
30. [4.6] $9x^{10} - 16y^{10}$ 31. [4.7] $4x^2 + 3x - 5$
32. [4.7] $2x^2 - 4x - 2 + \frac{17}{3x+2}$ 33. [4.8] $\frac{1}{5^3}$
34. [4.8] y^{-8} 35. [4.8] $\frac{1}{6^5}$ 36. [4.8] $\frac{y^5}{x^5}$
37. [4.8] $\frac{b^4}{16a^{12}}$ 38. [4.8] $\frac{c^3}{a^3b^3}$ 39. [4.8] 3.9×10^9
40. [4.8] 0.00000005   41. [4.8] 1.75×10^{17}
42. [4.8] 1.296×10^{22} 43. [4.8] 1.5×10^4
44. [2.6] $\{x \mid x > 13\}$
45. [3.1]

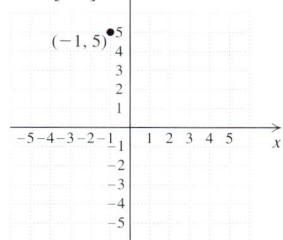

46. [1.5] $-\frac{7}{20}$ 47. [2.5] 64°, 32°, 84°
48. [4.4], [4.5] $V = l(l-2)(l-1) = l^3 - 3l^2 + 2l$
49. [2.2], [4.5] $\frac{100}{21}$

CHAPTER 5

TECHNOLOGY CONNECTION 5.1, P. 248

1. Let $y_1 = 8x^4 + 6x - 28x^3 - 21$ and $y_2 = (4x^3 + 3)(2x - 7)$. Note that the Y1- and Y2-columns of the table match regardless of how far we scroll up or down.

EXERCISE SET 5.1, P. 249

1. Answers may vary. $(10x)(x^2)$, $(5x^2)(2x)$, $(-2)(-5x^3)$
2. Answers may vary. $6x^3 = (6x)(x^2) = (3x^2)(2x) = (2x^2)(3x)$

3. Answers may vary. $(-3x^2)(2x^3)$, $(-x)(6x^4)$, $(2x^2)(-3x^3)$ **4.** Answers may vary.
$-15x^6 = (-3x^2)(5x^4) = (3x^3)(-5x^3) = (-15x)(x^5)$
5. Answers may vary. $(2x)(13x^4)$, $(13x^5)(2)$, $(-x^3)(-26x^2)$
6. Answers may vary. $25x^4 = (5x^2)(5x^2) = (x^3)(25x) = (-5x)(-5x^3)$ **7.** $x(x-6)$ **8.** $x(x+8)$
9. $4x(2x+1)$ **10.** $5x(x-2)$ **11.** $x^2(x+6)$
12. $x^2(4x^2+1)$ **13.** $8x^2(x^2-3)$ **14.** $5x^3(x^2+2)$
15. $2(x^2+x-4)$ **16.** $3(2x^2+x-5)$
17. $x^2(5x^4-10x+8)$ **18.** $x^3(10x^2+6x-3)$
19. $2x^2(x^6+2x^4-4x^2+5)$ **20.** $5(x^4-3x^3-5x-2)$
21. $x^2y^2(x^3y^3+x^2y+xy-1)$
22. $x^3y^3(x^6y^3-x^4y^2+xy+1)$
23. $\frac{1}{3}x^3(5x^3+4x^2+x+1)$
24. $\frac{1}{7}x(5x^6+3x^4-6x^2-1)$ **25.** $(y+3)(y+7)$
26. $b(b-5)+3(b-5) = (b-5)(b+3)$
27. $(x+3)(x^2-7)$
28. $3z^2(2z+9)+(2z+9) = (2z+9)(3z^2+1)$
29. $(y+8)(y^2+1)$
30. $x^2(x-7)-3(x-7) = (x-7)(x^2-3)$
31. $(x+3)(x^2+4)$
32. $6z^3+3z^2+2z+1 = 3z^2(2z+1)+(2z+1) = (2z+1)(3z^2+1)$
33. $(x+3)(2x^2+3)$
34. $3x^3+2x^2+3x+2 = x^2(3x+2)+(3x+2) = (3x+2)(x^2+1)$ **35.** $(3x-4)(3x^2+1)$
36. $10x^3-25x^2+4x-10 = 5x^2(2x-5)+2(2x-5) = (2x-5)(5x^2+2)$
37. $(5x-2)(x^2+1)$
38. $18x^3-21x^2+30x-35 = 3x^2(6x-7)+5(6x-7) = (6x-7)(3x^2+5)$
39. $(x+8)(x^2-3)$
40. $2x^3+12x^2-5x-30 = 2x^2(x+6)-5(x+6) = (x+6)(2x^2-5)$ **41.** $(w-7)(w^2+4)$
42. $y^3+8y^2-2y-16 = y^2(y+8)-2(y+8) = (y+8)(y^2-2)$ **43.** Not factorable by grouping
44. $p^3+p^2-3p+10 = p^2(p+1)-(3p-10)$, or $p^3-3p+p^2+10 = p(p^2-3)+p^2+10$
Not factorable by grouping
45. $(x-4)(2x^2-9)$
46. $20g^3-4g^2-25g+5 = 4g^2(5g-1)-5(5g-1) = (5g-1)(4g^2-5)$

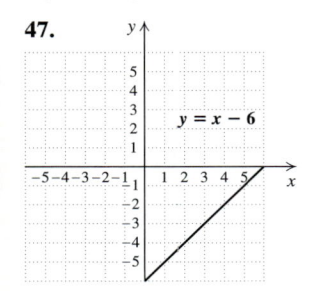

47.
Graph of $y = x - 6$

48. $4x - 8x + 16 \geq 6(x-2)$
$-4x + 16 \geq 6x - 12$
$-10x \geq -28$
$x \leq \frac{28}{10}$
$x \leq \frac{14}{5}$
$\{x \mid x \leq \frac{14}{5}\}$

49. 12 **50.** $A = \frac{p+q}{2}$ **51.** $y^2+12y+35$
$2A = p+q$
$2A - q = p$
52. $y^2+14y+49$ **53.** y^2-49 **54.** $y^2-14y+49$
55. ◆ One factor of $24x^4$ is 6. When we factor $24x^4$, we express it as a product of two monomials. **56.** ◆ Josh is correct, because answers can easily be checked by multiplying. **57.** ◆ Find the product of two binomials. For example, $(ax^2+b)(cx+d) = acx^3+adx^2+bcx+bd$. **58.** ◆ Yes; the opposite of a factor is also a factor so both can be correct.
59. $(2x^2+3)(2x^3+3)$
60. $x^6+x^4+x^2+1 = x^4(x^2+1)+(x^2+1)$
$= (x^2+1)(x^4+1)$
61. $(x^5+1)(x^7+1)$
62. $x^3+x^2-2x+2 = x^2(x+1)-2(x-1)$, or $x^3-2x+x^2+2 = x(x^2-2)+x^2+2$
Not factorable by grouping
63. $(x-1)(5x^4+x^2+3)$
64. $ax^2+2ax+3a+x^2+2x+3$
$= a(x^2+2x+3)+(x^2+2x+3)$
$= (x^2+2x+3)(a+1)$
65. ◆ Find the product of $5x^3y^2$ and a polynomial of degree 4 that has two or more terms and no common factors (other than 1). **66.** ◆ This is a good idea, because it is unlikely that Marlene will choose two replacement values that give the same value for nonequivalent expressions.

EXERCISE SET 5.2, PP. 255–256

1. $(x+2)(x+4)$ **2.** $(x+1)(x+6)$
3. $(x+1)(x+8)$ **4.** $(x+3)(x+4)$
5. $(y+4)(y+7)$ **6.** $(x-3)(x-3)$, or $(x-3)^2$
7. $(a+5)(a+6)$ **8.** $(x+2)(x+7)$
9. $(x-1)(x-4)$ **10.** $(b+1)(b+4)$
11. $(z-1)(z-7)$ **12.** $(a+2)(a-6)$
13. $(x-3)(x-5)$ **14.** $(d-2)(d-5)$
15. $(x+3)(x-5)$ **16.** $(y-1)(y-10)$
17. $(x-3)(x+5)$ **18.** $(x-6)(x+7)$
19. $3(y+4)(y-7)$ **20.** $2x^2-14x-36 = 2(x^2-7x-18) = 2(x+2)(x-9)$ **21.** $x(x+6)(x-7)$
22. $x^3-6x^2+16x = x(x^2-6x+16) = x(x+2)(x-8)$
23. $(x+5)(x-12)$ **24.** $(y+5)(y-9)$
25. $(x-6)(x+12)$ **26.** $-2x-99+x^2 = x^2-2x-99 = (x+9)(x-11)$ **27.** $5(b-3)(b+8)$
28. $c^4+c^3-56c^2 = c^2(c^2+c-56) = c^2(c-7)(c+8)$
29. $x^3(x+2)(x-1)$ **30.** $2a^2+4a-70 = 2(a^2+2a-35) = 2(a-5)(a+7)$ **31.** Prime
32. x^2+x+1. There are no factors of 1 whose sum is 1. This polynomial is not factorable into polynomials with integer coefficients. It is prime. **33.** Prime

34. $7 - 2p + p^2$. There are no factors of 7 whose sum is -2. This polynomial is not factorable into polynomials with integer coefficients. It is prime. **35.** $(x + 9)(x + 11)$
36. $(x + 10)(x + 10)$, or $(x + 10)^2$
37. $2x(x - 8)(x - 12)$
38. $3x^3 - 63x^2 - 300x = 3x(x^2 - 21x - 100) = 3x(x + 4)(x - 25)$ **39.** $4(x + 5)^2$ **40.** $(x + 3)(x - 24)$
41. $(y - 9)(y - 12)$ **42.** $(x - 9)(x - 16)$
43. $a^4(a - 6)(a + 15)$ **44.** $a^4 + a^3 - 132a^2 = a^2(a^2 + a - 132) = a^2(a - 11)(a + 12)$ **45.** $\left(x - \frac{1}{5}\right)^2$
46. $\left(t + \frac{1}{3}\right)\left(t + \frac{1}{3}\right)$, or $\left(t + \frac{1}{3}\right)^2$ **47.** $(16 - y)(7 + y)$
48. $(9 - x)(12 + x)$ **49.** $(t - 0.5)(t + 0.2)$
50. $(y + 0.2)(y - 0.4)$ **51.** $(p + 5q)(p - 2q)$
52. $(a - 3b)(a + b)$ **53.** Prime **54.** $(x - 8y)(x - 3y)$
55. $(s - 5t)(s + 3t)$ **56.** $(b + 10c)(b - 2c)$
57. $6a^8(a - 7)(a + 2)$ **58.** $7x^9 - 28x^8 - 35x^7 = 7x^7(x^2 - 4x - 5) = 7x^7(x + 1)(x - 5)$ **59.** $\frac{8}{3}$
60. $2x + 7 = 0$
$ 2x = -7$
$ x = -\frac{7}{2}$
61. $3x^2 + 22x + 24$
62. $49w^2 + 84w + 36$ **63.** 29,443
64. Solve: $4x + x + (x + 30) = 180$
$ x = 25$
The angles are $100°$, $25°$, and $55°$.
65. ◆ There is a finite number of pairs of numbers with the correct product, but there are infinitely many pairs with the correct sum. **66.** ◆ Common factors must be factored out at some point. If they are factored out first, it is easier to find the factorization of the expression because we are dealing with "smaller" terms. **67.** ◆ Since both constants are negative, the middle term will be negative so $(x - 17)(x - 18)$ cannot be a factorization of $x^2 + 35x + 306$. **68.** ◆ Although $x^3 - 8x^2 + 15x$ can be factored as $(x^2 - 5x)(x - 3)$, this is not a complete factorization of the polynomial since $x^2 - 5x = x(x - 5)$. The student should be advised *always* to look for a common factor first. **69.** $15, -15, 27, -27, 51, -51$
70. $a^2 + ba - 50$

Pairs of Factors Whose Product is -50	Sums of Factors
$-1, \ 50$	49
$1, -50$	-49
$-2, \ 25$	23
$2, -25$	-23
$-5, \ 10$	5
$5, -10$	-5

$a^2 + ba - 50$ can be factored if b is 49, -49, 23, -23, 5, or -5.

71. $\left(x + \frac{1}{2}\right)\left(x - \frac{1}{4}\right)$ **72.** $\left(x + \frac{3}{4}\right)\left(x - \frac{1}{4}\right)$
73. $\frac{1}{3}a(a - 3)(a + 2)$ **74.** $a^7 - \frac{25}{7}a^5 - \frac{30}{7}a^6 = a^5\left(a^2 - \frac{30}{7}a - \frac{25}{7}\right) = a^5(a - 5)\left(a + \frac{5}{7}\right)$
75. $(x^m + 4)(x^m + 7)$ **76.** $(t^n - 2)(t^n - 5)$
77. $(x + 1)(a + 2)(a + 1)$
78. $ ax^2 - 5x^2 + 8ax - 40x - (a - 5)9$
$= (a - 5)x^2 + (a - 5)8x - (a - 5)9$
$= (a - 5)(x^2 + 8x - 9)$
$= (a - 5)(x + 9)(x - 1)$
79. $2x^2(4 - \pi)$
80. Shaded area = Area of circle − Area of triangle = $\pi x^2 - \frac{1}{2}(2x)(x) = \pi x^2 - x^2 = x^2(\pi - 1)$

EXERCISE SET 5.3, PP. 264–265

1. $(3x + 4)(x - 1)$ **2.** $(2x - 1)(x + 4)$
3. $(5x + 9)(x - 2)$ **4.** $(3x + 5)(x - 3)$
5. $(2x - 3)(3x - 2)$ **6.** $(3x - 1)(2x - 7)$
7. $(3x + 1)(x + 1)$ **8.** $(7x + 1)(x + 2)$
9. $(2x + 5)(2x - 3)$ **10.** $(3x - 2)(3x + 4)$
11. $(5x + 2)(3x - 5)$ **12.** $(3x + 1)(x - 2)$
13. $(3x + 8)(3x - 2)$ **14.** $(2x + 1)(x - 1)$
15. $(2x - 1)(x - 2)$ **16.** $(6x - 5)(3x + 2)$
17. $(3x - 4)(4x - 5)$ **18.** $(5x + 3)(3x + 2)$
19. $2(7x - 1)(2x + 3)$ **20.** $(7x + 4)(5x + 2)$
21. $(3x + 4)(3x + 2)$ **22.** $4 - 13x + 6x^2$ is prime.
23. $(7 + 3x)^2$ **24.** $(5x + 4)^2$ **25.** $(x + 2)(24x - 1)$
26. $(8a + 3)(2a + 9)$ **27.** $(7x + 4)(5x - 11)$
28. $18t^2 + 24t - 10 = 2(9t^2 + 12t - 5) = 2(3t - 1)(3t + 5)$ **29.** Prime **30.** $(2x + 5)(x - 3)$
31. $4(3x - 2)(x + 3)$ **32.** $6x^2 + 33x + 15 = 3(2x^2 + 11x + 5) = 3(2x + 1)(x + 5)$
33. $6(5x - 9)(x + 1)$ **34.** $20x^2 - 25x + 5 = 5(4x^2 - 5x + 1) = 5(4x - 1)(x - 1)$
35. $3(2x + 1)(x + 5)$ **36.** $12x^2 + 28x - 24 = 4(3x^2 + 7x - 6) = 4(3x - 2)(x + 3)$
37. $3(2x + 3)(3x - 1)$ **38.** $4x + 1 + 3x^2 = 3x^2 + 4x + 1 = (3x + 1)(x + 1)$ **39.** $(y + 4)(y - 2)$
40. $x^2 + 5x + 2x + 10 = x(x + 5) + 2(x + 5) = (x + 5)(x + 2)$ **41.** $(x - 4)(x - 1)$
42. $a^2 + 5a - 2a - 10 = a(a + 5) - 2(a + 5) = (a + 5)(a - 2)$ **43.** $(3x + 2)(2x + 3)$
44. $3x^2 - 2x + 3x - 2 = x(3x - 2) + (3x - 2) = (3x - 2)(x + 1)$ **45.** $(3x - 4)(x - 4)$
46. $24 - 18y - 20y + 15y^2 = 6(4 - 3y) - 5y(4 - 3y) = (4 - 3y)(6 - 5y)$ **47.** $(7x - 8)(5x + 3)$
48. $8x^2 - 6x - 28x + 21 = 2x(4x - 3) - 7(4x - 3) = (4x - 3)(2x - 7)$ **49.** $(2x + 3)(2x - 3)$
50. $2x^4 - 6x^2 - 5x^2 + 15 = 2x^2(x^2 - 3) - 5(x^2 - 3) = (x^2 - 3)(2x^2 - 5)$ **51.** $(3x - 7)^2$

52. $25t^2 + 80t + 64 = 25t^2 + 40t + 40t + 64$
$= 5t(5t + 8) + 8(5t + 8)$
$= (5t + 8)(5t + 8)$, or
$(5t + 8)^2$

53. $(3t - 2)(6t + 5)$

54. $15x^2 - 25x - 10 = 5(3x^2 - 5x - 2)$
Factor $3x^2 - 5x - 2$ by grouping:
$3x^2 - 5x - 2 = 3x^2 - 6x + x - 2$
$= 3x(x - 2) + (x - 2)$
$= (x - 2)(3x + 1)$
Then $15x^2 - 25x - 10 = 5(x - 2)(3x + 1)$.

55. $7(2x - 1)(x - 2)$ **56.** $2x^2 + 6x - 14 = 2(x^2 + 3x - 7)$ **57.** $2x(3x - 5)(x + 1)$

58. $18x^3 + 21x^2 - 9x = 3x(6x^2 + 7x - 3)$
Factor $6x^2 + 7x - 3$ by grouping:
$6x^2 + 7x - 3 = 6x^2 + 9x - 2x - 3$
$= 3x(2x + 3) - (2x + 3)$
$= (2x + 3)(3x - 1)$
Then $18x^3 + 21x^2 - 9x = 3x(2x + 3)(3x - 1)$.

59. Prime

60. $89x + 64 + 25x^2 = 25x^2 + 25x + 64x + 64$
$= 25x(x + 1) + 64(x + 1)$
$= (x + 1)(25x + 64)$

61. $24x^3(3x - 2)(2x - 1)$

62. $168x^3 + 45x^2 + 3x = 3x(56x^2 + 15x + 1)$
Factor $56x^2 + 15x + 1$ by grouping:
$56x^2 + 15x + 1 = 56x^2 + 8x + 7x + 1$
$= 8x(7x + 1) + 7x + 1$
$= (7x + 1)(8x + 1)$
Then $168x^3 + 45x^2 + 3x = 3x(7x + 1)(8x + 1)$.

63. $2a^2(7a - 4)(5a - 2)$

64. $14t^4 - 19t^3 - 3t^2 = t^2(14t^2 - 19t - 3)$
Factor $14t^2 - 19t - 3$ by grouping:
$14t^2 - 19t - 3 = 14t^2 - 21t + 2t - 3$
$= 7t(2t - 3) + 2t - 3$
$= (2t - 3)(7t + 1)$
Then $14t^4 - 19t^3 - 3t^2 = t^2(2t - 3)(7t + 1)$.

65. $(4m - 5n)(3m + 4n)$

66. $6a^2 - ab - 15b^2 = 6a^2 - 10ab + 9ab - 15b^2$
$= 2a(3a - 5b) + 3b(3a - 5b)$
$= (3a - 5b)(2a + 3b)$

67. $(3p + 2q)(p - 6q)$

68. $9a^2 + 18ab + 8b^2 = 9a^2 + 6ab + 12ab + 8b^2$
$= 3a(3a + 2b) + 4b(3a + 2b)$
$= (3a + 2b)(3a + 4b)$

69. $2(5s - 3t)(s + t)$

70. $35p^2 + 34pq + 8q^2 = 35p^2 + 20pq + 14pq + 8q^2$
$= 5p(7p + 4q) + 2q(7p + 4q)$
$= (7p + 4q)(5p + 2q)$

71. $3(2a + 5b)(5a + 2b)$

72. $18x^2 - 6xy - 24y^2 = 6(3x^2 - xy - 4y^2)$
Factor $3x^2 - xy - 4y^2$ by grouping:
$3x^2 - xy - 4y^2 = 3x^2 - 4xy + 3xy - 4y^2$
$= x(3x - 4y) + y(3x - 4y)$
$= (3x - 4y)(x + y)$
Then $18x^2 - 6xy - 24y^2 = 6(3x - 4y)(x + y)$.

73. $5(3a - 4b)(a + b)$

74. $24a^2 - 34ab + 12b^2 = 2(12a^2 - 17ab + 6b^2)$
Factor $12a^2 - 17ab + 6b^2$ by grouping:
$12a^2 - 17ab + 6b^2 = 12a^2 - 8ab - 9ab + 6b^2$
$= 4a(3a - 2b) - 3b(3a - 2b)$
$= (3a - 2b)(4a - 3b)$
Then $24a^2 - 34ab + 12b^2 = 2(3a - 2b)(4a - 3b)$.

75. Prime **76.** $4x^2y + 10xy + 2y = 2y(2x^2 + 5x + 1)$

77. $(6xy - 5)(3xy + 2)$ **78.** $9a^2b^2 - 15ab - 2$ is prime.

79. $x^2(4x + 7)(2x + 5)$

80. $19x^3 - 3x^2 + 14x^4 = x^2(14x^2 + 19x - 3)$
Factor $14x^2 + 19x - 3$ by grouping:
$14x^2 + 19x - 3 = 14x^2 + 21x - 2x - 3$
$= 7x(2x + 3) - (2x + 3)$
$= (2x + 3)(7x - 1)$
Then $19x^3 - 3x^2 + 14x^4 = x^2(2x + 3)(7x - 1)$.

81. $(3a + 4)(3a + 2)$

82. $40a + 16 + 25a^2 = 25a^2 + 20a + 20a + 16$
$= 5a(5a + 4) + 4(5a + 4)$
$= (5a + 4)(5a + 4)$, or
$(5a + 4)^2$

83. 6369 km, 3949 mi

84. Solve: $x + (2x - 10) + (4x + 15) = 180$
$x = 25$
Then $2x - 10 = 2 \cdot 25 - 10$, or 40°.

85.

86. y^8 **87.** $9x^2 - 25$ **88.** $(4a - 3)^2 = 16a^2 - 24a + 9$

89. ◆ No; both $2x + 6$ and $2x + 8$ contain a factor of 2, so $2 \cdot 2$, or 4, must be factored out to reach the complete factorization. In other words, the largest common factor is 4, not 2. **90.** ◆ Answers will vary.

91. ◆ The student has incorrectly changed the sign of the middle term when factoring out the largest common factor. Thus, the signs in both terms of the final factorization are wrong. The number of points that should be deducted will vary.

92. ◆ The answer to Exercise 68 differs from the answer to Exercise 21 only in the variables.
93. $(4x^5 - 1)^2$ **94.** $(3x^5 + 2)^2$
95. $(10x^n + 3)(2x^n + 1)$ **96.** $-15x^{2m} + 26x^m - 8 = -(15x^{2m} - 26x^m + 8) = -(3x^m - 4)(5x^m - 2)$
97. $(x^{3a} - 1)(3x^{3a} + 1)$ **98.** $x^{2n+1} - 2x^{n+1} + x = x(x^{2n} - 2x^n + 1) = x(x^n - 1)^2$
99. $-2(a + 1)^n(a + 3)^2(a + 6)$
100. $7(t - 3)^{2n} + 5(t - 3)^n - 2 = [7(t - 3)^n - 2][(t - 3)^n + 1]$

EXERCISE SET 5.4, PP. 270–271

1. Yes
2. $x^2 - 16x + 64$
  (1) Two terms, x^2 and 64, are squares.
  (2) There is no minus sign before x^2 or 64.
  (3) Twice the product of the square roots, $2 \cdot x \cdot 8$, is $16x$, the opposite of the remaining term, $-16x$.
Thus, $x^2 - 16x + 64$ is a perfect-square trinomial.
3. No
4. $x^2 - 14x - 49$
  (1) Two terms, x^2 and 49, are squares.
  (2) There is a minus sign before 49, so $x^2 - 14x - 49$ is not a perfect-square trinomial.
5. No
6. $x^2 + 2x + 4$
  (1) Two terms, x^2 and 4, are squares.
  (2) There is no minus sign before x^2 or 4.
  (3) Twice the product of the square roots, $2 \cdot x \cdot 2$, is $4x$. This is neither the remaining term nor its opposite, so $x^2 + 2x + 4$ is not a perfect-square trinomial.
7. No
8. $36x^2 - 24x + 16$
  (1) Two terms, $36x^2$ and 16, are squares.
  (2) There is no minus sign before $36x^2$ or 16.
  (3) Twice the product of the square roots, $2 \cdot 6x \cdot 4$, is $48x$. This is neither the remaining term nor its opposite, so $36x^2 - 24x + 16$ is not a perfect-square trinomial.
9. $(x - 8)^2$ **10.** $(x - 7)^2$ **11.** $(x + 7)^2$ **12.** $(x + 8)^2$
13. $(x - 1)^2$ **14.** $5x^2 - 10x + 5 = 5(x^2 - 2x + 1) = 5(x - 1)^2$ **15.** $(2 + x)^2$ **16.** $(x - 2)^2$ **17.** $2(3x - 1)^2$
18. $(5x + 1)^2$ **19.** $(7 + 4y)^2$
20. $120m + 75 + 48m^2 = 3(16m^2 + 40m + 25) = 3(4m + 5)^2$ **21.** $x^3(x - 9)^2$ **22.** $2x^2 - 40x + 200 = 2(x^2 - 20x + 100) = 2(x - 10)^2$
23. $2x(x - 1)^2$ **24.** $x^3 - 24x^2 + 144x = x(x^2 + 24x + 144) = x(x + 12)^2$ **25.** $5(2x + 5)^2$
26. $12x^2 + 36x + 27 = 3(4x^2 + 12x + 9) = 3(2x + 3)^2$
27. $(7 - 3x)^2$ **28.** $(8 - 7x)^2$, or $(7x - 8)^2$

29. $5(y + 1)^2$ **30.** $2a^2 + 28a + 98 = 2(a^2 + 14a + 49) = 2(a + 7)^2$ **31.** $2(1 + 5x)^2$
32. $7 - 14a + 7a^2 = 7(1 - 2a + a^2) = 7(1 - a)^2$
(This result could also be expressed as $7(a - 1)^2$.)
33. $(2p + 3q)^2$ **34.** $(5m + 2n)^2$ **35.** $(a - 7b)^2$
36. $(x - 3y)^2$ **37.** $(8m + n)^2$ **38.** $(9p - q)^2$
39. $(4s - 5t)^2$ **40.** $36a^2 + 96ab + 64b^2 = 4(9a^2 + 24ab + 16b^2) = 4(3a + 4b)^2$ **41.** Yes
42. $x^2 - 36$
  (1) The first expression is a square: x^2
    The second expression is a square: $36 = 6^2$
  (2) The terms have different signs.
Thus, $x^2 - 36$ is a difference of squares, $x^2 - 6^2$.
43. No
44. $x^2 + 4$
  (1) The first expression is a square: x^2
    The second expression is a square: $4 = 2^2$
  (2) The terms do not have different signs.
Thus, $x^2 + 4$ is not a difference of squares.
45. No
46. $x^2 - 50y^2$ **47.** Yes
  (1) The expression $50y^2$ is not a square.
Thus, $x^2 - 50y^2$ is not a difference of squares.
48. $-1 + 36x^2$
  (1) The first expression is a square: $1 = 1^2$
    The second expression is a square: $36x^2 = (6x)^2$
  (2) The terms have different signs.
Thus, $-1 + 36x^2$ is a difference of squares, $(6x)^2 - 1^2$.
49. $(y + 2)(y - 2)$ **50.** $(x + 6)(x - 6)$
51. $(p + 3)(p - 3)$ **52.** $(q + 1)(q - 1)$
53. $(t + 7)(t - 7)$ **54.** $(m + 8)(m - 8)$
55. $(a + b)(a - b)$ **56.** $(p + q)(p - q)$
57. $(mn + 7)(mn - 7)$ **58.** $(5 + ab)(5 - ab)$
59. $2(10 + t)(10 - t)$ **60.** $(9 + w)(9 - w)$
61. $(4a + 3)(4a - 3)$ **62.** $(5x + 2)(5x - 2)$
63. $(2x + 5y)(2x - 5y)$ **64.** $(3a + 4b)(3a - 4b)$
65. $2(2x + 7)(2x - 7)$ **66.** $24x^2 - 54 = 6(4x^2 - 9) = 6(2x + 3)(2x - 3)$ **67.** $x(6 + 7x)(6 - 7x)$
68. $16x - 81x^3 = x(16 - 81x^2) = x(4 + 9x)(4 - 9x)$
69. $(7a^2 + 9)(7a^2 - 9)$ **70.** $(5a^2 + 3)(5a^2 - 3)$
71. $5(x^2 + 1)(x + 1)(x - 1)$
72. $(x^4 - 16) = (x^2 + 4)(x^2 - 4)$
$= (x^2 + 4)(x + 2)(x - 2)$
73. $4(x^2 + 4)(x + 2)(x - 2)$
74. $5x^4 - 80 = 5(x^4 - 16) = 5(x^2 + 4)(x^2 - 4) = 5(x^2 + 4)(x + 2)(x - 2)$
75. $(1 + y^4)(1 + y^2)(1 + y)(1 - y)$
76. $(x^8 - 1) = (x^4 + 1)(x^4 - 1)$
$= (x^4 + 1)(x^2 + 1)(x^2 - 1)$
$= (x^4 + 1)(x^2 + 1)(x + 1)(x - 1)$

77. $3x(x - 4)^2$ **78.** $2a^4 - 36a^3 + 162a^2 = 2a^2(a^2 - 18a + 81) = 2a^2(a - 9)^2$
79. $(x^6 + 4)(x^3 + 2)(x^3 - 2)$
80. $(x^8 - 81) = (x^4 + 9)(x^4 - 9)$
$= (x^4 + 9)(x^2 + 3)(x^2 - 3)$
81. $\left(y + \frac{1}{4}\right)\left(y - \frac{1}{4}\right)$ **82.** $\left(x + \frac{1}{5}\right)\left(x - \frac{1}{5}\right)$ **83.** $a^6(a - 1)^2$
84. $x^8 - 8x^7 + 16x^6 = x^6(x^2 - 8x + 16) = x^6(x - 4)^2$
85. $(4 + m^2n^2)(2 + mn)(2 - mn)$
86. $1 - a^4b^4 = (1 + a^2b^2)(1 - a^2b^2)$
$= (1 + a^2b^2)(1 + ab)(1 - ab)$
87. $s \geq 77$ **88.** Solve: $1.6a = 5$ **89.** $x^{12}y^{12}$
$a = 3.125\,L$
90. $25a^4b^6$

91.

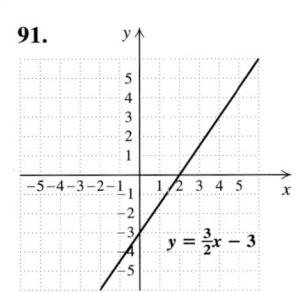

92. $3x - 5y = 30$

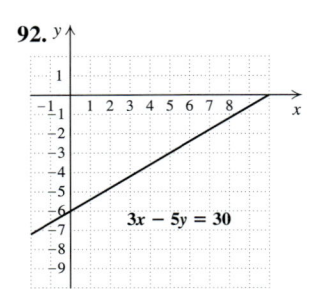

93. ◆ No; we reverse the procedures for squaring binomials and for multiplying sums and differences of terms in order to factor perfect-square trinomials and differences of squares, respectively.
94. ◆
$(x + 3)(x - 3)$
$= (x - 3)(x + 3)$ Using a commutative law
$= x^2 - 9$
Since $x^2 - 9$ and $x^2 + 9$ are not equivalent, the student's factorization of $x^2 + 9$ is incorrect. (Also it can be easily shown by multiplying that $(x + 3)(x - 3) \neq x^2 + 9$.) The student should recall that, if the greatest common factor has been removed, a sum of squares cannot be factored further.
95. ◆ Two terms must be squares. There must be no minus sign before either square. The remaining term must be twice the product of the square roots of the squares or must be the opposite of that product. **96.** ◆ First determine whether there are two expressions, both of which are squares. If this is the case, then determine whether the expressions have different signs. Both of these conditions must be satisfied in order for a polynomial to be a difference of squares. **97.** $(x + 1.5)(x - 1.5)$
98. $81x^2 + 216 = 27(3x^2 + 8)$ **99.** Prime
100. $x^8 - 2^8 = (x^4 + 2^4)(x^4 - 2^4)$
$= (x^4 + 2^4)(x^2 + 2^2)(x^2 - 2^2)$
$= (x^4 + 2^4)(x^2 + 2^2)(x + 2)(x - 2)$, or
$(x^4 + 16)(x^2 + 4)(x + 2)(x - 2)$
101. $3\left(x + \frac{1}{3}\right)\left(x - \frac{1}{3}\right)$

102. $18x^3 - \frac{8}{25}x = 2x\left(9x^2 - \frac{4}{25}\right) = 2x\left(3x + \frac{2}{5}\right)\left(3x - \frac{2}{5}\right)$
103. $p(0.7 + p)(0.7 - p)$
104. $0.64x^2 - 1.21 = (0.8x)^2 - (1.1)^2 =$
$(0.8x + 1.1)(0.8x - 1.1)$ **105.** $x(x + 6)(x^2 + 6x + 18)$
106. $(y - 5)^4 - z^8$
$= [(y - 5)^2 + z^4][(y - 5)^2 - z^4]$
$= [(y - 5)^2 + z^4][y - 5 + z^2][y - 5 - z^2]$
107. $\left(x + \frac{1}{x}\right)\left(x - \frac{1}{x}\right)$
108. $a^{2n} - 49b^{2n} = (a^n)^2 - (7b^n)^2 = (a^n + 7b^n)(a^n - 7b^n)$
109. $(9 + b^{2k})(3 + b^k)(3 - b^k)$
110. $x^4 - 8x^2 - 9 = (x^2 - 9)(x^2 + 1)$
$= (x + 3)(x - 3)(x^2 + 1)$
111. $(3b^n + 2)^2$
112. $16x^4 - 96x^2 + 144 = 16(x^4 - 6x^2 + 9)$
$= 16(x^2 - 3)^2$
113. $(y + 4)^2$ **114.** $49(x + 1)^2 - 42(x + 1) + 9 =$
$[7(x + 1) - 3]^2 = (7x + 7 - 3)^2 = (7x + 4)^2$
115. $(3x + 7)(3x - 7)^2$
116. $x^2(x + 1)^2 - (x^2 + 1)^2$
$= x^2(x^2 + 2x + 1) - (x^4 + 2x^2 + 1)$
$= x^4 + 2x^3 + x^2 - x^4 - 2x^2 - 1$
$= 2x^3 + x^2 - 2x^2 - 1$
$= (2x^3 - 2x^2) + (x^2 - 1)$
$= 2x^2(x - 1) + (x + 1)(x - 1)$
$= (x - 1)[2x^2 + (x + 1)]$
$= (x - 1)(2x^2 + x + 1)$
117. $(a + 4)(a - 2)$
118. $y^2 + 6y + 9 - x^2 - 8x - 16$
$= (y^2 + 6y + 9) - (x^2 + 8x + 16)$
$= (y + 3)^2 - (x + 4)^2$
$= [(y + 3) + (x + 4)][(y + 3) - (x + 4)]$
$= (y + 3 + x + 4)(y + 3 - x - 4)$
$= (y + x + 7)(y - x - 1)$
119. 9 **120.** For $c = a^2$, $2 \cdot a \cdot 3 = 24$. Then $a = 4$, so $c = 4^2 = 16$. **121.** 0, 2
122. $(x + 1)^2 - x^2$
$= [(x + 1) + x](x + 1 - x)$
$= 2x + 1$

EXERCISE SET 5.5, PP. 276–277

1. $5(x + 3)(x - 3)$ **2.** $10a^2 - 640 = 10(a^2 - 64) = 10(a + 8)(a - 8)$ **3.** $(a + 5)^2$ **4.** $(y - 7)^2$
5. $(2x - 3)(x - 4)$ **6.** $(2y + 5)(4y - 1)$
7. $x(x - 12)^2$ **8.** $x^3 - 18x^2 + 81x = x(x^2 - 18x + 81) = x(x - 9)^2$
9. $(x + 3)(x + 2)(x - 2)$ **10.** $x^3 - 5x^2 - 25x + 125 = x^2(x - 5) - 25(x - 5) = (x - 5)(x^2 - 25) = (x - 5)(x + 5)(x - 5) = (x + 5)(x - 5)^2$
11. $6(2x + 3)(2x - 3)$ **12.** $8x^2 - 98 = 2(4x^2 - 49) = 2(2x + 7)(2x - 7)$ **13.** $4x(x - 2)(5x + 9)$

14. $9x^3 + 12x^2 - 45x = 3x(3x^2 + 4x - 15) = 3x(x + 3)(3x - 5)$ **15.** Prime **16.** Prime
17. $a(a^2 + 8)(a + 8)$
18. $t^4 + 7t^2 - 3t^3 - 21t$
$= t(t^3 + 7t - 3t^2 - 21)$
$= t[t(t^2 + 7) - 3(t^2 + 7)]$
$= t(t^2 + 7)(t - 3)$
19. $x^3(x - 7)^2$ **20.** $2x^6 + 8x^5 + 8x^4 = 2x^4(x^2 + 4x + 4) = 2x^4(x + 2)^2$ **21.** $-2(x - 2)(x + 5)$
22. $45 - 3x - 6x^2 = -3(2x^2 + x - 15) = -3(2x - 5)(x + 3)$, or $3(5 - 2x)(3 + x)$ **23.** Prime
24. Prime **25.** $4(x^2 + 4)(x + 2)(x - 2)$
26. $5x^5 - 80x = 5x(x^4 - 16) = 5x(x^2 + 4)(x^2 - 4) = 5x(x^2 + 4)(x + 2)(x - 2)$
27. $(t^4 + 1)(t^2 + 1)(t + 1)(t - 1)$
28. $1 - n^8 = (1 + n^4)(1 - n^4)$
$= (1 + n^4)(1 + n^2)(1 - n^2)$
$= (1 + n^4)(1 + n^2)(1 + n)(1 - n)$
29. $x^3(x - 3)(x - 1)$ **30.** $x^6 - 2x^5 + 7x^4 = x^4(x^2 - 2x + 7)$ **31.** $(x + y)(x - y)$
32. $(pq + r)(pq - r)$ **33.** $12n^2(1 + 2n)$
34. $a(x^2 + y^2)$ **35.** $ab(b - a)$ **36.** $36mn - 9m^2n^2 = 9mn(4 - mn)$ **37.** $2\pi r(h + r)$ **38.** $10p^4q^4 + 35p^3q^3 + 10p^2q^2 = 5p^2q^2(2p^2q^2 + 7pq + 2)$ **39.** $(a + b)(2x + 1)$
40. $(a^3 + b)(5c - 1)$ **41.** $(x + 2)(x - 5 - y)$
42. $n^2 + 2n + np + 2p$ **43.** $(x + 1)(x + y)$
$= n(n + 2) + p(n + 2)$
$= (n + 2)(n + p)$
44. $2x^2 - 4x + xz - 2z$ **45.** $(a - 3)(a + y)$
$= 2x(x - 2) + z(x - 2)$
$= (x - 2)(2x + z)$
46. $x^2 + y^2 - 2xy$ **47.** $(2y - 1)(3y + p)$
$= x^2 - 2xy + y^2$
$= (x - y)^2$
48. $9c^2 + 6cd + d^2$ **49.** $(a - 2b)^2$, or $(2b - a)^2$
$= (3c)^2 + 2 \cdot 3c \cdot d + d^2$
$= (3c + d)^2$
50. $7p^4 - 7q^4$
$= 7(p^4 - q^4)$
$= 7(p^2 + q^2)(p^2 - q^2)$
$= 7(p^2 + q^2)(p + q)(p - q)$
51. $(4x + 3y)^2$ **52.** $(5z + y)^2$ **53.** $(2xy + 3z)^2$
54. $a^5 + 4a^4b - 5a^3b^2$
$= a^3(a^2 + 4ab - 5b^2)$
$= a^3(a + 5b)(a - b)$
55. $(a^2b^2 + 4)(ab + 2)(ab - 2)$ **56.** $(a - 2b)(a + b)$
57. $p(2p + q)^2$ **58.** $2mn - 360n^2 + m^2$
$= m^2 + 2mn - 360n^2$
$= (m + 20n)(m - 18n)$
59. $(3b + a)(b - 6a)$ **60.** $(mn - 8)(mn + 4)$
61. $(xy + 5)(xy + 3)$ **62.** $a^5b^2 + 3a^4b - 10a^3$
$= a^3(a^2b^2 + 3ab - 10)$
$= a^3(ab + 5)(ab - 2)$

63. $(pq + 6)(pq + 1)$ **64.** $(7x + 8y)^2$
65. $b^4(ab + 8)(ab - 4)$ **66.** $2s^6t^2 + 10s^3t^3 + 12t^4 = 2t^2(s^6 + 5s^3t + 6t^2) = 2t^2(s^3 + 3t)(s^3 + 2t)$
67. $x^4(x + 2y)(x - y)$ **68.** $a^2 + 2a^2bc + a^2b^2c^2 = a^2(1 + 2bc + b^2c^2) = a^2(1 + bc)^2$ **69.** $\left(6a - \frac{5}{4}\right)^2$
70. $\left(\frac{1}{9}x - \frac{4}{3}\right)^2$, or $\frac{1}{9}\left(\frac{1}{3}x - 4\right)^2$ **71.** $\left(\frac{1}{2}a + \frac{1}{3}b\right)^2$
72. $1 - 16x^{12}y^{12}$
$= (1 + 4x^6y^6)(1 - 4x^6y^6)$
$= (1 + 4x^6y^6)(1 + 2x^3y^3)(1 - 2x^3y^3)$
73. $a(b^2 + 9a^2)(b + 3a)(b - 3a)$ **74.** $(0.1x - 0.5y)^2$, or $0.01(x - 5y)^2$ **75.** $(w - 7)(w + 2)(w - 2)$
76. $y^3 + 8y^2 - y - 8 = y^2(y + 8) - (y + 8) = (y + 8)(y^2 - 1) = (y + 8)(y + 1)(y - 1)$

77.
$y = -4x + 7$		$y = -4x + 7$	
11 ? $-4(-1) + 7$		7 ? $-4 \cdot 0 + 7$	
$\qquad$ 4 + 7		$\qquad$ 0 + 7	
11 $\vert$ 11	TRUE	7 $\vert$ 7	TRUE

$y = -4x + 7$	
-5 ? $-4 \cdot 3 + 7$	
$\qquad$ $-12 + 7$	
-5 $\vert$ -5	TRUE

78. $2x - 7 = 0$
$2x = 7$
$x = \frac{7}{2}$

79. $-\frac{4}{3}$

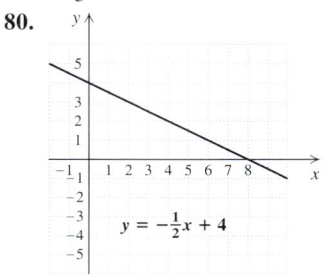

80.

$y = -\frac{1}{2}x + 4$

81. $X = \dfrac{A + 7}{a + b}$

82. $4(x - 9) - 2(x + 7) < 14$
$\quad 4x - 36 - 2x - 14 < 14$
$\qquad\qquad\qquad\quad 2x < 64$
$\qquad\qquad\qquad\quad\ x < 32$

The solution set is $\{x \mid x < 32\}$.
83. ◆ No; the degree of each binomial must be at least 1, so the degree of their product must be at least 4.
84. ◆ See page 273 of the text.

85. ◆ For $x = -3$:
$(x - 4)^2 = (-3 - 4)^2 = (-7)^2 = 49$
$(4 - x)^2 = [4 - (-3)]^2 = 7^2 = 49$
For $x = 1$:
$(x - 4)^2 = (1 - 4)^2 = (-3)^2 = 9$
$(4 - x)^2 = (4 - 1)^2 = 3^2 = 9$
In general, $(x - 4)^2 = [-(-x + 4)]^2 = [-(4 - x)]^2 = (-1)^2(4 - x)^2 = (4 - x)^2$.

86. ◆ Find the product of a binomial and a trinomial. One example is found as follows:
$(x + 1)(2x^2 - 3x + 4)$
$= 2x^3 - 3x^2 + 4x + 2x^2 - 3x + 4$
$= 2x^3 - x^2 + x + 4$

87. $-x(x^2 + 9)(x^2 - 2)$

88. $18 + a^3 - 9a - 2a^2$
$= a^3 - 2a^2 - 9a + 18$
$= a^2(a - 2) - 9(a - 2)$
$= (a - 2)(a^2 - 9)$
$= (a - 2)(a + 3)(a - 3)$

89. $3(a + 1)(a - 1)(a + 2)(a - 2)$

90. $x^4 - 7x^2 - 18$
$= (x^2 + 2)(x^2 - 9)$
$= (x^2 + 2)(x + 3)(x - 3)$

91. $(x - 1)(x + 2)(x - 2)$

92. $y^2(y + 1) - 4y(y + 1) - 21(y + 1)$
$= (y + 1)(y^2 - 4y - 21)$
$= (y + 1)(y - 7)(y + 3)$

93. $(y - 1)^3$

94. $6(x - 1)^2 + 7y(x - 1) - 3y^2$
$= [2(x - 1) + 3y][3(x - 1) - y]$
$= (2x + 3y - 2)(3x - y - 3)$

95. $[(y + 4) + x]^2$

96. $2(a + 3)^2 - (a + 3)(b - 2) - (b - 2)^2$
$= [2(a + 3) + (b - 2)][(a + 3) - (b - 2)]$
$= (2a + b + 4)(a - b + 5)$

97. $(x^4 + 16)(x^2 + 4)(x + 2)(x - 2)$

98. $6x^2 - xy - 15y^2$
$= 6 \cdot 2^2 - 2(-1) - 15(-1)^2$
$= 6 \cdot 4 - 2(-1) - 15 \cdot 1$
$= 24 + 2 - 15 = 11$
$(2x + 3y)(3x - 5y)$
$= [2 \cdot 2 + 3(-1)][3 \cdot 2 - 5(-1)]$
$= (4 - 3)(6 + 5)$
$= 1 \cdot 11 = 11$
Since the value of both expressions is 11, the factorization is probably correct.

99. $49 = 49$; probably correct

TECHNOLOGY CONNECTION 5.6, P. 282

1. $-4.65, 0.65$ 2. $-0.37, 5.37$ 3. $-8.98, -4.56$
4. No solution

EXERCISE SET 5.6, PP. 283–284

1. $-7, -6$ 2. $-1, -2$ 3. $3, -5$ 4. $-9, 3$
5. $\frac{9}{2}, -4$ 6. $\frac{5}{3}, -1$ 7. $\frac{7}{10}, -\frac{9}{4}$ 8. $\frac{7}{2}, -\frac{4}{3}$ 9. $0, -6$

10. $0, -9$ 11. $\frac{18}{11}, \frac{1}{21}$
12. $\left(\frac{1}{9} - 3x\right)\left(\frac{1}{5} + 2x\right) = 0$
$\frac{1}{9} - 3x = 0$ or $\frac{1}{5} + 2x = 0$
$-3x = -\frac{1}{9}$ or $2x = -\frac{1}{5}$
$x = \frac{1}{27}$ or $x = -\frac{1}{10}$

13. $0, -\frac{9}{2}$ 14. $0, -\frac{7}{4}$ 15. $50, 70$
16. $(1 - 0.05x)(1 - 0.3x) = 0$
$1 - 0.05x = 0$ or $1 - 0.3x = 0$
$-0.05x = -1$ or $-0.3x = -1$
$x = 20$ or $x = \frac{10}{3}$

17. $-3, \frac{5}{2}, 6$ 18. $4, -9, \frac{1}{3}$ 19. $1, 6$
20. $x^2 - 6x + 5 = 0$ 21. $-3, 7$
$(x - 1)(x - 5) = 0$
$x = 1$ or $x = 5$

22. $x^2 - 7x - 18 = 0$ 23. $-7, -2$
$(x + 2)(x - 9) = 0$
$x = -2$ or $x = 9$

24. $x^2 + 8x + 15 = 0$ 25. $0, -3$
$(x + 5)(x + 3) = 0$
$x = -5$ or $x = -3$

26. $x^2 + 8x = 0$ 27. $0, 9$
$x(x + 8) = 0$
$x = 0$ or $x = -8$

28. $x^2 + 4x = 0$ 29. $-10, 10$
$x(x + 4) = 0$
$x = 0$ or $x = -4$

30. $x^2 = 16$ 31. $-\frac{3}{2}, \frac{3}{2}$
$x^2 - 16 = 0$
$(x - 4)(x + 4) = 0$
$x = 4$ or $x = -4$

32. $9x^2 - 4 = 0$ 33. -5
$(3x - 2)(3x + 2) = 0$
$x = \frac{2}{3}$ or $x = -\frac{2}{3}$

34. $0 = 6x + x^2 + 9$ 35. 1
$0 = x^2 + 6x + 9$
$0 = (x + 3)(x + 3)$
$x = -3$

36. $x^2 + 16 = 8x$ 37. $0, \frac{4}{9}$ 38. $3x^2 = 7x$
$x^2 - 8x + 16 = 0$ $3x^2 - 7x = 0$
$(x - 4)(x - 4) = 0$ $x(3x - 7) = 0$
$x = 4$ $x = 0$ or $x = \frac{7}{3}$

39. $-\frac{5}{3}, 4$ 40. $6x^2 - 4x = 10$ 41. $-5, -1$
$6x^2 - 4x - 10 = 0$
$2(3x^2 - 2x - 5) = 0$
$2(3x - 5)(x + 1) = 0$
$x = \frac{5}{3}$ or $x = -1$

42.
$$12y^2 - 5y = 2$$
$$12y^2 - 5y - 2 = 0$$
$$(4y + 1)(3y - 2) = 0$$
$$y = -\tfrac{1}{4} \text{ or } y = \tfrac{2}{3}$$

43. $-3, 7$

44.
$$(x + 1)(x - 7) = -15$$
$$x^2 - 6x - 7 = -15$$
$$x^2 - 6x + 8 = 0$$
$$(x - 2)(x - 4) = 0$$
$$x = 2 \text{ or } x = 4$$

45. $-1, \tfrac{2}{3}$

46.
$$t(t - 5) = 14$$
$$t^2 - 5t = 14$$
$$t^2 - 5t - 14 = 0$$
$$(t - 7)(t + 2) = 0$$
$$t = 7 \text{ or } t = -2$$

47. $-\tfrac{5}{9}, \tfrac{5}{9}$

48.
$$36m^2 - 9 = 40$$
$$36m^2 - 49 = 0$$
$$(6m + 7)(6m - 7) = 0$$
$$m = -\tfrac{7}{6} \text{ or } m = \tfrac{7}{6}$$

49. $-1, \tfrac{6}{5}$

50.
$$(x + 3)(3x + 5) = 7$$
$$3x^2 + 14x + 15 = 7$$
$$3x^2 + 14x + 8 = 0$$
$$(3x + 2)(x + 4) = 0$$
$$x = -\tfrac{2}{3} \text{ or } x = -4$$

51. $-2, 9$

52.
$$3x^2 - 2x = 9 - 8x$$
$$3x^2 + 6x - 9 = 0$$
$$3(x - 1)(x + 3) = 0$$
$$x = 1 \text{ or } x = -3$$

53. $-\tfrac{5}{2}, \tfrac{4}{3}$

54.
$$(2t + 1)(4t - 1) = 14$$
$$8t^2 + 2t - 1 = 14$$
$$8t^2 + 2t - 15 = 0$$
$$(2t + 3)(4t - 5) = 0$$
$$t = -\tfrac{3}{2} \text{ or } t = \tfrac{5}{4}$$

55. $-3, 2$

56. The solutions of the equation are the first coordinates of the x-intercepts of the graph. From the graph we see that the x-intercepts are $(-1, 0)$ and $(4, 0)$, so the solutions of the equation are -1 and 4.

57. $(-4, 0), (1, 0)$

58. We let $y = 0$ and solve for x.
$$0 = x^2 - x - 6$$
$$0 = (x - 3)(x + 2)$$
$$x - 3 = 0 \text{ or } x + 2 = 0$$
$$x = 3 \text{ or } x = -2$$
The x-intercepts are $(3, 0)$ and $(-2, 0)$.

59. $(-3, 0), (5, 0)$

60. We let $y = 0$ and solve for x.
$$0 = x^2 + 2x - 8$$
$$0 = (x + 4)(x - 2)$$
$$x + 4 = 0 \text{ or } x - 2 = 0$$
$$x = -4 \text{ or } x = 2$$
The x-intercepts are $(-4, 0)$ and $(2, 0)$.

61. $\left(-\tfrac{5}{2}, 0\right), (2, 0)$

62. We let $y = 0$ and solve for x.
$$0 = 2x^2 + 3x - 9$$
$$0 = (2x - 3)(x + 3)$$
$$2x - 3 = 0 \text{ or } x + 3 = 0$$
$$2x = 3 \text{ or } x = -3$$
$$x = \tfrac{3}{2} \text{ or } x = -3$$
The x-intercepts are $\left(\tfrac{3}{2}, 0\right)$ and $(-3, 0)$.

63. $(a + b)^2$ **64.** $a^2 + b^2$ **65.** Let x represent the first integer; $x + (x + 1)$ **66.** Let x represent the number; $2x + 5 < 19$ **67.** Let n represent the number; $\tfrac{1}{2}n - 7 > 24$ **68.** Let n represent the number; $n - 3 \geq 34$ **69.** ◆ A quadratic polynomial is a second-degree algebraic expression. It has no equals sign. A quadratic equation is a second-degree equation equivalent to one of the form $ax^2 + bx + c = 0, a \neq 0$. It has an equals sign.   **70.** ◆ One solution of the equation is 0. Dividing both sides of the equation by x, leaving the solution $x = 3$, is equivalent to dividing by 0.   **71.** ◆ The graph has no x-intercepts.   **72.** ◆ No; when $ax^2 + bx + c$ is a perfect-square trinomial, then $ax^2 + bx + c = 0$ will have only one solution.   **73.** (a) $x^2 - x - 12 = 0$; (b) $x^2 + 7x + 12 = 0$; (c) $4x^2 - 4x + 1 = 0$; (d) $x^2 - 25 = 0$; (e) $40x^3 - 14x^2 + x = 0$

74.
$$16(x - 1) = x(x + 8)$$
$$16x - 16 = x^2 + 8x$$
$$0 = x^2 - 8x + 16$$
$$0 = (x - 4)(x - 4)$$
$$x = 4$$

75. $-5, 4$

76.
$$(t - 5)^2 = 2(5 - t)$$
$$t^2 - 10t + 25 = 10 - 2t$$
$$t^2 - 8t + 15 = 0$$
$$(t - 5)(t - 3) = 0$$
$$t = 5 \text{ or } t = 3$$

77. $-\tfrac{3}{5}, \tfrac{3}{5}$

78.
$$x^2 - \tfrac{25}{36} = 0$$
$$\left(x + \tfrac{5}{6}\right)\left(x - \tfrac{5}{6}\right) = 0$$
$$x = -\tfrac{5}{6} \text{ or } x = \tfrac{5}{6}$$

79. $-4, 4$

80.
$$\tfrac{27}{25}x^2 = \tfrac{1}{3}$$
$$\tfrac{27}{25}x^2 - \tfrac{1}{3} = 0$$
$$27\left(\tfrac{1}{25}x^2 - \tfrac{1}{81}\right) = 0$$
$$27\left(\tfrac{1}{5}x + \tfrac{1}{9}\right)\left(\tfrac{1}{5}x - \tfrac{1}{9}\right) = 0$$
$$x = -\tfrac{5}{9} \text{ or } x = \tfrac{5}{9}$$

81. (a) $2x^2 + 20x - 4 = 0$; (b) $x^2 - 3x - 18 = 0$; (c) $(x + 1)(5x - 5) = 0$; (d) $(2x + 8)(2x - 5) = 0$; (e) $4x^2 + 8x + 36 = 0$; (f) $9x^2 - 12x + 24 = 0$
82. ◆ Find the product of 7 binomials of the form $x + a$ where all of the constant terms of the binomials are different. **83.** ◆ Graph $y = x^2 + 3x - 4$ and $y = -6$ on the same set of axes. The first coordinates of the points of intersection of the two graphs are the solutions of $x^2 + 3x - 4 = -6$. **84.** $-3.45, 1.65$ **85.** $2.33, 6.77$ **86.** $-0.25, 0.88$ **87.** $-9.15, -4.59$ **88.** $4.55, -3.23$ **89.** $-6.75, -3.25$

EXERCISE SET 5.7, PP. 290–293

1. $-1, 2$ **2.** Let $x =$ the number (or numbers).
Solve: $x^2 - 6 = x$
$x = 3$ or $x = -2$
3. $3, 5$ **4.** Let $n =$ the number (or numbers).
Solve: $6n = 8 + n^2$
$n = 2$ or $n = 4$
5. 10 and 11
6. Solve: $x(x + 1) = 210$
The solutions of the equation are $x = 14$ or $x = -15$. We only need to check 14, since negative numbers do not make sense in the original problem. It checks.
The page numbers are 14 and 15.
7. 14 and 16, -16 and -14
8. Solve: $x(x + 2) = 255$
$x = 15$ or $x = -17$
The integers are 15 and 17 or -17 and -15.
9. Length: 12 cm; width: 7 cm
10. Solve: $(w + 4)w = 96$
$w = -12$ or $w = 8$
Only 8 checks in the original problem. The length is $8 + 4$, or 12 m, and the width is 8 m.
11. Height: 7 cm; base: 10 cm
12. Solve: $\frac{1}{2}(h + 10)(h) = 28$
$h = -14$ or $h = 4$
Only 4 checks in the original problem. The height is 4 cm, and the base is $4 + 10$, or 14 cm.
13. Foot: 7 ft; height: 12 ft
14. Let $h =$ the height, in m, and $\frac{1}{2}h =$ the base, in m.
Solve: $\frac{1}{2} \cdot \frac{1}{2}h \cdot h = 64$
$h = 16$ or $h = -16$
Only 16 checks in the original problem. The base is $\frac{1}{2} \cdot 16$, or 8 m, and the height is 16 m.
15. 25 ft

16. Let $a =$ the altitude from which the descent began, in feet.

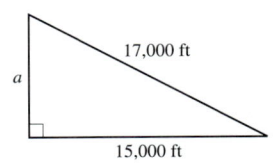

Solve: $a^2 + 15{,}000^2 = 17{,}000^2$
$a = -8000$ or $a = 8000$
Only 8000 checks in the original problem. The descent began from an altitude of 8000 ft.
17. 380 **18.** Solve: $14^2 - 14 = N$ **19.** 12
$N = 182$
20. Solve: $n^2 - n = 90$
$n = -9$ or $n = 10$
Only 10 checks in the original problem. There are 10 teams in the league.
21. 24 ft
22. Let $h =$ the vertical height to which each brace reaches, in feet.
Solve: $h^2 + 12^2 = 15^2$
$h = 9$ or $h = -9$
Only 9 checks in the original problem. Each brace reaches 9 ft vertically.
23. 66 **24.** Solve: $N = \frac{1}{2}(30^2 - 30)$ **25.** 25
$N = 435$
26. Solve: $190 = \frac{1}{2}(n^2 - n)$
$n = -19$ or $n = 20$
Only 20 checks in the original problem. Twenty people took part in the toast.
27. 4 m
28. We label the drawing. Let $x =$ the length of a side of the dining room, in ft.

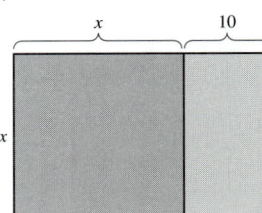

Solve: $x(x + 10) = 264$
$x = -22$ or $x = 12$
Only 12 checks in the original problem. The dining room is 12 ft by 12 ft, and the kitchen is 12 ft by 10 ft.
29. 20 ft **30.** Solve: $h = 48(1.5) - 16(1.5)^2$
$h = 36$ ft
31. At 1 sec and at 2 sec after it has been launched

32. Solve: $0 = 48t - 16t^2$

$t = 0$ or $t = 3$

Only 3 checks in the original problem. The rocket crashes into the ground 3 sec after it is launched.

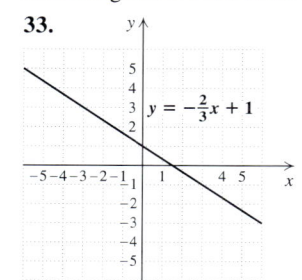

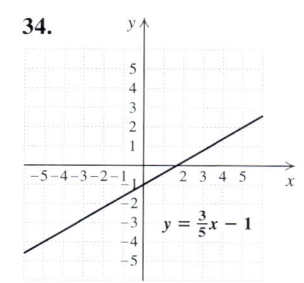

33. Graph of $y = -\frac{2}{3}x + 1$

34. Graph of $y = \frac{3}{5}x - 1$

35. 7 **36.** $m^6 n^6$ **37.** $-\frac{7}{24}$

38. $\frac{5}{8} - \frac{7}{3} \cdot \frac{9}{8} = \frac{5}{8} - \frac{21}{8} = -\frac{16}{8} = -2$

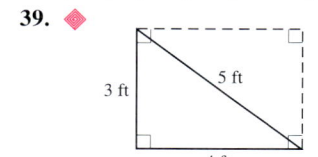

39. Right triangle with legs 3 ft and 4 ft, hypotenuse 5 ft.

40. Answers may vary. The area of a rectangle is 90 m². The length is 1 m greater than the width. Find the length and width. **41.** Answers may vary. The problem in Exercise 40 works here also. **42.** No; if a problem translates to a quadratic equation, $ax^2 + bx + c = 0$, we can solve it only if $ax^2 + bx + c$ can be factored.

43. 39 cm

44. Solve: $(40 + 2x)(20 + 2x) = 1500$

$x = -35$ or $x = 5$

Only 5 checks in the original problem. The walk is 5 ft wide.

45. 960 ft²

46. Let y = the ten's digit. Then $y + 4$ = the one's digit and $10y + y + 4$, or $11y + 4$, represents the number.

Solve: $11y + 4 + y(y + 4) = 58$

$y = -18$ or $y = 3$

Only 3 checks in the original problem. The number is 37.

47. 7 m

48. Let w = the width of the piece of cardboard, in cm. Then $2w$ = the length, in cm. The length and width of the base of the box are $2x - 8$ and $x - 8$, respectively, and its height is 4.

Solve: $(2x - 8)(x - 8)(4) = 616$

$w = 15$ or $w = -3$

Only 15 checks in the original problem. The original dimension of the cardboard are 15 cm by 2 · 15, or 30 cm.

49. 5 in.

50. Let s = the length of a side of the original square, in cm. Then $s + 5$ = the length of a side of the new square, in cm.

Solve: $(s + 5)^2 = 2\frac{1}{4} \cdot x^2$

$s = 10$ or $s = -2$

Only 10 checks in the original problem. The length of a side of the original square is 10 cm, so its area is 10 · 10, or 100 cm². The length of a side of the new square is 10 + 5, or 15 cm, and its area is 15 · 15, or 225 cm².

51. $1200

52. Let h and w represent the height and width of the old screen, in inches, respectively. Then since the new screen has the same ratio of height to width as the old screen, we let kh and kw represent the height and width of the new screen, in inches, respectively. We know that

$$h^2 + w^2 = 13.5^2, \text{ or } h^2 + w^2 = 182.25$$

and

$$(kh)^2 + (kw)^2 = 27^2.$$

Then

$$k^2 h^2 + k^2 w^2 = 729$$
$$k^2(h^2 + w^2) = 729$$
$$k^2(182.25) = 729 \quad \text{Substituting 182.25 for } h^2 + w^2$$
$$k^2 = 4$$
$$k^2 - 4 = 0$$
$$(k + 2)(k - 2) = 0$$

$k = -2$ or $k = 2$

Only 2 has meaning in the original problem. Then the dimensions of the new screen are $2h$ by $2w$, and its area is $2h \cdot 2w$, or $4hw$. The area of the old screen is hw, so the new screen is 4 times as large as the old screen.

REVIEW EXERCISES: CHAPTER 5, P. 295

1. [5.1] Answers may vary. $(12x)(3x^2)$, $(-9x^2)(-4x)$, $(6x)(6x^2)$ **2.** [5.1] Answers may vary. $(-4x^3)(5x^2)$, $(2x^4)(-10x)$, $(-5x)(4x^4)$ **3.** [5.1] $2x^3(x + 3)$
4. [5.1] $x(x - 3)$ **5.** [5.4] $(3x + 2)(3x - 2)$
6. [5.2] $(x + 6)(x - 2)$ **7.** [5.4] $(x + 7)^2$
8. [5.1] $3x(2x^2 + 4x + 1)$ **9.** [5.1] $(2x + 3)(3x^2 + 1)$
10. [5.3] $(3x - 1)(2x - 1)$
11. [5.4] $(x^2 + 9)(x + 3)(x - 3)$
12. [5.3] $3x(3x - 5)(x + 3)$ **13.** [5.4] $2(x + 5)(x - 5)$
14. [5.1] $(x^3 - 2)(x + 4)$
15. [5.4] $(4x^2y^2 + 1)(2xy + 1)(2xy - 1)$
16. [5.1] $4x^4(2x^2 - 8x + 1)$ **17.** [5.4] $3(2x + 5)^2$
18. [5.4] Prime **19.** [5.2] $x(x - 6)(x + 5)$
20. [5.4] $(2x + 5)(2x - 5)$ **21.** [5.4] $(3x - 5)^2$
22. [5.3] $2(3x + 4)(x - 6)$ **23.** [5.4] $(x - 3)^2$
24. [5.3] $(2x + 1)(x - 4)$ **25.** [5.4] $2(3x - 1)^2$
26. [5.4] $3(x + 3)(x - 3)$ **27.** [5.2] $(x - 5)(x - 3)$

28. [5.4] $(5x - 2)^2$ **29.** [5.2] $(xy + 4)(xy - 3)$
30. [5.4] $3(2a + 7b)^2$ **31.** [5.1] $(m + t)(m + 5)$
32. [5.4] $32(x^2 - 2y^2z^2)(x^2 + 2y^2z^2)$ **33.** [5.6] $1, -3$
34. [5.6] $-7, 5$ **35.** [5.6] $-\frac{1}{3}, \frac{1}{3}$ **36.** [5.6] $\frac{2}{3}, 1$
37. [5.6] $-4, \frac{3}{2}$ **38.** [5.6] $-2, 3$ **39.** [5.7] $-3, 4$
40. [5.6] $(-1, 0), \left(\frac{5}{2}, 0\right)$ **41.** [5.7] 40 cm **42.** [5.7] 24 m
43. [1.6] $\frac{1}{2}$
44. [3.2]

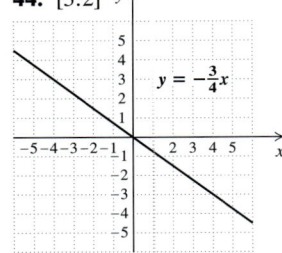

29. [3.2]

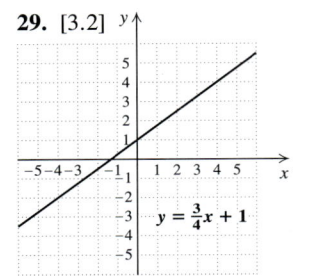

30. [4.1] $49a^6b^{10}$
31. [5.7] $l = 15, w = 3$
32. [5.2] $(a - 4)(a + 8)$

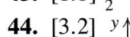

45. [4.1] m^2n^5 **46.** [2.7] Let x represent the number; $\frac{1}{2}x - 2 \geq 10$ **47.** [5.4] ◆ Answers may vary. Because Edith did not first factor out the largest common factor, 4, her factorization will not be "complete" until she removes a common factor of 2 from each binomial. Awarding 3 to 7 points would seem reasonable. **48.** [5.6] ◆ The equations solved in this chapter have an x^2-term (are quadratic), whereas those solved previously have no x^2-term (are linear). The principle of zero products is used to solve quadratic equations and is not used to solve linear equations. **49.** [5.7] $2\frac{1}{2}$ cm **50.** [5.7] $0, 2$
51. [5.7] $l = 12$ cm, $w = 6$ cm **52.** [5.6] No real solution
53. [5.6] $-3, 2, \frac{5}{2}$

TEST: CHAPTER 5, P. 296

1. [5.1] Answers may vary. $(2x^2)(4x^2), (8x)(x^3), (-4x^3)(-2x)$ **2.** [5.2] $(x - 5)(x - 2)$
3. [5.4] $(x - 5)^2$ **4.** [5.1] $2y^2(3 - 4y + 2y^2)$
5. [5.1] $(x^2 + 2)(x + 1)$ **6.** [5.1] $x(x - 5)$
7. [5.2] $x(x + 3)(x - 1)$ **8.** [5.3] $2(5x - 6)(x + 4)$
9. [5.4] $(2x + 3)(2x - 3)$ **10.** [5.2] $(x - 4)(x + 3)$
11. [5.3] $3m(2m + 1)(m + 1)$ **12.** [5.4] $3(w + 5)(w - 5)$
13. [5.4] $5(3x + 2)^2$ **14.** [5.4] $3(x^2 + 4)(x + 2)(x - 2)$
15. [5.4] $(7x - 6)^2$ **16.** [5.3] $(5x - 1)(x - 5)$
17. [5.1] $(x^3 - 3)(x + 2)$
18. [5.4] $5(4 + x^2)(2 + x)(2 - x)$
19. [5.3] $(2x - 5)(2x + 3)$ **20.** [5.3] $3t(2t + 5)(t - 1)$
21. [5.2] $3(m + 2n)(m - 5n)$ **22.** [5.6] $5, -4$
23. [5.6] $-5, \frac{3}{2}$ **24.** [5.6] $-4, 7$ **25.** [5.7] Length: 8 m; width: 6 m **26.** [5.7] 5 ft **27.** [1.6] -0.4
28. [2.7] $\{l \mid l < 13 \text{ cm}\}$

CHAPTER 6
TECHNOLOGY CONNECTION 6.1, P. 299
1.

2.

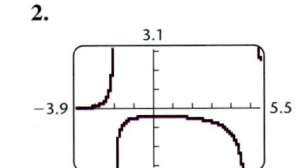

EXERCISE SET 6.1, PP. 304–305
1. 0 **2.** 0 **3.** -6 **4.** -7 **5.** 5
6. $4x - 12 = 0$ **7.** $-4, 7$
   $x = 3$
8. $p^2 - 7p + 10 = 0$ **9.** $-5, 5$
   $(p - 5)(p - 2) = 0$
   $p = 5$ or $p = 2$
10. $49 - x^2 = 0$
   $(7 + x)(7 - x) = 0$
   $x = -7$ or $x = 7$
11. $\dfrac{5a}{4b^2}$ **12.** $\dfrac{45x^3y^2}{9x^5y} = \dfrac{5y \cdot 9x^3y}{x^2 \cdot 9x^3y} = \dfrac{5y}{x^2}$
13. $\dfrac{5}{2xy^4}$
14. $\dfrac{12a^5b^6}{18a^3b} = \dfrac{2a^2b^5 \cdot 6a^3b}{3 \cdot 6a^3b} = \dfrac{2a^2b^5}{3}$ **15.** $\dfrac{3}{2}$
16. $\dfrac{14x - 7}{10x - 5} = \dfrac{7(2x - 1)}{5(2x - 1)} = \dfrac{7}{5}$ **17.** $\dfrac{a - 3}{a + 1}$
18. $\dfrac{a^2 + 5a + 6}{a^2 - 9} = \dfrac{(a + 3)(a + 2)}{(a + 3)(a - 3)} = \dfrac{a + 2}{a - 3}$
19. $\dfrac{3}{2x^3}$

20. $\dfrac{76a^5}{24a^3} = \dfrac{19a^2 \cdot 4a^3}{6 \cdot 4a^3}$

$= \dfrac{19a^2}{6} \cdot \dfrac{4a^3}{4a^3}$

$= \dfrac{19a^2}{6} \cdot 1$

$= \dfrac{19a^2}{6}$

Check: Let $a = 1$.

$\dfrac{76a^5}{24a^3} = \dfrac{76 \cdot 1^5}{24 \cdot 1^3} = \dfrac{76}{24} = \dfrac{19}{6}$

$\dfrac{19a^2}{6} = \dfrac{19 \cdot 1^2}{6} = \dfrac{19}{6}$

The answer is probably correct.

21. $\dfrac{y - 3}{2y}$

22. $\dfrac{4x - 12}{4x} = \dfrac{4(x - 3)}{4 \cdot x}$

$= \dfrac{4(x - 3)}{4 \cdot x}$

$= \dfrac{x - 3}{x}$

Check: Let $x = 2$.

$\dfrac{4x - 12}{4x} = \dfrac{4 \cdot 2 - 12}{4 \cdot 2} = \dfrac{-4}{8} = -\dfrac{1}{2}$

$\dfrac{x - 3}{x} = \dfrac{2 - 3}{2} = \dfrac{-1}{2} = -\dfrac{1}{2}$

The answer is probably correct.

23. $\dfrac{3(2a - 1)}{7(a - 1)}$

24. $\dfrac{3m^2 + 3m}{6m^2 + 9m} = \dfrac{3m(m + 1)}{3m(2m + 3)}$

$= \dfrac{3m}{3m} \cdot \dfrac{m + 1}{2m + 3}$

$= 1 \cdot \dfrac{m + 1}{2m + 3}$

$= \dfrac{m + 1}{2m + 3}$

Check: Let $m = 1$.

$\dfrac{3m^2 + 3m}{6m^2 + 9m} = \dfrac{3 \cdot 1^2 + 3 \cdot 1}{6 \cdot 1^2 + 9 \cdot 1} = \dfrac{6}{15} = \dfrac{2}{5}$

$\dfrac{m + 1}{2m + 3} = \dfrac{1 + 1}{2 \cdot 1 + 3} = \dfrac{2}{5}$

The answer is probably correct.

25. $\dfrac{t - 5}{t - 4}$

26. $\dfrac{a^2 - 4}{a^2 + 5a + 6} = \dfrac{(a + 2)(a - 2)}{(a + 2)(a + 3)}$

$= \dfrac{a + 2}{a + 2} \cdot \dfrac{a - 2}{a + 3}$

$= 1 \cdot \dfrac{a - 2}{a + 3}$

$= \dfrac{a - 2}{a + 3}$

Check: Let $a = 1$.

$\dfrac{a^2 - 4}{a^2 + 5a + 6} = \dfrac{1^2 - 4}{1^2 + 5 \cdot 1 + 6} = \dfrac{-3}{12} = -\dfrac{1}{4}$

$\dfrac{a - 2}{a + 3} = \dfrac{1 - 2}{1 + 3} = \dfrac{-1}{4} = -\dfrac{1}{4}$

The answer is probably correct.

27. $\dfrac{a - 4}{2(a + 4)}$

28. $\dfrac{2t^2 + 6t + 4}{4t^2 - 12t - 16} = \dfrac{2(t^2 + 3t + 2)}{4(t^2 - 3t - 4)}$

$= \dfrac{2(t + 2)(t + 1)}{2 \cdot 2(t - 4)(t + 1)}$

$= \dfrac{2(t + 1)}{2(t + 1)} \cdot \dfrac{t + 2}{2(t - 4)}$

$= 1 \cdot \dfrac{t + 2}{2(t - 4)}$

$= \dfrac{t + 2}{2(t - 4)}$

Check: Let $t = 1$.

$\dfrac{2t^2 + 6t + 4}{4t^2 - 12t - 16} = \dfrac{2 \cdot 1^2 + 6 \cdot 1 + 4}{4 \cdot 1^2 - 12 \cdot 1 - 16} = \dfrac{12}{-24} = -\dfrac{1}{2}$

$\dfrac{t + 2}{2(t - 4)} = \dfrac{1 + 2}{2(1 - 4)} = \dfrac{3}{-6} = -\dfrac{1}{2}$

The answer is probably correct.

29. $\dfrac{x + 4}{x - 4}$

30. $\dfrac{x^2 - 25}{x^2 - 10x + 25} = \dfrac{(x - 5)(x + 5)}{(x - 5)(x - 5)}$

$= \dfrac{x - 5}{x - 5} \cdot \dfrac{x + 5}{x - 5}$

$= 1 \cdot \dfrac{x + 5}{x - 5}$

$= \dfrac{x + 5}{x - 5}$

Check: Let $x = 2$.

$\dfrac{x^2 - 25}{x^2 - 10x + 25} = \dfrac{2^2 - 25}{2^2 - 10 \cdot 2 + 25} = \dfrac{-21}{9} = -\dfrac{7}{3}$

$\dfrac{x + 5}{x - 5} = \dfrac{2 + 5}{2 - 5} = \dfrac{7}{-3} = -\dfrac{7}{3}$

The answer is probably correct.

31. $t - 1$

32. $\dfrac{a^2 - 1}{a - 1} = \dfrac{(a - 1)(a + 1)}{a - 1}$

$= \dfrac{a - 1}{a - 1} \cdot \dfrac{a + 1}{1}$

$= 1 \cdot \dfrac{a + 1}{1}$

$= a + 1$

Check: Let $a = 2$.

$\dfrac{a^2 - 1}{a - 1} = \dfrac{2^2 - 1}{2 - 1} = \dfrac{3}{1} = 3$

$a + 1 = 2 + 1 = 3$

The answer is probably correct.

33. $\dfrac{y^2 + 4}{y + 2}$

34. $\dfrac{x^2 + 1}{x + 1}$ cannot be simplified.

Neither the numerator nor the denominator can be factored.

35. $\dfrac{1}{2}$

36. $\dfrac{6x^2 - 54}{4x^2 - 36} = \dfrac{2 \cdot 3(x^2 - 9)}{2 \cdot 2(x^2 - 9)}$

$= \dfrac{2(x^2 - 9)}{2(x^2 - 9)} \cdot \dfrac{3}{2}$

$= 1 \cdot \dfrac{3}{2}$

$= \dfrac{3}{2}$

Check: Let $x = 1$.

$\dfrac{6x^2 - 54}{4x^2 - 36} = \dfrac{6 \cdot 1^2 - 54}{4 \cdot 1^2 - 36} = \dfrac{-48}{-32} = \dfrac{3}{2}$

$\dfrac{3}{2} = \dfrac{3}{2}$

The answer is probably correct.

37. $\dfrac{5}{y + 6}$

38. $\dfrac{6t + 12}{t^2 - t - 6} = \dfrac{6(t + 2)}{(t - 3)(t + 2)}$

$= \dfrac{6}{t - 3} \cdot \dfrac{t + 2}{t + 2}$

$= \dfrac{6}{t - 3} \cdot 1$

$= \dfrac{6}{t - 3}$

Check: Let $t = 1$.

$\dfrac{6t + 12}{t^2 - t - 6} = \dfrac{6 \cdot 1 + 12}{1^2 - 1 - 6} = \dfrac{18}{-6} = -3$

$\dfrac{6}{t - 3} = \dfrac{6}{1 - 3} = \dfrac{6}{-2} = -3$

The answer is probably correct.

39. $\dfrac{y - 6}{y - 5}$

40. $\dfrac{a^2 - 10a + 21}{a^2 - 11a + 28} = \dfrac{(a - 7)(a - 3)}{(a - 7)(a - 4)}$

$= \dfrac{a - 7}{a - 7} \cdot \dfrac{a - 3}{a - 4}$

$= 1 \cdot \dfrac{a - 3}{a - 4}$

$= \dfrac{a - 3}{a - 4}$

Check: Let $a = 2$.

$\dfrac{a^2 - 10a + 21}{a^2 - 11a + 28} = \dfrac{2^2 - 10 \cdot 2 + 21}{2^2 - 11 \cdot 2 + 28} = \dfrac{5}{10} = \dfrac{1}{2}$

$\dfrac{a - 3}{a - 4} = \dfrac{2 - 3}{2 - 4} = \dfrac{-1}{-2} = \dfrac{1}{2}$

The answer is probably correct.

41. $\dfrac{a - 3}{a + 3}$

42. $\dfrac{t^2 - 4}{(t + 2)^2} = \dfrac{(t - 2)(t + 2)}{(t + 2)(t + 2)}$

$= \dfrac{t - 2}{t + 2} \cdot \dfrac{t + 2}{t + 2}$

$= \dfrac{t - 2}{t + 2} \cdot 1$

$= \dfrac{t - 2}{t + 2}$

Check: Let $t = 1$.

$\dfrac{t^2 - 4}{(t + 2)^2} = \dfrac{1^2 - 4}{(1 + 2)^2} = \dfrac{-3}{9} = -\dfrac{1}{3}$

$\dfrac{t - 2}{t + 2} = \dfrac{1 - 2}{1 + 2} = \dfrac{-1}{3} = -\dfrac{1}{3}$

The answer is probably correct.

43. -1

44. $\dfrac{6 - x}{x - 6} = \dfrac{-(x - 6)}{x - 6}$

$= \dfrac{-1}{1} \cdot \dfrac{x - 6}{x - 6}$

$= -1$

Check: Let $x = 3$.

$\dfrac{6 - x}{x - 6} = \dfrac{6 - 3}{3 - 6} = \dfrac{3}{-3} = -1$

$-1 = -1$

The answer is probably correct.

45. 1

46. $\dfrac{a-b}{b-a} = \dfrac{-1(-a+b)}{b-a}$
$= \dfrac{-1(b-a)}{b-a}$
$= -1 \cdot \dfrac{b-a}{b-a}$
$= -1 \cdot 1$
$= -1$

Check: Let $a = 2$ and $b = 1$.
$\dfrac{a-b}{b-a} = \dfrac{2-1}{1-2} = \dfrac{1}{-1} = -1$
$-1 = -1$

The answer is probably correct.

47. -5

48. $\dfrac{6t-12}{2-t} = \dfrac{-6(-t+2)}{2-t}$
$= \dfrac{-6(2-t)}{2-t}$
$= -6 \cdot \dfrac{2-t}{2-t}$
$= -6$

Check: Let $t = 3$.
$\dfrac{6t-12}{2-t} = \dfrac{6 \cdot 3 - 12}{2-3} = \dfrac{6}{-1} = -6$
$-6 = -6$

The answer is probably correct.

49. $-\dfrac{3}{2}$

50. $\dfrac{7a^2-7}{1-a} = \dfrac{7(a+1)(a-1)}{-(a-1)}$
$= \dfrac{7(a+1)}{-1} \cdot \dfrac{a-1}{a-1}$
$= -7(a+1)$

Check: Let $a = 2$.
$\dfrac{7a^2-7}{1-a} = \dfrac{7 \cdot 2^2 - 7}{1-2} = \dfrac{21}{-1} = -21$
$-7(a+1) = -7(2+1) = -7 \cdot 3 = -21$

The answer is probably correct.

51. $(x+1)(x+7)$ 52. $(x-2)(x-7)$

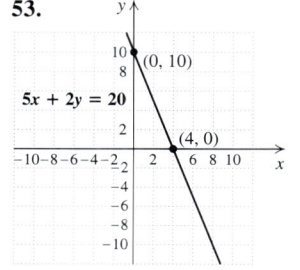

53. [Graph showing line $5x + 2y = 20$ with intercepts $(0, 10)$ and $(4, 0)$]

54. $2x - 4y = 8$
Find the y-intercept: $-4y = 8$
$y = -2$
The y-intercept is $(0, -2)$.
Find the x-intercept: $2x = 8$
$x = 4$

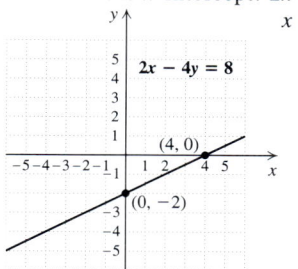

[Graph of $2x - 4y = 8$ showing intercepts $(4, 0)$ and $(0, -2)$]

The x-intercept is $(4, 0)$.

55. $\dfrac{5}{48}$ 56. $\dfrac{7}{9} - \dfrac{2}{3} \cdot \dfrac{6}{7} = \dfrac{7}{9} - \dfrac{4}{7} = \dfrac{49}{63} - \dfrac{36}{63} = \dfrac{13}{63}$

57. ◆ Form a rational expression that has factors of $x+3$ and $x-4$ in the denominator.   58. ◆ Canceling removes a factor equal to 1, allowing us to rewrite $a \cdot 1$ as a.

59. ◆ Although a rational expression has been simplified incorrectly, it is possible that there are one or more values of the variable(s) for which the two expressions are the same. For example, $\dfrac{x^2+x-2}{x^2+3x+2}$ could be simplified incorrectly as $\dfrac{x-1}{x+2}$, but evaluating the expressions for $x = 1$ gives 0 in each case. $\left(\text{The correct simplification is } \dfrac{x-1}{x+1}.\right)$

60. ◆ Show that $(a-b) + (b-a) = 0$.   61. $-x - 2y$

62. $\dfrac{(a-b)^2}{b^2-a^2} = \dfrac{(a-b)(a-b)}{-(a+b)(a-b)} = \dfrac{a-b}{-a-b}$, or $\dfrac{b-a}{a+b}$

63. $\dfrac{(t-1)(t-9)^2}{(t^2+9)(t+1)}$

64. $\dfrac{(t+2)^3(t^2+2t+1)(t+1)}{(t+1)^3(t^2+4t+4)(t+2)} =$
$\dfrac{(t+2)^3(t+1)^2(t+1)}{(t+1)^3(t+2)^2(t+2)} = 1$

65. $\dfrac{(x-y)^3}{(x+y)^2(x-5y)}$

66. $\dfrac{(x-1)(x^4-1)(x^2-1)}{(x^2+1)(x-1)^2(x^4-2x^2+1)} =$
$\dfrac{(x-1)(x^4-1)(x^2-1)}{(x^2+1)(x-1)^2(x^2-1)^2} =$
$\dfrac{(x-1)(x^2+1)(x^2-1)(x+1)(x-1)}{(x^2+1)(x-1)(x-1)(x^2-1)(x^2-1)} =$
$\dfrac{(x-1)(x^2+1)(x^2-1)(x+1)(x-1) \cdot 1}{(x^2+1)(x-1)(x-1)(x^2-1)(x+1)(x-1)} =$
$\dfrac{1}{x-1}$

A-114 INSTRUCTOR'S ANSWERS

67. $\dfrac{x^3 + 4}{(x^2 + 2)(x^3 + 2)}$

68. $\dfrac{10t^4 - 8t^3 + 15t - 12}{8 - 10t + 12t^2 - 15t^3}$
$= \dfrac{2t^3(5t - 4) + 3(5t - 4)}{2(4 - 5t) + 3t^2(4 - 5t)}$
$= \dfrac{(5t - 4)(2t^3 + 3)}{(4 - 5t)(2 + 3t^2)}$
$= \dfrac{(5t\!\!-\!\!4)(2t^3 + 3)}{(-1)(5t\!\!-\!\!4)(2 + 3t^2)}$
$= -\dfrac{2t^3 + 3}{2 + 3t^2}$, or $\dfrac{-2t^3 - 3}{2 + 3t^2}$, or $\dfrac{2t^3 + 3}{-2 - 3t^2}$

69. ◆ $\dfrac{5(2x + 5) - 25}{10} = \dfrac{10x + 25 - 25}{10}$
$= \dfrac{10x}{10}$
$= x$

You get the same number you selected. A person asked to select a number and then perform these operations would probably be surprised that the result is the original number.

EXERCISE SET 6.2, PP. 308–310

1. $\dfrac{7x(x - 3)}{5(2x + 1)}$ **2.** $\dfrac{3x(5x + 2)}{4(x - 1)}$ **3.** $\dfrac{(a - 4)(a + 2)}{(a + 6)(a + 6)}$
4. $\dfrac{(a + 3)(a + 3)}{(a + 6)(a - 1)}$ **5.** $\dfrac{(2x + 3)(x + 1)}{4(x - 5)}$
6. $\dfrac{(-5)(-6)}{(3x - 4)(5x + 6)}$ **7.** $\dfrac{(a - 5)(a + 2)}{(a^2 + 1)(a^2 - 1)}$
8. $\dfrac{(t + 3)(t + 3)}{(t^2 - 2)(t^2 - 2)}$ **9.** $\dfrac{(x + 1)(x - 1)}{(2 + x)(x + 1)}$
10. $\dfrac{(m^2 + 5)(m^2 - 4)}{(m + 8)(m^2 - 4)}$ **11.** $\dfrac{5a^2}{4}$
12. $\dfrac{10}{t^7} \cdot \dfrac{3t^2}{25t} = \dfrac{2 \cdot 5 \cdot 3 \cdot t^2}{t^2 \cdot t^5 \cdot 5 \cdot 5 \cdot t} = \dfrac{6}{5t^6}$ **13.** $\dfrac{2}{dc^2}$
14. $\dfrac{3x^2y}{2} \cdot \dfrac{4}{xy^3} = \dfrac{3 \cdot x \cdot x \cdot y \cdot 2 \cdot 2}{2 \cdot x \cdot y \cdot y^2} = \dfrac{6x}{y^2}$ **15.** $\dfrac{x + 2}{x - 2}$
16. $\dfrac{t^2}{t^2 - 4} \cdot \dfrac{t^2 - 5t + 6}{t^2 - 3t} = \dfrac{t \cdot t(t\!-\!3)(t\!-\!2)}{(t + 2)(t\!-\!2)(t)(t\!-\!3)} = \dfrac{t}{t + 2}$
17. $\dfrac{(a + 5)(a - 5)(2a - 5)}{(a - 3)(a - 1)(2a + 5)}$
18. $\dfrac{x + 3}{x^2 + 9} \cdot \dfrac{x^2 + 5x + 4}{x + 9} = \dfrac{(x + 3)(x + 4)(x + 1)}{(x^2 + 9)(x + 9)}$
19. $\dfrac{(a + 3)(a - 3)}{a(a + 4)}$
20. $\dfrac{x^2 + 10x - 11}{x^2 - 1} \cdot \dfrac{x + 1}{x + 11} = \dfrac{(x - 1)(x + 11)(x + 1)}{(x + 1)(x - 1)(x + 11)} = 1$
21. $\dfrac{2a}{a - 2}$

22. $\dfrac{5v + 5}{v - 2} \cdot \dfrac{v^2 - 4v + 4}{v^2 - 1} = \dfrac{5(v + 1)(v - 2)(v - 2)}{(v - 2)(v + 1)(v - 1)} = \dfrac{5(v - 2)}{v - 1}$

23. $\dfrac{t - 5}{t + 5}$

24. $\dfrac{x^2 + 5x + 4}{x^2 - 6x + 8} \cdot \dfrac{x^2 + 5x - 14}{x^2 + 8x + 7}$ **25.** $\dfrac{5(a + 6)}{a - 1}$
$= \dfrac{(x + 4)(x + 1)(x + 7)(x - 2)}{(x - 4)(x - 2)(x + 7)(x + 1)}$
$= \dfrac{x + 4}{x - 4}$

26. $\dfrac{2t^2 - 98}{4t^2 - 4} \cdot \dfrac{8t + 8}{16t - 112} =$
$\dfrac{2(t + 7)(t - 7)(8)(t + 1)}{2 \cdot 2(t + 1)(t - 1)(2)(8)(t - 7)} = \dfrac{t + 7}{4(t - 1)}$

27. $\dfrac{(x - 1)(x - 3)^3}{(x + 3)(x + 1)}$

28. $\dfrac{(x + 2)^5}{(x - 1)^3} \cdot \dfrac{x^2 - 1}{x^2 + 5x + 6}$
$= \dfrac{(x + 2)(x + 2)(x + 2)(x + 2)(x + 2)(x + 1)(x - 1)}{(x - 1)(x - 1)(x - 1)(x + 3)(x + 2)}$
$= \dfrac{(x + 2)^4(x + 1)}{(x - 1)^2(x + 3)}$

29. $\dfrac{t - 2}{t - 1}$

30. $\dfrac{(y + 4)^3}{(y + 2)^3} \cdot \dfrac{y^2 + 4y + 4}{y^2 + 8y + 16} =$
$\dfrac{(y + 4)(y + 4)(y + 4)(y + 2)(y + 2)}{(y + 2)(y + 2)(y + 2)(y + 4)(y + 4)} = \dfrac{y + 4}{y + 2}$

31. $\dfrac{9}{x}$ **32.** $\dfrac{x^2 + 4}{3 - x}$ **33.** $\dfrac{1}{a^3 - 8a}$ **34.** $\dfrac{a^2 - b^2}{7}$

35. $\dfrac{x^2 - 4x + 7}{x^2 + 2x - 5}$ **36.** $\dfrac{x^2 + 7xy - y^2}{x^2 - 3xy + y^2}$ **37.** $\dfrac{35}{18}$

38. $\dfrac{3}{8} \div \dfrac{5}{2} = \dfrac{3}{8} \cdot \dfrac{2}{5} = \dfrac{3 \cdot 2}{2 \cdot 4 \cdot 5} = \dfrac{3}{20}$ **39.** 4

40. $\dfrac{x}{4} \div \dfrac{5}{x} = \dfrac{x}{4} \cdot \dfrac{x}{5} = \dfrac{x^2}{20}$ **41.** $\dfrac{x^3}{y}$ **42.** $\dfrac{a^5}{b^4} \div \dfrac{a^3}{b} =$
$\dfrac{a^5}{b^4} \cdot \dfrac{b}{a^3} = \dfrac{a^2 \cdot a^3 \cdot b}{b \cdot b^3 \cdot a^3} = \dfrac{a^2}{b^3}$ **43.** $\dfrac{(a + 2)(a + 3)}{(a - 3)(a - 1)}$

44. $\dfrac{y + 2}{4} \div \dfrac{y}{2} = \dfrac{y + 2}{4} \cdot \dfrac{2}{y} = \dfrac{2(y + 2)}{2 \cdot 2 \cdot y} = \dfrac{y + 2}{2y}$

45. $\dfrac{(x - 1)^2}{x}$

46. $\dfrac{4y - 8}{y + 2} \div \dfrac{y - 2}{y^2 - 4} = \dfrac{4y - 8}{y + 2} \cdot \dfrac{y^2 - 4}{y - 2} =$
$\dfrac{4(y - 2)(y + 2)(y - 2)}{(y + 2)(y - 2)} = 4(y - 2)$

CHAPTER 6

47. $\dfrac{1}{2}$ **48.** $\dfrac{a}{a-b} \div \dfrac{b}{a-b} = \dfrac{a}{a-b} \cdot \dfrac{a-b}{b} = \dfrac{a}{b}$

49. $\dfrac{(y+3)(y^2+1)}{y+1}$

50. $(x^2 - 5x - 6) \div \dfrac{x^2-1}{x+6} = \dfrac{x^2-5x-6}{1} \cdot \dfrac{x+6}{x^2-1} = \dfrac{(x-6)(x+1)(x+6)}{(x+1)(x-1)} = \dfrac{(x-6)(x+6)}{x-1}$

51. $\dfrac{15}{8}$

52. $\dfrac{-4+2x}{8} \div \dfrac{x-2}{2} = \dfrac{-4+2x}{8} \cdot \dfrac{2}{x-2} = \dfrac{2(x-2) \cdot 2}{2 \cdot 2 \cdot 2 \cdot (x-2)} = \dfrac{1}{2}$

53. $\dfrac{15}{4}$

54. $\dfrac{-12+4x}{4} \div \dfrac{-6+2x}{6} = \dfrac{-12+4x}{4} \cdot \dfrac{6}{-6+2x} = \dfrac{4(-3+x) \cdot 2 \cdot 3}{4 \cdot 2(-3+x) \cdot 1} = 3$

55. $\dfrac{a-5}{3(a-1)}$

56. $\dfrac{t-3}{t+2} \cdot \dfrac{4t-12}{t+1} = \dfrac{t-3}{t+2} \cdot \dfrac{t+1}{4t-12} = \dfrac{(t-3)(t+1)}{(t+2)(4)(t-3)} = \dfrac{t+1}{4(t+2)}$ **57.** $2x+1$

58. $(a+7) \div \dfrac{3a^2+14a-49}{a^2+8a+7}$
$= \dfrac{a+7}{1} \cdot \dfrac{a^2+8a+7}{3a^2+14a-49}$
$= \dfrac{(a+7)(a+7)(a+1)}{(3a-7)(a+7)}$
$= \dfrac{(a+7)(a+1)}{3a-7}$

59. $\dfrac{(x+2)^2}{x}$

60. $\dfrac{x+y}{x-y} \div \dfrac{x^2+y}{x^2-y^2} = \dfrac{x+y}{x-y} \cdot \dfrac{x^2-y^2}{x^2+y} = \dfrac{(x+y)(x+y)(x-y)}{(x-y)(x^2+y)} = \dfrac{(x+y)^2}{x^2+y}$

61. $\dfrac{x-5}{x+3}$

62. $\dfrac{a^2+2a-3}{a^2+3a} \div \dfrac{a+1}{a} = \dfrac{a^2+2a-3}{a^2+3a} \cdot \dfrac{a}{a+1} = \dfrac{(a+3)(a-1)(a)}{a(a+3)(a+1)} = \dfrac{a-1}{a+1}$

63. $\dfrac{(t^2+4)(t+3)^2}{(t-3)^2(t^4+16)}$

64. $\dfrac{x^2-4}{x^2+2x+1} \div \dfrac{x^4+1}{x-1} = \dfrac{x^2-4}{x^2+2x+1} \cdot \dfrac{x-1}{x^4+1} = \dfrac{(x+2)(x-2)(x-1)}{(x+1)^2(x^4+1)}$

65. $\dfrac{1}{(c-5)^2}$

66. $\dfrac{1-z}{1+2z+z^2} \div (1-z) = \dfrac{1-z}{1+2z-z^2} \cdot \dfrac{1}{1-z} = \dfrac{(1-z) \cdot 1}{(1+2z-z^2)(1-z)} = \dfrac{1}{1+2z-z^2}$

67. $\dfrac{t+5}{t-5}$

68. $\dfrac{(y-3)^3}{(y+3)^3} \div \dfrac{(y-3)^2}{(y+3)^2} = \dfrac{(y-3)^3}{(y+3)^3} \cdot \dfrac{(y+3)^2}{(y-3)^2} = \dfrac{(y-3)(y-3)^2(y+3)^2}{(y+3)(y+3)^2(y-3)^2} = \dfrac{y-3}{y+3}$

69. 4 **70.** $2x^2+16$ **71.** $8x^3 - 11x^2 - 3x + 12$

72. $0.06y^3 - 0.09y^2 + 0.01y - 1$ **73.** $-\dfrac{37}{20}$

74. $\dfrac{5}{9} + \dfrac{2}{3} \cdot \dfrac{4}{5} = \dfrac{5}{9} + \dfrac{8}{15} = \dfrac{25}{45} + \dfrac{24}{45} = \dfrac{49}{45}$

75. ◆ Yes; consider the product $\dfrac{a}{b} \cdot \dfrac{c}{d} = \dfrac{ac}{bd}$. The reciprocal of the product is $\dfrac{bd}{ac}$. This is equal to the product of the reciprocals of the two original factors: $\dfrac{b}{a} \cdot \dfrac{d}{c} = \dfrac{bd}{ac}$.

76. ◆ Parentheses are required to ensure that numerators and denominators are multiplied correctly. For example, to indicate the product of $a+b$ and $c+d$, we must write $(a+b)(c+d)$ rather than $a+b(c+d)$ or $(a+b)c+d$.

77. ◆ The quotient is undefined because $\dfrac{x+3}{x-5}$ is undefined for $x=5$, $\dfrac{x-7}{x+1}$ is undefined for $x=-1$, and $\dfrac{x+1}{x-7}$ (the reciprocal of $\dfrac{x-7}{x+1}$) is undefined for $x=7$.

78. ◆ To divide we multiply by the reciprocal of the divisor.

79. $\dfrac{a}{(c-3d)(2a+5b)}$

80. $(x-2a) \div \dfrac{a^2x^2-4a^4}{a^2x+2a^3} = \dfrac{x-2a}{1} \cdot \dfrac{a^2x+2a^3}{a^2x^2-4a^4}$
$= \dfrac{(x-2a)(a^2)(x+2a)}{a^2(x+2a)(x-2a)}$
$= 1$

81. $\dfrac{1}{b^3(a-3b)}$

A-116 INSTRUCTOR'S ANSWERS

82. $\dfrac{3x^2 - 2xy - y^2}{x^2 - y^2} \div (3x^2 + 4xy + y^2)^2 =$
$\dfrac{3x^2 - 2xy - y^2}{x^2 - y^2} \cdot \dfrac{1}{(3x^2 + 4xy + y^2)^2} =$
$\dfrac{(3x + y)(x - y) \cdot 1}{(x + y)(x - y)(3x + y)(3x + y)(x + y)(x + y)} =$
$\dfrac{1}{(x + y)^3(3x + y)}$

83. $\dfrac{x^3 y^2}{4}$

84. $\dfrac{z^2 - 8z + 16}{z^2 + 8z + 16} \div \dfrac{(z - 4)^5}{(z + 4)^5} \div \dfrac{3z + 12}{z^2 - 16}$
$= \dfrac{(z - 4)^2}{(z + 4)^2} \cdot \dfrac{(z + 4)^5}{(z - 4)^5} \cdot \dfrac{(z + 4)(z - 4)}{3(z + 4)}$
$= \dfrac{(z - 4)^2(z + 4)^2(z + 4)^3(z + 4)(z - 4)}{(z + 4)^2(z - 4)^2(z - 4)(z - 4)^2(3)(z + 4)}$
$= \dfrac{(z + 4)^3}{3(z - 4)^2}$

85. $\dfrac{4}{x + 7}$

86. $\dfrac{3x + 3y + 3}{9x} \div \dfrac{x^2 + 2xy + y^2 - 1}{x^4 + x^2}$
$= \dfrac{3x + 3y + 3}{9x} \cdot \dfrac{x^4 + x^2}{x^2 + 2xy + y^2 - 1}$
$= \dfrac{3(x + y + 1)(x^2)(x^2 + 1)}{9x[(x + y) + 1][(x + y) - 1]}$
$= \dfrac{3(x + y + 1)(x)(x)(x^2 + 1)}{3 \cdot 3 \cdot x(x + y + 1)(x + y - 1)}$
$= \dfrac{3x(x + y + 1)}{3x(x + y + 1)} \cdot \dfrac{x(x^2 + 1)}{3(x + y - 1)}$
$= \dfrac{x(x^2 + 1)}{3(x + y - 1)}$

87. $\dfrac{(t - 1)(t - 9)(t - 9)}{(t^2 + 9)(t + 1)}$

88. $\dfrac{(t + 2)^3}{(t + 1)^3} \div \dfrac{t^2 + 4t + 4}{t^2 + 2t + 1} \cdot \dfrac{t + 1}{t + 2}$
$= \dfrac{(t + 2)^3}{(t + 1)^3} \cdot \dfrac{t^2 + 2t + 1}{t^2 + 4t + 4} \cdot \dfrac{t + 1}{t + 2}$
$= \dfrac{(t + 2)(t + 2)(t + 2)(t + 1)(t + 1)(t + 1)}{(t + 1)(t + 1)(t + 1)(t + 2)(t + 2)(t + 2)}$
$= 1$

89. $\dfrac{3(y + 2)^3}{y(y - 1)}$

90. $\dfrac{a^4 - 81b^4}{a^2c - 6abc + 9b^2c} \cdot \dfrac{a + 3b}{a^2 + 9b^2} \div \dfrac{a^2 + 6ab + 9b^2}{(a - 3b)^2}$
$= \dfrac{(a^2 + 9b^2)(a + 3b)(a - 3b)}{c(a - 3b)^2} \cdot \dfrac{a + 3b}{a^2 + 9b^2} \cdot \dfrac{(a - 3b)^2}{(a + 3b)^2}$
$= \dfrac{a - 3b}{c}$

EXERCISE SET 6.3, PP. 317–318

1. $\dfrac{13}{x}$ 2. $\dfrac{13}{a^2}$ 3. $\dfrac{3x + 1}{15}$ 4. $\dfrac{4a - 4}{7}$ 5. $\dfrac{6}{a + 3}$
6. $\dfrac{13}{x + 2}$ 7. $\dfrac{6}{a + 2}$ 8. $\dfrac{6}{x + 7}$ 9. $\dfrac{y + 4}{y}$
10. $\dfrac{5 + 3t}{4t} - \dfrac{2t + 1}{4t} = \dfrac{5 + 3t - 2t - 1}{4t} = \dfrac{t + 4}{4t}$
11. 11
12. $\dfrac{3a + 13}{a + 4} + \dfrac{2a + 7}{a + 4} = \dfrac{5a + 20}{a + 4} = \dfrac{5(a + 4)}{a + 4} = 5$
13. $\dfrac{7x + 5}{x + 1}$
14. $\dfrac{3a + 13}{a + 4} - \dfrac{2a + 7}{a + 4} = \dfrac{3a + 13 - 2a - 7}{a + 4} = \dfrac{a + 6}{a + 4}$
15. $a + 5$
16. $\dfrac{x^2}{x + 5} + \dfrac{7x + 10}{x + 5} = \dfrac{x^2 + 7x + 10}{x + 5} = \dfrac{(x + 5)(x + 2)}{x + 5} = x + 2$
17. $x - 4$
18. $\dfrac{a^2}{a + 3} - \dfrac{2a + 15}{a + 3} = \dfrac{a^2 - 2a - 15}{a + 3} = \dfrac{(a + 3)(a - 5)}{a + 3} = a - 5$
19. $t + 7$
20. $\dfrac{y^2 + 6y}{y + 2} + \dfrac{2y + 12}{y + 2} = \dfrac{y^2 + 8y + 12}{y + 2} =$
$\dfrac{(y + 6)(y + 2)}{y + 2} = y + 6$
21. $\dfrac{1}{x + 2}$
22. $\dfrac{-7}{x^2 - 4x + 3} + \dfrac{x + 4}{x^2 - 4x + 3} = \dfrac{x - 3}{x^2 - 4x + 3} =$
$\dfrac{x - 3}{(x - 3)(x - 1)} = \dfrac{1}{x - 1}$
23. $\dfrac{a + 1}{a + 6}$
24. $\dfrac{a^2 - 1}{a^2 - 7a + 12} - \dfrac{8}{a^2 - 7a + 12} = \dfrac{a^2 - 9}{a^2 - 7a + 12} =$
$\dfrac{(a + 3)(a - 3)}{(a - 4)(a - 3)} = \dfrac{a + 3}{a - 4}$
25. $\dfrac{t - 4}{t + 3}$
26. $\dfrac{y^2 - 7y}{y^2 + 8y + 16} + \dfrac{6y - 20}{y^2 + 8y + 16} = \dfrac{y^2 - y - 20}{y^2 + 8y + 16} =$
$\dfrac{(y - 5)(y + 4)}{(y + 4)(y + 4)} = \dfrac{y - 5}{y + 4}$
27. $\dfrac{x + 5}{x - 6}$

28. $\dfrac{2x^2+3}{x^2-6x+5} - \dfrac{(x^2-5x+9)}{x^2-6x+5} =$
$\dfrac{2x^2+3-x^2+5x-9}{x^2-6x+5} = \dfrac{x^2+5x-6}{x^2-6x+5} =$
$\dfrac{(x+6)(x-1)}{(x-5)(x-1)} = \dfrac{x+6}{x-5}$

29. $\dfrac{-5}{x-4}$

30. $\dfrac{3-2t}{t^2-5t+4} + \dfrac{2-3t}{t^2-5t+4} = \dfrac{5-5t}{t^2-5t+4} =$
$\dfrac{-5(-1+t)}{(t-4)(t-1)} = \dfrac{-5(t-1)}{(t-4)(t-1)} = \dfrac{-5}{t-4}$

31. $\dfrac{-1}{x-1}$

32. $\dfrac{5-3x}{x^2-2x+1} - \dfrac{x+1}{x^2-2x+1} = \dfrac{5-3x-x-1}{x^2-2x+1} =$
$\dfrac{4-4x}{x^2-2x+1} = \dfrac{-4(-1+x)}{(x-1)^2} = \dfrac{-4(x-1)}{(x-1)(x-1)} = \dfrac{-4}{x-1}$

33. 135

34. $10 = 2 \cdot 5$
$15 = 3 \cdot 5$
LCM $= 2 \cdot 3 \cdot 5$, or 30

35. 72

36. $12 = 2 \cdot 2 \cdot 3$
$15 = 3 \cdot 5$
LCM $= 2 \cdot 2 \cdot 3 \cdot 5$, or 60

37. 126

38. $8 = 2 \cdot 2 \cdot 2$
$36 = 2 \cdot 2 \cdot 3 \cdot 3$
$40 = 2 \cdot 2 \cdot 2 \cdot 5$
LCM $= 2 \cdot 2 \cdot 2 \cdot 3 \cdot 3 \cdot 5$, or 360

39. $12x^3$

40. $2a^2b = 2 \cdot a \cdot a \cdot b$
$8ab^2 = 2 \cdot 2 \cdot 2 \cdot a \cdot b \cdot b$
LCM $= 2 \cdot 2 \cdot 2 \cdot a \cdot a \cdot b \cdot b$, or $8a^2b^2$

41. $30a^4b^8$

42. $6a^2b^7 = 2 \cdot 3 \cdot a \cdot a \cdot b \cdot b \cdot b \cdot b \cdot b \cdot b \cdot b$
$9a^5b^2 = 3 \cdot 3 \cdot a \cdot a \cdot a \cdot a \cdot a \cdot b \cdot b$
LCM $= 2 \cdot 3 \cdot 3 \cdot a \cdot a \cdot a \cdot a \cdot a \cdot b \cdot b \cdot b \cdot b \cdot b \cdot b \cdot b$,
or $18a^5b^7$

43. $6(y-3)$

44. $4(x-1) = 2 \cdot 2 \cdot (x-1)$
$8(x-1) = 2 \cdot 2 \cdot 2 \cdot (x-1)$
LCM $= 2 \cdot 2 \cdot 2 \cdot (x-1)$, or $8(x-1)$

45. $(x+2)(x-2)(x+3)$

46. $x^2 + 3x + 2 = (x+2)(x+1)$
$x^2 - 4 = (x+2)(x-2)$
LCM $= (x+2)(x+1)(x-2)$

47. $t(t+2)^2(t-4)$

48. $y^3 - y^2 = y \cdot y(y-1)$
$y^4 - y^2 = y \cdot y(y+1)(y-1)$
LCM $= y \cdot y(y+1)(y-1)$
$= y^2(y+1)(y-1)$

49. $30x^2y^2z^3$

50. $8x^3z = 2 \cdot 2 \cdot 2 \cdot x \cdot x \cdot x \cdot z$
$12xy^2 = 2 \cdot 2 \cdot 3 \cdot x \cdot y \cdot y$
$4y^5z^2 = 2 \cdot 2 \cdot y \cdot y \cdot y \cdot y \cdot y \cdot z \cdot z$
LCM $= 2 \cdot 2 \cdot 2 \cdot 3 \cdot x \cdot x \cdot x \cdot y \cdot y \cdot y \cdot y \cdot y \cdot z \cdot z$
$= 24x^3y^5z^2$

51. $(a-1)^2(a+1)$

52. $x^2 - y^2 = (x+y)(x-y)$
$2x + 2y = 2(x+y)$
$x^2 + 2xy + y^2 = (x+y)(x+y)$
LCM $= 2(x+y)(x+y)(x-y) = 2(x+y)^2(x-y)$

53. $(m-2)^2(m-3)$

54. $2x^2 + 5x + 2 = (2x+1)(x+2)$
$2x^2 - x - 1 = (2x+1)(x-1)$
LCM $= (2x+1)(x+2)(x-1)$

55. $10v(v+3)(v+4)$

56. $12a^2 + 24a = 2 \cdot 2 \cdot 3a(a+2)$
$4a^2 + 20a + 24 = 2 \cdot 2(a+3)(a+2)$
LCM $= 2 \cdot 2 \cdot 3a(a+2)(a+3) = 12a(a+2)(a+3)$

57. $18x^3(x-2)^2(x+1)$

58. $x^5 - 4x^3 = x \cdot x \cdot x(x+2)(x-2)$
$x^3 + 4x^2 + 4x = x(x+2)(x+2)$
LCM $= x \cdot x \cdot x(x+2)(x+2)(x-2) = x^3(x+2)^2(x-2)$

59. $\dfrac{26}{12x^5}, \dfrac{x^2y}{12x^5}$

60. $10a^3 = 2 \cdot 5 \cdot a \cdot a \cdot a$
$5a^6 = 5 \cdot a \cdot a \cdot a \cdot a \cdot a \cdot a$
The LCD is $2 \cdot 5 \cdot a \cdot a \cdot a \cdot a \cdot a \cdot a$, or $10a^6$.
$\dfrac{3}{10a^3} \cdot \dfrac{a^3}{a^3} = \dfrac{3a^3}{10a^6}$ and
$\dfrac{b}{5a^6} \cdot \dfrac{2}{2} = \dfrac{2b}{10a^6}$

61. $\dfrac{12b}{8a^2b^2}, \dfrac{5a}{8a^2b^2}$

62. $3x^4y^2 = 3 \cdot x \cdot x \cdot x \cdot x \cdot y \cdot y$
$9xy^3 = 3 \cdot 3 \cdot x \cdot y \cdot y \cdot y$
The LCD is $3 \cdot 3 \cdot x \cdot x \cdot x \cdot x \cdot y \cdot y \cdot y$, or $9x^4y^3$.
$\dfrac{7}{3x^4y^2} \cdot \dfrac{3y}{3y} = \dfrac{21y}{9x^4y^3}$ and
$\dfrac{4}{9xy^3} \cdot \dfrac{x^3}{x^3} = \dfrac{4x^3}{9x^4y^3}$

63. $\dfrac{(x+3)(x+1)}{(x+3)(x+2)(x-2)}, \dfrac{(x-2)^2}{(x+3)(x+2)(x-2)}$

A-118 INSTRUCTOR'S ANSWERS

64. $x^2 - 9 = (x + 3)(x - 3)$
$x^2 + 11x + 24 = (x + 3)(x + 8)$
The LCD is $(x + 3)(x - 3)(x + 8)$

$$\frac{x-4}{x^2-9} = \frac{x-4}{(x+3)(x-3)} \cdot \frac{x+8}{x+8}$$
$$= \frac{(x-4)(x+8)}{(x+3)(x-3)(x+8)}$$

$$\frac{x+2}{x^2+11x+24} = \frac{x+2}{(x+3)(x+8)} \cdot \frac{x-3}{x-3}$$
$$= \frac{(x+2)(x-3)}{(x+3)(x+8)(x-3)}$$

65. $(x - 4)(x - 15)$
66. $(x + 12)(x - 3)$ 67. $x^2 - 9x + 18$ 68. $s^2 - \pi r^2$
69. $-\dfrac{1}{72}$ 70. $\dfrac{2}{15} - \dfrac{7}{20} = \dfrac{8}{60} - \dfrac{21}{60} = -\dfrac{13}{60}$

71. ◆ If the numbers have a common factor, their product contains that factor more than the greatest number of times it occurs in any one factorization.   72. ◆ No; an LCM is an LCD only when it is the LCM of a set of denominators.   73. ◆ No; if rational expressions have different denominators we must factor to find the LCD. We must also be able to factor in order to simplify after adding or subtracting rational expressions.   74. ◆ The binomial is a factor of the trinomial.   75. 0

76. $\dfrac{6x-1}{x-1} + \dfrac{3(2x+5)}{x-1} + \dfrac{3(2x-3)}{x-1}$
$= \dfrac{6x - 1 + 6x + 15 + 6x - 9}{x-1}$
$= \dfrac{18x + 5}{x-1}$

77. $\dfrac{30}{(x-3)(x+4)}$

78. $\dfrac{x^2}{3x^2 - 5x - 2} - \dfrac{2x}{3x+1} \cdot \dfrac{1}{x-2}$
$= \dfrac{x^2}{(3x+1)(x-2)} - \dfrac{2x}{(3x+1)(x-2)}$
$= \dfrac{x^2 - 2x}{(3x+1)(x-2)}$
$= \dfrac{x(x-2)}{(3x+1)(x-2)}$
$= \dfrac{x}{3x+1}$

79. 1440
80. $8x^2 - 8 = 8(x^2 - 1) = 2 \cdot 2 \cdot 2(x+1)(x-1)$
$6x^2 - 12x + 6 = 6(x^2 - 2x + 1) = 2 \cdot 3(x-1)(x-1)$
$10x - 10 = 10(x - 1) = 2 \cdot 5(x - 1)$
LCM $= 2 \cdot 2 \cdot 2 \cdot 3 \cdot 5(x+1)(x-1)(x-1) = 120(x+1)(x-1)^2$

$\left(\text{We could also express the LCM as}\right.$
$120(x + 1)(x - 1)(1 - x)$. It is not necessary to include both a factor and its opposite in the LCM since
$\left. \dfrac{a}{-b} = \dfrac{-a}{b} = -\dfrac{a}{b}. \right)$

81. 7:55 A.M.
82. The time it takes Kim and Jed to meet again at the starting place is the LCM of the times it takes them to complete one round of the course.
$6 = 2 \cdot 3$
$8 = 2 \cdot 2 \cdot 2$
LCM $= 2 \cdot 2 \cdot 2 \cdot 3$, or 24
It takes 24 min.
83. ◆ Evaluate both expressions for some value of the variable for which both are defined. If the results are the same, we can conclude that the answer is probably correct.
84. ◆ The LCD can be found regardless of the factorization selected. However, if the factorization selected contains the other factorizations, then no multiplication is required to find the LCD.

EXERCISE SET 6.4, PP. 325–326

1. $\dfrac{3x + 4}{x^2}$
2. LCD $= x^2$
$\dfrac{5}{x} + \dfrac{6}{x^2} = \dfrac{5}{x} \cdot \dfrac{x}{x} + \dfrac{6}{x^2} = \dfrac{5x + 6}{x^2}$
3. $\dfrac{5}{24r}$
4. LCD $= 18t$
$\dfrac{2}{9t} - \dfrac{11}{6t} = \dfrac{2}{9t} \cdot \dfrac{2}{2} - \dfrac{11}{6t} \cdot \dfrac{3}{3} = \dfrac{4 - 33}{18t} = \dfrac{-29}{18t}$
5. $\dfrac{7x + 3y}{x^2 y^2}$
6. LCD $= c^2 d^3$
$\dfrac{2}{c^2 d} + \dfrac{7}{cd^3} = \dfrac{2}{c^2 d} \cdot \dfrac{d^2}{d^2} + \dfrac{7}{cd^3} \cdot \dfrac{c}{c} = \dfrac{2d^2 + 7c}{c^2 d^3}$
7. $\dfrac{16 - 15t}{18t^3}$
8. LCD $= 3x^2 y^3$
$\dfrac{-2}{3xy^2} - \dfrac{6}{x^2 y^3} = \dfrac{-2}{3xy^2} \cdot \dfrac{xy}{xy} - \dfrac{6}{x^2 y^3} \cdot \dfrac{3}{3} = \dfrac{-2xy - 18}{3x^2 y^3}$
9. $\dfrac{5x + 9}{24}$

10. LCD = 18
$$\frac{x-4}{9} + \frac{x+5}{6} = \frac{x-4}{9} \cdot \frac{2}{2} + \frac{x+5}{6} \cdot \frac{3}{3}$$
$$= \frac{2(x-4) + 3(x+5)}{18}$$
$$= \frac{2x - 8 + 3x + 15}{18}$$
$$= \frac{5x + 7}{18}$$

11. $\dfrac{a+8}{4}$

12. $\left.\begin{array}{l}6 = 2 \cdot 3 \\ 3 = 3\end{array}\right\}$ LCD = 6
$$\frac{x-2}{6} - \frac{x+1}{3} = \frac{x-2}{6} - \frac{x+1}{3} \cdot \frac{2}{2}$$
$$= \frac{x-2}{6} - \frac{2x+2}{6}$$
$$= \frac{x-2 - (2x+2)}{6}$$
$$= \frac{x - 2 - 2x - 2}{6}$$
$$= \frac{-x-4}{6}, \text{ or } \frac{-(x+4)}{6}$$

13. $\dfrac{5a^2 + 7a - 3}{9a^2}$

14. $\left.\begin{array}{l}16a = 2 \cdot 2 \cdot 2 \cdot 2 \cdot a \\ 4a^2 = 2 \cdot 2 \cdot a \cdot a\end{array}\right\}$ LCD $= 2 \cdot 2 \cdot 2 \cdot 2 \cdot a \cdot a$, or $16a^2$
$$\frac{a+4}{16a} + \frac{3a+4}{4a^2} = \frac{a+4}{16a} \cdot \frac{a}{a} + \frac{3a+4}{4a^2} \cdot \frac{4}{4}$$
$$= \frac{a^2 + 4a}{16a^2} + \frac{12a + 16}{16a^2}$$
$$= \frac{a^2 + 16a + 16}{16a^2}$$

15. $\dfrac{-7x - 13}{4x}$

16. $\left.\begin{array}{l}3z = 3 \cdot z \\ 4z = 2 \cdot 2 \cdot z\end{array}\right\}$ LCD $= 2 \cdot 2 \cdot 3 \cdot z$, or $12z$
$$\frac{4z-9}{3z} - \frac{3z-8}{4z} = \frac{4z-9}{3z} \cdot \frac{4}{4} - \frac{3z-8}{4z} \cdot \frac{3}{3}$$
$$= \frac{16z - 36}{12z} - \frac{9z - 24}{12z}$$
$$= \frac{16z - 36 - (9z - 24)}{12z}$$
$$= \frac{16z - 36 - 9z + 24}{12z}$$
$$= \frac{7z - 12}{12z}$$

17. $\dfrac{c^2 + 3cd - d^2}{c^2 d^2}$

18. LCD $= x^2 y^2$ (See Exercise 5.)
$$\frac{x+y}{xy^2} + \frac{3x+y}{x^2 y} = \frac{x+y}{xy^2} \cdot \frac{x}{x} + \frac{3x+y}{x^2 y} \cdot \frac{y}{y}$$
$$= \frac{x(x+y) + y(3x+y)}{x^2 y^2}$$
$$= \frac{x^2 + xy + 3xy + y^2}{x^2 y^2}$$
$$= \frac{x^2 + 4xy + y^2}{x^2 y^2}$$

19. $\dfrac{3y^2 - 3xy - 6x^2}{2x^2 y^2}$

20. $\left.\begin{array}{l}3xt^2 = 3 \cdot x \cdot t \cdot t \\ x^2 t = x \cdot x \cdot t\end{array}\right\}$ LCD $= 3 \cdot x \cdot x \cdot t \cdot t$, or $3x^2 t^2$
$$\frac{4x + 2t}{3xt^2} - \frac{5x - 3t}{x^2 t}$$
$$= \frac{4x + 2t}{3xt^2} \cdot \frac{x}{x} - \frac{5x - 3t}{x^2 t} \cdot \frac{3t}{3t}$$
$$= \frac{4x^2 + 2tx}{3x^2 t^2} - \frac{15xt - 9t^2}{3x^2 t^2}$$
$$= \frac{4x^2 + 2tx - (15xt - 9t^2)}{3x^2 t^2}$$
$$= \frac{4x^2 + 2tx - 15xt + 9t^2}{3x^2 t^2}$$
$$= \frac{4x^2 - 13xt + 9t^2}{3x^2 t^2}$$
(Although $4x^2 - 13xt + 9t^2$ can be factored, doing so will not enable us to simplify the result further.)

21. $\dfrac{4x}{(x-1)(x+1)}$

22. The denominators cannot be factored, so the LCD is their product, $(x-2)(x+2)$.
$$\frac{3}{x-2} + \frac{3}{x+2} = \frac{3}{x-2} \cdot \frac{x+2}{x+2} + \frac{3}{x+2} \cdot \frac{x-2}{x-2}$$
$$= \frac{3(x+2) + 3(x-2)}{(x-2)(x+2)}$$
$$= \frac{3x + 6 + 3x - 6}{(x-2)(x+2)}$$
$$= \frac{6x}{(x-2)(x+2)}$$

23. $\dfrac{-z^2 + 5z}{(z-1)(z+1)}$

24. $\dfrac{5}{x+5} - \dfrac{3}{x-5}$ LCD $= (x+5)(x-5)$
$$= \frac{5}{x+5} \cdot \frac{x-5}{x-5} - \frac{3}{x-5} \cdot \frac{x+5}{x+5}$$

$$= \frac{5x-25}{(x+5)(x-5)} - \frac{3x+15}{(x+5)(x-5)}$$
$$= \frac{5x-25-(3x+15)}{(x+5)(x-5)}$$
$$= \frac{5x-25-3x-15}{(x+5)(x-5)}$$
$$= \frac{2x-40}{(x+5)(x-5)}$$

(Although $2x - 40$ can be factored, doing so will not enable us to simplify the result further.)

25. $\dfrac{11x+15}{4x(x+5)}$

26. $\left.\begin{array}{l} 3x = 3 \cdot x \\ x+1 = x+1 \end{array}\right\}$LCD $= 3x(x+1)$

$$\frac{3}{x+1} + \frac{2}{3x} = \frac{3}{x+1} \cdot \frac{3x}{3x} + \frac{2}{3x} \cdot \frac{x+1}{x+1}$$
$$= \frac{9x+2(x+1)}{3x(x+1)}$$
$$= \frac{9x+2x+2}{3x(x+1)}$$
$$= \frac{11x+2}{3x(x+1)}$$

27. $\dfrac{14-3x}{(x+2)(x-2)}$

28. $\dfrac{3}{2t^2-2t} - \dfrac{5}{2t-2}$

$$= \frac{3}{2t(t-1)} - \frac{5}{2(t-1)} \quad \text{LCD} = 2t(t-1)$$
$$= \frac{3}{2t(t-1)} - \frac{5}{2(t-1)} \cdot \frac{t}{t}$$
$$= \frac{3-5t}{2t(t-1)}$$

29. $\dfrac{x^2-x}{(x+5)(x-5)}$

30. $\dfrac{2x}{x^2-16} + \dfrac{x}{x-4}$

$$= \frac{2x}{(x+4)(x-4)} + \frac{x}{x-4} \quad \text{LCD} = (x+4)(x-4)$$
$$= \frac{2x}{(x+4)(x-4)} + \frac{x}{x-4} \cdot \frac{x+4}{x+4}$$
$$= \frac{2x+x(x+4)}{(x+4)(x-4)}$$
$$= \frac{2x+x^2+4x}{(x+4)(x-4)}$$
$$= \frac{x^2+6x}{(x+4)(x-4)}$$

(Although $x^2 + 6x$ can be factored, doing so will not enable us to simplify the result further.)

31. $\dfrac{4t-5}{4(t-3)}$

32. $\dfrac{6}{z+4} - \dfrac{2}{3z+12} = \dfrac{6}{z+4} - \dfrac{2}{3(z+4)}$

$$\text{LCD} = 3(z+4)$$
$$= \frac{6}{z+4} \cdot \frac{3}{3} - \frac{2}{3(z+4)}$$
$$= \frac{18}{3(z+4)} - \frac{2}{3(z+4)}$$
$$= \frac{16}{3(z+4)}$$

33. $\dfrac{2x+10}{(x+3)^2}$

34. $\dfrac{3}{x-1} + \dfrac{2}{(x-1)^2}$ LCD $= (x-1)^2$

$$= \frac{3}{x-1} \cdot \frac{x-1}{x-1} + \frac{2}{(x-1)^2}$$
$$= \frac{3(x-1)+2}{(x-1)^2}$$
$$= \frac{3x-3+2}{(x-1)^2}$$
$$= \frac{3x-1}{(x-1)^2}$$

35. $\dfrac{6-20x}{15x(x+1)}$

36. $\dfrac{2t}{t^2-9} - \dfrac{3}{t-3} = \dfrac{2t}{(t+3)(t-3)} - \dfrac{3}{t-3}$

$$\text{LCD} = (t+3)(t-3)$$
$$= \frac{2t}{(t+3)(t-3)} - \frac{3}{t-3} \cdot \frac{t+3}{t+3}$$
$$= \frac{2t-3(t+3)}{(t+3)(t-3)}$$
$$= \frac{2t-3t-9}{(t+3)(t-3)}$$
$$= \frac{-t-9}{(t+3)(t-3)}$$

37. $\dfrac{9a}{4(a-5)}$

38. $\dfrac{4a}{5a-10} + \dfrac{3a}{10a-20} = \dfrac{4a}{5(a-2)} + \dfrac{3a}{2 \cdot 5(a-2)}$

$$\text{LCD} = 2 \cdot 5(a-2)$$
$$= \frac{4a}{5(a-2)} \cdot \frac{2}{2} + \frac{3a}{2 \cdot 5(a-2)}$$
$$= \frac{8a+3a}{10(a-2)}$$
$$= \frac{11a}{10(a-2)}$$

39. $\dfrac{t^2+2ty-y^2}{(y-t)(y+t)}$

40. $\dfrac{a}{x+a} - \dfrac{a}{x-a}$ LCD $= (x+a)(x-a)$

$= \dfrac{a}{x+a} \cdot \dfrac{x-a}{x-a} - \dfrac{a}{x-a} \cdot \dfrac{x+a}{x+a}$

$= \dfrac{ax - a^2}{(x+a)(x-a)} - \dfrac{ax + a^2}{(x+a)(x-a)}$

$= \dfrac{ax - a^2 - (ax + a^2)}{(x+a)(x-a)}$

$= \dfrac{ax - a^2 - ax - a^2}{(x+a)(x-a)}$

$= \dfrac{-2a^2}{(x+a)(x-a)}$

41. $\dfrac{2x^2 - 10x + 25}{x(x-5)}$

42. $\dfrac{x+4}{x} + \dfrac{x}{x+4}$ LCD $= x(x+4)$

$= \dfrac{x+4}{x} \cdot \dfrac{x+4}{x+4} + \dfrac{x}{x+4} \cdot \dfrac{x}{x}$

$= \dfrac{(x+4)^2 + x^2}{x(x+4)}$

$= \dfrac{x^2 + 8x + 16 + x^2}{x(x+4)}$

$= \dfrac{2x^2 + 8x + 16}{x(x+4)}$

(Although $2x^2 + 8x + 16$ can be factored, doing so will not enable us to simplify the result further.)

43. $\dfrac{x-3}{(x+1)(x+3)}$

44. $\dfrac{x}{x^2 + 9x + 20} - \dfrac{4}{x^2 + 7x + 12}$

$= \dfrac{x}{(x+4)(x+5)} - \dfrac{4}{(x+3)(x+4)}$

LCD $= (x+3)(x+4)(x+5)$

$= \dfrac{x}{(x+4)(x+5)} \cdot \dfrac{x+3}{x+3} - \dfrac{4}{(x+3)(x+4)} \cdot \dfrac{x+5}{x+5}$

$= \dfrac{x(x+3) - 4(x+5)}{(x+3)(x+4)(x+5)}$

$= \dfrac{x^2 + 3x - 4x - 20}{(x+3)(x+4)(x+5)}$

$= \dfrac{x^2 - x - 20}{(x+3)(x+4)(x+5)}$

$= \dfrac{(x+4)(x-5)}{(x+3)(x+4)(x+5)}$

$= \dfrac{x-5}{(x+3)(x+5)}$

45. $\dfrac{x^2 + 5x + 1}{(x+1)^2(x+4)}$

46. $\dfrac{7}{a^2 + a - 2} + \dfrac{5}{a^2 - 4a + 3}$

$= \dfrac{7}{(a+2)(a-1)} + \dfrac{5}{(a-3)(a-1)}$

LCD $= (a+2)(a-1)(a-3)$

$= \dfrac{7(a-3) + 5(a+2)}{(a+2)(a-1)(a-3)}$

$= \dfrac{7a - 21 + 5a + 10}{(a+2)(a-1)(a-3)}$

$= \dfrac{12a - 11}{(a+2)(a-1)(a-3)}$

47. $\dfrac{x^2 - 48}{(x+7)(x+8)(x+6)}$

48. $\dfrac{-5}{x^2 + 17x + 16} - \dfrac{3}{x^2 + 9x + 8}$

$= \dfrac{-5}{(x+1)(x+16)} - \dfrac{3}{(x+1)(x+8)}$

LCD $= (x+1)(x+16)(x+8)$

$= \dfrac{-5(x+8) - 3(x+16)}{(x+1)(x+16)(x+8)}$

$= \dfrac{-5x - 40 - 3x - 48}{(x+1)(x+16)(x+8)}$

$= \dfrac{-8x - 88}{(x+1)(x+16)(x+8)}$

(Although $-8x - 88$ can be factored, doing so will not enable us to simplify the result further.)

49. $\dfrac{5x + 12}{(x+3)(x-3)(x+2)}$

50. $\dfrac{3z}{z^2 - 4x + 4} + \dfrac{10}{z^2 + z - 6}$

$= \dfrac{3z}{(z-2)^2} + \dfrac{10}{(z-2)(z+3)}$,

LCD $= (z-2)^2(z+3)$

$= \dfrac{3z}{(z-2)^2} \cdot \dfrac{z+3}{z+3} + \dfrac{10}{(z-2)(z+3)} \cdot \dfrac{z-2}{z-2}$

$= \dfrac{3z^2 + 9z + 10z - 20}{(z-2)^2(z+3)}$

$= \dfrac{3z^2 + 19z - 20}{(z-2)^2(z+3)}$

51. $\dfrac{11}{x-1}$

52. $\dfrac{x}{4} - \dfrac{3x-5}{-4} = \dfrac{x}{4} - \dfrac{3x-5}{-4} \cdot \dfrac{-1}{-1}$

$= \dfrac{x}{4} - \dfrac{-3x+5}{4}$

$= \dfrac{x - (-3x+5)}{4}$

$= \dfrac{x + 3x - 5}{4}$

$= \dfrac{4x - 5}{4}$

53. $t + 2$

54. $\dfrac{y^2}{y-3} + \dfrac{9}{3-y} = \dfrac{y^2}{y-3} + \dfrac{9}{3-y} \cdot \dfrac{-1}{-1}$

$= \dfrac{y^2}{y-3} + \dfrac{-9}{-3+y}$

$= \dfrac{y^2 - 9}{y - 3}$

$= \dfrac{(y+3)(y-3)}{y-3}$

$= y + 3$

55. 0

56. $\dfrac{b-7}{b^2-16} + \dfrac{7-b}{16-b^2} = \dfrac{b-7}{b^2-16} + \dfrac{7-b}{16-b^2} \cdot \dfrac{-1}{-1}$

$= \dfrac{b-7}{b^2-16} + \dfrac{b-7}{b^2-16}$

$= \dfrac{2b-14}{b^2-16}$

(Although both $2b - 14$ and $b^2 - 16$ can be factored, doing so will not enable us to simplify the result further.)

57. $\dfrac{p^2 + 7p + 1}{(p+5)(p-5)}$

58. $\dfrac{y+2}{y-7} + \dfrac{3-y}{49-y^2}$

$= \dfrac{y+2}{y-7} + \dfrac{3-y}{(7+y)(7-y)}$

$= \dfrac{y+2}{y-7} + \dfrac{3-y}{(7+y)(7-y)} \cdot \dfrac{-1}{-1}$

$= \dfrac{y+2}{y-7} + \dfrac{y-3}{(y+7)(y-7)}$ LCD $= (y+7)(y-7)$

$= \dfrac{y+2}{y-7} \cdot \dfrac{y+7}{y+7} + \dfrac{y-3}{(y+7)(y-7)}$

$= \dfrac{y^2 + 9y + 14 + y - 3}{(y+7)(y-7)}$

$= \dfrac{y^2 + 10y + 11}{(y+7)(y-7)}$

59. $\dfrac{9x+12}{(x+3)(x-3)}$

60. $\dfrac{8x}{16-x^2} - \dfrac{5}{x-4}$

$= \dfrac{8x}{(4+x)(4-x)} - \dfrac{5}{x-4}$

$= \dfrac{8x}{(4+x)(4-x)} - \dfrac{5}{x-4} \cdot \dfrac{-1}{-1}$

$= \dfrac{8x}{(4+x)(4-x)} - \dfrac{-5}{4-x}$ LCD $= (4+x)(4-x)$

$= \dfrac{8x}{(4+x)(4-x)} - \dfrac{-5}{4-x} \cdot \dfrac{4+x}{4+x}$

$= \dfrac{8x - (-5)(4+x)}{(4+x)(4-x)}$

$= \dfrac{8x + 20 + 5x}{(4+x)(4-x)}$

$= \dfrac{13x + 20}{(4+x)(4-x)}$, or $\dfrac{-13x - 20}{(x+4)(x-4)}$

61. $\dfrac{-3x^2 + 7x + 4}{3(x+2)(2-x)}$

62. $\dfrac{a}{a^2-1} + \dfrac{2a}{a-a^2} = \dfrac{a}{a^2-1} + \dfrac{2 \cdot a}{a(1-a)}$

$= \dfrac{a}{(a+1)(a-1)} + \dfrac{2}{1-a}$

$= \dfrac{a}{(a+1)(a-1)} + \dfrac{2}{1-a} \cdot \dfrac{-1}{-1}$

$= \dfrac{a}{(a+1)(a-1)} + \dfrac{-2}{a-1}$

LCD $= (a+1)(a-1)$

$= \dfrac{a}{(a+1)(a-1)} + \dfrac{-2}{a-1} \cdot \dfrac{a+1}{a+1}$

$= \dfrac{a - 2a - 2}{(a+1)(a-1)}$

$= \dfrac{-a - 2}{(a+1)(a-1)}$

63. $\dfrac{a-2}{(a+3)(a-3)}$

64. $\dfrac{4x}{x^2-y^2} - \dfrac{6}{y-x}$

$= \dfrac{4x}{(x+y)(x-y)} - \dfrac{6}{y-x}$

$= \dfrac{4x}{(x+y)(x-y)} - \dfrac{6}{y-x} \cdot \dfrac{-1}{-1}$

$= \dfrac{4x}{(x+y)(x-y)} - \dfrac{-6}{x-y}$ LCD $= (x+y)(x-y)$

$= \dfrac{4x}{(x+y)(x-y)} - \dfrac{-6}{x-y} \cdot \dfrac{x+y}{x+y}$

$= \dfrac{4x - (-6)(x+y)}{(x+y)(x-y)}$

$= \dfrac{4x + 6x + 6y}{(x+y)(x-y)}$

$= \dfrac{10x + 6y}{(x+y)(x-y)}$

(Although $10x + 6y$ can be factored, doing so will not enable us to simplify the result further.)

65. $\dfrac{2}{y(y-1)}$

66. $\dfrac{x+6}{4-x^2} - \dfrac{x+3}{x+2} + \dfrac{x-3}{2-x}$ LCD $= (2+x)(2-x)$

$= \dfrac{(x+6) - (x+3)(2-x) + (x-3)(2+x)}{(2+x)(2-x)}$

$= \dfrac{x+6 + x^2 + x - 6 + x^2 - x - 6}{(2+x)(2-x)}$

$= \dfrac{2x^2 + x - 6}{(2+x)(2-x)}$

$= \dfrac{(2x-3)(x+2)}{(2+x)(2-x)}$

$= \dfrac{2x-3}{2-x}$

67. $\dfrac{z-3}{2z-1}$

68. $\dfrac{1}{x+y} + \dfrac{1}{x-y} - \dfrac{2x}{x^2-y^2}$ LCD $= (x+y)(x-y)$

$= \dfrac{(x-y) + (x+y) - 2x}{(x+y)(x-y)}$

$= 0$

69. $\dfrac{2}{r+s}$

70. $\dfrac{3}{2c-1} - \dfrac{1}{c+2} - \dfrac{5}{2c^2+3c-2}$

$= \dfrac{3}{2c-1} - \dfrac{1}{c+2} - \dfrac{5}{(2c-1)(c+2)}$

   LCD $= (2c-1)(c+2)$

$= \dfrac{3}{2c-1} \cdot \dfrac{c+2}{c+2} - \dfrac{1}{c+2} \cdot \dfrac{2c-1}{2c-1} - \dfrac{5}{(2c-1)(c+2)}$

$= \dfrac{(3c+6) - (2c-1) - 5}{(2c-1)(c+2)}$

$= \dfrac{3c + 6 - 2c + 1 - 5}{(2c-1)(c+2)}$

$= \dfrac{c+2}{(2c-1)(c+2)}$

$= \dfrac{1}{2c-1}$

71. [graph of $y = \tfrac{1}{2}x - 5$]

72. [graph of $y = -\tfrac{1}{2}x - 5$]

73. [graph of $y = 3$]

74. [graph of $x = -5$]

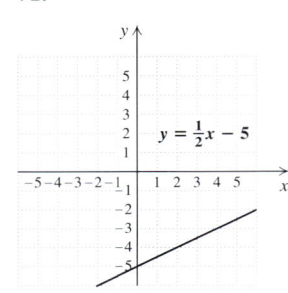

75. -8

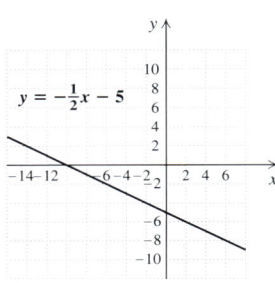

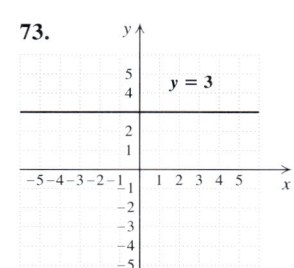

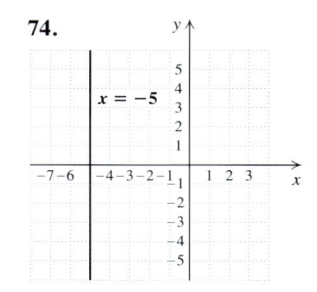

76. $2a + 8 = 13 - 4a$
$6a = 5$
$a = \dfrac{5}{6}$

77. $3, 5$

78. $x^2 - 7x - 18 = 0$
$(x - 9)(x + 2) = 0$
$x - 9 = 0$ or $x + 2 = 0$
$x = 9$ or $x = -2$

79. ◆ No; when adding, no sign changes are required so the result is the same regardless of parentheses. When subtracting, however, the sign of each term of the expression being subtracted must be changed and parentheses are needed to make sure this is done.   80. ◆ Their sum is zero. Another explanation is that

$-\left(\dfrac{1}{3-x}\right) = \dfrac{1}{-(3-x)} = \dfrac{1}{x-3}.$

81. ◆ Using the least common denominator usually reduces the complexity of computations and requires less simplification of the sum or difference.   82. ◆ If the denominators are the same, add the numerators and keep the same denominator. Simplify, if possible. If the denominators are different, follow the steps in the box on page 320 of the text.

83. Perimeter: $\dfrac{4x^2 + 18x}{(x+4)(x+5)}$; area: $\dfrac{x^2}{(x+4)(x+5)}$

84. $P = 2\left(\dfrac{3}{x+4}\right) + 2\left(\dfrac{2}{x-5}\right)$

$= \dfrac{6}{x+4} + \dfrac{4}{x-5}$ LCD $= (x+4)(x-5)$

$= \dfrac{6}{x+4} \cdot \dfrac{x-5}{x-5} + \dfrac{4}{x-5} \cdot \dfrac{x+4}{x+4}$

$= \dfrac{6x - 30 + 4x + 16}{(x+4)(x-5)}$

$= \dfrac{10x - 14}{(x+4)(x-5)}$, or $\dfrac{10x - 14}{x^2 - x - 20}$

$A = \left(\dfrac{3}{x+4}\right)\left(\dfrac{2}{x-5}\right) = \dfrac{6}{(x+4)(x-5)}$, or

$\dfrac{6}{x^2 - x - 20}$

A-124 INSTRUCTOR'S ANSWERS

85. $\dfrac{30}{(x-3)(x+4)}$

86. $\dfrac{x^2}{3x^2-5x-2} - \dfrac{2x}{3x+1} \cdot \dfrac{1}{x-2}$

$= \dfrac{x^2}{(3x+1)(x-2)} - \dfrac{2x}{(3x+1)(x-2)}$

$= \dfrac{x^2 - 2x}{(3x+1)(x-2)}$

$= \dfrac{x(x-2)}{(3x+1)(x-2)}$

$= \dfrac{x}{3x+1} \cdot \dfrac{x-2}{x-2}$

$= \dfrac{x}{3x+1}$

87. $\dfrac{x^2 + xy - x^3 + x^2y - xy^2 + y^3}{(x^2+y^2)(x+y)^2(x-y)}$

88. $\dfrac{1}{ay - 3a + 2xy - 6x} - \dfrac{xy + ay}{a^2 - 4x^2}\left(\dfrac{1}{y-3}\right)^2$

$= \dfrac{1}{ay - 3a + 2xy - 6x} - \dfrac{xy + ay}{(a^2 - 4x^2)(y-3)^2}$

$= \dfrac{1}{a(y-3) + 2x(y-3)} - \dfrac{xy + ay}{(a+2x)(a-2x)(y-3)^2}$

$= \dfrac{1}{(y-3)(a+2x)} - \dfrac{xy + ay}{(a+2x)(a-2x)(y-3)^2}$

LCD $= (y-3)^2(a+2x)(a-2x)$

$= \dfrac{1}{(y-3)(a+2x)} \cdot \dfrac{(y-3)(a-2x)}{(y-3)(a-2x)} -$

$\dfrac{xy + ay}{(y-3)^2(a+2x)(a-2x)}$

$= \dfrac{ay - 2xy - 3a + 6x - xy - ay}{(y-3)^2(a+2x)(a-2x)}$

$= \dfrac{-3xy - 3a + 6x}{(y-3)^2(a+2x)(a-2x)}$

89. $\dfrac{-x^2 - 3}{(2x-3)(x-3)}$

90. $\left(\dfrac{a}{a-b} + \dfrac{b}{a+b}\right)\left(\dfrac{1}{3a+b} + \dfrac{2a+6b}{9a^2-b^2}\right)$

$= \dfrac{a}{(a-b)(3a+b)} + \dfrac{a(2a+6b)}{(a-b)(9a^2-b^2)} +$

$\dfrac{b}{(a+b)(3a+b)} + \dfrac{b(2a+6b)}{(a+b)(9a^2-b^2)}$

$= \dfrac{a}{(a-b)(3a+b)} + \dfrac{2a^2 + 6ab}{(a-b)(3a+b)(3a-b)} +$

$\dfrac{b}{(a+b)(3a+b)} + \dfrac{2ab + 6b^2}{(a+b)(3a+b)(3a-b)}$

LCD $= (a-b)(a+b)(3a+b)(3a-b)$

$= [a(a+b)(3a-b) + (2a^2 + 6ab)(a+b) +$
$\quad b(a-b)(3a-b) + (2ab + 6b^2)(a-b)]/$
$[(a-b)(a+b)(3a+b)(3a-b)]$

$= (3a^3 + 2a^2b - ab^2 + 2a^3 + 8a^2b + 6ab^2 + b^3 -$
$\quad 4ab^2 + 3a^2b + 4ab^2 - 6b^3 + 2a^2)/$
$[(a-b)(a+b)(3a+b)(3a-b)]$

$= \dfrac{5a^3 + 15a^2b + 5ab^2 - 5b^3}{(a-b)(a+b)(3a+b)(3a-b)}$

$= \dfrac{5(a+b)(a^2 + 2ab - b^2)}{(a-b)(a+b)(3a+b)(3a-b)}$

$= \dfrac{5(a^2 + 2ab - b^2)}{(a-b)(3a+b)(3a-b)}$

91. Answers may vary; $\dfrac{a}{a-b} + \dfrac{3b}{b-a}$

EXERCISE SET 6.5, PP. 331–332

1. $\dfrac{2}{5}$
2. $\dfrac{1 - \frac{3}{4}}{1 + \frac{9}{16}} = \dfrac{1 - \frac{3}{4}}{1 + \frac{9}{16}} \cdot \dfrac{16}{16}$

$= \dfrac{16 - 12}{16 + 9} = \dfrac{4}{25}$

3. $\dfrac{18}{65}$
4. $\dfrac{1 + \frac{1}{5}}{1 - \frac{3}{5}} = \dfrac{\frac{6}{5}}{\frac{2}{5}} = \dfrac{6}{5} \cdot \dfrac{5}{2} = 3$
5. $\dfrac{4s^2}{9 + 3s^2}$

6. $\dfrac{\frac{1}{x} - 5}{\frac{1}{x} + 3} = \dfrac{\frac{1}{x} - 5}{\frac{1}{x} + 3} \cdot \dfrac{x}{x}$
7. $\dfrac{3x}{2x+1}$

$= \dfrac{1 - 5x}{1 + 3x}$

8. $\dfrac{\frac{4}{x} - \frac{1}{x^2}}{\frac{2}{x}} = \dfrac{\frac{4}{x} - \frac{1}{x^2}}{\frac{2}{x}} \cdot \dfrac{x^2}{x^2}$
9. $\dfrac{4a - 10}{a - 1}$

$= \dfrac{4x - 1}{2x}$

10. $\dfrac{\frac{a+4}{a^2}}{\frac{a-2}{3a}} = \dfrac{a+4}{a^2} \cdot \dfrac{3a}{a-2}$
11. $x - 4$

$= \dfrac{3(a+4)}{a(a-2)}$

$= \dfrac{3a + 12}{a^2 - 2a}$

12. $\dfrac{\dfrac{3}{x}+\dfrac{3}{8}}{\dfrac{x}{8}-\dfrac{3}{x}} = \dfrac{\dfrac{3}{x}+\dfrac{3}{8}}{\dfrac{x}{8}-\dfrac{3}{x}}\cdot\dfrac{8x}{8x}$

$=\dfrac{24+3x}{x^2-24}$

(Although the numerator can be factored, doing so will not enable us to simplify further.)

13. $\dfrac{1}{x}$

14. $\dfrac{\dfrac{1}{5}-\dfrac{1}{a}}{\dfrac{5-a}{5}} = \dfrac{\dfrac{1}{5}-\dfrac{1}{a}}{\dfrac{5-a}{5}}\cdot\dfrac{5a}{5a}$

$=\dfrac{a-5}{a(5-a)}$

$=\dfrac{a-5}{a(-1)(a-5)}$

$=-\dfrac{1}{a}$

15. $\dfrac{1+t^2}{t-t^2}$

16. $\dfrac{2+\dfrac{1}{x}}{2-\dfrac{1}{x^2}} = \dfrac{2+\dfrac{1}{x}}{2-\dfrac{1}{x^2}}\cdot\dfrac{x^2}{x^2}$

$=\dfrac{2x^2+x}{2x^2-1}$

(Although the numerator can be factored, doing so will not enable us to simplify further.)

17. $\dfrac{x}{x-y}$

18. $\dfrac{\dfrac{a^2}{a-3}}{\dfrac{2a}{a^2-9}} = \dfrac{a^2}{a-3}\cdot\dfrac{a^2-9}{2a}$

$=\dfrac{a\cdot a\cdot(a+3)(a-3)}{(a-3)\cdot 2\cdot a}$

$=\dfrac{a^2+3a}{2}$

19. $\dfrac{3a^2-4a}{2+3a^2}$

20. $\dfrac{\dfrac{5}{x^3}+\dfrac{1}{x^2}}{\dfrac{2}{x}-\dfrac{3}{x^2}} = \dfrac{\dfrac{5}{x^3}+\dfrac{1}{x^2}}{\dfrac{2}{x}-\dfrac{3}{x^2}}\cdot\dfrac{x^3}{x^3}$

$=\dfrac{5+x}{2x^2-3x}$

(Although the denominator can be factored, doing so will not enable us to simplify further.)

21. $\dfrac{60-15a^3}{126a^2+28a^3}$

22. $\dfrac{\dfrac{5}{4x^3}-\dfrac{3}{8x}}{\dfrac{3}{2x}+\dfrac{3}{4x^3}} = \dfrac{\dfrac{5}{4x^3}\cdot\dfrac{2}{2}-\dfrac{3}{8x}\cdot\dfrac{x^2}{x^2}}{\dfrac{3}{2x}\cdot\dfrac{2x^2}{2x^2}+\dfrac{3}{4x^3}}$

$=\dfrac{\dfrac{10-3x^2}{8x^3}}{\dfrac{6x^2+3}{4x^3}}$

$=\dfrac{10-3x^2}{8x^3}\cdot\dfrac{4x^3}{6x^2+3}$

$=\dfrac{4x^3(10-3x^2)}{2\cdot 4x^3\cdot 3(2x^2+1)}$

$=\dfrac{10-3x^2}{6(2x^2+1)}$, or

$=\dfrac{10-3x^2}{12x^2+6}$

23. $\dfrac{2xy-3y^3}{xy^3+30}$

24. $\dfrac{\dfrac{a}{6b^3}+\dfrac{4}{9b^2}}{\dfrac{5}{6b}-\dfrac{1}{9b^3}}$ LCM of the denominators is $18b^3$

$=\dfrac{\dfrac{a}{6b^3}+\dfrac{4}{9b^2}}{\dfrac{5}{6b}-\dfrac{1}{9b^3}}\cdot\dfrac{18b^3}{18b^3}$

$=\dfrac{3a+8b}{15b^2-2}$

25. $\dfrac{5a^2+2b^3}{5b^3-3a^2b^3}$

26. $\dfrac{\dfrac{2}{x^2y}+\dfrac{3}{xy^2}}{\dfrac{2}{xy^3}+\dfrac{1}{x^2y}} = \dfrac{\dfrac{2}{x^2y}+\dfrac{3}{xy^2}}{\dfrac{2}{xy^3}+\dfrac{1}{x^2y}}\cdot\dfrac{x^2y^3}{x^2y^3}$

$=\dfrac{2y^2+3xy}{2x+y^2}$

(Although the numerator can be factored, doing so will not enable us to simplify further.)

27. $\dfrac{2x^4-3x^2}{2x^4+3}$

28. $\dfrac{3-\dfrac{2}{a^4}}{2+\dfrac{3}{a^3}} = \dfrac{3-\dfrac{2}{a^4}}{2+\dfrac{3}{a^3}}\cdot\dfrac{a^4}{a^4}$

$=\dfrac{3a^4-2}{2a^4+3a}$

29. $\dfrac{t^2-2}{t^2+5}$

30. $\dfrac{x+\dfrac{3}{x}}{x-\dfrac{2}{x}} = \dfrac{x\cdot x+\dfrac{3}{x}\cdot x}{x\cdot x-\dfrac{2}{x}\cdot x}$

$$= \frac{\dfrac{x^2+3}{x}}{\dfrac{x^2-2}{x}}$$
$$= \frac{x^2+3}{x} \cdot \frac{x}{x^2-2}$$
$$= \frac{x(x^2+3)}{x(x^2-2)}$$
$$= \frac{x^2+3}{x^2-2}$$

31. $\dfrac{7a^2b^3 - 5a}{4b^2 + ab^2}$

32. $\dfrac{5 + \dfrac{3}{x^2y}}{3+x \over x^3y} = \dfrac{5 + \dfrac{3}{x^2y}}{\dfrac{3+x}{x^3y}} \cdot \dfrac{x^3y}{x^3y} = \dfrac{5x^3y + 3x}{3+x}$

33. $\dfrac{a-7}{3a+2a^2}$

34. $\dfrac{\dfrac{x+5}{x^2}}{\dfrac{2}{x} - \dfrac{3}{x^2}} = \dfrac{\dfrac{x+5}{x^2}}{\dfrac{2}{x} \cdot \dfrac{x}{x} - \dfrac{3}{x^2}}$
$$= \dfrac{\dfrac{x+5}{x^2}}{\dfrac{2x-3}{x^2}}$$
$$= \dfrac{x+5}{x^2} \cdot \dfrac{x^2}{2x-3}$$
$$= \dfrac{x^2(x+5)}{x^2(2x-3)}$$
$$= \dfrac{x+5}{2x-3}$$

35. $\dfrac{10b^2 + 2a}{ab - 10b^2}$

36. $\dfrac{x - 3 + \dfrac{2}{x}}{x - 4 + \dfrac{3}{x}} = \dfrac{x \cdot \dfrac{x}{x} - 3 \cdot \dfrac{x}{x} + \dfrac{2}{x}}{x \cdot \dfrac{x}{x} - 4 \cdot \dfrac{x}{x} + \dfrac{3}{x}}$
$$= \dfrac{\dfrac{x^2 - 3x + 2}{x}}{\dfrac{x^2 - 4x + 3}{x}}$$
$$= \dfrac{x^2 - 3x + 2}{x} \cdot \dfrac{x}{x^2 - 4x + 3}$$
$$= \dfrac{(x-2)(x-1)}{x} \cdot \dfrac{x}{(x-3)(x-1)}$$
$$= \dfrac{x(x-1)}{x(x-1)} \cdot \dfrac{x-2}{x-3}$$
$$= \dfrac{x-2}{x-3}$$

37. $\dfrac{20x^2 - 30x}{15x^2 + 105x - 12}$

38. $\dfrac{a + 5 - \dfrac{3}{a}}{a - 3 + \dfrac{5}{a}} = \dfrac{a + 5 - \dfrac{3}{a}}{a - 3 + \dfrac{5}{a}} \cdot \dfrac{a}{a}$
$$= \dfrac{a^2 + 5a - 3}{a^2 - 3a + 5}$$

39. -4

40. $(x-1)7 - (x+1)9 = 4(x+2)$
$7x - 7 - 9x - 9 = 4x + 8$
$-2x - 16 = 4x + 8$
$-6x = 24$
$x = -4$

41. $23x^4 + 50x^3 + 23x^2 - 163x + 41$

42. Solve: $w(w+3) = 10$
$w = -5$ or $w = 2$

Only 2 makes sense in the original problem. The width is 2 yd and the length is $2 + 3$, or 5 yd, so the perimeter is $2 \cdot 2 + 2 \cdot 5$, or 14 yd.

43. ◆ Since there is a single rational expression in the numerator and in the denominator, Method 2 would be used. **44.** ◆ Although either method could be used, Method 2 requires fewer steps. **45.** ◆ The distributive law is used in Step 3 to eliminate the rational expressions in the numerator and in the denominator. **46.** ◆ Factoring is used to find the LCD of all rational expressions within a complex rational expression when Method 1 is used. It is also used to find a common denominator in the numerator and in the denominator when Method 2 is used. Regardless of the method used, factoring might be used in the final step to simplify the result.

47. $\dfrac{(x-1)(3x-2)}{5x-3}$

48. $\dfrac{\dfrac{a}{b} - \dfrac{c}{d}}{\dfrac{b}{a} - \dfrac{d}{c}} = \dfrac{\dfrac{a}{b} \cdot \dfrac{d}{d} - \dfrac{c}{d} \cdot \dfrac{b}{b}}{\dfrac{b}{a} \cdot \dfrac{c}{c} - \dfrac{d}{c} \cdot \dfrac{a}{a}}$
$$= \dfrac{\dfrac{ad - bc}{bd}}{\dfrac{bc - ad}{ac}}$$
$$= \dfrac{ad - bc}{bd} \cdot \dfrac{ac}{bc - ad}$$
$$= \dfrac{-1(bc - ad)(ac)}{bd(bc - ad)}$$
$$= \dfrac{-1(\cancel{bc - ad})(ac)}{bd(\cancel{bc - ad})}$$
$$= -\dfrac{ac}{bd}$$

49. $\dfrac{ac}{bd}$

50. $\left[\dfrac{\dfrac{x+1}{x-1}+1}{\dfrac{x+1}{x-1}-1}\right]^5 = \left[\dfrac{\dfrac{x+1+x-1}{x-1}}{\dfrac{x+1-(x-1)}{x-1}}\right]^5$

$= \left[\dfrac{\dfrac{2x}{x-1}}{\dfrac{2}{x-1}}\right]^5$

$= \left[\dfrac{2x}{\cancel{x-1}} \cdot \dfrac{\cancel{x-1}}{2}\right]^5$

$= x^5$

51. $\dfrac{3x+2}{2x+1}$

52. $\dfrac{\dfrac{z}{1-\dfrac{z}{2+2z}} - 2z}{\dfrac{2z}{5z-2} - 3} = \dfrac{\dfrac{z}{\dfrac{2+2z-z}{2+2z}} - 2z}{\dfrac{2z-15z+6}{5z-2}}$

$= \dfrac{\dfrac{z}{\dfrac{2+z}{2+2z}} - 2z}{\dfrac{-13z+6}{5z-2}}$

$= \dfrac{\dfrac{z \cdot \dfrac{2+2z}{2+z}} - 2z}{\dfrac{-13z+6}{5z-2}}$

$= \dfrac{\dfrac{z(2+2z) - 2z(2+z)}{2+z}}{\dfrac{-13z+6}{5z-2}}$

$= \dfrac{\dfrac{2z+2z^2-4z-2z^2}{2+z}}{\dfrac{-13z+6}{5z-2}}$

$= \dfrac{\dfrac{-2z}{2+z}}{\dfrac{-13z+6}{5z-2}}$

$= \dfrac{-2z}{2+z} \cdot \dfrac{5z-2}{-13z+6}$

$= \dfrac{-2z(5z-2)}{(2+z)(-13z+6)}$

53. ◆ Evaluate both the original complex rational expression and its simplified form for a value of a (other than 0). If both expressions have the same value, the simplification is probably correct. If they have different values, the result is not correct.   **54.** 📈

EXERCISE SET 6.6, PP. 336–337

1. $\dfrac{6}{5}$

2. $\dfrac{3}{8} - \dfrac{4}{5} = \dfrac{x}{20}$, LCD = 40

$40\left(\dfrac{3}{8} - \dfrac{4}{5}\right) = 40 \cdot \dfrac{x}{20}$

$15 - 32 = 2x$

$-17 = 2x$

$-\dfrac{17}{2} = x$

This checks.

3. $\dfrac{40}{29}$

4. $\dfrac{2}{3} + \dfrac{5}{6} = \dfrac{1}{x}$, LCD = 6x

$6x\left(\dfrac{2}{3} + \dfrac{5}{6}\right) = 6x \cdot \dfrac{1}{x}$

$4x + 5x = 6$

$9x = 6$

$x = \dfrac{2}{3}$

This checks.

5. $\dfrac{40}{9}$

6. $\dfrac{1}{6} + \dfrac{1}{8} = \dfrac{1}{t}$, LCD = 24t

$24t\left(\dfrac{1}{6} + \dfrac{1}{8}\right) = 24t \cdot \dfrac{1}{t}$

$4t + 3t = 24$

$7t = 24$

$t = \dfrac{24}{7}$

This checks.

7. $-3, -1$

8. $x + \dfrac{4}{x} = -5$, LCD = x

$x\left(x + \dfrac{4}{x}\right) = x(-5)$

$x^2 + 4 = -5x$

$x^2 + 5x + 4 = 0$

$(x+4)(x+1) = 0$

$x + 4 = 0$ or $x + 1 = 0$

$x = -4$ or $x = -1$

Both of these check.

9. $-7, 7$

10.
$$\frac{x}{6} - \frac{6}{x} = 0, \quad \text{LCD} = 6x$$
$$6x\left(\frac{x}{6} - \frac{6}{x}\right) = 6x \cdot 0$$
$$x^2 - 36 = 0$$
$$(x+6)(x-6) = 0$$
$$x + 6 = 0 \quad \text{or} \quad x - 6 = 0$$
$$x = -6 \quad \text{or} \quad x = 6$$

Both of these check.

11. 2

12.
$$\frac{5}{x} = \frac{6}{x} - \frac{1}{3}, \quad \text{LCD} = 3x$$
$$3x \cdot \frac{5}{x} = 3x\left(\frac{6}{x} - \frac{1}{3}\right)$$
$$15 = 18 - x$$
$$-3 = -x$$
$$3 = x$$

This checks.

13. $\frac{23}{4}$

14.
$$\frac{5}{3x} + \frac{3}{x} = 1, \quad \text{LCD} = 3x$$
$$3x\left(\frac{5}{3x} + \frac{3}{x}\right) = 3x \cdot 1$$
$$5 + 9 = 3x$$
$$14 = 3x$$
$$\frac{14}{3} = x$$

This checks.

15. 5

16.
$$\frac{x-7}{x+2} = \frac{1}{4}, \quad \text{LCD} = 4(x+2)$$
$$4(x+2) \cdot \frac{x-7}{x+2} = 4(x+2) \cdot \frac{1}{4}$$
$$4(x-7) = x + 2$$
$$4x - 28 = x + 2$$
$$3x = 30$$
$$x = 10$$

This checks.

17. $-\frac{13}{2}$

18.
$$\frac{2}{x+1} = \frac{1}{x-2},$$
$$\text{LCD} = (x+1)(x-2)$$
$$(x+1)(x-2) \cdot \frac{2}{x+1} = (x+1)(x-2) \cdot \frac{1}{x-2}$$
$$2(x-2) = x + 1$$
$$2x - 4 = x + 1$$
$$x = 5$$

This checks.

19. 3

20.
$$\frac{x}{6} - \frac{x}{10} = \frac{1}{6}, \quad \text{LCD} = 30$$
$$30\left(\frac{x}{6} - \frac{x}{10}\right) = 30 \cdot \frac{1}{6}$$
$$5x - 3x = 5$$
$$2x = 5$$
$$x = \frac{5}{2}$$

This checks.

21. -2

22.
$$\frac{x+1}{3} - 1 = \frac{x-1}{2}, \quad \text{LCD} = 6$$
$$6\left(\frac{x+1}{3} - 1\right) = 6 \cdot \frac{x-1}{2}$$
$$2(x+1) - 6 = 3(x-1)$$
$$2x + 2 - 6 = 3x - 3$$
$$2x - 4 = 3x - 3$$
$$-1 = x$$

This checks.

23. No solution

24.
$$\frac{x-1}{x-5} = \frac{4}{x-5}, \quad \text{LCD} = x - 5$$
$$(x-5) \cdot \frac{x-1}{x-5} = (x-5) \cdot \frac{4}{x-5}$$
$$x - 1 = 4$$
$$x = 5$$

Check:
$$\frac{x-1}{x-5} = \frac{4}{x-5}$$
$$\frac{5-1}{5-5} \,?\, \frac{4}{5-5}$$
$$\frac{4}{0} \,\Big|\, \frac{4}{0} \quad \text{UNDEFINED}$$

The number 5 is not a solution of the original equation because it results in division by 0. The equation has no solution.

25. -10

26.
$$\frac{2}{x+3} = \frac{7}{x}, \quad \text{LCD} = x(x+3)$$
$$x(x+3) \cdot \frac{2}{x+3} = x(x+3) \cdot \frac{7}{x}$$
$$2x = 7(x+3)$$
$$2x = 7x + 21$$
$$-21 = 5x$$
$$-\frac{21}{5} = x$$

This checks.

27. $\frac{1}{5}$

28.
$$\frac{x-2}{x-3} = \frac{x-1}{x+1},$$
$$\text{LCD} = (x-3)(x+1)$$
$$(x-3)(x+1) \cdot \frac{x-2}{x-3} = (x-3)(x+1) \cdot \frac{x-1}{x+1}$$
$$(x+1)(x-2) = (x-3)(x-1)$$
$$x^2 - x - 2 = x^2 - 4x + 3$$
$$-x - 2 = -4x + 3$$
$$3x = 5$$
$$x = \frac{5}{3}$$

This checks.

29. No solution

30.
$$\frac{x}{x+4} - \frac{4}{x-4} = \frac{x^2+16}{x^2-16},$$
$$\text{LCD} = (x+4)(x-4)$$
$$(x+4)(x-4)\left(\frac{x}{x+4} - \frac{4}{x-4}\right) =$$
$$(x+4)(x-4) \cdot \frac{x^2+16}{(x+4)(x-4)}$$
$$x(x-4) - 4(x+4) = x^2 + 16$$
$$x^2 - 4x - 4x - 16 = x^2 + 16$$
$$x^2 - 8x - 16 = x^2 + 16$$
$$-8x - 16 = 16$$
$$-8x = 32$$
$$x = -4$$

The number -4 is not a solution of the original equation because it results in division by 0. The equation has no solution.

31. 2

32.
$$\frac{1}{x+3} + \frac{1}{x-3} = \frac{1}{x^2-9},$$
$$\text{LCD} = (x+3)(x-3)$$
$$(x+3)(x-3)\left(\frac{1}{x+3} + \frac{1}{x-3}\right) =$$
$$(x+3)(x-3) \cdot \frac{1}{(x+3)(x-3)}$$
$$(x-3) + (x+3) = 1$$
$$2x = 1$$
$$x = \frac{1}{2}$$

This checks.

33. No solution

34.
$$\frac{t+10}{7-t} = \frac{3}{t-7}$$
$$\frac{t+10}{7-t} = \frac{-1}{-1} \cdot \frac{3}{t-7}$$
$$\frac{t+10}{7-t} = \frac{-3}{7-t}, \quad \text{LCD} = 7-t$$

$$(7-t) \cdot \frac{t+10}{7-t} = (7-t) \cdot \frac{-3}{7-t}$$
$$t + 10 = -3$$
$$t = -13$$

This checks.

35. $\dfrac{25}{9}$ **36.** $(-4)^{-3} = \dfrac{1}{(-4)^3} = -\dfrac{1}{64}$

37. $a^{-6}b^{-15}$, or $\dfrac{1}{a^6b^{15}}$ **38.** x^8y^{12}

39. $\dfrac{3x - 16}{(x+3)(x-3)(x-2)}$

40. $\dfrac{2}{3x-12} + \dfrac{5}{x^2-16}$
$$= \frac{2}{3(x-4)} + \frac{5}{(x+4)(x-4)}$$
$$= \frac{2(x+4) + 5 \cdot 3}{3(x-4)(x+4)}$$
$$= \frac{2x + 8 + 15}{3(x-4)(x+4)}$$
$$= \frac{2x + 23}{3(x-4)(x+4)}$$

41. ◆ When adding rational expressions, we use the LCD to write an expression equivalent to the sum of the given expressions. When solving rational equations, we use the LCD to clear fractions and then proceed to find the value(s) of the variable for which the equation is true.

42. ◆ Since the denominators are the same, the numerators must be the same. Then $x = -2$, but this value of x makes the denominator 0, so the equation cannot have a solution.   **43.** ◆ If we multiply both sides of a rational equation by a variable expression in order to clear fractions, it is possible that the variable expression is equal to 0. Thus, an equivalent equation might not be produced.

44. ◆ Graph each side of the equation and find the first coordinate(s) of the point(s) of intersection of the graphs.

45. -2

46.
$$\frac{4}{y-2} - \frac{2y-3}{y^2-4} = \frac{5}{y+2},$$
$$\text{LCD} = (y+2)(y-2)$$
$$(y+2)(y-2)\left(\frac{4}{y-2} - \frac{2y-3}{(y+2)(y-2)}\right) =$$
$$(y+2)(y-2) \cdot \frac{5}{y+2}$$
$$4(y+2) - (2y-3) = 5(y-2)$$
$$4y + 8 - 2y + 3 = 5y - 10$$
$$2y + 11 = 5y - 10$$
$$21 = 3y$$
$$7 = y$$

This checks.

47. $-\dfrac{1}{6}$

48.
$$\frac{12-6x}{x^2-4} = \frac{3x}{x+2} - \frac{2x-3}{x-2},$$
$$\text{LCD} = (x+2)(x-2)$$

$$(x+2)(x-2) \cdot \frac{12-6x}{(x+2)(x-2)} =$$
$$(x+2)(x-2)\left(\frac{3x}{x+2} - \frac{2x-3}{x-2}\right)$$
$$12 - 6x =$$
$$3x(x-2) - (x+2)(2x-3)$$
$$12 - 6x =$$
$$3x^2 - 6x - 2x^2 - x + 6$$
$$0 = x^2 - x - 6$$
$$0 = (x-3)(x+2)$$

$x - 3 = 0 \quad \text{or} \quad x + 2 = 0$
$x = 3 \quad \text{or} \quad x = -2$

Only 3 checks, so the solution is 3.

49. $-1, 0$

50.
$$2 - \frac{a-2}{a+3} = \frac{a^2-4}{a+3}, \quad \text{LCD} = a+3$$

$$(a+3)\left(2 - \frac{a-2}{a+3}\right) = (a+3) \cdot \frac{a^2-4}{a+3}$$
$$2(a+3) - (a-2) = a^2 - 4$$
$$2a + 6 - a + 2 = a^2 - 4$$
$$0 = a^2 - a - 12$$
$$0 = (a-4)(a+3)$$

$a - 4 = 0 \quad \text{or} \quad a + 3 = 0$
$a = 4 \quad \text{or} \quad a = -3$

Only 4 checks, so the solution is 4.

51. 4

52.
$$\frac{3a-5}{a^2+4a+3} + \frac{2a+2}{a+3} = \frac{a-3}{a+1}$$
$$\frac{3a-5}{(a+3)(a+1)} + \frac{2a+2}{a+3} = \frac{a-3}{a+1},$$
$$\text{LCD} = (a+3)(a+1)$$

$$(a+3)(a+1)\left(\frac{3a-5}{(a+3)(a+1)} + \frac{2a+2}{a+3}\right) =$$
$$(a+3)(a+1) \cdot \frac{a-3}{a+1}$$
$$3a - 5 + (a+1)(2a+2) =$$
$$(a+3)(a-3)$$
$$3a - 5 + 2a^2 + 4a + 2 = a^2 - 9$$
$$a^2 + 7a + 6 = 0$$
$$(a+6)(a+1) = 0$$

$a = -6 \text{ or } a = -1$

Only -6 checks, so the solution is -6.

53. 〰️   **54.** 〰️

EXERCISE SET 6.7, PP. 344–348

1. $-1, 5$

2. Let $x =$ the number.   **3.** 1

Solve: $x - \dfrac{4}{x} = 3$

$x = 4 \text{ or } x = -1$

Both numbers check.

4. Let $x =$ the number.   **5.** $6\frac{6}{7}$ hr

Solve: $x + \dfrac{5}{x} = 6$

$x = 1 \text{ or } x = 5$

Both numbers check.

6. Let $t =$ the number of hours it takes to do the job, working together.

Solve: $\dfrac{t}{5} + \dfrac{t}{4} = 1$

$t = \dfrac{20}{9}$, or $2\dfrac{2}{9}$ hr

7. $25\frac{5}{7}$ min

8. Let $t =$ the number of hours it takes them to do the job, working together.

Solve: $\dfrac{t}{4} + \dfrac{t}{3} = 1$

$t = \dfrac{12}{7}$, or $1\dfrac{5}{7}$ hr

9. $20\frac{4}{7}$ hr

10. Let $t =$ the number of hours it takes to fill the tank when both pipes are working.

Solve: $\dfrac{t}{12} + \dfrac{t}{9} = 1$

$t = \dfrac{36}{7}$, or $5\dfrac{1}{7}$ hr

11. $11\frac{1}{9}$ min

12. Let $t =$ the number of hours it would take to construct the wall, working together.

Solve: $\dfrac{t}{6} + \dfrac{t}{8} = 1$

$t = \dfrac{24}{7}$, or $3\dfrac{3}{7}$ hr

13. $4\frac{4}{9}$ min

14. Let $t =$ the number of minutes it takes to mow the yard, working together.

Solve: $\dfrac{t}{40} + \dfrac{t}{50} = 1$

$t = \dfrac{200}{9}$, or $22\dfrac{2}{9}$ min

15.

Speed	Time
r	t
r + 40	t

Speed	Time
r	$\frac{300}{r}$
r + 40	$\frac{700}{r+40}$

Train: 30 mph; police car: 70 mph

16.

	Distance	Speed	Time
Harley	75	r	t
Lexus	120	r + 30	t

Solve: $\frac{75}{r} = \frac{120}{r+30}$
$r = 50$
Then $r + 30 = 80$.
The speed of Bill's Harley is 50 mph, and the speed of Hillary's Lexus is 80 mph.

17.

Speed	Time
r − 14	t
r	t

Speed	Time
r − 14	$\frac{330}{r-14}$
r	$\frac{400}{r}$

Passenger: 80 km/h; freight: 66 km/h

18.

	Distance	Speed	Time
Freight	390	r − 15	t
Passenger	480	r	t

Solve: $\frac{390}{r-15} = \frac{480}{r}$
$r = 80$
Then $r - 15 = 65$.
The speed of the passenger train is 80 km/h. The speed of the freight train is 65 km/h.

19. Ted: 14 km/h; Joni: 19 km/h

20.

	Distance	Speed	Time
Mark	42	r	t
Ellie	48	r + 5	t

Solve: $\frac{42}{r} = \frac{48}{r+5}$
$r = 35$
Then $r + 5 = 40$.
The speed of Mark's snowmobile is 35 km/h, and the speed of Ellie's snowmobile is 40 km/h.

21. 3 hr

22.

	Distance	Speed	Time
Tory	40	r	t
Emilio	100	r	t + 2

Solve: $\frac{40}{t} = \frac{100}{t+2}$
$t = \frac{4}{3}$, or $1\frac{1}{3}$ hr

23. 15 students per faculty member

24. $\frac{800 \text{ mi}}{50 \text{ gal}} = 16 \frac{\text{mi}}{\text{gal}}$

25. 1088 ft/sec

26. $\frac{4.6 \text{ km}}{2 \text{ hr}} = 2.3$ km/h

27. $\frac{15}{4}$ cm

28. $\frac{a}{b} = \frac{c}{d}$
$\frac{a}{9} = \frac{10}{7}$
$a = \frac{90}{7}$ cm

29. $b = 3$ m, $c = 5$ m, $d = 1$ m; $b = 10$ m, $c = 12$ m, $d = 8$ m

30. $\frac{a}{b} = \frac{c}{d}$
$\frac{18}{b} = \frac{b+3}{b-2}$
$18(b-2) = b(b+3)$
$18b - 36 = b^2 + 3b$
$0 = b^2 - 15b + 36$
$0 = (b-3)(b-12)$
$b = 3$ or $b = 12$
If $b = 3$ m, then $c = 3 + 3$, or 6 m and $d = 3 - 2$, or 1 m.
If $b = 12$ m, then $c = 12 + 3$, or 15 m and $d = 12 - 2$, or 10 m.

31. 582

32. Solve: $\dfrac{K}{42} = \dfrac{234}{14}$
$K = 702$ km

33. $21\dfrac{2}{3}$ cups

34. Solve: $\dfrac{H}{16} = \dfrac{1.2}{10}$
$H = 1.92$ grams

35. 10.5

36. Solve: $\dfrac{9}{a} = \dfrac{8}{6}$
$a = 6.75$

$\left(\text{One of the following proportions could also be used:}\right.$
$\left.\dfrac{a}{9} = \dfrac{6}{8}, \dfrac{9}{8} = \dfrac{a}{6}, \dfrac{8}{9} = \dfrac{6}{a}\right)$

37. $\dfrac{8}{3}$

38. Solve: $\dfrac{r}{10} = \dfrac{6}{8}$
$r = 7.5$

$\left(\text{One of the following proportions could also be used:}\right.$
$\dfrac{10}{r} = \dfrac{8}{6}, \dfrac{r}{6} = \dfrac{10}{8}, \dfrac{6}{r} = \dfrac{8}{10}, \dfrac{12}{16} = \dfrac{10}{r}, \dfrac{16}{12} = \dfrac{r}{12}, \dfrac{r}{16} = \dfrac{10}{16},$
$\left.\dfrac{12}{r} = \dfrac{16}{10}\right)$

39. 15 ft

40. $\dfrac{h}{6} = \dfrac{6}{8}$
$h = \dfrac{36}{8} = 4.5$ ft

$\left(\text{The proportion } \dfrac{6}{h} = \dfrac{8}{6} \text{ could also be used.}\right)$

41. 287   **42.** Solve: $\dfrac{318}{D} = \dfrac{56}{168}$
$D = 954$ **43.** 184

44. Solve: $\dfrac{184}{6} = \dfrac{1288}{D}$ **45.** No   **46.** Solve: $\dfrac{9}{144} = \dfrac{D}{320}$
$D = 42$ $D = 20$

47. 225

48. (a) Solve: $\dfrac{0.16}{1} = \dfrac{R}{12}$
$R = 1.92$ tons
(b) Solve: $\dfrac{0.16}{1} = \dfrac{A}{180}$
$A = 28.8$ lb

49. (a) 4.8 tons; (b) 48 lb

50. Let x = the numerator. Then $104 - x$ = the denominator.
Solve: $\dfrac{9}{17} = \dfrac{x}{104 - x}$
$x = 36$
Then $104 - x = 104 - 36 = 68$, and the desired ratio is $\dfrac{36}{68}$.

51. 1   **52.** $x^2 - 1$ **53.** $13y^3 - 14y^2 + 12y - 73$

54. Solve: $w + w + (w + 15) + (w + 15) = 642$
$w = 153$, so $w + 15 = 168$, and area $= 168(153) = 25{,}704$ ft^2

55. $t = \dfrac{n - ar}{s}$ **56.** $uv - av = m$
$v(u - a) = m$
$v = \dfrac{m}{u - a}$

57. ◆ No. If the workers work at different rates, two workers will complete a task in more than half the time of the faster person working alone but in less than half the slower person's time. This is illustrated in Example 2.

58. ◆ Answers may vary. Casey can paint a room in 7 hr. It takes Lee 5 hr to paint the same room. How long would it take them to paint the room working together?

59. ◆ Answers may vary. One motorboat travels 4 mph faster than another. While one boat travels 30 mi, the other travels 18 mi. Find their speeds.   **60.** ◆ Yes; corresponding angles have the same measure and corresponding sides are proportional. (The ratio of each pair of corresponding sides is equivalent to 1.)

61. Ann: 6 hr; Betty: 12 hr

62.

	Distance	Speed	Time
Upstream	24	$10 - r$	t
Downstream	24	$10 + r$	$5 - t$

From the rows of the table we get two equations:
$24 = (10 - r)t$
$24 = (10 + r)(5 - t)$

We solve each equation for t and set the results equal:
Solving $24 = (10 - r)t$ for t: $t = \dfrac{24}{10 - r}$
Solving $24 = (10 + r)(5 - t)$ for t: $t = 5 - \dfrac{24}{10 + r}$
Then $\dfrac{24}{10 - r} = 5 - \dfrac{24}{10 + r}$.
$r = -2$ or $r = 2$
Only 2 checks in the original problem. The speed of the current is 2 mph.

63. $27\dfrac{3}{11}$ minutes after 5:00

64. Find a second proportion:

$$\frac{A}{B} = \frac{C}{D} \quad \text{Given}$$

$$\frac{D}{A} \cdot \frac{A}{B} = \frac{D}{A} \cdot \frac{C}{D} \quad \text{Multiplying by } \frac{D}{A}$$

$$\frac{D}{B} = \frac{C}{A}$$

Find a third proportion:

$$\frac{A}{B} = \frac{C}{D} \quad \text{Given}$$

$$\frac{B}{C} \cdot \frac{A}{B} = \frac{B}{C} \cdot \frac{C}{D} \quad \text{Multiplying by } \frac{B}{C}$$

$$\frac{A}{C} = \frac{B}{D}$$

Find a fourth proportion:

$$\frac{A}{B} = \frac{C}{D} \quad \text{Given}$$

$$\frac{DB}{AC} \cdot \frac{A}{B} = \frac{DB}{AC} \cdot \frac{C}{D} \quad \text{Multiplying by } \frac{DB}{AC}$$

$$\frac{D}{C} = \frac{B}{A}$$

65. $9\frac{3}{13}$ days

66. Let t = the time it takes Michelle to wax the car alone. Then $t/2$ = Sal's time alone, and $t - 2$ = Kristen's time alone. In 1 hr they do $\frac{1}{t} + \frac{1}{\frac{t}{2}} + \frac{1}{t-2}$, or $\frac{1}{t} + \frac{2}{t} + \frac{1}{t-2}$ of the job working together. The entire job takes 1 hr and 20 min, or $\frac{4}{3}$ hr so we solve $\frac{4}{3}\left(\frac{1}{t} + \frac{2}{t} + \frac{1}{t-2}\right) = 1$. We get $t = \frac{4}{3}$ or $t = 6$. Only 6 makes sense in the original problem. Thus, working alone, it would take Michelle 6 hr, Sal 3 hr, and Kristen 4 hr to wax the car.

67. $30\frac{22}{31}$ hr

68. Let p = the width of the pond. We have similar triangles:

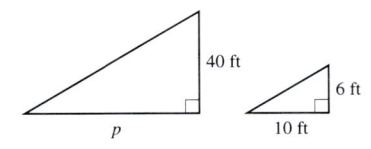

We write a proportion and solve it.

$$\frac{p}{10} = \frac{40}{6}$$

$$p = \frac{40 \cdot 10}{6} \quad \text{Multiplying by 10}$$

$$p = \frac{200}{3}, \text{ or } 66\frac{2}{3}$$

The pond is $66\frac{2}{3}$ ft wide.

69. 270

70. We organize the information in a table. Let r = the speed on the first part of the trip and t = the time driven at that speed.

	Distance	Speed	Time
First part	30	r	t
Second part	30	$r + 15$	$1 - t$

From the rows of the table we obtain two equations:

$30 = rt$
$30 = (r + 15)(1 - t)$

We solve each equation for t and set the results equal:

Solving $30 = rt$ for t: $t = \frac{30}{r}$

Solving $20 = (r + 15)(1 - t)$ for t: $t = 1 - \frac{20}{r + 15}$

Then $\frac{30}{r} = 1 - \frac{20}{r + 15}$.

$r = 45$ or $r = -10$

Only 45 checks in the original problem. The speed for the first 30 miles was 45 mph.

71. ◆
$$\frac{A + B}{B} = \frac{C + D}{D}$$

$$\frac{A}{B} + \frac{B}{B} = \frac{C}{D} + \frac{D}{D}$$

$$\frac{A}{B} + 1 = \frac{C}{D} + 1$$

$$\frac{A}{B} = \frac{C}{D}$$

The equations are equivalent.

72. ◆ No; consider similar right triangles with sides a, b, c and ka, kb, kc, for example, where $k \neq 1$. Their areas are $\frac{1}{2}ab$ and $\frac{1}{2} \cdot ka \cdot kb$, or $\frac{1}{2}k^2ab$. The perimeters of these triangles are $a + b + c$ and $ka + kb + kc$, or $k(a + b + c)$, respectively. The ratio of the areas is

$$\frac{\frac{1}{2}ab}{\frac{1}{2}k^2ab}, \text{ or } \frac{1}{k^2},$$

but the ratio of the perimeters is $\dfrac{a+b+c}{k(a+b+c)}$, or $\dfrac{1}{k}$. Since the ratios are not equal, the areas and perimeters are not proportional.

EXERCISE SET 6.8, PP. 352–354

1. $h = \dfrac{S}{2\pi r}$

2. $A = P(1+rt)$
$A = P + Prt$
$A - P = Prt$
$\dfrac{A-P}{Pr} = t$

3. $h = \dfrac{2A}{b}$

4. $s = \dfrac{1}{2}gt^2$
$2s = gt^2$
$\dfrac{2s}{t^2} = g$

5. $n = \dfrac{s}{180} + 2$, or $n = \dfrac{s+360}{180}$

6. $S = \dfrac{n}{2}(a+l)$
$2S = n(a+l)$
$2S = na + nl$
$2S - nl = na$
$\dfrac{2S-nl}{n} = a$

7. $b = \dfrac{3V - kB + 2kn}{k}$

8. $A = P + Prt$
$A = P(1+rt)$
$\dfrac{A}{1+rt} = P$

9. $r = \dfrac{L}{l-S}$

10. $T = mg - mf$
$T = m(g-f)$
$\dfrac{T}{g-f} = m$

11. $h = \dfrac{2A}{b_1 + b_2}$

12. $S = 2\pi r(r+h)$
$\dfrac{S}{2\pi r} = r + h$
$\dfrac{S}{2\pi r} - r = h$, or
$\dfrac{S - 2\pi r^2}{2\pi r} = h$

13. $a = \dfrac{d}{b-c}$

14. $mn + p = np$
$p = np - mn$
$p = n(p-m)$
$\dfrac{p}{p-m} = n$

15. $r = \dfrac{m}{s}$

16. $\dfrac{V}{m} = d$
$V = dm$
$\dfrac{V}{d} = m$

17. $d = \dfrac{c}{a+b}$

18. $\dfrac{m}{n} = p - q$
$m = n(p-q)$
$\dfrac{m}{p-q} = n$

19. $R = \dfrac{V_1 - V_2}{I}$

20. $\dfrac{M-g}{t} = r + s$
$M - g = t(r+s)$
$\dfrac{M-g}{r+s} = t$

21. $f = \dfrac{pq}{q+p}$

22. $\dfrac{1}{R} = \dfrac{1}{r_1} + \dfrac{1}{r_2}$
$Rr_1 r_2 \cdot \dfrac{1}{R} = Rr_1 r_2 \left(\dfrac{1}{r_1} + \dfrac{1}{r_2}\right)$
$r_1 r_2 = Rr_2 + Rr_1$
$r_1 r_2 = R(r_2 + r_1)$
$\dfrac{r_1 r_2}{r_2 + r_1} = R$

23. $t = \dfrac{v_2 - v_1}{a}$

24. $r = \dfrac{v^2 pL}{a}$
$ar = v^2 pL$
$\dfrac{ar}{v^2 L} = p$

25. $l = \dfrac{P - 2w}{2}$, or $l = \dfrac{P}{2} - w$

26. $\dfrac{a}{c} = n + bn$
$\dfrac{a}{c} = n(1+b)$
$\dfrac{a}{c(1+b)} = n$

27. $a = \dfrac{Q}{M(b-c)}$

28. $S = \dfrac{a+2b}{3b}$
$3bS = a + 2b$
$3bS - 2b = a$
$b(3S - 2) = a$
$b = \dfrac{a}{3S - 2}$

29. $a = \dfrac{b}{K-C}$

30. $C = \dfrac{5}{9}(F - 32)$
$9C = 5(F - 32)$
$9C = 5F - 160$
$9C + 160 = 5F$
$\dfrac{9C + 160}{5} = F$, or
$\dfrac{9}{5}C + 32 = F$

31. $r^3 = \dfrac{3V}{4\pi}$

32. $f = \dfrac{gm - t}{m}$
$fm = gm - t$
$fm + t = gm$
$\dfrac{fm + t}{m} = g$

33. $r = \dfrac{Sl - a}{S - l}$

34.
$$f = \frac{gm - t}{m}$$
$$fm = gm - t$$
$$fm - gm = -t$$
$$m(f - g) = -t$$
$$m = \frac{-t}{f - g}, \text{ or } \frac{t}{g - f}$$

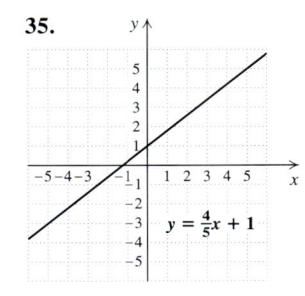

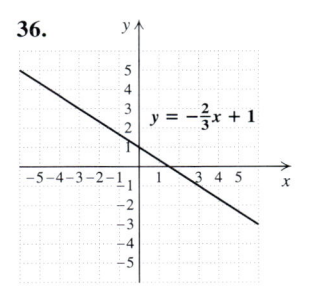

35. Graph of $y = \frac{4}{5}x + 1$

36. Graph of $y = -\frac{2}{3}x + 1$

37. $-\frac{3}{4}$

38. $3x + 4y = 24$
y-intercept: $4y = 24$
$y = 6$

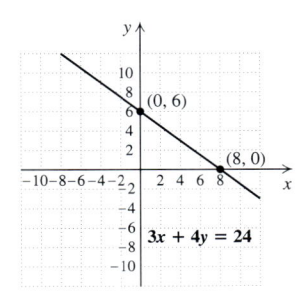

Graph of $3x + 4y = 24$ with points $(0, 6)$ and $(8, 0)$.

The y-intercept is $(0, 6)$.
x-intercept: $3x = 24$
$x = 8$
The x-intercept is $(8, 0)$.

39. $(x + 2)(x - 15)$

40. $-3x^3 - 5x^2 + 5$

41. ◆ It is slightly easier to solve $\frac{1}{p} + \frac{1}{q} = \frac{1}{f}$ for f because, when clearing fractions, the multiplication on the right $\left(pqf \cdot \frac{1}{f}\right)$ is easier to perform than the corresponding multiplication in the first equation $\left(25 \cdot 23 \cdot x \cdot \frac{1}{x}\right)$.

42. ◆ Someone might know the areas and heights of a number of triangles and want to know their bases. Solving for b would eliminate the need to perform the operations necessary to isolate b each time.

43. ◆ No; for $A \neq 0$, $B \neq 0$, $A = B$ is equivalent to $\frac{1}{A} = \frac{1}{B}$:
$$A = B$$
$$\frac{1}{AB} \cdot A = \frac{1}{AB} \cdot B$$
$$\frac{1}{B} = \frac{1}{A}$$

44. ◆ The formula $t = \frac{ab}{b + a}$ is especially useful when we know how long it takes each of two workers, working alone, to do a job and we want to find how long it will take them to do the job, working together.

45. $n_2 = \frac{n_1 p_2 R + p_1 p_2 n_1}{p_1 p_2 - p_1 R}$

46.
$$u = -F\left(E - \frac{P}{T}\right)$$
$$u = -EF + \frac{FP}{T}$$
$$T \cdot u = T\left(-EF + \frac{FP}{T}\right)$$
$$Tu = -EFT + FP$$
$$Tu + EFT = FP$$
$$T(u + EF) = FP$$
$$T = \frac{FP}{u + EF}$$

47. $v = \frac{Nbf_2 - bf_1 - df_1}{Nf_2 - 1}$

48. When $C = F$, we have
$$C = \frac{5}{9}(C - 32)$$
$$9C = 5(C - 32)$$
$$9C = 5C - 160$$
$$4C = -160$$
$$C = -40$$

At $-40°$, the Fahrenheit and Celsius readings are the same.

REVIEW EXERCISES: CHAPTER 6, PP. 356–357

1. [6.1] 0 **2.** [6.1] 6 **3.** [6.1] $-6, 6$ **4.** [6.1] $-6, 5$
5. [6.1] -2 **6.** [6.1] $\frac{x - 2}{x + 1}$ **7.** [6.1] $\frac{7x + 3}{x - 3}$
8. [6.1] $\frac{y - 5}{y + 5}$ **9.** [6.1] $-5(x + 2y)$ **10.** [6.2] $\frac{a - 6}{5}$
11. [6.2] $\frac{6}{2t - 1}$ **12.** [6.2] $-20t$ **13.** [6.2] $\frac{2x^2 - 2x}{x + 1}$
14. [6.2] $\frac{(x^2 + 1)(2x + 1)}{(x - 2)(x + 1)}$ **15.** [6.2] $\frac{(t + 4)^2}{t + 1}$
16. [6.3] $24a^5b^7$ **17.** [6.3] $x^4(x + 1)(x - 1)$

18. [6.3] $(y-2)(y+2)(y+1)$ **19.** [6.3] $\dfrac{-3x+18}{x+7}$
20. [6.4] -1 **21.** [6.3] $\dfrac{4}{x-4}$ **22.** [6.4] $\dfrac{x+5}{2x}$
23. [6.4] $\dfrac{2x+3}{x-2}$ **24.** [6.4] $\dfrac{2a}{a-1}$ **25.** [6.4] $d+c$
26. [6.4] $\dfrac{-x^2+x+26}{(x-5)(x+5)(x+1)}$ **27.** [6.4] $\dfrac{2(x-2)}{x+2}$
28. [6.4] $\dfrac{19x+8}{10x(x+2)}$ **29.** [6.5] $\dfrac{z}{1-z}$
30. [6.5] $\dfrac{2x^4y^2+x^3}{y+xy}$ **31.** [6.5] $c-d$ **32.** [6.6] 8
33. [6.6] $-\dfrac{1}{2}$ **34.** [6.6] 3, -5 **35.** [6.7] $5\tfrac{1}{7}$ hr
36. [6.7] Car: 105 km/h; train: 90 km/h **37.** [6.7] -2
38. [6.7] 80 **39.** [6.7] 6 **40.** [6.8] $s=\dfrac{rt}{r-t}$
41. [6.8] $C=\tfrac{5}{9}(F-32)$, or $C=\tfrac{5}{9}F-\dfrac{160}{9}$ **42.** [2.1] 11
43. [3.3] (3, 0), (0, −6)

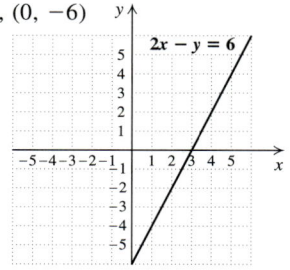

44. [5.2] $(x+12)(x-4)$
45. [4.3] $-2x^3+3x^2+12x-18$
46. [6.3], [6.6] ◆ A student should master factoring before beginning a study of rational equations because it is necessary to factor when finding the LCD of the rational expressions. It may also be necessary to factor to use the principle of zero products after fractions have been cleared.
47. [6.3] ◆ Although multiplying the denominators of the expressions being added results in a common denominator, it is often not the *least* common denominator. Using a common denominator other than the LCD makes the expressions more complicated, requires additional simplifying after the addition has been performed, and leaves more room for error.
48. [6.2] $\dfrac{5(a+3)^2}{a}$ **49.** [6.4] $\dfrac{10a}{(a-b)(b-c)}$

TEST: CHAPTER 6, P. 358

1. [6.1] 0 **2.** [6.1] -8 **3.** [6.1] $-7, 7$ **4.** [6.1] 1, 2
5. [6.1] $\dfrac{3x+7}{x+3}$ **6.** [6.2] $\dfrac{-2(a+5)}{3}$
7. [6.2] $\dfrac{(5y+1)(y+1)}{3y(y+2)}$

8. [6.2] $\dfrac{(2x+1)(2x-1)(x^2+1)}{(x-1)^2(x-2)}$
9. [6.2] $(x+3)(x-3)$
10. [6.3] $(y-3)(y+3)(y+7)$ **11.** [6.3] $\dfrac{23-3x}{x^3}$
12. [6.3] $\dfrac{8-2t}{t^2+1}$ **13.** [6.4] $\dfrac{-3}{x-3}$ **14.** [6.4] $\dfrac{2x-5}{x-3}$
15. [6.4] $\dfrac{8t-3}{t(t-1)}$ **16.** [6.4] $\dfrac{-x^2-7x-15}{(x+4)(x-4)(x+1)}$
17. [6.4] $\dfrac{x^2+2x-7}{(x-1)^2(x+1)}$ **18.** [6.5] $\dfrac{3y+1}{y}$
19. [6.5] $\dfrac{3a^2b^2-2a^3}{a^3b^2+2b^2}$ **20.** [6.6] 12 **21.** [6.6] $-3, 5$
22. [6.7] 12 min **23.** [6.7] $2\tfrac{1}{7}$ cups
24. [6.7] 45 km/h, 65 km/h **25.** [6.8] $t=\dfrac{d}{r+w}$
26. [2.1] $-\dfrac{3}{7}$
27. [3.3] (10, 0), (0, 4)

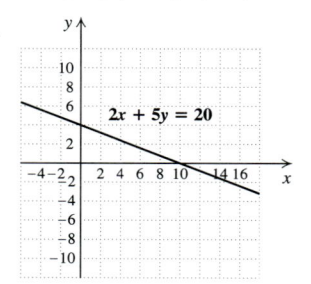

28. [5.2] $(x-9)(x+5)$ **29.** [4.3] $13x^2-29x+76$
30. [6.7] Reggie: 10 hr; Rema: 4 hr **31.** [6.5] $\dfrac{3a+2}{2a+1}$

CUMULATIVE REVIEW: 1–6, PP. 359–360

1. [1.2] $2b+a$ **2.** [1.4] $-3.1 > -3.15$ **3.** [1.8] 49
4. [1.8] $-8x+28$ **5.** [1.5] $-\dfrac{43}{8}$ **6.** [1.7] 1
7. [1.7] -6.2 **8.** [1.8] 8 **9.** [2.2] 10 **10.** [5.6] $-7, 7$
11. [2.1] $\tfrac{9}{2}$ **12.** [2.2] -2 **13.** [2.2] $\tfrac{8}{3}$
14. [5.6] $-10, -1$ **15.** [2.2] -8 **16.** [6.6] 1, 4
17. [2.6] $\{y|\ y \leq -\tfrac{2}{3}\}$ **18.** [6.6] -17 **19.** [5.6] $-4, \tfrac{1}{2}$
20. [2.6] $\{x|\ x > 43\}$ **21.** [6.6] 2 **22.** [5.6] $-\tfrac{7}{2}, 5$
23. [6.6] -13 **24.** [6.8] $t=\dfrac{4b}{A}$ **25.** [6.8] $n=\dfrac{tm}{t-m}$
26. [6.8] $c=\dfrac{a-b}{r}$ **27.** [1.6] $\tfrac{3}{2}x+2y-3z$
28. [4.2] $-4x^3-\tfrac{1}{7}x^2-2$

29. [3.2]

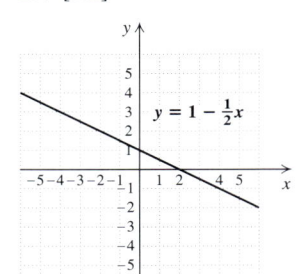

30. [3.3]

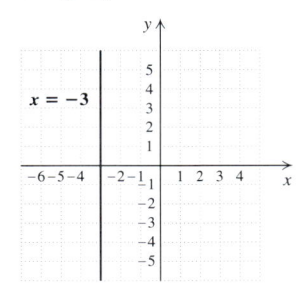

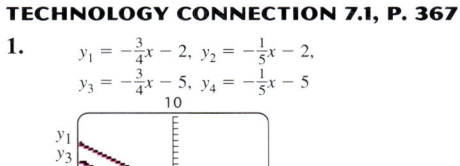

31. [3.3]

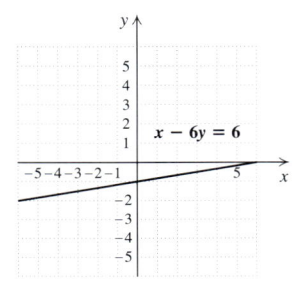

32. [3.5] -2 **33.** [4.8] x^{-2}, or $\dfrac{1}{x^2}$ **34.** [4.8] y^{-8}, or $\dfrac{1}{y^8}$
35. [4.1] $-4a^4b^{14}$ **36.** [4.3] $-y^3 - 2y^2 - 2y + 7$
37. [1.2] $12x + 16y + 4z$
38. [4.4] $2x^5 + x^3 - 6x^2 - x + 3$
39. [4.5] $36x^2 - 60xy + 25y^2$ **40.** [4.5] $2x^2 - x - 21$
41. [4.5] $4x^6 - 1$ **42.** [5.1] $2x(3 - x - 12x^3)$
43. [5.4] $(4x + 9)(4x - 9)$ **44.** [5.2] $(x - 6)(x - 4)$
45. [5.3] $(2x + 1)(4x + 3)$ **46.** [5.3] $2(3x - 2)(x - 4)$
47. [5.4] $2(x + 3)(x - 3)$ **48.** [5.4] $(4x + 5)^2$
49. [5.3] $(3x - 2)(x + 4)$ **50.** [5.1] $(x^3 - 3)(x + 2)$
51. [6.2] $\dfrac{y - 6}{2}$ **52.** [6.2] 1 **53.** [6.4] $\dfrac{a^2 + 7ab + b^2}{a^2 - b^2}$
54. [6.4] $\dfrac{-2x - 5}{x - 4}$ **55.** [6.5] $\dfrac{x}{x - 2}$ **56.** [6.5] $\dfrac{t + 2t^3}{t^3 - 2}$
57. [4.7] $5x^2 - 4x + 2 + \dfrac{2}{3x} + \dfrac{6}{x^2}$
58. [4.7] $15x^3 - 57x^2 + 177x - 529 + \dfrac{1605}{x + 3}$
59. [2.7] At most 225 **60.** [2.4] $3.60 **61.** [5.7] 14 ft
62. [2.5] $-278, -276$ **63.** [6.7] 30 min
64. [6.7] Harley: 40 km/h; Millie: 50 km/h
65. [2.5] 26 in. **66.** [4.3], [4.5] 12
67. [1.4], [2.2] $-144, 144$
68. [4.5] $16y^6 - y^4 + 6y^2 - 9$
69. [5.4] $2(a^{16} + 81b^{20})(a^8 + 9b^{10})(a^4 + 3b^5)(a^4 - 3b^5)$
70. [5.6] $4, -7, 12$ **71.** [1.4], [1.6] -7

CHAPTER 7

TECHNOLOGY CONNECTION 7.1, P. 367

1.
$y_1 = -\dfrac{3}{4}x - 2,\ y_2 = -\dfrac{1}{5}x - 2,$
$y_3 = -\dfrac{3}{4}x - 5,\ y_4 = -\dfrac{1}{5}x - 5$

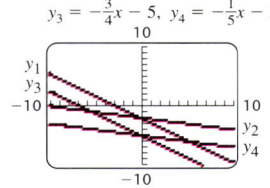

EXERCISE SET 7.1, PP. 367–369

1.

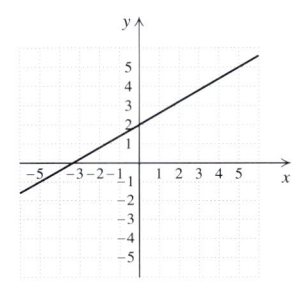

2. Plot $(0, -1)$; move up 2 units and right 5 units to $(5, 1)$. Draw a line through $(0, -1)$ and $(5, 1)$.

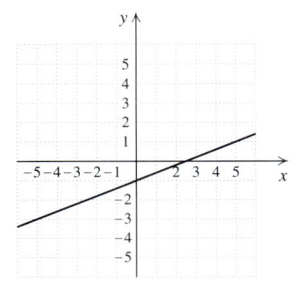

3.

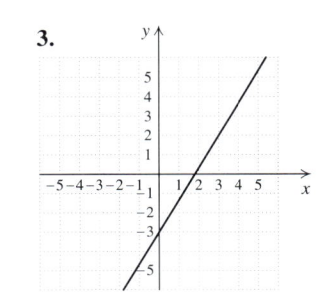

4. Plot $(0, 1)$; move up 5 units and right 2 units to $(2, 6)$. Draw a line through $(0, 1)$ and $(2, 6)$.

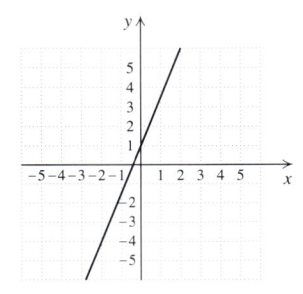

5.

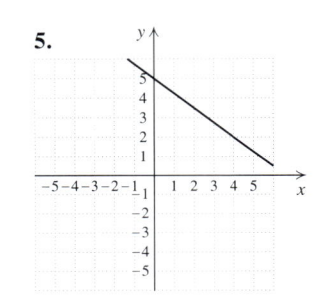

6. Plot (0, 6). Thinking of the slope as $\frac{-4}{5}$, move down 4 units and right 5 units to (5, 2). Draw a line through (0, 6) and (5, 2).

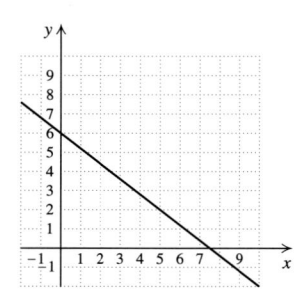

7.

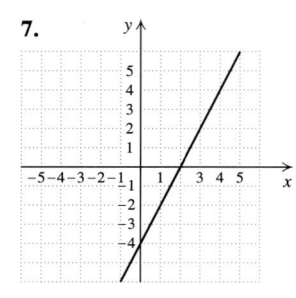

8. Plot (0, −3); move down 2 units and right 1 unit to (1, −5). Draw a line through (0, −3) and (1, −5).

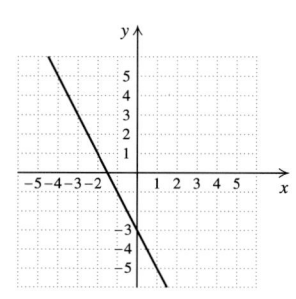

9.

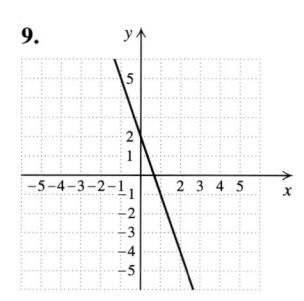

10. Plot (0, 4); move up 3 units and right 1 unit to (1, 7). Draw a line through (0, 4) and (1, 7).

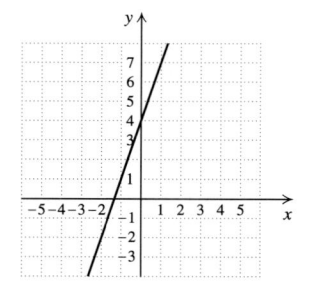

11. $\frac{3}{7}$; (0, 6) **12.** $-\frac{3}{8}$, (0, 7) **13.** $-\frac{5}{6}$; (0, 2)
14. $\frac{7}{2}$, (0, 4) **15.** $\frac{9}{4}$; (0, −7)
16. $y = \frac{2}{9}x - 1$
$y = \frac{2}{9}x + (-1)$
Slope: $\frac{2}{9}$, y-intercept: (0, −1)   **17.** $-\frac{2}{5}$; (0, 0)

18. $y = \frac{4}{3}x$
$y = \frac{4}{3}x + 0$
Slope: $\frac{4}{3}$, y-intercept: (0, 0)
19. 2; (0, 4)
20. $-5x + y = 5$
$y = 5x + 5$
Slope: 5, y-intercept: (0, 5)
21. $\frac{3}{4}$; (0, −3)
22. $3x - 2y = 18$
$-2y = -3x + 18$
$y = \frac{3}{2}x - 9$
Slope: $\frac{3}{2}$, y-intercept: (0, −9)
23. $\frac{1}{5}$; $\left(0, \frac{8}{5}\right)$
24. $x - 6y = 9$
$-6y = -x + 9$
$y = \frac{1}{6}x - \frac{3}{2}$
Slope: $\frac{1}{6}$, y-intercept: $\left(0, -\frac{3}{2}\right)$
25. 0; (0, 4)
26. $y - 3 = 5$
$y = 8$
$y = 0x + 8$
Slope: 0, y-intercept: (0, 8)
27. $y = 3x + 7$
28. $y = -4x - 2$ **29.** $y = \frac{7}{8}x - 1$ **30.** $y = \frac{5}{7}x + 4$
31. $y = -\frac{5}{3}x - 8$ **32.** $y = \frac{3}{4}x + 23$ **33.** $y = -2x + 3$
34. $y = 7x - 6$
35. 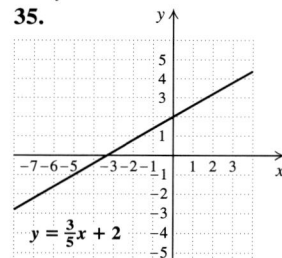
36. 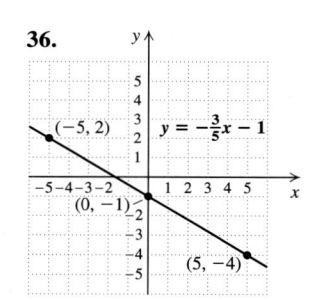
37. 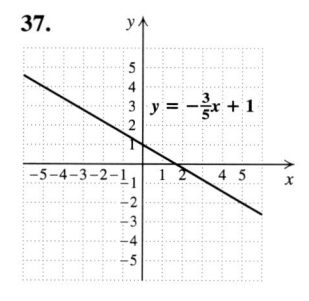
38.

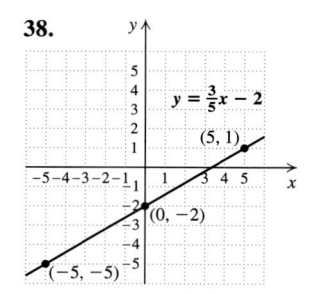

39. [graph: $y = \frac{5}{3}x + 3$]

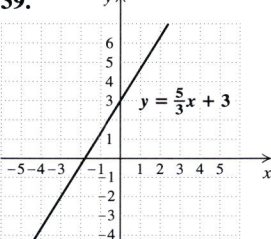

40. [graph: $y = \frac{5}{3}x - 2$ through (3, 3) and (−3, −7), (0, −2)]

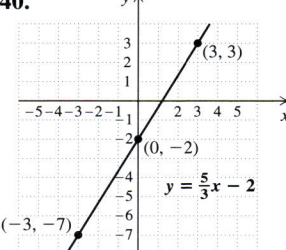

41. [graph: $y = -\frac{3}{2}x - 2$]

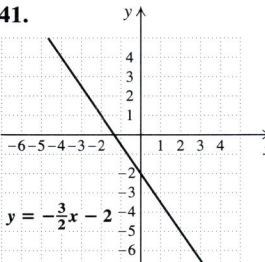

42. [graph: $y = -\frac{4}{3}x + 3$ through (0, 3), (3, −1), (6, −5)]

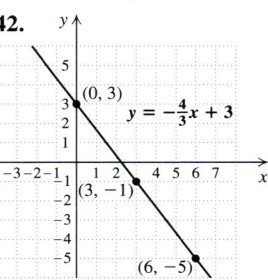

43. [graph: $2x + y = 1$]

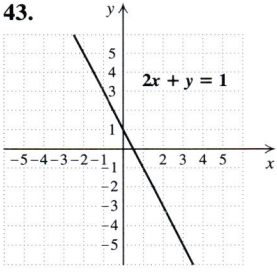

44. $3x + y = 2$
$y = -3x + 2$
[graph through (0, 2), (1, −1), (2, −4)]

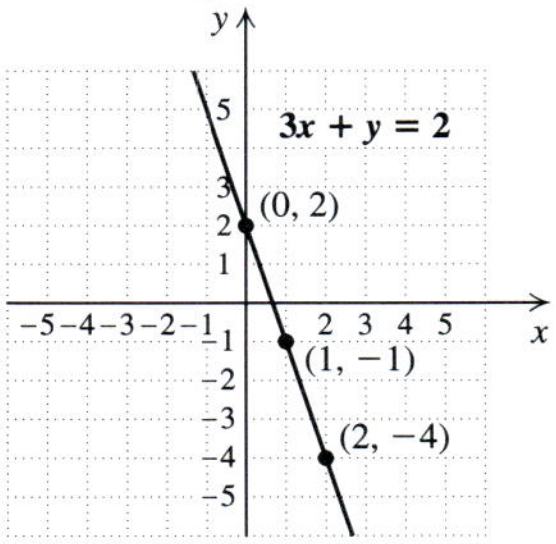

45. [graph: $3x - y = 4$]

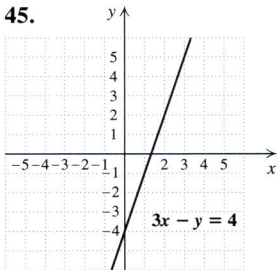

46. $2x - y = 5$
$y = 2x - 5$
[graph through (0, −5), (1, −3), (2, −1); labeled $2x - y = 5$]

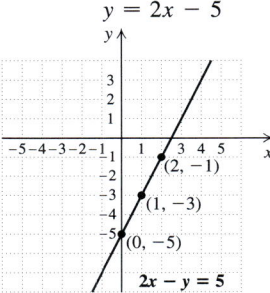

47. [graph: $2x + 3y = 9$]

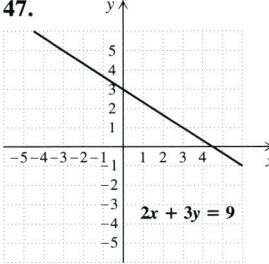

48. $4x + 5y = 15$
$y = -\frac{4}{5}x + 3$
[graph through (−5, 7), (0, 3), (5, −1); labeled $4x + 5y = 15$]

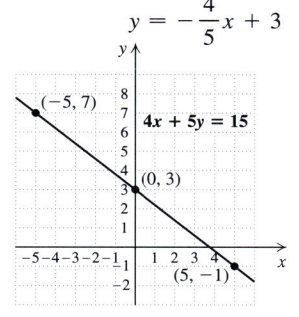

49. [graph: $x - 4y = 12$]

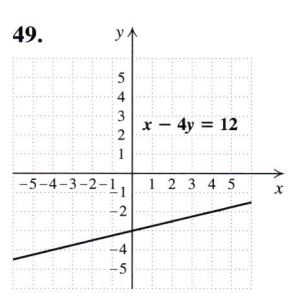

50. $x + 5y = 20$
$y = -\frac{1}{5}x + 4$
[graph through (−5, 5), (0, 4), (5, 3); labeled $x + 5y = 20$]

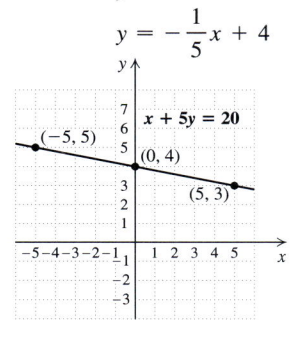

51. [graph: Cost vs. Amount of water used (in 10,000 gallons); $y = x + 9$; $1 per 10,000 gallons]

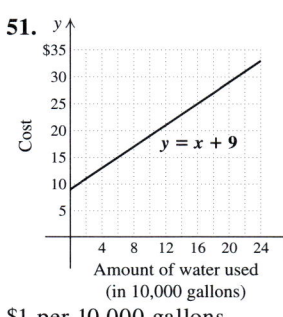

52. Two points on the graph are (0, 50) and (1, 70), so the y-intercept is (0, 50). Now we find the slope:
$$m = \frac{70 - 50}{1 - 0} = 20$$
Then the equation is $y = 20x + 50$.

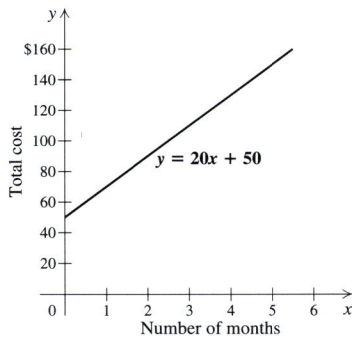

[graph: Total cost vs. Number of months; $y = 20x + 50$]

Since the slope is 20, the rate is $20 per month.

53. $y = 1.5x + 16$

54. Two points on the graph are (0, 4.95) and (1, 4.95 + 2.50), or (1, 7.45), so the y-intercept is (0, 4.95). Now we find the slope:
$$m = \frac{7.45 - 4.95}{1 - 0} = 2.5$$
Then the equation is $y = 2.5x + 4.95$.

55. Yes **56.** Since the slopes are different $\left(-\frac{5}{4} \text{ and } \frac{5}{4}\right)$, the graphs of the equations are not parallel. **57.** No
58. $y = -3x + 1$ represents a line with slope -3 and y-intercept $(0, 1)$. $6x + 2y = 8$, or $y = -3x + 4$ represents a line with slope -3 and y-intercept $(0, 4)$. Since both lines have slope -3 but different y-intercepts, their graphs are parallel. **59.** Yes **60.** $3x = 5y - 2$, or $y = \frac{3}{5}x + \frac{2}{5}$, represents a line with slope $\frac{3}{5}$ and y-intercept $\left(0, \frac{2}{5}\right)$. $10y = 4 - 6x$, or $y = -\frac{3}{5}x + \frac{2}{5}$, represents a line with slope $-\frac{3}{5}$ and y-intercept $\left(0, \frac{2}{5}\right)$. Since the lines have different slopes, their graphs are not parallel. **61.** $-3, 0$
62. $x^3 + 5x^2 - 14x = x(x^2 + 5x - 14)$
$ = x(x + 7)(x - 2)$

63. 13 and 15, -15 and -13
64. Let $x =$ the number. **65.** $\frac{53}{7}$
Solve: $x^2 - 11 = 10x$
$ x = 11 \text{ or } x = -1$

66. $2(5 + 2y) + 4y = 13$
$10 + 4y + 4y = 13$
$10 + 8y = 13$
$8y = 3$
$y = \frac{3}{8}$

67. ◆ Yes; think of the slope as $0/a$ for any nonzero value of a. **68.** ◆ No; the slope of a vertical line is undefined.
69. ◆ Some such circumstances include using an incorrect slope and/or y-intercept when drawing the graph.
70. ◆ Answers will vary. **71.** Yes **72.** $y + 3x = 10$, or $y = -3x + 10$ represents a line with slope -3. Since $-3 \cdot \frac{1}{3} = -1$, the graphs of the equations are perpendicular.
73. No **74.** $10 - 4y = 7x$, or $y = -\frac{7}{4}x + \frac{5}{2}$ represents a line with slope $-\frac{7}{4}$. $7y + 21 = 4x$, or $y = \frac{4}{7}x - 3$ represents a line with slope $\frac{4}{7}$. Since $-\frac{7}{4} \cdot \frac{4}{7} = -1$, the graphs of the equations are perpendicular. **75.** Yes
76. The slope of $y = -2x$ is -2. The slope of $x = \frac{1}{2}$ is undefined. Since the product of the slopes is not -1, the graphs of the equations are not perpendicular.
77. When $x = 0$, $y = b$, so $(0, b)$ is on the line. When $x = 1$, $y = m + b$, so $(1, m + b)$ is on the line. Then
$$\text{slope} = \frac{(m + b) - b}{1 - 0} = m.$$

78. Rewrite each equation in slope-intercept form.
$3x - 2y = 8$
$y = \frac{3}{2}x - 4$

The slope is $\frac{3}{2}$.

$2y + 3x = -4$
$y = -\frac{3}{2}x - 2$

The y-intercept is $(0, -2)$.

We write an equation of the line with slope $\frac{3}{2}$ and y-intercept $(0, -2)$:

$y = \frac{3}{2}x - 2$

79. $y = \frac{5}{2}x + 1$
80. ◆

$T = -\frac{3}{4}a + 165$

The slope is the rate at which the target heart rate decreases for each increase of 1 year in age.

EXERCISE SET 7.2, PP. 372–373

1. $y - 7 = 5(x - 3)$ **2.** $y - 0 = -2(x - (-3))$
3. $y - 4 = \frac{3}{4}(x - 2)$ **4.** $y - 2 = -1\left(x - \frac{1}{2}\right)$
5. $y - (-6) = 1 \cdot (x - 2)$ **6.** $y - (-2) = 6(x - 4)$
7. $y - 0 = -3(x - (-4))$ **8.** $y - 3 = -3(x - 0)$
9. $y - 6 = \frac{2}{3}(x - 5)$ **10.** $y - 7 = \frac{5}{6}(x - 2)$
11. $y = 2x + 2$ **12.** $y - 5 = 4(x - 1)$
 $y = 4x + 1$
13. $y = -x + 8$ **14.** $y - (-3) = 1 \cdot (x - 2)$
 $y = x - 5$
15. $y = \frac{1}{2}x + 4$ **16.** $y - (-4) = -\frac{1}{2}(x - 6)$
 $y = -\frac{1}{2}x - 1$
17. $y = -\frac{1}{3}x - 7$ **18.** $y - 7 = \frac{1}{5}(x - (-5))$
 $y = \frac{1}{5}x + 8$
19. $y = \frac{5}{4}x - 8$ **20.** $y - 8 = \frac{4}{3}(x - (-3))$
 $y = \frac{4}{3}x + 12$

21. $y = \frac{1}{4}x + \frac{5}{2}$

22. $m = \dfrac{5-16}{1-12} = \dfrac{-11}{-11} = 1$
$y - 16 = 1(x - 12)$
$y = x + 4$

23. $y = -\frac{1}{2}x + 4$

24. $m = \dfrac{2-0}{4-0} = \dfrac{2}{4} = \dfrac{1}{2}$
$y - 0 = \dfrac{1}{2}(x - 0)$
$y = \dfrac{1}{2}x$

25. $y = -\frac{3}{2}x + \frac{13}{2}$

26. $m = \dfrac{4-1}{-1-(-4)} = \dfrac{3}{3} = 1$
$y - 1 = 1(x - (-4))$
$y = x + 5$

27. $y = \frac{2}{5}x - 2$

28. $m = \dfrac{3-(-2)}{1-(-2)} = \dfrac{5}{3}$
$y - 3 = \dfrac{5}{3}(x - 1)$
$y = \dfrac{5}{3}x + \dfrac{4}{3}$

29. $y = \frac{3}{4}x - \frac{5}{2}$

30. $m = \dfrac{-3-5}{-1-(-3)} = \dfrac{-8}{2} = -4$
$y - 5 = -4(x - (-3))$
$y = -4x - 7$

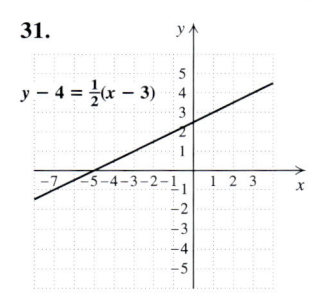

31. [graph of $y - 4 = \frac{1}{2}(x - 3)$]

32. $y - 2 = \frac{1}{3}(x - 5)$, $m = \frac{1}{3}$, or $\frac{-1}{-3}$
Plot (5, 2), move down 1 unit and left 3 units to (2, 1), and draw the line.

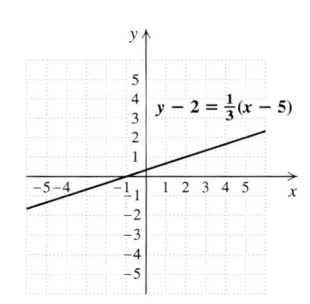

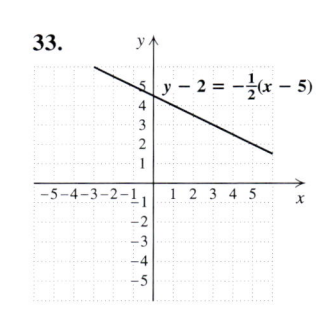

33. [graph of $y - 2 = \frac{1}{3}(x - 5)$ and $y - 2 = -\frac{1}{2}(x - 5)$]

34. $y - 1 = -\frac{1}{4}(x - 3)$, $m = -\frac{1}{4}$, or $\frac{1}{-4}$
Plot (3, 1), move up 1 unit and left 4 units to (−1, 2), and draw the line.

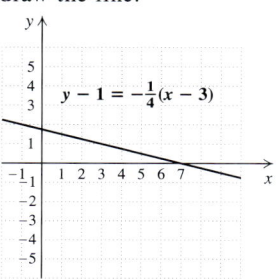

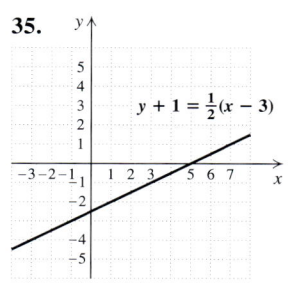

35. [graph of $y + 1 = \frac{1}{2}(x - 3)$]

36. $y - 2 = \frac{1}{3}(x + 5)$, $m = \frac{1}{3}$
Plot (−5, 2), move up 1 unit and right 3 units to (−2, 3), and draw the line.

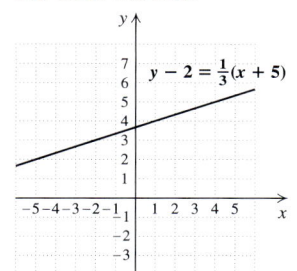

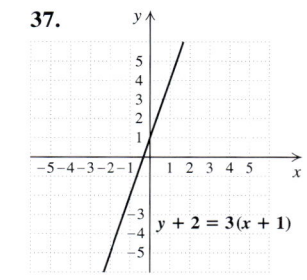

37. [graph of $y + 2 = 3(x + 1)$]

38. $y + 4 = 2(x + 1)$, $m = 2$, or $\frac{2}{1}$
Plot (−1, −4), move up 2 units and right 1 unit to (0, −2), and draw the line.

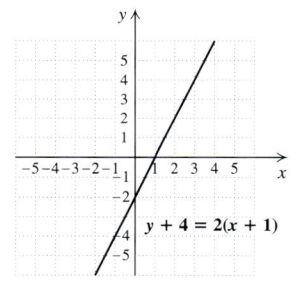

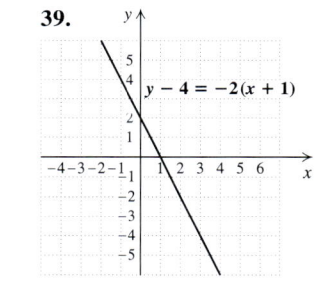

39. [graph of $y - 4 = -2(x + 1)$]

40. $y + 3 = -1(x - 4)$, $m = -1$, or $\frac{1}{-1}$
Plot (4, −3), move up 1 unit and left 1 unit to (3, −2), and draw the line.

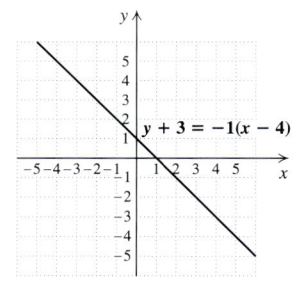

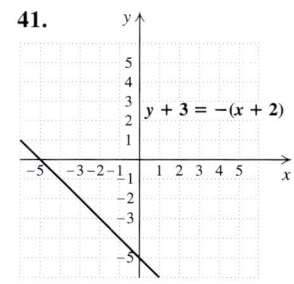

41. [graph of $y + 3 = -(x + 2)$]

22.
$$M = kE$$
$$28.6 = k(171.6)$$
$$\frac{1}{6} = k$$
$$M = \frac{1}{6}E \quad \text{Equation of variation}$$
$$M = \frac{1}{6} \cdot 110$$
$$M = 18\frac{1}{3} \text{ lb}$$

23. 66.88 lb

24.
$$P = \frac{k}{W}$$
$$660 = \frac{k}{1.6}$$
$$1056 = k$$
$$P = \frac{1056}{W} \quad \text{Equation of variation}$$
$$440 = \frac{1056}{W}$$
$$W = 2.4 \text{ ft}$$

25. 54 min

26. Note that 1 day = 24 hours and 30 days = 720 hours.
$$c = kn$$
$$14 = k \cdot 720$$
$$\frac{7}{360} = k$$
$$c = \frac{7}{360}n \quad \text{Equation of variation}$$
For 1 day: $c = \frac{7}{360} \cdot 24 \approx \$0.467$, or 46.7¢
For 1 hour: $c = \frac{7}{360} \cdot 1 \approx \$0.019$, or 1.9¢

27. 10 **28.** $c = \frac{k}{n}$ **29.** $b = \frac{2a-4}{3}$
$$17.50 = \frac{k}{9}$$
$$157.5 = k$$
$$c = \frac{157.5}{n}$$
$$31.50 = \frac{157.5}{n}$$
$$n = 5$$

30. $4m + 2n = p$ **31.** $3x + 12y$
$$4m = p - 2n$$
$$m = \frac{p - 2n}{4}$$

32. $(4a - 5b) - (7a - 6b)$
$= 4a - 5b - 7a + 6b$
$= -3a + b$

33.

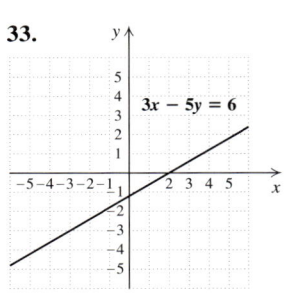

34.

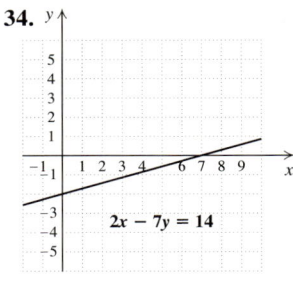

35. ◆ Neither; the cost of mailing a letter in the United States is not affected by the distance it travels.
36. ◆ Inverse variation; the higher the speed, the shorter the time. **37.** ◆ Direct variation; the greater the weight, the longer the cooking time. **38.** ◆ Inverse variation; the greater the average gain per play, the smaller the number of plays required. **39.** $P = kS, k = 8$
40. $C = kr$ **41.** $B = kN$ **42.** $A = kr^2$ **43.** $S = kv^6$
$k = 2\pi$ $k = \pi$
44. $p^2 = kt$ **45.** $I = \frac{k}{d^2}$ **46.** $D = \frac{k}{V}$
47. $V = kr^3, k = \frac{4}{3}\pi$ **48.** $P = kv^3$ **49.** ◆ If $a = kb$ and $b = hc$, then $a = k(hc) = (kh)c$, or $a = Kc$ where $K = kh$, so a varies directly as c.

REVIEW EXERCISES: CHAPTER 7, PP. 385–386
1. [7.1] -9; (0, 46) **2.** [7.1] -1; (0, 9)
3. [7.1] $\frac{1}{3}$; $\left(0, -\frac{2}{3}\right)$ **4.** [7.1] 0; (0, 8)
5. [7.1] $y = 2x - 4$ **6.** [7.1] $y = -\frac{3}{2}x + 1$
7. [7.1]   **8.** [7.1]

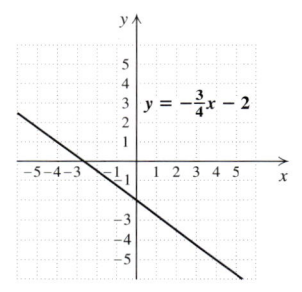

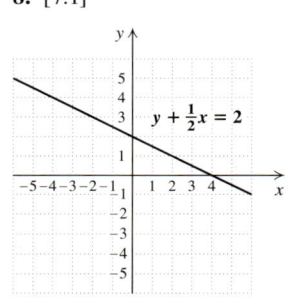

9. [7.2] 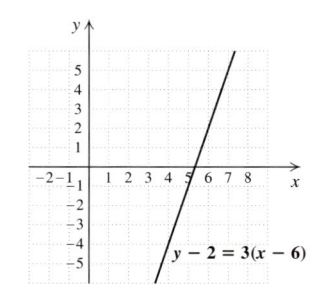   **10.** [7.2]

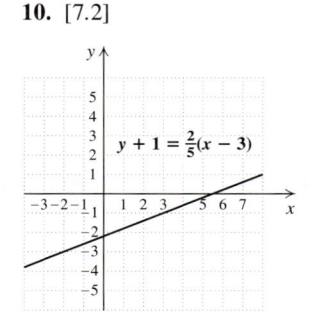

21. $y = \frac{1}{4}x + \frac{5}{2}$

22. $m = \dfrac{5 - 16}{1 - 12} = \dfrac{-11}{-11} = 1$
$y - 16 = 1(x - 12)$
$y = x + 4$

23. $y = -\frac{1}{2}x + 4$

24. $m = \dfrac{2 - 0}{4 - 0} = \dfrac{2}{4} = \dfrac{1}{2}$
$y - 0 = \dfrac{1}{2}(x - 0)$
$y = \dfrac{1}{2}x$

25. $y = -\frac{3}{2}x + \frac{13}{2}$

26. $m = \dfrac{4 - 1}{-1 - (-4)} = \dfrac{3}{3} = 1$
$y - 1 = 1(x - (-4))$
$y = x + 5$

27. $y = \frac{2}{5}x - 2$

28. $m = \dfrac{3 - (-2)}{1 - (-2)} = \dfrac{5}{3}$
$y - 3 = \dfrac{5}{3}(x - 1)$
$y = \dfrac{5}{3}x + \dfrac{4}{3}$

29. $y = \frac{3}{4}x - \frac{5}{2}$

30. $m = \dfrac{-3 - 5}{-1 - (-3)} = \dfrac{-8}{2} = -4$
$y - 5 = -4(x - (-3))$
$y = -4x - 7$

31. [graph of $y - 4 = \frac{1}{2}(x - 3)$]

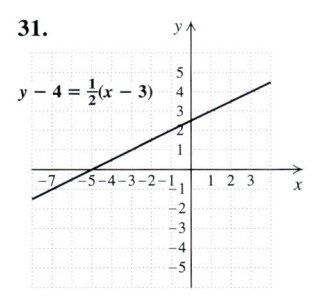

32. $y - 2 = \frac{1}{3}(x - 5)$, $m = \frac{1}{3}$, or $\frac{-1}{-3}$
Plot (5, 2), move down 1 unit and left 3 units to (2, 1), and draw the line.

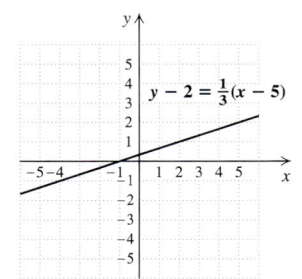

33. [graphs of $y - 2 = \frac{1}{3}(x - 5)$ and $y - 2 = -\frac{1}{2}(x - 5)$]

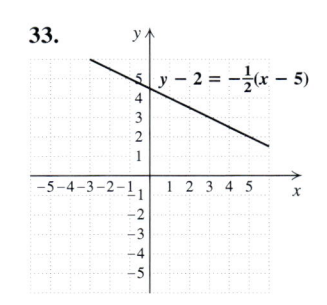

34. $y - 1 = -\frac{1}{4}(x - 3)$, $m = -\frac{1}{4}$, or $\frac{1}{-4}$
Plot (3, 1), move up 1 unit and left 4 units to (−1, 2), and draw the line.

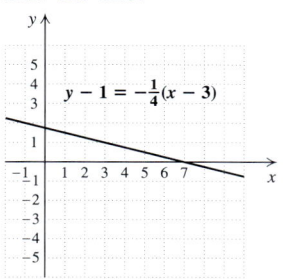

35. [graph of $y + 1 = \frac{1}{2}(x - 3)$]

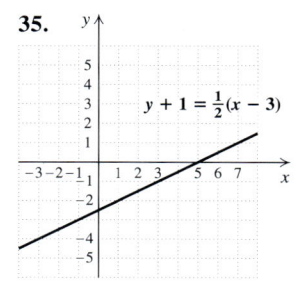

36. $y - 2 = \frac{1}{3}(x + 5)$, $m = \frac{1}{3}$
Plot (−5, 2), move up 1 unit and right 3 units to (−2, 3), and draw the line.

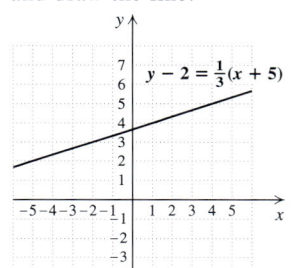

37. [graph of $y + 2 = 3(x + 1)$]

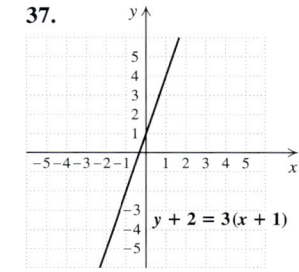

38. $y + 4 = 2(x + 1)$, $m = 2$, or $\frac{2}{1}$
Plot (−1, −4), move up 2 units and right 1 unit to (0, −2), and draw the line.

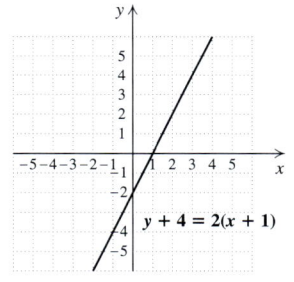

39. [graph of $y - 4 = -2(x + 1)$]

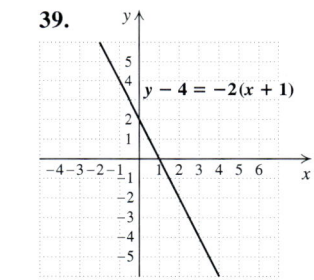

40. $y + 3 = -1(x - 4)$, $m = -1$, or $\frac{1}{-1}$
Plot (4, −3), move up 1 unit and left 1 unit to (3, −2), and draw the line.

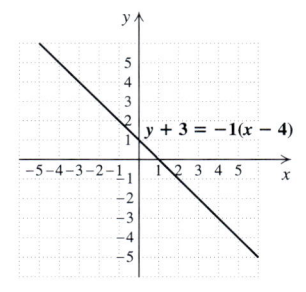

41. [graph of $y + 3 = -(x + 2)$]

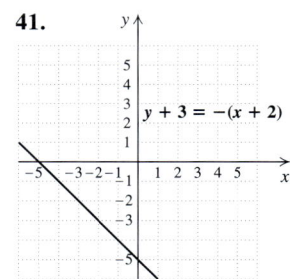

42. $y + 4 = 2(x + 2)$, $m = 2$, or $\frac{2}{1}$
Plot $(-2, -4)$, move up 2 units and right 1 unit to $(-1, -2)$, and draw the line.

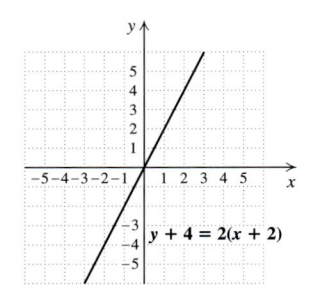

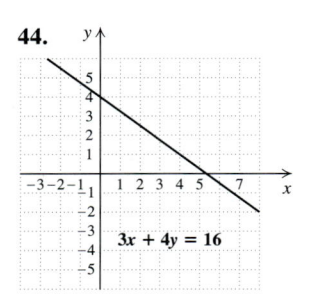

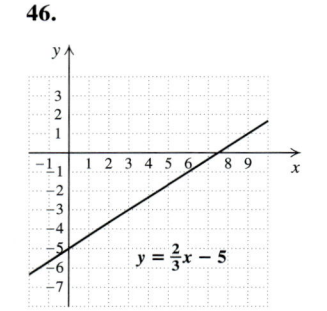

[Graph showing line $y + 4 = 2(x + 2)$]

43. [Graph showing line $5x - 2y = 5$]

44. [Graph showing line $3x + 4y = 16$]

45. [Graph showing line $y = -\frac{4}{3}x - 6$]

46. [Graph showing line $y = \frac{2}{3}x - 5$]

47. $\left\{x \mid x \geq -\frac{7}{4}\right\}$

48. $2(x - 4) > 9 - 5x$
$2x - 8 > 9 - 5x$
$7x > 17$
$x > \frac{17}{7}$

The solution set is $\left\{x \mid x > \frac{17}{7}\right\}$.

49. ◆ The equation of a horizontal line $y = b$ can be written in point-slope form:
$y - b = 0(x - x_1)$
The equation of a vertical line cannot be written in point-slope form because the slope of a vertical line is undefined.

50. ◆
1. Find the slope of the line using $m = \frac{y_2 - y_1}{x_2 - x_1}$. If the slope is undefined, the line is vertical and its equation is $x = x_1$ (or $x = x_2$ since $x_1 = x_2$). If the slope is defined, proceed to Step (2).
2. Substitute in the point-slope equation, $y - y_1 = m(x - x_1)$.
3. Solve for y.

51. ◆ $\frac{y - 1}{x - 4}$ is undefined for $x = 4$ since the denominator is $4 - 4$, or 0.   **52.** ◆ An infinite number of points (x_1, y_1) can be used in the point-slope form of an equation of any nonvertical line. However, such a line has exactly one slope and exactly one y-intercept, so it has only one slope-intercept equation.   **53.** $y = 3x + 17$

54. $x - 3y = 6$
$-3y = -x + 6$
$y = \frac{1}{3}x - 2$

The y-intercept is $(0, -2)$.
Then $m = \frac{-2 - (-1)}{0 - 5} = \frac{-1}{-5} = \frac{1}{5}$.

$y - (-1) = \frac{1}{5}(x - 5)$
$y + 1 = \frac{1}{5}x - 1$
$y = \frac{1}{5}x - 2$

55. $y = \frac{10}{3}x + \frac{25}{3}$

56. First we find the slope of the given line:
$4x - 8y = 12$
$-8y = -4x + 12$
$y = \frac{1}{2}x - \frac{3}{2}$

The slope is $\frac{1}{2}$.
Then we use the point-slope equation to find the equation of a line with slope $\frac{1}{2}$ containing the point $(-2, 0)$:

$y - 0 = \frac{1}{2}(x - (-2))$
$y = \frac{1}{2}(x + 2)$
$y = \frac{1}{2}x + 1$

57. ◆ Equations are entered on most graphers in slope-intercept form. Writing point-slope form in the modified form $y = m(x - x_1) + y_1$ or $y = mx - mx_1 + y_1$ better accommodates graphers.

EXERCISE SET 7.3, P. 377

1. No
2. $$\frac{-2x + 4y \leq -2}{-2 \cdot 5 + 4(-3) \;?\; -2}$$
   $$-10 - 12$$
   $$-22 \;|\; -2 \;\text{TRUE}$$
   $(5, -3)$ is a solution.
3. Yes
4. $$\frac{x + 0 \cdot y < 3}{-6 + 0 \cdot 5 \;?\; 3}$$
   $$-6 \;|\; 3 \;\text{TRUE}$$
   $(-6, 5)$ is a solution.

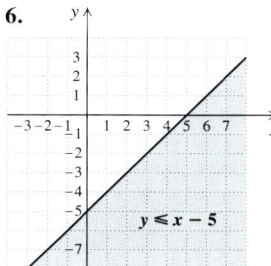

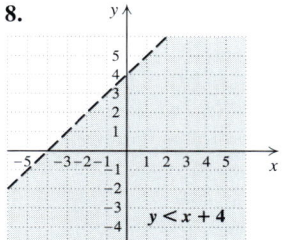

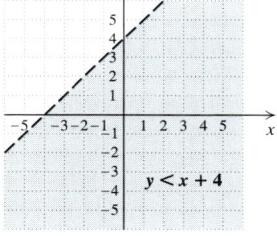

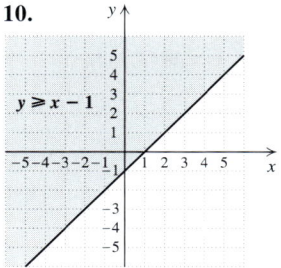

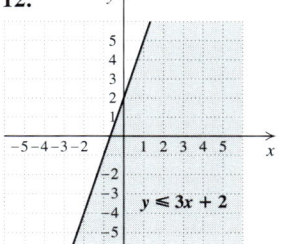

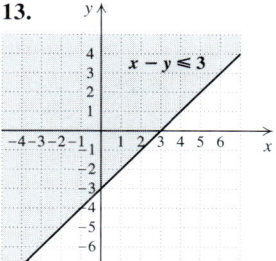

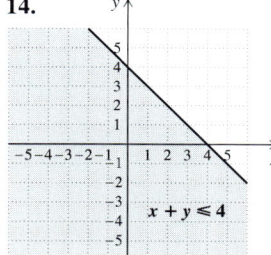

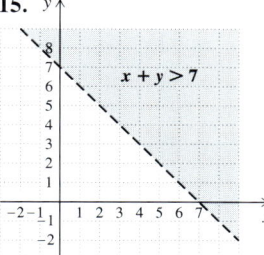

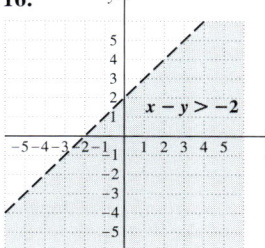

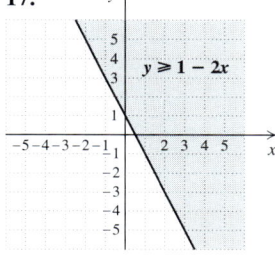

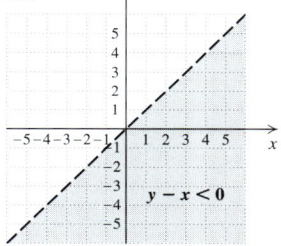

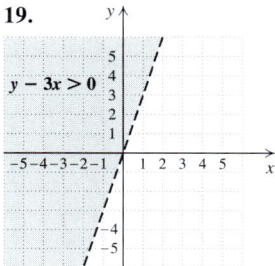

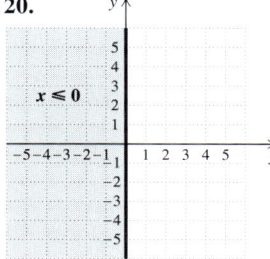

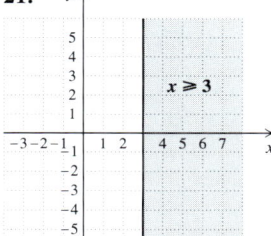

5. [graph: $y \leq x + 3$]
6. [graph: $y \leq x - 5$]
7. [graph: $y < x - 1$]
8. [graph: $y < x + 4$]
9. [graph: $y \geq x - 2$]
10. [graph: $y \geq x - 1$]
11. [graph: $y \leq 2x - 1$]
12. [graph: $y \leq 3x + 2$]
13. [graph: $x - y \leq 3$]
14. [graph: $x + y \leq 4$]
15. [graph: $x + y > 7$]
16. [graph: $x - y > -2$]
17. [graph: $y \geq 1 - 2x$]
18. [graph: $y - x < 0$]
19. [graph: $y - 3x > 0$]
20. [graph: $x \leq 0$]
21. [graph: $x \geq 3$]
22. [graph: $x > -4$]

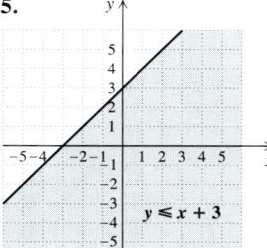

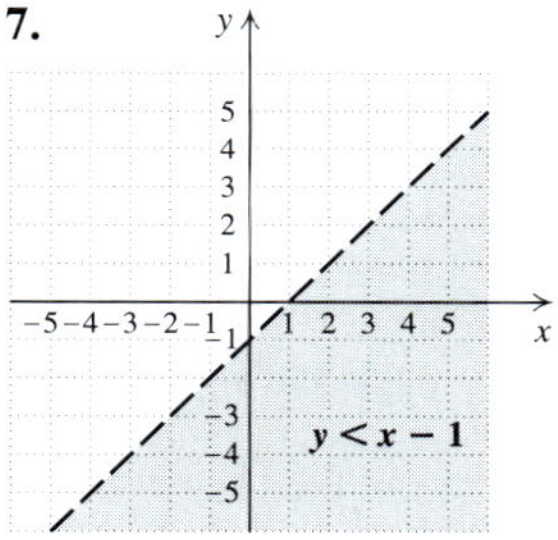

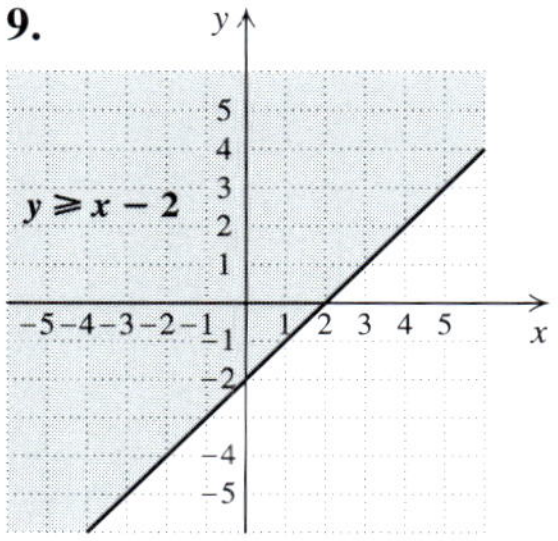

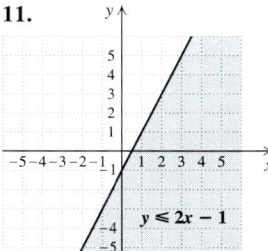

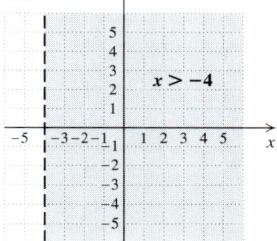

23. [graph: $y \le 3$]

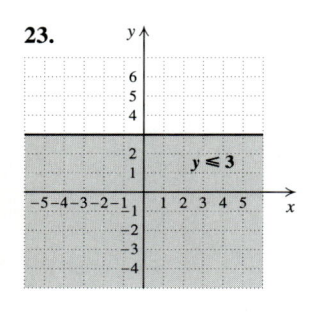

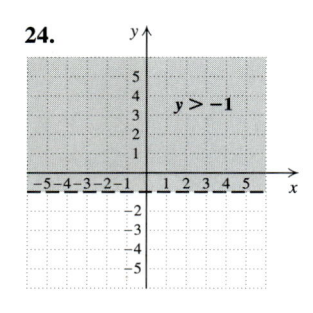

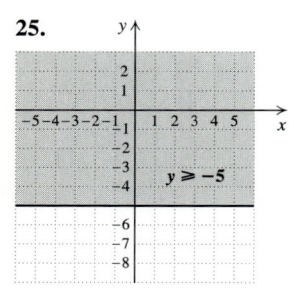

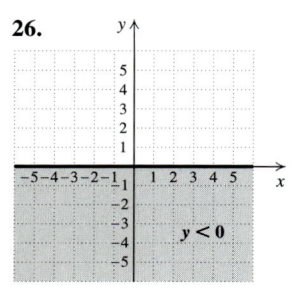

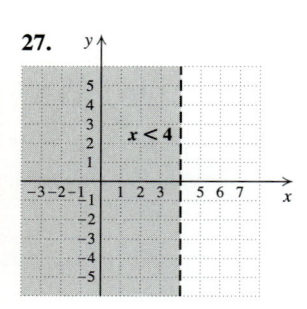

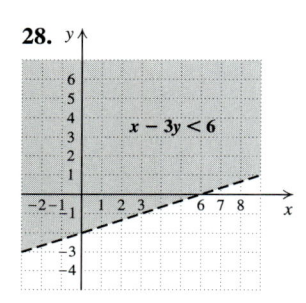

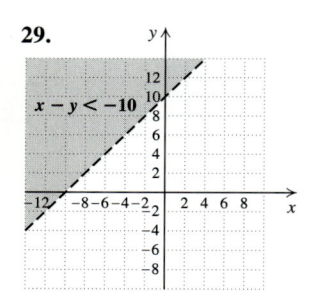

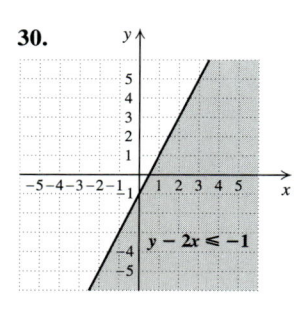

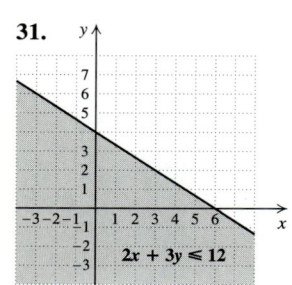

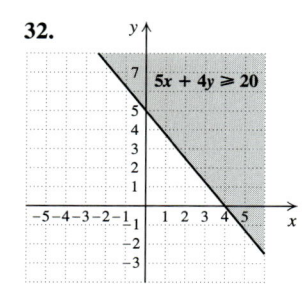

24. [graph: $y > -1$]

25. [graph: $y \ge -5$]

26. [graph: $y < 0$]

27. [graph: $x < 4$]

28. [graph: $x - 3y < 6$]

29. [graph: $x - y < -10$]

30. [graph: $y - 2x \le -1$]

31. [graph: $2x + 3y \le 12$]

32. [graph: $5x + 4y \ge 20$]

33. $2x^3 + 5x^2 - 4x - 3$ **34.** $(3x - 5)(3x + 5) = 9x^2 - 25$ **35.** $(a + 6)(3a^2 - 4)$

36. $\dfrac{x^2 - 9x + 14}{x^2 + 3x - 10} = \dfrac{(x - 2)(x - 7)}{(x - 2)(x + 5)} = \dfrac{x - 7}{x + 5}$ **37.** -18

38. $4x - 3(8 - x) = 19$
$4x - 24 + 3x = 19$
$7x - 24 = 19$
$7x = 43$
$x = \dfrac{43}{7}$

39. ◆ The point $(4.5, -1)$ is on the boundary line rather than in one of the regions created by the boundary line.

40. ◆ It is easy to do computations that involve 0.

41. ◆ If $b > 0$, then the y-intercept of $y = mx + b$ is on the positive y-axis and the graph of $y = mx + b$ lies "above" the origin. Using $(0, 0)$ as a test point, we have the false inequality $0 > b$ so the region above $y = mx + b$ is shaded.

If $b = 0$, the line $y = mx + b$ or $y = mx$ passes through the origin. Testing a point above the line, such as $(1, m + 1)$, we have the true inequality $m + 1 > m$ so the region above the line is shaded.

If $b < 0$, then the y-intercept of $y = mx + b$ is on the negative y-axis and the graph of $y = mx + b$ lies "below" the origin. Using $(0, 0)$ as a test point we get the true inequality $0 > b$ so the region above $y = mx + b$ is shaded.

Thus, we see that in any case the graph of any inequality of the form $y > mx + b$ is always shaded above the line $y = mx + b$.

42. ◆ See the procedure at the top of page 376 in the text.

43. $35c + 75a > 1000$

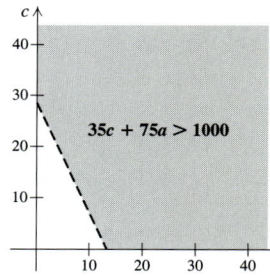

[graph: $35c + 75a > 1000$]

44. $2w + t \ge 60$. (Since w and t must also be nonnegative, we will show only the portion of the graph that is in the first quadrant.)

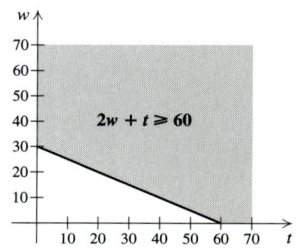

[graph: $2w + t \ge 60$]

45. $r + t \geq 17$

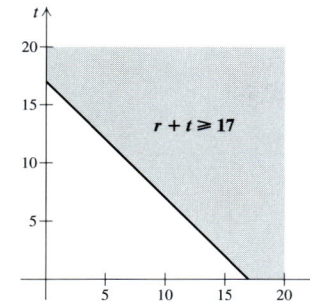

46. First find the equation of the line containing the points $(2, 0)$ and $(0, -2)$. The slope is

$$\frac{-2 - 0}{0 - 2} = \frac{-2}{-2} = 1.$$

We know that the y-intercept is $(0, -2)$, so we write the equation using slope-intercept form: $y = x - 2$. Since the line is dashed, the inequality symbol will be $<$ or $>$. To determine which, we substitute the coordinates of a point in the shaded region. We will use $(0, 0)$.

$$\begin{array}{c|c} y & x - 2 \\ \hline 0 \; ? \; 0 - 2 \\ 0 \; | \; -2 \end{array}$$

Since $0 > -2$ is true, the correct symbol is $>$. The inequality is $y > x - 2$.

47. $x \geq -2$ **48.** Graph $xy \leq 0$. From the principle of zero products, we know that $xy = 0$ when $x = 0$ or $y = 0$. Therefore, the graph contains the lines $x = 0$ and $y = 0$, or the y- and x-axes. Also, $xy < 0$ when x and y have different signs. This is the case for all points in the second quadrant (x is negative and y is positive) and in the fourth quadrant (x is positive and y is negative). Thus, we shade the second and fourth quadrants.

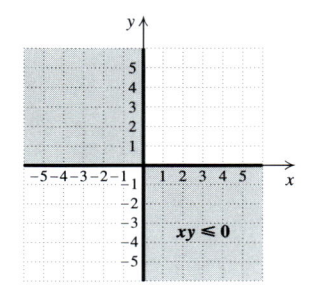

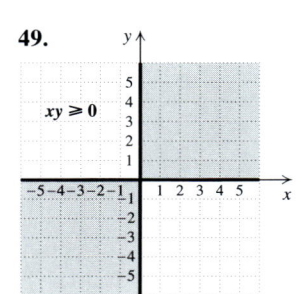

49.

EXERCISE SET 7.4, PP. 383–384

1. $y = 4x$ **2.** $y = kx$
$30 = k \cdot 8$
$3.75 = k$
$y = 3.75x$

3. $y = 1.75x$

4. $y = kx$
$0.8 = k(0.5)$
$1.6 = k$
$y = 1.6x$

5. $y = 3.2x$

6. $y = kx$
$630 = k \cdot 175$
$3.6 = k$
$y = 3.6x$

7. $y = \frac{2}{3}x$

8. $y = kx$
$500 = 60 \cdot x$
$\frac{25}{3} = x$
$y = \frac{25}{3}x$

9. $y = \frac{90}{x}$

10. $y = \frac{k}{x}$
$8 = \frac{k}{10}$
$80 = k$
$y = \frac{80}{x}$

11. $y = \frac{70}{x}$

12. $y = \frac{k}{x}$
$0.125 = \frac{k}{8}$
$1 = k$
$y = \frac{1}{x}$

13. $y = \frac{1}{x}$ **14.** $y = \frac{k}{x}$
$42 = \frac{k}{25}$
$1050 = k$
$y = \frac{1050}{x}$

15. $y = \frac{2100}{x}$

16. $y = \frac{k}{x}$
$0.2 = \frac{k}{0.3}$
$0.06 = k$
$y = \frac{0.06}{x}$

17. $183.75

18. $B = kT$
$6578 = k \cdot 2$
$3289 = k$
$B = 3289T$ Equation of variation
$B = 3289(5)$
$B = 16{,}445$

19. $22\frac{6}{7}$

20. $V = \frac{k}{P}$
$200 = \frac{k}{32}$
$6400 = k$
$V = \frac{6400}{P}$ Equation of variation
$V = \frac{6400}{20}$
$V = 320 \text{ cm}^3$

21. $3\frac{5}{9}$ amperes

22. $M = kE$
$28.6 = k(171.6)$
$\dfrac{1}{6} = k$

$M = \dfrac{1}{6}E$ Equation of variation

$M = \dfrac{1}{6} \cdot 110$

$M = 18\dfrac{1}{3}$ lb

23. 66.88 lb

24. $P = \dfrac{k}{W}$

$660 = \dfrac{k}{1.6}$

$1056 = k$

$P = \dfrac{1056}{W}$ Equation of variation

$440 = \dfrac{1056}{W}$

$W = 2.4$ ft

25. 54 min

26. Note that 1 day = 24 hours and 30 days = 720 hours.

$c = kn$
$14 = k \cdot 720$
$\dfrac{7}{360} = k$

$c = \dfrac{7}{360}n$ Equation of variation

For 1 day: $c = \dfrac{7}{360} \cdot 24 \approx \$0.467$, or 46.7¢

For 1 hour: $c = \dfrac{7}{360} \cdot 1 \approx \$0.019$, or 1.9¢

27. 10 **28.** $c = \dfrac{k}{n}$ **29.** $b = \dfrac{2a - 4}{3}$

$17.50 = \dfrac{k}{9}$

$157.5 = k$

$c = \dfrac{157.5}{n}$

$31.50 = \dfrac{157.5}{n}$

$n = 5$

30. $4m + 2n = p$ **31.** $3x + 12y$
$4m = p - 2n$
$m = \dfrac{p - 2n}{4}$

32. $(4a - 5b) - (7a - 6b)$
$= 4a - 5b - 7a + 6b$
$= -3a + b$

33. [graph of $3x - 5y = 6$]

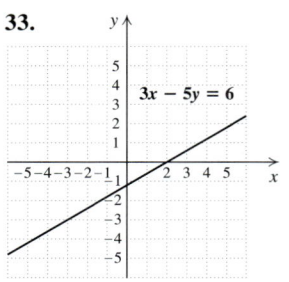

34. [graph of $2x - 7y = 14$]

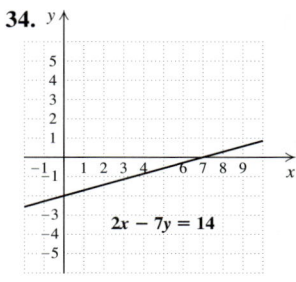

35. ◆ Neither; the cost of mailing a letter in the United States is not affected by the distance it travels.
36. ◆ Inverse variation; the higher the speed, the shorter the time. **37.** ◆ Direct variation; the greater the weight, the longer the cooking time. **38.** ◆ Inverse variation; the greater the average gain per play, the smaller the number of plays required. **39.** $P = kS, k = 8$
40. $C = kr$ **41.** $B = kN$ **42.** $A = kr^2$ **43.** $S = kv^6$
$k = 2\pi$ $k = \pi$
44. $p^2 = kt$ **45.** $I = \dfrac{k}{d^2}$ **46.** $D = \dfrac{k}{V}$
47. $V = kr^3, k = \dfrac{4}{3}\pi$ **48.** $P = kv^3$ **49.** ◆ If $a = kb$ and $b = hc$, then $a = k(hc) = (kh)c$, or $a = Kc$ where $K = kh$, so a varies directly as c.

REVIEW EXERCISES: CHAPTER 7, PP. 385–386

1. [7.1] -9; (0, 46) **2.** [7.1] -1; (0, 9)
3. [7.1] $\dfrac{1}{3}$; $\left(0, -\dfrac{2}{3}\right)$ **4.** [7.1] 0; (0, 8)
5. [7.1] $y = 2x - 4$ **6.** [7.1] $y = -\dfrac{3}{2}x + 1$
7. [7.1] **8.** [7.1]

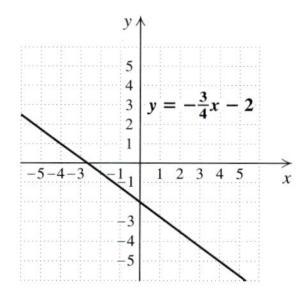

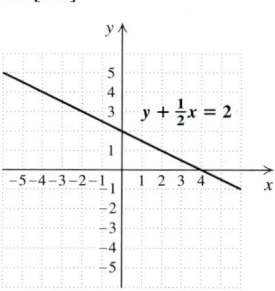

[graph of $y = -\dfrac{3}{4}x - 2$] [graph of $y + \dfrac{1}{2}x = 2$]

9. [7.2] 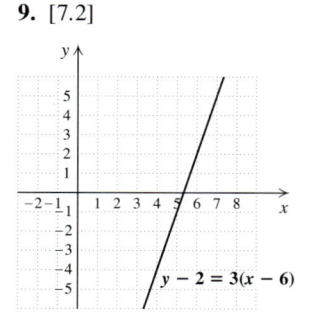 **10.** [7.2]

[graph of $y - 2 = 3(x - 6)$] [graph of $y + 1 = \dfrac{2}{5}(x - 3)$]

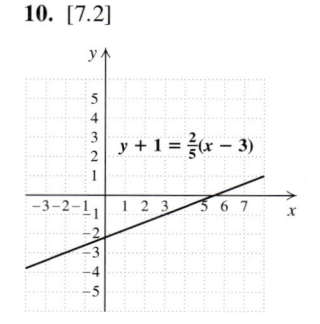

11. [7.1] (graph of $4y = 3x - 8$)

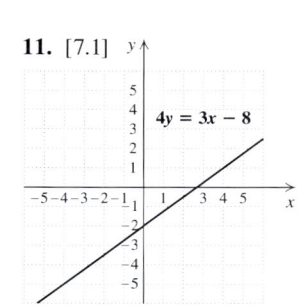

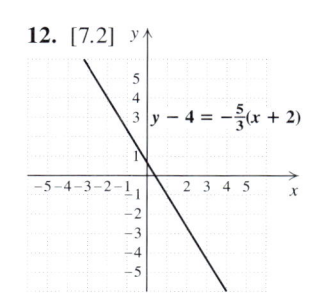

12. [7.2] (graph of $y - 4 = -\frac{5}{3}(x + 2)$)

13. [7.1] Yes **14.** [7.1] No **15.** [7.1] $y = 0.5x + 12$
16. [7.2] $y - 2 = 3(x - 1)$
17. [7.2] $y - (-5) = \frac{2}{3}(x - (-2))$ **18.** [7.2] $y = x + 2$
19. [7.2] $y = \frac{1}{2}x - 1$

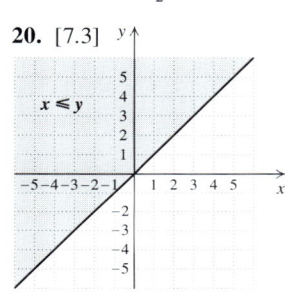

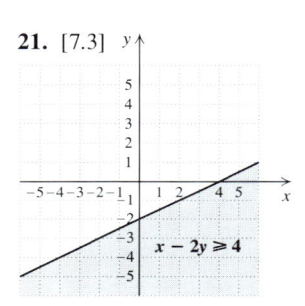

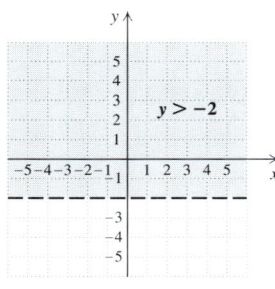

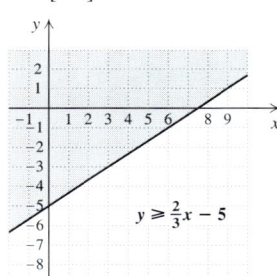

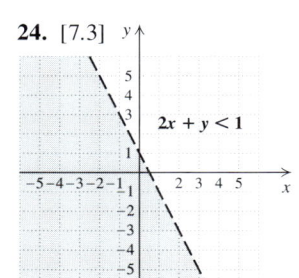

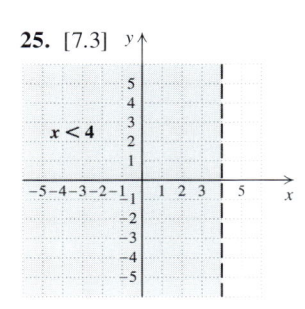

20. [7.3] (graph of $x \leq y$)

21. [7.3] (graph of $x - 2y \geq 4$)

22. [7.3] (graph of $y > -2$)

23. [7.3] (graph of $y \geq \frac{2}{3}x - 5$)

24. [7.3] (graph of $2x + y < 1$)

25. [7.3] (graph of $x < 4$)

26. [7.4] $y = \dfrac{243}{x}$ **27.** [7.4] 9.6 lb
28. [4.5] $\frac{1}{4}y^2 + 3y + 9$ **29.** [5.1] $(x - 1)(x^2 + 2)$
30. [4.3] $5x^3 - x^2 + 6x + 5$ **31.** [4.3] $2a + 28b$

32. [7.1], [7.2] Answers may vary. Some will consider it easier to convert the point–slope form $y - y_1 = m(x - x_1)$ to the form $y = mx - mx_1 + y_1$, or $y = mx + b$, where $b = -mx_1 + y_1$. Others will consider it easier to convert the slope–intercept form $y = mx + b$ to the form

$$y - 0 = m\left(x + \frac{b}{m}\right),$$

or $y - y_1 = m(x - x_1)$, where $y_1 = 0$ and $x_1 = -\dfrac{b}{m}$.

33. [7.3] ◆ The boundary line is part of the graph of a linear inequality $ax + by \leq c$ because the $\leq$ sign indicates that the graph of $ax + by < c$, as well as the graph of $ax + by = c$, form the solution set. The graph of $ax + by < c$ does not contain the graph of $ax + by = c$.

34. [7.1] $y = -x$ **35.** [7.1] $-\dfrac{c}{a}$; $(0, c)$, $(a, 0)$
36. [7.1] $y = -2x - 3$ **37.** [7.1], [7.2] $y = \frac{3}{2}x + \frac{7}{2}$

TEST: CHAPTER 7, P. 386

1. [7.1] $-\frac{1}{5}$; $(0, -7)$ **2.** [7.1] 0; $(0, -3)$
3. [7.1] 2, $\left(0, \frac{1}{4}\right)$ **4.** [7.1] $\frac{4}{3}$; $(0, -2)$
5. [7.1] $y = \frac{1}{2}x - 7$ **6.** [7.1] $y = -4x + 3$
7. [7.2] $y - 5 = 1(x - 3)$ **8.** [7.2] $y - 0 = -\frac{1}{2}(x - (-2))$
9. [7.2] $y = -3x + 4$ **10.** [7.2] $y = \frac{1}{4}x - 2$
11. [7.1] (graph of $2x + y = 5$) **12.** [7.1] (graph of $y = \frac{2}{3}x - 6$)

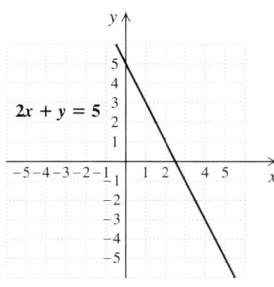

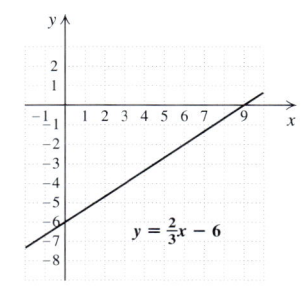

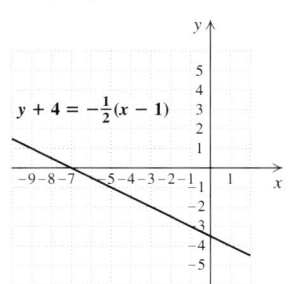

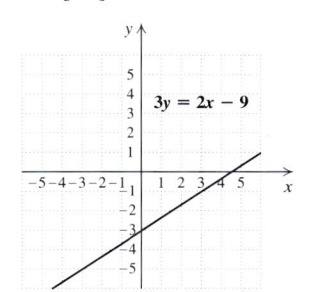

13. [7.2] (graph of $y + 4 = -\frac{1}{2}(x - 1)$) **14.** [7.1] (graph of $3y = 2x - 9$)

15. [7.1] Yes

16. [7.3] $y > x - 1$ (graph)

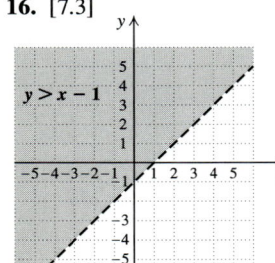

17. [7.3] $2x - y \le 4$ (graph)

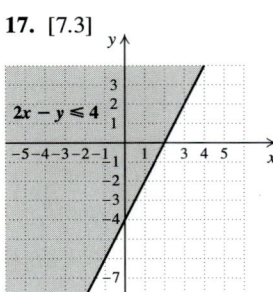

18. [7.3] $y < -2$ (graph)

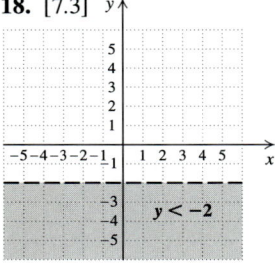

19. [7.4] $y = 4.5x$ **20.** [7.4] 18 min
21. [4.4] $-3y^7 + 9y^5 - 21y^4$ **22.** [4.5] $x^2 - 0.01$
23. [2.3] $x = \dfrac{5y + 2z}{3}$ **24.** [4.3] $8x$
25. [7.1], [7.2] $y = \dfrac{2}{3}x + \dfrac{11}{3}$ **26.** [7.1] $y = \dfrac{5}{2}x$

CHAPTER 8

TECHNOLOGY CONNECTION 8.1, P. 390
1.  (graph)   **2.** No. The lines are parallel.

EXERCISE SET 8.1, PP. 392–393
1. Yes

2.
$$\begin{array}{c|c} 5x - 2y = -5 \\ \hline 5 \cdot 1 - 2 \cdot 5 \ ? \ -5 \\ 5 - 10 \\ -5 \ | \ -5 \quad \text{TRUE} \end{array}$$

$$\begin{array}{c|c} 3x - 7y = -32 \\ \hline 3 \cdot 1 - 7 \cdot 5 \ ? \ -32 \\ 3 - 35 \\ -32 \ | \ -32 \quad \text{TRUE} \end{array}$$

(1, 5) is a solution of the system.

3. No

4.
$$\begin{array}{c|c} b + 2a = 2 \\ \hline -2 + 2 \cdot 2 \ ? \ 2 \\ -2 + 4 \\ 2 \ | \ 2 \quad \text{TRUE} \end{array} \qquad \begin{array}{c|c} b - a = -4 \\ \hline -2 - 2 \ ? \ -4 \\ -4 \ | \ -4 \quad \text{TRUE} \end{array}$$

$(2, -2)$ is a solution of the system.

5. Yes

6.
$$\begin{array}{c|c} 3r + s = -6 \\ \hline 3(-1) + (-3) \ ? \ -6 \\ -3 - 3 \\ -6 \ | \ -6 \quad \text{TRUE} \end{array}$$

$$\begin{array}{c|c} 2r = 1 + s \\ \hline 2(-1) \ ? \ 1 + (-3) \\ -2 \ | \ -2 \quad \text{TRUE} \end{array}$$

$(-1, -3)$ is a solution of the system.

7. (4, 2)

8. (graph, $x + y = 3$, $x - y = 1$, (2, 1)) (2, 1) checks.

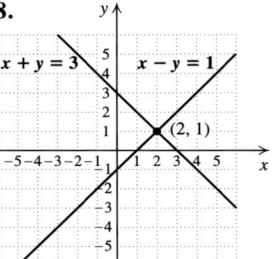

9. (2, 1)

10. (graph, $x + y = 4$, $y = 2x - 5$, (3, 1)) (3, 1) checks.

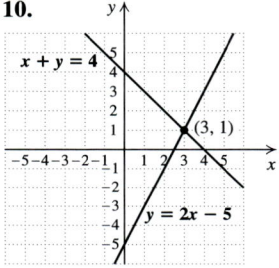

11. $(2, -2)$

12. (graph, $4x + 2y = 6$, $3x - y = 2$, (1, 1)) (1, 1) checks.

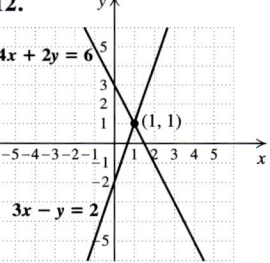

13. No solution

14. (graph, $6x + 12 = 2y$, $6 - y = -3x$) The equations represent the same line. There is an infinite number of solutions.

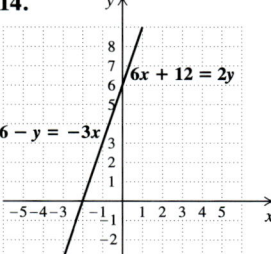

CHAPTERS 7–8   **A-149**

15. $(2, -1)$

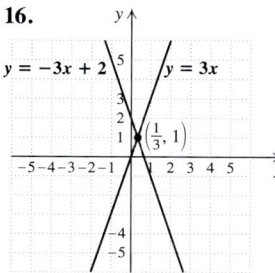

16. [graph: $y = -3x + 2$ and $y = 3x$ intersecting at $\left(\frac{1}{3}, 1\right)$]

$\left(\frac{1}{3}, 1\right)$ checks.

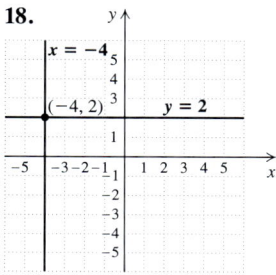

17. Infinite number of solutions

18. [graph: $x = -4$ and $y = 2$ intersecting at $(-4, 2)$]

$(-4, 2)$ checks.

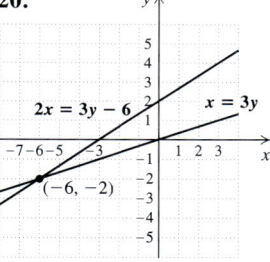

19. $(-3, -3)$

20. [graph: $2x = 3y - 6$ and $x = 3y$ intersecting at $(-6, -2)$]

$(-6, -2)$ checks.

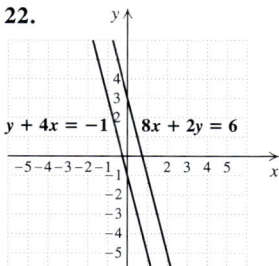

21. $(-5, 3)$

22. [graph: $y + 4x = -1$ and $8x + 2y = 6$, parallel lines]

The lines are parallel. There is no solution.

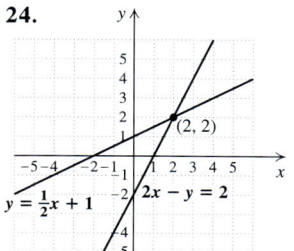

23. $(1, -3)$

24. [graph: $y = \frac{1}{2}x + 1$ and $2x - y = 2$ intersecting at $(2, 2)$]

$(2, 2)$ checks.

25. $(-8, -7)$

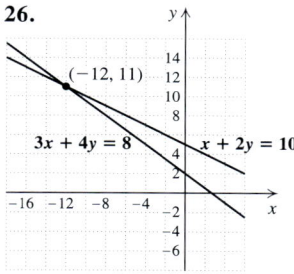

26. [graph: $3x + 4y = 8$ and $x + 2y = 10$ intersecting at $(-12, 11)$]

$(-12, 11)$ checks.

27. $(3, -4)$

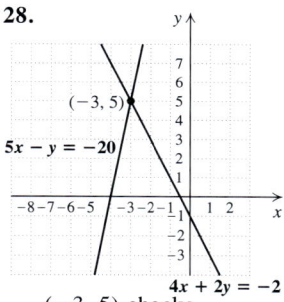

28. [graph: $5x - y = -20$ and $4x + 2y = -2$ intersecting at $(-3, 5)$]

$(-3, 5)$ checks.

29. $(6, 2)$

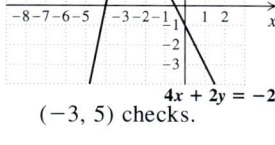

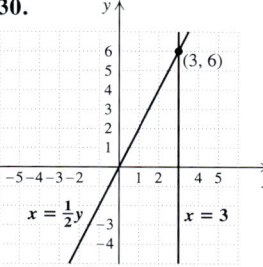

30. [graph: $x = \frac{1}{2}y$ and $x = 3$ intersecting at $(3, 6)$]

$(3, 6)$ checks.

31. $\dfrac{9x + 12}{(x + 4)(x - 4)}$

32. $\dfrac{2x^2 - x - 15}{x^2 - 9} = \dfrac{(2x + 5)(x - 3)}{(x + 3)(x - 3)} = \dfrac{2x + 5}{x + 3}$

33. Trinomial   **34.** Binomial   **35.** -2

36. $4x - 5(9 - 2x) = 7$
$4x - 45 + 10x = 7$
$14x = 52$
$x = \dfrac{26}{7}$

37. ◆ Slope-intercept form allows us to determine by observation of the equations whether their graphs are parallel or the same line. **38.** ◆ No; given a pair of linear equations there are only three possibilities: they intersect in exactly one point, they have no points of intersection, or they represent the same line and hence have an infinite number of points of intersection. **39.** ◆ Yes; since the graphs of the equations are the same, there is an infinite number of solutions and the system is consistent. **40.** ◆ We know that the first coordinate of the point of intersection is 2. Substitute 2 for x in either $y = 3x - 1$ or $y = 9 - 2x$ and find y, the second coordinate of the point of intersection. **41.** Exercises 14 and 17 **42.** The systems in Exercises 7–12, 14–21, and 23–30 are consistent. **43.** Exercises 13 and 22 **44.** The systems in Exercises 7–13, 15, 16, and 18–30 contain independent equations.
45. Answers may vary.
$2x - y = 8$,
$x + 3y = -10$
46. Answers may vary. Any equation with $(3, -2)$ as a solution and that is independent of $4x + 3y = 6$ will do. One such equation is $x + y = 1$. **47.** $A = 2, B = 2$
48.

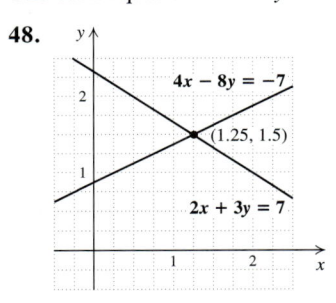

(1.25, 1.5) checks.
49. (a) Electric: $y = 100x$; gas: $y = 25x + 400$;
(b)

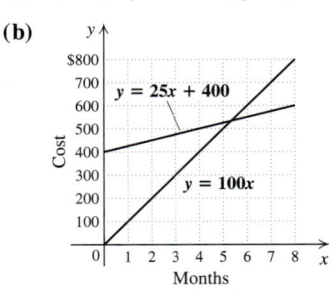

(c) $5\dfrac{1}{3}$ months

50. (a) Let x = the number of copies, up to 300, and y = the cost.
Copy card: $y = 18$
Per page: $y = 0.08x$
(b)

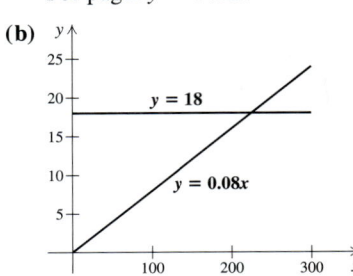

(c) The graphs appear to intersect at the point with first coordinate 225. The graph of $y = 18$ lies below the graph of $y = 0.08x$ to the right of this point, so the card is more economical when more than 225 copies are made.
51. (41.5, 17.1)

TECHNOLOGY CONNECTION 8.2, P. 397
1. $(2.5, -1.75)$

EXERCISE SET 8.2, PP. 399–401
1. $(2, -4)$
2. $x + y = 10$, (1)   **3.** $(2, 1)$
$x = y + 8$ (2)
Substitute $y + 8$ for x in (1).
$(y + 8) + y = 10$
$2y + 8 = 10$
$2y = 2$
$y = 1$
Substitute 1 for y in (2).
$x = 1 + 8 = 9$
The solution is $(9, 1)$.
4. $y = x - 3$, (1)   **5.** $(4, 3)$
$3x + y = 5$ (2)
Substitute $x - 3$ for y in (2).
$3x + (x - 3) = 5$
$4x - 3 = 5$
$4x = 8$
$x = 2$
Substitute 2 for x in (1).
$y = 2 - 3 = -1$
The solution is $(2, -1)$.

6. $y = 2x + 1$, (1)
$x + y = 4$ (2)
Substitute $2x + 1$ for y in (2).
$$x + (2x + 1) = 4$$
$$3x + 1 = 4$$
$$3x = 3$$
$$x = 1$$
Substitute 1 for x in (1).
$$y = 2 \cdot 1 + 1 = 3$$
The solution is $(1, 3)$.

7. $(1, -2)$

8. $r = -3s$, (1)
$r + 4s = 10$ (2)
Substitute $-3s$ for r in (2).
$$-3s + 4s = 10$$
$$s = 10$$
Substitute 10 for s in (1).
$$r = -3(10) = -30$$
The solution is $(-30, 10)$.

9. $(-2, 4)$

10. $x = y - 8$, (1)
$3x + 2y = 1$ (2)
Substitute $y - 8$ for x in (2).
$$3(y - 8) + 2y = 1$$
$$3y - 24 + 2y = 1$$
$$5y = 25$$
$$y = 5$$
Substitute 5 for y in (1).
$$x = 5 - 8 = -3$$
The solution is $(-3, 5)$.

11. No solution

12. $y = 3x - 1$, (1)
$6x - 2y = 2$ (2)
Substitute $3x - 1$ for y in (2).
$$6x - 2(3x - 1) = 2$$
$$6x - 6x + 2 = 2$$
$$2 = 2 \quad \text{True for all } x$$
Infinite number of solutions

13. $(-1, -3)$

14. $x - y = 6$, (1)
$x + y = -2$ (2)
Solve (1) for x.
$$x - y = 6 \quad (1)$$
$$x = y + 6 \quad (3)$$
Substitute $y + 6$ for x in (2).
$$(y + 6) + y = -2$$
$$2y + 6 = -2$$
$$2y = -8$$
$$y = -4$$
Substitute -4 for y in (3).
$$x = y + 6 = -4 + 6 = 2$$
The solution is $(2, -4)$.

15. $\left(\frac{17}{3}, \frac{2}{3}\right)$

16. $y - 2x = -6$, (1)
$2y - x = 5$ (2)
Solve (1) for y.
$$y - 2x = -6, \quad (1)$$
$$y = 2x - 6 \quad (3)$$
Substitute $2x - 6$ for y in (2).
$$2y - x = 5$$
$$2(2x - 6) - x = 5$$
$$4x - 12 - x = 5$$
$$3x - 12 = 5$$
$$3x = 17$$
$$x = \frac{17}{3}$$
Substitute $\frac{17}{3}$ for x in (3).
$$y = 2x - 6 = 2\left(\frac{17}{3}\right) - 6 = \frac{16}{3}$$
The solution is $\left(\frac{17}{3}, \frac{16}{3}\right)$.

17. Infinite number of solutions

18. $x - 4y = 3$, (1)
$2x - 6 = 8y$ (2)
Solve (1) for x.
$$x - 4y = 3$$
$$x = 4y + 3$$
Substitute $4y + 3$ for x in (2).
$$2(4y + 3) - 6 = 8y$$
$$8y + 6 - 6 = 8y$$
$$8y = 8y$$
The last equation is true for any choice of y, so there is an infinite number of solutions.

19. Infinite number of solutions

20. $y = 2x + 5$, (1)
$y = 2x - 5$ (2)
We substitute $2x + 5$ for y in (2).
$$2x + 5 = 2x - 5$$
$$-5 = 5$$
We obtain a false equation, so the system has no solution.

21. $(-12, 11)$

22. $2x + 3y = -2$, (1)
$2x - y = 9$ (2)
Solve (2) for y.
$$2x - y = 9 \quad (2)$$
$$2x - 9 = y \quad (3)$$
Substitute $2x - 9$ for y in (1).
$$2x + 3(2x - 9) = -2$$
$$2x + 6x - 27 = -2$$
$$8x = 25$$
$$x = \frac{25}{8}$$

Now substitute $\frac{25}{8}$ for x in (3).
$$y = 2x - 9 = 2\left(\frac{25}{8}\right) - 9 = \frac{25}{4} - \frac{36}{4} = -\frac{11}{4}$$
The solution is $\left(\frac{25}{8}, -\frac{11}{4}\right)$.

23. $(-2, 4)$
24. $x - y = -3$, (1)
$2x + 3y = -6$ (2)

Solve (1) for x.
$x - y = -3$ (1)
$x = y - 3$ (3)

Substitute $y - 3$ for x in (2).
$2(y - 3) + 3y = -6$
$2y - 6 + 3y = -6$
$5y = 0$
$y = 0$

Now substitute 0 for y in (3).
$x = y - 3 = 0 - 3 = -3$

The solution is $(-3, 0)$.

25. $(1, 2)$

26. $r - 2s = 0$, (1)
$4r - 3s = 15$ (2)

Solve (1) for r.
$r - 2s = 0$ (1)
$r = 2s$ (3)

Substitute $2s$ for r in (2).
$4(2s) - 3s = 15$
$8s - 3s = 15$
$5s = 15$
$s = 3$

Now substitute 3 for s in (3).
$r = 2s = 2 \cdot 3 = 6$

The solution is $(6, 3)$.

27. Infinite number of solutions
28. $x - 3y = 7$, (1)
$-4x + 12y = 28$ (2)

Solve (1) for x.
$x - 3y = 7$ (1)
$x = 3y + 7$ (3)

Substitute $3y + 7$ for x in (2).
$-4(3y + 7) + 12y = 28$
$-12y - 28 + 12y = 28$
$-28 = 28$

We obtain a false equation, so the system has no solution.

29. $(-7, -2)$
30. $x - 2y = 5$, (1)
$2y - 3x = 1$ (2)

Solve (1) for x.

$x - 2y = 5$
$x = 2y + 5$ (3)

Substitute $2y + 5$ for x in (2).
$2y - 3(2y + 5) = 1$
$2y - 6y - 15 = 1$
$-4y = 16$
$y = -4$

Next substitute -4 for y in (3).
$x = 2y + 5 = 2(-4) + 5 = -8 + 5 = -3$

The solution is $(-3, -4)$.

31. No solution
32. $2x = y - 3$, (1)
$2x = y + 5$ (2)

Solve (1) for y.
$2x = y - 3$ (1)
$2x + 3 = y$ (3)

Substitute $2x + 3$ for y in (2).
$2x = (2x + 3) + 5$
$2x = 2x + 8$
$0 = 8$

We obtain a false equation, so the system has no solution.

33. 23 and 26
34. Solve: $x + y = 56$,
$x = y + 2$,

where x and y are the numbers. The solution is $(29, 27)$, so the numbers are 29 and 27.

35. 36 and 22
36. Solve: $x + y = 66$,
$x - y = 12$,

where x and y are the numbers. The solution is $(39, 27)$, so the numbers are 39 and 27.

37. 28 and 12
38. Solve: $x - y = 18$,
$2y + 3x = 74$,

where x = the larger number and y = the smaller number. The solution is $(22, 4)$, so the numbers are 22 and 4.

39. 70°, 110°
40. Solve: $x + y = 180$,
$x = 3y - 8$,

where x = one angle and y = the other angle. The solution is $(133, 47)$, so the angles are 133° and 47°.

41. 62°, 28°
42. Solve: $x + y = 90$,
$$y = \frac{1}{2}x + 42,$$

where x and y represent the angles. The solution is $(32, 58)$, so the angles are 32° and 58°.

43. 380 mi, 270 mi

44. Solve: $2l + 2w = 1280$,
$w = l - 90$,
where l = the length and w = the width. The solution is (365, 275), so the length is 365 mi and the width is 275 mi.
45. 40 ft, 20 ft
46. Solve: $h = 4w$,
$h + w = 25$.

The solution is (20, 5), so the height is 20 ft and the width of the service zone is 5 ft.
47. 110 yd, 60 yd
48. Solve: $2l + 2w = 346\frac{2}{3}$,
$l = w + 66\frac{2}{3}$,

where l = the length and w = the width. The solution is $\left(120, 53\frac{1}{3}\right)$, so the length is 120 yd and the width is $53\frac{1}{3}$ yd.
49. $(3x - 2)(2x - 3)$ **50.** $4p^2 - p - 3 = (4p + 3)(p - 1)$
51. Prime **52.** $9a^2 - 25 = (3a + 5)(3a - 5)$ **53.** $-11y$
54. $4(2x + 3y) + 3(5x - 4y)$
$= 8x + 12y + 15x - 12y$
$= 23x$

55. ◆ This is not the best approach, in general. If the first equation has x alone on one side, for instance, or if the second equation has a variable alone on one side, solving for y in the first equation is inefficient. This procedure could also introduce fractions in the computations unnecessarily.
56. ◆ The substitution method always yields exact solutions. It might also yield the solution more quickly than the graphing method, particularly when a variable is alone on one side of an equation in the system.   **57.** ◆ Some systems can be solved more easily by graphing, if substitution requires extensive use of fractions.   **58.** ◆ Because the equations share the same coefficients of x and y but have different constant terms, their graphs have the same slope but different y-intercepts. Thus, the lines are parallel and the system has no solution.   **59.** (4.382, 4.328)
60. $\frac{1}{4}(a - b) = 2$ (1)
$\frac{1}{6}(a + b) = 1$ (2)

We first clear the fractions.
$a - b = 8$ (1a)
$a + b = 6$ (2a)

We solve Equation (1a) for a.
$a - b = 8$ (1a)
$a = b + 8$

We substitute $b + 8$ for a in Equation (2a) and solve for b.
$(b + 8) + b = 6$
$2b + 8 = 6$
$2b = -2$
$b = -1$

Next we substitute -1 for b in Equation (1a) and solve for a.
$a - b = 8$
$a - (-1) = 8$
$a + 1 = 8$
$a = 7$

Since $(7, -1)$ checks in both equations, it is the solution.
61. $(10, -2)$
62. $0.4x + 0.7y = 0.1$ (1)
$0.5x - 0.1y = 1.1$ (2)

We first multiply each equation by 10 to clear the decimals.
$4x + 7y = 1$ (1a)
$5x - y = 11$ (2a)

We solve Equation (2a) for y.
$5x - y = 11$ (2a)
$5x - 11 = y$ (3)

Substitute $5x - 11$ for y in Equation (1a) and solve for x.
$4x + 7y = 1$ (1a)
$4x + 7(5x - 11) = 1$ Substituting
$4x + 35x - 77 = 1$
$39x = 78$
$x = 2$

Next we substitute 2 for x in Equation (3) and compute y.
$y = 5x - 11 = 5 \cdot 2 - 11 = 10 - 11 = -1$

Since $(2, -1)$ checks in both equations, it is the solution.
63. $(2, -1, 3)$
64. $x + y + z = 180$,   (1)
$x = z - 70$,   (2)
$2y - z = 0$ (3)

Substitute $z - 70$ for x in (1).
$(z - 70) + y + z = 180$
$y + 2z = 250$ (4)

We now have a system of two equations in two variables.
$2y - z = 0$ (3)
$y + 2z = 250$ (4)

Solve (3) for z.
$2y - z = 0$ (3)
$2y = z$ (5)

Substitute $2y$ for z in (4).
$y + 2(2y) = 250$
$5y = 250$
$y = 50$

Substitute 50 for y in (5).
$z = 2y = 2 \cdot 50 = 100$

Substitute 100 for z in (2).
$x = z - 70 = 100 - 70 = 30$

The solution is (30, 50, 100).

65. $x - 2y = 6$, (1)
$3x + 2y = 4$ (2)

Solve (1) for $2y$.
$x - 2y = 6$
$x - 6 = 2y$ (3)

Substitute $x - 6$ for $2y$ in (2).
$3x + x - 6 = 4$
$4x = 10$
$x = \frac{5}{2}$

Substitute $\frac{5}{2}$ for x in (3).
$\frac{5}{2} - 6 = 2y$
$-\frac{7}{2} = 2y$
$-\frac{7}{4} = y$

The solution is $\left(\frac{5}{2}, -\frac{7}{4}\right)$.

It can be argued that neither procedure is easier to use than the other. The procedure in Example 3 requires the use of the distributive law after the first substitution while the procedure above does not. However, the procedure above requires the use of the multiplication principle in solving for the second variable while the procedure in Example 3 does not. Elsewhere, the two procedures require the same number of steps.

66. Answers may vary.
$2x + 3y = 5$,
$5x + 4y = 2$

EXERCISE SET 8.3, PP. 408–409

1. $(9, 3)$ **2.** $x - y = 7$ (1)
$x + y = 3$ (2)
$2x = 10$
$x = 5$

Substitute 5 for x in (2).
$5 + y = 3$
$y = -2$

The solution is $(5, -2)$.

3. $(-1, 7)$

4. $x + y = 6$ (1)
$-x + 3y = -2$ (2)
$4y = 4$
$y = 1$

Substitute 1 for y in Equation (1).
$x + 1 = 6$
$x = 5$

The solution is $(5, 1)$.

5. $(3, 0)$

6. $4x - y = 1$ (1)
$3x + y = 13$ (2)
$7x = 14$
$x = 2$

Substitute 2 for x in (2).
$3 \cdot 2 + y = 13$
$6 + y = 13$
$y = 7$

The solution is $(2, 7)$.

7. $(-1, 3)$

8. $7c + 5d = 18$ (1)
$c - 5d = -2$ (2)
$8c = 16$
$c = 2$

Substitute 2 for c in Equation (1).
$7 \cdot 2 + 5d = 18$
$14 + 5d = 18$
$5d = 4$
$d = \frac{4}{5}$

The solution is $\left(2, \frac{4}{5}\right)$.

9. $\left(-1, \frac{1}{5}\right)$

10. $3a - 3b = -15$ (1)
$-3a - 3b = -3$ (2)
$-6b = -18$
$b = 3$

Substitute 3 for b in Equation (1).
$3a - 3 \cdot 3 = -15$
$3a - 9 = -15$
$3a = -6$
$a = -2$

The solution is $(-2, 3)$.

11. Infinite number of solutions

12. $2x + 3y = 4$
$-2x - 3y = -4$
$0 = 0$

The equation $0 = 0$ is always true, so there is an infinite number of solutions.

13. $(-3, -5)$

14. $x + y = -7$, (1)
$3x + y = -9$ (2)
$-x - y = 7$ Multiplying (1) by -1
$3x + y = -9$
$2x = -2$
$x = -1$

Substitute -1 for x in (1).
$-1 + y = -7$
$y = -6$

The solution is $(-1, -6)$.

15. $(4, 5)$

16. $3x - y = 8$, (1)
$x + 2y = 5$ (2)

$6x - 2y = 16$ Multiplying (1) by 2
$\underline{x + 2y = 5}$
$7x = 21$
$x = 3$

Substitute 3 for x in (2).
$3 + 2y = 5$
$2y = 2$
$y = 1$

The solution is $(3, 1)$.

17. $(4, 1)$

18. $x - y = 7$, (1)
$4x - 5y = 25$ (2)

$-5x + 5y = -35$ Multiplying (1) by -5
$\underline{4x - 5y = 25}$
$-x = -10$
$x = 10$

Substitute 10 for x in (1).
$10 - y = 7$
$3 = y$

The solution is $(10, 3)$.

19. $(4, 3)$

20. $7p + 5q = 2$ (1)
$8p - 9q = 17$ (2)

$63p + 45q = 18$ Multiplying (1) by 9
$\underline{40p - 45q = 85}$ Multiplying (2) by 5
$103p = 103$
$p = 1$

Substitute 1 for p in Equation (1).
$7 \cdot 1 + 5q = 2$
$7 + 5q = 2$
$5q = -5$
$q = -1$

The solution is $(1, -1)$.

21. $(1, -1)$

22. $3x - 4y = 16$ (1)
$5x + 6y = 14$ (2)

$9x - 12y = 48$ Multiplying (1) by 3
$\underline{10x + 12y = 28}$ Multiplying (2) by 2
$19x = 76$
$x = 4$

Substitute 4 for x in Equation (1).
$3 \cdot 4 - 4y = 16$
$12 - 4y = 16$
$-4y = 4$
$y = -1$

The solution is $(4, -1)$.

23. $(-3, -1)$

24. $5a = 2b$,
$2a + 11 = 3b$

We write the equations in the form $Aa + Bb = C$.
$5a - 2b = 0$, (1)
$2a - 3b = -11$ (2)

$15a - 6b = 0$ Multiplying (1) by 3
$\underline{-4a + 6b = 22}$ Multiplying (2) by -2
$11a = 22$
$a = 2$

Substitute 2 for a in Equation (1).
$5 \cdot 2 - 2b = 0$
$10 - 2b = 0$
$-2b = -10$
$b = 5$

The solution is $(2, 5)$.

25. No solution

26. $2p + 5q = 9$, (1)
$3p - 2q = 4$ (2)

$4p + 10q = 18$ Multiplying (1) by 2
$\underline{15p - 10q = 20}$ Multiplying (2) by 5
$19p = 38$
$p = 2$

Substitute 2 for p in (1).
$2 \cdot 2 + 5q = 9$
$5q = 5$
$q = 1$

The solution is $(2, 1)$.

27. $\left(5, \frac{1}{2}\right)$

28. $m = 32 + n$,
$3m = 8n + 6$

We write the equations in the form $Am + Bn = C$.
$m - n = 32$, (1)
$3m - 8n = 6$ (2)

$-3m + 3n = -96$ Multiplying (1) by -3
$\underline{3m - 8n = 6}$
$-5n = -90$
$n = 18$

Substitute 18 for n in (1).
$m - 18 = 32$
$m = 50$

The solution is $(50, 18)$.

29. $(-2, 2)$

30. $2x + y = 13$ (1)
$4x + 2y = 23$ (2)

$-4x - 2y = -26$ Multiplying (1) by -2
$\underline{4x + 2y = 23}$ (2)
$0 = -3$

We get a false equation, so there is no solution.

31. $(2, -1)$

CHAPTER 8   A-155

32. $x - \frac{3}{2}y = 13$
$\frac{3}{2}x - y = 17$

We clear the fractions.

$2x - 3y = 26$, (1)
$3x - 2y = 34$ (2)

$\begin{array}{rl} 4x - 6y = & 52 \quad \text{Multiplying (1) by 2} \\ -9x + 6y = & -102 \quad \text{Multiplying (2) by } -3 \\ \hline -5x = & -50 \\ x = & 10 \end{array}$

Substitute 10 for x in (1).

$2 \cdot 10 - 3y = 26$
$-3y = 6$
$y = -2$

The solution is $(10, -2)$.

34. $1.8x - 2y = 0.9$,
$0.04x + 0.18y = 0.15$

Clear the decimals.

$18x - 20y = 9$, (1)
$4x + 18y = 15$ (2)

This is the system that results from clearing the fractions in Exercise 33. The solution is $\left(\frac{231}{202}, \frac{117}{202}\right)$.

35. 10 mi

36. Solve: $45.95 + 0.40m = c$,
$46.95 + 0.20m = c$,

where m represents the mileage and c the cost. We find $m = 5$, so the cost is the same when the cars are driven 5 miles.

37. $64°, 26°$

38. Solve: $x + y = 90$,
$x - y = 26$,

where x and y represent the measures of the larger angle and the smaller angle, respectively. The solution is $(58, 32)$, so the angles are $58°$ and $32°$.

39. $145°, 35°$

40. Solve: $x + y = 180$,
$x = 2y - 45$,

where x and y represent the angles. The solution is $(105, 75)$, so the angles are $105°$ and $75°$.

41. Oats: 11 acres; hay: 20 acres

42. Solve: $c + r = 820$,
$c = 140 + r$,

where c and r represent the number of acres of Chardonnay and Riesling grapes that should be planted, respectively. The solution is $(480, 340)$, so 480 acres of Chardonnay grapes and 340 acres of Riesling grapes should be planted.

43. 2 ft by 4 ft

44. Solve: $2l + 2w = 30$,
$l = 1.5w$,

33. $\left(\frac{231}{202}, \frac{117}{202}\right)$

where l and w represent the length and width of the garden, respectively. The solution is $(9, 6)$, so the garden is 9 ft by 6 ft.

45. a^3b^{-6}, or $\frac{a^3}{b^6}$ **46.** $\frac{9a^4}{4b^6}$ **47.** $(2x + 5)^2$

48. $9a^2 - 24a + 16 = (3a - 4)^2$ **49.** 30.1

50. $0.3(8) + 0.5(9) = 2.4 + 4.5 = 6.9$

51. ◆ Write an equation $Ax + By = C$. Then write a second equation $kAx + kBy = D$, where $D \neq kC$.

52. ◆ Write an equation $Ax + By = C$. Then write a second equation $kAx + kBy = kC$. **53.** ◆ No; only ordered pairs that satisfy the equation(s) of the system are solutions. **54.** ◆ The multiplication principle might be used to obtain a pair of terms that are opposites. The addition principle is used to eliminate a variable. Once a variable has been eliminated, the multiplication and addition principles are also used to solve for the remaining variable and, after a substitution, are used again to find the variable that was eliminated. In Example 4 each of these principles is used four times. **55.** $(-1, 1)$

56. $x + y = 7$, (1)
$3(y - x) = 9$ (2)

Multiply (1) by 3 and remove parentheses in (2) and then rewrite this equation in the form $Ax + By = C$. Then add.

$\begin{array}{rl} 3x + 3y = & 21 \\ -3x + 3y = & 9 \\ \hline 6y = & 30 \\ y = & 5 \end{array}$

Substitute 5 for y in (1).

$x + 5 = 7$
$x = 2$

The ordered pair $(2, 5)$ checks, so it is the solution.

57. $(0, 4)$

58. $2(5a - 5b) = 10$, (1)
$-5(2a + 6b) = 10$ (2)

Remove parentheses and add.

$\begin{array}{rl} 10a - 10b = & 10 \\ -10a - 30b = & 10 \\ \hline -40b = & 20 \\ b = & -\frac{1}{2} \end{array}$

Substitute $-\frac{1}{2}$ for b in Equation (2).

$-5\left(2a + 6\left(-\frac{1}{2}\right)\right) = 10$
$-5(2a - 3) = 10$
$-10a + 15 = 10$
$-10a = -5$
$a = \frac{1}{2}$

The ordered pair $\left(\frac{1}{2}, -\frac{1}{2}\right)$ checks, so it is the solution.

59. $\left(\dfrac{b-c}{1-a}, \dfrac{b-ac}{1-a}\right)$

60. $ax + by + c = 0,$
$ax + cy + b = 0$

Put both equations in the form $Ax + By = C$.

$ax + by = -c, \quad (1)$
$ax + cy = -b \quad (2)$

$\begin{aligned} ax + by &= -c & (1) \\ -ax - cy &= b & \text{Multiplying (2) by } -1 \\ \hline by - cy &= b - c \\ y(b - c) &= b - c \\ y &= 1 \end{aligned}$

Substitute 1 for y in (1).

$ax + b \cdot 1 = -c$
$ax = -b - c$
$x = \dfrac{-b - c}{a}$

The solution is $\left(\dfrac{-b-c}{a}, 1\right)$.

61. 12 rabbits, 23 pheasants

62. Solve: $x = 0.2y,$
$x + 20 = 0.52(y + 20),$

where x is Patrick's age and y is his mother's age. The solution is (6, 30), so Patrick is 6 years old, and his mother is 30

63. 45, 10

64. Solve: $h = \dfrac{2}{3}b,$
$h + 2 = \dfrac{4}{5}(b + 1),$

where b and h represent the original base and height of the triangle, respectively. The solution is (9, 6), so the original triangle has a base of 9 ft and a height of 6 ft.

EXERCISE SET 8.4, PP. 415–418

1. 175 cars, 85 trucks
2. Solve: $x + y = 45,$
$y = x + 23,$

where x represents the number of kilometers walked and y the number of kilometers driven. We find $x = 11$, so they walk 11 km.

3. Soda: $0.49; pizza: $1.50
4. Solve: $b + 2c = 2.39,$
$b + c = 1.69,$

where $b =$ the cost of one burger and $c =$ the cost of one piece of chicken. The solution is (0.99, 0.7), so one burger costs $0.99.

5. Hendersons: 10 bags; Savickis: 4 bags

6. Solve: $m = 3.5k,$
$m + k = 36,$

where m and k are the number of rolls of insulation required by the Mazzas and the Kranepools, respectively. The solution is (8, 28), so the Mazza's attic required 28 rolls of insulation and the Kranepool's attic required 8 rolls.

7. 8

8. Solve: $2t + f = 100,$
$t + f = 64,$

where $t =$ the number of two-point shots made and $f =$ the number of foul shots made. We find that $t = 36$ and $f = 28$, so Chamberlain made 36 two-point shots and 28 foul shots.

9. 5-cent: 340; 10-cent: 96

10. Solve: $x + y = 75,$
$1.25x + 1.50y = 104.25,$

where x and y represent the number of soft-serve and hard-pack cones ordered, respectively. The solution is (33, 42), so 33 soft-serve and 42 hard-pack cones were ordered.

11. 210 full-price; 225 half-price

12. Solve: $x + y = 490,$
$4x + 7y = 2905,$

where $x =$ the number of student tickets and $y =$ the number of other tickets sold. The solution is (175, 315), so 175 student tickets and 315 other tickets were sold.

13. 128 activity-card; 75 noncard

14. Solve: $x + y = 200,$
$2x + 3y = 530,$

where $x =$ the number of student tickets sold and $y =$ the number of adult tickets sold. The solution is (70, 130), so 70 student tickets and 130 adult tickets were sold.

15. Brazilian: 200 kg; Turkish: 100 kg

16. Solve: $x + y = 50,$
$x + 1.35y = 1.14(50),$

where x and y represent the number of pounds of sunflower seed and rolled oats to be used, respectively. The solution is (30, 20), so 30 lb of sunflower seed and 20 lb of rolled oats should be used.

17. Peanuts: 135 lb; Brazil nuts: 345 lb

18. Solve: $x + y = 10,$
$8x + 9y = 8.40(10),$

where x and y represent the number of kg of cashews and pecans in the mixture, respectively. The solution is (6, 4), so the mixture contains 6 kg of cashews and 4 kg of pecans.

19.

Type of Solution	50%-acid	80%-acid	68%-acid mix
Amount of Solution	x	y	200
Percent Acid	50%	80%	68%
Amount of Acid in Solution	$0.5x$	$0.8y$	136

80 mL of 50%; 120 mL of 80%

20. Solve: $x + y = 90$,
$0.12x + 0.3y = 0.2(90)$,

where x and y represent the number of ounces of Clear Shine and Sunstream in the mixture, respectively. The solution is (50, 40), so 50 oz of Clear Shine and 40 oz of Sunstream should be used.

21. 100 L of 28%; 200 L of 40%

22. Solve: $x + y = 200$,
$0.8x + 0.3y = 0.62(200)$,

where x and y represent the number of liters of 80%-base solution and 30%-base solution to be used in the mixture, respectively. The solution is (128, 72), so 128 L of 80%-base solution and 72 L of 30%-base solution should be used.

23. 4 gal of 87-octane; 8 gal of 93-octane

24. Solve: $x + y = 16$,
$10x + 15y = 180$,

where x and y represent the number of type A questions and type B questions Amy answered correctly, respectively. The solution is (12, 4), so 12 type A questions and 4 type B questions were answered, respectively.

25. 45 oz of three-fourths gold; 15 oz of five-twelfths gold

26. Solve: $d + q = 103$,
$0.1d + 0.25q = 15.25$,

where d and q represent the number of dimes and quarters in the mixture, respectively. The solution is (70, 33), so the collection contains 70 dimes and 33 quarters.

27. 10 nickels; 3 quarters

28. Solve: $x + y = 12$,
$1300x + 1850y = 18,350$,

where x represents the number of 1300-word pages and y represents the number of 1850-word pages in the document. The solution is (7, 5), so seven 1300-word pages and five 1850-word pages were used.

29. $6\frac{2}{3}$ lb of peanuts; $3\frac{1}{3}$ lb of fancy nuts

30. Solve: $9x + 7y = 19.70(16)$,
$3x + 5y = 19.825(8)$,

where x and y represent the price per gallon, in dollars, of inexpensive paint and expensive paint in a mixture, respectively. The solution is about (19.408, 20.075), so the price of the inexpensive paint is about $19.408 per gallon and the price of the expensive paint is about $20.075 per gallon.

31. $(5x + 9)(5x - 9)$ **32.** $36 - a^2 = (6 + a)(6 - a)$
33. $4(x^2 + 25)$

34. $\dfrac{x^2}{x+4} = \dfrac{16}{x+4}$, LCD $= x + 4$ **35.** 5

$(x+4) \cdot \dfrac{x^2}{x+4} = (x+4) \cdot \dfrac{16}{x+4}$

$x^2 = 16$
$x^2 - 16 = 0$
$(x+4)(x-4) = 0$
$x + 4 = 0$ or $x - 4 = 0$
$x = -4$ or $x = 4$

Only 4 checks.

36. $x^2 = 100$
$x^2 - 100 = 0$
$(x+10)(x-10) = 0$
$x + 10 = 0$ or $x - 10 = 0$
$x = -10$ or $x = 10$

37. ◆ Answers may vary. The attendance at a school play was 398. Admission was $3 each for adults and $2 each for children. The receipts were $1074. How many adults and how many children attended?   **38.** ◆ All four problems translate to one equation of the form $x + y =$ the total number of items, and a second equation of the form $ax + by =$ total points or total value or total pigment.

39. ◆ Answers will vary.   **40.** ◆ Although it makes sense to consider fractional parts of pounds or gallons, it does not make sense to talk about fractional parts of baskets or people.

41. $4\frac{4}{7}$ L

42. Solve: $8000 + x = y$,
$0.4(8000) + 0 \cdot x = 0.3y$,

where x is the amount of water to be added and y is the amount of 30% solution. We find that $x = 2666\frac{2}{3}$, so $2666\frac{2}{3}$ L of water should be added.

43. 2.5 gal

44. Solve: $100 + x = y$,
$4.6 = 0.032y$,

where x and y represent the number of liters of skim milk and 3.2% butterfat milk, respectively. We find that $x = 43.75$, so 43.75 L of skim milk should be added.

45. 10 $20-workers; 5 $25-workers

46. Solve: $x + y = 54{,}000$,
$0.06x + 0.065y = 3385$,

where x and y represent the amount invested at 6% and 6.5%, respectively. The solution is $(25{,}000, 29{,}000)$, so $25,000 is invested at 6% and $29,000 is invested at 6.5%.
47. 9%, 10.5%
48. Solve: $10x + y = 6(x + y)$,
$x = y + 1$,

where $x =$ the tens digit and $y =$ the ones digit. The solution is $(5, 4)$, so the number is 54.
49. 75
50. Solve: $x + y + z = 99$,
$x = y + 9.95$,
$z = x + 65.45$,

where x, y, and z represent the costs of the bat, ball, and glove, respectively. The solution is $(14.5, 4.55, 79.95)$, so the bat costs $14.50, the ball costs $4.55, and the glove costs $79.95.
51. Tweedledum: 120 lb; Tweedledee: 121 lb

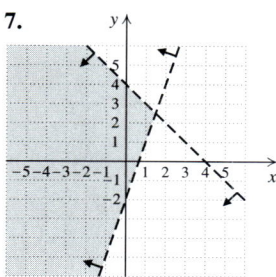

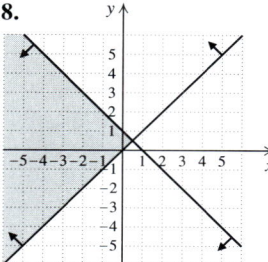

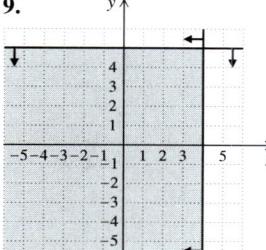

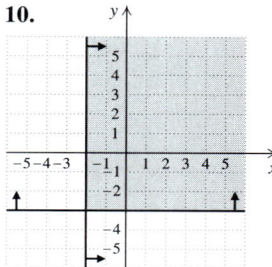

EXERCISE SET 8.5, PP. 420–421

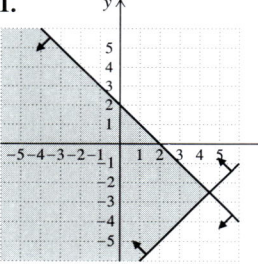

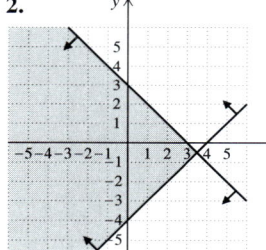

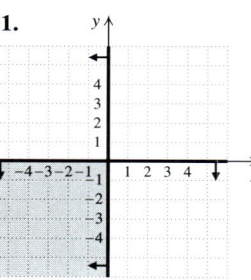

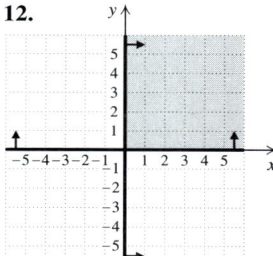

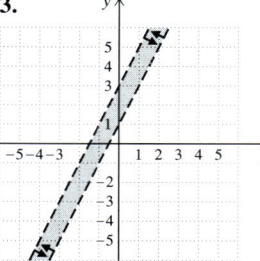

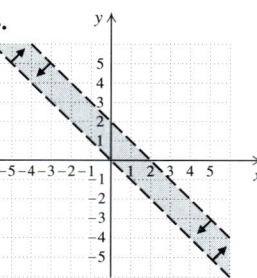

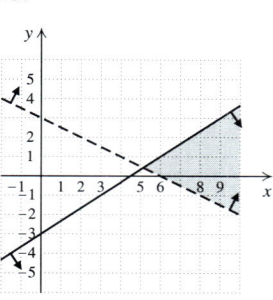

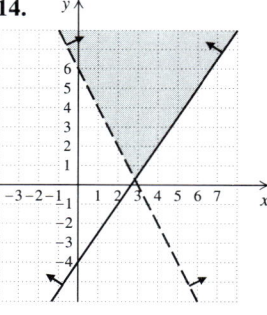

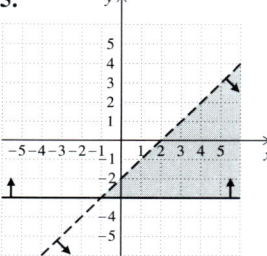

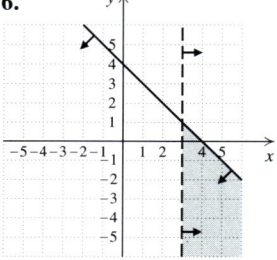

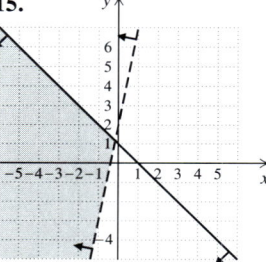

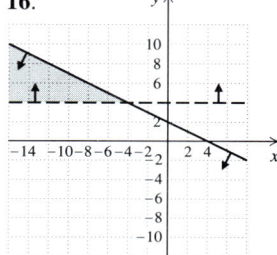

17. [graph] **18.** [graph]

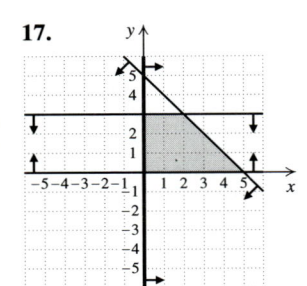

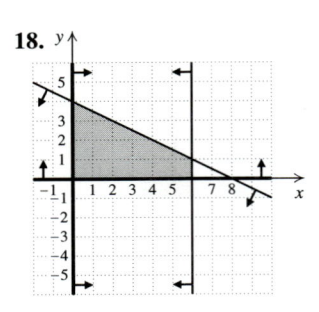

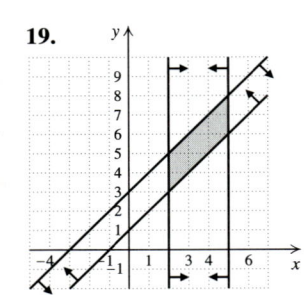

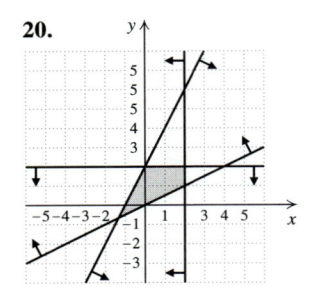

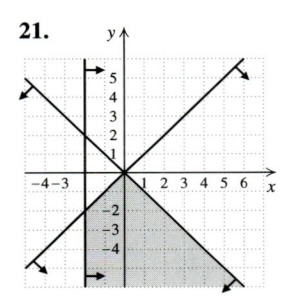

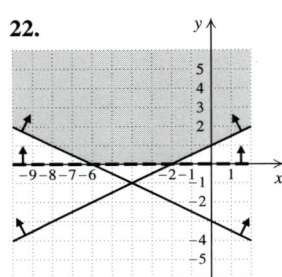

19. [graph] **20.** [graph]

21. [graph] **22.** [graph]

23. $\dfrac{13 - 2x}{3(x + 2)(x - 2)}$

24. $\dfrac{5}{4a^3 - 12a^2} - \dfrac{2}{a^2 - 4a + 3}$
$= \dfrac{5}{4a^2(a - 3)} - \dfrac{2}{(a - 3)(a - 1)}$,
LCD $= 4a^2(a - 3)(a - 1)$
$= \dfrac{5(a - 1) - 2(4a^2)}{4a^2(a - 3)(a - 1)}$
$= \dfrac{5a - 5 - 8a^2}{4a^2(a - 3)(a - 1)}$

25. 26
26. $4x^3 - 5x^2 = 4(-3)^3 - 5(-3)^2 = -108 - 45 = -153$
27. $(x - 5)^2$
28. $3a^2 - 18a + 27 = 3(a^2 - 6a + 9) = 3(a - 3)^2$
29. ◆ The point (0, 0) could not be used as a test point for the graph of $x - 2y < 0$, because (0, 0) is on the line $x - 2y = 0$. However, (0, 0) could be used as a test point for $-2x + y > 2$, since (0, 0) is not on the line $-2x + y = 2$.

30. ◆ Yes; the solution set includes the points on the solid line for $x < 3$ since these points are in the solution set of both inequalities. **31.** ◆ One condition under which a system of two linear inequalities will have no solution is that boundary lines of the graphs are parallel and the graphs have no points in common. Another is that the boundary lines of the graphs coincide, but at most one graph includes the boundary line, and the graphs have no non-boundary line points in common. **32.** ◆ Yes; assuming the solution set is nonempty, the darkest region is the region common to all of the inequalities in the system and represents the solution set.

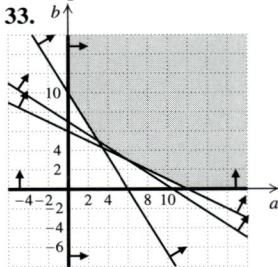

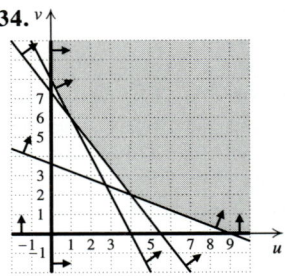

33. [graph] **34.** [graph]

REVIEW EXERCISES: CHAPTER 8, PP. 422–423

1. [8.1] Yes **2.** [8.1] No **3.** [8.1] (1, −2)
4. [8.1] (6, −2) **5.** [8.1] No solution
6. [8.1] Infinite number of solutions **7.** [8.2] (−5, 9)
8. [8.2] $\left(\dfrac{10}{3}, \dfrac{4}{3}\right)$ **9.** [8.2] No solution **10.** [8.2] (−3, 9)
11. [8.2] (4, 5) **12.** [8.2] Infinite number of solutions
13. [8.3] $\left(\dfrac{3}{2}, \dfrac{15}{4}\right)$ **14.** [8.3] (5, −3)
15. [8.3] Infinite number of solutions **16.** [8.3] (−2, 4)
17. [8.3] (3, 2) **18.** [8.3] (3, 2) **19.** [8.3] No solution
20. [8.3] (−4, 2) **21.** [8.2] 12 and 15
22. [8.2] $37\tfrac{1}{2}$ cm; $10\tfrac{1}{2}$ cm **23.** [8.4] 15
24. [8.4] 410 adults'; 415 children's
25. [8.4] Café Rich: $66\tfrac{2}{3}$ g; Café Light: $133\tfrac{1}{3}$ g
26. [8.5] **27.** [8.5]

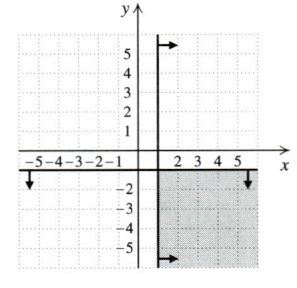

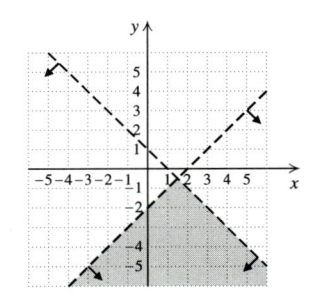

[graph] [graph]

28. [6.1] $\dfrac{a + 2}{2a + 1}$ **29.** [4.8], [6.1] $\dfrac{3ab^7}{2}$
30. [4.4] $15a^6b - 6a^3b^2$ **31.** [5.5] $9x(x - 2)^2$

32. [8.1] ◆ A solution of a system of two equations is an ordered pair that makes both equations true. The graph of an equation represents all ordered pairs that make that equation true. So in order for an ordered pair to make *both* equations true, it must be on both graphs.
33. [8.5] ◆ The solution sets of linear inequalities are regions, not lines. Thus the solution sets can intersect even if the boundary lines do not.
34. [8.1] $C = 1, D = 3$ **35.** [8.2] $(2, 1, -2)$
36. [8.2] $(2, 0)$ **37.** [8.4] 24   **38.** [8.4] $336

TEST: CHAPTER 8, PP. 423–424

1. [8.1] Yes   **2.** [8.1] $(1, 3)$ **3.** [8.1] No solution
4. [8.2] $(8, -2)$ **5.** [8.2] $(-1, 3)$
6. [8.2] Infinite number of solutions   **7.** [8.3] $(1, -5)$
8. [8.3] $(12, -6)$ **9.** [8.3] $(0, 1)$ **10.** [8.3] $(5, 1)$
11. [8.4] 40 L of 25%; 20 L of 40%   **12.** [8.2] 36°, 54°
13. [8.4] Oak: 15; pine: 3
14. [8.5]   **15.** [8.5]

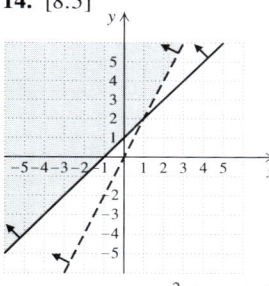

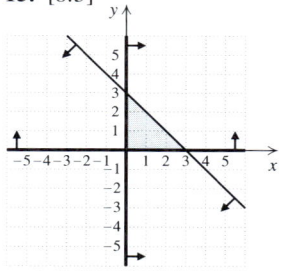

16. [6.4] $\dfrac{-x^2 + x + 17}{(x-4)(x+4)(x+1)}$ **17.** [4.2] 0
18. [2.2] $-\dfrac{15}{2}$ **19.** [5.6] $-3, 3$ **20.** [8.4] 9
21. [8.5]   **22.** [8.1] $C = -\dfrac{19}{2}, D = \dfrac{14}{3}$

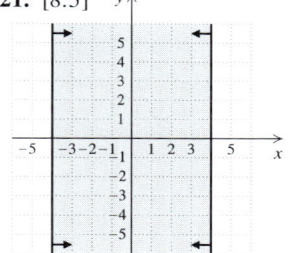

CHAPTER 9

TECHNOLOGY CONNECTION 9.1, P. 429

1. $y = \sqrt{x}$ **2.** $y = \sqrt{2x}$ 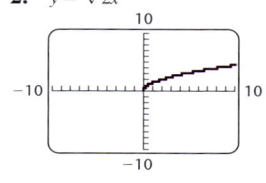
3. $y = \sqrt{x^2}$ 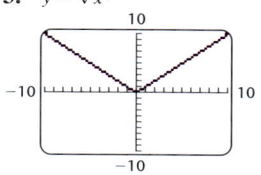   **4.** $y = \sqrt{(2x)^2}$ 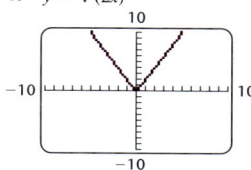
5. $y = \sqrt{x+4}$ 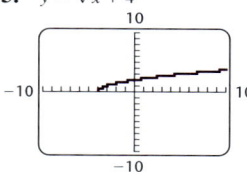   **6.** $y = \sqrt{6-x}$

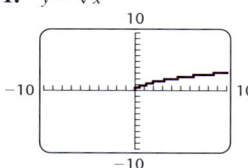

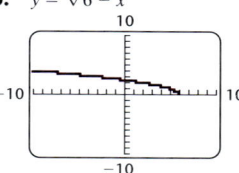

EXERCISE SET 9.1, PP. 430–431

1. $3, -3$ **2.** $2, -2$ **3.** $4, -4$ **4.** $1, -1$ **5.** $7, -7$
6. $11, -11$ **7.** $13, -13$ **8.** $12, -12$ **9.** 2   **10.** 3
11. -1 **12.** -5 **13.** 0   **14.** -9 **15.** -11 **16.** 19
17. 20   **18.** 21   **19.** 13   **20.** 12   **21.** -25 **22.** -30
23. $a - 4$ **24.** $t + 3$ **25.** $t^2 + 1$ **26.** $x^2 + 5$
27. $\dfrac{3}{x+2}$ **28.** $\dfrac{a}{a-b}$ **29.** Rational   **30.** Irrational
31. Irrational   **32.** Irrational   **33.** Irrational
34. Rational   **35.** Irrational   **36.** Irrational
37. Rational   **38.** Rational   **39.** Irrational
40. Irrational   **41.** 2.236   **42.** 2.449   **43.** 4.123
44. 4.359   **45.** 9.644   **46.** 6.557   **47.** t **48.** x
49. $5x$ **50.** $3a$ **51.** $7a$ **52.** $4x$ **53.** $17x$ **54.** $8ab$
55. (a) 15; (b) 14
56. (a) $N = 2.5\sqrt{49} = 2.5(7) = 17.5 \approx 18$
   (b) $N = 2.5\sqrt{53} \approx 2.5(7.280) \approx 18.2 \approx 19$
57. 0.864 sec   **58.** $T = 0.144\sqrt{25} = 0.144(5) = 0.72$ sec
59. $10x^9$ **60.** $(3a^3b^2)(5ab^7) = 15a^4b^9$ **61.** $-\dfrac{5}{4}$
62. $-3x + 5y = 15$ **63.** $y = 2x + 10$
   $5y = 3x + 15$
   $y = \dfrac{3}{5}x + 3$

The slope is $\dfrac{3}{5}$.

64. $m = -\dfrac{5}{4}$ (See Exercise 61.)
   $y - 4 = -\dfrac{5}{4}(x - (-3))$
   $y - 4 = -\dfrac{5}{4}(x + 3)$
   $y - 4 = -\dfrac{5}{4}x - \dfrac{15}{4}$
   $y = -\dfrac{5}{4}x + \dfrac{1}{4}$

A-162 INSTRUCTOR'S ANSWERS

65. ◆ $\sqrt{A^2}$ denotes the principal, or positive, square root of A^2. Thus, when A is negative $\sqrt{A^2} \neq A$. **66.** ◆ $\sqrt{12}$ is more exact, since 3.464101615 is an approximation of $\sqrt{12}$ while $\sqrt{12}$ denotes the exact value of the principal square root of 12. **67.** ◆ The square root of 10 is the principal, or positive, square root. A square root of 10 could refer to either the positive or the negative square root.
68. ◆ Zero is the only number that has only one square root. It is unique in this regard since $0 = -0$. **69.** 2
70. $\sqrt{3^2 + 4^2} = \sqrt{9 + 16} = \sqrt{25} = 5$ **71.** -6 and -5
72. 64; answers may vary. **73.** $-8, 8$ **74.** $\sqrt{y^2} = -5$ has no solution, because the principal square root must be positive. **75.** $-3, 3$ **76.** $t = -7$ or $t = 7$ **77.** $9a^3b^4$
78. $3a$ **79.** $\dfrac{2x^4}{y^3}$ **80.** $\dfrac{y^6}{90}$ **81.** $\dfrac{20}{m^8}$ **82.** $\dfrac{p}{60}$
83. (a) 1.7; (b) 2.2; (c) 2.6. Answers may vary.
84. $V = \dfrac{1087\sqrt{273 + 28}}{16.52} \approx 1141.6$ ft/sec
85. 1097.1 ft/sec
86. $V = \dfrac{1087\sqrt{273 - 10}}{16.52} \approx 1067.1$ ft/sec
87. 1270.8 ft/sec
88. ◆

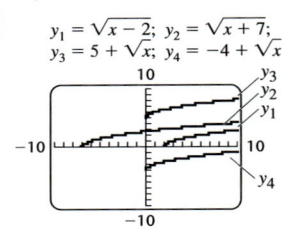

$y_1 = \sqrt{x-2}$; $y_2 = \sqrt{x+7}$;
$y_3 = 5 + \sqrt{x}$; $y_4 = -4 + \sqrt{x}$

EXERCISE SET 9.2, PP. 436–437

1. $\sqrt{35}$ **2.** $\sqrt{15}$ **3.** $\sqrt{12}$, or $2\sqrt{3}$ **4.** $\sqrt{18}$, or $3\sqrt{2}$
5. $\sqrt{\dfrac{3}{10}}$ **6.** $\sqrt{\dfrac{3}{8}} \sqrt{\dfrac{1}{5}} = \sqrt{\dfrac{3}{40}}$, or $\sqrt{\dfrac{1}{2}} \sqrt{\dfrac{3}{10}}$ **7.** 8
8. 18 **9.** $\sqrt{75}$, or $5\sqrt{3}$ **10.** $\sqrt{72}$, or $6\sqrt{2}$ **11.** $\sqrt{2x}$
12. $\sqrt{3a}$ **13.** $\sqrt{21x}$ **14.** $\sqrt{20x}$, or $2\sqrt{5x}$ **15.** $\sqrt{7xy}$
16. $\sqrt{10mn}$ **17.** $\sqrt{6ac}$ **18.** $\sqrt{3xyz}$ **19.** $2\sqrt{7}$
20. $\sqrt{12} = \sqrt{4 \cdot 3} = 2\sqrt{3}$ **21.** $2\sqrt{2}$
22. $\sqrt{45} = \sqrt{9 \cdot 5} = 3\sqrt{5}$ **23.** $10\sqrt{5}$
24. $\sqrt{200} = \sqrt{100 \cdot 2} = 10\sqrt{2}$ **25.** $3\sqrt{x}$
26. $\sqrt{4y} = 2\sqrt{y}$ **27.** $5\sqrt{3a}$
28. $\sqrt{40m} = \sqrt{4 \cdot 10m} = 2\sqrt{10m}$ **29.** $4\sqrt{a}$
30. $\sqrt{49b} = 7\sqrt{b}$ **31.** $8y$ **32.** $\sqrt{9x^2} = 3x$ **33.** $x\sqrt{13}$
34. $\sqrt{29t^2} = t\sqrt{29}$ **35.** $2t\sqrt{2}$
36. $\sqrt{125a^2} = \sqrt{25 \cdot a^2 \cdot 5} = 5a\sqrt{5}$ **37.** $4\sqrt{5}$
38. $\sqrt{98} = \sqrt{49 \cdot 2} = 7\sqrt{2}$ **39.** $12\sqrt{2y}$
40. $\sqrt{363p} = \sqrt{121 \cdot 3p} = 11\sqrt{3p}$ **41.** x^{10}
42. $\sqrt{x^{30}} = x^{15}$ **43.** x^6 **44.** $\sqrt{x^{16}} = x^8$ **45.** $x^2\sqrt{x}$
46. $\sqrt{x^3} = \sqrt{x^2 x} = x\sqrt{x}$ **47.** $t^9\sqrt{t}$
48. $\sqrt{p^{17}} = \sqrt{p^{16}p} = p^8\sqrt{p}$ **49.** $6m\sqrt{m}$

50. $\sqrt{250y^3} = \sqrt{25y^2(10y)} = 5y\sqrt{10y}$ **51.** $2a^2\sqrt{2a}$
52. $\sqrt{12b^7} = \sqrt{4b^6(3b)} = 2b^3\sqrt{3b}$ **53.** $2p^8\sqrt{26p}$
54. $\sqrt{90m^{23}} = \sqrt{9m^{22}(10m)} = 3m^{11}\sqrt{10m}$ **55.** $5\sqrt{3}$
56. $\sqrt{3} \sqrt{6} = \sqrt{3 \cdot 6} = \sqrt{3 \cdot 3 \cdot 2} = 3\sqrt{2}$ **57.** 9
58. $\sqrt{14} \cdot \sqrt{21} = \sqrt{14 \cdot 21} = \sqrt{2 \cdot 7 \cdot 3 \cdot 7} = 7\sqrt{6}$
59. $6\sqrt{xy}$
60. $\sqrt{5x} \sqrt{20y} = \sqrt{5x \cdot 20y} = \sqrt{5 \cdot x \cdot 2 \cdot 2 \cdot 5 \cdot y} = 2 \cdot 5\sqrt{xy} = 10\sqrt{xy}$ **61.** 10
62. $\sqrt{11} \sqrt{11x} = \sqrt{11 \cdot 11x} = 11\sqrt{x}$ **63.** $5b\sqrt{3}$
64. $\sqrt{6a} \sqrt{18a} = \sqrt{6a \cdot 18a} = \sqrt{2 \cdot 3 \cdot a \cdot 2 \cdot 3 \cdot 3 \cdot a} = 2 \cdot 3 \cdot a\sqrt{3} = 6a\sqrt{3}$ **65.** $7x$
66. $\sqrt{3a} \sqrt{3a} = \sqrt{3a \cdot 3a} = 3a$ **67.** $a\sqrt{bc}$
68. $\sqrt{xy} \sqrt{xz} = \sqrt{x^2yz} = x\sqrt{yz}$ **69.** $2x^3\sqrt{2}$
70. $\sqrt{15m^6} \sqrt{5m^2} = m^3\sqrt{15} \cdot m\sqrt{5} = m^4\sqrt{15 \cdot 5} = m^4\sqrt{3 \cdot 5 \cdot 5} = 5m^4\sqrt{3}$ **71.** $xy^3\sqrt{xy}$
72. $\sqrt{x^3y^2} \sqrt{xy} = y\sqrt{x^3}\sqrt{xy} = y\sqrt{x^4y} = x^2y\sqrt{y}$
73. $10ab^2\sqrt{5ab}$
74. $\sqrt{10xy^2} \sqrt{5x^2y^3} = y\sqrt{10x} \cdot x\sqrt{5y^3} = xy\sqrt{10x \cdot 5y^3} = xy\sqrt{2 \cdot 5 \cdot x \cdot 5 \cdot y^2 \cdot y} = xy \cdot 5 \cdot y\sqrt{2xy} = 5xy^2\sqrt{2xy}$
75. 20 mph, 54.8 mph
76. $r = 2\sqrt{5 \cdot 30} = 2\sqrt{150} = 2\sqrt{25 \cdot 6} = 2 \cdot 5\sqrt{6} \approx 10(2.449) \approx 24.49$, or 24.5 mph, to the nearest tenth.
$r = 2\sqrt{5 \cdot 70} = 2\sqrt{350} = 2\sqrt{25 \cdot 14} = 2 \cdot 5\sqrt{14} \approx 10(3.742) \approx 37.42$, or 37.4 mph, to the nearest tenth.
77. 168 km **78.** $(a - 5b)(a + 5b) = a^2 - 25b^2$
79. $3a^2b^5$ **80.** $\dfrac{12x^8y^6}{3x^2y^2} = 4x^6y^4$
81. ◆ Yes; we often use the rules for manipulating exponents "in reverse" when simplifying radical expressions. For example, we might write x^5 as $x^4 \cdot x$ or y^6 and $(y^3)^2$. **82.** ◆ $\sqrt{16x^4} = \sqrt{(4x^2)^2} = 4x^2$, but $\sqrt{4x^{16}} = \sqrt{(2x^8)^2} = 2x^8 \neq 2x^4$. **83.** ◆ In general, $\sqrt{a^2 - b^2} \neq \sqrt{a^2} - \sqrt{b^2}$. In this case, let $x = 13$. Then $\sqrt{x^2 - 25} = \sqrt{13^2 - 25} = \sqrt{169 - 25} = \sqrt{144} = 12$, but
$\sqrt{x^2} - \sqrt{25} = \sqrt{13^2} - \sqrt{25} = 13 - 5 = 8$.
84. ◆ $\sqrt{49} = 7$
$\sqrt{490} = \sqrt{49 \cdot 10} = 7\sqrt{10}$
$\sqrt{4900} = 70$
$\sqrt{49,000} = \sqrt{4900 \cdot 10} = 70\sqrt{10}$
$\sqrt{490,000} = 700$
Each is $\sqrt{10}$ times the one that precedes it.
85. 0.1 **86.** $\sqrt{0.25} = \sqrt{(0.5)^2} = 0.5$ **87.** 0.25
88. $\sqrt{0.000001} = \sqrt{(0.001)^2} = 0.001$ **89.** >
90. $\sqrt{450} = \sqrt{225 \cdot 2} = 15\sqrt{2}$, so $15\sqrt{2} = \sqrt{450}$.
91. >
92. $3\sqrt{11} = \sqrt{9} \sqrt{11} = \sqrt{99}$ and $7\sqrt{2} = \sqrt{49} \sqrt{2} = \sqrt{98}$, so $3\sqrt{11} > 7\sqrt{2}$.
93. <

94. $8^2 = 64$
$(\sqrt{15} + \sqrt{17})^2 = 15 + 2\sqrt{255} + 17 = 32 + 2\sqrt{255}$
Now $\sqrt{255} < \sqrt{256}$, or $\sqrt{255} < 16$, so
$2\sqrt{255} < 2 \cdot 16 = 32$. Then $32 + 2\sqrt{255} < 32 + 32$, or
$(\sqrt{15} + \sqrt{17})^2 < 64$. Thus, $8 > \sqrt{15} + \sqrt{17}$.
95. $18(x + 1)\sqrt{(x + 1)y}$
96. $\sqrt{18(x - 2)} \sqrt{20(x - 2)^3} = \sqrt{9 \cdot 2 \cdot 4 \cdot 5(x - 2)^4} = 3 \cdot 2(x - 2)^2\sqrt{2 \cdot 5} = 6(x - 2)^2\sqrt{10}$
97. $2x^3\sqrt{5x}$
98. $\sqrt{2^{109}} \sqrt{x^{306}} \sqrt{x^{11}} = \sqrt{2^{109}x^{317}} = \sqrt{2^{108} \cdot 2 \cdot x^{316} \cdot x} = 2^{54}x^{158}\sqrt{2x}$
99. x^{4n} **100.** $\sqrt{0.04x^{4n}} = \sqrt{(0.2x^{2n})^2} = 0.2x^{2n}$
101. $y^k\sqrt{y}$, where $k = \dfrac{n - 1}{2}$

EXERCISE SET 9.3, PP. 441–442

1. 2 **2.** $\dfrac{\sqrt{20}}{\sqrt{5}} = \sqrt{\dfrac{20}{5}} = \sqrt{4} = 2$ **3.** 2
4. $\dfrac{\sqrt{72}}{\sqrt{2}} = \sqrt{\dfrac{72}{2}} = \sqrt{36} = 6$ **5.** $\sqrt{5}$
6. $\dfrac{\sqrt{18}}{\sqrt{3}} = \sqrt{\dfrac{18}{3}} = \sqrt{6}$ **7.** $\dfrac{1}{4}$
8. $\dfrac{\sqrt{3}}{\sqrt{48}} = \sqrt{\dfrac{3}{48}} = \sqrt{\dfrac{1}{16}} = \dfrac{1}{4}$ **9.** $\dfrac{2}{5}$
10. $\dfrac{\sqrt{18}}{\sqrt{32}} = \sqrt{\dfrac{18}{32}} = \sqrt{\dfrac{9}{16}} = \dfrac{3}{4}$ **11.** 2
12. $\dfrac{\sqrt{18b}}{\sqrt{2b}} = \sqrt{\dfrac{18b}{2b}} = \sqrt{9} = 3$ **13.** $3y$
14. $\dfrac{\sqrt{48x^3}}{\sqrt{3x}} = \sqrt{\dfrac{48x^3}{3x}} = \sqrt{16x^2} = 4x$ **15.** $3x^2$
16. $\dfrac{\sqrt{20a^8}}{\sqrt{5a^2}} = \sqrt{\dfrac{20a^8}{5a^2}} = \sqrt{4a^6} = 2a^3$ **17.** $\dfrac{\sqrt{3}}{x^4}$
18. $\dfrac{\sqrt{6x^9}}{\sqrt{2x^3}} = \sqrt{\dfrac{6x^9}{2x^3}} = \sqrt{3x^6} = x^3\sqrt{3}$ **19.** $\dfrac{6}{5}$
20. $\sqrt{\dfrac{9}{49}} = \dfrac{\sqrt{9}}{\sqrt{49}} = \dfrac{3}{7}$ **21.** $\dfrac{7}{4}$ **22.** $\sqrt{\dfrac{1}{4}} = \dfrac{\sqrt{1}}{\sqrt{4}} = \dfrac{1}{2}$
23. $-\dfrac{1}{9}$ **24.** $-\sqrt{\dfrac{25}{49}} = -\dfrac{\sqrt{25}}{\sqrt{49}} = -\dfrac{5}{7}$ **25.** $\dfrac{2}{3}$
26. $\sqrt{\dfrac{81}{121}} = \dfrac{\sqrt{81}}{\sqrt{121}} = \dfrac{9}{11}$ **27.** $\dfrac{6}{a}$
28. $\sqrt{\dfrac{25}{x^2}} = \dfrac{\sqrt{25}}{\sqrt{x^2}} = \dfrac{5}{x}$ **29.** $\dfrac{3a}{25}$
30. $\sqrt{\dfrac{x^2y^2}{144}} = \dfrac{\sqrt{x^2y^2}}{\sqrt{144}} = \dfrac{xy}{12}$ **31.** $\dfrac{\sqrt{15}}{3}$

32. $\sqrt{\dfrac{2}{7}} = \dfrac{\sqrt{2}}{\sqrt{7}} \cdot \dfrac{\sqrt{7}}{\sqrt{7}} = \dfrac{\sqrt{14}}{7}$ **33.** $\dfrac{\sqrt{35}}{10}$
34. $\sqrt{\dfrac{1}{12}} = \dfrac{\sqrt{1}}{\sqrt{4}\sqrt{3}} = \dfrac{1}{2\sqrt{3}} \cdot \dfrac{\sqrt{3}}{\sqrt{3}} = \dfrac{\sqrt{3}}{6}$ **35.** $\dfrac{\sqrt{5}}{15}$
36. $\sqrt{\dfrac{7}{18}} = \dfrac{\sqrt{7}}{3\sqrt{2}} = \dfrac{\sqrt{7}}{3\sqrt{2}} \cdot \dfrac{\sqrt{2}}{\sqrt{2}} = \dfrac{\sqrt{14}}{6}$ **37.** $\dfrac{3\sqrt{5}}{5}$
38. $\dfrac{4}{\sqrt{3}} = \dfrac{4}{\sqrt{3}} \cdot \dfrac{\sqrt{3}}{\sqrt{3}} = \dfrac{4\sqrt{3}}{3}$ **39.** $\dfrac{2\sqrt{6}}{3}$
40. $\sqrt{\dfrac{12}{5}} = \dfrac{2\sqrt{3}}{\sqrt{5}} = \dfrac{2\sqrt{3}}{\sqrt{5}} \cdot \dfrac{\sqrt{5}}{\sqrt{5}} = \dfrac{2\sqrt{15}}{5}$ **41.** $\dfrac{\sqrt{3x}}{x}$
42. $\sqrt{\dfrac{2}{x}} = \dfrac{\sqrt{2}}{\sqrt{x}} \cdot \dfrac{\sqrt{x}}{\sqrt{x}} = \dfrac{\sqrt{2x}}{x}$ **43.** $\dfrac{\sqrt{21}}{3}$
44. $\dfrac{\sqrt{11}}{\sqrt{7}} = \dfrac{\sqrt{11}}{\sqrt{7}} \cdot \dfrac{\sqrt{7}}{\sqrt{7}} = \dfrac{\sqrt{77}}{7}$ **45.** $\dfrac{3\sqrt{2}}{4}$
46. $\dfrac{\sqrt{4}}{\sqrt{27}} = \dfrac{2}{3\sqrt{3}} \cdot \dfrac{\sqrt{3}}{\sqrt{3}} = \dfrac{2\sqrt{3}}{9}$ **47.** $\dfrac{\sqrt{42}}{14}$
48. $\dfrac{\sqrt{3}}{\sqrt{2}} = \dfrac{\sqrt{3}}{\sqrt{2}} \cdot \dfrac{\sqrt{2}}{\sqrt{2}} = \dfrac{\sqrt{6}}{2}$ **49.** $\dfrac{\sqrt{21}}{6}$
50. $\dfrac{\sqrt{5}}{\sqrt{18}} = \dfrac{\sqrt{5}}{3\sqrt{2}} \cdot \dfrac{\sqrt{2}}{\sqrt{2}} = \dfrac{\sqrt{10}}{6}$ **51.** $\dfrac{\sqrt{2x}}{8}$
52. $\dfrac{\sqrt{a}}{\sqrt{40}} = \dfrac{\sqrt{a}}{\sqrt{4}\sqrt{10}} = \dfrac{\sqrt{a}}{2\sqrt{10}} = \dfrac{\sqrt{a}}{2\sqrt{10}} \cdot \dfrac{\sqrt{10}}{\sqrt{10}} = \dfrac{\sqrt{10a}}{2 \cdot 10} = \dfrac{\sqrt{10a}}{20}$ **53.** $\dfrac{4y\sqrt{3}}{3}$
54. $\dfrac{8x}{\sqrt{5}} = \dfrac{8x}{\sqrt{5}} \cdot \dfrac{\sqrt{5}}{\sqrt{5}} = \dfrac{8x\sqrt{5}}{5}$ **55.** $\dfrac{\sqrt{3a}}{2}$
56. $\dfrac{\sqrt{3x}}{\sqrt{27}} = \sqrt{\dfrac{3x}{27}} = \sqrt{\dfrac{x}{9}} = \dfrac{\sqrt{x}}{3}$ **57.** $\dfrac{3\sqrt{10x}}{5x}$
58. $\dfrac{\sqrt{45}}{\sqrt{8a}} = \dfrac{3\sqrt{5}}{2\sqrt{2a}} = \dfrac{3\sqrt{5}}{2\sqrt{2a}} \cdot \dfrac{\sqrt{2a}}{\sqrt{2a}} = \dfrac{3\sqrt{10a}}{4a}$ **59.** $\dfrac{3\sqrt{6}}{8c}$
60. $\dfrac{\sqrt{7x^3}}{\sqrt{12x}} = \sqrt{\dfrac{7x^3}{12x}} = \sqrt{\dfrac{7x^2}{12}} = \dfrac{x\sqrt{7}}{2\sqrt{3}} = \dfrac{x\sqrt{7}}{2\sqrt{3}} \cdot \dfrac{\sqrt{3}}{\sqrt{3}} = \dfrac{x\sqrt{21}}{6}$ **61.** 1.57 sec, 3.14 sec, 8.8813 sec, 11.102 sec
62. $\dfrac{2}{3}$ in. $= \dfrac{1}{18}$ ft
Substitute $\dfrac{1}{18}$ for L and 3.14 for π in the formula.
$T = 2\pi\sqrt{\dfrac{L}{32}} \approx 2(3.14)\sqrt{\dfrac{\frac{1}{18}}{32}} \approx 6.28\sqrt{\dfrac{1}{18} \cdot \dfrac{1}{32}} \approx 6.28\sqrt{\dfrac{1}{576}} \approx 6.28\left(\dfrac{1}{24}\right) \approx 0.26$ sec
63. 1.19 sec

A-164 INSTRUCTOR'S ANSWERS

64. $T = 2\pi\sqrt{\dfrac{\frac{\pi^2}{32}}{32}} = 2\pi\sqrt{\dfrac{32}{\pi^2} \cdot \dfrac{1}{32}} = 2\pi\sqrt{\dfrac{1}{\pi^2}} = 2\pi\left(\dfrac{1}{\pi}\right) = 2$ sec

The time it takes the pendulum to swing from one side to the other and back is 2 sec, so it takes 1 sec to swing from one side to the other.

65. $(4, 2)$

66. $2x - 3y = 7$ (1)
    $2x + 3y = 9$ (2)
    $4x = 16$ Adding
    $x = 4$

   $2 \cdot 4 + 3y = 9$ Substituting in (2)
   $8 + 3y = 9$
   $3y = 1$
   $y = \dfrac{1}{3}$

The pair $\left(4, \dfrac{1}{3}\right)$ checks.

67. $9x^2 - 49$ 68. $(4a - 5b)(4a + 5b) = 16a^2 - 25b^2$
69. $21x - 9y$ 70. $17a + 9b - 3a - 15b = 14a - 6b$
71. ◆ Not necessarily; while it would probably be best to rewrite $\sqrt{2}/\sqrt{18}$ as $\sqrt{2/18}$ before simplifying, this isn't particularly helpful when simplifying an expression like $\sqrt{4}/\sqrt{9}$.   72. ◆ $\sqrt{2} \approx 1.414213562$; the long division $1.414213562/2$ is easier to perform than the long division $1/1.414213562$.   73. ◆ If division requires rationalizing the denominator, it is necessary to know how to multiply radical expressions.
74. ◆ 1. If necessary, rewrite the expression as $\sqrt{a}/\sqrt{b}$.
    2. Simplify the numerator and denominator, if possible, by taking the square roots of perfect square factors.
    3. Multiply by a form of 1 that produces an expression without a radical in the numerator.

75. $\dfrac{\sqrt{14}}{40}$ 76. $\sqrt{\dfrac{3}{1000}} = \dfrac{\sqrt{3}}{10\sqrt{10}} = \dfrac{\sqrt{3}}{10\sqrt{10}} \cdot \dfrac{\sqrt{10}}{\sqrt{10}} = \dfrac{\sqrt{30}}{100}$

77. $\dfrac{\sqrt{10x}}{4x^4}$

78. $\sqrt{\dfrac{3x^2y}{a^2x^5}} = \sqrt{\dfrac{3y}{a^2x^3}} = \dfrac{\sqrt{3y}}{ax\sqrt{x}} = \dfrac{\sqrt{3y}}{ax\sqrt{x}} \cdot \dfrac{\sqrt{x}}{\sqrt{x}} = \dfrac{\sqrt{3xy}}{ax^2}$

79. $\dfrac{\sqrt{10abc}}{5b^2c^5}$

80. $\sqrt{\dfrac{1}{5zw^2}} = \dfrac{1}{w\sqrt{5z}} = \dfrac{1}{w\sqrt{5z}} \cdot \dfrac{\sqrt{5z}}{\sqrt{5z}} = \dfrac{\sqrt{5z}}{5zw}$

81. $\dfrac{y-x}{xy}$, or $\dfrac{1}{x} - \dfrac{1}{y}$

82. $\sqrt{2 - \dfrac{4}{z^2} + \dfrac{2}{z^4}} = \sqrt{\dfrac{2z^4 - 4z^2 + 2}{z^4}} =$
$\sqrt{\dfrac{2(z^2 - 1)^2}{z^4}} = \dfrac{\sqrt{2}(z^2 - 1)}{z^2}$

An alternative method of simplifying this expression is shown below.

$\sqrt{2 - \dfrac{4}{z^2} + \dfrac{2}{z^4}} = \sqrt{2\left(1 - \dfrac{2}{z^2} + \dfrac{1}{z^4}\right)} =$
$\sqrt{2\left(1 - \dfrac{1}{z^2}\right)^2} = \left(1 - \dfrac{1}{z^2}\right)\sqrt{2}$

The two answers are equivalent.

EXERCISE SET 9.4, PP. 446–447
1. $11\sqrt{2}$ 2. $7\sqrt{3}$ 3. $3\sqrt{5}$ 4. $3\sqrt{2}$ 5. $13\sqrt{x}$
6. $12\sqrt{y}$ 7. $-2\sqrt{x}$ 8. $-8\sqrt{a}$ 9. $8\sqrt{2a}$ 10. $9\sqrt{6x}$
11. $8\sqrt{10y}$ 12. $11\sqrt{14y}$ 13. $11\sqrt{7}$ 14. $15\sqrt{5}$
15. $2\sqrt{2}$ 16. 0 17. $5\sqrt{3} + \sqrt{8} = 5\sqrt{3} + 2\sqrt{2}$ cannot be simplified further.
18. $2\sqrt{5} + \sqrt{45} = 2\sqrt{5} + \sqrt{9 \cdot 5} = 2\sqrt{5} + 3\sqrt{5} = 5\sqrt{5}$
19. $-2\sqrt{x}$ 20. $\sqrt{25a} - \sqrt{a} = 5\sqrt{a} - \sqrt{a} = 4\sqrt{a}$
21. $25\sqrt{2}$
22. $3\sqrt{12} + 2\sqrt{300} = 3\sqrt{4 \cdot 3} + 2\sqrt{100 \cdot 3}$ 23. $\sqrt{3}$
$= 3 \cdot 2\sqrt{3} + 2 \cdot 10\sqrt{3}$
$= 6\sqrt{3} + 20\sqrt{3}$
$= 26\sqrt{3}$
24. $7\sqrt{50} - 3\sqrt{2} = 7\sqrt{25 \cdot 2} - 3\sqrt{2}$ 25. $13\sqrt{2}$
$= 7 \cdot 5\sqrt{2} - 3\sqrt{2}$
$= 35\sqrt{2} - 3\sqrt{2}$
$= 32\sqrt{2}$
26. $\sqrt{45} + \sqrt{80} = \sqrt{9 \cdot 5} + \sqrt{16 \cdot 5} = 3\sqrt{5} + 4\sqrt{5} = 7\sqrt{5}$ 27. $11\sqrt{3} - 2\sqrt{2}$
28. $9\sqrt{8} - \sqrt{72} + \sqrt{98} = 9\sqrt{4 \cdot 2} - \sqrt{36 \cdot 2} + \sqrt{49 \cdot 2}$
$= 18\sqrt{2} - 6\sqrt{2} + 7\sqrt{2}$
$= 19\sqrt{2}$
29. $2\sqrt{2}$
30. $\sqrt{18} - 3\sqrt{8} + \sqrt{75}$ 31. $\sqrt{x}$
$= \sqrt{9 \cdot 2} - 3\sqrt{4 \cdot 2} + \sqrt{25 \cdot 3}$
$= 3\sqrt{2} - 6\sqrt{2} + 5\sqrt{3}$
$= -3\sqrt{2} + 5\sqrt{3}$
32. $\sqrt{16a} - 4\sqrt{a} + \sqrt{25a} = 4\sqrt{a} - 4\sqrt{a} + 5\sqrt{a} = 5\sqrt{a}$
33. $\sqrt{10} + \sqrt{14}$ 34. $\sqrt{10} + \sqrt{55}$ 35. $\sqrt{30} - 5\sqrt{2}$
36. $\sqrt{6}(\sqrt{15} - \sqrt{7}) = \sqrt{90} - \sqrt{42}$ 37. $22 + 9\sqrt{2}$
$= \sqrt{9 \cdot 10} - \sqrt{42}$
$= 3\sqrt{10} - \sqrt{42}$
38. $(5 + \sqrt{11})(3 + \sqrt{11}) = 15 + 5\sqrt{11} + 3\sqrt{11} + 11 = 26 + 8\sqrt{11}$ 39. $16 - 7\sqrt{6}$
40. $(\sqrt{10} + 4)(\sqrt{10} - 7) = 10 - 7\sqrt{10} + 4\sqrt{10} - 28 = -18 - 3\sqrt{10}$ 41. -44
42. $(1 + \sqrt{5})(1 - \sqrt{5}) = 1 - 5 = -4$ 43. 3
44. $(\sqrt{2} + \sqrt{6})(\sqrt{2} - \sqrt{6}) = 2 - 6 = -4$

45. $-1 - 2\sqrt{2}$

46. $(8 - \sqrt{7})(3 + 2\sqrt{7}) = 24 + 16\sqrt{7} - 3\sqrt{7} - 2 \cdot 7$
$= 24 + 16\sqrt{7} - 3\sqrt{7} - 14$
$= 10 + 13\sqrt{7}$

47. $52 + 14\sqrt{3}$

48. $(2 + \sqrt{5})^2 = 4 + 4\sqrt{5} + 5 = 9 + 4\sqrt{5}$

49. $13 - 4\sqrt{3}$

50. $(6 - 3\sqrt{5})^2 = 36 - 36\sqrt{5} + 9 \cdot 5$
$= 36 - 36\sqrt{5} + 45$
$= 81 - 36\sqrt{5}$

51. $x - 2\sqrt{10x} + 10$

52. $(\sqrt{a} - \sqrt{6})^2 = a - 2\sqrt{6a} + 6$ **53.** $\dfrac{45 - 9\sqrt{2}}{23}$

54. $\dfrac{2}{3 + \sqrt{5}} = \dfrac{2}{3 + \sqrt{5}} \cdot \dfrac{3 - \sqrt{5}}{3 - \sqrt{5}} = \dfrac{6 - 2\sqrt{5}}{9 - 5} =$
$\dfrac{6 - 2\sqrt{5}}{4} = \dfrac{2(3 - \sqrt{5})}{2 \cdot 2} = \dfrac{3 - \sqrt{5}}{2}$ **55.** $-4 - 2\sqrt{7}$

56. $\dfrac{3}{7 - \sqrt{2}} = \dfrac{3}{7 - \sqrt{2}} \cdot \dfrac{7 + \sqrt{2}}{7 + \sqrt{2}} = \dfrac{21 + 3\sqrt{2}}{49 - 2} =$
$\dfrac{21 + 3\sqrt{2}}{47}$ **57.** $3 - \sqrt{7}$, or $-\sqrt{7} + 3$

58. $\dfrac{6}{\sqrt{10} + 5} = \dfrac{6}{\sqrt{10} + 5} \cdot \dfrac{\sqrt{10} - 5}{\sqrt{10} - 5} = \dfrac{6\sqrt{10} - 30}{10 - 25} =$
$\dfrac{6\sqrt{10} - 30}{-15} = \dfrac{3(2\sqrt{10} - 10)}{3(-5)} = \dfrac{2\sqrt{10} - 10}{-5} =$
$-\dfrac{2\sqrt{10} - 10}{5}$ **59.** $-\dfrac{6 + 5\sqrt{6}}{19}$

60. $\dfrac{\sqrt{10}}{\sqrt{10} - 7} = \dfrac{\sqrt{10}}{\sqrt{10} - 7} \cdot \dfrac{\sqrt{10} + 7}{\sqrt{10} + 7} = \dfrac{10 + 7\sqrt{10}}{10 - 49} =$
$\dfrac{10 + 7\sqrt{10}}{-39} = -\dfrac{10 + 7\sqrt{10}}{39}$ **61.** $\dfrac{5 + \sqrt{15}}{2}$

62. $\dfrac{\sqrt{7}}{\sqrt{7} - \sqrt{5}} = \dfrac{\sqrt{7}}{\sqrt{7} - \sqrt{5}} \cdot \dfrac{\sqrt{7} + \sqrt{5}}{\sqrt{7} + \sqrt{5}} = \dfrac{7 + \sqrt{35}}{7 - 5} =$
$\dfrac{7 + \sqrt{35}}{2}$ **63.** $\dfrac{\sqrt{15} - 3}{2}$

64. $\dfrac{\sqrt{6}}{\sqrt{7} - \sqrt{6}} = \dfrac{\sqrt{6}}{\sqrt{7} - \sqrt{6}} \cdot \dfrac{\sqrt{7} + \sqrt{6}}{\sqrt{7} + \sqrt{6}} =$
$\dfrac{\sqrt{42} + 6}{7 - 6} = \sqrt{42} + 6$ **65.** $\dfrac{2\sqrt{7} + 2\sqrt{2}}{5}$

66. $\dfrac{6}{\sqrt{5} - \sqrt{3}} = \dfrac{6}{\sqrt{5} - \sqrt{3}} \cdot \dfrac{\sqrt{5} + \sqrt{3}}{\sqrt{5} + \sqrt{3}} =$
$\dfrac{6\sqrt{5} + 6\sqrt{3}}{5 - 3} = \dfrac{6\sqrt{5} + 6\sqrt{3}}{2} = 3\sqrt{5} + 3\sqrt{3}$

67. $11 + 2\sqrt{30}$

68. $\dfrac{\sqrt{10} - \sqrt{7}}{\sqrt{10} + \sqrt{7}} = \dfrac{\sqrt{10} - \sqrt{7}}{\sqrt{10} + \sqrt{7}} \cdot \dfrac{\sqrt{10} - \sqrt{7}}{\sqrt{10} - \sqrt{7}} =$
$\dfrac{10 - 2\sqrt{70} + 7}{10 - 7} = \dfrac{17 - 2\sqrt{70}}{3}$ **69.** $\dfrac{5}{11}$

70. $3(x - 4) - 2 = 8(2x + 3)$ **71.** $-1, 6$
$3x - 12 - 2 = 16x + 24$
$3x - 14 = 16x + 24$
$-13x = 38$
$x = -\dfrac{38}{13}$

72. $\quad x^2 + 10 = 7x$
$x^2 - 7x + 10 = 0$
$(x - 2)(x - 5) = 0$
$x = 2 \quad \text{or} \quad x = 5$

73. 1.6 L of Jolly Juice; 6.4 L of Real Squeeze
74. Solve $3 + 1.5h \geq 7.5$, where $h =$ the number of hours Thelma parks. The solution is $h \geq 3$, so Thelma parks for at least 3 hr. **75.** ◆ Multiply by a form of 1, using the conjugate of the numerator. **76.** ◆ It is important for the signs to differ to ensure that the product of the conjugates will be free of radicals. **77.** ◆ After multiplying to rationalize the denominator it might be necessary to add or subtract radical expressions in the numerator. **78.** ◆ If we use the rule $(A + B)^2 = A^2 + 2AB + B^2$, then we can find $(\sqrt{a} + \sqrt{b})^2$ without adding like radicals. However, the middle term, $2\sqrt{ab}$, is the result of adding $\sqrt{ab} + \sqrt{ab}$ in the FOIL method, from which the rule above is derived. **79.** $18\sqrt{2}$

80. $\sqrt{a^3b^6} - b\sqrt{a^5} + a\sqrt{a^7}$ **81.** $\dfrac{7\sqrt{x}}{2x}$
$= \sqrt{a^2b^6 \cdot a} - b\sqrt{a^4 \cdot a} + a\sqrt{a^6 \cdot a}$
$= ab^3\sqrt{a} - a^2b\sqrt{a} + a \cdot a^3\sqrt{a}$
$= ab^3\sqrt{a} - a^2b\sqrt{a} + a^4\sqrt{a}$
$= (ab^3 - a^2b + a^4)\sqrt{a}$

82. $\sqrt{ab^6} + b\sqrt{a^3} + a\sqrt{a} = \sqrt{b^6 \cdot a} + b\sqrt{a^2 \cdot a} + a\sqrt{a}$
$= b^3\sqrt{a} + ab\sqrt{a} + a\sqrt{a}$
$= (b^3 + ab + a)\sqrt{a}$

83. 0

84. $7x\sqrt{12xy^2} - 9y\sqrt{27x^3} + 5\sqrt{300x^3y^2}$
$= 7x\sqrt{4y^2 \cdot 3x} - 9y\sqrt{9x^2 \cdot 3x} + 5\sqrt{100x^2y^2 \cdot 3x}$
$= 7x \cdot 2y\sqrt{3x} - 9y \cdot 3x\sqrt{3x} + 5 \cdot 10xy\sqrt{3x}$
$= 14xy\sqrt{3x} - 27xy\sqrt{3x} + 50xy\sqrt{3x}$
$= (14xy - 27xy + 50xy)\sqrt{3x}$
$= 37xy\sqrt{3x}$

85. Any pair of numbers a, b such that $a = 0$ or $b = 0$.
86. $\sqrt{10} + \sqrt{50} = \sqrt{10} + \sqrt{10}\sqrt{5} = \sqrt{10}(1 + \sqrt{5})$
$\sqrt{10} + \sqrt{50} = \sqrt{10} + \sqrt{25 \cdot 2} = \sqrt{10} + 5\sqrt{2}$
$\sqrt{10} + \sqrt{50} = \sqrt{2}\sqrt{5} + \sqrt{2}\sqrt{25} =$
$\sqrt{2}(\sqrt{5} + \sqrt{25}) = \sqrt{2}(\sqrt{5} + 5), \quad \text{or} \quad \sqrt{2}(5 + \sqrt{5})$
All three are correct.

TECHNOLOGY CONNECTION 9.5, P. 449

1. $y_1 = 3\sqrt{x}$, $y_2 = \sqrt{x+32}$ 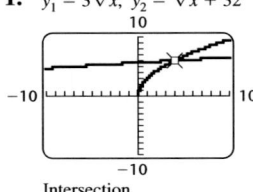
Intersection: $X = 4$, $Y = 6$

2. $y_1 = 3 + \sqrt{27-3x}$, $y_2 = x$ 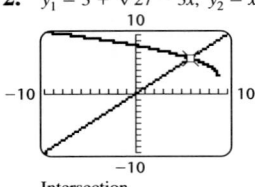
Intersection: $X = 6$, $Y = 6$

EXERCISE SET 9.5, PP. 451–453

1. 64

2. $\sqrt{x} = 7$
$x = 49$

3. 13

4. $\sqrt{x+4} = 11$
$x + 4 = 121$
$x = 117$

5. $\frac{77}{2}$

6. $\sqrt{2x+1} = 13$
$2x + 1 = 169$
$2x = 168$
$x = 84$

7. 5

8. $4 + \sqrt{y-3} = 11$
$\sqrt{y-3} = 7$
$y - 3 = 49$
$y = 52$

9. 3

10. $8 - 4\sqrt{5n} = 0$
$8 = 4\sqrt{5n}$
$2 = \sqrt{5n}$
$4 = 5n$
$\frac{4}{5} = n$

11. 3

12. $\sqrt{4x-5} = \sqrt{x+9}$
$4x - 5 = x + 9$
$3x = 14$
$x = \frac{14}{3}$

13. No solution

14. $\sqrt{x} = -9$
The principal square root of x cannot be negative. There is no solution.

15. No solution

16. $\sqrt{3x-5} = \sqrt{2x+1}$
$3x - 5 = 2x + 1$
$x = 6$

17. 9

18. $\sqrt{3x-2} = x - 4$
$3x - 2 = x^2 - 8x + 16$
$0 = x^2 - 11x + 18$
$0 = (x - 2)(x - 9)$
$x = 2$ or $x = 9$
Only 9 checks.

19. 7

20. $x - 9 = \sqrt{x-3}$
$x^2 - 18x + 81 = x - 3$
$x^2 - 19x + 84 = 0$
$(x - 12)(x - 7) = 0$
$x = 12$ or $x = 7$
Only 12 checks.

21. 5

22. $x - 1 = 6\sqrt{x-9}$
$x^2 - 2x + 1 = 36(x - 9)$
$x^2 - 2x + 1 = 36x - 324$
$x^2 - 38x + 325 = 0$
$(x - 13)(x - 25) = 0$
$x = 13$ or $x = 25$

23. 3

24. $\sqrt{22-x} = x - 2$
$22 - x = x^2 - 4x + 4$
$0 = x^2 - 3x - 18$
$0 = (x - 6)(x + 3)$
$x = 6$ or $x = -3$
Only 6 checks.

25. 0, 8

26. $1 + 2\sqrt{x-1} = x$
$2\sqrt{x-1} = x - 1$
$4(x - 1) = x^2 - 2x + 1$
$4x - 4 = x^2 - 2x + 1$
$0 = x^2 - 6x + 5$
$0 = (x - 1)(x - 5)$
$x = 1$ or $x = 5$

27. No solution

28. $\sqrt{x^2+5} - x + 2 = 0$
$\sqrt{x^2+5} = x - 2$
$x^2 + 5 = x^2 - 4x + 4$
$1 = -4x$
$-\frac{1}{4} = x$
$-\frac{1}{4}$ does not check. There is no solution.

29. 3

30. $\sqrt{(4x+5)(x+4)} = 2x + 5$
$(4x + 5)(x + 4) = 4x^2 + 20x + 25$
$4x^2 + 21x + 20 = 4x^2 + 20x + 25$
$x = 5$

31. $\frac{7}{3}$

32. $\sqrt{7-3x} = \sqrt{x-2}$
$7 - 3x = x - 2$
$-4x = -9$
$x = \frac{9}{4}$

33. 3

34. $x = 1 + \sqrt{1-x}$
$x - 1 = \sqrt{1-x}$
$x^2 - 2x + 1 = 1 - x$
$x^2 - x = 0$
$x(x - 1) = 0$
$x = 0$ or $x = 1$
Only 1 checks.

35. 36 m

36. $84 = 3.5\sqrt{h}$
$24 = \sqrt{h}$
$576 = h$
The altitude of the scout's eyes is 576 m.

37. 11,664 m

38. $99.4 = 3.5\sqrt{h}$
$28.4 = \sqrt{h}$
$806.56 = h$
The mast is 806.56 m high.

39. About 2.08 ft

40.
$$T = 2\pi\sqrt{\frac{L}{32}}$$
$$3 = 2(3.14)\sqrt{\frac{L}{32}} \quad \text{Substituting 3 for } T \text{ and 3.14 for } \pi$$
$$3 = 6.28\sqrt{\frac{L}{32}}$$
$$(3)^2 = \left(6.28\sqrt{\frac{L}{32}}\right)^2$$
$$9 = \frac{39.4384L}{32}$$
$$\frac{32 \cdot 9}{39.4384} = L$$
$$7.30 \approx L$$

The pendulum is about 7.30 ft long.

41. 80 ft, 320 ft

42.
$$48 = 2\sqrt{5L}$$
$$24 = \sqrt{5L}$$
$$576 = 5L$$
$$115.2 = L$$

The car will skid 115.2 ft at 48 mph.

$$60 = 2\sqrt{5L}$$
$$30 = \sqrt{5L}$$
$$900 = 5L$$
$$180 = L$$

The car will skid 180 ft at 60 mph.

43. 121

44. Let x represent the number.
Solve $\sqrt{2x} - 1 = 7$.
$x = 32$

45. -32 **46.** $(-5)^3 = -125$ **47.** $\dfrac{x^6}{3}$

48. $\dfrac{3}{x^2 - 9} \cdot \dfrac{x^2 - 6x + 9}{12} = \dfrac{3(x-3)(x-3)}{(x+3)(x-3)(3)(4)} = \dfrac{x-3}{4(x+3)}$

49. $x - 4$

50. $\dfrac{9}{a-1} - \dfrac{4}{1-a} = \dfrac{9}{a-1} - \dfrac{-4}{a-1} = \dfrac{9-(-4)}{a-1} = \dfrac{13}{a-1}$

51. ◆ One method is to write an equation of the form $\sqrt{ax} = b$, where $b < 0$.   **52.** ◆ A radical term would remain after the principle of squaring was used. We could still have solved the equation by isolating the radical at this point and proceeding using the techniques of this section.

53. ◆ The square of a number is equal to the square of its opposite. Thus, while squaring both sides of a radical equation allows us to find the solutions of the original equation, this procedure can also introduce numbers that are not solutions of the original equation.   **54.** ◆ Yes; if $a = b$, then by the multiplication principle we can multiply on the left by a n times and on the right by b n times and to produce an equivalent equation.   **55.** 9

56.
$$1 + \sqrt{x} = \sqrt{x+9}$$
$$1 + 2\sqrt{x} + x = x + 9$$
$$2\sqrt{x} = 8$$
$$\sqrt{x} = 4$$
$$x = 16$$

57. No solution

58.
$$\sqrt{y+8} - \sqrt{y} = 2$$
$$\sqrt{y+8} = \sqrt{y} + 2$$
$$y + 8 = y + 4\sqrt{y} + 4$$
$$4 = 4\sqrt{y}$$
$$1 = \sqrt{y}$$
$$1 = y$$

59. $-\dfrac{57}{16}$

60.
$$\sqrt{y+1} - \sqrt{2y-5} = \sqrt{y-2}$$
$$y + 1 - 2\sqrt{(y+1)(2y-5)} + 2y - 5 = y - 2$$
$$2y - 2 = 2\sqrt{2y^2 - 3y - 5}$$
$$y - 1 = \sqrt{2y^2 - 3y - 5}$$
$$y^2 - 2y + 1 = 2y^2 - 3y - 5$$
$$0 = y^2 - y - 6$$
$$0 = (y-3)(y+2)$$
$$y = 3 \quad \text{or} \quad y = -2$$

Only 3 checks.

61. 10

62.
$$x + (2-x)\sqrt{x} = 0$$
$$x = -(2-x)\sqrt{x}$$
$$x^2 = (4 - 4x + x^2)(x)$$
$$x^2 = 4x - 4x^2 + x^3$$
$$0 = x^3 - 5x^2 + 4x$$
$$0 = x(x^2 - 5x + 4)$$
$$0 = x(x-1)(x-4)$$
$$x = 0 \quad \text{or} \quad x = 1 \quad \text{or} \quad x = 4$$

Only 0 and 4 check.

63. 34.726 m

64.
$$A = \sqrt{1 + \sqrt{a/b}}$$
$$A^2 = 1 + \sqrt{a/b}$$
$$A^2 - 1 = \sqrt{a/b}$$
$$A^4 - 2A^2 + 1 = \dfrac{a}{b}$$
$$b = \dfrac{a}{A^4 - 2A^2 + 1}$$

65.

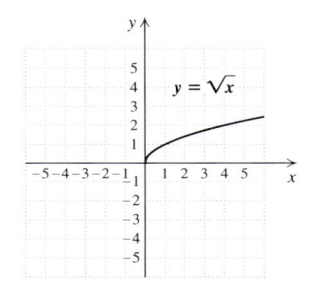

$y = \sqrt{x}$

66.

$y = \sqrt{x-2}$

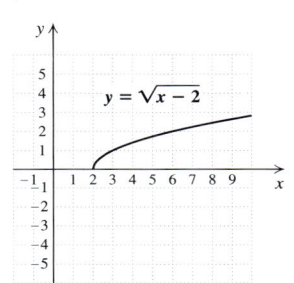

67. [graph of $y = \sqrt{x} - 3$]

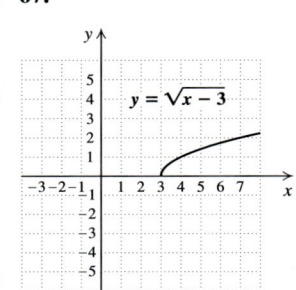

68. [graph of $y = \sqrt{x} + 1$]

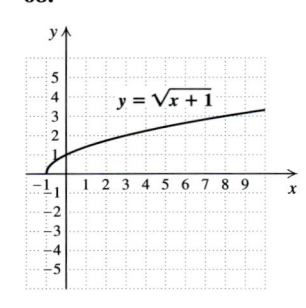

69. [graph showing $y = \sqrt{x-5}$ and $y = x - 7$ intersecting at $(9, 2)$]; 9

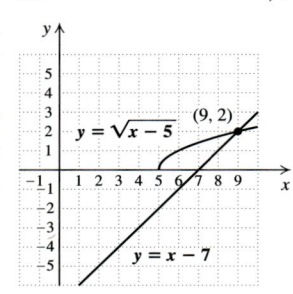

70. [graph showing $y = \sqrt{x} + 9$ and $y = 1 + \sqrt{x}$ intersecting at $(16, 5)$] The solution is 16.

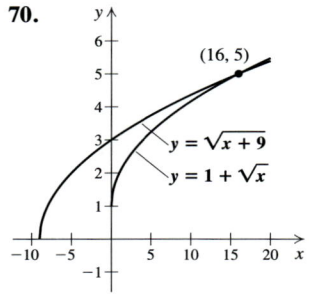

71. 1.57

72. Graph $y_1 = -\sqrt{x+3}$ and $y_2 = 2x - 1$ and then find the first coordinate(s) of the point(s) of intersection. The solution is about -0.32.

EXERCISE SET 9.6, PP. 457–460

1. 17 **2.** $3^2 + 5^2 = c^2$
$9 + 25 = c^2$
$34 = c^2$
$\sqrt{34} = c$
$5.831 \approx c$

3. $\sqrt{72} \approx 8.485$

4. $7^2 + 7^2 = c^2$
$49 + 49 = c^2$
$98 = c^2$
$\sqrt{98} = c$
$9.899 \approx c$

5. 12 **6.** $a^2 + 12^2 = 13^2$
$a^2 + 144 = 169$
$a^2 = 25$
$a = 5$

7. 6 **8.** $(\sqrt{5})^2 + b^2 = 6^2$
$5 + b^2 = 36$
$b^2 = 31$
$b = \sqrt{31}$
$b \approx 5.568$

9. 26

10. $5^2 + 12^2 = c^2$
$25 + 144 = c^2$
$169 = c^2$
$13 = c$

11. 12 **12.** $18^2 + b^2 = 30^2$
$324 + b^2 = 900$
$b^2 = 576$
$b = 24$

13. 2 **14.** $a^2 + 1^2 = (\sqrt{2})^2$
$a^2 + 1 = 2$
$a^2 = 1$
$a = 1$

15. $\sqrt{2} \approx 1.414$

16. $(\sqrt{3})^2 + (\sqrt{5})^2 = c^2$
$3 + 5 = c^2$
$8 = c^2$
$\sqrt{8} = c$
$2.828 \approx c$

17. 5 **18.** $5^2 + 5^2 = c^2$
$25 + 25 = c^2$
$50 = c^2$
$\sqrt{50} = c$
$7.071 \approx c$

19. $\sqrt{147} \approx 12.124$ m

20. [square with side 4 cm and diagonal d]
Solve $4^2 + 4^2 = d^2$.
$d = \sqrt{32}$ cm ≈ 5.657 cm

21. $\sqrt{208} \approx 14.422$ ft

22. [right triangle with legs 9 m and 13 m, hypotenuse w]
Solve $9^2 + 13^2 = w^2$.
$w = \sqrt{250}$ m ≈ 15.811 m

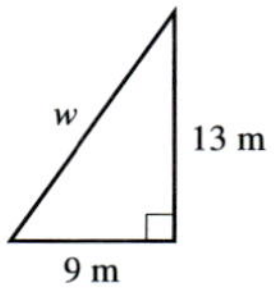

23. $\sqrt{12,500} \approx 111.803$ yd

24. [rectangle 130 yd by 100 yd with diagonal d]
Solve $100^2 + 130^2 = d^2$.
$d = \sqrt{26,900}$ yd ≈ 164.012 yd

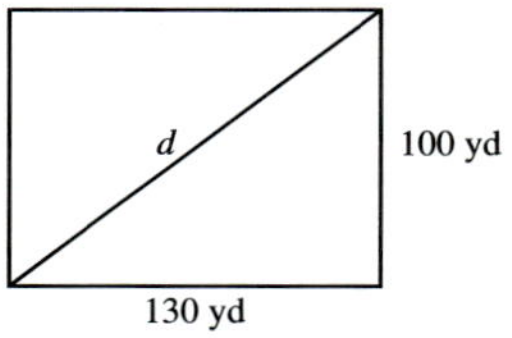

25. 240 ft

26. Solve $90^2 + 90^2 = d^2$.
$d = \sqrt{16,200}$ ft ≈ 127.279 ft

27. $\sqrt{2,600,000,000} \approx 50,990.195$ ft

28. Solve $5^2 + 7^2 = d^2$.
$d = \sqrt{74}$ km ≈ 8.602 km

29. $-\frac{1}{3}$

30. $m = \dfrac{-8 - (-3)}{0 - 8} = \dfrac{-5}{-8} = \dfrac{5}{8}$

31. $\{x \mid x > -25\}$

32. $-2x + 6 \geq 7x - 3$
$9 \geq 9x$
$1 \geq x$
$\{x \mid x \leq 1\}$

33. $81x^4$

34. $\left(-\dfrac{2}{5}\right)^3 = -\dfrac{8}{125}$

35. ◆ No; $a^2 + b^2 \neq (a + b)^2$.

36. ◆ No; consider an equilateral triangle with side s. If this were a right triangle, then it would hold that $s^2 + s^2 = s^2$, or $2s^2 = s^2$. This is true only for $s = 0$, so a right triangle cannot be equilateral.

37. ◆ Yes; consider an isosceles triangle with two sides of length s. Then a triangle whose third side has length $s^2 + s^2$, or $2s^2$, is an isosceles right triangle. **38.** ◆ No; consider the clapboard's height above ground level to be one leg of a right triangle. Then the length of the ladder is the hypotenuse of that triangle. Since the length of the hypotenuse must be greater than the length of a leg, a 28-ft ladder cannot be used to repair a clapboard that is 28 ft above ground level. **39.** Yes
40. Solve $s^2 + s^2 = (8\sqrt{2})^2$, where s is the length of a side of the square, in feet.
$$s = 8 \text{ ft}$$
41. $\sqrt{7} \approx 2.646$ m
42. Solve $x^2 + (x+1)^2 = (x+2)^2$, where x, $x+1$, and $x+2$ represent the lengths of the sides
We find $x = 3$, so the lengths of the sides are 3, 4, and 5.
43. $s\sqrt{3}$
44. Solve $\left(\dfrac{\sqrt{2}}{3}\right)^2 + \left(\dfrac{\sqrt{2}}{3}\right)^2 = c^2$, where c is the length of a diagonal of the square.
$$c = \dfrac{2}{3}$$
45. $h = \dfrac{a}{2}\sqrt{3}$
46.   **47.** $\sqrt{1525} \approx 39.051$ mi

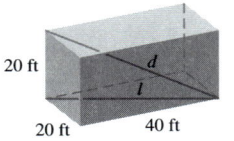

First find l^2:
$$20^2 + 40^2 = l^2$$
$$2000 = l^2$$
Then find d:
$$20^2 + l^2 = d^2$$
$$400 + 2000 = d^2$$
$$2400 = d^2$$
$$\sqrt{2400} = d$$
$$48.990 \approx d$$
The longest straight-line distance that can be measured is $\sqrt{2400}$ ft ≈ 48.990 ft.
48.

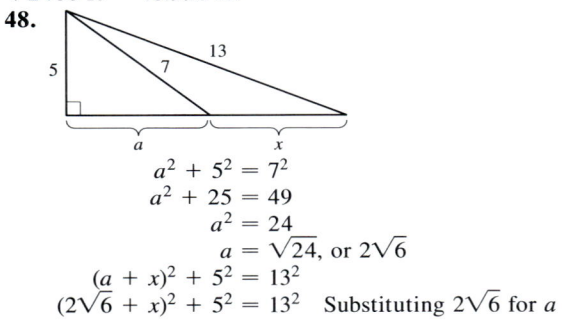

$$a^2 + 5^2 = 7^2$$
$$a^2 + 25 = 49$$
$$a^2 = 24$$
$$a = \sqrt{24}, \text{ or } 2\sqrt{6}$$
$$(a+x)^2 + 5^2 = 13^2$$
$$(2\sqrt{6} + x)^2 + 5^2 = 13^2 \quad \text{Substituting } 2\sqrt{6} \text{ for } a$$

$$(2\sqrt{6} + x)^2 + 25 = 169$$
$$(2\sqrt{6} + x)^2 = 144$$
$$2\sqrt{6} + x = 12 \quad \text{Taking the principal}$$
$$\qquad\qquad\qquad \text{square root}$$
$$x = 12 - 2\sqrt{6}$$
$$x \approx 7.101$$
49. 6
50. If the area of square $PQRS$ is 100 ft², then each side measures 10 ft. If A, B, C, and D are midpoints, then each of the segments PB, BQ, QC, CR, RD, DS, SA, and AP measures 5 ft. We can label the figure with additional information.

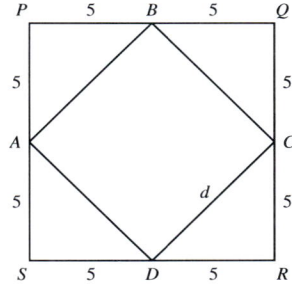

We label a side of the square $ABCD$ with d. Then we use the Pythagorean theorem.
$$5^2 + 5^2 = d^2$$
$$25 + 25 = d^2$$
$$50 = d^2$$
$$\sqrt{50} = d$$
If a side of square $ABCD$ is $\sqrt{50}$, then its area is $\sqrt{50} \cdot \sqrt{50}$, or 50 ft².
51. 640 acres

EXERCISE SET 9.7, PP. 465–466
1. -4 **2.** -2 **3.** 10 **4.** -3 **5.** -5 **6.** -2
7. 6 **8.** -7 **9.** 5 **10.** 3 **11.** 0 **12.** 1 **13.** -1
14. -3 **15.** Not a real number **16.** Not a real number
17. 10 **18.** 10 **19.** 5 **20.** 7 **21.** 2 **22.** 1 **23.** a
24. n **25.** $2\sqrt[3]{4}$ **26.** $\sqrt[3]{54} = \sqrt[3]{27 \cdot 2} = 3\sqrt[3]{2}$
27. $2\sqrt[4]{3}$ **28.** $\sqrt[5]{160} = \sqrt[5]{32 \cdot 5} = 2\sqrt[5]{5}$ **29.** $\dfrac{4}{5}$
30. $\dfrac{5}{3}$ **31.** $\dfrac{2}{3}$ **32.** $\dfrac{5}{4}$ **33.** $\dfrac{\sqrt[3]{17}}{2}$ **34.** $\dfrac{\sqrt[5]{11}}{2}$
35. $\dfrac{\sqrt[4]{13}}{3}$ **36.** $\dfrac{\sqrt[3]{10}}{3}$ **37.** 5 **38.** 3 **39.** 10 **40.** 5
41. 2 **42.** 2 **43.** 8 **44.** $8^{4/3} = (8^{1/3})^4 = 2^4 = 16$
45. 32 **46.** $9^{3/2} = (9^{1/2})^3 = 3^3 = 27$ **47.** 16
48. $32^{2/5} = (32^{1/5})^2 = 2^2 = 4$ **49.** 32
50. $16^{5/4} = (16^{1/4})^5 = 2^5 = 32$ **51.** 3125
52. $4^{-1/2} = \dfrac{1}{4^{1/2}} = \dfrac{1}{2}$ **53.** $\dfrac{1}{6}$ **54.** $32^{-1/5} = \dfrac{1}{32^{1/5}} = \dfrac{1}{2}$

55. $\frac{1}{4}$ **56.** $100^{-3/2} = \frac{1}{100^{3/2}} = \frac{1}{(\sqrt{100})^3} = \frac{1}{10^3} = \frac{1}{1000}$

57. $\frac{1}{27}$ **58.** $16^{-3/4} = \frac{1}{16^{3/4}} = \frac{1}{(\sqrt[4]{16})^3} = \frac{1}{2^3} = \frac{1}{8}$

59. $\frac{1}{243}$ **60.** $32^{-2/5} = \frac{1}{32^{2/5}} = \frac{1}{(\sqrt[5]{32})^2} = \frac{1}{2^2} = \frac{1}{4}$

61. $\frac{1}{4}$ **62.** $625^{-3/4} = \frac{1}{625^{3/4}} = \frac{1}{(\sqrt[4]{625})^3} = \frac{1}{5^3} = \frac{1}{125}$

63. $(x-5)^2$ **64.** $a^2 - 12a + 36 = (a-6)^2$

65. $(t+9)^2$ **66.** $\begin{aligned} y &= kx \\ 2.5 &= k \cdot 4 \\ 0.625 &= k \\ y &= 0.625x \quad \text{Equation of variation} \end{aligned}$

67. $\frac{-2}{a-5}$

68. $\frac{x}{x^2-9} \cdot \frac{x^2-4x+3}{3x^4} = \frac{x(x-3)(x-1)}{(x+3)(x-3)3 \cdot x \cdot x^3}$
$= \frac{x-1}{3x^3(x+3)}$

69. ◆ The product of an odd number of negative factors is negative. **70.** ◆ You would probably use $(\sqrt[n]{a})^m$, because it is easier to compute $(\sqrt{25})^3 = 5^3 = 125$ than $\sqrt{25^3} = \sqrt{15,625} = 125$. **71.** ◆ Yes; we can think of a^n as the product of n factors of $a^{1/n}$ and of b^n as the product of n factors of $b^{1/n}$. Then, if $a > b$, each factor of $a^{1/n}$ must be greater than each factor of $b^{1/n}$, or $a^{1/n} > b^{1/n}$.
72. ◆ $a^{-3/5}$ will be negative when $a < 0$. **73.** 6.310
74. 2.213 **75.** 31.623 **76.** 9.391 **77.** $x^{14/9}$
78. $a^{1/4}a^{3/2} = a^{1/4+6/4} = a^{7/4}$ **79.** $p^{2/15}$
80. $m^{-2/3}m^{3/4}m^{1/2} = m^{-2/3+3/4+1/2} = m^{-8/12+9/12+6/12} = m^{7/12}$

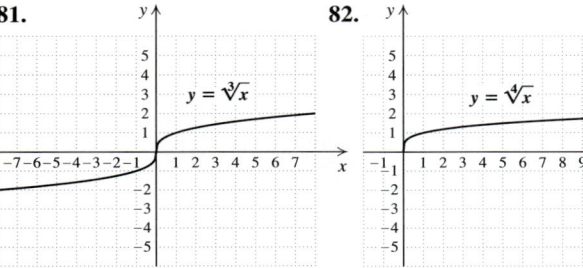

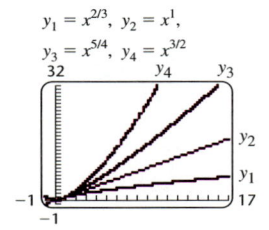

81.

82.

83.

REVIEW EXERCISES: CHAPTER 9, PP. 467–468

1. [9.1] 5, −5 **2.** [9.1] 14, −14 **3.** [9.1] 30, −30
4. [9.1] 15, −15 **5.** [9.1] −12 **6.** [9.1] 9 **7.** [9.1] 7
8. [9.1] −13 **9.** [9.1] $2x^2y$ **10.** [9.1] $\frac{a}{b}$
11. [9.1] Rational **12.** [9.1] Irrational
13. [9.1] Irrational **14.** [9.1] Rational **15.** [9.1] 1.732
16. [9.1] 9.950 **17.** [9.1] p **18.** [9.1] $7x$ **19.** [9.1] $4m$
20. [9.1] ac **21.** [9.2] $4\sqrt{3}$ **22.** [9.2] $7t\sqrt{2}$
23. [9.2] $4\sqrt{2p}$ **24.** [9.2] x^3 **25.** [9.2] $2a^6\sqrt{3a}$
26. [9.2] $6m^7\sqrt{m}$ **27.** [9.2] $\sqrt{33}$ **28.** [9.2] $2\sqrt{15}$
29. [9.2] $\sqrt{35xt}$ **30.** [9.2] $2a\sqrt{6}$ **31.** [9.2] $5xy\sqrt{2}$
32. [9.2] $10a^2b\sqrt{ab}$ **33.** [9.3] $\frac{\sqrt{7}}{3}$ **34.** [9.3] $\frac{y^4\sqrt{5}}{3}$
35. [9.3] $\frac{5}{8}$ **36.** [9.3] $\frac{2}{3}$ **37.** [9.3] $\frac{7}{t}$ **38.** [9.4] $13\sqrt{5}$
39. [9.4] $\sqrt{5}$ **40.** [9.4] $-3\sqrt{x}$ **41.** [9.4] $7 + 4\sqrt{3}$
42. [9.4] 1 **43.** [9.4] $-11 + 5\sqrt{7}$ **44.** [9.3] $\frac{\sqrt{2}}{2}$
45. [9.3] $\frac{\sqrt{10}}{4}$ **46.** [9.3] $\frac{\sqrt{5y}}{y}$ **47.** [9.3] $\frac{2\sqrt{3}}{3}$
48. [9.4] $8 - 4\sqrt{3}$ **49.** [9.4] $-7 - 3\sqrt{5}$ **50.** [9.5] 52
51. [9.5] No solution **52.** [9.5] 4 **53.** [9.5] 0, 3
54. [9.5] 405 ft **55.** [9.6] 20 **56.** [9.6] $\sqrt{3} \approx 1.732$
57. [9.6] $\sqrt{10,625} \approx 103.078$ ft **58.** [9.7] 2
59. [9.7] Not a real number **60.** [9.7] −3
61. [9.7] $2\sqrt[4]{2}$ **62.** [9.7] 10 **63.** [9.7] $\frac{1}{3}$ **64.** [9.7] 64
65. [9.7] $\frac{1}{27}$ **66.** [8.3] (13, −3) **67.** [2.6] $\{x| x < -\frac{17}{2}\}$
68. [7.1] $\frac{2}{5}$; $(0, -\frac{14}{5})$ **69.** [6.2] $\frac{(x+1)(x+2)}{x(x-2)}$
70. [9.5] ◆ He could be mistakenly assuming that negative solutions are not possible because principal square roots are never negative. He could also be assuming that substituting a negative value for the variable in a radicand always produces a negative radicand.
71. [9.4] ◆ Some radical terms that are like terms may not appear to be so until they are in simplified form.
72. [9.1] 2 **73.** [9.1] No solution
74. [9.4] $(x + \sqrt{5})(x - \sqrt{5})$
75. [9.5] $b = \sqrt{A^2 - a^2}$ or $b = -\sqrt{A^2 - a^2}$

TEST: CHAPTER 9, P. 469

1. [9.1] 9, −9 **2.** [9.1] 8 **3.** [9.1] −5 **4.** [9.1] $x + 4$
5. [9.1] Irrational **6.** [9.1] Rational **7.** [9.1] 9.327
8. [9.1] 2.646 **9.** [9.1] a **10.** [9.1] $6y$ **11.** [9.2] $2\sqrt{6}$
12. [9.2] $3x^3\sqrt{3}$ **13.** [9.2] $2t^2\sqrt{t}$ **14.** [9.2] $\sqrt{30}$
15. [9.2] $5\sqrt{2}$ **16.** [9.2] $\sqrt{14xy}$ **17.** [9.2] $4t$
18. [9.2] $3ab^2\sqrt{2}$ **19.** [9.3] $\frac{3}{4}$ **20.** [9.3] $\frac{\sqrt{7}}{4y}$

21. [9.3] $\dfrac{3}{2}$ **22.** [9.3] $\dfrac{12}{a}$ **23.** [9.4] $-6\sqrt{2}$
24. [9.4] $7\sqrt{3}$ **25.** [9.4] $21 - 8\sqrt{5}$ **26.** [9.4] 11
27. [9.3] $\dfrac{\sqrt{10}}{5}$ **28.** [9.3] $\dfrac{2x\sqrt{y}}{y}$ **29.** [9.4] $\dfrac{40 + 10\sqrt{5}}{11}$
30. [9.6] $\sqrt{80} \approx 8.944$ cm **31.** [9.5] 48
32. [9.5] $-2, 2$ **33.** [9.5] About 5000 m **34.** [9.7] 2
35. [9.7] -1 **36.** [9.7] -4 **37.** [9.7] Not a real number
38. [9.7] 3 **39.** [9.7] $\dfrac{1}{3}$ **40.** [9.7] 1000 **41.** [9.7] $\dfrac{1}{32}$
42. [1.8] -27 **43.** [7.1] $y = -\dfrac{1}{2}x - 1$
44. [6.2] $\dfrac{(a+1)^2}{a}$ **45.** [6.4] $\dfrac{x+4}{3x-5}$ **46.** [9.5] -3
47. [9.2] y^{8n}

CHAPTER 10

TECHNOLOGY CONNECTION 10.1, P. 473

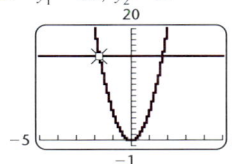

1. $y_1 = 5x^2, y_2 = 15$
Intersection X = −1.732051, Y = 15

$y_1 = 5x^2, y_2 = 15$
Intersection X = 1.7320508, Y = 15

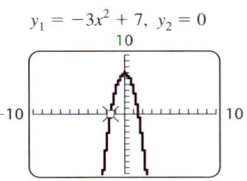

$y_1 = -3x^2 + 7, y_2 = 0$
Intersection X = −1.527525, Y = 0

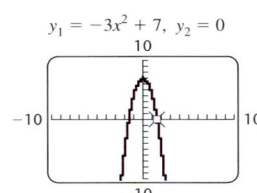

$y_1 = -3x^2 + 7, y_2 = 0$
Intersection X = 1.5275252, Y = 0

EXERCISE SET 10.1, PP. 475–476

1. $10, -10$ **2.** $x^2 = 25$ **3.** $6, -6$
$\qquad x = \sqrt{25}$ or $x = -\sqrt{25}$
$\qquad x = 5$ or $x = -5$
4. $a^2 = 49$
$\quad a = \sqrt{49}$ or $a = -\sqrt{49}$
$\quad a = 7$ or $a = -7$
5. $\sqrt{23}, -\sqrt{23}$
6. $n^2 = 13$ **7.** $\sqrt{10}, -\sqrt{10}$
$\quad n = \sqrt{13}$ or $n = -\sqrt{13}$
8. $5x^2 = 70$ **9.** $2\sqrt{2}, -2\sqrt{2}$
$\quad x^2 = 14$
$\quad x = \sqrt{14}$ or $x = -\sqrt{14}$

10. $7t^2 = 140$ **11.** $\dfrac{2}{3}, -\dfrac{2}{3}$
$\quad t^2 = 20$
$\quad t = \sqrt{20}$ or $t = -\sqrt{20}$
$\quad t = 2\sqrt{5}$ or $t = -2\sqrt{5}$
12. $25 - 4a^2 = 0$ **13.** $\dfrac{2\sqrt{5}}{7}, -\dfrac{2\sqrt{5}}{7}$
$\quad 25 = 4a^2$
$\quad \dfrac{25}{4} = a^2$
$\quad a = \sqrt{\dfrac{25}{4}}$ or $a = -\sqrt{\dfrac{25}{4}}$
$\quad a = \dfrac{5}{2}$ or $a = -\dfrac{5}{2}$
14. $4y^2 - 3 = 9$ **15.** $2\sqrt{6}, -2\sqrt{6}$
$\quad 4y^2 = 12$
$\quad y^2 = 3$
$\quad y = \sqrt{3}$ or $y = -\sqrt{3}$
16. $25x^2 - 35 = 0$
$\quad 25x^2 = 35$
$\quad x^2 = \dfrac{35}{25}$
$\quad x = \sqrt{\dfrac{35}{25}}$ or $x = -\sqrt{\dfrac{35}{25}}$
$\quad x = \sqrt{\dfrac{7}{5}}$ or $x = -\sqrt{\dfrac{7}{5}}$
If you wish to rationalize denominators, these answers can also be written $\dfrac{\sqrt{35}}{5}$ and $-\dfrac{\sqrt{35}}{5}$.
17. $4, -6$ **18.** $(x - 2)^2 = 49$
$\qquad\qquad x - 2 = 7$ or $x - 2 = -7$
$\qquad\qquad x = 9$ or $x = -5$
19. $13, -5$ **20.** $(x + 3)^2 = 36$
$\qquad\qquad x + 3 = 6$ or $x + 3 = -6$
$\qquad\qquad x = 3$ or $x = -9$
21. $-3 \pm \sqrt{6}$
22. $(m - 4)^2 = 21$
$\quad m - 4 = \sqrt{21}$ or $m - 4 = -\sqrt{21}$
$\quad m = 4 + \sqrt{21}$ or $m = 4 - \sqrt{21}$
$\quad m = 4 \pm \sqrt{21}$
23. $21, 5$
24. $(a + 13)^2 = 8$
$\quad a + 13 = \sqrt{8}$ or $a + 13 = -\sqrt{8}$
$\quad a + 13 = 2\sqrt{2}$ or $a + 13 = -2\sqrt{2}$
$\quad a = -13 + 2\sqrt{2}$ or $a = -13 - 2\sqrt{2}$
$\quad a = -13 \pm 2\sqrt{2}$
25. $5 \pm \sqrt{14}$

26. $(x-7)^2 = 12$
$x - 7 = \sqrt{12}$ or $x - 7 = -\sqrt{12}$
$x - 7 = 2\sqrt{3}$ or $x - 7 = -2\sqrt{3}$
$x = 7 + 2\sqrt{3}$ or $x = 7 - 2\sqrt{3}$
$x = 7 \pm 2\sqrt{3}$

27. $3, -7$

28. $(x + 9)^2 = 34$
$x + 9 = \sqrt{34}$ or $x + 9 = -\sqrt{34}$
$x = -9 + \sqrt{34}$ or $x = -9 - \sqrt{34}$
$x = -9 \pm \sqrt{34}$

29. $\dfrac{3 \pm \sqrt{17}}{4}$

30. $\left(x + \dfrac{3}{2}\right)^2 = \dfrac{7}{2}$
$x + \dfrac{3}{2} = \sqrt{\dfrac{7}{2}}$ or $x + \dfrac{3}{2} = -\sqrt{\dfrac{7}{2}}$
$x = -\dfrac{3}{2} + \sqrt{\dfrac{7}{2}}$ or $x = -\dfrac{3}{2} - \sqrt{\dfrac{7}{2}}$
$x = -\dfrac{3}{2} + \dfrac{\sqrt{14}}{2}$ or $x = -\dfrac{3}{2} - \dfrac{\sqrt{14}}{2}$
$x = \dfrac{-3 + \sqrt{14}}{2}$ or $x = \dfrac{-3 - \sqrt{14}}{2}$
$x = \dfrac{-3 \pm \sqrt{14}}{2}$

31. $5, -15$

32. $x^2 - 6x + 9 = 64$
$(x - 3)^2 = 64$
$x - 3 = 8$ or $x - 3 = -8$
$x = 11$ or $x = -5$

33. $5, 3$

34. $y^2 + 14y + 49 = 4$
$(y + 7)^2 = 4$
$y + 7 = 2$ or $y + 7 = -2$
$y = -5$ or $y = -9$

35. $-3 \pm \sqrt{13}$

36. $m^2 - 2m + 1 = 5$
$(m - 1)^2 = 5$
$m - 1 = \sqrt{5}$ or $m - 1 = -\sqrt{5}$
$m = 1 + \sqrt{5}$ or $m = 1 - \sqrt{5}$
$m = 1 \pm \sqrt{5}$

37. $6 \pm 3\sqrt{2}$

38. $x^2 + 4x + 4 = 12$
$(x + 2)^2 = 12$
$x + 2 = \sqrt{12}$ or $x + 2 = -\sqrt{12}$
$x = -2 + \sqrt{12}$ or $x = -2 - \sqrt{12}$
$x = -2 + 2\sqrt{3}$ or $x = -2 - 2\sqrt{3}$
$x = -2 \pm 2\sqrt{3}$

39. Yes

40. $y - 4x = 6$ $x + 4y = 8$
$y = 4x + 6$ $4y = -x + 8$
$\qquad\qquad\qquad y = -\dfrac{1}{4}x + 2$

Since the slopes are different, the equations do not represent parallel lines.

41. Yes

42. $2x - 3y = 12$ $3y - 2x = 24$
$-3y = -2x + 12$ $3y = 2x + 24$
$y = \dfrac{2}{3}x - 4$ $y = \dfrac{2}{3}x + 8$

Since the slopes are the same and the y-intercepts are different, the equations represent parallel lines.

43. $x^2 - \dfrac{3}{4}x + \dfrac{9}{64}$ **44.** $\left(t + \dfrac{5}{4}\right)^2 = t^2 + \dfrac{5}{2}t + \dfrac{25}{16}$

45. ◆ The solutions will be rational numbers because each is the solution of a linear equation of the form $bx + a = 0$.

46. ◆ By the principle of square roots, if $x^2 = k$, then $x = \sqrt{k}$ or $x = -\sqrt{k}$. Thus -9 is also a solution.

47. ◆ Answers may vary. One such equation is $x^2 + 2x + 1 = 5$. In order to solve this equation using the principle of zero products it would be necessary to factor $x^2 + 2x - 4$ as $(x + 1 - \sqrt{5})(x + 1 + \sqrt{5})$.

48. ◆ She does not recognize that the $\pm$ sign yields two solutions, one in which the radical is added to 3 and the other in which the radical is subtracted from 3.

49. $\dfrac{-7 \pm \sqrt{7}}{6}$

50. $x^2 - 5x + \dfrac{25}{4} = \dfrac{13}{4}$
$\left(x - \dfrac{5}{2}\right)^2 = \dfrac{13}{4}$
$x - \dfrac{5}{2} = \dfrac{\sqrt{13}}{2}$ or $x - \dfrac{5}{2} = -\dfrac{\sqrt{13}}{2}$
$x = \dfrac{5}{2} + \dfrac{\sqrt{13}}{2}$ or $x = \dfrac{5}{2} - \dfrac{\sqrt{13}}{2}$
$x = \dfrac{5 \pm \sqrt{13}}{2}$

51. $\dfrac{3 \pm \sqrt{17}}{4}$

52. $t^2 + 3t + \dfrac{9}{4} = \dfrac{49}{4}$
$\left(t + \dfrac{3}{2}\right)^2 = \dfrac{49}{4}$
$t + \dfrac{3}{2} = \dfrac{7}{2}$ or $t + \dfrac{3}{2} = -\dfrac{7}{2}$
$t = \dfrac{4}{2}$ or $t = -\dfrac{10}{2}$
$t = 2$ or $t = -5$

53. 1.85, −4.35
54. $a^2 − 3.8a + 3.61 = 27.04$
$(a − 1.9)^2 = 27.04$
$a − 1.9 = 5.2$ or $a − 1.9 = −5.2$
$a = 7.1$ or $a = −3.3$
55. −5, −1
56. From the graph we see that when $y = 9$, then $x = −6$ or $x = 0$. Thus, the solutions of $(x + 3)^2 = 9$ are −6 and 0.
57. $d = \sqrt{\dfrac{kMm}{f}}$

EXERCISE SET 10.2, PP. 480–481

1. $x^2 + 8x + 16$ **2.** $x^2 + 4x$
$\left(\dfrac{4}{2}\right)^2 = 2^2 = 4$
$x^2 + 4x + 4$ is the square of $x + 2$.
3. $x^2 − 14x + 49$ **4.** $x^2 − 12x$
$\left(\dfrac{-12}{2}\right)^2 = (−6)^2 = 36$
$x^2 − 12x + 36$ is the square of $x − 6$.
5. $x^2 − 3x + \dfrac{9}{4}$ **6.** $x^2 − x$
$\left(\dfrac{-1}{2}\right)^2 = \dfrac{1}{4}$
$x^2 − x + \dfrac{1}{4}$ is the square of $x − \dfrac{1}{2}$.
7. $t^2 + t + \dfrac{1}{4}$ **8.** $y^2 − 9y$
$\left(\dfrac{-9}{2}\right)^2 = \dfrac{81}{4}$
$y^2 − 9y + \dfrac{81}{4}$ is the square of $y − \dfrac{9}{2}$.
9. $x^2 + \dfrac{5}{4}x + \dfrac{25}{64}$ **10.** $x^2 + \dfrac{4}{3}x$
$\left(\dfrac{1}{2}\cdot\dfrac{4}{3}\right)^2 = \left(\dfrac{2}{3}\right)^2 = \dfrac{4}{9}$
$x^2 + \dfrac{4}{3}x + \dfrac{4}{9}$ is the square of $x + \dfrac{2}{3}$.
11. $m^2 − \dfrac{9}{2}m + \dfrac{81}{16}$ **12.** $r^2 − \dfrac{2}{5}r$
$\left[\dfrac{1}{2}\left(-\dfrac{2}{5}\right)\right]^2 = \left(-\dfrac{1}{5}\right)^2 = \dfrac{1}{25}$
$r^2 − \dfrac{2}{5}r + \dfrac{1}{25}$ is the square of $r − \dfrac{1}{5}$.
13. 3, 5 **14.** $x^2 − 6x − 7 = 0$
$x^2 − 6x = 7$
$x^2 − 6x + 9 = 7 + 9$
$(x − 3)^2 = 16$
$x − 3 = 4$ or $x − 3 = −4$
$x = 7$ or $x = −1$
15. −21, −1 **16.** $x^2 + 14x + 40 = 0$
$x^2 + 14x = −40$
$x^2 + 14x + 49 = −40 + 49$
$(x + 7)^2 = 9$
$x + 7 = 3$ or $x + 7 = −3$
$x = −4$ or $x = −10$
17. $1 \pm \sqrt{6}$
18. $3x^2 − 12x − 33 = 0$
$x^2 − 4x − 11 = 0$
$x^2 − 4x = 11$
$x^2 − 4x + 4 = 11 + 4$
$(x − 2)^2 = 15$
$x − 2 = \sqrt{15}$ or $x − 2 = −\sqrt{15}$
$x = 2 + \sqrt{15}$ or $x = 2 − \sqrt{15}$
$x = 2 \pm \sqrt{15}$
19. $11 \pm \sqrt{19}$
20. $x^2 − 18x + 74 = 0$
$x^2 − 18x = −74$
$x^2 − 18x + 81 = −74 + 81$
$(x − 9)^2 = 7$
$x − 9 = \sqrt{7}$ or $x − 9 = −\sqrt{7}$
$x = 9 + \sqrt{7}$ or $x = 9 − \sqrt{7}$
$x = 9 \pm \sqrt{7}$
21. $−5 \pm \sqrt{29}$
22. $x^2 − 7x − 2 = 0$
$x^2 − 7x = 2$
$x^2 − 7x + \dfrac{49}{4} = 2 + \dfrac{49}{4}$
$\left(x − \dfrac{7}{2}\right)^2 = \dfrac{8}{4} + \dfrac{49}{4} = \dfrac{57}{4}$
$x − \dfrac{7}{2} = \dfrac{\sqrt{57}}{2}$ or $x − \dfrac{7}{2} = −\dfrac{\sqrt{57}}{2}$
$x = \dfrac{7 + \sqrt{57}}{2}$ or $x = \dfrac{7 − \sqrt{57}}{2}$
$x = \dfrac{7 \pm \sqrt{57}}{2}$
23. $\dfrac{−5 \pm \sqrt{33}}{2}$
24. $2x^2 + 6x − 56 = 0$
$x^2 + 3x − 28 = 0$
$x^2 + 3x = 28$
$x^2 + 3x + \dfrac{9}{4} = 28 + \dfrac{9}{4}$
$\left(x + \dfrac{3}{2}\right)^2 = \dfrac{121}{4}$
$x + \dfrac{3}{2} = \dfrac{11}{2}$ or $x + \dfrac{3}{2} = −\dfrac{11}{2}$
$x = 4$ or $x = −7$
25. $\dfrac{−3 \pm \sqrt{41}}{4}$

26. $x^2 - \dfrac{3}{2}x - 2 = 0$

$x^2 - \dfrac{3}{2}x = 2$

$x^2 - \dfrac{3}{2}x + \dfrac{9}{16} = 2 + \dfrac{9}{16}$

$\left(x - \dfrac{3}{4}\right)^2 = \dfrac{41}{16}$

$x - \dfrac{3}{4} = \dfrac{\sqrt{41}}{4}$ or $x - \dfrac{3}{4} = -\dfrac{\sqrt{41}}{4}$

$x = \dfrac{3}{4} + \dfrac{\sqrt{41}}{4}$ or $x = \dfrac{3}{4} - \dfrac{\sqrt{41}}{4}$

$x = \dfrac{3 \pm \sqrt{41}}{4}$

27. $\dfrac{-3 \pm \sqrt{137}}{4}$

28. $2x^2 - 3x - 8 = 0$

$x^2 - \dfrac{3}{2}x - 4 = 0$

$x^2 - \dfrac{3}{2}x = 4$

$x^2 - \dfrac{3}{2}x + \dfrac{9}{16} = 4 + \dfrac{9}{16}$

$\left(x - \dfrac{3}{4}\right)^2 = \dfrac{73}{16}$

$x - \dfrac{3}{4} = \dfrac{\sqrt{73}}{4}$ or $x - \dfrac{3}{4} = -\dfrac{\sqrt{73}}{4}$

$x = \dfrac{3}{4} + \dfrac{\sqrt{73}}{4}$ or $x = \dfrac{3}{4} - \dfrac{\sqrt{73}}{4}$

$x = \dfrac{3 \pm \sqrt{73}}{4}$

29. $\dfrac{-2 \pm \sqrt{7}}{3}$

30. $3x^2 - 4x - 3 = 0$

$x^2 - \dfrac{4}{3}x - 1 = 0$

$x^2 - \dfrac{4}{3}x = 1$

$x^2 - \dfrac{4}{3}x + \dfrac{4}{9} = 1 + \dfrac{4}{9}$

$\left(x - \dfrac{2}{3}\right)^2 = \dfrac{13}{9}$

$x - \dfrac{2}{3} = \dfrac{\sqrt{13}}{3}$ or $x - \dfrac{2}{3} = -\dfrac{\sqrt{13}}{3}$

$x = \dfrac{2}{3} + \dfrac{\sqrt{13}}{3}$ or $x = \dfrac{2}{3} - \dfrac{\sqrt{13}}{3}$

$x = \dfrac{2 \pm \sqrt{13}}{3}$

31. $\dfrac{5 \pm \sqrt{97}}{4}$

32. $2x^2 = 5 + 9x$

$2x^2 - 9x - 5 = 0$

$x^2 - \dfrac{9}{2}x - \dfrac{5}{2} = 0$

$x^2 - \dfrac{9}{2}x = \dfrac{5}{2}$

$x^2 - \dfrac{9}{2}x + \dfrac{81}{16} = \dfrac{5}{2} + \dfrac{81}{16}$

$\left(x - \dfrac{9}{4}\right)^2 = \dfrac{121}{16}$

$x - \dfrac{9}{4} = \dfrac{11}{4}$ or $x - \dfrac{9}{4} = -\dfrac{11}{4}$

$x = 5$ or $x = -\dfrac{1}{2}$

33. $-\dfrac{7}{2}, \dfrac{1}{2}$

34. $6x^2 + 11x = 10$

$x^2 + \dfrac{11}{6}x = \dfrac{5}{3}$

$x^2 + \dfrac{11}{6}x + \dfrac{121}{144} = \dfrac{5}{3} + \dfrac{121}{144}$

$\left(x + \dfrac{11}{12}\right)^2 = \dfrac{361}{144}$

$x + \dfrac{11}{12} = \dfrac{19}{12}$ or $x + \dfrac{11}{12} = -\dfrac{19}{12}$

$x = \dfrac{2}{3}$ or $x = -\dfrac{5}{2}$

35. $\left(\dfrac{2}{3}, \dfrac{17}{3}\right)$

36. $2x + 3y = 8$, (1)
$x = y - 6$ (2)

Substitute $y - 6$ for x in (1).

$2(y - 6) + 3y = 8$
$2y - 12 + 3y = 8$
$5y = 20$
$y = 4$

Substitute 4 for y in (2).

$x = 4 - 6 = -2$

The solution is $(-2, 4)$.

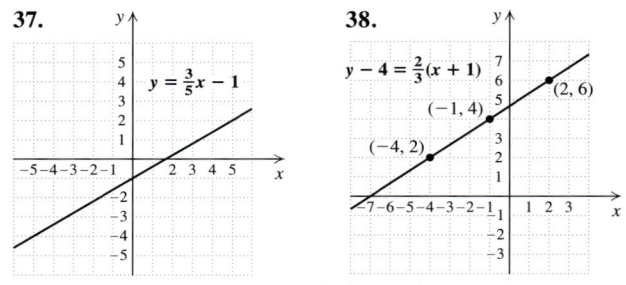

37. [graph of $y = \dfrac{3}{5}x - 1$]

38. [graph of $y - 4 = \dfrac{2}{3}(x + 1)$ through $(-4, 2)$, $(-1, 4)$, $(2, 6)$]

39. $2\sqrt{21}$ **40.** $\sqrt{90} = \sqrt{9 \cdot 10} = 3\sqrt{10}$

41. ◆ Yes; we add the same number on both sides of an equation every time we complete the square.

42. ◆ Given an equation that cannot be solved using the principle of zero products or the principle of square roots, completing the square enables us to express the equation in an equivalent form that can be solved using the principle of square roots. **43.** ◆ No; the solutions are of the form $x = \frac{g \pm \sqrt{h}}{k}$. They are opposites only when $g = 0$ (that is, when the quadratic equation has no x-term.) **44.** ◆ For an equation $ax^2 + bx + c = 0$, the coefficient of the x-term determines if the number added to complete the square is a whole number or a fraction. If b is an even number, then a whole number is added. Otherwise, a fraction is added.
45. $14, -14$ **46.** $x^2 + bx + 36$
$$\left(\frac{b}{2}\right)^2 = 36$$
$$\frac{b^2}{4} = 36$$
$$b^2 = 144$$
$$b = 12 \quad \text{or} \quad b = -12$$
47. $\pm 10\sqrt{2}$ **48.** $x^2 + bx + 45$
$$\left(\frac{b}{2}\right)^2 = 45$$
$$\frac{b^2}{4} = 45$$
$$b^2 = 180$$
$$b = \sqrt{180} \quad \text{or} \quad b = -\sqrt{180}$$
$$b = 6\sqrt{5} \quad \text{or} \quad b = -6\sqrt{5}$$
49. $\pm 8\sqrt{3}$ **50.** $4x^2 + bx + 16$
$$4\left(x^2 + \frac{b}{4}x + 4\right)$$
$$\left(\frac{b/4}{2}\right)^2 = 4$$
$$\left(\frac{b}{8}\right)^2 = 4$$
$$\frac{b^2}{64} = 4$$
$$b^2 = 256$$
$$b = 16 \quad \text{or} \quad b = -16$$
51. $-0.39, -7.61$ **52.** $7.41, 4.59$
53. $0.37, -5.37$ **54.** $7.27, -0.27$
55. $3.71, -1.21$ **56.** $5, -0.5$

TECHNOLOGY CONNECTION 10.3, P. 485
1. An error message appears.
2. The graph has no x-intercepts, so there is no value of x for which $x^2 + x + 1 = 0$ or, equivalently, for which $x^2 + x = -1$.

EXERCISE SET 10.3, PP. 488–491
1. $-2, 9$ **2.** $x^2 + 4x = 21$ **3.** 3
$$x^2 + 4x - 21 = 0$$
$$(x + 7)(x - 3) = 0$$
$$x = -7 \quad \text{or} \quad x = 3$$
4. $x^2 = 8x - 16$ **5.** $-\frac{4}{3}, -1$
$$x^2 - 8x + 16 = 0$$
$$(x - 4)(x - 4) = 0$$
$$x = 4$$
6. $3y^2 + 2y - 8 = 0$ **7.** $-\frac{1}{2}, \frac{7}{2}$
$$(3y - 4)(y + 2) = 0$$
$$y = \frac{4}{3} \quad \text{or} \quad y = -2$$
8. $4x^2 + 4x = 15$ **9.** $-8, 8$
$$4x^2 + 4x - 15 = 0$$
$$(2x - 3)(2x + 5) = 0$$
$$x = \frac{3}{2} \quad \text{or} \quad x = -\frac{5}{2}$$
10. $x^2 - 4 = 0$ **11.** $-2 \pm \sqrt{11}$
$$(x + 2)(x - 2) = 0$$
$$x = -2 \quad \text{or} \quad x = 2$$
12. $x^2 + 2x - 2 = 0$ **13.** $5 \pm \sqrt{3}$
$$x = \frac{-2 \pm \sqrt{2^2 - 4 \cdot 1 \cdot (-2)}}{2 \cdot 1}$$
$$x = \frac{-2 \pm \sqrt{12}}{2} = \frac{-2 \pm 2\sqrt{3}}{2}$$
$$x = -1 \pm \sqrt{3}$$
14. $y^2 + 6y - 1 = 0$ **15.** $-1 \pm \sqrt{7}$
$$y = \frac{-6 \pm \sqrt{6^2 - 4 \cdot 1 \cdot (-1)}}{2 \cdot 1}$$
$$y = \frac{-6 \pm \sqrt{40}}{2} = \frac{-6 \pm 2\sqrt{10}}{2}$$
$$y = -3 \pm \sqrt{10}$$
16. $x^2 - 4x + 4 = 5$ **17.** $\frac{-2 \pm \sqrt{10}}{3}$
$$x^2 - 4x - 1 = 0$$
$$x = \frac{-(-4) \pm \sqrt{(-4)^2 - 4 \cdot 1 \cdot (-1)}}{2 \cdot 1}$$
$$x = \frac{4 \pm \sqrt{20}}{2} = \frac{4 \pm 2\sqrt{5}}{2}$$
$$x = 2 \pm \sqrt{5}$$
18. $3x^2 - 8x + 2 = 0$
$$x = \frac{-(-8) \pm \sqrt{(-8)^2 - 4 \cdot 3 \cdot 2}}{2 \cdot 3}$$
$$x = \frac{8 \pm \sqrt{40}}{6} = \frac{8 \pm 2\sqrt{10}}{6}$$
$$x = \frac{4 \pm \sqrt{10}}{3}$$
19. $\frac{5 \pm \sqrt{33}}{4}$

INSTRUCTOR'S ANSWERS

20. $2x^2 + 2x = 3$
$2x^2 + 2x - 3 = 0$
$x = \dfrac{-2 \pm \sqrt{2^2 - 4 \cdot 2 \cdot (-3)}}{2 \cdot 2}$
$x = \dfrac{-2 \pm \sqrt{28}}{4} = \dfrac{-2 \pm 2\sqrt{7}}{4}$
$x = \dfrac{-1 \pm \sqrt{7}}{2}$

21. $\dfrac{-1 \pm \sqrt{2}}{2}$

22. $4y^2 - 4y - 1 = 0$
$y = \dfrac{-(-4) \pm \sqrt{(-4)^2 - 4 \cdot 4 \cdot (-1)}}{2 \cdot 4}$
$y = \dfrac{4 \pm \sqrt{32}}{8} = \dfrac{4 \pm 4\sqrt{2}}{8}$
$y = \dfrac{1 \pm \sqrt{2}}{2}$

23. No real-number solutions

24. $4y^2 + 2y + 3 = 0$
$y = \dfrac{-2 \pm \sqrt{2^2 - 4 \cdot 4 \cdot 3}}{2 \cdot 4}$
$y = \dfrac{-2 \pm \sqrt{-44}}{8}$

Since the radicand, -44, is negative, there are no real-number solutions.

25. $\dfrac{5 \pm \sqrt{73}}{6}$

26. $2x^2 + 3x = 1$
$2x^2 + 3x - 1 = 0$
$x = \dfrac{-3 \pm \sqrt{3^2 - 4 \cdot 2 \cdot (-1)}}{2 \cdot 2}$
$x = \dfrac{-3 \pm \sqrt{17}}{4}$

27. $\dfrac{3 \pm \sqrt{29}}{2}$

28. $5m^2 = 3 + 11m$
$5m^2 - 11m - 3 = 0$
$m = \dfrac{-(-11) \pm \sqrt{(-11)^2 - 4 \cdot 5 \cdot (-3)}}{2 \cdot 5}$
$m = \dfrac{11 \pm \sqrt{181}}{10}$

29. $0, \dfrac{3}{2}$

30. $7x^2 + 2 = 6x$
$7x^2 - 6x + 2 = 0$
$x = \dfrac{-(-6) \pm \sqrt{(-6)^2 - 4 \cdot 7 \cdot 2}}{2 \cdot 7}$
$x = \dfrac{6 \pm \sqrt{-20}}{14}$

Since the radicand, -20, is negative, there are no real-number solutions.

31. No real-number solutions

32. $15t^2 + 10t = 0$
$5t(3t + 2) = 0$
$t = 0 \quad or \quad t = -\dfrac{2}{3}$

33. $\pm 3\sqrt{2}$

34. $5t^2 = 100$
$t^2 = 20$
$t = \sqrt{20} \quad or \quad t = -\sqrt{20}$
$t = 2\sqrt{5} \quad or \quad t = -2\sqrt{5}$

35. $5.317, -1.317$

36. $x^2 + 2x - 2 = 0$
$x = \dfrac{-2 \pm \sqrt{2^2 - 4 \cdot 1 \cdot (-2)}}{2 \cdot 1}$
$x = \dfrac{-2 \pm \sqrt{12}}{2} = \dfrac{-2 \pm 2\sqrt{3}}{2}$
$x = -1 \pm \sqrt{3}$
$\sqrt{3} \approx 1.732$, so
$-1 + \sqrt{3} \approx -1 + 1.732 \quad or \quad -1 - \sqrt{3} \approx -1 - 1.732$
$\approx 0.732 \quad or \quad \approx -2.732$

37. $6.162, -0.162$

38. $y^2 + 10y + 22 = 0$
$y = \dfrac{-10 \pm \sqrt{10^2 - 4 \cdot 1 \cdot 22}}{2 \cdot 1}$
$= \dfrac{-10 \pm \sqrt{12}}{2} = \dfrac{-10 \pm 2\sqrt{3}}{2}$
$= -5 \pm \sqrt{3}$
$\sqrt{3} \approx 1.732$, so
$-5 + \sqrt{3} \approx -5 + 1.732 \quad or \quad -5 - \sqrt{3} \approx -5 - 1.732$
$\approx -3.268 \quad or \quad \approx -6.732$

39. $0.207, -1.207$

40. $4x^2 = 4x + 1$
$4x^2 - 4x - 1 = 0$
$x = \dfrac{-(-4) \pm \sqrt{(-4)^2 - 4 \cdot 4 \cdot (-1)}}{2 \cdot 4}$
$= \dfrac{4 \pm \sqrt{32}}{8} = \dfrac{4 \pm 4\sqrt{2}}{8}$
$= \dfrac{1 \pm \sqrt{2}}{2}$
$\sqrt{2} \approx 1.414$, so
$\dfrac{1 + \sqrt{2}}{2} \approx \dfrac{1 + 1.414}{2} \quad or \quad \dfrac{1 - \sqrt{2}}{2} \approx \dfrac{1 - 1.414}{2}$
$\approx \dfrac{2.414}{2} \quad or \quad \approx \dfrac{-0.414}{2}$
$\approx 1.207 \quad or \quad \approx -0.207$

41. 10

42. Solve $20 = \dfrac{n^2 - 3n}{2}$
$n = 8 \quad or \quad n = -5$
Only 8 checks in the original problem.

43. About 7.95 sec

44. Solve $1136 = 16t^2$
$t \approx 8.43 \quad or \quad t \approx -8.43$
The solution that checks in the original problem is about 8.43 sec.

45. About 6.61 sec
46. Solve $175 = 16t^2$.
$$t \approx 3.31 \quad or \quad t \approx -3.31$$
The solution that checks in the original problem is approximately 3.31 sec.
47. 7 ft, 24 ft
48. Solve $x^2 + (x + 14)^2 = 26^2$, where x is the length of the shorter leg and $x + 14$ is the length of the longer leg, in yards.
$$x = -24 \quad or \quad x = 10$$
Only 10 checks in the original problem. The lengths of the legs are 10 yd and $10 + 14$, or 24 yd.
49. 10 cm, 6 cm
50. Solve $(w + 3)w = 70$, where $w = $ the width and $w + 3 = $ the length of the rectangle, in meters.
$$w = -10 \quad or \quad w = 7$$
Only 7 checks in the original problem. The length is $7 + 3$, or 10 m, and the width is 7 m.
51. 15 m by 20 m
52.

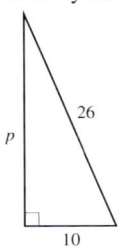

Solve $10^2 + p^2 = 26^2$
$$p = 24 \quad or \quad p = -24$$
Only 24 checks in the original problem. The wire reaches 24 ft up the pole.
53. 4 m, 6.5 m
54. Solve $15.5 = \frac{1}{2}(x + 1.2)(x)$, where x represents the length of the shorter leg and $x + 1.2$ represents the length of the longer leg, in centimeters.
$$x = -6.2 \quad or \quad x = 5$$
Only 5 checks in the original problem. The legs are 5 cm and 6.2 cm.
55. 5.58 in., 3.58 in.
56. Solve $(x + 3)x = 15$, where $x = $ the width and $x + 3 = $ the length of the rectangle, in feet.
$$x \approx 2.65 \quad or \quad x \approx -5.65$$
Only 2.65 checks in the original problem. The length is about $2.65 + 3$, or 5.65 ft, and the width is 2.65 ft.
57. 5.66 m, 2.83 m
58. Solve $2x \cdot x = 20$, where x represents the width and $2x$ represents the length of the rectangle, in centimeters.
$$x \approx 3.16 \quad or \quad x \approx -3.16$$
Only 3.16 checks in the original problem. The length is $2(3.16)$, or 6.32 cm, and the width is 3.16 cm.

59. 20%
60. Solve $3610 = 2560(1 + r)^2$.
$$r = 0.1875 \quad or \quad r = -2.1875$$
Only 0.1875 checks in the original problem, so the interest rate is 18.75%.
61. 4%
62. Solve $7290 = 6250(1 + r)^2$.
$$r = 0.08 \quad or \quad r = -2.08$$
Only 0.08 checks in the original problem, so the interest rate is 8%.
63. 17.84 ft
64. Solve $20{,}000 = 3.14\left(\dfrac{d}{2}\right)^2$, where d is the width (diameter) of the oil slick.
$$d \approx 159.62 \quad or \quad d \approx -159.62$$
Only 159.62 makes sense in this problem. The oil slick is about 160 m wide.
65. $\frac{6}{5}$ **66.** $\dfrac{3}{x - 4} = \dfrac{2}{x + 4}$, $LCD = (x - 4)(x + 4)$
$$3(x + 4) = 2(x - 4)$$
$$3x + 12 = 2x - 8$$
$$x = -20$$
67. $3\sqrt{10}$ **68.** $\sqrt{9000x^{10}} = \sqrt{900x^{10} \cdot 10} = 30x^5\sqrt{10}$
69. -1 **70.** $(-1)^{64} = 1$ **71.** ◆ The quadratic formula would not be the easiest way to solve a quadratic equation when the equation can be solved by factoring or by using the principle of square roots.   **72.** ◆ Yes; the quadratic formula is derived by solving $ax^2 + bx + c = 0$ by completing the square.   **73.** ◆ Answers may vary. The length of a rectangle is 3 ft greater than the width. The area is 50 ft². Find the length and width.   **74.** ◆ No; because of the plus or minus sign in the quadratic formula, any irrational solutions that arise must occur in pairs. (No more than two solutions exist for any quadratic equation.)
75. $-\frac{4}{3}, 0$
76. $5x + x(x - 7) = 0$ **77.** $\dfrac{7 \pm \sqrt{69}}{10}$
$5x + x^2 - 7x = 0$
$x^2 - 2x = 0$
$x(x - 2) = 0$
$x = 0 \quad or \quad x = 2$
78. $3 - x(x - 3) = 4$
$3 - x^2 + 3x = 4$
$0 = x^2 - 3x + 1$
$$x = \dfrac{-(-3) \pm \sqrt{(-3)^2 - 4 \cdot 1 \cdot 1}}{2 \cdot 1}$$
$$x = \dfrac{3 \pm \sqrt{5}}{2}$$
79. $\dfrac{-2 \pm \sqrt{10}}{2}$

80. $(y+4)(y+3) = 15$
$y^2 + 7y + 12 = 15$
$y^2 + 7y - 3 = 0$
$y = \dfrac{-7 \pm \sqrt{7^2 - 4 \cdot 1 \cdot (-3)}}{2 \cdot 1}$
$y = \dfrac{-7 \pm \sqrt{61}}{2}$

81. $\pm \sqrt{7}$

82. $\dfrac{x^2}{x+3} - \dfrac{5}{x+3} = 0,\quad \text{LCM is } x+3$
$(x+3)\left(\dfrac{x^2}{x+3} - \dfrac{5}{x+3}\right) = (x+3) \cdot 0$
$x^2 - 5 = 0$
$x^2 = 5$
$x = \sqrt{5} \quad \text{or} \quad x = -\sqrt{5}$

83. $2 \pm \sqrt{34}$

84. $\dfrac{1}{x} + \dfrac{1}{x+1} = \dfrac{1}{3},\quad \text{LCM is } 3x(x+1)$
$3x(x+1)\left(\dfrac{1}{x} + \dfrac{1}{x+1}\right) = 3x(x+1) \cdot \dfrac{1}{3}$
$3(x+1) + 3x = x(x+1)$
$3x + 3 + 3x = x^2 + x$
$0 = x^2 - 5x - 3$
$x = \dfrac{-(-5) \pm \sqrt{(-5)^2 - 4 \cdot 1 \cdot (-3)}}{2 \cdot 1}$
$x = \dfrac{5 \pm \sqrt{37}}{2}$

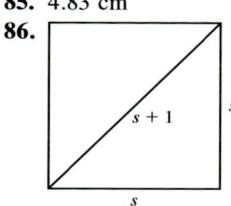

85. 4.83 cm

86.
Solve $s^2 + s^2 = (s+1)^2$, where $s =$ the length of a side of the square.
$s = 1 + \sqrt{2}$ or $s = 1 - \sqrt{2}$
Only $s = 1 + \sqrt{2}$ checks in the original problem. The area of the square is $(1 + \sqrt{2})^2 = 3 + 2\sqrt{2} \approx 5.828$ square units.

87. 7.5 ft from the bottom

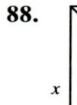

88.
Solve $x^2 + x^2 = (x+3)^2$.
$x = 3 \pm 3\sqrt{2}$

Only $3 + 3\sqrt{2}$ makes sense in the original problem. The length of a side of the square is $3 + 3\sqrt{2}$ cm, or about 7.2 cm.

89. 5.5%

90. Solve $3000 = P(1 + 0.0575)^2$.
$P \approx \$2682.63$

91. About 19 in. by 44 in.   **92.**

EXERCISE SET 10.4, PP. 493–494

1. i **2.** $6i$ **3.** $3i$ **4.** $9i$ **5.** $5\sqrt{2}i$
6. $\sqrt{-44} = \sqrt{-1 \cdot 4 \cdot 11} = 2i\sqrt{11}$, or $2\sqrt{11}i$
7. $-2\sqrt{5}i$ **8.** $-\sqrt{-45} = -\sqrt{-1 \cdot 9 \cdot 5} = -3i\sqrt{5}$, or $-3\sqrt{5}i$ **9.** $-3\sqrt{2}i$ **10.** $-\sqrt{-28} = -\sqrt{-1 \cdot 4 \cdot 7} = -2i\sqrt{7}$, or $-2\sqrt{7}i$ **11.** $4 + 7i$ **12.** $7 + 2i$
13. $7 + 4i$ **14.** $-8 - 6i$ **15.** $3 - 7\sqrt{2}i$
16. $-2 + \sqrt{-125} = -2 + \sqrt{-1 \cdot 25 \cdot 5} = -2 + 5i\sqrt{5}$
17. $\pm 3i$
18. $x^2 + 4 = 0$
$x^2 = -4$
$x = \sqrt{-4} \quad \text{or} \quad x = -\sqrt{-4}$
$x = 2i \quad \text{or} \quad x = -2i$
$x = \pm 2i$
19. $\pm 2\sqrt{7}i$
20. $x^2 = -48$
$x = \sqrt{-48} \quad \text{or} \quad x = -\sqrt{-48}$
$x = 4i\sqrt{3} \quad \text{or} \quad x = -4i\sqrt{3}$
$x = \pm 4i\sqrt{3}$
21. $2 \pm \sqrt{2}i$
22. $x^2 + 4x + 5 = 0$
$x = \dfrac{-4 \pm \sqrt{4^2 - 4 \cdot 1 \cdot 5}}{2 \cdot 1}$
$x = \dfrac{-4 \pm \sqrt{-4}}{2} = \dfrac{-4 \pm 2i}{2}$
$x = -2 \pm i$
23. $3 \pm 5i$
24. $(x+1)^2 = -16$
$x + 1 = \sqrt{-16} \quad \text{or} \quad x + 1 = -\sqrt{-16}$
$x + 1 = 4i \quad \text{or} \quad x + 1 = -4i$
$x = -1 + 4i \quad \text{or} \quad x = -1 - 4i$
$x = -1 \pm 4i$
25. $-1 \pm i$
26. $x^2 + 5 = 2x$
$x^2 - 2x + 5 = 0$
$x = \dfrac{-(-2) \pm \sqrt{(-2)^2 - 4 \cdot 1 \cdot 5}}{2 \cdot 1}$
$x = \dfrac{2 \pm \sqrt{-16}}{2} = \dfrac{2 \pm 4i}{2}$
$x = 1 \pm 2i$
27. $2 \pm \sqrt{3}i$

28. $x^2 + 7 + 4x = 0$
$x^2 + 4x + 7 = 0$
$x = \dfrac{-4 \pm \sqrt{4^2 - 4 \cdot 1 \cdot 7}}{2 \cdot 1}$
$x = \dfrac{-4 \pm \sqrt{-12}}{2} = \dfrac{-4 \pm 2i\sqrt{3}}{2}$
$x = \dfrac{-4}{2} \pm \dfrac{2\sqrt{3}}{2}i = -2 \pm \sqrt{3}i$

29. $-\dfrac{3}{2} \pm \dfrac{1}{2}i$

30. $4y^2 + 3y + 2 = 0$
$y = \dfrac{-3 \pm \sqrt{3^2 - 4 \cdot 4 \cdot 2}}{2 \cdot 4}$
$y = \dfrac{-3 \pm \sqrt{-23}}{8} = \dfrac{-3 \pm i\sqrt{23}}{8}$
$y = -\dfrac{3}{8} \pm \dfrac{\sqrt{23}}{8}i$

31. $-\dfrac{1}{3} \pm \dfrac{\sqrt{2}}{3}i$

32. $4p^2 + 3 = 6p$
$4p^2 - 6p + 3 = 0$
$p = \dfrac{-(-6) \pm \sqrt{(-6)^2 - 4 \cdot 4 \cdot 3}}{2 \cdot 4}$
$p = \dfrac{6 \pm \sqrt{-12}}{8} = \dfrac{6 \pm 2i\sqrt{3}}{8}$
$p = \dfrac{6}{8} \pm \dfrac{2\sqrt{3}}{8}i = \dfrac{3}{4} \pm \dfrac{\sqrt{3}}{4}i$

33. [graph of $y = \tfrac{3}{5}x - 2$]

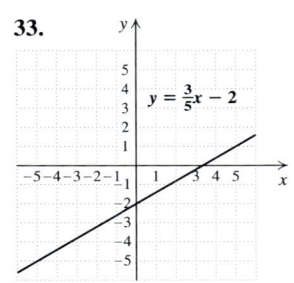

34. [graph of $2x - 3y = 10$]

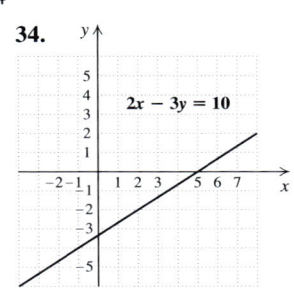

35. [graph of $y = -4$]

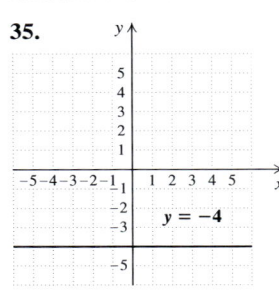

36. [graph of $x = 2$]

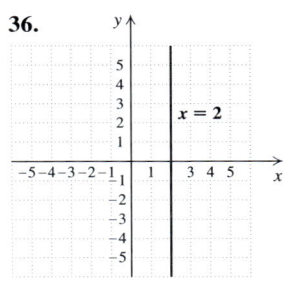

37. -17 **38.** $-1(-4)^2 = -1 \cdot 16 = -16$ **39.** ◆ An equation of the form $x^2 = c$ has imaginary-number solutions when $c < 0$. **40.** ◆ If $b^2 - 4ac < 0$, the radicand in the quadratic formula is negative and the solutions are imaginary. If $b^2 - 4ac \geq 0$, the radicand is nonnegative and the solution(s) are real numbers.

41. ◆ For $b^2 - 4ac \neq 0$, $\pm\sqrt{b^2 - 4ac}$ yields either two imaginary numbers or two real numbers. Thus, it is not possible for a quadratic equation (with real-number coefficients) to have one imaginary-number solution and one real-number solution. (If imaginary-number coefficients are allowed, then a quadratic equation can have one imaginary-number solution and one real-number solution. One example is $x^2 - (5 + i)x + 5i = 0$ with solutions 5 and i.) **42.** ◆ Yes; after completing the square and factoring we have an equation of the form $(x - a)^2 = k$. If $k < 0$, the method of completing the square yields imaginary-number solutions.

43. $-2 \pm i$

44. $(p + 5)^2 + (p + 1)^2 = 0$
$p^2 + 10p + 25 + p^2 + 2p + 1 = 0$
$2p^2 + 12p + 26 = 0$
$p^2 + 6p + 13 = 0$
$p = \dfrac{-6 \pm \sqrt{6^2 - 4 \cdot 1 \cdot 13}}{2 \cdot 1}$
$p = \dfrac{-6 \pm \sqrt{-16}}{2} = \dfrac{-6 \pm 4i}{2}$
$p = -3 \pm 2i$

45. $-1 \pm \sqrt{19}i$

46. $\dfrac{1}{a - 1} - \dfrac{2}{a - 1} = 3a$
$-\dfrac{1}{a - 1} = 3a$
$-1 = 3a^2 - 3a$
$0 = 3a^2 - 3a + 1$
$a = \dfrac{-(-3) \pm \sqrt{(-3)^2 - 4 \cdot 3 \cdot 1}}{2 \cdot 3}$
$a = \dfrac{3 \pm \sqrt{-3}}{6} = \dfrac{3 \pm i\sqrt{3}}{6}$
$a = \dfrac{3}{6} \pm \dfrac{\sqrt{3}}{6}i = \dfrac{1}{2} \pm \dfrac{\sqrt{3}}{6}i$

47. $y = x^2 + 3x + 4$ [graph] No x-intercepts exist.

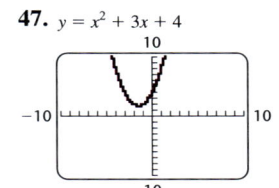

$y_1 = x^2 + 2$, $y_2 = 2x$ [graph] No intersection exists.

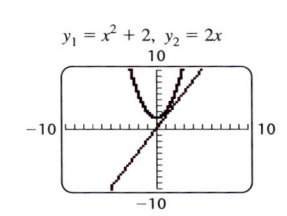

EXERCISE SET 10.5, PP. 499–501

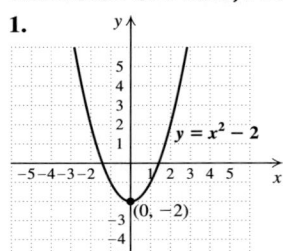

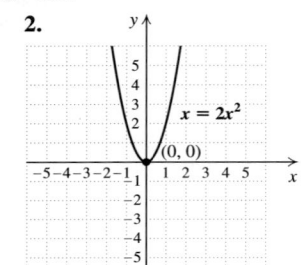

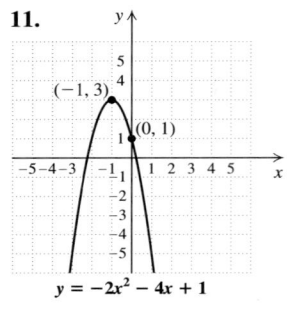

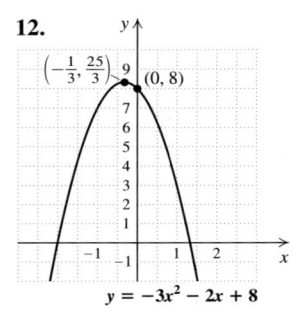

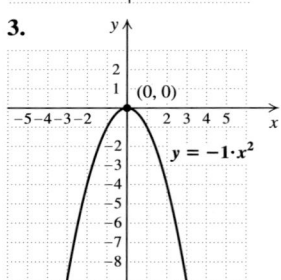

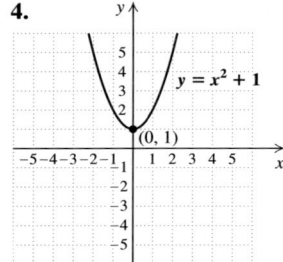

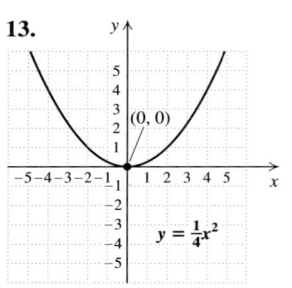

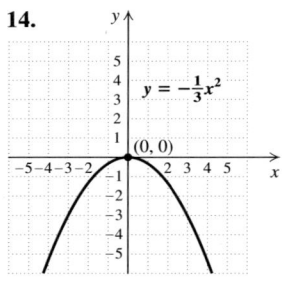

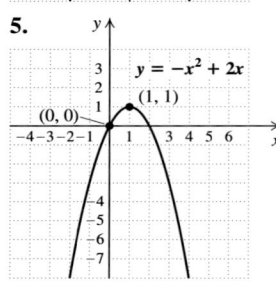

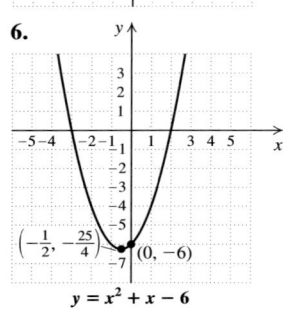

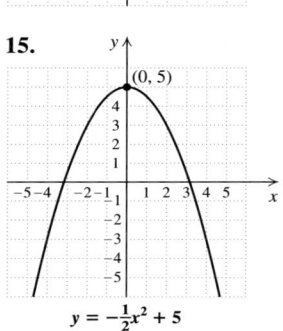

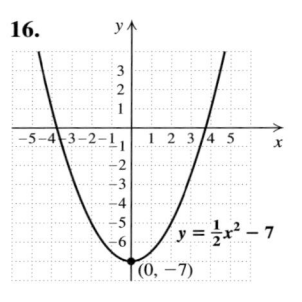

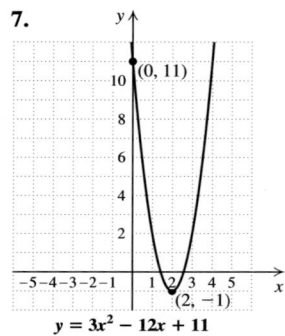

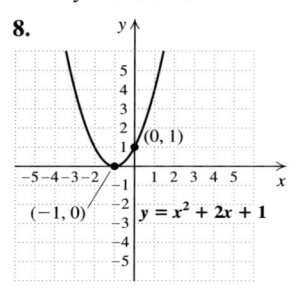

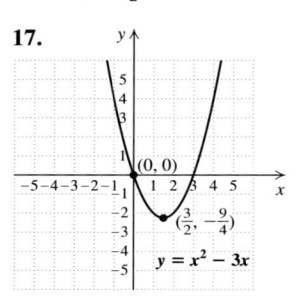

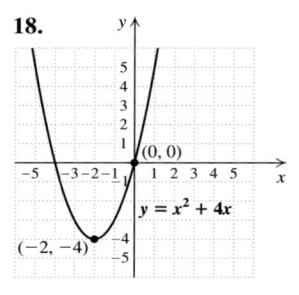

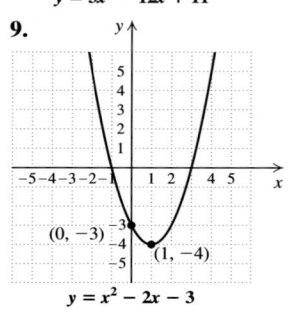

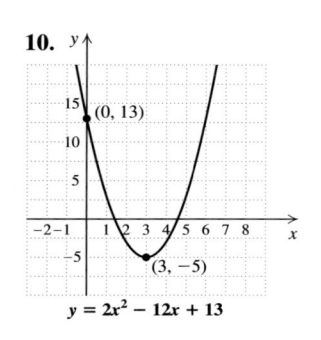

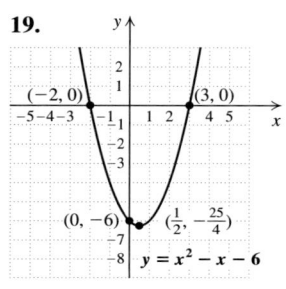

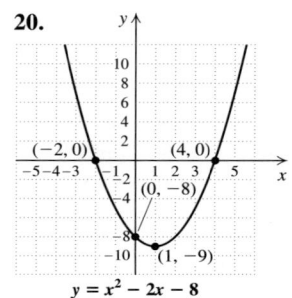

1. $y = x^2 - 2$; vertex $(0, -2)$

2. $x = 2x^2$ [graph labeled $(0,0)$]

3. $y = -1 \cdot x^2$; vertex $(0, 0)$

4. $y = x^2 + 1$; vertex $(0, 1)$

5. $y = -x^2 + 2x$; vertex $(1, 1)$; $(0, 0)$

6. $y = x^2 + x - 6$; vertex $\left(-\frac{1}{2}, -\frac{25}{4}\right)$; $(0, -6)$

7. $y = 3x^2 - 12x + 11$; vertex $(2, -1)$; $(0, 11)$

8. $y = x^2 + 2x + 1$; vertex $(-1, 0)$; $(0, 1)$

9. $y = x^2 - 2x - 3$; vertex $(1, -4)$; $(0, -3)$

10. $y = 2x^2 - 12x + 13$; vertex $(3, -5)$; $(0, 13)$

11. $y = -2x^2 - 4x + 1$; vertex $(-1, 3)$; $(0, 1)$

12. $y = -3x^2 - 2x + 8$; vertex $\left(-\frac{1}{3}, \frac{25}{3}\right)$; $(0, 8)$

13. $y = \frac{1}{4}x^2$; vertex $(0, 0)$

14. $y = -\frac{1}{3}x^2$; vertex $(0, 0)$

15. $y = -\frac{1}{2}x^2 + 5$; vertex $(0, 5)$

16. $y = \frac{1}{2}x^2 - 7$; vertex $(0, -7)$

17. $y = x^2 - 3x$; vertex $\left(\frac{3}{2}, -\frac{9}{4}\right)$; $(0, 0)$

18. $y = x^2 + 4x$; vertex $(-2, -4)$; $(0, 0)$

19. $y = x^2 - x - 6$; $(-2, 0)$, $(3, 0)$; vertex $\left(\frac{1}{2}, -\frac{25}{4}\right)$; $(0, -6)$

20. $y = x^2 - 2x - 8$; $(-2, 0)$, $(4, 0)$; vertex $(1, -9)$; $(0, -8)$

21. [graph: $y = 2x^2 - 5x$, vertex $(5/4, -25/8)$, intercepts $(0,0)$, $(5/2, 0)$]

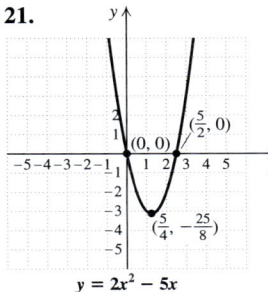

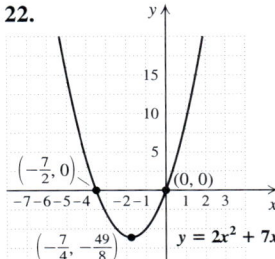

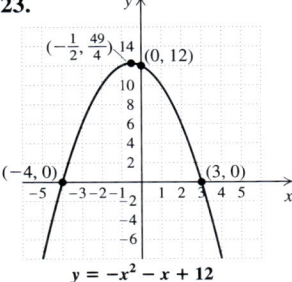

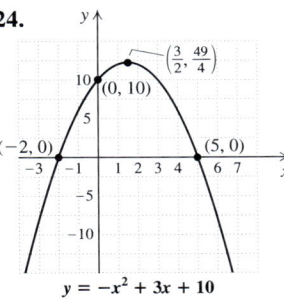

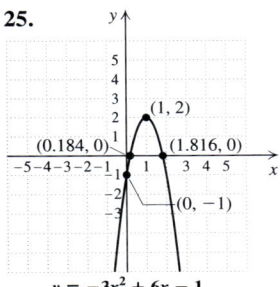

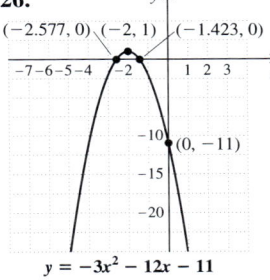

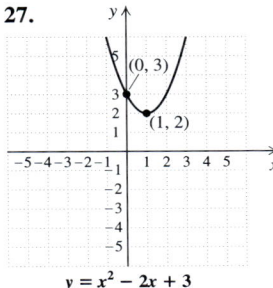

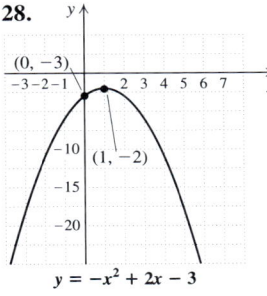

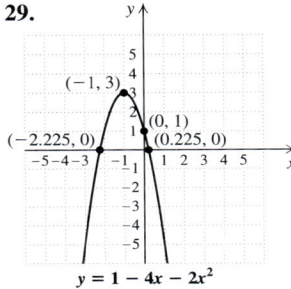

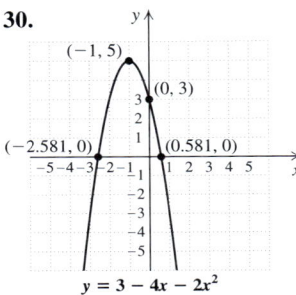

22. [graph: $y = 2x^2 + 7x$, vertex $(-7/4, -49/8)$, intercepts $(-7/2, 0)$, $(0,0)$]

23. [graph: $y = -x^2 - x + 12$, vertex $(-1/2, 49/4)$, intercepts $(-4, 0)$, $(3, 0)$, $(0, 12)$]

24. [graph: $y = -x^2 + 3x + 10$, vertex $(3/2, 49/4)$, intercepts $(-2, 0)$, $(5, 0)$, $(0, 10)$]

25. [graph: $y = -3x^2 + 6x - 1$, vertex $(1, 2)$, intercepts $(0.184, 0)$, $(1.816, 0)$, $(0, -1)$]

26. [graph: $y = -3x^2 - 12x - 11$, vertex $(-2, 1)$, intercepts $(-2.577, 0)$, $(-1.423, 0)$, $(0, -11)$]

27. [graph: $y = x^2 - 2x + 3$, vertex $(1, 2)$, $(0, 3)$]

28. [graph: $y = -x^2 + 2x - 3$, vertex $(1, -2)$, $(0, -3)$]

29. [graph: $y = 1 - 4x - 2x^2$, vertex $(-1, 3)$, intercepts $(-2.225, 0)$, $(0.225, 0)$, $(0, 1)$]

30. [graph: $y = 3 - 4x - 2x^2$, vertex $(-1, 5)$, intercepts $(-2.581, 0)$, $(0.581, 0)$, $(0, 3)$]

31. $\sqrt{432} \approx 20.78$ ft

32. Solve $94^2 + 50^2 = d^2$, where $d = $ the length of a diagonal, in feet.
$$d = \sqrt{11{,}336} \quad or \quad d = -\sqrt{11{,}336}$$
Only $\sqrt{11{,}336}$ makes sense in the original problem. The length of the diagonal is $\sqrt{11{,}336} \approx 106.47$ ft.

33. $y = -\dfrac{5}{3}x + \dfrac{11}{3}$

34. $8 - 6x = 2y$
$4 - 3x = y$
$m = -3$

35. -3

36. $3(-2)^4 + 5(-2)^2 = 3 \cdot 16 + 5 \cdot 4 = 48 + 20 = 68$

37. ◆ $|a|$ affects the width of the graph. As $|a|$ increases the graph becomes narrower.

38. ◆ Average the values of the x-coordinates of the x-intercepts to find the x-coordinate of the vertex. Then substitute this value for x in the equation of the parabola to find the y-coordinate of the vertex.

39. ◆ Yes; when $x = 0$, $y = a \cdot 0^2 + b \cdot 0 + c$ so a y-intercept always exists. It is $(0, c)$. If $a = 0$, then the equation becomes a linear equation, $y = bx + c$, whose y-intercept is also $(0, c)$.

40. ◆ Write an equation of the form $y = k(x - r_1)(x - r_2)$.

41.
$$\dfrac{\dfrac{-b - \sqrt{b^2 - 4ac}}{2a} + \dfrac{-b + \sqrt{b^2 - 4ac}}{2a}}{2} = \dfrac{\dfrac{-2b}{2a}}{2} = \dfrac{-2b}{2a} \cdot \dfrac{1}{2} = -\dfrac{b}{2a}$$

42. (a) We substitute 128 for H and solve for t:
$$128 = -16t^2 + 96t$$
$$16t^2 - 96t + 128 = 0$$
$$16(t^2 - 6t + 8) = 0$$
$$16(t - 2)(t - 4) = 0$$
$t - 2 = 0 \quad or \quad t - 4 = 0$
$t = 2 \quad or \quad t = 4$

The projectile is 128 ft from the ground 2 sec after launch and again 4 sec after launch. The graph confirms this.

(b) We find the first coordinate of the vertex of the function $H = -16t^2 + 96t$:
$$-\dfrac{b}{2a} = -\dfrac{96}{2(-16)} = -\dfrac{96}{-32} = -(-3) = 3$$

The projectile reaches its maximum height 3 sec after launch. The graph confirms this.

(c) We substitute 0 for H and solve for t:
$$0 = -16t^2 + 96t$$
$$0 = -16t(t - 6)$$
$-16t = 0 \quad or \quad t - 6 = 0$
$t = 0 \quad or \quad t = 6$

At $t = 0$ sec the projectile has not yet been launched. Thus, we use $t = 6$. The projectile returns to the ground 6 sec after launch. The graph confirms this.

43. (a) 56.25 ft, 120 ft, 206.25 ft, 276.25 ft, 356.25 ft, 600 ft;
(b) [graph of $d = r + 0.05r^2$, Stopping distance (in feet) vs Speed (in miles per hour)]

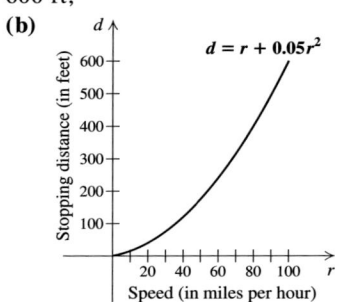

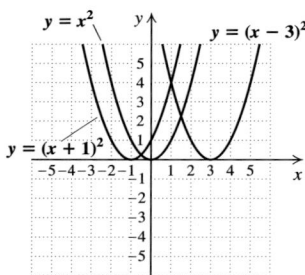

44. [graphs of $y = x^2$, $y = (x-3)^2$, $y = (x+1)^2$]

We can move the graph of $y = x^2$ to the right h units if $h \geq 0$ or to the left $|h|$ units if $h < 0$ to obtain the graph of $y = (x - h)^2$.

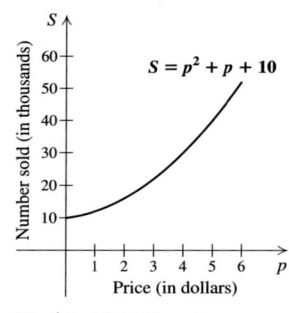

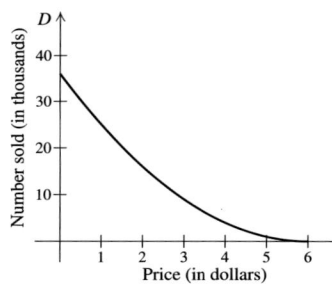

45. [graph of $S = p^2 + p + 10$, Number sold (in thousands) vs Price (in dollars)]

46. [graph of D, Number sold (in thousands) vs Price (in dollars)]

47. $2; 16,000 units

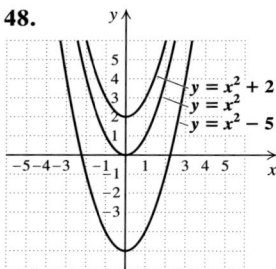

48. [graphs of $y = x^2 + 2$, $y = x^2$, $y = x^2 - 5$]

We can move the graph of $y = x^2$ up k units if $k \geq 0$ or down $|k|$ units if $k < 0$ to obtain the graph of $y = x^2 + k$.

49. 2.2361

EXERCISE SET 10.6, PP. 509–511

1. Yes **2.** Yes **3.** Yes **4.** No **5.** No **6.** Yes
7. Yes **8.** Yes **9.** 9, 12, 2
10. $g(0) = 0 - 6 = -6$ **11.** $-21, 15, 42$
$g(6) = 6 - 6 = 0$
$g(13) = 13 - 6 = 7$
12. $f(6) = -4 \cdot 6 = -24$
$f\left(-\frac{1}{2}\right) = -4\left(-\frac{1}{2}\right) = 2$
$f(20) = -4 \cdot 20 = -80$
13. 7, -17, 24.1
14. $h(4) = 19$
$h(-6) = 19$
$h(12) = 19$
15. 0, 5, 2
16. $P(0) = 3 \cdot 0^2 - 2 \cdot 0 = 0$
$P(-2) = 3(-2)^2 - 2(-2) = 12 + 4 = 16$
$P(3) = 3 \cdot 3^2 - 2 \cdot 3 = 27 - 6 = 21$
17. 6, 1, $\frac{13}{4}$
18. $f(-3) = |-3| - 2 = 3 - 2 = 1$
$f(93) = |93| - 2 = 93 - 2 = 91$
$f(-100) = |-100| - 2 = 100 - 2 = 98$
19. 4, -122, 3
20. $h(0) = 0^4 - 3 = 0 - 3 = -3$
$h(-1) = (-1)^4 - 3 = 1 - 3 = -2$
$h(3) = 3^4 - 3 = 81 - 3 = 78$
21. (a) 159.48 cm; (b) 153.98 cm
22. (a) $M(30) = 2.89(30) + 70.64 = 157.34$ cm
(b) $M(35) = 2.89(35) + 70.64 = 171.79$ cm
23. $1\frac{20}{33}$ atm, $1\frac{10}{11}$ atm, $4\frac{1}{33}$ atm
24. $T(5) = 10 \cdot 5 + 20 = 50 + 20 = 70°$ C
$T(20) = 10 \cdot 20 + 20 = 200 + 20 = 220°$ C
$T(1000) = 10 \cdot 1000 + 20 = 10{,}000 + 20 = 10{,}020°$ C
25. 1.792 cm, 2.8 cm, 11.2 cm
26. $C(62) = \frac{5}{9}(62 - 32) = \frac{5}{9} \cdot 30 = \frac{50}{3} = 16\frac{2}{3}°$
$C(77) = \frac{5}{9}(77 - 32) = \frac{5}{9} \cdot 45 = 25°$
$C(23) = \frac{5}{9}(23 - 32) = \frac{5}{9}(-9) = -5°$

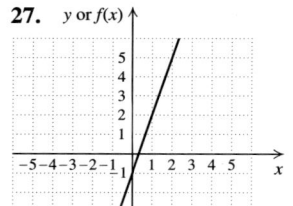

27. [graph of $f(x) = 3x - 1$]

28. [graph of $g(x) = 2x + 5$]

29. [graph: $g(x) = -2x + 3$]

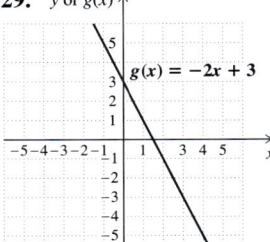

30. [graph: $f(x) = -\frac{1}{2}x + 2$]

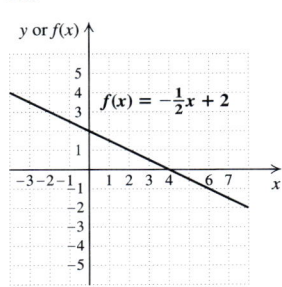

31. [graph: $f(x) = \frac{1}{2}x + 1$]

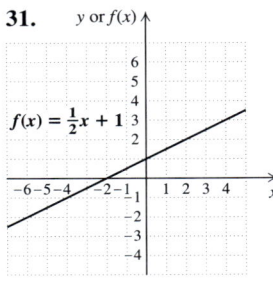

32. [graph: $f(x) = -\frac{3}{4}x - 2$]

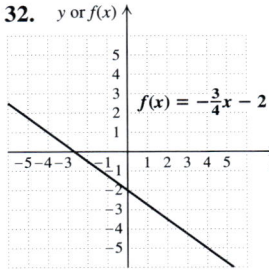

33. [graph: $g(x) = 2|x|$]

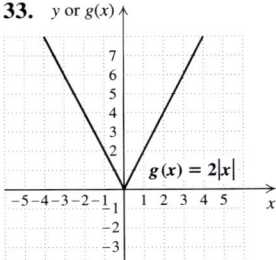

34. [graph: $h(x) = -|x|$]

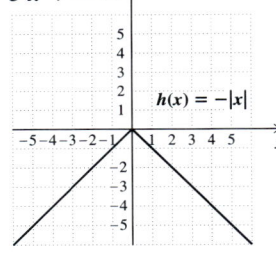

35. [graph: $g(x) = x^2$]

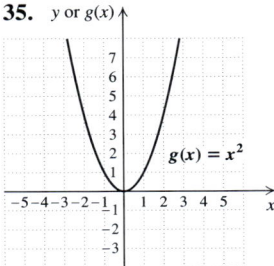

36. [graph: $f(x) = x^2 - 1$]

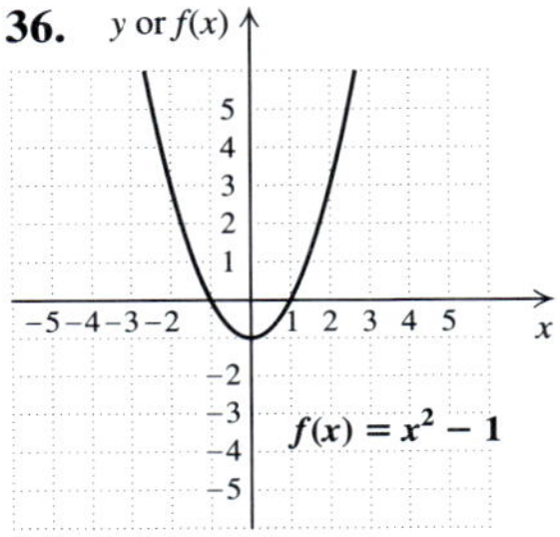

37. [graph: $f(x) = x^2 - x - 2$]

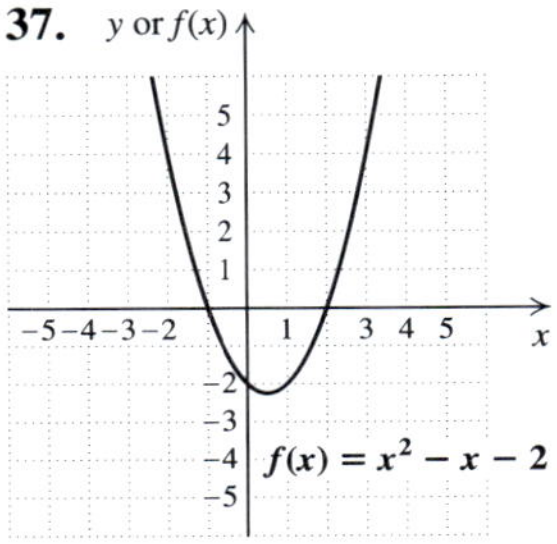

38. [graph: $g(x) = x^2 + 6x + 5$]

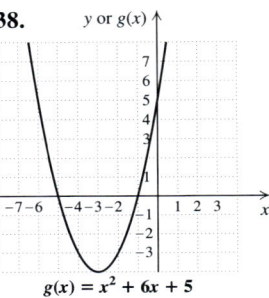

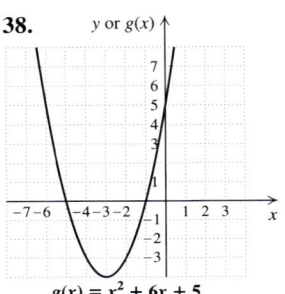

39. Yes **40.** The graph is not that of a function because a vertical line, say $x = 1$, crosses the graph at more than one point. **41.** No **42.** The graph is not that of function because a vertical line, say $x = 3$, crosses the graph at more than one point. **43.** Yes **44.** The graph is that of a function because no vertical line can cross the graph at more than one point. **45.** No **46.** Since the slopes are the same ($m = 0$), and the y-intercepts are different, the equations represent parallel lines. **47.** No solution
48. $x - 3y = 12$, (1)
$\quad\;\; 3x - 9y = 6$ (2)

Solve (1) for x.

$\quad x = 3y + 2$

Substitute $3y + 2$ for x in (2).

$\quad 3(3y + 2) - 9y = 6$
$\quad\quad 9y + 6 - 9y = 6$
$\quad\quad\quad\quad\quad\;\; 6 = 6$ True for all values of x

There are an infinite number of solutions.

49. ◆ No; since each input has exactly one output, the number of outputs cannot exceed the number of inputs.
50. ◆ Yes; it is possible for two or more inputs to be matched to the same output. **51.** ◆ Answers will vary.
52. ◆ One definition of "function" is "the proper or characteristic action of a person, living thing, manufactured or created thing." In mathematics a function acts in a prescribed, or characteristic, manner on an input to produce the corresponding output.

53. [graph: $g(x) = x^3$]

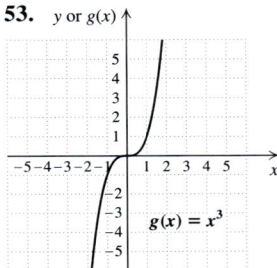

54. [graph: $f(x) = 2 + \sqrt{x}$]

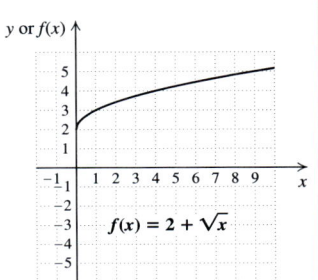

55. [graph: $f(x) = |x| + x$]

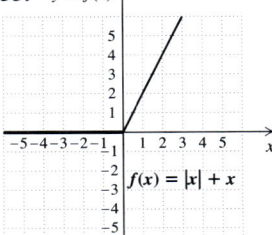

56. [graph: $g(x) = |x| - x$]

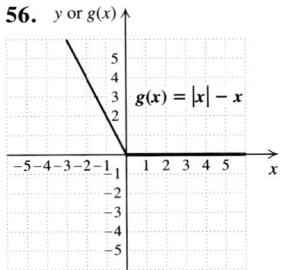

57. Answers may vary.

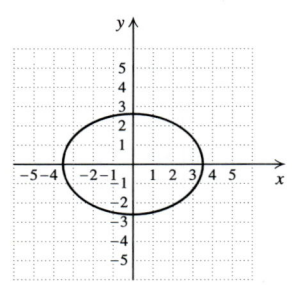

58. $f(-1) = -7$ gives us the ordered pair $(-1, -7)$. $f(3) = 8$ gives us the ordered pair $(3, 8)$. The slope of the line determined by these points is
$$m = \frac{-7 - 8}{-1 - 3} = \frac{-15}{-4} = \frac{15}{4}.$$
We use the point-slope equation to find the equation of the line with slope $\frac{15}{4}$ and containing the point $(3, 8)$.
$$y - y_1 = m(x - x_1)$$
$$y - 8 = \frac{15}{4}(x - 3)$$
$$y - 8 = \frac{15}{4}x - \frac{45}{4}$$
$$y = \frac{15}{4}x - \frac{13}{4}$$

59. $g(x) = x^2 - 4$

60. $f(x) = 3x + 5$

The domain is the set $\{0, 1, 2, 3\}$.
$$f(0) = 3 \cdot 0 + 5 = 0 + 5 = 5$$
$$f(1) = 3 \cdot 1 + 5 = 3 + 5 = 8$$
$$f(2) = 3 \cdot 2 + 5 = 6 + 5 = 11$$
$$f(3) = 3 \cdot 3 + 5 = 9 + 5 = 14$$
The range is the set $\{5, 8, 11, 14\}$.

61. $\{-5, -4, -1, 4\}$

62. $h(x) = |x| - x$

The domain is the set $\{-1, 0, 1, 2, 3, 4, 5, 6, 7, 8, 9, 10, 11, 12, 13, 14, 15, 16, 17, 18, 19\}$.
$$h(-1) = |-1| - (-1) = 1 + 1 = 2$$
$$h(0) = |0| - 0 = 0 - 0 = 0$$
$$h(1) = |1| - 1 = 1 - 1 = 0$$
$$h(2) = |2| - 2 = 2 - 2 = 0$$
$$h(3) = |3| - 3 = 3 - 3 = 0$$
$$\vdots$$
$$h(19) = |19| - 19 = 19 - 19 = 0$$
The range is the set $\{0, 2\}$.

63. $\{-7, 0, 1, 2, 9\}$ **64.**

REVIEW EXERCISES: CHAPTER 10, PP. 512–513

1. [10.2] $1, -\frac{5}{3}$ **2.** [10.2] $\frac{3 \pm \sqrt{5}}{2}$ **3.** [10.1] $\pm\sqrt{3}$

4. [10.3] $\frac{3}{5}, 1$ **5.** [10.3] $1 \pm \sqrt{11}$ **6.** [10.4] $\pm 8i$

7. [10.3] $\frac{1}{3}, -2$ **8.** [10.1] $-8 \pm \sqrt{13}$ **9.** [10.4] $3 \pm 2i$

10. [10.3] $\frac{1 \pm \sqrt{10}}{3}$ **11.** [10.1] 0   **12.** [10.3] $-3 \pm 3\sqrt{2}$

13. [10.4] $-\frac{1}{2} \pm \frac{\sqrt{3}}{2}i$ **14.** [10.3] $\frac{2 \pm \sqrt{3}}{2}$ **15.** [10.1] 1

16. [10.1] $\pm 2\sqrt{2}$ **17.** [10.4] $\frac{3}{10} \pm \frac{\sqrt{71}}{10}i$

18. [10.3] $\frac{5}{3}, -\frac{7}{2}$ **19.** [10.3] 2.618, 0.382

20. [10.3] $-0.134, -1.866$ **21.** [10.3] 1.7 m, 4.7 m

22. [10.3] 5%   **23.** [10.3] 12 m, 9 m   **24.** [10.3] 6.3 sec

25. [10.4] $2 - 8i$ **26.** [10.4] $2\sqrt{6}i$

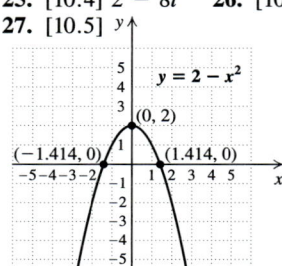

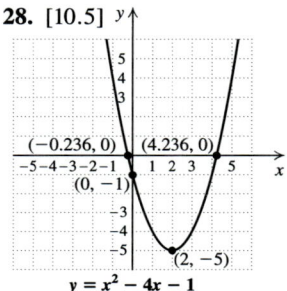

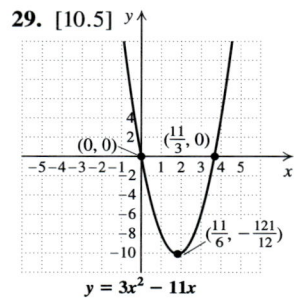

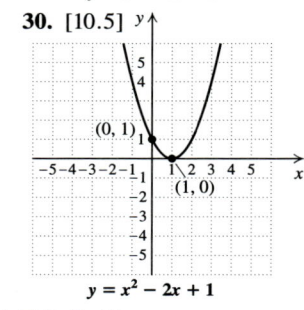

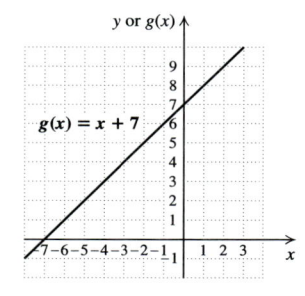

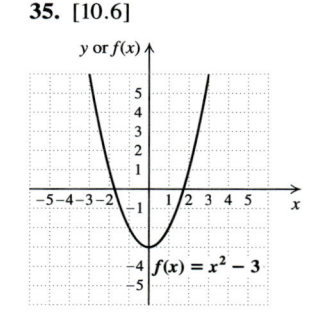

27. [10.5] $y = 2 - x^2$

28. [10.5] $y = x^2 - 4x - 1$

29. [10.5] $y = 3x^2 - 11x$

30. [10.5] $y = x^2 - 2x + 1$

31. [10.6] $-1, -7, 2$ **32.** [10.6] $0, 0, 19$

33. [10.6] 2700

34. [10.6] $g(x) = x + 7$

35. [10.6] $f(x) = x^2 - 3$

36. [10.6] *[graph of h(x) = 3|x|]*

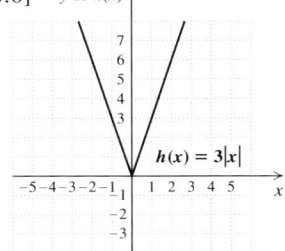

37. [10.6] No **38.** [10.6] Yes **39.** [7.1] Yes
40. [9.2] $2p^3\sqrt{21}$ **41.** [9.4] 7 **42.** [7.2] $y = -\frac{4}{3}x + \frac{7}{3}$
43. [10.5] ◆ The graph can be drawn more accurately and with fewer points plotted when the general shape is known.
44. [10.5] ◆ If the radicand is 0, the quadratic formula becomes $x = -b/(2a)$; thus there is only one x-intercept. If the radicand is negative, there are no real-number solutions and thus no x-intercepts. If the radicand is positive, then

$$x = \frac{-b + \sqrt{b^2 - 4ac}}{2a}$$

or

$$x = \frac{-b - \sqrt{b^2 - 4ac}}{2a}$$

so there must be two x-intercepts.
45. [10.3] 31 and 32; -32 and -31
46. [10.2] $b = 14$ or -14 **47.** [10.3] 25
48. [10.3] $s = 5\sqrt{\pi}$

TEST: CHAPTER 10, PP. 513–514

1. [10.1] $\pm\sqrt{5}$ **2.** [10.3] $0, \frac{7}{3}$ **3.** [10.3] $\frac{1 \pm \sqrt{13}}{2}$
4. [10.3] $\frac{5}{3}, -3$ **5.** [10.3] $-8, 6$ **6.** [10.4] $\pm 7i$
7. [10.1] $2 \pm \sqrt{5}$ **8.** [10.1] 2 **9.** [10.4] $2 \pm 2i$
10. [10.3] $\frac{3 \pm \sqrt{37}}{2}$ **11.** [10.3] $-2 \pm \sqrt{14}$
12. [10.3] $\frac{7 \pm \sqrt{37}}{6}$ **13.** [10.2] $2 \pm \sqrt{14}$
14. [10.3] 4.702, -1.702 **15.** [10.3] 6.5 m, 2.5 m
16. [10.3] 11 **17.** [10.4] $7i$ **18.** [10.4] $3 - 4\sqrt{2}i$
19. [10.5] *[graph of $y = -x^2 + x - 5$, vertex $(\frac{1}{2}, -\frac{19}{4})$, y-intercept $(0, -5)$]*
20. [10.5] *[graph of $y = x^2 + 2x - 15$, x-intercepts $(-5, 0)$ and $(3, 0)$, vertex $(-1, -16)$, y-intercept $(0, -15)$]*

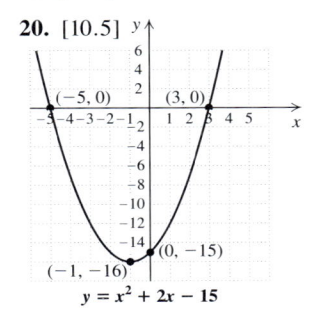

21. [10.6] $1, \frac{3}{2}, 2$ **22.** [10.6] $1, 3, -3$
23. [10.6] 25.98 min
24. [10.6] *[graph of $h(x) = x - 4$]* **25.** [10.6] *[graph of $g(x) = x^2 - 4$]*

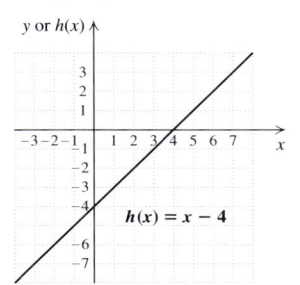

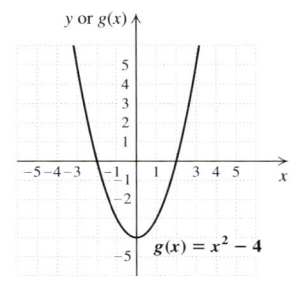

26. [10.6] Yes **27.** [10.6] No **28.** [7.2] $y = -3x - 11$
29. [8.2] $(2, 4)$ **30.** [9.4] $-\sqrt{3}$
31. [7.1] *[graph of $y = -\frac{1}{3}x + 2$]*

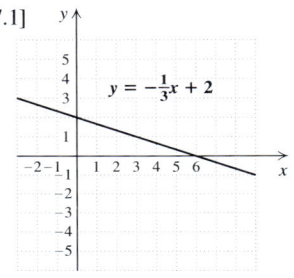

32. [10.3] $5 + 5\sqrt{2} \approx 12.071$ ft **33.** [10.3] $1 \pm \sqrt{5}$

CUMULATIVE REVIEW: 1–10, PP. 515–517

1. [1.8] x^3 **2.** [1.8] 54 **3.** [1.2] $(6x)y, x(6y)$; there are other answers. **4.** [6.3] 240 **5.** [1.4] 7 **6.** [1.5] 9
7. [1.6] 15 **8.** [1.7] $-\frac{3}{20}$ **9.** [1.8] 4
10. [1.8] $-2m - 4$ **11.** [2.1] -9
12. [2.6] $\{x | x < -9\}$ **13.** [2.2] 7 **14.** [5.6] 3, 5
15. [8.2] $(1, 2)$ **16.** [6.6] $-\frac{3}{5}$ **17.** [8.3] $(6, 7)$
18. [5.6] $3, -2$ **19.** [10.3] $\frac{-3 \pm \sqrt{29}}{2}$ **20.** [9.5] 2
21. [2.6] $\{x | x \geq -1\}$ **22.** [2.2] 8 **23.** [2.2] -3
24. [8.3] $(15, 0)$ **25.** [10.4] $-1 \pm 2i$ **26.** [10.1] $\pm\sqrt{10}$
27. [10.1] $3 \pm \sqrt{6}$ **28.** [6.6] $\frac{2}{9}$ **29.** [5.6] $\frac{5}{4}, -\frac{4}{3}$
30. [6.6] No solution **31.** [9.5] 12 **32.** [6.8] $t = \frac{4s + 3}{A}$
33. [6.8] $m = \frac{tn}{t + n}$ **34.** [4.8] 2.73×10^{-5}
35. [4.8] 2.0×10^{14} **36.** [4.8] x^{-4} **37.** [4.8] y^7
38. [4.1] $4y^{12}$ **39.** [4.2] $10x^3 + 3x - 3$
40. [4.3] $7x^3 - 2x^2 + 4x - 17$ **41.** [4.3] $8x^2 - 4x - 6$
42. [4.4] $-8y^4 + 6y^3 - 2y^2$
43. [4.4] $6t^3 - 17t^2 + 16t - 6$ **44.** [4.5] $t^2 - \frac{1}{16}$
45. [4.5] $9m^2 - 12m + 4$

46. [4.6] $15x^2y^3 + x^2y^2 + 5xy^2 + 7$
47. [4.6] $x^4 - 0.04y^2$ **48.** [4.6] $9p^2 + 24pq^2 + 16q^4$
49. [6.2] $\dfrac{2}{x+3}$ **50.** [6.2] $\dfrac{3a(a-1)}{2(a+1)}$
51. [6.4] $\dfrac{27x-4}{5x(3x-1)}$ **52.** [6.4] $\dfrac{-x^2+x+2}{(x+4)(x-4)(x-5)}$
53. [4.7] $x^2 + 9x + 16 + \dfrac{35}{x-2}$
54. [5.4] $(7x+2)(7x-2)$ **55.** [5.1] $(x-8)(x^2-5)$
56. [5.4] $x(4x+1)(4x-1)$ **57.** [5.4] $(m-4)^2$
58. [5.3] $(5x-2)(3x+4)$ **59.** [5.1] $2x(9x^4+2x^3-5)$
60. [5.3] $(3p+2)(2p-3)$ **61.** [5.1] $(c-3b)(2a+d)$
62. [5.4] $(3x+5y)^2$ **63.** [5.5] $3(t^2+6)(t+2)(t-2)$
64. [5.4] $(7ab+3)(7ab-3)$ **65.** [6.5] $-\dfrac{42}{5}$ **66.** [9.1] 7
67. [9.7] -5 **68.** [9.2] $8x$ **69.** [9.2] $\sqrt{a^2-b^2}$
70. [9.2] $8a^2b\sqrt{3ab}$ **71.** [9.7] $\tfrac{1}{4}$ **72.** [9.2] $9xy\sqrt{3x}$
73. [9.3] $\dfrac{10}{9}$ **74.** [9.4] 7 **75.** [9.4] $16\sqrt{3}$
76. [9.3] $\dfrac{2\sqrt{10}}{5}$ **77.** [9.6] 40 cm

78. [7.1] **79.** [3.3]

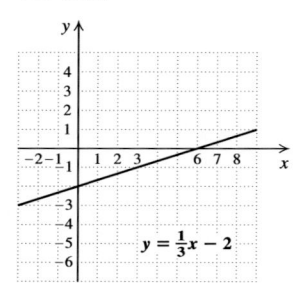

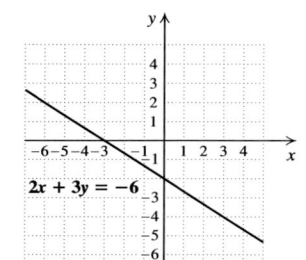

80. [3.3] 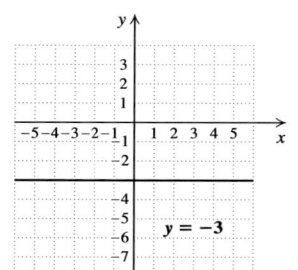 **81.** [7.3]

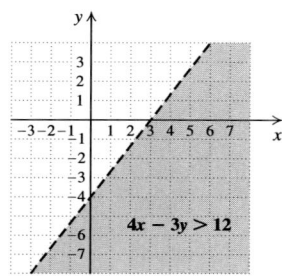

82. [10.5] 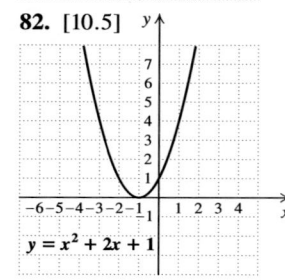 **83.** [7.3]

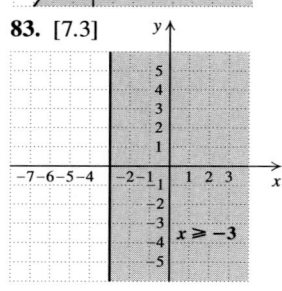

84. [10.2] $\dfrac{2 \pm \sqrt{6}}{3}$ **85.** [10.3] $1.207, -0.207$
86. [10.5]

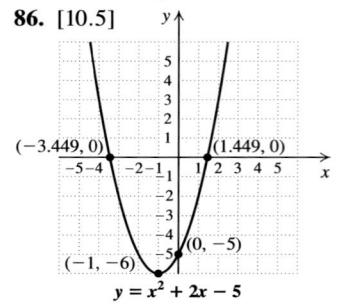

87. [2.4], [2.5] 48.2 million **88.** [2.7] $\{x \mid x > 5\}$
89. [5.7] 14, 16; $-16, -14$ **90.** [10.3] 12 m
91. [2.5] 6090
92. [8.4] $1.10 per lb: 14 lb; $0.80 per lb: 28 lb
93. [6.7] 24 min
94. [3.4] $225

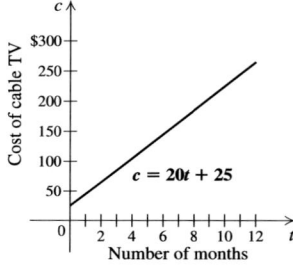

95. [10.6] $-4, 0, 0$ **96.** [7.1] Parallel
97. [8.5]

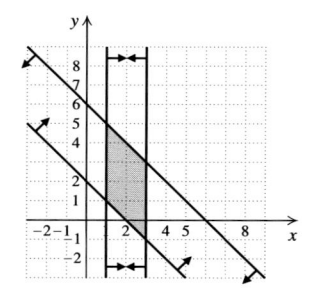

98. [7.1] 2, $(0, -8)$ **99.** [3.5] 15
100. [7.2] $y + 3 = -\tfrac{1}{2}(x-1)$ **101.** [10.4] $-5i$
102. [10.2] $30, -30$ **103.** [9.6] $\dfrac{\sqrt{6}}{3}$ **104.** [4.5] Yes
105. [6.1] No **106.** [4.5] No **107.** [9.1] No
108. [9.1] Yes

Index

A

Absolute value, 29
  and square roots, 428, 467
Addition
  associative law of, 11, 59
  commutative law of, 10, 59
  of exponents, 176, 181, 231, 238
  of fractional expressions, 20
  of polynomials, 192, 217
  of radical expressions, 443
  of rational expressions, 310, 320, 355
  of real numbers, 32, 33
Addition principle
  for equations, 65, 115
  for inequalities, 102, 115
Additive inverse, 37, 38. *See also* Opposite.
Algebraic expressions, 3
  evaluating, 2
  LCM of, 312
  terms, 13
  translating to, 4
Algebraic fractions, 298. *See also* Rational expressions.
Angles
  complementary, 99
  right, 288
  supplementary, 99
Applications, *see* Applied problems; Index of Applications
Applied problems, 3, 6–9, 23, 30, 34, 36, 37, 41, 43, 59, 61, 78, 81, 87–89, 91–100, 109–114, 116–118, 120–123, 125–128, 134, 136, 137, 146–156, 163, 166–174, 187–192, 195, 196, 198, 199, 206, 214, 215, 219–222, 228, 234, 236, 240, 241, 255, 265, 271, 285–293, 295, 296, 309, 318, 331, 338–348, 357–360, 368, 369, 377, 380, 382–386, 393, 398–401, 407–418, 423, 424, 429–431, 437, 442, 446, 450, 452, 455–460, 468, 469, 485–491, 500, 501, 510, 512–514, 516. *See also Index of Applications.*
Approximately equal to ($\approx$), 27
Approximating
  higher roots, 465
  square roots, 427
Area
  circle, 82
  parallelogram, 7, 59
  rectangle, 3, 59
  square, 51
  surface, right circular cylinder, 216
  trapezoid, 83
  triangle, 3, 59
Associative laws, 11, 59
Axes of graphs, 123
Axis, 123
Axis of symmetry, 494

B

Bar graph, 120
Base, 51
Binomials, 185
  as divisors, 225
  product of, 208
    FOIL method, 208
    sum and difference of two terms, 209, 238
    squares of, 210, 238
Braces, 52
Brackets, 52

C

Calculator
  approximating higher roots on, 465
  approximating square roots on, 427
Canceling, 19, 301
Carry out, 91, 116
Changing the sign, 39, 193
Checking
  by evaluating, 204, 248, 302
  factorizations, 245, 246, 248
  in problem-solving process, 91, 116
  quotients, 204, 225
  solutions of equations, 66, 278, 333, 390, 448, 473
  solutions of inequalities, 374
Circle
  area, 82
    of a sector, 82
  circumference, 79
Circle graph, 121
Circumference, 79
Clearing decimals, 73, 106, 406
Clearing fractions, 73, 332, 406
Closed set, 522
Coefficients, 68, 185
  leading, 186
Collaborative Corner, 15, 58, 77, 84, 90, 114, 128, 138, 156, 183, 207, 222, 237, 256, 272, 293, 319, 349, 354, 369, 378, 394, 418, 432, 453, 460, 476, 481, 501
Collecting like terms, *see* Combining like terms
Combining like terms, 35, 186
  in equation solving, 72
Common denominators, 20. *See also* Least common denominator.
Common factor, 14, 245
Commutative laws, 10, 59
Complementary angles, 99
Completing the square, 477
  solving equations by, 478

Complex numbers, 492
Complex rational, or fractional, expressions, 327
   simplifying, 327, 329, 355, 356
Composite number, 17
Compound interest, 83, 183, 490
Conjugates, 445
Consistent system of equations, 390, 391, 421
Constant, 2
   of proportionality, 379, 381
   variation, 379
Coordinates, 123
   finding, 124
   on the globe, 128
Correspondence, 502. *See also* Function.
Cube
   surface area, 81
   volume, 51
Cube root, 461
Cubes, factoring sums or differences, 523

D

Decimal notation
   converting from/to percent notation, 85, 86
   converting from/to scientific notation, 232, 233
   for irrational numbers, 27, 428
   for rational numbers, 26
   repeating, 26
   terminating, 26
Decimals, clearing, 73, 106, 406
Degree
   of a polynomial, 186
   of term, 185, 216
Demand, 501
Denominator, 17
   common, 20
   least common, 312, 313, 355
   with opposite factors, 323
   rationalizing, 439, 444
Dependent equations, 391, 421
Descending order, 187
Diagonals, number of, 485
Diagram, set, 520, 521
Difference, *see* Subtraction
Differences of cubes, 523
Differences of squares, 209, 268
   factoring, 268, 294
Direct variation, 378, 385

Distance, 7
Distributive law, 12, 59
   and combining like terms, 35
   and factoring, 13, 245
Division
   using exponents, 177, 181, 231, 238
   of fractional expressions, 21
   of polynomials, 223–227
   of radical expressions, 438, 463, 467
   of rational expressions, 307, 355
   of real numbers, 46
   and reciprocals, 21, 307
   using scientific notation, 233
   of square roots, 438, 467
   by zero, 46, 48, 59
Domain, 502

E

Element of a set, 519
Elimination method, solving systems of equations, 402
Empty set, 521
Equations, 5. *See also* Formulas.
   dependent, 391, 421
   of direct variation, 378, 385
   equivalent, 65
   fractional, 332
   graphs of, *see* Graphing
   independent, 391, 421
   of inverse variation, 381, 385
   linear, 130
   containing parentheses, 74
   point–slope, 370, 385
   proportion, 342
   quadratic, 278, 472
   radical, 447
   rational, 332
   slope–intercept, 364, 385
   solutions, 5, 64, 129
   solving, *see* Solving equations
   systems of, 388
   translating to, 6, 146
   of variation, 378
Equilateral triangle, 459
Equilibrium point, 501
Equivalent equations, 65
Equivalent expressions, 9
Equivalent inequalities, 102
Evaluating expressions, 2
   and checking, 204, 248, 302
   polynomials, 187, 215

Even roots, 462
Exponential notation, 51, 52. *See also* Exponents.
Exponents, 51
   adding, 176, 181, 231, 238
   definitions for, 181, 231, 238
   dividing using, 177, 181, 231, 238
   multiplying, 179, 181, 231, 238
   multiplying using, 176, 181, 231, 238
   negative, 229, 231, 238
   one as, 181, 231, 238
   raising a power to a power, 179, 181, 231, 238
   raising a product to a power, 180, 181, 231, 238
   raising a quotient to a power, 181, 231, 238
   rational, 463, 464, 467
   rules for, 181, 238
   subtracting, 177, 181, 231, 238
   zero as, 178, 181, 238
Expressions
   algebraic, 3
   equivalent, 9
   evaluating, 2, 3
   fractional, 298
   radical, 427
   rational, 298
   simplifying, *see* Simplifying
   terms of, 13
   value of, 2
   variable, 2

F

Factoring, 13
   checking by evaluating, 245, 246, 248
   common factor, 14, 245
   completely, 269
   numbers, 16
   polynomials, 244, 270, 273, 294
      with a common factor, 245
      differences of cubes, 523
      differences of squares, 268, 294
      by grouping, 247
      monomials, 244
      perfect-square trinomials, 266, 294
      squares of binomials, 266, 294
      sums of cubes, 523
      trinomials, 250–255, 257–264

radical expressions, 433
solving equations by, 279
Factorization, 14. *See also*
Factoring.
prime, 17
Factors, 13. *See also* Factoring.
opposites, 303, 323
Familiarize, 91, 116
First coordinate, 123
Five-step process for problem
solving, 91, 116
FOIL method, 208
and factoring, 257
Formulas, 78. *See also* Equations.
compound-interest, 83, 183, 490
functions as, 503
quadratic, 483, 511
simple interest, 82
solving for given letter, 78, 80,
116, 350, 356
Fractional equations, 332
Fractional expressions, 298. *See also*
Fractional notation; Rational
expressions.
addition of, 20
complex, 327
division of, 21
multiplication of, 17
simplifying, 18
subtraction of, 20
Fractional notation, 17. *See also*
Fractional expressions;
Fractions; Rational
expressions.
converting to percent notation, 86
for one, 17
for rational numbers, 25, 26
simplifying, 18
Fractions. *See also* Fractional
expressions; Fractional
notation; Rational
expressions.
addition of, 20
algebraic, 298. *See also* Rational
expressions.
clearing, 73, 332
converting to decimal notation, 27
division of, 21
improper, 20
multiplication of, 17
subtraction of, 20
Function, 502
domain, 502

as a formula, 503
graphs of, 505
inputs, 502
"machine," 505
notation, 504
outputs, 502
range, 502
vertical-line test, 507, 512

G

Globe, coordinates on, 128
Grade, 163
Graph. *See also* Graphing.
bar, 120
circle, 121
line, 122
nonlinear, 135
and rate, 149
Grapher, 135
intersect, 336
range, 135
regression, 501
root, 282
simultaneous mode, 367
table, 148
trace, 148
value, 188
Y-VARS, 505
Zdecimal, 299
zoom, 141
Graphing
equations, 130
horizontal line, 142
using intercepts, 139
linear, 130, 131, 169
and point–slope form, 372
and problem solving, 146
quadratic, 494–499, 512
slope, 362, 365, 372
using y-intercept and slope,
362, 365
vertical line, 142
x-intercept, 139
y-intercept, 134, 139
functions, 505
inequalities, 101
linear, 373, 376, 385
in one variable, 101, 376
systems of, 419
numbers, 26, 27
points on a plane, 123
solving equations by, 336
solving problems with, 146

systems of equations, 389
Greater than ($>$), 28
Greater than or equal to ($\geq$), 29
Greatest common factor, 245
Grouping
factoring by, 247, 262
symbols, 52

H

Higher roots, 461
approximating, 465
Horizontal lines, 142
slope, 162, 169
Hypotenuse, 288

I

i, 491
Identity property
of one, 18, 59
of zero, 33, 59
Imaginary number, 492
Improper fraction, 20
Inconsistent system of equations,
390, 391, 421
Independent equations, 391, 421
Indeterminate, 48
Index, 461
Inequalities, 29
addition principle for, 102, 115
equivalent, 102
graphs of, 101
linear, 373, 376, 385
in one variable, 101
linear, 373
system of, 419
multiplication principle for, 104,
115
solution set, 103
solutions of, 101
solving, *see* Solving inequalities
translating to, 109
Inputs, 502
Integers, 24
Intercepts, 134, 139, 281, 496, 512
graphs using, 139
Interest
compound, 83, 183, 490
simple, 82
Intersect, on a grapher, 336
Intersection, 521
Inverse
additive, 37, 38. *See also* Opposite
of a number.

Inverse (*continued*)
　multiplicative, 21
Inverse variation, 381, 385
Irrational numbers, 27, 429
　decimal notation, 27, 428
　graphing, 27
Isosceles triangle, 459

L

Largest common factor, 245
Law of opposites, 38, 59
LCD, *see* Least common denominator
LCM, *see* Least common multiple
Leading coefficient, 186
Leading term, 186
Least common denominator, 312, 313, 355
Least common multiple, 312
Legs of a right triangle, 288
Less than ($<$), 28
Less than or equal to ($\leq$), 29
Like radicals, 443
Like terms, 35, 54, 186, 217
　combining (or collecting), 35, 186
Line. *See also* Lines.
　horizontal, 142
　point–slope equation, 370, 385
　slope, 160, 169
　slope–intercept equation, 364, 385
　vertical, 142
Line graph, 122
Linear equations, 130
　graphing, 130, 169
　in one variable, 143
Linear inequalities, 373
　systems of, 419
Lines
　parallel, 367, 385, 390, 391, 421
　perpendicular, 369

M

Magic number, 222
Member of a set, 519
Mixture problems, 412
Monomials, 184
　as divisors, 223
　factoring, 244
　and multiplying, 200
Motion problems, 340
Multiple, least common, 312
Multiplication. *See also* Multiplying.
　associative law of, 11, 59
　commutative law of, 10, 59
　using exponents, 176, 181, 231, 238
　of fractional expressions, 17
　of monomials, 200
　of polynomials, 200–204, 207–212, 218
　of radical expressions, 433, 435, 444, 463, 467
　of rational expressions, 305, 355
　of real numbers, 44–46
　using scientific notation, 233
　of square roots, 433, 435, 467
　by zero, 44, 59
Multiplication principle
　for equations, 67, 115
　for inequalities, 104, 115
Multiplicative inverse, 21
Multiplicative property of zero, 44, 59
Multiplying. *See also* Multiplication.
　exponents, 179, 181, 238
　by 1, 18, 59
　by -1, 54, 59

N

nth roots, 462
　and rational exponents, 464
Natural numbers, 16
Negative exponents, 229, 464, 467
Negative integers, 24
Negative one, property of, 54, 59
Nonlinear graphs, 135
Notation
　exponential, 51, 52
　fractional, 17
　function, 504
　percent, 85, 116
　scientific, 232
　set, 103, 519
Number line
　addition on, 32
　and graphing, 26, 27
　order on, 28
Numbers
　complex, 492
　composite, 17
　factoring, 16
　graphing, 26, 27
　imaginary, 492
　integers, 24
　irrational, 27, 427
　natural, 16
　opposites of, 24, 37, 38, 47
　order of, 28
　prime, 16
　rational, 25, 26
　real, 27
　signs of, 39
　whole, 16
Numerator, 17

O

Odd roots, 462
One
　as exponent, 181, 231, 238
　fractional notation for, 17
　identity property of, 18, 59
　removing a factor equal to, 18, 19
Operations, order of, 52, 59
Opposite. *See also* Additive inverse.
　and changing the sign, 39
　factors, 303, 323
　of a number, 24, 37, 38, 47
　of a polynomial, 193
　in subtraction, 39
　of a sum, 55, 59
Opposites, law of, 38, 59
Order
　descending, 187
　on number line, 28
　of operations, 52, 59
Ordered pairs, 123
Origin, 123
Outputs, 502

P

Pairs, ordered, 123
Parabolas, 494
　axis of symmetry, 494
　intercepts, 496, 512
　vertex, 494, 497, 512
Parallel lines, 367, 385, 390, 391, 421
Parallelogram, area, 7, 59
Parentheses, 52
　in equations, 74
　in inequalities, 106
　removing, 13, 54, 55
Percent notation, 85, 116
　converting to/from decimal notation, 85, 86
Percent problems, 86
Perfect square radicand, 427
Perfect-square trinomial, 265
　factoring, 266, 294

Perimeter, rectangle, 82
Perpendicular lines, 369
Pi (π), 27
Pie chart, 121
Plotting points, 123
Point–slope equation, 370, 385
Points, coordinates of, 123
Polygon, number of diagonals, 485
Polynomials, 184
   addition of, 192, 217
   additive inverse of, 193
   binomials, 185
   coefficients in, 185
   combining like terms (or collecting similar terms), 186
   degree of, 186
   in descending order, 187
   division of, 223–227
   evaluating, 187, 215
   factoring, *see* Factoring, polynomials
   leading term, 186
   least common multiple of, 315
   monomials, 184
   multiplication of, 200–204, 207–212, 218
   opposite of, 193
   prime, 254
   quadratic, 494
   in several variables, 215
   subtraction of, 194, 217
   terms of, 185
   trinomials, 185
   value of, 187
Positive integers, 24
Positive rational exponents, 464, 467
Positive square root, 426
Power, 51. *See also* Exponents.
   raising to a power, 179, 181, 231, 238
   rule, 179, 181, 231, 238
   square root of, 435
Prime factorization, 17
   and LCM, 312
Prime numbers, 16
Prime polynomial, 254
Principal square root, 426
Principle of square roots, 472, 511
Principle of squaring, 447, 467
Principle of zero products, 278, 294
Problem solving. *See also* Applied problems.
   five-step process, 91, 116

   and graphing, 146
   other tips, 97
Problems, applied, *see* Applied Problems; *Index of Applications*
Product. *See also* Multiplication; Multiplying.
   raising to a power, 180, 181, 231, 238
   of square roots, 433, 435, 444, 467
   of sums and differences, 209, 238
   of two binomials, 208
Product rule
   for exponential notation, 176, 181, 231, 238
   for radicals, 433, 463, 467
Profit, total, 199
Properties of exponents, 181, 238
Property of -1, 54, 59
Proportion, 342
Proportional, 342, 379
Proportionality, constant of, 379, 381
Pure imaginary numbers, 492
Pythagorean theorem, 288, 294, 454, 467

Q

Quadrants, 124
Quadratic equations, 278
   complex-number solutions, 492
   graphs of, 494–499, 512
   solving, 511
      by completing the square, 478
      by factoring, 279
      using principle of square roots, 472, 474
      using quadratic formula, 483
   in standard form, 482
Quadratic formula, 483, 511
Quadratic polynomial, 494
Quotient, raising to a power, 181, 231, 238
Quotient rule
   for exponential notation, 177, 181, 231, 238
   for radicals, 438, 463, 467

R

Radical equations, 447
Radical expressions, 429. *See also* Square roots.
   adding, 443

   conjugates, 445
   dividing, 438, 463, 467
   in equations, 447
   factoring, 433
   index, 461
   multiplying, 433, 435, 444, 463, 467
   perfect square radicand, 427
   radicand, 427
   rationalizing denominators, 439, 444
   simplifying, 433, 467
   subtracting, 443
Radical sign, 426
Radicals, like, 443
Radicand, 427, 461
   negative, 427
   perfect square, 427
Raising a power to a power, 179, 181, 231, 238
Raising a product to a power, 180, 181, 231, 238
Raising a quotient to a power, 181, 231, 238
Range
   of a function, 502
   on a grapher, 135
Rate, 149
   of change, 151
Ratio, 342. *See also* Proportion.
   of two integers, 25
Rational equations, 332
Rational exponents, 463, 464, 467
Rational expressions, 298. *See also* Fractional expressions; Rational numbers.
   addition, 310, 320, 355
   complex, 327
   division, 307, 355
   multiplying, 305, 355
   with opposite factors, 303, 323
   reciprocals, 306
   simplifying, 300
   subtraction, 311, 320, 355
   undefined, 298
Rational numbers, 25, 26. *See also* Fractional expressions; Rational expressions.
Rationalizing denominators, 439, 444
Real-number system, 27
Real numbers, 28
   addition, 32, 33

Real numbers (*continued*)
  division, 46
  graphing, 26, 27
  multiplication, 44–46
  order, 28
  subtraction, 39
Reciprocals, 21, 47, 229, 306
Rectangle
  area, 3, 59
  perimeter, 82
Regression, 501
Remainder, 225
Removing a factor equal to one, 18, 19, 300
Removing parentheses, 13, 54, 55
Repeating decimals, 26
Reversing the sign, 39
Right angle, 288
Right circular cylinder, surface area, 216
Right triangles, 288, 454
Rise, 160
Root, on a grapher, 282
Roots
  cube, 461
  even, 462
  higher, 461
  nth, 462
  odd, 462
  square, *see* Square roots; Radical expressions
Roster notation, 519
Rules
  for exponents, 181, 238
  for order of operations, 52, 59
Run, 160

S

Scientific notation, 232, 238
  on a calculator, 233
    converting from/to decimal notation, 232, 233
    dividing using, 233
    multiplying using, 233
Second coordinate, 123
Sector of a circle, area, 82
Set. *See also* Sets.
  of integers, 24
  of rational numbers, 26
  of real numbers, 28
Set-builder notation, 103, 519
Sets, 24. *See also* Set.
  closed, 522

diagram, 520, 521
element of, 519
empty, 521
intersection of, 521
member of, 519
naming, 519
notation, 103, 519
solution, 103
subset, 520
union of, 521
Several variables, polynomials in, 215
Sign, changing, 39, 193
Signs of numbers, 39
Similar terms, 35, 54, 186, 217. *See also* Like terms.
Similar triangles, 343
Simple interest, 82
Simplifying
  complex rational expressions, 327, 329, 355, 356
  fractional expressions, 18
  radical expressions, 433, 469
  rational expressions, 300
  removing parentheses, 13, 54, 55
Simultaneous mode, 367
Slant, *see* Slope
Slope, 160, 169, 364
  applications, 162
  and graphing, 362, 365, 372
  of horizontal line, 162, 169
  of parallel lines, 367, 385
  of perpendicular lines, 369
  point–slope equation, 370, 385
  slope–intercept equation, 364, 385
  of vertical line, 102, 169
Slope–intercept equation, 364, 385
Solution set, 103
Solutions
  of equations, 5, 64, 129
    complex-number, 492
  of inequalities, 101
  of systems of equations, 388
  of systems of inequalities, 419
Solving equations, 5, 64, 75, 115. *See also* Solving formulas.
  using the addition principle, 65, 115
  clearing decimals, 73
  clearing fractions, 73
  combining like terms, 72
  containing parentheses, 74
  by factoring, 279

fractional, 332, 355
  by graphing, 336
  using multiplication principle, 67, 115
  with parentheses, 74
  using principle of zero products, 278
  using principles together, 71
  procedure, 75, 115
  quadratic, 511
    by completing the square, 478
    by factoring, 278
    using principle of square roots, 472, 474
    using quadratic formula, 483
  with radicals, 447
  rational, 332, 355
  systems of, 406, 422
    by elimination method, 402
    by graphing, 389
    by substitution method, 395
Solving formulas, 78, 80, 116, 350, 356
Solving inequalities
  using addition principle, 102, 115
  using multiplication principle, 104, 114
  using principles together, 105
  solution set, 102
  systems of, 419
Speed, 151
Sphere, volume, 353
Square, area, 51
Square of a binomial, 210, 238
Square, completing, 477
Square, perfect, 427
Square roots, 426. *See also* Radical expressions.
  and absolute value, 428, 467
  adding, 443
  approximating, 429
  dividing, 438, 467
  irrational, 427
  multiplying, 433, 435, 444, 467
  of negative numbers, 427
  positive, 426
  of powers, 435
  principal, 426
  principle of, 472, 511
  rational, 427
  simplifying, 433, 467
Squares, differences of, 209, 268
  factoring, 268, 294

Squaring, principle of, 447, 467
Standard form of a quadratic equation, 482
State answer, problem-solving process, 91, 116
Subset, 520
Substituting, 3. *See also* Evaluating expressions.
Substitution method, solving systems of equations, 395
Subtraction
  by adding the opposite, 39
  of exponents, 177, 181, 231, 238
  of fractional expressions, 20
  of polynomials, 194, 217
  of radical expressions, 443
  of rational expressions, 311, 320, 355
  of real numbers, 39
Sum
  of cubes, factoring, 523
  opposite of, 55, 59
Sum and difference of two terms, product of, 209, 238
Supplementary angles, 99
Supply, 500
Surface area, right circular cylinder, 216
Symmetry, axis of, 494
Systems of equations, 388
  consistent, 390, 391, 421
  dependent equations, 391, 421
  inconsistent, 390, 391, 421
  independent equations, 391, 421
  solution of, 388
  solving, 406, 422
    by elimination method, 402
    by graphing, 389
    by substitution method, 395
Systems of linear inequalities, 419

T

Table, on a grapher, 148
Target heart rate, 369
Technology Connection, 135, 141, 148, 188, 195, 204, 233, 248, 282, 367, 390, 397, 428, 449, 473, 485, 505
Temperature conversion, 354
Terminating decimal, 26
Terms, 13, 184
  coefficients of, 185
  combining (or collecting) like, 35, 186
  degrees of, 185, 216
  leading, 186
  like, 35, 54
  of polynomials, 185
  similar, 35, 54
Total profit, 199
Total value problems, 410
Trace, 148
Translating
  to algebraic expressions, 4
  to equations, 6, 146
  to inequalities, 109
  in problem-solving process, 91, 116
Trapezoid, area, 83
Triangle
  area, 3, 59
  equilateral, 459
  isosceles, 459
  right, 288, 454
  similar, 343
Trinomial, 185
  factoring, 250–255, 257–264
Trinomial, perfect-square, 265
  factoring, 266, 294

U

Undefined rational expression, 298
Undefined slope, 162, 169
Union, 521

V

Value
  of an expression, 2
  on a grapher, 188
  of a polynomial, 187
Variable, 2
Variable expression, 2
Variation
  constant, 379
  direct, 378, 385
  equation of, 378
  inverse, 381, 385
Vertex, 494, 497, 512
Vertical lines, 142
  slope, 162, 169
Vertical-line test, 507, 512
Volume
  cube, 51
  sphere, 353

W

Whole numbers, 16
Wind chill temperature, 453
Window, 135
Work principle, 340, 356

X

x-intercept, 139, 281, 496, 512

Y

y-intercept, 134, 364, 496, 512
  and graphing, 139, 362, 365
Y-VARS, on a grapher, 505

Z

Zdecimal, on a grapher, 299
Zero
  division by, 46, 48, 59
  as exponent, 178, 181, 231, 238
  identity property, 33, 59
  multiplicative property, 44, 59
Zero products, principle of, 278, 294
Zoom, 141

INDEX OF APPLICATIONS

In addition to the applications highlighted below, there are other applied problems and examples of problem solving in the text. An extensive list of their locations can be found under the heading "Applied problems" in the index at the back of the book.

ASTRONOMY
Lunar weight, 383
Orbit time, 7
Weight on Mars, 347, 383
Weight on the moon, 347

BIOLOGY
Cricket chirps and temperature, 99
Horticulture, 417
Speed of a snake, 346
Weight of a fish, 83
Weight gain, 112
Zoology, 7

BUSINESS
Allocating resources, 154
Audio sales, 98
Blending coffees, 412
Car rentals, 99
Catering, 417
Coffee blends, 416
Computers, 236
Consumer's demand, 501
Cost of self-employment, 89
Dairy farming, 418
Depreciation of an office machine, 148
Earnings, 113
Electrician visits, 112
Farming, 409
Fax machines, 345
Food preparation, 154
Harvesting, 345
Hours worked, 8
Ice cream cones, 416
Light bulbs, 347
Manufacturing, 383
Mixed nuts, 416
Mixing juice, 446
Packaging, 23
Paid admissions, 411, 416
Painting, 417
Parking spaces, 430
Parking-lot arrival spaces, 429
Payroll, 418
Peanut sales, 384
Planting grapes, 409
Price of printing, 136
Printing, 417
Production, 416
Profits and losses, 36
Programming, 348
Pumping time, 383
Retail sales, 87
Salaries, 156
Seed mix, 416
Seller's supply, 500
Selling vehicles, 415
Total cost, 191
Total profit, 199
Total revenue, 191
Truck rentals, 93, 99, 111
Value of an office machine, 134
Van rentals, 111
Vinyl phonograph singles, 136
Volunteer work, 113
Wages, 383
Wages and commissions, 154, 155
Waxing, 348
Work time, 7
Zoo admissions, 416

CHEMISTRY
Acid mixtures, 416
Automobile maintenance, 417
Chemistry, 417
Gas volume, 383
Melting butter, 112
Metallurgy, 417
Octane rating, 417, 418
Ski wax, 113
Temperature conversions, 510

CONSTRUCTION/ENGINEERING
Architecture, 167, 297, 377
Carpentry, 167, 344
Construction, 167, 291, 344, 500
Electrical current, 383
Electrical power, 81
Elevators, 377
Engineering, 167
Furnace output, 81
Insulation, 416
Masonry, 345, 458
Plumbing, 345, 458, 489
Rain gutter design, 293
Road design, 290
Roadway design, 288
Roofing, 292
Surveying, 166, 167, 459
Water intake, 345
Wiring, 348

CONSUMER APPLICATIONS
Banquet costs, 110
Car rental rates, 407
Car rentals, 408
Chartering a boat, 383
Commuting costs, 446
Copying costs, 154, 393
Cost of cable TV, 368
Cost of clothes, 113
Cost of a FedEx delivery, 154
Cost of a road call, 154
Cost of road service, 112
Cost of television, 383
Cost of water, 368
Credit card bills, 36

(continued)

Deducting sales tax, 89
Energy use, 214
Food prices, 415
Insurance-covered repairs, 112
Internet service, 368
Music-club purchases, 100
Net weight, 113
Parking costs, 111
Parking fees, 113, 114, 154
Phone costs, 111
Price of a CD player, 98
Price of a movie ticket, 113
Price of a textbook, 98
Real-estate depreciation, 154
Reducing utility costs, 393
Refrigerator size, 368
Selling a home, 95
Sporting-goods prices, 418
Telephone bills, 36
Tipping, 89
Toll charges, 113
Truck rentals, 146, 153
Turkey servings, 383
Value of computer software, 136
Van rentals, 153
Well drilling, 111

ECONOMICS
Account balance, 36
Budgeting, 100
Compounding interest, 83
Equilibrium point, 501
Interest compounded annually, 183, 222
Investing, 418
Investments, 490, 491
Loan interest, 99
Loan repayment, 43
Savings interest, 99
Spending on health, education, and defense, 126
Stock growth, 36
Stock prices, 37
Use of tax dollars, 121, 125
Yearly depreciation, 222

ENVIRONMENT
Altitude and temperature, 83
Barometric pressure, 36
Changes in elevation, 41, 43
Deer population, 347
Distance from a storm, 78
Ecology, 384
Elevation extremes, 43
Environmental science, 490
Estimating populations, 244
Fish population, 347
Fuel consumption, 8
Household waste, 416
Lake level, 34
Melting snow, 510
Moose population, 347
Peak elevation, 36
Pond depth, 112
Record temperature drop, 43, 137
Returnable bottles, 416
River discharge, 236
Sorting recyclables, 338
Sorting solid waste, 125
Thunderstorm distance, 504
Underwater elevation, 43
Water contamination, 236

GEOMETRY
Angles of a triangle, 98
Angles in a triangle, 96
Angles in a pentagon, 100
Angles in a quadrilateral, 100
Area of a circle, 191
Area of a garden, 290
Area of a parallelogram, 7
Area of a rectangle, 112, 113, 489, 490
Area of a sector, 82
Area of a square, 490
Area of a trapezoid, 83, 351
Area of a triangle, 100, 113, 409
Baseball, 458
Calculator design, 290
Changing elevations, 452
Circumference, 191
Circumference of a circle, 79
Complementary angles, 99, 400, 408
Cordless telephones, 458, 459
Court dimensions, 500
Diagonals in a polygon, 485
Dimensions of a closed box, 292
Dimensions of a garden, 285
Dimensions of an open box, 292
Dimensions of a sail, 286, 290
Dimensions of a square, 491
Dimensions of a state, 99
Dimensions of a triangle, 290
Distances, 348
Flagpoles, 490
Framing, 409
Gardening, 94, 409, 490
Geometry, 37, 346, 384
Guy wire, 291, 458, 489
Hancock Building dimensions, 98, 99
Home maintenance, 457
Perimeter of a pool, 113
Perimeter of a rectangle, 100, 112, 113, 400
Perimeter of a triangle, 98, 113
Pool sidewalk, 292
Ranching, 460
Reach of a ladder, 291, 455
Right triangle geometry, 288
Right triangles, 487, 489
Sighting to the horizon, 450, 452

Similar triangles, 243, 347
Soccer fields, 458
Softball dimensions, 456
Supplementary angles, 99, 398, 400, 409
Surface area of a cube, 81
Surface area of a right circular cylinder, 216
Surface area of a silo, 220
Typing paper, 99

HEALTH/LIFE SCIENCES
Absorption of ibuprofen, 81
Aerobic exercise, 369
Blood alcohol level, 125
Body temperature, 112
Calorie content, 89
Dietary restrictions, 109
Dosage size, 83
Driving under the influence, 120
Exercise, 167
Exercise and pulse rate, 122
Fat content, 89
Fat content in foods, 113
Fruit servings, 111
Hair growth, 154
Hemoglobin, 346
Increasing life expectancy, 136
Infant health, 88
Kissing and colds, 88
Lung capacity, 219
Medical dosage, 187, 188
Nutrition, 79
Nutritional standards, 114
Young's rule, 352

MISCELLANEOUS
Age, 409
Answering questions, 383
Caged rabbits and pheasants, 409
Coin value, 417
College course load, 111
FBI recruiting, 88
Grade average, 111
High-fives, 291
Mowing, 345
Number of handshakes, 291
Page numbers, 92, 98, 285, 290
Paint colors, 413
Raking, 345
Shoveling, 345
Television screens, 293
Test scores, 100, 417
Toasting, 291
Quiz average, 111

PHYSICS
Acoustics, 384
Altitude of a launched object, 219
Density, 384
Firecrackers, 347
Free-fall record, 489
Free-fall time, 489
Free-falling objects, 486
Gravitational force, 476
Height of a projectile, 500
Height of a rocket, 292
Lighting, 384
Musical tones, 383
Period of a swinging pendulum, 442, 452
Pressure at sea depth, 510
Speed of sound, 346, 431
Temperature as a function of depth, 510
Wavelength of a musical note, 81
Wind energy, 384
Work formula, 350

SOCIAL SCIENCES
Changing attitudes, 127
Coordinates on the globe, 128
Electrons, 348
Gettysburg Address, 100
Intelligence quotient, 350
Literature, 418
Memorizing words, 190
Predicting heights, 510
Votes for president, 88

SPORTS/HOBBIES/ENTERTAINMENT
Baking, 346
Basketball scores, 410
Basketball scoring, 99, 416
Boating, 348
Chess rating, 83
Enlarged strike zone, 491
Football, 400
Games, 8
Games in a league, 291
Games in a league's schedule, 287
Games in a sports league, 187
Golfing, 37
Hang time, 430
Hiking, 91
Hockey wins and losses, 377
Left-handed bowlers, 88
Olympic softball, 7
Path of the Olympic arrow, 192
Physical education, 291
Race time, 99
Racquetball, 400, 460
Running, 98
Sailing, 293
Sewing a quilt, 348
Size of a league schedule, 81
Skiing, 163
Skydiving, 190, 191

(continued)

Sled-dog racing, 98
Soccer, 400
Sports card values, 37
Track records, 112
Winning percentage, 100
Yardage gained, 36

STATISTICS/DEMOGRAPHICS
Change in tuition and fees, 164
Changing populations, 127
Cheese consumption, 127
Coffee harvest, 346
College size, 81
Cost of college, 136, 137
Daily accidents, 191
Health care costs, 88
Home remodeling, 98
Junk mail, 88
Lotteries and education, 88
Meat consumption, 127
Number of U.S. farms, 164
Orange juice consumption, 236
Tea consumption, 137
Two-person households, 236
U.S. defense outlays, 164
U.S. households and answering machines, 516
U.S. population, 164
Where college spending money goes, 509
White office paper vs. newsprint, 221
World birth rate vs. death rate, 30

TRANSPORTATION
Aviation, 155, 291, 458
Bicycle speed, 346
Boat speed, 346
Commuting, 348
Distance driven, 7, 460
Driving distances, 437
Driving and hiking, 415
Driving speed, 341, 345
Flight delays, 236
Mileage, 242
Miles driven, 347
Navigation, 166
Snowmobile speed, 346
Speed of a skidding car, 437, 452
Speed of travel, 345
Stopping distance, 500
Tractor speed, 346
Train schedules and speeds, 151
Train speed, 346
Train speeds, 345
Travel to work, 8
Walking speed, 346